P9-CKW-793

FUNDAMENTALS OF PHOTONICS

FUNDAMENTALS OF PHOTONICS

BAHAA E. A. SALEH
Department of Electrical and Computer Engineering
University of Wisconsin — Madison
Madison, Wisconsin

MALVIN CARL TEICH
Department of Electrical Engineering
Columbia University
New York, New York

A WILEY-INTERSCIENCE PUBLICATION
JOHN WILEY & SONS, INC.
NEW YORK / CHICHESTER / BRISBANE / TORONTO / SINGAPORE

Library of Congress Cataloging in Publication Data:

Saleh, Bahaa E. A., 1944–
Fundamentals of photonics/Bahaa E. A. Saleh, Malvin Carl Teich.
p. cm.—(Wiley series in pure and applied optics)
"A Wiley-Interscience publication."
Includes bibliographical references and index.
ISBN 0-471-83965-5
1. Photonics. I. Teich, Malvin Carl. II. Title. III. Series.

TA1520.S24 1991
621.36—dc20 90-44694
CIP

Printed in the United States of America

10 9 8

PREFACE

Optics is an old and venerable subject involving the generation, propagation, and detection of light. Three major developments, which have been achieved in the last thirty years, are responsible for the rejuvenation of optics and for its increasing importance in modern technology: the invention of the laser, the fabrication of low-loss optical fibers, and the introduction of semiconductor optical devices. As a result of these developments, new disciplines have emerged and new terms describing these disciplines have come into use: **electro-optics**, **optoelectronics**, **quantum electronics**, **quantum optics**, and **lightwave technology**. Although there is a lack of complete agreement about the precise usages of these terms, there is a general consensus regarding their meanings.

Photonics

Electro-optics is generally reserved for optical devices in which electrical effects play a role (lasers, and electro-optic modulators and switches, for example). *Optoelectronics*, on the other hand, typically refers to devices and systems that are essentially electronic in nature but involve light (examples are light-emitting diodes, liquid-crystal display devices, and array photodetectors). The term *quantum electronics* is used in connection with devices and systems that rely principally on the interaction of light with matter (lasers and nonlinear optical devices used for optical amplification and wave mixing serve as examples). Studies of the quantum and coherence properties of light lie within the realm of *quantum optics*. The term *lightwave technology* has been used to describe devices and systems that are used in optical communications and optical signal processing.

In recent years, the term **photonics** has come into use. This term, which was coined in analogy with electronics, reflects the growing tie between optics and electronics forged by the increasing role that semiconductor materials and devices play in optical systems. *Electronics* involves the control of electric-charge flow (in vacuum or in matter); *photonics* involves the control of photons (in free space or in matter). The two disciplines clearly overlap since electrons often control the flow of photons and, conversely, photons control the flow of electrons. The term *photonics* also reflects the importance of the photon nature of light in describing the operation of many optical devices.

Scope

This book provides an introduction to the fundamentals of photonics. The term *photonics* is used broadly to encompass all of the aforementioned areas, including the following:

- The *generation* of coherent light by lasers, and incoherent light by luminescence sources such as light-emitting diodes.
- The *transmission* of light in free space, through conventional optical components such as lenses, apertures, and imaging systems, and through waveguides such as optical fibers.
- The *modulation*, switching, and scanning of light by the use of electrically, acoustically, or optically controlled devices.
- The *amplification* and *frequency conversion* of light by the use of wave interactions in nonlinear materials.
- The *detection* of light.

These areas have found ever-increasing applications in optical communications, signal processing, computing, sensing, display, printing, and energy transport.

Approach and Presentation

The underpinnings of photonics are provided in a number of chapters that offer concise introductions to:

- The four theories of light (each successively more advanced than the preceding): ray optics, wave optics, electromagnetic optics, and photon optics.
- The theory of interaction of light with matter.
- The theory of semiconductor materials and their optical properties.

These chapters serve as basic building blocks that are used in other chapters to describe the *generation* of light (by lasers and light-emitting diodes); the *transmission* of light (by optical beams, diffraction, imaging, optical waveguides, and optical fibers); the *modulation* and switching of light (by the use of electro-optic, acousto-optic, and nonlinear-optic devices); and the *detection* of light (by means of photodetectors). Many applications and examples of real systems are provided so that the book is a blend of theory and practice. The final chapter is devoted to the study of fiber-optic communications, which provides an especially rich example in which the generation, transmission, modulation, and detection of light are all part of a single photonic system used for the transmission of information.

The theories of light are presented at progressively increasing levels of difficulty. Thus light is described first as rays, then scalar waves, then electromagnetic waves, and finally, photons. Each of these descriptions has its domain of applicability. Our approach is to draw from the simplest theory that adequately describes the phenomenon or intended application. Ray optics is therefore used to describe imaging systems and the confinement of light in waveguides and optical resonators. Scalar wave theory provides a description of optical beams, which are essential for the understanding of lasers, and of Fourier optics, which is useful for describing coherent optical systems and holography. Electromagnetic theory provides the basis for the polarization and dispersion of light, and the optics of guided waves, fibers, and resonators. Photon optics serves to describe the interactions of light with matter, explaining such processes as light generation and detection, and light mixing in nonlinear media.

Intended Audience

Fundamentals of Photonics is meant to serve as:

- An introductory textbook for students in electrical engineering or applied physics at the senior or first-year graduate level.
- A self-contained work for self-study.
- A text for programs of continuing professional development offered by industry, universities, and professional societies.

The reader is assumed to have a background in engineering or applied physics, including courses in modern physics, electricity and magnetism, and wave motion. Some knowledge of linear systems and elementary quantum mechanics is helpful but not essential. Our intent has been to provide an introduction to photonics that emphasizes the concepts governing applications of current interest. The book should, therefore, not be considered as a compendium that encompasses all photonic devices and systems. Indeed, some areas of photonics are not included at all, and many of the individual chapters could easily have been expanded into separate monographs.

Organization

The book consists of four parts: **Optics and Fiber Optics** (Chapters 1 to 10), **Quantum Electronics** (Chapters 11 to 14), **Optoelectronics** (Chapters 15 to 17), and **Electro-Optics and Lightwave Technology** (Chapters 18 to 22). The form of the book is modular so that it can be used by readers with different needs; it also provides instructors an opportunity to select topics for different courses. Essential material from one chapter is often briefly summarized in another to make each chapter as self-contained as possible. For example, at the beginning of Chapter 22 (Fiber-Optic Communications), relevant material from earlier chapters that describe fibers, light sources, and detectors is briefly reviewed. This places the important features of the various components at the disposal of the reader before the chapter proceeds with a discussion of the design and performance of the overall communication system that makes use of these components.

Recognizing the different degrees of mathematical sophistication of the intended readership, we have endeavored to present difficult concepts in two steps: at an introductory level providing physical insight and motivation, followed by a more advanced analysis. This approach is exemplified by the treatment in Chapter 18 (Electro-Optics) in which the subject is first presented using scalar notation, and then treated again using tensor notation.

Commonly accepted notation and symbols have been used wherever possible. Because of the broad spectrum of topics covered, however, there are a good number of symbols that have multiple meanings; a list of symbols is provided at the end of the book to help clarify symbol usage. Important equations are highlighted by boxes to simplify future retrieval. Sections dealing with material of a more advanced nature are indicated by asterisks and may be omitted if desired. Summaries are provided throughout the chapters at points where a recapitulation is deemed useful because of the involved nature of the material.

Representative Courses

The chapters of this book may be combined in various ways for use in semester or quarter courses. Representative examples of such courses are provided below. Some of these courses may be offered as part of a sequence. Other selections may also be made to suit the particular objectives of instructors and students.

Optics

Background: Chapter 1 (Ray Optics) and Chapter 2 (Wave Optics)

Chapter 3 (Beam Optics)

Chapter 4 (Fourier Optics)

Chapter 5 (Electromagnetic Optics)

Chapter 6 (Polarization and Crystal Optics)

Chapter 7 (Guided-Wave Optics)

Chapter 10 (Statistical Optics)

Optical Information Processing

Background: Chapter 1 (Ray Optics) and Chapter 2 (Wave Optics)

Chapter 4 (Fourier Optics)

Chapter 10 (Statistical Optics)

Chapter 18 (Electro-Optics)

Chapter 20 (Acousto-Optics)

Chapter 21 (Photonic Switching and Computing)

Lasers* or *Quantum Electronics

Background: Chapter 1 (Ray Optics); Chapter 2 (Wave Optics); and Chapter 15 (Photons in Semiconductors, Section 15.1)

Chapter 3 (Beam Optics)

Chapter 9 (Resonator Optics)

Chapter 11 (Photon Optics)

Chapter 12 (Photons and Atoms)

Chapter 13 (Laser Amplifiers)

Chapter 14 (Lasers)

Chapter 15 (Photons in Semiconductors, Section 15.2)

Chapter 16 (Semiconductor Photon Sources, Sections 16.2 and 16.3)

Optoelectronics

Background: Chapter 6 (Polarization and Crystal Optics); Chapter 11 (Photon Optics, Sections 11.1A and 11.2); Chapter 12 (Photons and Atoms, Sections 12.1 and 12.2); Chapter 13 (Laser Amplifiers, Section 13.1); Chapter 14 (Lasers, Sections 14.1 and 14.2); and Chapter 15 (Photons in Semiconductors, Section 15.1)

Chapter 15 (Photons in Semiconductors, Section 15.2)

Chapter 16 (Semiconductor Photon Sources)

Chapter 17 (Semiconductor Photon Detectors)

Chapter 18 (Electro-Optics)

Chapter 21 (Photonic Switching and Computing, Sections 21.1 to 21.3)

Chapter 22 (Fiber-Optic Communications)

Optical Electronics and Communications

Background: Chapter 1 (Ray Optics); Chapter 2 (Wave Optics); and Chapter 15 (Photons in Semiconductors, Section 15.1)

Chapter 9 (Resonator Optics, Section 9.1)

Chapter 11 (Photon Optics, Sections 11.1 and 11.2)

Chapter 12 (Photons and Atoms)
Chapter 13 (Laser Amplifiers)
Chapter 14 (Lasers, Sections 14.1 and 14.2)
Chapter 15 (Photons in Semiconductors, Section 15.2)
Chapter 16 (Semiconductor Photon Sources)
Chapter 17 (Semiconductor Photon Detectors)
Chapter 22 (Fiber-Optic Communications)

Lightwave Devices

Background: Chapter 5 (Electromagnetic Optics); Chapter 9 (Resonator Optics, Section 9.1); Chapter 11 (Photon Optics, Sections 11.1A and 11.2); Chapter 12 (Photons and Atoms, Sections 12.1 and 12.2); and Chapter 15 (Photons in Semiconductors)
Chapter 6 (Polarization and Crystal Optics)
Chapter 7 (Guided-Wave Optics)
Chapter 8 (Fiber Optics)
Chapter 16 (Semiconductor Photon Sources)
Chapter 17 (Semiconductor Photon Detectors)
Chapter 18 (Electro-Optics)
Chapter 19 (Nonlinear Optics)
Chapter 20 (Acousto-Optics)

Fiber-Optic Communications* or *Lightwave Systems

Background: Chapter 5 (Electromagnetic Optics); Chapter 6 (Polarization and Crystal Optics); Chapter 9 (Resonator Optics, Section 9.1); Chapter 11 (Photon Optics, Sections 11.1A and 11.2); and Chapter 12 (Photons and Atoms, Sections 12.1 and 12.2)
Chapter 7 (Guided-Wave Optics)
Chapter 8 (Fiber Optics)
Chapter 15 (Photons in Semiconductors, Section 15.2)
Chapter 16 (Semiconductor Photon Sources)
Chapter 17 (Semiconductor Photon Detectors)
Chapter 21 (Photonic Switching and Computing, Sections 21.1 to 21.3)
Chapter 22 (Fiber-Optic Communications)

Problems, Reading Lists, and Appendices

A set of problems is provided at the end of each chapter. Problems are numbered in accordance with the chapter sections to which they apply. Quite often, problems deal with ideas or applications not mentioned in the text, analytical derivations, and numerical computations designed to illustrate the magnitudes of important quantities. Problems marked with asterisks are of a more advanced nature. A number of exercises also appear within the text of each chapter to help the reader develop a better understanding of (or to introduce an extension of) the material.

Appendices summarize the properties of one- and two-dimensional Fourier transforms, linear-systems theory, and modes of linear systems (which are important in polarization devices, optical waveguides, and resonators); these are called upon at appropriate points throughout the book. Each chapter ends with a reading list that includes a selection of important books, review articles, and a few classic papers of special significance.

Acknowledgments

We are grateful to many colleagues for reading portions of the text and providing helpful comments: Govind P. Agrawal, David H. Auston, Rasheed Azzam, Nikolai G. Basov, Franco Cerrina, Emmanuel Desurvire, Paul Diament, Eric Fossum, Robert J. Keyes, Robert H. Kingston, Rodney Loudon, Leonard Mandel, Leon McCaughan, Richard M. Osgood, Jan Peřina, Robert H. Rediker, Arthur L. Schawlow, S. R. Seshadri, Henry Stark, Ferrel G. Stremler, John A. Tataronis, Charles H. Townes, Patrick R. Trischitta, Wen I. Wang, and Edward S. Yang.

We are especially indebted to John Whinnery and Emil Wolf for providing us with many suggestions that greatly improved the presentation.

Several colleagues used portions of the notes in their classes and provided us with invaluable feedback. These include Etan Bourkoff at Johns Hopkins University (now at the University of South Carolina), Mark O. Freeman at the University of Colorado, George C. Papen at the University of Illinois, and Paul R. Prucnal at Princeton University.

Many of our students and former students contributed to this material in various ways over the years and we owe them a great debt of thanks: Gaetano L. Aiello, Mohamad Asi, Richard Campos, Buddy Christyono, Andrew H. Cordes, Andrew David, Ernesto Fontenla, Evan Goldstein, Matthew E. Hansen, Dean U. Hekel, Conor Heneghan, Adam Heyman, Bradley M. Jost, David A. Landgraf, Kanghua Lu, Ben Nathanson, Winslow L. Sargeant, Michael T. Schmidt, Raul E. Sequeira, David Small, Kraisin Songwatana, Nikola S. Subotic, Jeffrey A. Tobin, and Emily M. True. Our thanks also go to the legions of unnamed students who, through a combination of vigilance and the desire to understand the material, found countless errors.

We particularly appreciate the many contributions and help of those students who were intimately involved with the preparation of this book at its various stages of completion: Niraj Agrawal, Suzanne Keilson, Todd Larchuk, Guifang Li, and Philip Tham.

We are grateful for the assistance given to us by a number of colleagues in the course of collecting the photographs used at the beginnings of the chapters: E. Scott Barr, Nicolaas Bloembergen, Martin Carey, Marjorie Graham, Margaret Harrison, Ann Kottner, G. Thomas Holmes, John Howard, Theodore H. Maiman, Edward Palik, Martin Parker, Aleksandr M. Prokhorov, Jarus Quinn, Lesley M. Richmond, Claudia Schüler, Patrick R. Trischitta, J. Michael Vaughan, and Emil Wolf. Specific photo credits are as follows: AIP Meggers Gallery of Nobel Laureates (Gabor, Townes, Basov, Prokhorov, W. L. Bragg); AIP Niels Bohr Library (Rayleigh, Frauenhofer, Maxwell, Planck, Bohr, Einstein in Chapter 12, W. H. Bragg); Archives de l'Académie des Sciences de Paris (Fabry); The Astrophysical Journal (Perot); AT & T Bell Laboratories (Shockley, Brattain, Bardeen); Bettmann Archives (Young, Gauss, Tyndall); Bibliothèque Nationale de Paris (Fermat, Fourier, Poisson); Burndy Library (Newton, Huygens); Deutsches Museum (Hertz); ETH Bibliothek (Einstein in Chapter 11); Bruce Fritz (Saleh); Harvard University (Bloembergen); Heidelberg University (Pockels); Kelvin Museum of the University of Glasgow (Kerr); Theodore H. Maiman (Maiman); Princeton University (von Neumann); Smithsonian Institution (Fresnel); Stanford University (Schawlow); Emil Wolf (Born, Wolf). Corning Incorporated kindly provided the photograph used at the beginning of Chapter 8. We are grateful to GE for the use of their logotype, which is a registered trademark of the General Electric Company, at the beginning of Chapter 16. The IBM logo at the beginning of Chapter 16 is being used with special permission from IBM. The right-most logotype at the beginning of Chapter 16 was supplied courtesy of Lincoln Laboratory, Massachusetts Institute of Technology. AT & T Bell Laboratories kindly permitted us use of the diagram at the beginning of Chapter 22.

We greatly appreciate the continued support provided to us by the National Science Foundation, the Center for Telecommunications Research, and the Joint Services Electronics Program through the Columbia Radiation Laboratory.

Finally, we extend our sincere thanks to our editors, George Telecki and Bea Shube, for their guidance and suggestions throughout the course of preparation of this book.

BAHAA E. A. SALEH

Madison, Wisconsin

MALVIN CARL TEICH

New York, New York

April 3, 1991

CONTENTS

CHAPTER 1

RAY OPTICS 1

1.1 Postulates of Ray Optics 3
1.2 Simple Optical Components 6
1.3 Graded-Index Optics 18
1.4 Matrix Optics 26
Reading List 37
Problems 39

CHAPTER 2

WAVE OPTICS 41

2.1 Postulates of Wave Optics 43
2.2 Monochromatic Waves 44
2.3 Relation Between Wave Optics and Ray Optics 52
2.4 Simple Optical Components 53
2.5 Interference 63
2.6 Polychromatic Light 72
Reading List 77
Problems 78

CHAPTER 3

BEAM OPTICS 80

3.1 The Gaussian Beam 81
3.2 Transmission Through Optical Components 92
3.3 Hermite – Gaussian Beams 100
3.4 Laguerre – Gaussian and Bessel Beams 104
Reading List 106
Problems 106

CHAPTER 4
FOURIER OPTICS 108

4.1 Propagation of Light in Free Space 111
4.2 Optical Fourier Transform 121
4.3 Diffraction of Light 127
4.4 Image Formation 135
4.5 Holography 143
Reading List 151
Problems 153

CHAPTER 5
ELECTROMAGNETIC OPTICS 157

5.1 Electromagnetic Theory of Light 159
5.2 Dielectric Media 162
5.3 Monochromatic Electromagnetic Waves 167
5.4 Elementary Electromagnetic Waves 169
5.5 Absorption and Dispersion 174
5.6 Pulse Propagation in Dispersive Media 182
Reading List 191
Problems 191

CHAPTER 6
POLARIZATION AND CRYSTAL OPTICS 193

6.1 Polarization of Light 195
6.2 Reflection and Refraction 203
6.3 Optics of Anisotropic Media 210
6.4 Optical Activity and Faraday Effect 223
6.5 Optics of Liquid Crystals 227
6.6 Polarization Devices 230
Reading List 234
Problems 235

CHAPTER 7
GUIDED-WAVE OPTICS 238

7.1 Planar-Mirror Waveguides 240
7.2 Planar Dielectric Waveguides 248
7.3 Two-Dimensional Waveguides 258
7.4 Optical Coupling in Waveguides 261
Reading List 269
Problems 270

CHAPTER 8

FIBER OPTICS 272

8.1 Step-Index Fibers 274
8.2 Graded-Index Fibers 287
8.3 Attenuation and Dispersion 296
Reading List 306
Problems 307

CHAPTER 9

RESONATOR OPTICS 310

9.1 Planar-Mirror Resonators 312
9.2 Spherical-Mirror Resonators 327
Reading List 339
Problems 340

CHAPTER 10

STATISTICAL OPTICS 342

10.1 Statistical Properties of Random Light 344
10.2 Interference of Partially Coherent Light 360
10.3 Transmission of Partially Coherent Light Through Optical Systems 366
10.4 Partial Polarization 376
Reading List 380
Problems 381

CHAPTER 11

PHOTON OPTICS 384

11.1 The Photon 386
11.2 Photon Streams 398
11.3 Quantum States of Light 411
Reading List 416
Problems 418

CHAPTER 12

PHOTONS AND ATOMS 423

12.1 Atoms, Molecules, and Solids 424
12.2 Interactions of Photons with Atoms 434
12.3 Thermal Light 450
12.4 Luminescence Light 454
Reading List 457
Problems 458

CHAPTER 13
LASER AMPLIFIERS 460

13.1 The Laser Amplifier 463
13.2 Amplifier Power Source 468
13.3 Amplifier Nonlinearity and Gain Saturation 480
13.4 Amplifier Noise 488
Reading List 489
Problems 491

CHAPTER 14
LASERS 494

14.1 Theory of Laser Oscillation 496
14.2 Characteristics of the Laser Output 503
14.3 Pulsed Lasers 522
Reading List 536
Problems 538

CHAPTER 15
PHOTONS IN SEMICONDUCTORS 542

15.1 Semiconductors 544
15.2 Interactions of Photons with Electrons and Holes 573
Reading List 588
Problems 590

CHAPTER 16
SEMICONDUCTOR PHOTON SOURCES 592

16.1 Light-Emitting Diodes 594
16.2 Semiconductor Laser Amplifiers 609
16.3 Semiconductor Injection Lasers 619
Reading List 638
Problems 640

CHAPTER 17
SEMICONDUCTOR PHOTON DETECTORS 644

17.1 Properties of Semiconductor Photodetectors 648
17.2 Photoconductors 654
17.3 Photodiodes 657
17.4 Avalanche Photodiodes 666

17.5 Noise in Photodetectors 673
Reading List 691
Problems 692

CHAPTER 18
ELECTRO-OPTICS 696

18.1 Principles of Electro-Optics 698
18.2 Electro-Optics of Anisotropic Media 712
18.3 Electro-Optics of Liquid Crystals 721
18.4 Photorefractive Materials 729
Reading List 733
Problems 735

CHAPTER 19
NONLINEAR OPTICS 737

19.1 Nonlinear Optical Media 739
19.2 Second-Order Nonlinear Optics 743
19.3 Third-Order Nonlinear Optics 751
19.4 Coupled-Wave Theory of Three-Wave Mixing 762
19.5 Coupled-Wave Theory of Four-Wave Mixing 774
19.6 Anisotropic Nonlinear Media 779
19.7 Dispersive Nonlinear Media 782
19.8 Optical Solitons 786
Reading List 793
Problems 796

CHAPTER 20
ACOUSTO-OPTICS 799

20.1 Interaction of Light and Sound 802
20.2 Acousto-Optic Devices 815
20.3 Acousto-Optics of Anisotropic Media 825
Reading List 830
Problems 830

CHAPTER 21
PHOTONIC SWITCHING AND COMPUTING 832

21.1 Photonic Switches 833
21.2 All-Optical Switches 840
21.3 Bistable Optical Devices 843
21.4 Optical Interconnections 855

21.5 Optical Computing 862
Reading List 870
Problems 872

CHAPTER 22
FIBER-OPTIC COMMUNICATIONS 874

22.1 Components of the Optical Fiber Link 876
22.2 Modulation, Multiplexing, and Coupling 887
22.3 System Performance 893
22.4 Receiver Sensitivity 903
22.5 Coherent Optical Communications 907
Reading List 913
Problems 915

APPENDIX A
FOURIER TRANSFORM 918

A.1 One-Dimensional Fourier Transform 918
A.2 Time Duration and Spectral Width 921
A.3 Two-Dimensional Fourier Transform 924
Reading List 927

APPENDIX B
LINEAR SYSTEMS 928

B.1 One-Dimensional Linear Systems 928
B.2 Two-Dimensional Linear Systems 931

APPENDIX C
MODES OF LINEAR SYSTEMS 934

SYMBOLS 937

INDEX 949

FUNDAMENTALS OF PHOTONICS

CHAPTER

1

RAY OPTICS

1.1 POSTULATES OF RAY OPTICS

1.2 SIMPLE OPTICAL COMPONENTS
- A. Mirrors
- B. Planar Boundaries
- C. Spherical Boundaries and Lenses
- D. Light Guides

1.3 GRADED-INDEX OPTICS
- A. The Ray Equation
- B. Graded-Index Optical Components
- *C. The Eikonal Equation

1.4 MATRIX OPTICS
- A. The Ray-Transfer Matrix
- B. Matrices of Simple Optical Components
- C. Matrices of Cascaded Optical Components
- D. Periodic Optical Systems

Sir Isaac Newton (1642–1727) set forth a theory of optics in which light emissions consist of collections of corpuscles that propagate rectilinearly.

Pierre de Fermat (1601–1665) developed the principle that light travels along the path of least time.

Light is an electromagnetic wave phenomenon described by the same theoretical principles that govern all forms of electromagnetic radiation. Electromagnetic radiation propagates in the form of two mutually coupled *vector* waves, an electric-field wave and a magnetic-field wave. Nevertheless, it is possible to describe many optical phenomena using a *scalar* wave theory in which light is described by a single scalar wavefunction. This approximate way of treating light is called scalar wave optics, or simply **wave optics**.

When light waves propagate through and around objects whose dimensions are much greater than the wavelength, the wave nature of light is not readily discerned, so that its behavior can be adequately described by rays obeying a set of geometrical rules. This model of light is called **ray optics**. Strictly speaking, ray optics is the limit of wave optics when the wavelength is infinitesimally small.

Thus the electromagnetic theory of light (**electromagnetic optics**) encompasses wave optics, which, in turn, encompasses ray optics, as illustrated in Fig. 1.0-1. Ray optics and wave optics provide approximate models of light which derive their validity from their successes in producing results that approximate those based on rigorous electromagnetic theory.

Although electromagnetic optics provides the most complete treatment of light within the confines of **classical optics**, there are certain optical phenomena that are characteristically quantum mechanical in nature and cannot be explained classically. These phenomena are described by a quantum electromagnetic theory known as **quantum electrodynamics**. For optical phenomena, this theory is also referred to as **quantum optics**.

Historically, optical theory developed roughly in the following sequence: (1) ray optics; → (2) wave optics; → (3) electromagnetic optics; → (4) quantum optics. Not

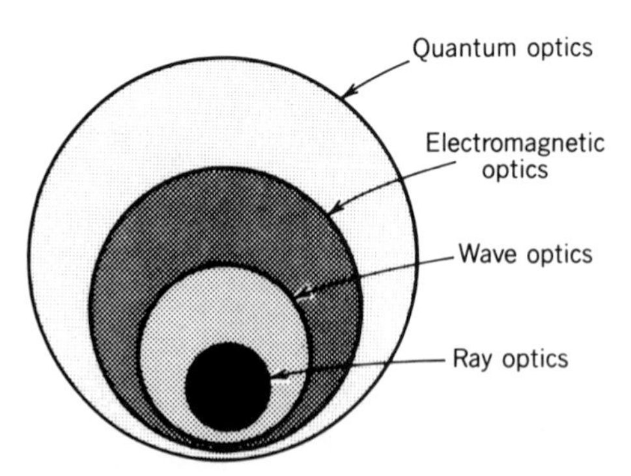

Figure 1.0-1 The theory of quantum optics provides an explanation of virtually all optical phenomena. The electromagnetic theory of light (electromagnetic optics) provides the most complete treatment of light within the confines of classical optics. Wave optics is a scalar approximation of electromagnetic optics. Ray optics is the limit of wave optics when the wavelength is very short.

surprisingly, these models are progressively more difficult and sophisticated, having being developed to provide explanations for the outcomes of successively more complex and precise optical experiments.

For pedagogical reasons, the chapters in this book follow the historical order noted above. Each model of light begins with a set of postulates (provided without proof), from which a large body of results are generated. The postulates of each model are then shown to follow naturally from the next-higher-level model. In this chapter we begin with ray optics.

Ray Optics

Ray optics is the simplest theory of light. Light is described by rays that travel in different optical media in accordance with a set of geometrical rules. Ray optics is therefore also called **geometrical optics**. Ray optics is an approximate theory. Although it adequately describes most of our daily experiences with light, there are many phenomena that ray optics does not adequately describe (as amply attested to by the remaining chapters of this book).

Ray optics is concerned with the *location* and *direction* of light rays. It is therefore useful in studying *image formation*—the collection of rays from each point of an object and their redirection by an optical component onto a corresponding point of an image. Ray optics permits us to determine conditions under which light is guided within a given medium, such as a glass fiber. In isotropic media, optical rays point in the direction of the flow of *optical energy*. Ray bundles can be constructed in which the density of rays is proportional to the density of light energy. When light is generated isotropically from a point source, for example, the energy associated with the rays in a given cone is proportional to the solid angle of the cone. Rays may be traced through an optical system to determine the optical energy crossing a given area.

This chapter begins with a set of postulates from which the simple rules that govern the propagation of light rays through optical media are derived. In Sec. 1.2 these rules are applied to simple optical components such as mirrors and planar or spherical boundaries between different optical media. Ray propagation in inhomogeneous (graded-index) optical media is examined in Sec. 1.3. Graded-index optics is the basis of a technology that has become an important part of modern optics.

Optical components are often centered about an optical axis, around which the rays travel at small inclinations. Such rays are called **paraxial rays**. This assumption is the basis of **paraxial optics**. The change in the position and inclination of a paraxial ray as it travels through an optical system can be efficiently described by the use of a 2×2-matrix algebra. Section 1.4 is devoted to this algebraic tool, called **matrix optics**.

1.1 POSTULATES OF RAY OPTICS

Postulates of Ray Optics

- Light travels in the form of rays. The rays are emitted by light sources and can be observed when they reach an optical detector.
- An optical medium is characterized by a quantity $n \geq 1$, called the **refractive index**. The refractive index is the ratio of the speed of light in free space c_o to that in the medium c. Therefore, the time taken by light to travel a distance d equals $d/c = nd/c_o$. It is thus proportional to the product nd, known as the **optical path length**.

- In an inhomogeneous medium the refractive index $n(\mathbf{r})$ is a function of the position $\mathbf{r} = (x, y, z)$. The optical path length along a given path between two points A and B is therefore

$$\text{Optical path length} = \int_A^B n(\mathbf{r})\, ds,$$

where ds is the differential element of length along the path. The time taken by light to travel from A to B is proportional to the optical path length.

- **Fermat's Principle**. Optical rays traveling between two points, A and B, follow a path such that the time of travel (or the optical path length) between the two points is an extremum relative to neighboring paths. An extremum means that the rate of change is zero, i.e.,

$$\delta \int_A^B n(\mathbf{r})\, ds = 0.$$

The extremum may be a minimum, a maximum, or a point of inflection. It is, however, usually a minimum, in which case

light rays travel along the path of least time.

Sometimes the minimum time is shared by more than one path, which are then all followed simultaneously by the rays.

In this chapter we use the postulates of ray optics to determine the rules governing the propagation of light rays, their reflection and refraction at the boundaries between different media, and their transmission through various optical components. A wealth of results applicable to numerous optical systems are obtained without the need for any other assumptions or rules regarding the nature of light.

Propagation in a Homogeneous Medium

In a homogeneous medium the refractive index is the same everywhere, and so is the speed of light. The path of minimum time, required by Fermat's principle, is therefore also the path of minimum distance. The principle of the *path of minimum distance* is known as **Hero's principle**. The path of minimum distance between two points is a straight line so that *in a homogeneous medium, light rays travel in straight lines* (Fig. 1.1-1).

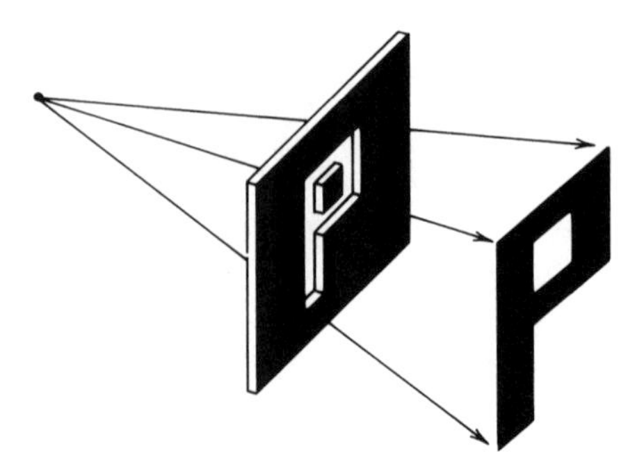

Figure 1.1-1 Light rays travel in straight lines. Shadows are perfect projections of stops.

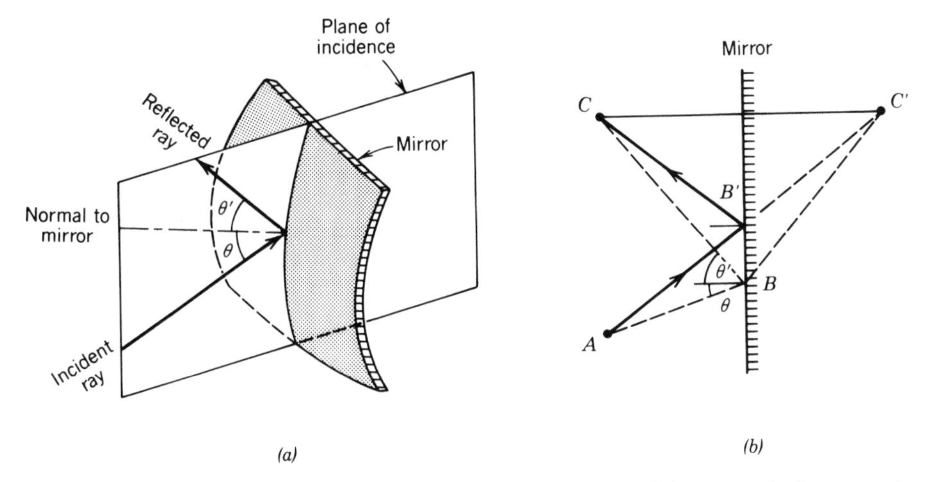

Figure 1.1-2 (*a*) Reflection from the surface of a curved mirror. (*b*) Geometrical construction to prove the law of reflection.

Reflection from a Mirror

Mirrors are made of certain highly polished metallic surfaces, or metallic or dielectric films deposited on a substrate such as glass. Light reflects from mirrors in accordance with the law of reflection:

The reflected ray lies in the plane of incidence;
the angle of reflection equals the angle of incidence.

The plane of incidence is the plane formed by the incident ray and the normal to the mirror at the point of incidence. The angles of incidence and reflection, θ and θ', are defined in Fig. 1.1-2(*a*). To prove the law of reflection we simply use Hero's principle. Examine a ray that travels from point A to point C after reflection from the planar mirror in Fig. 1.1-2(*b*). According to Hero's principle the distance $\overline{AB} + \overline{BC}$ must be minimum. If C' is a mirror image of C, then $\overline{BC} = \overline{BC'}$, so that $\overline{AB} + \overline{BC'}$ must be a minimum. This occurs when $\overline{ABC'}$ is a straight line, i.e., when B coincides with B' and $\theta = \theta'$.

Reflection and Refraction at the Boundary Between Two Media

At the boundary between two media of refractive indices n_1 and n_2 an incident ray is split into two—a reflected ray and a refracted (or transmitted) ray (Fig. 1.1-3). The

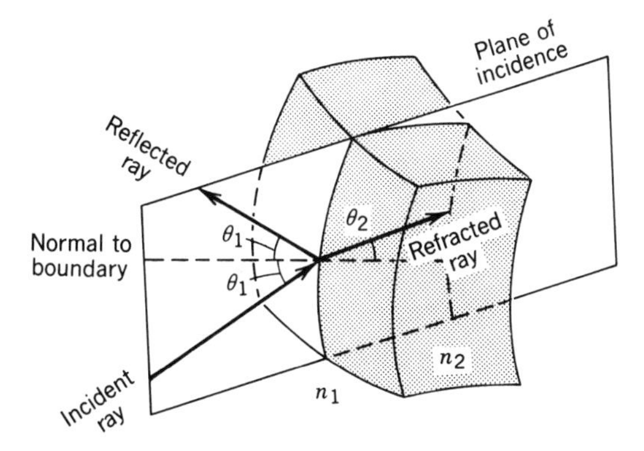

Figure 1.1-3 Reflection and refraction at the boundary between two media.

reflected ray obeys the law of reflection. The refracted ray obeys the law of refraction:

The refracted ray lies in the plane of incidence; the angle of refraction θ_2 is related to the angle of incidence θ_1 by Snell's law,

$$n_1 \sin \theta_1 = n_2 \sin \theta_2. \tag{1.1-1}$$

Snell's Law

EXERCISE 1.1-1

Proof of Snell's Law. The proof of Snell's law is an exercise in the application of Fermat's principle. Referring to Fig. 1.1-4, we seek to minimize the optical path length $n_1\overline{AB} + n_2\overline{BC}$ between points A and C. We therefore have the following optimization problem: Find θ_1 and θ_2 that minimize $n_1 d_1 \sec \theta_1 + n_2 d_2 \sec \theta_2$, subject to the condition $d_1 \tan \theta_1 + d_2 \tan \theta_2 = d$. Show that the solution of this constrained minimization problem yields Snell's law.

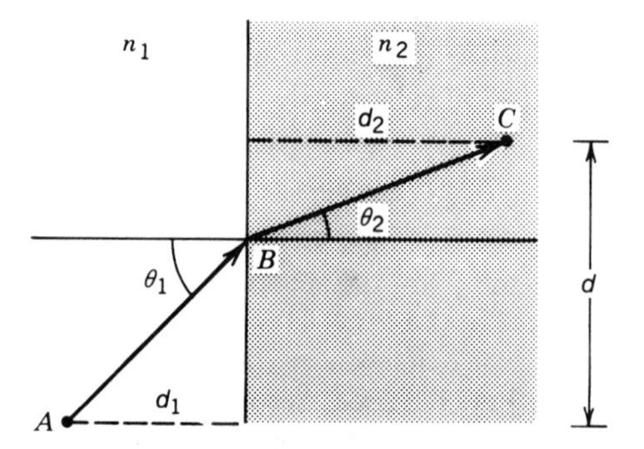

Figure 1.1-4 Construction to prove Snell's law.

The three simple rules—propagation in straight lines and the laws of reflection and refraction—are applied in Sec. 1.2 to several geometrical configurations of mirrors and transparent optical components, without further recourse to Fermat's principle.

1.2 SIMPLE OPTICAL COMPONENTS

A. Mirrors

Planar Mirrors

A planar mirror reflects the rays originating from a point P_1 such that the reflected rays appear to originate from a point P_2 behind the mirror, called the image (Fig. 1.2-1).

Paraboloidal Mirrors

The surface of a paraboloidal mirror is a paraboloid of revolution. It has the useful property of focusing all incident rays parallel to its axis to a single point called the **focus**. The distance $\overline{PF} = f$ defined in Fig. 1.2-2 is called the **focal length**. Paraboloidal

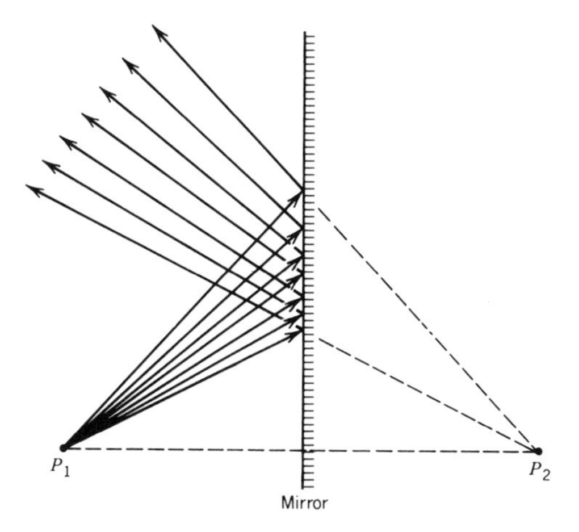

Figure 1.2-1 Reflection from a planar mirror.

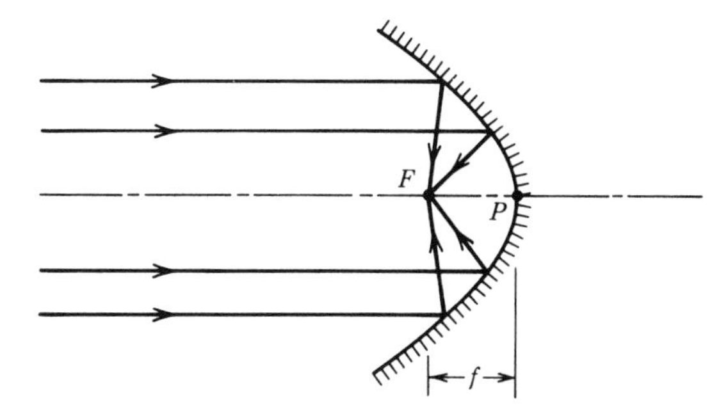

Figure 1.2-2 Focusing of light by a paraboloidal mirror.

mirrors are often used as light-collecting elements in telescopes. They are also used for making parallel beams of light from point sources such as in flashlights.

Elliptical Mirrors

An elliptical mirror reflects all the rays emitted from one of its two foci, e.g., P_1, and images them onto the other focus, P_2 (Fig. 1.2-3). The distances traveled by the light from P_1 to P_2 along any of the paths are all equal, in accordance with Hero's principle.

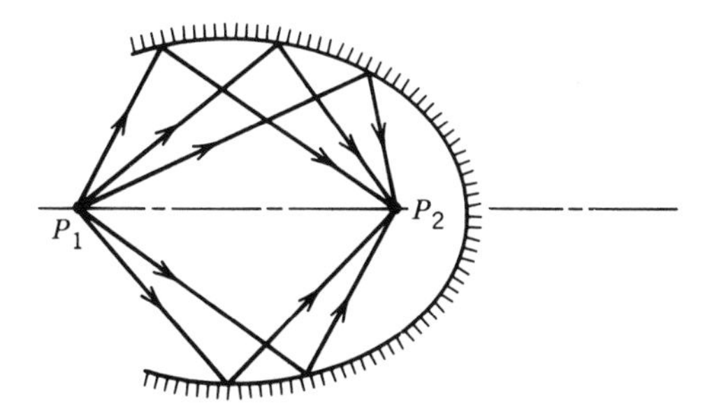

Figure 1.2-3 Reflection from an elliptical mirror.

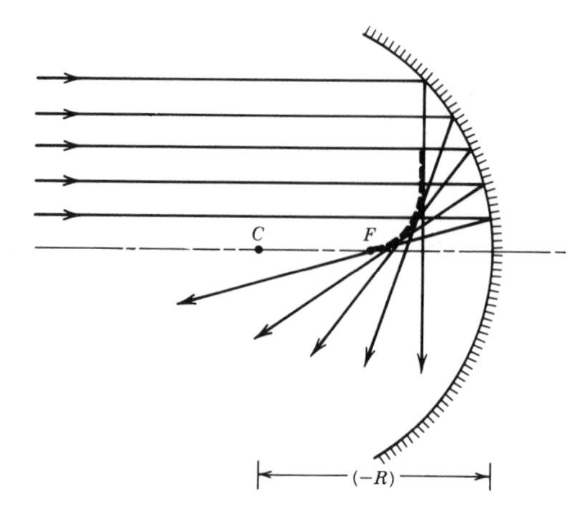

Figure 1.2-4 Reflection of parallel rays from a concave spherical mirror.

Spherical Mirrors

A spherical mirror is easier to fabricate than a paraboloidal or an elliptical mirror. However, it has neither the focusing property of the paraboloidal mirror nor the imaging property of the elliptical mirror. As illustrated in Fig. 1.2-4, parallel rays meet the axis at different points; their envelope (the dashed curve) is called the caustic curve. Nevertheless, parallel rays close to the axis are approximately focused onto a single point F at distance $(-R)/2$ from the mirror center C. By convention, R is negative for concave mirrors and positive for convex mirrors.

Paraxial Rays Reflected from Spherical Mirrors

Rays that make small angles (such that $\sin\theta \approx \theta$) with the mirror's axis are called **paraxial rays**. In the **paraxial approximation**, where only paraxial rays are considered, a spherical mirror has a focusing property like that of the paraboloidal mirror *and* an imaging property like that of the elliptical mirror. The body of rules that results from this approximation forms **paraxial optics**, also called first-order optics or Gaussian optics.

A spherical mirror of radius R therefore acts like a paraboloidal mirror of focal length $f = R/2$. This is in fact plausible since at points near the axis, a parabola can be approximated by a circle with radius equal to the parabola's radius of curvature (Fig. 1.2-5).

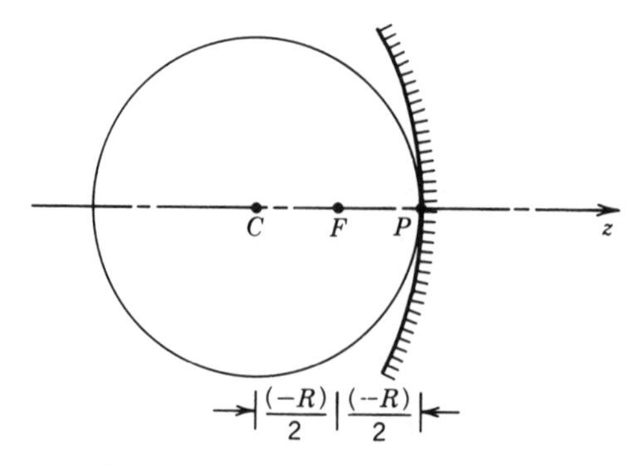

Figure 1.2-5 A spherical mirror approximates a paraboloidal mirror for paraxial rays.

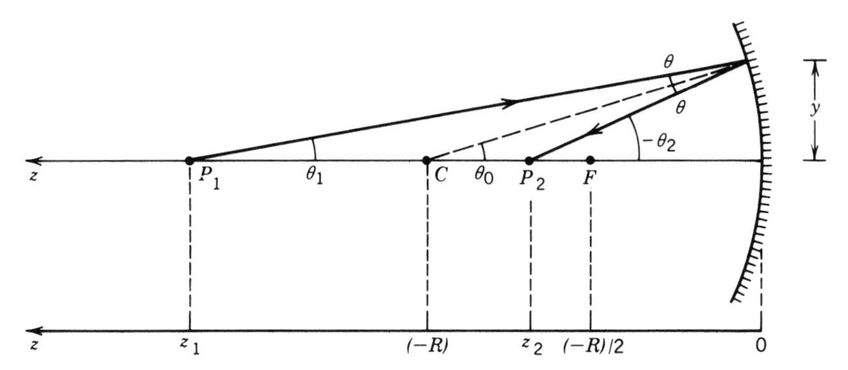

Figure 1.2-6 Reflection of paraxial rays from a concave spherical mirror of radius $R < 0$.

All paraxial rays originating from each point on the axis of a spherical mirror are reflected and focused onto a single corresponding point on the axis. This can be seen (Fig. 1.2-6) by examining a ray emitted at an angle θ_1 from a point P_1 at a distance z_1 away from a concave mirror of radius R, and reflecting at angle $(-\theta_2)$ to meet the axis at a point P_2 a distance z_2 away from the mirror. The angle θ_2 is negative since the ray is traveling downward. Since $\theta_1 = \theta_0 - \theta$ and $(-\theta_2) = \theta_0 + \theta$, it follows that $(-\theta_2) + \theta_1 = 2\theta_0$. If θ_0 is sufficiently small, the approximation $\tan\theta_0 \approx \theta_0$ may be used, so that $\theta_0 \approx y/(-R)$, from which

$$(-\theta_2) + \theta_1 \approx \frac{2y}{-R}, \tag{1.2-1}$$

where y is the height of the point at which the reflection occurs. Recall that R is negative since the mirror is concave. Similarly, if θ_1 and θ_2 are small, $\theta_1 \approx y/z_1$, $(-\theta_2) \approx y/z_2$, and (1.2-1) yields $y/z_1 + y/z_2 \approx 2y/(-R)$, from which

$$\frac{1}{z_1} + \frac{1}{z_2} \approx \frac{2}{-R}. \tag{1.2-2}$$

This relation hold regardless of y (i.e., regardless of θ_1) as long as the approximation is valid. This means that all paraxial rays originating at point P_1 arrive at P_2. The distances z_1 and z_2 are measured in a coordinate system in which the z axis points to the left. Points of negative z therefore lie to the right of the mirror.

According to (1.2-2), rays that are emitted from a point very far out on the z axis $(z_1 = \infty)$ are focused to a point F at a distance $z_2 = (-R)/2$. This means that within the paraxial approximation, all rays coming from infinity (parallel to the mirror's axis) are focused to a point at a distance

$$\boxed{f = \frac{-R}{2},} \tag{1.2-3}$$

Focal Length of a Spherical Mirror

which is called the mirror's focal length. Equation (1.2-2) is usually written in the form

$$\frac{1}{z_1} + \frac{1}{z_2} = \frac{1}{f}, \qquad (1.2\text{-}4)$$

Imaging Equation (Paraxial Rays)

known as the imaging equation. Both the incident and the reflected rays must be paraxial for this equation to be valid.

EXERCISE 1.2-1

Image Formation by a Spherical Mirror. Show that within the paraxial approximation, rays originating from a point $P_1 = (y_1, z_1)$ are reflected to a point $P_2 = (y_2, z_2)$, where z_1 and z_2 satisfy (1.2-4) and $y_2 = -y_1 z_2 / z_1$ (Fig. 1.2-7). This means that rays from each point in the plane $z = z_1$ meet at a single corresponding point in the plane $z = z_2$, so that the mirror acts as an image-forming system with magnification $-z_2/z_1$. Negative magnification means that the image is inverted.

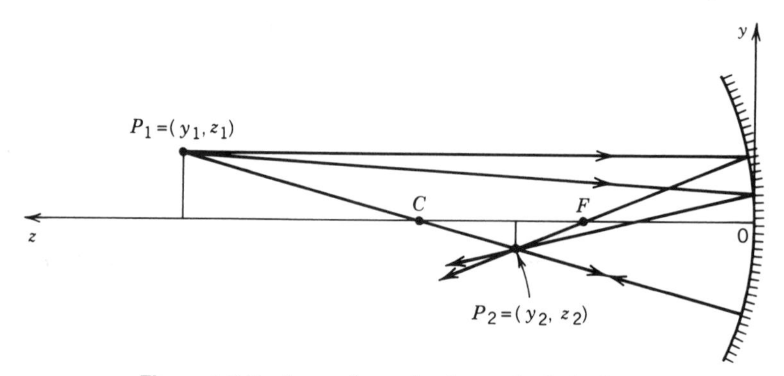

Figure 1.2-7 Image formation by a spherical mirror.

B. Planar Boundaries

The relation between the angles of refraction and incidence, θ_2 and θ_1, at a planar boundary between two media of refractive indices n_1 and n_2 is governed by Snell's law (1.1-1). This relation is plotted in Fig. 1.2-8 for two cases:

- *External Refraction* ($n_1 < n_2$). When the ray is incident from the medium of smaller refractive index, $\theta_2 < \theta_1$ and the refracted ray bends away from the boundary.
- *Internal Refraction* ($n_1 > n_2$). If the incident ray is in a medium of higher refractive index, $\theta_2 > \theta_1$ and the refracted ray bends toward the boundary.

In both cases, when the angles are small (i.e., the rays are paraxial), the relation between θ_2 and θ_1 is approximately linear, $n_1\theta_1 \approx n_2\theta_2$, or $\theta_2 \approx (n_1/n_2)\theta_1$.

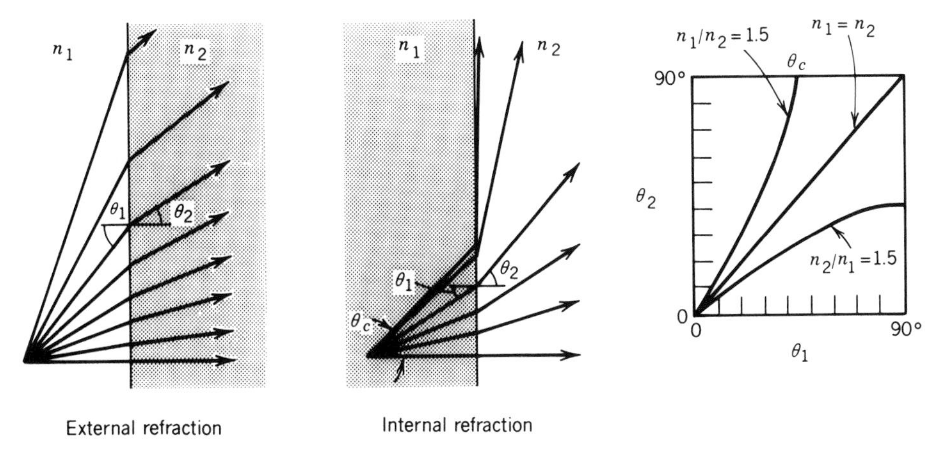

Figure 1.2-8 Relation between the angles of refraction and incidence.

Total Internal Reflection

For internal refraction ($n_1 > n_2$), the angle of refraction is greater than the angle of incidence, $\theta_2 > \theta_1$, so that as θ_1 increases, θ_2 reaches 90° first (see Fig. 1.2-8). This occurs when $\theta_1 = \theta_c$ (the **critical angle**), with $n_1 \sin\theta_c = n_2$, so that

$$\theta_c = \sin^{-1}\frac{n_2}{n_1}. \tag{1.2-5}$$

Critical Angle

When $\theta_1 > \theta_c$, Snell's law (1.1-1) cannot be satisfied and refraction does not occur. The incident ray is totally reflected as if the surface were a perfect mirror [Fig. 1.2-9(*a*)]. The phenomenon of total internal reflection is the basis of many optical devices and systems, such as reflecting prisms [see Fig. 1.2-9(*b*)] and optical fibers (see Sec. 1.2D).

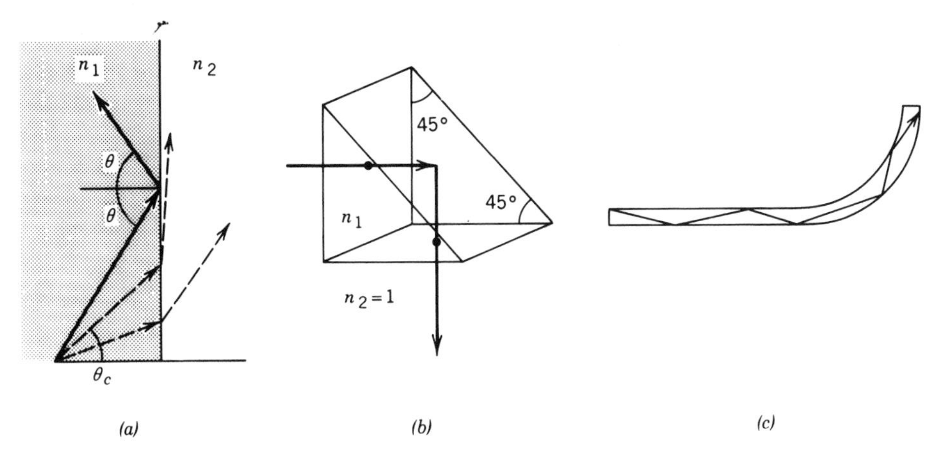

Figure 1.2-9 (*a*) Total internal reflection at a planar boundary. (*b*) The reflecting prism. If $n_1 > \sqrt{2}$ and $n_2 = 1$ (air), then $\theta_c < 45°$; since $\theta_1 = 45°$, the ray is totally reflected. (*c*) Rays are guided by total internal reflection from the internal surface of an optical fiber.

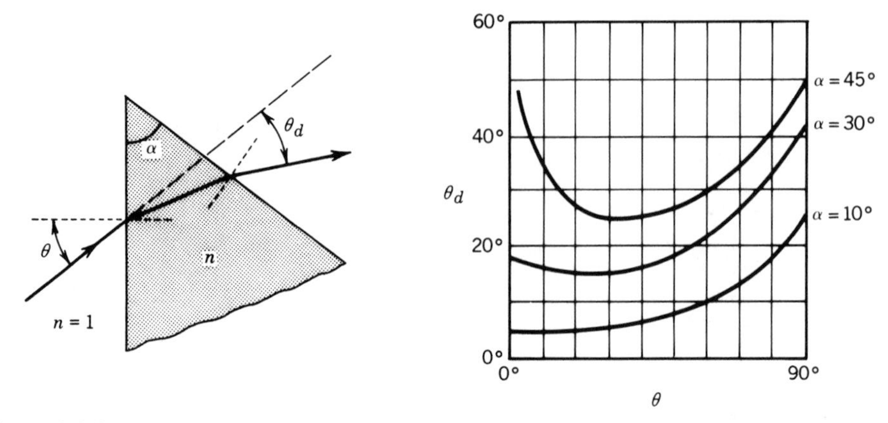

Figure 1.2-10 Ray deflection by a prism. The angle of deflection θ_d as a function of the angle of incidence θ for different apex angles α when $n = 1.5$. When both α and θ are small $\theta_d \approx (n - 1)\alpha$, which is approximately independent of θ. When $\alpha = 45°$ and $\theta = 0°$, total internal reflection occurs, as illustrated in Fig. 1.2-9(*b*).

Prisms

A prism of apex angle α and refractive index n (Fig. 1.2-10) deflects a ray incident at an angle θ by an angle

$$\theta_d = \theta - \alpha + \sin^{-1}\left[(n^2 - \sin^2\theta)^{1/2}\sin\alpha - \sin\theta\cos\alpha\right]. \tag{1.2-6}$$

This may be shown by using Snell's law twice at the two refracting surfaces of the prism. When α is very small (thin prism) and θ is also very small (paraxial approximation), (1.2-6) is approximated by

$$\theta_d \approx (n - 1)\alpha. \tag{1.2-7}$$

Beamsplitters

The beamsplitter is an optical component that splits the incident light beam into a reflected beam and a transmitted beam, as illustrated in Fig. 1.2-11. Beamsplitters are also frequently used to combine two light beams into one [Fig. 1.2-11(*c*)]. Beamsplitters are often constructed by depositing a thin semitransparent metallic or dielectric film on a glass substrate. A thin glass plate or a prism can also serve as a beamsplitter.

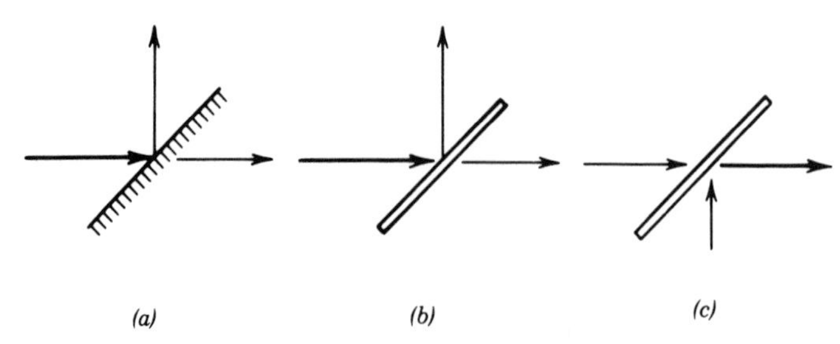

Figure 1.2-11 Beamsplitters and combiners: (*a*) partially reflective mirror; (*b*) thin glass plate; (*c*) beam combiner.

C. Spherical Boundaries and Lenses

We now examine the refraction of rays from a spherical boundary of radius R between two media of refractive indices n_1 and n_2. By convention, R is positive for a convex boundary and negative for a concave boundary. By using Snell's law, and considering only paraxial rays making small angles with the axis of the system so that $\tan\theta \approx \theta$, the following properties may be shown to hold:

- A ray making an angle θ_1 with the z axis and meeting the boundary at a point of height y [see Fig. 1.2-12(*a*)] refracts and changes direction so that the refracted ray makes an angle θ_2 with the z axis,

$$\theta_2 \approx \frac{n_1}{n_2}\theta_1 - \frac{n_2 - n_1}{n_2 R} y. \tag{1.2-8}$$

- All paraxial rays originating from a point $P_1 = (y_1, z_1)$ in the $z = z_1$ plane meet at a point $P_2 = (y_2, z_2)$ in the $z = z_2$ plane, where

$$\frac{n_1}{z_1} + \frac{n_2}{z_2} \approx \frac{n_2 - n_1}{R} \tag{1.2-9}$$

and

$$y_2 = -\frac{n_1}{n_2}\frac{z_2}{z_1} y_1. \tag{1.2-10}$$

The $z = z_1$ and $z = z_2$ planes are said to be conjugate planes. Every point in the first plane has a corresponding point (image) in the second with magnification

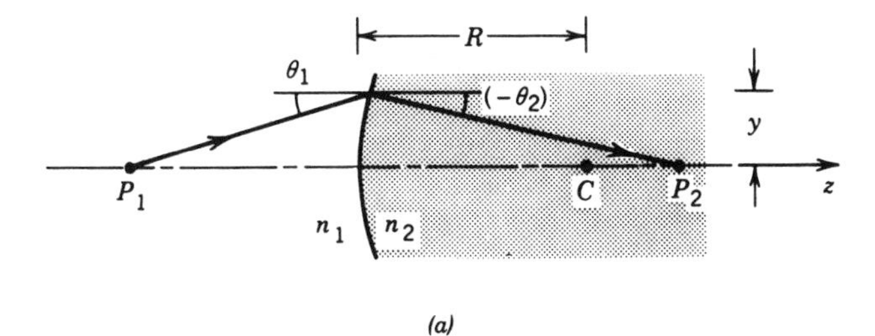

(*a*)

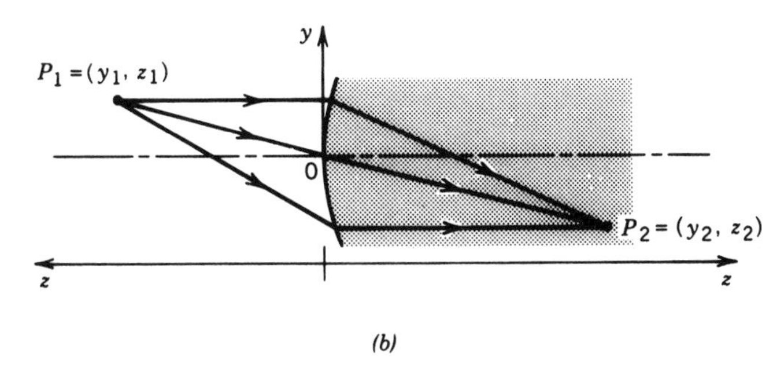

(*b*)

Figure 1.2-12 Refraction at a convex spherical boundary ($R > 0$).

$-(n_1/n_2)(z_2/z_1)$. Again, negative magnification means that the image is inverted. By convention P_1 is measured in a coordinate system pointing to the left and P_2 in a coordinate system pointing to the right (e.g., if P_2 lies to the left of the boundary, then z_2 would be negative).

The similarities between these properties and those of the spherical mirror are evident. It is important to remember that the image formation properties described above are approximate. They hold only for paraxial rays. Rays of large angles do not obey these paraxial laws; the deviation results in image distortion called **aberration**.

EXERCISE 1.2-2

Image Formation. Derive (1.2-8). Prove that paraxial rays originating from P_1 pass through P_2 when (1.2-9) and (1.2-10) are satisfied.

EXERCISE 1.2-3

Aberration-Free Imaging Surface. Determine the equation of a convex aspherical (nonspherical) surface between media of refractive indices n_1 and n_2 such that all rays (not necessarily paraxial) from an axial point P_1 at a distance z_1 to the left of the surface are imaged onto an axial point P_2 at a distance z_2 to the right of the surface [Fig. 1.2-12(a)]. *Hint:* In accordance with Fermat's principle the optical path lengths between the two points must be equal for all paths.

Lenses

A spherical lens is bounded by two spherical surfaces. It is, therefore, defined completely by the radii R_1 and R_2 of its two surfaces, its thickness Δ, and the refractive index n of the material (Fig. 1.2-13). A glass lens in air can be regarded as a combination of two spherical boundaries, air-to-glass and glass-to-air.

A ray crossing the first surface at height y and angle θ_1 with the z axis [Fig. 1.2-14(a)] is traced by applying (1.2-8) at the first surface to obtain the inclination angle θ of the refracted ray, which we extend until it meets the second surface. We then use (1.2-8) once more with θ replacing θ_1 to obtain the inclination angle θ_2 of the ray after refraction from the second surface. The results are in general complicated. When the lens is thin, however, it can be assumed that the incident ray emerges from the lens at

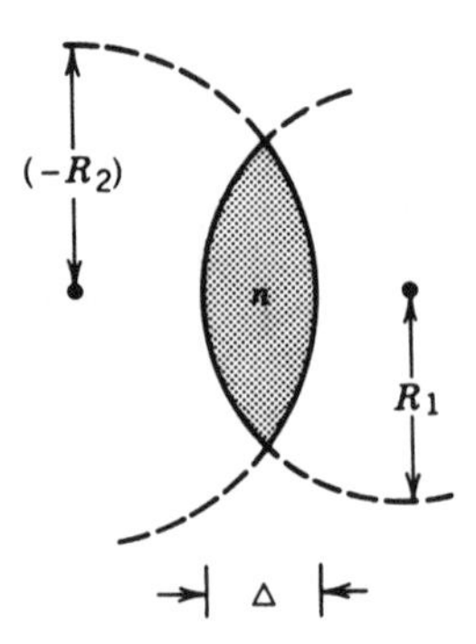

Figure 1.2-13 A biconvex spherical lens.

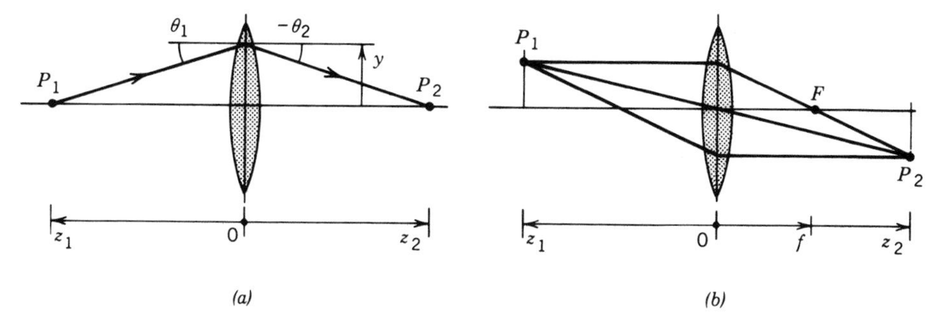

Figure 1.2-14 (*a*) Ray bending by a thin lens. (*b*) Image formation by a thin lens.

about the same height y at which it enters. Under this assumption, the following relations follow:

- The angles of the refracted and incident rays are related by

$$\theta_2 = \theta_1 - \frac{y}{f}, \tag{1.2-11}$$

where f, called the **focal length**, is given by

$$\boxed{\frac{1}{f} = (n-1)\left(\frac{1}{R_1} - \frac{1}{R_2}\right).} \tag{1.2-12}$$

Focal Length of a Thin Spherical Lens

- All rays originating from a point $P_1 = (y_1, z_1)$ meet at a point $P_2 = (y_2, z_2)$ [Fig. 1.2-14(*b*)], where

$$\boxed{\frac{1}{z_1} + \frac{1}{z_2} = \frac{1}{f}} \tag{1.2-13}$$

Imaging Equation

and

$$\boxed{y_2 = -\frac{z_2}{z_1} y_1.} \tag{1.2-14}$$

Magnification

This means that each point in the $z = z_1$ plane is imaged onto a corresponding point in the $z = z_2$ plane with the magnification factor $-z_2/z_1$. The focal length f of a lens therefore completely determines its effect on paraxial rays.

As indicated earlier, P_1 and P_2 are measured in coordinate systems pointing to the left and right, respectively, and the radii of curvatures R_1 and R_2 are positive for convex surfaces and negative for concave surfaces. For the biconvex lens shown in Fig. 1.2-13, R_1 is positive and R_2 is negative, so that the two terms of (1.2-12) add and provide a positive f.

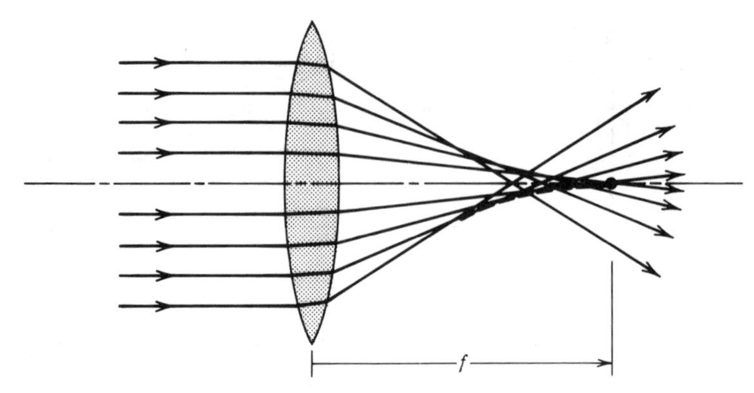

Figure 1.2-15 Nonparaxial rays do not meet at the paraxial focus. The dashed envelope of the refracted rays is called the caustic curve.

EXERCISE 1.2-4

Proof of the Thin Lens Formulas. Using (1.2-8), prove (1.2-11), (1.2-12), and (1.2-13).

It is emphasized once more that the foregoing relations hold only for paraxial rays. The deviations of nonparaxial rays from these relations result in aberrations, as illustrated in Fig. 1.2-15.

D. Light Guides

Light may be guided from one location to another by use of a set of lenses or mirrors, as illustrated schematically in Fig. 1.2-16. Since refractive elements (such as lenses) are usually partially reflective and since mirrors are partially absorptive, the cumulative loss of optical power will be significant when the number of guiding elements is large. Components in which these effects are minimized can be fabricated (e.g., antireflection coated lenses), but the system is generally cumbersome and costly.

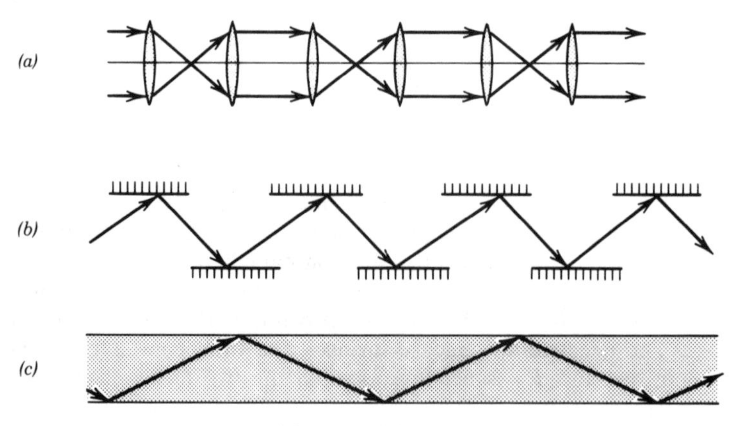

Figure 1.2-16 Guiding light: (a) lenses; (b) mirrors; (c) total internal reflection.

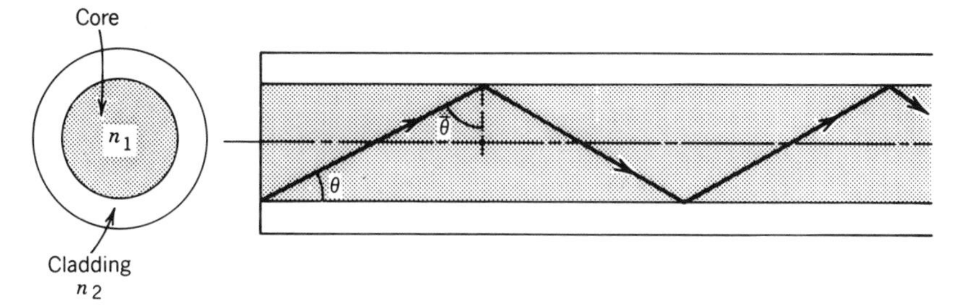

Figure 1.2-17 The optical fiber. Light rays are guided by multiple total internal reflections.

An ideal mechanism for guiding light is that of total internal reflection at the boundary between two media of different refractive indices. Rays are reflected repeatedly without undergoing refraction. Glass fibers of high chemical purity are used to guide light for tens of kilometers with relatively low loss of optical power.

An optical fiber is a light conduit made of two concentric glass (or plastic) cylinders (Fig. 1.2-17). The inner, called the core, has a refractive index n_1, and the outer, called the cladding, has a slightly smaller refractive index, $n_2 < n_1$. Light rays traveling in the core are totally reflected from the cladding if their angle of incidence is greater than the critical angle, $\bar{\theta} > \theta_c = \sin^{-1}(n_2/n_1)$. The rays making an angle $\theta = 90° - \bar{\theta}$ with the optical axis are therefore confined in the fiber core if $\theta < \bar{\theta}_c$, where $\bar{\theta}_c = 90° - \theta_c = \cos^{-1}(n_2/n_1)$. Optical fibers are used in optical communication systems (see Chaps. 8 and 22). Some important properties of optical fibers are derived in Exercise 1.2-5.

Trapping of Light in Media of High Refractive Index

It is often difficult for light originating inside a medium of large refractive index to be extracted into air, especially if the surfaces of the medium are parallel. This occurs since certain rays undergo multiple total internal reflections without ever refracting into air. The principle is illustrated in Exercise 1.2-6.

EXERCISE 1.2-5

Numerical Aperture and Angle of Acceptance of an Optical Fiber. An optical fiber is illuminated by light from a source (e.g., a light-emitting diode, LED). The refractive indices of the core and cladding of the fiber are n_1 and n_2, respectively, and the refractive index of air is 1 (Fig. 1.2-18). Show that the angle θ_a of the cone of rays accepted by the

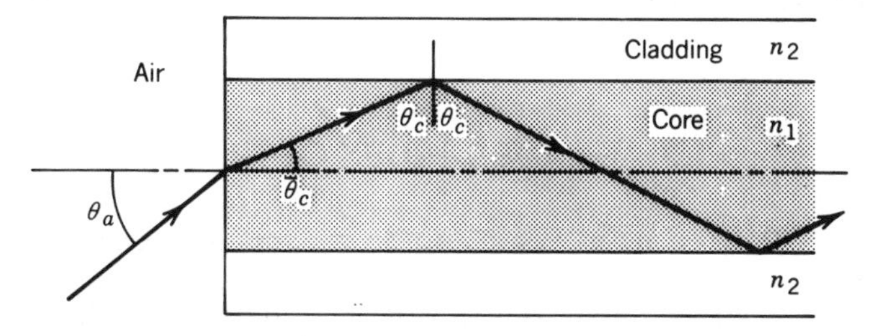

Figure 1.2-18 Acceptance angle of an optical fiber.

fiber (transmitted through the fiber without undergoing refraction at the cladding) is given by

$$\text{NA} = \sin\theta_a = \left(n_1^2 - n_2^2\right)^{1/2}. \tag{1.2-15}$$

Numerical Aperture of an Optical Fiber

The parameter $\text{NA} = \sin\theta_a$ is known as the **numerical aperture** of the fiber. Calculate the numerical aperture and acceptance angle for a silica glass fiber with $n_1 = 1.475$ and $n_2 = 1.460$.

EXERCISE 1.2-6

Light Trapped in a Light-Emitting Diode

(a) Assume that light is generated in all directions inside a material of refractive index n cut in the shape of a parallelepiped (Fig. 1.2-19). The material is surrounded by air with refractive index 1. This process occurs in light-emitting diodes (see Chap. 16). What is the angle of the cone of light rays (inside the material) that will emerge from each face? What happens to the other rays? What is the numerical value of this angle for GaAs ($n = 3.6$)?

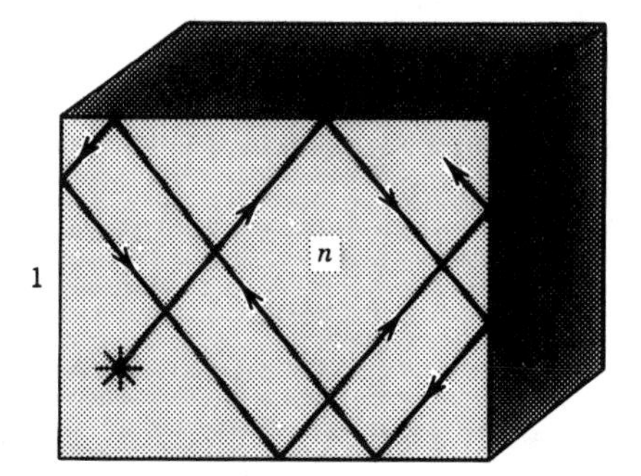

Figure 1.2-19 Trapping of light in a parallelepiped of high refractive index.

(b) Assume that when light is generated isotropically the amount of optical power associated with the rays in a given cone is proportional to the solid angle of the cone. Show that the ratio of the optical power that is extracted from the material to the total generated optical power is $3[1 - (1 - 1/n^2)^{1/2}]$, provided that $n > \sqrt{2}$. What is the numerical value of this ratio for GaAs?

1.3 GRADED-INDEX OPTICS

A graded-index (GRIN) material has a refractive index that varies with position in accordance with a continuous function $n(\mathbf{r})$. These materials are often fabricated by adding impurities (dopants) of controlled concentrations. In a GRIN medium the optical rays follow curved trajectories, instead of straight lines. By appropriate choice of $n(\mathbf{r})$, a GRIN plate can have the same effect on light rays as a conventional optical component, such as a prism or a lens.

A. The Ray Equation

To determine the trajectories of light rays in an inhomogeneous medium with refractive index $n(\mathbf{r})$, we use Fermat's principle,

$$\delta \int_A^B n(\mathbf{r})\, ds = 0,$$

where ds is a differential length along the ray trajectory between A and B. If the trajectory is described by the functions $x(s)$, $y(s)$, and $z(s)$, where s is the length of the trajectory (Fig. 1.3-1), then using the calculus of variations it can be shown† that $x(s)$, $y(s)$, and $z(s)$ must satisfy three partial differential equations,

$$\frac{d}{ds}\left(n\frac{dx}{ds}\right) = \frac{\partial n}{\partial x}, \qquad \frac{d}{ds}\left(n\frac{dy}{ds}\right) = \frac{\partial n}{\partial y}, \qquad \frac{d}{ds}\left(n\frac{dz}{ds}\right) = \frac{\partial n}{\partial z}. \tag{1.3-1}$$

By defining the vector $\mathbf{r}(s)$, whose components are $x(s)$, $y(s)$, and $z(s)$, (1.3-1) may be written in the compact vector form

$$\boxed{\frac{d}{ds}\left(n\frac{d\mathbf{r}}{ds}\right) = \nabla n,} \tag{1.3-2}$$

Ray Equation

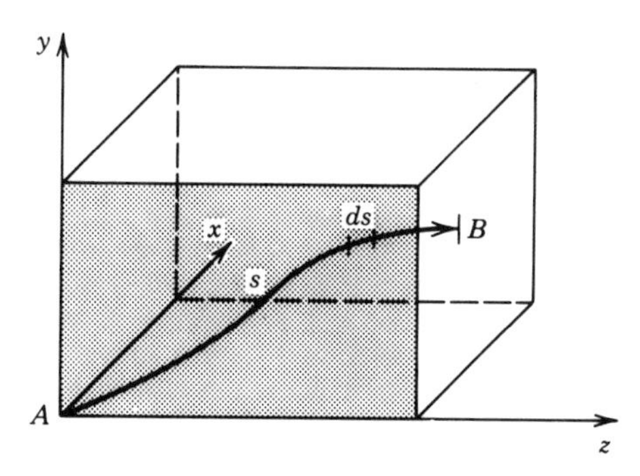

Figure 1.3-1 The ray trajectory is described parametrically by three functions $x(s)$, $y(s)$, and $z(s)$, or by two functions $x(z)$ and $y(z)$.

---

†This derivation is beyond the scope of this book; see, e.g., R. Weinstock, *Calculus of Variation*, Dover, New York, 1974.

Figure 1.3-2 Trajectory of a paraxial ray in a graded-index medium.

where ∇n, the gradient of n, is a vector with Cartesian components $\partial n/\partial x$, $\partial n/\partial y$, and $\partial n/\partial z$. Equation (1.3-2) is known as the **ray equation**.

One approach to solving the ray equation is to describe the trajectory by two functions $x(z)$ and $y(z)$, write $ds = dz[1 + (dx/dz)^2 + (dy/dz)^2]^{1/2}$, and substitute in (1.3-2) to obtain two partial differential equations for $x(z)$ and $y(z)$. The algebra is generally not trivial, but it simplifies considerably when the paraxial approximation is used.

The Paraxial Ray Equation

In the paraxial approximation, the trajectory is almost parallel to the z axis, so that $ds \approx dz$ (Fig. 1.3-2). The ray equations (1.3-1) then simplify to

$$\frac{d}{dz}\left(n\frac{dx}{dz}\right) \approx \frac{\partial n}{\partial x}, \qquad \frac{d}{dz}\left(n\frac{dy}{dz}\right) \approx \frac{\partial n}{\partial y}. \tag{1.3-3}$$

Paraxial Ray Equations

Given $n = n(x, y, z)$, these two partial differential equations may be solved for the trajectory $x(z)$ and $y(z)$.

In the limiting case of a homogeneous medium for which n is independent of x, y, z, (1.3-3) gives $d^2x/d^2z = 0$ and $d^2y/d^2z = 0$, from which it follows that x and y are linear functions of z, so that the trajectories are straight lines. More interesting cases will be examined subsequently.

B. Graded-Index Optical Components

Graded-Index Slab

Consider a slab of material whose refractive index $n = n(y)$ is uniform in the x and z directions but varies continuously in the y direction (Fig. 1.3-3). The trajectories of

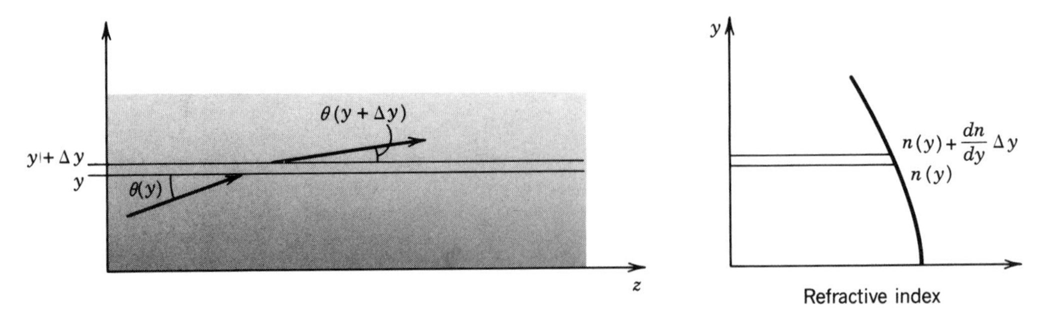

Figure 1.3-3 Refraction in a graded-index slab.

paraxial rays in the y–z plane are described by the paraxial ray equation

$$\frac{d}{dz}\left(n\frac{dy}{dz}\right) = \frac{dn}{dy}, \tag{1.3-4}$$

from which

$$\frac{d^2y}{dz^2} = \frac{1}{n}\frac{dn}{dy}. \tag{1.3-5}$$

Given $n(y)$ and the initial conditions (y and dy/dz at $z = 0$), (1.3-5) can be solved for the function $y(z)$, which describes the ray trajectories.

Derivation of the Paraxial Ray Equation in a Graded-Index Slab Using Snell's Law

Equation (1.3-5) may also be derived by the direct use of Snell's law (Fig. 1.3-3). Let $\theta(y) \approx dy/dz$ be the angle that the ray makes with the z-axis at the position (y, z). After traveling through a layer of width Δy the ray changes its angle to $\theta(y + \Delta y)$. The two angles are related by Snell's law,

$$n(y)\cos\theta(y) = n(y+\Delta y)\cos\theta(y+\Delta y)$$
$$= \left[n(y) + \frac{dn}{dy}\Delta y\right]\left[\cos\theta(y) - \frac{d\theta}{dy}\Delta y \sin\theta(y)\right],$$

where we have applied the expansion $f(y + \Delta y) = f(y) + (df/dy)\Delta y$ to the function $f(y) = \cos\theta(y)$. In the limit $\Delta y \to 0$, we obtain the differential equation

$$\frac{dn}{dy} = n\tan\theta\frac{d\theta}{dy}. \tag{1.3-6}$$

For paraxial rays θ is very small so that $\tan\theta \approx \theta$. Substituting $\theta = dy/dz$ in (1.3-6), we obtain (1.3-5).

EXAMPLE 1.3-1. ***Slab with Parabolic Index Profile.*** An important particular distribution for the graded refractive index is

$$n^2(y) = n_0^2(1 - \alpha^2 y^2). \tag{1.3-7}$$

This is a symmetric function of y that has its maximum value at $y = 0$ (Fig. 1.3-4). A glass slab with this profile is known by the trade name SELFOC. Usually, α is chosen to be sufficiently small so that $\alpha^2 y^2 \ll 1$ for all y of interest. Under this condition, $n(y) = n_0(1 - \alpha^2 y^2)^{1/2} \approx n_0(1 - \frac{1}{2}\alpha^2 y^2)$; i.e., $n(y)$ is a parabolic distribution. Also, because

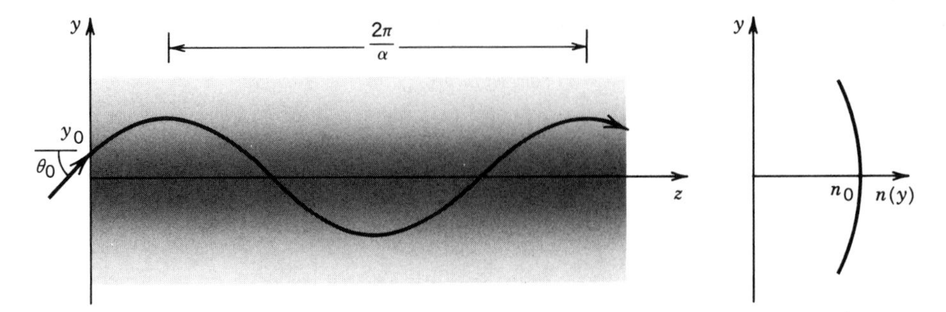

Figure 1.3-4 Trajectory of a ray in a GRIN slab of parabolic index profile (SELFOC).

$n(y) - n_0 \ll n_0$, the fractional change of the refractive index is very small. Taking the derivative of (1.3-7), the right-hand side of (1.3-5) is $(1/n)\,dn/dy = -(n_0/n)^2\alpha^2 y \approx -\alpha^2 y$, so that (1.3-5) becomes

$$\frac{d^2y}{dz^2} \approx -\alpha^2 y. \tag{1.3-8}$$

The solutions of this equation are harmonic functions with period $2\pi/\alpha$. Assuming an initial position $y(0) = y_0$ and an initial slope $dy/dz = \theta_0$ at $z = 0$,

$$y(z) = y_0 \cos \alpha z + \frac{\theta_0}{\alpha} \sin \alpha z, \tag{1.3-9}$$

from which the slope of the trajectory is

$$\theta(z) = \frac{dy}{dz} = -y_0\alpha \sin \alpha z + \theta_0 \cos \alpha z. \tag{1.3-10}$$

The ray oscillates about the center of the slab with a period $2\pi/\alpha$ known as the **pitch**, as illustrated in Fig. 1.3-4.

The maximum excursion of the ray is $y_{\max} = [y_0^2 + (\theta_0/\alpha)^2]^{1/2}$ and the maximum angle is $\theta_{\max} = \alpha y_{\max}$. The validity of this approximate analysis is ensured if $\theta_{\max} \ll 1$. If $2y_{\max}$ is smaller than the width of the slab, the ray remains confined and the slab serves as a light guide. Figure 1.3-5 shows the trajectories of a number of rays transmitted through a SELFOC slab. Note that all rays have the same pitch. This GRIN slab may be used as a lens, as demonstrated in Exercise 1.3-1.

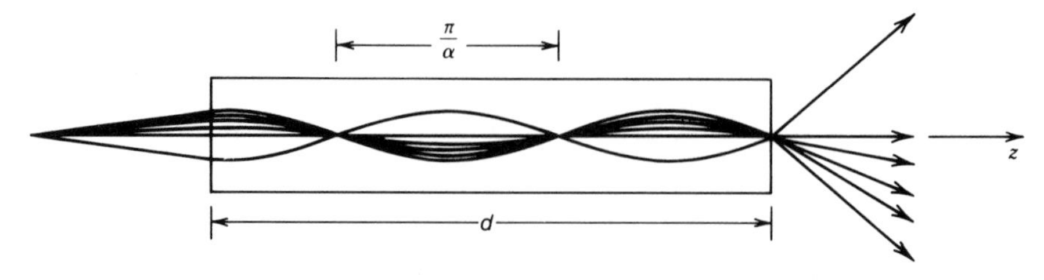

Figure 1.3-5 Trajectories of rays in a SELFOC slab.

EXERCISE 1.3-1

The GRIN Slab as a Lens. Show that a SELFOC slab of length $d < \pi/2\alpha$ and refractive index given by (1.3-7) acts as a cylindrical lens (a lens with focusing power in the y–z plane) of focal length

$$f \approx \frac{1}{n_0 \alpha \sin \alpha d}. \qquad (1.3\text{-}11)$$

Show that the principal point (defined in Fig. 1.3-6) lies at a distance from the slab edge $\overline{AH} \approx (1/n_0\alpha)\tan(\alpha d/2)$. Sketch the ray trajectories in the special cases $d = \pi/\alpha$ and $\pi/2\alpha$.

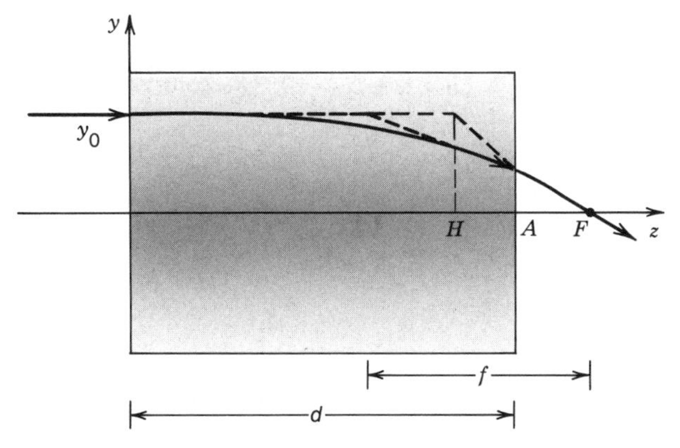

Figure 1.3-6 The SELFOC slab used as a lens; F is the focal point and H is the principal point.

Graded-Index Fibers

A graded-index fiber is a glass cylinder with a refractive index n that varies as a function of the radial distance from its axis. In the paraxial approximation, the ray trajectories are governed by the paraxial ray equations (1.3-3). Consider, for example, the distribution

$$n^2 = n_0^2\left[1 - \alpha^2\left(x^2 + y^2\right)\right]. \qquad (1.3\text{-}12)$$

Substituting (1.3-12) into (1.3-3) and assuming that $\alpha^2(x^2 + y^2) \ll 1$ for all x and y of interest, we obtain

$$\frac{d^2x}{dz^2} \approx -\alpha^2 x, \qquad \frac{d^2y}{dz^2} \approx -\alpha^2 y. \qquad (1.3\text{-}13)$$

Both x and y are therefore harmonic functions of z with period $2\pi/\alpha$. The initial positions (x_0, y_0) and angles ($\theta_{x0} = dx/dz$ and $\theta_{y0} = dy/dz$) at $z = 0$ determine the amplitudes and phases of these harmonic functions. Because of the circular symmetry,

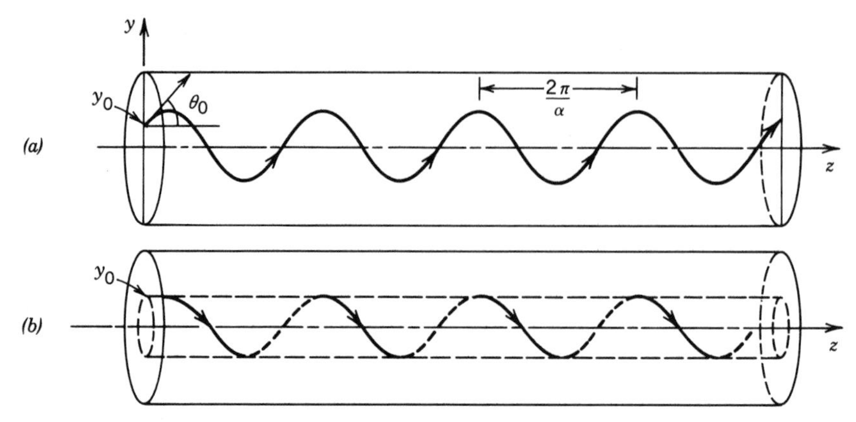

Figure 1.3-7 (*a*) Meridional and (*b*) helical rays in a graded-index fiber with parabolic index profile.

there is no loss of generality in choosing $x_0 = 0$. The solution of (1.3-13) is then

$$x(z) = \frac{\theta_{x0}}{\alpha} \sin \alpha z$$
$$y(z) = \frac{\theta_{y0}}{\alpha} \sin \alpha z + y_0 \cos \alpha z. \tag{1.3-14}$$

If $\theta_{x0} = 0$, i.e., the incident ray lies in a meridional plane (a plane passing through the axis of the cylinder, in this case the y–z plane), the ray continues to lie in that plane following a sinusoidal trajectory similar to that in the GRIN slab [Fig. 1.3-7(*a*)].

On the other hand, if $\theta_{y0} = 0$, and $\theta_{x0} = \alpha y_0$, then

$$x(z) = y_0 \sin \alpha z$$
$$y(z) = y_0 \cos \alpha z, \tag{1.3-15}$$

so that the ray follows a helical trajectory lying on the surface of a cylinder of radius y_0 [Fig. 1.3-7(*b*)]. In both cases the ray remains confined within the fiber, so that the fiber serves as a light guide. Other helical patterns are generated with different incident rays.

Graded-index fibers and their use in optical communications are discussed in Chaps. 8 and 22.

EXERCISE 1.3-2

Numerical Aperture of the Graded-Index Fiber. Consider a graded-index fiber with the index profile in (1.3-12) and radius a. A ray is incident from air into the fiber at its center, making an angle θ_0 with the fiber axis (see Fig. 1.3-8). Show, in the paraxial

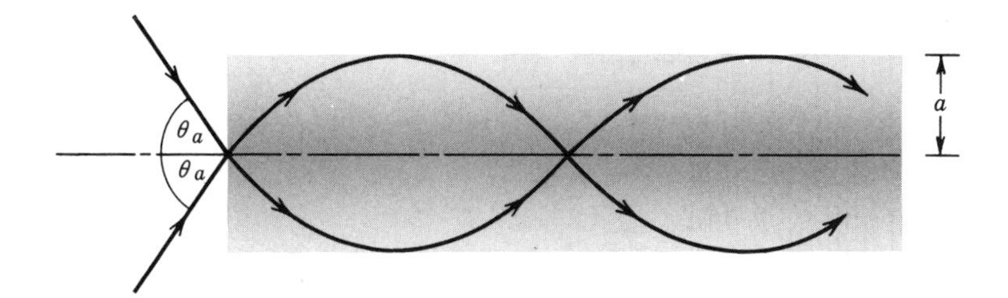

Figure 1.3-8 Acceptance angle of a graded-index optical fiber.

approximation, that the numerical aperture is

$$\mathrm{NA} = \sin\theta_a \approx n_0 a\alpha, \tag{1.3-16}$$

Numerical Aperture (Graded-Index Fiber)

where θ_a is the maximum angle θ_0 for which the ray trajectory is confined within the fiber. Compare this to the numerical aperture of a step-index fiber such as the one discussed in Exercise 1.2-5. To make the comparison fair, take the refractive indices of the core and cladding of the step-index fiber to be $n_1 = n_0$ and $n_2 = n_0(1 - \alpha^2 a^2)^{1/2} \approx n_0(1 - \frac{1}{2}\alpha^2 a^2)$.

*C. The Eikonal Equation

The ray trajectories are often characterized by the surfaces to which they are normal. Let $S(\mathbf{r})$ be a scalar function such that its equilevel surfaces, $S(\mathbf{r}) = \text{constant}$, are everywhere normal to the rays (Fig. 1.3-9). If $S(\mathbf{r})$ is known, the ray trajectories can readily be constructed since the normal to the equilevel surfaces at a position $\mathbf{r}$ is in the direction of the gradient vector $\nabla S(\mathbf{r})$. The function $S(\mathbf{r})$, called the **eikonal**, is akin to the potential function $V(\mathbf{r})$ in electrostatics; the role of the optical rays is played by the lines of electric field $\mathbf{E} = -\nabla V$.

To satisfy Fermat's principle (which is the main postulate of ray optics) the eikonal $S(\mathbf{r})$ must satisfy a partial differential equation known as the **eikonal equation**,

$$\left(\frac{\partial S}{\partial x}\right)^2 + \left(\frac{\partial S}{\partial y}\right)^2 + \left(\frac{\partial S}{\partial z}\right)^2 = n^2, \tag{1.3-17}$$

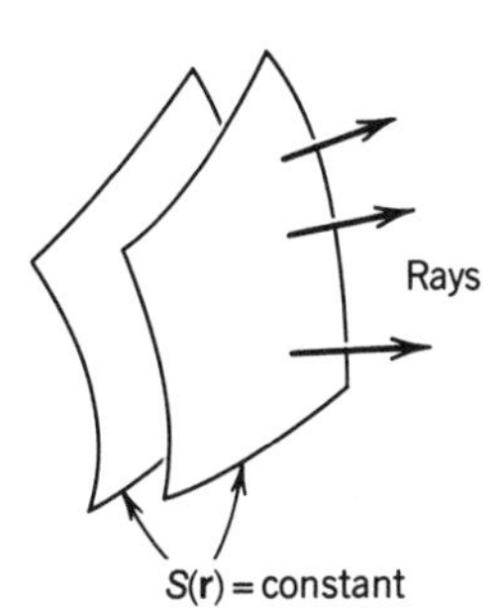

Figure 1.3-9 Ray trajectories are normal to the surfaces of constant $S(\mathbf{r})$.

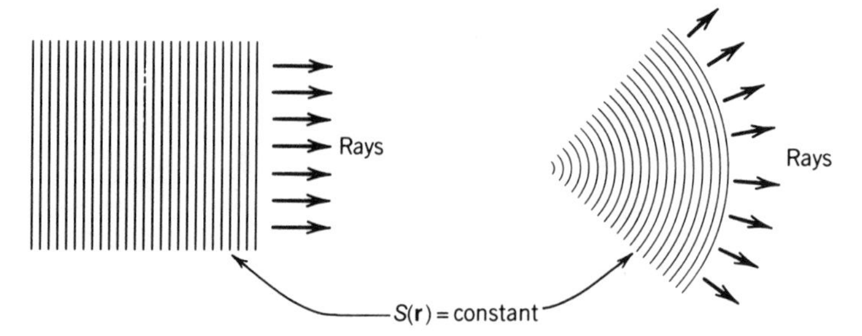

Figure 1.3-10 Rays and surfaces of constant $S(\mathbf{r})$ in a homogeneous medium.

which is usually written in the vector form

$$|\nabla S|^2 = n^2, \tag{1.3-18}$$

Eikonal Equation

where $|\nabla S|^2 = \nabla S \cdot \nabla S$. The proof of the eikonal equation from Fermat's principle is a mathematical exercise that lies beyond the scope of this book.† Fermat's principle (and the ray equation) can also be shown to follow from the eikonal equation. Therefore, either the eikonal equation or Fermat's principle may be regarded as the principal postulate of ray optics.

Integrating the eikonal equation (1.3-18) along a ray trajectory between two points A and B gives

$$S(\mathbf{r}_B) - S(\mathbf{r}_A) = \int_A^B |\nabla S|\, ds = \int_A^B n\, ds = \text{optical path length between } A \text{ and } B.$$

This means that the difference $S(\mathbf{r}_B) - S(\mathbf{r}_A)$ represents the optical path length between A and B. In the electrostatics analogy, the optical path length plays the role of the potential difference.

To determine the ray trajectories in an inhomogeneous medium of refractive index $n(\mathbf{r})$, we can either solve the ray equation (1.3-2), as we have done earlier, or solve the eikonal equation for $S(\mathbf{r})$, from which we calculate the gradient ∇S.

If the medium is homogeneous, i.e., $n(\mathbf{r})$ is constant, the magnitude of ∇S is constant, so that the wavefront normals (rays) must be straight lines. The surfaces $S(\mathbf{r}) =$ constant may be parallel planes or concentric spheres, as illustrated in Fig. 1.3-10.

1.4 MATRIX OPTICS

Matrix optics is a technique for tracing paraxial rays. The rays are assumed to travel only within a single plane, so that the formalism is applicable to systems with planar geometry and to meridional rays in circularly symmetric systems.

A ray is described by its position and its angle with respect to the optical axis. These variables are altered as the ray travels through the system. In the paraxial approximation, the position and angle at the input and output planes of an optical system are

†See, e.g., M. Born and E. Wolf, *Principles of Optics*, Pergamon Press, New York, 6th ed. 1980.

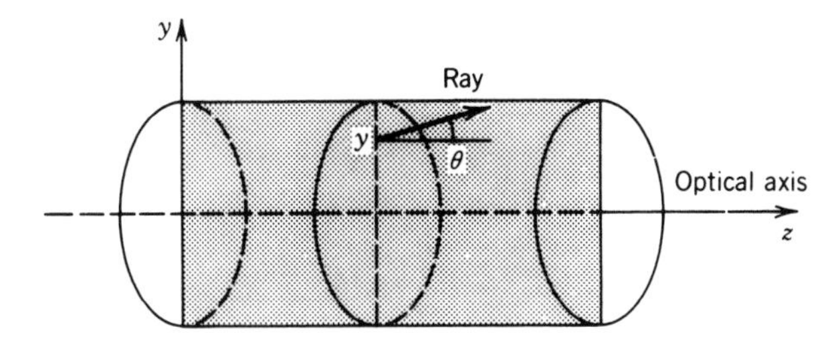

Figure 1.4-1 A ray is characterized by its coordinate y and its angle θ.

related by two *linear* algebraic equations. As a result, the optical system is described by a 2×2 matrix called the ray-transfer matrix.

The convenience of using matrix methods lies in the fact that the ray-transfer matrix of a cascade of optical components (or systems) is a product of the ray-transfer matrices of the individual components (or systems). Matrix optics therefore provides a formal mechanism for describing complex optical systems in the paraxial approximation.

A. The Ray-Transfer Matrix

Consider a circularly symmetric optical system formed by a succession of refracting and reflecting surfaces all centered about the same axis (optical axis). The z axis lies along the optical axis and points in the general direction in which the rays travel. Consider rays in a plane containing the optical axis, say the y–z plane. We proceed to trace a ray as it travels through the system, i.e., as it crosses the transverse planes at different axial distances. A ray crossing the transverse plane at z is completely characterized by the coordinate y of its crossing point and the angle θ (Fig. 1.4-1).

An optical system is a set of optical components placed between two transverse planes at z_1 and z_2, referred to as the input and output planes, respectively. The system is characterized completely by its effect on an incoming ray of arbitrary position and direction (y_1, θ_1). It steers the ray so that it has new position and direction (y_2, θ_2) at the output plane (Fig. 1.4-2).

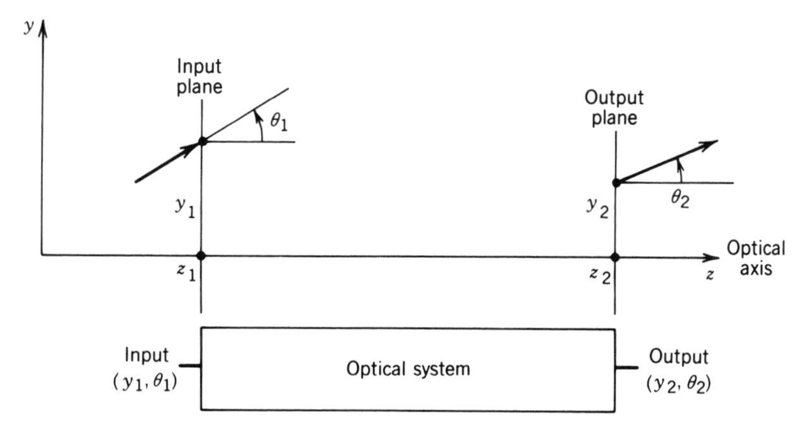

Figure 1.4-2 A ray enters an optical system at position y_1 and angle θ_1 and leaves at position y_2 and angle θ_2.

In the paraxial approximation, when all angles are sufficiently small so that $\sin\theta \approx \theta$, the relation between (y_2, θ_2) and (y_1, θ_1) is linear and can generally be written in the form

$$y_2 = Ay_1 + B\theta_1 \tag{1.4-1}$$

$$\theta_2 = Cy_1 + D\theta_1, \tag{1.4-2}$$

where A, B, C and D are real numbers. Equations (1.4-1) and (1.4-2) may be conveniently written in matrix form as

$$\begin{bmatrix} y_2 \\ \theta_2 \end{bmatrix} = \begin{bmatrix} A & B \\ C & D \end{bmatrix} \begin{bmatrix} y_1 \\ \theta_1 \end{bmatrix}.$$

The matrix **M**, whose elements are A, B, C, D, characterizes the optical system completely since it permits (y_2, θ_2) to be determined for any (y_1, θ_1). It is known as the **ray-transfer matrix**.

EXERCISE 1.4-1

Special Forms of the Ray-Transfer Matrix. Consider the following situations in which one of the four elements of the ray-transfer matrix vanishes:

(a) Show that if $A = 0$, all rays that enter the system at the same angle leave at the same position, so that parallel rays in the input are focused to a single point at the output.

(b) What are the special features of each of the systems for which $B = 0$, $C = 0$, or $D = 0$?

B. Matrices of Simple Optical Components

Free-Space Propagation

Since rays travel in free space along straight lines, a ray traversing a distance d is altered in accordance with $y_2 = y_1 + \theta_1 d$ and $\theta_2 = \theta_1$. The ray-transfer matrix is therefore

$$\mathbf{M} = \begin{bmatrix} 1 & d \\ 0 & 1 \end{bmatrix}. \tag{1.4-3}$$

Refraction at a Planar Boundary

At a planar boundary between two media of refractive indices n_1 and n_2, the ray angle changes in accordance with Snell's law $n_1 \sin\theta_1 = n_2 \sin\theta_2$. In the paraxial approximation, $n_1\theta_1 \approx n_2\theta_2$. The position of the ray is not altered, $y_2 = y_1$. The ray-transfer

matrix is

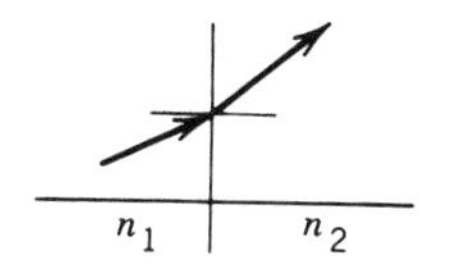

$$\mathbf{M} = \begin{bmatrix} 1 & 0 \\ 0 & \frac{n_1}{n_2} \end{bmatrix}. \tag{1.4-4}$$

Refraction at a Spherical Boundary

The relation between θ_1 and θ_2 for paraxial rays refracted at a spherical boundary between two media is provided in (1.2-8). The ray height is not altered, $y_2 \approx y_1$. The ray-transfer matrix is

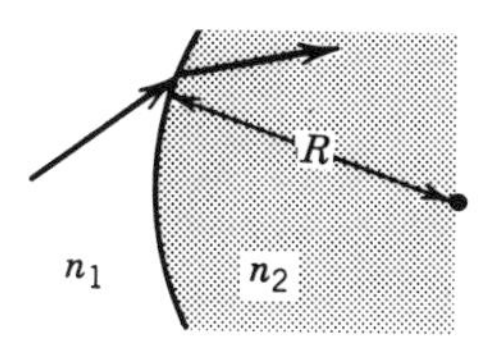

Convex, $R > 0$; concave, $R < 0$

$$\mathbf{M} = \begin{bmatrix} 1 & 0 \\ -\frac{(n_2 - n_1)}{n_2 R} & \frac{n_1}{n_2} \end{bmatrix}. \tag{1.4-5}$$

Transmission Through a Thin Lens

The relation between θ_1 and θ_2 for paraxial rays transmitted through a thin lens of focal length f is given in (1.2-11). Since the height remains unchanged ($y_2 = y_1$),

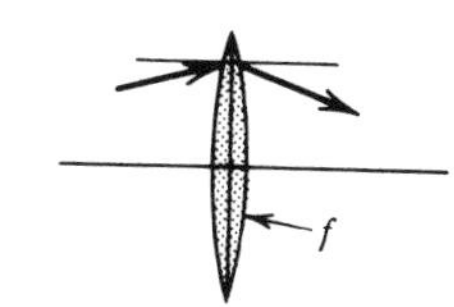

Convex, $f > 0$; concave, $f < 0$

$$\mathbf{M} = \begin{bmatrix} 1 & 0 \\ -\frac{1}{f} & 1 \end{bmatrix}. \tag{1.4-6}$$

Reflection from a Planar Mirror

Upon reflection from a planar mirror, the ray position is not altered, $y_2 = y_1$. Adopting the convention that the z axis points in the general direction of travel of the rays, i.e., toward the mirror for the incident rays and away from it for the reflected rays, we conclude that $\theta_2 = \theta_1$. The ray-transfer matrix is therefore the identity matrix

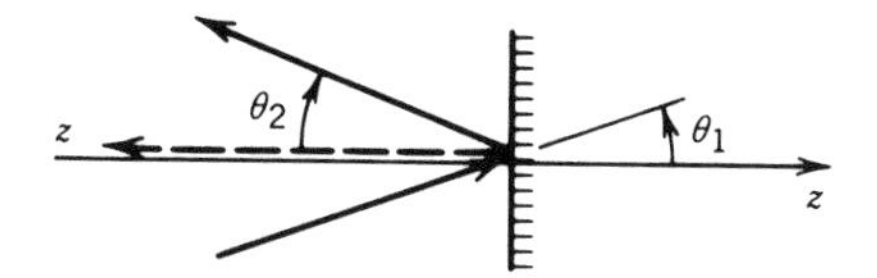

$$\mathbf{M} = \begin{bmatrix} 1 & 0 \\ 0 & 1 \end{bmatrix}. \tag{1.4-7}$$

Reflection from a Spherical Mirror

Using (1.2-1), and the convention that the z axis follows the general direction of the rays as they reflect from mirrors, we similarly obtain

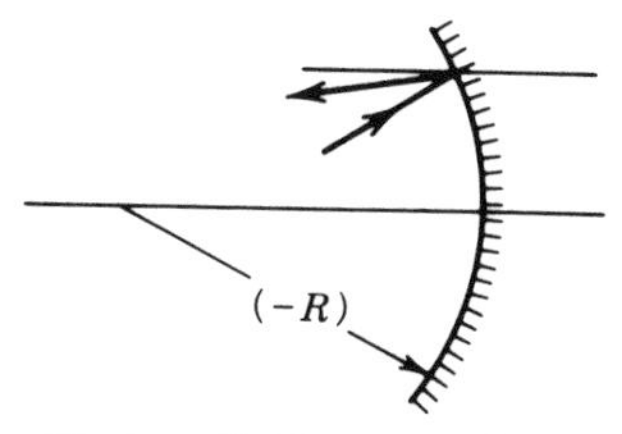

Concave, $R < 0$; convex, $R > 0$

$$\mathbf{M} = \begin{bmatrix} 1 & 0 \\ \frac{2}{R} & 1 \end{bmatrix}. \tag{1.4-8}$$

Note the similarity between the ray-transfer matrices of a spherical mirror (1.4-8) and a thin lens (1.4-6). A mirror with radius of curvature R bends rays in a manner that is identical to that of a thin lens with focal length $f = -R/2$.

C. Matrices of Cascaded Optical Components

A cascade of optical components whose ray-transfer matrices are $\mathbf{M}_1, \mathbf{M}_2, \ldots, \mathbf{M}_N$ is equivalent to a single optical component of ray-transfer matrix

M_1 M_2 ... M_N

$$\mathbf{M} = \mathbf{M}_N \cdots \mathbf{M}_2 \mathbf{M}_1. \tag{1.4-9}$$

Note the order of matrix multiplication: The matrix of the system that is crossed by the rays first is placed to the right, so that it operates on the column matrix of the incident ray first.

EXERCISE 1.4-2

A Set of Parallel Transparent Plates. Consider a set of N parallel planar transparent plates of refractive indices $n_1, n_2, \ldots, n_N$ and thicknesses $d_1, d_2, \ldots, d_N$, placed in air $(n = 1)$ normal to the z axis. Show that the ray-transfer matrix is

1 n_1 n_2 n_N 1 z d_1 d_2 d_N

$$\mathbf{M} = \begin{bmatrix} 1 & \sum_{i=1}^{N} \frac{d_i}{n_i} \\ 0 & 1 \end{bmatrix}. \tag{1.4-10}$$

Note that the order of placing the plates does not affect the overall ray-transfer matrix. What is the ray-transfer matrix of an inhomogeneous transparent plate of thickness d_0 and refractive index $n(z)$?

EXERCISE 1.4-3

A Gap Followed by a Thin Lens. Show that the ray-transfer matrix of a distance d of free space followed by a lens of focal length f is

f d

$$\mathbf{M} = \begin{bmatrix} 1 & d \\ -\frac{1}{f} & 1 - \frac{d}{f} \end{bmatrix}. \tag{1.4-11}$$

EXERCISE 1.4-4

Imaging with a Thin Lens. Derive an expression for the ray-transfer matrix of a system comprised of free space/thin lens/free space, as shown in Fig. 1.4-3. Show that if the imaging condition ($1/d_1 + 1/d_2 = 1/f$) is satisfied, all rays originating from a single point in the input plane reach the output plane at the single point y_2, regardless of their angles. Also show that if $d_2 = f$, all parallel incident rays are focused by the lens onto a single point in the output plane.

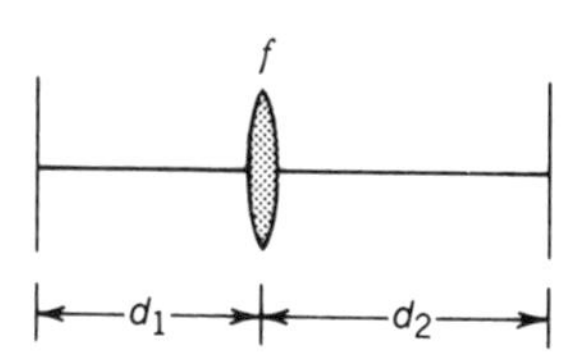

Figure 1.4-3 Single-lens imaging system.

EXERCISE 1.4-5

Imaging with a Thick Lens. Consider a glass lens of refractive index n, thickness d, and two spherical surfaces of equal radii R (Fig. 1.4-4). Determine the ray-transfer matrix of the system between the two planes at distances d_1 and d_2 from the vertices of the lens. The lens is placed in air (refractive index = 1). Show that the system is an imaging system (i.e., the input and output planes are conjugate) if

$$\frac{1}{z_1} + \frac{1}{z_2} = \frac{1}{f} \qquad \text{or} \qquad s_1 s_2 = f^2, \tag{1.4-12}$$

where

$$z_1 = d_1 + h \qquad s_1 = z_1 - f \tag{1.4-13}$$

$$z_2 = d_2 + h, \qquad s_2 = z_2 - f \tag{1.4-14}$$

and

$$h = \frac{(n-1)fd}{nR} \tag{1.4-15}$$

$$\frac{1}{f} = \frac{(n-1)}{R}\left[2 - \frac{n-1}{n}\frac{d}{R}\right]. \tag{1.4-16}$$

The points F_1 and F_2 are known as the front and back **focal points**, respectively. The points P_1 and P_2 are known as the first and second **principal points**, respectively. Show the importance of these points by tracing the trajectories of rays that are incident parallel to the optical axis.

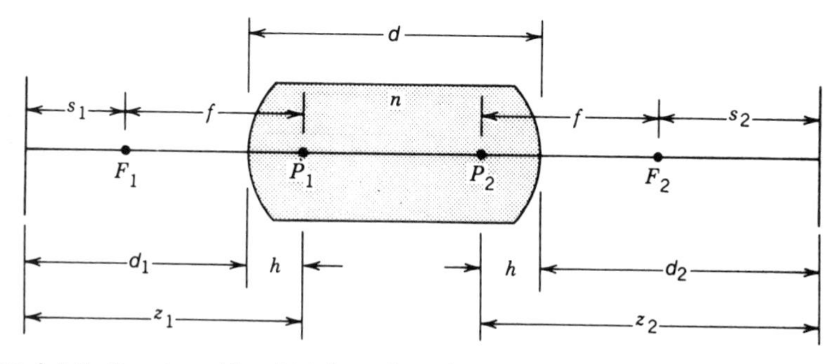

Figure 1.4-4 Imaging with a thick lens. P_1 and P_2 are the principal points and F_1 and F_2 are the focal points.

D. Periodic Optical Systems

A periodic optical system is a cascade of identical unit systems. An example is a sequence of equally spaced identical relay lenses used to guide light, as shown in Fig. 1.2-16(a). Another example is the reflection of light between two parallel mirrors forming an optical resonator (see Chap. 9); in that case, the ray traverses the same unit system (a round trip of reflections) repeatedly. A homogeneous medium, such as a glass fiber, may be considered as a periodic system if it is divided into contiguous identical segments of equal length. A general theory of ray propagation in periodic optical systems will now be formulated using matrix methods.

Difference Equation for the Ray Position

A periodic system is composed of a cascade of identical unit systems (stages), each with a ray-transfer matrix (A, B, C, D), as shown in Fig. 1.4-5. A ray enters the system with initial position y_0 and slope θ_0. To determine the position and slope (y_m, θ_m) of the ray at the exit of the mth stage, we apply the $ABCD$ matrix m times,

$$\begin{bmatrix} y_m \\ \theta_m \end{bmatrix} = \begin{bmatrix} A & B \\ C & D \end{bmatrix}^m \begin{bmatrix} y_0 \\ \theta_0 \end{bmatrix}.$$

We can also apply the relations

$$y_{m+1} = Ay_m + B\theta_m \tag{1.4-17}$$

$$\theta_{m+1} = Cy_m + D\theta_m \tag{1.4-18}$$

iteratively to determine (y_1, θ_1) from (y_0, θ_0), then (y_2, θ_2) from (y_1, θ_1), and so on, using a computer.

It is of interest to derive equations that govern the dynamics of the position y_m, $m = 0, 1, \ldots,$ irrespective of the angle θ_m. This is achieved by eliminating θ_m from (1.4-17) and (1.4-18). From (1.4-17)

$$\theta_m = \frac{y_{m+1} - Ay_m}{B}. \tag{1.4-19}$$

Replacing m with $m + 1$ in (1.4-19) yields

$$\theta_{m+1} = \frac{y_{m+2} - Ay_{m+1}}{B}. \tag{1.4-20}$$

Substituting (1.4-19) and (1.4-20) into (1.4-18) gives

$$\boxed{y_{m+2} = 2by_{m+1} - F^2 y_m,} \tag{1.4-21}$$

Recurrence Relation for Ray Position

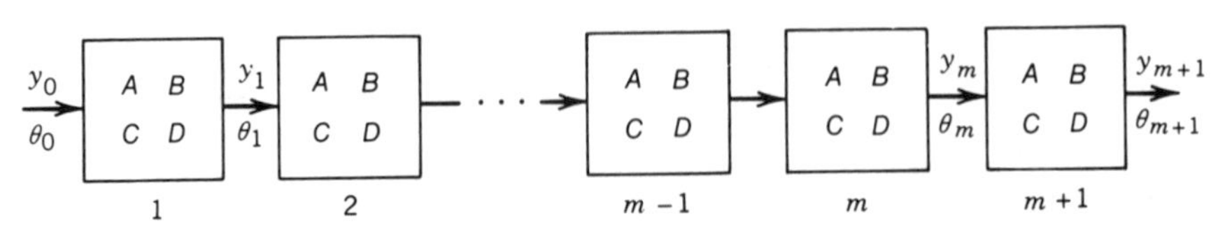

Figure 1.4-5 A cascade of identical optical components.

where

$$b = \frac{A + D}{2} \tag{1.4-22}$$

$$F^2 = AD - BC = \det[\mathbf{M}], \tag{1.4-23}$$

and det[**M**] is the determinant of **M**.

Equation (1.4-21) is a linear difference equation governing the ray position y_m. It can be solved iteratively on a computer by computing y_2 from y_0 and y_1, then y_3 from y_1 and y_2, and so on. y_1 may be computed from y_0 and θ_0 by use of (1.4-17) with $m = 0$.

It is useful, however, to derive an explicit expression for y_m by solving the difference equation (1.4-21). As in linear differential equations, a solution satisfying a linear difference equation and the initial conditions is a unique solution. It is therefore appropriate to make a judicious guess for the solution of (1.4-21). We use a trial solution of the geometric form

$$y_m = y_0 h^m, \tag{1.4-24}$$

where h is a constant. Substituting (1.4-24) into (1.4-21) immediately shows that the trial solution is suitable provided that h satisfies the quadratic algebraic equation

$$h^2 - 2bh + F^2 = 0, \tag{1.4-25}$$

from which

$$h = b \pm j(F^2 - b^2)^{1/2}. \tag{1.4-26}$$

The results can be presented in a more compact form by defining the variable

$$\varphi = \cos^{-1}\frac{b}{F}, \tag{1.4-27}$$

so that $b = F\cos\varphi$, $(F^2 - b^2)^{1/2} = F\sin\varphi$, and therefore $h = F(\cos\varphi \pm j\sin\varphi) = F\exp(\pm j\varphi)$, whereupon (1.4-24) becomes $y_m = y_0 F^m \exp(\pm jm\varphi)$.

A general solution may be constructed from the two solutions with positive and negative signs by forming their linear combination. The sum of the two exponential functions can always be written as a harmonic (circular) function, so that

$$y_m = y_{\max} F^m \sin(m\varphi + \varphi_0), \tag{1.4-28}$$

where $y_{\max}$ and φ_0 are constants to be determined from the initial conditions y_0 and y_1. In particular, $y_{\max} = y_0/\sin\varphi_0$.

The parameter F is related to the determinant of the ray-transfer matrix of the unit system by $F = \det^{1/2}[\mathbf{M}]$. It can be shown that regardless of the unit system, $\det[\mathbf{M}] = n_1/n_2$, where n_1 and n_2 are the refractive indices of the initial and final sections of the unit system. This general result is easily verified for the ray-transfer matrices of all the optical components considered in this section. Since the determinant of a product of two matrices is the product of their determinants, it follows that the relation $\det[\mathbf{M}] = n_1/n_2$ is applicable to any cascade of these optical components. For example, if $\det[\mathbf{M}_1] = n_1/n_2$ and $\det[\mathbf{M}_2] = n_2/n_3$, then $\det[\mathbf{M}_2\mathbf{M}_1] = (n_2/n_3)(n_1/n_2) = n_1/n_3$.

In most applications $n_1 = n_2$, so that det[**M**] $= 1$ and $F = 1$, in which case the solution for the ray position is

$$y_m = y_{\max} \sin(m\varphi + \varphi_0). \tag{1.4-29}$$

Ray Position in a Periodic System

We shall assume henceforth that $F = 1$.

Condition for a Harmonic Trajectory

For y_m to be a harmonic (instead of hyperbolic) function, $\varphi = \cos^{-1} b$ must be real. This requires that

$$|b| \le 1 \qquad \text{or} \qquad \frac{|A + D|}{2} \le 1. \tag{1.4-30}$$

Condition for a Stable Solution

If, instead, $|b| > 1$, φ is then imaginary and the solution is a hyperbolic function (cosh or sinh), which increases without bound, as illustrated in Fig. 1.4-6(a). A harmonic solution ensures that y_m is bounded for all m, with a maximum value of $y_{\max}$. The bound $|b| \le 1$ therefore provides a condition of **stability** (boundedness) of the ray trajectory.

The ray angle corresponding to (1.4-29) is also a harmonic function $\theta_m = \theta_{\max} \sin(m\varphi + \varphi_1)$, where $\theta_{\max}$ and φ_1 are constants. This can be shown by use of (1.4-19) and trigonometric identities. The maximum angle $\theta_{\max}$ must be sufficiently small so that the paraxial approximation, which underlies this analysis, is applicable.

Condition for a Periodic Trajectory

The harmonic function (1.4-29) is periodic in m if it is possible to find an integer s such that $y_{m+s} = y_m$ for all m. The smallest such integer is the period. The ray then

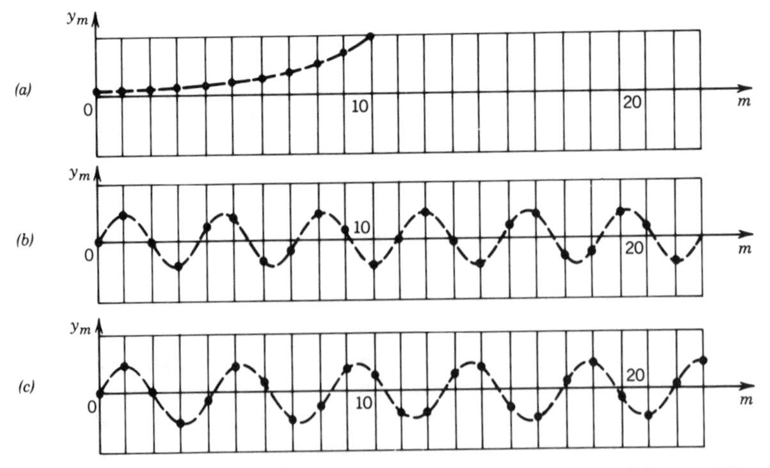

Figure 1.4-6 Examples of trajectories in periodic optical systems: (a) unstable trajectory ($b > 1$); (b) stable and periodic trajectory ($\varphi = 6\pi/11$; period = 11 stages); (c) stable but nonperiodic trajectory ($\varphi = 1.5$).

retraces its path after s stages. This condition is satisfied if $s\varphi = 2\pi q$, where q is an integer. Thus the necessary and sufficient condition for a periodic trajectory is that $\varphi/2\pi$ is a rational number q/s. If $\varphi = 6\pi/11$, for example, then $\varphi/2\pi = \frac{3}{11}$ and the trajectory is periodic with period $s = 11$ stages. This case is illustrated in Fig. 1.4-6(*b*).

Summary

A paraxial ray traveling through a cascade of identical unit optical systems, each with a ray-transfer matrix with elements (A, B, C, D) such that $AD - BC = 1$, follows a harmonic (and therefore bounded) trajectory if the condition $|(A + D)/2| \leq 1$, called the stability condition, is satisfied. The position at the mth stage is then $y_m = y_{max} \sin(m\varphi + \varphi_0)$, $m = 0, 1, 2, \ldots$, where $\varphi = \cos^{-1}[(A + D)/2]$. The constants y_{max} and φ_0 are determined from the initial positions y_0 and $y_1 = Ay_0 + B\theta_0$, where θ_0 is the initial ray inclination. The ray angles are related to the positions by $\theta_m = (y_{m+1} - Ay_m)/B$ and follow a harmonic function $\theta_m = \theta_{max} \sin(m\varphi + \varphi_1)$. For the paraxial approximation to be valid, $\theta_{max} \ll 1$. The ray trajectory is periodic with period s if $\varphi/2\pi$ is a rational number q/s.

EXAMPLE 1.4-1. ***A Sequence of Equally Spaced Identical Lenses.*** A set of identical lenses of focal length f separated by distance d, as shown in Fig. 1.4-7, may be used to relay light between two locations. The unit system, a distance d of free space followed by a lens, has a ray-transfer matrix given by (1.4-11); $A = 1$, $B = d$, $C = -1/f$, $D = 1 - d/f$.

The parameter $b = (A + D)/2 = 1 - d/2f$ and the determinant is unity. The condition for a stable ray trajectory, $|b| \leq 1$ or $-1 \leq b \leq 1$, is therefore

$$0 \leq d \leq 4f, \tag{1.4-31}$$

so that the spacing between the lenses must be smaller than four times the focal length. Under this condition the positions of paraxial rays obey the harmonic function

$$y_m = y_{max} \sin(m\varphi + \varphi_0), \qquad \varphi = \cos^{-1}\left(1 - \frac{d}{2f}\right). \tag{1.4-32}$$

When $d = 2f$, $\varphi = \pi/2$ and $\varphi/2\pi = \frac{1}{4}$, so that the trajectory of an arbitrary ray is

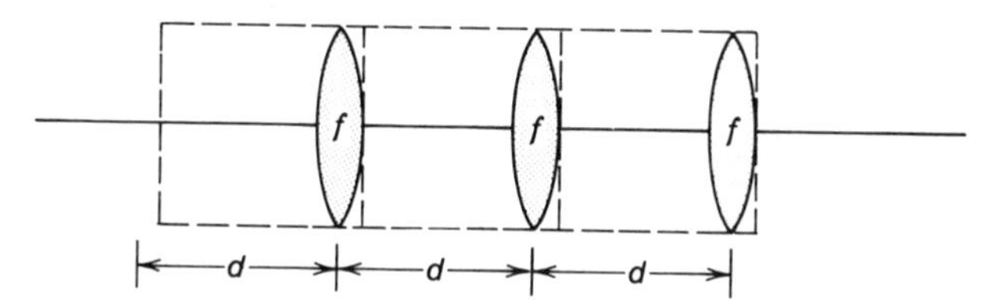

Figure 1.4-7 A periodic sequence of lenses.

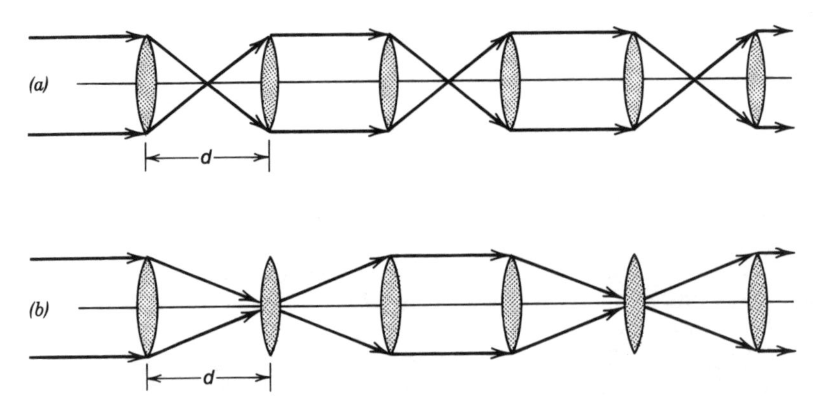

Figure 1.4-8 Examples of stable ray trajectories in a periodic lens system: (*a*) $d = 2f$; (*b*) $d = f$.

periodic with period equal to four stages. When $d = f$, $\varphi = \pi/3$ and $\varphi/2\pi = \frac{1}{6}$, so that the ray trajectory is periodic and retraces itself each six stages. These cases are illustrated in Fig. 1.4-8.

EXERCISE 1.4-6

A Periodic Set of Pairs of Different Lenses. Examine the trajectories of paraxial rays through a periodic system composed of a set of lenses with alternating focal lengths f_1 and

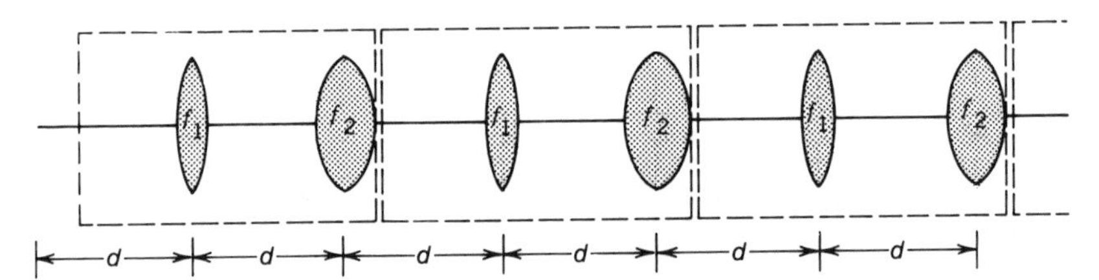

Figure 1.4-9 A periodic sequence of lens pairs.

f_2 as shown in Fig. 1.4-9. Show that the ray trajectory is bounded (stable) if

$$0 \le \left(1 - \frac{d}{2f_1}\right)\left(1 - \frac{d}{2f_2}\right) \le 1. \tag{1.4-33}$$

EXERCISE 1.4-7

An Optical Resonator. Paraxial rays are reflected repeatedly between two spherical mirrors of radii R_1 and R_2 separated by a distance d (Fig. 1.4-10). Regarding this as a periodic system whose unit system is a single round trip between the mirrors, determine the condition of stability of the ray trajectory. Optical resonators will be studied in detail in Chap. 9.

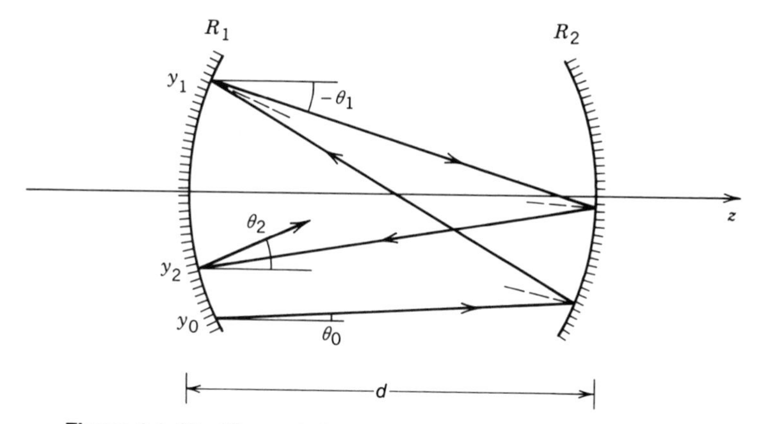

Figure 1.4-10 The optical resonator as a periodic optical system.

READING LIST

General

P. P. Banerjee and T. Poon, *Principles of Applied Optics*, Aksen Associates, Pacific Palisades, CA, 1991.

B. D. Guenther, *Modern Optics*, Wiley, New York, 1990.

J. R. Meyer-Arendt, *Introduction to Classical and Modern Optics*, Prentice-Hall, Englewood Cliffs, NJ, 1972, 3rd ed. 1989.

J. Strong, *Procedures in Applied Optics*, Marcel Dekker, New York, 1989.

D. Malacara, *Optics*, Academic Press, New York, 1988.

K. D. Möller, *Optics*, University Science Books, Mill Valley, CA, 1988.

F. G. Smith and J. H. Thomson, *Optics*, Wiley, New York, 1971, 2nd ed. 1988.

W. T. Welford, *Optics*, Oxford University Press, New York, 1976, 3rd ed. 1988.

R. W. Wood, *Physical Optics*, Macmillan, New York, 3rd ed. 1934; Reprinted by the Optical Society of America, Washington, DC, 1988.

E. Hecht and A. Zajac, *Optics*, Addison-Wesley, Reading, MA, 1974, 2nd ed. 1987.

F. L. Pedrotti and L. S. Pedrotti, *Introduction to Optics*, Prentice-Hall, Englewood Cliffs, NJ, 1987.

M. V. Klein and T. E. Furtak, *Optics*, Wiley, New York, 1982, 2nd ed. 1986.

M. Young, *Optics and Lasers: An Engineering Physics Approach*, Springer-Verlag, New York, 1977, 3rd ed. 1986.

K. Iizuka, *Engineering Optics*, Springer-Verlag, New York, 1985.

H. Haken, *Light*, North-Holland, Amsterdam, vol. 1, 1981; vol. 2, 1985.

Research & Education Association, *The Optics Problem Solver*, New York, 1981.

W. H. A. Fincham and M. H. Freeman, *Optics*, Butterworth, London, 9th ed. 1980.

M. Born and E. Wolf, *Principles of Optics*, Pergamon Press, New York, 1959, 6th ed. 1980.

A. K. Ghatak and K. Thyagarajan, *Contemporary Optics*, Plenum Press, New York, 1978.

E. W. Marchand, *Gradient-Index Optics*, Academic Press, New York, 1978.

F. P. Carlson, *Introduction to Applied Optics for Engineers*, Academic Press, New York, 1977.

F. A. Jenkins and H. E. White, *Fundamentals of Optics*, McGraw-Hill, New York, 1937, 4th ed. 1976.

A. Nussbaum and R. A. Phillips, *Contemporary Optics for Scientists and Engineers*, Prentice-Hall, Englewood Cliffs, NJ, 1976.

R. W. Ditchburn, *Light*, Academic Press, New York, 3rd ed. 1976.
G. F. Lothian, *Optics and Its Uses*, Van Nostrand Reinhold, New York, 1976.
J. P. Mathieu, *Optics*, Pergamon Press, New York, 1975.
E. Hecht, *Schaum's Outline of Theory and Problems of Optics*, McGraw-Hill, New York, 1975.
R. S. Longhurst, *Geometrical and Physical Optics*, Longman, Inc., New York, 3rd ed. 1973.
C. S. Williams and O. A. Becklund, *Optics: A Short Course for Engineers and Scientists*, Wiley-Interscience, New York, 1972.
A. K. Ghatak, *An Introduction to Modern Optics*, McGraw-Hill, New York, 1971.
M. V. Klein, *Optics*, Wiley, New York, 1970.
D. W. Tenquist, R. M. Whittle, and J. Yarwood, *University Optics*, Gordon and Breach, New York, 1970.
G. R. Fowles, *Introduction to Modern Optics*, Holt, Rinehart and Winston, New York, 1968.
E. B. Brown, *Modern Optics*, Reinhold, New York, 1966.
J. M. Stone, *Radiation and Optics*, McGraw-Hill, New York, 1963.
J. Strong, *Concepts of Classical Optics*, W. H. Freeman, San Francisco, 1958.
B. Rossi, *Optics*, Addison-Wesley, Reading, MA, 1957.
A. Sommerfeld, *Optics*, Academic Press, New York, 1954.

Geometrical Optics

W. T. Welford and R. Winston, *The Optics of Nonimaging Concentrators*, Academic Press, New York, 1978.
O. N. Stavroudis, *The Optics of Rays, Wavefronts and Caustics*, Academic Press, New York, 1972.
H. G. Zimmer, *Geometrical Optics*, Springer-Verlag, New York, 1970.
G. A. Fry, *Geometrical Optics*, Chilton, Philadelphia, 1969.
A. Nussbaum, *Geometric Optics: An Introduction*, Addison-Wesley, Reading, MA, 1968.
R. K. Luneburg, *Mathematical Theory of Optics*, University of California Press, Berkeley, CA, 1964.

Optical System Design

D. C. O'Shea, *Elements of Modern Optical Design*, Wiley, New York, 1985.
R. Kingslake, *Optical System Design*, Academic Press, New York, 1983.
L. Levi, *Applied Optics: A Guide to Optical System Design*, Wiley, New York, vol. 1, 1968; vol. 2, 1980.
W. J. Smith, *Modern Optical Engineering*, McGraw-Hill, New York, 1966.

Matrix Optics

A. Gerrard and J. M. Burch, *Introduction to Matrix Methods in Optics*, Wiley, New York, 1974.
W. Brouwer, *Matrix Methods in Optical Instrument Design*, W. A. Benjamin, New York, 1974.
J. W. Blaker, *Geometric Optics: The Matrix Theory*, Marcel Dekker, New York, 1971.

Popular and Historical

R. Kingslake, *A History of the Photographic Lens*, Academic Press, Orlando, 1989.
M. I. Sobel, *Light*, University of Chicago Press, Chicago, 1987.
A. I. Sabra, *Theories of Light from Descartes to Newton*, Cambridge University Press, New York, 1981.
I. Newton, *Opticks*, Dover, New York, 1979 (originally published 1704).
V. Ronchi, *The Nature of Light*, Harvard University Press, Cambridge, MA, 1971.
S. Tolansky, *Revolution in Optics*, Penguin, Baltimore, 1968.
A. C. S. Van Heel and C. H. F. Velzel, *What Is Light?*, McGraw-Hill, New York, 1968.
L. Basford and J. Pick, *The Rays of Light*, Samson, Low, Marsten & Co., London, 1966.

S. Tolansky, *Curiosities of Light Rays and Light Waves*, American Elsevier, New York, 1965.
W. H. Bragg, *The Universe of Light*, Dover, New York, 1959.
E. Ruchardt, *Light, Visible and Invisible*, University of Michigan Press, Ann Arbor, MI, 1958.

PROBLEMS

1.2-1 **Transmission through Planar Plates.** (a) Use Snell's law to show that a ray entering a planar plate of width d and refractive index n_1 (placed in air; $n \approx 1$) emerges parallel to its initial direction. The ray need not be paraxial. Derive an expression for the lateral displacement of the ray as a function of the angle of incidence θ. Explain your results in terms of Fermat's principle.
(b) If the plate is, instead, made of a stack of N parallel layers of thicknesses $d_1, d_2, \ldots, d_N$ and refractive indices $n_1, n_2, \ldots, n_N$, show that the transmitted ray is parallel to the incident ray. If θ_m is the angle of the ray in the mth layer, show that $n_m \sin\theta_m = \sin\theta$, $m = 1, 2, \ldots$.

1.2-2 **Lens in Water.** Determine the focal length f of a biconvex lens with radii 20 cm and 30 cm and refractive index $n = 1.5$. What is the focal length when the lens is immersed in water ($n = \frac{4}{3}$)?

1.2-3 **Numerical Aperture of a Claddless Fiber.** Determine the numerical aperture and the acceptance angle of an optical fiber if the refractive index of the core is $n_1 = 1.46$ and the cladding is stripped out (replaced with air $n_2 \approx 1$).

1.2-4 **Fiber Coupling Spheres.** Tiny glass balls are often used as lenses to couple light into and out of optical fibers. The fiber end is located at a distance f from the sphere. For a sphere of radius $a = 1$ mm and refractive index $n = 1.8$, determine f such that a ray parallel to the optical axis at a distance $y = 0.7$ mm is focused onto the fiber, as illustrated in Fig. P1.2-4.

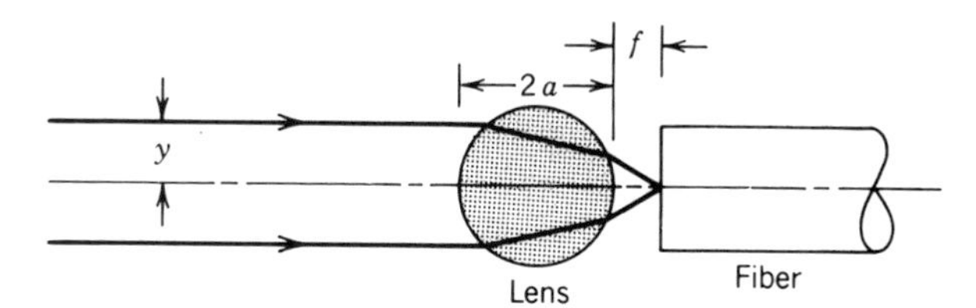

Figure P1.2-4 Focusing light into an optical fiber with a spherical glass ball.

1.2-5 **Extraction of Light from a High-Refractive-Index Medium.** Assume that light is generated isotropically in all directions inside a material of refractive index $n = 3.7$ cut in the shape of a parallelepiped and placed in air ($n = 1$) (see Exercise 1.2-6).
(a) If a reflective material acting as a perfect mirror is coated on all sides except the front side, determine the percentage of light that may be extracted from the front side.
(b) If another transparent material of refractive index $n = 1.4$ is placed on the front side, would that help extract some of the trapped light?

1.3-1 **Axially Graded Plate.** A plate of thickness d is oriented normal to the z-axis. The refractive index $n(z)$ is graded in the z direction. Show that a ray entering the plate from air at an incidence angle θ_0 in the y–z plane makes an angle $\theta(z)$ at position z in the medium given by $n(z)\sin\theta(z) = \sin\theta_0$. Show that the ray emerges into air parallel to the original incident ray. *Hint*: You may use the results of Problem 1.2-1. Show that the ray position $y(z)$ inside the plate obeys the differential equation $(dy/dz)^2 = (n^2/\sin^2\theta - 1)^{-1}$.

1.3-2 **Ray Trajectories in GRIN Fibers.** Consider a graded-index optical fiber with cylindrical symmetry about the z axis and refractive index $n(\rho)$, $\rho = (x^2 + y^2)^{1/2}$. Let (ρ, ϕ, z) be the position vector in a cylindrical coordinate system. Rewrite the paraxial ray equations, (1.3-3), in a cylindrical system and derive differential equations for ρ and ϕ as functions of z.

1.4-1 **Ray-Transfer Matrix of a Lens System.** Determine the ray-transfer matrix for an optical system made of a thin convex lens of focal length f and a thin concave lens of focal length $-f$ separated by a distance f. Discuss the imaging properties of this composite lens.

1.4-2 **Ray-Transfer Matrix of a GRIN Plate.** Determine the ray-transfer matrix of a SELFOC plate [i.e., a graded-index material with parabolic refractive index $n(y) \approx n_0(1 - \frac{1}{2}\alpha^2 y^2)$] of width d.

1.4-3 **The GRIN Plate as a Periodic System.** Consider the trajectories of paraxial rays inside a SELFOC plate normal to the z axis. This system may be regarded as a periodic system made of a sequence of identical contiguous plates of thickness d each. Using the result of Problem 1.4-2, determine the stability condition of the ray trajectory. Is this condition dependent on the choice of d?

1.4-4 **4 × 4 Ray-Transfer Matrix for Skewed Rays.** Matrix methods may be generalized to describe skewed paraxial rays in circularly symmetric systems, and to astigmatic (non-circularly symmetric) systems. A ray crossing the plane $z = 0$ is generally characterized by four variables—the coordinates (x, y) of its position in the plane, and the angles (θ_x, θ_y) that its projections in the x–z and y–z planes make with the z axis. The emerging ray is also characterized by four variables linearly related to the initial four variables. The optical system may then be characterized completely, within the paraxial approximation, by a 4 × 4 matrix.
(a) Determine the 4 × 4 ray-transfer matrix of a distance d in free space.
(b) Determine the 4 × 4 ray-transfer matrix of a thin cylindrical lens with focal length f oriented in the y direction (Fig. P1.4-4). The cylindrical lens has focal length f for rays in the y–z plane, and no focusing power for rays in the x–z plane.

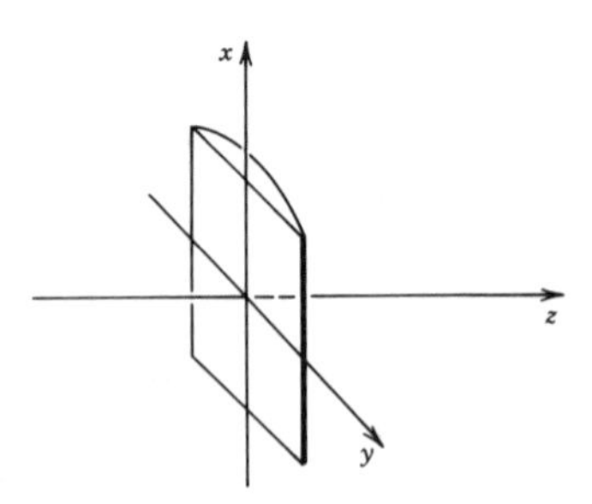

Figure P1.4-4 Cylindrical lens.

CHAPTER

2

WAVE OPTICS

2.1 POSTULATES OF WAVE OPTICS

2.2 MONOCHROMATIC WAVES

- A. Complex Representation and the Helmholtz Equation
- B. Elementary Waves
- C. Paraxial Waves

*2.3 RELATION BETWEEN WAVE OPTICS AND RAY OPTICS

2.4 SIMPLE OPTICAL COMPONENTS

- A. Reflection and Refraction
- B. Transmission Through Optical Components
- C. Graded-Index Optical Components

2.5 INTERFERENCE

- A. Interference of Two Waves
- B. Multiple-Wave Interference

2.6 POLYCHROMATIC LIGHT

- A. Fourier Decomposition
- B. Light Beating

Christiaan Huygens (1629–1695) advanced several new concepts concerning the propagation of light waves.

Thomas Young (1773–1829) championed the wave theory of light and discovered the principle of optical interference.

Light propagates in the form of waves. In free space, light waves travel with a constant speed $c_o = 3.0 \times 10^8$ m/s (30 cm/ns or 0.3 mm/ps). The range of optical wavelengths contains three bands— ultraviolet (10 to 390 nm), visible (390 to 760 nm), and infrared (760 nm to 1 mm). The corresponding range of optical frequencies stretches from 3×10^{11} Hz to 3×10^{16} Hz, as illustrated in Fig. 2.0-1.

The **wave theory** of light encompasses the ray theory (Fig. 2.0-2). Strictly speaking, ray optics is the limit of wave optics when the wavelength is infinitesimally short. However, the wavelength need not actually be equal to zero for the ray-optics theory to be useful. As long as the light waves propagate through and around objects whose dimensions are much greater than the wavelength, the ray theory suffices for describing most phenomena. Because the wavelength of visible light is much shorter than the dimensions of the visible objects encountered in our daily lives, manifestations of the wave nature of light are not apparent without careful observation.

In this chapter, light is described by a scalar function, called the wavefunction, which obeys the wave equation. The precise physical meaning of the wavefunction is

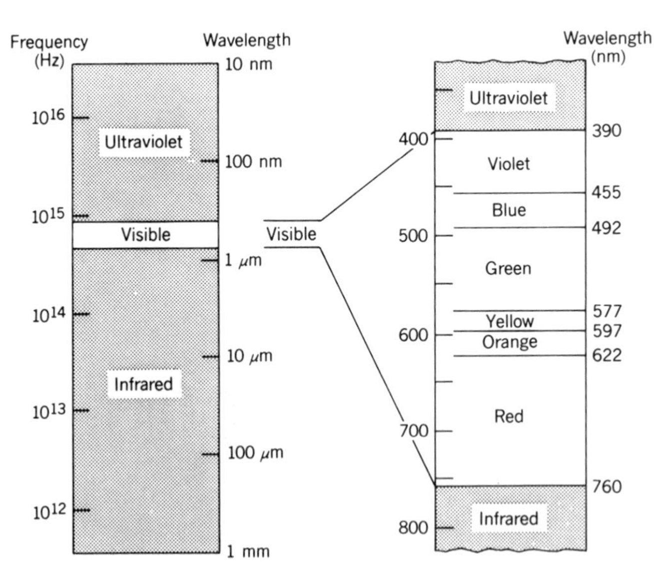

Figure 2.0-1 Optical frequencies and wavelengths.

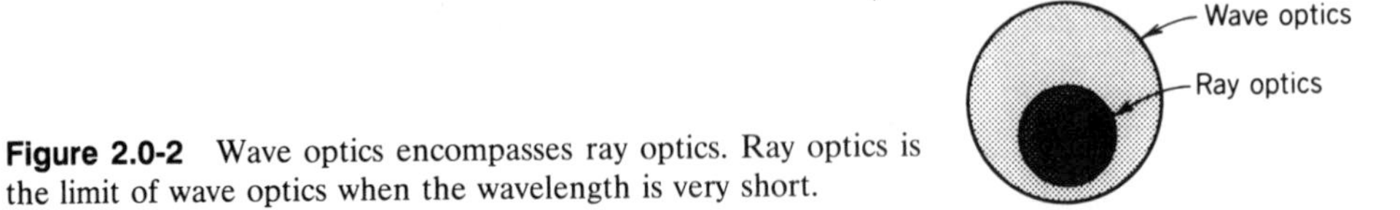

Figure 2.0-2 Wave optics encompasses ray optics. Ray optics is the limit of wave optics when the wavelength is very short.

not specified; it suffices to say at this point that it may represent any of the components of the electric or magnetic fields (as described in Chap. 5, which covers the electromagnetic theory of light). This, and a relation between the optical power density and the wavefunction, constitute the postulates of the scalar wave model of light, hereafter called **wave optics**. The consequences of these simple postulates are many and far reaching. Wave optics constitutes a basis for describing a host of optical phenomena that fall outside the confines of ray optics, including interference and diffraction, as demonstrated in this and the following two chapters.

Wave optics has its limitations. It is not capable of providing a complete picture of the reflection and refraction of light at the boundaries between dielectric materials, nor of explaining those optical phenomena that require a vector formulation, such as polarization effects. In Chap. 5 the electromagnetic theory of light is presented and the conditions under which scalar wave optics provides a good approximation to certain electromagnetic phenomena are elucidated.

This chapter begins with the postulates of wave optics (Sec. 2.1). In Secs. 2.2 to 2.5 we consider monochromatic waves, and polychromatic light is discussed in Sec. 2.6. Elementary waves, such as the plane wave and the spherical wave, are introduced in Sec. 2.2. Section 2.3 establishes that ray optics can be derived from wave optics. The transmission of optical waves through simple optical components such as mirrors, prisms, lenses, and gratings is examined in Sec. 2.4. Interference, an important manifestation of the wave nature of light, is the subject of Secs. 2.5 and 2.6.

2.1 POSTULATES OF WAVE OPTICS

The Wave Equation

Light propagates in the form of waves. In free space, light waves travel with speed c_o. A homogeneous transparent medium such as glass is characterized by a single constant, its refractive index n (≥ 1). In a medium of refractive index n, light waves travel with a reduced speed

$$c = \frac{c_o}{n}. \tag{2.1-1}$$

Speed of Light in a Medium

An optical wave is described mathematically by a real function of position $\mathbf{r} = (x, y, z)$ and time t, denoted $u(\mathbf{r}, t)$ and known as the **wavefunction**. It satisfies the wave equation,

$$\nabla^2 u - \frac{1}{c^2}\frac{\partial^2 u}{\partial t^2} = 0, \tag{2.1-2}$$

The Wave Equation

where ∇^2 is the Laplacian operator, $\nabla^2 = \partial^2/\partial x^2 + \partial^2/\partial y^2 + \partial^2/\partial z^2$. Any function satisfying (2.1-2) represents a possible optical wave.

Because the wave equation is linear, the **principle of superposition** applies; i.e., if $u_1(\mathbf{r}, t)$ and $u_2(\mathbf{r}, t)$ represent optical waves, then $u(\mathbf{r}, t) = u_1(\mathbf{r}, t) + u_2(\mathbf{r}, t)$ also represents a possible optical wave.

At the boundary between two different media, the wavefunction changes in a way that depends on the refractive indices. However, the laws that govern this change depend on the physical significance assigned to the wavefunction (i.e., the component of the electromagnetic field it represents), as discussed in Chap. 5.

The wave equation is approximately applicable to media with position-dependent refractive indices, provided that the variation is slow within distances of a wavelength. The medium is then said to be locally homogeneous. For such media, n in (2.1-1) and c in (2.1-2) are simply replaced by position-dependent functions $n(\mathbf{r})$ and $c(\mathbf{r})$, respectively.

Intensity, Power, and Energy

The optical **intensity** $I(\mathbf{r}, t)$, defined as the optical power per unit area (units of watts/cm^2), is proportional to the average of the squared wavefunction,

$$I(\mathbf{r}, t) = 2\langle u^2(\mathbf{r}, t)\rangle. \tag{2.1-3}$$

Optical Intensity

The operation $\langle \cdot \rangle$ denotes averaging over a time interval that is much longer than the time of an optical cycle, but much shorter than any other time of interest (the duration of a pulse of light, for example). The duration of an optical cycle is extremely short: 2×10^{-15} s = 2 fs for light of wavelength 600 nm, as an example. This concept is explained further in Sec. 2.6.

Although the physical meaning of the wavefunction $u(\mathbf{r}, t)$ has not been specified, (2.1-3) represents its connection with a physically measurable quantity—the optical intensity. There is some arbitrariness in the definition of the wavefunction and its relation to the intensity. Equation (2.1-3) could have, for example, been written without the factor 2 and the wavefunction scaled by a factor $\sqrt{2}$, so that the intensity remains the same. The choice of the factor 2 will later prove convenient, however.

The optical **power** $P(t)$ (units of watts) flowing into an area A normal to the direction of propagation of light is the integrated intensity

$$P(t) = \int_A I(\mathbf{r}, t)\, dA. \tag{2.1-4}$$

The optical **energy** (units of joules) collected in a given time interval is the time integral of the optical power over the time interval.

2.2 MONOCHROMATIC WAVES

A monochromatic wave is represented by a wavefunction with harmonic time dependence,

$$u(\mathbf{r}, t) = a(\mathbf{r}) \cos[2\pi\nu t + \varphi(\mathbf{r})], \tag{2.2-1}$$

as shown in Fig. 2.2-1(a), where

$a(\mathbf{r})$ = amplitude
$\varphi(\mathbf{r})$ = phase
ν = frequency (cycles/s or Hz)
$\omega = 2\pi\nu$ = angular frequency (radians/s).

Both the amplitude and the phase are generally position dependent, but the wavefunc-

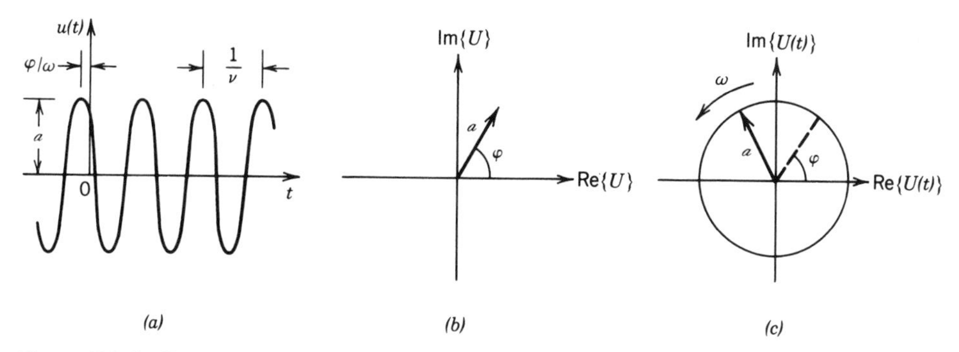

Figure 2.2-1 Representations of a monochromatic wave at a fixed position **r**: (*a*) the wavefunction $u(t)$ is a harmonic function of time; (*b*) the complex amplitude $U = a \exp(j\varphi)$ is a fixed phasor; (*c*) the complex wavefunction $U(t) = U \exp(j2\pi\nu t)$ is a phasor rotating with angular velocity $\omega = 2\pi\nu$ radians/s.

tion is a harmonic function of time with frequency ν at all positions. The frequency of optical waves lies in the range 3×10^{11} to 3×10^{16} Hz, as depicted in Fig. 2.0-1.

A. Complex Representation and the Helmholtz Equation

Complex Wavefunction

It is convenient to represent the real wavefunction $u(\mathbf{r}, t)$ in (2.2-1) in terms of a complex function

$$U(\mathbf{r}, t) = a(\mathbf{r}) \exp[j\varphi(\mathbf{r})] \exp(j2\pi\nu t), \tag{2.2-2}$$

so that

$$u(\mathbf{r}, t) = \operatorname{Re}\{U(\mathbf{r}, t)\} = \tfrac{1}{2}[U(\mathbf{r}, t) + U^*(\mathbf{r}, t)]. \tag{2.2-3}$$

The function $U(\mathbf{r}, t)$, known as the **complex wavefunction**, describes the wave completely; the wavefunction $u(\mathbf{r}, t)$ is simply its real part. Like the wavefunction $u(\mathbf{r}, t)$, the complex wavefunction $U(\mathbf{r}, t)$ must also satisfy the wave equation,

$$\boxed{\nabla^2 U - \frac{1}{c^2} \frac{\partial^2 U}{\partial t^2} = 0.} \tag{2.2-4}$$

The Wave Equation

The two functions satisfy the same boundary conditions.

Complex Amplitude

Equation (2.2-2) may be written in the form

$$U(\mathbf{r}, t) = U(\mathbf{r}) \exp(j2\pi\nu t), \tag{2.2-5}$$

where the time-independent factor $U(\mathbf{r}) = a(\mathbf{r}) \exp[j\varphi(\mathbf{r})]$ is referred to as the **complex**

amplitude. The wavefunction $u(\mathbf{r}, t)$ is therefore related to the complex amplitude by

$$u(\mathbf{r}, t) = \mathrm{Re}\{U(\mathbf{r}) \exp(j2\pi\nu t)\} = \tfrac{1}{2}[U(\mathbf{r}) \exp(j2\pi\nu t) + U^*(\mathbf{r}) \exp(-j2\pi\nu t)]. \tag{2.2-6}$$

At a given position $\mathbf{r}$, the complex amplitude $U(\mathbf{r})$ is a complex variable [depicted in Fig. 2.2-1(b)] whose magnitude $|U(\mathbf{r})| = a(\mathbf{r})$ is the amplitude of the wave and whose argument $\arg\{U(\mathbf{r})\} = \varphi(\mathbf{r})$ is the phase. The complex wavefunction $U(\mathbf{r}, t)$ is represented graphically by a phasor rotating with angular velocity $\omega = 2\pi\nu$ radians/s [Fig. 2.2-1(c)]. Its initial value at $t = 0$ is the complex amplitude $U(\mathbf{r})$.

The Helmholtz Equation

Substituting $U(\mathbf{r}, t) = U(\mathbf{r}) \exp(j2\pi\nu t)$ into the wave equation (2.2-4), we obtain the differential equation

$$(\nabla^2 + k^2)U(\mathbf{r}) = 0, \tag{2.2-7}$$

Helmholtz Equation

called the **Helmholtz equation**, where

$$k = \frac{2\pi\nu}{c} = \frac{\omega}{c} \tag{2.2-8}$$

Wavenumber

is referred to as the **wavenumber**.

Optical Intensity

The optical intensity is determined by use of (2.1-3). When

$$\begin{aligned} 2u^2(\mathbf{r}, t) &= 2a^2(\mathbf{r}) \cos^2[2\pi\nu t + \varphi(r)] \\ &= |U(\mathbf{r})|^2\{1 + \cos(2[2\pi\nu t + \varphi(\mathbf{r})])\} \end{aligned} \tag{2.2-9}$$

is averaged over a time longer than an optical period, $1/\nu$, the second term of (2.2-9) vanishes, so that

$$I(\mathbf{r}) = |U(\mathbf{r})|^2. \tag{2.2-10}$$

Optical Intensity

Thus *the optical intensity of a monochromatic wave is the absolute square of its complex amplitude*. The intensity of a monochromatic wave does *not* vary with time.

Wavefronts

The wavefronts are the surfaces of equal phase, $\varphi(\mathbf{r}) =$ constant. The constants are often taken to be multiples of 2π, $\varphi(\mathbf{r}) = 2\pi q$, where q is an integer. The wavefront normal at position $\mathbf{r}$ is parallel to the gradient vector $\nabla\varphi(\mathbf{r})$ (a vector with components $\partial\varphi/\partial x$, $\partial\varphi/\partial y$, and $\partial\varphi/\partial z$ in a Cartesian coordinate system). It represents the direction at which the rate of change of the phase is maximum.

Summary

- A monochromatic wave of frequency ν is described by a *complex wavefunction* $U(\mathbf{r}, t) = U(\mathbf{r})\exp(j2\pi\nu t)$, which satisfies the wave equation.
- The *complex amplitude* $U(\mathbf{r})$ satisfies the Helmholtz equation; its magnitude $|U(\mathbf{r})|$ and argument $\arg\{U(\mathbf{r})\}$ are the *amplitude* and *phase* of the wave, respectively. The optical *intensity* is $I(\mathbf{r}) = |U(\mathbf{r})|^2$. The *wavefronts* are the surfaces of constant phase, $\varphi(\mathbf{r}) = \arg\{U(\mathbf{r})\} = 2\pi q$ (q = integer).
- The *wavefunction* $u(\mathbf{r}, t)$ is the real part of the complex wavefunction, $u(\mathbf{r}, t) = \mathrm{Re}\{U(\mathbf{r}, t)\}$. The wavefunction also satisfies the wave equation.

B. Elementary Waves

The simplest solutions of the Helmholtz equation in a homogeneous medium are the plane wave and the spherical wave.

The Plane Wave

The plane wave has complex amplitude

$$U(\mathbf{r}) = A\exp(-j\mathbf{k}\cdot\mathbf{r}) = A\exp\left[-j(k_x x + k_y y + k_z z)\right], \qquad (2.2\text{-}11)$$

where A is a complex constant called the **complex envelope** and $\mathbf{k} = (k_x, k_y, k_z)$ is called the **wavevector**. For (2.2-11) to satisfy the Helmholtz equation (2.2-7), $k_x^2 + k_y^2 + k_z^2 = k^2$, so that the magnitude of the wavevector $\mathbf{k}$ is the wavenumber k.

Since the phase $\arg\{U(\mathbf{r})\} = \arg\{A\} - \mathbf{k}\cdot\mathbf{r}$, the wavefronts obey $\mathbf{k}\cdot\mathbf{r} = k_x x + k_y y + k_z z = 2\pi q + \arg\{A\}$ (q = integer). This is the equation describing parallel planes perpendicular to the wavevector $\mathbf{k}$ (hence the name "plane wave"). These planes are separated by a distance $\lambda = 2\pi/k$, so that

$$\boxed{\lambda = \frac{c}{\nu},} \qquad (2.2\text{-}12)$$

Wavelength

where λ is called the **wavelength**. The plane wave has a constant intensity $I(\mathbf{r}) = |A|^2$ everywhere in space so that it carries infinite power. This wave is clearly an idealization since it exists everywhere and at all times.

If the z axis is taken in the direction of the wavevector $\mathbf{k}$, then $U(\mathbf{r}) = A\exp(-jkz)$ and the corresponding wavefunction obtained from (2.2-6) is

$$u(\mathbf{r}, t) = |A|\cos\left[2\pi\nu t - kz + \arg\{A\}\right] = |A|\cos\left[2\pi\nu(t - z/c) + \arg\{A\}\right]. \qquad (2.2\text{-}13)$$

The wavefunction is therefore periodic in time with period $1/\nu$, and periodic in space with period $2\pi/k$, which is equal to the wavelength λ (see Fig. 2.2-2). Since the phase of the complex wavefunction, $\arg\{U(\mathbf{r}, t)\} = 2\pi\nu(t - z/c) + \arg\{A\}$, varies with time

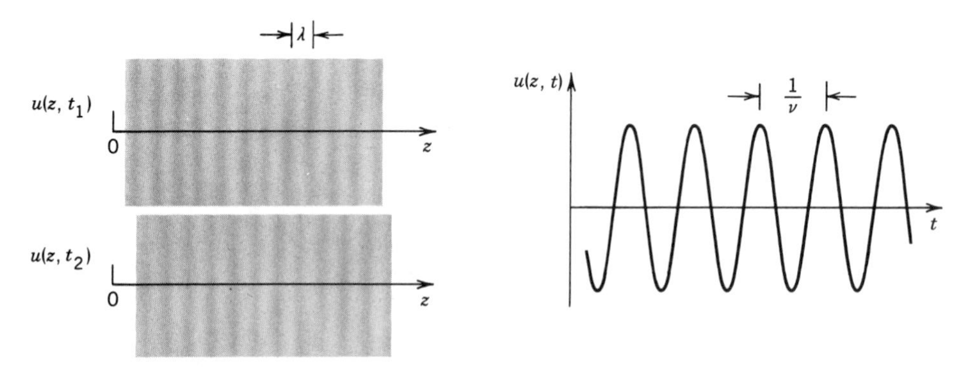

Figure 2.2-2 A plane wave traveling in the z direction is a periodic function of z with spatial period λ and a periodic function of t with temporal period $1/\nu$.

and position as a function of the variable $t - z/c$ (see Fig. 2.2-2), c is called the **phase velocity** of the wave.

In a medium of refractive index n, the phase velocity $c = c_o/n$ and the wavelength $\lambda = c/\nu = c_o/n\nu$, so that $\lambda = \lambda_o/n$ where $\lambda_o = c_o/\nu$ is the wavelength in free space. For a given frequency ν, the wavelength in the medium is reduced relative to that in free space by the factor n. As a consequence, the wavenumber $k = 2\pi/\lambda$ increases relative to that in free space ($k_o = 2\pi/\lambda_o$) by the factor n. In summary: *As a monochromatic wave propagates through media of different refractive indices its frequency remains the same, but its velocity, wavelength, and wavenumber are altered*:

$$c = \frac{c_o}{n}, \qquad \lambda = \frac{\lambda_o}{n}, \qquad k = nk_o. \tag{2.2-14}$$

The wavelengths shown in Fig. 2.0-1 are in free space ($n = 1$).

The Spherical Wave

Another simple solution of the Helmholtz equation is the spherical wave

$$U(\mathbf{r}) = \frac{A}{r}\exp(-jkr), \tag{2.2-15}$$

where r is the distance from the origin and $k = 2\pi\nu/c = \omega/c$ is the wavenumber. The intensity $I(\mathbf{r}) = |A|^2/r^2$ is inversely proportional to the square of the distance. Taking $\arg\{A\} = 0$ for simplicity, the wavefronts are the surfaces $kr = 2\pi q$ or $r = q\lambda$, where q is an integer. These are concentric spheres separated by a radial distance $\lambda = 2\pi/k$ that advance radially at the phase velocity c (Fig. 2.2-3).

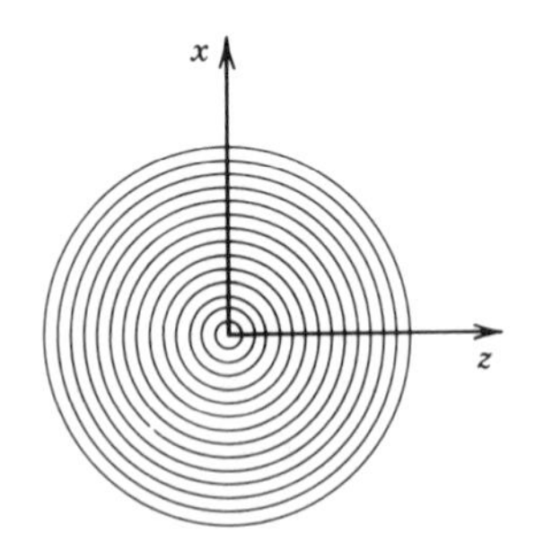

Figure 2.2-3 Cross section of the wavefronts of a spherical wave.

A spherical wave originating at the position $\mathbf{r}_0$ has a complex amplitude $U(\mathbf{r}) = (A/|\mathbf{r}-\mathbf{r}_0|)\exp(-jk|\mathbf{r}-\mathbf{r}_0|)$. Its wavefronts are spheres centered about $\mathbf{r}_0$. A wave with complex amplitude $U(\mathbf{r}) = (A/r)\exp(+jkr)$ is a spherical wave traveling inwardly (toward the origin) instead of outwardly (away from the origin).

Fresnel Approximation of the Spherical Wave; The Paraboloidal Wave

Let us examine a spherical wave originating at $\mathbf{r} = \mathbf{0}$ at points $\mathbf{r} = (x, y, z)$ sufficiently close to the z axis but far from the origin, so that $(x^2 + y^2)^{1/2} \ll z$. The paraxial approximation of ray optics (see Sec. 1.2) would be applicable were these points the endpoints of rays beginning at the origin. Denoting $\theta^2 = (x^2 + y^2)/z^2 \ll 1$, we use an approximation based on the Taylor series expansion

$$r = (x^2 + y^2 + z^2)^{1/2} = z(1 + \theta^2)^{1/2} = z\left(1 + \frac{\theta^2}{2} - \frac{\theta^4}{8} + \cdots\right)$$

$$\approx z\left(1 + \frac{\theta^2}{2}\right) = z + \frac{x^2 + y^2}{2z}.$$

Substituting $r = z + (x^2 + y^2)/2z$ into the phase, and $r = z$ into the magnitude of $U(\mathbf{r})$ in (2.2-15), we obtain

$$U(\mathbf{r}) \approx \frac{A}{z}\exp(-jkz)\exp\left[-jk\frac{x^2 + y^2}{2z}\right]. \tag{2.2-16}$$

Fresnel Approximation of a Spherical Wave

A more accurate value of r was used in the phase since the sensitivity to errors of the phase is greater. This is called the **Fresnel approximation**. It plays an important role in simplifying the theory of transmission of optical waves through apertures (**diffraction**), as discussed in Chap. 4.

The complex amplitude in (2.2-16) may be viewed as representing a plane wave $A\exp(-jkz)$ modulated by the factor $(1/z)\exp[-jk(x^2 + y^2)/2z]$, which involves a phase $k(x^2 + y^2)/2z$. This phase factor serves to bend the planar wavefronts of the plane wave into paraboloidal surfaces (Fig. 2.2-4), since the equation of a paraboloid of revolution is $(x^2 + y^2)/z =$ constant. Thus the spherical wave is approximated by a **paraboloidal wave**. When z becomes very large, the phase in (2.2-16) approaches kz and the magnitude varies slowly with z, so that the spherical wave eventually resembles the plane wave $\exp(-jkz)$, as illustrated in Fig. 2.2-4.

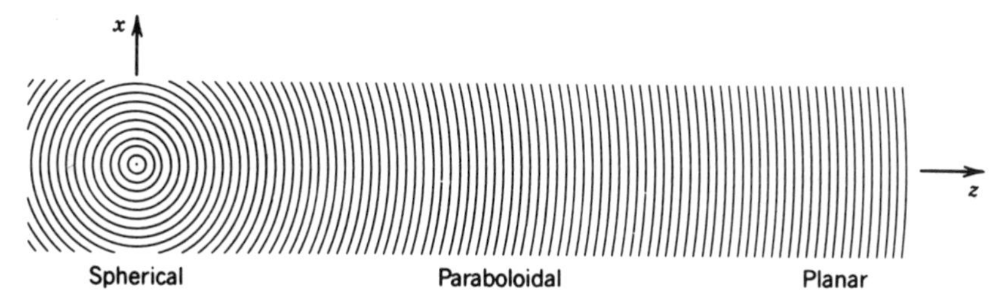

Figure 2.2-4 A spherical wave may be approximated at points near the z axis and sufficiently far from the origin by a paraboloidal wave. For very far points, the spherical wave approaches the plane wave.

The condition of validity of the Fresnel approximation is *not* simply that $\theta^2 \ll 1$. Although the third term of the series expansion, $\theta^4/8$, may be very small in comparison with the second and first terms, when multiplied by kz it may become comparable to π. The approximation is therefore valid when $kz\theta^4/8 \ll \pi$, or $(x^2 + y^2)^2 \ll 4z^3\lambda$. For points (x, y) lying within a circle of radius a centered about the z axis, the validity condition is $a^4 \ll 4z^3\lambda$ or

$$\frac{N_F\theta_m^2}{4} \ll 1, \tag{2.2-17}$$

where $\theta_m = a/z$ is the maximum angle and

$$\boxed{N_F = \frac{a^2}{\lambda z}} \tag{2.2-18}$$

Fresnel Number

is known as the **Fresnel number.**

EXERCISE 2.2-1

Validity of the Fresnel Approximation. Determine the radius of a circle within which a spherical wave of wavelength $\lambda = 633$ nm, originating at a distance 1 m away, may be approximated by a paraboloidal wave. Determine the maximum angle θ_m and the Fresnel number N_F.

C. Paraxial Waves

A wave is said to be paraxial if its wavefront normals are paraxial rays. One way of constructing a paraxial wave is to start with a plane wave $A\exp(-jkz)$, regard it as a "carrier" wave, and modify or "modulate" its complex envelope A, making it a slowly varying function of position $A(\mathbf{r})$ so that the complex amplitude of the modulated wave becomes

$$U(\mathbf{r}) = A(\mathbf{r})\exp(-jkz). \tag{2.2-19}$$

The variation of $A(\mathbf{r})$ with position must be slow within the distance of a wavelength $\lambda = 2\pi/k$, so that the wave approximately maintains its underlying plane-wave nature.

The wavefunction $u(\mathbf{r}, t) = |A(\mathbf{r})|\cos[2\pi\nu t - kz + \arg\{A(\mathbf{r})\}]$ of a paraxial wave is sketched in Fig. 2.2-5(a) as a function of z at $t = 0$ and $x = y = 0$. This is a sinusoidal function of z with amplitude $|A(0, 0, z)|$ and phase $\arg\{A(0, 0, z)\}$ that vary slowly with z. Since the change of the phase $\arg\{A(x, y, z)\}$ is small within the distance of a wavelength, the planar wavefronts, $kz = 2\pi q$, of the carrier plane wave bend only slightly, so that their normals are paraxial rays [Fig. 2.2-5(b)].

The Paraxial Helmholtz Equation

For the paraxial wave (2.2-19) to satisfy the Helmholtz equation (2.2-7), the complex envelope $A(\mathbf{r})$ must satisfy another partial differential equation obtained by substituting (2.2-19) into (2.2-7). The assumption that $A(\mathbf{r})$ varies slowly with respect to z signifies that within a distance $\Delta z = \lambda$, the change ΔA is much smaller than A itself;

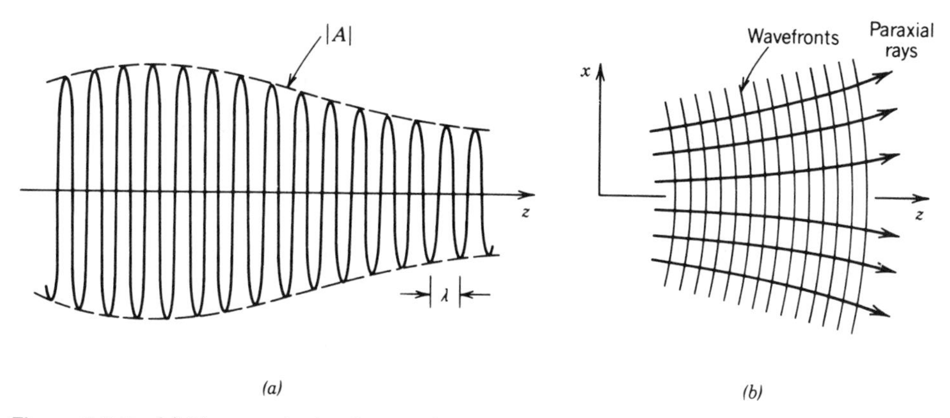

Figure 2.2-5 (a) The magnitude of a paraxial wave as a function of the axial distance z. (b) The wavefronts and wavefront normals of a paraxial wave.

i.e., $\Delta A \ll A$. This inequality of complex variables applies to the magnitudes of the real and imaginary parts separately. Since $\Delta A = (\partial A/\partial z)\,\Delta z = (\partial A/\partial z)\lambda$, it follows that $\partial A/\partial z \ll A/\lambda = Ak/2\pi$, and therefore

$$\frac{\partial A}{\partial z} \ll kA. \tag{2.2-20}$$

Similarly, the derivative $\partial A/\partial z$ varies slowly within the distance λ, so that $\partial^2 A/\partial^2 z \ll k\,\partial A/\partial z$, and therefore

$$\frac{\partial^2 A}{\partial z^2} \ll k^2 A. \tag{2.2-21}$$

Substituting (2.2-19) into (2.2-7) and neglecting $\partial^2 A/\partial z^2$ in comparison with $k\,\partial A/\partial z$ or $k^2 A$, we obtain

$$\nabla_T^2 A - j2k\frac{\partial A}{\partial z} = 0, \tag{2.2-22}$$

Paraxial Helmholtz Equation

where $\nabla_T^2 = \partial^2/\partial x^2 + \partial^2/\partial y^2$ is the transverse Laplacian operator.

Equation (2.2-22) is the **slowly varying envelope approximation** of the Helmholtz equation. We shall simply call it the **paraxial Helmholtz equation**. It is a partial differential equation that resembles the Schrödinger equation of quantum physics.

The simplest solution of the paraxial Helmholtz equation is the paraboloidal wave (Exercise 2.2-2), which is the paraxial approximation of the spherical wave. The most interesting and useful solution, however, is the **Gaussian beam**, to which Chap. 3 is devoted.

EXERCISE 2.2-2

The Paraboloidal Wave and the Gaussian Beam. Verify that a paraboloidal wave with the complex envelope $A(\mathbf{r}) = (A_0/z)\exp[-jk(x^2 + y^2)/2z]$ [see (2.2-16)] satisfies the paraxial Helmholtz equation (2.2-22). Show that the wave with complex amplitude $A(\mathbf{r}) =$

$[A/q(z)]\exp[-jk(x^2+y^2)/2q(z)]$, where $q(z) = z + jz_0$ and z_0 is a constant, also satisfies the paraxial Helmholtz equation. This wave, called the Gaussian beam, is the subject of Chap. 3. Sketch the intensity of the Gaussian beam in the plane $z = 0$.

*2.3 RELATION BETWEEN WAVE OPTICS AND RAY OPTICS

We proceed to show that ray optics is the limit of wave optics when the wavelength $\lambda_o \to 0$. Consider a monochromatic wave of free-space wavelength λ_o in a medium with refractive index $n(\mathbf{r})$ that varies sufficiently slowly with position so that the medium may be regarded as locally homogeneous. We write the complex amplitude in the form

$$U(\mathbf{r}) = \mathscr{a}(\mathbf{r}) \exp[-jk_o S(\mathbf{r})], \tag{2.3-1}$$

where $\mathscr{a}(\mathbf{r})$ is its magnitude, $-k_o S(\mathbf{r})$ its phase, and $k_o = 2\pi/\lambda_o$ is the wavenumber. We assume that $\mathscr{a}(\mathbf{r})$ varies sufficiently slowly with $\mathbf{r}$, so that it may be regarded as constant within the distance of a wavelength λ_o.

The wavefronts are the surfaces $S(\mathbf{r})$ = constant and the wavefront normals point in the direction of the gradient ∇S. In the neighborhood of a given position $\mathbf{r}_0$, the wave can be locally regarded as a plane wave with amplitude $\mathscr{a}(\mathbf{r}_0)$ and wavevector $\mathbf{k}$ with magnitude $k = n(\mathbf{r}_0)k_o$ and direction parallel to the gradient vector ∇S at $\mathbf{r}_0$. A different neighborhood exhibits a local plane wave of different amplitude and different wavevector.

In Chap. 1 it was shown that the optical rays are normal to the equilevel surfaces of a function $S(\mathbf{r})$ called the eikonal (see Sec. 1.3C). We therefore associate the local wavevectors (wavefront normals) in wave optics with the rays of ray optics and recognize that the function $S(\mathbf{r})$, which is proportional to the phase of the wave, is nothing but the eikonal of ray optics (Fig. 2.3-1). This association has a formal mathematical basis, as will be demonstrated subsequently. With this analogy, ray optics can serve to determine the approximate effects of optical components on the wavefront normals, as illustrated in Fig. 2.3-1.

The Eikonal Equation

Substituting (2.3-1) into the Helmholtz equation, (2.2-7) provides

$$k_o^2\left[n^2 - |\nabla S|^2\right]\mathscr{a} + \nabla^2\mathscr{a} - jk_o[2\nabla S \cdot \nabla\mathscr{a} + \mathscr{a}\nabla^2 S] = 0, \tag{2.3-2}$$

where $\mathscr{a} = \mathscr{a}(\mathbf{r})$ and $S = S(\mathbf{r})$. The real and imaginary parts of the left-hand side of (2.3-2) must both vanish. Equating the real part to zero and using $k_o = 2\pi/\lambda_o$, we obtain

$$|\nabla S|^2 = n^2 + \left(\frac{\lambda_o}{2\pi}\right)^2 \nabla^2\mathscr{a}/\mathscr{a}. \tag{2.3-3}$$

The assumption that $\mathscr{a}$ varies slowly over the distance λ_o means that $\lambda_o^2\nabla^2\mathscr{a}/\mathscr{a} \ll 1$,

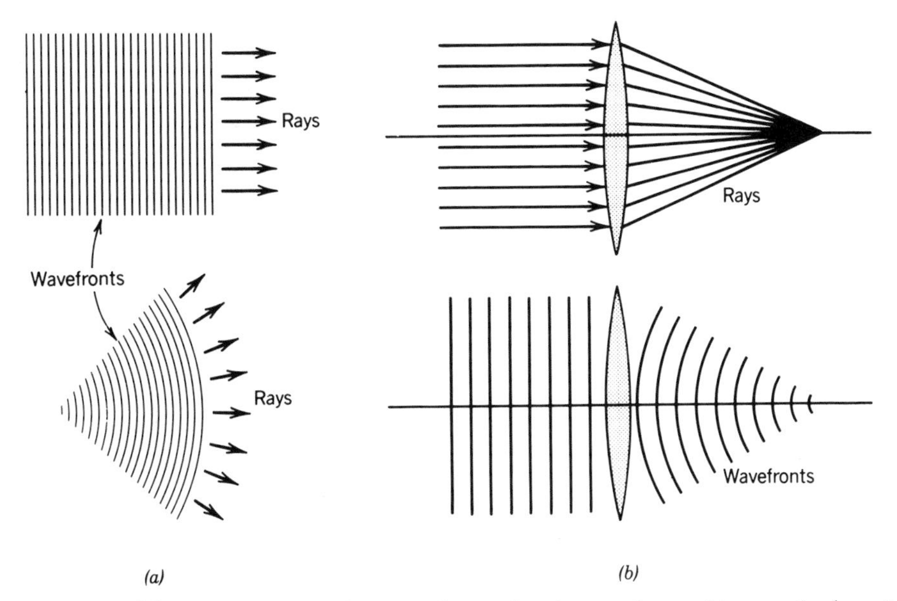

Figure 2.3-1 (*a*) The rays of ray optics are orthogonal to the wavefronts of wave optics (see also Fig. 1.3-10). (*b*) The effect of a lens on rays and wavefronts.

so that the second term of the right-hand side may be neglected in the limit $\lambda_o \to 0$ and

$$|\nabla S|^2 \approx n^2. \tag{2.3-4}$$

Eikonal Equation

This is the eikonal equation (1.3-18), which may be regarded as the main postulate of ray optics (Fermat's principle can be derived from the eikonal equation, and vice versa).

In conclusion: The scalar function $S(\mathbf{r})$, which is proportional to the phase in wave optics, is the eikonal of ray optics. This is also consistent with the observation that in ray optics $S(\mathbf{r}_B) - S(\mathbf{r}_A)$ equals the optical path length between the points $\mathbf{r}_A$ and $\mathbf{r}_B$.

The eikonal equation is the limit of the Helmholtz equation when $\lambda_o \to 0$. Given $n(\mathbf{r})$ we may use the eikonal equation to determine $S(r)$. By equating the imaginary part of (2.3-2) to zero, we obtain a relation between a and S, thereby permitting us to determine the wavefunction.

2.4 SIMPLE OPTICAL COMPONENTS

In this section we examine the effects of optical components, such as mirrors, transparent plates, prisms, and lenses, on optical waves.

A. Reflection and Refraction

Reflection from a Planar Mirror

A plane wave of wavevector $\mathbf{k}_1$ is incident onto a planar mirror located in free space in the $z = 0$ plane. A reflected plane wave of wavevector $\mathbf{k}_2$ is created. The angles of

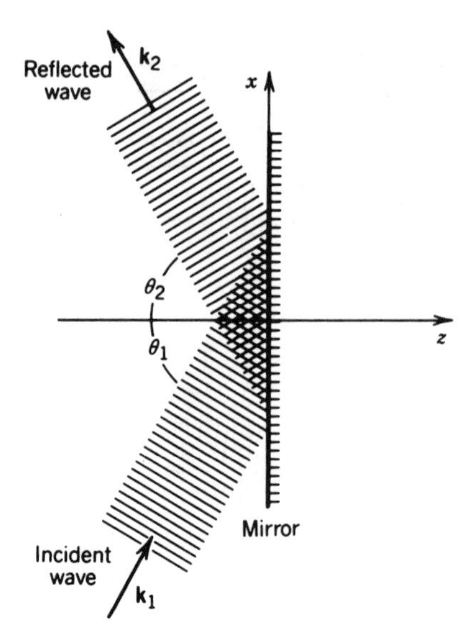

Figure 2.4-1 Reflection of a plane wave from a planar mirror. Phase matching at the surface of the mirror requires that the angles of incidence and reflection be equal.

incidence and reflection are θ_1 and θ_2, as illustrated in Fig. 2.4-1. The sum of the two waves satisfies the Helmholtz equation if $k_1 = k_2 = k_o$. Certain boundary conditions must be satisfied at the surface of the mirror. Since these conditions are the same at all points (x, y), it is necessary that the wavefronts of the two waves match, i.e., the phases must be equal,

$$\mathbf{k}_1 \cdot \mathbf{r} = \mathbf{k}_2 \cdot \mathbf{r} \qquad \text{for all } \mathbf{r} = (x, y, 0), \tag{2.4-1}$$

or differ by a constant.

Substituting $\mathbf{r} = (x, y, 0)$, $\mathbf{k}_1 = (k_o \sin\theta_1, 0, k_o \cos\theta_1)$, and $\mathbf{k}_2 = (k_o \sin\theta_2, 0, -k_o \cos\theta_2)$ into (2.4-1), we obtain $k_o \sin(\theta_1)x = k_o \sin(\theta_2)x$, from which $\theta_1 = \theta_2$, so that the angles of incidence and reflection must be equal. Thus the law of reflection of optical rays is applicable to the wavevectors of plane waves.

Reflection and Refraction at a Planar Dielectric Boundary

We now consider a plane wave of wavevector $\mathbf{k}_1$ incident on a planar boundary between two homogeneous media of refractive indices n_1 and n_2. The boundary lies in the $z = 0$ plane (Fig. 2.4-2). Refracted and reflected plane waves of wavevectors $\mathbf{k}_2$ and $\mathbf{k}_3$ emerge. The combination of the three waves satisfies the Helmholtz equation everywhere if each of the waves has the appropriate wavenumber in the medium in which it propagates ($k_1 = k_3 = n_1 k_o$ and $k_2 = n_2 k_o$).

Since the boundary conditions are invariant to x and y, it is necessary that the wavefronts of the three waves match, i.e., the phases must be equal,

$$\mathbf{k}_1 \cdot \mathbf{r} = \mathbf{k}_2 \cdot \mathbf{r} = \mathbf{k}_3 \cdot \mathbf{r} \qquad \text{for all } \mathbf{r} = (x, y, 0), \tag{2.4-2}$$

or differ by constants. Since $\mathbf{k}_1 = (n_1 k_o \sin\theta_1, 0, n_1 k_o \cos\theta_1)$, $\mathbf{k}_3 = (n_1 k_o \sin\theta_3, 0, -n_1 k_o \cos\theta_3)$, and $\mathbf{k}_2 = (n_2 k_o \sin\theta_2, 0, n_2 k_o \cos\theta_2)$, where θ_1, θ_2, and θ_3 are the angles of incidence, refraction, and reflection, respectively, it follows from (2.4-2) that $\theta_1 = \theta_3$ and $n_1 \sin\theta_1 = n_2 \sin\theta_2$. These are the laws of reflection and refraction (Snell's law) of ray optics, now applicable to the wavevectors.

It is not possible to determine the amplitudes of the reflected and refracted waves using the scalar wave theory of light since the boundary conditions are not completely specified in this theory. This will be achieved in Sec. 6.2 using the electromagnetic theory of light (Chaps. 5 and 6).

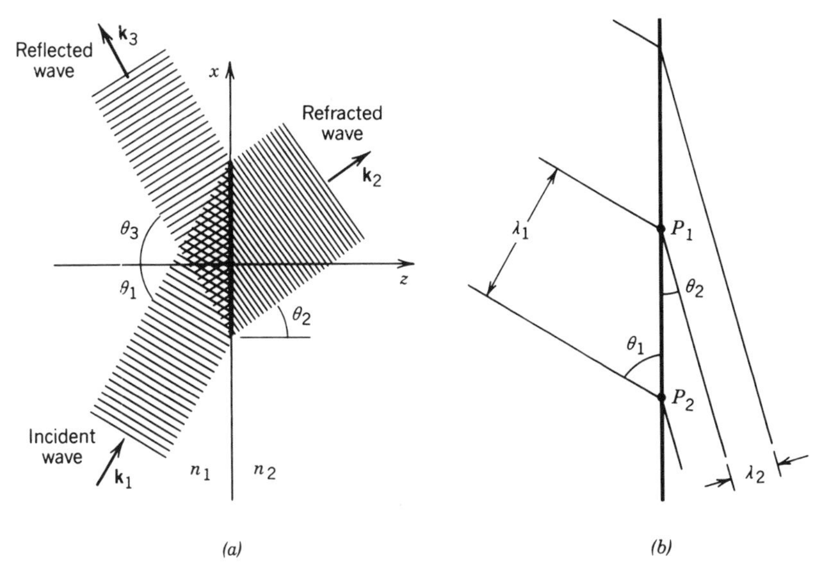

Figure 2.4-2 (*a*) Reflection and refraction of a plane wave at a dielectric boundary. (*b*) Matching the wavefronts at the boundary; the distance $\overline{P_1P_2}$ for the incident wave, $\lambda_1/\sin\theta_1 = \lambda_o/n_1 \sin\theta_1$, equals that for the refracted wave, $\lambda_2/\sin\theta_2 = \lambda_o/n_2 \sin\theta_2$, from which Snell's law follows.

B. Transmission Through Optical Components

We now proceed to examine the transmission of optical waves through transparent optical components such as plates, prisms, and lenses. The effect of reflection at the surfaces of these components will be ignored, since it cannot be properly accounted for using the scalar wave-optics model of light. The effect of absorption in the material is also ignored and relegated to Sec. 5.5. The main emphasis here is on the phase shift introduced by these components and on the associated wavefront bending.

Transmission Through a Transparent Plate

Consider first the transmission of a plane wave through a transparent plate of refractive index n and thickness d surrounded by free space. The surfaces of the plate are the planes $z = 0$ and $z = d$ and the incident wave travels in the z direction (Fig. 2.4-3). Let $U(x, y, z)$ be the complex amplitude of the wave. Since external and internal reflections are ignored, $U(x, y, z)$ is assumed to be continuous at the bound-

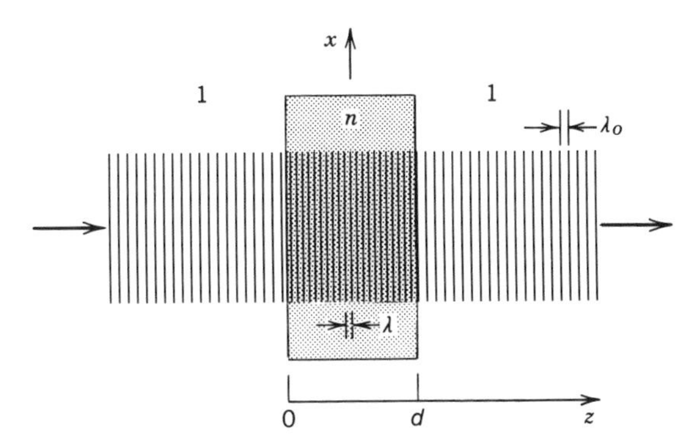

Figure 2.4-3 Transmission of a plane wave through a transparent plate.

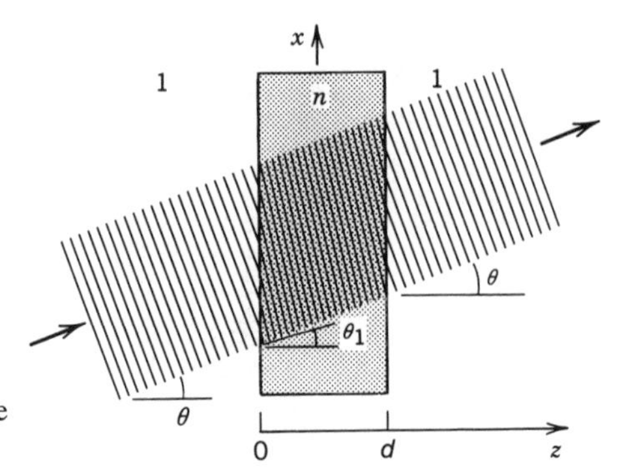

Figure 2.4-4 Transmission of an oblique plane wave through a thin transparent plate.

aries. The ratio $t(x, y) = U(x, y, d)/U(x, y, 0)$ therefore represents the **complex amplitude transmittance** of the plate. The effect of reflection is considered in Sec. 6.2 and the effect of multiple internal reflections within the plate is examined in Sec. 9.1.

The incident plane wave continues to propagate inside the plate as a plane wave with wavenumber nk_o, so that $U(x, y, z)$ is proportional to $\exp(-jnk_o z)$. Thus $U(x, y, d)/U(x, y, 0) = \exp(-jnk_o d)$, so that

$$t(x, y) = \exp(-jnk_o d), \tag{2.4-3}$$

Complex Amplitude Transmittance of a Transparent Plate

i.e., the plate introduces a phase shift $nk_o d = 2\pi(d/\lambda)$.

If the incident plane wave makes an angle θ with the z axis and has wavevector $\mathbf{k}$ (Fig. 2.4-4), the refracted and transmitted waves are also plane waves with wavevectors $\mathbf{k}_1$ and $\mathbf{k}$ and angles θ_1 and θ, respectively, where θ_1 and θ are related by Snell's law, $\sin\theta = n\sin\theta_1$. The complex amplitude $U(x, y, z)$ inside the plate is now proportional to $\exp(-j\mathbf{k}_1 \cdot \mathbf{r}) = \exp[-jnk_o(z\cos\theta_1 + x\sin\theta_1)]$, so that the complex amplitude transmittance of the plate $U(x, y, d)/U(x, y, 0)$ is

$$t(x, y) = \exp\left[-jnk_o(d\cos\theta_1 + x\sin\theta_1)\right].$$

If the angle of incidence θ is small (i.e., the incident wave is *paraxial*), then $\theta_1 \approx \theta/n$ is also small and the approximations $\sin\theta \approx \theta$ and $\cos\theta \approx 1 - \frac{1}{2}\theta^2$ yield $t(x, y) \approx \exp(-jnk_o d)\exp(jk_o\theta^2 d/2n - jk_o\theta x)$. If the plate is sufficiently *thin* and the angle θ is sufficiently small such that $k_o\theta^2 d/2n \ll 2\pi$ [or $(d/\lambda_o)\theta^2/2n \ll 1$] and if $(x/\lambda)\theta \ll 1$ for all values of x of interest, then the transmittance of the plate may be approximated by (2.4-3). Under these conditions the transmittance of the plate is approximately independent of the angle θ.

Thin Transparent Plate of Varying Thickness

We now determine the amplitude transmittance of a thin transparent plate whose thickness $d(x, y)$ varies smoothly as a function of x and y, assuming that the incident wave is an arbitrary *paraxial* wave. The plate lies between the planes $z = 0$ and $z = d_0$, which are regarded as the boundaries of the optical component (Fig. 2.4-5).

In the vicinity of the position $(x, y, 0)$ the incident paraxial wave may be regarded locally as a plane wave traveling along a direction making a small angle with the z axis. It crosses a thin plate of width $d(x, y)$ surrounded by thin layers of air of total width $d_0 - d(x, y)$. In accordance with the approximate relation (2.4-3), the local transmit-

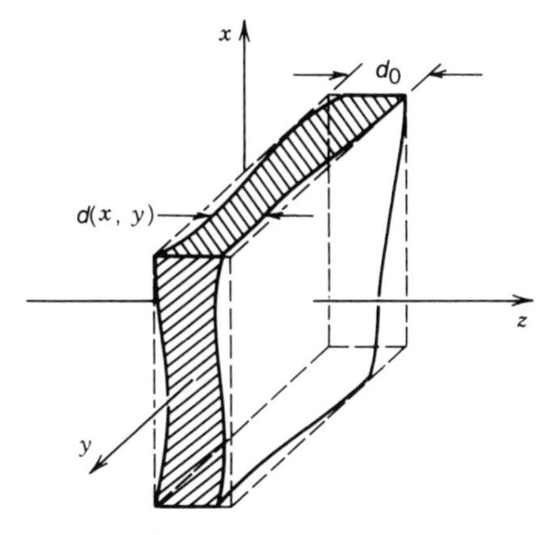

Figure 2.4-5 A transparent plate of varying thickness.

tance is the product of the transmittances of a thin layer of air of thickness $d_0 - d(x, y)$ and a thin layer of material of thickness $d(x, y)$, so that $t(x, y) \approx \exp[-jnk_o d(x, y)] \exp\{-jk_o[d_0 - d(x, y)]\}$, from which

$$t(x, y) \approx h_0 \exp[-j(n - 1)k_o d(x, y)], \tag{2.4-4}$$

Transmittance of a Variable-Thickness Plate

where $h_0 = \exp(-jk_o d_0)$ is a constant phase factor. This relation is valid in the paraxial approximation (all angles θ are small) and when the thickness d_0 is sufficiently small so that $(d_0/\lambda)\theta^2/2n \ll 1$ at all points (x, y) for which $(x/\lambda)\theta \ll 1$ and $(y/\lambda)\theta \ll 1$.

EXERCISE 2.4-1

Transmission Through a Prism. Use (2.4-4) to show that the complex amplitude transmittance of a thin inverted prism with small angle $\alpha \ll 1$ and width d_0 (Fig. 2.4-6) is $t(x, y) = h_0 \exp[-j(n - 1)k_o \alpha x]$, where $h_0 = \exp(-jk_o d_0)$. What is the effect of the

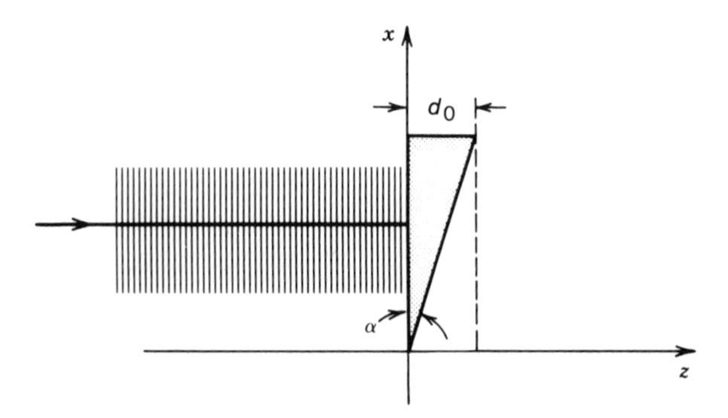

Figure 2.4-6 Transmission of a plane wave through a thin prism.

prism on an incident plane wave traveling in the z direction? Compare your results with the results obtained in the ray-optics model [see (1.2-7)].

Thin Lens

The general expression (2.4-4) for the complex amplitude transmittance of a thin transparent plate of variable thickness is now applied to the planoconvex thin lens shown in Fig. 2.4-7. Since the lens is the cap of a sphere of radius R, the thickness at the point (x, y) is $d(x,y) = d_0 - \overline{PQ} = d_0 - (R - \overline{QC})$, or

$$d(x,y) = d_0 - \left\{R - \left[R^2 - (x^2+y^2)\right]^{1/2}\right\}. \tag{2.4-5}$$

This expression may be simplified by considering only points for which x and y are sufficiently small in comparison with R so that $x^2 + y^2 \ll R^2$. In this case

$$\left[R^2 - (x^2+y^2)\right]^{1/2} = R\left(1 - \frac{x^2+y^2}{R^2}\right)^{1/2} \approx R\left(1 - \frac{x^2+y^2}{2R^2}\right),$$

and (2.4-5) gives

$$d(x,y) \approx d_0 - \frac{x^2+y^2}{2R}.$$

Upon substitution into (2.4-4) we obtain

$$t(x,y) \approx h_0 \exp\left[jk_o \frac{x^2+y^2}{2f}\right], \tag{2.4-6}$$

Transmittance of a Thin Lens

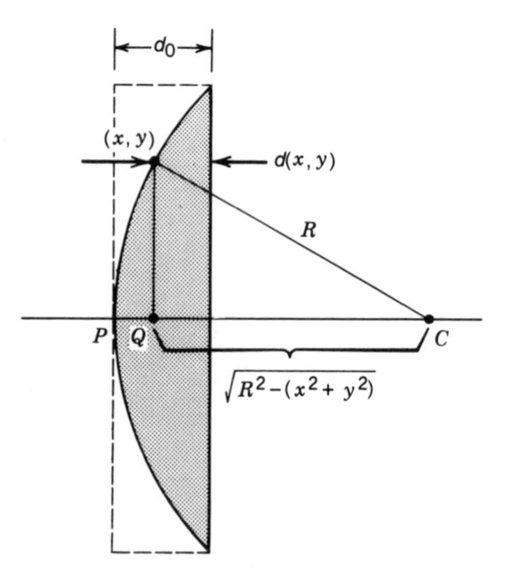

Figure 2.4-7 A planoconvex lens.

where

$$f = \frac{R}{n-1} \tag{2.4-7}$$

is the focal length of the lens (see Sec. 1.2C) and $h_0 = \exp(-jnk_o d_0)$ is a constant phase factor that is usually of no significance.

Since the lens imparts to the incident wave a phase proportional to $x^2 + y^2$, it bends the planar wavefronts of a plane wave, transforming it into a paraboloidal wave centered at a distance f from the lens, as demonstrated in Exercise 2.4-3.

EXERCISE 2.4-2

Double-Convex Lens. Show that the complex amplitude transmittance of the double-convex lens shown in Fig. 2.4-8 is given by (2.4-6) with

$$\frac{1}{f} = (n-1)\left(\frac{1}{R_1} - \frac{1}{R_2}\right). \tag{2.4-8}$$

You may prove this either by using the general formula (2.4-4) or by regarding the double-convex lens as a cascade of two planoconvex lenses. Recall that, by convention, the radius of a convex/concave surface is positive/negative, i.e., R_1 is positive and R_2 is negative for the lens in Fig. 2.4-8. The parameter f is recognized as the focal length of the lens [see (1.2-12)].

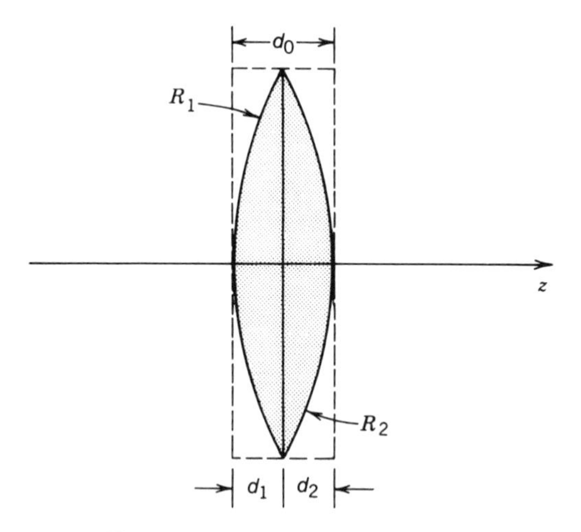

Figure 2.4-8 A double-convex lens.

EXERCISE 2.4-3

Focusing of a Plane Wave by a Thin Lens. Show that when a plane wave is transmitted through a thin lens of focal length f in a direction parallel to the axis of the

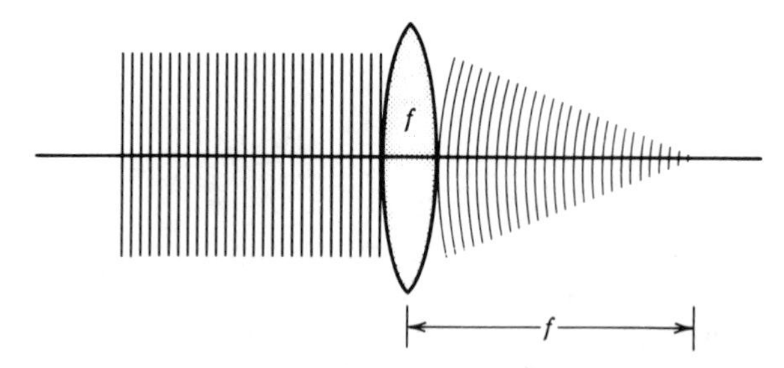

Figure 2.4-9 A thin lens transforms a plane wave into a paraboloidal wave.

lens, it is converted into a paraboloidal wave (the Fresnel approximation of a spherical wave) centered about a point at a distance f from the lens, as illustrated in Fig. 2.4-9. What is the effect of the lens on a plane wave incident at a small angle θ?

EXERCISE 2.4-4

Imaging Property of a Lens. Show that a paraboloidal wave centered at the point P_1 (Fig. 2.4-10) is converted by a lens of focal length f into a paraboloidal wave centered about P_2, where $1/z_1 + 1/z_2 = 1/f$.

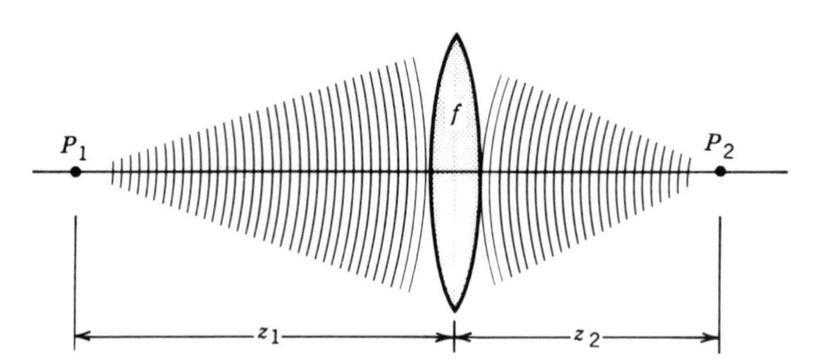

Figure 2.4-10 A lens transforms a paraboloidal wave into another paraboloidal wave. The two waves are centered at distances satisfying the imaging equation.

Diffraction Gratings

A **diffraction grating** is an optical component that serves to periodically modulate the phase or the amplitude of the incident wave. It can be made of a transparent plate with periodically varying thickness or periodically graded refractive index (see Sec. 2.4C). Repetitive arrays of diffracting elements such as apertures, obstacles, or absorbing elements can also be used (see Sec. 4.3). Reflection diffraction gratings are often fabricated by use of periodically ruled thin films of aluminum that have been evaporated onto a glass substrate.

Consider here a diffraction grating made of a thin transparent plate placed in the $z = 0$ plane whose thickness varies periodically in the x direction with period Λ (Fig. 2.4-11). As will be demonstrated in Exercise 2.4-5, this plate converts an incident plane wave of wavelength $\lambda \ll \Lambda$, traveling at a small angle θ_i with respect to the z axis, into

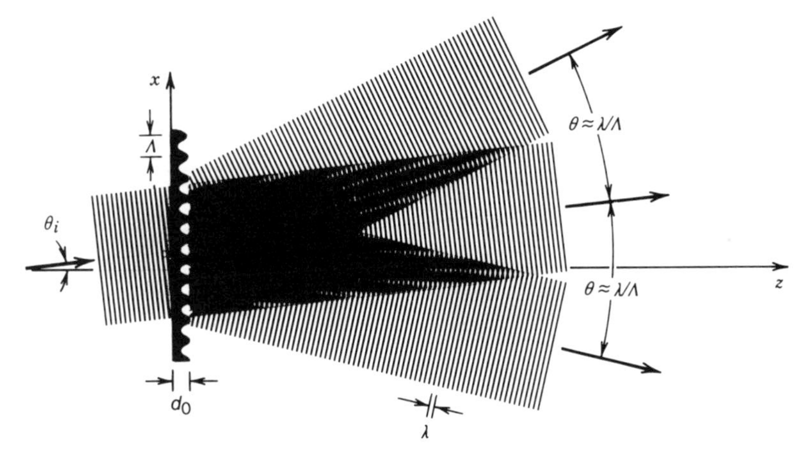

Figure 2.4-11 A thin transparent plate with periodically varying thickness serves as a diffraction grating. It splits an incident plane wave into multiple plane waves traveling in different directions.

several plane waves at small angles

$$\theta_q = \theta_i + q\frac{\lambda}{\Lambda}, \tag{2.4-9}$$

Grating Equation

$q = 0, \pm 1, \pm 2, \ldots,$ with the z axis, where q is called the diffraction order. The diffracted waves are separated by an angle $\theta = \lambda/\Lambda$, as shown schematically in Fig. 2.4-11.

EXERCISE 2.4-5

Transmission Through a Diffraction Grating

(a) The thickness of a thin transparent plate varies sinusoidally in the x direction, $d(x, y) = \frac{1}{2}d_0[1 + \cos(2\pi x/\Lambda)]$, as illustrated in Fig. 2.4-11. Show that the complex amplitude transmittance is $t(x, y) = h_0 \exp[-j\frac{1}{2}(n - 1)k_o d_0 \cos(2\pi x/\Lambda)]$ where $h_0 = \exp[-j\frac{1}{2}(n + 1)k_o d_0]$.

(b) Show that an incident plane wave traveling at a small angle θ_i with the z direction is transmitted in the form of a sum of plane waves traveling at angles θ_q given by (2.4-9). *Hint*: Expand the periodic function $t(x, y)$ in a Fourier series.

Equation (2.4-9) is valid only in the paraxial approximation (when all angles are small). This approximation is applicable when the period Λ is much greater than the wavelength λ. A more general analysis of thin diffraction gratings, without the use of the paraxial approximation, shows that the incident plane wave is converted into

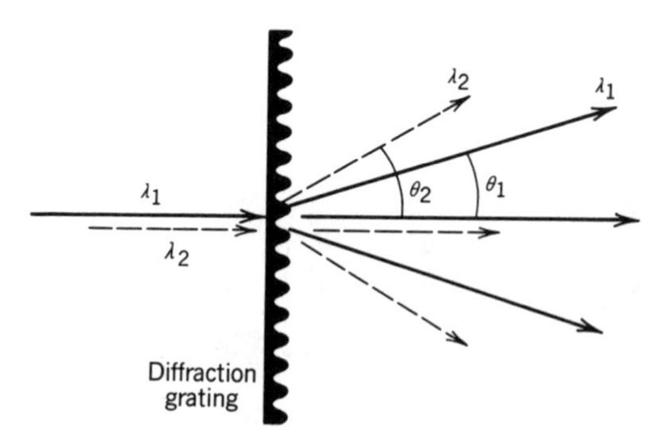

Figure 2.4-12 A diffraction grating directs two waves of wavelengths λ_1 and λ_2 into two directions θ_1 and θ_2. It therefore serves as a spectrum analyzer or a spectrometer.

several plane waves at angles θ_q satisfying[†]

$$\sin \theta_q = \sin \theta_i + q \frac{\lambda}{\Lambda}. \tag{2.4-10}$$

Diffraction gratings are used as filters and spectrum analyzers. Since the angles θ_q are dependent on the wavelength λ (and therefore on the frequency ν), an incident polychromatic wave is separated by the grating into its spectral components (Fig. 2.4-12). Diffraction gratings have found numerous applications in spectroscopy.

C. Graded-Index Optical Components

The effect of a prism, lens, or diffraction grating lies in the phase shift it imparts to the incident wave, which serves to bend the wavefront in some prescribed manner. This phase shift is controlled by the variation of the thickness of the material with the transverse distance from the optical axis (linearly, quadratically, or periodically, in the cases of the prism, lens, and diffraction grating, respectively). The same phase shift may instead be introduced by a transparent planar plate of fixed width but with varying refractive index.

The complex amplitude transmittance of a thin transparent planar plate of width d_0 and graded refractive index $n(x, y)$ is

$$t(x, y) = \exp[-jn(x, y) k_o d_0]. \tag{2.4-11}$$

Transmittance of a Graded-Index Thin Plate

By selecting the appropriate variation of $n(x, y)$ with x and y, the action of any constant-index thin optical component can be reproduced, as demonstrated in Exercise 2.4-6.

[†]See, e.g., E. Hecht and A. Zajac, *Optics*, Addison-Wesley, Reading, MA, 1974.

EXERCISE 2.4-6

Graded-Index Lens. Show that a thin plate (Fig. 2.4-13) of uniform thickness d_0 and quadratically graded refractive index $n(x, y) = n_0[1 - \frac{1}{2}\alpha^2(x^2 + y^2)]$, where $\alpha d_0 \ll 1$, acts as a lens of focal length $f = 1/n_0\alpha^2 d_0$ (see Exercise 1.3-1).

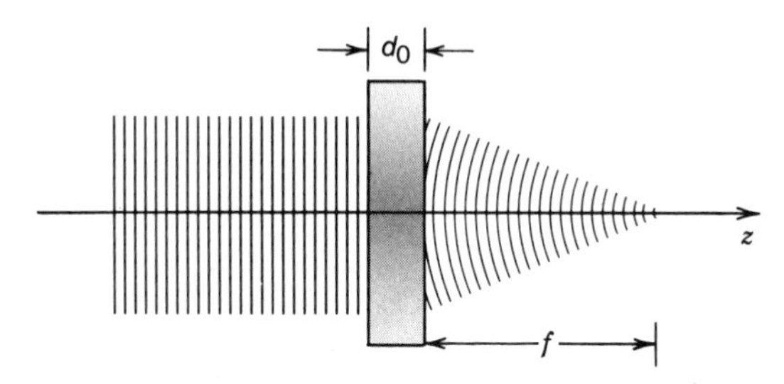

Figure 2.4-13 A graded-index plate acts as a lens.

2.5 INTERFERENCE

When two or more optical waves are present simultaneously in the same region of space, the total wavefunction is the sum of the individual wavefunctions. This basic principle of superposition follows from the linearity of the wave equation. For monochromatic waves of the same frequency, the superposition principle is also applicable to the complex amplitudes. This is consistent with the linearity of the Helmholtz equation.

The superposition principle does not apply to the optical intensity. The intensity of the superposition of two or more waves is not necessarily the sum of their intensities. The difference is attributed to the interference between these waves. Interference cannot be explained on the basis of ray optics since it is dependent on the phase relationship between the superposed waves.

In this section we examine the interference between two or more monochromatic waves of the same frequency. The interference of waves of different frequencies is discussed in Sec. 2.6.

A. Interference of Two Waves

When two monochromatic waves of complex amplitudes $U_1(\mathbf{r})$ and $U_2(\mathbf{r})$ are superposed, the result is a monochromatic wave of the same frequency and complex amplitude

$$U(\mathbf{r}) = U_1(\mathbf{r}) + U_2(\mathbf{r}). \tag{2.5-1}$$

In accordance with (2.2-10), the intensities of the constituent waves are $I_1 = |U_1|^2$ and $I_2 = |U_2|^2$ and the intensity of the total wave is

$$I = |U|^2 = |U_1 + U_2|^2 = |U_1|^2 + |U_2|^2 + U_1^* U_2 + U_1 U_2^*. \tag{2.5-2}$$

The explicit dependence on **r** has been omitted for convenience. Substituting

$$U_1 = I_1^{1/2} \exp(j\varphi_1) \qquad \text{and} \qquad U_2 = I_2^{1/2} \exp(j\varphi_2) \tag{2.5-3}$$

into (2.5-2), where φ_1 and φ_2 are the phases of the two waves, we obtain

$$\boxed{I = I_1 + I_2 + 2(I_1 I_2)^{1/2} \cos\varphi,} \tag{2.5-4}$$

Interference Equation

with

$$\varphi = \varphi_2 - \varphi_1. \tag{2.5-5}$$

This relation, called the **interference equation**, can also be seen from the geometry of the phasor diagram in Fig. 2.5-1(*a*), which demonstrates that the magnitude of the phasor U is sensitive to the phase difference φ, not only to the magnitudes of the constituent phasors.

The intensity of the sum of the two waves is *not* the sum of their intensities [Fig. 2.5-1(*b*)]; an additional term, attributed to **interference** between the two waves, is present in (2.5-4). This term may be positive or negative, corresponding to constructive or destructive interference, respectively. If $I_1 = I_2 = I_0$, for example, then $I = 2I_0(1 + \cos\varphi) = 4I_0\cos^2(\varphi/2)$, so that for $\varphi = 0$, $I = 4I_0$ (i.e., the total intensity is four times the intensity of each of the superposed waves). For $\varphi = \pi$, the superposed waves cancel one another and the total intensity $I = 0$. When $\varphi = \pi/2$ or $3\pi/2$, the interference term vanishes and $I = 2I_0$, i.e., the total intensity is the sum of the constituent intensities. The strong dependence of the intensity I on the phase difference φ permits us to measure phase differences by detecting light intensity. This principle is used in numerous optical systems.

Interference is not observed under ordinary lighting conditions since the random fluctuations of the phases φ_1 and φ_2 cause the phase difference φ to assume random values, which are uniformly distributed between 0 and 2π, so that the average of $\cos\varphi = 0$ and the interference term is washed out. Light with such randomness is said to be *partially coherent* and Chap. 10 is devoted to its study. We limit ourselves here to the study of *coherent* light.

Interference is accompanied by a spatial redistribution of the optical intensity without violation of power conservation. For example, the two waves may have uniform intensities I_1 and I_2 in some plane, but as a result of a position-dependent phase

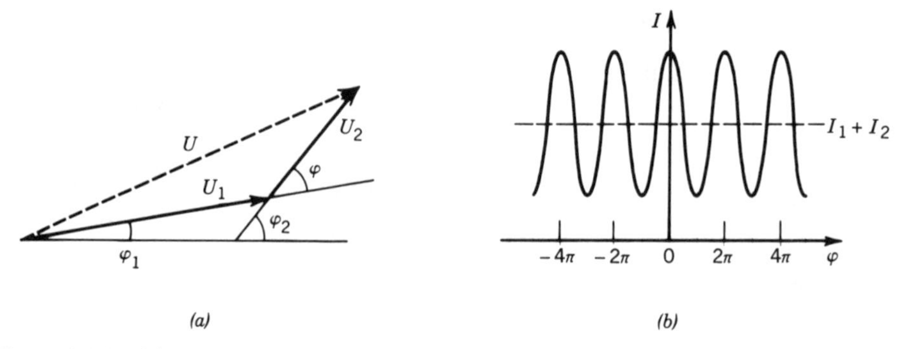

Figure 2.5-1 (*a*) Phasor diagram for the superposition of two waves of intensities I_1 and I_2 and phase difference $\varphi = \varphi_2 - \varphi_1$. (*b*) Dependence of the total intensity I on the phase difference φ.

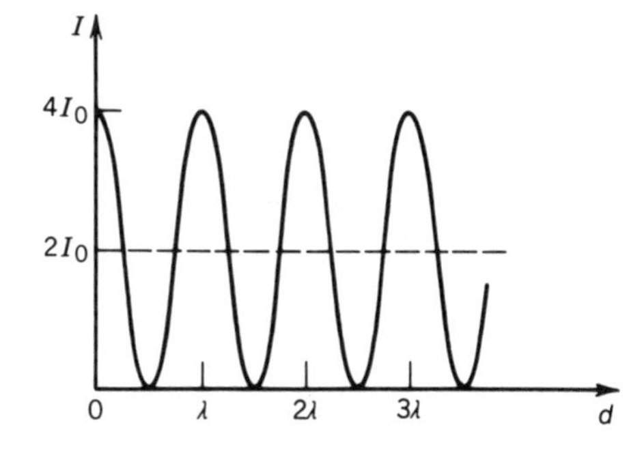

Figure 2.5-2 Dependence of the intensity I of the superposition of two waves, each of intensity I_0, on the delay distance d. When the delay distance is a multiple of λ, the interference is constructive; when it is an odd multiple of $\lambda/2$, the interference is destructive.

difference φ, the total intensity may be smaller than $I_1 + I_2$ at some positions and greater than $I_1 + I_2$ at others, with the total power (integral of the intensity) conserved.

Interferometers

Consider the superposition of two plane waves, each of intensity I_0, propagating in the z direction, and assume that one wave is delayed by a distance d with respect to the other so that $U_1 = I_0^{1/2}\exp(-jkz)$ and $U_2 = I_0^{1/2}\exp[-jk(z - d)]$. The intensity I of the sum of these two waves can be determined by substituting $I_1 = I_2 = I_0$ and $\varphi = kd = 2\pi d/\lambda$ into the interference equation (2.5-4),

$$I = 2I_0\left[1 + \cos\left(2\pi\frac{d}{\lambda}\right)\right]. \tag{2.5-6}$$

The dependence of I on the delay d is sketched in Fig. 2.5-2. If the delay is an integer multiple of λ, complete constructive interference occurs and the total intensity $I = 4I_0$. On the other hand, if d is an odd integer multiple of $\lambda/2$, complete destructive interference occurs and $I = 0$. The average intensity is the sum of the two intensities $2I_0$.

An **interferometer** is an optical instrument that splits a wave into two waves using a beamsplitter, delays them by unequal distances, redirects them using mirrors, recombines them using another (or the same) beamsplitter, and detects the intensity of their superposition. Three important examples, the **Mach–Zehnder interferometer**, the **Michelson interferometer**, and the **Sagnac interferometer**, are illustrated in Fig. 2.5-3.

Since the intensity I is sensitive to the phase $\varphi = 2\pi d/\lambda = 2\pi n d/\lambda_o = 2\pi n\nu d/c_o$, where d is the difference between the distances traveled by the two waves, the interferometer can be used to measure small variations of the distance d, the refractive index n, or the wavelength λ_o (or frequency ν). For example, if $d/\lambda_o = 10^4$, a change $\Delta n = 10^{-4}$ of the refractive index corresponds to a phase change $\Delta\varphi = 2\pi$. Also, the phase φ changes by a full 2π if d changes by a wavelength λ. An incremental change of the frequency $\Delta\nu = c/d$ has the same effect. Interferometers can serve as spectrometers, which measure the spectrum of polychromatic light (see Sec. 10.2B). In the Sagnac interferometer the optical paths are identical but opposite, so that rotation of the interferometer results in a phase shift φ proportional to the angular velocity of rotation. This system is therefore often used as a gyroscope.†

Interference of Two Oblique Plane Waves

Consider now the interference of two plane waves of equal intensities—one propagating in the z direction $U_1 = I_0^{1/2}\exp(-jkz)$, and the other at an angle θ with the z axis

†See, e.g., E. Hecht and A. Zajac, *Optics*, Addison-Wesley, Reading, MA, 1974.

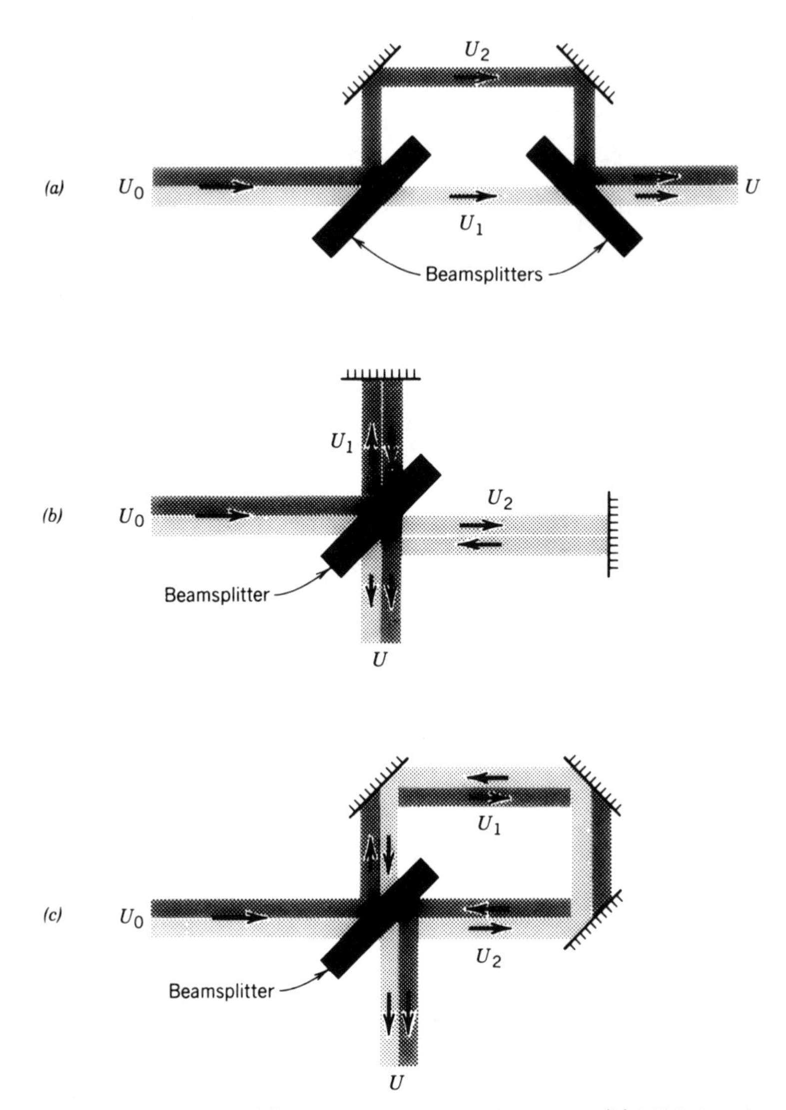

Figure 2.5-3 Interferometers: (*a*) Mach–Zehnder interferometer; (*b*) Michelson interferometer; (*c*) Sagnac interferometer. A wave U_0 is split into two waves U_1 and U_2. After traveling through different paths, the waves are recombined into a superposition wave $U = U_1 + U_2$ whose intensity is recorded. The waves are split and recombined using beamsplitters. In the Sagnac interferometer the two waves travel through the same path in opposite directions.

in the x-z plane, $U_2 = I_0^{1/2} \exp[-j(k \cos \theta z + k \sin \theta x)]$, as illustrated in Fig. 2.5-4. At the $z = 0$ plane the two waves have a phase difference $\varphi = kx \sin \theta$, so that the interference equation (2.5-4) yields the total intensity:

$$I = 2I_0[1 + \cos(k \sin \theta x)]. \tag{2.5-7}$$

This pattern varies sinusoidally with x, with period $2\pi/k \sin \theta = \lambda/\sin \theta$, as shown in Fig. 2.5-4. If $\theta = 30°$, for example, the period is 2λ. This suggests a method of printing a sinusoidal pattern of high resolution for use as a diffraction grating. It also suggests a method of monitoring the angle of arrival θ of a wave by mixing it with a reference

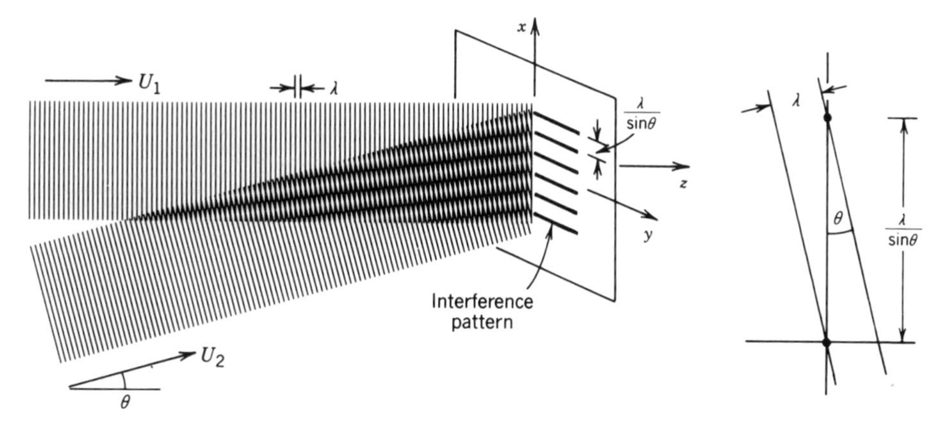

Figure 2.5-4 The interference of two plane waves at an angle θ results in a sinusoidal intensity pattern of period $\lambda/\sin\theta$.

wave and recording the resultant intensity distribution. As discussed in Sec. 4.5, this is the principle behind holography.

EXERCISE 2.5-1

Interference of a Plane Wave and a Spherical Wave. A plane wave of complex amplitude $A_1 \exp(-jkz)$ and a spherical wave approximated by the paraboloidal wave of complex amplitude $(A_2/z)\exp(-jkz)\exp[-jk(x^2+y^2)/2z]$ [see (2.2-16)], interfere in the $z = d$ plane. Derive an expression for the total intensity $I(x, y, d)$. Verify that the locus of points of zero intensity is a set of concentric rings, as illustrated in Fig. 2.5-5.

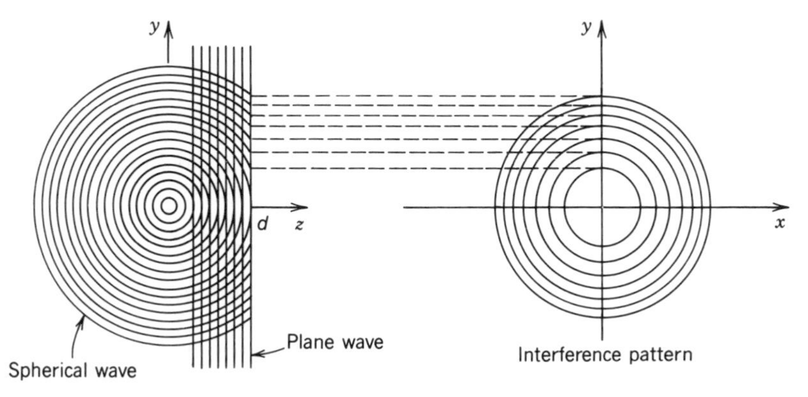

Figure 2.5-5 The interference of a plane wave and a spherical wave creates a pattern of concentric rings (illustrated at the plane $z = d$).

EXERCISE 2.5-2

Interference of Two Spherical Waves. Two spherical waves of equal intensity I_0, originating at the points $(a, 0, 0)$ and $(-a, 0, 0)$ interfere in the plane $z = d$ as illustrated in Fig. 2.5-6. The system is similar to that used by Thomas Young in his celebrated double-slit

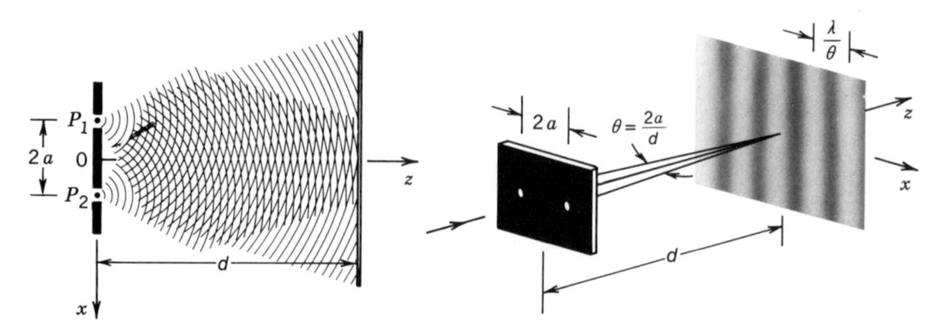

Figure 2.5-6 Interference of two spherical waves of equal intensities originating at the points P_1 and P_2. The two waves can be obtained by permitting a plane wave to impinge on two pinholes in a screen. The light intensity at an observation plane a large distance d away takes the form of a sinusoidal pattern with period $\approx \lambda/\theta$.

experiment in which he demonstrated interference. Use the paraboloidal approximation for the spherical waves to show that the detected intensity is

$$I(x, y, d) = 2I_0\left(1 + \cos\frac{2\pi x\theta}{\lambda}\right), \tag{2.5-8}$$

where $\theta = 2a/d$ is approximately the angle subtended by the centers of the two waves at the observation plane. The intensity pattern is periodic with period λ/θ.

B. Multiple-Wave Interference

When M monochromatic waves of complex amplitudes $U_1, U_2, \ldots, U_M$ and the same frequency are added, the result is a monochromatic wave with complex amplitude $U = U_1 + U_2 + \cdots + U_M$. Knowing the intensities of the individual waves, $I_1, I_2, \ldots, I_M$, is not sufficient to determine the total intensity $I = |U|^2$ since the relative phases must also be known. The role played by the phase is dramatically illustrated by the following examples.

Interference of M Waves of Equal Amplitudes and Equal Phase Differences
We first examine the interference of M waves with complex amplitudes

$$U_m = I_0^{1/2} \exp[j(m-1)\varphi], \qquad m = 1, 2, \ldots, M. \tag{2.5-9}$$

The waves have equal intensities I_0, and phase difference φ between successive waves, as illustrated in Fig. 2.5-7(a). To derive an expression for the intensity of the superposition, it is convenient to introduce $h = \exp(j\varphi)$, and write $U_m = I_0^{1/2} h^{m-1}$. The complex amplitude of the superposed wave is then

$$U = I_0^{1/2}(1 + h + h^2 + \cdots + h^{M-1}) = I_0^{1/2}\frac{1 - h^M}{1 - h}$$

$$= I_0^{1/2}\frac{1 - \exp(jM\varphi)}{1 - \exp(j\varphi)},$$

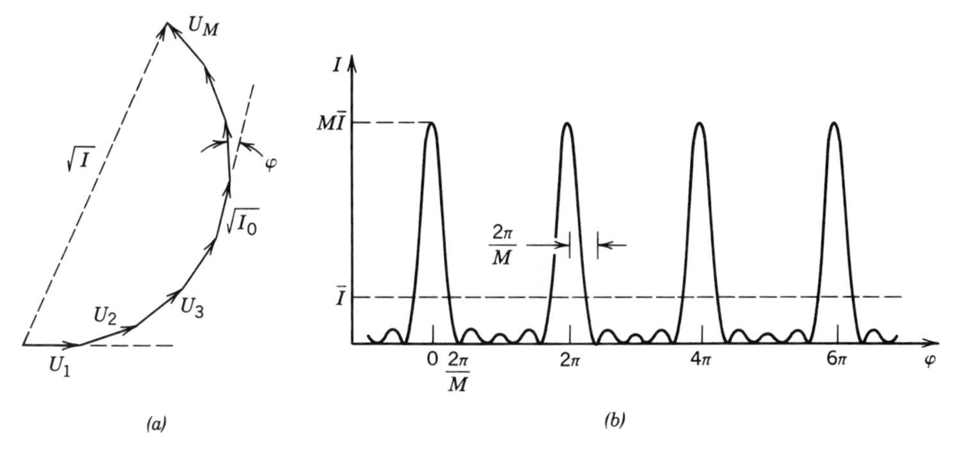

Figure 2.5-7 (*a*) The sum of M phasors of equal magnitudes and equal phase differences. (*b*) The intensity I as a function of φ. The peak intensity occurs when all the phasors are aligned; it is M times greater than the mean intensity $\bar{I} = MI_0$. In this graph $M = 5$.

and the corresponding intensity is

$$I = |U|^2 = I_0 \left| \frac{\exp(-jM\varphi/2) - \exp(jM\varphi/2)}{\exp(-j\varphi/2) - \exp(j\varphi/2)} \right|^2,$$

from which

$$I = I_0 \frac{\sin^2(M\varphi/2)}{\sin^2(\varphi/2)}. \qquad (2.5\text{-}10)$$

Interference of M Waves

The intensity I is strongly dependent on the phase difference φ, as illustrated in Fig. 2.5-7(*b*) for $M = 5$. When $\varphi = 2\pi q$, where q is an integer, all the phasors are aligned so that the amplitude of the total wave is M times that of an individual component, and the intensity reaches its peak value of $M^2 I_0$. The mean intensity averaged over a uniform distribution of φ is $\bar{I} = (1/2\pi)\int_0^{2\pi} I\, d\varphi = MI_0$, which is the result obtained in the absence of interference. The peak intensity is therefore M times greater than the mean intensity. If M is large, the sensitivity to the phase can be dramatic since the peak intensity will be much greater than the mean intensity. For a phase difference φ slightly different from $2\pi q$, the intensity drops sharply. In particular, when it is $2\pi/M$ the intensity is zero. A comparison of Fig. 2.5-7(*b*) for $M = 5$ with Fig. 2.5-2 for $M = 2$ is instructive.

EXERCISE 2.5-3

Bragg Reflection. Light is reflected at an angle θ from M parallel reflecting planes separated by a distance d as shown in Fig. 2.5-8. Assume that only a small fraction of the light is reflected from each plane, so that the amplitudes of the M reflected waves are

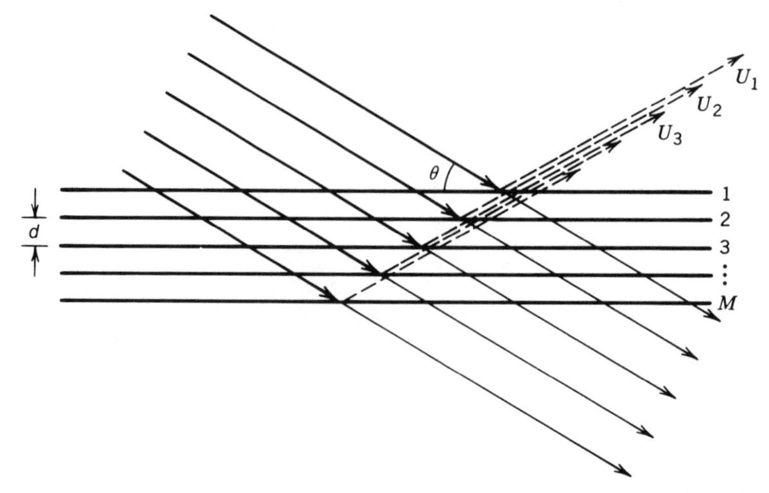

Figure 2.5-8 Reflection of a plane wave from M planes separated from each other by a distance d. The reflected waves interfere constructively and yield maximum intensity when the angle θ is the Bragg angle.

approximately equal. Show that the reflected waves have a phase difference $\varphi = k(2d \sin \theta)$ and that the angle θ at which the intensity of the total reflected light is maximum satisfies

$$\sin \theta = \frac{\lambda}{2d}. \tag{2.5-11}$$

Bragg Angle

This angle is known as the Bragg angle. Such reflections are encountered when x-ray waves are reflected from atomic planes in crystalline structures. It also occurs when light is reflected from a periodic structure created by an acoustic wave (see Chap. 20).

Interference of an Infinite Number of Waves of Progressively Smaller Amplitudes and Equal Phase Differences

We now examine the superposition of an infinite number of waves with equal phase differences and with amplitudes that decrease at a geometric rate,

$$U_1 = I_0^{1/2}, \quad U_2 = hU_1, \quad U_3 = hU_2 = h^2 U_1, \quad \ldots, \tag{2.5-12}$$

where $h = \mathscr{r} e^{j\varphi}$, $|h| = \mathscr{r} < 1$, and I_0 is the intensity of the initial wave. The amplitude of the mth wave is smaller than that of the $(m-1)$st wave by the factor $\mathscr{r}$ and the phase differs by φ. The phasor diagram is shown in Fig. 2.5-9(a).

The superposition wave has a complex amplitude

$$\begin{aligned} U &= U_1 + U_2 + U_3 + \cdots \\ &= I_0^{1/2}(1 + h + h^2 + \cdots) \\ &= \frac{I_0^{1/2}}{1 - h} = \frac{I_0^{1/2}}{1 - \mathscr{r} e^{j\varphi}}. \end{aligned} \tag{2.5-13}$$

The intensity $I = |U|^2 = I_0/|1 - re^{j\varphi}|^2 = I_0/[(1 - r\cos\varphi)^2 + r^2\sin^2\varphi]$, from which

$$I = \frac{I_0}{(1-r)^2 + 4r\sin^2(\varphi/2)}. \tag{2.5-14}$$

It is convenient to write this equation in the form

$$I = \frac{I_{\max}}{1 + (2\mathcal{F}/\pi)^2 \sin^2(\varphi/2)}, \tag{2.5-15}$$

Intensity of an Infinite Number of Waves

where

$$I_{\max} = \frac{I_0}{(1-r)^2} \tag{2.5-16}$$

and the quantity

$$\mathcal{F} = \frac{\pi r^{1/2}}{1-r} \tag{2.5-17}$$

Finesse

is a parameter called the **finesse**.

The intensity I is a periodic function of φ with period 2π, as illustrated in Fig. 2.5-9(*b*). It reaches its maximum value $I_{\max}$ when $\varphi = 2\pi q$, where q is an integer. This occurs when the phasors align to form a straight line. (This result is not unlike that displayed in Fig. 2.5-7(*b*) for the interference of M waves of equal amplitudes and equal phase differences.) When the finesse $\mathcal{F}$ is large (i.e., the factor r is close to 1), I becomes a sharply peaked function of φ. Consider, for example, values of φ near the

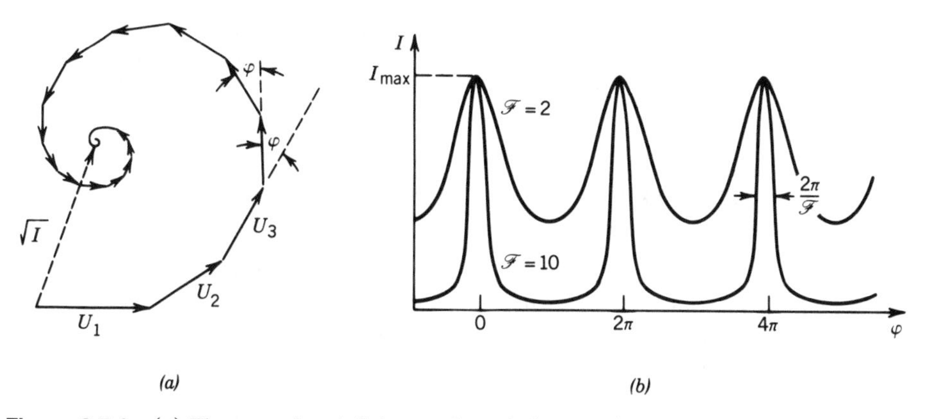

Figure 2.5-9 (*a*) The sum of an infinite number of phasors whose magnitudes are successively reduced at a geometric rate and whose phase differences φ are equal. (*b*) Dependence of the intensity I on the phase difference φ for two values of $\mathcal{F}$. Peak values occur at $\varphi = 2\pi q$. The width (FWHM) of each peak is approximately $2\pi/\mathcal{F}$ when $\mathcal{F} \gg 1$. The sharpness of the peaks increases with increasing $\mathcal{F}$.

$\varphi = 0$ peak. For $|\varphi| \ll 1$, $\sin(\varphi/2) \approx \varphi/2$ and (2.5-15) may be approximated by

$$I \approx \frac{I_{\max}}{1 + (\mathscr{F}/\pi)^2 \varphi^2}. \tag{2.5-18}$$

The intensity I decreases to half its peak valued when $\varphi = \pi/\mathscr{F}$, so that the full width at half maximum (FWHM) of the peak is

$$\Delta\varphi = \frac{2\pi}{\mathscr{F}}. \tag{2.5-19}$$

Width of Interference Pattern

If $\mathscr{F} \gg 1$, $\Delta\varphi \ll 2\pi$ and the assumption that $\varphi \ll 1$ is applicable. The finesse $\mathscr{F}$ is therefore the ratio between the period 2π and the FWHM of the interference pattern. It is a measure of the sharpness of the interference function, i.e., the sensitivity of the intensity to deviations of φ from the values $2\pi q$ corresponding to the peaks.

The Fabry–Perot interferometer is a useful device based on this principle. It consists of two parallel mirrors within which light undergoes multiple reflections. In the course of each round trip, the light suffers a fixed amplitude reduction $\mathscr{r}$ and a phase shift $\varphi = k2d = 4\pi\nu d/c$, where d is the mirror separation. The total light intensity depends on the phase shift φ in accordance with (2.5-15). Because the phase shift φ is proportional to the optical frequency ν, the intensity transmission of the device exhibits spectral characteristics with peaks at resonance frequencies separated by $c/2d$. The width of these resonances is $(c/2d)/\mathscr{F}$, where the finesse $\mathscr{F}$ is governed by losses (since it is related to the attenuation factor $\mathscr{r}$). The Fabry–Perot interferometer serves as a spectrum analyzer and as an optical resonator, which is one of the essential components of a laser. Optical resonators are discussed in Chap. 9.

2.6 POLYCHROMATIC LIGHT

Since the wavefunction of monochromatic light is a harmonic function of time that extends over all time (from $-\infty$ to ∞), it is an idealization that cannot be met in reality. This section is devoted to polychromatic waves of finite time duration, including optical pulses.

A. Fourier Decomposition

A polychromatic wave can be expanded as a sum of monochromatic waves by the use of Fourier methods. Since we already know how monochromatic waves are transmitted through optical components, we can determine the effect of optical systems on polychromatic light by using the principle of superposition.

An arbitrary function of time, such as the wavefunction $u(\mathbf{r}, t)$ at a fixed position $\mathbf{r}$, can be analyzed as a superposition integral of harmonic functions of different frequencies, amplitudes, and phases,

$$u(\mathbf{r}, t) = \int_{-\infty}^{\infty} U_\nu(\mathbf{r}) \exp(j2\pi\nu t)\, d\nu, \tag{2.6-1}$$

where $U_\nu(\mathbf{r})$ is determined by carrying out the **Fourier transform**

$$U_\nu(\mathbf{r}) = \int_{-\infty}^{\infty} u(\mathbf{r}, t) \exp(-j2\pi\nu t)\, dt. \tag{2.6-2}$$

A review of the Fourier transform and its properties is presented in Appendix A.

Complex Representation

Since $u(\mathbf{r}, t)$ is real, $U_\nu(\mathbf{r})$ must be a symmetric function of ν, i.e., $U_{-\nu}(\mathbf{r}) = U_\nu^*(\mathbf{r})$. The integral in (2.6-1) may therefore be simplified by use of the relation

$$\int_{-\infty}^{0} U_\nu(\mathbf{r}) \exp(j2\pi\nu t)\, d\nu = \int_{0}^{\infty} U_{-\nu}(\mathbf{r}) \exp(-j2\pi\nu t)\, d\nu$$
$$= \int_{0}^{\infty} U_\nu^*(\mathbf{r}) \exp(-j2\pi\nu t)\, d\nu,$$

so that $u(\mathbf{r}, t)$ is the sum of a complex function and its conjugate,

$$u(\mathbf{r}, t) = \int_{0}^{\infty} \left[U_\nu(\mathbf{r}) \exp(j2\pi\nu t) + U_\nu^*(\mathbf{r}) \exp(-j2\pi\nu t) \right] d\nu. \tag{2.6-3}$$

As in the case of monochromatic light (Sec. 2.2A), the complex wavefunction is defined as twice the first term in (2.6-3),

$$U(\mathbf{r}, t) = 2\int_{0}^{\infty} U_\nu(\mathbf{r}) \exp(j2\pi\nu t)\, d\nu, \tag{2.6-4}$$

so that its real part is the wavefunction

$$u(\mathbf{r}, t) = \mathrm{Re}\{U(\mathbf{r}, t)\} = \tfrac{1}{2}[U(\mathbf{r}, t) + U^*(\mathbf{r}, t)], \tag{2.6-5}$$

as in (2.2-3). The complex wavefunction (also called the **complex analytic signal**) is therefore obtained from the wavefunction by a process of three steps: (1) determine its Fourier transform; (2) eliminate negative frequencies and multiply by 2; and (3) determine the inverse Fourier transform. Since each of its Fourier components satisfies the wave equation, the complex wavefunction $U(\mathbf{r}, t)$ itself satisfies the wave equation.

The magnitudes of the Fourier transforms of the wavefunction and the complex wavefunction of a **quasi-monochromatic** wave are illustrated in Fig. 2.6-1. A quasi-monochromatic wave has Fourier components with frequencies confined within a narrow band of width $\Delta\nu$ surrounding a central frequency ν_0, such that $\Delta\nu \ll \nu_0$.

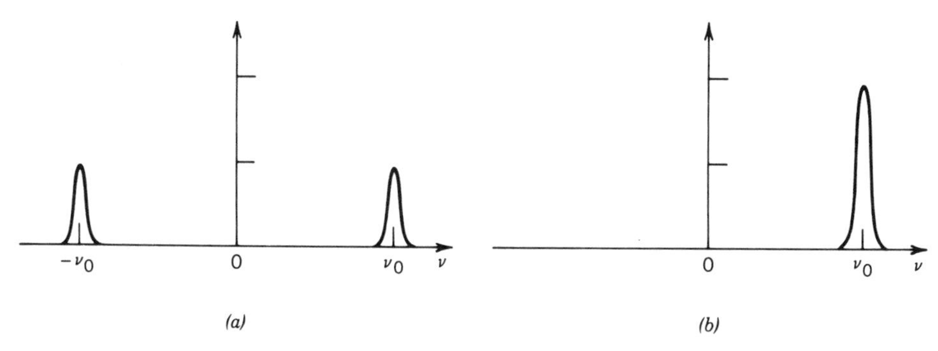

Figure 2.6-1 (*a*) The magnitude of the Fourier transform of the wavefunction. (*b*) The magnitude of the Fourier transform of the corresponding complex wavefunction.

Intensity of a Polychromatic Wave

The intensity is related to the wavefunction by

$$\begin{aligned} I(\mathbf{r},t) &= 2\langle u^2(\mathbf{r},t)\rangle \\ &= 2\langle\{\tfrac{1}{2}[U(\mathbf{r},t) + U^*(\mathbf{r},t)]\}^2\rangle \\ &= \tfrac{1}{2}\langle U^2(\mathbf{r},t)\rangle + \tfrac{1}{2}\langle U^{*2}(\mathbf{r},t)\rangle + \langle U(\mathbf{r},t)U^*(\mathbf{r},t)\rangle. \end{aligned} \tag{2.6-6}$$

If the wave is quasi-monochromatic with central frequency ν_0 and spectral width $\Delta\nu \ll \nu_0$, the average $\langle \cdot \rangle$ is taken over a time interval much longer than the time of an optical cycle $1/\nu_0$ but much shorter than $1/\Delta\nu$ (see Sec. 2.1). Since $U(\mathbf{r}, t)$ is given by (2.6-4), the term U^2 in (2.6-6) has components oscillating at frequencies $\approx 2\nu_0$. Similarly, the components of U^{*2} have frequencies $\approx -2\nu_0$. These terms are washed out by the averaging operation. The third term contains only frequency differences of the order of $\Delta\nu \ll \nu_0$. It therefore varies slowly and is unaffected by the time-averaging operation. Thus the third term survives and the light intensity is given by

$$I(\mathbf{r},t) = |U(\mathbf{r},t)|^2. \tag{2.6-7}$$

Optical Intensity of Quasi-Monochromatic Light

The intensity of a quasi-monochromatic wave is therefore given by the squared-absolute-value of its complex wavefunction. The simplicity of this result is, in fact, the rationale for introducing the concept of the complex wavefunction.

The Pulsed Plane Wave

As an example, consider a polychromatic wave each of whose monochromatic components is a plane wave traveling in the z direction with speed c. The complex wavefunction is the superposition integral

$$U(\mathbf{r},t) = \int_0^\infty A_\nu \exp(-jkz)\exp(j2\pi\nu t)\,d\nu = \int_0^\infty A_\nu \exp\left[j2\pi\nu\left(t - \frac{z}{c}\right)\right]d\nu, \tag{2.6-8}$$

where A_ν is the complex envelope of the component of frequency ν and wavenumber $k = 2\pi\nu/c$.

Assuming that the speed $c = c_o/n$ is independent of the frequency ν, (2.6-8) may be written in the form

$$U(\mathbf{r},t) = a\left(t - \frac{z}{c}\right), \tag{2.6-9}$$

where

$$a(t) = \int_0^\infty A_\nu \exp(j2\pi\nu t)\,d\nu. \tag{2.6-10}$$

Since A_ν may be arbitrarily chosen, (2.6-9) represents a valid wave, regardless of the function $a(\cdot)$ (provided that d^2a/dt^2 exists). Indeed, it can be easily verified that $U(\mathbf{r}, t) = a(t - z/c)$ satisfies the wave equation for an arbitrary form of $a(t)$.

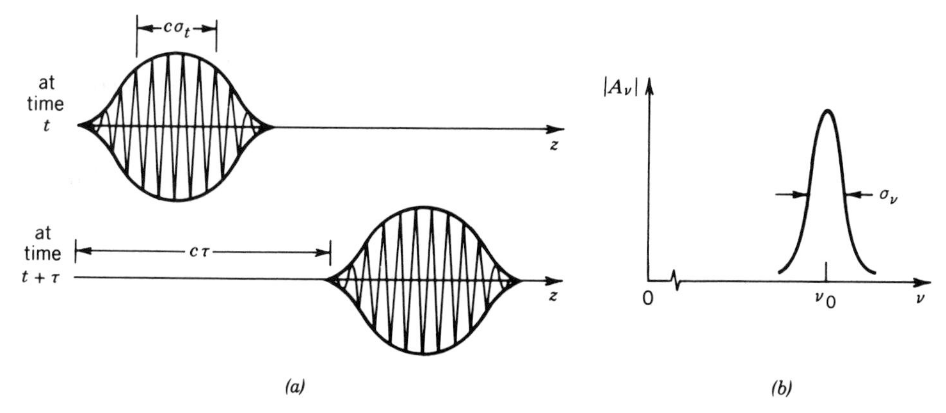

Figure 2.6-2 (a) The wavefunction $u(\mathbf{r}, t) = \text{Re}\{a(t - z/c)\}$ of a pulsed plane wave of time duration σ_t at times t and $t + \tau$. The pulse travels with speed c and occupies a distance $\sigma_z = c\sigma_t$. (b) The magnitude $|A_\nu|$ of the Fourier transform of the wavefunction is centered at ν_0 and has a width σ_ν.

If $a(t)$ is of finite duration σ_t, for example, then the wave is a plane-wave pulse of light (a **wavepacket**) traveling in the z direction. At any time, the wavepacket extends over a distance $\sigma_z = c\sigma_t$ (Fig. 2.6-2). A pulse of duration $\sigma_t = 1$ ps, for example, extends over a distance of 0.3 mm. If the pulse intensity is Gaussian with rms width $\sigma_t = 1$ ps, its spectral bandwidth is $\sigma_\nu = 1/4\pi\sigma_t \approx 80$ GHz (see Appendix A, Sec. A.2). If the central frequency ν_0 is 5×10^{14} Hz (corresponding to $\lambda = 0.6\ \mu$m), the condition of quasi-monochromaticity is clearly satisfied.

The propagation of optical pulses through media with frequency-dependent refractive indices (i.e., with a frequency-dependent speed of light $c = c_o/n$) is discussed in Sec. 5.6.

B. Light Beating

The dependence of the intensity of a polychromatic wave on time may be attributed to interference among the monochromatic components that constitute the wave. This concept is now demonstrated by means of two examples: interference between two monochromatic waves and interference among a finite number of monochromatic waves.

Interference Between Two Monochromatic Waves

An optical wave composed of two monochromatic waves of frequencies ν_1 and ν_2 and intensities I_1 and I_2 has a complex wavefunction at some point in space

$$U(t) = I_1^{1/2} \exp(j2\pi\nu_1 t) + I_2^{1/2} \exp(j2\pi\nu_2 t), \tag{2.6-11}$$

where the phases are assumed to be zero. The $\mathbf{r}$ dependence has been suppressed for notational convenience. The intensity of the total wave is determined by use of the interference equation (2.5-4),

$$I(t) = I_1 + I_2 + 2(I_1 I_2)^{1/2} \cos[2\pi(\nu_2 - \nu_1)t]. \tag{2.6-12}$$

The intensity therefore varies sinusoidally at the difference frequency $|\nu_2 - \nu_1|$, which is called the "beat frequency." The effect is called **light beating** or **light mixing**.

Equation (2.6-12) is analogous to (2.5-7), which describes the "spatial" interference of two monochromatic waves of the same frequency but different directions. This can be understood from the phasor diagram in Fig. 2.5-1. The two phasors U_1 and U_2 rotate at angular frequencies $\omega_1 = 2\pi\nu_1$ and $\omega_2 = 2\pi\nu_2$, so that the difference angle $\varphi = \varphi_2 - \varphi_1$ is $2\pi(\nu_2 - \nu_1)t$, in accord with (2.6-12).

Beating occurs in electronics when the sum of two sinusoidal signals drives a nonlinear (e.g., quadratic) device called a mixer and produces signals at the sum and difference frequencies. It is used in heterodyne radio receivers. In optics, the nonlinearity results from the squared-absolute-value relation between the optical intensity and the complex wavefunction. Only the difference frequency is detected in this case. The use of optical beating in optical heterodyne receivers is discussed in Sec. 22.5. Other forms of optical mixing make use of nonlinear media to generate optical frequency differences and sums, as described in Chap. 19.

EXERCISE 2.6-1

Optical Doppler Radar. As a result of the **Doppler effect** a monochromatic optical wave of frequency ν reflected from an object moving with velocity v undergoes a frequency shift $\Delta\nu = \pm(2v/c)\nu$, depending on whether the object is moving toward (+) or away (−) from the observer. Assuming that the original and reflected waves are superimposed, derive an expression for the intensity of the resultant wave. Suggest a method for measuring the velocity of a target using such an arrangement. If one of the mirrors of a Michelson interferometer moves with velocity v, use (2.5-6) to show that the beat frequency is $(2v/c)\nu$.

Interference of M Monochromatic Waves

The interference of a large number of monochromatic waves with equal intensities, equal phases, and equally spaced frequencies can result in the generation of narrow pulses of light. Consider an odd number $M = 2L + 1$ waves, each with intensity I_0 and zero phase, and with frequencies

$$\nu_q = \nu_0 + q\nu_F, \qquad q = -L, \dots, 0, \dots, L,$$

centered about the frequency ν_0 and spaced by the frequency $\nu_F \ll \nu_0$. At a given position, the total wave has a complex wavefunction

$$U(t) = I_0^{1/2} \sum_{q=-L}^{L} \exp[j2\pi(\nu_0 + q\nu_F)t]. \tag{2.6-13}$$

This is the sum of M phasors of equal magnitudes and phases differing by $\varphi = 2\pi\nu_F t$. Using the result of the analysis for an identical situation provided in (2.5-10) and Fig. 2.5-7, the intensity becomes

$$I(t) = |U(t)|^2 = I_0 \frac{\sin^2(M\pi t/T_F)}{\sin^2(\pi t/T_F)}. \tag{2.6-14}$$

As illustrated in Fig. 2.6-3 the intensity $I(t)$ is a periodic sequence of pulses with period $T_F = 1/\nu_F$, peak intensity $M^2 I_0$, and mean intensity MI_0. The peak intensity is M times greater than the mean intensity. The width of each pulse is approximately

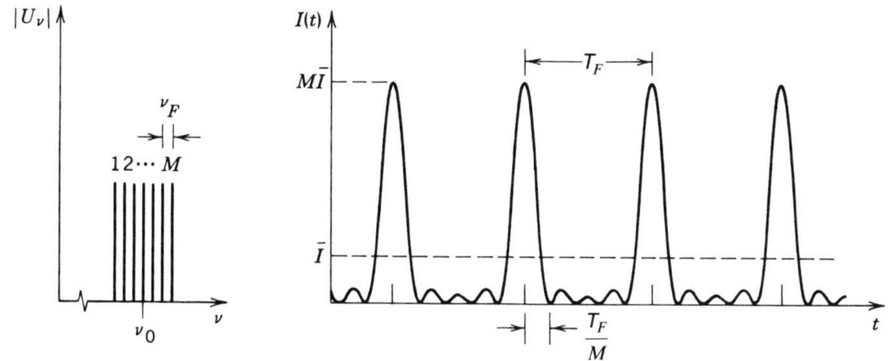

Figure 2.6-3 Time dependence of the intensity of a polychromatic wave composed of a sum of M monochromatic waves, of equal intensities, equal phases, and frequencies differing by ν_F. The intensity is a periodic train of pulses of period $T_F = 1/\nu_F$ with a peak M times greater than the mean. The duration of each pulse is M times smaller than the period. This should be compared with Fig. 2.5-7.

T_F/M. For large M, these pulses can be very narrow. If $\nu_F = 1$ GHz, for example, then $T_F = 1$ ns. If $M = 1000$, pulses of 1-ps width are generated.

This example provides a dramatic demonstration of how M monochromatic waves may cooperate to produce a train of very narrow pulses. In Chap. 14 we shall see that the modes of a laser can be "phase locked" in the fashion described above to produce narrow laser pulses.

READING LIST

General

See the general list in Chapter 1.

S. G. Lipson and H. Lipson, *Optical Physics*, Cambridge University Press, London, 1969, 2nd ed. 1981.

H. D. Young, *Fundamentals of Waves, Optics, and Modern Physics*, McGraw-Hill, New York, 2nd ed. 1976.

J. R. Pierce, *Almost All About Waves*, MIT Press, Cambridge, MA, 1974.

R. H. Webb, *Elementary Wave Optics*, Academic Press, New York, 1969.

D. H. Towne, *Wave Phenomena*, Addison-Wesley, Reading, MA, 1967.

C. Curry, *Wave Optics*, Edward Arnold, London, 1957.

Interferometry

J. M. Vaughan, *The Fabry-Perot Interferometer*, Adam Hilger, Bristol, England, 1989.

S. Tolansky, *An Introduction to Interferometry*, Wiley, New York, 1973.

A. H. Cook, *Interference of Electromagnetic Waves*, Clarendon Press, Oxford, 1971.

J. Dyson, *Interferometry as a Measuring Tool*, Machinery Publishing, Brighton, 1970.

W. H. Steel, *Interferometry*, Cambridge University Press, London, 1967.

M. Françon, *Optical Interferometry*, Academic Press, New York, 1966.

Spectroscopy

J. E. Chamberlain, *The Principles of Interferometric Spectroscopy*, Wiley, New York, 1979.

R. J. Bell, *Introductory Fourier Transform Spectroscopy*, Academic Press, New York, 1972.

J. F. James and R. S. Sternberg, *Design of Optical Spectrometers*, Chapman & Hall, London, 1969.

Diffraction Gratings

M. C. Hutley, *Diffraction Gratings*, Academic Press, New York, 1982.

R. Petit, ed., *Electromagnetic Theory of Gratings*, Springer-Verlag, New York, 1980.

S. P. Davis, *Diffraction Gratings and Spectrographs*, Holt, Rinehart and Winston, New York, 1970.

E. G. Loewen, *Diffraction Grating Handbook*, Bausch & Lomb, Rochester, NY, 1970.

Popular and Historical

J. Z. Buchwald, *The Rise of the Wave Theory of Light: Optical Theory and Experiment in the Early Nineteenth Century*, University of Chicago Press, Chicago, 1989.

W. E. Kock, *Sound Waves and Light Waves*, Doubleday (Anchor Books), Garden City, NY, 1965.

C. Huygens, *Treatise on Light*, Dover, New York, 1962 (originally published in 1690).

PROBLEMS

2.2-1 **Spherical Waves.** Use a spherical coordinate system to verify that the complex amplitude of the spherical wave (2.2-15) satisfies the Helmholtz equation (2.2-7).

2.2-2 **Intensity of a Spherical Wave.** Derive an expression for the intensity I of a spherical wave at a distance r from its center in terms of the optical power P. What is the intensity at $r = 1$ m for $P = 100$ W?

2.2-3 **Cylindrical Waves.** Derive expressions for the complex amplitude and intensity of a monochromatic wave whose wavefronts are cylinders centered about the y axis.

2.2-4 **Paraxial Helmholtz Equation.** Derive the paraxial Helmholtz equation (2.2-22) using the approximations in (2.2-20) and (2.2-21).

2.2-5 **Conjugate Waves.** Compare a monochromatic wave with complex amplitude $U(\mathbf{r})$ to a monochromatic wave of the same frequency but with complex amplitude $U^*(\mathbf{r})$, with respect to intensity, wavefronts, and wavefront normals. Use the plane wave $U(\mathbf{r}) = A\exp[-jk(x+y)/\sqrt{2}]$ and the spherical wave $U(\mathbf{r}) = (A/r)\exp(-jkr)$ as examples.

2.3-1 **Wave in a GRIN Slab.** Sketch the wavefronts of a wave traveling in the graded-index SELFOC slab described in Example 1.3-1.

2.4-1 **Reflection of a Spherical Wave from a Planar Mirror.** A spherical wave is reflected from a planar mirror sufficiently far from the wave origin so that the Fresnel approximation is satisfied. By regarding the spherical wave locally as a plane wave with slowly varying direction, use the law of reflection of plane waves to determine the nature of the reflected wave.

2.4-2 **Optical Path Length.** A plane wave travels in a direction normal to a thin plate made of N thin parallel layers of thicknesses d_q and refractive indices n_q, $q = 1, 2, \ldots, N$. If all reflections are ignored, determine the complex amplitude transmittance of the plate. If the plate is replaced with a distance d of free space, what should d be so that the same complex amplitude transmittance is obtained? Show that this distance is the optical path length defined in Sec. 1.1.

2.4-3 **Diffraction Grating.** Repeat Exercise 2.4-5 for a thin transparent plate whose thickness $d(x, y)$ is a square (instead of sinusoidal) periodic function of x of period

$\Lambda \gg \lambda$. Show that the angle θ between the diffracted waves is still given by $\theta \approx \lambda/\Lambda$. If a plane wave is incident in a direction normal to the grating, determine the amplitudes of the different diffracted plane waves.

2.4-4 **Reflectance of a Spherical Mirror.** Show that the complex amplitude reflectance $r(x, y)$ (the ratio of the complex amplitudes of the reflected and incident waves) of a thin spherical mirror of radius R is given by $r(x, y) = h_0 \exp[-jk_o(x^2 + y^2)/R]$, where h_0 is a constant. Compare this to the complex amplitude transmittance of a lens of focal length $f = -R/2$.

2.5-1 **Standing Waves.** Derive an expression for the intensity I of the superposition of two plane waves of wavelength λ traveling in opposite directions along the z axis. Sketch I versus z.

2.5-2 **Fringe Visibility.** The visibility of an interference pattern such as that described by (2.5-4) and plotted in Fig. 2.5-1 is defined as the ratio $\mathscr{V} = (I_{max} - I_{min})/(I_{max} + I_{min})$, where I_{max} and I_{min} are the maximum and minimum values of I. Derive an expression for $\mathscr{V}$ as a function of the ratio I_1/I_2 of the two interfering waves and determine the ratio I_1/I_2 for which the visibility is maximum.

2.5-3 **Michelson Interferometer.** If one of the mirrors of the Michelson interferometer (Fig. 2.5-3(b)) is misaligned by a small angle $\Delta\theta$, describe the shape of the interference pattern in the detector plane. What happens to this pattern as the other mirror moves?

2.6-1 **Pulsed Spherical Wave.** (a) Show that a pulsed spherical wave has a complex wavefunction of the form $U(\mathbf{r}, t) = (1/r)a(t - r/c)$, where $a(t)$ is an arbitrary function. (b) An ultrashort optical pulse has a complex wavefunction with central frequency corresponding to a wavelength $\lambda_o = 585$ nm and a Gaussian envelope of rms width $\sigma_t = 6$ fs (1 fs $= 10^{-15}$ s). How many optical cycles are contained within the pulse width? If the pulse propagates in free space as a spherical wave initiated at the origin at $t = 0$, describe the spatial distribution of the intensity as a function of the radial distance at time $t = 1$ ps.

CHAPTER

3

BEAM OPTICS

3.1 THE GAUSSIAN BEAM
- A. Complex Amplitude
- B. Properties

3.2 TRANSMISSION THROUGH OPTICAL COMPONENTS
- A. Transmission Through a Thin Lens
- B. Beam Shaping
- C. Reflection from a Spherical Mirror
- *D. Transmission Through an Arbitrary Optical System

3.3 HERMITE - GAUSSIAN BEAMS

*3.4 LAGUERRE - GAUSSIAN AND BESSEL BEAMS

The Gaussian beam is named after the great mathematician **Karl Friedrich Gauss (1777–1855)**.

Lord Rayleigh (John W. Strutt) (1842–1919) contributed to many areas of optics, including scattering, diffraction, radiation, and image formation. The depth of focus of the Gaussian beam is named after him.

Can light be spatially confined and transported in free space without angular spread? Although the wave nature of light precludes the existence of such an idealization, light can take the form of beams that come as close as possible to spatially localized and nondiverging waves.

A plane wave and a spherical wave represent the two opposite extremes of angular and spatial confinement. The wavefront normals (rays) of a plane wave are parallel to the direction of the wave so that there is no angular spread, but the energy extends spatially over the entire space. The spherical wave, on the other hand, originates from a single point, but its wavefront normals (rays) diverge in all directions.

Waves with wavefront normals making small angles with the z axis are called paraxial waves. They must satisfy the paraxial Helmholtz equation derived in Sec. 2.2C. An important solution of this equation that exhibits the characteristics of an optical beam is a wave called the **Gaussian beam**. The beam power is principally concentrated within a small cylinder surrounding the beam axis. The intensity distribution in any transverse plane is a circularly symmetric Gaussian function centered about the beam axis. The width of this function is minimum at the beam waist and grows gradually in both directions. The wavefronts are approximately planar near the beam waist, but they gradually curve and become approximately spherical far from the waist. The angular divergence of the wavefront normals is the minimum permitted by the wave equation for a given beam width. The wavefront normals are therefore much like a thin pencil of rays. Under ideal conditions, the light from a laser takes the form of a Gaussian beam.

An expression for the complex amplitude of the Gaussian beam is derived in Sec. 3.1 and a detailed discussion of its physical properties (intensity, power, beam radius, angular divergence, depth of focus, and phase) is provided. The shaping of Gaussian beams (focusing, relaying, collimating, and expanding) by the use of various optical components is the subject of Sec. 3.2. A family of optical beams called Hermite–Gaussian beams, of which the Gaussian beam is a member, is introduced in Sec. 3.3. Laguerre–Gaussian and Bessel beams are discussed in Sec. 3.4.

3.1 THE GAUSSIAN BEAM

A. Complex Amplitude

The concept of paraxial waves was introduced in Sec. 2.2C. A paraxial wave is a plane wave e^{-jkz} (with wavenumber $k = 2\pi/\lambda$ and wavelength λ) modulated by a complex envelope $A(\mathbf{r})$ that is a slowly varying function of position (see Fig. 2.2-5). The complex amplitude is

$$U(\mathbf{r}) = A(\mathbf{r}) \exp(-jkz). \tag{3.1-1}$$

The envelope is assumed to be approximately constant within a neighborhood of size λ, so that the wave is locally like a plane wave with wavefront normals that are paraxial rays.

For the complex amplitude $U(\mathbf{r})$ to satisfy the Helmholtz equation, $\nabla^2 U + k^2 U = 0$, the complex envelope $A(\mathbf{r})$ must satisfy the paraxial Helmholtz equation (2.2-22)

$$\nabla_T^2 A - j2k \frac{\partial A}{\partial z} = 0, \tag{3.1-2}$$

where $\nabla_T^2 = \partial^2/\partial x^2 + \partial^2/\partial y^2$ is the transverse part of the Laplacian operator. One simple solution to the paraxial Helmholtz equation provides the paraboloidal wave for which

$$A(\mathbf{r}) = \frac{A_1}{z} \exp\left(-jk \frac{\rho^2}{2z}\right), \qquad \rho^2 = x^2 + y^2 \tag{3.1-3}$$

(see Exercise 2.2-2) where A_1 is a constant. The paraboloidal wave is the paraxial approximation of the spherical wave $U(r) = (A_1/r)\exp(-jkr)$ when x and y are much smaller than z (see Sec. 2.2B).

Another solution of the paraxial Helmholtz equation provides the Gaussian beam. It is obtained from the paraboloidal wave by use of a simple transformation. Since the complex envelope of the paraboloidal wave (3.1-3) is a solution of the paraxial Helmholtz equation (3.1-2), a shifted version of it, with $z - \xi$ replacing z where ξ is a constant,

$$A(\mathbf{r}) = \frac{A_1}{q(z)} \exp\left[-jk \frac{\rho^2}{2q(z)}\right], \qquad q(z) = z - \xi, \tag{3.1-4}$$

is also a solution. This provides a paraboloidal wave centered about the point $z = \xi$ instead of $z = 0$. When ξ is complex, (3.1-4) remains a solution of (3.1-2), but it acquires dramatically different properties. In particular, when ξ is purely imaginary, say $\xi = -jz_0$ where z_0 is real, (3.1-4) gives rise to the complex envelope of the Gaussian beam

$$A(\mathbf{r}) = \frac{A_1}{q(z)} \exp\left[-jk \frac{\rho^2}{2q(z)}\right], \qquad q(z) = z + jz_0. \tag{3.1-5}$$

Complex Envelope

The parameter z_0 is known as the **Rayleigh range**.

To separate the amplitude and phase of this complex envelope, we write the complex function $1/q(z) = 1/(z + jz_0)$ in terms of its real and imaginary parts by defining two new real functions $R(z)$ and $W(z)$, such that

$$\frac{1}{q(z)} = \frac{1}{R(z)} - j\frac{\lambda}{\pi W^2(z)}. \tag{3.1-6}$$

It will be shown subsequently that $W(z)$ and $R(z)$ are measures of the beam width and wavefront radius of curvature, respectively. Expressions for $W(z)$ and $R(z)$ as functions of z and z_0 are provided in (3.1-8) and (3.1-9). Substituting (3.1-6) into (3.1-5) and using (3.1-1), an expression for the complex amplitude $U(\mathbf{r})$ of the Gaussian beam is obtained:

$$U(\mathbf{r}) = A_0 \frac{W_0}{W(z)} \exp\left[-\frac{\rho^2}{W^2(z)}\right] \exp\left[-jkz - jk\frac{\rho^2}{2R(z)} + j\zeta(z)\right]$$

(3.1-7)
Gaussian-Beam Complex Amplitude

$$W(z) = W_0\left[1 + \left(\frac{z}{z_0}\right)^2\right]^{1/2} \quad (3.1\text{-}8)$$

$$R(z) = z\left[1 + \left(\frac{z_0}{z}\right)^2\right] \quad (3.1\text{-}9)$$

$$\zeta(z) = \tan^{-1}\frac{z}{z_0} \quad (3.1\text{-}10)$$

$$W_0 = \left(\frac{\lambda z_0}{\pi}\right)^{1/2}. \quad (3.1\text{-}11)$$

Beam Parameters

A new constant $A_0 = A_1/jz_0$ has been defined for convenience.

The expression for the complex amplitude of the Gaussian beam is central to this chapter. It contains two parameters, A_0 and z_0, which are determined from the boundary conditions. All other parameters are related to the Rayleigh range z_0 and the wavelength λ by (3.1-8) to (3.1-11).

B. Properties

Equations (3.1-7) to (3.1-11) will now be used to determine the properties of the Gaussian beam.

Intensity

The optical intensity $I(\mathbf{r}) = |U(\mathbf{r})|^2$ is a function of the axial and radial distances z and $\rho = (x^2 + y^2)^{1/2}$,

$$I(\rho, z) = I_0\left[\frac{W_0}{W(z)}\right]^2 \exp\left[-\frac{2\rho^2}{W^2(z)}\right], \quad (3.1\text{-}12)$$

where $I_0 = |A_0|^2$. At each value of z the intensity is a Gaussian function of the radial distance ρ. This is why the wave is called a Gaussian beam. The Gaussian function has its peak at $\rho = 0$ (on axis) and drops monotonically with increasing ρ. The width $W(z)$ of the Gaussian distribution increases with the axial distance z as illustrated in Fig. 3.1-1.

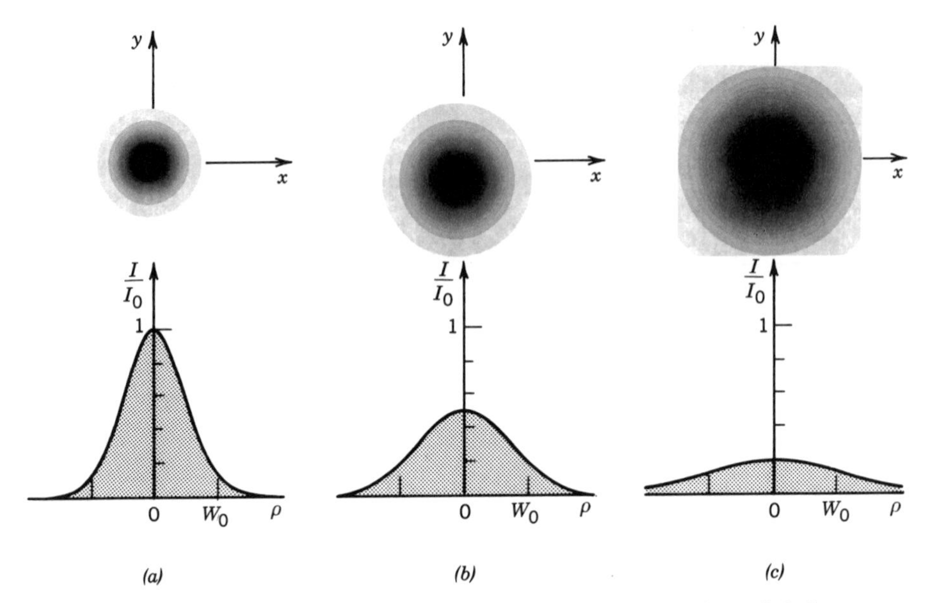

Figure 3.1-1 The normalized beam intensity I/I_0 as a function of the radial distance ρ at different axial distances: (a) $z = 0$; (b) $z = z_0$; (c) $z = 2z_0$.

On the beam axis ($\rho = 0$) the intensity

$$I(0, z) = I_0 \left[\frac{W_0}{W(z)} \right]^2 = \frac{I_0}{1 + (z/z_0)^2} \tag{3.1-13}$$

has its maximum value I_0 at $z = 0$ and drops gradually with increasing z, reaching half its peak value at $z = \pm z_0$ (Fig. 3.1-2). When $|z| \gg z_0$, $I(0, z) \approx I_0 z_0^2/z^2$, so that the intensity decreases with the distance in accordance with an inverse-square law, as for spherical and paraboloidal waves. The overall peak intensity $I(0, 0) = I_0$ occurs at the beam center ($z = 0, \rho = 0$).

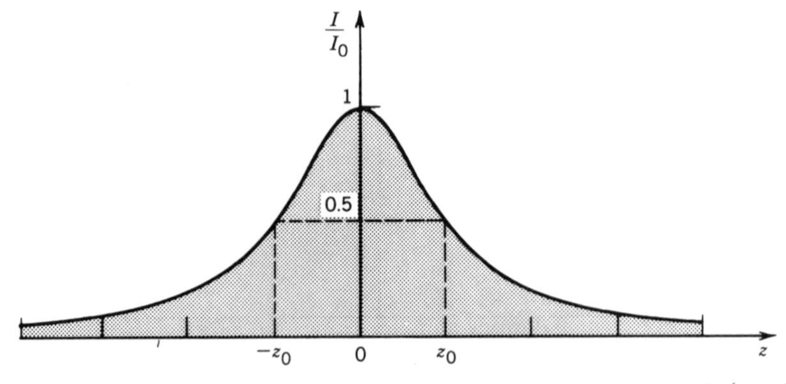

Figure 3.1-2 The normalized beam intensity I/I_0 at points on the beam axis ($\rho = 0$) as a function of z.

Power

The total optical power carried by the beam is the integral of the optical intensity over a transverse plane (say at a distance z),

$$P = \int_0^\infty I(\rho, z) 2\pi\rho \, d\rho,$$

which gives

$$P = \tfrac{1}{2} I_0 (\pi W_0^2). \tag{3.1-14}$$

The result is independent of z, as expected. Thus the beam power is one-half the peak intensity times the beam area. Since beams are often described by their power P, it is useful to express I_0 in terms of P using (3.1-14) and to rewrite (3.1-12) in the form

$$I(\rho, z) = \frac{2P}{\pi W^2(z)} \exp\left[-\frac{2\rho^2}{W^2(z)}\right]. \tag{3.1-15}$$

Beam Intensity

The ratio of the power carried within a circle of radius ρ_0 in the transverse plane at position z to the total power is

$$\frac{1}{P}\int_0^{\rho_0} I(\rho, z) 2\pi\rho \, d\rho = 1 - \exp\left[-\frac{2\rho_0^2}{W^2(z)}\right]. \tag{3.1-16}$$

The power contained within a circle of radius $\rho_0 = W(z)$ is approximately 86% of the total power. About 99% of the power is contained within a circle of radius $1.5W(z)$.

Beam Radius

Within any transverse plane, the beam intensity assumes its peak value on the beam axis, and drops by the factor $1/e^2 \approx 0.135$ at the radial distance $\rho = W(z)$. Since 86% of the power is carried within a circle of radius $W(z)$, we regard $W(z)$ as the beam radius (also called the beam width). The rms width of the intensity distribution is $\sigma = \frac{1}{2}W(z)$ (see Appendix A, Sec. A.2, for the different definitions of width).

The dependence of the beam radius on z is governed by (3.1-8),

$$W(z) = W_0\left[1 + \left(\frac{z}{z_0}\right)^2\right]^{1/2}. \tag{3.1-17}$$

Beam Radius

It assumes its minimum value W_0 in the plane $z = 0$, called the beam waist. Thus W_0 is the **waist radius**. The waist diameter $2W_0$ is called the **spot size**. The beam radius increases gradually with z, reaching $\sqrt{2}\,W_0$ at $z = z_0$, and continues increasing monotonically with z (Fig. 3.1-3). For $z \gg z_0$ the first term of (3.1-17) may be neglected, resulting in the linear relation

$$W(z) \approx \frac{W_0}{z_0} z = \theta_0 z, \tag{3.1-18}$$

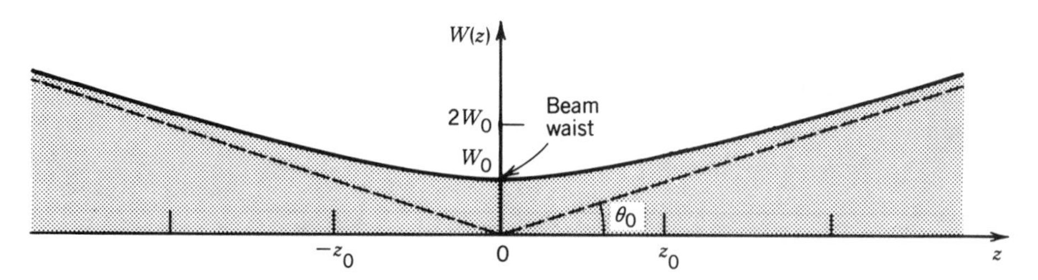

Figure 3.1-3 The beam radius $W(z)$ has its minimum value W_0 at the waist ($z = 0$), reaches $\sqrt{2}\, W_0$ at $z = \pm z_0$, and increases linearly with z for large z.

where $\theta_0 = W_0/z_0$. Using (3.1-11), we can also write

$$\theta_0 = \frac{\lambda}{\pi W_0}. \tag{3.1-19}$$

Beam Divergence

Far from the beam center, when $z \gg z_0$, the beam radius increases approximately linearly with z, defining a cone with half-angle θ_0. About 86% of the beam power is confined within this cone. The angular divergence of the beam is therefore defined by the angle

$$\boxed{\theta_0 = \frac{2}{\pi} \frac{\lambda}{2W_0}.} \tag{3.1-20}$$

Divergence Angle

The beam divergence is directly proportional to the ratio between the wavelength λ and the beam-waist diameter $2W_0$. If the waist is squeezed, the beam diverges. To obtain a highly directional beam, therefore, a short wavelength and a fat beam waist should be used.

Depth of Focus

Since the beam has its minimum width at $z = 0$, as shown in Fig. 3.1-3, it achieves its best focus at the plane $z = 0$. In either direction, the beam gradually grows "out of focus." The axial distance within which the beam radius lies within a factor $\sqrt{2}$ of its minimum value (i.e., its area lies within a factor of 2 of its minimum) is known as the **depth of focus** or **confocal parameter** (Fig. 3.1-4). It can be seen from (3.1-17) that the

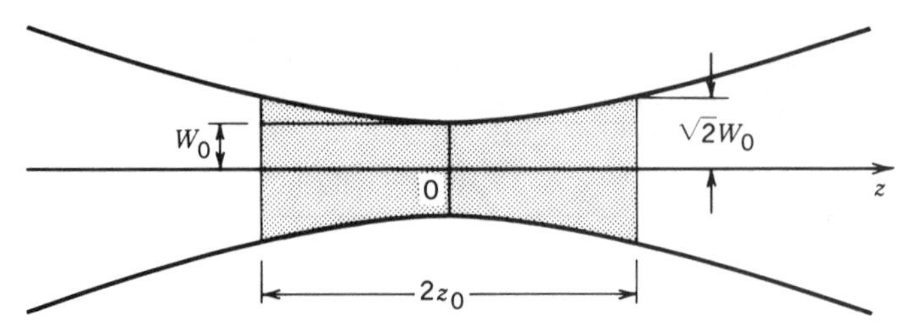

Figure 3.1-4 The depth of focus of a Gaussian beam.

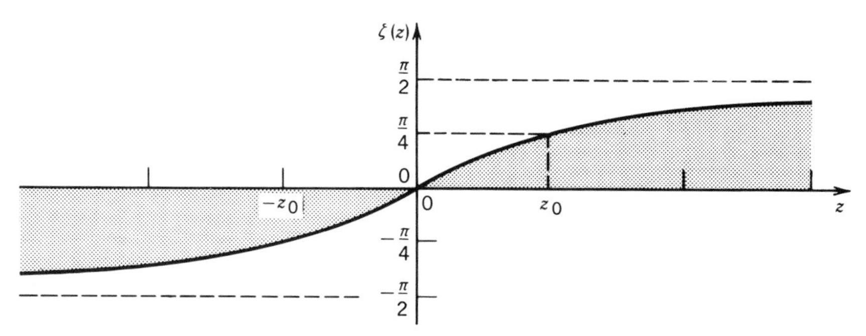

Figure 3.1-5 $\zeta(z)$ is the phase retardation of the Gaussian beam relative to a uniform plane wave at points on the beam axis.

depth of focus is twice the Rayleigh range,

$$2z_0 = \frac{2\pi W_0^2}{\lambda}. \tag{3.1-21}$$

Depth of Focus

The depth of focus is directly proportional to the area of the beam at its waist, and inversely proportional to the wavelength. Thus when a beam is focused to a small spot size, the depth of focus is short and the plane of focus must be located with greater accuracy. A small spot size and a long depth of focus cannot be obtained simultaneously unless the wavelength of the light is short. For $\lambda = 633$ nm (the wavelength of a He–Ne laser line), for example, a spot size $2W_0 = 2$ cm corresponds to a depth of focus $2z_0 \approx 1$ km. A much smaller spot size of 20 μm corresponds to a much shorter depth of focus of 1 mm.

Phase

The phase of the Gaussian beam is, from (3.1-7),

$$\varphi(\rho, z) = kz - \zeta(z) + \frac{k\rho^2}{2R(z)}. \tag{3.1-22}$$

On the beam axis ($\rho = 0$) the phase

$$\varphi(0, z) = kz - \zeta(z) \tag{3.1-23}$$

comprises two components. The first, kz, is the phase of a plane wave. The second represents a phase retardation $\zeta(z)$ given by (3.1-10) which ranges from $-\pi/2$ at $z = -\infty$ to $+\pi/2$ at $z = \infty$, as illustrated in Fig. 3.1-5. This phase retardation corresponds to an excess delay of the wavefront in comparison with a plane wave or a spherical wave (see also Fig. 3.1-8). The total accumulated excess retardation as the wave travels from $z = -\infty$ to $z = \infty$ is π. This phenomenon is known as the **Guoy effect.**[†]

Wavefronts

The third component in (3.1-22) is responsible for wavefront bending. It represents the deviation of the phase at off-axis points in a given transverse plane from that at the

[†]See, for example, A. E. Siegman, *Lasers*, University Science Books, Mill Valley, CA, 1986.

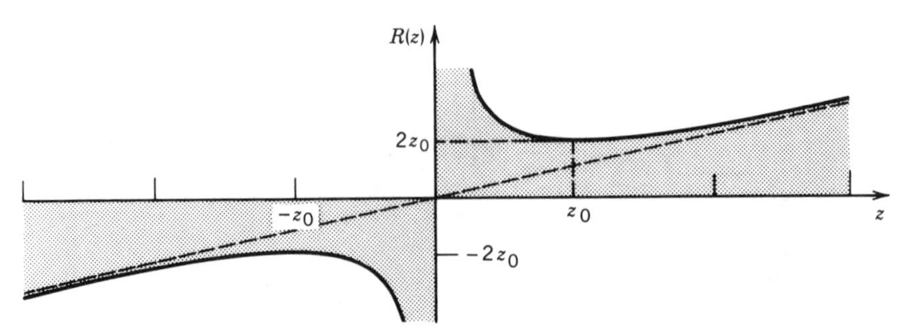

Figure 3.1-6 The radius of curvature $R(z)$ of the wavefronts of a Gaussian beam. The dashed line is the radius of curvature of a spherical wave.

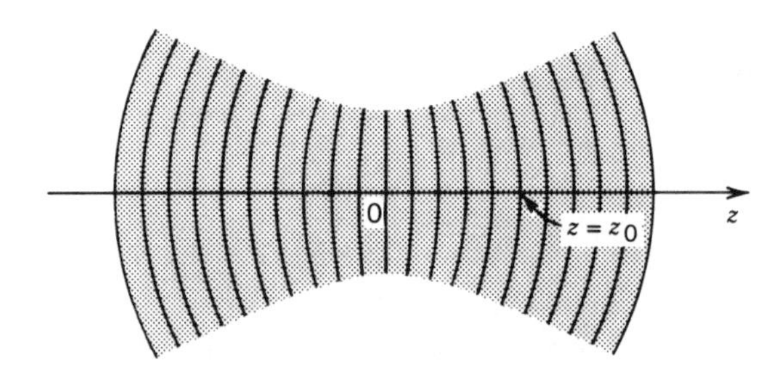

Figure 3.1-7 Wavefronts of a Gaussian beam.

axial point. The surfaces of constant phase satisfy $k[z + \rho^2/2R(z)] - \zeta(z) = 2\pi q$. Since $\zeta(z)$ and $R(z)$ are relatively slowly varying, they are approximately constant at points within the beam radius on each wavefront. We may therefore write $z + \rho^2/2R = q\lambda + \zeta\lambda/2\pi$, where $R = R(z)$ and $\zeta = \zeta(z)$. This is precisely the equation of a paraboloidal surface of radius of curvature R. Thus $R(z)$, plotted in Fig. 3.1-6, is the radius of curvature of the wavefront at position z on the beam axis.

As illustrated in Fig. 3.1-6, the radius of curvature $R(z)$ is infinite at $z = 0$, corresponding to planar wavefronts. It decreases to a minimum value of $2z_0$ at $z = z_0$. This is the point at which the wavefront has the greatest curvature (Fig. 3.1-7). The radius of curvature subsequently increases with further increase of z until $R(z) \approx z$ for $z \gg z_0$. The wavefront is then approximately the same as that of a spherical wave. For negative z the wavefronts follow an identical pattern, except for a change in sign. We have adopted the convention that a diverging wavefront has a positive radius of curvature, whereas a converging wavefront has a negative radius of curvature.

Summary: Properties of the Gaussian Beam at Special Points

- *At the plane $z = z_0$.* At an axial distance z_0 from the beam waist, the wave has the following properties:
  - (i) The beam radius is $\sqrt{2}$ times greater than the radius at the beam waist, and the area is larger by a factor of 2.
  - (ii) The intensity on the beam axis is $\frac{1}{2}$ the peak intensity.
  - (iii) The phase on the beam axis is retarded by an angle $\pi/4$ relative to the phase of a plane wave.
  - (iv) The radius of curvature of the wavefront is the smallest, so that the wavefront has the greatest curvature ($R = 2z_0$).

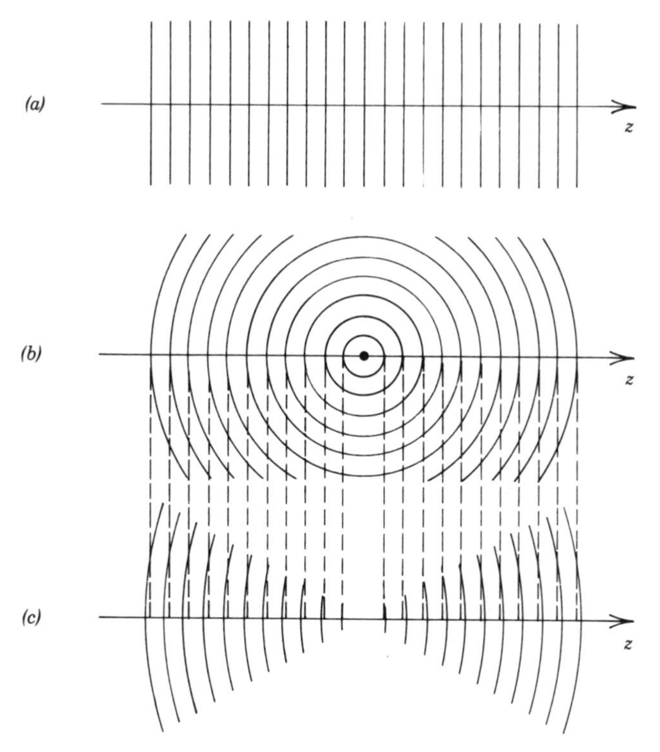

Figure 3.1-8 Wavefronts of (*a*) a uniform plane wave; (*b*) a spherical wave; (*c*) a Gaussian beam. At points near the beam center, the Gaussian beam resembles a plane wave. At large z the beam behaves like a spherical wave except that the phase is retarded by 90° (shown in this diagram by a quarter of the distance between two adjacent wavefronts).

- *Near the Beam Center.* At points for which $|z| \ll z_0$ and $\rho \ll W_0$, $\exp[-\rho^2/W^2(z)] \approx \exp(-\rho^2/W_0^2) \approx 1$, so that the beam intensity is approximately constant. Also, $R(z) \approx z_0^2/z$ and $\zeta(z) \approx 0$, so that the phase $k[z + \rho^2/2R(z)] = kz(1 + \rho^2/2z_0^2) \approx kz$. As a result, the wavefronts are approximately planar. The Gaussian beam may therefore be approximated near its center by a plane wave.
- *Far from the Beam Waist.* At points within the beam-waist radius ($\rho < W_0$) but far from the beam waist ($z \gg z_0$) the wave is approximately like a spherical wave. Since $W(z) \approx W_0 z/z_0 \gg W_0$ and $\rho < W_0$, $\exp[-\rho^2/W^2(z)] \approx 1$, so that the beam intensity is approximately uniform. Since $R(z) \approx z$ the wavefronts are approximately spherical. Thus, except for an excess phase $\zeta(z) \approx \pi/2$, the complex amplitude of the Gaussian beam approaches that of the paraboloidal wave, which in turn approaches that of the spherical wave in the paraxial approximation (Fig. 3.1-8).

EXERCISE 3.1-1

Parameters of a Gaussian Laser Beam. A 1-mW He–Ne laser produces a Gaussian beam of wavelength $\lambda = 633$ nm and a spot size $2W_0 = 0.1$ mm.

(a) Determine the angular divergence of the beam, its depth of focus, and its diameter at $z = 3.5 \times 10^5$ km (approximately the distance to the moon).

(b) What is the radius of curvature of the wavefront at $z = 0$, $z = z_0$, and $z = 2z_0$?

(c) What is the optical intensity (in W/cm^2) at the beam center ($z = 0, \rho = 0$) and at the axial point $z = z_0$? Compare this with the intensity at $z = z_0$ of a 100-W spherical wave produced by a small isotropically emitting light source located at $z = 0$.

EXERCISE 3.1-2

Validity of the Paraxial Approximation for a Gaussian Beam. The complex envelope $A(\mathbf{r})$ of a Gaussian beam is an exact solution of the paraxial Helmholtz equation (3.1-2), but its corresponding complex amplitude $U(\mathbf{r}) = A(\mathbf{r})\exp(-jkz)$ is only an approximate solution of the Helmholtz equation (2.2-7). This is because the paraxial Helmholtz equation is itself approximate. The approximation is satisfactory if the condition (2.2-20) is satisfied. Show that if the divergence angle θ_0 of a Gaussian beam is small ($\theta_0 \ll 1$), the condition (2.2-20) for the validity of the paraxial Helmholtz equation is satisfied.

Parameters Required to Characterize a Gaussian Beam

Assuming that the wavelength λ is known, how many parameters are required to describe a plane wave, a spherical wave, and a Gaussian beam? The plane wave is completely specified by its complex amplitude and direction. The spherical wave is specified by its amplitude and the location of its origin. The Gaussian beam, in contrast, is characterized by more parameters—its peak amplitude [the parameter A_0 in (3.1-7)], its direction (the beam axis), the location of its waist, *and* one additional parameter: the waist radius W_0 *or* the Rayleigh range z_0, for example. Thus, if the beam peak amplitude and the axis are known, two additional parameters are necessary.

If the complex number $q(z) = z + jz_0$ is known, the distance z to the beam waist and the Rayleigh range z_0 are readily identified as the real and imaginary parts of $q(z)$. As an example, if the q-parameter is $3 + j4$ cm at some point on the beam axis, we conclude that the beam waist lies at a distance $z = 3$ cm to the left of that point and that the depth of focus is $2z_0 = 8$ cm. The waist radius W_0 may be determined by use of (3.1-11). The **q-parameter** $q(z)$ is therefore sufficient for characterizing a Gaussian beam of known peak amplitude and beam axis. The linear dependence of the q-parameter on z permits us to readily determine q at all points, given q at a single point. If $q(z) = q_1$ and $q(z + d) = q_2$, then $q_2 = q_1 + d$. In the present example, at $z = 13$ cm, $q = 13 + j4$.

If the beam width $W(z)$ and the radius of curvature $R(z)$ are known at an arbitrary point on the axis, the beam can be identified completely by solving (3.1-8), (3.1-9), and (3.1-11) for z, z_0, and W_0. Alternatively, the q-parameter may be determined from $W(z)$ and $R(z)$ using the relation, $1/q(z) = 1/R(z) - j\lambda/[\pi W^2(z)]$, from which the beam is identified.

EXERCISE 3.1-3

Determination of a Beam with Given Width and Curvature. Assuming that the width W and the radius of curvature R of a Gaussian beam are known at some point on the beam axis (Fig. 3.1-9), show that the beam waist is located at a distance

$$z = \frac{R}{1 + \left(\lambda R/\pi W^2\right)^2} \tag{3.1-24}$$

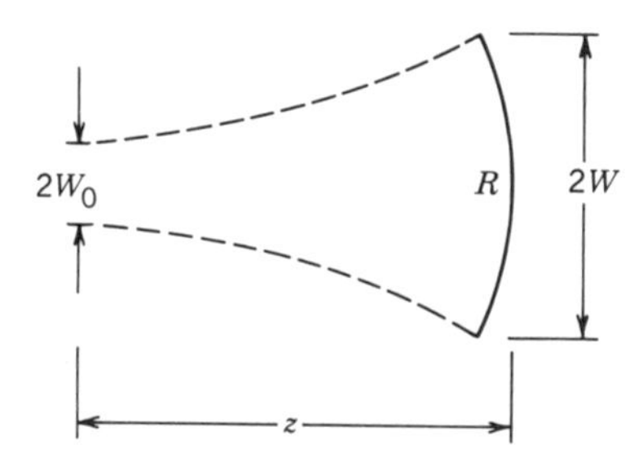

Figure 3.1-9 Given W and R, determine z and W_0.

to the left and the waist radius is

$$W_0 = \frac{W}{\left[1 + \left(\pi W^2/\lambda R\right)^2\right]^{1/2}}. \tag{3.1-25}$$

EXERCISE 3.1-4

Determination of the Width and Curvature at One Point Given the Width and Curvature at Another Point. Assume that the radius of curvature and the width of a Gaussian beam of wavelength $\lambda = 1$ μm at some point on the beam axis are $R_1 = 1$ m and $W_1 = 1$ mm, respectively (Fig. 3.1-10). Determine the beam width and the radius of curvature at a distance $d = 10$ cm to the right.

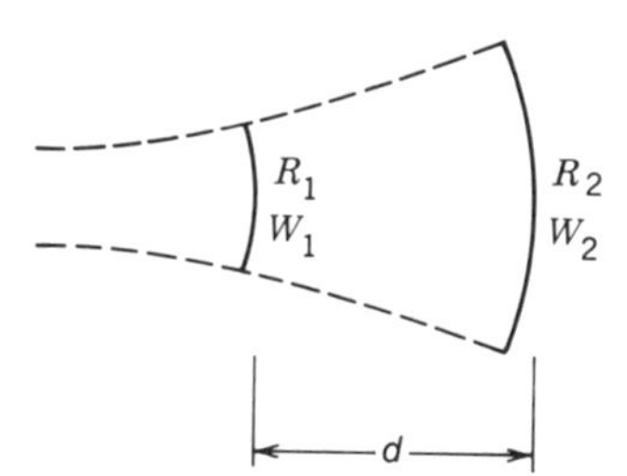

Figure 3.1-10 Given R_1, W_1, and d, determine R_2 and W_2.

EXERCISE 3.1-5

Identification of a Beam with Known Curvatures at Two Points. A Gaussian beam has radii of curvature R_1 and R_2 at two points on the beam axis separated by a distance d, as illustrated in Fig. 3.1-11. Verify that the location of the beam center and its depth of

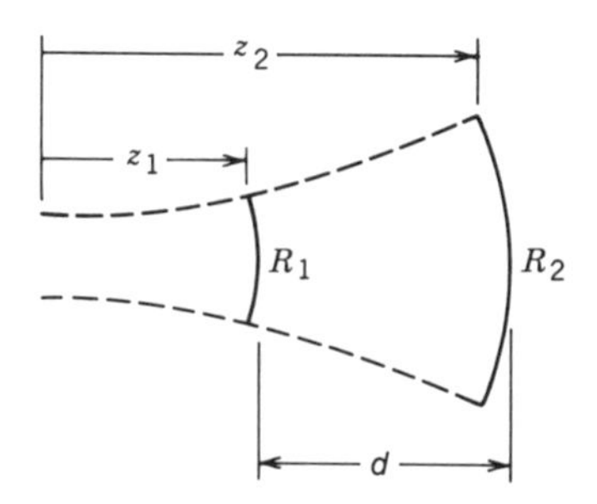

Figure 3.1-11 Given R_1, R_2, and d, determine z_1, z_2, z_0, and W_0.

focus may be determined from the relations

$$z_1 = \frac{-d(R_2 - d)}{R_2 - R_1 - 2d} \tag{3.1-26}$$

$$z_0^2 = \frac{-d(R_1 + d)(R_2 - d)(R_2 - R_1 - d)}{(R_2 - R_1 - 2d)^2} \tag{3.1-27}$$

$$W_0 = \left(\frac{\lambda z_0}{\pi}\right)^{1/2}.$$

3.2 TRANSMISSION THROUGH OPTICAL COMPONENTS

The effects of different optical components on a Gaussian beam are discussed in this section. We show that if a Gaussian beam is transmitted through a set of circularly symmetric optical components aligned with the beam axis, *the Gaussian beam remains a Gaussian beam* as long as the overall system maintains the paraxial nature of the wave. Only the beam waist and curvature are altered so that the beam is only reshaped. The results of this section are important in the design of optical instruments in which Gaussian beams are used.

A. Transmission Through a Thin Lens

The complex amplitude transmittance of a thin lens of focal length f is proportional to $\exp(jk\rho^2/2f)$ (see Sec. 2.4B). When a Gaussian beam crosses the lens its complex amplitude, given in (3.1-7), is multiplied by this phase factor. As a result, its wavefront is bent, but the beam radius is not altered.

A Gaussian beam centered at $z = 0$ with waist radius W_0 is transmitted through a thin lens located at a distance z, as illustrated in Fig. 3.2-1. The phase at the plane of the lens is $kz + k\rho^2/2R - \zeta$, where $R = R(z)$ and $\zeta = \zeta(z)$ are given by (3.1-9) and (3.1-10), respectively. The phase of the transmitted wave is altered to

$$kz + k\frac{\rho^2}{2R} - \zeta - k\frac{\rho^2}{2f} = kz + k\frac{\rho^2}{2R'} - \zeta, \tag{3.2-1}$$

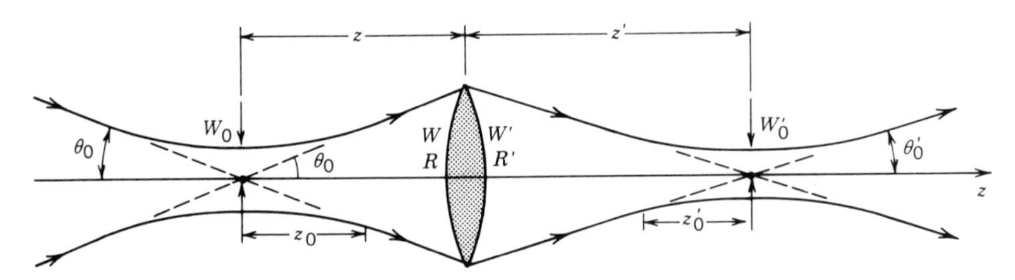

Figure 3.2-1 Transmission of a Gaussian beam through a thin lens.

where

$$\frac{1}{R'} = \frac{1}{R} - \frac{1}{f}. \tag{3.2-2}$$

We conclude that the transmitted wave is itself a Gaussian beam with width $W' = W$ and radius of curvature R', where R' satisfies the imaging equation $1/R - 1/R' = 1/f$. Note that R is positive since the wavefront of the incident beam is diverging and R' is negative since the wavefront of the transmitted beam is converging.

The parameters of the emerging beam may be determined by referring to Exercise 3.1-3, in which the parameters of a Gaussian beam were determined from its width and curvature at a given point. By use of (3.1-25) and (3.1-24) the waist radius of the new beam is

$$W_0' = \frac{W}{\left[1 + \left(\pi W^2/\lambda R'\right)^2\right]^{1/2}}, \tag{3.2-3}$$

and the center is located a distance

$$-z' = \frac{R'}{1 + \left(\lambda R'/\pi W^2\right)^2} \tag{3.2-4}$$

from the lens. A minus sign is used in (3.2-4) since the waist lies to the right of the lens. Substituting $R = z[1 + (z_0/z)^2]$ and $W = W_0[1 + (z/z_0)^2]^{1/2}$ into (3.2-2) to (3.2-4), the following expressions, which relate the parameters of the two beams, are obtained (Fig. 3.2-1):

Waist radius	$W_0' = MW_0$	(3.2-5)
Waist location	$(z' - f) = M^2(z - f)$	(3.2-6)
Depth of focus	$2z_0' = M^2(2z_0)$	(3.2-7)
Divergence	$2\theta_0' = \dfrac{2\theta_0}{M}$	(3.2-8)
Magnification	$M = \dfrac{M_r}{(1 + r^2)^{1/2}}$	(3.2-9)
$r = \dfrac{z_0}{z - f}$,	$M_r = \left\lvert\dfrac{f}{z - f}\right\rvert.$	(3.2-9a)

Parameter Transformation by a Lens

The magnification factor M plays an important role. The beam waist is magnified by M, the beam depth of focus is magnified by M^2, and the angular divergence is minified by the factor M.

Limit of Ray Optics

Consider the limiting case in which $(z - f) \gg z_0$, so that the lens is well outside the depth of focus of the incident beam (Fig. 3.2-2). The beam may then be approximated by a spherical wave, and the parameter $r \ll 1$ so that $M \approx M_r$ [see (3.2-9a)]. Thus

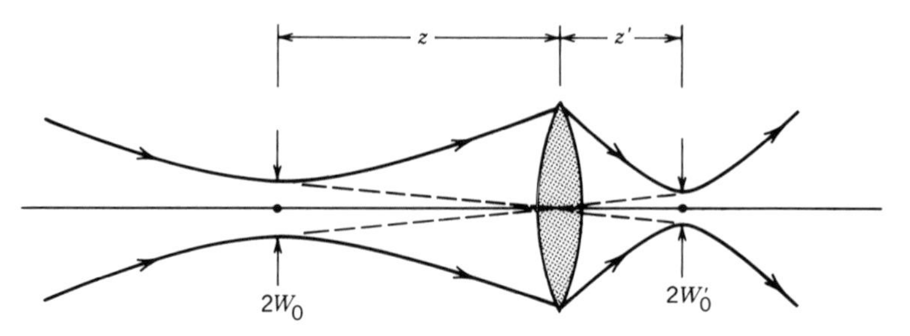

Figure 3.2-2 Beam imaging in the ray-optics limit.

(3.2-5) to (3.2-9a) reduce to

$$W_0' \approx MW_0 \tag{3.2-10}$$

$$\frac{1}{z'} + \frac{1}{z} \approx \frac{1}{f} \tag{3.2-11}$$

$$M \approx M_r = \left| \frac{f}{z - f} \right|. \tag{3.2-12}$$

Equations (3.2-10) to (3.2-12) are precisely the relations provided by ray optics for the location and size of a patch of light of diameter $2W_0$ located a distance z to the left of a thin lens (see Sec. 1.2C). The magnification factor M_r is that based on ray optics. Since (3.2-9) provides that $M < M_r$, the maximum magnification attainable is the ray-optics magnification M_r. As r^2 increases, the deviation from ray optics grows and the magnification decreases. Equations (3.2-10) to (3.2-12) also correspond to the results obtained from wave optics for the focusing of a spherical wave in the paraxial approximation (see Sec. 2.4B).

B. Beam Shaping

A lens, or sequence of lenses, may be used to reshape a Gaussian beam without compromising its Gaussian nature.

Beam Focusing

If a lens is placed at the waist of a Gaussian beam, as shown in Fig. 3.2-3, the parameters of the transmitted Gaussian beam are determined by substituting $z = 0$ in

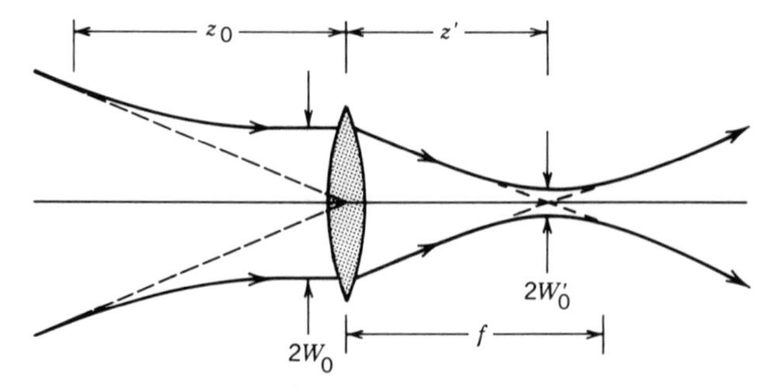

Figure 3.2-3 Focusing a beam with a lens at the beam waist.

(3.2-5) to (3.2-9a). The transmitted beam is then focused to a waist radius W_0' at a distance z' given by

$$W_0' = \frac{W_0}{\left[1 + (z_0/f)^2\right]^{1/2}} \tag{3.2-13}$$

$$z' = \frac{f}{1 + (f/z_0)^2}. \tag{3.2-14}$$

If the depth of focus of the incident beam $2z_0$ is much longer than the focal length f of the lens (Fig. 3.2-4), then $W_0' \approx (f/z_0)W_0$. Using $z_0 = \pi W_0^2/\lambda$, we obtain

$$W_0' \approx \frac{\lambda}{\pi W_0} f = \theta_0 f \tag{3.2-15}$$

$$z' \approx f. \tag{3.2-16}$$

The transmitted beam is then focused at the lens' focal plane as would be expected for parallel rays incident on a lens. This occurs because the incident Gaussian beam is well approximated by a plane wave at its waist. The spot size expected from ray optics is, of course, zero. In wave optics, however, the focused waist radius W_0' is directly proportional to the wavelength and the focal length, and inversely proportional to the radius of the incident beam. In the limit $\lambda \to 0$, the spot size does indeed approach zero in accordance with ray optics.

In many applications, such as laser scanning, laser printing, and laser fusion, it is desirable to generate the smallest possible spot size. It is clear from (3.2-15) that this may be achieved by use of the shortest possible wavelength, the thickest incident beam, and the shortest focal length. Since the lens should intercept the incident beam, its diameter D must be at least $2W_0$. Assuming that $D = 2W_0$, the diameter of the focused spot is given by

$$2W_0' \approx \frac{4}{\pi}\lambda F_{\#}, \qquad F_{\#} = \frac{f}{D}, \tag{3.2-17}$$

Focused Spot Size

where $F_{\#}$ is the F-number of the lens. A microscope objective with small F-number is often used. Since (3.2-15) and (3.2-16) are approximate, their validity must always be confirmed before use.

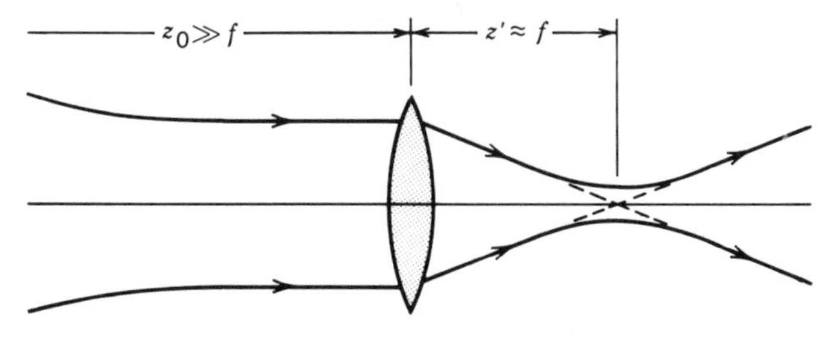

Figure 3.2-4 Focusing a collimated beam.

EXERCISE 3.2-1

Beam Relaying. A Gaussian beam of radius W_0 and wavelength λ is repeatedly focused by a sequence of identical lenses, each of focal length f and separated by distance d (Fig. 3.2-5). The focused waist radius is equal to the incident waist radius, i.e., $W_0' = W_0$. Using (3.2-6), (3.2-9), and (3.2-9a) show that this condition can arise only if the inequality $d \leq 4f$ is satisfied. Note that this is the same condition of ray confinement for a sequence of lenses derived in Sec. 1.4D using ray optics.

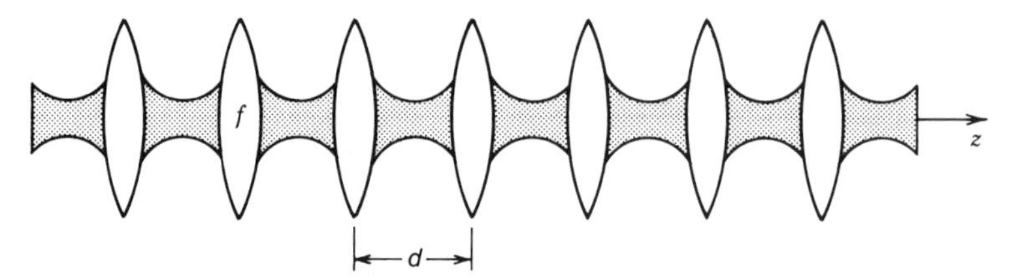

Figure 3.2-5 Beam relaying.

EXERCISE 3.2-2

Beam Collimation. A Gaussian beam is transmitted through a thin lens of focal length f.

(a) Show that the locations of the waists of the incident and transmitted beams, z and z', are related by

$$\frac{z'}{f} - 1 = \frac{z/f - 1}{(z/f - 1)^2 + (z_0/f)^2}. \tag{3.2-18}$$

This relation is plotted in Fig. 3.2-6.

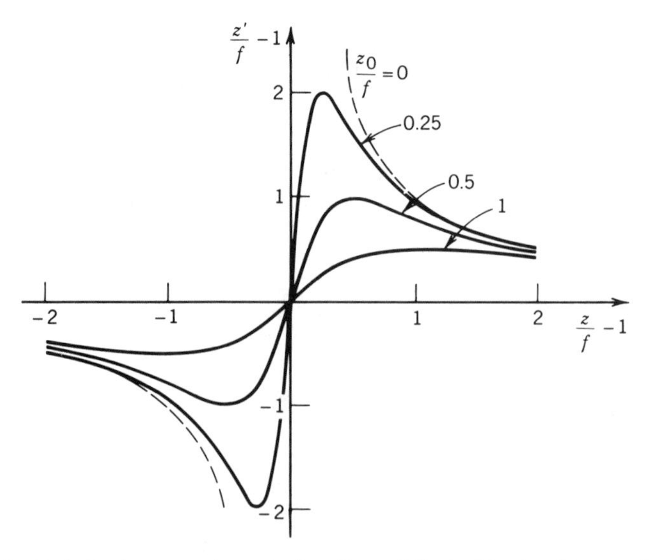

Figure 3.2-6 Relation between the waist locations of the incident and transmitted beams.

(b) The beam is collimated by making the location of the new waist z' as distant as possible from the lens. This is achieved by using the smallest ratio z_0/f (short depth of focus and long focal length). For a given ratio z_0/f, show that the optimal value of z for collimation is $z = f + z_0$.

(c) If $\lambda = 1\ \mu\text{m}$, $z_0 = 1$ cm and $f = 50$ cm, determine the optimal value of z for collimation, and the corresponding magnification M, distance z', and width W_0' of the collimated beam.

EXERCISE 3.2-3

Beam Expansion. A Gaussian beam is expanded and collimated using two lenses of focal lengths f_1 and f_2, as illustrated in Fig. 3.2-7. Parameters of the initial beam (W_0, z_0) are modified by the first lens to (W_0'', z_0'') and subsequently altered by the second lens to (W_0', z_0'). The first lens, which has a short focal length, serves to reduce the depth of focus $2z_0''$ of the beam. This prepares it for collimation by the second lens, which has a long focal length. The system functions as an inverse Keplerian telescope.

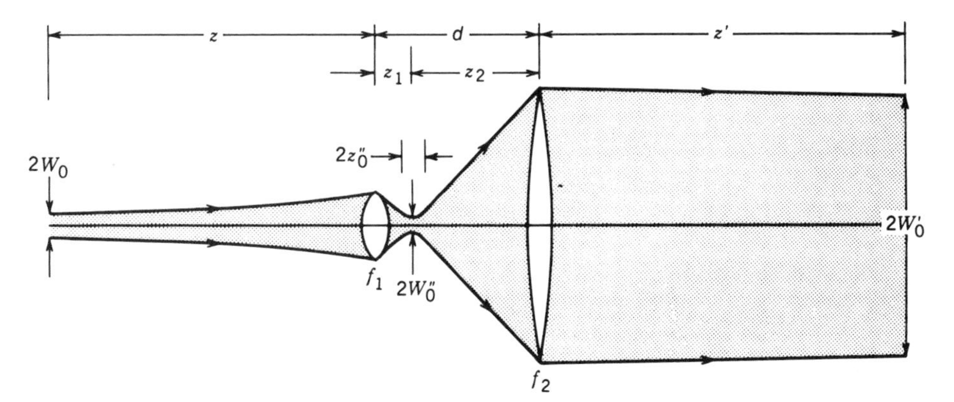

Figure 3.2-7 Beam expansion using a two-lens system.

(a) Assuming that $f_1 \ll z$ and $z - f_1 \gg z_0$, use the results of Exercise 3.2-2 to determine the optimal distance d between the lenses such that the distance z' to the waist of the final beam is as large as possible.

(b) Determine an expression for the overall magnification $M = W_0'/W_0$ of the system.

C. Reflection from a Spherical Mirror

We now examine the reflection of a Gaussian beam from a spherical mirror. Since the complex amplitude reflectance of the mirror is proportional to $\exp(-jk\rho^2/R)$, where by convention $R > 0$ for convex mirrors and $R < 0$ for concave mirrors, the action of the mirror on a Gaussian beam of width W_1 and radius of curvature R_1 is to reflect the beam and to modify its phase by the factor $-k\rho^2/R$, keeping its radius unaltered. Thus the reflected beam remains Gaussian, with parameters W_2 and R_2 given by

$$W_2 = W_1 \tag{3.2-19}$$

$$\frac{1}{R_2} = \frac{1}{R_1} + \frac{2}{R}. \tag{3.2-20}$$

Equation (3.2-20) is the same as (3.2-2) if $f = -R/2$. Thus the Gaussian beam is

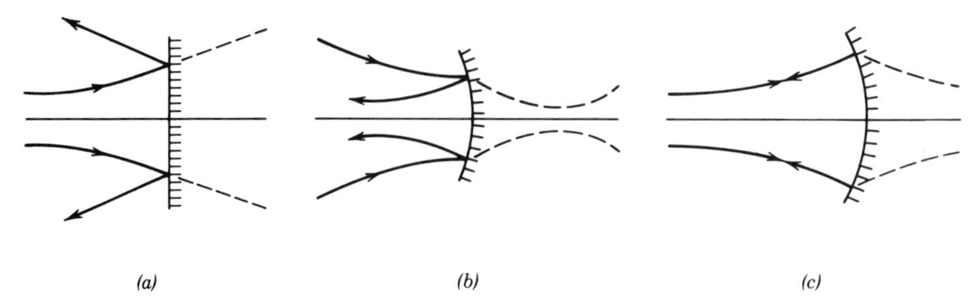

Figure 3.2-8 Reflection of a Gaussian beam of curvature R_1 from a mirror of curvature R: (a) $R = \infty$; (b) $R_1 = \infty$; (c) $R_1 = -R$. The dashed curves show the effects of replacing the mirror by a lens of focal length $f = -R/2$.

modified in precisely the same way as by the lens, except for a reversal of the direction of propagation.

Three special cases (illustrated in Fig. 3.2-8) are of interest:

- If the *mirror is planar*, i.e., $R = \infty$, then $R_2 = R_1$, so that the mirror reverses the direction of the beam without altering its curvature, as illustrated in Fig. 3.2-8(a).
- If $R_1 = \infty$, i.e., the *beam waist lies on the mirror*, then $R_2 = R/2$. If the mirror is concave ($R < 0$), $R_2 < 0$, so that the reflected beam acquires a negative curvature and the wavefronts converge. The mirror then focuses the beam to a smaller spot size, as illustrated in Fig. 3.2-8(b).
- If $R_1 = -R$, i.e., the incident *beam has the same curvature as the mirror*, then $R_2 = R$. The wavefronts of both the incident and reflected waves coincide with the mirror and the wave retraces its path as shown in Fig. 3.2-8(c). This is expected since the wavefront normals are also normal to the mirror, so that the mirror reflects the wave back onto itself. In the illustration in Fig. 3.2-8(c) the mirror is concave ($R < 0$); the incident wave is diverging ($R_1 > 0$) and the reflected wave is converging ($R_2 < 0$).

EXERCISE 3.2-4

Variable-Reflectance Mirrors. A spherical mirror of radius R has a variable intensity reflectance characterized by $\mathscr{R}(\rho) = |\mathscr{r}(\rho)|^2 = \exp(-2\rho^2/W_m^2)$, which is a Gaussian function of the radial distance ρ. The reflectance is unity on axis and falls by a factor $1/e^2$ when $\rho = W_m$. Determine the effect of the mirror on a Gaussian beam with radius of curvature R_1 and beam radius W_1 at the mirror.

---

*D. Transmission Through an Arbitrary Optical System

In the paraxial approximation, an optical system is completely characterized by the 2×2 ray-transfer matrix relating the position and inclination of the transmitted ray to those of the incident ray (see Sec. 1.4). We now consider how an arbitrary paraxial optical system, characterized by a matrix $\mathbf{M}$ of elements (A, B, C, D), modifies a Gaussian beam (Fig. 3.2-9).

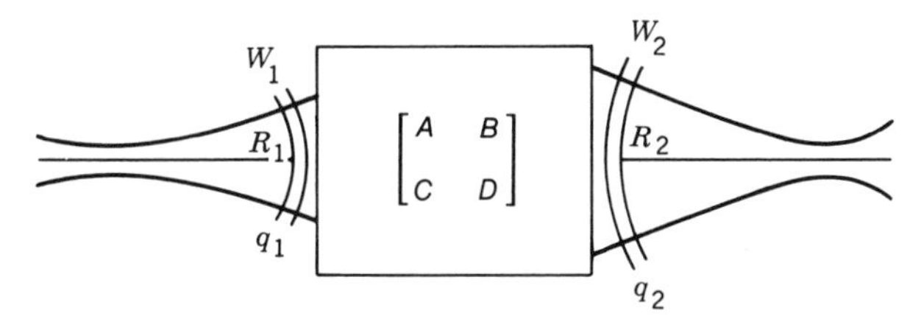

Figure 3.2-9 Modification of a Gaussian beam by an arbitrary paraxial system described by an *ABCD* matrix.

The ABCD Law

The q-parameters, q_1 and q_2, of the incident and transmitted Gaussian beams at the input and output planes of a paraxial optical system described by the (A, B, C, D) matrix are related by

$$q_2 = \frac{Aq_1 + B}{Cq_1 + D}. \tag{3.2-21}$$

The *ABCD* Law

Because the q parameter identifies the width W and curvature R of the Gaussian beam (see Exercise 3.1-3), this simple law, called the ***ABCD* law**, governs the effect of an arbitrary paraxial system on the Gaussian beam. The *ABCD* law will be proved by verification in special cases, and its generality will ultimately be established by induction.

Transmission Through Free Space

When the optical system is a distance d of free space (or of any homogeneous medium), the elements of the ray-transfer matrix **M** are $A = 1, B = d, C = 0, D = 1$. Since $q = z + jz_0$ in free space, the q-parameter is modified by the optical system in accordance with $q_2 = q_1 + d = (1 \cdot q_1 + d)/(0 \cdot q_1 + 1)$, so that the *ABCD* law applies.

Transmission Through a Thin Optical Component

An arbitrary thin optical component does not affect the ray position, so that

$$y_2 = y_1, \tag{3.2-22}$$

but does alter the angle in accordance with

$$\theta_2 = Cy_1 + D\theta_1, \tag{3.2-23}$$

as illustrated in Fig. 3.2-10. Thus $A = 1$ and $B = 0$, but C and D are arbitrary. In all of the thin optical components described in Sec. 1.4B, however, $D = n_1/n_2$. Since the

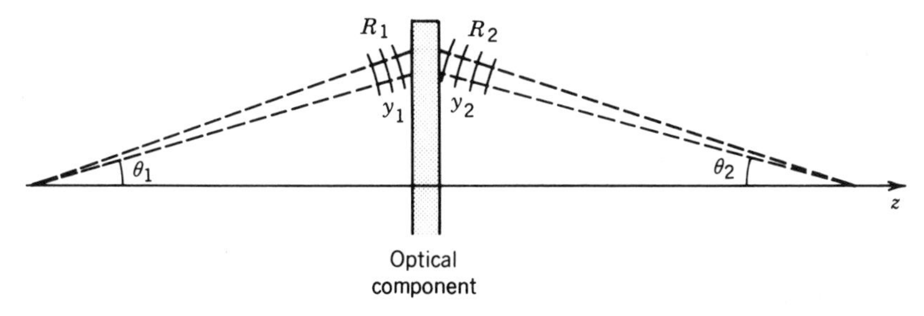

Figure 3.2-10 Modification of a Gaussian beam by a thin optical component.

optical component is thin, the beam width does not change, i.e.,

$$W_2 = W_1. \tag{3.2-24}$$

If the input and output beams are approximated by spherical waves of radii R_1 and R_2 at the input and output planes of the component, respectively, then in the paraxial approximation (small θ_1 and θ_2), $\theta_1 \approx y_1/R_1$ and $\theta_2 \approx y_2/R_2$. Substituting into (3.2-23), and using (3.2-22), we obtain

$$\frac{1}{R_2} = C + \frac{D}{R_1}. \tag{3.2-25}$$

Using (3.1-6), which is the expression for q as a function of R and W, and noting that $D = n_1/n_2 = \lambda_2/\lambda_1$, (3.2-24) and (3.2-25) can be combined into a single equation,

$$\frac{1}{q_2} = C + \frac{D}{q_1}, \tag{3.2-26}$$

from which $q_2 = (1 \cdot q_1 + 0)/(Cq_1 + D)$, so that the *ABCD* law also applies.

Invariance of the ABCD Law to Cascading

If the *ABCD* law is applicable to each of two optical systems with matrices $\mathbf{M}_i = (A_i, B_i, C_i, D_i)$, $i = 1, 2$, it must also apply to a system comprising their cascade (a system with matrix $\mathbf{M} = \mathbf{M}_2\mathbf{M}_1$). This may be shown by straightforward substitution.

Generality of the ABCD Law

Since the *ABCD* law applies to thin optical components and to propagation in a homogeneous medium, it also applies to any combination thereof. All of the paraxial optical systems of interest are combinations of propagation in homogeneous media and thin optical components such as thin lenses and mirrors. We therefore conclude that the *ABCD* law is applicable to all these systems. Since an inhomogeneous continuously varying medium may be regarded as a cascade of incremental thin elements followed by incremental distances, we conclude that the *ABCD* law applies to these systems as well, provided that all rays (wavefront normals) remain paraxial.

EXERCISE 3.2-5

Transmission of a Gaussian Beam Through a Transparent Plate. Use the *ABCD* law to examine the transmission of a Gaussian beam from air, through a transparent plate of refractive index n and thickness d, and again into air. Assume that the beam axis is normal to the plate.

3.3 HERMITE–GAUSSIAN BEAMS

The Gaussian beam is not the only beam-like solution of the paraxial Helmholtz equation (3.1-2). There are may other solutions including beams with non-Gaussian intensity distributions. Of particular interest are solutions that share the paraboloidal wavefronts of the Gaussian beam, but exhibit different intensity distributions. Beams of paraboloidal wavefronts are of importance since they match the curvatures of spherical mirrors of large radius. They can therefore reflect between two spherical mirrors that form a resonator, without being altered. Such self-reproducing waves are called the **modes** of the resonator. The optics of resonators is discussed in Chap. 9.

Consider a Gaussian beam of complex envelope

$$A_G(x, y, z) = \frac{A_1}{q(z)} \exp\left[-jk\frac{x^2 + y^2}{2q(z)}\right], \tag{3.3-1}$$

where $q(z) = z + jz_0$. The beam radius $W(z)$ is given by (3.1-8) and the wavefront radius of curvature $R(z)$ is given by (3.1-9). Consider a second wave whose complex envelope is a modulated version of the Gaussian beam,

$$A(x, y, z) = \mathscr{X}\left[\sqrt{2}\frac{x}{W(z)}\right]\mathscr{Y}\left[\sqrt{2}\frac{y}{W(z)}\right]\exp[j\mathscr{Z}(z)]A_G(x, y, z), \tag{3.3-2}$$

where $\mathscr{X}(\cdot)$, $\mathscr{Y}(\cdot)$, and $\mathscr{Z}(\cdot)$ are real functions. This wave, if it exists, has the following two properties:

- The phase is the same as that of the underlying Gaussian wave, except for an excess phase $\mathscr{Z}(z)$ that is independent of x and y. If $\mathscr{Z}(z)$ is a slowly varying function of z, the two waves have paraboloidal wavefronts with the same radius of curvature $R(z)$. These two waves are therefore focused by thin lenses and mirrors in precisely the same manner.
- The magnitude

$$A_0\mathscr{X}\left[\sqrt{2}\frac{x}{W(z)}\right]\mathscr{Y}\left[\sqrt{2}\frac{y}{W(z)}\right]\left[\frac{W_0}{W(z)}\right]\exp\left[-\frac{x^2 + y^2}{W^2(z)}\right],$$

  where $A_0 = A_1/jz_0$, is a function of $x/W(z)$ and $y/W(z)$ whose widths in the x and y directions vary with z in accordance with the same scaling factor $W(z)$. As z increases, the intensity distribution in the transverse plane remains fixed, except for a magnification factor $W(z)$. This distribution is a Gaussian function modulated in the x and y directions by the functions $\mathscr{X}^2(\cdot)$ and $\mathscr{Y}^2(\cdot)$.

The modulated wave therefore represents a beam of non-Gaussian intensity distribution, but with the same wavefronts and angular divergence as the Gaussian beam.

The existence of this wave is assured if three real functions $\mathscr{X}(\cdot)$, $\mathscr{Y}(\cdot)$, and $\mathscr{Z}(z)$ can be found such that (3.3-2) satisfies the paraxial Helmholtz equation (3.1-2). Substituting (3.3-2) into (3.1-2), using the fact that A_G itself satisfies (3.1-2), and defining two new variables $u = \sqrt{2}\,x/W(z)$ and $v = \sqrt{2}\,y/W(z)$, we obtain

$$\frac{1}{\mathscr{X}}\left(\frac{\partial^2\mathscr{X}}{\partial u^2} - 2u\frac{\partial\mathscr{X}}{\partial u}\right) + \frac{1}{\mathscr{Y}}\left(\frac{\partial^2\mathscr{Y}}{\partial v^2} - 2v\frac{\partial\mathscr{Y}}{\partial v}\right) + kW^2(z)\frac{\partial\mathscr{Z}}{\partial z} = 0. \tag{3.3-3}$$

Since the left-hand side of this equation is the sum of three terms, each of which is a function of a single independent variable, u, v, or z, respectively, each of these terms must be constant. Equating the first term to the constant $-2\mu_1$ and the second to $-2\mu_2$, the third must be equal to $2(\mu_1 + \mu_2)$. This technique of "separation of variables" permits us to reduce the partial differential equation (3.3-3) into three ordinary differential equations for $\mathscr{X}(u)$, $\mathscr{Y}(v)$, and $\mathscr{Z}(z)$, respectively:

$$-\frac{1}{2}\frac{d^2\mathscr{X}}{du^2} + u\frac{d\mathscr{X}}{du} = \mu_1\mathscr{X} \tag{3.3-4a}$$

$$-\frac{1}{2}\frac{d^2\mathscr{Y}}{dv^2} + v\frac{d\mathscr{Y}}{dv} = \mu_2\mathscr{Y} \tag{3.3-4b}$$

$$z_0\left[1 + \left(\frac{z}{z_0}\right)^2\right]\frac{d\mathscr{Z}}{dz} = \mu_1 + \mu_2, \tag{3.3-4c}$$

where we have used the expression for $W(z)$ given in (3.1-8) and (3.1-11).

Equation (3.3-4a) represents an eigenvalue problem whose eigenvalues are $\mu_1 = l$, where $l = 0, 1, 2, \ldots$ and whose eigenfunctions are the **Hermite polynomials** $\mathscr{X}(u) = H_l(u)$, $l = 0, 1, 2, \ldots$. These polynomials are defined by the recurrence relation

$$H_{l+1}(u) = 2uH_l(u) - 2lH_{l-1}(u) \tag{3.3-5}$$

and

$$H_0(u) = 1, \qquad H_1(u) = 2u. \tag{3.3-6}$$

Thus

$$H_2(u) = 4u^2 - 2, \qquad H_3(u) = 8u^3 - 12u, \qquad \ldots. \tag{3.3-7}$$

Similarly, the solutions of (3.3-4b) are $\mu_2 = m$ and $\mathscr{Y}(v) = H_m(v)$, where $m = 0, 1, 2, \ldots$. There is therefore a family of solutions labeled by the indices (l, m).

Substituting $\mu_1 = l$ and $\mu_2 = m$ in (3.3-4c), and integrating, we obtain

$$\mathscr{Z}(z) = (l + m)\zeta(z), \tag{3.3-8}$$

where $\zeta(z) = \tan^{-1}(z/z_0)$. The excess phase $\mathscr{Z}(z)$ varies slowly between $-(l + m)\pi/2$ and $(l + m)\pi/2$, as z varies between $-\infty$ and ∞ (see Fig. 3.1-5).

We finally substitute into (3.3-2) to obtain an expression for the complex envelope of the beam labeled by the indices (l, m). Rearranging terms and multiplying by $\exp(-jkz)$ provides the complex amplitude

$$U_{l,m}(x, y, z) = A_{l,m}\left[\frac{W_0}{W(z)}\right]G_l\left[\frac{\sqrt{2}x}{W(z)}\right]G_m\left[\frac{\sqrt{2}y}{W(z)}\right] \times \exp\left[-jkz - jk\frac{x^2 + y^2}{2R(z)} + j(l + m + 1)\zeta(z)\right], \tag{3.3-9}$$

Hermite–Gaussian Beam Complex Amplitude

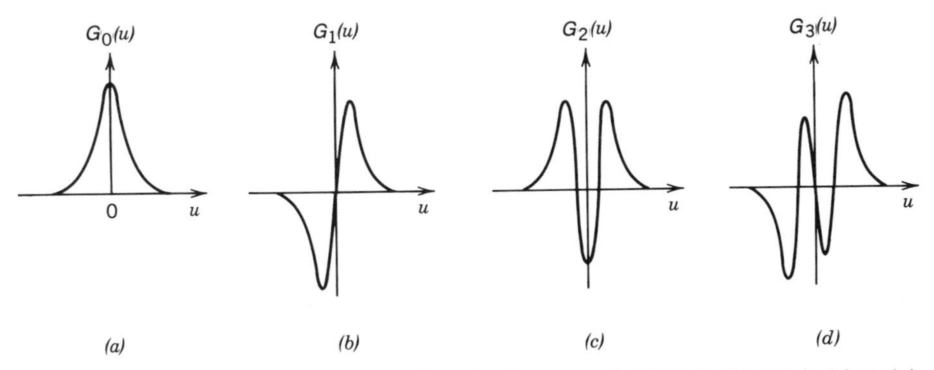

Figure 3.3-1 Several low-order Hermite-Gaussian functions: (*a*) $G_0(u)$; (*b*) $G_1(u)$; (*c*) $G_2(u)$; (*d*) $G_3(u)$.

where

$$G_l(u) = H_l(u) \exp\left(\frac{-u^2}{2}\right), \qquad l = 0, 1, 2, \ldots, \tag{3.3-10}$$

is known as the **Hermite–Gaussian function** of order l, and $A_{l,m}$ is a constant.

Since $H_0(u) = 1$, the Hermite–Gaussian function of order 0 is simply the Gaussian function. $G_1(u) = 2u \exp(-u^2/2)$ is an odd function, $G_2(u) = (4u^2 - 2)\exp(-u^2/2)$ is even, $G_3(u) = (8u^3 - 12u)\exp(-u^2/2)$ is odd, and so on. These functions are shown in Fig. 3.3-1.

An optical wave with complex amplitude given by (3.3-9) is known as the Hermite–Gaussian beam of order (l, m). The Hermite–Gaussian beam of order $(0, 0)$ is the Gaussian beam.

Intensity Distribution

The optical intensity of the (l, m) Hermite–Gaussian beam is

$$I_{l,m}(x, y, z) = |A_{l,m}|^2 \left[\frac{W_0}{W(z)}\right]^2 G_l^2\left[\frac{\sqrt{2}\,x}{W(z)}\right] G_m^2\left[\frac{\sqrt{2}\,y}{W(z)}\right]. \tag{3.3-11}$$

Figure 3.3-2 illustrates the dependence of the intensity on the normalized transverse distances $u = \sqrt{2}\,x/W(z)$ and $v = \sqrt{2}\,y/W(z)$ for several values of l and m. Beams of higher order have larger widths than those of lower order as is evident from Fig. 3.3-1.

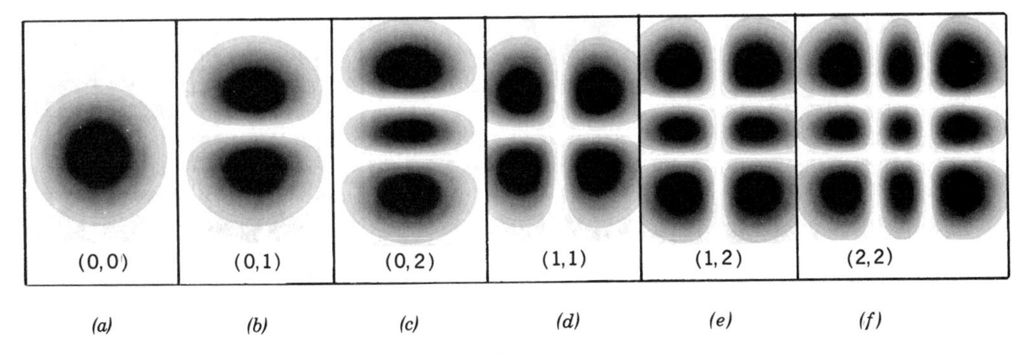

Figure 3.3-2 Intensity distributions of several low-order Hermite–Gaussian beams in the transverse plane. The order (l, m) is indicated in each case.

Regardless of the order, however, the width of the beam is proportional to $W(z)$, so that as z increases the intensity pattern is magnified by the factor $W(z)/W_0$ but otherwise maintains its profile. Among the family of Hermite–Gaussian beams, the only circularly symmetric member is the Gaussian beam.

EXERCISE 3.3-1

The Donut Beam. A wave is a superposition of two Hermite–Gaussian beams of orders $(1,0)$ and $(0,1)$ of equal intensities. The two beams have independent and random phases so that their intensities add with no interference. Show that the total intensity is a donut-shaped circularly symmetric function. Assuming that $W_0 = 1$ mm, determine the radius of the circle of peak intensity and the radii of the two circles of $1/e^2$ times the peak intensity at the beam waist.

---

*3.4 LAGUERRE – GAUSSIAN AND BESSEL BEAMS

Laguerre – Gaussian Beams

The Hermite–Gaussian beams form a complete set of solutions to the paraxial Helmholtz equation. Any other solution can be written as a superposition of these beams. But this family is not the only one. Another complete set of solutions, known as **Laguerre–Gaussian beams**, may be obtained by writing the paraxial Helmholtz equation in cylindrical coordinates (ρ, ϕ, z) and using separation of variables in ρ and ϕ, instead of x and y. The lowest-order Laguerre–Gaussian beam is the Gaussian beam.

Bessel Beams

In the search for beamlike waves, it is natural to examine the possibility of the existence of waves with planar wavefronts but with nonuniform intensity distributions in the transverse plane. Consider a wave with the complex amplitude

$$U(\mathbf{r}) = A(x, y)e^{-j\beta z}. \tag{3.4-1}$$

For this wave to satisfy the Helmholtz equation, $\nabla^2 U + k^2 U = 0$, $A(x, y)$ must satisfy

$$\nabla_T^2 A + k_T^2 A = 0, \tag{3.4-2}$$

where $k_T^2 + \beta^2 = k^2$ and $\nabla_T^2 = \partial^2/\partial x^2 + \partial^2/\partial y^2$ is the transverse Laplacian operator. Equation (3.4-2), known as the two-dimensional Helmholtz equation, may be solved using the method of separation of variables. Using polar coordinates ($x = \rho \cos \phi$, $y = \rho \sin \phi$), the result is

$$A(x, y) = A_m J_m(k_T \rho) e^{jm\phi}, \qquad m = 0, \pm 1, \pm 2, \ldots, \tag{3.4-3}$$

where $J_m(\cdot)$ is the Bessel function of the first kind and mth order, and A_m is a constant. Solutions of (3.4-3) that are singular at $\rho = 0$ are not included.

For $m = 0$, the wave has a complex amplitude

$$U(\mathbf{r}) = A_0 J_0(k_T \rho) e^{-j\beta z} \tag{3.4-4}$$

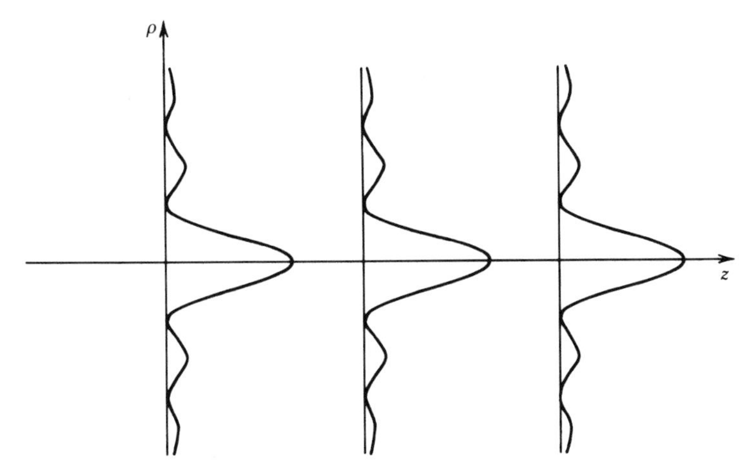

Figure 3.4-1 The intensity distribution of the Bessel beam in the transverse plane is independent of z; the beam does not diverge.

and therefore has planar wavefronts. The wavefront normals (rays) are all parallel to the z axis. The intensity distribution $I(\rho, \phi, z) = |A_0|^2 J_0^2(k_T \rho)$ is circularly symmetric, varies with ρ as illustrated in Fig. 3.4-1, and is independent of z, so that there is no spread of the optical power. This wave is called the **Bessel beam**.

It is interesting to compare the Bessel beam to the Gaussian beam. Whereas the complex amplitude of the Bessel beam is an *exact* solution of the Helmholtz equation, the complex amplitude of the Gaussian beam is only an approximate solution (its complex envelope is an exact solution of the paraxial Helmholtz equation, however). The intensity distribution of these two beams are compared in Fig. 3.4-2. The asymptotic behavior of these distributions in the limit of large radial distances is significantly different. Whereas the intensity of the Gaussian beam decreases exponentially in proportionality to $\exp[-2\rho^2/W^2(z)]$, the intensity of the Bessel beam is proportional to $J_0^2(k_T \rho) \approx (2/\pi k_T \rho)\cos^2(k_T \rho - \pi/4)$, which is an oscillatory function with slowly decaying magnitude. Whereas the rms width of the Gaussian beam, $\sigma = \frac{1}{2}W(z)$, is finite, the rms width of the Bessel beam is *infinite* at all z (see

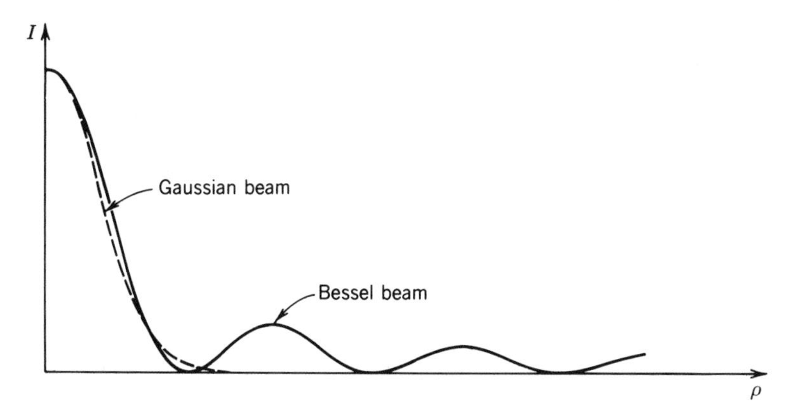

Figure 3.4-2 Comparison of the radial intensity distributions of a Gaussian beam and a Bessel beam. Parameters are selected such that the peak intensities and $1/e^2$ widths are identical in both cases.

Appendix A, Sec. A.2 for the definition of rms width). There is a tradeoff between the minimum beam size and the divergence. Thus although the divergence of the Bessel beam is zero, its rms width is infinite. The generation of Bessel beams requires special schemes.† Since Gaussian beams are the modes of spherical resonators, they are created naturally by lasers.

READING LIST

Books with Chapters on Optical Beams

A. Yariv, *Quantum Electronics*, Wiley, New York, 1967, 3rd ed. 1989.

J. T. Verdeyen, *Laser Electronics*, Prentice-Hall, Englewood Cliffs, NJ, 1981, 2nd ed. 1989.

P. W. Milonni and J. H. Eberly, *Lasers*, Wiley, New York, 1988.

W. Witteman, *The Laser*, Springer-Verlag, New York, 1987.

A. E. Siegman, *Lasers*, University Science Books, Mill Valley, CA, 1986.

K. Shimoda, *Introduction to Laser Physics*, Springer-Verlag, New York, 2nd ed. 1986.

S. Solimeno, B. Crosignani, and P. DiPorto, *Guiding, Diffraction and Confinement of Optical Radiation*, Academic Press, New York, 1986.

A. Yariv, *Optical Electronics*, Holt, Rinehart and Winston, New York, 1971, 3rd ed. 1985.

D. C. O'Shea, *Elements of Modern Optical Design*, Wiley, New York, 1985.

D. Marcuse, *Light Transmission Optics*, Van Nostrand Reinhold, New York, 1972, 2nd ed. 1982.

M. S. Sodha and A. K. Ghatak, *Inhomogeneous Optical Waveguides*, Plenum Press, New York, 1977.

J. A. Arnaud, *Beam and Fiber Optics*, Academic Press, New York, 1976.

A. E. Siegman, *An Introduction to Lasers and Masers*, McGraw-Hill, New York, 1971.

Special Journal Issue

Special issue on propagation and scattering of beam fields, *Journal of the Optical Society of America A*, vol. 3, no. 4, 1986.

Articles

H. Kogelnik and T. Li, Laser Beams and Resonators, *Proceedings of the IEEE*, vol. 54, pp. 1312–1329, 1966.

A. G. Fox and T. Li, Resonant Modes in a Maser Interferometer, *Bell System Technical Journal*, vol. 40, pp. 453–488, 1961.

G. D. Boyd and J. P. Gordon, Confocal Multimode Resonator for Millimeter Through Optical Wavelength Masers, *Bell System Technical Journal*, vol. 40, pp. 489–508, 1961.

PROBLEMS

3.1-1 **Beam Parameters.** The light from a Nd:YAG laser at wavelength 1.06 μm is a Gaussian beam of 1-W optical power and beam divergence $2\theta_0 = 1$ mrad. Determine the beam waist radius, the depth of focus, the maximum intensity, and the intensity on the beam axis at a distance $z = 100$ cm from the beam waist.

3.1-2 **Beam Identification by Two Widths.** A Gaussian beam of wavelength $\lambda_o = 10.6\ \mu$m (emitted by a CO_2 laser) has widths $W_1 = 1.699$ mm and $W_2 = 3.38$ mm at two points separated by a distance $d = 10$ cm. Determine the location of the waist and the waist radius.

†See P. W. Milonni and J. H. Eberly, *Lasers*, Wiley, New York, 1988, Sec. 14.14.

3.1-3 **The Elliptic Gaussian Beam.** The paraxial Helmholtz equation admits a Gaussian beam with intensity $I(x, y, 0) = |A_0|^2 \exp[-2(x^2/W_{0x}^2 + y^2/W_{0y}^2)]$ in the $z = 0$ plane, with beam waist radii W_{0x} and W_{0y} in the x and y-directions respectively. The contours of constant intensity are therefore ellipses instead of circles. Write expressions for the beam depth of focus, angular divergence, and radii of curvature in the x and y directions, as functions of W_{0x}, W_{0y}, and the wavelength λ. If $W_{0x} = 2W_{0y}$, sketch the shape of the beam spot in the $z = 0$ plane and in the far field (z much greater than the depths of focus in both transverse directions).

3.2-1 **Beam Focusing.** An argon-ion laser produces a Gaussian beam of wavelength $\lambda = 488$ nm and waist radius $W_0 = 0.5$ mm. Design a single-lens optical system for focusing the light to a spot of diameter 100 μm. What is the shortest focal-length lens that may be used?

3.2-2 **Spot Size.** A Gaussian beam of Rayleigh range $z_0 = 50$ cm and wavelength $\lambda = 488$ nm is converted into a Gaussian beam of waist radius W_0' using a lens of focal length $f = 5$ cm at a distance z from its waist, as illustrated in Fig. 3.2-2. Write a computer program to plot W_0' as a function of z. Verify that in the limit $z - f \gg z_0$, (3.2-10) and (3.2-12) hold; and in the limit $z \ll z_0$ (3.2-13) holds.

3.2-3 **Beam Refraction.** A Gaussian beam is incident from air ($n = 1$) into a medium with a planar boundary and refractive index $n = 1.5$. The beam axis is normal to the boundary and the beam waist lies at the boundary. Sketch the transmitted beam. If the angular divergence of the beam in air is 1 mrad, what is the angular divergence in the medium?

*3.2-4 **Transmission of a Gaussian Beam Through a Graded-Index Slab.** The *ABCD* matrix of a SELFOC graded-index slab with quadratic refractive index (see Sec. 1.3B) $n(y) \approx n_0(1 - \frac{1}{2}\alpha^2 y^2)$ and length d is: $A = \cos \alpha d$, $B = (1/\alpha) \sin \alpha d$, $C = -\alpha \sin \alpha d$, $D = \cos \alpha d$ for paraxial rays along the z direction. A Gaussian beam of wavelength λ_o, waist radius W_0 in free space, and axis in the z direction enters the slab at its waist. Use the *ABCD* law to determine an expression for the beam width in the y direction as a function of d. Sketch the shape of the beam as it travels through the medium.

3.3-1 **Power Confinement in Hermite–Gaussian Beams.** Determine the ratio of the power contained within a circle of radius $W(z)$ in the transverse plane to the total power in the Hermite–Gaussian beams of orders $(0,0)$, $(1,0)$, $(0,1)$, and $(1,1)$. What is the ratio of the power contained within a circle of radius $W(z)/10$ to the total power for the $(0,0)$ and $(1,1)$ Hermite–Gaussian beams?

3.3-2 **Superposition of Two Beams.** Sketch the intensity of a superposition of the $(1,0)$ and $(1,0)$ Hermite–Gaussian beams assuming that the complex coefficients $A_{1,0}$ and $A_{0,1}$ in (3.3-9) are equal.

3.3-3 **Axial Phase.** Consider the Hermite–Gaussian beams of all orders (l, m) and Rayleigh range $z_0 = 30$ cm in a medium of refractive index $n = 1$. Determine the frequencies within the band $\nu = 10^{14} \pm 2 \times 10^9$ Hz for which the phase retardation between the planes $z = -z_0$ and $z = z_0$ is an integer multiple of π on the beam axis. These frequencies are the modes of a resonator made of two spherical mirrors placed at the $z = \pm z_0$ planes, as described in Sec. 9.2D.

CHAPTER

4

FOURIER OPTICS

4.1 PROPAGATION OF LIGHT IN FREE SPACE
- A. Correspondence Between the Spatial Harmonic Function and the Plane Wave
- B. Transfer Function of Free Space
- C. Impulse-Response Function of Free Space

4.2 OPTICAL FOURIER TRANSFORM
- A. Fourier Transform in the Far Field
- B. Fourier Transform Using a Lens

4.3 DIFFRACTION OF LIGHT
- A. Fraunhofer Diffraction
- *B. Fresnel Diffraction

4.4 IMAGE FORMATION
- A. Ray-Optics Description of Image Formation
- B. Spatial Filtering
- C. Single-Lens Imaging System

4.5 HOLOGRAPHY

Josef von Frauenhofer (1787–1826) developed diffraction gratings and contributed to the understanding of light diffraction. His epitaph reads "*Approximavit sidera*; he brought the stars nearer."

Jean-Baptiste Joseph Fourier (1768–1830) recognized that periodic functions can be considered as sums of sinusoids. Harmonic analysis is the basis of Fourier optics.

Dennis Gabor (1900–1979) made the first hologram in 1947. He received the Nobel Prize in 1971.

Fourier optics provides a description of the propagation of light waves based on harmonic analysis (the Fourier transform) and linear systems. The methods of harmonic analysis have proven to be useful in describing signals and systems in many disciplines. Harmonic analysis is based on the expansion of an arbitrary function of time $f(t)$ as a superposition (a sum or an integral) of harmonic functions of time of different frequencies (see Appendix A, Sec. A.1). The harmonic function $F(\nu)\exp(j2\pi\nu t)$, which has frequency ν and complex amplitude $F(\nu)$, is the building block of the theory. Several of these functions, each with its own value of $F(\nu)$, are added to construct the function $f(t)$, as illustrated in Fig. 4.0-1. The complex amplitude $F(\nu)$, as a function of frequency, is called the Fourier transform of $f(t)$. This approach is useful for the description of linear systems (see Appendix B, Sec. B.1). If the response of the system to each harmonic function is known, the response to an arbitrary input function is readily determined by the use of harmonic analysis at the input and superposition at the output.

An arbitrary function $f(x, y)$ of the two variables x and y, representing the spatial coordinates in a plane, may similarly be written as a superposition of harmonic functions of x and y of the form $F(\nu_x, \nu_y)\exp[-j2\pi(\nu_x x + \nu_y y)]$, where $F(\nu_x, \nu_y)$ is the complex amplitude and ν_x and ν_y are the **spatial frequencies** (cycles per unit length; typically cycles/mm) in the x and y directions, respectively.† The harmonic function $F(\nu_x, \nu_y)\exp[-j2\pi(\nu_x x + \nu_y y)]$ is the two-dimensional building block of the theory. It can be used to generate an arbitrary function of two variables $f(x, y)$, as illustrated in Fig. 4.0-2 (see Appendix A, Sec. A.3).

The plane wave $U(x, y, z) = A\exp[-j(k_x x + k_y y + k_z z)]$ plays an important role in wave optics. The coefficients (k_x, k_y, k_z) are components of the wavevector $\mathbf{k}$ and A is a complex constant. At points in an arbitrary plane, $U(x, y, z)$ is a spatial harmonic function. In the $z = 0$ plane, for example, $U(x, y, 0)$ is the harmonic function $f(x, y) = A\exp[-j2\pi(\nu_x x + \nu_y y)]$, where $\nu_x = k_x/2\pi$ and $\nu_y = k_y/2\pi$ are the spatial frequencies (cycles/mm) and k_x and k_y are the spatial angular frequencies (radians/mm). There is a one-to-one correspondence between the plane wave $U(x, y, z)$ and the spatial harmonic function $f(x, y) = U(x, y, 0)$, provided that the spatial frequency does not exceed the inverse wavelength $1/\lambda$. Since an arbitrary function $f(x,y)$ can be analyzed as a superposition of harmonic functions, an arbitrary traveling

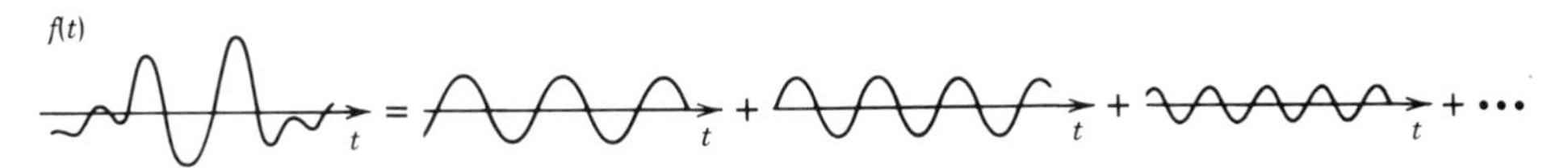

Figure 4.0-1 An arbitrary function $f(t)$ may be analyzed as a sum of harmonic functions of different frequencies and complex amplitudes.

†The spatial harmonic function is defined with a minus sign in the exponent, in contrast to the plus sign used in the definition of the temporal harmonic function (see Appendix A, Sec. A.3). These signs match those of a forward-traveling plane wave.

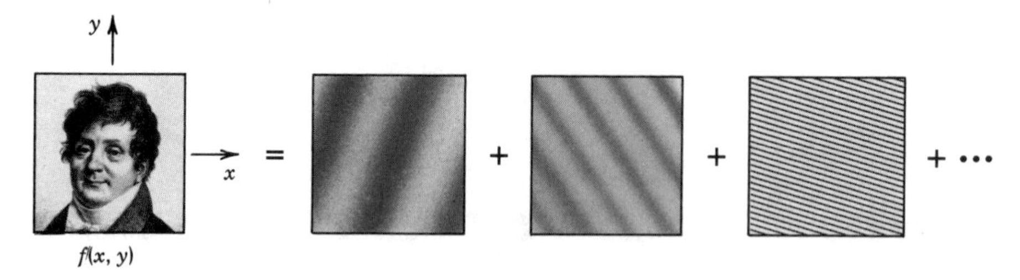

Figure 4.0-2 An arbitrary function $f(x, y)$ may be analyzed as a sum of harmonic functions of different spatial frequencies and complex amplitudes.

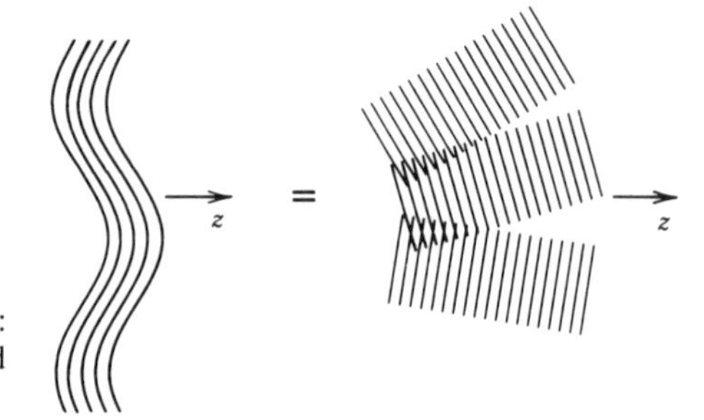

Figure 4.0-3 The principle of Fourier optics: an arbitrary wave in free space can be analyzed as a superposition of plane waves.

wave $U(x, y, z)$ may be analyzed as a sum of plane waves (Fig. 4.0-3). The plane wave is the building block used to construct a wave of arbitrary complexity. Furthermore, if it is known how a linear optical system modifies plane waves, the principle of superposition can be used to determine the effect of the system on an arbitrary wave.

Because of the important role Fourier analysis plays in describing linear systems, it is useful to describe the propagation of light through linear optical components, including free space, using a linear-system approach. The complex amplitudes in two planes normal to the optic (z) axis are regarded as the input and output of the system (Fig. 4.0-4). A linear system may be characterized by either its **impulse-response function** (the response of the system to an impulse, or a point, at the input) or by its **transfer function** (the response to spatial harmonic functions), as described in Appendix B.

The chapter begins with a Fourier description of the propagation of light in free space (Sec. 4.1). The transfer function and impulse-response function of the free-space

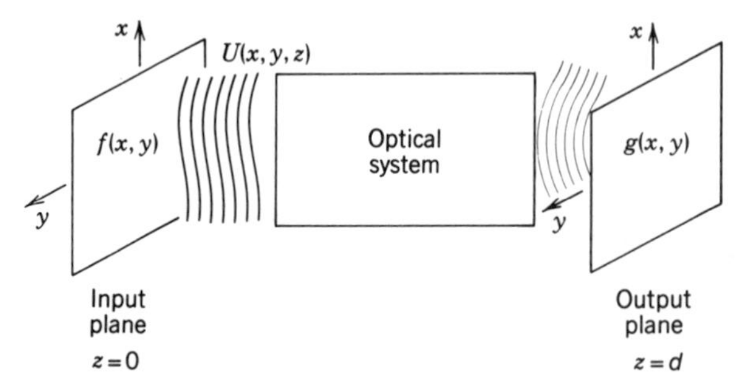

Figure 4.0-4 The transmission of an optical wave $U(x, y, z)$ through an optical system between an input plane $z = 0$ and an output plane $z = d$. This is regarded as a linear system whose input and output are the functions $f(x, y) = U(x, y, 0)$ and $g(x, y) = U(x, y, d)$, respectively.

propagation system are determined. In Sec. 4.2 we show that a lens may perform the operation of the spatial Fourier transform. The transmission of light through apertures is discussed in Sec. 4.3; this is a Fourier-optics approach to the diffraction of light. Section 4.4 is devoted to image formation and spatial filtering. Finally, an introduction to holography, the recording and reconstruction of optical waves, is presented in Sec. 4.5. Knowledge of the basic properties of the Fourier transform and linear systems in one and two dimensions (reviewed in Appendices A and B) is necessary for understanding this chapter.

4.1 PROPAGATION OF LIGHT IN FREE SPACE

A. Correspondence Between the Spatial Harmonic Function and the Plane Wave

Consider a plane wave of complex amplitude $U(x, y, z) = A \exp[-j(k_x x + k_y y + k_z z)]$ with wavevector $\mathbf{k} = (k_x, k_y, k_z)$, wavelength λ, wavenumber $k = (k_x^2 + k_y^2 + k_z^2)^{1/2} = 2\pi/\lambda$, and complex envelope A. The vector $\mathbf{k}$ makes angles $\theta_x = \sin^{-1}(k_x/k)$ and $\theta_y = \sin^{-1}(k_y/k)$ with the y–z and x–z planes, respectively, as illustrated in Fig. 4.1-1. The complex amplitude in the $z = 0$ plane, $U(x, y, 0)$, is a spatial harmonic function $f(x, y) = A \exp[-j2\pi(\nu_x x + \nu_y y)]$ with spatial frequencies $\nu_x = k_x/2\pi$ and $\nu_y = k_y/2\pi$ (cycles/mm). The angles of the wavevector are therefore related to the spatial frequencies of the harmonic function by

$$\theta_x = \sin^{-1} \lambda\nu_x$$
$$\theta_y = \sin^{-1} \lambda\nu_y. \tag{4.1-1}$$

Correspondence Between Spatial Frequencies and Angles

Recognizing $\Lambda_x = 1/\nu_x$ and $\Lambda_y = 1/\nu_y$ as the periods of the harmonic function in the x and y directions, we see that the angles $\theta_x = \sin^{-1}(\lambda/\Lambda_x)$ and $\theta_y = \sin^{-1}(\lambda/\Lambda_y)$ are governed by the ratios of the wavelength of light to the period of the harmonic function in each direction. These geometrical relations follow from matching the wavefronts of the wave to the periodic pattern of the harmonic function in the $z = 0$ plane, as illustrated in Fig. 4.1-1.

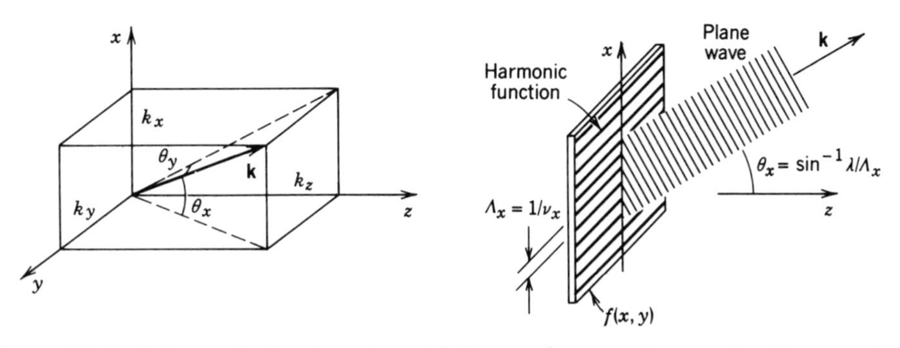

Figure 4.1-1 A harmonic function of spatial frequencies ν_x and ν_y at the plane $z = 0$ is consistent with a plane wave traveling at angles $\theta_x = \sin^{-1} \lambda\nu_x$ and $\theta_y = \sin^{-1} \lambda\nu_y$.

If $k_x \ll k$ and $k_y \ll k$, so that the wavevector $\mathbf{k}$ is paraxial, the angles θ_x and θ_y are small ($\sin\theta_x \approx \theta_x$ and $\sin\theta_y \approx \theta_y$) and

$$\theta_x \approx \lambda\nu_x$$
$$\theta_y \approx \lambda\nu_y. \qquad (4.1\text{-}2)$$

Spatial Frequencies and Angles (Paraxial Approximation)

Thus the angles of inclination of the wavevector are directly proportional to the spatial frequencies of the corresponding harmonic function.

Apparently, there is a one-to-one correspondence between the plane wave $U(x, y, z)$ and the harmonic function $f(x, y)$. Given one, the other can be readily determined (if the wavelength λ is known). Given the wave $U(x, y, z)$, the harmonic function $f(x, y)$ is obtained by sampling in the $z = 0$ plane, $f(x, y) = U(x, y, 0)$. Given the harmonic function $f(x, y)$, on the other hand, the wave $U(x, y, z)$ is constructed by using the relation $U(x, y, z) = f(x, y)\exp(-jk_z z)$ with

$$k_z = \pm\left(k^2 - k_x^2 - k_y^2\right)^{1/2}, \qquad k = 2\pi/\lambda. \qquad (4.1\text{-}3)$$

A condition of validity of this correspondence is that $k_x^2 + k_y^2 < k^2$, so that k_z is real. This condition implies that $\lambda\nu_x < 1$ and $\lambda\nu_y < 1$, so that the angles θ_x and θ_y defined by (4.1-1) exist. The $+$ and $-$ signs in (4.1-3) represent waves traveling in the forward and backward directions, respectively. We shall be concerned with forward waves only.

Spatial Spectral Analysis

When a plane wave of unity amplitude traveling in the z direction is transmitted through a thin optical element with complex amplitude transmittance $f(x, y) = \exp[-j2\pi(\nu_x x + \nu_y y)]$ the wave is modulated by the harmonic function, so that $U(x, y, 0) = f(x, y)$. The incident wave is then converted into a plane wave with a wavevector at angles $\theta_x = \sin^{-1}\lambda\nu_x$ and $\theta_y = \sin^{-1}\lambda\nu_y$ (see Fig. 4.1-2). The optical element is a diffraction grating which acts like a prism (see Exercise 2.4-5).

If the transmittance of the optical element $f(x, y)$ is the sum of several harmonic functions of different spatial frequencies, the transmitted optical wave is also the sum of an equal number of plane waves dispersed into different directions; each spatial frequency is mapped into a corresponding direction, in accordance with (4.1-1). The

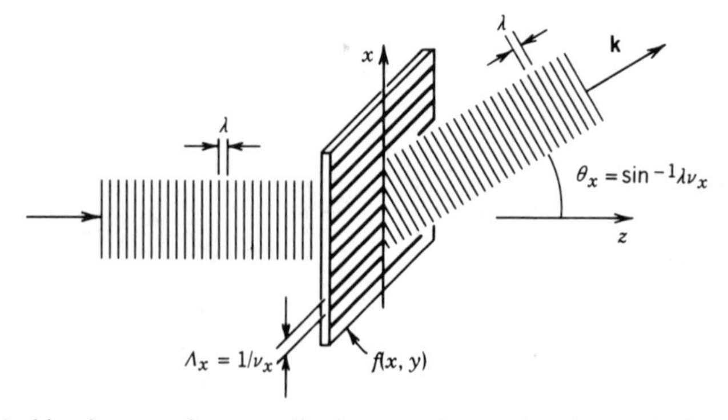

Figure 4.1-2 A thin element whose amplitude transmittance is a harmonic function of spatial frequency ν_x (period $\Lambda_x = 1/\nu_x$) bends a plane wave of wavelength λ by an angle $\theta_x = \sin^{-1}\lambda\nu_x = \sin^{-1}(\lambda/\Lambda_x)$.

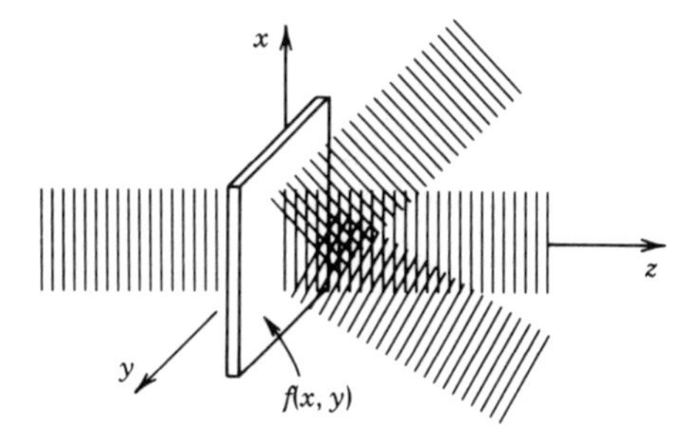

Figure 4.1-3 A thin optical element of amplitude transmittance $f(x, y)$ decomposes an incident plane wave into many plane waves. The plane wave traveling at the angles $\theta_x = \sin^{-1} \lambda \nu_x$ and $\theta_y = \sin^{-1} \lambda \nu_y$ has a complex envelope $F(\nu_x, \nu_y)$, the Fourier transform of $f(x, y)$.

amplitude of each wave is proportional to the amplitude of the corresponding harmonic component of $f(x, y)$.

More generally, if $f(x, y)$ is a superposition integral of harmonic functions,

$$f(x, y) = \iint_{-\infty}^{\infty} F(\nu_x, \nu_y) \exp\left[-j2\pi(\nu_x x + \nu_y y)\right] d\nu_x \, d\nu_y, \tag{4.1-4}$$

with frequencies (ν_x, ν_y) and amplitudes $F(\nu_x, \nu_y)$, the transmitted wave $U(x, y, z)$ is the superposition of plane waves,

$$U(x, y, z) = \iint_{-\infty}^{\infty} F(\nu_x, \nu_y) \exp\left[-j(2\pi\nu_x x + 2\pi\nu_y y)\right] \exp(-jk_z z) \, d\nu_x \, d\nu_y,$$

with complex envelopes $F(\nu_x, \nu_y)$, where $k_z = (k^2 - k_x^2 - k_y^2)^{1/2} = 2\pi(1/\lambda^2 - \nu_x^2 - \nu_y^2)^{1/2}$. Note that $F(\nu_x, \nu_y)$ is the Fourier transform of $f(x, y)$ (see Appendix A, Sec. A.3).

Since an arbitrary function may be Fourier analyzed as a superposition integral of the form (4.1-4), the light transmitted through a thin optical element of arbitrary transmittance may be written as a superposition of plane waves (see Fig. 4.1-3), provided that $\nu_x^2 + \nu_y^2 < 1/\lambda^2$. This process of "spatial spectral analysis" is akin to the angular dispersion of different temporal-frequency components (wavelengths) provided by a prism. Free-space propagation serves as a natural "spatial prism," sensitive to the spatial instead of the temporal frequencies of the optical wave.

Amplitude Modulation

Consider a transparency with complex amplitude transmittance $f_0(x, y)$. If the Fourier transform $F_0(\nu_x, \nu_y)$ extends over widths $\pm\Delta\nu_x$ and $\pm\Delta\nu_y$ in the x and y directions, the transparency will deflect an incident plane wave by angles θ_x and θ_y in the range $\pm\sin^{-1}(\lambda \, \Delta\nu_x)$ and $\pm\sin^{-1}(\lambda \, \Delta\nu_y)$, respectively.

Consider a second transparency of complex amplitude transmittance $f(x, y) = f_0(x, y) \exp[-j2\pi(\nu_{x0} x + \nu_{y0} y)]$, where $f_0(x, y)$ is slowly varying compared to $\exp[-j2\pi(\nu_{x0} x + \nu_{y0} y)]$ so that $\Delta\nu_x \ll \nu_{x0}$ and $\Delta\nu_y \ll \nu_{y0}$. We may regard $f(x, y)$ as an amplitude-modulated function with a carrier frequency ν_{x0} and ν_{y0} and modulation function $f_0(x, y)$. The Fourier transform of $f(x, y)$ is $F_0(\nu_x - \nu_{x0}, \nu_y - \nu_{y0})$, in accordance with the frequency-shifting property of the Fourier transform (see Appendix A). The transparency will deflect a plane wave to directions centered about the angles $\theta_{x0} = \sin^{-1} \lambda \nu_{x0}$ and $\theta_{y0} = \sin^{-1} \lambda \nu_{y0}$ (Fig. 4.1-4). This can also be readily seen by regarding $f(x, y)$ as a transparency of transmittance $f_0(x, y)$ in contact with a grating or prism of transmittance $\exp[-j2\pi(\nu_{x0} x + \nu_{y0} y)]$ that provides the angular deflection θ_{x0} and θ_{y0}.

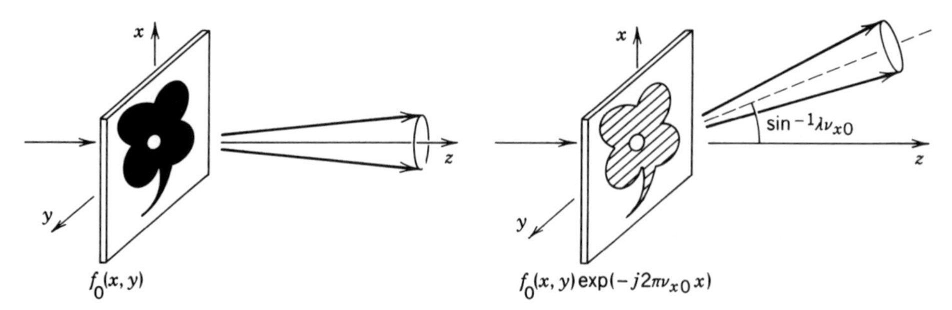

Figure 4.1-4 Deflection of light by the transparencies $f_0(x, y)$ and $f_0(x, y)\exp(-j2\pi\nu_{x0}x)$. The "carrier" harmonic function $\exp(-j2\pi\nu_{x0}x)$ acts as a prism that deflects the wave by an angle $\theta_{x0} = \sin^{-1}\lambda\nu_{x0}$.

This idea may be used to record two images $f_1(x, y)$ and $f_2(x, y)$ on the same transparency using the *spatial-frequency multiplexing* scheme $f(x, y) = f_1(x, y)\exp[-j2\pi(\nu_{x1}x + \nu_{y1}y)] + f_2(x, y)\exp[-j2\pi(\nu_{x2}x + \nu_{y2}y)]$. The two images may be easily separated by illuminating the transparency with a plane wave, whereupon the two images are deflected at different angles and are thus separated. This principle will prove useful in holography (Sec. 4.5), where it is often desired to separate two images recorded on the same transparency.

Frequency Modulation

We now examine the transmission of a plane wave through a transparency made of a "collage" of several regions, the transmittance of each of which is a harmonic function of some spatial frequency, as illustrated in Fig. 4.1-5. If the dimensions of each region are much greater than the period, each region acts as a grating or a prism that deflects the wave in some direction, so that different portions of the incident wavefront are deflected into different directions. This principle may be used to create maps of optical interconnections, which may be used in optical computing applications, as described in Sec. 21.5.

A transparency may also have a harmonic transmittance with a spatial frequency that varies continuously and slowly with position (in comparison with λ), much as the frequency of a frequency-modulated (FM) signal varies slowly with time. Consider, for example, the phase function $f(x, y) = \exp[-j2\pi\phi(x, y)]$, where $\phi(x, y)$ is a continuous slowly varying function of x and y. In the neighborhood of a point (x_0, y_0), we may use the Taylor's series expansion $\phi(x, y) \approx \phi(x_0, y_0) + (x - x_0)\nu_x + (y - y_0)\nu_y$, where the derivatives $\nu_x = \partial\phi/\partial x$ and $\nu_y = \partial\phi/\partial y$ are evaluated at the position (x_0, y_0). The local variation of $f(x, y)$ with x and y is therefore proportional to the quantity $\exp[-j2\pi(\nu_x x + \nu_y y)]$, which is a harmonic function with spatial frequencies

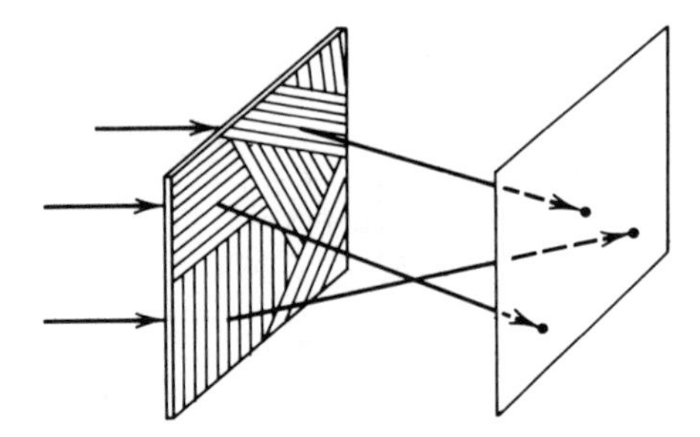

Figure 4.1-5 Deflection of light by a transparency made of several harmonic functions (phase gratings) of different spatial frequencies.

$\nu_x = \partial\phi/\partial x$ and $\nu_y = \partial\phi/\partial y$. Since the derivatives $\partial\phi/\partial x$ and $\partial\phi/\partial y$ vary with x and y, so do the spatial frequencies. The transparency $f(x, y) = \exp[-j2\pi\phi(x, y)]$ therefore deflects the portion of the wave at the position (x, y) by the position-dependent angles $\theta_x = \sin^{-1}(\lambda\, \partial\phi/\partial x)$ and $\theta_y = \sin^{-1}(\lambda\, \partial\phi/\partial y)$.

EXAMPLE 4.1-1. ***Scanning.*** A thin transparency with complex amplitude transmittance $f(x, y) = \exp(j\pi x^2/\lambda f)$ introduces a phase shift $2\pi\phi(x, y)$ where $\phi(x, y) = -x^2/2\lambda f$, so that the wave is deflected at the position (x, y) by the angles $\theta_x = \sin^{-1}(\lambda\, \partial\phi/\partial x) = \sin^{-1}(-x/f)$ and $\theta_y = 0$. If $|x/f| \ll 1$, $\theta_x \approx -x/f$ and the deflection angle θ_x is directly proportional to the transverse distance x. This transparency may be used to deflect a narrow beam of light. If the transparency is moved at a uniform speed, the beam is deflected by a linearly increasing angle as illustrated in Fig. 4.1-6.

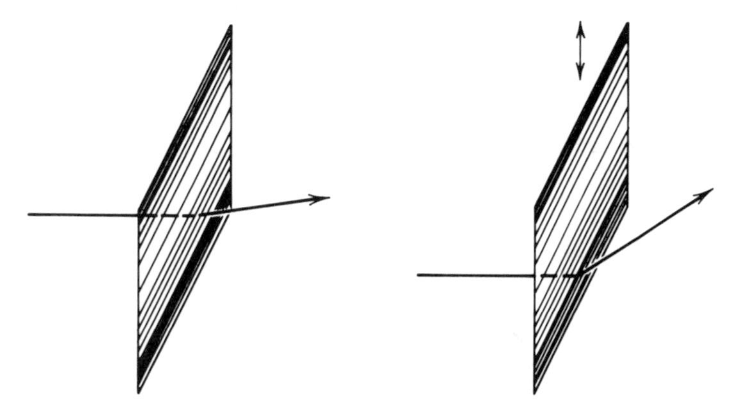

Figure 4.1-6 Using a frequency-modulated transparency to scan an optical beam.

EXAMPLE 4.1-2. ***Imaging.*** If the transparency in Example 4.1-1 is illuminated by a plane wave, each part of the wave is deflected by a different angle and as a result the wavefront is altered. The local wavevector at position x bends by an angle $-x/f$ so that all wavevectors meet at a single point on the optical axis a distance f from the transparency, as illustrated in Fig. 4.1-7. The transparency acts as a cylindrical lens with focal length f. Similarly, a transparency with the transmittance $f(x, y) = \exp[j\pi(x^2 + y^2)/\lambda f]$ acts as a

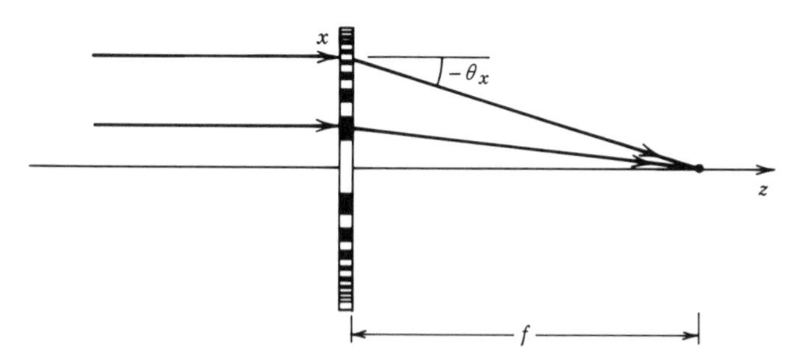

Figure 4.1-7 A transparency with transmittance $f(x, y) = \exp(j\pi x^2/\lambda f)$ bends the wave at position x by an angle $\theta_x \approx -x/f$ so that it acts as a cylindrical lens with focal length f.

spherical lens with focal length f. Indeed, this is the expression for the transmittance of a thin lens [see (2.4-6)].

EXERCISE 4.1-1

The Fresnel Zone Plate

(a) Use harmonic analysis near the position x to show that a transparency with complex amplitude transmittance

$$f(x,y)=\begin{cases}1, & \text{if } \cos\left(\pi\dfrac{x^2}{\lambda f}\right)>0\\ 0, & \text{otherwise}\end{cases}$$

acts as a cylindrical lens with multiple focal lengths.

(b) A circularly symmetric transparency of complex amplitude transmittance

$$f(x,y)=\begin{cases}1, & \text{if } \cos\left(\pi\dfrac{x^2+y^2}{\lambda f}\right)>0\\ 0, & \text{otherwise}\end{cases}$$

is known as a Fresnel zone plate (see Fig. 4.1-8). Show that it acts as a spherical lens with multiple focal lengths.

Figure 4.1-8 The Fresnel zone plate.

B. Transfer Function of Free Space

We now examine the propagation of a monochromatic optical wave of wavelength λ and complex amplitude $U(x, y, z)$ in the free space between the planes $z = 0$ and $z = d$, called the input and output planes, respectively (see Fig. 4.1-9). Given the complex amplitude of the wave at the input plane, $f(x, y) = U(x, y, 0)$, we shall determine the complex amplitude at the output plane, $g(x, y) = U(x, y, d)$.

We regard $f(x, y)$ and $g(x, y)$ as the input and output of a linear system. The system is linear since the Helmholtz equation, which $U(x, y, z)$ must satisfy, is linear. The system is shift-invariant because of the invariance of free space to displacement of the coordinate system. A linear shift-invariant system is characterized by its impulse

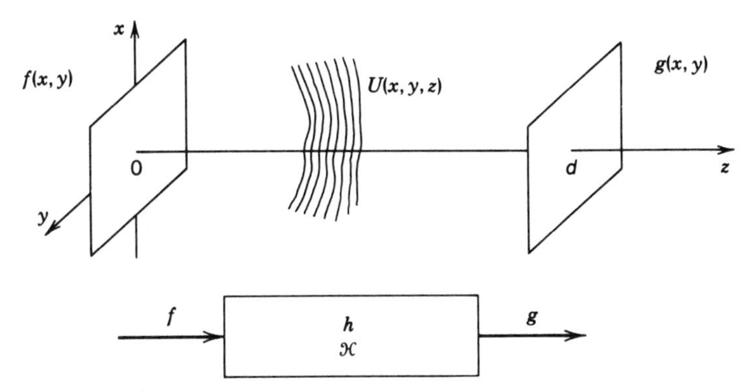

Figure 4.1-9 Propagation of light between two planes is regarded as a linear system whose input and output are the complex amplitudes of the wave in the two planes.

response function $h(x, y)$ or by its transfer function $\mathcal{H}(\nu_x, \nu_y)$, as explained in Appendix B, Sec. B.2. We now proceed to determine expressions for these functions.

The transfer function $\mathcal{H}(\nu_x, \nu_y)$ is the factor by which an input spatial harmonic function of frequencies ν_x and ν_y is multiplied to yield the output harmonic function. We therefore consider a harmonic input function $f(x, y) = A \exp[-j2\pi(\nu_x x + \nu_y y)]$. As explained earlier, this corresponds to a plane wave $U(x, y, z) = A \exp[-j(k_x x + k_y y + k_z z)]$ where $k_x = 2\pi\nu_x$, $k_y = 2\pi\nu_y$, and

$$k_z = \left(k^2 - k_x^2 - k_y^2\right)^{1/2} = 2\pi\left(\frac{1}{\lambda^2} - \nu_x^2 - \nu_y^2\right)^{1/2}. \tag{4.1-5}$$

The output $g(x, y) = A \exp[-j(k_x x + k_y y + k_z d)]$, so that we can write $\mathcal{H}(\nu_x, \nu_y) = g(x, y)/f(x, y) = \exp(-jk_z d)$, from which

$$\mathcal{H}(\nu_x, \nu_y) = \exp\left[-j2\pi\left(\frac{1}{\lambda^2} - \nu_x^2 - \nu_y^2\right)^{1/2} d\right]. \tag{4.1-6}$$

Transfer Function of Free Space

The transfer function $\mathcal{H}(\nu_x, \nu_y)$ is therefore a circularly symmetric complex function of the spatial frequencies ν_x and ν_y. Its magnitude and phase are sketched in Fig. 4.1-10.

For spatial frequencies for which $\nu_x^2 + \nu_y^2 \leq 1/\lambda^2$ (i.e., frequencies lying within a circle of radius $1/\lambda$) the magnitude $|\mathcal{H}(\nu_x, \nu_y)| = 1$ and the phase $\arg\{\mathcal{H}(\nu_x, \nu_y)\}$ is a function of ν_x and ν_y. A harmonic function with such frequencies therefore undergoes a spatial phase shift as it propagates, but its magnitude is not altered.

At higher spatial frequencies, $\nu_x^2 + \nu_y^2 > 1/\lambda^2$, the quantity under the square root in (4.1-6) is negative so that the exponent is real and the transfer function $\exp[-2\pi(\nu_x^2 + \nu_y^2 - 1/\lambda^2)^{1/2} d]$ represents an attenuation factor.† The wave is then called an **evanescent wave**. When $\nu_\rho = (\nu_x^2 + \nu_y^2)^{1/2}$ exceeds $1/\lambda$ slightly, i.e., $\nu_\rho \approx 1/\lambda$, the attenuation factor is $\exp[-2\pi(\nu_\rho^2 - 1/\lambda^2)^{1/2} d] = \exp[-2\pi(\nu_\rho - 1/\lambda)^{1/2}(\nu_\rho + 1/\lambda)^{1/2} d] \approx \exp[-2\pi(\nu_\rho - 1/\lambda)^{1/2}(2d^2/\lambda)^{1/2}]$, which equals $\exp(-2\pi)$ when $(\nu_\rho - 1/\lambda) \approx \lambda/2d^2$, or $(\nu_\rho - 1/\lambda)/(1/\lambda) \approx \frac{1}{2}(\lambda/d)^2$. For $d \gg \lambda$ the attenuation factor drops sharply when the spatial frequency slightly exceeds $1/\lambda$, as illustrated in Fig. 4.1-10.

†The − sign in (4.1-3) was used since the + sign would have resulted in an exponentially growing function, which is physically unacceptable.

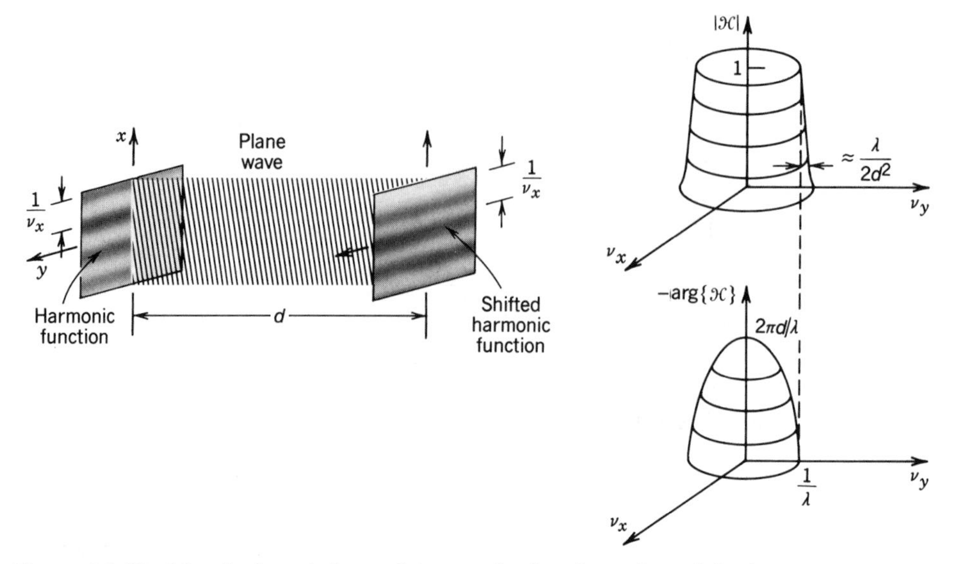

Figure 4.1-10 Magnitude and phase of the transfer function $\mathcal{H}(\nu_x, \nu_y)$ for free-space propagation between two planes separated by a distance d.

We may therefore regard $1/\lambda$ as the cutoff spatial frequency (the spatial bandwidth) of the system. Thus

the spatial bandwidth of light propagation
in free space is approximately $1/\lambda$ *cycles/mm.*

Features contained in spatial frequencies greater than $1/\lambda$ (corresponding to details of size finer than λ) cannot be transmitted by an optical wave of wavelength λ over distances much greater than λ.

Fresnel Approximation

The expression for the transfer function in (4.1-6) may be simplified if the input function $f(x, y)$ contains only spatial frequencies that are much smaller than the cutoff frequency $1/\lambda$, so that $\nu_x^2 + \nu_y^2 \ll 1/\lambda^2$. The plane-wave components of the propagating light then make small angles $\theta_x \approx \lambda\nu_x$ and $\theta_y \approx \lambda\nu_y$ corresponding to paraxial rays.

Denoting $\theta^2 = \theta_x^2 + \theta_y^2 \approx \lambda^2(\nu_x^2 + \nu_y^2)$, where θ is the angle with the optical axis, the phase factor in (4.1-6) is

$$2\pi\left(\frac{1}{\lambda^2} - \nu_x^2 - \nu_y^2\right)^{1/2} d = 2\pi\frac{d}{\lambda}(1 - \theta^2)^{1/2}$$

$$= 2\pi\frac{d}{\lambda}\left(1 - \frac{\theta^2}{2} + \frac{\theta^4}{8} - \cdots\right). \qquad (4.1\text{-}7)$$

Neglecting the third and higher terms of this expansion, (4.1-6) may be approximated by

$$\mathcal{H}(\nu_x, \nu_y) \approx \mathcal{H}_0 \exp\left[j\pi\lambda d\left(\nu_x^2 + \nu_y^2\right)\right], \qquad (4.1\text{-}8)$$

Transfer Function
of Free Space
(Fresnel Approximation)

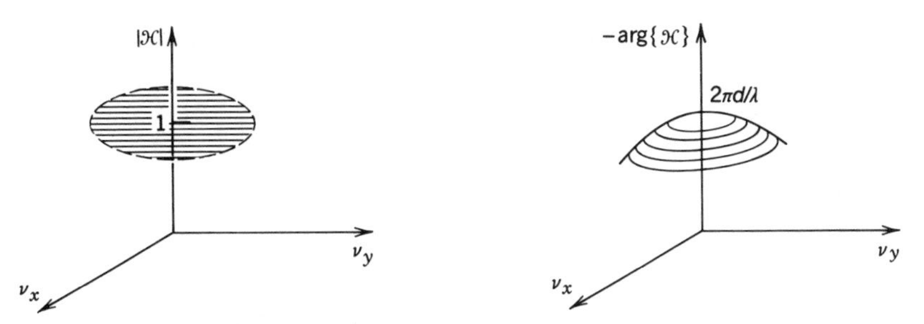

Figure 4.1-11 The transfer function of free-space propagation for low spatial frequencies (much less than $1/\lambda$ cycles/mm) has a constant magnitude and a quadratic phase.

where $\mathcal{H}_0 = \exp(-jkd)$. In this approximation, the phase is a quadratic function of ν_x and ν_y, as illustrated in Fig. 4.1-11. This approximation is known as the **Fresnel approximation.**

The condition of validity of the Fresnel approximation is that the third term in (4.1-7) is much smaller than π for all θ. This is equivalent to

$$\frac{\theta^4 d}{4\lambda} \ll 1. \tag{4.1-9}$$

If a is the largest radial distance in the output plane, the largest angle $\theta_m \approx a/d$, and (4.1-9) may be written in the form

$$\frac{N_F \theta_m^2}{4} \ll 1, \tag{4.1-10}$$

Condition of Validity of Fresnel Approximation

where $N_F = a^2/\lambda d$ is the Fresnel number. For example, if $a = 1$ cm, $d = 100$ cm, and $\lambda = 0.5\ \mu$m, then $\theta_m = 10^{-2}$ radian, $N_F = 200$, and $N_F\theta^2/4 = 5 \times 10^{-3}$. In this case the Fresnel approximation is applicable.

Input–Output Relation

Given the input function $f(x, y)$, the output function $g(x, y)$ may be determined as follows: (1) We determine the Fourier transform

$$F(\nu_x, \nu_y) = \iint_{-\infty}^{\infty} f(x, y) \exp\left[j2\pi(\nu_x x + \nu_y y)\right] dx\, dy, \tag{4.1-11}$$

which represents the complex envelopes of the plane-wave components in the input plane; (2) the product $\mathcal{H}(\nu_x, \nu_y)F(\nu_x, \nu_y)$ gives the complex envelopes of the plane-wave components in the output plane; and (3) the complex amplitude in the output plane is the sum of the contributions of these plane waves,

$$g(x, y) = \iint_{-\infty}^{\infty} \mathcal{H}(\nu_x, \nu_y)F(\nu_x, \nu_y) \exp\left[-j2\pi(\nu_x x + \nu_y y)\right] d\nu_x\, d\nu_y.$$

Using the Fresnel approximation for $\mathcal{H}(\nu_x, \nu_y)$, which is given by (4.1-8), we have

$$g(x, y) = \mathcal{H}_0 \iint_{-\infty}^{\infty} F(\nu_x, \nu_y) \exp\left[j\pi\lambda d\left(\nu_x^2 + \nu_y^2\right)\right] \exp\left[-j2\pi(\nu_x x + \nu_y y)\right] d\nu_x \, d\nu_y. \quad (4.1\text{-}12)$$

Equations (4.1-12) and (4.1-11) serve to relate the output function $g(x, y)$ to the input function $f(x, y)$.

C. Impulse-Response Function of Free Space

The impulse-response function $h(x, y)$ of the system of free-space propagation is the response $g(x, y)$ when the input $f(x, y)$ is a point at the origin $(0, 0)$. It is the inverse Fourier transform of the transfer function $\mathcal{H}(\nu_x, \nu_y)$. Using Sec. A.3 and Table A.1-1 in Appendix A and $k = 2\pi/\lambda$, the inverse Fourier transform of (4.1-8) is

$$h(x, y) \approx h_0 \exp\left[-jk\frac{x^2 + y^2}{2d}\right], \quad (4.1\text{-}13)$$

Impulse-Response Function of Free Space (Fresnel Approximation)

where $h_0 = (j/\lambda d)\exp(-jkd)$. This function is proportional to the complex amplitude at the $z = d$ plane of a parabolodial wave centered about the origin $(0, 0)$ [see (2.2-16)]. Thus each point in the input plane generates a paraboloidal wave; all such waves are superposed at the output plane.

Free-Space Propagation as a Convolution

An alternative procedure for relating the complex amplitudes $f(x, y)$ and $g(x, y)$ is to regard $f(x, y)$ as a superposition of different points (delta functions), each producing a paraboloidal wave. The wave originating at the point (x', y') has an amplitude $f(x', y')$ and is centered about (x', y') so that it generates a wave with amplitude $f(x', y')h(x - x', y - y')$ at the point (x, y) in the output plane. The sum of these contributions is the two-dimensional convolution

$$g(x, y) = \iint_{-\infty}^{\infty} f(x', y')h(x - x', y - y')\, dx'\, dy',$$

which, in the Fresnel approximation, becomes

$$g(x, y) = h_0 \iint_{-\infty}^{\infty} f(x', y') \exp\left[-j\pi\frac{(x - x')^2 + (y - y')^2}{\lambda d}\right] dx'\, dy', \quad (4.1\text{-}14)$$

where $h_0 = (j/\lambda d)\exp(-jkd)$.

In summary: Within the Fresnel approximation, there are two approaches to determining the complex amplitude $g(x, y)$ in the output plane, given the complex amplitude $f(x, y)$ in the input plane: (1) Equation (4.1-14) is based on a space-domain

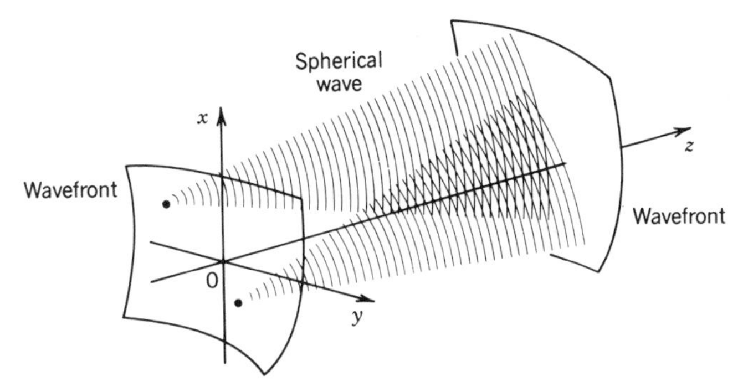

Figure 4.1-12 The Huygens–Fresnel principle. Each point on a wavefront generates a spherical wave.

approach in which the input wave is expanded in terms of paraboloidal elementary waves; and (2) Equation (4.1-12) is a frequency-domain approach in which the input wave is expanded as a sum of plane waves.

EXERCISE 4.1-2

Gaussian Beams Revisited. If the function $f(x, y) = A \exp[-(x^2 + y^2)/W_0^2]$ represents the complex amplitude of an optical wave $U(x, y, z)$ in the plane $z = 0$, show that $U(x, y, z)$ is the Gaussian beam discussed in Chap. 3, (3.1-7). Use both the space- and frequency-domain methods.

Huygens – Fresnel Principle

The **Huygens–Fresnel principle** states that each point on a wavefront generates a spherical wave (Fig. 4.1-12). The envelope of these secondary waves constitutes a new wavefront. Their superposition constitutes the wave in another plane. The system's impulse-response function for propagation between the planes $z = 0$ and $z = d$ is

$$h(x, y) \propto \frac{1}{r} \exp(-jkr), \qquad r = \left(x^2 + y^2 + d^2\right)^{1/2}. \tag{4.1-15}$$

In the paraxial approximation, the spherical wave given by (4.1-15) is approximated by the paraboloidal wave in (4.1-13) (see Sec. 2.2B). Our derivation of the impulse response function is therefore consistent with the Huygens–Fresnel principle.

4.2 OPTICAL FOURIER TRANSFORM

As has been shown in Sec. 4.1, the propagation of light in free space is described conveniently by Fourier analysis. If the complex amplitude of a monochromatic wave of wavelength λ in the $z = 0$ plane is a function $f(x, y)$ composed of harmonic components of different spatial frequencies, each harmonic component corresponds to a plane wave: The plane wave traveling at angles $\theta_x = \sin^{-1} \lambda\nu_x, \theta_y = \sin^{-1} \lambda\nu_y$ corresponds to the components with spatial frequencies ν_x and ν_y and has an amplitude

$F(\nu_x, \nu_y)$, the Fourier transform of $f(x, y)$. This suggests that light can be used to compute the Fourier transform of a two-dimensional function $f(x, y)$, simply by making a transparency with amplitude transmittance $f(x, y)$ through which a uniform plane wave of unity magnitude is transmitted.

Because each of the plane waves has an infinite extent and therefore overlaps with the other plane waves, however, it is necessary to find a method of separating these waves. It will be shown that at a sufficiently long distance, only a single plane wave contributes to the total amplitude at each point in the output plane, so that the Fourier components are eventually separated naturally. A more practical approach is to use a lens to focus each of the plane waves into a single point.

A. Fourier Transform in the Far Field

We now proceed to show that if the propagation distance d is sufficiently long, the only plane wave that contributes to the complex amplitude at a point (x, y) in the output plane is the wave with direction making angles $\theta_x \approx x/d$ and $\theta_y \approx y/d$ with the optical axis (see Fig. 4.2-1). This is the wave with wavevector components $k_x \approx (x/d)k$ and $k_y \approx (y/d)k$ and amplitude $F(\nu_x, \nu_y)$ with $\nu_x = x/\lambda d$, and $\nu_y = y/\lambda d$. The complex amplitudes $g(x, y)$ and $f(x, y)$ of the wave at the $z = d$ and $z = 0$ planes are related by

$$g(x, y) \approx h_0 F\left(\frac{x}{\lambda d}, \frac{y}{\lambda d}\right), \tag{4.2-1}$$

Free-Space Propagation
(Fraunhofer Approximation)

where $F(\nu_x, \nu_y)$ is the Fourier transform of $f(x, y)$ and $h_0 = (j/\lambda d)\exp(-jkd)$. Contributions of all other waves cancel out as a result of destructive interference. This approximation is known as the **Fraunhofer approximation**. Two proofs of (4.2-1) are provided.

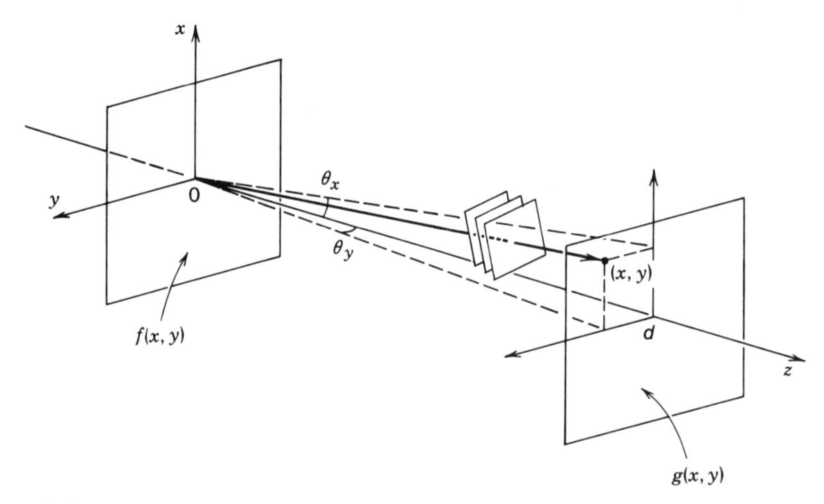

Figure 4.2-1 When the distance d is sufficiently long, the complex amplitude at point (x, y) in the $z = d$ plane is proportional to the complex amplitude of the plane-wave component with angles $\theta_x \approx x/d \approx \lambda\nu_x$ and $\theta_y \approx y/d \approx \lambda\nu_y$, i.e., to the Fourier transform $F(\nu_x, \nu_y)$ of $f(x, y)$, with $\nu_x = x/\lambda d$ and $\nu_y = y/\lambda d$.

Proof 1. We begin with the relation between $g(x, y)$ and $f(x, y)$ in (4.1-14). The phase in the argument of the exponent is $(\pi/\lambda d)[(x - x')^2 + (y - y')^2] = (\pi/\lambda d)[(x^2 + y^2) + (x'^2 + y'^2) - 2(xx' + yy')]$. If $f(x, y)$ is confined to a small area of radius b, and if the distance d is sufficiently large so that the Fresnel number $N_F' = b^2/\lambda d$ is small,

$$\boxed{N_F' \ll 1,} \tag{4.2-2}$$

Condition of Validity of Fraunhofer Approximation

then the phase factor $(\pi/\lambda d)(x'^2 + y'^2) \le \pi(b^2/\lambda d)$ is negligible and (4.1-14) may be approximated by

$$g(x, y) = h_0 \exp\left(-j\pi\frac{x^2 + y^2}{\lambda d}\right) \iint_{-\infty}^{\infty} f(x', y') \exp\left(j2\pi\frac{xx' + yy'}{\lambda d}\right) dx'\, dy'. \tag{4.2-3}$$

The factors $x/\lambda d$ and $y/\lambda d$ may be regarded as the frequencies $\nu_x = x/\lambda d$ and $\nu_y = y/\lambda d$, so that

$$g(x, y) = h_0 \exp\left(-j\pi\frac{x^2 + y^2}{\lambda d}\right) F\left(\frac{x}{\lambda d}, \frac{y}{\lambda d}\right), \tag{4.2-4}$$

where $F(\nu_x, \nu_y)$ is the Fourier transform of $f(x, y)$. The phase factor given by $\exp[-j\pi(x^2 + y^2)/\lambda d]$ in (4.2-4) may also be neglected and (4.2-1) obtained if we also limit our interest to points in the output plane within a circle of radius a centered about the z-axis so that $\pi(x^2 + y^2)/\lambda d \le \pi a^2/\lambda d \ll \pi$. This is applicable when the Fresnel number $N_F = a^2/\lambda d \ll 1$.

The **Fraunhofer approximation** is therefore valid whenever the Fresnel numbers N_F and N_F' are small. The Fraunhofer approximation is more difficult to satisfy than the Fresnel approximation, which requires that $N_F\theta_m^2/4 \ll 1$ [see (4.1-10)]. Since $\theta_m \ll 1$ in the paraxial approximation, it is possible to satisfy the Fresnel condition $N_F\theta_m^2/4 \ll 1$ for Fresnel numbers N_F not necessarily $\ll 1$.

Summary

In the Fraunhofer approximation, the complex amplitude $g(x, y)$ of a wave of wavelength λ in the $z = d$ plane is proportional to the Fourier transform $F(\nu_x, \nu_y)$ of the complex amplitude $f(x, y)$ in the $z = 0$ plane, evaluated at the spatial frequencies $\nu_x = x/\lambda d$ and $\nu_y = y/\lambda d$. The approximation is valid if $f(x, y)$ is confined to a circle of radius b satisfying, $b^2/\lambda d \ll 1$, and at points in the output plane within a circle of radius a satisfying $a^2/\lambda d \ll 1$.

EXERCISE 4.2-1

Conditions of Validity of the Fresnel and Fraunhofer Approximations: A Comparison. Demonstrate that the Fraunhofer approximation is more restrictive than the Fresnel approximation by taking $\lambda = 0.5\ \mu m$, assuming that the object points (x, y) lie within a circle of radius $b = 1$ cm, and determining the range of distances d for which the two approximations are applicable.

****Proof* 2.* The complex amplitude $g(x, y)$ in (4.1-12) is expressed as an integral of plane waves of different frequencies. If d is sufficiently large so that the phase in the integrand is much greater than 2π, it can be shown using the method of stationary phase[†] that only one value of ν_x contributes to the integral. This is the value for which the derivative of the phase $\pi\lambda d\nu_x^2 - 2\pi\nu_x x$ with respect to ν_x vanishes; i.e., $\nu_x = x/\lambda d$. Similarly, the only value of ν_y that contributes to the integral is $\nu_y = y/\lambda d$. This proves the assertion that only one plane wave contributes to the far field at a given point.

B. Fourier Transform Using a Lens

The plane-wave components that constitute a wave may also be separated by use of a lens. A thin spherical lens transforms a plane wave into a paraboloidal wave focused to a point in the lens focal plane (see Sec. 2.4 and Exercise 2.4-3). If the plane wave arrives at small angles θ_x and θ_y, the paraboloidal wave is centered about the point $(\theta_x f, \theta_y f)$, where f is the focal length (see Fig. 4.2-2). The lens therefore maps each direction (θ_x, θ_y) into a single point $(\theta_x f, \theta_y f)$ in the focal plane and thus separates the contributions of the different plane waves.

In reference to the optical system shown in Fig. 4.2-3, let $f(x, y)$ be the complex amplitude of the optical wave in the $z = 0$ plane. Light is decomposed into plane waves, with the wave traveling at small angles $\theta_x = \lambda\nu_x$ and $\theta_y = \lambda\nu_y$ having a complex amplitude proportional to the Fourier transform $F(\nu_x, \nu_y)$. This wave is focused by the lens into a point (x, y) in the focal plane where $x = \theta_x f = \lambda f \nu_x$, $y = \theta_y f = \lambda f \nu_y$. The complex amplitude at point (x, y) in the output plane is therefore proportional to the Fourier transform of $f(x, y)$ evaluated at $\nu_x = x/\lambda f$ and $\nu_y = y/\lambda f$, so that

$$g(x, y) \propto F\left(\frac{x}{\lambda f}, \frac{y}{\lambda f}\right). \tag{4.2-5}$$

To determine the proportionality factor in (4.2-5), we analyze the input function $f(x, y)$ into its Fourier components and trace the plane wave corresponding to each component through the optical system. We then superpose the contributions of these waves at the output plane to obtain $g(x, y)$. All these waves will be assumed to be

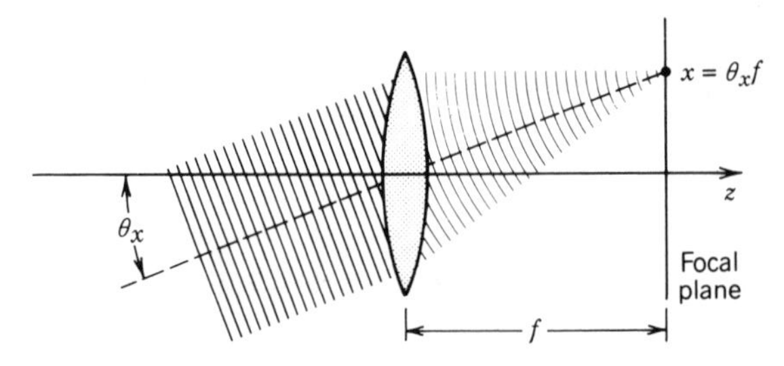

Figure 4.2-2 Focusing of a plane wave into a point. A direction (θ_x, θ_y) is mapped into a point $(x, y) = (\theta_x f, \theta_y f)$.

[†]See, e.g., Appendix III in M. Born and E. Wolf, *Principles of Optics*, Pergamon Press, New York, 6th ed. 1980.

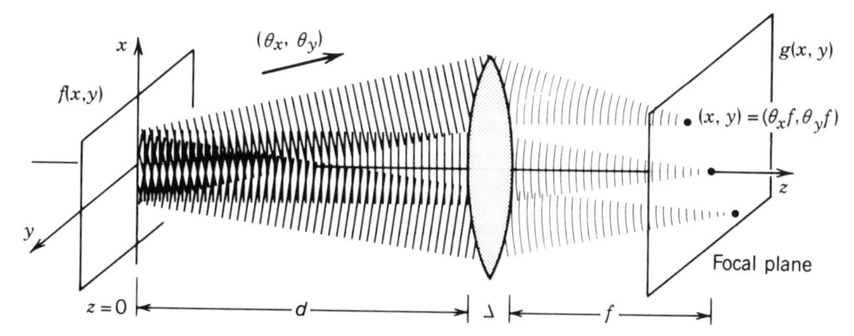

Figure 4.2-3 Focusing of the plane waves associated with the harmonic Fourier components of the input function $f(x, y)$ into points in the focal plane. The amplitude of the plane wave with direction $(\theta_x, \theta_y) = (\lambda\nu_x, \lambda\nu_y)$ is proportional to the Fourier transform $F(\nu_x, \nu_y)$ and is focused at the point $(x, y) = (\theta_x f, \theta_y f) = (\lambda f \nu_x, \lambda f \nu_y)$.

paraxial and the Fresnel approximation will be used. The procedure takes the following four steps.

1. The plane wave with angles $\theta_x = \lambda\nu_x$ and $\theta_y = \lambda\nu_y$ has a complex amplitude $U(x,y,0) = F(\nu_x, \nu_y)\exp[-j2\pi(\nu_x x + \nu_y y)]$ in the $z = 0$ plane and $U(x, y, d) = \mathcal{H}(\nu_x, \nu_y)F(\nu_x, \nu_y)\exp[-j2\pi(\nu_x x + \nu_y y)]$ in the $z = d$ plane, immediately before crossing the lens, where $\mathcal{H}(\nu_x, \nu_y) = \mathcal{H}_0 \exp[j\pi\lambda d(\nu_x^2 + \nu_y^2)]$ is the transfer function of a distance d of free space and $\mathcal{H}_0 = \exp(-jkd)$.
2. Upon crossing the lens, the complex amplitude is multiplied by the lens phase factor $\exp[j\pi(x^2 + y^2)/\lambda f]$ [the phase factor $\exp(-jk\Delta)$, where Δ is the width of the lens, has been ignored]. Thus

$$U(x, y, d + \Delta) = \mathcal{H}_0 \exp\left(j\pi \frac{x^2 + y^2}{\lambda f}\right)$$

$$\times \exp\left[j\pi\lambda d\left(\nu_x^2 + \nu_y^2\right)\right] F(\nu_x, \nu_y) \exp\left[-j2\pi(\nu_x x + \nu_y y)\right].$$

This expression is simplified by writing $-2\nu_x x + x^2/\lambda f = (x^2 - 2\nu_x \lambda f x)/\lambda f = [(x - x_0)^2 - x_0^2]/\lambda f$, with $x_0 = \lambda\nu_x f$; a similar relation for y is written with $y_0 = \lambda\nu_y f$, so that

$$U(x, y, d + \Delta) = A(\nu_x, \nu_y) \exp\left[j\pi \frac{(x - x_0)^2 + (y - y_0)^2}{\lambda f}\right], \quad (4.2\text{-}6)$$

where

$$A(\nu_x, \nu_y) = \mathcal{H}_0 \exp\left[j\pi\lambda(d - f)\left(\nu_x^2 + \nu_y^2\right)\right] F(\nu_x, \nu_y). \quad (4.2\text{-}7)$$

Equation (4.2-6) is recognized as the complex amplitude of a paraboloidal wave converging toward the point (x_0, y_0) in the lens focal plane, $z = d + \Delta + f$.
3. We now examine the propagation in the free space between the lens and the output plane to determine $U(x, y, d + \Delta + f)$. We apply (4.1-14) to (4.2-6), use

the relation $\int \exp[j2\pi(x - x_0)x'/\lambda f]\,dx' = \lambda f \delta(x - x_0)$, and obtain

$$U(x, y, d + \Delta + f) = h_0(\lambda f)^2 A(\nu_x, \nu_y)\delta(x - x_0)\delta(y - y_0)$$

where $h_0 = (j/\lambda f)\exp(-jkf)$. Indeed, the plane wave is focused into a single point at $x_0 = \lambda\nu_x f$ and $y_0 = \lambda\nu_y f$.

4. The last step is to integrate over all the plane waves (all ν_x and ν_y). By virtue of the sifting property of the delta function, $\delta(x - x_0) = \delta(x - \lambda f \nu_x) = (1/\lambda f)\delta(\nu_x - x/\lambda f)$, this integral gives $g(x, y) = h_0 A(x/\lambda f, y/\lambda f)$. Substituting from (4.2-7) we finally obtain

$$g(x, y) = h_l \exp\left[j\pi \frac{(x^2 + y^2)(d - f)}{\lambda f^2}\right] F\left(\frac{x}{\lambda f}, \frac{y}{\lambda f}\right), \tag{4.2-8}$$

where $h_l = \mathcal{H}_0 h_0 = (j/\lambda f)\exp[-jk(d + f)]$. Thus the coefficient of proportionality in (4.2-5) contains a phase factor that is a quadratic function of x and y.

Since $|h_l| = 1/\lambda f$ it follows from (4.2-8) that the optical intensity at the output plane is

$$I(x, y) = \frac{1}{(\lambda f)^2}\left|F\left(\frac{x}{\lambda f}, \frac{y}{\lambda f}\right)\right|^2. \tag{4.2-9}$$

The intensity of light at the output plane (the back focal plane of the lens) is therefore proportional to the squared absolute value of the Fourier transform of the complex amplitude of the wave at the input plane, regardless of the distance d.

The phase factor in (4.2-8) vanishes if $d = f$, so that

$$g(x, y) = h_l F\left(\frac{x}{\lambda f}, \frac{y}{\lambda f}\right), \tag{4.2-10}$$

Fourier Transform Property of a Lens

where $h_l = (j/\lambda f)\exp(-j2kf)$. This geometry is shown in Fig. 4.2-4.

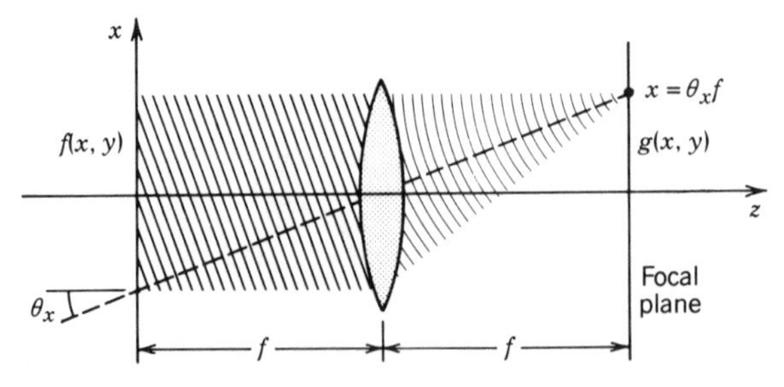

Figure 4.2-4 Fourier transform system. The Fourier component of $f(x, y)$ with spatial frequencies ν_x and ν_y generates a plane wave at angles $\theta_x = \lambda\nu_x$ and $\theta_y = \lambda\nu_y$ and is focused by the lens to the point $(x, y) = (f\theta_x, f\theta_y) = (\lambda f\nu_x, \lambda f\nu_y)$ so that $g(x, y)$ is proportional to the Fourier transform $F(x/\lambda f, y/\lambda f)$.

In summary: The complex amplitude of light at a point (x, y) in the back focal plane of a lens of focal length f is proportional to the Fourier transform of the complex amplitude in the front focal plane evaluated at the frequencies $\nu_x = x/\lambda f, \nu_y = y/\lambda f$. This relation is valid in the Fresnel approximation. Without the lens, the Fourier transformation is obtained only in the Fraunhofer approximation, which is more restrictive.

EXERCISE 4.2-2

The Inverse Fourier Transform. Verify that the optical system in Fig. 4.2-4 performs the inverse Fourier transform operation if the coordinate system in the front focal plane is inverted, i.e., $(x, y) \to (-x, -y)$.

4.3 DIFFRACTION OF LIGHT

When an optical wave is transmitted through an aperture in an opaque screen and travels some distance in free space, its intensity distribution is called the diffraction pattern. If light were treated as rays, the diffraction pattern would be a shadow of the aperture. Because of the wave nature of light, however, the diffraction pattern may deviate slightly or substantially from the aperture shadow, depending on the distance between the aperture and observation plane, the wavelength, and the dimensions of the aperture. An example is illustrated in Fig. 4.3-1. It is difficult to determine exactly the manner in which the screen modifies the incident wave, but the propagation in free space beyond the aperture is always governed by the laws described earlier in this chapter.

The simplest theory of diffraction is based on the *assumption* that the incident wave is transmitted without change at points within the aperture, but is reduced to zero at points on the back side of the opaque part of the screen. If $U(x, y)$ and $f(x, y)$ are the complex amplitudes of the wave immediately to the left and right of the screen (Fig. 4.3-2), then in accordance with this assumption,

$$f(x, y) = U(x, y)p(x, y), \qquad (4.3\text{-}1)$$

Figure 4.3-1 Diffraction pattern of the teeth of a saw. (From M. Cagnet, M. Françon, and J. C. Thrierr, *Atlas of Optical Phenomena*, Springer-Verlag, Berlin, 1962.)

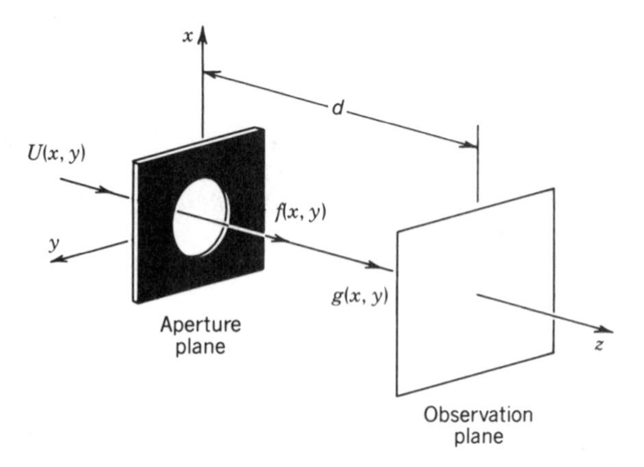

Figure 4.3-2 A wave $U(x, y)$ is transmitted through an aperture of amplitude transmittance $p(x, y)$, generating a wave of complex amplitude $f(x, y) = U(x, y)p(x, y)$. After propagation a distance d in free space the complex amplitude is $g(x, y)$ and the diffraction pattern is the intensity $I(x, y) = |g(x, y)|^2$.

where

$$p(x, y) = \begin{cases} 1 & \text{inside the aperture} \\ 0, & \text{outside the aperture} \end{cases} \tag{4.3-2}$$

is called the **aperture function**.

Given $f(x, y)$, the complex amplitude $g(x, y)$ at an observation plane a distance d from the screen may be determined using the methods described in Secs. 4.1 and 4.2. The diffraction pattern $I(x, y) = |g(x, y)|^2$ is known as **Fraunhofer diffraction** or **Fresnel diffraction**, depending on whether free-space propagation is described using the Fraunhofer approximation or the Fresnel approximation, respectively.

Although this approach gives reasonably accurate results in most cases, it is not exact. The validity and self-consistency of the assumption that the complex amplitude $f(x, y)$ vanishes at points outside the aperture on the back of the screen are questionable since the transmitted wave propagates in all directions and reaches those points. A theory of diffraction based on the exact solution of the Helmholtz equation under the boundary conditions imposed by the aperture is mathematically difficult. Only a few geometrical structures have yielded exact solutions. However, different diffraction theories have been developed using a variety of assumptions, leading to results with varying accuracies. Rigorous diffraction theory is beyond the scope of this book.

A. Fraunhofer Diffraction

Fraunhofer diffraction is the theory of transmission of light through apertures under the assumption that the incident wave is multiplied by the aperture function and using the Fraunhofer approximation to determine the propagation of light in the free space beyond the aperture. The Fraunhofer approximation is valid if the propagation distance d between the aperture and observation planes is sufficiently large so that the Fresnel number $N_F' = b^2/\lambda d \ll 1$, where b is the largest radial distance within the aperture.

Assuming that the incident wave is a plane wave of intensity I_i traveling in the z direction so that $U(x, y) = I_i^{1/2}$, then $f(x, y) = I_i^{1/2}p(x, y)$. In the Fraunhofer approximation [see (4.2-1)],

$$g(x, y) \approx I_i^{1/2} h_0 P\left(\frac{x}{\lambda d}, \frac{y}{\lambda d}\right), \tag{4.3-3}$$

where

$$P(\nu_x, \nu_y) = \iint_{-\infty}^{\infty} p(x, y) \exp\left[j2\pi(\nu_x x + \nu_y y)\right] dx\, dy$$

is the Fourier transform of $p(x, y)$ and $h_0 = (j/\lambda d)\exp(-jkd)$. The diffraction pattern is therefore

$$I(x, y) = \frac{I_i}{(\lambda d)^2} \left| P\left(\frac{x}{\lambda d}, \frac{y}{\lambda d}\right)\right|^2. \tag{4.3-4}$$

In summary: The Fraunhofer diffraction pattern at the point (x, y) is proportional to the squared magnitude of the Fourier transform of the aperture function $p(x, y)$ evaluated at the spatial frequencies $\nu_x = x/\lambda d$ and $\nu_y = y/\lambda d$.

EXERCISE 4.3-1

Fraunhofer Diffraction from a Rectangular Aperture. Verify that the Fraunhofer diffraction pattern from a rectangular aperture, of height and width D_x and D_y respectively, observed at a distance d is

$$I(x, y) = I_o \operatorname{sinc}^2 \frac{D_x x}{\lambda d} \operatorname{sinc}^2 \frac{D_y y}{\lambda d}, \tag{4.3-5}$$

where $I_o = (D_x D_y/\lambda d)^2 I_i$ is the peak intensity and $\operatorname{sinc}(x) = \sin(\pi x)/(\pi x)$. Verify that the first zeros of this pattern occur at $x = \pm\lambda d/D_x$ and $y = \pm\lambda d/D_y$, so that the angular divergence of the diffracted light is given by

$$\theta_x = \frac{\lambda}{D_x}, \qquad \theta_y = \frac{\lambda}{D_y}. \tag{4.3-6}$$

If $D_y < D_x$, the diffraction pattern is wider in the y direction than in the x direction, as illustrated in Fig. 4.3-3.

EXERCISE 4.3-2

Fraunhofer Diffraction from a Circular Aperture. Verify that the Fraunhofer diffraction pattern from a circular aperture of diameter D (Fig. 4.3-4) is

$$I(x, y) = I_o\left[\frac{2J_1(\pi D\rho/\lambda d)}{\pi D\rho/\lambda d}\right]^2, \qquad \rho = \left(x^2 + y^2\right)^{1/2}, \tag{4.3-7}$$

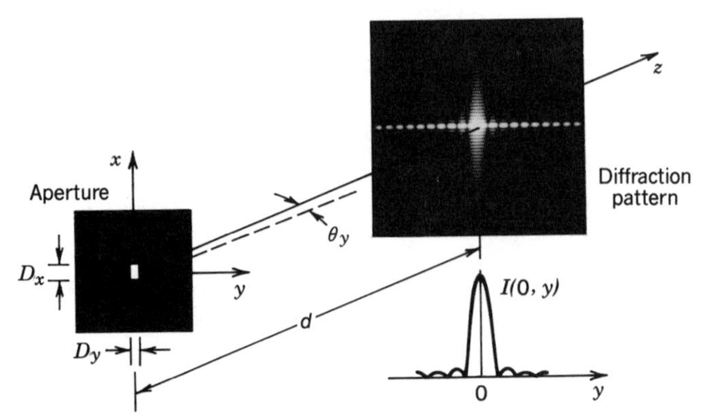

Figure 4.3-3 Fraunhofer diffraction from a rectangular aperture. The central lobe of the pattern has half-angular widths $\theta_x = \lambda/D_x$ and $\theta_y = \lambda/D_y$.

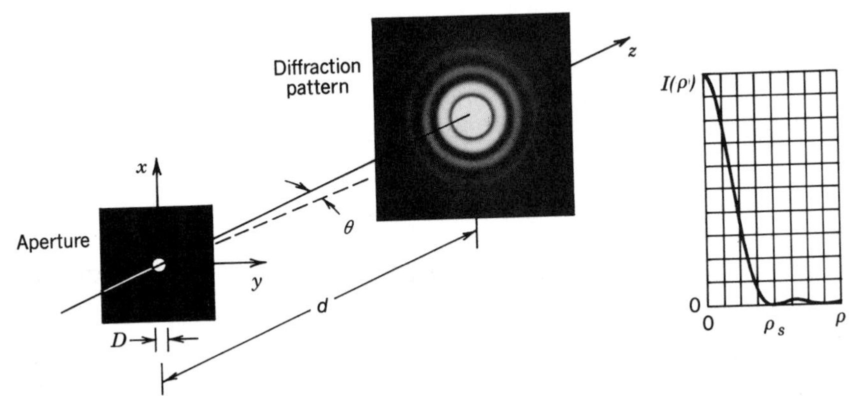

Figure 4.3-4 The Fraunhofer diffraction pattern from a circular aperture produces the Airy pattern with the radius of the central disk subtending an angle $\theta = 1.22\lambda/D$.

where $I_o = (\pi D^2/4\lambda d)^2 I_i$ is the peak intensity and $J_1(\cdot)$ is the Bessel function of order 1. The Fourier transform of circularly symmetric functions is discussed in Appendix A, Sec. A.3. The circularly symmetric pattern (4.3-7), known as the **Airy pattern**, consists of a central disc surrounded by rings. Verify that the radius of the central disk, known as the **Airy disk**, is $\rho_s = 1.22\lambda d/D$ and subtends an angle

$$\theta = 1.22\frac{\lambda}{D}. \tag{4.3-8}$$

Half-Angle Subtended by the Airy Disk

The Fraunhofer approximation is valid for distances d that are usually extremely large. They are satisfied in applications of long-distance free-space optical communication such as laser radar (lidar) and satellite communication. However, as shown in Sec. 4.2B, if a lens of focal length f is used to focus the diffracted light, the intensity pattern in the focal plane is proportional to the squared magnitude of the Fourier transform of

$p(x, y)$ evaluated at $\nu_x = x/\lambda f$ and $\nu_y = y/\lambda f$. The observed pattern is therefore identical to that obtained from (4.3-4), with the distance d replaced by the focal length f.

EXERCISE 4.3-3

Spot Size of a Focused Optical Beam. A beam of light is focused using a lens of focal length f with a circular aperture of diameter D (Fig. 4.3-5). If the beam is approximated by a plane wave at points within the aperture, verify that the pattern of the focused spot is

$$I(x, y) = I_o \left[\frac{2J_1(\pi D\rho/\lambda f)}{\pi D\rho/\lambda f} \right]^2, \qquad \rho = \left(x^2 + y^2\right)^{1/2}, \tag{4.3-9}$$

where I_o is the peak intensity. Compare the radius of the focused spot,

$$\rho_s = 1.22\lambda \frac{f}{D}, \tag{4.3-10}$$

to the spot size obtained when a Gaussian beam of waist radius W_0 is focused by an ideal lens of infinite aperture [see (3.2-15)].

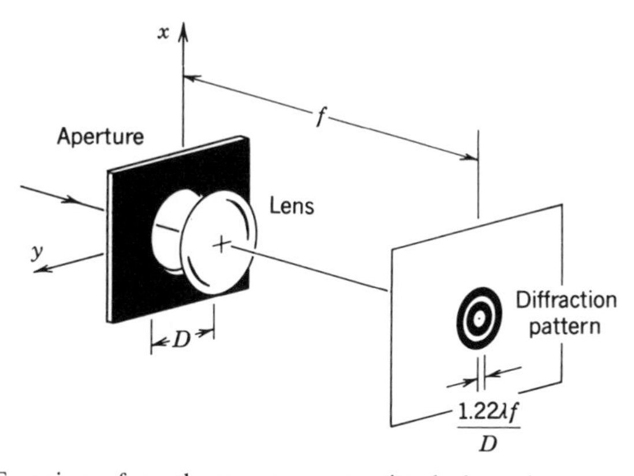

Figure 4.3-5 Focusing of a plane wave transmitted through a circular aperture of diameter D.

*B. Fresnel Diffraction

The theory of Fresnel diffraction is based on the assumption that the incident wave is multiplied by the aperture function $p(x, y)$ and propagates in free space in accordance with the Fresnel approximation. If the incident wave is a plane wave traveling in the z-direction with intensity I_i, the complex amplitude immediately after the aperture is $f(x, y) = I_i^{1/2} p(x, y)$. Using (4.1-14), the diffraction pattern $I(x, y) = |g(x, y)|^2$ at a

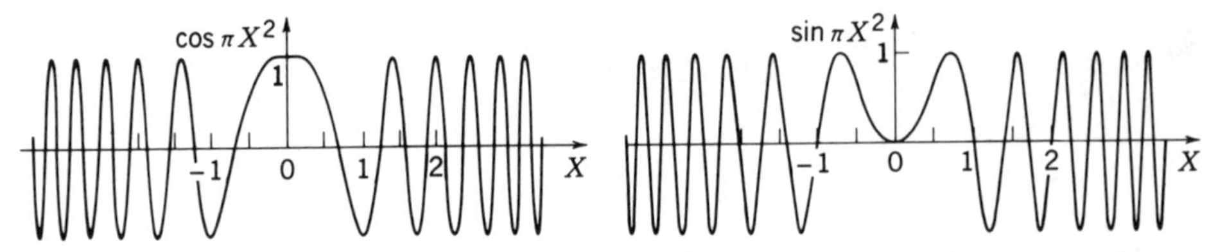

Figure 4.3-6 The real and imaginary parts of $\exp(-j\pi X^2)$.

distance d is

$$I(x, y) = \frac{I_i}{(\lambda d)^2} \left| \iint_{-\infty}^{\infty} p(x', y') \exp\left[-j\pi \frac{(x - x')^2 + (y - y')^2}{\lambda d} \right] dx'\, dy' \right|^2. \quad (4.3\text{-}11)$$

It is convenient to normalize all distances using $(\lambda d)^{1/2}$ as a unit of distance, so that $X = x/(\lambda d)^{1/2}$ and $X' = x'/(\lambda d)^{1/2}$ are the normalized distances (and similarly for y and y'). Equation (4.3-11) then gives

$$I(X, Y) = I_i \left| \iint_{-\infty}^{\infty} p(X', Y') \exp\left\{ -j\pi \left[(X - X')^2 + (Y - Y')^2 \right] \right\} dX'\, dY' \right|^2. \quad (4.3\text{-}12)$$

The integral in (4.3-12) is the convolution of $p(X, Y)$ and $\exp[-j\pi(X^2 + Y^2)]$. The real and imaginary parts of $\exp(-j\pi X^2)$, $\cos \pi X^2$ and $\sin \pi X^2$, are plotted in Fig. 4.3-6. They oscillate at an increasing frequency and their first lobes lie in the intervals $|X| < 1/\sqrt{2}$ and $|X| < 1$, respectively. The total area under the function $\exp(-j\pi X^2)$ is 1, with the main contribution to the area coming from the first few lobes, since subsequent lobes cancel out. If a is the radius of the aperture, the radius of the normalized function $p(X, Y)$ is $a/(\lambda d)^{1/2}$. The result of the convolution, which depends on the relative size of the two functions, is therefore governed by the Fresnel number $N_F = a^2/\lambda d$.

If the Fresnel number is large, the normalized width of the aperture $a/(\lambda d)^{1/2}$ is much greater than the width of the main lobe, and the convolution yields approximately the wider function $p(X, Y)$. Under this condition the Fresnel diffraction pattern is a shadow of the aperture, as would be expected from ray optics. Note that ray optics is applicable in the limit $\lambda \to 0$, which corresponds to the limit $N_F \to \infty$. In the opposite limit, when N_F is small, the Fraunhofer approximation becomes applicable and the Fraunhofer diffraction pattern is obtained.

EXAMPLE 4.3-1. ***Fresnel Diffraction from a Slit.*** Assume that the aperture is a slit of width $D = 2a$, so that $p(x, y) = 1$ when $|x| \le a$, and 0 elsewhere. The normalized coordinate is $X = x/(\lambda d)^{1/2}$ and

$$p(X, Y) = \begin{cases} 1, & |X| \le \dfrac{a}{(\lambda d)^{1/2}} = N_F^{1/2} \\ 0, & \text{elsewhere}, \end{cases} \quad (4.3\text{-}13)$$

where $N_F = a^2/\lambda d$ is the Fresnel number. Substituting into (4.3-12), we obtain $I(X, Y) = I_i |g(X)|^2$, where

$$g(X) = \int_{-\sqrt{N_F}}^{\sqrt{N_F}} \exp\left[-j\pi (X - X')^2 \right] dX' = \int_{X - \sqrt{N_F}}^{X + \sqrt{N_F}} \exp(-j\pi X'^2)\, dX'. \quad (4.3\text{-}14)$$

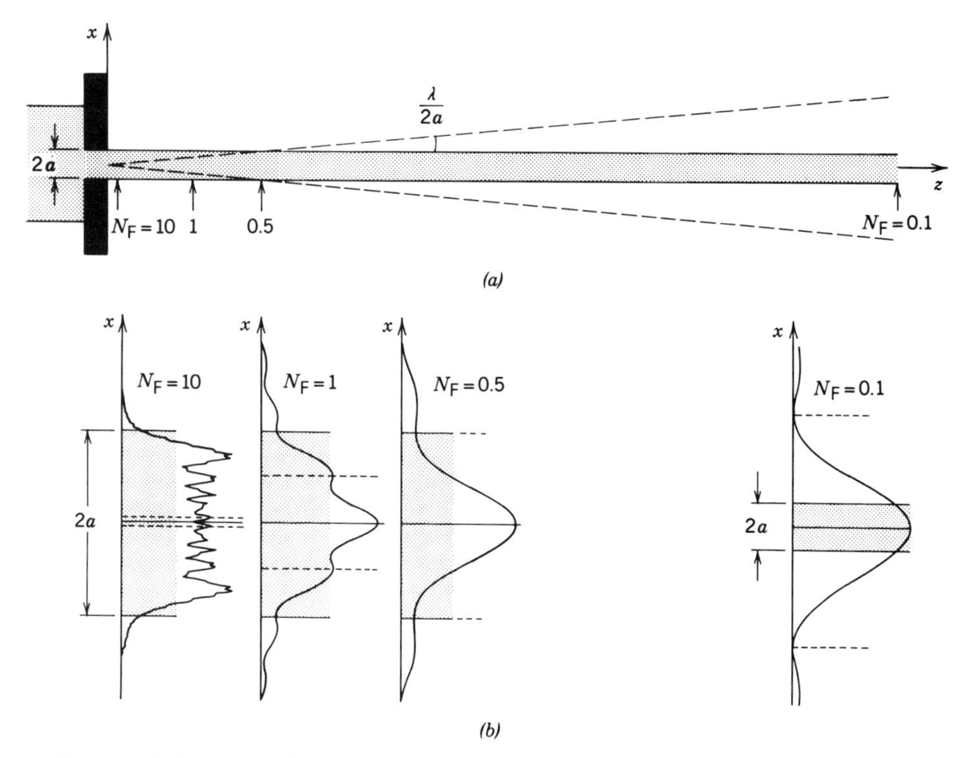

Figure 4.3-7 Fresnel diffraction from a slit of width $D = 2a$. (a) Shaded area is the geometrical shadow of the aperture. The dashed line is the width of the Fraunhofer diffracted beam. (b) Diffraction pattern at four axial positions marked by the arrows in (a) and corresponding to the Fresnel numbers $N_F = 10, 1, 0.5$, and 0.1. The shaded area represents the geometrical shadow of the slit. The dashed lines at $|x| = (\lambda/D)d$ represent the width of the Fraunhofer pattern in the far field. Where the dashed lines coincide with the edges of the geometrical shadow, the Fresnel number $N_F = a^2/\lambda d = 0.5$.

This integral is usually written in terms of the Fresnel integrals

$$C(x) = \int_0^x \cos\frac{\pi\alpha^2}{2}\,d\alpha, \qquad S(x) = \int_0^x \sin\frac{\pi\alpha^2}{2}\,d\alpha,$$

which are available in the standard computer mathematical libraries.

The complex function $g(X)$ may also be evaluated using Fourier-transform techniques. Since $g(x)$ is the convolution of a rectangular function of width $N_F^{1/2}$ and $\exp(-j\pi X^2)$, its Fourier transform $G(\nu_x) \propto \operatorname{sinc}(N_F^{1/2}\nu_x)\exp(j\pi\nu_x^2)$ (see Table A.1-1 in Appendix A). Thus $g(X)$ may be computed by determining the inverse Fourier transform of $G(\nu_x)$. If $N_F \gg 1$, the width of $\operatorname{sinc}(N_F^{1/2}\nu_x)$ is much narrower than the width of the first lobe of $\exp(j\pi\nu_x^2)$ (see Fig. 4.3-6) so that $G(\nu_x) \approx \operatorname{sinc}(N_F^{1/2}\nu_x)$ and $g(X)$ is the rectangular function representing the aperture shadow.

The diffraction pattern from a slit is plotted in Fig. 4.3-7 for different Fresnel numbers corresponding to different distances d from the aperture. At very small distances (very large N_F), the diffraction pattern is a perfect shadow of the slit. As the distance increases (N_F decreases), the wave nature of light is exhibited in the form of small oscillations around the edges of the aperture (see also the diffraction pattern in Fig. 4.3-1). For very small N_F, the Fraunhofer pattern described by (4.3-5) is obtained. This is a sinc function with the first zero subtending an angle $\lambda/D = \lambda/2a$.

EXAMPLE 4.3-2. ***Fresnel Diffraction from a Gaussian Aperture.*** If the aperture function $p(x, y)$ is the Gaussian function $p(x, y) = \exp[-(x^2 + y^2)/W_0^2]$, the Fresnel diffraction equation (4.3-11) may be evaluated exactly by finding the convolution of

$\exp[-(x^2 + y^2)/W_0^2]$ with $h_0 \exp[-j\pi(x^2 + y^2)/\lambda d]$ using, for example, Fourier transform techniques (see Appendix A). The resultant diffraction pattern is

$$I(x, y) = I_i \left[\frac{W_0}{W(d)} \right]^2 \exp\left[-2\frac{x^2 + y^2}{W^2(d)} \right],$$

where $W^2(d) = W_0^2 + \theta_0^2 d^2$ and $\theta_0 = \lambda/\pi W_0$.

The diffraction pattern is a Gaussian function of $1/e^2$ half-width $W(d)$. For small d, $W(d) \approx W_0$; but as d increases, $W(d)$ increases and approaches $W(d) \approx \theta_0 d$ when d is sufficiently large for the Fraunhofer approximation to be applicable, so that the angle subtended by the Fraunhofer diffraction pattern is θ_0. These results are illustrated in Fig. 4.3-8, which is analogous to the illustration in Fig. 4.3-7 for diffraction from a slit. The wave diffracted from a Gaussian aperture is the Gaussian beam described in detail in Chap. 3.

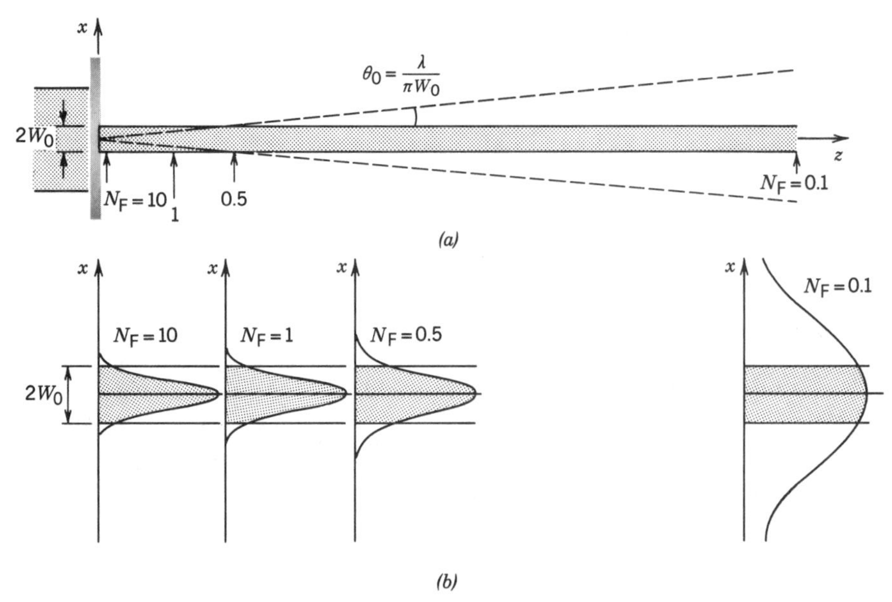

Figure 4.3-8 Fresnel diffraction pattern for a Gaussian aperture of radius W_0 at distances d such that the parameter $(\pi/2)W_0^2/\lambda d$, which is analogous to the Fresnel number N_F in Fig. 4.3-7, is 10, 1, 0.5, and 0.1. These values correspond to $W(d)/W_0 =$ 1.001, 1.118, 1.414, and 5.099, respectively. The diffraction pattern is Gaussian at all distances.

Summary

In the order of increasing distance from the aperture, the diffraction pattern is:

1. A shadow of the aperture.
2. A Fresnel diffraction pattern, which is the convolution of the normalized aperture function with $\exp[-j\pi(X^2 + Y^2)]$.
3. A Fraunhofer diffraction pattern, which is the squared-absolute value of the Fourier transform of the aperture function. The far field has an angular divergence proportional to λ/D, where D is the diameter of the aperture.

4.4 IMAGE FORMATION

An ideal image formation system is an optical system that replicates the distribution of light in one plane, the object plane, into another, the image plane. Since the optical transmission process is never perfect, the image is never an exact replica of the object. Aside from image magnification, there is also blur resulting from imperfect focusing and from the diffraction of optical waves. This section is devoted to the description of image formation systems and their fidelity. Methods of linear systems, such as the impulse-response function and the transfer function (Appendix B), are used to characterize image formation. A simple ray-optics approach is presented first, then a treatment based on wave optics is subsequently developed.

A. Ray-Optics Description of Image Formation

Consider an imaging system using a lens of focal length f at distances d_1 and d_2 from the object and image planes, respectively, as shown in Fig. 4.4-1. When $1/d_1 + 1/d_2 = 1/f$, the system is focused so that paraxial rays emitted from each point in the object plane reach a single corresponding point in the image plane. Within the ray theory of light, the imaging is "ideal," with each point of the object producing a single point of the image. The impulse-response function of the system is an impulse function.

Suppose now that the system is not in focus, as illustrated in Fig. 4.4-2, and assume that the focusing error is

$$\epsilon = \frac{1}{d_2} + \frac{1}{d_1} - \frac{1}{f}. \tag{4.4-1}$$

A point in the object plane generates a patch of light in the image plane that is a shadow of the lens aperture. The distribution of this patch is the system's impulse-response function. For simplicity, we shall consider an object point lying on the optical axis and determine the distribution of light $h(x, y)$ it generates in the image plane.

Assume that the plane of the focused image lies at a distance d_{2o} satisfying the imaging equation $1/d_{2o} + 1/d_1 = 1/f$. The shadow of a point on the edge of the aperture at a radial distance ρ is a point in the image plane with radial distance ρ_s where the ratio $\rho_s/\rho = (d_{2o} - d_2)/d_{2o} = 1 - d_2/d_{2o} = 1 - d_2(1/f - 1/d_1) = 1 - d_2(1/d_2 - \epsilon) = \epsilon d_2$. If $p(x, y)$ is the aperture function, also called the **pupil**

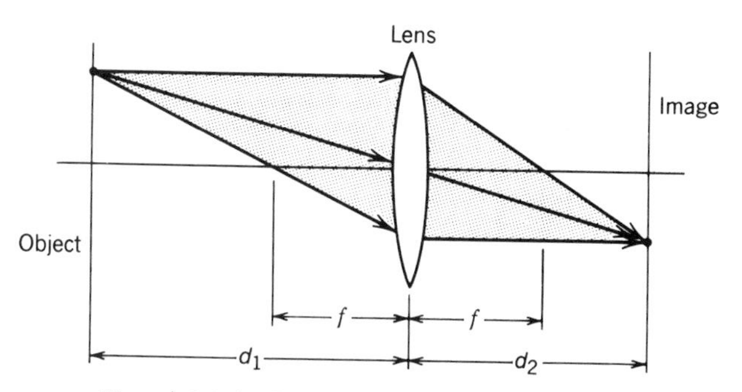

Figure 4.4-1 Rays in a focused imaging system.

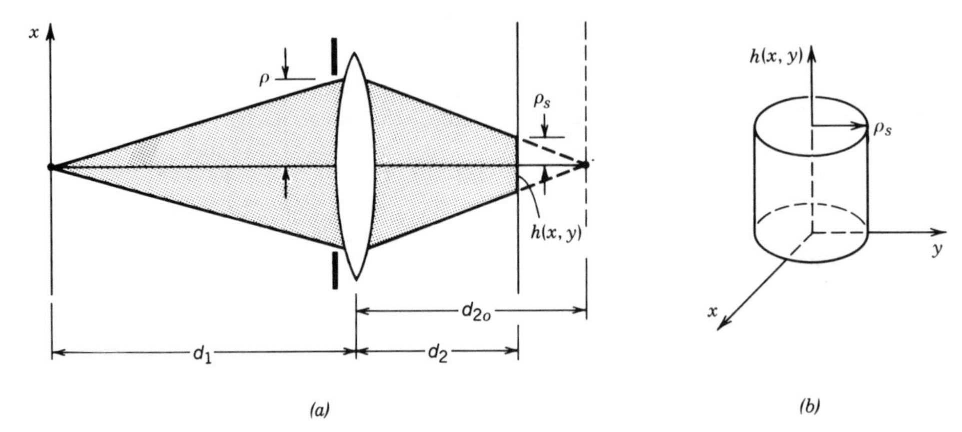

(a) (b)

Figure 4.4-2 (*a*) Rays in a defocused imaging system. (*b*) The impulse-response function of an imaging system with a circular aperture of diameter D is a circle of radius $\rho_s = \epsilon d_2 D/2$, where ϵ is the focusing error.

function [$p(x, y) = 1$ for points inside the aperture, and 0 elsewhere], then $h(x, y)$ is a scaled version of $p(x, y)$ magnified by a factor $\rho_s/\rho = \epsilon d_2$, so that

$$h(x, y) \propto p\left(\frac{x}{\epsilon d_2}, \frac{y}{\epsilon d_2}\right). \tag{4.4-2}$$

Impulse-Response Function of a Defocused System (Ray-Optics Theory)

As an example, a circular aperture of diameter D corresponds to an impulse-response function confined to a circle of radius

$$\rho_s = \frac{\epsilon d_2 D}{2}, \tag{4.4-3}$$

Radius of Blur Spot

as illustrated in Fig. 4.4-2. The radius ρ_s of this "blur spot" is an inverse measure of resolving power and image quality. A small value of ρ_s means that the system is capable of resolving fine details. Since ρ_s is proportional to the aperture diameter D, the image quality may be improved by use of a small aperture. A small aperture corresponds to a reduced sensitivity of the system to focusing errors, so that it corresponds to an increased "depth of focus."

B. Spatial Filtering

Consider now the two-lens imaging system illustrated in Fig. 4.4-3. This system, called the **4-*f* system**, serves as a focused imaging system with unity magnification, as can be easily verified by ray tracing.

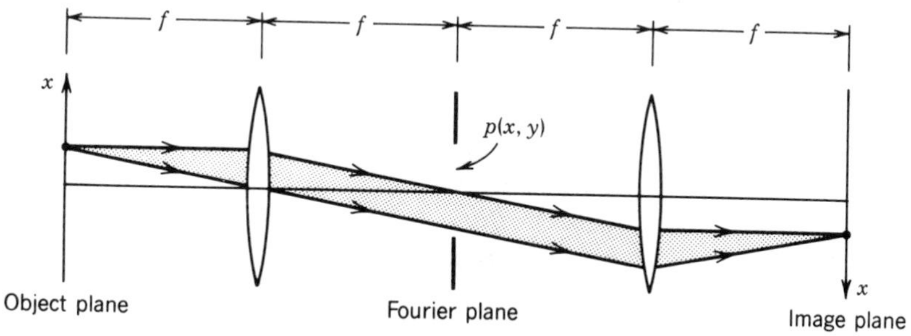

Figure 4.4-3 The 4–f imaging system. If an inverted coordinate system is used in the image plane, the magnification is unity.

The analysis of wave propagation through this system becomes simple if we recognize it as a cascade of two Fourier-transforming subsystems. The first subsystem (between the object plane and the Fourier plane) performs a Fourier transform, and the second (between the Fourier plane and the image plane) performs an inverse Fourier transform since the coordinate system in the image plane is inverted (see Exercise 4.2-2). As a result, in the absence of an aperture the image is a perfect replica of the object.

Let $f(x, y)$ be the complex amplitude transmittance of a transparency placed in the object plane and illuminated by a plane wave $\exp(-jkz)$ traveling in the z direction, as illustrated in Fig. 4.4-4, and let $g(x, y)$ be the complex amplitude in the image plane. The first lens system analyzes $f(x, y)$ into its spatial Fourier transform and separates its Fourier components so that each point in the Fourier plane corresponds to a single spatial frequency. These components are then recombined by the second lens system and the object distribution is perfectly reconstructed.

The 4–f imaging system can be used as a spatial filter in which the image $g(x, y)$ is a filtered version of the object $f(x, y)$. Since the Fourier components of $f(x, y)$ are available in the Fourier plane, a mask may be used to adjust them selectively, blocking some components and transmitting others, as illustrated in Fig. 4.4-5. The Fourier component of $f(x, y)$ at the spatial frequency (ν_x, ν_y) is located in the Fourier plane at the point $x = \lambda f \nu_x$, $y = \lambda f \nu_y$. To implement a filter of transfer function $\mathcal{H}(\nu_x, \nu_y)$, the

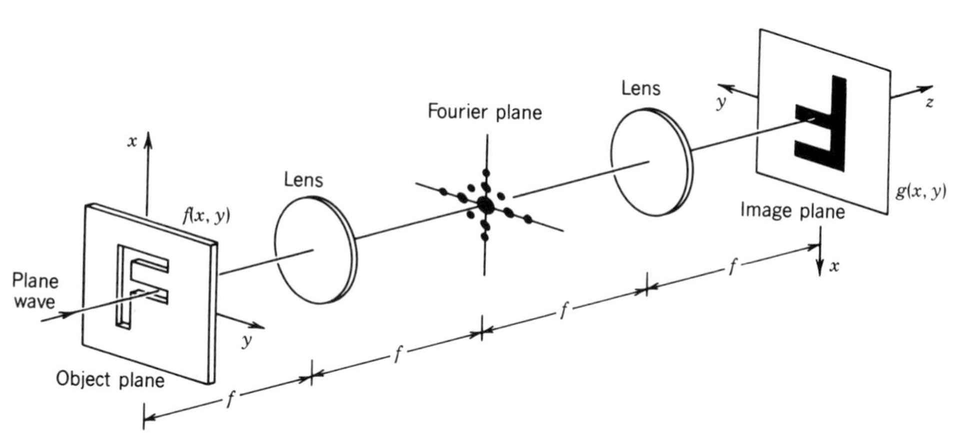

Figure 4.4-4 The 4–f system performs a Fourier transform followed by an inverse Fourier transform, so that the image is a perfect replica of the object.

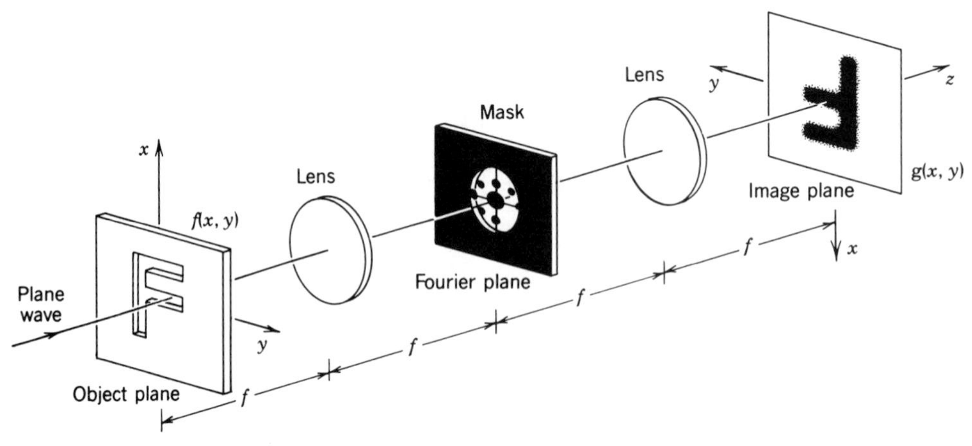

Figure 4.4-5 Spatial filtering. The transparencies in the object and Fourier planes have complex amplitude transmittances $f(x, y)$ and $p(x, y)$. A plane wave traveling in the z direction is modulated by the object transparency, Fourier transformed by the first lens, multiplied by the transmittance of the mask in the Fourier plane and inverse Fourier transformed by the second lens. As a result, the complex amplitude in the image plane $g(x, y)$ is a filtered version of $f(x, y)$. The system has a transfer function $\mathcal{H}(\nu_x, \nu_y) = p(\lambda f \nu_x, \lambda f \nu_y)$.

complex amplitude transmittance $p(x, y)$ of the mask must be proportional to $\mathcal{H}(x/\lambda f, y/\lambda f)$. Thus the transfer function of the filter realized by a mask of transmittance $p(x, y)$ is

$$\mathcal{H}(\nu_x, \nu_y) = p(\lambda f \nu_x, \lambda f \nu_y), \tag{4.4-4}$$

Transfer Function of the 4-f Spatial Filter With Mask Transmittance $p(x, y)$

where we have ignored the phase factor $j \exp(-j2kf)$ associated with each Fourier transform operation [the argument of h_l in (4.2-10)]. The Fourier transforms $G(\nu_x, \nu_y)$ and $F(\nu_x, \nu_y)$ of $g(x, y)$ and $f(x, y)$ are related by $G(\nu_x, \nu_y) = \mathcal{H}(\nu_x, \nu_y) F(\nu_x, \nu_y)$.

This is a rather simple result. *The transfer function has the same shape as the pupil function*. The corresponding impulse-response function $h(x, y)$ is the inverse Fourier transform of $\mathcal{H}(\nu_x, \nu_y)$,

$$h(x, y) = \frac{1}{(\lambda f)^2} P\left(\frac{x}{\lambda f}, \frac{y}{\lambda f}\right), \tag{4.4-5}$$

where $P(\nu_x, \nu_y)$ is the Fourier transform of $p(x, y)$.

Examples of Spatial Filters

- The ideal circularly symmetric *low-pass filter* has a transfer function $\mathcal{H}(\nu_x, \nu_y) = 1$, $\nu_x^2 + \nu_y^2 < \nu_s^2$ and $\mathcal{H}(\nu_x, \nu_y) = 0$, otherwise. It passes spatial frequencies that are smaller than the cutoff frequency ν_s and blocks higher frequencies. This filter is implemented by a mask in the form of a circular aperture of diameter D, with $D/2 = \nu_s \lambda f$. For example, if $D = 2$ cm, $\lambda = 1$ μm, and $f = 100$ cm, the cutoff

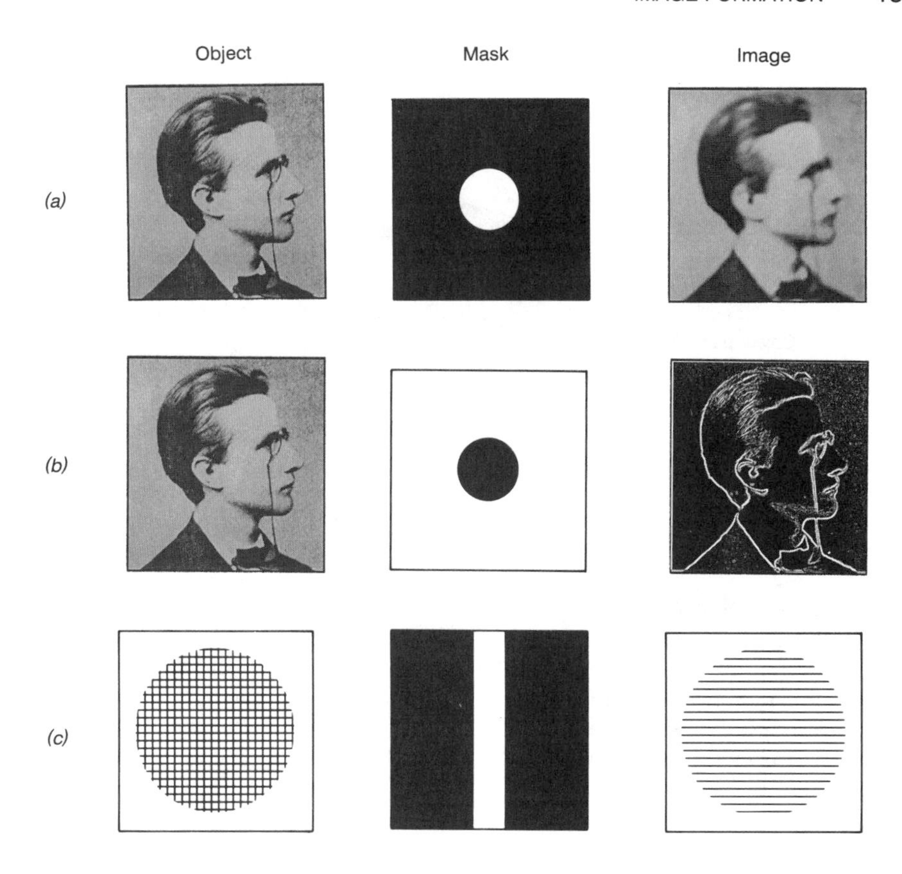

Figure 4.4-6 Examples of object, mask, and filtered image for three spatial filters: (*a*) low-pass filter; (*b*) high-pass filter; (*c*) vertical-pass filter. Black means the transmittance is zero and white means the transmittance is unity.

frequency (spatial bandwidth) $\nu_s = D/2\lambda f = 10$ lines/mm. This filter eliminates spatial frequencies that are greater than 10 lines/mm, so that the smallest size of discernible detail in the filtered image is approximately 0.1 mm.

- The *high-pass filter* is the complement of the low-pass filter. It blocks low frequencies and transmits high frequencies. The mask is a clear transparency with an opaque central circle. The filter output is high at regions of large rate of change and small at regions of smooth or slow variation of the object. The filter is therefore useful for edge enhancement in image-processing applications.
- The *vertical-pass filter* blocks horizontal frequencies and transmits vertical frequencies. Only variations in the x direction are transmitted. If the mask is a vertical slit of width D, the highest transmitted frequency is $\nu_y = (D/2)/\lambda f$.

Examples of these filters and their effects on images are illustrated in Fig. 4.4-6.

C. Single-Lens Imaging System

We now consider image formation in the single-lens imaging system shown in Fig. 4.4-7 using a wave-optics approach. We first determine the impulse-response function, and then derive the transfer function.

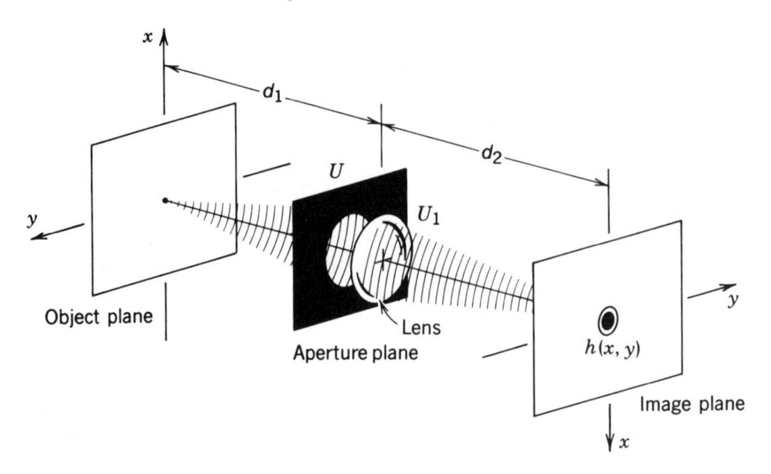

Figure 4.4-7 Single-lens imaging system.

Impulse-Response Function

To determine the impulse-response function we consider an object composed of a single point (an impulse) on the optical axis at the point $(0, 0)$, and follow the emitted optical wave as it travels to the image plane. The resultant complex amplitude is the impulse-response function $h(x, y)$.

An impulse in the object plane produces in the aperture plane a spherical wave approximated by [see (4.1-13)]

$$U(x, y) \approx h_1 \exp\left[-jk\frac{x^2 + y^2}{2d_1}\right], \tag{4.4-6}$$

where $h_1 = (j/\lambda d_1)\exp(-jkd_1)$. Upon crossing the aperture and the lens, $U(x, y)$ is multiplied by the pupil function $p(x, y)$ and the lens quadratic phase factor $\exp[jk(x^2 + y^2)/2f]$, becoming

$$U_1(x, y) = U(x, y)\exp\left(jk\frac{x^2 + y^2}{2f}\right)p(x, y). \tag{4.4-7}$$

The resultant field $U_1(x, y)$ then propagates in free space a distance d_2. In accordance with (4.1-14) it produces the amplitude

$$h(x, y) = h_2 \iint_{-\infty}^{\infty} U_1(x', y')\exp\left[-j\pi\frac{(x - x')^2 + (y - y')^2}{\lambda d_2}\right]dx'\,dy', \tag{4.4-8}$$

where $h_2 = (j/\lambda d_2)\exp(-jkd_2)$. Substituting from (4.4-6) and (4.4-7) into (4.4-8) and casting the integrals as a Fourier transform, we obtain

$$h(x, y) = h_1 h_2 \exp\left(-j\pi\frac{x^2 + y^2}{\lambda d_2}\right)P_1\left(\frac{x}{\lambda d_2}, \frac{y}{\lambda d_2}\right), \tag{4.4-9}$$

where $P_1(\nu_x, \nu_y)$ is the Fourier transform of the function

$$p_1(x, y) = p(x, y)\exp\left(-j\pi\epsilon\frac{x^2 + y^2}{\lambda}\right), \tag{4.4-10}$$

known as the **generalized pupil function**. The factor ϵ is the focusing error given by (4.4-1).

For a high-quality imaging system, the impulse-response function is a narrow function, extending only over a small range of values of x and y. If the phase factor $\pi(x^2 + y^2)/\lambda d_2$ in (4.4-9) is much smaller than 1 for all x and y within this range, it can be neglected, so that

$$h(x, y) = h_0 P_1\left(\frac{x}{\lambda d_2}, \frac{y}{\lambda d_2}\right), \tag{4.4-11}$$

Impulse-Response Function

where $h_0 = h_1 h_2$ is a constant of magnitude $(1/\lambda d_1)(1/\lambda d_2)$. It follows that the system's impulse-response function is proportional to the Fourier transform of the generalized pupil function $p_1(x, y)$ evaluated at $\nu_x = x/\lambda d_2$ and $\nu_y = y/\lambda d_2$.

If the system is focused ($\epsilon = 0$), then $p_1(x, y) = p(x, y)$, and

$$h(x, y) \approx h_0 P\left(\frac{x}{\lambda d_2}, \frac{y}{\lambda d_2}\right), \tag{4.4-12}$$

where $P(\nu_x, \nu_y)$ is the Fourier transform of $p(x, y)$. This result is similar to the corresponding result in (4.4-5) for the 4-f system.

EXAMPLE 4.4-1. ***Impulse-Response Function of a Focused Imaging System with a Circular Aperture.*** If the aperture is a circle of diameter D so that $p(x, y) = 1$ if $\rho = (x^2 + y^2)^{1/2} \le D/2$, and zero otherwise, then the impulse-response function is

$$h(x, y) = h(0,0)\frac{2J_1(\pi D\rho/\lambda d_2)}{\pi D\rho/\lambda d_2}, \qquad \rho = \left(x^2 + y^2\right)^{1/2}, \tag{4.4-13}$$

and $|h(0,0)| = (\pi D^2/4\lambda^2 d_1 d_2)$. This is a circularly symmetric function whose cross section is shown in Fig. 4.4-8. It drops to zero at a radius $\rho_s = 1.22\lambda d_2/D$ and oscillates slightly before it vanishes. The radius ρ_s is therefore a measure of the size of the blur circle. If the system is focused at ∞, $d_1 = \infty$, $d_2 = f$, and $\rho_s = 1.22\lambda F_{\#}$, where $F_{\#} = f/D$ is the lens

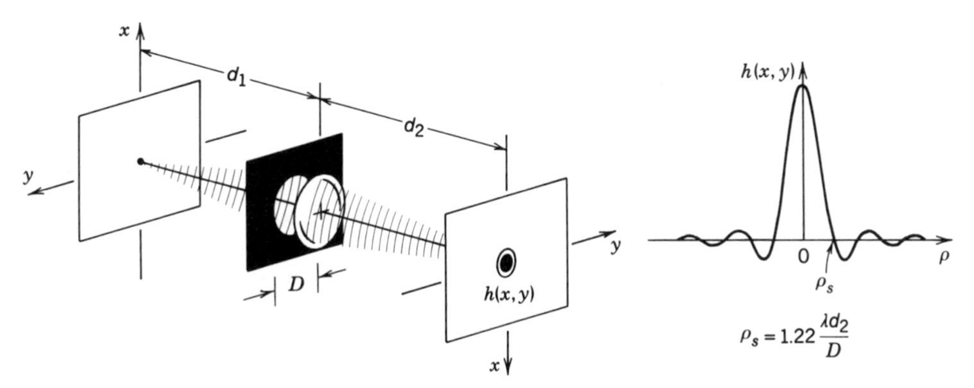

Figure 4.4-8 Impulse-response function of an imaging system with a circular aperture.

F-number. Thus systems of smaller $F_{\#}$ (larger apertures) have better image quality. This assumes, of course, that the larger lens does not introduce geometrical aberrations.

Transfer Function

The transfer function of a linear system can only be defined when the system is shift invariant (see Appendix B). Evidently, the single-lens imaging system is not shift invariant since a shift Δ of a point in the object plane is accompanied by a *different* shift $M\Delta$ in the image plane, where $M = -d_2/d_1$ is the magnification.

The image is different from the object in two ways. First, the image is a magnified replica of the object, i.e., the point (x, y) of the object is located at a new point (Mx, My) in the image. Second, every point is smeared into a patch as a result of defocusing or diffraction. We can therefore think of image formation as a cascade of two systems—a system of ideal magnification followed by a system of blur, as depicted in Fig. 4.4-9. By its nature, the magnification system is shift variant. For points near the optical axis, the blur system is approximately shift invariant and therefore can be described by a transfer function.

The transfer function $\mathcal{H}(\nu_x, \nu_y)$ of the blur system is determined by obtaining the Fourier transform of the impulse-response function $h(x, y)$ in (4.4-11). The result is

$$\mathcal{H}(\nu_x, \nu_y) \approx p_1(\lambda d_2 \nu_x, \lambda d_2 \nu_y), \tag{4.4-14}$$

Transfer Function

where $p_1(x, y)$ is the generalized pupil function and we have ignored a constant phase factor $\exp(-jkd_1)\exp(-jkd_2)$. If the system is focused, then

$$\mathcal{H}(\nu_x, \nu_y) \approx p(\lambda d_2 \nu_x, \lambda d_2 \nu_y), \tag{4.4-15}$$

where $p(x, y)$ is the pupil function. This result is identical to that obtained for the 4-f imaging system [see (4.4-4)]. If the aperture is a circle of diameter D, for example, then

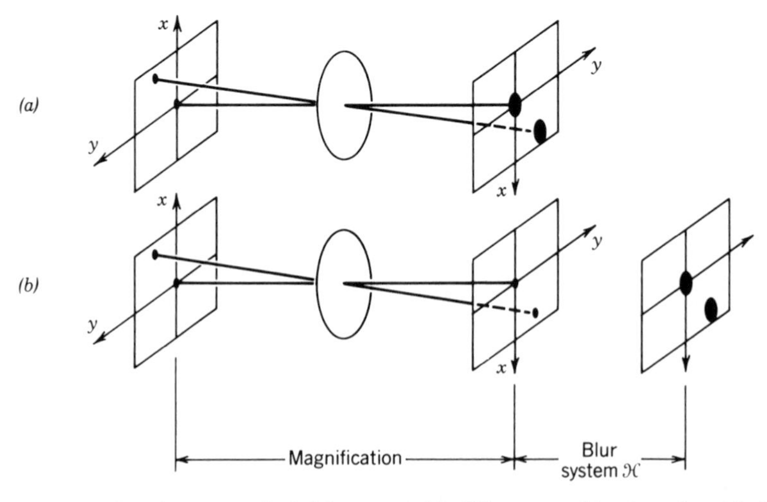

Figure 4.4-9 The imaging system in (a) is regarded in (b) as a combination of an ideal imaging system with only magnification, followed by shift-invariant blur in which each point is blurred into a patch with a distribution equal to the impulse-response function.

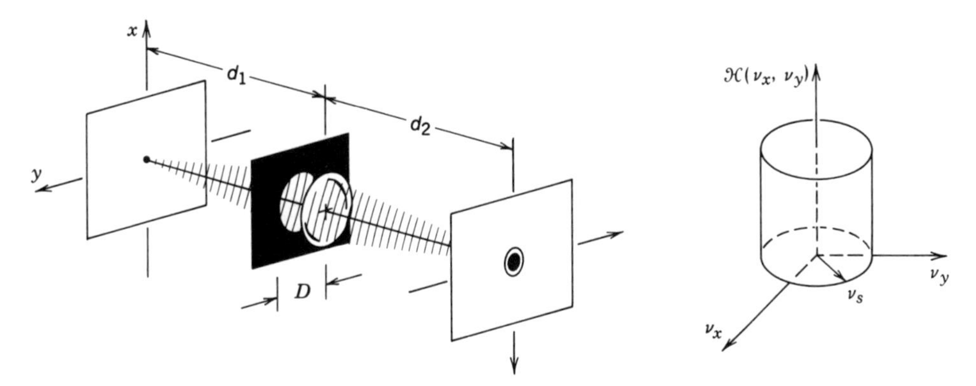

Figure 4.4-10 Transfer function of a focused imaging system with a circular aperture of diameter D. The system has a spatial bandwidth $\nu_s = D/2\lambda d_2$.

the transfer function is constant within a circle of radius ν_s, where

$$\nu_s = \frac{D}{2\lambda d_2}, \tag{4.4-16}$$

and vanishes elsewhere, as illustrated in Fig. 4.4-10.

If the lens is focused at infinity, i.e., $d_2 = f$,

$$\boxed{\nu_s = \frac{1}{2\lambda F_{\#}},} \tag{4.4-17}$$

Spatial Bandwidth

where $F_{\#} = f/D$ is the lens F-number. For example, for an F-2 lens ($F_{\#} = f/D = 2$) and for $\lambda = 0.5$ μm, $\nu_s = 500$ lines/mm. The frequency ν_s is the spatial bandwidth, i.e., the highest spatial frequency that the imaging system can transmit.

4.5 HOLOGRAPHY

Holography involves the recording and reconstruction of optical waves. A **hologram** is a transparency containing a coded record of the optical wave.

Consider a monochromatic optical wave whose complex amplitude in some plane, say the $z = 0$ plane, is $U_o(x, y)$. If, somehow, a thin optical element (call it a transparency) with complex amplitude transmittance $\mathscr{t}(x, y)$ equal to $U_o(x, y)$ were able to be made, it would provide a complete record of the wave. The wave could then be reconstructed simply by illuminating the transparency with a uniform plane wave of unit amplitude traveling in the z direction. The transmitted wave would have a complex amplitude in the $z = 0$ plane $U(x, y) = 1 \cdot \mathscr{t}(x, y) = U_o(x, y)$. The original wave would then be reproduced at all points in the $z = 0$ plane, and therefore reconstructed everywhere in the space $z > 0$.

As an example, we know that a uniform plane wave traveling at an angle θ with respect to the z axis in the x-z plane has a complex amplitude $U_o(x, y) = \exp[-jk \sin\theta\, x]$. A record of this wave would be a transparency with complex amplitude transmittance $\mathscr{t}(x, y) = \exp[-jk \sin\theta\, x]$. Such a transparency acts as a prism that bends an incident plane wave $\exp(-jkz)$ by an angle θ (see Sec. 2.4B), thus reproducing the original wave.

The question is how to make a transparency $t(x, y)$ from the original wave $U_o(x, y)$. One key impediment is that optical detectors, including the photographic emulsions used to make transparencies, are responsive to the optical intensity, $|U_o(x, y)|^2$, and are therefore insensitive to the phase $\arg\{U_o(x, y)\}$. Phase information is obviously important and cannot be disregarded, however. For example, if the phase of the oblique wave $U_o(x, y) = \exp[-jk \sin\theta\, x]$ were not recorded, neither would the direction of travel of the wave. To record the phase of $U_o(x, y)$, a code must be found that transforms phase into intensity. The recorded information could then be optically decoded in order to reconstruct the wave.

The Holographic Code

The holographic code is based on mixing the original wave (hereafter called the **object wave**) U_o with a known **reference wave** U_r and recording their interference pattern in the $z = 0$ plane. The intensity of the sum of the two waves is photographically recorded and a transparency of complex amplitude transmittance t, proportional to the intensity, is made [Fig. 4.5-1(a)]. The transmittance is therefore given by

$$
\begin{aligned}
t \propto |U_o + U_r|^2 &= |U_r|^2 + |U_o|^2 + U_r^* U_o + U_r U_o^*, \\
&= I_r + I_o + U_r^* U_o + U_r U_o^*, \\
&= I_r + I_o + 2(I_r I_o)^{1/2} \cos[\arg\{U_r\} - \arg\{U_o\}], \qquad (4.5\text{-}1)
\end{aligned}
$$

where I_r and I_o are, respectively, the intensities of the reference wave and the object wave in the $z = 0$ plane.

The transparency, called a **hologram**, clearly carries coded information pertinent to the magnitude and phase of the wave U_o. In fact, as an interference pattern the transmittance t is highly sensitive to the difference between the phases of the two waves, as was shown in Sec. 2.5.

To decode the information in the hologram and reconstruct the object wave, the reference wave U_r is again used to illuminate the hologram [Fig. 4.5-1(b)]. The result is

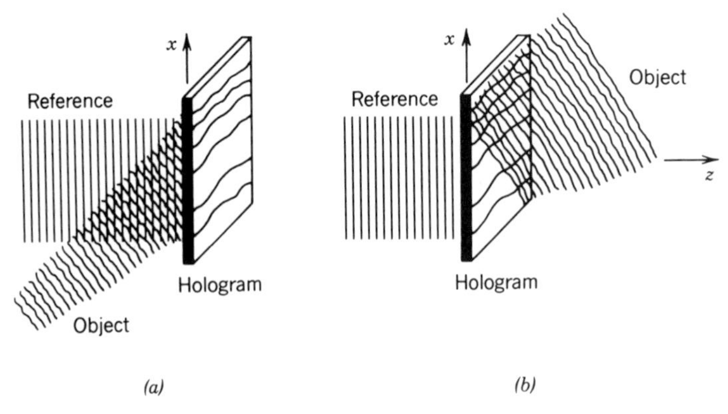

Figure 4.5-1 (a) A hologram is a transparency on which the interference pattern between the original wave (object wave) and a reference wave is recorded. (b) The original wave is reconstructed by illuminating the hologram with the reference wave.

a wave with complex amplitude

$$U = t U_r \propto U_r I_r + U_r I_o + I_r U_o + U_r^2 U_o^* \tag{4.5-2}$$

in the hologram plane $z = 0$. The third term on the right-hand side is the original wave multiplied by the intensity I_r of the reference wave. If I_r is uniform (independent of x and y), this term constitutes the desired reconstructed wave. But it must be separated from the other three terms. The fourth term is a conjugated version of the original wave modulated by U_r^2. The first two terms represent the reference wave, modulated by the sum of the intensities of the two waves.

If the reference wave is selected to be a uniform plane wave propagating along the z axis, $I_r^{1/2} \exp(-jkz)$, then in the $z = 0$ plane $U_r(x, y) = I_r^{1/2}$ is a constant independent of x and y. Dividing (4.5-2) by $U_r = I_r^{1/2}$ gives

$$U(x, y) \propto I_r + I_o(x, y) + I_r^{1/2} U_o(x, y) + I_r^{1/2} U_o^*(x, y). \tag{4.5-3}$$

Reconstructed Wave in Plane of Hologram

The significance of the various terms in (4.5-3), and the methods of extracting the original wave (the third term), are clarified by means of a number of examples.

EXAMPLE 4.5-1. ***Hologram of an Oblique Plane Wave.*** If the object wave is an oblique plane wave at angle θ [Fig. 4.5-2(a)], $U_o(x, y) = I_o^{1/2} \exp(-jk \sin\theta\, x)$, then (4.5-3) gives $U(x, y) \propto I_r + I_o + (I_r I_o)^{1/2} \exp(-jk \sin\theta\, x) + (I_r I_o)^{1/2} \exp(jk \sin\theta\, x)$. Since the first two terms are constant, they correspond to a wave propagating in the z direction (the continuance of the reference wave). The third term corresponds to the original object wave, whereas the fourth term represents the **conjugate wave**, a plane wave traveling at an angle $-\theta$. The object wave is therefore separable from the other waves. In fact, this hologram is nothing but a recording of the interference pattern formed from two oblique plane waves at an angle θ (Sec. 2.5A). It serves as a sinusoidal diffraction grating that splits an incident reference wave into three waves at angles 0, θ, and $-\theta$ [see Fig. 4.5-2(b) and Sec. 2.4B].

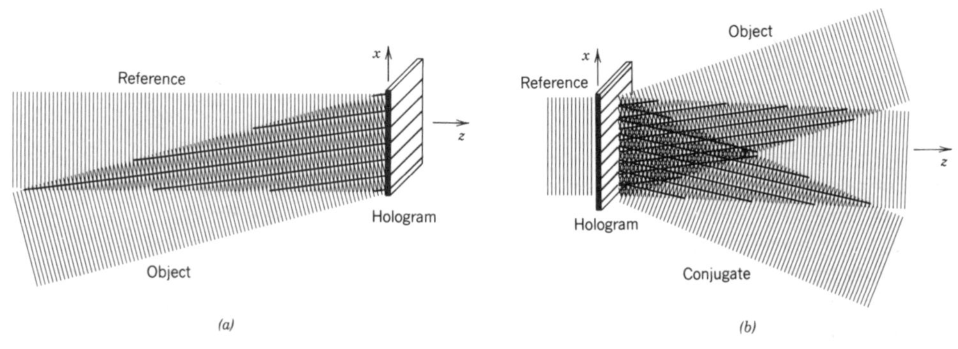

Figure 4.5-2 The hologram of an oblique plane wave is a sinusoidal diffraction grating: (a) recording; (b) reconstruction.

EXAMPLE 4.5-2. *Hologram of a Point Source.* Here the object wave is a spherical wave originating at the point $\mathbf{r}_0 = (0, 0, -d)$, as illustrated in Fig. 4.5-3, so that $U_o(x, y) \propto \exp(-jk|\mathbf{r} - \mathbf{r}_0|)/|\mathbf{r} - \mathbf{r}_0|$, where $\mathbf{r} = (x, y, 0)$. The first term of (4.5-3) corresponds to a plane wave traveling in the z direction, whereas the third is proportional to the amplitude of the original spherical wave originating at $(0, 0, -d)$. The fourth term is proportional to the amplitude of the conjugate wave $U_o^*(x, y) \propto \exp(jk|\mathbf{r} - \mathbf{r}_0|)/|\mathbf{r} - \mathbf{r}_0|$, which is a converging spherical wave centered at the point $(0, 0, d)$. The second term is proportional to $1/|\mathbf{r} - \mathbf{r}_0|^2$ and its corresponding wave therefore travels in the z direction with very small angular spread since its intensity varies slowly in the transverse plane.

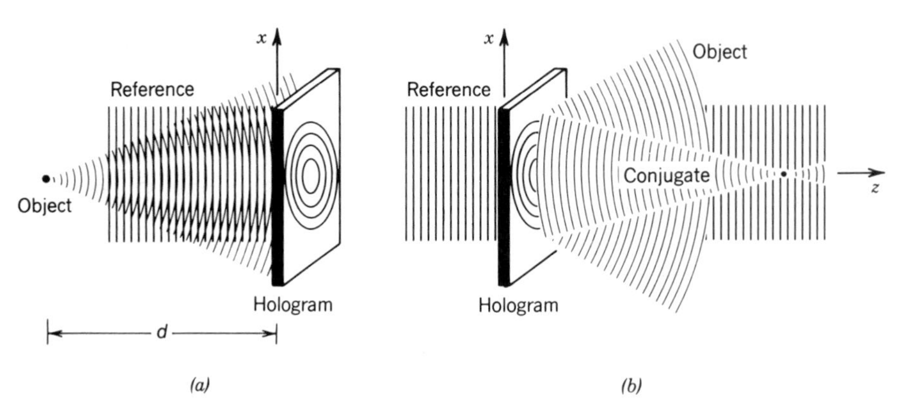

Figure 4.5-3 Hologram of a spherical wave originating from a point source: (*a*) recording; (*b*) reconstruction. The conjugate wave forms a real image of the point.

Off-Axis Holography

One means of separating the four components of the reconstructed wave is to ensure that they vary at well-separated spatial frequencies, so that they have well-separated directions. This form of spatial frequency multiplexing (see Sec. 4.1A) is assured if the object and reference waves are offset so that they arrive from well-separated directions.

Assume that the object wave has a complex amplitude $U_o(x, y) = f(x, y)\exp(-jk\sin\theta\, x)$. This is a wave of complex envelope $f(x, y)$ modulated by a phase factor equal to that introduced by a prism with deflection angle θ. It is assumed that $f(x, y)$ varies slowly so that its maximum spatial frequency ν_s corresponds to an angle $\theta_s = \sin^{-1}\lambda\nu_s$ much smaller than θ. The object wave therefore has directions centered about the angle θ, as illustrated in Fig. 4.5-4. Equation (4.5-3) gives

$$U(x, y) \propto I_r + |f(x, y)|^2 + I_r^{1/2} f(x, y)\exp(-jk\sin\theta\, x) + I_r^{1/2} f^*(x, y)\exp(+jk\sin\theta\, x).$$

The third term is evidently a replica of the object wave, which arrives from a direction at an angle θ. The presence of the phase factor $\exp(jk\sin\theta\, x)$ in the fourth term indicates that it is deflected in the $-\theta$ direction. The first term corresponds to a plane wave traveling in the z direction. The second term, usually known as the **ambiguity term**, corresponds to a nonuniform plane wave in directions within a cone of small angle $2\theta_s$ around the z direction. The offset of the directions of the object and reference waves results in a natural angular separation of the object and conjugate waves from each other and from the other two waves if $\theta > 3\theta_s$, thus allowing the original wave to be recovered unambiguously.

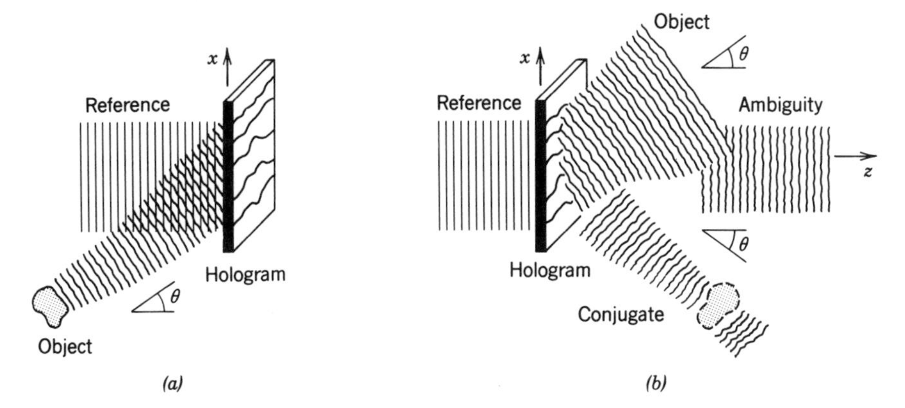

Figure 4.5-4 Hologram of an off-axis object wave: (*a*) recording; (*b*) reconstruction. The object wave is separated from both the reference and conjugate waves.

An alternative method of reducing the effect of the ambiguity wave is to make the intensity of the reference wave much greater than that of the object wave. The ambiguity wave [second term of (4.5-3)] is then much smaller than the other terms since it involves only object waves; it is therefore relatively negligible.

Fourier-Transform Holography

The Fourier transform $F(\nu_x, \nu_y)$ of a function $f(x, y)$ may be computed optically by use of a lens (see Sec. 4.2). If $f(x, y)$ is the complex amplitude in one focal plane of the lens, then $F(x/\lambda f, y/\lambda f)$ is the complex amplitude in the other focal plane, where f is the focal length of the lens and λ is the wavelength. Since the Fourier transform is usually a complex-valued function, it cannot be recorded directly.

The Fourier transform $F(x/\lambda f, y/\lambda f)$ may be recorded holographically by regarding it as an object wave, $U_o(x, y) = F(x/\lambda f, y/\lambda f)$, mixing it with a reference wave $U_r(x, y)$, and recording the superposition as a hologram [Fig. 4.5-5(*a*)]. Reconstruction is achieved by illumination of the hologram with the reference wave as usual. The reconstructed wave may be inverse Fourier transformed using a lens so that the original function $f(x, y)$ is recovered [Fig. 4.5-5(*b*)].

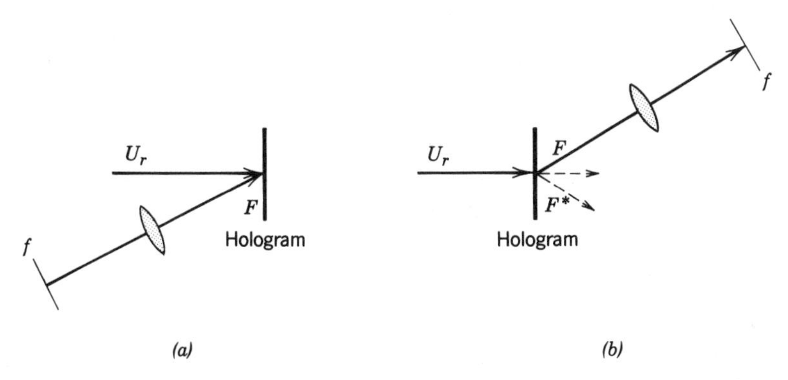

Figure 4.5-5 Hologram of a wave whose complex amplitude represents the Fourier transform of a function $f(x, y)$: (*a*) recording; (*b*) reconstruction.

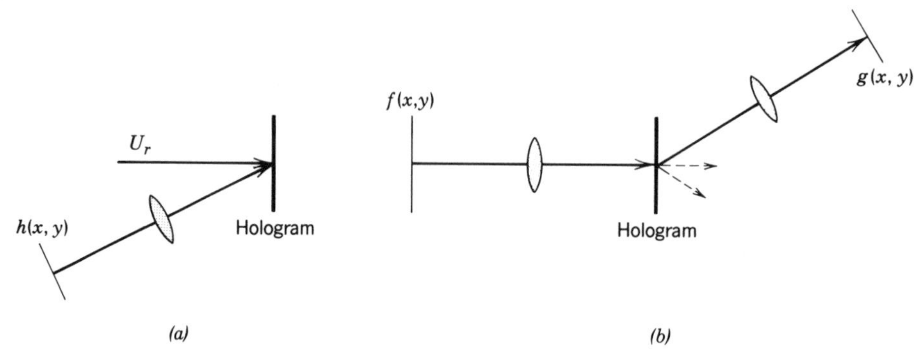

Figure 4.5-6 The Vander Lugt holographic filter. (*a*) A hologram of the Fourier transform of $h(x, y)$ is recorded. (*b*) The Fourier transform of $f(x, y)$ is transmitted through the hologram and inverse Fourier transformed by a lens. The result is a function $g(x, y)$ proportional to the convolution of $f(x, y)$ and $h(x, y)$. The overall process provides a spatial filter with impulse-response function $h(x, y)$.

Holographic Spatial Filters

A spatial filter of transfer function $\mathcal{H}(\nu_x, \nu_y)$ may be implemented by use of a 4-f optical system with a mask of complex amplitude transmittance $p(x, y) = \mathcal{H}(x/\lambda f, y/\lambda f)$ placed in the Fourier plane (see Sec. 4.4B). Since the transfer function $\mathcal{H}(\nu_x, \nu_y)$ is usually complex-valued, the mask transmittance $p(x, y)$ has a phase component and is difficult to fabricate using conventional printing techniques. If the filter impulse-response function $h(x, y)$ is real-valued, however, a Fourier-transform hologram of $h(x, y)$ may be created by holographically recording the Fourier transform $U_o(x, y) = \mathcal{H}(x/\lambda f, y/\lambda f)$.

Using the Fourier transform of the input $f(x, y)$ as a reference, $U_r(x, y) = F(x/\lambda f, y/\lambda f)$, the hologram constructs the wave

$$U_r(x, y)U_o(x, y) = F(x/\lambda f, y/\lambda f)\mathcal{H}(x/\lambda f, y/\lambda f).$$

The inverse Fourier transform of the reconstructed object wave, obtained with a lens of focal length f as illustrated in Fig. 4.5-6(*b*), therefore yields a complex amplitude $g(x, y)$ with a Fourier transform $G(\nu_x, \nu_y) = \mathcal{H}(\nu_x, \nu_y)F(\nu_x, \nu_y)$. Thus $g(x, y)$ is the convolution of $f(x, y)$ with $h(x, y)$. The overall system, known as the **Vander Lugt filter**, performs the operation of convolution, which is the basis of spatial filtering.

If the conjugate wave $U_r(x, y)U_o^*(x, y) = F(x/\lambda f, y/\lambda f)\mathcal{H}^*(x/\lambda f, y/\lambda f)$ is, instead, inverse Fourier transformed, the correlation, instead of the convolution, of the functions $f(x, y)$ and $h(x, y)$ is obtained. The operation of correlation is useful in image-processing applications, including pattern recognition.

The Holographic Apparatus

An essential condition for the successful fabrication of a hologram is the availability of a monochromatic light source with minimal phase fluctuations. The presence of phase fluctuations results in the random shifting of the interference pattern and the washing out of the hologram. For this reason, a coherent light source (usually a laser) is a necessary part of the apparatus. The coherence requirements for the interference of light waves are discussed in Chap. 10.

Figure 4.5-7 illustrates a typical experimental configuration used to record a hologram and reconstruct the optical wave scattered from the surface of a physical object. Using a beamsplitter, laser light is split into two portions, one is used as the reference wave, whereas the other is scattered from the object to form the object wave. The

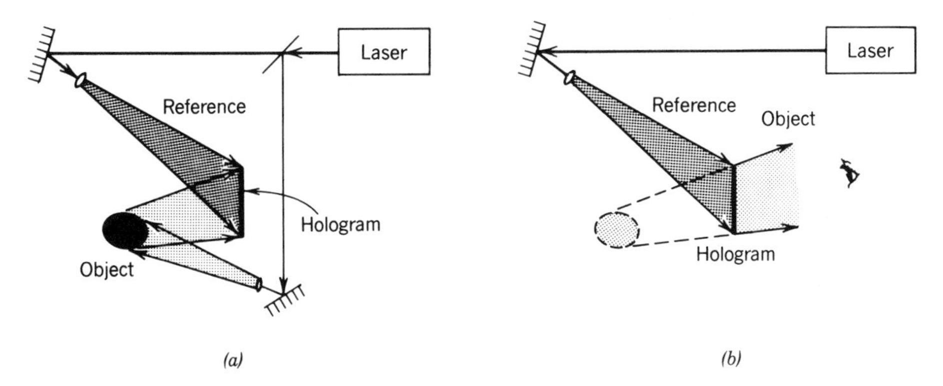

Figure 4.5-7 Holographic recording (*a*) and reconstruction (*b*).

optical path difference between the two waves should be as small as possible to ensure that the two beams maintain a nonrandom phase difference [the term $\arg\{U_r\} - \arg\{U_o\}$ in (4.5-1)].

Since the interference pattern forming the hologram is composed of fine lines separated by distances of the order of $\lambda/\sin\theta$, where θ is the angular offset between the reference and object waves, the photographic film must be of high resolution and the system must not vibrate during the exposure. The larger θ, the smaller the distances between the hologram lines, and the more stringent these requirements are. The object wave is reconstructed when the recorded hologram is illuminated with the reference wave, so that a viewer sees the object as if it were actually there, with its three-dimensional character preserved.

Volume Holography

It has been assumed so far that the hologram is a thin planar transparency on which the interference pattern of the object and reference waves is recorded. We now consider recording the hologram in a relatively thick medium and show that this offers an advantage. Consider the simple case when the object and reference waves are plane waves with wavevectors $\mathbf{k}_r$ and $\mathbf{k}_o$. The recording medium extends between the planes $z = 0$ and $z = \Delta$, as illustrated in Fig. 4.5-8. The interference pattern is now a function of x, y, and z:

$$
\begin{aligned}
I(x, y, z) &= |I_r^{1/2} \exp(-j\mathbf{k}_r \cdot \mathbf{r}) + I_o^{1/2} \exp(-j\mathbf{k}_o \cdot \mathbf{r})|^2 \\
&= I_r + I_o + 2(I_r I_o)^{1/2} \cos(\mathbf{k}_o \cdot \mathbf{r} - \mathbf{k}_r \cdot \mathbf{r}) \\
&= I_r + I_o + 2(I_r I_o)^{1/2} \cos(\mathbf{k}_g \cdot \mathbf{r}),
\end{aligned}
$$

where $\mathbf{k}_g = \mathbf{k}_o - \mathbf{k}_r$. This is a sinusoidal pattern of period $\Lambda = 2\pi/|\mathbf{k}_g|$ and with the surfaces of constant intensity normal to the vector $\mathbf{k}_g$.

For example, if the reference wave points in the z direction and the object wave makes an angle θ with the z axis, $|\mathbf{k}_g| = 2k\sin(\theta/2)$ and the period is

$$
\Lambda = \frac{\lambda}{2\sin(\theta/2)}, \tag{4.5-4}
$$

as illustrated in Fig. 4.5-8.

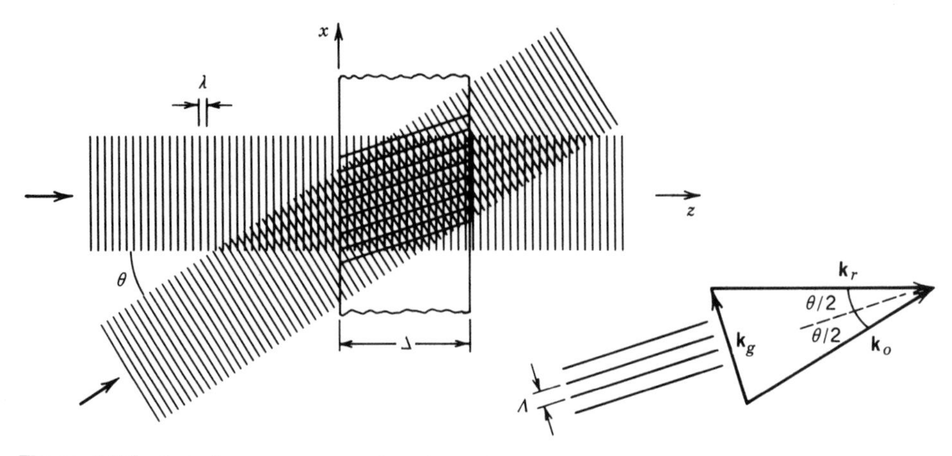

Figure 4.5-8 Interference pattern when the reference and object waves are plane waves. Since $|\mathbf{k}_r| = |\mathbf{k}_o| = 2\pi/\lambda$ and $|\mathbf{k}_g| = 2\pi/\Lambda$, from the geometry of the vector diagram $2\pi/\Lambda = 2(2\pi/\lambda)\sin(\theta/2)$, so that $\Lambda = \lambda/2\sin(\theta/2)$.

If recorded in an emulsion, this pattern serves as a thick diffraction grating, a **volume hologram**. The vector $\mathbf{k}_g$ is called the **grating vector**. When illuminated with the reference wave as illustrated in Fig. 4.5-9, the parallel planes of the grating reflect the wave only when the Bragg condition $\sin\phi = \lambda/2\Lambda$ is satisfied, where ϕ is the angle between the planes of the grating and the incident reference wave (see Exercise 2.5-3). In our case $\phi = \theta/2$, so that $\sin(\theta/2) = \lambda/2\Lambda$. In view of (4.5-4), the Bragg condition is indeed satisfied, so that the reference wave is indeed reflected. As evident from the geometry, the reflected wave is an extension of the object wave, so that the reconstruction process is successful.

Suppose now that the hologram is illuminated with a reference wave of different wavelength λ'. Evidently, the Bragg condition, $\sin(\theta/2) = \lambda'/2\Lambda$, will not be satisfied and the wave will not be reflected. It follows that the object wave is reconstructed only if the wavelength of the reconstruction source matches that of the recording source. If light with a broad spectrum (white light) is used as a reconstruction source, only the "correct" wavelength would be reflected and the reconstruction process would be successful.

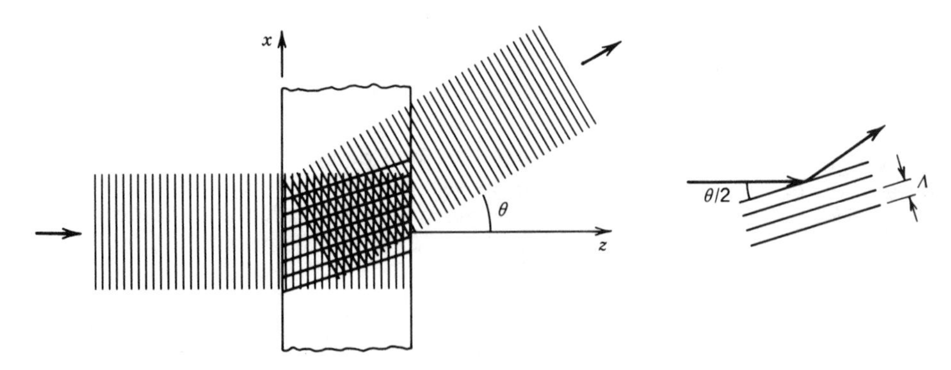

Figure 4.5-9 The reference wave is Bragg reflected from the thick hologram and the object wave is reconstructed.

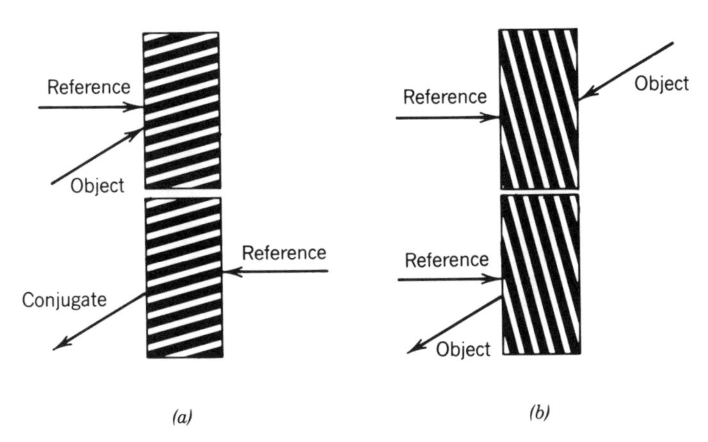

Figure 4.5-10 Two geometries for recording and reconstruction of a volume hologram. (*a*) This hologram is reconstructed by use of a reversed reference wave; the reconstructed wave is a conjugate wave traveling in a direction opposite to the original object wave. (*b*) A reflection hologram is recorded with the reference and object waves arriving from opposite sides; the object wave is reconstructed by reflection from the grating.

Although the recording process must be done with monochromatic light, the reconstruction can be achieved with white light. This provides a clear advantage in many applications of holography. Other geometries for recording and reconstruction of a volume hologram are illustrated in Fig. 4.5-10.

Another type of hologram that may be viewed with white light is the **rainbow hologram**. This hologram is recorded through a narrow slit so that the reconstructed image, of course, also appears as if seen through a slit. However, if the wavelength of reconstruction differs from the recording wavelength, the reconstructed wave will appear to be coming from a displaced slit since a magnification effect will be introduced. If white light is used for reconstruction, the reconstructed wave appears as the object seen through many displaced slits, each with a different wavelength (color). The result is a rainbow of images seen through parallel slits. Each slit displays the object with parallax effect in the direction of the slit, but not in the orthogonal direction. Rainbow holograms have many commercial uses as displays.

READING LIST

Fourier Optics and Optical Signal Processing

G. Reynolds, J. B. DeVelis, G. B. Parrent, and B. J. Thompson, *The New Physical Optics Notebook*: *Tutorials in Fourier Optics*, SPIE–The International Society for Optical Engineering, Bellingham, WA, and American Institute of Physics, New York, 1989.

J. L. Horner, ed., *Optical Signal Processing*, Academic Press, San Diego, CA, 1987.

F. T. S. Yu, *White-Light Optical Signal Processing*, Wiley, New York, 1985.

E. G. Steward, *Fourier Optics*: *An Introduction*, Halsted Press, New York, 1983.

P. M. Duffieux, *Fourier Transform and Its Applications to Optics*, Wiley, New York, 2nd ed. 1983.

F. T. S. Yu, *Optical Information Processing*, Wiley, New York, 1983.

H. Stark, ed., *Applications of Optical Fourier Transforms*, Academic Press, New York, 1982.

S. H. Lee, ed., *Optical Information Processing Fundamentals*, Springer-Verlag, New York, 1981.

J. D. Gaskill, *Linear Systems, Fourier Transforms and Optics*, Wiley, New York, 1978.

F. P. Carlson, *Introduction to Applied Optics for Engineers*, Academic Press, New York, 1978.

D. Casasent, ed., *Optical Data Processing: Applications*, Springer-Verlag, New York, 1978.

W. E. Kock, G. W. Stroke, and Yu. E. Nesterikhin, *Optical Information Processing*, Plenum Press, New York, 1976.

G. Harburn, C. A. Taylor, and T. R. Welberry, *Atlas of Optical Transforms*, Cornell University Press, Ithaca, NY, 1975.

T. Cathey, *Optical Information Processing and Holography*, Wiley, New York, 1974.

H. S. Lipson, ed., *Optical Transforms*, Academic Press, New York, 1972.

M. Cagnet, M. Françon, and S. Mallick, *Atlas of Optical Phenomena*, Springer-Verlag, New York, 1971.

A. R. Shulman, *Optical Data Processing*, Wiley, New York, 1970.

J. W. Goodman, *Introduction to Fourier Optics*, McGraw-Hill, New York, 1968.

A. Papoulis, *Systems and Transforms with Applications in Optics*, McGraw-Hill, New York, 1968.

G. W. Stroke, *An Introduction to Coherent Optics and Holography*, Academic Press, New York, 1966.

L. Mertz, *Transformations in Optics*, Wiley, New York, 1965.

C. A. Taylor and H. Lipson, *Optical Transforms*, Cornell University Press, Ithaca, NY, 1964.

E. L. O'Neill, *Introduction to Statistical Optics*, Addison-Wesley, Reading, MA, 1963.

Diffraction

S. Solimeno, B. Crosignani, and P. DiPorto, *Guiding, Diffraction, and Confinement of Optical Radiation*, Academic Press, New York, 1986.

J. M. Cowley, *Diffraction Physics*, North-Holland, New York, 1981, 3rd ed. 1984.

M. Françon, *Diffraction: Coherence in Optics*, Pergamon Press, New York, 1966.

Image Formation

C. S. Williams and O. A. Becklund, *Introduction to the Optical Transfer Function*, Wiley, New York, 1989.

M. Françon, *Optical Image Formation and Processing*, Academic Press, New York, 1979.

J. C. Dainty and R. Shaw, *Image Science*, Academic Press, New York, 1974.

K. R. Barnes, *The Optical Transfer Function*, Elsevier, New York, 1971.

E. H. Linfoot, *Fourier Methods in Optical Image Evaluation*, Focal Press, New York, 1964.

Holography

G. Saxby, *Practical Holography*, Prentice-Hall, Englewood Cliffs, NJ, 1989.

J. E. Kasper, *Complete Book of Holograms: How They Work and How to Make Them*, Wiley, New York, 1987.

W. Schumann, J.-P. Zurcher, and D. Cuche, *Holography and Deformation Analysis*, Springer-Verlag, New York, 1985.

N. Abramson, *The Making and Evaluation of Holograms*, Academic Press, New York, 1981.

Y. I. Ostrovsky, M. M. Butusov, and G. V. Ostrovskaya, *Interferometry by Holography*, Springer-Verlag, New York, 1980.

H. J. Caulfield, ed., *Handbook of Optical Holography*, Academic Press, New York, 1979.

W. Schumann and M. Dubas, *Holographic Interferometry*, Springer-Verlag, New York, 1979.

C. M. Vest, *Holographic Interferometry*, Wiley, New York, 1979.

G. Bally, ed., *Holography in Medicine and Biology*, Springer-Verlag, New York, 1979.

L. M. Soroko, *Holography and Coherent Optics*, Plenum Press, New York, 1978.

R. J. Collier, C. B. Burckhardt, and L. H. Lin, *Optical Holography*, Academic Press, New York, 1971, paperback edition 1977.

H. M. Smith, *Principles of Holography*, Wiley, New York, 1969, 2nd ed. 1975.

M. Françon, *Holography*, Academic Press, New York, 1974.

H. J. Caulfield and L. Sun, *The Applications of Holography*, Wiley-Interscience, New York, 1970.

J. B. DeVelis and G. O. Reynolds, *Theory and Applications of Holography*, Addison-Wesley, Reading, MA, 1967.

PROBLEMS

4.1-1 **Correspondence Between Harmonic Functions and Plane Waves.** The complex amplitudes of a monochromatic wave of wavelength λ in the $z = 0$ and $z = d$ planes are $f(x, y)$ and $g(x, y)$, respectively. Assuming that $d = 10^4\lambda$, use harmonic analysis to determine $g(x, y)$ in the following cases:
(a) $f(x, y) = 1$;
(b) $f(x, y) = \exp[(-j\pi/\lambda)(x + y)]$;
(c) $f(x, y) = \cos(\pi x/2\lambda)$;
(d) $f(x, y) = \cos^2(\pi y/2\lambda)$;
(e) $f(x, y) = \sum_m \mathrm{rect}[(x/10\lambda) - 2m]$, $m = 0, \pm 1, \pm 2, \ldots$, where $\mathrm{rect}(x) = 1$ if $|x| \leq \frac{1}{2}$ and 0, otherwise.
Describe the physical nature of the wave in each case.

4.1-2 In Problem 4.1-1, if $f(x, y)$ is a circularly symmetric function with a maximum spatial frequency of 200 lines/mm, determine the angle of the cone within which the wave directions are confined. Assume that $\lambda = 633$ nm.

4.1-3 **Logarithmic Interconnection Map.** A transparency of amplitude transmittance $t(x, y) = \exp[-j2\pi\phi(x)]$ is illuminated with a uniform plane wave of wavelength $\lambda = 1$ μm. The transmitted light is focused by an adjacent lens of focal length $f = 100$ cm. What must $\phi(x)$ be so that the ray that hits the transparency at position x is deflected and focused to a position $x' = \ln(x)$ for all $x > 0$? (Note that x and x' are measured in millimeters.) If the lens is removed, how should $\phi(x)$ be modified so that the system performs the same function? This system may be used to perform a logarithmic coordinate transformation, as discussed in Chap. 21.

4.2-1 **Proof of the Lens Fourier-Transform Property.** (a) Show that the convolution of $f(x)$ and $\exp(-j\pi x^2/\lambda d)$ may be obtained in three steps: Multiply $f(x)$ by $\exp(-j\pi x^2/\lambda d)$; evaluate the Fourier transform of the product at the frequency $\nu_x = x/\lambda d$; and multiply the result by $\exp(-j\pi x^2/\lambda d)$.
(b) The Fourier transform system in Fig. 4.2-4 is a cascade of three systems—propagation a distance f in free space, transmission through a lens of focal length f, and propagation a distance f in free space. Noting that propagation a distance d in free space is equivalent to convolution with $\exp(-j\pi x^2/\lambda d)$ [see (4.1-14)], and using the result in (a), derive the lens' Fourier transform equation (4.2-10). For simplicity ignore the y dependence.

4.2-2 **Fourier Transform of Line Functions.** A transparency of amplitude transmittance $t(x, y)$ is illuminated with a plane wave of wavelength $\lambda = 1$ μm and focused with a lens of focal length $f = 100$ cm. Sketch the intensity distribution in the plane of the transparency and in the lens focal plane in the following cases (all distances are measured in mm):
(a) $t(x, y) = \delta(x - y)$;
(b) $t(x, y) = \delta(x + a) + \delta(x - a)$, $a = 1$ mm;
(c) $t(x, y) = \delta(x + a) + j\delta(x - a)$, $a = 1$ mm,
where $\delta(\cdot)$ is the delta function (see Appendix A, Sec. A.1).

4.2-3 **Design of an Optical Fourier-Transform System.** A lens is used to display the Fourier transform of a two-dimensional function with spatial frequencies between 20 and 200 lines/mm. If the wavelength of light is $\lambda = 488$ nm, what should be the focal length of the lens so that the highest and lowest spatial frequencies are separated by a distance of 9 cm in the Fourier plane?

4.3-1 **Fraunhofer Diffraction from a Diffraction Grating.** Derive an expression for the Fraunhofer diffraction pattern for an aperture made of $M = 2L + 1$ parallel slits of infinitesimal widths separated by equal distances $a = 10\lambda$,

$$p(x, y) = \sum_{m=-L}^{L} \delta(x - ma).$$

Sketch the pattern as a function of the observation angle $\theta = x/d$, where d is the observation distance.

4.3-2 **Fraunhofer Diffraction with an Oblique Incident Wave.** The diffraction pattern from an aperture with aperture function $p(x, y)$ is $(1/\lambda d)^2 |P(x/\lambda d, y/\lambda d)|^2$, where $P(\nu_x, \nu_y)$ is the Fourier transform of $p(x, y)$ and d is the distance between the aperture and observation planes. What is the diffraction pattern when the direction of the incident wave makes a small angle $\theta_x \ll 1$, with the z-axis in the x-z plane?

*4.3-3 **Fresnel Diffraction from Two Pinholes.** Show that the Fresnel diffraction pattern from two pinholes separated by a distance $2a$, i.e., $p(x, y) = [\delta(x - a) + \delta(x + a)]\delta(y)$, at an observation distance d is the periodic pattern, $I(x, y) = (2/\lambda d)^2 \cos^2(2\pi ax/\lambda d)$.

*4.3-4 **Relation Between Fresnel and Fraunhofer Diffraction.** Show that the Fresnel diffraction pattern of the aperture function $p(x, y)$ is equal to the Fraunhofer diffraction pattern of the aperture function $p(x, y)\exp[-j\pi(x^2 + y^2)/\lambda d]$.

4.4-1 **Blurring a Sinusoidal Grating.** An object $f(x, y) = \cos^2(2\pi x/a)$ is imaged by a defocused single-lens imaging system whose impulse-response function $h(x, y) = 1$ within a square of width D, and $= 0$ elsewhere. Derive an expression for the distribution of the image $g(x, 0)$ in the x direction. Derive an expression for the contrast of the image in terms of the ratio D/a. The contrast = (max − min)/(max + min), where max and min are the maximum and minimum values of $g(x, 0)$.

4.4-2 **Image of a Phase Object.** An imaging system has an impulse-response function $h(x,y) = \text{rect}(x)\delta(y)$. If the input wave is

$$f(x, y) = \begin{cases} \exp\left(j\dfrac{\pi}{2}\right) & \text{for } x > 0 \\ \exp\left(-j\dfrac{\pi}{2}\right) & \text{for } x \le 0, \end{cases}$$

determine and sketch the intensity $|g(x, y)|^2$ of the output wave $g(x, y)$. Verify that even though the intensity of the input wave $|f(x, y)|^2 = 1$, the intensity of the output wave is not uniform.

4.4-3 **Optical Spatial Filtering.** Consider the spatial filtering system shown in Fig. 4.4-5 with $f = 1000$ mm. The system is illuminated with a uniform plane wave of unit amplitude and wavelength $\lambda = 10^{-3}$ mm. The input transparency has amplitude transmittance $f(x, y)$ and the mask has amplitude transmittance $p(x, y)$. Write an expression relating the complex amplitude $g(x, y)$ of light in the image plane to $f(x, y)$ and $p(x, y)$. Assuming that all distances are measured in mm, sketch $g(x, 0)$

in the following cases:
(a) $f(x, y) = \delta(x - 5)$ and $p(x, y) = \mathrm{rect}(x)$;
(b) $f(x, y) = \mathrm{rect}(x)$ and $p(x, y) = \mathrm{sinc}(x)$.
Determine $p(x, y)$ such that $g(x, y) = \nabla_T^2 f(x, y)$, where $\nabla_T^2 = \partial^2/\partial x^2 + \partial^2/\partial y^2$ is the transverse Laplacian operator.

4.4-4 **Optical Cross-Correlation.** Show how a spatial filter may be used to perform the operation of cross-correlation (defined in Appendix A) between two images described by the real-valued functions $f_1(x, y)$ and $f_2(x, y)$. Under what conditions would the complex amplitude transmittances of the masks and transparencies used be real-valued?

*4.4-5 **Impulse-Response Function of a Severely Defocused System.** Using wave optics, show that the impulse-response function of a severely defocused imaging system (one for which the defocusing error ϵ is very large) may be approximated by $h(x, y) = p(x/\epsilon d_2, y/\epsilon d_2)$, where $p(x, y)$ is the pupil function. *Hint:* Use the method of stationary phase described on page 124 (proof 2) to evaluate the integral that results from the use of (4.4-11) and (4.4-10). Note that this is the same result predicted by the ray theory of light [see (4.4-2)].

4.4-6 **Two-Point Resolution.** (a) Consider the single-lens imaging system discussed in Sec. 4.4C. Assuming a square aperture of width D, unit magnification, and perfect focus, write an expression for the impulse-response function $h(x, y)$.
(b) Determine the response of the system to an object consisting of two points separated by a distance b, i.e.,

$$f(x, y) = \delta(x)\delta(y) + \delta(x - b)\delta(y).$$

(c) If $\lambda d_2/D = 0.1$ mm, sketch the magnitude of the image $g(x, 0)$ as a function of x when the points are separated by a distance $b = 0.5$, 1, and 2 mm. What is the minimum separation between the two points such that the image remains discernible as two spots instead of a single spot, i.e., has two peaks.

4.4-7 **Ring Aperture.** (a) A focused single-lens imaging system, with magnification $M = 1$ and focal length $f = 100$ cm has an aperture in the form of a ring

$$p(x, y) = \begin{cases} 1, & a \le \left(x^2 + y^2\right)^{1/2} \le b, \\ 0, & \text{otherwise}, \end{cases}$$

where $a = 5$ mm and $b = 6$ mm. Determine the transfer function $H(\nu_x, \nu_y)$ of the system and sketch its cross section $H(\nu_x, 0)$. The wavelength $\lambda = 1\ \mu$m.
(b) If the image plane is now moved closer to the lens so that its distance from the lens becomes $d_2 = 25$ cm, with the distance between the object plane and the lens d_1 as in (a), use the ray-optics approximation to determine the impulse-response function of the imaging system $h(x, y)$ and sketch $h(x, 0)$.

4.5-1 **Holography with a Spherical Reference Wave.** The choice of a uniform plane wave as a reference wave is not essential to holography; other waves can be used. Assuming that the reference wave is a spherical wave centered about the point $(0, 0, -d)$, determine the hologram pattern and examine the reconstructed wave when:
(a) the object wave is a plane wave traveling at an angle θ_x;
(b) the object wave is a spherical wave centered at $(-x_0, 0, -d_1)$.
Approximate spherical waves by paraboloidal waves.

4.5-2 **Optical Correlation.** A transparency with an amplitude transmittance given by $f(x, y) = f_1(x - a, y) + f_2(x + a, y)$ is Fourier transformed by a lens and the intensity is recorded on a transparency (hologram). The hologram is subsequently illuminated with a reference wave and the reconstructed wave is Fourier transformed with a lens to generate the function $g(x, y)$. Derive an expression relating $g(x, y)$ to $f_1(x, y)$ and $f_2(x, y)$. Show how the correlation of the two functions $f_1(x,y)$ and $f_2(x, y)$ may be determined with this system.

CHAPTER

5

ELECTROMAGNETIC OPTICS

5.1 ELECTROMAGNETIC THEORY OF LIGHT

5.2 DIELECTRIC MEDIA

- A. Linear, Nondispersive, Homogeneous, and Isotropic Media
- B. Nonlinear, Dispersive, Inhomogeneous, or Anisotropic Media

5.3 MONOCHROMATIC ELECTROMAGNETIC WAVES

5.4 ELEMENTARY ELECTROMAGNETIC WAVES

- A. Plane, Spherical, and Gaussian Electromagnetic Waves
- B. Relation Between Electromagnetic Optics and Scalar Wave Optics

5.5 ABSORPTION AND DISPERSION

- A. Absorption
- B. Dispersion
- C. The Resonant Medium

5.6 PULSE PROPAGATION IN DISPERSIVE MEDIA

James Clerk Maxwell (1831–1879) advanced the theory that light is an electromagnetic wave phenomenon.

Light is an electromagnetic wave phenomenon described by the same theoretical principles that govern all forms of electromagnetic radiation. Optical frequencies occupy a band of the electromagnetic spectrum that extends from the infrared through the visible to the ultraviolet (Fig. 5.0-1). Because the wavelength of light is relatively short (between 10 nm and 1 mm), the techniques used for generating, transmitting, and detecting optical waves have traditionally differed from those used for electromagnetic waves of longer wavelength. However, the recent miniaturization of optical components (e.g., optical waveguides and integrated-optical devices) has caused these differences to become less significant.

Electromagnetic radiation propagates in the form of two mutually coupled *vector* waves, an electric-field wave and a magnetic-field wave. The wave optics theory described in Chap. 2 is an approximation of the electromagnetic theory, in which light is described by a single *scalar* function of position and time (the wavefunction). This approximation is adequate for paraxial waves under certain conditions. As shown in Chap. 2, the ray optics approximation provides a further simplification valid in the limit of short wavelengths. Thus electromagnetic optics encompasses wave optics, which, in turn, encompasses ray optics (Fig. 5.0-2).

This chapter provides a brief review of the aspects of electromagnetic theory that are of importance in optics. The basic principles of the theory—Maxwell's equations—are provided in Sec. 5.1, whereas Sec. 5.2 covers the electromagnetic properties of dielectric media. These two sections may be regarded as the postulates of electromagnetic optics, i.e., the set of rules on which the remaining sections are based. In Sec. 5.3 we provide a restatement of these rules for the important special case of monochromatic light. Elementary electromagnetic waves (plane waves, spherical waves, and Gaussian beams) are introduced as examples in Sec. 5.4. Dispersive media, which exhibit wavelength-dependent absorption coefficients and refractive indices, are discussed in Sec. 5.5. Section 5.6 is devoted to the propagation of light pulses in dispersive

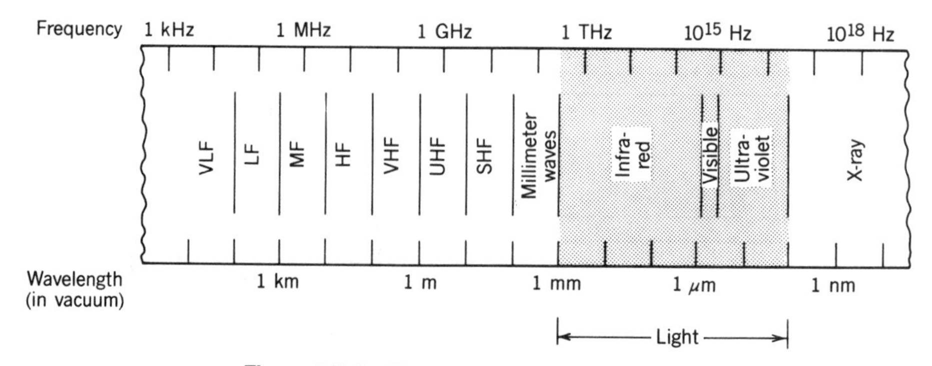

Figure 5.0-1 The electromagnetic spectrum.

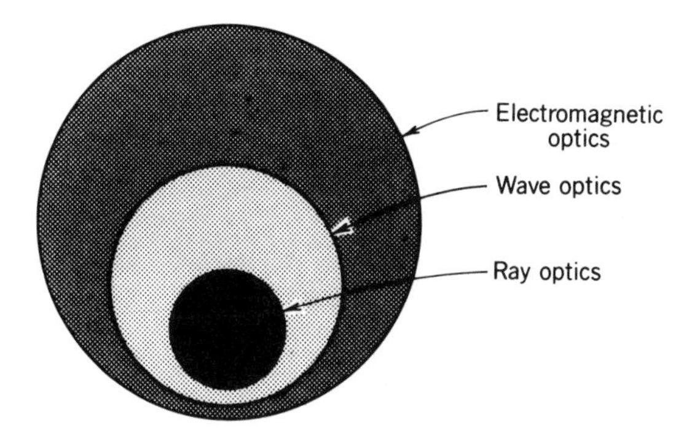

Figure 5.0-2 Wave optics is the scalar approximation of electromagnetic optics. Ray optics is the limit of wave optics when the wavelength is very short.

media. Chapter 6 covers the polarization of light and the optics of anisotropic media, and Chap. 19 is devoted to the electromagnetic optics of nonlinear media.

5.1 ELECTROMAGNETIC THEORY OF LIGHT

An electromagnetic field is described by two related vector fields: the **electric field** $\mathscr{E}(\mathbf{r}, t)$ and the **magnetic field** $\mathscr{H}(\mathbf{r}, t)$. Both are *vector* functions of position and time. In general, six scalar functions of position and time are therefore required to describe light in free space. Fortunately, these functions are related since they must satisfy a set of coupled partial differential equations known as Maxwell's equations.

Maxwell's Equations in Free Space

The electric and magnetic fields in free space satisfy the following partial differential equations, known as **Maxwell's equations**:

$$\nabla \times \mathscr{H} = \epsilon_o \frac{\partial \mathscr{E}}{\partial t} \tag{5.1-1}$$

$$\nabla \times \mathscr{E} = -\mu_o \frac{\partial \mathscr{H}}{\partial t} \tag{5.1-2}$$

$$\nabla \cdot \mathscr{E} = 0 \tag{5.1-3}$$

$$\nabla \cdot \mathscr{H} = 0, \tag{5.1-4}$$

Maxwell's Equations (Free Space)

where the constants $\epsilon_o \approx (1/36\pi) \times 10^{-9}$ and $\mu_o = 4\pi \times 10^{-7}$ (MKS units) are, respectively, the **electric permittivity** and the **magnetic permeability** of free space; and $\nabla \cdot$ and $\nabla \times$ are the divergence and the curl operations.[†]

[†]In a Cartesian coordinate system $\nabla \cdot \mathscr{E} = \partial \mathscr{E}_x/\partial x + \partial \mathscr{E}_y/\partial y + \partial \mathscr{E}_z/\partial z$ and $\nabla \times \mathscr{E}$ is a vector with Cartesian components $(\partial \mathscr{E}_z/\partial y - \partial \mathscr{E}_y/\partial z)$, $(\partial \mathscr{E}_x/\partial z - \partial \mathscr{E}_z/\partial x)$, and $(\partial \mathscr{E}_y/\partial x - \partial \mathscr{E}_x/\partial y)$.

The Wave Equation

A necessary condition for $\mathscr{E}$ and $\mathscr{H}$ to satisfy Maxwell's equations is that each of their components satisfy the wave equation

$$\nabla^2 u - \frac{1}{c_o^2} \frac{\partial^2 u}{\partial t^2} = 0, \tag{5.1-5}$$

The Wave Equation

where

$$c_o = \frac{1}{(\epsilon_o \mu_o)^{1/2}} \approx 3 \times 10^8 \text{ m/s} \tag{5.1-6}$$

Speed of Light (Free Space)

is the speed of light, and the scalar function u represents any of the three components $(\mathscr{E}_x, \mathscr{E}_y, \mathscr{E}_z)$ of $\mathscr{E}$, or the three components $(\mathscr{H}_x, \mathscr{H}_y, \mathscr{H}_z)$ of $\mathscr{H}$. The wave equation may be derived from Maxwell's equations by applying the curl operation $\nabla \times$ to (5.1-2), using the vector identity $\nabla \times (\nabla \times \mathscr{E}) = \nabla(\nabla \cdot \mathscr{E}) - \nabla^2 \mathscr{E}$, and then using (5.1-1) and (5.1-3) to show that each component of $\mathscr{E}$ satisfies the wave equation. A similar procedure is followed for $\mathscr{H}$.

Since Maxwell's equations and the wave equation are linear, the principle of superposition applies; i.e., if two sets of electric and magnetic fields are solutions to these equations, their sum is also a solution.

The connection between electromagnetic optics and wave optics is now eminently clear. The wave equation, which is the basis of wave optics, is embedded in the structure of electromagnetic theory; and the speed of light is related to the electromagnetic constants ϵ_o and μ_o by (5.1-6).

Maxwell's Equations in a Medium

In a medium in which there are no free electric charges or currents, two more vector fields need to be defined—the **electric flux density** (also called the electric displacement) $\mathscr{D}(\mathbf{r}, t)$ and the **magnetic flux density** $\mathscr{B}(\mathbf{r}, t)$. Maxwell's equations relate the four fields $\mathscr{E}$, $\mathscr{H}$, $\mathscr{D}$, and $\mathscr{B}$, by

$$\nabla \times \mathscr{H} = \frac{\partial \mathscr{D}}{\partial t} \tag{5.1-7}$$

$$\nabla \times \mathscr{E} = -\frac{\partial \mathscr{B}}{\partial t} \tag{5.1-8}$$

$$\nabla \cdot \mathscr{D} = 0 \tag{5.1-9}$$

$$\nabla \cdot \mathscr{B} = 0. \tag{5.1-10}$$

Maxwell's Equations (Source-Free Medium)

The relation between the electric flux density $\mathscr{D}$ and the electric field $\mathscr{E}$ depends on the electric properties of the medium. Similarly, the relation between the magnetic flux density $\mathscr{B}$ and the magnetic field $\mathscr{H}$ depends on the magnetic properties of the medium. Two equations help define these relations:

$$\mathscr{D} = \epsilon_o \mathscr{E} + \mathscr{P} \tag{5.1-11}$$

$$\mathscr{B} = \mu_o \mathscr{H} + \mu_o \mathscr{M}, \tag{5.1-12}$$

in which $\mathscr{P}$ is the **polarization density** and $\mathscr{M}$ is the **magnetization density**. In a dielectric medium, the polarization density is the macroscopic sum of the electric dipole moments that the electric field induces. The magnetization density is similarly defined.

The vector fields $\mathscr{P}$ and $\mathscr{M}$ are, in turn, related to the electric and magnetic fields $\mathscr{E}$ and $\mathscr{H}$ by relations that depend on the electric and magnetic properties of the medium, respectively, as will be described subsequently. Once the medium is known, an equation relating $\mathscr{P}$ and $\mathscr{E}$, and another relating $\mathscr{M}$ and $\mathscr{H}$ are established. When substituted in Maxwell's equations, we are left with equations governing only the two vector fields $\mathscr{E}$ and $\mathscr{H}$.

In free space, $\mathscr{P} = \mathscr{M} = 0$, so that $\mathscr{D} = \epsilon_o \mathscr{E}$ and $\mathscr{B} = \mu_o \mathscr{H}$; the free-space Maxwell's equations, (5.1-1) to (5.1-4), are then recovered. In a nonmagnetic medium $\mathscr{M} = 0$. Throughout this book, unless otherwise stated, it is assumed that the medium is nonmagnetic ($\mathscr{M} = 0$). Equation (5.1-12) is then replaced by

$$\mathscr{B} = \mu_o \mathscr{H}. \tag{5.1-13}$$

Boundary Conditions

In a homogeneous medium, all components of the fields $\mathscr{E}$, $\mathscr{H}$, $\mathscr{D}$, and $\mathscr{B}$ are continuous functions of position. At the boundary between two dielectric media and in the absence of free electric charges and currents, the tangential components of the electric and magnetic fields $\mathscr{E}$ and $\mathscr{H}$ and the normal components of the electric and magnetic flux densities $\mathscr{D}$ and $\mathscr{B}$ must be continuous (Fig. 5.1-1).

Intensity and Power

The flow of electromagnetic power is governed by the vector

$$\mathscr{S} = \mathscr{E} \times \mathscr{H}, \tag{5.1-14}$$

known as the **Poynting vector**. The direction of power flow is along the direction of the Poynting vector, i.e., is orthogonal to both $\mathscr{E}$ and $\mathscr{H}$. The optical intensity I (power flow across a unit area normal to the vector $\mathscr{S}$)† is the magnitude of the time-averaged

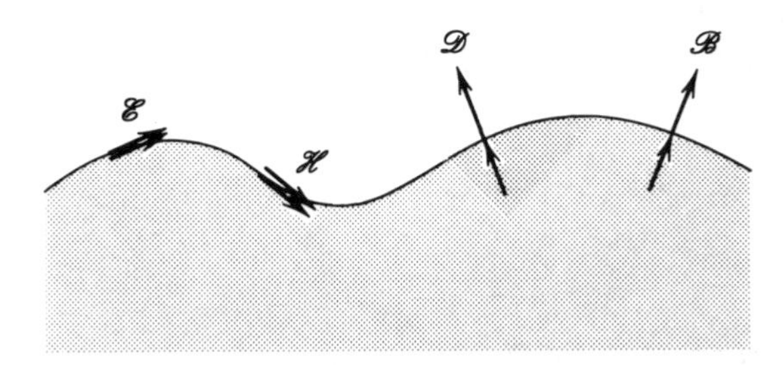

Figure 5.1-1 Tangential components of $\mathscr{E}$ and $\mathscr{H}$ and normal components of $\mathscr{D}$ and $\mathscr{B}$ are continuous at the boundaries between different media without free electric charges and currents.

†For a discussion of this interpretation, see M. Born and E. Wolf, *Principles of Optics*, Pergamon Press, New York, 6th ed. 1980, pp. 9–10; and E. Wolf, Coherence and Radiometry, *Journal of the Optical Society of America*, vol. 68, pp. 6–17, 1978.

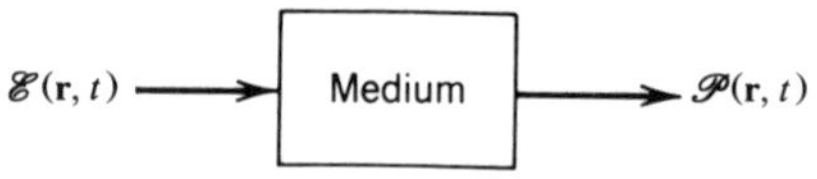

Figure 5.2-1 The dielectric medium responds to an applied electric field $\mathscr{E}$ and creates a polarization density $\mathscr{P}$.

$\mathscr{E}(\mathbf{r}, t)$ → Medium → $\mathscr{P}(\mathbf{r}, t)$

Poynting vector $\langle \mathscr{S} \rangle$. The average is taken over times that are long compared to an optical cycle, but short compared to other times of interest.

5.2 DIELECTRIC MEDIA

The nature of the dielectric medium is exhibited in the relation between the polarization density $\mathscr{P}$ and the electric field $\mathscr{E}$, called the **medium equation** (Fig. 5.2-1). It is useful to think of the $\mathscr{P}$–$\mathscr{E}$ relation as a system in which $\mathscr{E}$ is regarded as an applied input and $\mathscr{P}$ as the output or response. Note that $\mathscr{E} = \mathscr{E}(\mathbf{r}, t)$ and $\mathscr{P} = \mathscr{P}(\mathbf{r}, t)$ are functions of position and time.

Definitions

- A dielectric medium is said to be *linear* if the vector field $\mathscr{P}(\mathbf{r}, t)$ is linearly related to the vector field $\mathscr{E}(\mathbf{r}, t)$. The principle of superposition then applies.
- The medium is said to be *nondispersive* if its response is instantaneous; i.e., $\mathscr{P}$ at time t is determined by $\mathscr{E}$ at the same time t and not by prior values of $\mathscr{E}$. Nondispersiveness is clearly an idealization since any physical system, however fast it may be, has a finite response time.
- The medium is said to be *homogeneous* if the relation between $\mathscr{P}$ and $\mathscr{E}$ is independent of the position $\mathbf{r}$.
- The medium is called *isotropic* if the relation between the vectors $\mathscr{P}$ and $\mathscr{E}$ is independent of the direction of the vector $\mathscr{E}$, so that the medium looks the same from all directions. The vectors $\mathscr{P}$ and $\mathscr{E}$ must then be parallel.
- The medium is said to be *spatially nondispersive* if the relation between $\mathscr{P}$ and $\mathscr{E}$ is local; i.e., $\mathscr{P}$ at each position $\mathbf{r}$ is influenced only by $\mathscr{E}$ at the same position. In this chapter the medium is always assumed to be spatially nondispersive.

A. Linear, Nondispersive, Homogeneous, and Isotropic Media

Let us first consider the simplest case of *linear*, *nondispersive*, *homogeneous*, and *isotropic* media. The vectors $\mathscr{P}$ and $\mathscr{E}$ at any position and time are parallel and proportional, so that

$$\boxed{\mathscr{P} = \epsilon_o \chi \mathscr{E},} \tag{5.2-1}$$

where χ is a scalar constant called the **electric susceptibility** (Fig. 5.2-2).

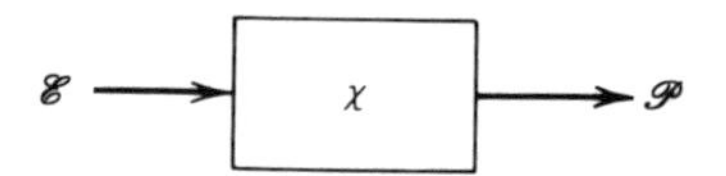

Figure 5.2-2 A linear, homogenous, isotropic, and nondispersive medium is characterized completely by one constant, the electric susceptibility χ.

Substituting (5.2-1) in (5.1-11), it follows that $\mathscr{D}$ and $\mathscr{E}$ are also parallel and proportional,

$$\mathscr{D} = \epsilon \mathscr{E}, \tag{5.2-2}$$

where

$$\epsilon = \epsilon_o(1 + \chi) \tag{5.2-3}$$

is another scalar constant, the **electric permittivity** of the medium. The radio ϵ/ϵ_o is the **relative permittivity** or **dielectric constant.**

Under these conditions, Maxwell's equations simplify to

$$\nabla \times \mathscr{H} = \epsilon \frac{\partial \mathscr{E}}{\partial t} \tag{5.2-4}$$

$$\nabla \times \mathscr{E} = -\mu_o \frac{\partial \mathscr{H}}{\partial t} \tag{5.2-5}$$

$$\nabla \cdot \mathscr{E} = 0 \tag{5.2-6}$$

$$\nabla \cdot \mathscr{H} = 0. \tag{5.2-7}$$

Maxwell's Equations (Linear, Homogeneous, Isotropic, Nondispersive, Source-Free Medium)

We are now left with two related vector fields, $\mathscr{E}(\mathbf{r}, t)$ and $\mathscr{H}(\mathbf{r}, t)$ that satisfy equations identical to Maxwell's equations in free space with ϵ_o replaced by ϵ. Each of the components of $\mathscr{E}$ and $\mathscr{H}$ therefore satisfies the wave equation

$$\nabla^2 u - \frac{1}{c^2}\frac{\partial^2 u}{\partial t^2} = 0, \tag{5.2-8}$$

Wave Equation

with a speed $c = 1/(\epsilon\mu_o)^{1/2}$. The different components of the electric and magnetic fields propagate in the form of waves of speed

$$c = \frac{c_o}{n}, \tag{5.2-9}$$

Speed of Light (In a Medium)

where

$$n = \left(\frac{\epsilon}{\epsilon_o}\right)^{1/2} = (1 + \chi)^{1/2} \tag{5.2-10}$$

Refractive Index

and

$$c_o = \frac{1}{(\epsilon_o \mu_o)^{1/2}} \tag{5.2-11}$$

is the speed of light in free space. The constant n is the ratio of the speed of light in free space to that in the medium. It therefore represents the refractive index of the medium.

The refractive index is the square root of the dielectric constant.

This is another point of connection between scalar wave optics (Chap. 2) and electromagnetic optics. Other connections are discussed in Sec. 5.4B.

B. Nonlinear, Dispersive, Inhomogeneous, or Anisotropic Media

We now consider media for which one or more of the properties of linearity, nondispersiveness, homogeneity, and isotropy are not satisfied.

Inhomogeneous Media

In an inhomogeneous dielectric medium (such as a graded-index medium) that is linear, nondispersive, and isotropic, the simple proportionality relations $\mathscr{P} = \epsilon_o \chi \mathscr{E}$, and $\mathscr{D} = \epsilon \mathscr{E}$ remain valid, but the coefficients χ and ϵ are functions of position, $\chi = \chi(\mathbf{r})$ and $\epsilon = \epsilon(\mathbf{r})$ (Fig. 5.2-3). Likewise, the refractive index $n = n(\mathbf{r})$ is position dependent.

For locally homogeneous media, in which $\epsilon(\mathbf{r})$ varies sufficiently slowly so that it can be assumed constant within a distance of a wavelength, the wave equation is modified to

$$\boxed{\nabla^2 \mathscr{E} - \frac{1}{c^2(\mathbf{r})} \frac{\partial^2 \mathscr{E}}{\partial t^2} = 0} \tag{5.2-12}$$

Wave Equation (Inhomogeneous Medium)

where $c(\mathbf{r}) = c_o/n(\mathbf{r})$ is a spatially varying speed and $n(\mathbf{r}) = [\epsilon(\mathbf{r})/\epsilon_o]^{1/2}$ is the refractive index at position $\mathbf{r}$. This relation, which was provided as one of the postulates of wave optics (Sec. 2.1), will now be shown to be a consequence of Maxwell's equations.

Beginning with Maxwell's equations (5.1-7) to (5.1-10) and noting that $\epsilon = \epsilon(\mathbf{r})$ is position dependent, we apply the curl operation $\nabla \times$ to both sides of (5.1-8) and use Maxwell's equation (5.1-7) to write

$$\nabla \times (\nabla \times \mathscr{E}) = \nabla(\nabla \cdot \mathscr{E}) - \nabla^2 \mathscr{E} = -\mu_o \frac{\partial^2 \mathscr{D}}{\partial t^2}. \tag{5.2-13}$$

Maxwell's equation (5.1-9) gives $\nabla \cdot \epsilon \mathscr{E} = 0$ and the identity $\nabla \cdot \epsilon \mathscr{E} = \epsilon \nabla \cdot \mathscr{E} + \mathscr{E} \cdot \nabla \epsilon$

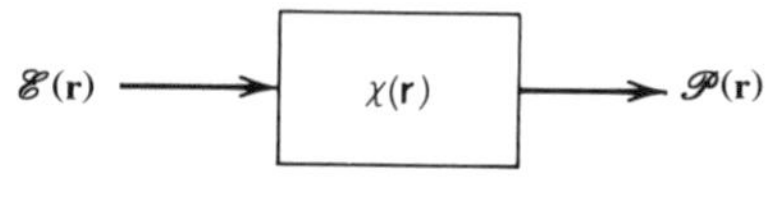

Figure 5.2-3 An inhomogeneous (but linear, nondispersive, and isotropic) medium is characterized by a position dependent susceptibility $\chi(\mathbf{r})$.

permits us to obtain $\nabla \cdot \mathscr{E} = -(1/\epsilon)\nabla\epsilon \cdot \mathscr{E}$, which when substituted in (5.2-13) yields

$$\nabla^2 \mathscr{E} - \frac{1}{c^2(\mathbf{r})} \frac{\partial^2 \mathscr{E}}{\partial t^2} + \nabla\left(\frac{1}{\epsilon} \nabla\epsilon \cdot \mathscr{E}\right) = 0, \tag{5.2-14}$$

where $c(\mathbf{r}) = 1/[\mu_o \epsilon(\mathbf{r})]^{1/2} = c_o/n(\mathbf{r})$. If $\epsilon(\mathbf{r})$ varies in space at a much slower rate than $\mathscr{E}(\mathbf{r}, t)$; i.e., $\epsilon(\mathbf{r})$ does not vary significantly within a wavelength distance, the third term in (5.2-14) may be neglected in comparison with the first, so that (5.2-12) is approximately applicable.

Anisotropic Media

In an anisotropic dielectric medium, the relation between the vectors $\mathscr{P}$ and $\mathscr{E}$ depends on the direction of the vector $\mathscr{E}$, and these two vectors are not necessarily parallel. If the medium is linear, nondispersive, and homogeneous, each component of $\mathscr{P}$ is a linear combination of the three components of $\mathscr{E}$

$$\mathscr{P}_i = \sum_j \epsilon_o \chi_{ij} \mathscr{E}_j, \tag{5.2-15}$$

where the indices $i, j = 1, 2, 3$ denote the x, y, and z components.

The dielectric properties of the medium are described by an array $\{\chi_{ij}\}$ of 3×3 constants known as the **susceptibility tensor** (Fig. 5.2-4). A similar relation between $\mathscr{D}$ and $\mathscr{E}$ applies:

$$\mathscr{D}_i = \sum_j \epsilon_{ij} \mathscr{E}_j, \tag{5.2-16}$$

where $\{\epsilon_{ij}\}$ are elements of the **electric permittivity tensor**. The optical properties of anisotropic media are examined in Chap. 6.

Dispersive Media

The relation between $\mathscr{P}$ and $\mathscr{E}$ is a dynamic relation with "memory" rather than an instantaneous relation. The vector $\mathscr{E}$ "creates" the vector $\mathscr{P}$ by inducing oscillation of the bound electrons in the atoms of the medium, which collectively produce the polarization density. A time delay between this cause and effect (or input and output)

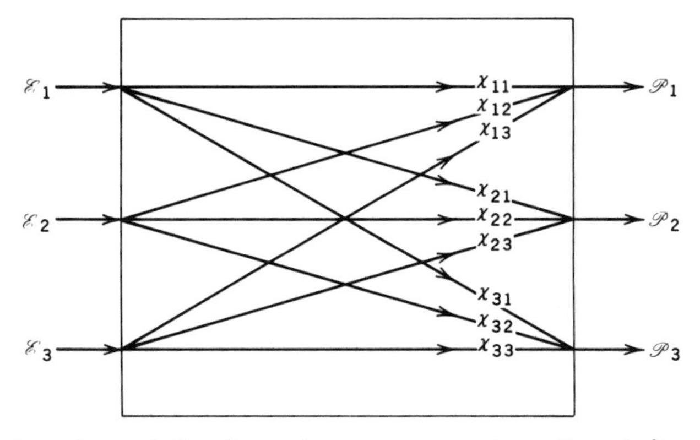

Figure 5.2-4 An anisotropic (but linear, homogeneous, and nondispersive) medium is characterized completely by nine constants, elements of the susceptibility tensor χ_{ij}. Each of the components of $\mathscr{P}$ is a weighted superposition of the three components of $\mathscr{E}$.

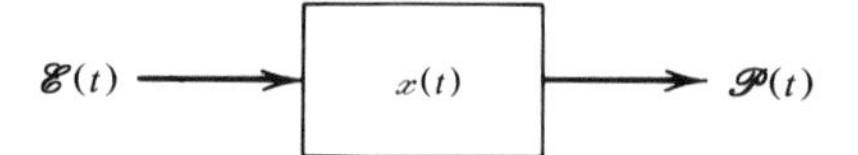

Figure 5.2-5 In a dispersive (but linear, homogeneous, and isotropic) medium, the relation between $\mathscr{P}(t)$ and $\mathscr{E}(t)$ is governed by a dynamic linear system described by an impulse-response function $\epsilon_o x(t)$ corresponding to a frequency dependent susceptibility $\chi(\nu)$.

is exhibited. When this time is extremely short in comparison with other times of interest, however, the response may be regarded as instantaneous, so that the medium is approximately nondispersive. For simplicity, we shall limit this discussion to dispersive media that are linear, homogeneous, and isotropic.

The dynamic relation between $\mathscr{P}(t)$ and $\mathscr{E}(t)$ may be described by a linear differential equation; for example, $a_1\, d^2\mathscr{P}/dt^2 + a_2\, d\mathscr{P}/dt + a_3\mathscr{P} = \mathscr{E}$, where a_1, a_2, and a_3 are constants. This equation is similar to that describing the response of a harmonic oscillator to a driving force. More generally, a linear dynamic relation may be described by the methods of linear systems (see Appendix B).

A linear system is characterized by its response to an impulse. An impulse of electric field of magnitude $\delta(t)$ at time $t = 0$ induces a time-dispersed polarization density of magnitude $\epsilon_o x(t)$, where $x(t)$ is a scalar function of time beginning at $t = 0$ and lasting for some duration. Since the medium is linear, an arbitrary electric field $\mathscr{E}(t)$ induces a polarization density that is a superposition of the effects of $\mathscr{E}(t')$ at all $t' \leq t$, i.e., a convolution (see Appendix A)

$$\mathscr{P}(t) = \epsilon_o \int_{-\infty}^{\infty} x(t - t')\mathscr{E}(t')\, dt'. \tag{5.2-17}$$

The dielectric medium is therefore described completely by the impulse-response function $\epsilon_o x(t)$.

Dynamic linear systems are also described by their transfer function (which governs the response to harmonic inputs). The transfer function is the Fourier transform of the impulse-response function. In our case the transfer function at frequency ν is $\epsilon_o \chi(\nu)$, where $\chi(\nu)$, the Fourier transform of $x(t)$, is a frequency-dependent susceptibility (Fig. 5.2-5). This concept is discussed in Sec. 5.3.

Nonlinear Media

In a nonlinear dielectric medium, the relation between $\mathscr{P}$ and $\mathscr{E}$ is nonlinear. If the medium is homogeneous, isotropic, and nondispersive, then $\mathscr{P}$ is some nonlinear function of $\mathscr{E}$, $\mathscr{P} = \Psi(\mathscr{E})$, at every position and time; for example, $\mathscr{P} = a_1\mathscr{E} + a_2\mathscr{E}^2 + a_3\mathscr{E}^3$, where a_1, a_2, and a_3 are constants. The wave equation (5.2-8) is not applicable to electromagnetic waves in nonlinear media. However, Maxwell's equations can be used to derive a *nonlinear* partial differential equation that these waves obey.

Operating on Maxwell's equation (5.1-8) with the curl operator $\nabla \times$, using the relation $\mathscr{B} = \mu_o \mathscr{H}$, and substituting from Maxwell's equation (5.1-7), we obtain $\nabla \times (\nabla \times \mathscr{E}) = -\mu_o\, \partial^2\mathscr{D}/\partial t^2$. Using the relation $\mathscr{D} = \epsilon_o\mathscr{E} + \mathscr{P}$ and the vector identity $\nabla \times (\nabla \times \mathscr{E}) = \nabla(\nabla \cdot \mathscr{E}) - \nabla^2\mathscr{E}$, we write

$$\nabla(\nabla \cdot \mathscr{E}) - \nabla^2\mathscr{E} = -\epsilon_o\mu_o \frac{\partial^2\mathscr{E}}{\partial t^2} - \mu_o \frac{\partial^2\mathscr{P}}{\partial t^2}. \tag{5.2-18}$$

For a homogeneous and isotropic medium $\mathscr{D} = \epsilon\mathscr{E}$, so that from Maxwell's equation, $\nabla \cdot \mathscr{D} = 0$, we conclude that $\nabla \cdot \mathscr{E} = 0$. Substituting $\nabla \cdot \mathscr{E} = 0$ and $\epsilon_o\mu_o = 1/c_o^2$ into

(5.2-18), we obtain

$$\nabla^2 \mathscr{E} - \frac{1}{c_o^2}\frac{\partial^2 \mathscr{E}}{\partial t^2} = \mu_o \frac{\partial^2 \mathscr{P}}{\partial t^2}. \tag{5.2-19}$$

Wave Equation (Homogeneous and Isotropic Medium)

Equation (5.2-19) is applicable to all homogeneous and isotropic dielectric media. If, in addition, the medium is nondispersive, $\mathscr{P} = \Psi(\mathscr{E})$ and therefore (5.2-19) yields a nonlinear partial differential equation for the electric field $\mathscr{E}$,

$$\nabla^2 \mathscr{E} - \frac{1}{c_o^2}\frac{\partial^2 \mathscr{E}}{\partial t^2} = \mu_o \frac{\partial^2 \Psi(\mathscr{E})}{\partial t^2}. \tag{5.2-20}$$

The nonlinearity of the wave equation implies that the principle of superposition is no longer applicable.

Most optical media are approximately linear, unless the optical intensity is very large, as in the case of focused laser beams. Nonlinear optical media are discussed in Chap. 19.

5.3 MONOCHROMATIC ELECTROMAGNETIC WAVES

When the electromagnetic wave is monochromatic, all components of the electric and magnetic fields are harmonic functions of time of the same frequency. These components are expressed in terms of their complex amplitudes as was done in Sec. 2.2A,

$$\mathscr{E}(\mathbf{r}, t) = \operatorname{Re}\{\mathbf{E}(\mathbf{r}) \exp(j\omega t)\}$$

$$\mathscr{H}(\mathbf{r}, t) = \operatorname{Re}\{\mathbf{H}(\mathbf{r}) \exp(j\omega t)\}, \tag{5.3-1}$$

where $\mathbf{E}(\mathbf{r})$ and $\mathbf{H}(\mathbf{r})$ are the complex amplitudes of the electric and magnetic fields, respectively, $\omega = 2\pi\nu$ is the angular frequency, and ν is the frequency. The complex amplitudes $\mathbf{P}$, $\mathbf{D}$, and $\mathbf{B}$ of the real functions $\mathscr{P}$, $\mathscr{D}$, and $\mathscr{B}$ are similarly defined. The relations between these complex amplitudes that follow from Maxwell's equations and the medium equations will now be determined.

Maxwell's Equations

Substituting $\partial/\partial t = j\omega$ in Maxwell's equations (5.1-7) to (5.1-10), we obtain

$$\nabla \times \mathbf{H} = j\omega \mathbf{D} \tag{5.3-2}$$

$$\nabla \times \mathbf{E} = -j\omega \mathbf{B} \tag{5.3-3}$$

$$\nabla \cdot \mathbf{D} = 0 \tag{5.3-4}$$

$$\nabla \cdot \mathbf{B} = 0. \tag{5.3-5}$$

Maxwell's Equations (Source-Free Medium; Monochromatic Light)

Equations (5.1-11) and (5.1-13) similarly provide

$$\mathbf{D} = \epsilon_o \mathbf{E} + \mathbf{P} \tag{5.3-6}$$

$$\mathbf{B} = \mu_o \mathbf{H}. \tag{5.3-7}$$

Optical Intensity and Power

The flow of electromagnetic power is governed by the time average of the Poynting vector $\mathscr{S} = \mathscr{E} \times \mathscr{H}$. In terms of the complex amplitudes,

$$\mathscr{S} = \operatorname{Re}\{\mathbf{E}e^{j\omega t}\} \times \operatorname{Re}\{\mathbf{H}e^{j\omega t}\} = \tfrac{1}{2}(\mathbf{E}e^{j\omega t} + \mathbf{E}^* e^{-j\omega t}) \times \tfrac{1}{2}(\mathbf{H}e^{j\omega t} + \mathbf{H}^* e^{-j\omega t})$$

$$= \tfrac{1}{4}(\mathbf{E} \times \mathbf{H}^* + \mathbf{E}^* \times \mathbf{H} + \mathbf{E} \times \mathbf{H}e^{j2\omega t} + \mathbf{E}^* \times \mathbf{H}^* e^{-j2\omega t}).$$

The terms containing $e^{j2\omega t}$ and $e^{-j2\omega t}$ are washed out by the averaging process so that

$$\langle \mathscr{S} \rangle = \tfrac{1}{4}(\mathbf{E} \times \mathbf{H}^* + \mathbf{E}^* \times \mathbf{H}) = \tfrac{1}{2}(\mathbf{S} + \mathbf{S}^*) = \operatorname{Re}\{\mathbf{S}\}, \tag{5.3-8}$$

where

$$\mathbf{S} = \tfrac{1}{2}\mathbf{E} \times \mathbf{H}^* \tag{5.3-9}$$

is regarded as a "complex Poynting vector." The optical intensity is the magnitude of the vector $\operatorname{Re}\{\mathbf{S}\}$.

Linear, Nondispersive, Homogeneous, and Isotropic Media

With the medium equations

$$\mathbf{D} = \epsilon \mathbf{E} \quad \text{and} \quad \mathbf{B} = \mu_o \mathbf{H}, \tag{5.3-10}$$

Maxwell's equations, (5.3-2) to (5.3-5), become

$$\nabla \times \mathbf{H} = j\omega\epsilon \mathbf{E} \tag{5.3-11}$$

$$\nabla \times \mathbf{E} = -j\omega\mu_o \mathbf{H} \tag{5.3-12}$$

$$\nabla \cdot \mathbf{E} = 0 \tag{5.3-13}$$

$$\nabla \cdot \mathbf{H} = 0. \tag{5.3-14}$$

Maxwell's Equations (Monochromatic Light; Linear, Homogeneous, Isotropic, Nondispersive, Source-Free Medium)

Since the components of $\mathscr{E}$ and $\mathscr{H}$ satisfy the wave equation [with $c = c_o/n$ and $n = (\epsilon/\epsilon_o)^{1/2}$], the components of $\mathbf{E}$ and $\mathbf{H}$ must satisfy the Helmholtz equation

$$\nabla^2 U + k^2 U = 0, \qquad k = \omega(\epsilon\mu_o)^{1/2} = nk_o, \tag{5.3-15}$$

Helmholtz Equation

where the scalar function $U = U(\mathbf{r})$ represents any of the six components of the vectors $\mathbf{E}$ and $\mathbf{H}$, and $k_o = \omega/c_o$.

Inhomogeneous Media

In an inhomogeneous medium, Maxwell's equations (5.3-11) to (5.3-14) remain applicable, but $\epsilon = \epsilon(\mathbf{r})$ is now position dependent. For locally homogeneous media in which $\epsilon(\mathbf{r})$ varies slowly with respect to the wavelength, the Helmholtz equation (5.3-15) is approximately valid with $k = n(\mathbf{r})k_o$ and $n(\mathbf{r}) = [\epsilon(\mathbf{r})/\epsilon_o]^{1/2}$.

Dispersive Media

In a dispersive medium $\mathscr{P}(t)$ and $\mathscr{E}(t)$ are related by the dynamic relation in (5.2-17). To determine the corresponding relation between the complex amplitudes $\mathbf{P}$ and $\mathbf{E}$, we substitute (5.3-1) into (5.2-17) and equate the coefficients of $e^{j\omega t}$. The result is

$$\boxed{\mathbf{P} = \epsilon_o \chi(\nu)\mathbf{E},} \tag{5.3-16}$$

where

$$\chi(\nu) = \int_{-\infty}^{\infty} x(t) \exp(-j2\pi\nu t)\, dt \tag{5.3-17}$$

is the Fourier transform of $x(t)$. This can also be seen if we invoke the convolution theorem (convolution in the time domain is equivalent to multiplication in the frequency domain; see Secs. A.1 and B.1 of Appendices A and B), and recognize $\mathbf{E}$ and $\mathbf{P}$ as the components of $\mathscr{E}$ and $\mathscr{P}$ of frequency ν. The function $\epsilon_o\chi(\nu)$ may be regarded as the transfer function of the linear system that relates $\mathscr{P}(t)$ to $\mathscr{E}(t)$.

The relation between $\mathscr{D}$ and $\mathscr{E}$ is similar,

$$\mathbf{D} = \epsilon(\nu)\mathbf{E}, \tag{5.3-18}$$

where

$$\epsilon(\nu) = \epsilon_o[1 + \chi(\nu)]. \tag{5.3-19}$$

The only difference between the idealized nondispersive medium and the dispersive medium is that in the latter the susceptibility χ and the permittivity ϵ are frequency dependent. The Helmholtz equation (5.3-15) is applicable to dispersive media with the wavenumber

$$k = \omega\left[\epsilon(\nu)\mu_o\right]^{1/2} = n(\nu)k_o, \tag{5.3-20}$$

where the refractive index $n(\nu) = [\epsilon(\nu)/\epsilon_o]^{1/2}$ is now frequency dependent. If $\chi(\nu)$, $\epsilon(\nu)$, and $n(\nu)$ are approximately constant within the frequency band of interest, the medium may be treated as approximately nondispersive. Dispersive media are discussed further in Sec. 5.5.

5.4 ELEMENTARY ELECTROMAGNETIC WAVES

A. Plane, Spherical, and Gaussian Electromagnetic Waves

Three important examples of monochromatic electromagnetic waves are introduced in this section—the plane wave, the spherical wave, and the Gaussian beam. The medium is assumed linear, homogeneous, and isotropic.

The Transverse Electromagnetic (TEM) Plane Wave

Consider a monochromatic electromagnetic wave whose electric and magnetic field components are plane waves of wavevector **k** (see Sec. 2.2B), so that

$$\mathbf{E}(\mathbf{r}) = \mathbf{E}_0 \exp(-j\mathbf{k}\cdot\mathbf{r}) \tag{5.4-1}$$

$$\mathbf{H}(\mathbf{r}) = \mathbf{H}_0 \exp(-j\mathbf{k}\cdot\mathbf{r}), \tag{5.4-2}$$

where $\mathbf{E}_0$ and $\mathbf{H}_0$ are constant vectors. Each of these components satisfies the Helmholtz equation if the magnitude of **k** is $k = nk_o$, where n is the refractive index of the medium.

We now examine the conditions $\mathbf{E}_0$ and $\mathbf{H}_0$ must satisfy so that Maxwell's equations are satisfied. Substituting (5.4-1) and (5.4-2) into Maxwell's equations (5.3-11) and (5.3-12), we obtain

$$\mathbf{k} \times \mathbf{H}_0 = -\omega\epsilon\mathbf{E}_0 \tag{5.4-3}$$

$$\mathbf{k} \times \mathbf{E}_0 = \omega\mu_o\mathbf{H}_0. \tag{5.4-4}$$

The other two Maxwell's equations are satisfied identically since the divergence of a uniform plane wave is zero.

It follows from (5.4-3) that **E** is normal to both **k** and **H**. Equation (5.4-4) similarly implies that **H** is normal to both **k** and **E**. Thus **E**, **H**, and **k** must by mutually orthogonal (Fig. 5.4-1). Since **E** and **H** lie in a plane normal to the direction of propagation **k**, the wave is called a **transverse electromagnetic (TEM)** wave.

In accordance with (5.4-3) the magnitudes H_0 and E_0 are related by $H_0 = (\omega\epsilon/k)E_0$. Similarly, (5.4-4) yields $H_0 = (k/\omega\mu_o)E_0$. For these two equations to be consistent $\omega\epsilon/k = k/\omega\mu_o$, or $k = \omega(\epsilon\mu_o)^{1/2} = \omega/c = n\omega/c_o = nk_o$. This is, in fact, the condition for the wave to satisfy the Helmholtz equation. The ratio between the amplitudes of the electric and magnetic fields is therefore $E_0/H_0 = \omega\mu_o/k = \mu_o c_o/n = (\mu_o/\epsilon_o)^{1/2}/n$, or

$$\frac{E_0}{H_0} = \eta, \tag{5.4-5}$$

where

$$\boxed{\eta = \frac{\eta_o}{n}} \tag{5.4-6}$$

Impedance of the Medium

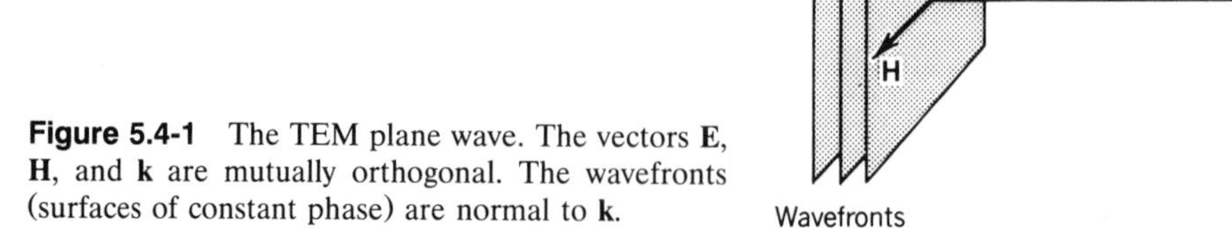

Figure 5.4-1 The TEM plane wave. The vectors **E**, **H**, and **k** are mutually orthogonal. The wavefronts (surfaces of constant phase) are normal to **k**.

is known as the **impedance** of the medium and

$$\eta_o = \left(\frac{\mu_o}{\epsilon_o}\right)^{1/2} \approx 120\pi \approx 377\ \Omega \tag{5.4-7}$$

Impedance of Free Space

is the impedance of free space.

The complex Poynting vector $\mathbf{S} = \frac{1}{2}\mathbf{E} \times \mathbf{H}^*$ is parallel to the wavevector $\mathbf{k}$, so that the power flows along a direction normal to the wavefronts. The magnitude of the Poynting vector $\mathbf{S}$ is $\frac{1}{2}E_0 H_0^* = |E_0|^2/2\eta$, so that the intensity is

$$I = \frac{|E_0|^2}{2\eta}. \tag{5.4-8}$$

Intensity

The intensity of the TEM wave is therefore proportional to the squared absolute value of the complex envelope of the electric field. For example, an intensity of 10 W/cm^2 in free space corresponds to an electric field of $\approx$ 87 V/cm. Note the similarity between (5.4-8) and the relation $I = |U|^2$, which is applicable to scalar waves (Sec. 2.2A).

The Spherical Wave

An example of an electromagnetic wave with features resembling the scalar spherical wave discussed in Sec. 2.2B is the field radiated by an oscillating electric dipole. This wave is constructed from an auxiliary vector field

$$\mathbf{A}(\mathbf{r}) = A_0 U(\mathbf{r})\hat{\mathbf{x}}, \tag{5.4-9}$$

where

$$U(\mathbf{r}) = \frac{1}{r}\exp(-jkr) \tag{5.4-10}$$

represents a scalar spherical wave originating at $r = 0$, $\hat{\mathbf{x}}$ is a unit vector in the x direction, and A_0 is a constant. Because $U(\mathbf{r})$ satisfies the Helmholtz equation (as we know from scalar wave optics), $\mathbf{A}(\mathbf{r})$ also satisfies the Helmholtz equation, $\nabla^2\mathbf{A} + k^2\mathbf{A} = 0$.

We now define the magnetic field

$$\mathbf{H} = \frac{1}{\mu_o}\nabla \times \mathbf{A} \tag{5.4-11}$$

and determine the corresponding electric field by using Maxwell's equation (5.3-11),

$$\mathbf{E} = \frac{1}{j\omega\epsilon}\nabla \times \mathbf{H}. \tag{5.4-12}$$

These fields satisfy the other three Maxwell's equations. The form of (5.4-11) and (5.4-12) ensures that $\nabla \cdot \mathbf{H} = 0$ and $\nabla \cdot \mathbf{E} = 0$, since the divergence of the curl of any vector field vanishes. Because $\mathbf{A}(\mathbf{r})$ satisfies the Helmholtz equation, it can be shown that the remaining Maxwell's equation ($\nabla \times \mathbf{E} = -j\omega\mu_o \mathbf{H}$) is also satisfied. Thus (5.4-9) to (5.4-12) define a valid electromagnetic wave. The vector $\mathbf{A}$ is known in electromagnetic theory as the **vector potential**. Its introduction often facilitates the solution of Maxwell's equation.

To obtain explicit expressions for $\mathbf{E}$ and $\mathbf{H}$ the curl operations in (5.4-11) and (5.4-12) must be evaluated. This can be conveniently accomplished by use of the spherical coordinates (r, θ, ϕ) defined in Fig. 5.4-2(a). For points at distances from the origin much greater than a wavelength ($r \gg \lambda$, or $kr \gg 2\pi$), these expressions are approximated by

$$\mathbf{E}(\mathbf{r}) \approx E_0 \sin\theta \, U(\mathbf{r}) \, \hat{\boldsymbol{\theta}} \tag{5.4-13}$$

$$\mathbf{H}(\mathbf{r}) \approx H_0 \sin\theta \, U(\mathbf{r}) \, \hat{\boldsymbol{\phi}}, \tag{5.4-14}$$

where $E_0 = (jk/\mu_o)A_0$, $H_0 = E_0/\eta$, $\theta = \cos^{-1}(x/r)$, and $\hat{\boldsymbol{\theta}}$ and $\hat{\boldsymbol{\phi}}$ are unit vectors in spherical coordinates. Thus the wavefronts are spherical and the electric and magnetic fields are orthogonal to one another and to the radial direction $\hat{\mathbf{r}}$, as illustrated in Fig. 5.4-2(b). However, unlike the scalar spherical wave, the magnitude of this vector wave varies as $\sin\theta$. At points near the z axis and far from the origin, $\theta \approx \pi/2$ and $\phi \approx \pi/2$, so that the wavefront normals are almost parallel to the z axis (corresponding to paraxial rays) and $\sin\theta \approx 1$.

In a Cartesian coordinate system $\hat{\boldsymbol{\theta}} = -\sin\theta \, \hat{\mathbf{x}} + \cos\theta\cos\phi \, \hat{\mathbf{y}} + \cos\theta\sin\phi \, \hat{\mathbf{z}} \approx -\hat{\mathbf{x}} + (x/z)(y/z)\hat{\mathbf{y}} + (x/z)\hat{\mathbf{z}} \approx -\hat{\mathbf{x}} + (x/z)\hat{\mathbf{z}}$, so that

$$\mathbf{E}(\mathbf{r}) \approx E_0\left(-\hat{\mathbf{x}} + \frac{x}{z}\hat{\mathbf{z}}\right)U(\mathbf{r}), \tag{5.4-15}$$

where $U(\mathbf{r})$ is the paraxial approximation of the spherical wave (the paraboloidal wave

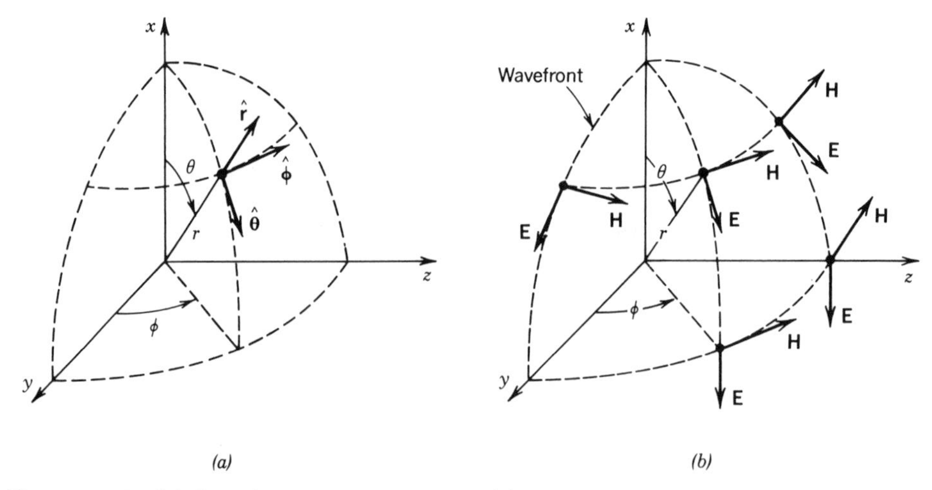

Figure 5.4-2 (a) Spherical coordinate system. (b) Electric and magnetic field vectors and wavefronts of the electromagnetic field radiated by an oscillating electric dipole at distances $r \gg \lambda$.

discussed in Sec. 2.2B). For very large z, the term (x/z) in (5.4-15) may also be neglected, so that

$$\mathbf{E}(\mathbf{r}) \approx -E_0 U(r)\,\hat{\mathbf{x}} \tag{5.4-16}$$

$$\mathbf{H}(\mathbf{r}) \approx H_0 U(\mathbf{r})\,\hat{\mathbf{y}}. \tag{5.4-17}$$

Under this approximation $U(\mathbf{r})$ approaches a plane wave $(1/z)e^{-jkz}$, so that we ultimately have a TEM plane wave.

The Gaussian Beam

As discussed in Sec. 3.1, a scalar Gaussian beam is obtained from a paraboloidal wave (the paraxial approximation to the spherical wave) by replacing the coordinate z with $z + jz_0$, where z_0 is a real constant. The same transformation can be applied to the electromagnetic spherical wave. Replacing z in (5.4-15) with $z + jz_0$, we obtain

$$\mathbf{E}(\mathbf{r}) = E_0\left(-\hat{\mathbf{x}} + \frac{x}{z + jz_0}\hat{\mathbf{z}}\right)U(\mathbf{r}), \tag{5.4-18}$$

where $U(\mathbf{r})$ now represents the scalar complex amplitude of a Gaussian beam [given by (3.1-7)]. Figure 5.4-3 illustrates the wavefronts of the Gaussian beam and the **E**-field lines determined from (5.4-18).

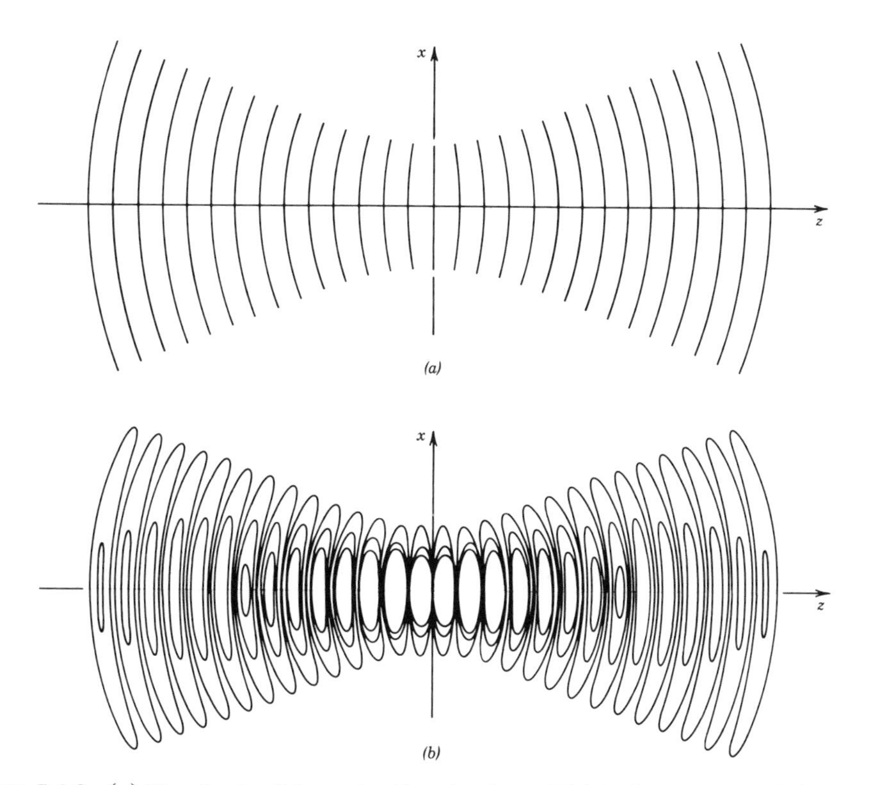

Figure 5.4-3 (*a*) Wavefronts of the scalar Gaussian beam $U(\mathbf{r})$ in the x–z plane. (*b*) Electric field lines of the electromagnetic Gaussian beam in the x–z plane. (After H. A. Haus, *Waves and Fields in Optoelectronics*, Prentice-Hall, Englewood Cliffs, NJ, 1984.)

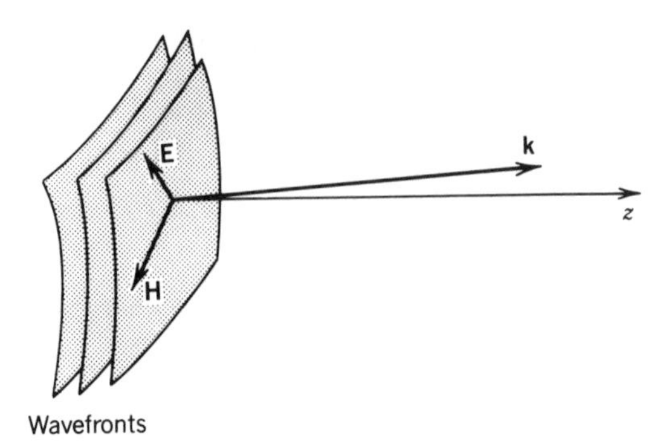

Figure 5.4-4 Paraxial electromagnetic wave.

B. Relation Between Electromagnetic Optics and Scalar Wave Optics

A paraxial scalar wave is a wave whose wavefront normals make small angles with the optical axis (see Sec. 2.2C). The wave behaves locally as a plane wave with the complex envelope and the direction of propagation varying slowly with the position.

The same idea is applicable to electromagnetic waves in isotropic media. A paraxial electromagnetic wave is locally approximated by a TEM plane wave. At each point, the vectors **E** and **H** lie in a plane tangential to the wavefront surfaces, i.e., normal to the wavevector **k** (Fig. 5.4-4). The optical power flows along the direction $\mathbf{E} \times \mathbf{H}$, which is parallel to **k** and approximately parallel to the optical axis; the intensity $I \approx |E|^2/2\eta$.

A scalar wave of complex amplitude $U = E/(2\eta)^{1/2}$ may be associated with the paraxial electromagnetic wave so that the two waves have the same intensity $I = |U|^2 = |E|^2/2\eta$ and the same wavefronts. The scalar description of light is an adequate approximation for solving problems of interference, diffraction, and propagation of paraxial waves, when polarization is not a factor. Take, for example, a Gaussian beam with very small divergence angle. Most questions regarding the intensity, focusing by lenses, reflection from mirrors, or interference may be addressed satisfactorily by use of the scalar theory (wave optics).

Note, however, that U and E do not satisfy the same boundary conditions. For example, if the electric field is tangential to the boundary between two dielectric media, E is continuous, but $U = E/(2\eta)^{1/2}$ is discontinuous since $\eta = \eta_o/n$ changes at the boundary. Problems involving reflection and refraction at dielectric boundaries cannot be addressed completely within the scalar wave theory. Similarly, problems involving the transmission of light through dielectric waveguides require an analysis based on the rigorous electromagnetic theory, as discussed in Chap. 7.

5.5 ABSORPTION AND DISPERSION

A. Absorption

The dielectric media discussed so far have been assumed to be totally transparent, i.e., not to absorb light. Glass is approximately transparent in the visible region of the optical spectrum, but it absorbs ultraviolet and infrared light. In those bands optical components are generally made of other materials (e.g., quartz and magnesium fluoride in the ultraviolet, and calcium fluoride and germanium in the infrared). Figure 5.5-1 shows the spectral windows within which selected materials are transparent.

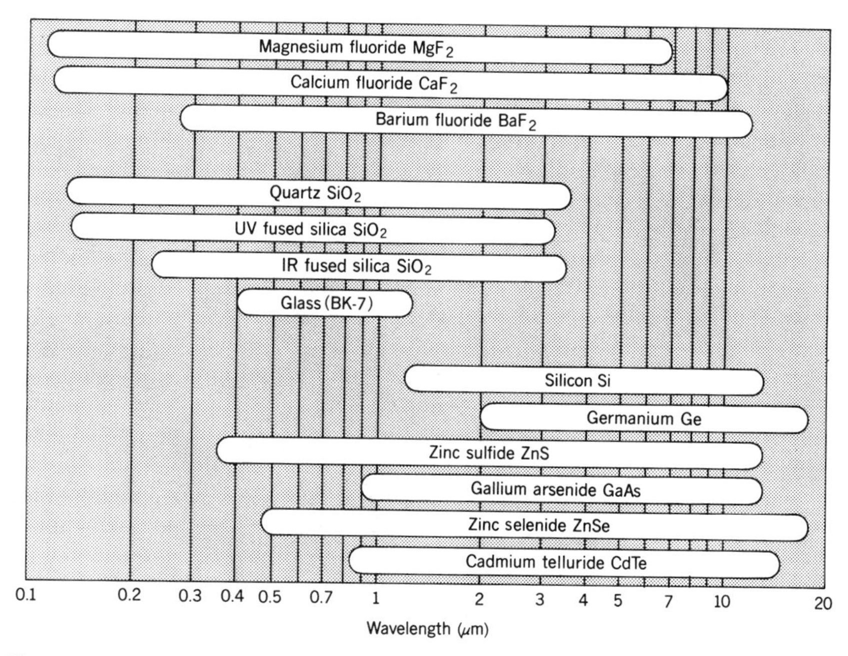

Figure 5.5-1 The spectral bands within which selected optical materials transmit light.

Dielectric materials that absorb light are often represented phenomenologically by a complex susceptibility,

$$\chi = \chi' + j\chi'', \tag{5.5-1}$$

corresponding to a complex permittivity $\epsilon = \epsilon_o(1 + \chi)$. The Helmholtz equation, $\nabla^2 U + k^2 U = 0$, remains applicable, but

$$k = \omega(\epsilon\mu_o)^{1/2} = (1 + \chi)^{1/2} k_o = (1 + \chi' + j\chi'')^{1/2} k_o \tag{5.5-2}$$

is now complex-valued ($k_o = \omega / c_o$ is the wavenumber in free space). A plane wave traveling in this medium in the z-direction is described by the complex amplitude $U = A \exp(-jkz)$. Since k is complex, both the magnitude and phase of U vary with z.

It is useful to write k in terms of its real and imaginary parts, $k = \beta - j\frac{1}{2}\alpha$, where β and α are real. Using (5.5-2), we obtain

$$\beta - j\tfrac{1}{2}\alpha = k_o(1 + \chi' + j\chi'')^{1/2}. \tag{5.5-3}$$

Equation (5.5-3) relates β and α to the susceptibility components χ' and χ''. Since $\exp(-jkz) = \exp(-\frac{1}{2}\alpha z)\exp(-j\beta z)$, the intensity of the wave is attenuated by the factor $|\exp(-jkz)|^2 = \exp(-\alpha z)$, so that the coefficient α represents the **absorption coefficient** (also called the **attenuation coefficient** or the **extinction coefficient**). We shall see in Chap. 13 that in certain media used in lasers, α is negative so that the medium amplifies instead of attenuates light.

Since the parameter β is the rate at which the phase changes with z, it is the propagation constant. The medium therefore has an effective refractive index n

defined by

$$\beta = nk_o, \tag{5.5-4}$$

and the wave travels with a phase velocity $c = c_o/n$.

Substituting (5.5-4) into (5.5-3) we obtain an equation relating the refractive index n and the absorption coefficient α to the real and imaginary parts of the susceptibility χ' and χ'',

$$n - j\frac{\alpha}{2k_o} = (1 + \chi' + j\chi'')^{1/2}. \tag{5.5-5}$$

Absorption Coefficient and Refractive Index

Weakly Absorbing Media

In a medium for which $\chi' \ll 1$ and $\chi'' \ll 1$ (a weakly absorbing gas, for example), $(1 + \chi' + j\chi'')^{1/2} \approx 1 + \frac{1}{2}(\chi' + j\chi'')$, so that (5.5-5) yields

$$n \approx 1 + \tfrac{1}{2}\chi' \tag{5.5-6}$$

$$\alpha \approx -k_o\chi''. \tag{5.5-7}$$

Weakly Absorptive Medium

The refractive index is then linearly related to the real part of the susceptibility, whereas the absorption coefficient is proportional to the imaginary part. For an absorptive medium χ'' is negative and α is positive. For an amplifying medium χ'' is positive and α is negative.

EXERCISE 5.5-1

Weakly Absorbing Medium. A nonabsorptive medium of refractive index n_0 is host to impurities with susceptibility $\chi = \chi' + j\chi''$, where $\chi' \ll 1$ and $\chi'' \ll 1$. Determine the total susceptibility and show that the refractive index and absorption coefficient are given approximately by

$$n \approx n_0 + \frac{\chi'}{2n_0} \tag{5.5-8}$$

$$\alpha \approx -\frac{k_o\chi''}{n_0}. \tag{5.5-9}$$

B. Dispersion

Dispersive media are characterized by a frequency-dependent (and therefore wavelength-dependent) susceptibility $\chi(\nu)$, refractive index $n(\nu)$, and speed of light $c(\nu) = c_o/n(\nu)$. The wavelength dependence of the refractive index of selected materials is shown in Fig. 5.5-2.

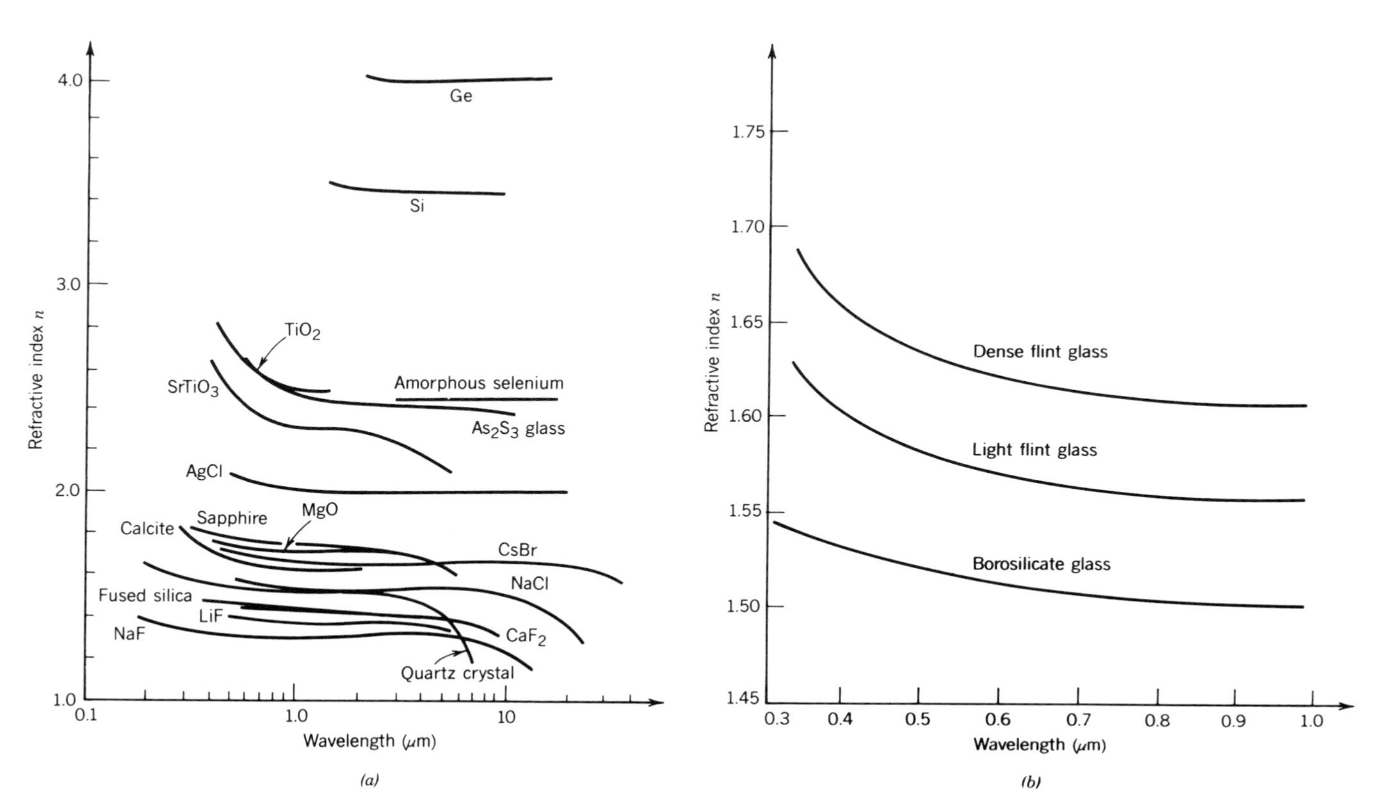

Figure 5.5-2 Wavelength dependence of the refractive index of (*a*) selected crystalline solids (from W. L. Wolfe and G. J. Zissis, Eds., *The Infrared Handbook*, Environmental Research Institute of Michigan, Ann Arbor, MI, 1978); (*b*) selected glasses (from W. D. Kingery, H. K. Bowen, and D. R. Uhlmann, *Introduction to Ceramics*, Wiley, New York, 1976).

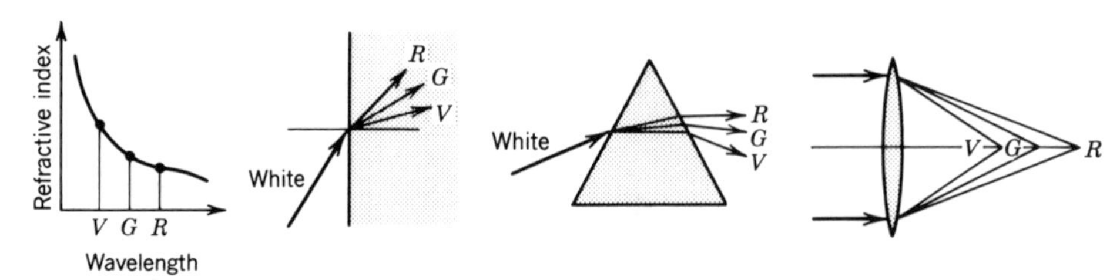

Figure 5.5-3 Optical components made of dispersive media refract waves of different wavelengths (e.g., V = violet, G = green, and R = red) by different angles.

Optical components such as prisms and lenses made of dispersive materials refract the waves of different wavelengths by different angles, thus dispersing polychromatic light, which comprises different wavelengths, into different directions. This accounts for the wavelength-resolving power of refracting surfaces and for the wavelength-dependent focusing powers of lenses, which is responsible for chromatic aberration in imaging systems. These effects are illustrated schematically in Fig. 5.5-3.

Since the speed of light in the dispersive medium is frequency dependent, each of the frequency components that constitute a short pulse of light undergoes a different time delay. If the distance of propagation through the medium is long (as in the case of light transmission through optical fibers), the pulse is dispersed in time and its width broadens, as illustrated in Fig. 5.5-4.

Measures of Dispersion

There are several measures of material dispersion. Dispersion in the glass optical components used with white light (light with a broad spectrum covering the visible band) is usually measured by the $\mathcal{V}$-number $\mathcal{V} = (n_D - 1)/(n_F - n_C)$, where n_F, n_D, and n_C are the refractive indices at three standard wavelengths (blue 486.1 nm, yellow 589.2 nm, and red 656.3 nm, respectively). For flint glass $\mathcal{V} \approx 38$, and for fused silica $\mathcal{V} \approx 68$.

One measure of dispersion near a specified wavelength λ_o is the magnitude of the derivative $dn/d\lambda_o$ at this wavelength. This measure is appropriate for prisms. Since the ray deflection angle θ_d in the prism is a function of n [see (1.2-6)], the angular dispersion $d\theta_d/d\lambda_o = (d\theta_d/dn)(dn/d\lambda_o)$ is a product of the material dispersion factor $dn/d\lambda_o$ and a factor $d\theta_d/dn$, which depends on the geometry and refractive index.

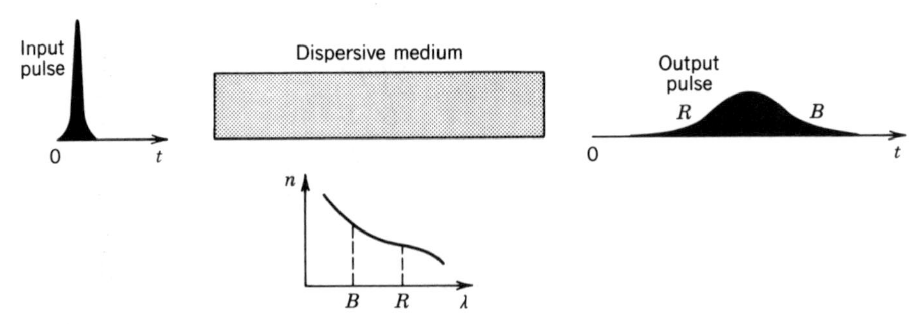

Figure 5.5-4 A dispersive medium broadens a pulse of light because the different frequency components that constitute the pulse travel at different velocities. In this illustration, the low-frequency component (long wavelength, denoted R) travels faster than the high-frequency component (short wavelength, denoted B) and arrives earlier.

The first and second derivatives $dn/d\lambda_o$ and $d^2n/d\lambda_o^2$ govern the effect of material dispersion on pulse propagation. It will be shown in Sec. 5.6 that a pulse of light of free-space wavelength λ_o travels with a velocity $v = c_o/N$, called the **group velocity**, where $N = n - \lambda_o\, dn/d\lambda_o$ is called the **group index**. As a result of the dependence of the group velocity itself on the wavelength, the pulse is broadened at a rate $|D_\lambda|\sigma_\lambda$ seconds per unit distance, where σ_λ is the spectral width of the light, and $D_\lambda = -(\lambda_o/c_o)\, d^2n/d\lambda_o^2$ is called the **dispersion coefficient**. For applications of pulse propagation in optical fibers D_λ is often measured in units of ps/km-nm (picoseconds of temporal spread per kilometer of optical fiber length per nanometer of spectral width; see Sec. 8.3B).

Absorption and Dispersion: The Kramers–Kronig Relations

Dispersion and absorption are intimately related. A dispersive material (with wavelength-dependent refractive index) must also be absorptive and the absorption coefficient must be wavelength dependent. This relation between the absorption coefficient and the refractive index has its origin in underlying relations between the real and imaginary parts of the susceptibility, $\chi'(\nu)$ and $\chi''(\nu)$, called the Kramers–Kronig relations:

$$\chi'(\nu) = \frac{2}{\pi}\int_0^\infty \frac{s\chi''(s)}{s^2 - \nu^2}\, ds \qquad (5.5\text{-}10)$$

$$\chi''(\nu) = \frac{2}{\pi}\int_0^\infty \frac{\nu\chi'(s)}{\nu^2 - s^2}\, ds. \qquad (5.5\text{-}11)$$

Kramers–Kronig Relations

These relations permit us to determine either the real or the imaginary component of the susceptibility, if the other is known for all ν. As a consequence of (5.5-5), the refractive index $n(\nu)$ is also related to the absorption coefficient $\alpha(\nu)$, so that if one is known for all ν, the other may be determined.

The Kramers–Kronig relations may be derived using a system's approach (see Appendix B, Sec. B.1). The system that relates the polarization density $\mathscr{P}(t)$ to the applied electric field $\mathscr{E}(t)$ is a linear shift-invariant system with transfer function $\epsilon_o\chi(\nu)$. Since $\mathscr{E}(t)$ and $\mathscr{P}(t)$ are real, $\chi(\nu)$ must be symmetric, $\chi(-\nu) = \chi^*(\nu)$. Since the system is causal (as all physical systems are), the real and imaginary parts of the transfer function $\epsilon_o\chi(\nu)$ must be related by the Kramers–Kronig relations (B.1-6) and (B.1-7), from which (5.5-10) and (5.5-11) follow.

C. The Resonant Medium

Consider a dielectric medium for which the dynamic relation between the polarization density and the electric field is described by the linear second-order differential equation

$$\frac{d^2\mathscr{P}}{dt^2} + \sigma\frac{d\mathscr{P}}{dt} + \omega_0^2\mathscr{P} = \omega_0^2\epsilon_o\chi_0\mathscr{E}, \qquad (5.5\text{-}12)$$

Resonant Dielectric Medium

where σ, ω_0, and χ_0 are constants.

This relation arises when the motion of each bound charge in the medium is modeled phenomenologically by a classical harmonic oscillator, with the displacement x and the applied force $\mathscr{F}$ related by a linear second-order differential equation,

$$\frac{d^2x}{dt^2} + \sigma \frac{dx}{dt} + \omega_0^2 x = \frac{\mathscr{F}}{m}. \tag{5.5-13}$$

Here m is the mass of the bound charge, $\omega_0 = (\kappa/m)^{1/2}$ is its resonance angular frequency, κ is the elastic constant, and σ is the damping coefficient. The force $\mathscr{F} = e\mathscr{E}$, and the polarization density $\mathscr{P} = Nex$, where e is the electron charge and N is the number of charges per unit volume. Therefore $\mathscr{P}$ and $\mathscr{E}$ are, respectively, proportional to x and $\mathscr{F}$, so that (5.5-13) yields (5.5-12) with $\chi_0 = e^2N/m\epsilon_o\omega_0^2$.

The dielectric medium is completely characterized by its response to harmonic (monochromatic) fields. Substituting $\mathscr{E}(t) = \mathrm{Re}\{E \exp(j\omega t)\}$ and $\mathscr{P}(t) = \mathrm{Re}\{P \exp(j\omega t)\}$ into (5.5-12) and equating coefficients of $\exp(j\omega t)$, we obtain

$$\left(-\omega^2 + j\sigma\omega + \omega_0^2\right)P = \omega_0^2 \epsilon_o \chi_0 E, \tag{5.5-14}$$

from which $P = \epsilon_o[\chi_0\omega_0^2/(\omega_0^2 - \omega^2 + j\sigma\omega)]E$. We write this relation in the form $P = \epsilon_o\chi(\nu)E$ and substitute $\omega = 2\pi\nu$ to obtain an expression for the frequency-dependent susceptibility,

$$\chi(\nu) = \chi_0 \frac{\nu_0^2}{\nu_0^2 - \nu^2 + j\nu\,\Delta\nu}, \tag{5.5-15}$$

Susceptibility of a Resonant Medium

where $\nu_0 = \omega_0/2\pi$ is the resonance frequency, and $\Delta\nu = \sigma/2\pi$.

The real and imaginary parts of $\chi(\nu)$,

$$\chi'(\nu) = \chi_0 \frac{\nu_0^2(\nu_0^2 - \nu^2)}{\left(\nu_0^2 - \nu^2\right)^2 + (\nu\,\Delta\nu)^2} \tag{5.5-16}$$

$$\chi''(\nu) = -\chi_0 \frac{\nu_0^2 \nu\,\Delta\nu}{\left(\nu_0^2 - \nu^2\right)^2 + (\nu\,\Delta\nu)^2} \tag{5.5-17}$$

are plotted in Fig. 5.5-5. At frequencies well below resonance ($\nu \ll \nu_0$), $\chi'(\nu) \approx \chi_0$ and $\chi''(\nu) \approx 0$, so that χ_0 is the low-frequency susceptibility. At frequencies well above resonance ($\nu \gg \nu_0$), $\chi'(\nu) \approx \chi''(\nu) \approx 0$ and the medium acts like free space. At resonance ($\nu = \nu_0$), $\chi'(\nu_0) = 0$ and $-\chi''(\nu)$ reaches its peak value of $(\nu_0/\Delta\nu)\chi_0$. Usually, ν_0 is much greater than $\Delta\nu$ so that the peak value of $-\chi''(\nu)$ is much greater than the low-frequency value χ_0.

We are often interested in the behavior of $\chi(\nu)$ near resonance, where $\nu \approx \nu_0$. We may then use the approximation $(\nu_0^2 - \nu^2) = (\nu_0 + \nu)(\nu_0 - \nu) \approx 2\nu_0(\nu_0 - \nu)$ in the real part of the denominator of (5.5-15), and replace ν with ν_0 in the imaginary part to obtain

$$\chi(\nu) \approx \chi_0 \frac{\nu_0/2}{(\nu_0 - \nu) + j\,\Delta\nu/2}, \tag{5.5-18}$$

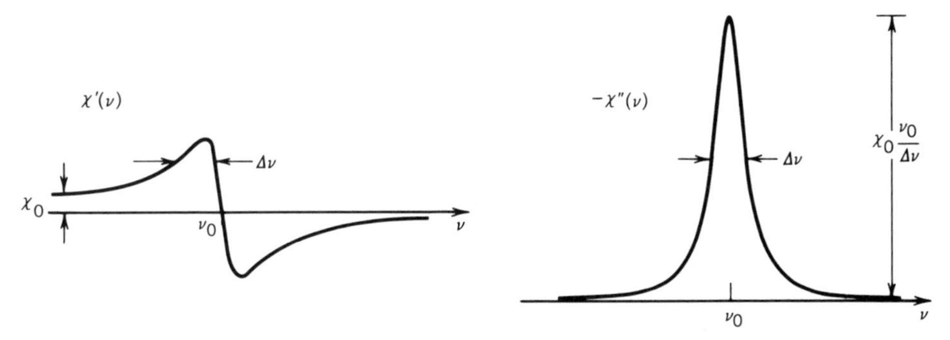

Figure 5.5-5 The real and imaginary parts of the susceptibility of a resonant dielectric medium. The real part $\chi'(\nu)$ is positive below resonance, zero at resonance, and negative above resonance. The imaginary part $\chi''(\nu)$ is negative so that $-\chi''(\nu)$ is positive and has a peak value $(\nu_0/\Delta\nu)\chi_0$ at $\nu = \nu_0$.

from which

$$\chi''(\nu) = -\chi_0 \frac{\nu_0 \, \Delta\nu}{4} \frac{1}{(\nu_0 - \nu)^2 + (\Delta\nu/2)^2} \tag{5.5-19}$$

$$\chi'(\nu) = 2\frac{\nu - \nu_0}{\Delta\nu}\chi''(\nu). \tag{5.5-20}$$

Susceptibility Near Resonance

In accordance with (5.5-19), $\chi''(\nu)$ drops to one-half its peak value when $|\nu - \nu_0| = \Delta\nu/2$. The parameter $\Delta\nu$ therefore represents the full-width half-maximum (FWHM) value of $\chi''(\nu)$.

If the resonant atoms are placed in a host medium of refractive index n_0, and if they are sufficiently dilute so that $\chi'(\nu)$ and $\chi''(\nu)$ are small, the overall absorption coefficient and refractive index are

$$n(\nu) \approx n_0 + \frac{\chi'(\nu)}{2n_0} \tag{5.5-21}$$

$$\alpha(\nu) \approx -\left(\frac{2\pi\nu}{n_0 c_o}\right)\chi''(\nu) \tag{5.5-22}$$

(see Exercise 5.5-1). The dependence of these coefficients on ν is illustrated in Fig. 5.5-6.

Media with Multiple Resonances

A typical dielectric medium contains multiple resonances, corresponding to different lattice and electronic vibrations. The overall susceptibility is the sum of contributions from these resonances. Whereas the imaginary part of the susceptibility is confined near the resonance frequency, the real part contributes at *all* frequencies near and *below* resonance, as Fig. 5.5-5 illustrates. This is exhibited in the frequency dependence of the absorption coefficient and the refractive index, as illustrated in Fig. 5.5-7. Absorption and dispersion are strongest near the resonance frequencies. Away from the resonance frequencies, the refractive index is constant and the medium is approxi-

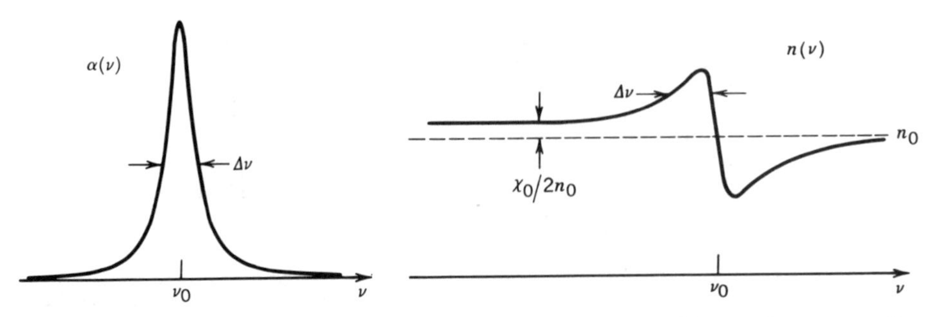

Figure 5.5-6 The absorption coefficient $\alpha(\nu)$ and refractive index $n(\nu)$ of a dielectric medium of refractive index n_0 with dilute atoms of resonance frequency ν_0.

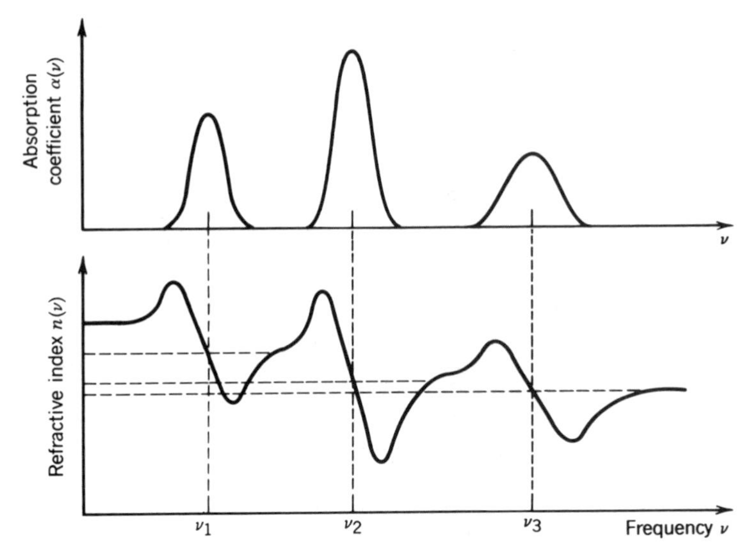

Figure 5.5-7 Absorption coefficient $\alpha(\nu)$ and refractive index $n(\nu)$ of a medium with three resonances.

mately nondispersive and nonabsorptive. Each resonance contributes a constant value to the refractive index at all frequencies smaller than its resonance frequency.

Other complex processes can also contribute to the absorption coefficient and the refractive index, so that different patterns of wavelength dependence can be exhibited. Figure 5.5-8 shows an example of the wavelength dependence of the absorption coefficient and refractive index for a dielectric material that is essentially transparent to light at visible wavelengths. In the visible band, the refractive index varies slightly because of proximity to ultraviolet absorption. In this band the refractive index is a decreasing function of wavelength. The rate of decrease is greater at shorter wavelengths, so that the material is more dispersive at short wavelengths.

5.6 PULSE PROPAGATION IN DISPERSIVE MEDIA

The study of pulse propagation in dispersive media is important in many applications, including the transmission of optical pulses through the glass fibers used in optical communication systems (as will become clear in Chaps. 8 and 22). The dispersive

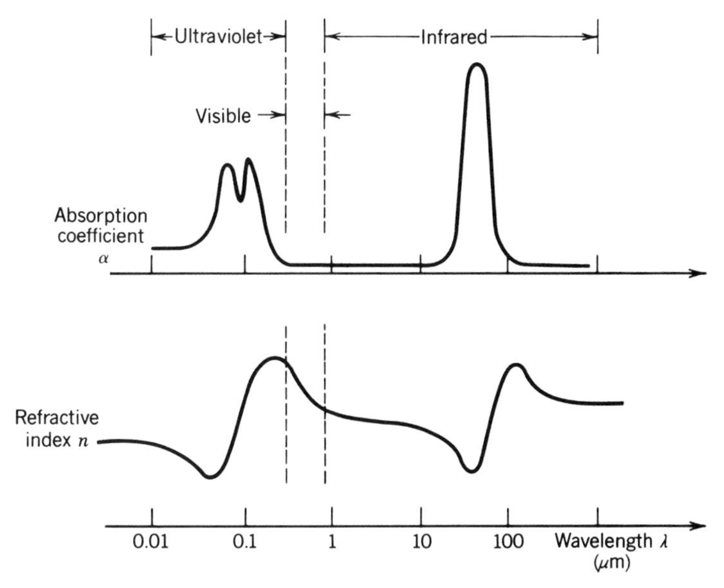

Figure 5.5-8 Typical wavelength dependence of the absorption coefficient and the refractive index of a dielectric material exhibiting resonance absorptions in the ultraviolet and infrared bands and low absorption in the visible band.

medium is characterized by a frequency-dependent refractive index, absorption coefficient, and phase velocity, so that monochromatic waves of different frequencies travel in the medium at different velocities and undergo different attenuations. Since a pulse of light is the sum of many monochromatic waves, each of which is modified differently, the pulse is delayed and broadened (dispersed in time) and its shape is altered. In this section we determine the velocity of a pulse, the rate at which it spreads in time, and the changes in its shape, as it travels through a dispersive medium.

Consider a pulsed plane wave traveling in the z direction in a linear, homogeneous, and isotropic medium with absorption coefficient $\alpha(\nu)$, refractive index $n(\nu)$, and propagation constant $\beta(\nu) = 2\pi\nu n(\nu)/c_o$. The complex wavefunction is

$$U(z,t) = \mathscr{A}(z,t)\exp[j(2\pi\nu_0 t - \beta_0 z)], \tag{5.6-1}$$

where ν_0 is the central frequency, $\beta_0 = \beta(\nu_0)$ is the central wavenumber, and $\mathscr{A}(z,t)$ is the complex envelope of the pulse, assumed to be slowly varying in comparison with the central frequency ν_0 (Fig. 5.6-1). The complex envelope $\mathscr{A}(0,t)$ in the plane $z = 0$ is assumed to be a known function, and we wish to determine $\mathscr{A}(z,t)$ at a distance z in the medium.

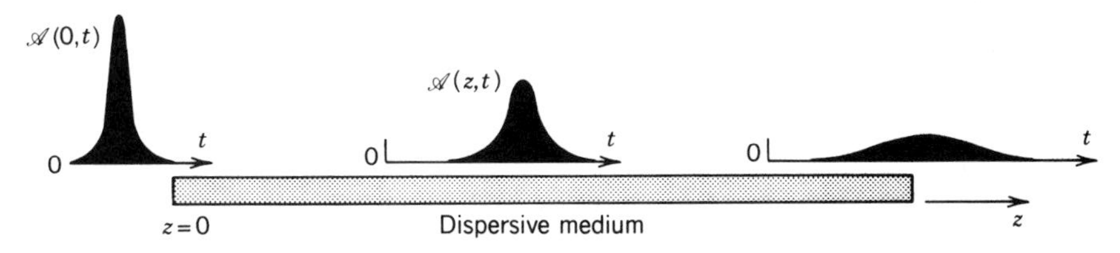

Figure 5.6-1 Broadening of the complex envelope of a pulse as a result of propagation in a dispersive medium.

Linear-System Description

The incident pulse $\mathscr{A}(0,t)$ and the transmitted pulse $\mathscr{A}(z,t)$ may be regarded as the input and output of a linear system using the techniques described in Appendix B, Sec. B.1. We aim at developing a procedure for determining $\mathscr{A}(z,t)$ from $\mathscr{A}(0,t)$.

Suppose first that the complex envelope $\mathscr{A}(0,t)$ is itself a harmonic function $\mathscr{A}(0,t) = A(0,f)\exp(j2\pi ft)$ with frequency f, so that the wave is monochromatic with frequency $\nu = f + \nu_0$. The complex wavefunction then varies with z in accordance with $U(z,t) = U(0,t)\exp[-\frac{1}{2}\alpha(f+\nu_0)z - j\beta(f+\nu_0)z]$. Using (5.6-1), $A(z,f) = A(0,f)\exp\{-\frac{1}{2}\alpha(f+\nu_0)z - j[\beta(f+\nu_0) - \beta(\nu_0)]z\}$, from which

$$A(z,f) = A(0,f)\mathscr{H}(f), \tag{5.6-2}$$

where

$$\mathscr{H}(f) = \exp\{-\tfrac{1}{2}\alpha(f+\nu_0)z - j[\beta(f+\nu_0) - \beta(\nu_0)]z\}. \tag{5.6-3}$$

Transfer Function

The factor $\mathscr{H}(f)$ is therefore the transfer function of the linear system whose input and output are the time functions $\mathscr{A}(0,t)$ and $\mathscr{A}(z,t)$ (see Appendix B, Sec. B.1).

We now describe a systematic procedure for determining the output $\mathscr{A}(z,t)$ from the input $\mathscr{A}(0,t)$ for an arbitrary dispersive medium. The complex envelope $\mathscr{A}(z,t)$ of an arbitrary pulse can always be decomposed as a superposition of harmonic functions by using the Fourier-transform relations,

$$\mathscr{A}(z,t) = \int_{-\infty}^{\infty} A(z,f)\exp(j2\pi ft)\,df \tag{5.6-4a}$$

$$A(z,f) = \int_{-\infty}^{\infty} \mathscr{A}(z,t)\exp(-j2\pi ft)\,dt. \tag{5.6-4b}$$

Starting with $\mathscr{A}(0,t)$, we determine the Fourier transform $A(0,f)$ by use of (5.6-4b) at $z=0$, then we use (5.6-2) and (5.6-3) to determine $A(z,f)$, from which $\mathscr{A}(z,t)$ is finally composed by using the inverse Fourier transform in (5.6-4a).

This procedure may be simplified by use of the convolution theorem (see Appendix A, Sec. A.1), which provides an explicit expression for $\mathscr{A}(z,t)$ as the convolution $\mathscr{A}(0,t)$ with $h(t)$,

$$\mathscr{A}(z,t) = \int_{-\infty}^{\infty} \mathscr{A}(0,t')h(t-t')\,dt', \tag{5.6-5}$$

where $h(t)$, the impulse-response function, is the inverse Fourier transform of $\mathscr{H}(f)$.

The Slowly Varying Envelope Approximation

Since $\mathscr{A}(z,t)$ is slowly varying in comparison with the central frequency ν_0, the Fourier transform $A(z,f)$ is a narrow function of f with width $\Delta\nu \ll \nu_0$. Such pulses are often called **wavepackets**. To simplify the analysis, we assume that within the frequency range $\Delta\nu$ centered about ν_0, the attenuation coefficient $\alpha(\nu)$ is approximately constant $\alpha(\nu) = \alpha$, and the propagation constant $\beta(\nu) = n(\nu)(2\pi\nu/c_o)$ varies only slightly and gradually with ν, so that it can be approximated by the first three

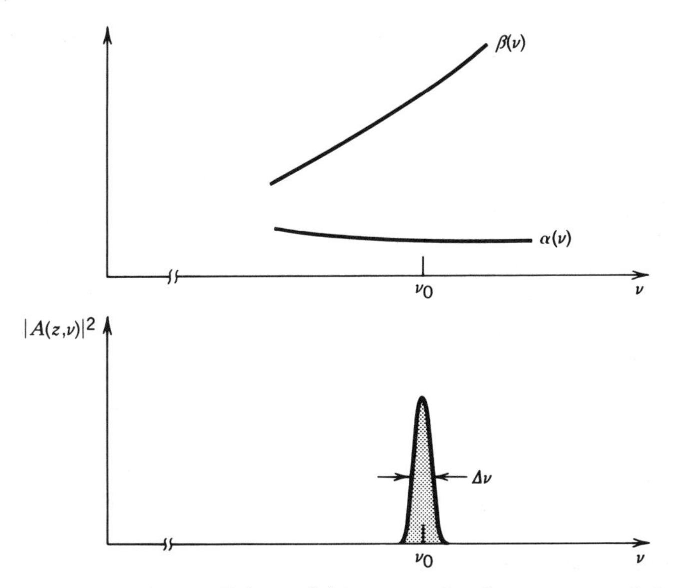

Figure 5.6-2 The attenuation coefficient $\alpha(\nu)$ is assumed to be constant and the propagation constant $\beta(\nu)$ is assumed to be a slowly varying function of ν within the spectral width $\Delta\nu$.

terms of a Taylor series expansion

$$\beta(\nu_0 + f) \approx \beta(\nu_0) + f\frac{d\beta}{d\nu} + \frac{1}{2}f^2\frac{d^2\beta}{d\nu^2}. \tag{5.6-6}$$

Figure 5.6-2 illustrates these functions.

Substituting (5.6-6) into (5.6-3) an approximate expression for the transfer function $\mathcal{H}(f)$ is obtained,

$$\mathcal{H}(f) \approx \mathcal{H}_0 \exp(-j2\pi f\tau_d)\exp(-j\pi D_\nu z f^2), \tag{5.6-7}$$

Approximate Transfer Function

where $\mathcal{H}_0 = e^{-\alpha z/2}$, $\tau_d = z/v$,

$$\frac{1}{v} = \frac{1}{2\pi}\frac{d\beta}{d\nu} = \frac{d\beta}{d\omega} \tag{5.6-8}$$

Group Velocity

and

$$D_\nu = \frac{1}{2\pi}\frac{d^2\beta}{d\nu^2} = 2\pi\frac{d^2\beta}{d\omega^2} = \frac{d}{d\nu}\left(\frac{1}{v}\right). \tag{5.6-9}$$

Dispersion Coefficient

The constants v and D_ν, called the **group velocity** and the **dispersion coefficient**, respectively, are important parameters that characterize the dispersive medium, as we shall see subsequently.

Group Velocity

If the dispersion coefficient is sufficiently small, the third term in the expansion (5.6-6) may also be neglected and $\mathcal{H}(f) \approx \mathcal{H}_0 \exp(-j2\pi f \tau_d)$. The system is then equivalent to an attenuation factor $\mathcal{H}_0 = e^{-\alpha z/2}$ and a time delay $\tau_d = z/v$ (see Appendix A, Sec. A.1, the delay property of the Fourier transform), so that $\mathscr{A}(z,t) = e^{-\alpha z/2}\mathscr{A}(0, t - \tau_d)$. In this approximation the pulse travels at the group velocity v, its intensity is attenuated by the factor $e^{-\alpha z}$, but its initial shape is not altered. By comparison, in an ideal (lossless and nondispersive) medium, $\alpha = 0$ and $\beta(\nu) = 2\pi\nu/c$, so that $v = c$; the pulse envelope travels at the speed of light in the medium and its height and shape are not altered.

Dispersion Coefficient

Since the group velocity $v = 2\pi/(d\beta/d\nu)$ is itself frequency dependent, different frequency components of the pulse undergo different delays $\tau_d = z/v$. As a result, the pulse spreads and its shape is altered. Two identical pulses of central frequencies ν and $\nu + \delta\nu$ suffer a differential delay

$$\delta\tau = \frac{d\tau_d}{d\nu}\delta\nu = \frac{d}{d\nu}\left(\frac{z}{v}\right)\delta\nu = D_\nu z\, \delta\nu.$$

If $D_\nu > 0$ (**normal dispersion**), the travel time for the higher-frequency component is longer than the travel time for the lower-frequency component. Thus shorter-wavelength components are slower, as illustrated schematically in Fig. 5.5-4. Normal dispersion occurs in glass in the visible band. At longer wavelengths, however, $D_\nu < 0$ (**anomalous dispersion**), so that the shorter-wavelength components are faster.

If the pulse has a spectral width σ_ν (Hz), then

$$\boxed{\sigma_\tau = |D_\nu| \sigma_\nu z} \tag{5.6-10}$$

is an estimate of the spread of its temporal width. The dispersion coefficient D_ν is therefore a measure of the pulse time broadening per unit spectral width per unit distance (s/m-Hz).

The shape of the transmitted pulse may be determined using the approximate transfer function (5.6-7). The corresponding impulse-response function $h(t)$ is obtained by taking the inverse Fourier transform,

$$\boxed{h(t) \approx \mathcal{H}_0 \frac{1}{(j|D_\nu|z)^{1/2}} \exp\left[j\pi \frac{(t-\tau_d)^2}{D_\nu z}\right].} \tag{5.6-11}$$

Impulse-Response Function

This may be shown by noting that the Fourier transform of $\exp(j\pi t^2)$ is $\sqrt{j}\,\exp(-j\pi f^2)$ and using the scaling and delay properties of the Fourier transform (see Appendix A, Sec. A.1 and Table A.1-1). The complex envelope $\mathscr{A}(z,t)$ may be obtained by convolving the initial complex envelope $\mathscr{A}(0,t)$ with the impulse-response function $h(t)$, as in (5.6-5).

Gaussian Pulses

As an example, assume that the complex envelope of the incident wave is a Gaussian pulse $\mathscr{A}(0, t) = \exp(-t^2/\tau_0^2)$ with $1/e$ half-width τ_0. The result of the convolution integral (5.6-5), when (5.6-11) is used and $\alpha = 0$, is

$$\mathscr{A}(z, t) = \left[\frac{q(0)}{q(z)}\right]^{1/2} \exp\left[j\pi \frac{(t - \tau_d)^2}{D_\nu q(z)}\right], \tag{5.6-12}$$

where

$$q(z) = z + jz_0, \qquad z_0 = \frac{\pi\tau_0^2}{-D_\nu}, \qquad \tau_d = \frac{z}{v}. \tag{5.6-13}$$

The intensity $|\mathscr{A}(z, t)|^2 = |q(0)/q(z)| \exp[-\pi(t - \tau_d)^2 \operatorname{Im}\{1/D_\nu q(z)\}]$ is a Gaussian function

$$|\mathscr{A}(z, t)|^2 = \frac{\tau_0}{\tau(z)} \exp\left[-\frac{2(t - \tau_d)^2}{\tau^2(z)}\right] \tag{5.6-14}$$

centered about the delay time $\tau_d = z/v$ and of width

$$\tau(z) = \tau_0\left[1 + \left(\frac{z}{z_0}\right)^2\right]^{1/2}. \tag{5.6-15}$$

Width Broadening of a Gaussian Pulse

The variation of $\tau(z)$ with z is illustrated in Fig. 5.6-3. In the limit $z \gg z_0$,

$$\tau(z) \approx \tau_0 \frac{z}{|z_0|} = |D_\nu| \frac{z}{\pi\tau_0}, \tag{5.6-16}$$

so that the pulse width increases linearly with z. The width of the transmitted pulse is

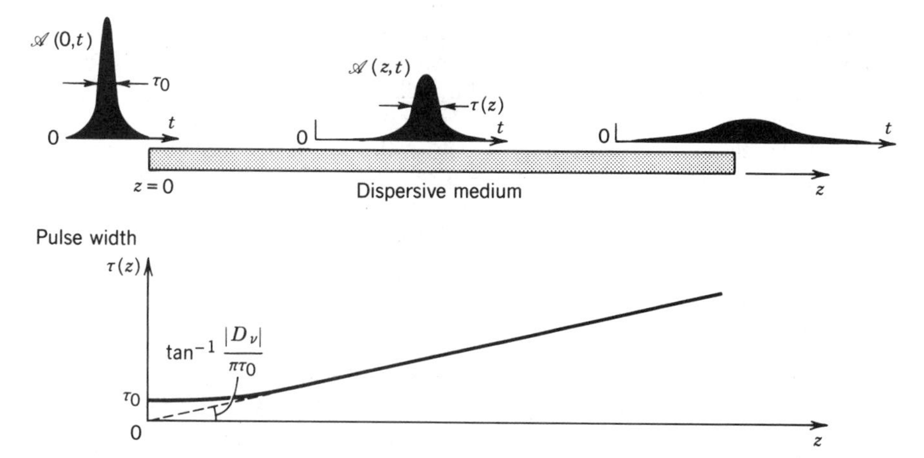

Figure 5.6-3 Gaussian pulse spreading as a function of distance. For large distances, the width increases at the rate $|D_\nu|/\pi\tau_0$, which is inversely proportional to the initial width τ_0.

then inversely proportional to the initial width τ_0. This is expected since a narrow pulse has a broad spectrum corresponding to a more pronounced dispersion. If $\sigma_\nu = 1/\pi\tau_0$ is interpreted as the spectral width of the initial pulse, then $\tau(z) = |D_\nu|\sigma_\nu z$, which is the same expression as in (5.6-10).

*Analogy Between Pulse Dispersion and Fresnel Diffraction

Expression (5.6-11) for the impulse-response function indicates that after traveling a distance z in a dispersive medium, an impulse at $t = 0$ spreads and becomes proportional to $\exp(j\pi t^2/D_\nu z)$, where the delay τ_d has been ignored. This is mathematically analogous to Fresnel diffraction, for which a point at $x = y = 0$ creates a paraboloidal wave proportional to $\exp[-j\pi(x^2 + y^2)/\lambda z]$ (see Sec. 4.1C). With the correspondences x (or y) $\leftrightarrow t$ and $\lambda \leftrightarrow -D_\nu$, the approximate temporal spread of a pulse is analogous to the Fresnel diffraction of a "spatial pulse" (an aperture function). The dispersion coefficient $-D_\nu$ for temporal dispersion is analogous to the wavelength for diffraction ("spatial dispersion"). The analogy holds because the Fresnel approximation and the dispersion approximation both make use of Taylor-series approximations carried to the quadratic term.

The temporal dispersion of a Gaussian pulse in a dispersive medium, for example, is analogous to the diffraction of a Gaussian beam in free space. The width of the beam is $W(z) = W_0[1 + (z/z_0)^2]^{1/2}$, where $z_0 = \pi W_0^2/\lambda$ [see (3.1-8) and (3.1-11)], which is analogous to the width in (5.6-15), $\tau(z) = \tau_0[1 + (z/z_0)^2]^{1/2}$, where $z_0 = \pi\tau_0^2/(-D_\nu)$.

*Pulse Compression in a Dispersive Medium by Chirping

The analogy between the diffraction of a Gaussian beam and the dispersion of a Gaussian pulse can be carried further. Since the spatial width of a Gaussian beam can be reduced by use of a focusing lens (see Sec. 3.2), could the temporal width of a Gaussian pulse be compressed by use of an analogous system?

A lens of focal length f introduces a phase factor $\exp[j\pi(x^2 + y^2)/\lambda f]$ (see Sec. 3.2A), which bends the wavefronts so that a beam of initial width W_0 is focused near the focal plane to a smaller width $W_0' = W_0/[1 + (z_0/f)^2]^{1/2}$, where $z_0 = \pi W_0^2/\lambda$ [see (3.2-13)]. Similarly, if the Gaussian pulse is multiplied by the phase factor $\exp(-j\pi t^2/D_\nu f)$, a pulse of initial width τ_0 would be compressed to a width $\tau_0' = \tau_0/[1 + (z_0/f)^2]^{1/2}$, after propagating a distance $\approx f$ in a dispersive medium with dispersion coefficient D_ν, where $z_0 = -\pi\tau_0^2/D_\nu$. Clearly, the pulse would be broadened again if it travels farther.

The phase factor $\exp(-j\pi t^2/D_\nu f)$ may be regarded as a frequency modulation of the initial pulse $\exp(-t^2/\tau_0^2)\exp(j2\pi\nu_0 t)$. The instantaneous frequency of the modulated pulse ($1/2\pi$ times the derivative of the phase) is $\nu_0 - t/D_\nu f$. Under conditions of normal dispersion, $D_\nu > 0$, the instantaneous frequency decreases linearly as a function of time. The pulse is said to be chirped.

The process of pulse compression is depicted in Fig. 5.6-4. The high-frequency components of the chirped pulse appear before the low frequency components. In a medium with normal dispersion, the travel time of the high-frequency components is longer than that of the low-frequency components. These two effects are balanced at a certain propagation distance at which the pulse is compressed to a minimum width.

*Differential Equation Governing Pulse Propagation

We now use the transfer function $\mathcal{H}(f)$ in (5.6-7) to generate a differential equation governing the envelope $\mathscr{A}(z, t)$. Substituting (5.6-7) into (5.6-2), we obtain $A(z, f) = A(0, f)\exp(-\alpha z/2 - j2\pi fz/v - j\pi D_\nu zf^2)$. Taking the derivative with respect to z, we obtain the differential equation $(d/dz)A(z, f) = (-\alpha/2 - j2\pi f/v - j\pi D_\nu f^2)A(z, f)$. Taking the inverse Fourier transform of both sides, and noting that the inverse Fourier transforms of $A(z, f)$, $j2\pi fA(z, f)$, and $(j2\pi f)^2 A(z, f)$ are $\mathscr{A}(z, t)$, $\partial\mathscr{A}(z, t)/\partial t$, and

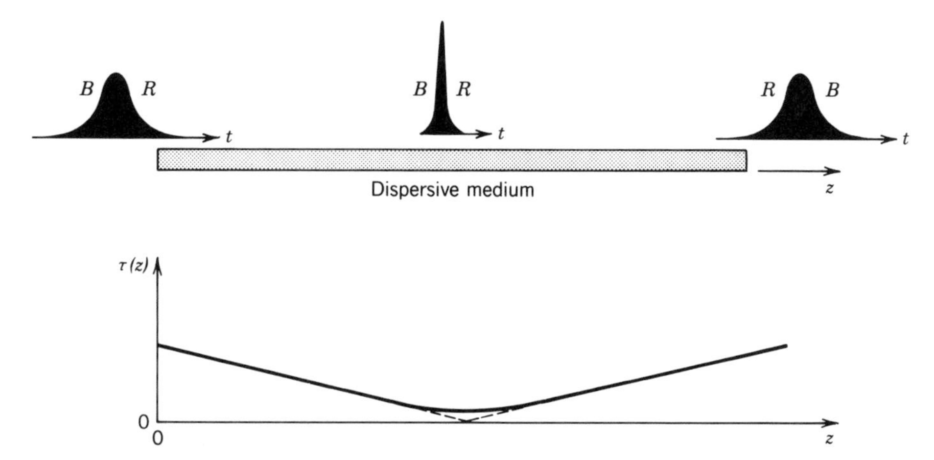

Figure 5.6-4 Compression of a chirped pulse in a medium with normal dispersion. The low frequency (marked R) occurs after the high frequency (marked B) in the initial pulse, but it catches up since it travels faster. Upon further propagation, the pulse spreads again as the R component arrives earlier than the B component.

$\partial^2\mathscr{A}(z,t)/\partial t^2$, respectively, we obtain a partial differential equation for $\mathscr{A} = \mathscr{A}(z,t)$:

$$\left(\frac{\partial}{\partial z} + \frac{1}{v}\frac{\partial}{\partial t}\right)\mathscr{A} + \frac{\alpha}{2}\mathscr{A} - j\frac{D_\nu}{4\pi}\frac{\partial^2\mathscr{A}}{\partial t^2} = 0. \tag{5.6-17}$$

Slowly Varying Envelope Wave Equation in a Dispersive Medium

The Gaussian pulse (5.6-12) is clearly a solution to this equation. Assuming that $\alpha = 0$ and using a coordinate system moving with velocity v, (5.6-17) simplifies to

$$\frac{\partial^2\mathscr{A}}{\partial t^2} + j\frac{4\pi}{D_\nu}\frac{\partial\mathscr{A}}{\partial z} = 0. \tag{5.6-18}$$

Equation (5.6-18) is analogous to the paraxial Helmholtz equation (2.2-22), confirming the analogy between dispersion in time and diffraction in space.

Wavelength Dependence of Group Velocity and Dispersion Coefficient

Since the group velocity v and the dispersion coefficient D_ν are the most important parameters governing pulse propagation in dispersive media, it is useful to examine their dependence on the wavelength. Substituting $\beta = n2\pi\nu/c_o = n2\pi/\lambda_o$ and $\nu = c_o/\lambda_o$ in the definitions (5.6-8) and (5.6-9) yields

$$v = \frac{c_o}{N}, \qquad N = n - \lambda_o\frac{dn}{d\lambda_o} \tag{5.6-19}$$

Group Velocity and Group Index

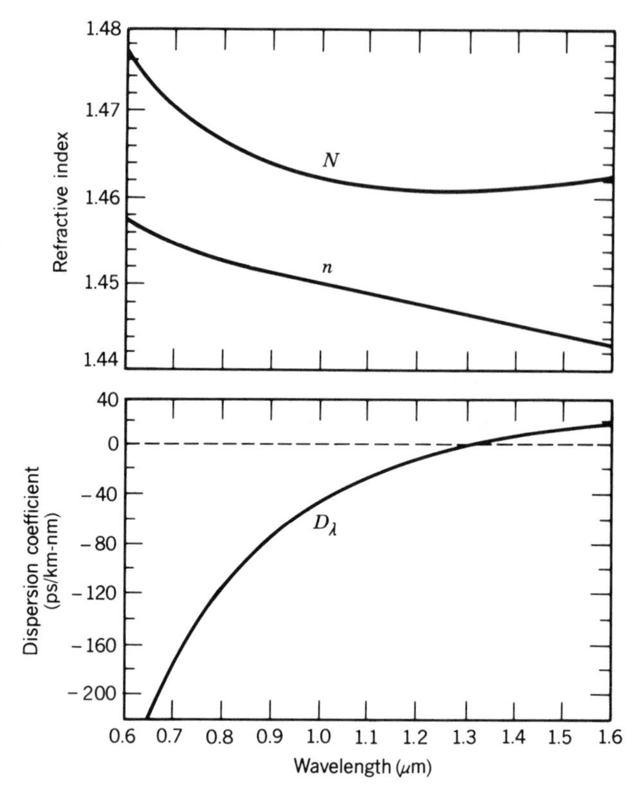

Figure 5.6-5 Wavelength dependence of optical parameters of fused silica: the refractive index n, the group index $N = c_o/v$, and the dispersion coefficient D_λ. At $\lambda_o = 1.312\ \mu$m, n has a point of inflection, the group velocity v is maximum, the group index N is minimum, and the dispersion coefficient D_λ vanishes. At this wavelength the pulse broadening is minimal.

and

$$D_\nu = \frac{\lambda_o^3}{c_o^2}\frac{d^2 n}{d\lambda_o^2}. \qquad (5.6\text{-}20)$$

Dispersion Coefficient (s / m-Hz)

The parameter N is often called the **group index**.

It is also common to define a dispersion coefficient D_λ in terms of the wavelength instead of the frequency by use of the relation $D_\lambda\, d\lambda = D_\nu\, d\nu$, which gives $D_\lambda = D_\nu\, d\nu/d\lambda_o = D_\nu(-c_o/\lambda_o^2)$, and†

$$D_\lambda = -\frac{\lambda_o}{c_o}\frac{d^2 n}{d\lambda_o^2}. \qquad (5.6\text{-}21)$$

Dispersion Coefficient (s / m-nm)

The pulse broadening for a source of spectral width σ_λ is, in analogy with (5.6-10), $\sigma_\tau = |D_\lambda|\sigma_\lambda z$.

†Another dispersion coefficient $M = -D_\lambda$ is also widely used in the literature.

In fiber-optics applications, D_λ is usually given in units of ps/km-nm, where the pulse broadening is measured in picoseconds, the length of the medium in kilometers, and the source spectral width in nanometers. The wavelength dependence of n, N, and D_λ for silica glass are illustrated in Fig. 5.6-5. For $\lambda_o < 1.312$ μm, $D_\lambda < 0$ ($D_\nu > 0$; normal dispersion). For $\lambda_o > 1.312$ μm, $D_\lambda > 0$, so that the dispersion is anomalous. Near $\lambda_o = 1.312$ μm, the dispersion coefficient vanishes. This property is significant in the design of light-transmission systems based on the use of optical pulses, as will become clear in Secs. 8.3, 19.8, and 22.1.

READING LIST

See also the list of general books on optics in Chapter 1.

E. D. Palik, ed., *Handbook of Optical Constants of Solids II*, Academic Press, Orlando, FL, 1991.

D. K. Cheng, *Field and Wave Electromagnetics*, Addison-Wesley, Reading, MA, 1983, 2nd ed. 1989.

W. H. Hayt, *Engineering Electromagnetics*, McGraw-Hill, New York, 1958, 5th ed. 1989.

H. A. Haus and J. R. Melcher, *Electromagnetic Fields and Energy*, Prentice-Hall, Englewood Cliffs, NJ, 1989.

P. Lorrain, D. Corson, and F. Lorrain, *Electromagnetic Fields and Waves*, W. H. Freeman, New York, 1970, 3rd ed. 1988.

J. A. Kong, *Electromagnetic Wave Theory*, Wiley, New York, 1986.

F. A. Hopf and G. I. Stegeman, *Applied Classical Electrodynamics*, Vol. I, *Linear Optics*, Wiley, New York, 1985.

H. A. Haus, *Waves and Fields in Optoelectronics*, Prentice-Hall, Englewood Cliffs, NJ, 1984.

S. Ramo, J. R. Whinnery, and T. Van Duzer, *Fields and Waves in Communication Electronics*, Wiley, New York, 1965, 2nd ed. 1984.

L. D. Landau, E. M. Lifshitz, and L. P. Pitaevskii, *Electrodynamics of Continuous Media*, Pergamon Press, New York, first English ed. 1960, 2nd ed. 1984.

H. C. Chen, *Theory of Electromagnetic Waves*, McGraw-Hill, New York, 1983.

J. D. Jackson, *Classical Electrodynamics*, Wiley, New York, 1962, 2nd ed. 1975.

L. Brillouin, *Wave Propagation and Group Velocity*, Academic Press, New York, 1960.

C. L. Andrews, *Optics of the Electromagnetic Spectrum*, Prentice-Hall, Englewood Cliffs, NJ, 1960.

PROBLEMS

5.1-1 **An Electromagnetic Wave.** An electromagnetic wave in free space has an electric field $\mathscr{E} = f(t - z/c_o)\hat{\mathbf{x}}$, where $\hat{\mathbf{x}}$ is a unit vector in the x direction, $f(t) = \exp(-t^2/\tau^2)\exp(j2\pi\nu_0 t)$, and τ is a constant. Describe the physical nature of this wave and determine an expression for the magnetic field vector.

5.2-1 **Dielectric Media.** Identify the media described by the following equations, regarding linearity, dispersiveness, spatial dispersiveness, and homogeneity.
(a) $\mathscr{P} = \epsilon_o \chi \mathscr{E} - a\nabla \times \mathscr{E}$,
(b) $\mathscr{P} + a\mathscr{P}^2 = \epsilon_o \mathscr{E}$,
(c) $a_1 \partial^2\mathscr{P}/\partial t^2 + a_2 \partial\mathscr{P}/\partial t + \mathscr{P} = \epsilon_o \chi \mathscr{E}$,
(d) $\mathscr{P} = \epsilon_o\{a_1 + a_2 \exp[-(x^2 + y^2)]\}\mathscr{E}$,
where χ, a, a_1, and a_2 are constants.

5.3-1 **Traveling Standing Wave.** The complex amplitude of the electric field of a monochromatic electromagnetic wave of wavelength λ_o traveling in free space is $\mathbf{E}(\mathbf{r}) = E_0 \sin\beta y \exp(-j\beta z)\hat{\mathbf{x}}$. (a) Determine a relation between β and λ_o.

(b) Derive an expression for the magnetic field vector $\mathbf{H}(\mathbf{r})$. (c) Determine the direction of flow of optical power. (d) This wave may be regarded as the sum of two TEM plane waves. Determine their directions of propagation.

5.4-1 **Electric Field of Focused Light.** (a) 1 W of optical power is focused uniformly on a flat target of size $0.1 \times 0.1\ \text{mm}^2$ placed in free space. Determine the peak value of the electric field E_0 (V/m). Assume that the optical wave is approximated as a TEM plane wave within the area of the target. (b) Determine the electric field at the center of a Gaussian beam (a point on the beam axis at the beam waist) if the beam power is 1 W and the beam waist radius $W_0 = 0.1$ mm. Refer to Sec. 3.1.

5.5-1 **Conductivity and Absorption.** In a medium with an electric current density $\mathcal{J}$, Maxwell's equation (5.2-4) is modified to $\nabla \times \mathcal{H} = \mathcal{J} + \epsilon\, \partial \mathcal{E}/\partial t$, with the other equations unaltered. If the medium is described by Ohm's law, $\mathcal{J} = \sigma \mathcal{E}$, where σ is the conductivity, show that the Helmholtz equation, (5.3-15), is applicable with a complex-valued k. Show that a plane wave traveling in this medium is attenuated, and determine an expression for the attenuation coefficient α.

5.5-2 **Dispersion in a Medium with Sharp Absorption Band.** Consider a resonant medium for which the susceptibility $\chi(\nu)$ is given by (5.5-15) with $\Delta\nu \approx 0$. Determine an expression for the refractive index $n(\nu)$ using (5.5-5) and plot it as a function of ν. Explain the physical significance of the result.

5.5-3 **Dispersion in a Medium with Two Absorption Bands.** Solid materials that could be used for making optical fibers typically exhibit strong absorption in the blue or ultraviolet region and strong absorption in the middle infrared region. Modeling the material as having two narrow resonant absorptions with $\Delta\nu \approx 0$ at wavelengths λ_{o1} and λ_{o2}, use the results of Problem 5.5-2 to sketch the wavelength dependence of the refractive index. Assume that the parameter χ_0 is the same for both resonances.

5.6-1 **Amplitude-Modulated Wave in a Dispersive Medium.** An amplitude-modulated wave with complex wavefunction $a(t) = [1 + m\cos(2\pi f_s t)]\exp(j2\pi\nu_0 t)$ at $z = 0$, where $f_s \ll \nu_0$, travels a distance z through a dispersive medium of propagation constant $\beta(\nu)$ and negligible attenuation. If $\beta(\nu_0) = \beta_0$, $\beta(\nu_0 - f_s) = \beta_1$, and $\beta(\nu_0 + f_s) = \beta_2$, derive an expression for the complex envelope of the transmitted wave as a function of β_0, β_1, β_2, and z. Show that at certain distances z the wave is amplitude modulated with no phase modulation.

5.6-2 **Group Velocity in a Resonant Medium.** Determine an expression for the group velocity v of a resonant medium with refractive index given by (5.5-21), (5.5-19), and (5.5-20). Plot v as a function of the frequency ν.

5.6-3 **Pulse broadening in an Optical Fiber.** A Gaussian pulse of width $\tau_0 = 100$ ps travels a distance of 1 km through an optical fiber made of fused silica with the characteristics shown in Fig. 5.6-5. Estimate the time delay τ_d and the width of the received pulse if the wavelength is (a) 0.8 μm; (b) 1.312 μm; (c) 1.55 μm.

CHAPTER

6

POLARIZATION AND CRYSTAL OPTICS

6.1 POLARIZATION OF LIGHT
- A. Polarization
- B. Matrix Representation

6.2 REFLECTION AND REFRACTION

6.3 OPTICS OF ANISOTROPIC MEDIA
- A. Refractive Indices
- B. Propagation Along a Principal Axis
- C. Propagation in an Arbitrary Direction
- D. Rays, Wavefronts, and Energy Transport
- E. Double Refraction

6.4 OPTICAL ACTIVITY AND FARADAY EFFECT
- A. Optical Activity
- B. Faraday Effect

6.5 OPTICS OF LIQUID CRYSTALS

6.6 POLARIZATION DEVICES
- A. Polarizers
- B. Wave Retarders
- C. Polarization Rotators

Augustin Jean Fresnel (1788–1827) advanced a theory of light in which waves exhibit transverse vibrations. The equations describing the partial reflection and refraction of light are named after him. Fresnel also made important contributions to the theory of light diffraction.

The polarization of light is determined by the time course of the *direction* of the electric-field vector $\mathscr{E}(\mathbf{r}, t)$. For monochromatic light, the three components of $\mathscr{E}(\mathbf{r}, t)$ vary sinusoidally with time with amplitudes and phases that are generally different, so that at each position $\mathbf{r}$ the endpoint of the vector $\mathscr{E}(\mathbf{r}, t)$ moves in a plane and traces an ellipse, as illustrated in Fig. 6.0-1(a). The plane, the orientation, and the shape of the ellipse generally vary with position.

In paraxial optics, however, light propagates along directions that lie within a narrow cone centered about the optical axis (the z axis). Waves are approximately transverse electromagnetic (TEM) and the electric-field vector therefore lies approximately in the transverse plane (the x-y plane), as illustrated in Fig. 6.0-1(b). If the medium is isotropic, the polarization ellipse is approximately the same everywhere, as illustrated in Fig. 6.0-1(b). The wave is said to be **elliptically polarized**.

The orientation and ellipticity of the ellipse determine the state of polarization of the optical wave, whereas the size of the ellipse is determined by the optical intensity. When the ellipse degenerates into a straight line or becomes a circle, the wave is said to be **linearly polarized** or **circularly polarized**, respectively.

Polarization plays an important role in the interaction of light with matter as attested to by the following examples:

- The amount of light reflected at the boundary between two materials depends on the polarization of the incident wave.
- The amount of light absorbed by certain materials is polarization dependent.
- Light scattering from matter is generally polarization sensitive.

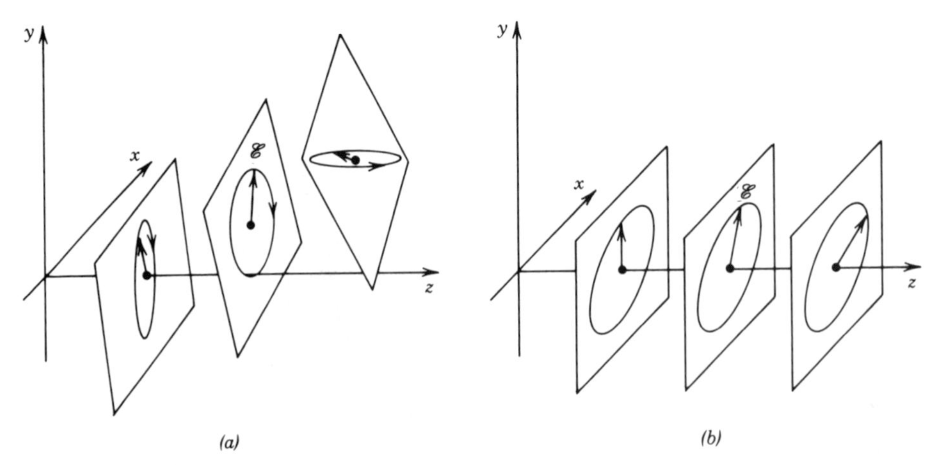

Figure 6.0-1 Time course of the electric field vector at several positions: (a) arbitrary wave; (b) paraxial wave or plane wave traveling in the z direction.

- The refractive index of anisotropic materials depends on the polarization. Waves with different polarizations therefore travel at different velocities and undergo different phase shifts, so that the polarization ellipse is modified as the wave advances (e.g., linearly polarized light can be transformed into circularly polarized light). This property is used in the design of many optical devices.
- So-called optically active materials have the natural ability to rotate the polarization plane of linearly polarized light. In the presence of a magnetic field, most materials rotate the polarization. When arranged in certain configurations, liquid crystals also act as polarization rotators.

This chapter is devoted to elementary polarization phenomena and a number of their applications. Elliptically polarized light is introduced in Sec. 6.1 using a matrix formalism that is convenient for describing polarization devices. Section 6.2 describes the effect of polarization on the reflection and refraction of light at the boundaries between dielectric media. The propagation of light through anisotropic media (crystals), optically active media, and liquid crystals are the subjects of Secs. 6.3, 6.4, and 6.5, respectively. Finally, basic polarization devices (polarizers, retarders, and rotators) are discussed in Sec. 6.6.

6.1 POLARIZATION OF LIGHT

A. Polarization

Consider a monochromatic plane wave of frequency ν traveling in the z direction with velocity c. The electric field lies in the x-y plane and is generally described by

$$\mathscr{E}(z,t) = \operatorname{Re}\left\{\mathbf{A}\exp\left[j2\pi\nu\left(t-\frac{z}{c}\right)\right]\right\}, \tag{6.1-1}$$

where the complex envelope

$$\mathbf{A} = A_x\hat{\mathbf{x}} + A_y\hat{\mathbf{y}}, \tag{6.1-2}$$

is a vector with complex components A_x and A_y. To describe the polarization of this wave, we trace the endpoint of the vector $\mathscr{E}(z,t)$ at each position z as a function of time.

The Polarization Ellipse

Expressing A_x and A_y in terms of their magnitudes and phases, $A_x = a_x\exp(j\varphi_x)$ and $A_y = a_y\exp(j\varphi_y)$, and substituting into (6.1-2) and (6.1-1), we obtain

$$\mathscr{E}(z,t) = \mathscr{E}_x\hat{\mathbf{x}} + \mathscr{E}_y\hat{\mathbf{y}}, \tag{6.1-3}$$

where

$$\mathscr{E}_x = a_x\cos\left[2\pi\nu\left(t-\frac{z}{c}\right)+\varphi_x\right] \tag{6.1-4a}$$

$$\mathscr{E}_y = a_y\cos\left[2\pi\nu\left(t-\frac{z}{c}\right)+\varphi_y\right] \tag{6.1-4b}$$

are the x and y components of the electric-field vector $\mathscr{E}(z,t)$. The components $\mathscr{E}_x$ and $\mathscr{E}_y$ are periodic functions of $t - z/c$ oscillating at frequency ν. Equations (6.1-4) are the parametric equations of the ellipse,

$$\frac{\mathscr{E}_x^2}{a_x^2} + \frac{\mathscr{E}_y^2}{a_y^2} - 2\cos\varphi \frac{\mathscr{E}_x \mathscr{E}_y}{a_x a_y} = \sin^2\varphi, \tag{6.1-5}$$

where $\varphi = \varphi_y - \varphi_x$ is the phase difference.

At a fixed value of z, the tip of the electric-field vector rotates periodically in the x–y plane, tracing out this ellipse. At a fixed time t, the locus of the tip of the electric-field vector follows a helical trajectory in space lying on the surface of an elliptical cylinder (see Fig. 6.1-1). The electric field rotates as the wave advances, repeating its motion periodically for each distance corresponding to a wavelength $\lambda = c/\nu$.

The state of polarization of the wave is determined by the shape of the ellipse (the direction of the major axis and the ellipticity, the ratio of the minor to the major axis of the ellipse). The shape of the ellipse therefore depends on two parameters—the ratio of the magnitudes a_y/a_x and the phase difference $\varphi = \varphi_y - \varphi_x$. The size of the ellipse, on the other hand, determines the intensity of the wave $I = (a_x^2 + a_y^2)/2\eta$, where η is the impedance of the medium.

Linearly Polarized Light

If one of the components vanishes ($a_x = 0$, for example), the light is linearly polarized in the direction of the other component (the y direction). The wave is also linearly polarized if the phase difference $\varphi = 0$ or π, since (6.1-4) gives $\mathscr{E}_y = \pm(a_y/a_x)\mathscr{E}_x$, which is the equation of a straight line of slope $\pm a_y/a_x$ (the + and − signs correspond to $\varphi = 0$ or π, respectively). In these cases the elliptical cylinder in Fig. 6.1-1(*b*) collapses into a plane as illustrated in Fig. 6.1-2. The wave is therefore also said to have **planar polarization**. If $a_x = a_y$, for example, the plane of polarization makes an angle 45° with the x axis. If $a_x = 0$, the plane of polarization is the y–z plane.

Circularly Polarized Light

If $\varphi = \pm\pi/2$ and $a_x = a_y = a_0$, (6.1-4) gives $\mathscr{E}_x = a_0 \cos[2\pi\nu(t - z/c) + \varphi_x]$ and $\mathscr{E}_y = \mp a_0 \sin[2\pi\nu(t - z/c) + \varphi_x]$, from which $\mathscr{E}_x^2 + \mathscr{E}_y^2 = a_0^2$, which is the equation of a circle. The elliptical cylinder in Fig. 6.1-1(*b*) becomes a circular cylinder and the wave is said to be circularly polarized. In the case $\varphi = +\pi/2$, the electric field at a fixed position z rotates in a clockwise direction when viewed from the direction toward which the wave is approaching. The light is then said to be **right circularly polarized.** The case $\varphi = -\pi/2$ corresponds to counterclockwise rotation and **left circularly**

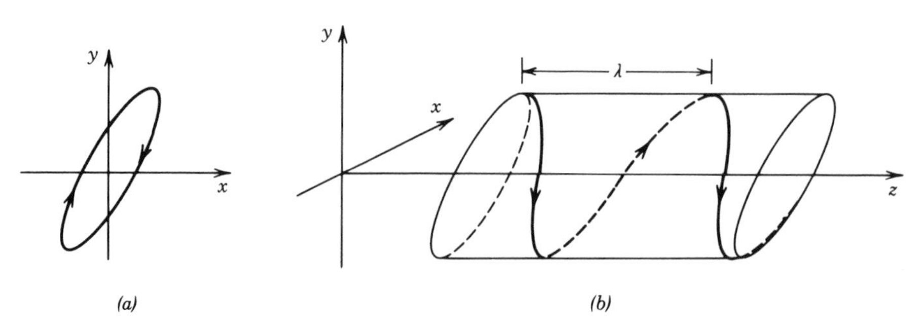

Figure 6.1-1 (*a*) Rotation of the endpoint of the electric-field vector in the x-y plane at a fixed position z. (*b*) Snapshot of the trajectory of the endpoint of the electric-field vector at a fixed time t.

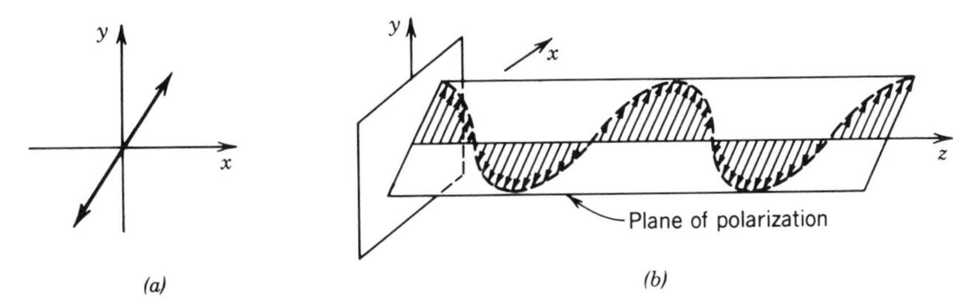

Figure 6.1-2 Linearly polarized light. (a) Time course at a fixed position z. (b) A snapshot (fixed time t).

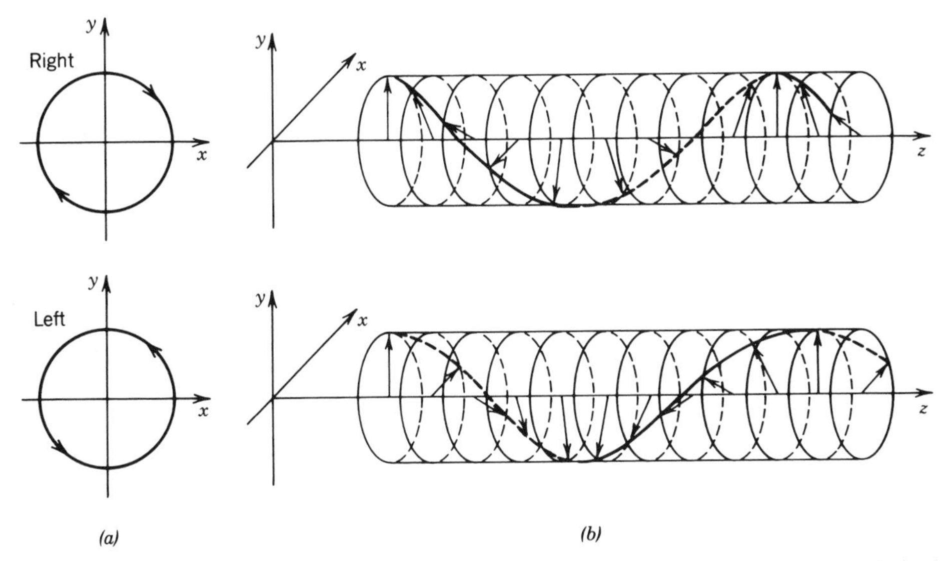

Figure 6.1-3 Trajectories of the endpoint of the electric-field vector of a circularly polarized plane wave. (a) Time course at a fixed position z. (b) A snapshot (fixed time t). The sense of rotation in (a) is opposite that in (b) because the traveling wave depends on $t - z/c$.

polarized light.† In the right circular case, a snapshot of the lines traced by the endpoints of the electric-field vectors at different positions is a right-handed helix (like a right-handed screw pointing in the direction of the wave), as illustrated in Fig. 6.1-3. For left circular polarization, a left-handed helix is followed.

B. Matrix Representation

The Jones Vector

A monochromatic plane wave of frequency ν traveling in the z direction is completely characterized by the complex envelopes $A_x = a_x \exp(j\varphi_x)$ and $A_y = a_y \exp(j\varphi_y)$ of the x and y components of the electric field. It is convenient to write these complex

†This convention is used in most textbooks of optics. The opposite designation is used in the engineering literature: in the case of right (left) circularly polarized light, the electric-field vector at a fixed position rotates counterclockwise (clockwise) when viewed from the direction toward which the wave is approaching.

TABLE 6.1-1 Jones Vectors

Polarization	Jones vector
Linearly polarized wave, in x direction	$\begin{bmatrix} 1 \\ 0 \end{bmatrix}$ 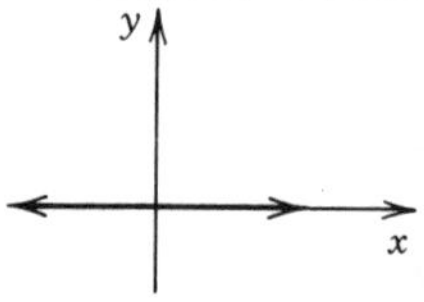
Linearly polarized wave, plane of polarization making angle θ with x axis	$\begin{bmatrix} \cos\theta \\ \sin\theta \end{bmatrix}$ 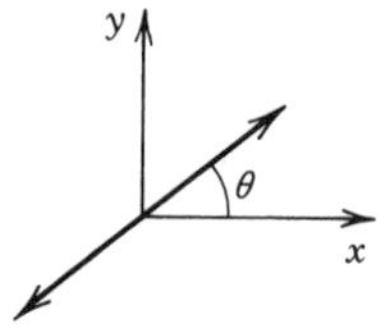
Right circularly polarized	$\frac{1}{\sqrt{2}}\begin{bmatrix} 1 \\ j \end{bmatrix}$ 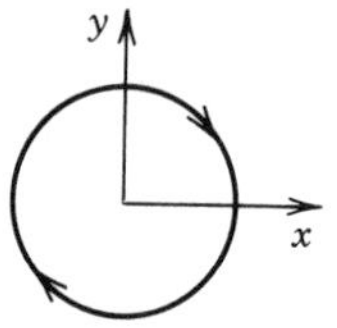
Left circularly polarized	$\frac{1}{\sqrt{2}}\begin{bmatrix} 1 \\ -j \end{bmatrix}$ 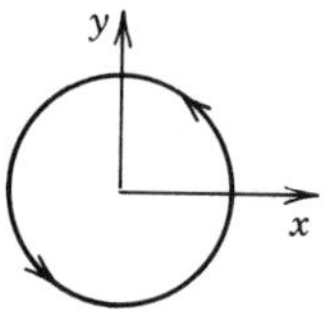

quantities in the form of a column matrix

$$\mathbf{J} = \begin{bmatrix} A_x \\ A_y \end{bmatrix}, \tag{6.1-6}$$

known as the **Jones vector**. Given the Jones vector, we can determine the total intensity of the wave, $I = (|A_x|^2 + |A_y|^2)/2\eta$, and use the ratio $a_y/a_x = |A_y|/|A_x|$ and the phase difference $\varphi = \varphi_y - \varphi_x = \arg\{A_y\} - \arg\{A_x\}$ to determine the orientation and shape of the polarization ellipse.

The Jones vectors for some special polarization states are provided in Table 6.1-1. The intensity in each case has been normalized so that $|A_x|^2 + |A_y|^2 = 1$ and the phase of the x component $\varphi_x = 0$.

Orthogonal Polarizations

Two polarization states represented by the Jones vectors $\mathbf{J}_1$ and $\mathbf{J}_2$ are said to be orthogonal if the inner product between $\mathbf{J}_1$ and $\mathbf{J}_2$ is zero. The inner product is defined by

$$(\mathbf{J}_1, \mathbf{J}_2) = A_{1x}A_{2x}^* + A_{1y}A_{2y}^*, \tag{6.1-7}$$

where A_{1x} and A_{1y} are the elements of $\mathbf{J}_1$ and A_{2x} and A_{2y} are the elements of $\mathbf{J}_2$. An example of orthogonal Jones vectors are the linearly polarized waves in the x and y directions. Another example is the right and left circularly polarized waves.

Expansion of Arbitrary Polarization as a Superposition of Two Orthogonal Polarizations

An arbitrary Jones vector **J** can always be analyzed as a weighted superposition of two orthogonal Jones vectors (say $\mathbf{J}_1$ and $\mathbf{J}_2$), called the expansion basis, $\mathbf{J} = \alpha_1\mathbf{J}_1 + \alpha_2\mathbf{J}_2$. If $\mathbf{J}_1$ and $\mathbf{J}_2$ are normalized such that $(\mathbf{J}_1, \mathbf{J}_1) = (\mathbf{J}_2, \mathbf{J}_2) = 1$, the expansion weights are the inner products $\alpha_1 = (\mathbf{J}, \mathbf{J}_1)$ and $\alpha_2 = (\mathbf{J}, \mathbf{J}_2)$. Using the x and y linearly polarized vectors $\begin{bmatrix}1\\0\end{bmatrix}$ and $\begin{bmatrix}0\\1\end{bmatrix}$, for example, as an expansion basis, the expansion weights for a Jones vector of components A_x and A_y are simply $\alpha_1 = A_x$ and $\alpha_2 = A_y$. Similarly, if the right and left circularly polarized waves $(1/\sqrt{2})\begin{bmatrix}1\\j\end{bmatrix}$ and $(1/\sqrt{2})\begin{bmatrix}1\\-j\end{bmatrix}$ are used as an expansion basis, the expansion weights are $\alpha_1 = (1/\sqrt{2})(A_x - jA_y)$ and $\alpha_2 = (1/\sqrt{2})(A_x + jA_y)$.

EXERCISE 6.1-1

Linearly Polarized Wave as a Sum of Right and Left Circularly Polarized Waves. Show that the linearly polarized wave with plane of polarization making an angle θ with the x axis is equivalent to a superposition of right and left circularly polarized waves with weights $(1/\sqrt{2})e^{-j\theta}$ and $(1/\sqrt{2})e^{j\theta}$, respectively.

Matrix Representation of Polarization Devices

Consider the transmission of a plane wave of arbitrary polarization through an optical system that maintains the plane-wave nature of the wave, but alters its polarization, as illustrated schematically in Fig. 6.1-4. The system is assumed to be linear, so that the principle of superposition of optical fields is obeyed. Two examples of such systems are the reflection of light from a planar boundary between two media, and the transmission of light through a plate with anisotropic optical properties.

The complex envelopes of the two electric-field components of the input (incident) wave, A_{1x} and A_{1y}, and those of the output (transmitted or reflected) wave, A_{2x} and A_{2y}, are in general related by the weighted superpositions

$$\begin{aligned} A_{2x} &= T_{11}A_{1x} + T_{12}A_{1y} \\ A_{2y} &= T_{21}A_{1x} + T_{22}A_{1y}, \end{aligned} \tag{6.1-8}$$

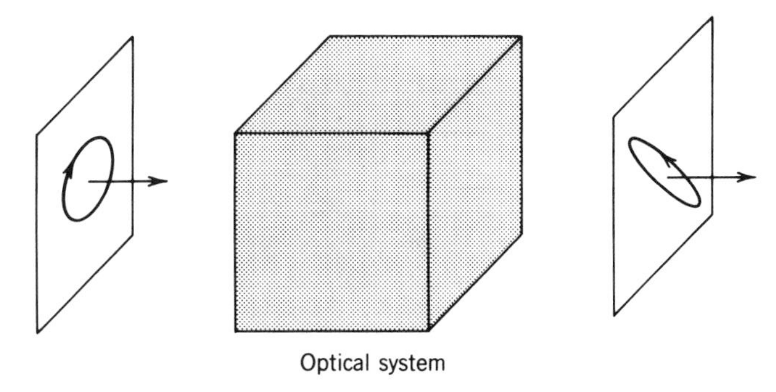

Figure 6.1-4 An optical system that alters the polarization of a plane wave.

where T_{11}, T_{12}, T_{21}, and T_{22} are constants describing the device. Equations (6.1-8) are general relations that all linear optical polarization devices must satisfy.

The linear relations in (6.1-8) may conveniently be written in matrix notation by defining a 2×2 matrix $\mathbf{T}$ with elements T_{11}, T_{12}, T_{21}, and T_{22} so that

$$\begin{bmatrix} A_{2x} \\ A_{2y} \end{bmatrix} = \begin{bmatrix} T_{11} & T_{12} \\ T_{21} & T_{22} \end{bmatrix} \begin{bmatrix} A_{1x} \\ A_{1y} \end{bmatrix}. \qquad (6.1\text{-}9)$$

If the input and output waves are described by the Jones vectors $\mathbf{J}_1$ and $\mathbf{J}_2$, respectively, then (6.1-9) may be written in the compact matrix form

$$\mathbf{J}_2 = \mathbf{T}\mathbf{J}_1. \qquad (6.1\text{-}10)$$

The matrix $\mathbf{T}$, called the **Jones matrix**, describes the optical system, whereas the vectors $\mathbf{J}_1$ and $\mathbf{J}_2$ describe the input and output waves.

The structure of the Jones matrix $\mathbf{T}$ of a given optical system determines its effect on the polarization state and intensity of the incident wave. The following is a list of the Jones matrices of some systems with simple characteristics. Physical devices that have such characteristics will be discussed subsequently in this chapter.

Linear Polarizers. The system represented by the Jones matrix

$$\mathbf{T} = \begin{bmatrix} 1 & 0 \\ 0 & 0 \end{bmatrix} \qquad (6.1\text{-}11)$$

Linear Polarizer along x Direction

transforms a wave of components (A_{1x}, A_{1y}) into a wave of components $(A_{1x}, 0)$, thus polarizing the wave along the x direction, as illustrated in Fig. 6.1-5. The system is a **linear polarizer** with transmission axis pointing in the x direction.

Wave Retarders. The system represented by the matrix

$$\mathbf{T} = \begin{bmatrix} 1 & 0 \\ 0 & \exp(-j\Gamma) \end{bmatrix} \qquad (6.1\text{-}12)$$

Wave-Retarder (Fast Axis along x Direction)

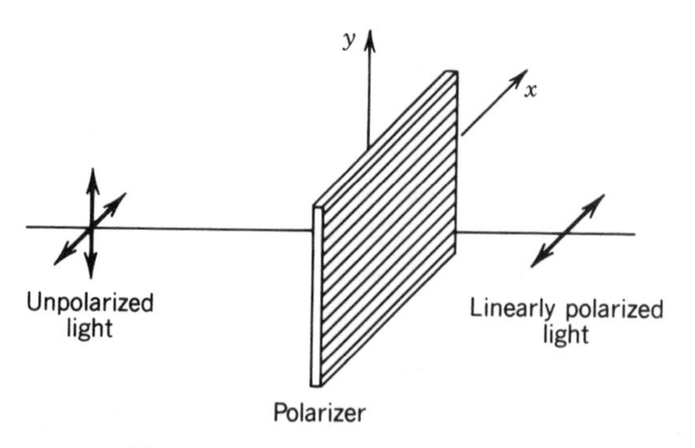

Figure 6.1-5 The linear polarizer.

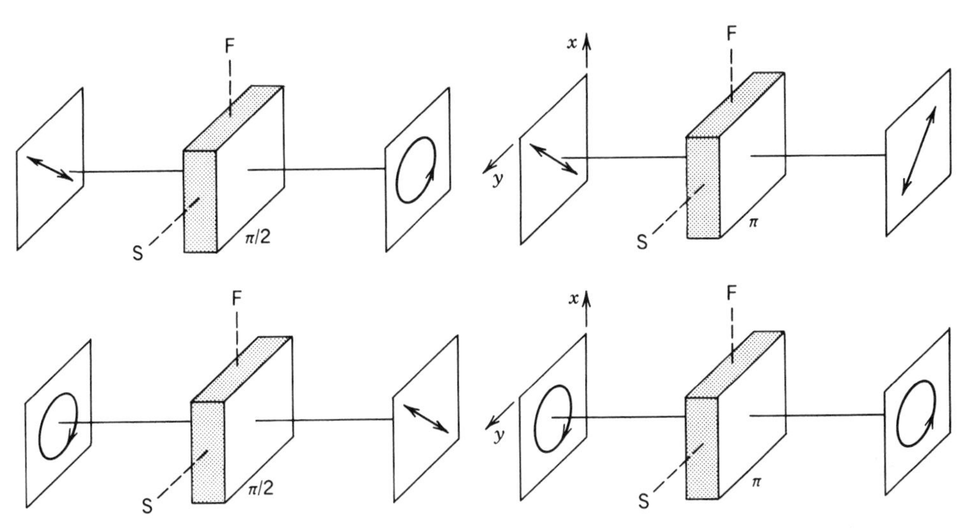

Figure 6.1-6 Operations of the quarter-wave ($\pi/2$) retarder and the half-wave (π) retarder. F and S represent the fast and slow axes of the retarder, respectively.

transforms a wave with field components (A_{1x}, A_{1y}) into another with components $(A_{1x}, e^{-j\Gamma}A_{1y})$, thus delaying the y component by a phase Γ, leaving the x component unchanged. It is therefore called a **wave retarder**. The x and y axes are called the fast and slow axes of the retarder, respectively. By simple application of matrix algebra, the following properties, illustrated in Fig. 6.1-6, may be shown:

- When $\Gamma = \pi/2$, the retarder (then called a **quarter-wave retarder**) converts linearly polarized light $\begin{bmatrix} 1 \\ 1 \end{bmatrix}$ into left circularly polarized light $\begin{bmatrix} 1 \\ -j \end{bmatrix}$, and converts right circularly polarized light $\begin{bmatrix} 1 \\ j \end{bmatrix}$ into linearly polarized light $\begin{bmatrix} 1 \\ 1 \end{bmatrix}$.
- When $\Gamma = \pi$, the retarder (then called a **half-wave retarder**) converts linearly polarized light $\begin{bmatrix} 1 \\ 1 \end{bmatrix}$ into linearly polarized light $\begin{bmatrix} 1 \\ -1 \end{bmatrix}$, thus rotating the plane of polarization by 90°. The half-wave retarder converts right circularly polarized light $\begin{bmatrix} 1 \\ j \end{bmatrix}$ into left circularly polarized light $\begin{bmatrix} 1 \\ -j \end{bmatrix}$.

Polarization Rotators. The Jones matrix

$$\mathbf{T} = \begin{bmatrix} \cos\theta & -\sin\theta \\ \sin\theta & \cos\theta \end{bmatrix} \tag{6.1-13}$$

Polarization Rotator

represents a device that converts a linearly polarized wave $\begin{bmatrix} \cos\theta_1 \\ \sin\theta_1 \end{bmatrix}$ into a linearly polarized wave $\begin{bmatrix} \cos\theta_2 \\ \sin\theta_2 \end{bmatrix}$ where $\theta_2 = \theta_1 + \theta$. It therefore rotates the plane of polarization of a linearly polarized wave by an angle θ. The device is called a **polarization rotator**.

Cascaded Polarization Devices

The action of cascaded optical systems on polarized light may be conveniently determined by using conventional matrix multiplication formulas. A system characterized by the Jones matrix $\mathbf{T}_1$ followed by another characterized by $\mathbf{T}_2$ are equivalent to a single system characterized by the product matrix $\mathbf{T} = \mathbf{T}_2\mathbf{T}_1$. The matrix of the system through which light is transmitted first should appear to the right in the matrix product since it applies on the input Jones vector first.

EXERCISE 6.1-2

Cascaded Wave Retarders. Show that two cascaded quarter-wave retarders with parallel fast axes are equivalent to a half-wave retarder. What if the fast axes are orthogonal?

Coordinate Transformation

Elements of the Jones vectors and Jones matrices depend on the choice of the coordinate system. If these elements are known in one coordinate system, they can be determined in another coordinate system by using matrix methods. If $\mathbf{J}$ is the Jones vector in the x-y coordinate system, then in a new coordinate system x'-y', with the x' direction making an angle θ with the x direction, the Jones vector $\mathbf{J}'$ is given by

$$\mathbf{J}' = \mathbf{R}(\theta)\mathbf{J}, \tag{6.1-14}$$

where $\mathbf{R}(\theta)$ is the matrix

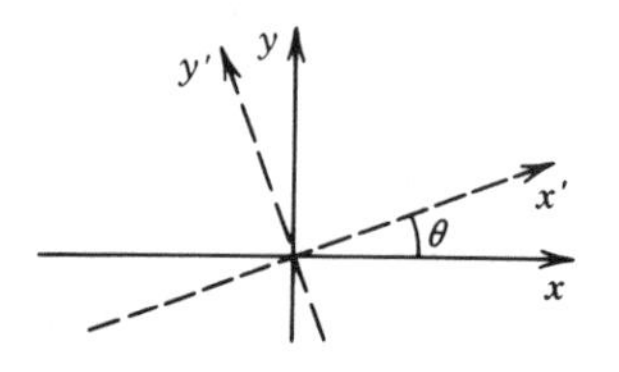

$$\mathbf{R}(\theta) = \begin{bmatrix} \cos\theta & \sin\theta \\ -\sin\theta & \cos\theta \end{bmatrix}. \tag{6.1-15}$$

Coordinate Transformation Matrix

This can be shown by relating the components of the electric field in the two coordinate systems.

The Jones matrix $\mathbf{T}$, which represents an optical system, is similarly transformed into $\mathbf{T}'$, in accordance with the matrix relations

$$\mathbf{T}' = \mathbf{R}(\theta)\mathbf{T}\mathbf{R}(-\theta) \tag{6.1-16}$$

$$\mathbf{T} = \mathbf{R}(-\theta)\mathbf{T}'\mathbf{R}(\theta), \tag{6.1-17}$$

where $\mathbf{R}(-\theta)$ is given by (6.1-15) with $-\theta$ replacing θ. The matrix $\mathbf{R}(-\theta)$ is the inverse of $\mathbf{R}(\theta)$, so that $\mathbf{R}(-\theta)\mathbf{R}(\theta)$ is a unit matrix. Equation (6.1-16) can be shown by using the relation $\mathbf{J}_2 = \mathbf{T}\mathbf{J}_1$ and the transformation $\mathbf{J}_2' = \mathbf{R}(\theta)\mathbf{J}_2 = \mathbf{R}(\theta)\mathbf{T}\mathbf{J}_1$. Since $\mathbf{J}_1 = \mathbf{R}(-\theta)\mathbf{J}_1'$, $\mathbf{J}_2' = \mathbf{R}(\theta)\mathbf{T}\mathbf{R}(-\theta)\mathbf{J}_1'$; since $\mathbf{J}_2' = \mathbf{T}'\mathbf{J}_1'$, (6.1-16) follows.

EXERCISE 6.1-3

Jones Matrix of a Polarizer. Show that the Jones matrix of a linear polarizer with a transmission axis making an angle θ with the x axis is

$$\mathbf{T} = \begin{bmatrix} \cos^2\theta & \sin\theta\cos\theta \\ \sin\theta\cos\theta & \sin^2\theta \end{bmatrix}. \tag{6.1-18}$$

Linear Polarizer at Angle θ

Derive (6.1-18) using (6.1-17), (6.1-15), and (6.1-11).

Normal Modes

The normal modes of a polarization system are the states of polarization that are not changed when the wave is transmitted through the system. These states have Jones vectors satisfying

$$\mathbf{T}\mathbf{J} = \mu\mathbf{J}, \tag{6.1-19}$$

where μ is a constant. The normal modes are therefore the eigenvectors of the Jones matrix $\mathbf{T}$, and the values of μ are the corresponding eigenvalues. Since the matrix $\mathbf{T}$ is of size 2×2 there are only two independent normal modes, $\mathbf{T}\mathbf{J}_1 = \mu_1\mathbf{J}_1$ and $\mathbf{T}\mathbf{J}_2 = \mu_2\mathbf{J}_2$. If the matrix $\mathbf{T}$ is Hermitian, i.e., $T_{12} = T_{21}^*$, the normal modes are orthogonal, $(\mathbf{J}_1, \mathbf{J}_2) = 0$. The normal modes are usually used as an expansion basis, so that an arbitrary input wave $\mathbf{J}$ may be expanded as a superposition of normal modes, $\mathbf{J} = \alpha_1\mathbf{J}_1 + \alpha_2\mathbf{J}_2$. The response of the system may be easily evaluated since $\mathbf{T}\mathbf{J} = \mathbf{T}(\alpha_1\mathbf{J}_1 + \alpha_2\mathbf{J}_2) = \alpha_1\mathbf{T}\mathbf{J}_1 + \alpha_2\mathbf{T}\mathbf{J}_2 = \alpha_1\mu_1\mathbf{J}_1 + \alpha_2\mu_2\mathbf{J}_2$ (see Appendix C).

EXERCISE 6.1-4

Normal Modes of Simple Polarization Systems

(a) Show that the normal modes of the linear polarizer are linearly polarized waves.
(b) Show that the normal modes of the wave retarder are linearly polarized waves.
(c) Show that the normal modes of the polarization rotator are right and left circularly polarized waves.

What are the eigenvalues of the systems above?

6.2 REFLECTION AND REFRACTION

In this section we examine the reflection and refraction of a monochromatic plane wave of arbitrary polarization incident at a planar boundary between two dielectric media. The media are assumed to be linear, homogeneous, isotropic, nondispersive, and nonmagnetic; the refractive indices are n_1 and n_2. The incident, refracted, and

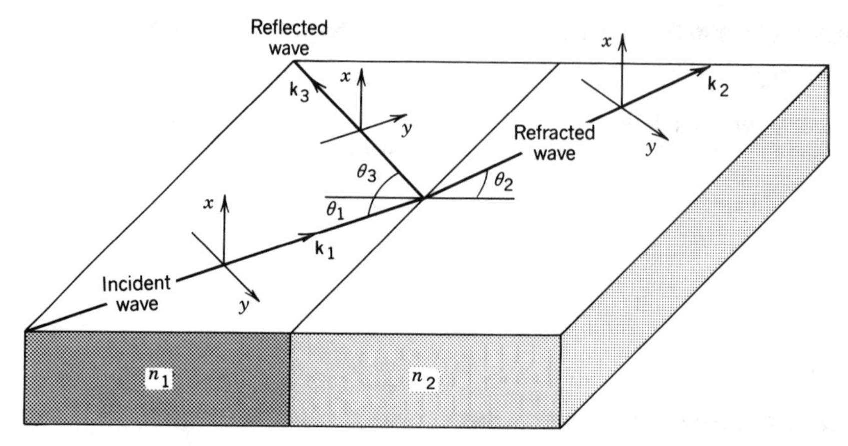

Figure 6.2-1 Reflection and refraction at the boundary between two dielectric media.

reflected waves are labeled with the subscripts 1, 2, and 3, respectively, as illustrated in Fig. 6.2-1.

As shown in Sec. 2.4A, the wavefronts of these waves are matched at the boundary if the angles of reflection and incidence are equal, $\theta_3 = \theta_1$, and the angles of refraction and incidence satisfy Snell's law,

$$n_1 \sin \theta_1 = n_2 \sin \theta_2. \tag{6.2-1}$$

To relate the amplitudes and polarizations of the three waves we associate with each wave an x-y coordinate system in a plane normal to the direction of propagation (Fig. 6.2-1). The electric-field envelopes of these waves are described by Jones vectors

$$\mathbf{J}_1 = \begin{bmatrix} A_{1x} \\ A_{1y} \end{bmatrix}, \qquad \mathbf{J}_2 = \begin{bmatrix} A_{2x} \\ A_{2y} \end{bmatrix}, \qquad \mathbf{J}_3 = \begin{bmatrix} A_{3x} \\ A_{3y} \end{bmatrix}.$$

We proceed to determine the relations between $\mathbf{J}_2$ and $\mathbf{J}_1$ and between $\mathbf{J}_3$ and $\mathbf{J}_1$. These relations are written in the matrix form $\mathbf{J}_2 = \mathbf{t}\mathbf{J}_1$, and $\mathbf{J}_3 = \mathbf{r}\mathbf{J}_1$, where $\mathbf{t}$ and $\mathbf{r}$ are 2×2 Jones matrices describing the transmission and reflection of the wave, respectively.

Elements of the transmission and reflection matrices may be determined by using the boundary conditions required by electromagnetic theory (tangential components of $\mathbf{E}$ and $\mathbf{H}$ and normal components of $\mathbf{D}$ and $\mathbf{B}$ are continuous at the boundary). The magnetic field associated with each wave is orthogonal to the electric field and their magnitudes are related by the characteristic impedances, η_o/n_1 for the incident and reflected waves, and η_o/n_2 for the transmitted wave, where $\eta_o = (\mu_o/\epsilon_o)^{1/2}$. The result is a set of equations that are solved to obtain relations between the components of the electric fields of the three waves.

The algebraic steps involved are reduced substantially if we observe that the two normal modes for this system are linearly polarized waves with polarization along the x and y directions. This may be proved if we show that an incident, a reflected, and a refracted wave with their electric field vectors pointing in the x direction are self-consistent with the boundary conditions, and similarly for three waves linearly polarized in the y direction. This is indeed the case. The x and y polarized waves are therefore separable and independent.

The x-polarized mode is called the **transverse electric (TE)** polarization or the **orthogonal** polarization, since the electric fields are orthogonal to the plane of incidence. The y-polarized mode is called the **transverse magnetic (TM)** polarization since the magnetic field is orthogonal to the plane of incidence, or the **parallel** polarization since the electric fields are parallel to the plane of incidence. The orthogonal and parallel polarizations are also called the s and p polarizations (s for the German *senkrecht*, meaning "perpendicular").

The independence of the x and y polarizations implies that the Jones matrices **t** and **r** are diagonal,

$$\mathbf{t} = \begin{bmatrix} t_x & 0 \\ 0 & t_y \end{bmatrix}, \qquad \mathbf{r} = \begin{bmatrix} r_x & 0 \\ 0 & r_y \end{bmatrix},$$

so that

$$E_{2x} = t_x E_{1x}, \qquad E_{2y} = t_y E_{1y} \tag{6.2-2}$$

$$E_{3x} = r_x E_{1x}, \qquad E_{3y} = r_y E_{1y}. \tag{6.2-3}$$

The coefficients t_x and t_y are the complex amplitude transmittances for the TE and TM polarizations, respectively, and similarly for the complex amplitude reflectances r_x and r_y.

Applying the boundary conditions to the TE and TM polarizations separately gives the following expressions for the reflection and transmission coefficients, known as the **Fresnel equations**:

$$r_x = \frac{n_1 \cos\theta_1 - n_2 \cos\theta_2}{n_1 \cos\theta_1 + n_2 \cos\theta_2} \tag{6.2-4}$$

$$t_x = 1 + r_x \tag{6.2-5}$$

Fresnel Equations (TE Polarization)

$$r_y = \frac{n_2 \cos\theta_1 - n_1 \cos\theta_2}{n_2 \cos\theta_1 + n_1 \cos\theta_2} \tag{6.2-6}$$

$$t_y = \frac{n_1}{n_2}(1 + r_y). \tag{6.2-7}$$

Fresnel Equations (TM Polarization)

Given n_1, n_2, and θ_1, the reflection coefficients can be determined by first determining θ_2 using Snell's law, (6.2-1), from which

$$\cos\theta_2 = \left(1 - \sin^2\theta_2\right)^{1/2} = \left[1 - \left(\frac{n_1}{n_2}\right)^2 \sin^2\theta_1\right]^{1/2}. \tag{6.2-8}$$

Since the quantities under the square roots in (6.2-8) can be negative, the reflection and transmission coefficients are in general complex. The magnitudes $|r_x|$ and $|r_y|$ and the phase shifts $\varphi_x = \arg\{r_x\}$ and $\varphi_y = \arg\{r_y\}$ are plotted as functions of the angle of incidence θ_1 in Figs. 6.2-2 to 6.2-5 for each of the two polarizations for external reflection ($n_1 < n_2$) and internal reflection ($n_1 > n_2$).

TE Polarization

The reflection coefficient r_x for the TE-polarized wave is given by (6.2-4).

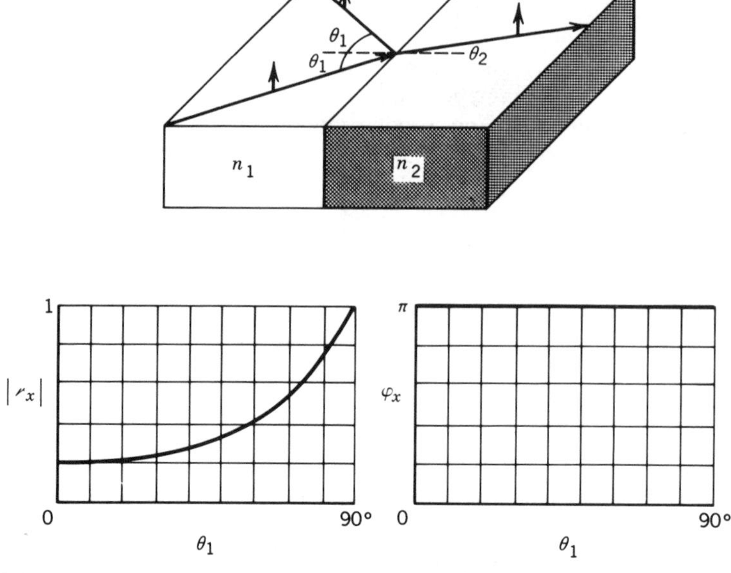

Figure 6.2-2 Magnitude and phase of the reflection coefficient as a function of the angle of incidence for external reflection of the TE polarized wave ($n_2/n_1 = 1.5$).

- *External Reflection* ($n_1 < n_2$). The reflection coefficient r_x is always real and negative, corresponding to a phase shift $\varphi_x = \pi$. The magnitude $|r_x| = (n_2 - n_1)/(n_1 + n_2)$ at $\theta_1 = 0$ (normal incidence) and increases to unity at $\theta_1 = 90°$ (grazing incidence).
- *Internal Reflection* ($n_1 > n_2$). For small θ_1 the reflection coefficient is real and positive. Its magnitude is $(n_1 - n_2)/(n_1 + n_2)$ when $\theta_1 = 0°$, increasing gradually

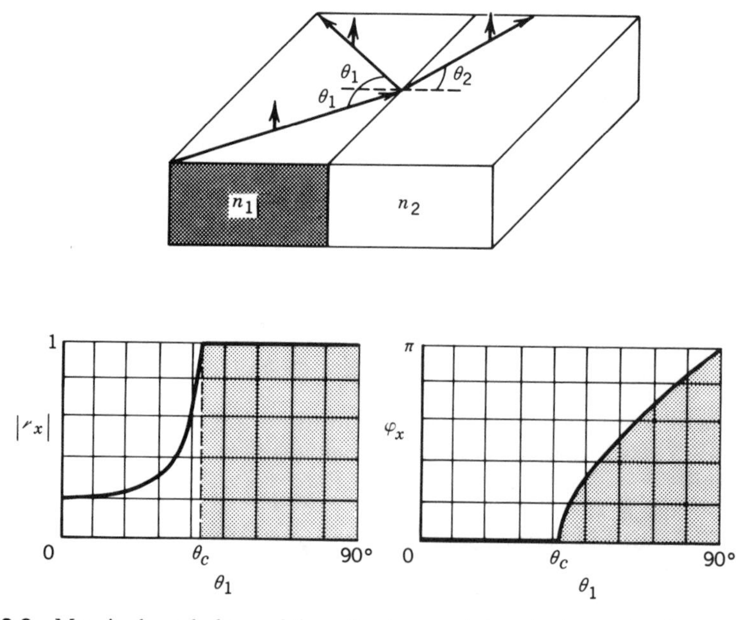

Figure 6.2-3 Magnitude and phase of the reflection coefficient for internal reflection of the TE wave ($n_1/n_2 = 1.5$).

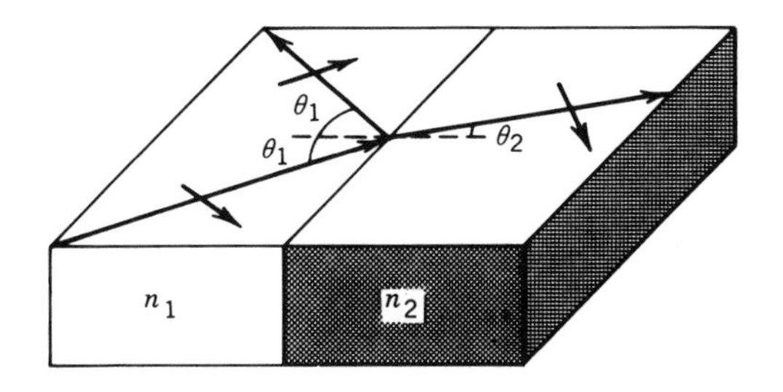

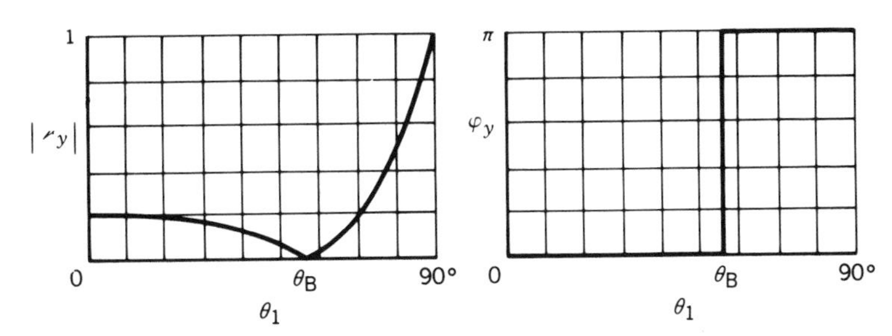

Figure 6.2-4 Magnitude and phase of the reflection coefficient for external reflection of the TM wave ($n_2/n_1 = 1.5$).

to unity when θ_1 equals the critical angle $\theta_c = \sin^{-1}(n_2/n_1)$. For $\theta_1 > \theta_c$, the magnitude of r_x remains unity, corresponding to total internal reflection. This may be shown by using (6.2-8) to write† $\cos\theta_2 = -[1 - \sin^2\theta_1/\sin^2\theta_c]^{1/2} = -j[\sin^2\theta_1/\sin^2\theta_c - 1]^{1/2}$, and substituting into (6.2-6). Total internal reflection is accompanied by a phase shift $\varphi_x = \arg\{r_x\}$ given by

$$\tan\frac{\varphi_x}{2} = \frac{\left(\sin^2\theta_1 - \sin^2\theta_c\right)^{1/2}}{\cos\theta_1}. \tag{6.2-9}$$

TE Reflection Phase Shift

The phase shift φ_x increases from 0 at $\theta_1 = \theta_c$ to π at $\theta_1 = 90°$, as illustrated in Fig. 6.2-3.

TM Polarization

The dependence of the reflection coefficient r_y on θ_1 in (6.2-6) is similarly examined for external and internal reflections:

- *External Reflection* ($n_1 < n_2$). The reflection coefficient is real. It decreases from a positive value of $(n_2 - n_1)/(n_2 + n_1)$ at normal incidence until it vanishes at an angle $\theta_1 = \theta_B$,

$$\theta_B = \tan^{-1}\frac{n_2}{n_1}, \tag{6.2-10}$$

Brewster Angle

†The choice of the minus sign for the square root is consistent with the derivation that leads to the Fresnel equations.

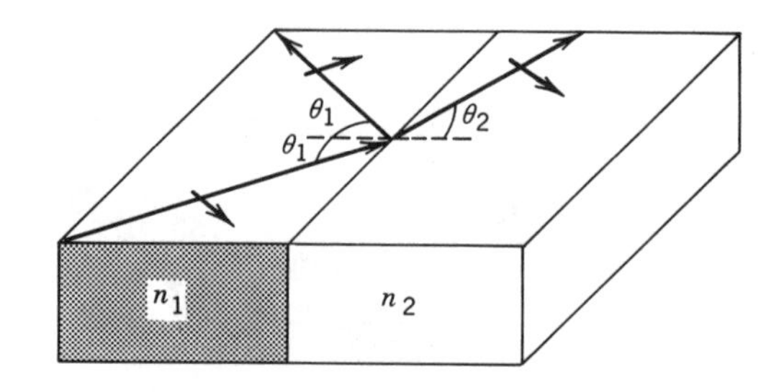

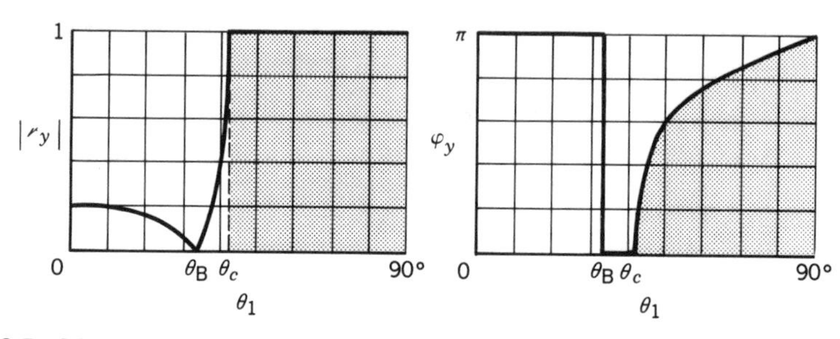

Figure 6.2-5 Magnitude and phase of the reflection coefficient for internal reflection of the TM wave ($n_1/n_2 = 1.5$).

known as the **Brewster angle**. For $\theta_1 > \theta_B$, r_y reverses sign and its magnitude increases gradually approaching unity at $\theta_1 = 90°$. The property that the TM wave is not reflected at the Brewster angle is used in making polarizers (see Sec. 6.6).

- *Internal Reflection* ($n_1 > n_2$). At $\theta_1 = 0°$, r_y is negative and has magnitude $(n_1 - n_2)/(n_1 + n_2)$. As θ_1 increases the magnitude drops until it vanishes at the Brewster angle $\theta_B = \tan^{-1}(n_2/n_1)$. As θ_1 increases beyond θ_B, r_y becomes positive and increases until it reaches unity at the critical angle θ_c. For $\theta_1 > \theta_c$ the wave undergoes total internal reflection accompanied by a phase shift $\varphi_y = \arg\{r_y\}$ given by

$$\tan\frac{\varphi_y}{2} = \frac{\left(\sin^2\theta_1 - \sin^2\theta_c\right)^{1/2}}{\cos\theta_1 \sin^2\theta_c}. \qquad (6.2\text{-}11)$$

TM Reflection Phase Shift

EXERCISE 6.2-1

Brewster Windows. At what angle is a TM-polarized beam of light transmitted through a glass plate of refractive index $n = 1.5$ placed in air ($n = 1$) without suffering reflection losses at either surface? These plates, known as Brewster windows, are used in lasers (Fig. 6.2-6; see Sec. 14.2D).

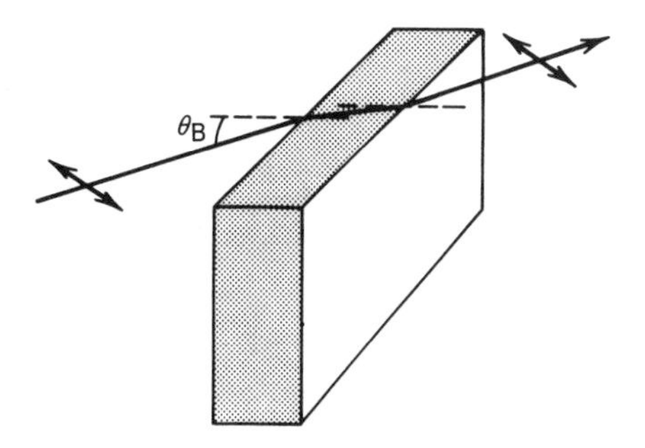

Figure 6.2-6 The Brewster window transmits TM-polarized light with no reflection loss.

Power Reflectance and Transmittance

The reflection and transmission coefficients r and t are ratios of the complex amplitudes. The power reflectance $\mathscr{R}$ and transmittance $\mathscr{T}$ are defined as the ratios of power flow (along a direction normal to the boundary) of the reflected and transmitted waves to that of the incident wave. Because the reflected and incident waves propagate in the same medium and make the same angle with the normal to the surface,

$$\mathscr{R} = |r|^2. \tag{6.2-12}$$

Conservation of power requires that

$$\mathscr{T} = 1 - \mathscr{R}. \tag{6.2-13}$$

Note, however, that $\mathscr{T} = [n_2 \cos\theta_2 / n_1 \cos\theta_1]|t|^2$ which is *not* generally equal to $|t|^2$ since the power travels at different angles. It follows that for both TE and TM polarizations, and for both external and internal reflection, the reflectance at normal incidence is

$$\boxed{\mathscr{R} = \left(\frac{n_1 - n_2}{n_1 + n_2}\right)^2.} \tag{6.2-14}$$

Power Reflectance at Normal Incidence

At a boundary between glass ($n = 1.5$) and air ($n = 1$), for example, $\mathscr{R} = 0.04$, so that 4% of the light is reflected at normal incidence. At the boundary between GaAs ($n = 3.6$) and air ($n = 1$), $\mathscr{R} \approx 0.32$, so that 32% of the light is reflected at normal incidence. The reflectance can be much greater or much less at oblique angles as illustrated in Fig. 6.2-7.

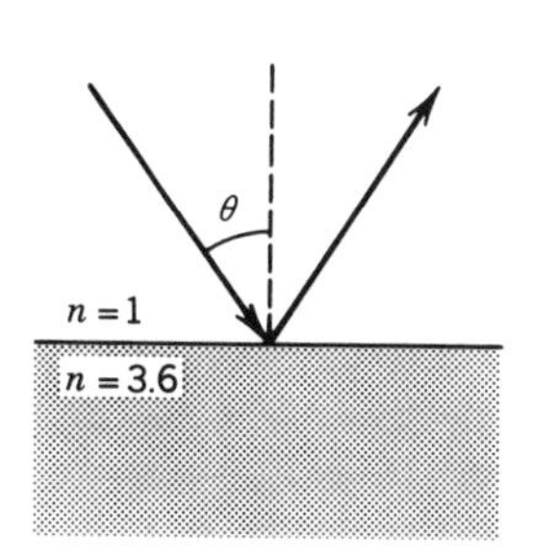

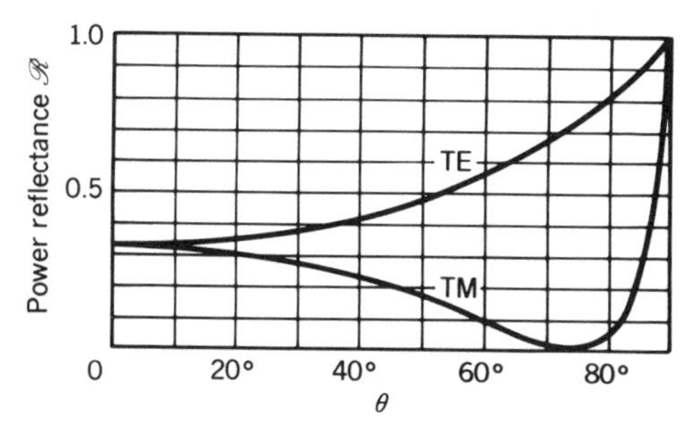

Figure 6.2-7 Power reflectance of TE and TM polarization plane waves at the boundary between air ($n = 1$) and GaAs ($n = 3.6$) as a function of the angle of incidence θ.

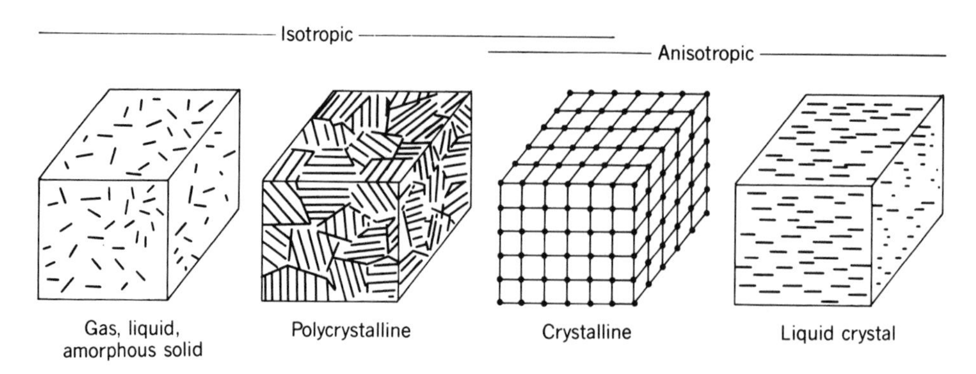

Figure 6.3-1 Positional and orientational order in different kinds of materials.

6.3 OPTICS OF ANISOTROPIC MEDIA

A dielectric medium is said to be anisotropic if its macroscopic optical properties depend on direction. The macroscopic properties of matter are of course governed by the microscopic properties: the shape and orientation of the individual molecules and the organization of their centers in space. The following is a description of the positional and orientational types of order inherent in several kinds of optical materials (see Fig. 6.3-1).

- If the molecules are located in space at totally random positions and are themselves isotropic or are oriented along totally random directions, the medium is isotropic. *Gases*, *liquids*, and *amorphous solids* are isotropic.
- If the molecules are anisotropic and their orientations are not totally random, the medium is anisotropic, even if the positions are totally random. This is the case for *liquid crystals*, which have orientational order but lack complete positional order.
- If the molecules are organized in space according to regular periodic patterns and are oriented in the same direction, as in *crystals*, the medium is in general anisotropic.
- *Polycrystalline materials* have a structure in the form of disjointed crystalline grains that are randomly oriented relative to each other. The grains are themselves generally anisotropic, but their averaged macroscopic behavior is isotropic.

A. Refractive Indices

Permittivity Tensor

In a linear anisotropic dielectric medium (a crystal, for example), each component of the electric flux density **D** is a linear combination of the three components of the electric field

$$D_i = \sum_j \epsilon_{ij} E_j, \tag{6.3-1}$$

where $i, j = 1, 2, 3$ indicate the x, y, and z components, respectively (see Sec. 5.2B). The dielectric properties of the medium are therefore characterized by a 3×3 array of nine coefficients $\{\epsilon_{ij}\}$ forming a tensor of second rank known as the **electric permittivity tensor** and denoted by the symbol $\boldsymbol{\epsilon}$. Equation (6.3-1) is usually written in the symbolic form $\mathbf{D} = \boldsymbol{\epsilon}\mathbf{E}$. The electric permittivity tensor is symmetrical, $\epsilon_{ij} = \epsilon_{ji}$, and is therefore characterized by only six independent numbers. For crystals of certain symmetries, some of these six coefficients vanish and some are related, so that even fewer coefficients are necessary.

Principal Axes and Principal Refractive Indices

Elements of the permittivity tensor depend on the choice of the coordinate system relative to the crystal structure. A coordinate system can always be found for which the off-diagonal elements of ϵ_{ij} vanish, so that

$$D_1 = \epsilon_1 E_1, \qquad D_2 = \epsilon_2 E_2, \qquad D_3 = \epsilon_3 E_3, \tag{6.3-2}$$

where $\epsilon_1 = \epsilon_{11}$, $\epsilon_2 = \epsilon_{22}$, and $\epsilon_3 = \epsilon_{33}$. These are the directions for which **E** and **D** are parallel. For example, if **E** points in the x direction, **D** must also point in the x direction. This coordinate system defines the **principal axes** and principal planes of the crystal. Throughout the remainder of this chapter, the coordinate system x, y, z (denoted also by the numbers 1, 2, 3) will be assumed to lie along the crystal's principal axes. The permittivities ϵ_1, ϵ_2, and ϵ_3 correspond to refractive indices

$$n_1 = \left(\frac{\epsilon_1}{\epsilon_o}\right)^{1/2}, \qquad n_2 = \left(\frac{\epsilon_2}{\epsilon_o}\right)^{1/2}, \qquad n_3 = \left(\frac{\epsilon_3}{\epsilon_o}\right)^{1/2}, \tag{6.3-3}$$

known as the **principal refractive indices** (ϵ_o is the permittivity of free space).

Biaxial, Uniaxial, and Isotropic Crystals

In crystals with certain symmetries two of the refractive indices are equal ($n_1 = n_2$) and the crystals are called **uniaxial** crystals. The indices are usually denoted $n_1 = n_2 = n_o$ and $n_3 = n_e$. For reasons to become clear later, n_o and n_e are called the **ordinary** and **extraordinary** indices, respectively. The crystal is said to be **positive uniaxial** if $n_e > n_o$, and **negative uniaxial** if $n_e < n_o$. The z axis of a uniaxial crystal is called the **optic axis**. In other crystals (those with cubic unit cells, for example) the three indices are equal and the medium is optically isotropic. Media for which the three principal indices are different are called **biaxial**.

Impermeability Tensor

The relation between **D** and **E** can be inverted and written in the form $\mathbf{E} = \boldsymbol{\epsilon}^{-1}\mathbf{D}$, where $\boldsymbol{\epsilon}^{-1}$ is the inverse of the tensor $\boldsymbol{\epsilon}$. It is also useful to define the tensor $\boldsymbol{\eta} = \epsilon_o \boldsymbol{\epsilon}^{-1}$ called the electric **impermeability tensor** (not to be confused with the impedance of the medium), so that $\epsilon_o \mathbf{E} = \boldsymbol{\eta}\mathbf{D}$. Since $\boldsymbol{\epsilon}$ is symmetrical, $\boldsymbol{\eta}$ is also symmetrical. Both tensors $\boldsymbol{\epsilon}$ and $\boldsymbol{\eta}$ share the same principal axes (directions for which **E** and **D** are parallel). In the principal coordinate system, $\boldsymbol{\eta}$ is diagonal with principal values $\epsilon_o/\epsilon_1 = 1/n_1^2$, $\epsilon_o/\epsilon_2 = 1/n_2^2$, and $\epsilon_o/\epsilon_3 = 1/n_3^2$. Either of the tensors $\boldsymbol{\epsilon}$ or $\boldsymbol{\eta}$ describes the optical properties of the crystal completely.

Geometrical Representation of Vectors and Tensors

A *vector* describes a physical variable with magnitude and direction (the electric field **E**, for example). It is represented *geometrically* by an arrow pointing in that direction with length proportional to the magnitude of the vector [Fig. 6.3-2(*a*)]. The vector is represented *numerically* by three numbers: its projections on the three axes of some coordinate system. These (components) are dependent on the choice of the coordinate system. However, the magnitude and direction of the vector in the physical space are independent of the choice of the coordinate system.

A second-rank *tensor* is a rule that relates two vectors. It is represented *numerically* in a given coordinate system by nine numbers. When the coordinate system is changed,

Figure 6.3-2 Geometrical representation of a vector (a) and a symmetrical tensor (b).

another set of nine numbers is obtained, but the physical nature of the rule is not changed. A useful *geometrical* representation of a symmetrical second-rank tensor (the dielectric tensor $\boldsymbol{\epsilon}$, for example) is a quadratic surface (an ellipsoid) defined by [Fig. 6.3-2(b)]

$$\sum_{ij} \epsilon_{ij} x_i x_j = 1, \tag{6.3-4}$$

known as the **quadric representation**. This surface is invariant to the choice of the coordinate system, so that if the coordinate system is rotated, both x_i and ϵ_{ij} are altered but the ellipsoid remains intact. In the principal coordinate system ϵ_{ij} is diagonal and the ellipsoid has a particularly simple form,

$$\epsilon_1 x_1^2 + \epsilon_2 x_2^2 + \epsilon_3 x_3^2 = 1. \tag{6.3-5}$$

The ellipsoid carries all information about the tensor (six degrees of freedom). Its principal axes are those of the tensor, and its axes have half-lengths $\epsilon_1^{-1/2}$, $\epsilon_2^{-1/2}$, and $\epsilon_3^{-1/2}$.

The Index Ellipsoid

The **index ellipsoid** (also called the **optical indicatrix**) is the quadric representation of the electric impermeability tensor $\boldsymbol{\eta} = \epsilon_o \boldsymbol{\epsilon}^{-1}$,

$$\sum_{ij} \eta_{ij} x_i x_j = 1. \tag{6.3-6}$$

Using the principal axes as a coordinate system, the index ellipsoid is described by

$$\boxed{\frac{x_1^2}{n_1^2} + \frac{x_2^2}{n_2^2} + \frac{x_3^2}{n_3^2} = 1,} \tag{6.3-7}$$

The Index Ellipsoid

where $1/n_1^2$, $1/n_2^2$, and $1/n_3^2$ are the principal values of $\boldsymbol{\eta}$.

The optical properties of the crystal (the directions of the principal axes and the values of the principal refractive indices) are therefore described completely by the index ellipsoid (Fig. 6.3-3). The index ellipsoid of a uniaxial crystal is an ellipsoid of revolution and that of an optically isotropic medium is a sphere.

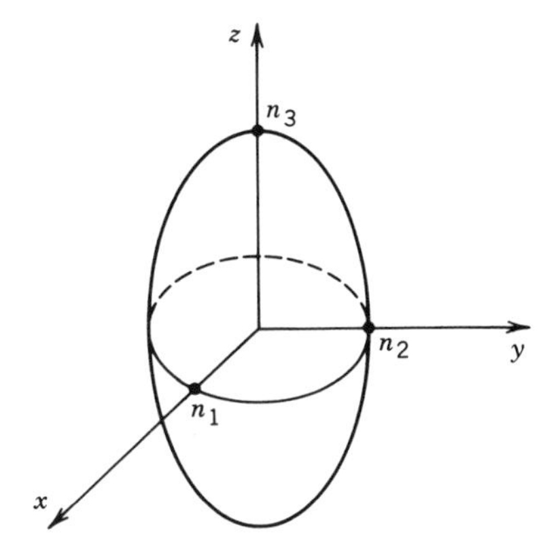

Figure 6.3-3 The index ellipsoid. The coordinates (x, y, z) are the principal axes and (n_1, n_2, n_3) are the principal refractive indices of the crystal.

B. Propagation Along a Principal Axis

The rules that govern the propagation of light in crystals under general conditions are rather complicated. However, they become relatively simple if the light is a plane wave traveling along one of the principal axes of the crystal. We begin with this case.

Normal Modes

Let x–y–z be a coordinate system in the directions of the principal axes of a crystal. A plane wave traveling in the z direction and linearly polarized in the x direction travels with phase velocity c_o/n_1 (wave number $k = n_1 k_o$) without changing its polarization. The reason is that the electric field then has only one component E_1 in the x direction, so that $\mathbf{D}$ is also in the x direction, $D_1 = \epsilon_1 E_1$, and the wave equation derived from Maxwell's equations will have a velocity $(\mu_o \epsilon_1)^{-1/2} = c_o/n_1$. A wave with linear polarization along the y direction similarly travels with phase velocity c_o/n_2 and "experiences" a refractive index n_2. Thus the normal modes for propagation in the z direction are the linearly polarized waves in the x and y directions. Other cases in which the wave propagates along one of the principal axes and is linearly polarized along another are treated similarly, as illustrated in Fig. 6.3-4.

Polarization Along an Arbitrary Direction

What if the wave travels along one principal axis (the z axis, for example) and is linearly polarized along an arbitrary direction in the x-y plane? This case can be

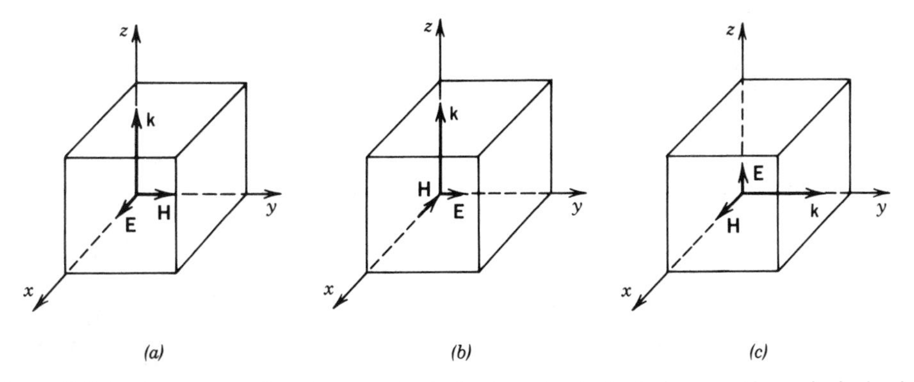

Figure 6.3-4 A wave traveling along a principal axis and polarized along another principal axis has a phase velocity c_o/n_1, c_o/n_2, or c_o/n_3, if the electric field vector points in the x, y, or z directions, respectively. (*a*) $k = n_1 k_o$; (*b*) $k = n_2 k_o$; (*c*) $k = n_3 k_o$.

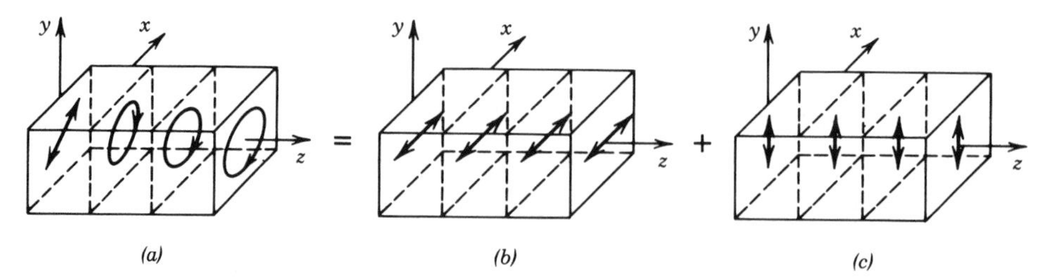

Figure 6.3-5 A linearly polarized wave at 45° in the $z = 0$ plane is analyzed as a superposition of two linearly polarized components in the x and y directions (normal modes), which travel at velocities c_o/n_1 and c_o/n_2. As a result of phase retardation, the wave is converted into an elliptically polarized wave.

addressed by analyzing the wave as a sum of the normal modes, the linearly polarized waves in the x and y directions. Since these two components travel with different velocities, c_o/n_1 and c_o/n_2, they undergo different phase shifts, $\varphi_x = n_1 k_o d$ and $\varphi_y = n_2 k_o d$, after propagating a distance d. Their phase retardation is therefore $\varphi = \varphi_y - \varphi_x = (n_2 - n_1) k_o d$. When the two components are combined, they form an elliptically polarized wave, as explained in Sec. 6.1 and illustrated in Fig. 6.3-5. The crystal can therefore be used as a **wave retarder**—a device in which two orthogonal polarizations travel at different phase velocities, so that one is retarded with respect to the other.

C. Propagation in an Arbitrary Direction

We now consider the general case of a plane wave traveling in an anisotropic crystal in an arbitrary direction defined by the unit vector $\hat{\mathbf{u}}$. The analysis is lengthy but the final results are simple. We will show that the two normal modes are linearly polarized waves. The refractive indices n_a and n_b and the directions of polarization of these modes may be determined by use of the following procedure based on the index ellipsoid. An analysis leading to a proof of this procedure will be subsequently provided.

Index-Ellipsoid Construction for Determining the Normal Modes

The following is a geometrical construction for determining the polarizations and refractive indices n_a and n_b of the normal modes of a wave traveling in the direction of the unit vector $\hat{\mathbf{u}}$ in an anisotropic material with the index ellipsoid $x_1^2/n_1^2 + x_2^2/n_2^2 + x_3^2/n_3^2 = 1$, illustrated in Fig. 6.3-6.

- Draw a plane passing through the origin of the index ellipsoid, normal to $\hat{\mathbf{u}}$. The intersection of the plane with the ellipsoid is an ellipse, called the index ellipse.
- The half-lengths of the major and minor axes of the index ellipse are the refractive indices n_a and n_b of the two normal modes.
- The directions of the major and minor axes of the index ellipse are the directions of the vectors $\mathbf{D}_a$ and $\mathbf{D}_b$ for the normal modes. These directions are orthogonal.
- The vectors $\mathbf{E}_a$ and $\mathbf{E}_b$ may be determined from $\mathbf{D}_a$ and $\mathbf{D}_b$ by use of (6.3-2).

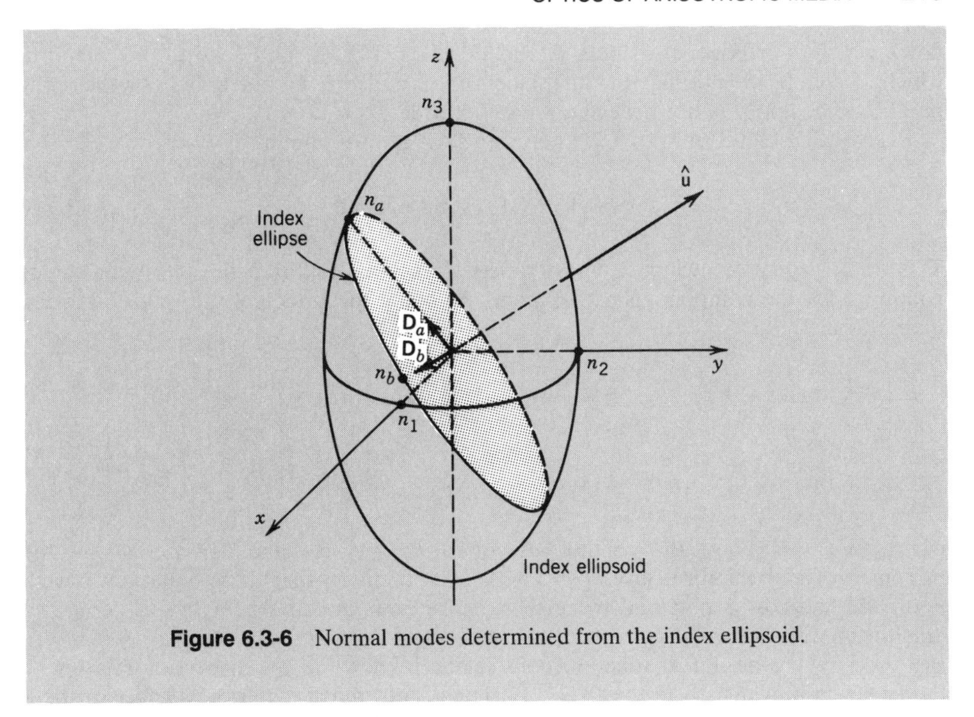

Figure 6.3-6 Normal modes determined from the index ellipsoid.

The Dispersion Relation

To determine the normal modes for a plane wave traveling in the direction $\hat{\mathbf{u}}$, we use Maxwell's equations (5.3-2) to (5.3-5) and the medium equation $\mathbf{D} = \boldsymbol{\epsilon}\mathbf{E}$. Since all fields are assumed to vary with the position $\mathbf{r}$ as $\exp(-j\mathbf{k}\cdot\mathbf{r})$, where $\mathbf{k} = k\hat{\mathbf{u}}$, Maxwell's equations (5.3-2) and (5.3-3) reduce to

$$\mathbf{k} \times \mathbf{H} = -\omega\mathbf{D} \tag{6.3-8}$$

$$\mathbf{k} \times \mathbf{E} = \omega\mu_o\mathbf{H}. \tag{6.3-9}$$

It follows from (6.3-8) that $\mathbf{D}$ is normal to both $\mathbf{k}$ and $\mathbf{H}$. Equation (6.3-9) similarly indicates that $\mathbf{H}$ is normal to both $\mathbf{k}$ and $\mathbf{E}$. These geometrical conditions are illustrated in Fig. 6.3-7, which also shows the Poynting vector $\mathbf{S} = \frac{1}{2}\mathbf{E} \times \mathbf{H}^*$ (direction of power

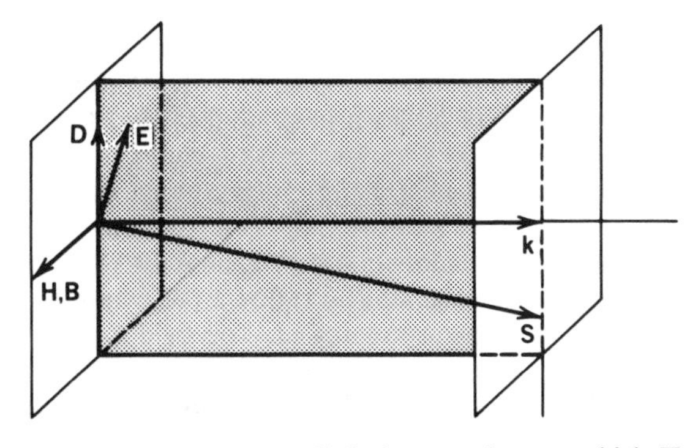

Figure 6.3-7 The vectors **D**, **E**, **k**, and **S** all lie in one plane to which **H** and **B** are normal. $\mathbf{D} \perp \mathbf{k}$ and $\mathbf{E} \perp \mathbf{S}$.

flow), which is orthogonal to both **E** and **H**. Thus **D**, **E**, **k**, and **S** lie in one plane to which **H** and **B** are normal. In this plane $\mathbf{D} \perp \mathbf{k}$ and $\mathbf{S} \perp \mathbf{E}$; but **D** is not necessarily parallel to **E**, and **S** is not necessarily parallel to **k**.

Substituting (6.3-8) into (6.3-9) and using $\mathbf{D} = \boldsymbol{\epsilon}\mathbf{E}$, we obtain

$$\mathbf{k} \times (\mathbf{k} \times \mathbf{E}) + \omega^2 \mu_o \boldsymbol{\epsilon} \mathbf{E} = \mathbf{0}. \tag{6.3-10}$$

This vector equation, which **E** must satisfy, translates to three linear homogeneous equations for the components E_1, E_2, and E_3 along the principal axes, written in the matrix form

$$\begin{bmatrix} n_1^2 k_o^2 - k_2^2 - k_3^2 & k_1 k_2 & k_1 k_3 \\ k_2 k_1 & n_2^2 k_o^2 - k_1^2 - k_3^2 & k_2 k_3 \\ k_3 k_1 & k_3 k_2 & n_3^2 k_o^2 - k_1^2 - k_2^2 \end{bmatrix} \begin{bmatrix} E_1 \\ E_2 \\ E_3 \end{bmatrix} = \begin{bmatrix} 0 \\ 0 \\ 0 \end{bmatrix}, \tag{6.3-11}$$

where (k_1, k_2, k_3) are the components of **k**, $k_o = \omega / c_o$, and (n_1, n_2, n_3) are the principal refractive indices given by (6.3-3). The condition that these equations have a nontrivial solution is obtained by setting the determinant of the matrix to zero. The result is an equation relating ω to k_1, k_2, and k_3 of the form $\omega = \omega(k_1, k_2, k_3)$, where $\omega(k_1, k_2, k_3)$ is a nonlinear function. This relation, known as the **dispersion relation**, is the equation of a surface in the k_1, k_2, k_3 space, known as the **normal surface** or the **k surface**. The intersection of the direction $\hat{\mathbf{u}}$ with the **k** surface determines the vector **k** whose magnitude $k = n\omega / c_o$ provides the refractive index n. There are two intersections corresponding to the two normal modes of each direction.

The **k** surface is a centrosymmetric surface made of two sheets, each corresponding to a solution (a normal mode). It can be shown that the **k** surface intersects each of the principal planes in an ellipse and a circle, as illustrated in Fig. 6.3-8. For biaxial crystals $(n_1 < n_2 < n_3)$, the two sheets meet at four points defining two optic axes. In the uniaxial case $(n_1 = n_2 = n_o,\ n_3 = n_e)$, the two sheets become a sphere and an ellipsoid of revolution meeting at only two points defining a single optic axis, the z axis. In the isotropic case $(n_1 = n_2 = n_3 = n)$, the two sheets degenerate into one sphere.

The intersection of the direction $\hat{\mathbf{u}} = (u_1, u_2, u_3)$ with the **k** surface corresponds to a wavenumber k satisfying

$$\sum_{j=1,2,3} \frac{u_j^2 k^2}{k^2 - n_j^2 k_o^2} = 1. \tag{6.3-12}$$

This is a fourth-order equation in k (or second order in k^2). It has four solutions $\pm k_a$ and $\pm k_b$, of which only the two positive values are meaningful, since the negative values represent a reversed direction of propagation. The problem is therefore solved: the wave numbers of the normal modes are k_a and k_b and the refractive indices are $n_a = k_a / k_o$ and $n_b = k_b / k_o$.

To determine the directions of polarization of the two normal modes, we determine the components $(k_1, k_2, k_3) = (ku_1, ku_2, ku_3)$ and the elements of the matrix in (6.3-11) for each of the two wavenumbers $k = k_a$ and k_b. We then solve two of the three equations in (6.3-11) to determine the ratios E_1/E_3 and E_2/E_3, from which we determine the direction of the corresponding electric field **E**.

*Proof of the Index-Ellipsoid Construction for Determining the Normal Modes

Since we already know that **D** lies in a plane normal to $\hat{\mathbf{u}}$, it is convenient to aim at finding **D** of the normal modes by rewriting (6.3-10) in terms of **D**. Using $\mathbf{E} = \boldsymbol{\epsilon}^{-1}\mathbf{D}$,

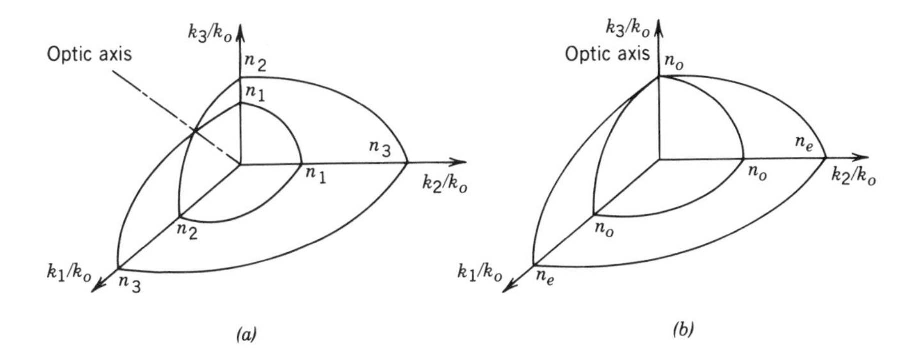

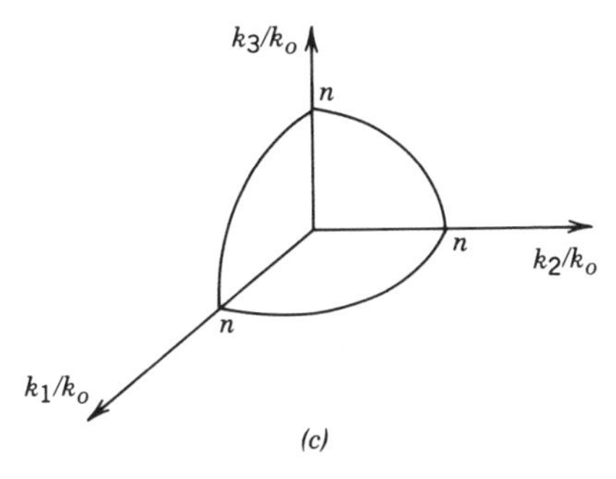

Figure 6.3-8 One octant of the **k** surface for (*a*) a biaxial crystal ($n_1 < n_2 < n_3$); (*b*) a uniaxial crystal ($n_1 = n_2 = n_o$, $n_3 = n_e$); and (*c*) an isotropic crystal ($n_1 = n_2 = n_3 = n$).

$\boldsymbol{\eta} = \epsilon_o \boldsymbol{\epsilon}^{-1}$, $\mathbf{k} = k\hat{\mathbf{u}}$, $n = k/k_o$, and $k_o^2 = \omega^2 \mu_o \epsilon_o$, (6.3-10) gives

$$-\hat{\mathbf{u}} \times (\hat{\mathbf{u}} \times \boldsymbol{\eta}\mathbf{D}) = \frac{1}{n^2}\mathbf{D}. \tag{6.3-13}$$

For each of the indices n_a and n_b of the normal modes, we determine the corresponding vector **D** by solving (6.3-13).

The operation $-\hat{\mathbf{u}} \times (\hat{\mathbf{u}} \times \boldsymbol{\eta}\mathbf{D})$ may be interpreted as a projection of the vector $\boldsymbol{\eta}\mathbf{D}$ onto a plane normal to $\hat{\mathbf{u}}$. We may therefore write (6.3-13) in the form

$$\mathbf{P}_u \boldsymbol{\eta} \mathbf{D} = \frac{1}{n^2}\mathbf{D}, \tag{6.3-14}$$

where $\mathbf{P}_u$ is an operator representing the projection operation. Equation (6.3-14) is an eigenvalue equation for the operator $\mathbf{P}_u\boldsymbol{\eta}$, with $1/n^2$ the eigenvalue and **D** the eigenvector. There are two eigenvalues, $1/n_a^2$ and $1/n_b^2$, and two corresponding eigenvectors, $\mathbf{D}_a$ and $\mathbf{D}_b$, representing the two normal modes.

The eigenvalue problem (6.3-14) has a simple geometrical interpretation. The tensor $\boldsymbol{\eta}$ is represented geometrically by its quadric representation—the index ellipsoid. The operator $\mathbf{P}_u\boldsymbol{\eta}$ represents projection onto a plane normal to $\hat{\mathbf{u}}$. Solving the eigenvalue problem in (6.3-14) is equivalent to finding the principal axes of the ellipse formed by the intersection of the plane normal to $\hat{\mathbf{u}}$ with the index ellipsoid. This proves the validity of the geometrical construction described earlier for using the index ellipsoid to determine the normal modes.

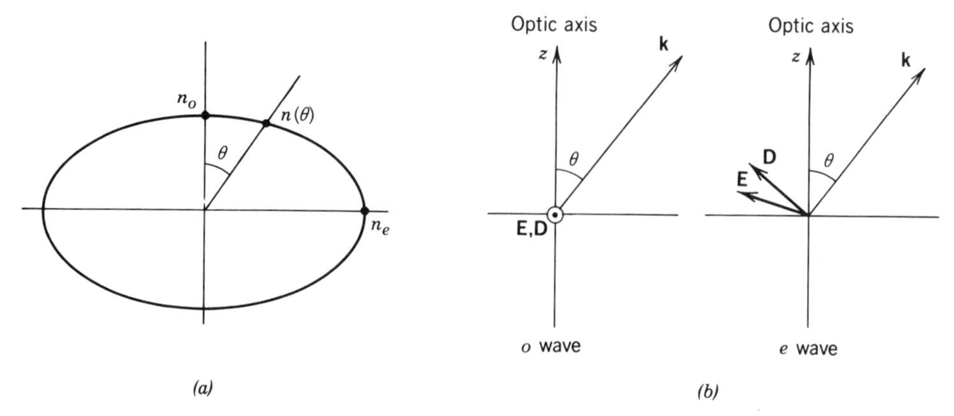

Figure 6.3-9 (*a*) Variation of the refractive index $n(\theta)$ of the extraordinary wave with θ (the angle between the direction of propagation and the optic axis). (*b*) The **E** and **D** vectors for the ordinary wave (*o* wave) and the extraordinary wave (*e* wave). The circle with a dot at the center signifies that the direction of the vector is out of the plane of the paper, toward the reader.

Special Case: Uniaxial Crystals

In uniaxial crystals ($n_1 = n_2 = n_o$ and $n_3 = n_e$) the index ellipsoid is an ellipsoid of revolution. For a wave traveling at an angle θ with the optic axis the index ellipse has half-lengths n_o and $n(\theta)$, where

$$\frac{1}{n^2(\theta)} = \frac{\cos^2\theta}{n_o^2} + \frac{\sin^2\theta}{n_e^2}, \tag{6.3-15}$$

Refractive Index of the Extraordinary Wave

so that the normal modes have refractive indices $n_a = n_o$ and $n_b = n(\theta)$. The first mode, called the **ordinary wave**, has a refractive index n_o regardless of θ. The second mode, called the **extraordinary wave**, has a refractive index $n(\theta)$ varying from n_o when $\theta = 0°$, to n_e when $\theta = 90°$, in accordance with the ellipse shown in Fig. 6.3-9(*a*). The vector **D** of the ordinary wave is normal to the plane defined by the optic axis (z axis) and the direction of wave propagation **k**, and the vectors **D** and **E** are parallel. The extraordinary wave, on the other hand, has a vector **D** in the k–z plane, which is normal to **k**, and **E** is not parallel to **D**. These vectors are illustrated in Fig. 6.3-9(*b*).

D. Rays, Wavefronts, and Energy Transport

The nature of waves in anisotropic media is best explained by examining the **k** surface $\omega = \omega(k_1, k_2, k_3)$ obtained by equating the determinant of the matrix in (6.3-11) to zero as illustrated in Fig. 6.3-8. The **k** surface describes the variation of the phase velocity $c = \omega/k$ with the direction $\hat{\mathbf{u}}$. The distance from the origin to the **k** surface in the direction of $\hat{\mathbf{u}}$ is therefore inversely proportional to the phase velocity.

The group velocity may also be determined from the **k** surface. In analogy with the group velocity $v = d\omega/dk$, which describes the velocity with which light pulses (wavepackets) travel (see Sec. 5.6), the group velocity for *rays* (localized beams, or spatial wavepackets) is the vector $\mathbf{v} = \nabla_k \omega(\mathbf{k})$, the gradient of ω with respect to **k**. Since the **k** surface is the surface $\omega(k_1, k_2, k_3) = \text{constant}$, **v** must be normal to the **k** surface. Thus rays travel along directions normal to the **k** surface.

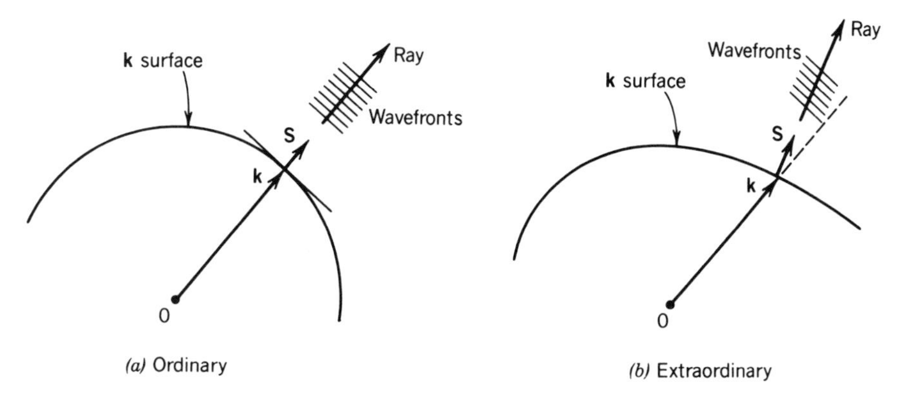

Figure 6.3-10 Rays and wavefronts for (a) spherical **k** surface, and (b) nonspherical **k** surface.

The Poynting vector $\mathbf{S} = \frac{1}{2}\mathbf{E} \times \mathbf{H}^*$ is also normal to the **k** surface. This can be shown by assuming a fixed ω and two vectors **k** and $\mathbf{k} + \Delta\mathbf{k}$ lying on the **k** surface. By taking the differential of (6.3-9) and (6.3-8) and using certain vector identities, it can be shown that $\Delta\mathbf{k} \cdot \mathbf{S} = 0$, so that **S** is normal to the **k** surface. Consequently, **S** is also parallel to the group velocity vector **v**. The wavefronts are perpendicular to the wavevector **k** (since the phase of the wave is $\mathbf{k} \cdot \mathbf{r}$). The wavefront normals are therefore parallel to the wavevector **k**.

If the **k** surface is a sphere, as in isotropic media, for example, the vectors **v**, **S**, and **k** are all parallel, indicating that rays are parallel to the wavefront normal **k** and energy flows in the same direction, as illustrated in Fig. 6.3-10(a). On the other hand, if the **k** surface is not normal to the wavevector **k**, as illustrated in Fig. 6.3-10(b), the rays and the direction of energy transport are not orthogonal to the wavefronts. Rays then have the "extraordinary" property of traveling at an oblique angle with their wavefronts [Fig. 6.3-10(b)].

Special Case: Uniaxial Crystals

In uniaxial crystals ($n_1 = n_2 = n_o$ and $n_3 = n_e$), the equation of the **k** surface $\omega = \omega(k_1, k_2, k_3)$ simplifies to

$$\left(k^2 - n_o^2 k_o^2\right)\left(\frac{k_1^2 + k_2^2}{n_e^2} + \frac{k_3^2}{n_o^2} - k_o^2\right) = 0, \tag{6.3-16}$$

which has two solutions: a sphere,

$$k = n_o k_o, \tag{6.3-17}$$

and an ellipsoid of revolution,

$$\frac{k_1^2 + k_2^2}{n_e^2} + \frac{k_3^2}{n_o^2} = k_o^2. \tag{6.3-18}$$

Because of symmetry about the z axis (optic axis), there is no loss of generality in assuming that the vector **k** lies in the y–z plane. Its direction is then characterized by the angle θ with the optic axis. It is therefore convenient to draw the k-surfaces only in the y–z plane—a circle and an ellipse, as shown in Fig. 6.3-11.

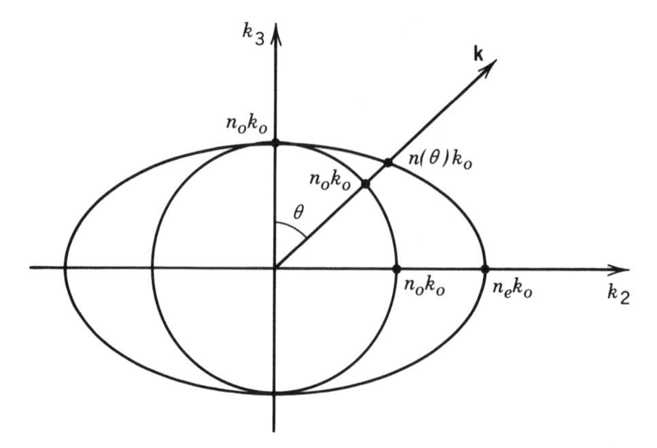

Figure 6.3-11 Intersection of the **k** surface with the y–z plane for a uniaxial crystal.

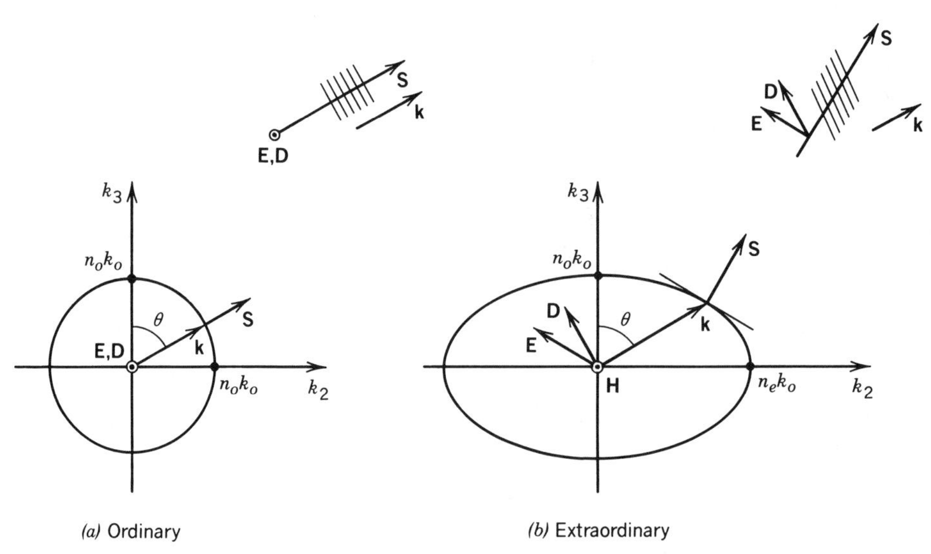

Figure 6.3-12 The normal modes for a plane wave traveling in a direction **k** at an angle θ with the optic axis z of a uniaxial crystal are: (a) An ordinary wave of refractive index n_o polarized in a direction normal to the k–z plane. (b) An extraordinary wave of refractive index $n(\theta)$ [given by (6.3-15)] polarized in the k–z plane along a direction tangential to the ellipse (the **k** surface) at the point of its intersection with **k**. This wave is "extraordinary" in the following ways: **D** is not parallel to **E** but both lie in the k–z plane; **S** is not parallel to **k** so that power does not flow along the direction of **k**; rays are not normal to wavefronts and the wave travels "sideways."

Given the direction $\hat{\mathbf{u}}$ of the vector **k**, the wavenumber k is determined by finding the intersection with the **k** surfaces. The two solutions define the two normal modes, the ordinary and extraordinary waves. The ordinary wave has a wavenumber $k = n_o k_o$ regardless of direction, whereas the extraordinary wave has a wavenumber $n(\theta)k_o$, where $n(\theta)$ is given by (6.3-15), confirming earlier results obtained from the index-ellipsoid geometrical construction. The directions of rays, wavefronts, energy flow, and field vectors **E** and **D** for the ordinary and extraordinary waves in a uniaxial crystal are illustrated in Fig. 6.3-12.

E. Double Refraction

Refraction of Plane Waves

We now examine the refraction of a plane wave at the boundary between an isotropic medium (say air, $n = 1$) and an anisotropic medium (a crystal). The key principle is that the wavefronts of the incident wave and the refracted wave must be matched at the boundary. Because the anisotropic medium supports two modes of distinctly different phase velocities, one expects that for each incident wave there are two refracted waves with two different directions and different polarizations. The effect is called **double refraction** or **birefringence**.

The phase-matching condition requires that

$$k_o \sin \theta_1 = k \sin \theta, \qquad (6.3\text{-}19)$$

where θ_1 and θ are the angles of incidence and refraction. In an anisotropic medium, however, the wave number $k = n(\theta)k_o$ is itself a function of θ, so that

$$\sin \theta_1 = n(\theta) \sin \theta, \qquad (6.3\text{-}20)$$

a modified Snell's law. To solve (6.3-19), we draw the intersection of the **k** surface with the plane of incidence and search for an angle θ for which (6.3-19) is satisfied. Two solutions, corresponding to the two normal modes, are expected. The polarization state of the incident light governs the distribution of energy among the two refracted waves.

Take, for example, a uniaxial crystal and a plane of incidence parallel to the optic axis. The **k** surfaces intersect the plane of incidence in a circle and an ellipse (Fig. 6.3-13). The two refracted waves that satisfy the phase-matching condition are:

- An ordinary wave of orthogonal polarization (TE) at an angle $\theta = \theta_o$ for which

$$\sin \theta_1 = n_o \sin \theta_o;$$

- An extraordinary wave of parallel polarization (TM) at an angle $\theta = \theta_e$, for which

$$\sin \theta_1 = n(\theta_e) \sin \theta_e,$$

where $n(\theta)$ is given by (6.3-15).

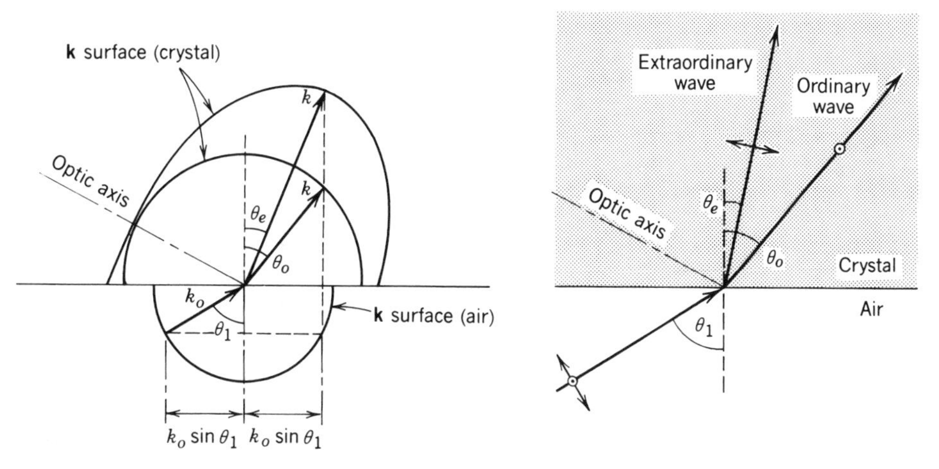

Figure 6.3-13 Determination of the angles of refraction by matching projections of the **k** vectors in air and in a uniaxial crystal.

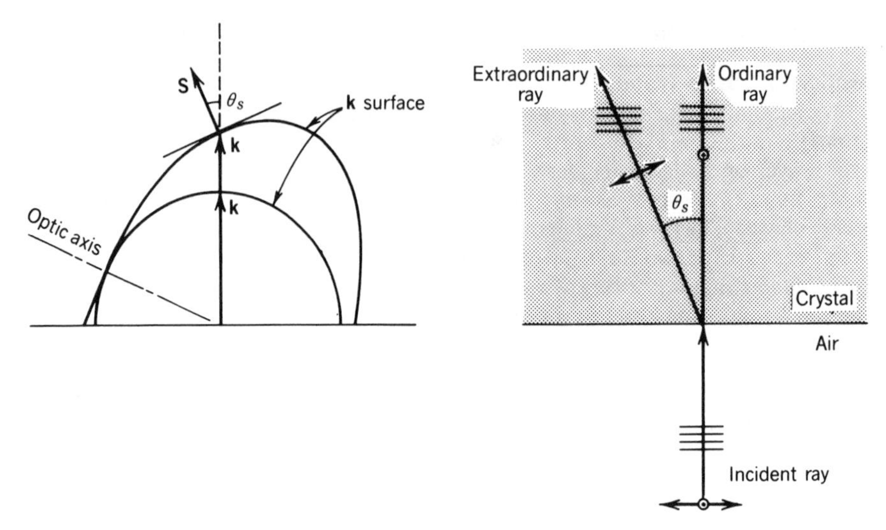

Figure 6.3-14 Double refraction at normal incidence.

If the incident wave carries the two polarizations, the two refracted waves will emerge.

Refraction of Rays

The previous analysis dealt with the refraction of plane waves. The refraction of rays is different since rays in an anisotropic medium do not necessarily travel in a direction normal to the wavefronts. In air, before entering the crystal, the wavefronts are normal to the rays. The refracted wave must have a wavevector satisfying the phase-matching condition, so that Snell's law (6.3-20) applies, with the angle of refraction θ determining the direction of **k**. Since the direction of **k** is not the direction of the ray, Snell's law is not applicable to rays.

An example that dramatizes the deviation from Snell's law is that of normal incidence at a uniaxial crystal whose optic axis is neither parallel nor perpendicular to the crystal boundary. The incident wave has a **k** vector normal to the boundary. To ensure phase matching, the refracted waves must also have wavevectors in the same direction. Intersections with the **k** surface yield two points corresponding to two waves. The ordinary ray is parallel to **k**. But the extraordinary ray points in the direction of the normal to the **k** surface, at an angle θ_s with the normal to the crystal boundary, as illustrated in Fig. 6.3-14. Thus normal incidence creates oblique refraction. Note, however, that the principle of phase matching is still maintained; wavefronts of both

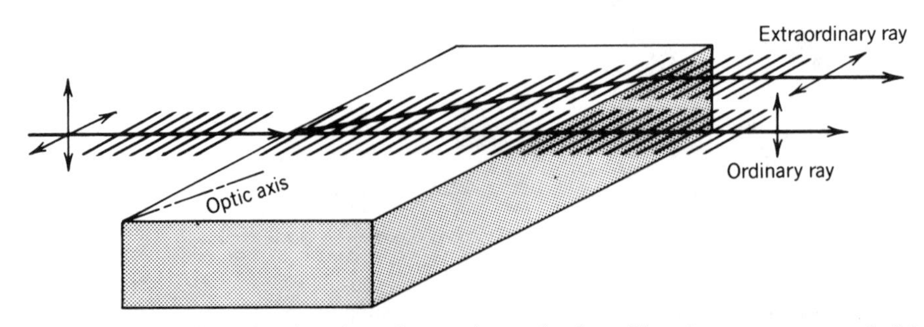

Figure 6.3-15 Double refraction through an anisotropic plate. The plate serves as a polarizing beamsplitter.

refracted rays are parallel to the crystal boundary and to the wavefront of the incident ray.

When light rays are transmitted through a plate of anisotropic material as described above, the two rays refracted at the first surface refract at the second surface, creating two laterally separated rays with orthogonal polarizations, as illustrated in Fig. 6.3-15.

6.4 OPTICAL ACTIVITY AND FARADAY EFFECT

A. Optical Activity

Certain materials act naturally as polarization rotators, a property known as optical activity. Their normal modes are circularly polarized, instead of linearly polarized waves; the waves with right- and left-circular polarizations travel at different phase velocities. Optical activity is found in materials in which the molecules have an inherently helical character. Examples are quartz, selenium, tellurium, and tellurium oxide (TeO_2). Many organic materials exhibit optical activity. The rotatory power and the sense of rotation are also sensitive to the chemical structure and concentration of solutions (this effect has been used, for example, to measure sugar content in solutions).

It will be shown subsequently that an optically active medium with right- and left-circular-polarization phase velocities c_o/n_+ and c_o/n_- acts as a polarization rotator with an angle of rotation $\pi(n_- - n_+)d/\lambda_o$ proportional to the distance d. The rotatory power (angle per unit length) of the optically active medium is therefore

$$\rho = \frac{\pi(n_- - n_+)}{\lambda_o}. \tag{6.4-1}$$

Rotatory Power

The direction of rotation of the polarization plane is in the same sense as that of the circularly polarized component of the greater phase velocity (smaller refractive index). If $n_+ < n_-$, ρ is positive and the rotation is in the same direction as the electric field vector of the right circularly polarized wave [clockwise when viewed from the direction toward which the wave is approaching, as illustrated in Fig. 6.4-1(*a*)].

The optically active medium is a spatially dispersive medium since the relation between $\mathbf{D}(\mathbf{r})$ and $\mathbf{E}(\mathbf{r})$ is not local. $\mathbf{D}(\mathbf{r})$ at position $\mathbf{r}$ is determined not only by $\mathbf{E}(\mathbf{r})$, but also by $\mathbf{E}(\mathbf{r}')$ at points $\mathbf{r}'$ in the immediate vicinity of $\mathbf{r}$ [since it is dependent on the derivatives in $\nabla \times \mathbf{E}(\mathbf{r})$]. Spatial dispersiveness is analogous to temporal dispersiveness, which is caused by the noninstantaneous response of the medium (see Sec. 5.2B).

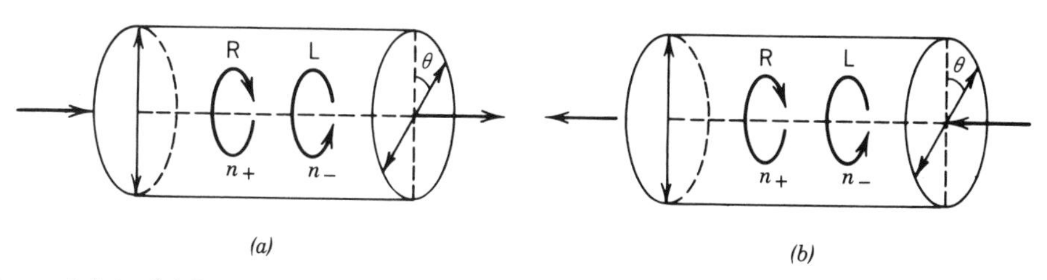

Figure 6.4-1 (*a*) Rotation of the plane of polarization in an optically active medium is a result of the difference in the velocities of the two circular polarizations. In this illustration, the right circularly polarized wave (R) is faster than the left circularly polarized wave (L), i.e., $n_+ < n_-$, so that ρ is positive. (*b*) If the wave in (*a*) is reflected after traversing the medium, the plane of polarization rotates in the opposite direction and the wave retraces itself.

Equation (6.4-1) may be obtained by decomposing the linearly polarized wave into a sum of right and left circularly polarized waves of equal amplitudes (see Exercise 6.1-1),

$$\begin{bmatrix} \cos\theta \\ \sin\theta \end{bmatrix} = \frac{1}{2}e^{-j\theta}\begin{bmatrix} 1 \\ j \end{bmatrix} + \frac{1}{2}e^{j\theta}\begin{bmatrix} 1 \\ -j \end{bmatrix},$$

where θ is the initial angle of the polarization plane. After a distance d of propagation in the medium, phase shifts $\varphi_+ = 2\pi n_+ d/\lambda_o$ and $\varphi_- = 2\pi n_- d/\lambda_o$, respectively, are encountered by the right and left circularly polarized waves, so that the new Jones vector is

$$\frac{1}{2}e^{-j\theta}e^{-j\varphi_+}\begin{bmatrix} 1 \\ j \end{bmatrix} + \frac{1}{2}e^{j\theta}e^{-j\varphi_-}\begin{bmatrix} 1 \\ -j \end{bmatrix} = e^{-j\varphi_o}\begin{bmatrix} \cos\left(\theta - \dfrac{\varphi}{2}\right) \\ \sin\left(\theta - \dfrac{\varphi}{2}\right) \end{bmatrix},$$

where $\varphi_o = \frac{1}{2}(\varphi_+ + \varphi_-)$ and $\varphi = \varphi_- - \varphi_+ = 2\pi(n_- - n_+)d/\lambda_o$. This Jones vector represents a linearly polarized wave with the plane of polarization rotated by an angle $\varphi/2 = \pi(n_- - n_+)d/\lambda_o$, as indicated above.

Medium Equations

We now show that a dielectric medium characterized by the medium equation

$$\mathbf{D} = \epsilon\mathbf{E} + \epsilon_o \xi\, j\omega\mathbf{B} = \epsilon\mathbf{E} - \epsilon_o \xi \nabla \times \mathbf{E}, \qquad (6.4\text{-}2)$$

where ξ is a constant, is optically active. This medium relation arises in molecular structures with a helical character. In these structures, a time-varying magnetic flux density $\mathbf{B}$ induces a circulating current that sets up an electric dipole moment (and hence polarization) proportional to $j\omega\mathbf{B} = -\nabla \times \mathbf{E}$, which is responsible for the last term in (6.4-2).

The optically active medium is a spatially dispersive medium since the relation between $\mathbf{D}(\mathbf{r})$ and $\mathbf{E}(\mathbf{r})$ is not local. $\mathbf{D}(\mathbf{r})$ at position $\mathbf{r}$ is determined not only by $\mathbf{E}(\mathbf{r})$, but also by $\mathbf{E}(\mathbf{r}')$ at points $\mathbf{r}'$ in the immediate vicinity of $\mathbf{r}$ [since it is dependent on the derivatives in $\nabla \times \mathbf{E}(\mathbf{r})$]. Spatial dispersiveness is analogous to temporal dispersiveness, which is caused by the noninstantaneous response of the medium (see Sec. 5.2B).

We proceed to show that the two normal modes of a medium satisfying (6.4-2) are circularly polarized waves and we determine the velocities c_o/n_+ and c_o/n_- in terms of the constant ξ.

Normal Modes of the Optically Active Medium

Consider the propagation of a plane wave $\mathbf{E}(\mathbf{r}) = \mathbf{E}\exp(-j\mathbf{k}\cdot\mathbf{r})$ in a medium satisfying (6.4-2). Setting $\mathbf{D}(\mathbf{r}) = \mathbf{D}\exp(-j\mathbf{k}\cdot\mathbf{r})$, (6.4-2) yields

$$\mathbf{D} = \epsilon\mathbf{E} + j\epsilon_o \mathbf{G} \times \mathbf{E}, \qquad (6.4\text{-}3)$$

where

$$\mathbf{G} = \xi\mathbf{k} \qquad (6.4\text{-}4)$$

is known as the **gyration vector**. Clearly, the vector $\mathbf{D}$ is not parallel to $\mathbf{E}$ since the vector $\mathbf{G} \times \mathbf{E}$ in (6.4-3) is perpendicular to $\mathbf{E}$. The relation between $\mathbf{D}$ and $\mathbf{E}$ is therefore dependent on the wavevector $\mathbf{k}$, which is not surprising since the medium is spatially dispersive. (This is analogous to the dependence of the dielectric properties of a temporally dispersive medium on ω.)

For simplicity, we assume that $\boldsymbol{\epsilon}$ has uniaxial symmetry (with indices n_o and n_e), use the principal axes of the tensor $\boldsymbol{\epsilon}$ as a coordinate system, and consider only waves propagating along the optic axis. The first term in (6.4-3) then corresponds to propagation of an ordinary wave of refractive index n_o.

To prove that the normal modes are circularly polarized, consider the two circularly polarized waves of electric-field vectors $\mathbf{E} = (E_0, \pm jE_0, 0)$ and wavevector $\mathbf{k} = (0, 0, k)$. The + and − signs correspond to right and left circularly polarized cases, respectively. Substituting in (6.4-3), we obtain $\mathbf{D} = (D_0, \pm jD_0, 0)$, where $D_0 = \epsilon_o(n_o^2 \pm G)E_0$. It follows that $\mathbf{D} = \epsilon_o n_\pm^2 \mathbf{E}$, where

$$n_\pm = \left(n_o^2 \pm G\right)^{1/2}, \tag{6.4-5}$$

so that for either of the two circularly polarized waves the vector $\mathbf{D}$ is parallel to the vector $\mathbf{E}$. Equation (6.3-10) is satisfied if the wavenumber $k = n_\pm k_0$. Thus the right and left circularly polarized waves propagate, without change of their state of polarization, with refractive indices n_+ and n_-, respectively. They *are* the normal modes for this medium.

EXERCISE 6.4-1

Rotatory Power of an Optically Active Medium. Show that if $G \ll n_o$, the rotatory power of an optically active medium (rotation of the polarization plane per unit length) is approximately given by

$$\rho \approx -\frac{\pi G}{\lambda_o n_o}. \tag{6.4-6}$$

The rotatory power is strongly dependent on the wavelength. Since G is proportional to k, as indicated by (6.4-4), it is inversely proportional to the wavelength λ_o. Thus the rotatory power in (6.4-6) is inversely proportional to λ_o^2. In addition, the refractive index n_o is itself wavelength dependent. The rotatory power ρ of quartz is ≈ 31 deg/mm at $\lambda_o = 500$ nm and ≈ 22 deg/mm at 600 nm; for silver thiogallate ($AgGaS_2$) ρ is ≈ 700 deg/mm at 490 nm and ≈ 500 deg/mm at 500 nm.

B. Faraday Effect

Certain materials act as polarization rotators when placed in a static magnetic field, a property known as the Faraday effect. The angle of rotation is proportional to the distance, and the rotatory power ρ (angle per unit length) is proportional to the component B of the magnetic flux density in the direction of wave propagation,

$$\rho = VB, \tag{6.4-7}$$

where V is known as the **Verdet constant.**

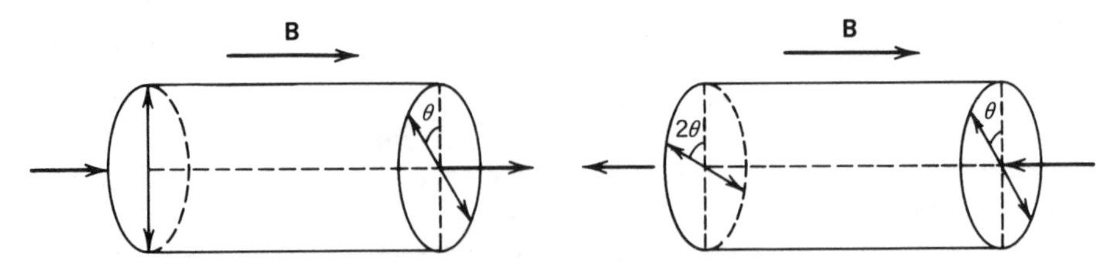

Figure 6.4-2 Polarization rotation in a medium exhibiting the Faraday effect. The sense of rotation is invariant to the direction of travel of the wave.

The sense of rotation is governed by the direction of the magnetic field: for $V > 0$, the rotation is in the direction of a right-handed screw pointing in the direction of the magnetic field. In contradistinction to optical activity, the sense of rotation does not reverse with the reversal of the direction of propagation of the wave (Fig. 6.4-2). When a wave travels through a Faraday rotator, reflects back onto itself, and travels once more through the rotator in the opposite direction, it undergoes twice the rotation.

The medium equation for materials exhibiting the Faraday effect is

$$\mathbf{D} = \boldsymbol{\epsilon}\mathbf{E} + j\epsilon_o\gamma\mathbf{B} \times \mathbf{E}, \tag{6.4-8}$$

where $\mathbf{B}$ is the magnetic flux density and γ is a constant of the medium that is called the **magnetogyration coefficient**. This relation originates from the interaction of the static magnetic field $\mathbf{B}$ with the motion of electrons in the molecules under the influence of the optical electric field $\mathbf{E}$.

To establish an analogy between the Faraday effect and optical activity (6.4-8) is written as

$$\mathbf{D} = \boldsymbol{\epsilon}\mathbf{E} + j\epsilon_o\mathbf{G} \times \mathbf{E}, \tag{6.4-9}$$

where

$$\mathbf{G} = \gamma\mathbf{B}. \tag{6.4-10}$$

Equation (6.4-9) is identical to (6.4-3) with the vector $\mathbf{G} = \gamma\mathbf{B}$ in Faraday rotators playing the role of the gyration vector $\mathbf{G} = \xi\mathbf{k}$ in optically active media. Note that in the Faraday effect $\mathbf{G}$ is independent of $\mathbf{k}$, so that reversal of the direction of propagation does not reverse the sense of rotation of the polarization plane. This property can be used to make optical isolators, as explained in Sec. 6.6.

With this analogy, and using (6.4-6), we conclude that the rotatory power of the Faraday medium is $\rho \approx -\pi G/\lambda_o n_o = -\pi\gamma B/\lambda_o n_o$, from which the Verdet constant (the rotatory power per unit magnetic flux density) is

$$\boxed{V \approx -\frac{\pi\gamma}{\lambda_o n_o}.} \tag{6.4-11}$$

Clearly, the Verdet constant is a function of the wavelength λ_o.

Materials that exhibit the Faraday effect include glasses, yttrium–iron–garnet (YIG), terbium–gallium–garnet (TGG), and terbium–aluminum–garnet (TbAlG). The Verdet constant V of TbAlG is $V = -1.16$ min/cm-Oe at $\lambda_o = 500$ nm.

6.5 OPTICS OF LIQUID CRYSTALS

Liquid Crystals

The liquid-crystal state is a state of matter in which the elongated (typically cigar-shaped) molecules have orientational order (like crystals) but lack positional order (like liquids). There are three types (phases) of liquid crystals, as illustrated in Fig. 6.5-1:

- In **nematic** liquid crystals the molecules tend to be parallel but their positions are random.
- In **smectic** liquid crystals the molecules are parallel, but their centers are stacked in parallel layers within which they have random positions, so that they have positional order in only one dimension.
- The **cholesteric** phase is a distorted form of the nematic phase in which the orientation undergoes helical rotation about an axis.

Liquid crystallinity is a *fluid* state of matter. The molecules change orientation when subjected to a force. For example, when a thin layer of liquid crystal is placed between two parallel glass plates the molecular orientation is changed if the plates are rubbed; the molecules orient themselves along the direction of rubbing.

Twisted nematic liquid crystals are nematic liquid crystals on which a twist, similar to the twist that exists naturally in the cholesteric phase, is imposed by external forces (for example, by placing a thin layer of the liquid crystal material between two glass plates polished in perpendicular directions as shown in Fig. 6.5-2). Because twisted nematic liquid crystals have enjoyed the greatest number of applications in photonics (in liquid-crystal displays, for example), this section is devoted to their optical properties. The electro-optic properties of twisted nematic liquid crystals, and their use as optical modulators and switches, are described in Chap. 18.

Optical Properties of Twisted Nematic Liquid Crystals

The twisted nematic liquid crystal is an optically *inhomogeneous anisotropic medium* that acts locally as a uniaxial crystal, with the optic axis parallel to the molecular

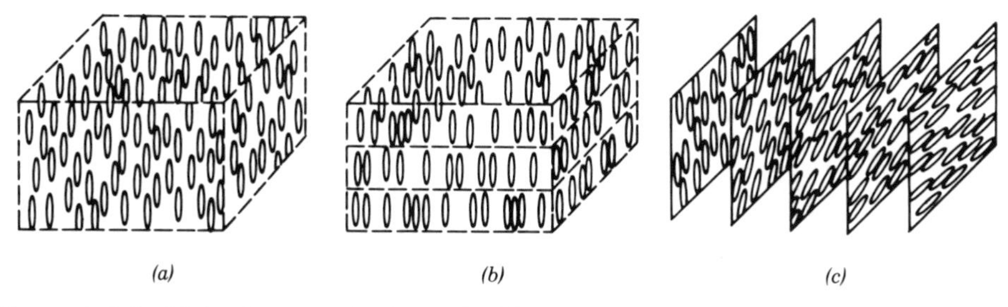

Figure 6.5-1 Molecular organizations of different types of liquid crystals: (a) nematic; (b) smectic; (c) cholesteric.

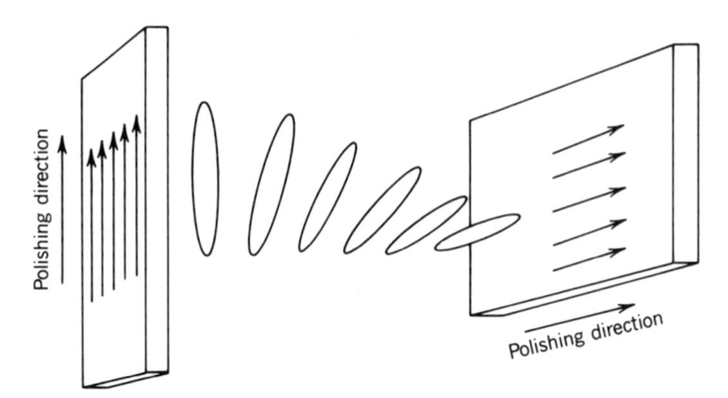

Figure 6.5-2 Molecular orientations of the twisted nematic liquid crystal.

direction. The optical properties are conveniently studied by dividing the material into thin layers perpendicular to the axis of twist, each of which acts as a uniaxial crystal, with the optic axis rotating gradually in a helical fashion (Fig. 6.5-3). The cumulative effects of these layers on the transmitted wave is determined. We proceed to show that under certain conditions the twisted nematic liquid crystal acts as a polarization rotator, with the polarization plane rotating in alignment with the molecular twist.

Consider the propagation of light along the axis of twist (the z axis) of a twisted nematic liquid crystal and assume that the twist angle varies linearly with z,

$$\theta = \alpha z, \tag{6.5-1}$$

where α is the twist coefficient (degrees per unit length). The optic axis is therefore parallel to the x-y plane and makes an angle θ with the x direction. The ordinary and extraordinary indices are n_o and n_e (typically, $n_e > n_o$), and the phase retardation coefficient (retardation per unit length) is

$$\beta = (n_e - n_o)k_o. \tag{6.5-2}$$

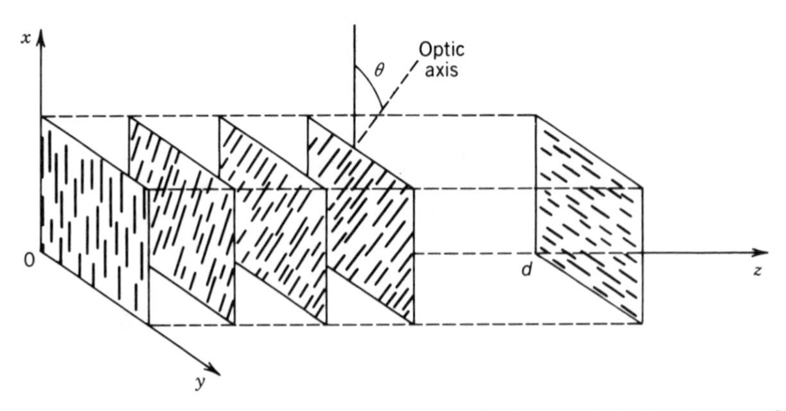

Figure 6.5-3 Propagation of light in a twisted nematic liquid crystal. In this diagram the angle of twist is 90°.

The liquid crystal cell is described completely by the twist coefficient α and the retardation coefficient β.

In practice, β is much greater than α, so that many cycles of phase retardation are introduced before the optic axis rotates appreciably. We show below that if the incident wave at $z = 0$ is linearly polarized in the x direction, then when $\beta \gg \alpha$, the wave maintains its linearly polarized state, but the plane of polarization rotates in alignment with the molecular twist, so that the angle of rotation is $\theta = \alpha z$ and the total rotation in a crystal of length d is the angle of twist αd. The liquid crystal cell then serves as a polarization rotator with rotatory power α. The polarization rotation property of the twisted nematic liquid crystal is useful for making display devices, as explained in Sec. 18.3.

Proof. We proceed to show that the twisted nematic liquid crystal acts as a polarization rotator if $\beta \gg \alpha$. We divide the width d of the cell into N incremental layers of equal widths $\Delta z = d/N$. The mth layer located at distance $z = z_m = m\,\Delta z$, $m = 1, 2, \ldots, N$, is a wave retarder whose slow axis (the optic axis) makes an angle $\theta_m = m\Delta\theta$ with the x axis, where $\Delta\theta = \alpha\Delta z$. It therefore has a Jones matrix

$$\mathbf{T}_m = \mathbf{R}(-\theta_m)\mathbf{T}_r\mathbf{R}(\theta_m), \tag{6.5-3}$$

where

$$\mathbf{T}_r = \begin{bmatrix} \exp(-jn_e k_o \Delta z) & 0 \\ 0 & \exp(-jn_o k_o \Delta z) \end{bmatrix} \tag{6.5-4}$$

is the Jones matrix of a retarder with axis in the x direction and $\mathbf{R}(\theta)$ is the coordinate rotation matrix in (6.1-15) [see (6.1-17)].

It is convenient to rewrite $\mathbf{T}_r$ in terms of the phase retardation coefficient $\beta = (n_e - n_o)k_o$,

$$\mathbf{T}_r = \exp(-j\varphi\,\Delta z)\begin{bmatrix} \exp\left(-j\beta\dfrac{\Delta z}{2}\right) & 0 \\ 0 & \exp\left(j\beta\dfrac{\Delta z}{2}\right) \end{bmatrix}, \tag{6.5-5}$$

where $\varphi = (n_o + n_e)k_o/2$. Since multiplying the Jones vector by a constant phase factor does not affect the state of polarization, we shall simply ignore the prefactor $\exp(-j\varphi\,\Delta z)$ in (6.5-5).

The overall Jones matrix of the device is the product

$$\mathbf{T} = \prod_{m=1}^{N} \mathbf{T}_m = \prod_{m=1}^{N} \mathbf{R}(-\theta_m)\mathbf{T}_r\mathbf{R}(\theta_m). \tag{6.5-6}$$

Using (6.5-3) and noting that $\mathbf{R}(\theta_m)\mathbf{R}(-\theta_{m-1}) = \mathbf{R}(\theta_m - \theta_{m-1}) = \mathbf{R}(\Delta\theta)$, we obtain

$$\mathbf{T} = \mathbf{R}(-\theta_N)\left[\mathbf{T}_r\mathbf{R}(\Delta\theta)\right]^{N-1}\mathbf{T}_r\mathbf{R}(\theta_1). \tag{6.5-7}$$

Substituting from (6.5-5) and (6.1-15)

$$\mathbf{T}_r\mathbf{R}(\Delta\theta) = \begin{bmatrix} \exp\left(-j\beta\frac{\Delta z}{2}\right) & 0 \\ 0 & \exp\left(j\beta\frac{\Delta z}{2}\right) \end{bmatrix} \begin{bmatrix} \cos\alpha\Delta z & \sin\alpha\Delta z \\ -\sin\alpha\Delta z & \cos\alpha\Delta z \end{bmatrix}. \quad (6.5\text{-}8)$$

Using (6.5-7) and (6.5-8), the Jones matrix $\mathbf{T}$ of the device can, in principle, be determined in terms of the parameters α, β, and $d = N\Delta z$.

When $\alpha \ll \beta$, we can assume that the incremental rotation matrix $\mathbf{R}(\Delta\theta)$ is approximately an identity matrix and obtain

$$\mathbf{T} \approx \mathbf{R}(-\theta_N)[\mathbf{T}_r]^N\mathbf{R}(\theta_1) = \mathbf{R}(-\alpha N\Delta z)\begin{bmatrix} \exp\left(-j\beta\frac{\Delta z}{2}\right) & 0 \\ 0 & \exp\left(j\beta\frac{\Delta z}{2}\right) \end{bmatrix}^N$$

$$= \mathbf{R}(-\alpha N\Delta z)\begin{bmatrix} \exp\left(-j\beta N\frac{\Delta z}{2}\right) & 0 \\ 0 & \exp\left(j\beta N\frac{\Delta z}{2}\right) \end{bmatrix}.$$

In the limit as $N \to \infty$, $\Delta z \to 0$, and $N\Delta z \to d$,

$$\mathbf{T} = \mathbf{R}(-\alpha d)\begin{bmatrix} \exp\left(-j\beta\frac{d}{2}\right) & 0 \\ 0 & \exp\left(j\beta\frac{d}{2}\right) \end{bmatrix}. \quad (6.5\text{-}9)$$

This Jones matrix represents a wave retarder of retardation βd with the slow axis along the x direction, followed by a polarization rotator with rotation angle αd. If the original wave is linearly polarized along the x direction the wave retarder provides only a phase shift; the device then simply rotates the polarization by an angle αd equal to the twist angle.

6.6 POLARIZATION DEVICES

This section is a brief description of a number of devices that are used to modify the state of polarization of light. The basic principles of most of these devices have been discussed earlier in this chapter.

A. Polarizers

A polarizer is a device that transmits the component of the electric field in the direction of its transmission axis and blocks the orthogonal component. This preferential treatment of the two components of the electric field is achieved by selective absorption, selective reflection from an isotropic medium, or selective reflection/refraction at the boundary of an anisotropic medium.

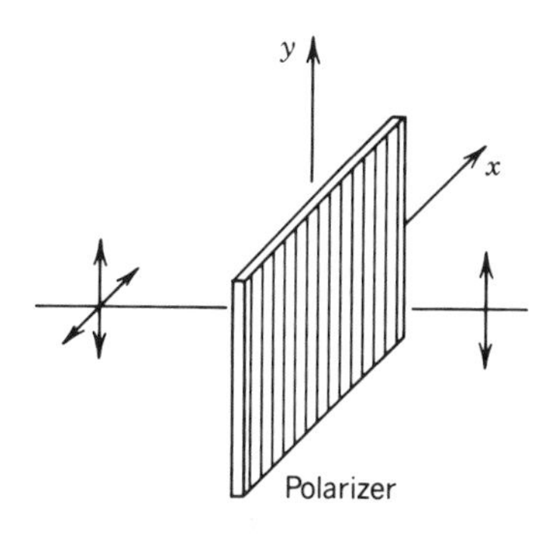

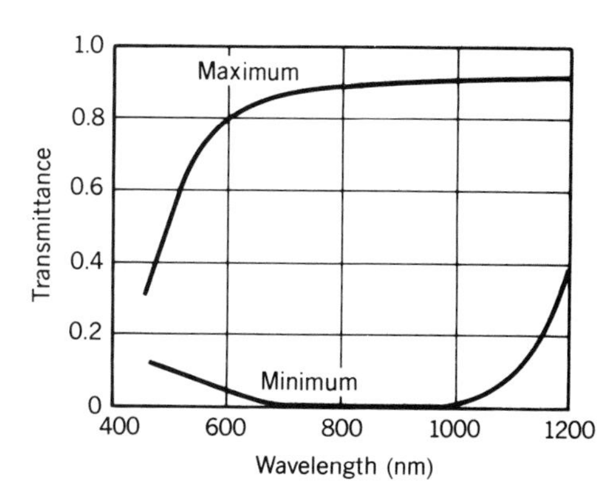

Figure 6.6-1 Power transmittances of a typical dichroic polarizer with the polarization plane of the light aligned for maximum and minimum transmittance.

Polarization by Selective Absorption (Dichroism)

The absorption of light by certain anisotropic materials, called **dichroic materials**, depends on the direction of the electric field (Fig. 6.6-1). These materials have anisotropic molecular structures whose response is sensitive to the direction of the applied field. The most common dichroic material is the Polaroid H-sheet (basically a sheet of polyvinyl alcohol heated and stretched in a certain direction then impregnated with iodine atoms).

Polarization by Selective Reflection

The reflection of light from the boundary between two dielectric isotropic materials is polarization dependent (see Sec. 6.2). At the Brewster angle of incidence, light of TM polarization is not reflected (i.e., is totally refracted). At this angle, only the TE component of the incident light is reflected, so that the reflector serves as a polarizer (Fig. 6.6-2).

Polarization by Selective Refraction in Anisotropic Media (Polarizing Beamsplitters)

When light refracts at the surface of an anisotropic crystal the two polarizations refract at different angles and are spatially separated (see Sec. 6.3E and Fig. 6.3-15). This is an excellent way of obtaining polarized light from unpolarized light. The device usually

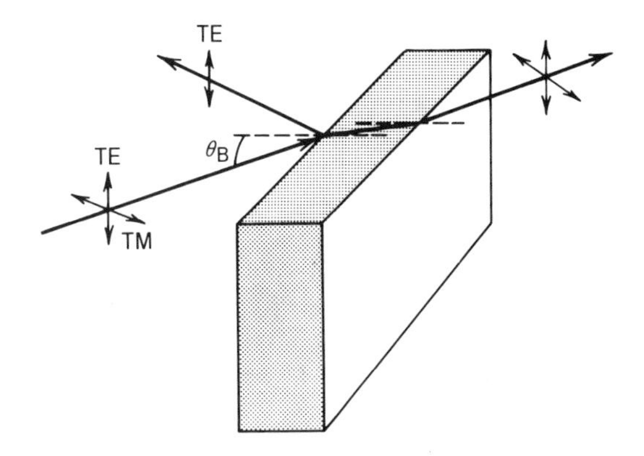

Figure 6.6-2 Brewster-angle polarizer.

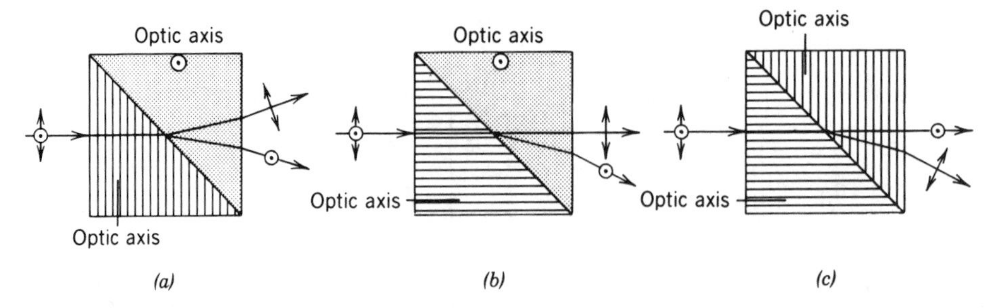

Figure 6.6-3 Polarizing prisms: (*a*) Wollaston prism; (*b*) Rochon prism; (*c*) Sénarmont prism. The directions and polarizations of the exiting waves differ in the three cases. In this illustration, the crystals are negative uniaxial (e.g., calcite).

takes the form of two cemented prisms made of anisotropic (uniaxial) crystals in different orientations, as illustrated by the examples in Fig. 6.6-3. These prisms serve as **polarizing beamsplitters**.

B. Wave Retarders

The wave retarder is characterized by its retardation Γ and its fast and slow axes (see Sec. 6.1B). The normal modes are linearly polarized waves polarized in the directions of the axes, and the velocities are different. Upon transmission through the retarder, a relative phase shift Γ between these modes ensues.

Wave retarders are often made of anisotropic materials. As explained in Sec. 6.3B, when light travels along a principal axis of a crystal (say the z axis), the normal modes are linearly polarized waves pointing along the two other principal axes (x and y axes). The two modes travel with the principal refractive indices n_1 and n_2. If $n_1 < n_2$, the x axis is the fast axis. If the plate has a thickness d, the phase retardation is $\Gamma = (n_2 - n_1)k_o d = 2\pi(n_2 - n_1)d/\lambda_o$. The retardation is directly proportional to the thickness d and inversely proportional to the wavelength λ_o (note, however, that $n_2 - n_1$ itself is wavelength dependent).

The refractive indices of mica, for example, are 1.599 and 1.594 at $\lambda_o = 633$ nm, so that $\Gamma/d \approx 15.8\pi$ rad/mm. A 63.3-μm thin sheet is a half-wave retarder ($\Gamma \approx \pi$).

Control of Light Intensity by Use of a Wave Retarder and Two Polarizers

The power (or intensity) transmittance of a system constructed from a wave retarder of retardation Γ placed between two crossed polarizers, at 45° with respect to the retarder's axes, as shown in Fig. 6.6-4, is

$$\mathscr{T} = \sin^2\frac{\Gamma}{2}. \tag{6.6-1}$$

This may be obtained by use of Jones matrices or by examining the polarization ellipse of the retarded light as a function of Γ and determining the component in the direction of the output polarizer, as illustrated in Fig. 6.6-4. If $\Gamma = 0$, no light is transmitted since the polarizers are orthogonal. If $\Gamma = \pi$, all the light is transmitted since the retarder rotates the polarization 90°, making it match the transmission axis of the second polarizer.

The intensity of the transmitted light can be controlled by altering the retardation Γ (for example, by changing the indices n_1 and n_2). This is the basic principle underlying the electro-optic modulators discussed in Chap. 18.

Furthermore, since Γ depends on d, slight variations in the thickness of a sample can be monitored by examining the pattern of the transmitted light. Also since Γ is

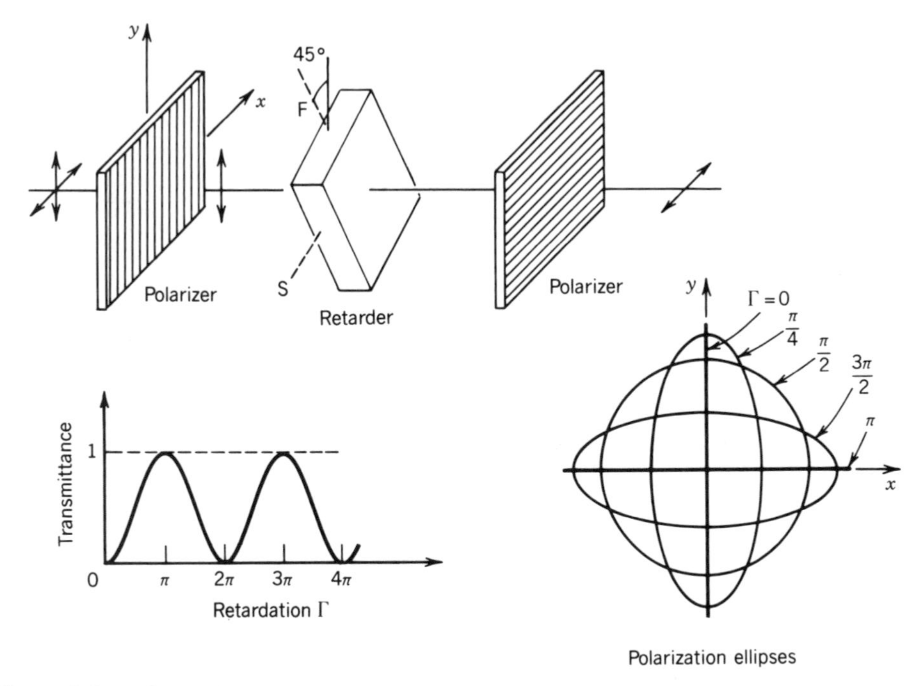

Figure 6.6-4 Controlling light intensity by use of a wave retarder with variable retardation Γ between two crossed polarizers.

wavelength dependent, the transmittance of the system is frequency sensitive. The system therefore serves as a filter, but the selectivity is not very sharp. Other configurations using wave retarders and polarizers can be used to construct narrowband transmission filters.

C. Polarization Rotators

A polarization rotator rotates the plane of polarization of linearly polarized light by a fixed angle, maintaining its linearly polarized nature. Optically active media and materials exhibiting the Faraday effect act as polarization rotators, as shown in Sec. 6.4. The twisted nematic liquid crystal also acts as a polarization rotator under certain conditions, as shown in Sec. 6.5.

If a polarization rotator is placed between two polarizers, the amount of transmitted light depends on the rotation angle. The intensity of light can be controlled (modulated) if the angle of rotation is controlled by some external means (e.g., by varying the magnetic flux density applied to a Faraday rotator, or by changing the molecular orientation of a liquid crystal by means of an applied electric field). Electro-optic modulation of light and liquid-crystal display devices are discussed in Chap. 18.

Optical Isolators

An optical isolator is a device that transmits light in only one direction, thus acting as a "one-way valve." Optical isolators are useful in preventing reflected light from returning back to the source. This type of feedback can have deleterious effects on the operation of certain light sources (semiconductor lasers, for example).

A system made of a polarizing beamsplitter followed by a quarter-wave retarder acts as an isolator. Light traveling in the forward direction is polarized by the cube, then circularly polarized by the retarder. Upon reflection from a mirror beyond the retarder,

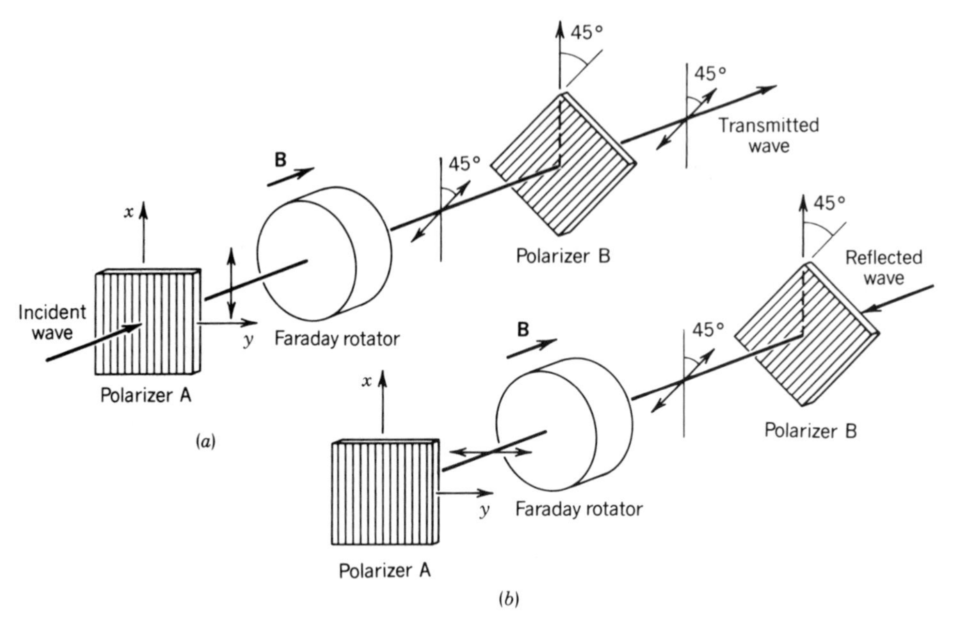

Figure 6.6-5 An optical isolator using a Faraday rotator transmits light in one direction, as in (*a*), and blocks light in the opposite direction, as in (*b*).

the sense of rotation is reversed (left to right, or vice versa), so that upon transmission back through the retarder it becomes polarized in the orthogonal direction and is therefore blocked by the polarizing cube (see Problem 6.1-6). Although this type of isolator can offer attenuation of the backward wave up to 30 dB (0.1%), it operates only over a narrow wavelength range.

A Faraday rotator placed between two polarizers making a 45° angle with each other can also be used as an optical isolator. The magnetic flux density applied to the rotator is adjusted so that the polarization is rotated by 45° in the direction of a right-handed screw pointing in the z direction [Fig. 6.6-5(*a*)]. Light traveling from left to right crosses polarizer A, rotates 45°, and is transmitted through polarizer B. However, light traveling in the opposite direction [Fig. 6.6-5(*b*)], although it crosses polarizer B, rotates an additional 45° and is blocked by polarizer A. A Faraday rotator cannot be replaced by an optically active or liquid-crystal polarization rotator since, in those devices, the sense of rotation is such that the polarization of the reflected wave retraces that of the incident wave and is therefore transmitted back through the polarizers to the source. Faraday-rotator isolators made of yttrium–iron–garnet (YIG) or terbium–gallium–garnet (TGG), for example, can offer an attenuation of the backward wave up to 90 dB, over a relatively wide wavelength range.

READING LIST

General

See also the list of general books on optics in Chapter 1.

D. S. Kliger, J. W. Lewis, and C. E. Randall, *Polarized Light in Optics and Spectroscopy*, Academic Press, Boston, 1990.

J. F. Nye, *Physical Properties of Crystals: Their Representation by Tensors and Matrices*, Oxford University Press, New York, 1967, 2nd ed. 1984.

R. M. A. Azzam and N. M. Bashara, *Ellipsometry and Polarized Light*, North-Holland, Amsterdam, 1977.

W. Swindell, *Polarized Light*, Dowden, Hutchinson & Ross, Stroudsburg, PA, 1975.

B. A. Robson, *The Theory of Polarization Phenomena*, Clarendon Press, Oxford, 1974.

D. Clarke and J. F. Grainger, *Polarized Light and Optical Measurement*, Pergamon Press, Oxford, 1971.

P. Gay, *An Introduction to Crystal Optics*, Longmans, London, 1967.

L. Velluz, M. LeGrand, and M. Grosjean, *Optical Circular Dichroism*, Academic Press, New York, 1965.

E. A. Wood, *Crystals and Light: An Introduction to Optical Crystallography*, Van Nostrand, Princeton, NJ, 1964.

W. A. Shurcliff and S. S. Ballard, *Polarized Light*, Van Nostrand, Princeton, NJ, 1964.

W. A. Shurcliff, *Polarized Light: Production and Use*, Harvard University Press, Cambridge, MA, 1962.

Books on Liquid Crystals

J. L. Ericksen and D. Kinderlehrer, eds., *Theory and Applications of Liquid Crystals*, Springer-Verlag, New York, 1987.

L. M. Blinov, *Electro-Optical and Magneto-Optical Properties of Liquid Crystals*, Wiley, New York, 1983.

W. H. de Jeu, *Physical Properties of Liquid Crystalline Materials*, Gordon and Breach, New York, 1980.

P.-G. de Gennes, *The Physics of Liquid Crystals*, Clarendon Press, Oxford, 1974, 1979.

S. Chandrasekhar, *Liquid Crystals*, Cambridge University Press, New York, 1977.

G. Meier, E. Sackmann, and J. G. Grabmaier, *Applications of Liquid Crystals*, Springer-Verlag, Berlin, 1975.

Articles

J. M. Bennett and H. E. Bennett, Polarization, in *Handbook of Optics*, W. G. Driscoll, ed., McGraw-Hill, New York, 1978.

V. M. Agranovich and V. L. Ginzburg, Crystal Optics with Spatial Dispersion, in *Progress in Optics*, vol. 9, E. Wolf, ed., North-Holland, Amsterdam, 1971.

PROBLEMS

6.1-1 **Orthogonal Polarizations.** Show that if two elliptically polarized states are orthogonal, the major axes of their ellipses are perpendicular and the senses of rotation are opposite.

6.1-2 **Rotating a Polarization Rotator.** Show that the Jones matrix of a polarization rotator is invariant to rotation of the coordinate system.

6.1-3 **The Half-Wave Retarder.** Linearly polarized light is transmitted through a half-wave retarder. If the polarization plane makes an angle θ with the fast axis of the retarder, show that the transmitted light is linearly polarized at an angle $-\theta$, i.e., rotates by an angle 2θ. Why is the half-wave retarder not equivalent to a polarization rotator?

6.1-4 **Wave Retarders in Tandem.** Write down the Jones matrices for:
(a) A $\pi/2$ wave retarder with the fast axis at 0°.
(b) A π wave retarder with the fast axis at 45°.
(c) A $\pi/2$ wave retarder with the fast axis at 90°.
If these three retarders are placed in tandem, show that the resulting device introduces a 90° rotation with a phase shift $\pi/2$.

6.1-5 **Reflection of Circularly Polarized Light.** Show that circularly polarized light changes handedness (right becomes left, and vice versa) upon reflection from a mirror.

6.1-6 **Optical Isolators.** An optical isolator transmits light traveling in one direction and blocks it in the opposite direction. Show that isolation of the light reflected by a planar mirror may be achieved by using a combination of a linear polarizer and a quarter-wave retarder with axes at 45° with respect to the transmission axis of the polarizer.

6.2-1 **Reflectance of Glass.** A plane wave is incident from air ($n = 1$) onto a glass plate ($n = 1.5$) at an angle of incidence 45°. Determine the intensity reflectances of the TE and TM waves. What is the average reflectance for unpolarized light (light carrying TE and TM waves of equal intensities)?

6.2-2 **Refraction at the Brewster Angle.** Show that at the Brewster angle of incidence the directions of the reflected and refracted waves are orthogonal. The electric field of the refracted TM wave is then parallel to the direction of the reflected wave.

6.2-3 **Retardation Associated with Total Internal Reflection.** Determine the phase retardation between the TE and TM waves introduced by total internal reflection at the boundary between glass ($n = 1.5$) and air ($n = 1$) at an angle of incidence $\theta = 1.2\theta_c$, where θ_c is the critical angle.

6.2-4 **Goos–Hänchen Shift.** Two TE plane waves undergo total internal reflection at angles θ and $\theta + d\theta$, where $d\theta$ is an incremental angle. If the phase retardation introduced between the reflected waves is written in the form $d\varphi = \xi\, d\theta$, determine an expression for the coefficient ξ. Sketch the interference patterns of the two incident waves and the two reflected waves and verify that they are shifted by a lateral distance proportional to ξ. When the incident wave is a beam (composed of many plane-wave components), the reflected beam is displaced laterally by a distance proportional to ξ. This effect is known as the Goos–Hänchen effect.

6.2-5 **Reflection from an Absorptive Medium.** Use Maxwell's equations and appropriate boundary conditions to show that the complex amplitude reflectance at the boundary between free space and a medium with refractive index n and absorption coefficient α at normal incidence is $r = [(n - j\alpha c/2\omega) - 1]/[(n - j\alpha c/2\omega) + 1]$.

6.3-1 **Maximum Retardation in Quartz.** Quartz is a positive uniaxial crystal with $n_e = 1.553$ and $n_o = 1.544$. (a) Determine the retardation per mm at $\lambda_o = 633$ nm when the crystal is oriented such that retardation is maximized. (b) At what thickness(es) does the crystal act as a quarter-wave retarder?

6.3-2 **Maximum Extraordinary Effect.** Determine the direction of propagation for which the angle between the wavevector **k** and the Poynting vector **S** (also the direction of ray propagation) in quartz ($n_e = 1.553$ and $n_o = 1.544$) is maximum.

6.3-3 **Double Refraction.** A plane wave is incident from free space onto a quartz crystal ($n_e = 1.553$ and $n_o = 1.544$) at an angle of incidence 30°. The optic axis is in the plane of incidence and is perpendicular to the direction of the incident wave. Determine the directions of the wavevectors and the rays of the two refracted waves.

6.3-4 **Lateral Shift in Double Refraction.** What is the optimum geometry for maximizing the lateral shift between the refracted ordinary and extraordinary beams in a positive uniaxial crystal? Indicate all pertinent angles and directions.

6.3-5 **Transmission Through a $LiNbO_3$ Plate.** Examine the transmission of an unpolarized He–Ne laser beam ($\lambda_o = 633$ nm) through a $LiNbO_3$ ($n_e = 2.29$, $n_o = 2.20$)

plate of thickness 1 cm, cut such that its optic axis makes an angle 45° with the normal to the plate. Determine the lateral shift and the retardation between the ordinary and extraordinary beams.

*6.3-6 **Conical Refraction.** When the wavevector **k** points along an optic axis of a biaxial crystal an unusual situation occurs. The two sheets of the **k** surface meet and the surface can be approximated by a conical surface. A ray is incident normal to the surface of a biaxial crystal with one of its optic axes also normal to the surface. Show that multiple refraction occurs with the refracted rays forming a cone. This effect is known as conical refraction. What happens when the conical rays refract from the parallel surface of the crystal into air?

6.6-1 **Circular Dichroism.** Determine the Jones matrix for a device that converts light with any state of polarization into right circularly polarized light. Certain materials have different absorption coefficients for right and left circularly polarized light, a property known as **circular dichroism**.

6.6-2 **Polarization Rotation by a Sequence of Linear Polarizers.** A wave that is linearly polarized in the x direction is transmitted through a sequence of N linear polarizers whose transmission axes are inclined by angles $m\theta$ ($m = 1, 2, \ldots, N$; $\theta = \pi/2N$) with respect to the x axis. Show that the transmitted light is linearly polarized in the y direction but its amplitude is reduced by the factor $\cos^N\theta$. What happens in the limit $N \to \infty$? *Hint*: Use Jones matrices and note that

$$\mathbf{R}[(m+1)\theta]\mathbf{R}(-m\theta) = \mathbf{R}(\theta),$$

where $\mathbf{R}(\theta)$ is the coordinate transformation matrix.

CHAPTER

7

GUIDED-WAVE OPTICS

7.1 PLANAR-MIRROR WAVEGUIDES

7.2 PLANAR DIELECTRIC WAVEGUIDES
- A. Waveguide Modes
- B. Field Distributions
- C. Group Velocities

7.3 TWO-DIMENSIONAL WAVEGUIDES

7.4 OPTICAL COUPLING IN WAVEGUIDES
- A. Input Couplers
- B. Coupling Between Waveguides

John Tyndall (1820–1893) was the first to demonstrate total internal reflection, which is the basis of guided-wave optics.

Conventional optical instruments make use of light that is transmitted between different locations in the form of beams that are collimated, relayed, focused, or scanned by mirrors, lenses, and prisms. Optical beams diffract and broaden, but they can be refocused by the use of lenses and mirrors. Although such beams are easily obstructed or scattered by various objects, this form of free-space transmission of light is the basis of most optical systems. There is, however, a relatively new technology for transmitting light through dielectric conduits, **guided-wave optics.** It has been developed to provide long-distance light transmission without the use of relay lenses. Guided-wave optics has important applications in directing light to awkward places, in establishing secure communications, and in the fabrication of miniaturized optical and optoelectronic devices requiring the confinement of light.

The basic concept of optical confinement is quite simple. A medium of one refractive index imbedded in a medium of lower refractive index acts as a light "trap" within which optical rays remain confined by multiple total internal reflections at the boundaries. Because this effect facilitates the confinement of light generated inside a medium of high refractive index (see Exercise 1.2-6), it can be exploited in making light conduits—guides that transport light from one location to another. An **optical waveguide** is a light conduit consisting of a slab, strip, or cylinder of dielectric material surrounded by another dielectric material of lower refractive index (Fig. 7.0-1). The light is transported through the inner medium without radiating into the surrounding medium. The most widely used of these waveguides is the optical fiber, which is made of two concentric cylinders of low-loss dielectric material such as glass (see Chap. 8).

Integrated optics is the technology of integrating various optical devices and components for the generation, focusing, splitting, combining, isolation, polarization, coupling, switching, modulation and detection of light, all on a single substrate (chip). Optical waveguides provide the connections between these components. Such chips (Fig. 7.0-2) are optical versions of electronic integrated circuits. Integrated optics has as its goal the miniaturization of optics in much the same way that integrated circuits have miniaturized electronics.

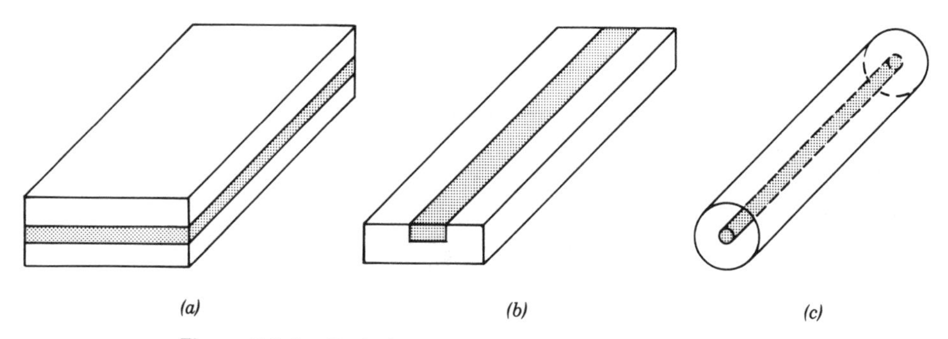

Figure 7.0-1 Optical waveguides: (*a*) slab; (*b*) strip; (*c*) fiber.

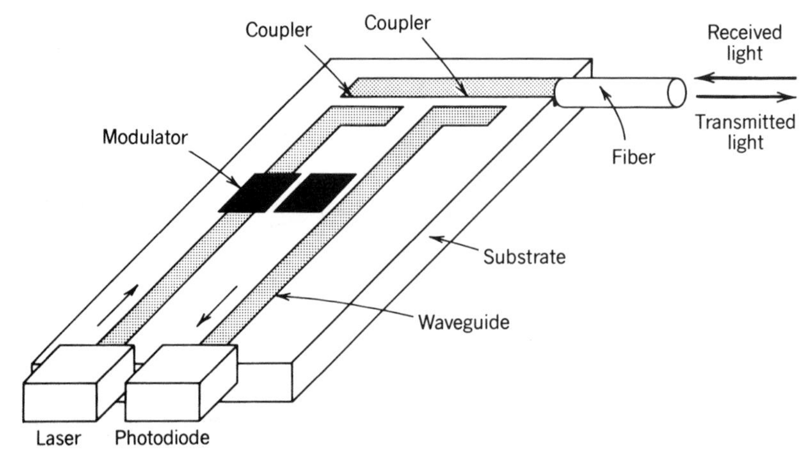

Figure 7.0-2 An example of an integrated-optic device used as an optical receiver/transmitter. Received light is coupled into a waveguide and directed to a photodiode where it is detected. Light from a laser is guided, modulated, and coupled into a fiber.

The basic theory of optical waveguides is presented in this and the following chapters. This chapter deals with rectangular waveguides which are used extensively in integrated optics. Cylindrical waveguides, which are used to make optical fibers, are the subject of Chap. 8. Integrated-optic devices (such as semiconductor lasers and detectors, modulators, and switches) are considered in the chapters that deal specifically with those devices. Fiber-optic communication systems are discussed in detail in Chap. 22.

7.1 PLANAR-MIRROR WAVEGUIDES

In this section we examine wave propagation in a waveguide made of two parallel infinite planar *mirrors* separated by a distance d (Fig. 7.1-1). The mirrors are assumed ideal; i.e., they reflect light without loss. A ray of light making an angle θ with the mirrors (say in the y–z plane) reflects and bounces between the mirrors without loss of energy. The ray is thus guided along the z direction. This seemingly perfect waveguide is *not* used in practical applications, mainly because of the difficulty and cost of fabricating low-loss mirrors. Nevertheless, this section is devoted to the study of this simple waveguide as a pedagogical introduction to the dielectric waveguide to be

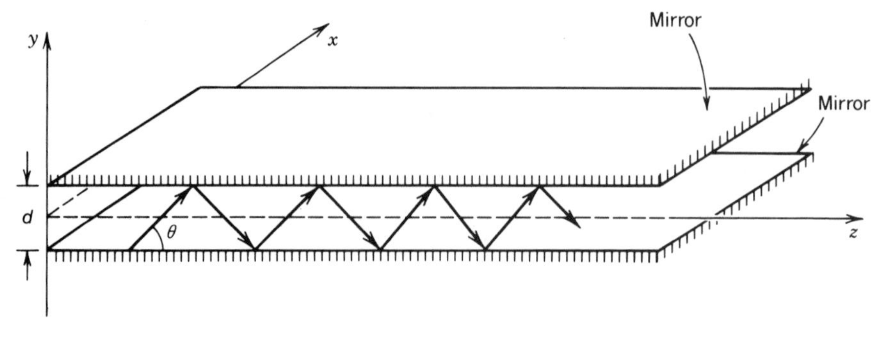

Figure 7.1-1 Planar-mirror waveguide.

examined subsequently in Sec. 7.2 and to the optical resonator, which is the subject of Chap. 9.

Waveguide Modes

The ray-optics picture of light guidance by multiple reflections does not explain a number of important effects that require the use of electromagnetic theory. A simple approach to carrying out an electromagnetic analysis is to associate with each optical ray a transverse electromagnetic (TEM) plane wave. The total electromagnetic field is the sum of these plane waves.

Consider a monochromatic TEM plane wave of wavelength $\lambda = \lambda_o/n$, wavenumber $k = nk_o$, and phase velocity $c = c_o/n$, where n is the refractive index of the medium between the mirrors. The wave is polarized in the x direction and its wavevector lies in the y–z plane at an angle θ with the z axis (Fig. 7.1-1). Like the optical ray, the wave reflects from the upper mirror, travels at an angle $-\theta$, reflects from the lower mirror, and travels once more at an angle θ, and so on. Since the electric field is parallel to the mirror, each reflection is accompanied by a phase shift π, but the amplitude and polarization are not changed. The π phase shift ensures that the sum of each wave and its own reflection vanishes so that the total field is zero at the mirrors. At each point within the waveguide we have TEM waves traveling in the upward direction at an angle θ and others traveling in the downward direction at an angle $-\theta$; all waves are polarized in the x direction.

We now impose a self-consistency condition by requiring that as the wave reflects twice, it reproduces itself [see Fig. 7.1-2(a)], so that we have only two distinct plane waves. Fields that satisfy this condition are called eigenmodes or simply modes of the waveguide (see Appendix C). *Modes are fields that maintain the same transverse distribution and polarization at all distances along the waveguide axis.* We shall see that self-consistency guarantees this shape invariance. In reference to Fig. 7.1-2, the phase shift encountered by the original wave in traveling from A to B must be equal to, or

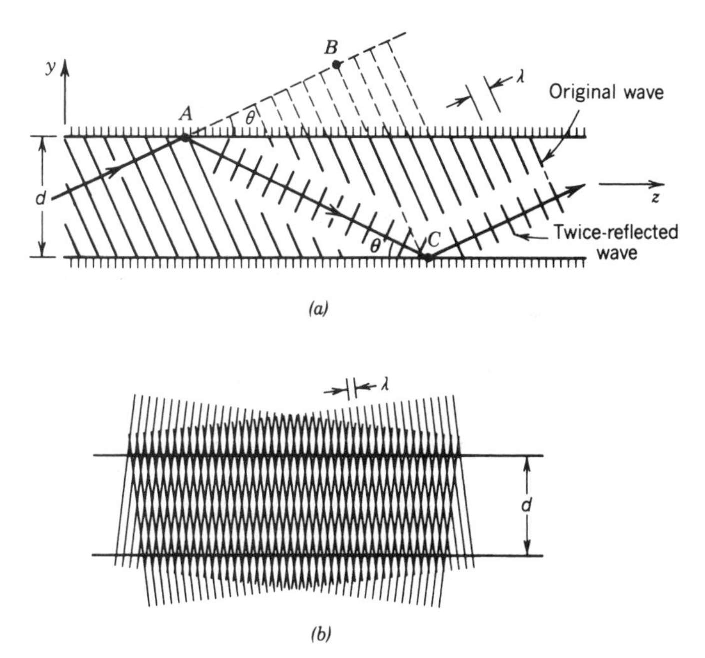

Figure 7.1-2 (a) Condition of self-consistency: as a wave reflects twice it duplicates itself. (b) At angles for which self-consistency is satisfied, the two waves interfere and create a pattern that does not change with z.

different by an integer multiple of 2π, from that encountered when the wave reflects, travels from A to C, and reflects once more. Accounting for a phase shift of π at each reflection, we have $2\pi\overline{AC}/\lambda - 2\pi - 2\pi\overline{AB}/\lambda = 2\pi q$, where $q = 0, 1, 2, \ldots$. Since $\overline{AC} - \overline{AB} = 2d \sin\theta$, where d is the distance between the mirrors, $2\pi(2d\sin\theta)/\lambda = 2\pi(q+1)$, and

$$\frac{2\pi}{\lambda} 2d \sin\theta = 2\pi m, \qquad m = 1, 2, \ldots, \tag{7.1-1}$$

where $m = q + 1$. The self-consistency condition is therefore satisfied only for certain bounce angles $\theta = \theta_m$ satisfying

$$\sin\theta_m = m\frac{\lambda}{2d}, \qquad m = 1, 2, \ldots. \tag{7.1-2}$$

Bounce Angles

Each integer m corresponds to a bounce angle θ_m, and the corresponding field is called the mth mode. The $m = 1$ mode has the smallest angle $\theta_1 = \sin^{-1}(\lambda/2d)$; modes with larger m are composed of more oblique plane-wave components.

When the self-consistency condition is satisfied, the phases of the upward and downward plane waves at points on the z axis differ by half the round-trip phase shift $q\pi$, $q = 0, 1, \ldots$, or $(m-1)\pi$, $m = 1, 2, \ldots$, so that they add for odd m and subtract for even m.

Since the y component of the propagation constant is $k_y = nk_o \sin\theta$, it is quantized to the values $k_{ym} = nk_o \sin\theta_m = (2\pi/\lambda)\sin\theta_m$. Using (7.1-2), we obtain

$$k_{ym} = m\frac{\pi}{d}, \qquad m = 1, 2, 3, \ldots, \tag{7.1-3}$$

Transverse Component of the Wavevector

so that the k_{ym} are spaced by π/d. Equation (7.1-3) states that the phase shift encountered when a wave travels a distance $2d$ (one round trip) in the y direction, with propagation constant k_{ym}, must be a multiple of 2π.

Propagation Constants

The guided wave is composed of two distinct plane waves traveling at angles $\pm\theta$ with the z axis in the y–z plane. Their wavevectors have components $(0, k_y, k_z)$ and $(0, -k_y, k_z)$. Their sum or difference therefore varies with z as $\exp(-jk_z z)$, so that the propagation constant of the guided wave is $\beta = k_z = k\cos\theta$. Thus β is quantized to the values $\beta_m = k\cos\theta_m$, from which $\beta_m^2 = k^2(1 - \sin^2\theta_m)$. Using (7.1-2), we obtain

$$\beta_m^2 = k^2 - \frac{m^2\pi^2}{d^2}. \tag{7.1-4}$$

Propagation Constants

Higher-order (more oblique) modes travel with smaller propagation constants. The values of θ_m, k_{ym}, and β_m for the different modes are illustrated in Fig. 7.1-3.

Field Distributions

The complex amplitude of the total field in the waveguide is the superposition of the two bouncing TEM plane waves. If $A_m \exp(-jk_{ym}y - j\beta_m z)$ is the upward wave, then $e^{j(m-1)\pi}A_m \exp(+jk_{ym}y - j\beta_m z)$ must be the downward wave [at $y = 0$, the two waves

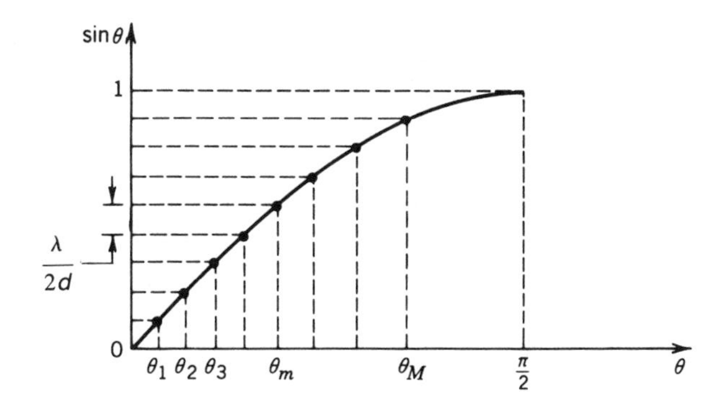

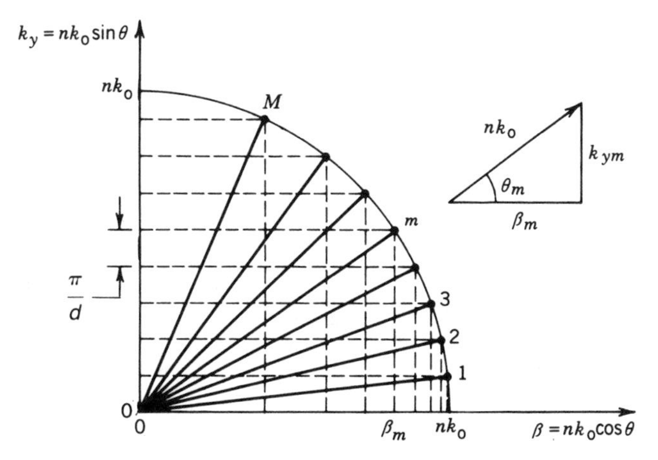

Figure 7.1-3 The bounce angles θ_m and the wavevector components of the modes of a planar-mirror waveguide (indicated by dots). The transverse components $k_{ym} = k \sin \theta_m$ are spaced uniformly at multiples of π/d, but the bounce angles θ_m and the propagation constants β_m are not equally spaced. Mode $m = 1$ has the smallest bounce angle and the largest propagation constant.

differ by a phase shift $(m - 1)\pi$]. There are therefore symmetric modes, for which the two plane-wave components are added, and antisymmetric modes, for which they are subtracted. The total field turns out to be $E_x(y, z) = 2A_m \cos(k_{ym} y) \exp(-j\beta_m z)$ for odd modes and $2jA_m \sin(k_{ym} y) \exp(-j\beta_m z)$ for even modes.

Using (7.1-3) we write the complex amplitude of the electric field in the form

$$E_x(y, z) = a_m u_m(y) \exp(-j\beta_m z), \tag{7.1-5}$$

where

$$u_m(y) = \begin{cases} \sqrt{\dfrac{2}{d}} \cos \dfrac{m\pi y}{d}, & m = 1, 3, 5, \ldots \\ \sqrt{\dfrac{2}{d}} \sin \dfrac{m\pi y}{d}, & m = 2, 4, 6, \ldots, \end{cases} \tag{7.1-6}$$

and $a_m = \sqrt{2d}\, A_m$ and $j\sqrt{2d}\, A_m$, for odd and even m, respectively. The functions $u_m(y)$ have been normalized to satisfy

$$\int_{-d/2}^{d/2} u_m^2(y)\, dy = 1. \tag{7.1-7}$$

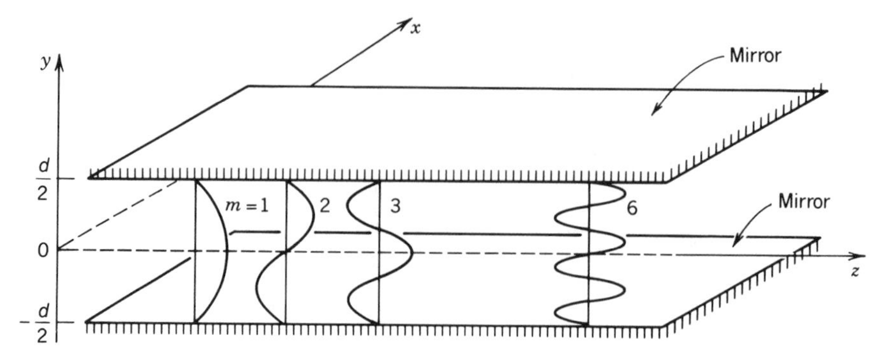

Figure 7.1-4 Field distributions of the modes of a planar-mirror waveguide.

Thus a_m is the amplitude of mode m. It can be shown that the functions $u_m(y)$ also satisfy

$$\int_{-d/2}^{d/2} u_m(y) u_l(y)\, dy = 0, \qquad l \neq m, \tag{7.1-8}$$

i.e., they are orthogonal in the $[-d/2, d/2]$ interval.

The transverse distributions $u_m(y)$ are plotted in Fig. 7.1-4. Each mode can be viewed as a standing wave in the y direction, traveling in the z direction. Modes of large m vary in the transverse plane at a greater rate k_y and travel with a smaller propagation constant β. The field vanishes at $y = \pm d/2$ for all modes, so that the boundary conditions at the surface of the mirrors are always satisfied.

Since we assumed that the bouncing TEM plane wave is polarized in the x direction, the total electric field is also in the x direction and the guided wave is a transverse-electric (TE) wave. Transverse magnetic (TM) waves may be treated similarly, as will be discussed later.

EXERCISE 7.1-1

Optical Power. Show that the optical power flow in the z direction associated with the TE mode $E_x(y, z) = a_m u_m(y) \exp(-j\beta_m z)$ is $(|a_m|^2/2\eta) \cos\theta_m$ where $\eta = \eta_o/n$ and $\eta_o = (\mu_o/\epsilon_o)^{1/2}$ is the impedance of free space.

Number of Modes

Since $\sin\theta_m = m\lambda/2d$, $m = 1, 2, \ldots$ and for $\sin\theta_m < 1$, the maximum allowed value of m is the greatest integer smaller than $(\lambda/2d)^{-1}$,

$$M \doteq \frac{2d}{\lambda}. \tag{7.1-9}$$

Number of Modes

The symbol $\doteq$ denotes that $2d/\lambda$ is reduced to the nearest integer. For example, when $2d/\lambda = 0.9$, 1, or 1.1, $M = 0$, 0, and 1, respectively. Thus M is the number of

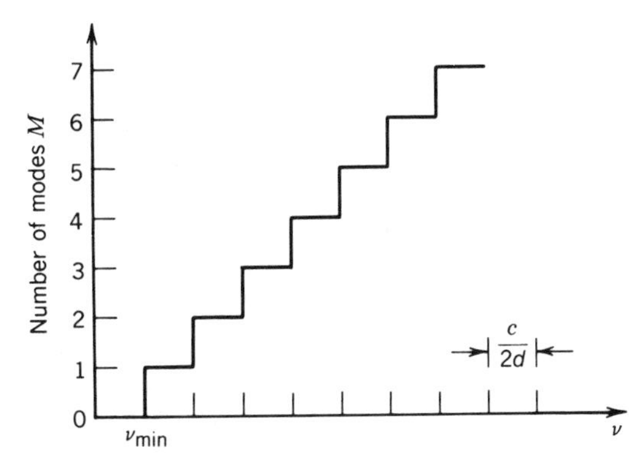

Figure 7.1-5 Number of modes as a function of frequency ν. The cutoff frequency is $\nu_{min} = c/2d$. As ν increases by $c/2d$, the number of modes M is incremented by one.

modes of the waveguide. Light can be transmitted through the waveguide in one, two, or many modes. The actual number of modes that carry optical power depends on the source of excitation, but the maximum number is M.

The number of modes increases with increasing ratio of the mirror separation to the wavelength. If $2d/\lambda \le 1$, $M = 0$, indicating that the self-consistency condition cannot be met and the waveguide cannot support any modes. The wavelength $\lambda_{max} = 2d$ is called the **cutoff wavelength** of the waveguide. It is the longest wavelength that can be guided by the structure. It corresponds to the **cutoff frequency** $\nu_{min} = c/2d$, the lowest frequency of light that can be guided by the waveguide. If $1 < 2d/\lambda \le 2$ (i.e., $d \le \lambda < 2d$), only one mode is allowed. The structure is said to be a **single-mode waveguide**. If $d = 5$ μm, for example, the waveguide has a cutoff wavelength $\lambda_{max} = 10$ μm; it supports a single mode for $5\ \mu\text{m} \le \lambda < 10\ \mu\text{m}$, and more modes for $\lambda < 5\ \mu$m. Equation (7.1-9) can also be written in terms of the frequency ν, $M \doteq \nu/(c/2d)$, so that the number of modes increases with the frequency ν, as illustrated in Fig. 7.1-5.

Group Velocities

A pulse of light (wavepacket) of angular frequency centered at ω and propagation constant β travels with a velocity $v = d\omega/d\beta$, known as the group velocity (see Sec. 5.6). The propagation constant of mode m is given by (7.1-4) from which $\beta_m^2 = (\omega/c)^2 - m^2\pi^2/d^2$, which is an explicit relation between β_m and ω known as the **dispersion relation**. Taking the derivative and assuming that c is independent of ω (i.e., ignoring dispersion in the waveguide material), we obtain $2\beta_m d\beta_m/d\omega = 2\omega/c^2$, so that $d\omega/d\beta_m = c^2\beta_m/\omega = c^2 k \cos\theta_m/\omega = c\cos\theta_m$, from which the group velocity of mode m is

$$v_m = c\cos\theta_m. \tag{7.1-10}$$

Group Velocity

Thus different modes have different group velocities. More oblique modes travel with a smaller group velocity since they are delayed by the longer path of the zigzaging process.

Equation (7.1-10) may also be obtained geometrically by examining the plane wave as it bounces between the mirrors and determining the distance advanced in the z direction and the time taken by the zigzaging process. For the trip from the bottom

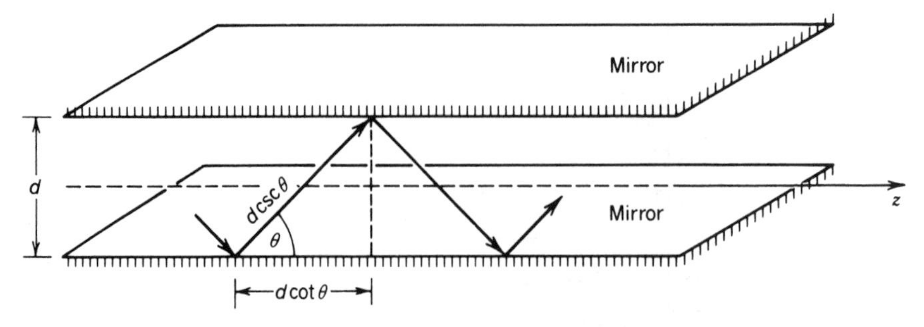

Figure 7.1-6 A plane wave bouncing at an angle θ advances in the z direction a distance $d \cot \theta$ in a time $d \csc \theta / c$. The velocity is $c \cos \theta$.

mirror to the top mirror (Fig. 7.1-6) we have

$$v = \frac{\text{distance}}{\text{time}} = \frac{d \cot \theta}{d \csc \theta / c} = c \cos \theta. \tag{7.1-11}$$

TM Modes

The modes considered so far have been TE modes (electric field in the x direction). TM modes (magnetic field in the x direction) can also be supported by the mirror waveguide. They can be studied by means of a TEM plane wave with the magnetic field in the x direction, traveling at an angle θ and reflecting from the two mirrors (Fig. 7.1-7). The electric-field complex amplitude then has components in the y and z directions. Since the z component is parallel to the mirror, it must behave like the x component of the TE mode (i.e., undergo a phase shift π at each reflection and vanish at the mirror). When the self-consistency condition is applied to this component the result is mathematically identical to that of the TE case. The angles θ, the transverse wavevector components k_y, and the propagation constants β of the TM modes associated with this component are identical to those of the TE modes. There are $M \doteq 2d/\lambda$ TM modes (and a total of $2M$ modes) supported by the waveguide.

As previously, the z component of the electric-field complex amplitude of mode m is the sum of an upward plane wave $A_m \exp(-jk_{ym} y) \exp(-j\beta_m z)$ and a downward plane wave $e^{j(m-1)\pi} A_m \exp(jk_{ym} y) \exp(-j\beta_m z)$, with equal amplitudes and phase shift $(m-1)\pi$, so that

$$E_z(y, z) = \begin{cases} a_m \sqrt{\dfrac{2}{d}} \cos \dfrac{m\pi y}{d} \exp(-j\beta_m z), & m = 1, 3, 5, \ldots \\ a_m \sqrt{\dfrac{2}{d}} \sin \dfrac{m\pi y}{d} \exp(-j\beta_m z), & m = 2, 4, 6, \ldots, \end{cases} \tag{7.1-12}$$

where $a_m = \sqrt{2d}\, A_m$ and $j\sqrt{2d}\, A_m$ for odd and even m, respectively. Since the electric-field vector of a TEM plane wave is normal to its direction of propagation, it

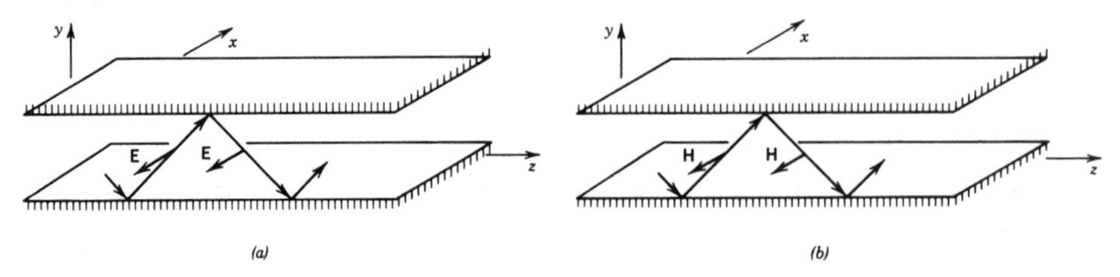

Figure 7.1-7 Polarization: (*a*) TE; (*b*) TM.

makes an angle $\pi/2 + \theta_m$ with the z axis for the upward wave, and $\pi/2 - \theta_m$ for the downward wave.

The y components of the electric field of these waves are

$$A_m \cot\theta_m \exp(-jk_{ym}y)\exp(-j\beta_m z) \quad \text{and} \quad e^{jm\pi} A_m \cot\theta_m \exp(jk_{ym}y)\exp(-j\beta_m z),$$

so that

$$E_y(y,z) = \begin{cases} a_m\sqrt{\dfrac{2}{d}}\cot\theta_m \cos\dfrac{m\pi y}{d}\exp(-j\beta_m z), & m = 1,3,5,\ldots \\ a_m\sqrt{\dfrac{2}{d}}\cot\theta_m \sin\dfrac{m\pi y}{d}\exp(-j\beta_m z), & m = 2,4,6,\ldots. \end{cases} \qquad (7.1\text{-}13)$$

Satisfaction of the boundary conditions is assured because $E_z(y,z)$ vanishes at the mirrors. The magnetic field component $H_x(y,z)$ may be similarly determined by noting that the ratio of the electric to the magnetic fields of a TEM wave is the impedance of the medium η. The resultant fields $E_y(y,z)$, $E_z(y,z)$, and $H_x(y,z)$ do, of course, satisfy Maxwell's equations.

Multimode Fields

It should not be thought that for light to be guided by the mirrors, it must have the distribution of one of the modes. In fact, a field satisfying the boundary conditions

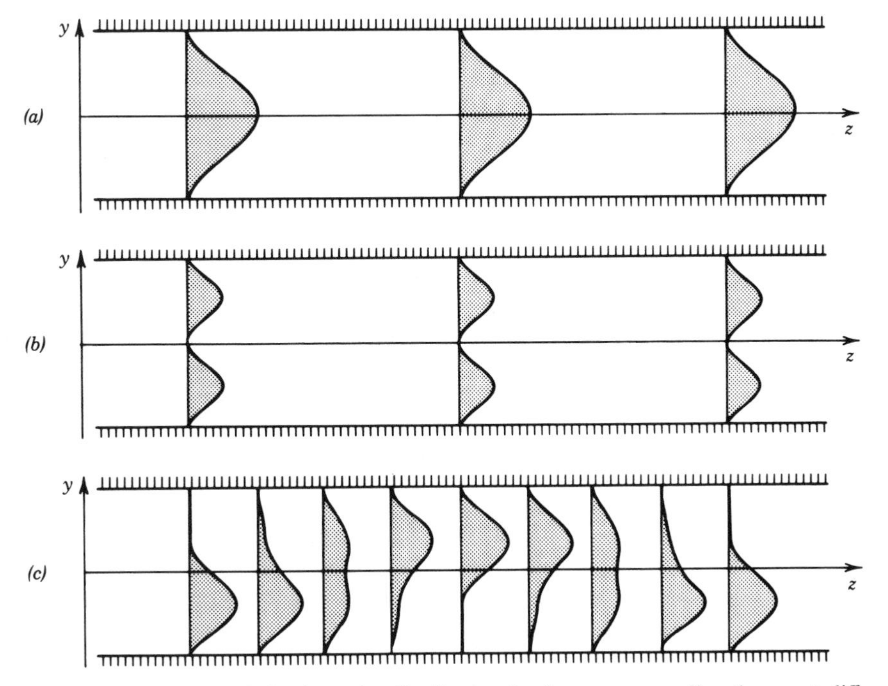

Figure 7.1-8 Variation of the intensity distribution in the transverse direction y at different axial distances z. (*a*) The electric-field complex amplitude in mode 1 is $E(y,z) = u_1(y)\exp(-j\beta_1 z)$, where $u_1(y) = \sqrt{2/d}\cos(\pi y/d)$. The intensity does not vary with z. (*b*) The complex amplitude in mode 2 is $E(y,z) = u_2(y)\exp(-j\beta_2 z)$, where $u_2(y) = \sqrt{2/d}\sin(2\pi y/d)$. The intensity does not vary with z. (*c*) The complex amplitude in a mixture of modes 1 and 2, $E(y,z) = u_1(y)\exp(-j\beta_1 z) + u_2(y)\exp(-j\beta_2 z)$. Since $\beta_1 \neq \beta_2$, the intensity distribution changes with z.

(vanishing at the mirrors) but otherwise having an arbitrary distribution in the transverse plane *can* be guided by the waveguide. The optical power, however, is divided among the modes. Since different modes travel with different propagation constants and different group velocities, the field changes its transverse distribution as it travels through the waveguide. Figure 7.1-8 illustrates how the transverse intensity distribution of a single mode is invariant to propagation, whereas the multimode distribution varies with z.

An arbitrary field polarized in the x direction and satisfying the boundary conditions can be written as a weighted superposition of the TE modes,

$$E_x(y, z) = \sum_{m=0}^{M} a_m u_m(y) \exp(-j\beta_m z), \tag{7.1-14}$$

where a_m, the superposition weights, are the amplitudes of the different modes.

EXERCISE 7.1-2

Optical Power in a Multimode Field. Show that the optical power flow in the z direction associated with the multimode field in (7.1-14) is the sum of the powers $(|a_m|^2/2\eta)\cos\theta_m$ carried by each of the modes.

7.2 PLANAR DIELECTRIC WAVEGUIDES

A planar dielectric waveguide is a slab of dielectric material surrounded by media of lower refractive indices. The light is guided inside the slab by total internal reflection. In thin-film devices the slab is called the "film" and the upper and lower media are called the "cover" and the "substrate," respectively. The inner medium and outer media may also be called the "core" and the "cladding" of the waveguide, respectively. In this section we study the propagation of light in a symmetric planar dielectric waveguide made of a slab of width d and refractive index n_1 surrounded by a cladding of smaller refractive index n_2, as illustrated in Fig. 7.2-1. All materials are assumed to be lossless.

Light rays making angles θ with the z axis, in the y–z plane, undergo multiple total internal reflections at the slab boundaries, provided that θ is smaller than the complement of the critical angle $\bar{\theta}_c = \pi/2 - \sin^{-1}(n_2/n_1) = \cos^{-1}(n_2/n_1)$ [see page 11 and Figs. 6.2-3 and 6.2-5]. They travel in the z direction by bouncing between the slab surfaces without loss of power. Rays making larger angles refract, losing a portion of their power at each reflection, and eventually vanish.

To determine the waveguide modes, a formal approach may be pursued by developing solutions to Maxwell's equations in the inner and outer media with the appropriate boundary conditions imposed (see Problem 7.2-4). We shall instead write the solution in terms of TEM plane waves bouncing between the surfaces of the slab. By imposing the self-consistency condition, we determine the bounce angles of the waveguide modes, from which the propagation constants, field distributions, and group velocities are determined. The analysis is analogous to that used in the previous section for the planar-mirror waveguide.

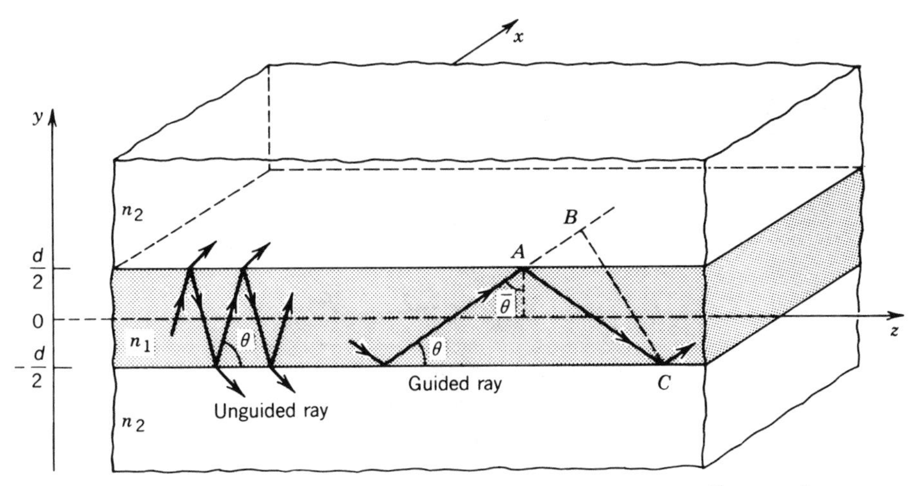

Figure 7.2-1 Planar dielectric waveguide. Rays making an angle $\theta < \bar{\theta}_c = \cos^{-1}(n_2/n_1)$ are guided by total internal reflection.

A. Waveguide Modes

Assume that the field in the slab is in the form of a monochromatic TEM plane wave of wavelength $\lambda = \lambda_o/n_1$ bouncing back and forth at an angle θ smaller than the complementary critical angle $\bar{\theta}_c$. The wave travels with a phase velocity $c_1 = c_o/n_1$, has a wavenumber $n_1 k_o$, and has wavevector components $k_x = 0$, $k_y = n_1 k_o \sin\theta$, and $k_z = n_1 k_o \cos\theta$. To determine the modes we impose the self-consistency condition that a wave reproduces itself after each round trip.

In one round trip, the twice-reflected wave lags behind the original wave by a distance $\overline{AC} - \overline{AB} = 2d\sin\theta$, as in Fig. 7.1-2. There is also a phase φ_r introduced by each internal reflection at the dielectric boundary (see Sec. 6.2). For self-consistency, the phase shift between the two waves must be zero or a multiple of 2π,

$$\frac{2\pi}{\lambda}2d\sin\theta - 2\varphi_r = 2\pi m, \qquad m = 0, 1, 2, \ldots \tag{7.2-1}$$

or

$$2k_y d - 2\varphi_r = 2\pi m. \tag{7.2-2}$$

The only difference between this condition and the corresponding condition in the mirror waveguide, (7.1-1) and (7.1-3), is that the phase shift π introduced by the mirror is replaced here by the phase shift φ_r introduced at the dielectric boundary.

The reflection phase shift φ_r is a function of the angle θ. It also depends on the polarization of the incident wave, TE or TM. In the TE case (the electric field is in the x direction), substituting $\theta_1 = \pi/2 - \theta$ and $\theta_c = \pi/2 - \bar{\theta}_c$ in (6.2-9) gives

$$\tan\frac{\varphi_r}{2} = \left(\frac{\sin^2\bar{\theta}_c}{\sin^2\theta} - 1\right)^{1/2}, \tag{7.2-3}$$

so that φ_r varies from π to 0 as θ varies from 0 to $\bar{\theta}_c$. Rewriting (7.2-1) in the form

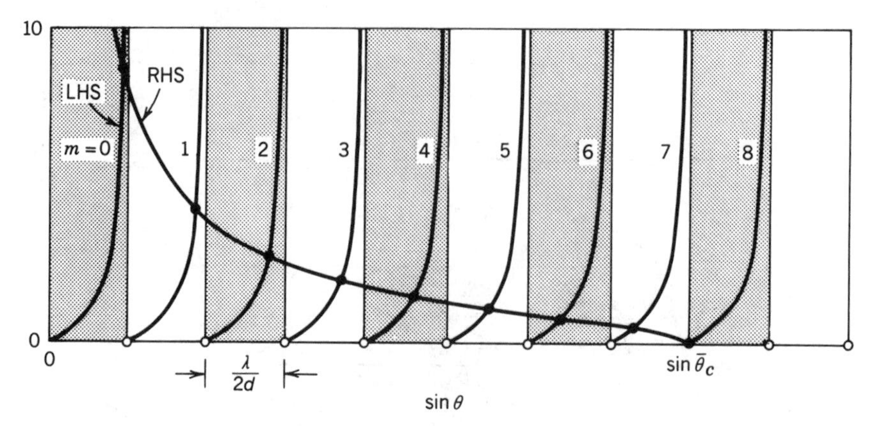

Figure 7.2-2 Graphical solution of (7.2-4) to determine the bounce angles θ_m of the modes of a planar dielectric waveguide. The RHS and LHS of (7.2-4) are plotted versus $\sin\theta$. The intersection points, marked by filled circles, determine $\sin\theta_m$. Each branch of the tan or cot function in the LHS corresponds to a mode. In this plot $\sin\bar{\theta}_c = 8(\lambda/2d)$ and the number of modes is $M = 9$. The open circles mark $\sin\theta_m = m\lambda/2d$, which provide the bounce angles of the modes of a planar-mirror waveguide of the same dimensions.

$\tan(\pi d\sin\theta/\lambda - m\pi/2) = \tan(\varphi_r/2)$ and using (7.2-3), we obtain

$$\tan\left(\pi\frac{d}{\lambda}\sin\theta - m\frac{\pi}{2}\right) = \left(\frac{\sin^2\bar{\theta}_c}{\sin^2\theta} - 1\right)^{1/2}. \tag{7.2-4}$$

Self-Consistency Condition (TE Modes)

This is a transcendental equation in one variable, $\sin\theta$. Its solutions yield the bounce angles θ_m of the modes. A graphical solution is instructive. The right- and left-hand sides of (7.2-4) are plotted in Fig. 7.2-2 as functions of $\sin\theta$. Solutions are given by the intersection points. The right-hand side (RHS), $\tan(\varphi_r/2)$, is a monotonic decreasing function of $\sin\theta$ which reaches 0 when $\sin\theta = \sin\bar{\theta}_c$. The left-hand side (LHS), generates two families of curves, $\tan[(\pi d/\lambda)\sin\theta]$ and $\cot[(\pi d/\lambda)\sin\theta]$, when m is even and odd, respectively. The intersection points determine the angles θ_m of the modes. The bounce angles of the modes of a *mirror* waveguide of mirror separation d may be obtained from this diagram by using $\varphi_r = \pi$ or, equivalently, $\tan(\varphi_r/2) = \infty$. For comparison, these angles are marked by open circles.

The angles θ_m lie between 0 and $\bar{\theta}_c$. They correspond to wavevectors with components $(0, n_1k_o\sin\theta_m, n_1k_o\cos\theta_m)$. The z components are the propagation constants

$$\beta_m = n_1k_o\cos\theta_m. \tag{7.2-5}$$

Propagation Constants

Since $\cos\theta_m$ lies between 1 and $\cos\bar{\theta}_c = n_2/n_1$, β_m lies between n_2k_o and n_1k_o, as illustrated in Fig. 7.2-3.

The bounce angles θ_m and the propagation constants β_m of TM modes can be found by using the same equation (7.2-1), but with the phase shift φ_r given by (6.2-11). Similar results are obtained.

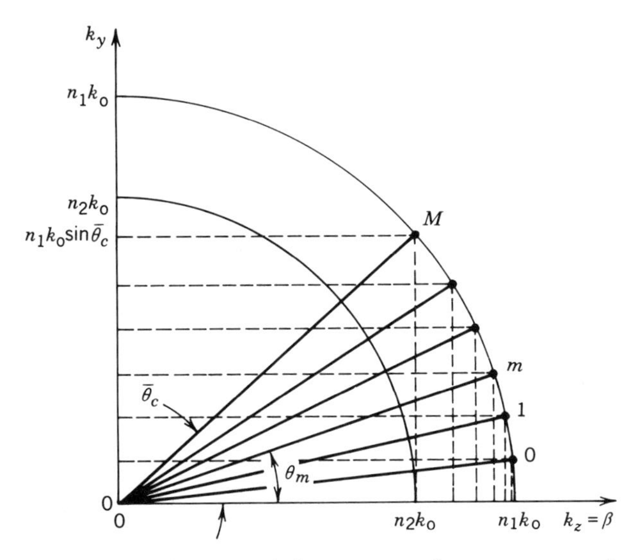

Figure 7.2-3 The bounce angles θ_m and the corresponding components k_z and k_y of the wavevector of the waveguide modes are indicated by dots. The angles θ_m lie between 0 and $\bar{\theta}_c$, and the propagation constants β_m lie between $n_2 k_o$ and $n_1 k_o$. These results should be compared with those shown in Fig. 7.1-3 for the planar-mirror waveguide.

Number of Modes

To determine the number of TE modes supported by the dielectric waveguide we examine the diagram in Fig. 7.2-2. The abscissa is divided into equal intervals of width $\lambda/2d$, each of which contains a mode marked by a filled circle. This extends over angles for which $\sin\theta \leq \sin\bar{\theta}_c$. The number of TE modes is therefore the smallest integer greater than $\sin\bar{\theta}_c/(\lambda/2d)$, so that

$$M \doteq \frac{\sin\bar{\theta}_c}{\lambda/2d}. \tag{7.2-6}$$

The symbol $\doteq$ denotes that $\sin\bar{\theta}_c/(\lambda/2d)$ is increased to the nearest integer. For example, if $\sin\bar{\theta}_c/(\lambda/2d) = 0.9$, 1, or 1.1, $M = 1$, 2, and 2, respectively. Substituting $\cos\bar{\theta}_c = n_2/n_1$ into (7.2-6), we obtain

$$\boxed{M \doteq 2\frac{d}{\lambda_o}\text{NA},} \tag{7.2-7}$$

Number of TE Modes

where

$$\boxed{\text{NA} = \left(n_1^2 - n_2^2\right)^{1/2}} \tag{7.2-8}$$

Numerical Aperture

is the numerical aperture of the waveguide (the NA is the sine of the angle of acceptance of rays from air into the slab; see Exercise 1.2-5). A similar expression can

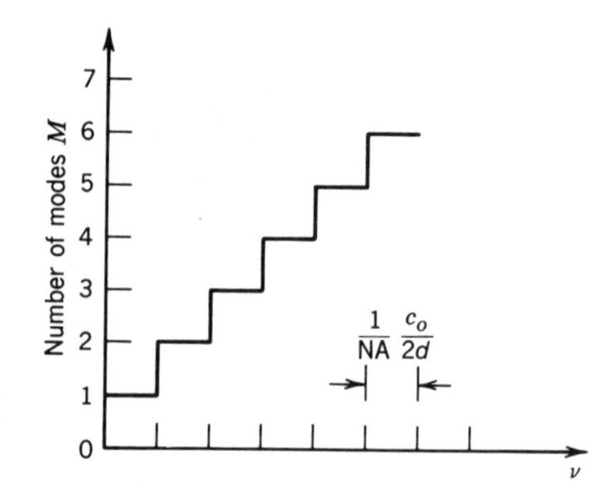

Figure 7.2-4 Number of TE modes as a function of frequency. Compare with Fig. 7.1-5 for the planar-mirror waveguide.

be obtained for the TM modes. If $d/\lambda_o = 10$, $n_1 = 1.47$, and $n_2 = 1.46$, for example, then $\bar{\theta}_c = 6.7°$, NA $= 0.171$, and $M = 4$ TE modes.

When $\lambda/2d > \sin\bar{\theta}_c$ or $(2d/\lambda_o)\text{NA} < 1$, only one mode is allowed. The waveguide is then a **single-mode waveguide**. This occurs when the slab is sufficiently thin or the wavelength is sufficiently long. Unlike the mirror waveguide, the dielectric waveguide has no absolute cutoff wavelength (or cutoff frequency). In a dielectric waveguide there is at least one TE mode, since the fundamental mode $m = 0$ is always allowed. Each of the modes $m = 1, 2, \ldots$ has its own cutoff wavelength, however.

The number of modes may also be written as a function of frequency,

$$M \doteq \frac{\text{NA}}{(c_o/2d)}\nu.$$

The relation is illustrated in Fig. 7.2-4. M is incremented by 1 as ν increases by $(c_o/2d)/\text{NA}$. Identical expressions for the number of TM modes may be derived similarly.

EXAMPLE 7.2-1. ***Modes in an AlGaAs Waveguide.*** A waveguide is made by sandwiching a layer of $Al_xGa_{1-x}As$ between two layers of $Al_yGa_{1-y}As$. By changing the concentrations x, y of Al in these compounds their refractive indices are controlled. If x and y are chosen such that at an operating wavelength $\lambda_o = 0.9\ \mu\text{m}$, $n_1 = 3.5$, and $n_1 - n_2 = 0.05$, then for a thickness $d = 10\ \mu\text{m}$ there are $M = 14$ TE modes. For $d < 0.76\ \mu\text{m}$, only a single mode is allowed.

B. Field Distributions

We now determine the field distributions of the TE modes.

Internal Field

The field inside the slab is composed of two TEM plane waves traveling at angles θ_m and $-\theta_m$ with the z axis with wavevector components $(0, \pm n_1 k_o \sin\theta_m, n_1 k_o \cos\theta_m)$. They have the same amplitude and a phase shift $m\pi$ (half that of a round trip) at the center of the slab. The electric-field complex amplitude is therefore $E_x(y, z) =$

$a_m u_m(y)\exp(-j\beta_m z)$, where $\beta_m = n_1 k_o \cos\theta_m$ is the propagation constant, a_m is a constant,

$$u_m(y) \propto \begin{cases} \cos\left(\dfrac{2\pi\sin\theta_m}{\lambda}y\right), & m = 0, 2, 4, \ldots \\ \sin\left(\dfrac{2\pi\sin\theta_m}{\lambda}y\right), & m = 1, 3, 5, \ldots, \end{cases} \qquad -\frac{d}{2} \le y \le \frac{d}{2}, \tag{7.2-9}$$

and $\lambda = \lambda_o/n_1$. Note that although the field is harmonic, it does not vanish at the slab boundary. As m increases, $\sin\theta_m$ increases, so that higher-order modes vary more rapidly with y.

External Field

The external field must match the internal field at all boundary points $y = \pm d/2$. It must therefore vary with z as $\exp(-j\beta_m z)$. Substituting $E_x(y, z) = a_m u_m(y)\exp(-j\beta_m z)$ into the Helmholtz equation $(\nabla^2 + n_2^2 k_o^2)E_x(y, z) = 0$, we obtain

$$\frac{d^2 u_m}{dy^2} - \gamma_m^2 u_m = 0, \tag{7.2-10}$$

where

$$\gamma_m^2 = \beta_m^2 - n_2^2 k_o^2. \tag{7.2-11}$$

Since $\beta_m > n_2 k_o$ for guided modes (see Fig. 7.2-3), $\gamma_m^2 > 0$, so that (7.2-10) is satisfied by the exponential functions $\exp(-\gamma_m y)$ and $\exp(\gamma_m y)$. Since the field must decay away from the slab, we choose $\exp(-\gamma_m y)$ in the upper medium and $\exp(\gamma_m y)$ in the lower medium,

$$u_m(y) \propto \begin{cases} \exp(-\gamma_m y), & y > \dfrac{d}{2} \\ \exp(\gamma_m y), & y < -\dfrac{d}{2}. \end{cases} \tag{7.2-12}$$

The decay rate γ_m is known as the **extinction coefficient**. The wave is said to be an **evanescent wave**. Substituting $\beta_m = n_1 k_o \cos\theta_m$ and $\cos\bar{\theta}_c = n_2/n_1$, into (7.2-11), we obtain

$$\gamma_m = n_2 k_o \left(\frac{\cos^2\theta_m}{\cos^2\bar{\theta}_c} - 1\right)^{1/2}. \tag{7.2-13}$$

Extinction Coefficient

As the mode number m increases, θ_m increases, and γ_m decreases. Higher-order modes therefore penetrate deeper into the cover and substrate.

To determine the proportionality constants in (7.2-9) and (7.2-12), we match the internal and external fields at $y = d/2$ and use the normalization

$$\int_{-\infty}^{\infty} u_m^2(y)\, dy = 1. \tag{7.2-14}$$

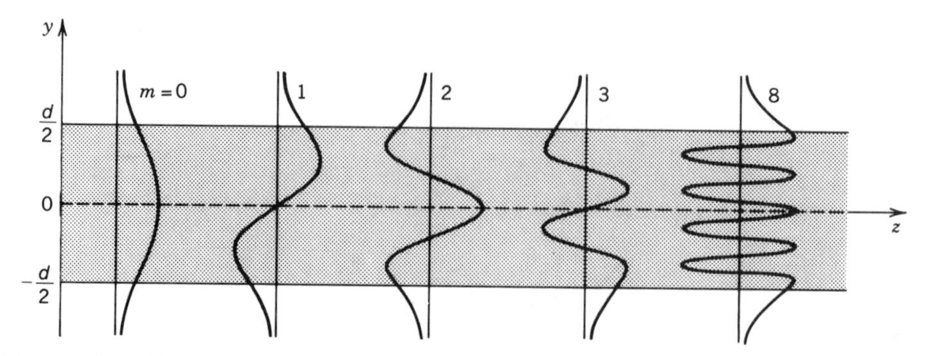

Figure 7.2-5 Field distributions for TE guided modes in a dielectric waveguide. These results should be compared with those shown in Fig. 7.1-4 for the planar-mirror waveguide.

This gives an expression for $u_m(y)$ valid for all y. These functions are illustrated in Fig. 7.2-5. As in the mirror waveguide, all of the $u_m(y)$ are orthogonal, i.e.,

$$\int_{-\infty}^{\infty} u_m(y) u_l(y)\, dy = 0, \qquad l \neq m. \tag{7.2-15}$$

An arbitrary TE field in the dielectric waveguide can be written as a superposition of these modes:

$$E_x(y, z) = \sum_m a_m u_m(y) \exp(-j\beta_m z), \tag{7.2-16}$$

where a_m is the amplitude of mode m.

EXERCISE 7.2-1

Confinement Factor. The power confinement factor is the ratio of power in the slab to the total power

$$\Gamma_m = \frac{\int_0^{d/2} u_m^2(y)\, dy}{\int_0^{\infty} u_m^2(y)\, dy}. \tag{7.2-17}$$

Derive an expression for Γ_m as a function of the angle θ_m and the ratio d/λ. Demonstrate that the lowest-order mode (smallest θ_m) has the highest power confinement factor.

The field distributions of the TM modes may be similarly determined (Fig. 7.2-6). Since it is parallel to the slab boundary, the z component of the electric field behaves similarly to the x component of the TE electric field. The analysis may start by determining $E_z(y, z)$. Using the properties of the constituent TEM waves, the other components $E_y(y, z)$ and $H_x(y, z)$ may readily be determined, as was done for mirror waveguides. Alternatively, Maxwell's equations may be used to determine these fields.

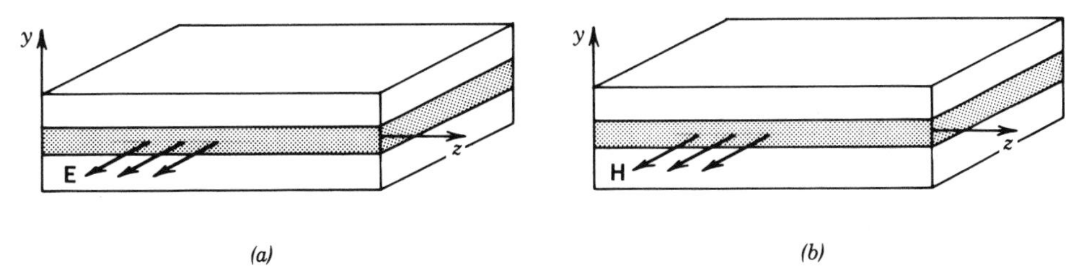

Figure 7.2-6 (*a*) TE and (*b*) TM modes in a dielectric planar waveguide.

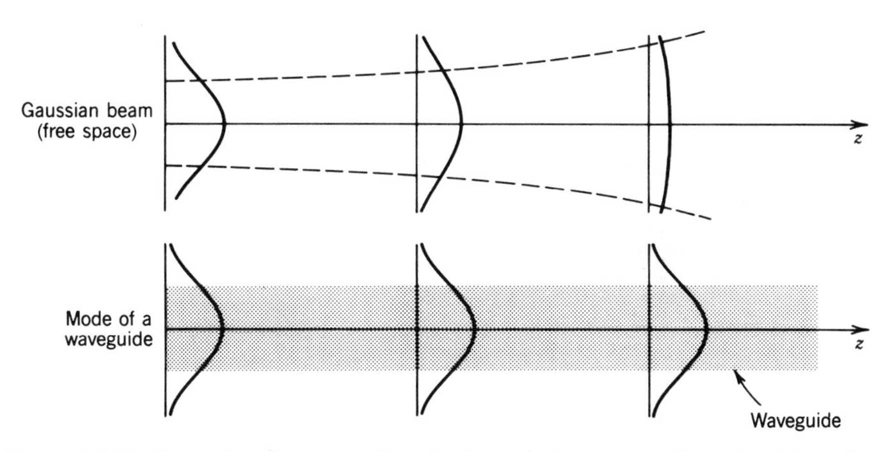

Figure 7.2-7 Comparison between a Gaussian beam in free space and a waveguide mode.

The field distribution of the lowest-order TE mode ($m = 0$) is similar in shape to that of the Gaussian beam (see Chap. 3). However, unlike the Gaussian beam, guided light does not spread in the transverse direction as it propagates in the axial direction (see Fig. 7.2-7). In a waveguide, the tendency of light to diffract is compensated by the guiding action of the medium.

C. Group Velocities

To determine the group velocity $v = d\omega/d\beta$ for each of the guided modes, we examine the dependence of the propagation constant β on the frequency ω by writing the self-consistency equation (7.2-2) in terms of β and ω. Since $k_y^2 = (\omega/c_1)^2 - \beta^2$, (7.2-2) gives

$$2d\left[\left(\frac{\omega}{c_1}\right)^2 - \beta^2\right]^{1/2} = 2\varphi_r + 2\pi m. \tag{7.2-18}$$

Since $\cos\theta = \beta/(\omega/c_1)$ and $\cos\bar{\theta}_c = n_2/n_1 = c_1/c_2$ (7.2-3) becomes

$$\tan^2\frac{\varphi_r}{2} = \frac{\beta^2 - \omega^2/c_2^2}{\omega^2/c_1^2 - \beta^2}. \tag{7.2-19}$$

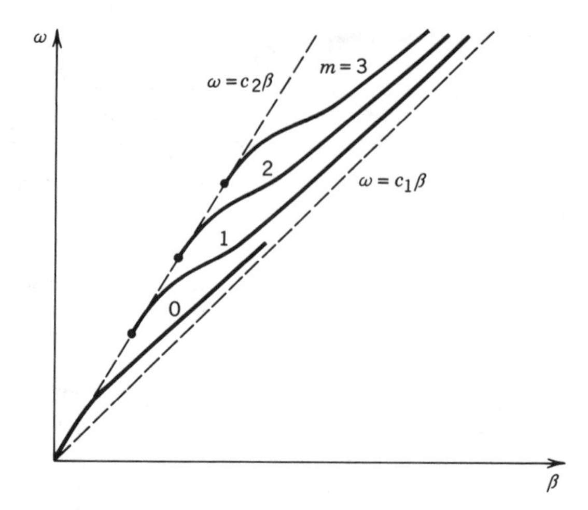

Figure 7.2-8 Schematic of the dispersion relation: angular frequency ω versus propagation constant β, for the different TE modes $m = 0, 1, 2, \ldots$. The group velocity is the slope $v = d\omega/d\beta$. As ω increases the group velocity for each mode decreases from approximately $c_2 = c_o/n_2$ to approximately $c_1 = c_o/n_1$. For $M \gg 1$, at a fixed ω, the group velocities of the different modes extend from approximately c_1 for $m = 0$ to approximately c_2 for $m = M$.

Substituting (7.2-19) into (7.2-18) we obtain

$$\tan^2\left\{\frac{d}{2}\left[\left(\frac{\omega}{c_1}\right)^2 - \beta^2\right]^{1/2} - \frac{m\pi}{2}\right\} = \frac{\beta^2 - \omega^2/c_2^2}{\omega^2/c_1^2 - \beta^2}. \tag{7.2-20}$$

The self-consistency condition therefore establishes a relation between β and ω, the **dispersion relation**. This relation is plotted schematically in Fig. 7.2-8 for the different modes $m = 0, 1, \ldots$.

The group velocities lie between c_1 and c_2 (the phase velocities in the slab and substrate). At a given ω, the lowest-order mode (the least oblique mode, $m = 0$) travels with a group velocity closest to c_1. The most oblique mode $m = M$ has a group velocity $\approx c_2$. This is not surprising. A large portion of the energy carried by the most oblique mode travels in the substrate where the velocity is c_2. Figure 7.2-9 provides a sketch of the group velocities v_m as a function of the mode angle θ_m.

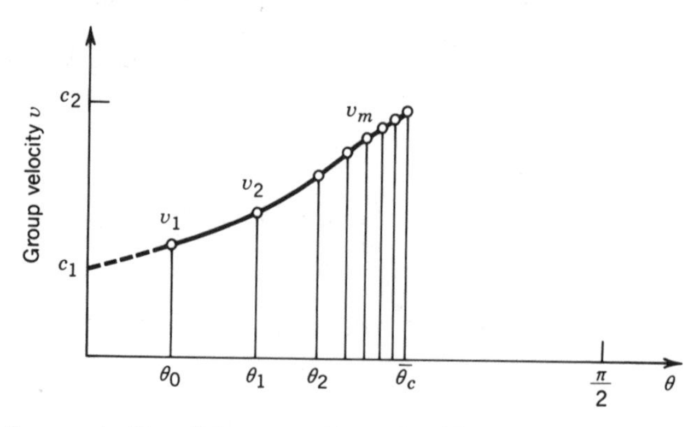

Figure 7.2-9 Group velocities of the waveguide modes. The least oblique mode travels with the smallest group velocity $\approx c_1 = c_o/n_1$. The most oblique mode has a group velocity $\approx c_2 = c_o/n_2$.

EXERCISE 7.2-2

Transit Time. Show that the maximum disparity between the times taken by the different modes of a planar dielectric waveguide to travel a distance L is

$$\sigma_\tau = \frac{L}{c_1}\Delta, \tag{7.2-21}$$

where $\Delta = (n_1 - n_2)/n_1$. If $n_1 - n_2 = 0.03$, at what distance is the disparity $\sigma_\tau = 1$ ns? Compare this to the case of a mirror waveguide with $n = 1$ and $d/\lambda = 10$. Use (7.1-11), (7.1-2), and (7.1-9).

By taking the total derivative of (7.2-18) with respect to β, we obtain

$$\frac{2d}{2k_y}\left(\frac{2\omega}{c_1^2}\frac{d\omega}{d\beta} - 2\beta\right) = 2\frac{\partial\varphi_r}{\partial\beta} + 2\frac{\partial\varphi_r}{\partial\omega}\frac{d\omega}{d\beta}.$$

Substituting $d\omega/d\beta = v$, $k_y/(\omega/c_1) = \sin\theta$, and $k_y/\beta = \tan\theta$ and introducing the new parameters

$$\Delta z = \frac{\partial\varphi_r}{\partial\beta}, \qquad \Delta\tau = -\frac{\partial\varphi_r}{\partial\omega}, \tag{7.2-22}$$

we obtain

$$v = \frac{d\cot\theta + \Delta z}{d\csc\theta/c_1 + \Delta\tau}. \tag{7.2-23}$$

As we recall from (7.1-11) and Fig. 7.1-6 for the planar-mirror waveguide, $d\cot\theta$ is the distance traveled in the z direction as a ray travels once between the two boundaries. This takes a time $d\csc\theta/c_1$. The ratio $d\cot\theta/(d\csc\theta/c_1) = c_1\cos\theta$ yields the group velocity for the mirror waveguide. The expression (7.2-23) for the group velocity in a dielectric waveguide indicates that the ray travels an additional distance $\Delta z = \partial\varphi_r/\partial\beta$, a trip that lasts a time $\Delta\tau = -\partial\varphi_r/\partial\omega$. We can think of this as an effective penetration of the ray into the cladding, or as an effective lateral shift of the ray, as shown in Fig. 7.2-10. The penetration of a ray undergoing total internal reflection is known as the **Goos–Hänchen effect** (see Problem 6.2-4). Using (7.2-22) it

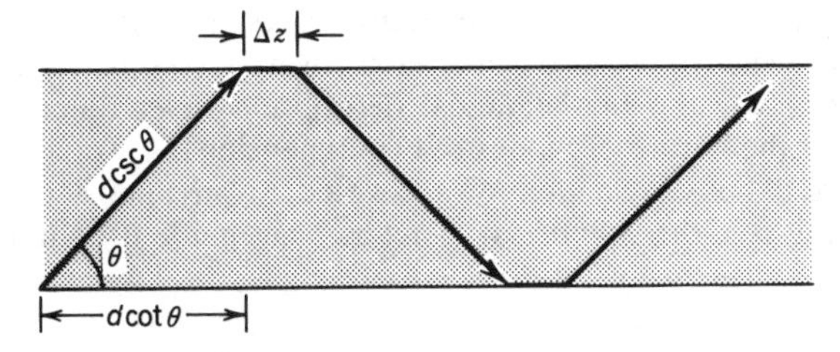

Figure 7.2-10 A ray model that replaces the reflection phase shift with an additional distance Δz traveled at velocity $c_1/\cos\theta$.

can be shown that $\Delta z/\Delta\tau = \omega/\beta = c_1/\cos\theta$. Therefore, more oblique modes travel this lateral distance at a faster speed than less oblique modes. This is responsible for the overall group velocity of more oblique modes being larger (contrary to the case of the mirror waveguide).

EXERCISE 7.2-3

The Asymmetric Planar Waveguide. Examine the TE field in an asymmetric planar waveguide consisting of a dielectric slab of width d and refractive index n_1 placed on a substrate of lower refractive index n_2 and covered with a medium of refractive index $n_3 < n_2 < n_1$, as illustrated in Fig. 7.2-11.

(a) Determine an expression for the maximum inclination angle θ of plane waves undergoing total internal reflection, and the corresponding numerical aperture NA of the waveguide.
(b) Write an expression for the self-consistency condition, similar to (7.2-4).
(c) Determine an approximate expression for the number of modes M (valid when M is very large).

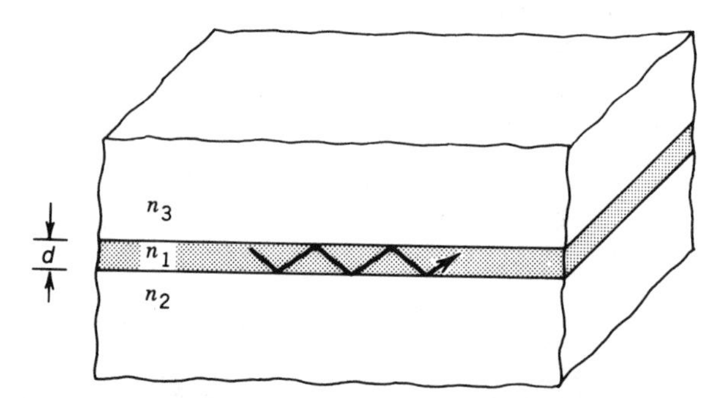

Figure 7.2-11 Asymmetric planar waveguide.

7.3 TWO-DIMENSIONAL WAVEGUIDES

The planar-mirror waveguide and the planar dielectric waveguide studied in the preceding two sections confine light in one transverse direction (the y direction) while guiding it along the z direction. Two-dimensional waveguides confine light in the two transverse directions (the x and y directions). The principle of operation and the underlying modal structure of two-dimensional waveguides is basically the same as planar waveguides; only the mathematical description is lengthier. This section is a brief description of the nature of modes in two-dimensional waveguides. Details can be found in specialized books. Chapter 8 is devoted to an important example of two-dimensional waveguides, the cylindrical dielectric waveguide used in optical fibers.

Rectangular Mirror Waveguide

The simplest generalization of the planar waveguide is the rectangular waveguide (Fig. 7.3-1). If the walls of the guide are mirrors, then, as in the planar case, light is guided

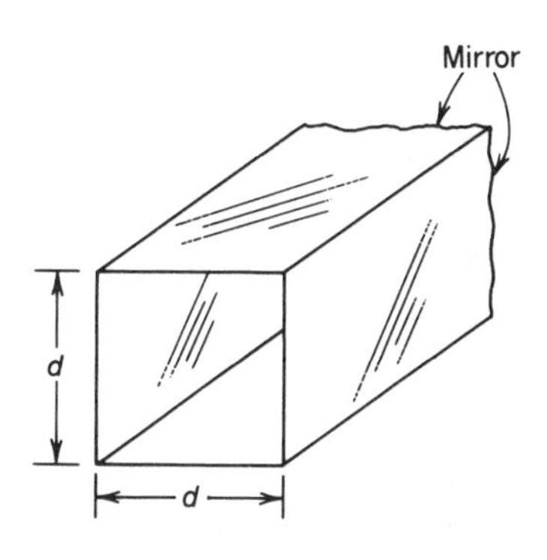

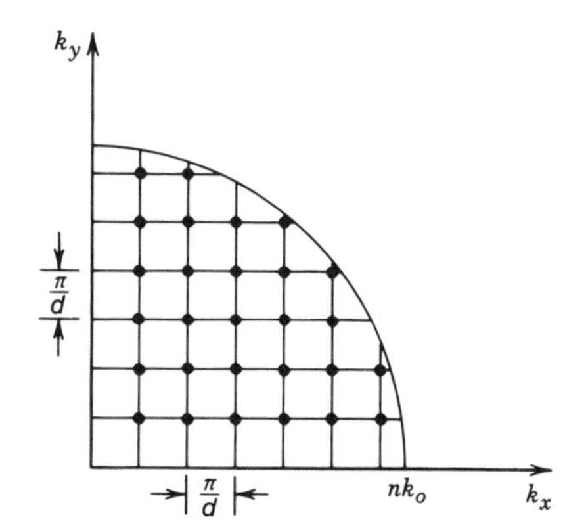

Figure 7.3-1 Modes of a rectangular mirror waveguide are characterized by a finite number of discrete values of k_x and k_y represented by dots.

by multiple reflections at all angles. For simplicity, we assume that the cross section of the guide is a square of width d. If a plane wave of wavevector (k_x, k_y, k_z) and its multiple reflections are to exist self-consistently inside the guide, it must satisfy the conditions:

$$2k_x d = 2\pi m_x, \qquad m_x = 1, 2, \ldots$$
$$2k_y d = 2\pi m_y, \qquad m_y = 1, 2, \ldots, \tag{7.3-1}$$

which are obvious generalizations of (7.1-3).

The propagation constant $\beta = k_z$ can be determined from k_x and k_y by using the relation $k_x^2 + k_y^2 + \beta^2 = n^2 k_o^2$. The three components of the wavevector therefore have discrete values, yielding a finite number of modes. Each mode is identified by two indices m_x and m_y (instead of one index m). All positive integer values of m_x and m_y are allowed as long as $k_x^2 + k_y^2 \le n^2 k_o^2$, as illustrated in Fig. 7.3-1.

The number of modes M can be easily determined by counting the number of dots within a quarter circle of radius nk_o in the k_x–k_y diagram (Fig. 7.3-1). If this number is large, it may be approximated by the ratio of the area $\pi(nk_o)^2/4$ to the area of a unit cell $(\pi/d)^2$,

$$M \approx \frac{\pi}{4}\left(\frac{2d}{\lambda}\right)^2. \tag{7.3-2}$$

Since there are two polarizations per mode, the total number of modes is actually $2M$. Comparing this to the number of modes in a one-dimensional mirror waveguide, $M \approx 2d/\lambda$, we see that increase of the dimensionality yields approximately the square of the number of modes. The number of modes is a measure of the degrees of freedom. When we add a second dimension we simply multiply the number of degrees of freedom.

The field distributions associated with these modes are generalizations of those in the planar case. Patterns such as those in Fig. 7.1-4 are obtained in each of the x and y directions depending on the mode indices m_x and m_y.

Rectangular Dielectric Waveguide

A dielectric cylinder of refractive index n_1 with square cross section of width d is embedded in a medium of slightly lower refractive index n_2. The waveguide modes can

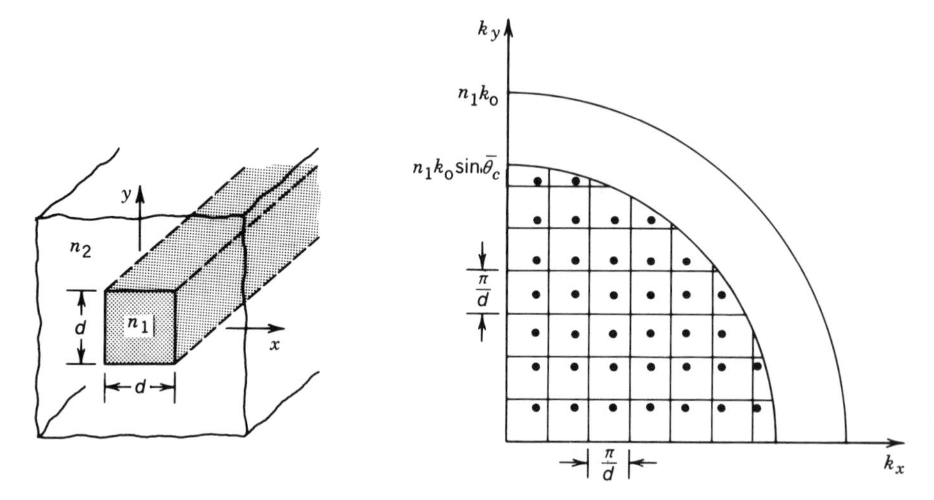

Figure 7.3-2 Geometry of a rectangular dielectric waveguide. The values of k_x and k_y for the waveguide modes are marked by dots.

be determined using a similar theory. Components of the wavevector (k_x, k_y, k_z) must satisfy the condition $k_x^2 + k_y^2 \leq n_1^2 k_o^2 \sin^2 \bar{\theta}_c$, where $\bar{\theta}_c = \cos^{-1}(n_2/n_1)$, so that k_x and k_y lie in the area shown in Fig. 7.3-2. The values of k_x and k_y for the different modes can be obtained from a self-consistency condition in which the phase shifts at the dielectric boundary are included, as was done in the planar case.

Unlike the mirror waveguide, k_x and k_y of the modes are not uniformly spaced. However, two consecutive values of k_x (or k_y) are separated by an *average* value of π/d (the same as for the mirror waveguide). The number of modes can therefore be approximated by counting the number of dots in the inner circle in the k_x–k_y diagram of Fig. 7.3-2, assuming an average spacing of π/d. The result is $M \approx (\pi/4)(n_1 k_o \sin \bar{\theta}_c)^2/(\pi/d)^2$, from which

$$M \approx \frac{\pi}{4}\left(\frac{2d}{\lambda_o}\right)^2 \mathrm{NA}^2, \tag{7.3-3}$$

Number of TE Modes

with $\mathrm{NA} = (n_1^2 - n_2^2)^{1/2}$ being the numerical aperture. The approximation is good when M is large. There is also an identical number M of TM modes. Compare this expression with that for the planar dielectric waveguide (7.2-7).

Geometries of Channel Waveguides

Useful geometries for waveguides include the strip, the embedded-strip, the rib or ridge, and the strip-loaded waveguides illustrated in Fig. 7.3-3. The exact analysis for some of these geometries is not easy, and approximations are usually used. The reader is referred to specialized books for further readings on this topic.

The waveguide may be fabricated in different configurations as illustrated in Fig. 7.3-4 for the embedded-strip geometry. S bends are used to offset the propagation axis. The Y branch plays the role of a beamsplitter or combiner. Two Y branches may be used to make a Mach–Zehnder interferometer. Two waveguides in close proximity (or

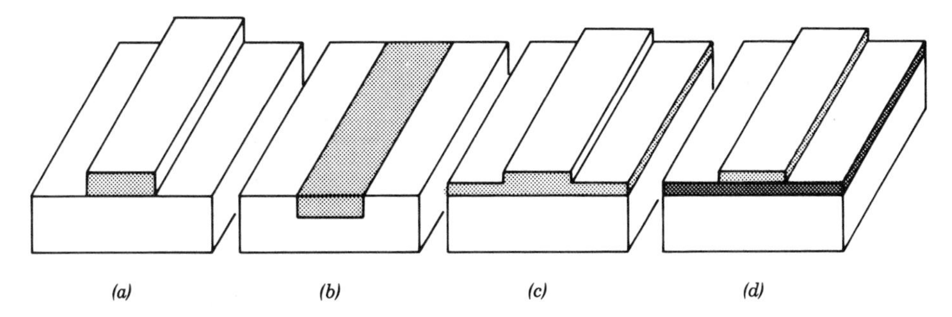

Figure 7.3-3 Various types of waveguide geometries: (*a*) strip; (*b*) embedded strip; (*c*) rib or ridge; (*d*) strip loaded. The darker the shading, the higher the refractive index.

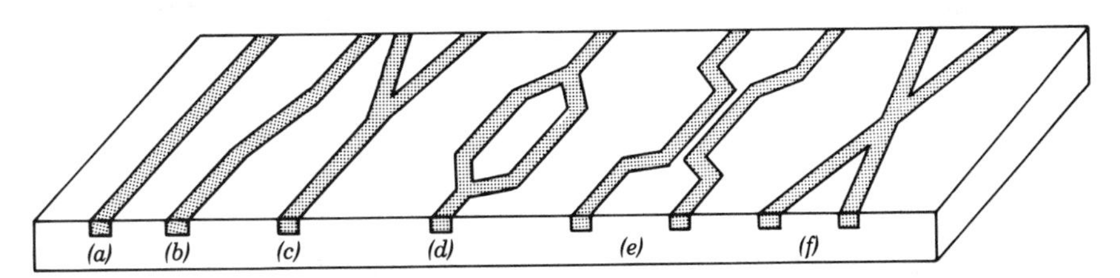

Figure 7.3-4 Different configurations for waveguides: (*a*) straight; (*b*) S bend; (*c*) Y branch; (*d*) Mach–Zehnder; (*e*) directional coupler; (*f*) intersection.

intersecting) can exchange power and may be used as directional couplers, as we shall see in the next section.

The most advanced technology for fabricating waveguides is Ti:$LiNbO_3$. An embedded-strip waveguide is fabricated by diffusing titanium into a lithium niobate substrate to raise its refractive index in the region of the strip. GaAs strip waveguides are made by using layers of GaAs and AlGaAs of lower refractive index. Glass waveguides are made by ion exchange. As we shall see in Chaps. 18 and 21, these waveguides are used to make a number of optical devices, e.g., light modulators and switches.

7.4 OPTICAL COUPLING IN WAVEGUIDES

A. Input Couplers

Mode Excitation

As was shown in previous sections, light propagates in a waveguide in the form of modes. The complex amplitude of the optical field is generally a superposition of these modes,

$$E(y, z) = \sum_m a_m u_m(y) \exp(-j\beta_m z), \tag{7.4-1}$$

where a_m is the amplitude, $u_m(y)$ is the transverse distribution (which is assumed to be real), and β_m is the propagation constant of mode m.

The amplitudes of the different modes depend on the nature of the light source used to "excite" the waveguide. If the source has a distribution that matches perfectly that of a specific mode, only that mode is excited. A source of arbitrary distribution

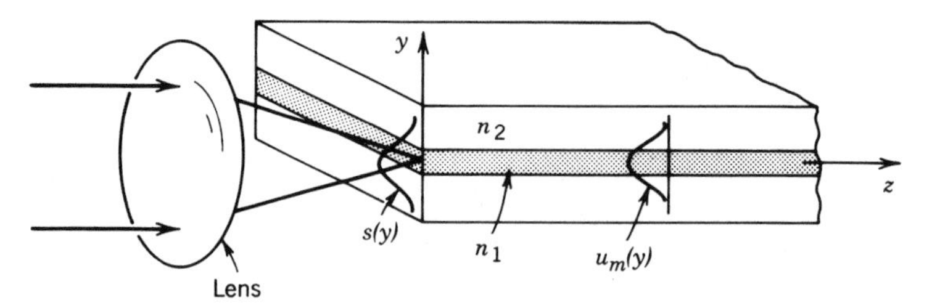

Figure 7.4-1 Coupling an optical beam into a waveguide.

$s(y)$ excites different modes by different amounts. The fraction of power transferred from the source to mode m depends on the degree of similarity between $s(y)$ and $u_m(y)$. We can write $s(y)$ as an expansion (a weighted superposition) of the orthogonal functions $u_m(y)$, i.e.,

$$s(y) = \sum_m a_m u_m(y), \tag{7.4-2}$$

where the coefficient a_l, the amplitude of the excited mode l, is

$$a_l = \int_{-\infty}^{\infty} s(y) u_l(y)\, dy. \tag{7.4-3}$$

This expression can be derived by multiplying both sides of (7.4-2) by $u_l(y)$, integrating with respect to y, and using the orthogonality equation $\int_{-\infty}^{\infty} u_l(y) u_m(y)\, dy = 0$ for $l \neq m$ along with the normalization condition. The coefficient a_l represents the degree of similarity (or correlation) between the source distribution $s(y)$ and the mode distribution $u_l(y)$.

Input Couplers

Light may be coupled into a waveguide by directly focusing it at one end (Fig. 7.4-1). To excite a given mode, the transverse distribution of the incident light $s(y)$ should match that of the mode. The polarization of the incident light must also match that of the desired mode. Because of the small dimensions of the waveguide slab, focusing and alignment are usually difficult and the coupling is inefficient.

In a multimode waveguide, the amount of coupling can be assessed by using a ray-optics approach (Fig. 7.4-2). The guided rays within the waveguide are confined to

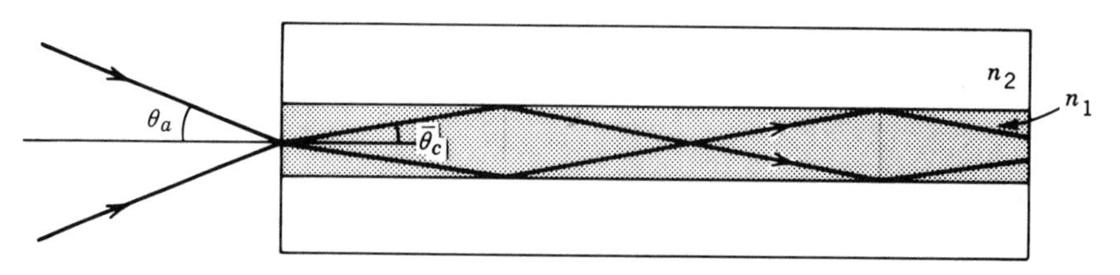

Figure 7.4-2 Focusing rays into a multimode waveguide.

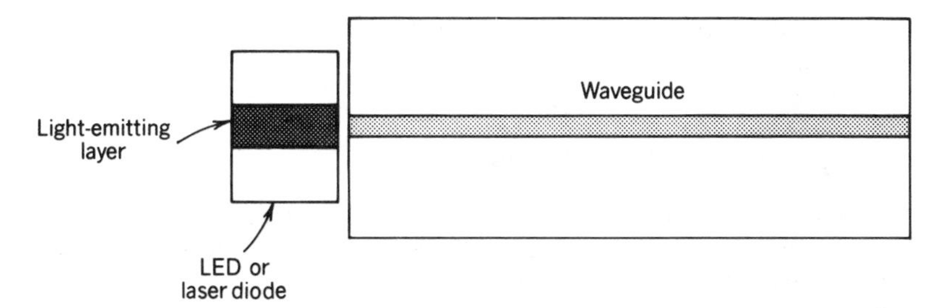

Figure 7.4-3 End butt coupling a light-emitting diode or a laser diode to a waveguide.

an angle $\bar{\theta}_c = \cos^{-1}(n_2/n_1)$. Because of refraction of the incident rays, this corresponds to an external angle θ_a satisfying NA $= \sin\theta_a = n_1 \sin\bar{\theta}_c = n_1[1 - (n_2/n_1)^2]^{1/2} = (n_1^2 - n_2^2)^{1/2}$, where NA is the numerical aperture of the waveguide (see Exercise 1.2-5). For maximum coupling efficiency the incident light should be focused to an angle not greater than θ_a.

Light may also be coupled from a semiconductor source (a light-emitting diode or a laser diode) into a waveguide simply by aligning the ends of the source and the waveguide while leaving a small space that is selected for maximum coupling (Fig. 7.4-3). In light-emitting diodes, light originates from within a narrow semiconductor junction and is emitted in all directions. In a laser diode, the emitted light is itself confined in a waveguide of its own (light-emitting diodes and laser diodes are described in Chap. 16). Other methods of coupling light into a waveguide include the use of a prism, a diffraction grating, or another waveguide.

The Prism Coupler

Optical power may be coupled into or out of a slab waveguide by use of a prism. A prism of refractive index $n_p > n_2$ is placed at a distance d_p from the slab of a waveguide of refractive indices n_1 and n_2, as illustrated in Fig. 7.4-4. An optical wave is incident into the prism such that it undergoes total internal reflection within the prism at an angle θ_p. The incident and reflected waves form a wave traveling in the z direction with a propagation constant $\beta_p = n_p k_o \cos\theta_p$. The transverse field distribution extends outside the prism and decays exponentially in the space separating the prism and the slab. If the distance d_p is sufficiently small, the wave is coupled to a mode of the slab waveguide with a matching propagation constant $\beta_m \approx \beta_p$. If an appropriate interaction distance is selected, power can be coupled into the slab waveguide, so that the prism acts as an input coupler. The operation may be reversed

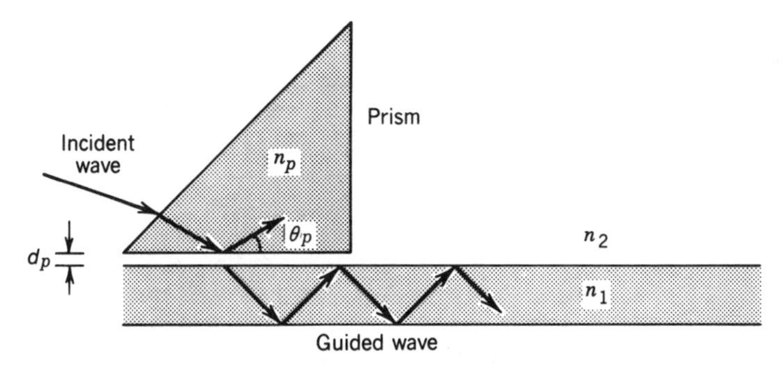

Figure 7.4-4 The prism coupler.

to make an output coupler, which extracts light from the slab waveguide into free space.

B. Coupling Between Waveguides

If two waveguides are sufficiently close such that their fields overlap, light can be coupled from one into the other. Optical power can be transferred between the waveguides, an effect that can be used to make optical couplers and switches. The basic principle of waveguide coupling is presented here; couplers and switches are discussed in Chaps. 21 and 22. Consider two parallel planar waveguides made of two slabs of widths d, separation $2a$, and refractive indices n_1 and n_2 embedded in a medium of refractive index n slightly smaller than n_1 and n_2, as illustrated in Fig. 7.4-5. Each of the waveguides is assumed to be single-mode. The separation between the waveguides is such that the optical field outside the slab of one waveguide (in the absence of the other) overlaps slightly with the slab of the other waveguide.

The formal approach to studying the propagation of light in this structure is to write Maxwell's equations in the different regions and use the boundary conditions to determine the modes of the overall system. These modes are different from those of each of the waveguides in isolation. An exact analysis is difficult and is beyond the scope of this book. However, for weak coupling, a simplified approximate theory, known as coupled-mode theory, is usually satisfactory.

The coupled-mode theory assumes that the modes of each of the waveguides, in the absence of the other, remain approximately the same, say $u_1(y)\exp(-j\beta_1 z)$ and $u_2(y)\exp(-j\beta_2 z)$, and that coupling modifies the *amplitudes* of these modes without affecting their transverse spatial distributions or their propagation constants. The amplitudes of the modes of waveguides 1 and 2 are therefore functions of z, $a_1(z)$ and $a_2(z)$. The theory aims at determining $a_1(z)$ and $a_2(z)$ under appropriate boundary conditions.

Coupling can be regarded as a scattering effect. The field of waveguide 1 is scattered from waveguide 2, creating a source of light that changes the amplitude of the field in waveguide 2. The field of waveguide 2 has a similar effect on waveguide 1. An analysis of this mutual interaction leads to two coupled differential equations that govern the variation of the amplitudes $a_1(z)$ and $a_2(z)$.

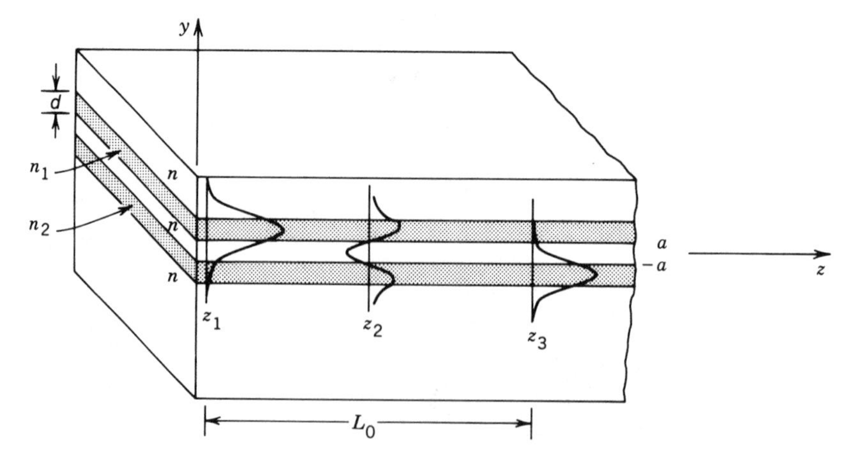

Figure 7.4-5 Coupling between two parallel planar waveguides. At z_1 light is mostly in waveguide 1, at z_2 it is divided equally between the two waveguides, and at z_3 it is mostly in waveguide 2.

It can be shown (see the derivation at the end of this section) that the amplitudes $a_1(z)$ and $a_2(z)$ are governed by two coupled first-order differential equations

$$\frac{da_1}{dz} = -j\mathcal{C}_{21}\exp(j\,\Delta\beta\,z)a_2(z) \tag{7.4-4a}$$

$$\frac{da_2}{dz} = -j\mathcal{C}_{12}\exp(-j\,\Delta\beta\,z)a_1(z), \tag{7.4-4b}$$

Coupled-Mode Equations

where

$$\Delta\beta = \beta_1 - \beta_2 \tag{7.4-5}$$

is the phase mismatch per unit length and

$$\mathcal{C}_{21} = \frac{1}{2}(n_2^2 - n^2)\frac{k_o^2}{\beta_1}\int_a^{a+d} u_1(y)u_2(y)\,dy,$$

$$\mathcal{C}_{12} = \frac{1}{2}(n_1^2 - n^2)\frac{k_o^2}{\beta_2}\int_{-a-d}^{-a} u_2(y)u_1(y)\,dy \tag{7.4-6}$$

are coupling coefficients.

We see from (7.4-4) that the rate of variation of a_1 is proportional to a_2, and vice versa. The coefficient of proportionality is the product of the coupling coefficient and the phase mismatch factor $\exp(j\,\Delta\beta\,z)$.

Assuming that the amplitude of light entering waveguide 1 is $a_1(0)$ and that no light enters waveguide 2, $a_2(0) = 0$, then (7.4-4) can be solved under these boundary conditions, yielding the harmonic solution

$$a_1(z) = a_1(0)\exp\left(+\frac{j\,\Delta\beta\,z}{2}\right)\left(\cos\gamma z - j\frac{\Delta\beta}{2\gamma}\sin\gamma z\right) \tag{7.4-7a}$$

$$a_2(z) = a_1(0)\frac{\mathcal{C}_{12}}{j\gamma}\exp\left(-j\frac{\Delta\beta\,z}{2}\right)\sin\gamma z, \tag{7.4-7b}$$

where

$$\gamma^2 = \left(\frac{\Delta\beta}{2}\right)^2 + \mathcal{C}^2 \tag{7.4-8}$$

and

$$\mathcal{C} = (\mathcal{C}_{12}\mathcal{C}_{21})^{1/2}. \tag{7.4-9}$$

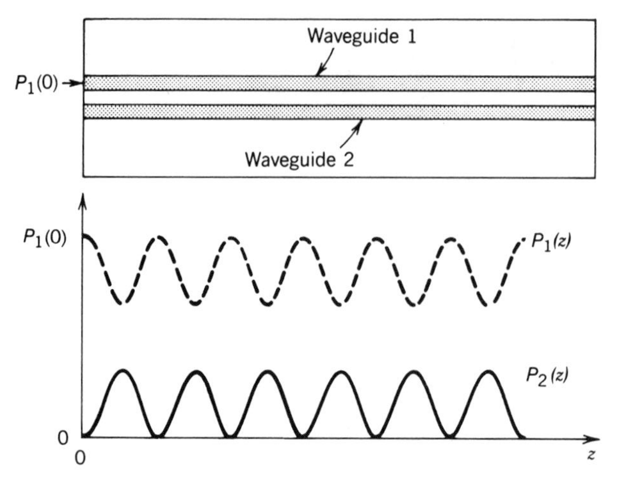

Figure 7.4-6 Periodic exchange of power between guides 1 and 2.

The optical powers $P_1(z) \propto |a_1(z)|^2$ and $P_2(z) \propto |a_2(z)|^2$ are therefore

$$P_1(z) = P_1(0)\left[\cos^2 \gamma z + \left(\frac{\Delta\beta}{2\gamma}\right)^2 \sin^2 \gamma z\right] \tag{7.4-10a}$$

$$P_2(z) = P_1(0)\frac{|\mathcal{C}_{12}|^2}{\gamma^2} \sin^2 \gamma z. \tag{7.4-10b}$$

Thus power is exchanged periodically between the two guides as illustrated in Fig. 7.4-6. The period is $2\pi/\gamma$. Power conservation requires that $\mathcal{C}_{12} = \mathcal{C}_{21} = \mathcal{C}$.

When the guides are identical, i.e., $n_1 = n_2$, $\beta_1 = \beta_2$, and $\Delta\beta = 0$, the two guided waves are said to be phase matched. Equations (7.4-10a, b) then simplify to

$$P_1(z) = P_1(0) \cos^2 \mathcal{C} z \tag{7.4-11a}$$

$$P_2(z) = P_1(0) \sin^2 \mathcal{C} z. \tag{7.4-11b}$$

The exchange of power between the waveguides can then be complete, as illustrated in Fig. 7.4-7.

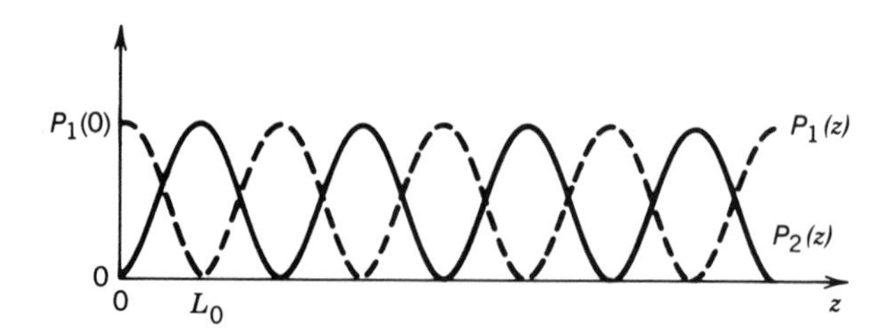

Figure 7.4-7 Exchange of power between guides 1 and 2 in the phase-matched case.

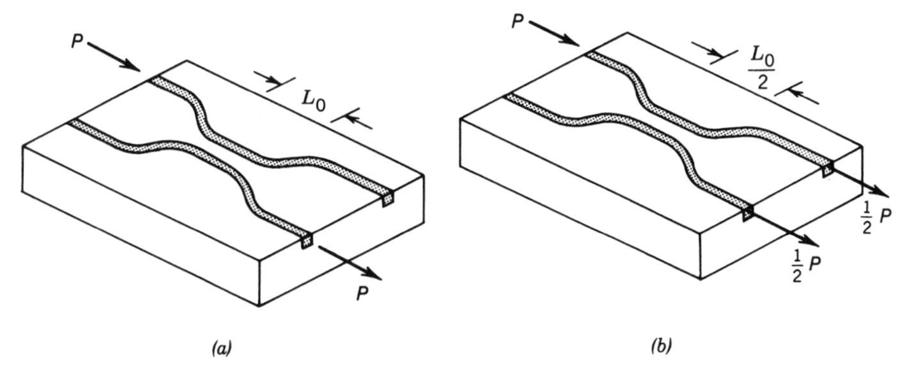

Figure 7.4-8 Optical couplers: (*a*) switching of power from one waveguide to another; (*b*) a 3-dB coupler.

We thus have a device for coupling desired fractions of optical power from one waveguide to another. At a distance $z = L_0 = \pi/2\mathcal{C}$, called the transfer distance, the power is transferred completely from waveguide 1 to waveguide 2 [Fig. 7.4-8(a)]. At a distance $L_0/2$, half the power is transferred, so that the device acts as a 3-dB coupler, i.e., a 50/50 beamsplitter [Fig. 7.4-8(b)].

Switching by Control of Phase Mismatch

A waveguide coupler of fixed length, $L_0 = \pi/2\mathcal{C}$, for example, changes its power-transfer ratio if a small phase mismatch $\Delta\beta$ is introduced. Using (7.4-10b) and (7.4-8), the power-transfer ratio $\mathcal{T} = P_2(L_0)/P_1(0)$ may be written as a function of $\Delta\beta$,

$$\mathcal{T} = \left(\frac{\pi}{2}\right)^2 \operatorname{sinc}^2\left\{\frac{1}{2}\left[1 + \left(\frac{\Delta\beta\, L_0}{\pi}\right)^2\right]^{1/2}\right\}, \tag{7.4-12}$$

Power-Transfer Ratio

where $\operatorname{sinc}(x) = \sin(\pi x)/(\pi x)$. Figure 7.4-9 illustrates the dependence of the power-transfer ratio $\mathcal{T}$ on the mismatch parameter $\Delta\beta\, L_0$. The ratio has a maximum value of unity at $\Delta\beta\, L_0 = 0$, decreases with increasing $\Delta\beta\, L_0$, and then vanishes when $\Delta\beta\, L_0 = \sqrt{3}\,\pi$.

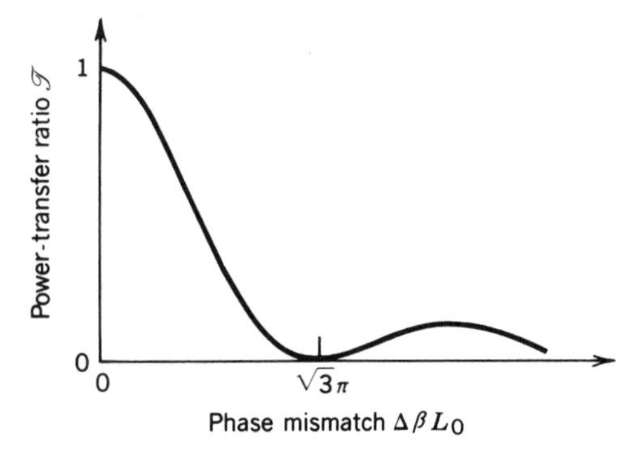

Figure 7.4-9 Dependence of the power transfer ratio $\mathcal{T} = P_2(L_0)/P_1(0)$ on the phase mismatch parameter $\Delta\beta L_0$. The waveguide length is chosen such that for $\Delta\beta = 0$ (the phase-matched case), maximum power is transferred to waveguide 2, i.e., $\mathcal{T} = 1$.

The dependence of the transferred power on the phase mismatch can be utilized in making electrically activated directional couplers. If the mismatch $\Delta\beta\, L_0$ is switched between 0 and $\sqrt{3}\,\pi$, the light is switched from waveguide 2 to waveguide 1. Electrical control of $\Delta\beta$ can be achieved if the material of the waveguides is electro-optic (i.e., if its refractive index can be altered by applying an electric field). Such a device will be studied in Chaps. 18 and 21 in connection with electro-optic switches.

*Derivation of the Coupled Wave Equations

We now derive the differential equations (7.4-4) that govern the amplitudes $a_1(z)$ and $a_2(z)$ of the coupled modes. When the two waveguides are not interacting they carry optical fields whose complex amplitudes are of the form

$$E_1(y,z) = a_1 u_1(y)\exp(-j\beta_1 z) \tag{7.4-13a}$$

$$E_2(y,z) = a_2 u_2(y)\exp(-j\beta_2 z). \tag{7.4-13b}$$

The amplitudes a_1 and a_2 are then constant. In the presence of coupling, we assume that the amplitudes a_1 and a_2 become functions of z but the transverse functions $u_1(y)$ and $u_2(y)$, and the propagation constants β_1 and β_2, are not altered. The amplitudes a_1 and a_2 are assumed to be slowly varying functions of z in comparison with the distance β^{-1} (the inverse of the propagation constant, β_1 or β_2, which is of the order of magnitude of the wavelength of light).

The presence of waveguide 2 is regarded as a perturbation of the medium outside waveguide 1 in the form of a slab of refractive index $n_2 - n$ and width d at a distance $2a$. The excess refractive index $(n_2 - n)$ and the field E_2 correspond to an excess polarization density $P = (\epsilon_2 - \epsilon)E_2 = \epsilon_o(n_2^2 - n^2)E_2$, which creates a source of optical radiation into waveguide 1 [see (5.2-19)] $\mathscr{S}_1 = -\mu_o\, \partial^2\mathscr{P}/\partial t^2$ with complex amplitude

$$\begin{aligned} S_1 = \mu_o\omega^2 P = \mu_o\omega^2\epsilon_o(n_2^2 - n^2)E_2 = (n_2^2 - n^2)k_o^2 E_2 \\ = (k_2^2 - k^2)E_2. \end{aligned} \tag{7.4-14}$$

Here ϵ_2 and ϵ are the permittivities associated with the refractive indices n_2 and n, and $k_2 = n_2 k_o$. This source is present only in the slab of waveguide 2.

To determine the effect of such a source on the field in waveguide 1, we write the Helmholtz equation in the presence of a source as

$$\nabla^2 E_1 + k_1^2 E_1 = -S_1 = -(k_2^2 - k^2)E_2. \tag{7.4-15a}$$

We similarly write the Helmholtz equation for the wave in waveguide 2 with a source generated as a result of the field in waveguide 1,

$$\nabla^2 E_2 + k_2^2 E_2 = -S_2 = -(k_1^2 - k^2)E_1, \tag{7.4-15b}$$

where $k_1 = n_1 k_o$. Equations (7.4-15a, b) are two coupled partial differential equations which we solve to determine E_1 and E_2. This type of perturbation analysis is valid only for weakly coupled waveguides.

We now write $E_1(y,z) = a_1(z)e_1(y,z)$ and $E_2(y,z) = a_2(z)e_2(y,z)$, where $e_1(y,z) = u_1(y)\exp(-j\beta_1 z)$ and $e_2(y,z) = u_2(y)\exp(-j\beta_2 z)$ and note that e_1 and e_2 must satisfy the Helmholtz equations,

$$\nabla^2 e_1 + k_1^2 e_1 = 0 \tag{7.4-16a}$$

$$\nabla^2 e_2 + k_2^2 e_2 = 0, \tag{7.4-16b}$$

where $k_1 = n_1 k_o$ and $k_2 = n_2 k_o$ for points inside the slabs of waveguides 1 and 2, respectively, and $k_1 = k_2 = nk_o$ elsewhere. Substituting $E_1 = a_1 e_1$ into (7.4-15a), we obtain

$$\frac{d^2 a_1}{dz^2} e_1 + 2\frac{da_1}{dz}\frac{de_1}{dz} = -\left(k_2^2 - k^2\right) a_2 e_2. \tag{7.4-17}$$

Noting that a_1 varies slowly, whereas e_1 varies rapidly with z, we neglect the first term of (7.4-17) compared to the second. The ratio between these terms is $[(d\Psi/dz)e_1]/[2\Psi de_1/dz] = [(d\Psi/dz)e_1]/[2\Psi(-j\beta_1 e_1)] = j(d\Psi/\Psi)/2\beta_1\, dz$ where $\Psi = da_1/dz$. The approximation is valid if $d\Psi/\Psi \ll \beta_1\, dz$, i.e., if the variation in $a_1(z)$ is slow in comparison with the length β_1^{-1}.

We now substitute for $e_1 = u_1 \exp(-j\beta_1 z)$ and $e_2 = u_2 \exp(-j\beta_2 z)$ into (7.4-17), after neglecting its first term, to obtain

$$2\frac{da_1}{dz}(-j\beta_1)u_1(y)e^{-j\beta_1 z} = -\left(k_2^2 - k^2\right) a_2 u_2(y) e^{-j\beta_2 z}. \tag{7.4-18}$$

Multiplying both sides of (7.4-18) by $u_1(y)$, integrating with respect to y, and using the fact that $u_1^2(y)$ is normalized so that its integral is unity, we obtain

$$\frac{da_1}{dz} e^{-j\beta_1 z} = -j\mathcal{C}_{21} a_2(z) e^{-j\beta_2 z}, \tag{7.4-19}$$

where $\mathcal{C}_{21}$ is given by (7.4-6). A similar equation is obtained by repeating the procedure for waveguide 2. These equations yield the coupled differential equations (7.4-4).

READING LIST

Books

T. Tamir, ed., *Guided-Wave Optoelectronics*, Springer-Verlag, New York, 2nd ed. 1990.

H. Nishihara, M. Haruna, and T. Suhara, *Optical Integrated Circuits*, McGraw-Hill, New York, 1989.

P. Yeh, *Optical Waves in Layered Media*, Wiley, New York, 1988.

L. D. Hutcheson, ed., *Integrated Optical Circuits and Components*, Marcel Dekker, New York, 1987.

D. L. Lee, *Electromagnetic Principles of Integrated Optics*, Wiley, New York, 1986.

S. Solimeno, B. Crosignani, and P. DiPorto, *Guiding, Diffraction, and Confinement of Optical Radiation*, Academic Press, Orlando, FL, 1986.

H. Nolting and R. Ulrich, eds., *Integrated Optics*, Springer-Verlag, New York, 1985.

R. G. Hunsperger, *Integrated Optics: Theory and Technology*, Springer-Verlag, New York, 1982, 2nd ed. 1984.

K. Iga, Y. Kokubun, and M. Oikawa, *Fundamentals of Microoptics*, Academic Press, Tokyo, 1984.

H. Huang, *Coupled Mode Theory as Applied to Microwave and Optical Transmission*, VNU Science Press, Utrecht, The Netherlands, 1984.

S. Martellucci and A. N. Chester, eds., *Integrated Optics: Physics and Applications*, Plenum Press, New York, 1983.

D. Marcuse, *Light Transmission Optics*, Van Nostrand-Reinhold, New York, 2nd ed. 1982.

T. Tamir, ed., *Integrated Optics*, Springer-Verlag, New York, 1979, 2nd ed. 1982.

M. J. Adams, *An Introduction to Optical Waveguides*, Wiley, New York, 1981.

G. H. Owyang, *Foundations of Optical Waveguides*, Elsevier/North-Holland, New York, 1981.

D. B. Ostrowsky, ed., *Fiber and Integrated Optics*, Plenum Press, New York, 1979.

M. S. Sodha and A. K. Ghatak, *Inhomogeneous Optical Waveguides*, Plenum Press, New York, 1977.

M. K. Barnoski, *Introduction to Integrated Optics*, Plenum Press, New York, 1974.

D. Marcuse, *Theory of Dielectric Optical Waveguides*, Academic Press, New York, 1974.

N. S. Kapany and J. J. Burke, *Optical Waveguides*, Academic Press, New York, 1972.

Special Journal Issues

Special issue on integrated optics, *Journal of Lightwave Technology*, vol. 6, no. 6, 1988.

Special section on integrated optics and optoelectronics, *Proceedings of the IEEE*, vol. 75, no. 11, 1987.

Special issue on integrated optics, *IEEE Journal of Quantum Electronics*, vol. QE-22, no. 6, 1986.

Joint special issue on optical guided-wave technology, *IEEE Journal of Quantum Electronics*, vol. QE-18, no. 4, 1982.

Special issue on integrated optics, *IEEE Journal of Quantum Electronics*, vol. QE-13, no. 4, 1977.

Articles

W. J. Tomlinson and S. K. Korotky, Integrated Optics: Basic Concepts and Techniques, in *Optical Fiber Telecommunications II*, S. E. Miller and I. P. Kaminow, eds., Academic Press, New York, 1988.

J. Viljanen, M. Maklin, and M. Leppihalme, Ion-Exchanged Integrated Waveguide Structures, *IEEE Circuits and Devices Magazine*, vol. 1, no. 2, pp. 13–16, 1985.

R. C. Alferness, Guided-Wave Devices for Optical Communication, *IEEE Journal of Quantum Electronics*, vol. QE-17, pp. 946–959, 1981.

R. Olshansky, Propagation in Glass Optical Waveguides, *Reviews of Modern Physics*, vol. 51, pp. 341–368, 1979.

P. K. Tien, Integrated Optics and New Wave Phenomena in Optical Waveguides, *Reviews of Modern Physics*, vol. 49, pp. 361–420, 1977.

H. Kogelnik, An Introduction to Integrated Optics, *IEEE Transactions on Microwave Theory and Techniques*, vol. MTT-23, pp. 2–20, 1975.

PROBLEMS

7.1-1 **Field Distribution.** (a) Show that a single TEM plane wave $E_x(y, z) = A\exp(-jk_y y)\exp(-j\beta z)$ cannot satisfy the boundary conditions, $E_x(\pm d/2, z) = 0$ at all z, in the mirror waveguide illustrated in Fig. 7.1-1.
(b) Show that the sum of two TEM plane waves written as $E_x(y, z) = A_1\exp(-jk_{y1} y)\exp(-j\beta_1 z) + A_2\exp(-jk_{y2} y)\exp(-j\beta_2 z)$ does satisfy the boundary conditions if $A_1 = \pm A_2$, $\beta_1 = \beta_2$, and $k_{y1} = -k_{y2} = m\pi/d$, $m = 1, 2, \ldots$.

7.1-2 **Modal Dispersion.** Light of wavelength $\lambda_o = 0.633\ \mu\text{m}$ is transmitted through a mirror waveguide of mirror separation $d = 10\ \mu\text{m}$ and $n = 1$. Determine the number of TE and TM modes. Determine the group velocities of the fastest and the slowest mode. If a narrow pulse of light is carried by all modes for a distance of 1 m in the waveguide, how much does the pulse spread as a result of the differences of the group velocities?

7.2-1 **Parameters of a Dielectric Waveguide.** Light of free-space wavelength $\lambda_o = 0.87\ \mu\text{m}$ is guided by a thin planar film of width $d = 2\ \mu\text{m}$ and refractive index $n_1 = 1.6$ surrounded by a medium of refractive index $n_2 = 1.4$.
(a) Determine the critical angle θ_c and its complement $\bar{\theta}_c$, the numerical aperture NA, and the maximum acceptance angle for light originating in air ($n = 1$).
(b) Determine the number of TE modes.
(c) Determine the bounce angle θ and the group velocity v of the $m = 0$ TE mode.

7.2-2 **Effect of Cladding.** Repeat Problem 7.2-1 if the thin film is suspended in air ($n_2 = 1$). Compare the results.

7.2-3 **Field Distribution.** The transverse distribution $u_m(y)$ of the electric-field complex amplitude of a TE mode in a slab waveguide is given by (7.2-9) and (7.2-12). Derive an expression for the ratio of the proportionality constants. Plot the distribution of the $m = 0$ TE mode for a slab waveguide with parameters $n_1 = 1.48$, $n_2 = 1.46$, $d = 0.5\ \mu\text{m}$, and $\lambda_o = 0.85\ \mu\text{m}$, and determine its confinement factor (percentage of power in the slab).

7.2-4 **Derivation of the Field Distributions Using Maxwell's Equations.** Assuming that the electric field in a symmetric dielectric waveguide is harmonic within the slab and exponential outside the slab and has a propagation constant β in both media, we may write $E_x(y, z) = u(y)e^{-j\beta z}$, where

$$u(y) = \begin{cases} A\cos(k_y y + \varphi), & -d/2 \le y \le d/2, \\ B\exp(-\gamma y), & y > d/2, \\ B\exp(\gamma y), & y < -d/2. \end{cases}$$

For the Helmholtz equation to be satisfied, $k_y^2 + \beta^2 = n_1^2 k_o^2$ and $-\gamma^2 + \beta^2 = n_2^2 k_o^2$. Use Maxwell's equations to derive expressions for $H_y(y, z)$ and $H_z(y, z)$. Show that the boundary conditions are satisfied if β, γ, and k_y take the values β_m, γ_m, and k_{ym} derived in the text and verify the self-consistency condition (7.2-4).

7.2-5 **Single-Mode Waveguide.** What is the largest thickness d of a planar symmetric dielectric waveguide with refractive indices $n_1 = 1.50$ and $n_2 = 1.46$ for which there is only one TE mode at $\lambda_o = 1.3\ \mu\text{m}$? What is the number of modes if a waveguide with this thickness is used at $\lambda_o = 0.85\ \mu\text{m}$ instead?

7.2-6 **Mode Cutoff.** Show that the cutoff condition for TE mode $m > 0$ in a symmetric slab waveguide with $n_1 \approx n_2$ is approximately $\lambda_o^2 \approx 8n_1\Delta n d^2/m^2$, where $\Delta n = n_1 - n_2$.

7.2-7 **TM Modes.** Derive an expression for the bounce angles of the TM modes similar to (7.2-4). Use a computer to generate a plot similar to Fig. 7.2-2 for TM modes in a waveguide with $\sin\bar{\theta}_c = 0.3$ and $\lambda/2d = 0.1$. What is the number of TM modes?

7.3-1 **Modes of a Rectangular Dielectric Waveguide.** A rectangular dielectric waveguide has a square cross section of area $10^{-2}\ \text{mm}^2$ and numerical aperture NA = 0.1. Use (7.3-3) to plot the number of TE modes as a function of frequency ν. Compare your results with Fig. 7.2-4.

7.4-1 **Coupling Coefficient Between Two Slabs.** (a) Use (7.4-6) to determine the coupling coefficient between two *identical* slab waveguides of width $d = 0.5\ \mu\text{m}$, spacing $2a = 1.0\ \mu\text{m}$, refractive indices $n_1 = n_2 = 1.48$, in a medium of refractive index $n = 1.46$, at $\lambda_o = 0.85\ \mu\text{m}$. Assume that both guides are operating in the $m = 0$ TE mode and use the results of Problem 7.2-3 to determine the transverse distributions.
(b) Determine the length of the guides so that the device acts as a 3-dB coupler.

CHAPTER

8

FIBER OPTICS

8.1 STEP-INDEX FIBERS
- A. Guided Rays
- B. Guided Waves
- C. Single-Mode Fibers

8.2 GRADED-INDEX FIBERS
- A. Guided Waves
- B. Propagation Constants and Velocities

8.3 ATTENUATION AND DISPERSION
- A. Attenuation
- B. Dispersion
- C. Pulse Propagation

Dramatic improvements in the development of low-loss materials for optical fibers are responsible for the commercial viability of fiber-optic communications. **Corning** Incorporated pioneered the development and manufacture of ultra-low-loss glass fibers.

An optical fiber is a cylindrical dielectric waveguide made of low-loss materials such as silica glass. It has a central **core** in which the light is guided, embedded in an outer **cladding** of slightly lower refractive index (Fig. 8.0-1). Light rays incident on the core–cladding boundary at angles greater than the critical angle undergo total internal reflection and are guided through the core without refraction. Rays of greater inclination to the fiber axis lose part of their power into the cladding at each reflection and are not guided.

As a result of recent technological advances in fabrication, light can be guided through 1 km of glass fiber with a loss as low as ≈ 0.16 dB ($\approx 3.6\,\%$). Optical fibers are replacing copper coaxial cables as the preferred transmission medium for electromagnetic waves, thereby revolutionizing terrestrial communications. Applications range from long-distance telephone and data communications to computer communications in a local area network.

In this chapter we introduce the principles of light transmission in optical fibers. These principles are essentially the same as those that apply in planar dielectric waveguides (Chap. 7), except for the cylindrical geometry. In both types of waveguide light propagates in the form of modes. Each mode travels along the axis of the waveguide with a distinct propagation constant and group velocity, maintaining its transverse spatial distribution and its polarization. In planar waveguides, we found that each mode was the sum of the multiple reflections of a TEM wave bouncing within the slab in the direction of an optical ray at a certain bounce angle. This approach is approximately applicable to cylindrical waveguides as well. When the core diameter is small, only a single mode is permitted and the fiber is said to be a **single-mode fiber**. Fibers with large core diameters are **multimode fibers**.

One of the difficulties associated with light propagation in multimode fibers arises from the differences among the group velocities of the modes. This results in a variety of travel times so that light pulses are broadened as they travel through the fiber. This effect, called **modal dispersion**, limits the speed at which adjacent pulses can be sent without overlapping and therefore the speed at which a fiber-optic communication system can operate.

Modal dispersion can be reduced by grading the refractive index of the fiber core from a maximum value at its center to a minimum value at the core–cladding boundary. The fiber is then called a **graded-index fiber**, whereas conventional fibers

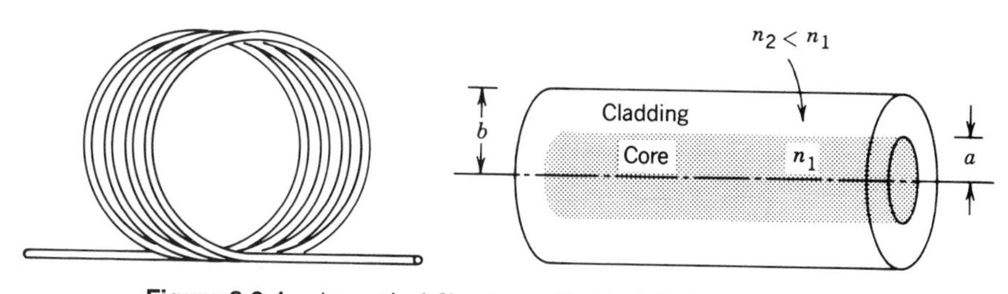

Figure 8.0-1 An optical fiber is a cylindrical dielectric waveguide.

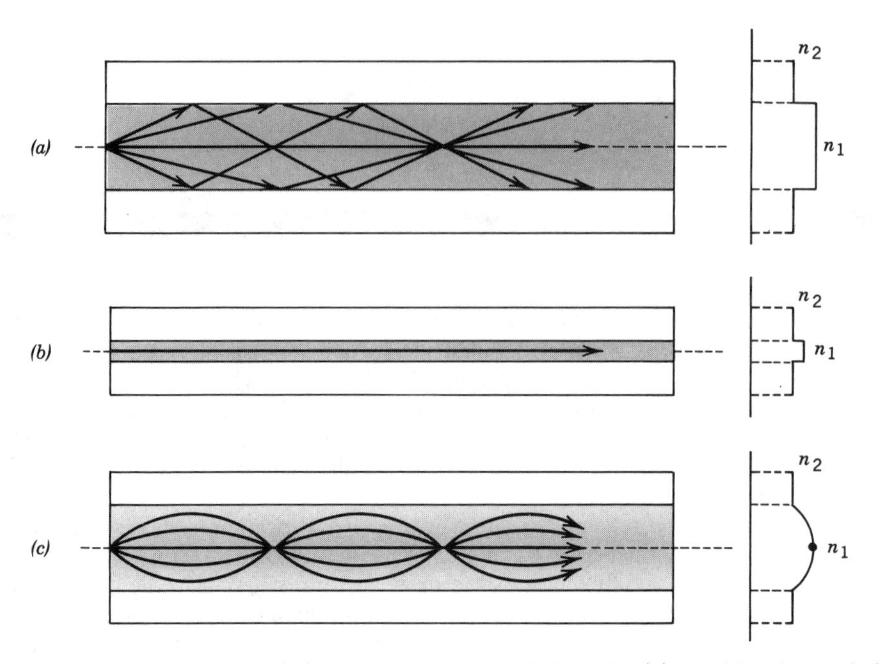

Figure 8.0-2 Geometry, refractive-index profile, and typical rays in: (*a*) a multimode step-index fiber, (*b*) a single-mode step-index fiber, and (*c*) a multimode graded-index fiber.

with constant refractive indices in the core and the cladding are called **step-index fibers**. In a graded-index fiber the velocity increases with distance from the core axis (since the refractive index decreases). Although rays of greater inclination to the fiber axis must travel farther, they travel faster, so that the travel times of the different rays are equalized. Optical fibers are therefore classified as step-index or graded-index, and multimode or single-mode, as illustrated in Fig. 8.0-2.

This chapter emphasizes the nature of optical modes and their group velocities in step-index and graded-index fibers. These topics are presented in Secs. 8.1 and 8.2, respectively. The optical properties of the fiber material (which is usually fused silica), including its attenuation and the effects of material, modal, and waveguide dispersion on the transmission of light pulses, are discussed in Sec. 8.3. Optical fibers are revisited in Chap. 22, which is devoted to their use in lightwave communication systems.

8.1 STEP-INDEX FIBERS

A step-index fiber is a cylindrical dielectric waveguide specified by its core and cladding refractive indices, n_1 and n_2, and the radii a and b (see Fig. 8.0-1). Examples of standard core and cladding diameters $2a/2b$ are $8/125$, $50/125$, $62.5/125$, $85/125$, $100/140$ (units of μm). The refractive indices differ only slightly, so that the fractional refractive-index change

$$\Delta = \frac{n_1 - n_2}{n_1} \tag{8.1-1}$$

is small ($\Delta \ll 1$).

Almost all fibers currently used in optical communication systems are made of fused silica glass (SiO_2) of high chemical purity. Slight changes in the refractive index are made by the addition of low concentrations of doping materials (titanium, germanium, or boron, for example). The refractive index n_1 is in the range from 1.44 to 1.46, depending on the wavelength, and Δ typically lies between 0.001 and 0.02.

A. Guided Rays

An optical ray is guided by total internal reflections within the fiber core if its angle of incidence on the core–cladding boundary is greater than the critical angle $\theta_c = \sin^{-1}(n_2/n_1)$, and remains so as the ray bounces.

Meridional Rays

The guiding condition is simple to see for meridional rays (rays in planes passing through the fiber axis), as illustrated in Fig. 8.1-1. These rays intersect the fiber axis and reflect in the same plane without changing their angle of incidence, as if they were in a planar waveguide. Meridional rays are guided if their angle θ with the fiber axis is smaller than the complement of the critical angle $\bar{\theta}_c = \pi/2 - \theta_c = \cos^{-1}(n_2/n_1)$. Since $n_1 \approx n_2$, $\bar{\theta}_c$ is usually small and the guided rays are approximately paraxial.

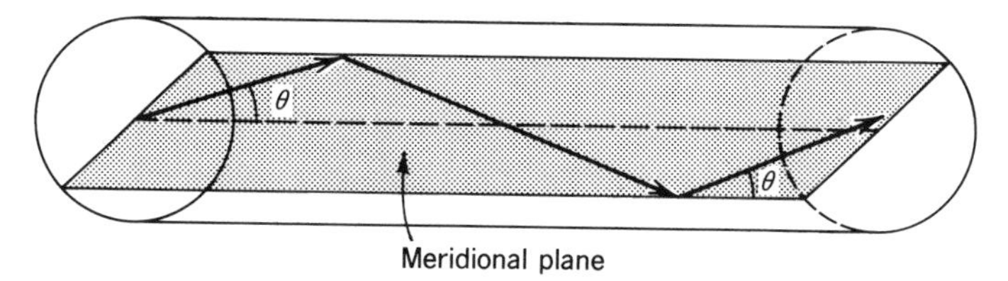

Figure 8.1-1 The trajectory of a meridional ray lies in a plane passing through the fiber axis. The ray is guided if $\theta < \bar{\theta}_c = \cos^{-1}(n_1/n_2)$.

Skewed Rays

An arbitrary ray is identified by its plane of incidence, a plane parallel to the fiber axis and passing through the ray, and by the angle with that axis, as illustrated in Fig. 8.1-2. The plane of incidence intersects the core–cladding cylindrical boundary at an angle ϕ with the normal to the boundary and lies at a distance R from the fiber axis. The ray is identified by its angle θ with the fiber axis and by the angle ϕ of its plane. When $\phi \neq 0$ ($R \neq 0$) the ray is said to be skewed. For meridional rays $\phi = 0$ and $R = 0$.

A skewed ray reflects repeatedly into planes that make the same angle ϕ with the core–cladding boundary, and follows a helical trajectory confined within a cylindrical shell of radii R and a, as illustrated in Fig. 8.1-2. The projection of the trajectory onto the transverse (x–y) plane is a regular polygon, not necessarily closed. It can be shown that the condition for a skewed ray to always undergo total internal reflection is that its angle θ with the z axis be smaller than $\bar{\theta}_c$.

Numerical Aperture

A ray incident from air into the fiber becomes a guided ray if upon refraction into the core it makes an angle θ with the fiber axis smaller than $\bar{\theta}_c$. Applying Snell's law at the air–core boundary, the angle θ_a in air corresponding to $\bar{\theta}_c$ in the core is given by the relation $1 \cdot \sin\theta_a = n_1 \sin\bar{\theta}_c$, from which (see Fig. 8.1-3 and Exercise 1.2-5) $\sin\theta_a = n_1(1 - \cos^2\bar{\theta}_c)^{1/2} = n_1[1 - (n_2/n_1)^2]^{1/2} = (n_1^2 - n_2^2)^{1/2}$. Therefore

$$\theta_a = \sin^{-1}\mathrm{NA}, \tag{8.1-2}$$

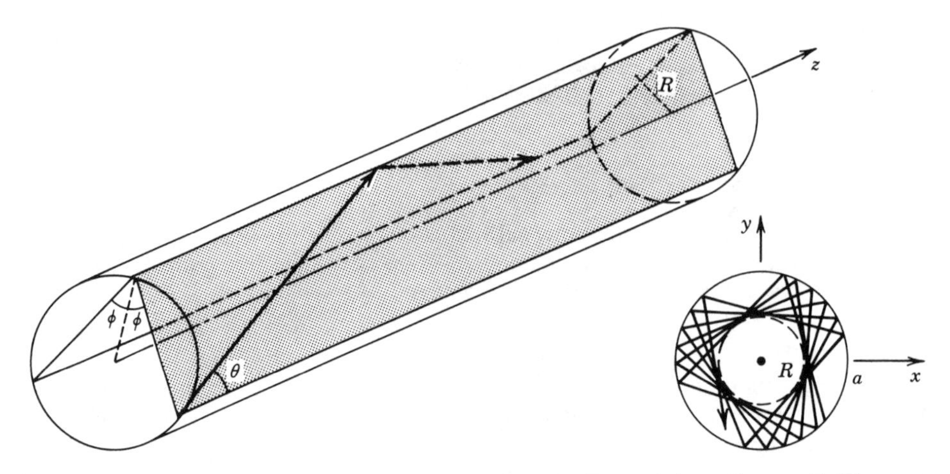

Figure 8.1-2 A skewed ray lies in a plane offset from the fiber axis by a distance R. The ray is identified by the angles θ and ϕ. It follows a helical trajectory confined within a cylindrical shell of radii R and a. The projection of the ray on the transverse plane is a regular polygon that is not necessarily closed.

where

$$\mathrm{NA} = \left(n_1^2 - n_2^2\right)^{1/2} \approx n_1(2\Delta)^{1/2} \tag{8.1-3}$$

Numerical Aperture

is the numerical aperture of the fiber. Thus θ_a is the acceptance angle of the fiber. It

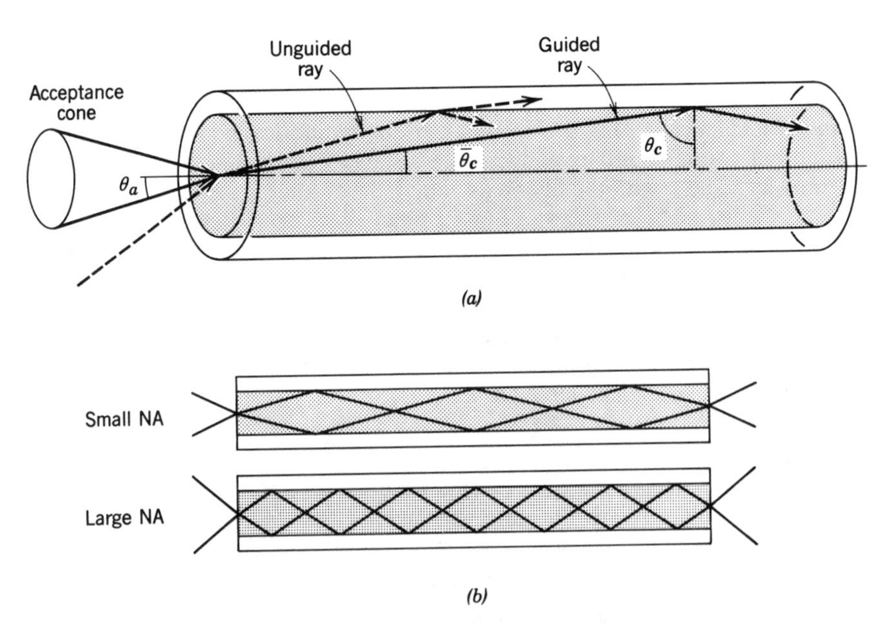

Figure 8.1-3 (a) The acceptance angle θ_a of a fiber. Rays within the acceptance cone are guided by total internal reflection. The numerical aperture NA $= \sin\theta_a$. (b) The light-gathering capacity of a large NA fiber is greater than that of a small NA fiber. The angles θ_a and $\bar{\theta}_c$ are typically quite small; they are exaggerated here for clarity.

determines the cone of external rays that are guided by the fiber. Rays incident at angles greater than θ_a are refracted into the fiber but are guided only for a short distance. The numerical aperture therefore describes the light-gathering capacity of the fiber.

When the guided rays arrive at the other end of the fiber, they are refracted into a cone of angle θ_a. Thus the acceptance angle is a crucial parameter for the design of systems for coupling light into or out of the fiber.

EXAMPLE 8.1-1. ***Cladded and Uncladded Fibers.*** In a silica glass fiber with $n_1 = 1.46$ and $\Delta = (n_1 - n_2)/n_1 = 0.01$, the complementary critical angle $\bar{\theta}_c = \cos^{-1}(n_2/n_1) = 8.1°$, and the acceptance angle $\theta_a = 11.9°$, corresponding to a numerical aperture NA = 0.206. By comparison, an uncladded silica glass fiber ($n_1 = 1.46$, $n_2 = 1$) has $\bar{\theta}_c = 46.8°$, $\theta_a = 90°$, and NA = 1. Rays incident from *all* directions are guided by the uncladded fiber since they reflect within a cone of angle $\bar{\theta}_c = 46.8°$ inside the core. Although its light-gathering capacity is high, the uncladded fiber is not a suitable optical waveguide because of the large number of modes it supports, as will be shown subsequently.

B. Guided Waves

In this section we examine the propagation of monochromatic light in step-index fibers using electromagnetic theory. We aim at determining the electric and magnetic fields of guided waves that satisfy Maxwell's equations and the boundary conditions imposed by the cylindrical dielectric core and cladding. As in all waveguides, there are certain special solutions, called modes (see Appendix C), each of which has a distinct propagation constant, a characteristic field distribution in the transverse plane, and two independent polarization states.

Spatial Distributions

Each of the components of the electric and magnetic fields must satisfy the Helmholtz equation, $\nabla^2 U + n^2 k_o^2 U = 0$, where $n = n_1$ in the core ($r < a$) and $n = n_2$ in the cladding ($r > a$) and $k_o = 2\pi/\lambda_o$ (see Sec. 5.3). We assume that the radius b of the cladding is sufficiently large that it can safely be assumed to be infinite when examining guided light in the core and near the core–cladding boundary. In a cylindrical coordinate system (see Fig. 8.1-4) the Helmholtz equation is

$$\frac{\partial^2 U}{\partial r^2} + \frac{1}{r}\frac{\partial U}{\partial r} + \frac{1}{r^2}\frac{\partial^2 U}{\partial \phi^2} + \frac{\partial^2 U}{\partial z^2} + n^2 k_o^2 U = 0, \qquad (8.1\text{-}4)$$

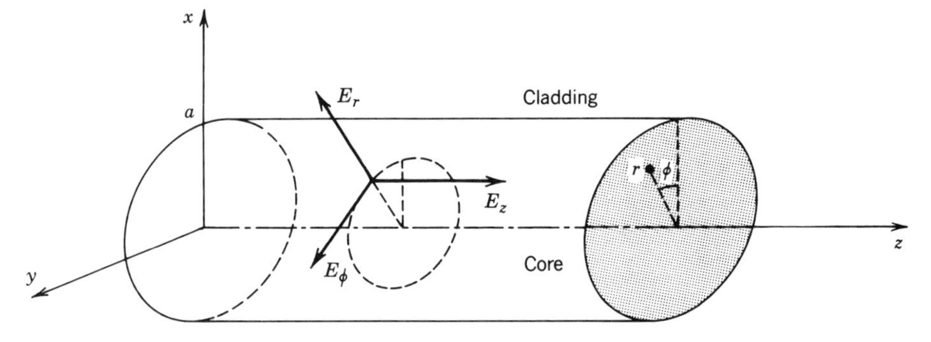

Figure 8.1-4 Cylindrical coordinate system.

where the complex amplitude $U = U(r, \phi, z)$ represents any of the Cartesian components of the electric or magnetic fields or the axial components E_z and H_z in cylindrical coordinates.

We are interested in solutions that take the form of waves traveling in the z direction with a propagation constant β, so that the z dependence of U is of the form $e^{-j\beta z}$. Since U must be a periodic function of the angle ϕ with period 2π, we assume that the dependence on ϕ is harmonic, $e^{-jl\phi}$, where l is an integer. Substituting

$$U(r, \phi, z) = u(r) e^{-jl\phi} e^{-j\beta z}, \qquad l = 0, \pm 1, \pm 2, \ldots, \tag{8.1-5}$$

into (8.1-4), an ordinary differential equation for $u(r)$ is obtained:

$$\frac{d^2u}{dr^2} + \frac{1}{r}\frac{du}{dr} + \left(n^2k_o^2 - \beta^2 - \frac{l^2}{r^2}\right)u = 0. \tag{8.1-6}$$

As in Sec. 7.2B, the wave is guided (or bound) if the propagation constant is smaller than the wavenumber in the core ($\beta < n_1 k_o$) and greater than the wavenumber in the cladding ($\beta > n_2 k_o$). It is therefore convenient to define

$$k_T^2 = n_1^2 k_o^2 - \beta^2 \tag{8.1-7a}$$

and

$$\gamma^2 = \beta^2 - n_2^2 k_o^2, \tag{8.1-7b}$$

so that for guided waves k_T^2 and γ^2 are positive and k_T and γ are real. Equation (8.1-6) may then be written in the core and cladding separately:

$$\frac{d^2u}{dr^2} + \frac{1}{r}\frac{du}{dr} + \left(k_T^2 - \frac{l^2}{r^2}\right)u = 0, \qquad r < a \text{ (core)}, \tag{8.1-8a}$$

$$\frac{d^2u}{dr^2} + \frac{1}{r}\frac{du}{dr} - \left(\gamma^2 + \frac{l^2}{r^2}\right)u = 0, \qquad r > a \text{ (cladding)}. \tag{8.1-8b}$$

Equations (8.1-8) are well-known differential equations whose solutions are the family of Bessel functions. Excluding functions that approach ∞ at $r = 0$ in the core or at $r \to \infty$ in the cladding, we obtain the bounded solutions:

$$u(r) \propto \begin{cases} J_l(k_T r), & r < a \text{ (core)} \\ K_l(\gamma r), & r > a \text{ (cladding)}, \end{cases} \tag{8.1-9}$$

where $J_l(x)$ is the Bessel function of the first kind and order l, and $K_l(x)$ is the modified Bessel function of the second kind and order l. The function $J_l(x)$ oscillates like the sine or cosine functions but with a decaying amplitude. In the limit $x \gg 1$,

$$J_l(x) \approx \left(\frac{2}{\pi x}\right)^{1/2} \cos\left[x - \left(l + \tfrac{1}{2}\right)\frac{\pi}{2}\right], \qquad x \gg 1. \tag{8.1-10a}$$

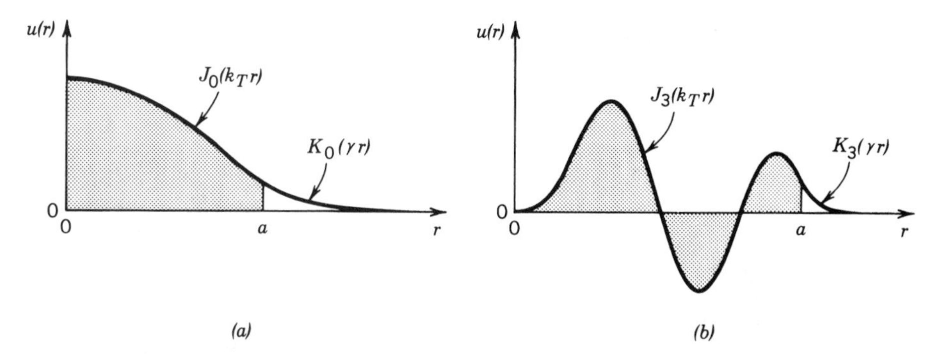

Figure 8.1-5 Examples of the radial distribution $u(r)$ given by (8.1-9) for (a) $l = 0$ and (b) $l = 3$. The shaded areas represent the fiber core and the unshaded areas the cladding. The parameters k_T and γ and the two proportionality constants in (8.1-9) have been selected such that $u(r)$ is continuous and has a continuous derivative at $r = a$. Larger values of k_T and γ lead to a greater number of oscillations in $u(r)$.

In the same limit, $K_l(x)$ decays with increasing x at an exponential rate,

$$K_l(x) \approx \left(\frac{\pi}{2x}\right)^{1/2}\left(1 + \frac{4l^2 - 1}{8x}\right)\exp(-x), \qquad x \gg 1. \tag{8.1-10b}$$

Two examples of the radial distribution $u(r)$ are shown in Fig. 8.1-5.

The parameters k_T and γ determine the rate of change of $u(r)$ in the core and in the cladding, respectively. A large value of k_T means faster oscillation of the radial distribution in the core. A large value of γ means faster decay and smaller penetration of the wave into the cladding. As can be seen from (8.1-7), the sum of the squares of k_T and γ is a constant,

$$k_T^2 + \gamma^2 = \left(n_1^2 - n_2^2\right)k_o^2 = \mathrm{NA}^2 \cdot k_o^2, \tag{8.1-11}$$

so that as k_T increases, γ decreases and the field penetrates deeper into the cladding. As k_T exceeds $\mathrm{NA} \cdot k_o$, γ becomes imaginary and the wave ceases to be bound to the core.

The V Parameter

It is convenient to normalize k_T and γ by defining

$$X = k_T a, \qquad Y = \gamma a. \tag{8.1-12}$$

In view of (8.1-11),

$$X^2 + Y^2 = V^2, \tag{8.1-13}$$

where $V = \mathrm{NA} \cdot k_o a$, from which

$$V = 2\pi \frac{a}{\lambda_o} \mathrm{NA}. \tag{8.1-14}$$

V Parameter

As we shall see shortly, V is an important parameter that governs the number of modes of the fiber and their propagation constants. It is called the **fiber parameter** or ***V* parameter**. It is important to remember that for the wave to be guided, X must be smaller than V.

Modes

We now consider the boundary conditions. We begin by writing the axial components of the electric- and magnetic-field complex amplitudes E_z and H_z in the form of (8.1-5). The condition that these components must be continuous at the core–cladding boundary $r = a$ establishes a relation between the coefficients of proportionality in (8.1-9), so that we have only one unknown for E_z and one unknown for H_z. With the help of Maxwell's equations, $j\omega\epsilon_o n^2 \mathbf{E} = \nabla \times \mathbf{H}$ and $-j\omega\mu_o \mathbf{H} = \nabla \times \mathbf{E}$, the remaining four components E_ϕ, H_ϕ, E_r, and H_r are determined in terms of E_z and H_z. Continuity of E_ϕ and H_ϕ at $r = a$ yields two more equations. One equation relates the two unknown coefficients of proportionality in E_z and H_z; the other equation gives a condition that the propagation constant β must satisfy. This condition, called the **characteristic equation** or **dispersion relation**, is an equation for β with the ratio a/λ_o and the fiber indices n_1, n_2 as known parameters.

For each azimuthal index l, the characteristic equation has multiple solutions yielding discrete propagation constants β_{lm}, $m = 1, 2, \dots$, each solution representing a mode. The corresponding values of k_T and γ, which govern the spatial distributions in the core and in the cladding, respectively, are determined by use of (8.1-7) and are denoted k_{Tlm} and γ_{lm}. A mode is therefore described by the indices l and m characterizing its azimuthal and radial distributions, respectively. The function $u(r)$ depends on both l and m; $l = 0$ corresponds to meridional rays. There are two independent configurations of the $\mathbf{E}$ and $\mathbf{H}$ vectors for each mode, corresponding to two states of polarization. The classification and labeling of these configurations are generally quite involved (see specialized books in the reading list for more details).

Characteristic Equation for the Weakly Guiding Fiber

Most fibers are weakly guiding (i.e., $n_1 \approx n_2$ or $\Delta \ll 1$) so that the guided rays are paraxial (i.e., approximately parallel to the fiber axis). The longitudinal components of the electric and magnetic fields are then much weaker than the transverse components and the guided waves are approximately transverse electromagnetic (TEM). The linear polarization in the x and y directions then form orthogonal states of polarization. The linearly polarized (l, m) mode is usually denoted as the LP_{lm} mode. The two polarizations of mode (l, m) travel with the same propagation constant and have the same spatial distribution.

For weakly guiding fibers the characteristic equation obtained using the procedure outlined earlier turns out to be approximately equivalent to the conditions that the scalar function $u(r)$ in (8.1-9) is continuous and has a continuous derivative at $r = a$. These two conditions are satisfied if

$$\frac{(k_T a) J_l'(k_T a)}{J_l(k_T a)} = \frac{(\gamma a) K_l'(\gamma a)}{K_l(\gamma a)}. \tag{8.1-15}$$

The derivatives J_l' and K_l' of the Bessel functions satisfy the identities

$$J_l'(x) = \pm J_{l \mp 1}(x) \mp l \frac{J_l(x)}{x}$$

$$K_l'(x) = -K_{l \mp 1}(x) \mp l \frac{K_l(x)}{x}.$$

Substituting these identities into (8.1-15) and using the normalized parameters $X = k_T a$ and $Y = \gamma a$, we obtain the characteristic equation

$$X \frac{J_{l \pm 1}(X)}{J_l(X)} = \pm Y \frac{K_{l \pm 1}(Y)}{K_l(Y)}. \tag{8.1-16}$$

Characteristic Equation

$$X^2 + Y^2 = V^2$$

Given V and l, the characteristic equation contains a single unknown variable X (since $Y^2 = V^2 - X^2$). Note that $J_{-l}(x) = (-1)^l J_l(x)$ and $K_{-l}(x) = K_l(x)$, so that if l is replaced with $-l$, the equation remains unchanged.

The characteristic equation may be solved graphically by plotting its right- and left-hand sides (RHS and LHS) versus X and finding the intersections. As illustrated in Fig. 8.1-6 for $l = 0$, the LHS has multiple branches and the RHS drops monotonically with increase of X until it vanishes at $X = V$ $(Y = 0)$. There are therefore multiple intersections in the interval $0 < X \leq V$. Each intersection point corresponds to a fiber mode with a distinct value of X. These values are denoted X_{lm}, $m = 1, 2, \ldots, M_l$ in order of increasing X. Once the X_{lm} are found, the corresponding transverse propagation constants k_{Tlm}, the decay parameters γ_{lm}, the propagation constants β_{lm}, and the radial distribution functions $u_{lm}(r)$ may be readily determined by use of (8.1-12), (8.1-7), and (8.1-9). The graph in Fig. 8.1-6 is similar to that in Fig. 7.2-2, which governs the modes of a planar dielectric waveguide.

Each mode has a distinct radial distribution. The radial distributions $u(r)$ shown in Fig. 8.1-5, for example, correspond to the LP_{01} mode ($l = 0, m = 1$) in a fiber with $V = 5$; and the LP_{34} mode ($l = 3, m = 4$) in a fiber with $V = 25$. Since the (l, m) and $(-l, m)$ modes have the same propagation constant, it is interesting to examine the spatial distribution of their superposition (with equal weights). The complex amplitude of the sum is proportional to $u_{lm}(r) \cos l\phi \exp(-j\beta_{lm} z)$. The intensity, which is proportional to $u_{lm}^2(r) \cos^2 l\phi$, is illustrated in Fig. 8.1-7 for the LP_{01} and LP_{34} modes (the same modes for which $u(r)$ is shown in Fig. 8.1-5).

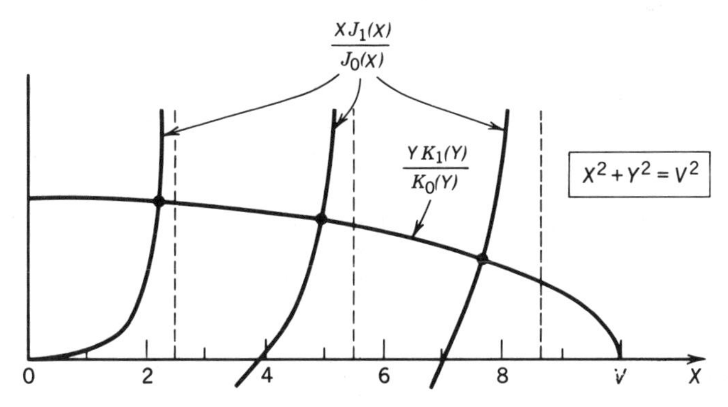

Figure 8.1-6 Graphical construction for solving the characteristic equation (8.1-16). The left- and right-hand sides are plotted as functions of X. The intersection points are the solutions. The LHS has multiple branches intersecting the abscissa at the roots of $J_{l \pm 1}(X)$. The RHS intersects each branch once and meets the abscissa at $X = V$. The number of modes therefore equals the number of roots of $J_{l \pm 1}(X)$ that are smaller than V. In this plot $l = 0$, $V = 10$, and either the $-$ or $+$ signs in (8.1-16) may be taken.

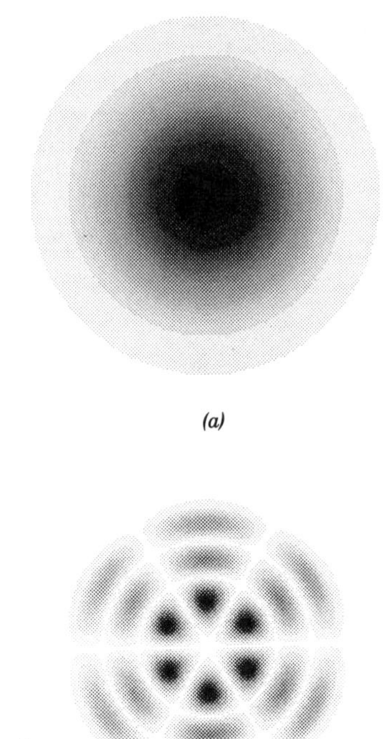

Figure 8.1-7 Distributions of the intensity of the (a) LP_{01} and (b) LP_{34} modes in the transverse plane, assuming an azimuthal cos $l\phi$ dependence. The fundamental LP_{01} mode has a distribution similar to that of the Gaussian beam discussed in Chap. 3.

Mode Cutoff and Number of Modes

It is evident from the graphical construction in Fig. 8.1-6 that as V increases, the number of intersections (modes) increases since the LHS of the characteristic equation (8.1-16) is independent of V, whereas the RHS moves to the right as V increases. Considering the minus signs in the characteristic equation, branches of the LHS intersect the abscissa when $J_{l-1}(X) = 0$. These roots are denoted by x_{lm}, $m = 1, 2, \ldots$. The number of modes M_l is therefore equal to the number of roots of $J_{l-1}(X)$ that are smaller than V. The (l, m) mode is allowed if $V > x_{lm}$. The mode reaches its cutoff point when $V = x_{lm}$. As V decreases further, the $(l, m - 1)$ mode also reaches its cutoff point when a new root is reached, and so on. The smallest root of $J_{l-1}(X)$ is $x_{01} = 0$ for $l = 0$ and the next smallest is $x_{11} = 2.405$ for $l = 1$. When $V < 2.405$, all modes with the exception of the fundamental LP_{01} mode are cut off. The fiber then operates as a single-mode waveguide. A plot of the number of modes M_l as a function of V is therefore a staircase function increasing by unity at each of the roots x_{lm} of the Bessel function $J_{l-1}(X)$. Some of these roots are listed in Table 8.1-1.

TABLE 8.1-1 Cutoff V Parameter for the LP_{0m} and LP_{1m} Modes[a]

l	m: 1	2	3
0	0	3.832	7.016
1	2.405	5.520	8.654

[a]The cutoffs of the $l = 0$ modes occur at the roots of $J_{-1}(X) = -J_1(X)$. The $l = 1$ modes are cut off at the roots of $J_0(X)$, and so on.

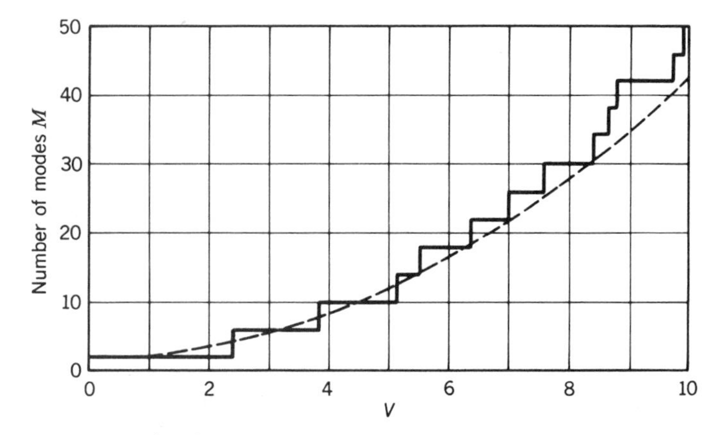

Figure 8.1-8 Total number of modes M versus the fiber parameter $V = 2\pi(a/\lambda_o)\text{NA}$. Included in the count are two helical polarities for each mode with $l > 0$ and two polarizations per mode. For $V < 2.405$, there is only one mode, the fundamental LP_{01} mode with two polarizations. The dashed curve is the relation $M = 4V^2/\pi^2 + 2$, which provides an approximate formula for the number of modes when $V \gg 1$.

A composite count of the total number of modes M (for all l) is shown in Fig. 8.1-8 as a function of V. This is a staircase function with jumps at the roots of $J_{l-1}(x)$. Each root must be counted twice since for each mode of azimuthal index $l > 0$ there is a corresponding mode $-l$ that is identical except for an opposite polarity of the angle ϕ (corresponding to rays with helical trajectories of opposite senses) as can be seen by using the plus signs in the characteristic equation. In addition, each mode has two states of polarization and must therefore be counted twice.

Number of Modes (Fibers with Large V Parameter)

For fibers with large V parameters, there are a large number of roots of $J_l(X)$ in the interval $0 < X < V$. Since $J_l(X)$ is approximated by the sinusoidal function in (8.1-10a) when $X \gg 1$, its roots x_{lm} are approximately given by $x_{lm} - (l + \frac{1}{2})(\pi/2) = (2m - 1)(\pi/2)$, i.e., $x_{lm} = (l + 2m - \frac{1}{2})\pi/2$, so that the cutoff points of modes (l, m), which are the roots of $J_{l\pm1}(X)$, are

$$x_{lm} \approx \left(l + 2m - \frac{1}{2} \pm 1\right)\frac{\pi}{2} \approx (l + 2m)\frac{\pi}{2}, \qquad l = 0, 1, \ldots; \quad m \gg 1, \quad (8.1\text{-}17)$$

when m is large.

For a fixed l, these roots are spaced uniformly at a distance π, so that the number of roots M_l satisfies $(l + 2M_l)\pi/2 = V$, from which $M_l \approx V/\pi - l/2$. Thus M_l drops linearly with increasing l, beginning with $M_l \approx V/\pi$ for $l = 0$ and ending at $M_l = 0$ when $l = l_{\max}$, where $l_{\max} = 2V/\pi$, as illustrated in Fig. 8.1-9. Thus the total number of modes is $M \approx \sum_{l=0}^{l_{\max}} M_l = \sum_{l=0}^{l_{\max}} (V/\pi - l/2)$.

Since the number of terms in this sum is assumed large, it may be readily evaluated by approximating it as the area of the triangle in Fig. 8.1-9, $M \approx \frac{1}{2}(2V/\pi)(V/\pi) = V^2/\pi^2$. Allowing for two degrees of freedom for positive and negative l and two polarizations for each index (l, m), we obtain

$$\boxed{M \approx \frac{4}{\pi^2}V^2.} \qquad (8.1\text{-}18)$$

Number of Modes
($V \gg 1$)

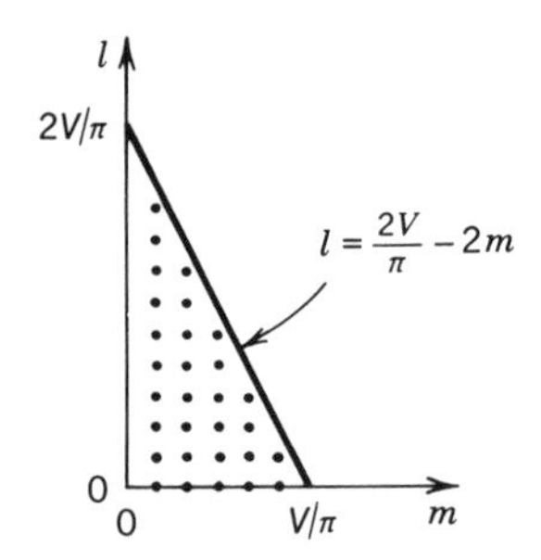

Figure 8.1-9 The indices of guided modes extend from $m = 1$ to $m \approx V/\pi - l/2$ and from $l = 0$ to $\approx 2V/\pi$.

This expression for M is analogous to that for the rectangular waveguide (7.3-3). Note that (8.1-18) is valid only for large V. This approximate number is compared to the exact number obtained from the characteristic equation in Fig. 8.1-8.

EXAMPLE 8.1-2. *Approximate Number of Modes.* A silica fiber with $n_1 = 1.452$ and $\Delta = 0.01$ has a numerical aperture $\mathrm{NA} = (n_1^2 - n_2^2)^{1/2} \approx n_1(2\Delta)^{1/2} \approx 0.205$. If $\lambda_o = 0.85$ μm and the core radius $a = 25$ μm, the V parameter is $V = 2\pi(a/\lambda_o)\mathrm{NA} \approx 37.9$. There are therefore approximately $M \approx 4V^2/\pi^2 \approx 585$ modes. If the cladding is stripped away so that the core is in direct contact with air, $n_2 = 1$ and $\mathrm{NA} = 1$. The V parameter is then $V = 184.8$ and more than 13,800 modes are allowed.

Propagation Constants (Fibers with Large V Parameter)

As mentioned earlier, the propagation constants can be determined by solving the characteristic equation (8.1-16) for the X_{lm} and using (8.1-7a) and (8.1-12) to obtain $\beta_{lm} = (n_1^2 k_o^2 - X_{lm}^2/a^2)^{1/2}$. A number of approximate formulas for X_{lm} applicable in certain limits are available in the literature, but there are no explicit exact formulas.

If $V \gg 1$, the crudest approximation is to assume that the X_{lm} are equal to the cutoff values x_{lm}. This is equivalent to assuming that the branches in Fig. 8.1-6 are approximately vertical lines, so that $X_{lm} \approx x_{lm}$. Since $V \gg 1$, the majority of the roots would be large and the approximation in (8.1-17) may be used to obtain

$$\beta_{lm} \approx \left[n_1^2 k_o^2 - (l + 2m)^2 \frac{\pi^2}{4a^2}\right]^{1/2}. \tag{8.1-19}$$

Since

$$M \approx \frac{4}{\pi^2} V^2 = \frac{4}{\pi^2} \mathrm{NA}^2 \cdot a^2 k_o^2 \approx \frac{4}{\pi^2}\left(2n_1^2\Delta\right) k_o^2 a^2, \tag{8.1-20}$$

(8.1-19) and (8.1-20) give

$$\beta_{lm} \approx n_1 k_o \left[1 - 2\frac{(l + 2m)^2}{M}\Delta\right]^{1/2}. \tag{8.1-21}$$

Because Δ is small we use the approximation $(1 + \delta)^{1/2} \approx 1 + \delta/2$ for $|\delta| \ll 1$, and

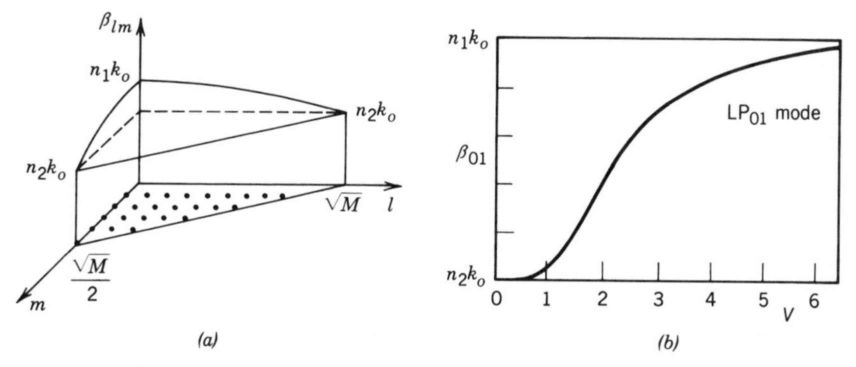

Figure 8.1-10 (*a*) Approximate propagation constants β_{lm} of the modes of a fiber with large V parameter as functions of the mode indices l and m. (*b*) Exact propagation constant β_{01} of the fundamental LP_{01} modes as a function of the V parameter. For $V \gg 1$, $\beta_{01} \approx n_1 k_o$.

obtain

$$\beta_{lm} \approx n_1 k_o \left[1 - \frac{(l + 2m)^2}{M} \Delta \right]. \tag{8.1-22}$$

Propagation Constants
$l = 0, 1, \ldots, \sqrt{M}$
$m = 1, 2, \ldots, (\sqrt{M} - l)/2$
$(V \gg 1)$

Since $l + 2m$ varies between 2 and $\approx 2V/\pi = \sqrt{M}$ (see Fig. 8.1-9), β_{lm} varies approximately between $n_1 k_o$ and $n_1 k_o(1 - \Delta) \approx n_2 k_o$, as illustrated in Fig. 8.1-10.

Group Velocities (Fibers with Large V Parameter)

To determine the group velocity, $v_{lm} = d\omega/d\beta_{lm}$, of the (l, m) mode we express β_{lm} as an explicit function of ω by substituting $n_1 k_o = \omega/c_1$ and $M = (4/\pi^2)(2n_1^2\Delta)k_o^2 a^2 = (8/\pi^2)a^2\omega^2\Delta/c_1^2$ into (8.1-22) and assume that c_1 and Δ are independent of ω. The derivative $d\omega/d\beta_{lm}$ gives

$$v_{lm} \approx c_1 \left[1 + \frac{(l + 2m)^2}{M} \Delta \right]^{-1}.$$

Since $\Delta \ll 1$, the approximate expansion $(1 + \delta)^{-1} \approx 1 - \delta$ when $|\delta| \ll 1$, gives

$$v_{lm} \approx c_1 \left[1 - \frac{(l + 2m)^2}{M} \Delta \right]. \tag{8.1-23}$$

Group Velocities
$(V \gg 1)$

Because the minimum and maximum values of $(l + 2m)$ are 2 and $\sqrt{M}$, respectively, and since $M \gg 1$, the group velocity varies approximately between c_1 and $c_1(1 - \Delta) = c_1(n_2/n_1)$. Thus the group velocities of the low-order modes are approximately equal to the phase velocity of the core material, and those of the high-order modes are smaller.

The fractional group-velocity change between the fastest and the slowest mode is roughly equal to Δ, the fractional refractive index change of the fiber. Fibers with large Δ, although endowed with a large NA and therefore large light-gathering capacity, also have a large number of modes, large modal dispersion, and consequently high pulse spreading rates. These effects are particularly severe if the cladding is removed altogether.

C. Single-Mode Fibers

As discussed earlier, a fiber with core radius a and numerical aperture NA operates as a single-mode fiber in the fundamental LP_{01} mode if $V = 2\pi(a/\lambda_o)\text{NA} < 2.405$ (see Table 8.1-1 on page 282). Single-mode operation is therefore achieved by using a small core diameter and small numerical aperture (making n_2 close to n_1), or by operating at a sufficiently long wavelength. The fundamental mode has a bell-shaped spatial distribution similar to the Gaussian distribution [see Figs. 8.1-5(a) and 8.1-7(a)] and a propagation constant β that depends on V as illustrated in Fig. 8.1-10(b). This mode provides the highest confinement of light power within the core.

EXAMPLE 8.1-3. *Single-Mode Operation.* A silica glass fiber with $n_1 = 1.447$ and $\Delta = 0.01$ (NA = 0.205) operates at $\lambda_o = 1.3$ μm as a single-mode fiber if $V = 2\pi(a/\lambda_o)\text{NA} < 2.405$, i.e., if the core diameter $2a < 4.86$ μm. If Δ is reduced to 0.0025, single-mode operation requires a diameter $2a < 9.72$ μm.

There are numerous advantages of using single-mode fibers in optical communication systems. As explained earlier, the modes of a multimode fiber travel at different group velocities and therefore undergo different time delays, so that a short-duration pulse of multimode light is delayed by different amounts and therefore spreads in time. Quantitative measures of modal dispersion are determined in Sec. 8.3B. In a single-mode fiber, on the other hand, there is only one mode with one group velocity, so that a short pulse of light arrives without delay distortion. As explained in Sec. 8.3B, other dispersion effects result in pulse spreading in single-mode fibers, but these are significantly smaller than modal dispersion.

As also shown in Sec. 8.3, the rate of power attenuation is lower in a single-mode fiber than in a multimode fiber. This, together with the smaller pulse spreading rate, permits substantially higher data rates to be transmitted by single-mode fibers in comparison with the maximum rates feasible with multimode fibers. This topic is discussed in Chap. 22.

Another difficulty with multimode fibers is caused by the random interference of the modes. As a result of uncontrollable imperfections, strains, and temperature fluctuations, each mode undergoes a random phase shift so that the sum of the complex amplitudes of the modes has a random intensity. This randomness is a form of noise known as **modal noise** or **speckle**. This effect is similar to the fading of radio signals due to multiple-path transmission. In a single-mode fiber there is only one path and therefore no modal noise.

Because of their small size and small numerical apertures, single-mode fibers are more compatible with integrated-optics technology. However, such features make them more difficult to manufacture and work with because of the reduced allowable mechanical tolerances for splicing or joining with demountable connectors and for coupling optical power into the fiber.

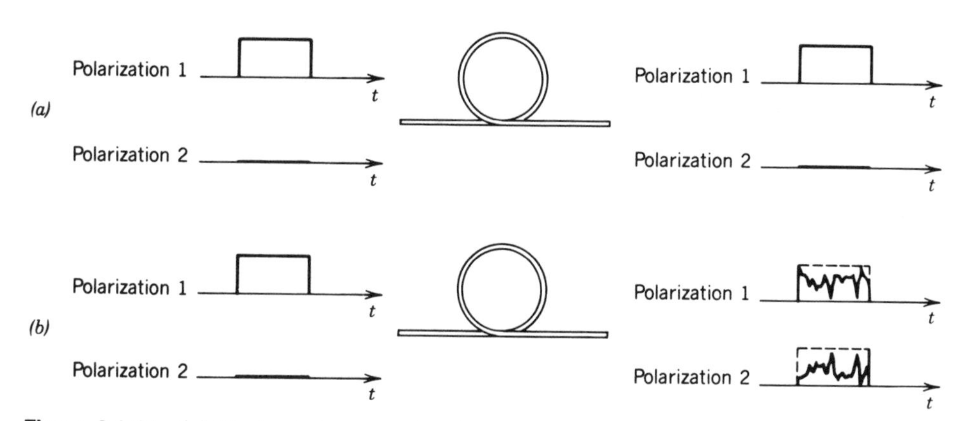

Figure 8.1-11 (*a*) Ideal polarization-maintaining fiber. (*b*) Random transfer of power between two polarizations.

Polarization-Maintaining Fibers

In a fiber with circular cross section, each mode has two independent states of polarization with the same propagation constant. Thus the fundamental LP_{01} mode in a single-mode weakly guiding fiber may be polarized in the x or y direction with the two orthogonal polarizations having the same propagation constant and the same group velocity.

In principle, there is no exchange of power between the two polarization components. If the power of the light source is delivered into one polarization only, the power received remains in that polarization. In practice, however, slight random imperfections or uncontrollable strains in the fiber result in random power transfer between the two polarizations. This coupling is facilitated since the two polarizations have the same propagation constant and their phases are therefore matched. Thus linearly polarized light at the fiber input is transformed into elliptically polarized light at the output. As a result of fluctuations of strain, temperature, or source wavelength, the ellipticity of the received light fluctuates randomly with time. Nevertheless, the total power remains fixed (Fig. 8.1-11). If we are interested only in transmitting light power, this randomization of the power division between the two polarization components poses no difficulty, provided that the total power is collected.

In many areas related to fiber optics, e.g., coherent optical communications, integrated-optic devices, and optical sensors based on interferometric techniques, the fiber is used to transmit the complex amplitude of a specific polarization (magnitude and phase). For these applications, polarization-maintaining fibers are necessary. To make a polarization-maintaining fiber the circular symmetry of the conventional fiber must be removed, by using fibers with elliptical cross sections or stress-induced anisotropy of the refractive index, for example. This eliminates the polarization degeneracy, i.e., makes the propagation constants of the two polarizations different. The coupling efficiency is then reduced as a result of the introduction of phase mismatch.

8.2 GRADED-INDEX FIBERS

Index grading is an ingenious method for reducing the pulse spreading caused by the differences in the group velocities of the modes of a multimode fiber. The core of a graded-index fiber has a varying refractive index, highest in the center and decreasing gradually to its lowest value at the cladding. The phase velocity of light is therefore minimum at the center and increases gradually with the radial distance. Rays of the

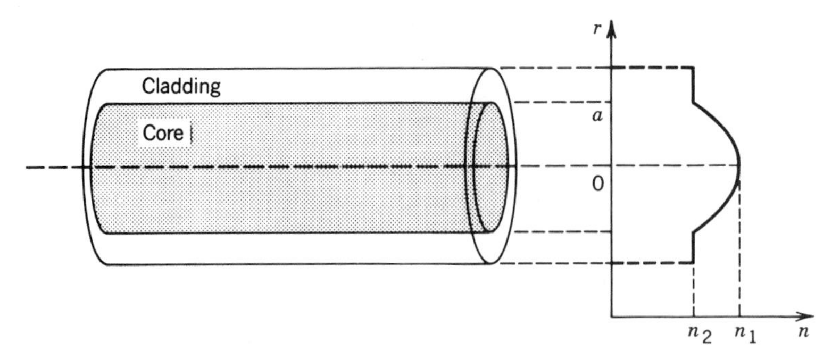

Figure 8.2-1 Geometry and refractive-index profile of a graded-index fiber.

most axial mode travel the shortest distance at the smallest phase velocity. Rays of the most oblique mode zigzag at a greater angle and travel a longer distance, mostly in a medium where the phase velocity is high. Thus the disparities in distances are compensated by opposite disparities in phase velocities. As a consequence, the differences in the group velocities and the travel times are expected to be reduced. In this section we examine the propagation of light in graded-index fibers.

The core refractive index is a function $n(r)$ of the radial position r and the cladding refractive index is a constant n_2. The highest value of $n(r)$ is $n(0) = n_1$ and the lowest value occurs at the core radius $r = a$, $n(a) = n_2$, as illustrated in Fig. 8.2-1.

A versatile refractive-index profile is the power-law function

$$n^2(r) = n_1^2\left[1 - 2\left(\frac{r}{a}\right)^p \Delta\right], \qquad r \le a, \tag{8.2-1}$$

where

$$\Delta = \frac{n_1^2 - n_2^2}{2n_1^2} \approx \frac{n_1 - n_2}{n_1}, \tag{8.2-2}$$

and p, called the **grade profile parameter**, determines the steepness of the profile. This function drops from n_1 at $r = 0$ to n_2 at $r = a$. For $p = 1$, $n^2(r)$ is linear, and for $p = 2$ it is quadratic. As $p \to \infty$, $n^2(r)$ approaches a step function, as illustrated in Fig. 8.2-2. Thus the step-index fiber is a special case of the graded-index fiber with $p = \infty$.

Guided Rays

The transmission of light rays in a graded-index medium with parabolic-index profile was discussed in Sec. 1.3. Rays in meridional planes follow oscillatory planar trajecto-

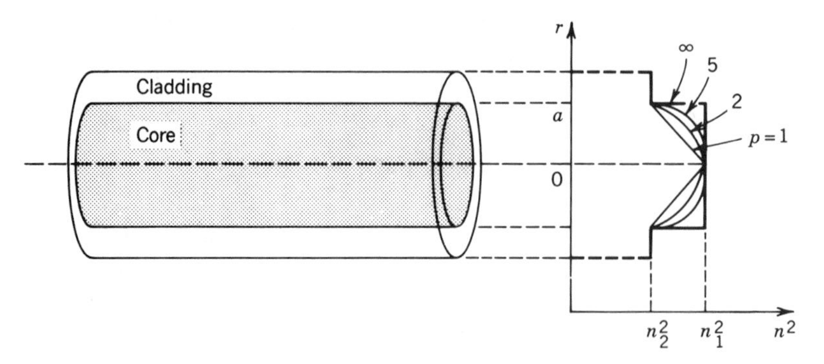

Figure 8.2-2 Power-law refractive-index profile $n^2(r)$ for different values of p.

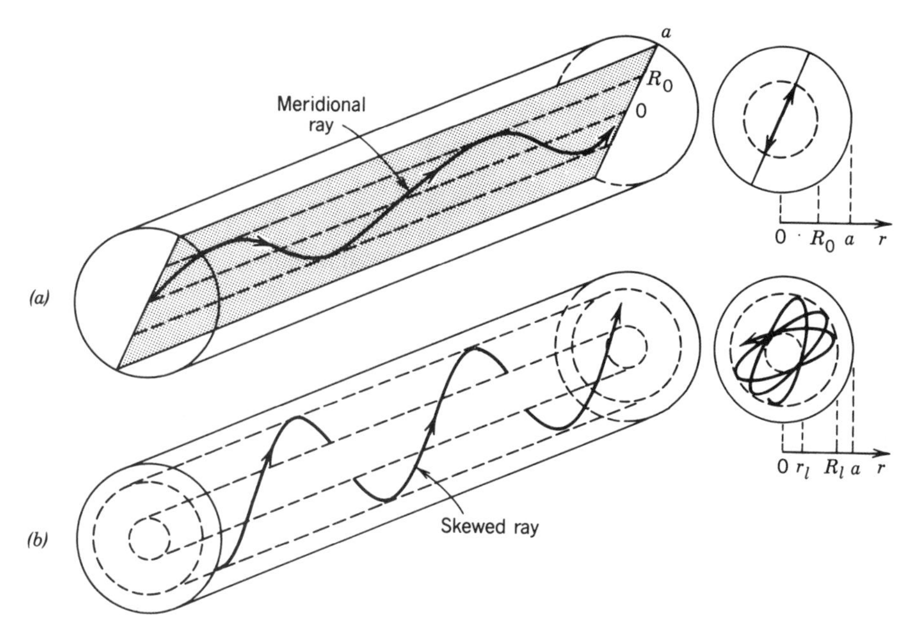

Figure 8.2-3 Guided rays in the core of a graded-index fiber. (a) A meridional ray confined to a meridional plane inside a cylinder of radius R_0. (b) A skewed ray follows a helical trajectory confined within two cylindrical shells of radii r_l and R_l.

ries, whereas skewed rays follow helical trajectories with the turning points forming cylindrical caustic surfaces, as illustrated in Fig. 8.2-3. Guided rays are confined within the core and do not reach the cladding.

A. Guided Waves

The modes of the graded-index fiber may be determined by writing the Helmholtz equation (8.1-4) with $n = n(r)$, solving for the spatial distributions of the field components, and using Maxwell's equations and the boundary conditions to obtain the characteristic equation as was done in the step-index case. This procedure is in general difficult.

In this section we use instead an approximate approach based on picturing the field distribution as a quasi-plane wave traveling within the core, approximately along the trajectory of the optical ray. A quasi-plane wave is a wave that is locally identical to a plane wave, but changes its direction and amplitude slowly as it travels. This approach permits us to maintain the simplicity of rays optics but retain the phase associated with the wave, so that we can use the self-consistency condition to determine the propagation constants of the guided modes (as was done in the planar waveguide in Sec. 7.2). This approximate technique, called the WKB (Wentzel–Kramers–Brillouin) method, is applicable only to fibers with a large number of modes (large V parameter).

Quasi-Plane Waves

Consider a solution of the Helmholtz equation (8.1-4) in the form of a quasi-plane wave (see Sec. 2.3)

$$U(\mathbf{r}) = \mathscr{a}(\mathbf{r}) \exp[-jk_o S(\mathbf{r})], \tag{8.2-3}$$

where $\mathscr{a}(\mathbf{r})$ and $S(\mathbf{r})$ are real functions of position that are slowly varying in comparison with the wavelength $\lambda_o = 2\pi/k_o$. We know from Sec. 2.3 that $S(\mathbf{r})$ approximately satisfies the eikonal equation $|\nabla S|^2 \approx n^2$, and that the rays travel in the direction of the gradient ∇S. If we take $k_o S(\mathbf{r}) = k_o s(r) + l\phi + \beta z$, where $s(r)$ is a slowly varying function of r, the eikonal equation gives

$$\left(k_o \frac{ds}{dr}\right)^2 + \beta^2 + \frac{l^2}{r^2} = n^2(r) k_o^2. \tag{8.2-4}$$

The local spatial frequency of the wave in the radial direction is the partial derivative of the phase $k_o S(\mathbf{r})$ with respect to r,

$$k_r = k_o \frac{ds}{dr}, \tag{8.2-5}$$

so that (8.2-3) becomes

$$U(r) = a(r) \exp\left(-j \int_0^r k_r \, dr\right) e^{-jl\phi} e^{-j\beta z}, \tag{8.2-6}$$

Quasi-Plane Wave

and (8.2-4) gives

$$k_r^2 = n^2(r) k_o^2 - \beta^2 - \frac{l^2}{r^2}. \tag{8.2-7}$$

Defining $k_\phi = l/r$, i.e., $\exp(-jl\phi) = \exp(-jk_\phi r\phi)$, and $k_z = \beta$, we find that (8.2-7) gives $k_r^2 + k_\phi^2 + k_z^2 = n^2(r)k_o^2$. The quasi-plane wave therefore has a local wavevector $\mathbf{k}$ with magnitude $n(r)k_o$ and cylindrical-coordinate components (k_r, k_ϕ, k_z). Since $n(r)$ and k_ϕ are functions of r, k_r is also generally position dependent. The direction of $\mathbf{k}$ changes slowly with r (see Fig. 8.2-4) following a helical trajectory similar to that of the skewed ray shown earlier in Fig. 8.2-3(b).

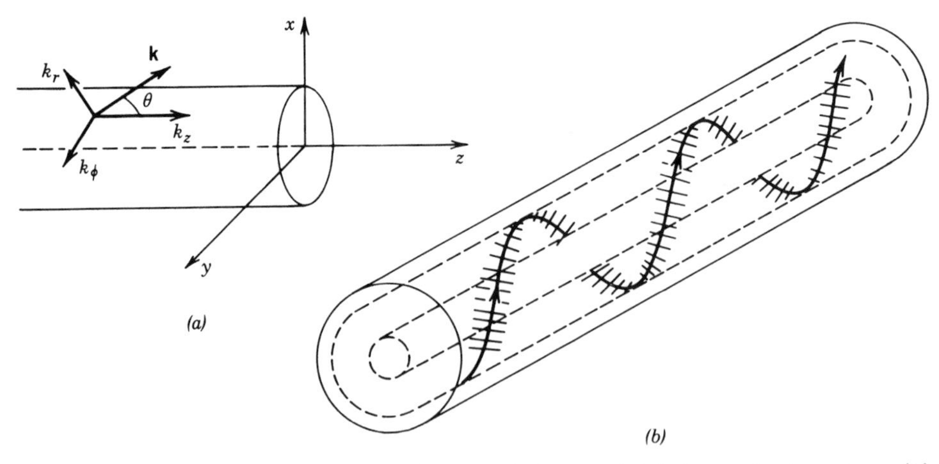

Figure 8.2-4 (a) The wavevector $\mathbf{k} = (k_r, k_\phi, k_z)$ in a cylindrical coordinate system. (b) Quasi-plane wave following the direction of a ray.

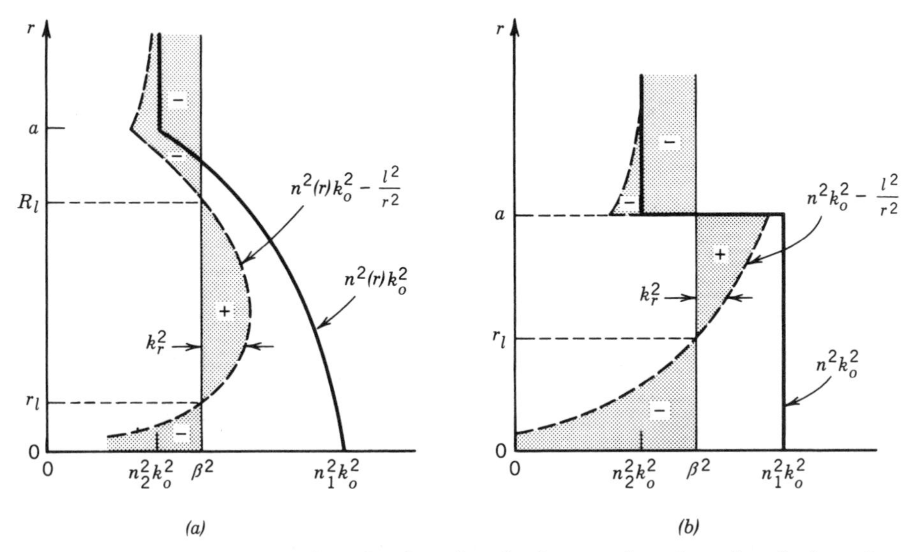

Figure 8.2-5 Dependence of $n^2(r)k_o^2$, $n^2(r)k_o^2 - l^2/r^2$, and $k_r^2 = n^2(r)k_o^2 - l^2/r^2 - \beta^2$ on the position r. At any r, k_r^2 is the width of the shaded area with the + and − signs denoting positive and negative k_r^2. (*a*) Graded-index fiber; k_r^2 is positive in the region $r_l < r < R_l$. (*b*) Step-index fiber; k_r^2 is positive in the region $r_l < r < a$.

To determine the region of the core within which the wave is bound, we determine the values of r for which k_r is real, or $k_r^2 > 0$. For a given l and β we plot $k_r^2 = [n^2(r)k_o^2 - l^2/r^2 - \beta^2]$ as a function of r. The term $n^2(r)k_o^2$ is first plotted as a function of r [the thick continuous curve in Fig. 8.2-5(*a*)]. The term l^2/r^2 is then subtracted, yielding the dashed curve. The value of β^2 is marked by the thin continuous vertical line. It follows that k_r^2 is represented by the difference between the dashed line and the thin continuous line, i.e., by the shaded area. Regions where k_r^2 is positive or negative are indicated by the + or − signs, respectively. Thus k_r is real in the region $r_l < r < R_l$, where

$$n^2(r)k_o^2 - \frac{l^2}{r^2} - \beta^2 = 0, \qquad r = r_l \quad \text{and} \quad r = R_l. \tag{8.2-8}$$

It follows that the wave is basically confined within a cylindrical shell of radii r_l and R_l just like the helical ray trajectory shown in Fig. 8.2-3(*b*).

These results are also applicable to the step-index fiber in which $n(r) = n_1$ for $r < a$, and $n(r) = n_2$ for $r > a$. In this case the quasi-plane wave is guided in the core by reflecting from the core–cladding boundary at $r = a$. As illustrated in Fig. 8.2-5(*b*), the region of confinement is $r_l < r < a$, where

$$n_1^2k_o^2 - \frac{l^2}{r_l^2} - \beta^2 = 0. \tag{8.2-9}$$

The wave bounces back and forth helically like the skewed ray shown in Fig. 8.1-2. In the cladding ($r > a$) and near the center of the core ($r < r_l$), k_r^2 is negative so that k_r is imaginary, and the wave therefore decays exponentially. Note that r_l depends on β. For large β (or large l), r_l is large; i.e., the wave is confined to a thin cylindrical shell near the edge of the core.

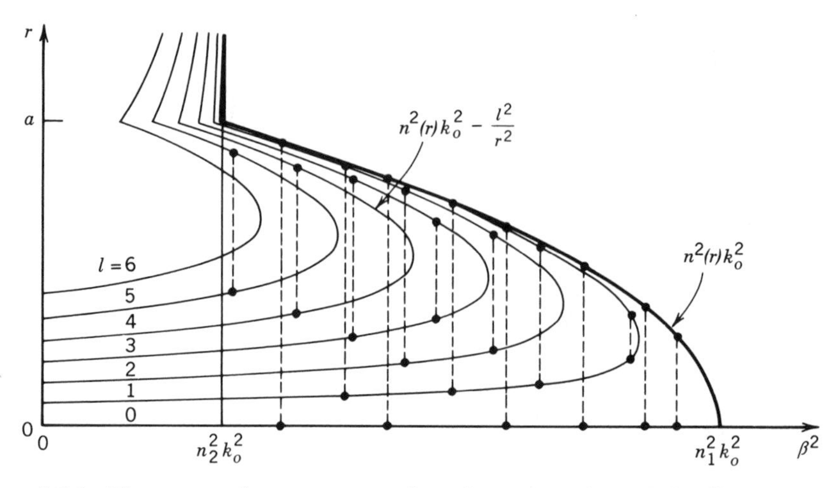

Figure 8.2-6 The propagation constants and confinement regions of the fiber modes. Each curve corresponds to an index l. In this plot $l = 0, 1, \ldots, 6$. Each mode (representing a certain value of m) is marked schematically by two dots connected by a dashed vertical line. The ordinates of the dots mark the radii r_l and R_l of the cylindrical shell within which the mode is confined. Values on the abscissa are the squared propagation constants β^2 of the mode.

Modes

The modes of the fiber are determined by imposing the self-consistency condition that the wave reproduce itself after one helical period of traveling between r_l and R_l and back. The azimuthal path length corresponding to an angle 2π must correspond to a multiple of 2π phase shift, i.e., $k_\phi 2\pi r = 2\pi l$; $l = 0, \pm 1, \pm 2, \ldots$. This condition is evidently satisfied since $k_\phi = l/r$. In addition, the radial round-trip path length must correspond to a phase shift equal to an integer multiple of 2π,

$$2\int_{r_l}^{R_l} k_r \, dr = 2\pi m, \qquad m = 1, 2, \ldots, M_l. \tag{8.2-10}$$

This condition, which is analogous to the self-consistency condition (7.2-2) for planar waveguides, provides the characteristic equation from which the propagation constants β_{lm} of the modes are determined. These values are marked schematically in Fig. 8.2-6; the mode $m = 1$ has the largest value of β (approximately $n_1 k_o$) and $m = M_l$ has the smallest value (approximately $n_2 k_o$).

Number of Modes

The total number of modes can be determined by adding the number of modes M_l for $l = 0, 1, \ldots, l_{\max}$. We shall address this problem using a different procedure. We first determine the number q_β of modes with propagation constants greater than a given value β. For each l, the number of modes $M_l(\beta)$ with propagation constant greater than β is the number of multiples of 2π the integral in (8.2-10) yields, i.e.,

$$M_l(\beta) = \frac{1}{\pi}\int_{r_l}^{R_l} k_r \, dr = \frac{1}{\pi}\int_{r_l}^{R_l} \left[n^2(r)k_o^2 - \frac{l^2}{r^2} - \beta^2 \right]^{1/2} dr, \tag{8.2-11}$$

where r_l and R_l are the radii of confinement corresponding to the propagation constant β as given by (8.2-8). Clearly, r_l and R_l depend on β.

The total number of modes with propagation constant greater than β is therefore

$$q_\beta = 4 \sum_{l=0}^{l_{\max}(\beta)} M_l(\beta), \tag{8.2-12}$$

where $l_{\max}(\beta)$ is the maximum value of l that yields a bound mode with propagation constants greater than β, i.e., for which the peak value of the function $n^2(r)k_o^2 - l^2/r^2$ is greater than β^2. The grand total number of modes M is q_β for $\beta = n_2 k_o$. The factor of 4 in (8.2-12) accounts for the two possible polarizations and the two possible polarities of the angle ϕ, corresponding to positive or negative helical trajectories for each (l, m). If the number of modes is sufficiently large, we can replace the summation in (8.2-12) by an integral,

$$q_\beta \approx 4 \int_0^{l_{\max}(\beta)} M_l(\beta)\, dl. \tag{8.2-13}$$

For fibers with a power-law refractive-index profile, we substitute (8.2-1) into (8.2-11), and the result into (8.2-13), and evaluate the integral to obtain

$$q_\beta \approx M \left[\frac{1 - (\beta/n_1 k_o)^2}{2\Delta} \right]^{(p+2)/p}, \tag{8.2-14}$$

where

$$M \approx \frac{p}{p+2} n_1^2 k_o^2 a^2 \Delta = \frac{p}{p+2} \frac{V^2}{2}. \tag{8.2-15}$$

Here $\Delta = (n_1 - n_2)/n_1$ and $V = 2\pi(a/\lambda_o)\mathrm{NA}$ is the fiber V parameter. Since $q_\beta \approx M$ at $\beta = n_2 k_o$, M is indeed the total number of modes.

For step-index fibers ($p = \infty$),

$$q_\beta \approx M \left[\frac{1 - (\beta/n_1 k_o)^2}{2\Delta} \right] \tag{8.2-16}$$

and

$$\boxed{M \approx \frac{V^2}{2}.} \tag{8.2-17}$$

Number of Modes (Step-Index Fiber) $V = 2\pi(a/\lambda_o)\mathrm{NA}$

This expression for M is nearly the same as $M \approx 4V^2/\pi^2 \approx 0.41V^2$ in (8.1-18), which was obtained in Sec. 8.1 using a different approximation.

B. Propagation Constants and Velocities

Propagation Constants

The propagation constant β_q of mode q is obtained by inverting (8.2-14),

$$\beta_q \approx n_1 k_o \left[1 - 2\left(\frac{q}{M}\right)^{p/(p+2)} \Delta\right]^{1/2}, \qquad q = 1, 2, \ldots, M, \tag{8.2-18}$$

where the index q_β has been replaced by q, and β replaced by β_q. Since $\Delta \ll 1$, the approximation $(1 + \delta)^{1/2} \approx 1 + \frac{1}{2}\delta$ (when $|\delta| \ll 1$) can be applied to (8.2-18), yielding

$$\beta_q \approx n_1 k_o \left[1 - \left(\frac{q}{M}\right)^{p/(p+2)} \Delta\right]. \tag{8.2-19}$$

Propagation Constants
$q = 1, 2, \ldots, M$

The propagation constant β_q therefore decreases from $\approx n_1 k_o$ (at $q = 1$) to $n_2 k_o$ (at $q = M$), as illustrated in Fig. 8.2-7.

In the step-index fiber ($p = \infty$),

$$\beta_q \approx n_1 k_o \left(1 - \frac{q}{M}\Delta\right). \tag{8.2-20}$$

Propagation Constants
(Step-Index Fiber)
$q = 1, 2, \ldots, M$

This expression is identical to (8.1-22) if the index $q = 1, 2, \ldots, M$ is replaced by $(l + 2m)^2$, where $l = 0, 1, \ldots, \sqrt{M}$; $m = 1, 2, \ldots, \sqrt{M}/2 - l/2$.

Group Velocities

To determine the group velocity $v_q = d\omega/d\beta_q$, we write β_q as a function of ω by substituting (8.2-15) into (8.2-19), substituting $n_1 k_o = \omega/c_1$ into the result, and evaluating $v_q = (d\beta_q/d\omega)^{-1}$. With the help of the approximation $(1 + \delta)^{-1} \approx 1 - \delta$ when

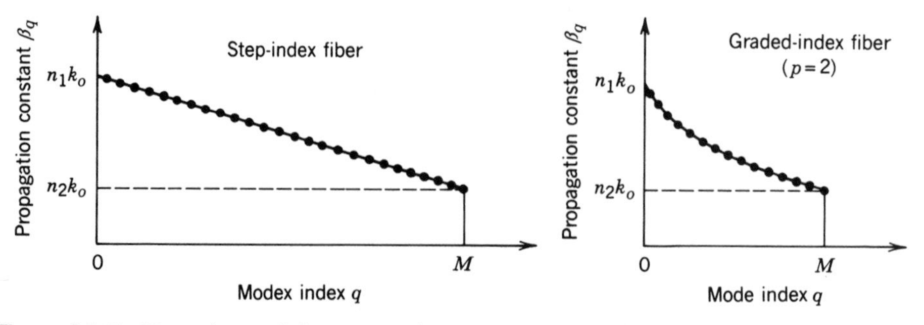

Figure 8.2-7 Dependence of the propagation constants β_q on the mode index $q = 1, 2, \ldots, M$.

$|\delta| \ll 1$, and assuming that c_1 and Δ are independent of ω (i.e., ignoring material dispersion), we obtain

$$v_q \approx c_1\left[1 - \frac{p-2}{p+2}\left(\frac{q}{M}\right)^{p/(p+2)}\Delta\right]. \tag{8.2-21}$$

Group Velocities
$q = 1, 2, \ldots, M$

For the step-index fiber ($p = \infty$)

$$v_q \approx c_1\left(1 - \frac{q}{M}\Delta\right). \tag{8.2-22}$$

Group Velocities
(Step-Index Fiber)
$q = 1, 2, \ldots, M$

The group velocity varies from approximately c_1 to $c_1(1 - \Delta)$. This reproduces the result obtained in (8.1-23).

Optimal Index Profile

Equation (8.2-21) indicates that the grade profile parameter $p = 2$ yields a group velocity $v_q \approx c_1$ for all q, so that all modes travel at approximately the same velocity c_1. The advantage of the graded-index fiber for multimode transmission is now apparent.

To determine the group velocity with better accuracy, we repeat the derivation of v_q from (8.2-18), taking three terms in the Taylor's expansion $(1 + \delta)^{1/2} \approx 1 + \delta/2 - \delta^2/8$, instead of two. For $p = 2$, the result is

$$v_q = c_1\left(1 - \frac{q}{M}\frac{\Delta^2}{2}\right). \tag{8.2-23}$$

Group Velocities
($p = 2$)
$q = 1, \ldots, M$

Thus the group velocities vary from approximately c_1 at $q = 1$ to approximately $c_1(1 - \Delta^2/2)$ at $q = M$. In comparison with the step-index fiber, for which the group velocity ranges between c_1 and $c_1(1 - \Delta)$, the fractional velocity difference for the parabolically graded fiber is $\Delta^2/2$ instead of Δ for the step-index fiber (Fig. 8.2-8). Under ideal conditions, the graded-index fiber therefore reduces the group velocity

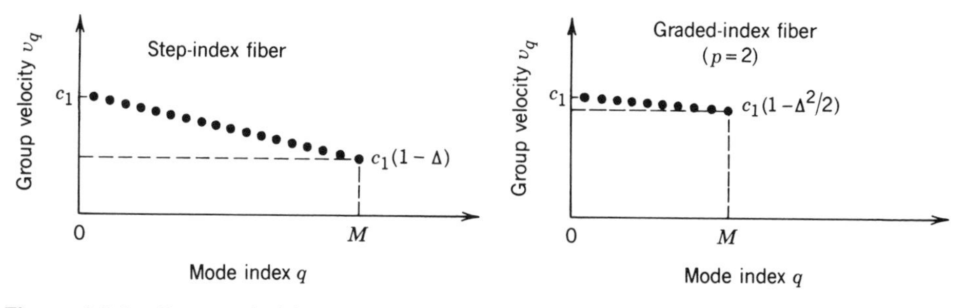

Figure 8.2-8 Group velocities v_q of the modes of a step-index fiber ($p = \infty$) and an optimal graded-index fiber ($p = 2$).

difference by a factor $\Delta/2$, thus realizing its intended purpose of equalizing the mode velocities. Since the analysis leading to (8.2-23) is based on a number of approximations, however, this improvement factor is only a rough estimate; indeed it is not fully attained in practice.

For $p = 2$, the number of modes M given by (8.2-15) becomes

$$\boxed{M \approx \frac{V^2}{4}.} \tag{8.2-24}$$

Number of Modes
(Graded-Index Fiber, $p = 2$)
$V = 2\pi(a/\lambda_o)\mathrm{NA}$

Comparing this with (8.2-17), we see that the number of modes in an optimal graded-index fiber is approximately one-half the number of modes in a step-index fiber of the same parameters n_1, n_2, and a.

8.3 ATTENUATION AND DISPERSION

Attenuation and dispersion limit the performance of the optical-fiber medium as a data transmission channel. Attenuation limits the magnitude of the optical power transmitted, whereas dispersion limits the rate at which data may be transmitted through the fiber, since it governs the temporal spreading of the optical pulses carrying the data.

A. Attenuation

The Attenuation Coefficient

Light traveling through an optical fiber exhibits a power that decreases exponentially with the distance as a result of absorption and scattering. The attenuation coefficient $\boldsymbol{\alpha}$ is usually defined in units of dB/km,

$$\boldsymbol{\alpha} = \frac{1}{L} 10 \log_{10} \frac{1}{\mathcal{T}}, \tag{8.3-1}$$

where $\mathcal{T} = P(L)/P(0)$ is the power transmission ratio (ratio of transmitted to incident power) for a fiber of length L km. The relation between $\boldsymbol{\alpha}$ and $\mathcal{T}$ is illustrated in Fig. 8.3-1 for $L = 1$ km. A 3-dB attenuation, for example, corresponds to $\mathcal{T} = 0.5$, while 10 dB is equivalent to $\mathcal{T} = 0.1$ and 20 dB corresponds to $\mathcal{T} = 0.01$, and so on.

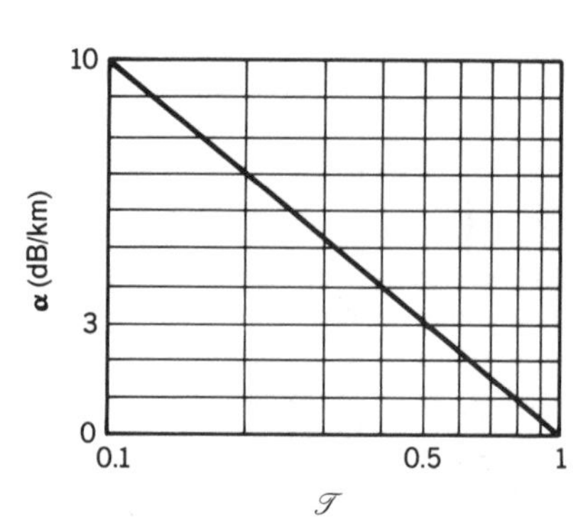

Figure 8.3-1 Relation between transmittance $\mathcal{T}$ and attenuation coefficient $\boldsymbol{\alpha}$ in dB units.

Losses in dB units are additive, whereas the transmission ratios are multiplicative. Thus for a propagation distance of z kilometers, the loss is $\boldsymbol{\alpha} z$ decibels and the power transmission ratio is

$$\frac{P(z)}{P(0)} = 10^{-\boldsymbol{\alpha} z/10} \approx e^{-0.23\boldsymbol{\alpha} z}. \qquad (\boldsymbol{\alpha} \text{ in dB/km}) \tag{8.3-2}$$

Note that if the attenuation coefficient is measured in km^{-1} units, instead of in dB/km, then

$$P(z)/P(0) = e^{-\alpha z} \tag{8.3-3}$$

where $\alpha \approx 0.23\boldsymbol{\alpha}$. Throughout this section $\boldsymbol{\alpha}$ is taken in dB/km units so that (8.3-2) applies. Elsewhere in the book, however, we use α to denote the attenuation coefficient (m^{-1} or cm^{-1}) in which case the power attenuation is described by (8.3-3).

Absorption

The attenuation coefficient of fused silica glass (SiO_2) is strongly dependent on wavelength, as illustrated in Fig. 8.3-2. This material has two strong absorption bands: a middle-infrared absorption band resulting from vibrational transitions and an ultraviolet absorption band due to electronic and molecular transitions. There is a window bounded by the tails of these bands in which there is essentially no intrinsic absorption. This window occupies the near-infrared region.

Scattering

Rayleigh scattering is another intrinsic effect that contributes to the attenuation of light in glass. The random localized variations of the molecular positions in glass create random inhomogeneities of the refractive index that act as tiny scattering centers. The amplitude of the scattered field is proportional to ω^2.† The scattered intensity is therefore proportional to ω^4 or to $1/\lambda_o^4$, so that short wavelengths are scattered more than long wavelengths. Thus blue light is scattered more than red (a similar effect, the

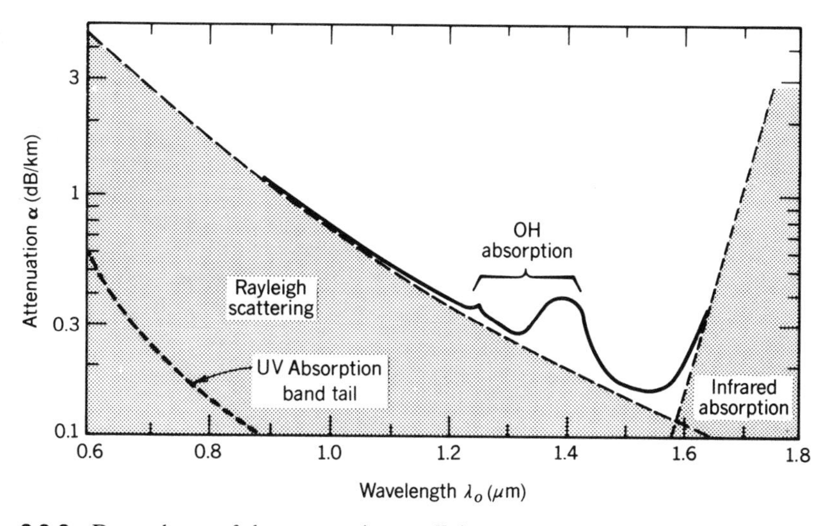

Figure 8.3-2 Dependence of the attenuation coefficient $\boldsymbol{\alpha}$ of silica glass on the wavelength λ_o. There is a local minimum at 1.3 μm ($\boldsymbol{\alpha} \approx 0.3$ dB/km) and an absolute minimum at 1.55 μm ($\boldsymbol{\alpha} \approx 0.16$ dB/km).

†The scattering medium creates a polarization density $\mathscr{P}$ which corresponds to a source of radiation proportional to $d^2\mathscr{P}/dt^2 = -\omega^2\mathscr{P}$; see (5.2-19).

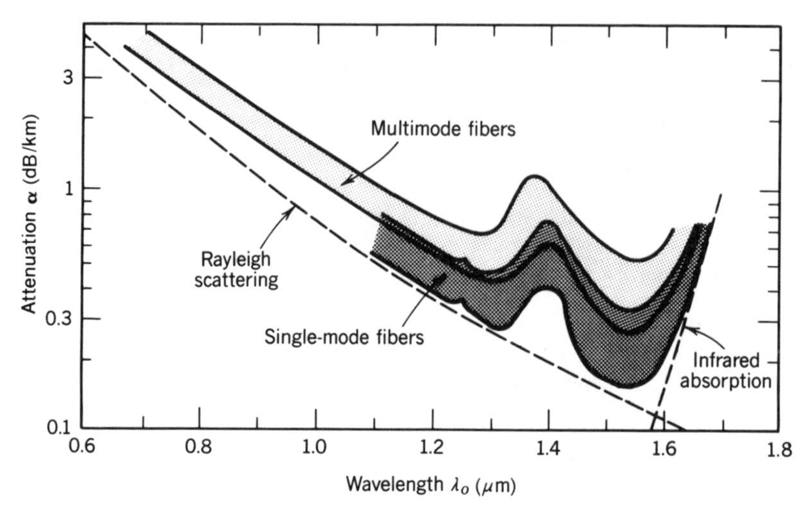

Figure 8.3-3 Ranges of attenuation coefficients of silica glass single-mode and multimode fibers.

scattering of sunlight from tiny atmospheric molecules, is the reason the sky appears blue). The attenuation caused by Rayleigh scattering therefore decreases with wavelength as $1/\lambda_o^4$, a relation known as **Rayleigh's inverse fourth-power law**. In the visible band, Rayleigh scattering is more significant than the tail of the ultraviolet absorption band, but it becomes negligible in comparison with infrared absorption for wavelengths greater than 1.6 μm.

The transparent window in silica glass is therefore bounded by Rayleigh scattering on the short-wavelength side and by infrared absorption on the long-wavelength side (as indicated by the dashed lines in Fig. 8.3-2).

Extrinsic Effects

In addition to these intrinsic effects there are extrinsic absorption bands due to impurities, mainly OH vibrations associated with water vapor dissolved in the glass and metallic-ion impurities. Recent progress in the technology of fabricating glass fibers has made it possible to remove most metal impurities, but OH impurities are difficult to eliminate. Wavelengths at which glass fibers are used for optical communication are selected to avoid these absorption bands. Light-scattering losses may also be accentuated when dopants are added for the purpose of index grading, for example.

The attenuation coefficient of guided light in glass fibers depends on the absorption and scattering in the core and cladding materials. Since each mode has a different penetration depth into the cladding so that rays travel different effective distances, the attenuation coefficient is mode dependent. It is generally higher for higher-order modes. Single-mode fibers therefore typically have smaller attenuation coefficients than multimode fibers (Fig. 8.3-3). Losses are also introduced by small random variations in the geometry of the fiber and by bends.

B. Dispersion

When a short pulse of light travels through an optical fiber its power is "dispersed" in time so that the pulse spreads into a wider time interval. There are four sources of dispersion in optical fibers: modal dispersion, material dispersion, waveguide dispersion, and nonlinear dispersion.

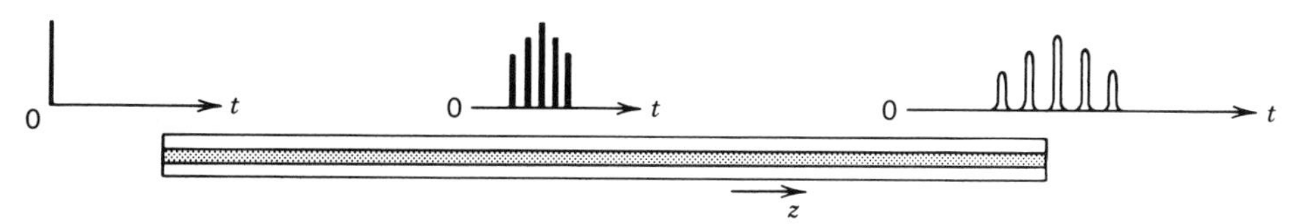

Figure 8.3-4 Pulse spreading caused by modal dispersion.

Modal Dispersion

Modal dispersion occurs in multimode fibers as a result of the differences in the group velocities of the modes. A single impulse of light entering an M-mode fiber at $z = 0$ spreads into M pulses with the differential delay increasing as a function of z. For a fiber of length L, the time delays encountered by the different modes are $\tau_q = L/v_q$, $q = 1, \ldots, M$, where v_q is the group velocity of mode q. If $v_{\min}$ and $v_{\max}$ are the smallest and largest group velocities, the received pulse spreads over a time interval $L/v_{\min} - L/v_{\max}$. Since the modes are generally not excited equally, the overall shape of the received pulse is a smooth profile, as illustrated in Fig. 8.3-4. An estimate of the overall rms pulse width is $\sigma_\tau = \frac{1}{2}(L/v_{\min} - L/v_{\max})$. This width represents the response time of the fiber.

In a step-index fiber with a large number of modes, $v_{\min} \approx c_1(1 - \Delta)$ and $v_{\max} \approx c_1$ (see Sec. 8.1B and Fig. 8.2-8). Since $(1 - \Delta)^{-1} \approx 1 + \Delta$, the response time is

$$\sigma_\tau \approx \frac{L}{c_1}\frac{\Delta}{2}, \tag{8.3-4}$$

Response Time (Multimode Step-Index Fiber)

i.e., it is a fraction $\Delta/2$ of the delay time L/c_1.

Modal dispersion is much smaller in graded-index fibers than in step-index fibers since the group velocities are equalized and the differences between the delay times $\tau_q = L/v_q$ of the modes are reduced. It was shown in Sec. 8.2B and in Fig. 8.2-8 that in a graded-index fiber with a large number of modes and with an optimal index profile, $v_{\max} \approx c_1$ and $v_{\min} \approx c_1(1 - \Delta^2/2)$. The response time is therefore

$$\sigma_\tau \approx \frac{L}{c_1}\frac{\Delta^2}{4}, \tag{8.3-5}$$

Response Time (Graded-Index Fiber)

which is a factor of $\Delta/2$ smaller than that in a step-index fiber.

EXAMPLE 8.3-1. ***Multimode Pulse Broadening Rate.*** In a step-index fiber with $\Delta = 0.01$ and $n = 1.46$, pulses spread at a rate of approximately $\sigma_\tau/L = \Delta/2c_1 = n_1\Delta/2c_o \approx 24$ ns/km. In a 100-km fiber, therefore, an impulse spreads to a width of ≈ 2.4 μs. If the same fiber is optimally index graded, the pulse broadening rate is approximately $n_1\Delta^2/4c_o \approx 122$ ps/km, which is substantially reduced.

The pulse broadening arising from modal dispersion is proportional to the fiber length L in both step-index and graded-index fibers. This dependence, however, does not necessarily hold when the fibers are longer than a certain critical length because of mode coupling. Coupling occurs between modes of approximately the same propagation constants as a result of small imperfections in the fiber (random irregularities of the fiber surface, or inhomogeneities of the refractive index) which permit the optical power to be exchanged between the modes. Under certain conditions, the response time σ_τ of mode-coupled fibers is proportional to L for small L and to $L^{1/2}$ when a critical length is exceeded, so that pulses are broadened at a slower rate†.

Material Dispersion

Glass is a dispersive medium; i.e, its refractive index is a function of wavelength. As discussed in Sec. 5.6, an optical pulse travels in a dispersive medium of refractive index n with a group velocity $v = c_o/N$, where $N = n - \lambda_o \, dn/d\lambda_o$. Since the pulse is a wavepacket, composed of a spectrum of components of different wavelengths each traveling at a different group velocity, its width spreads. The temporal width of an optical impulse of spectral width σ_λ (nm), after traveling a distance L, is $\sigma_\tau = |(d/d\lambda_o)(L/v)|\sigma_\lambda = |(d/d\lambda_o)(LN/c_o)|\sigma_\lambda$, from which

$$\boxed{\sigma_\tau = |D_\lambda|\sigma_\lambda L,} \tag{8.3-6}$$

Response Time (Material Dispersion)

where

$$D_\lambda = -\frac{\lambda_o}{c_o}\frac{d^2n}{d\lambda_o^2} \tag{8.3-7}$$

is the material dispersion coefficient [see (5.6-21)]. The response time increases linearly with the distance L. Usually, L is measured in km, σ_τ in ps, and σ_λ in nm, so that D_λ has units of ps/km-nm. This type of dispersion is called **material dispersion** (as opposed to modal dispersion).

The wavelength dependence of the dispersion coefficient D_λ for silica glass is shown in Fig. 8.3-5. At wavelengths shorter than 1.3 μm the dispersion coefficient is negative,

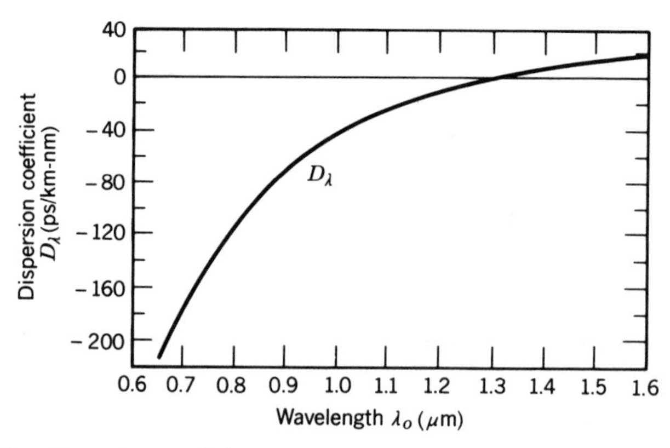

Figure 8.3-5 The dispersion coefficient D_λ of silica glass as a function of wavelength λ_o (see also Fig. 5.6-5).

†See, e.g., J. E. Midwinter, *Optical Fibers for Transmission*, Wiley, New York, 1979.

so that wavepackets of long wavelength travel faster than those of short wavelength. At a wavelength $\lambda_o = 0.87$ μm, the dispersion coefficient D_λ is approximately -80 ps/km-nm. At $\lambda_o = 1.55$ μm, $D_\lambda \approx +17$ ps/km-nm. At $\lambda_o \approx 1.312$ μm the dispersion coefficient vanishes, so that σ_τ in (8.3-6) vanishes. A more precise expression for σ_τ that incorporates the spread of the spectral width σ_λ about $\lambda_o = 1.312$ μm yields a very small, but nonzero, width.

EXAMPLE 8.3-2. ***Pulse Broadening Associated with Material Dispersion.*** The dispersion coefficient $D_\lambda \approx -80$ ps/km-nm at $\lambda_o \approx 0.87$ μm. For a source of linewidth $\sigma_\lambda = 50$ nm (from an LED, for example) the pulse spreading rate in a single-mode fiber with no other sources of dispersion is $|D_\lambda|\sigma_\lambda = 4$ ns/km. An impulse of light traveling a distance $L = 100$ km in the fiber is therefore broadened to a width $\sigma_\tau = |D_\lambda|\sigma_\lambda L = 0.4$ μs. The response time of the fiber is then 0.4 μs. An impulse of narrower linewidth $\sigma_\lambda = 2$ nm (from a laser diode, for example) operating near 1.3 μm, where the dispersion coefficient is 1 ps/km-nm, spreads at a rate of only 2 ps/km. A 100-km fiber thus has a substantially shorter response time, $\sigma_\tau = 0.2$ ns.

Waveguide Dispersion

The group velocities of the modes depend on the wavelength even if material dispersion is negligible. This dependence, known as **waveguide dispersion**, results from the dependence of the field distribution in the fiber on the ratio between the core radius and the wavelength (a/λ_o). If this ratio is altered, by altering λ_o, the relative portions of optical power in the core and cladding are modified. Since the phase velocities in the core and cladding are different, the group velocity of the mode is altered. Waveguide dispersion is particularly important in single-mode fibers, where modal dispersion is not exhibited, and at wavelengths for which material dispersion is small (near $\lambda_o = 1.3$ μm in silica glass).

As discussed in Sec. 8.1B, the group velocity $v = (d\beta/d\omega)^{-1}$ and the propagation constant β are determined from the characteristic equation, which is governed by the fiber V parameter $V = 2\pi(a/\lambda_o)\mathrm{NA} = (a \cdot \mathrm{NA}/c_o)\omega$. In the absence of material dispersion (i.e., when NA is independent of ω), V is directly proportional to ω, so that

$$\frac{1}{v} = \frac{d\beta}{d\omega} = \frac{d\beta}{dV}\frac{dV}{d\omega} = \frac{a \cdot \mathrm{NA}}{c_o}\frac{d\beta}{dV}. \tag{8.3-8}$$

The pulse broadening associated with a source of spectral width σ_λ is related to the time delay L/v by $\sigma_\tau = |(d/d\lambda_o)(L/v)|\sigma_\lambda$. Thus

$$\sigma_\tau = |D_w|\sigma_\lambda L, \tag{8.3-9}$$

where

$$D_w = \frac{d}{d\lambda_o}\left(\frac{1}{v}\right) = -\frac{\omega}{\lambda_o}\frac{d}{d\omega}\left(\frac{1}{v}\right) \tag{8.3-10}$$

is the waveguide dispersion coefficient. Substituting (8.3-8) into (8.3-10) we obtain

$$D_w = -\left(\frac{1}{2\pi c_o}\right) V^2 \frac{d^2\beta}{dV^2}. \tag{8.3-11}$$

Thus the group velocity is inversely proportional to $d\beta/dV$ and the dispersion coefficient is proportional to $V^2\, d^2\beta/dV^2$. The dependence of β on V is shown in Fig. 8.1-10(*b*) for the fundamental LP_{01} mode. Since β varies nonlinearly with V, the waveguide dispersion coefficient D_w is itself a function of V and is therefore also a function of the wavelength.† The dependence of D_w on λ_o may be controlled by altering the radius of the core or the index grading profile for graded-index fibers.

Combined Material and Waveguide Dispersion

The combined effects of material dispersion and waveguide dispersion (referred to here as **chromatic dispersion**) may be determined by including the wavelength dependence of the refractive indices, n_1 and n_2 and therefore NA, when determining $d\beta/d\omega$ from the characteristic equation. Although generally smaller than material dispersion, waveguide dispersion does shift the wavelength at which the total chromatic dispersion is minimum.

Since chromatic dispersion limits the performance of single-mode fibers, more advanced fiber designs aim at reducing this effect by using graded-index cores with refractive-index profiles selected such that the wavelength at which waveguide dispersion compensates material dispersion is shifted to the wavelength at which the fiber is to be used. **Dispersion-shifted fibers** have been successfully made by using a linearly tapered core refractive index and a reduced core radius, as illustrated in Fig. 8.3-6(*a*). This technique can be used to shift the zero-chromatic-dispersion wavelength from 1.3 μm to 1.55 μm, where the fiber has its lowest attenuation. Note, however, that the process of index grading itself introduces losses since dopants are used. Other grading profiles have been developed for which the chromatic dispersion vanishes at two wavelengths and is reduced for wavelengths between. These fibers, called **dispersion-flattened**, have been implemented by using a quadruple-clad layered grading, as illustrated in Fig. 8.3-6(*b*).

Combined Material and Modal Dispersion

The effect of material dispersion on pulse broadening in multimode fibers may be determined by returning to the original equations for the propagation constants β_q of the modes and determining the group velocities $v_q = (d\beta_q/d\omega)^{-1}$ with n_1 and n_2 being functions of ω. Consider, for example, the propagation constants of a graded-index fiber with a large number of modes, which are given by (8.2-19) and (8.2-15). Although n_1 and n_2 are dependent on ω, it is reasonable to assume that the ratio $\Delta = (n_1 - n_2)/n_1$ is approximately independent of ω. Using this approximation and evaluating $v_q = (d\beta_q/d\omega)^{-1}$, we obtain

$$v_q \approx \frac{c_o}{N_1}\left[1 - \frac{p-2}{p+2}\left(\frac{q}{M}\right)^{p/(p+2)} \Delta\right], \tag{8.3-12}$$

where $N_1 = (d/d\omega)(\omega n_1) = n_1 - \lambda_o(dn_1/d\lambda_o)$ is the group index of the core material. Under this approximation, the earlier expression (8.2-21) for v_q remains the same, except that the refractive index n_1 is replaced with the group index N_1. For a step-index fiber ($p = \infty$), the group velocities of the modes vary from c_o/N_1 to

† For more details on this topic, see the reading list, particularly the articles by Gloge.

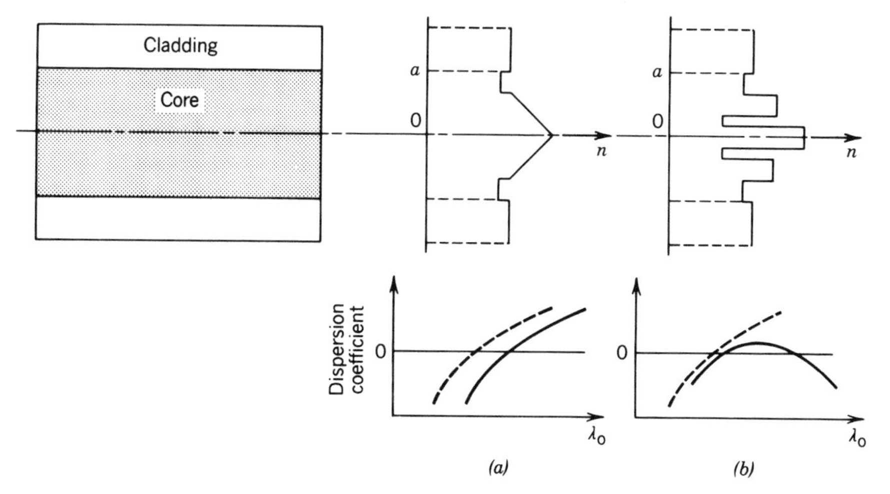

Figure 8.3-6 Refractive-index profiles and schematic wavelength dependences of the material dispersion coefficient (dashed curves) and the combined material and waveguide dispersion coefficients (solid curves) for (a) dispersion-shifted and (b) dispersion-flattened fibers.

$(c_o/N_1)(1 - \Delta)$, so that the response time is

$$\boxed{\sigma_\tau \approx \frac{L}{(c_o/N_1)} \frac{\Delta}{2}.} \tag{8.3-13}$$

Response Time
(Multimode Step-Index Fiber
with Material Dispersion)

This should be compared with (8.3-4) when there is no material dispersion.

EXERCISE 8.3-1

Optimal Grade Profile Parameter. Use (8.2-19) and (8.2-15) to derive the following expression for the group velocity v_q when both n_1 and Δ are wavelength dependent:

$$v_q \approx \frac{c_o}{N_1}\left[1 - \frac{p - 2 - p_s}{p + 2}\left(\frac{q}{M}\right)^{p/(p+2)} \Delta\right], \qquad q = 1, 2, \ldots, M, \tag{8.3-14}$$

where $p_s = 2(n_1/N_1)(\omega/\Delta)\, d\Delta/d\omega$. What is the optimal value of the grade profile parameter p for minimizing modal dispersion?

Nonlinear Dispersion

Yet another dispersion effect occurs when the intensity of light in the core is sufficiently high, since the refractive indices then become intensity dependent and the material exhibits nonlinear behavior. The high-intensity parts of an optical pulse undergo phase shifts different from the low-intensity parts, so that the frequency is shifted by different amounts. Because of material dispersion, the group velocities are modified, and consequently the pulse shape is altered. Under certain conditions, nonlinear dispersion can compensate material dispersion, so that the pulse travels without altering its temporal profile. The guided wave is then known as a solitary wave, or a soliton. Nonlinear optics is introduced in Chap. 19 and optical solitons are discussed in Sec. 19.8.

C. Pulse Propagation

As described in the previous sections, the propagation of pulses in optical fibers is governed by attenuation and several types of dispersion. The following is a summary and recapitulation of these effects, ignoring nonlinear dispersion.

An optical pulse of power $\tau_0^{-1}p(t/\tau_0)$ and short duration τ_0, where $p(t)$ is a function which has unit duration and unit area, is transmitted through a multimode fiber of length L. The received optical power may be written in the form of a sum

$$P(t) \propto \sum_{q=1}^{M} \exp(-0.23\boldsymbol{\alpha}_q L)\sigma_q^{-1} p\left(\frac{t-\tau_q}{\sigma_q}\right), \tag{8.3-15}$$

where M is the number of modes, the subscript q refers to mode q, $\boldsymbol{\alpha}_q$ is the attenuation coefficient (dB/km), $\tau_q = L/v_q$ is the delay time, v_q is the group velocity, and $\sigma_q > \tau_0$ is the width of the pulse associated with mode q. In writing (8.3-15), we have implicitly assumed that the incident optical power is distributed equally among the M modes of the fiber. It has also been assumed that the pulse shape $p(t)$ is not altered; it is only delayed by times τ_q and broadened to widths σ_q as a result of propagation. As was shown in Sec. 5.6, an initial pulse with a Gaussian profile is indeed broadened without altering its Gaussian nature.

The received pulse is thus composed of M pulses of widths σ_q centered at time delays τ_q, as illustrated in Fig. 8.3-7. The composite pulse has an overall width σ_τ which represents the overall response time of the fiber.

We therefore identify two basic types of dispersion: **intermodal** and **intramodal**. Intermodal, or simply modal, dispersion is the delay distortion caused by the disparity among the delay times τ_q of the modes. The time difference $\frac{1}{2}(\tau_{\max} - \tau_{\min})$ between the longest and shortest delay constitutes modal dispersion. It is given by (8.3-4) and (8.3-5) for step-index and graded-index fibers with a large number of modes, respectively. Material dispersion has some effect on modal dispersion since it affects the delay times. For example, (8.3-13) gives the modal dispersion of a multimode fiber with material dispersion. Modal dispersion is directly proportional to the fiber length L, except for long fibers, in which mode coupling plays a role, whereupon it becomes proportional to $L^{1/2}$.

Intramodal dispersion is the broadening of the pulses associated with the individual modes. It is caused by a combination of material dispersion and waveguide dispersion

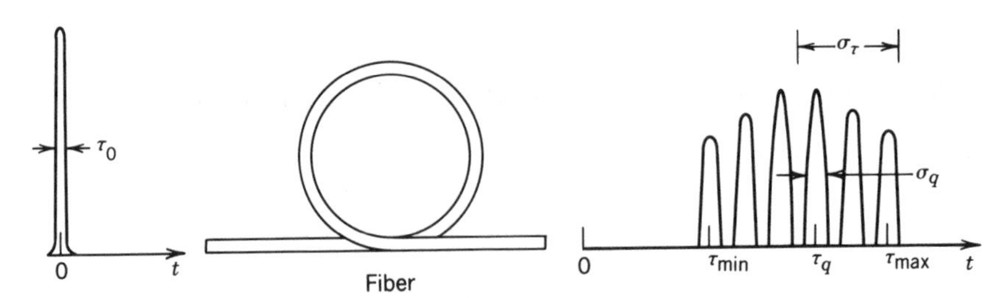

Figure 8.3-7 Response of a multimode fiber to a single pulse.

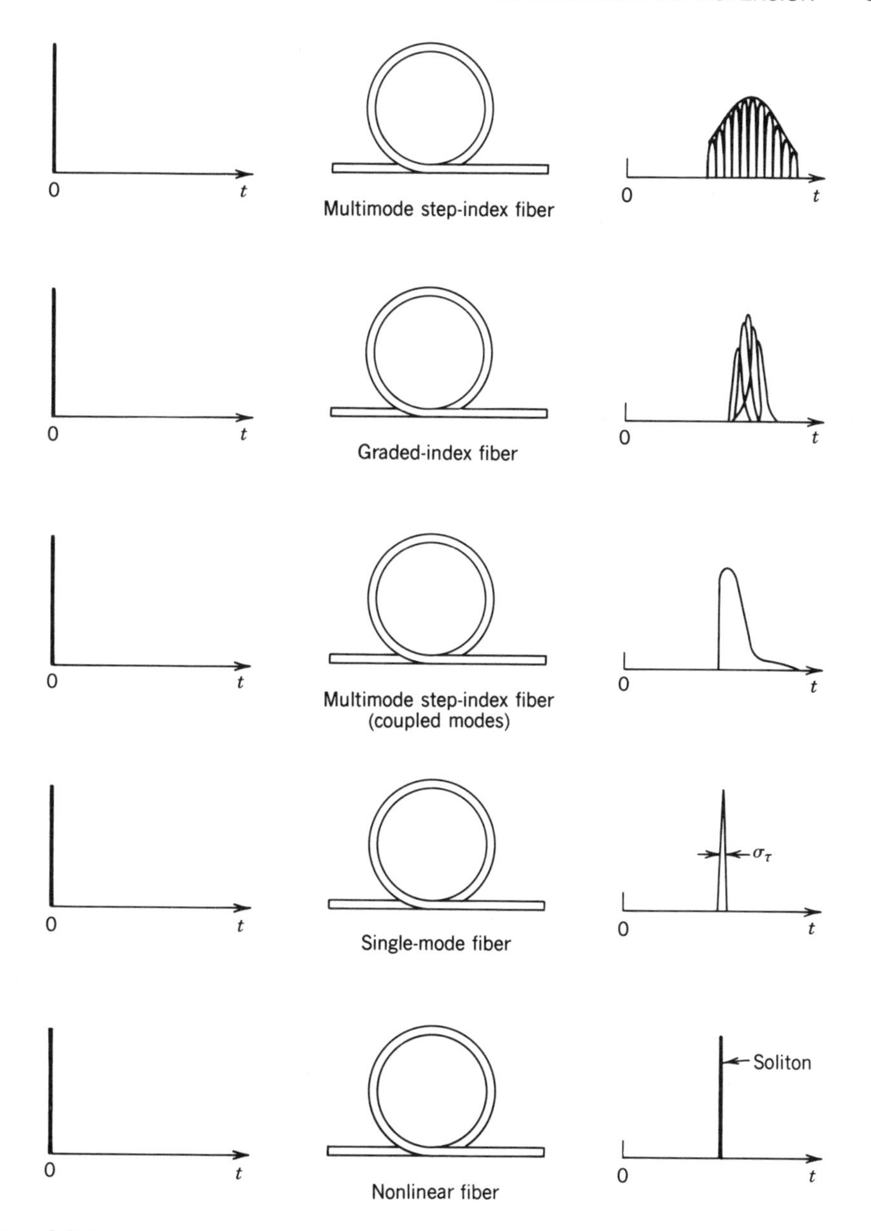

Figure 8.3-8 Broadening of a short optical pulse after transmission through different types of fibers. The width of the transmitted pulse is governed by modal dispersion in multimode (step-index and graded-index) fibers. In single-mode fibers the pulse width is determined by material dispersion and waveguide dispersion. Under certain conditions an intense pulse, called a soliton, can travel through a nonlinear fiber without broadening. This is a result of a balance between material dispersion and self-phase modulation (the dependence of the refractive index on the light intensity).

resulting from the finite spectral width of the initial optical pulse. The width σ_q is given by

$$\sigma_q^2 \approx \tau_0^2 + \left(D_q \sigma_\lambda L\right)^2, \tag{8.3-16}$$

where D_q is a dispersion coefficient representing the combined effects of material and waveguide dispersion for mode q. Material dispersion is usually more significant. For a very short initial width τ_0, (8.3-16) gives

$$\sigma_q \approx D_q \sigma_\lambda L. \tag{8.3-17}$$

Figure 8.3-8 is a schematic illustration in which the profiles of pulses traveling through different types of fibers are compared. In multimode step-index fibers, the modal dispersion $\frac{1}{2}(\tau_{\max} - \tau_{\min})$ is usually much greater than the material/waveguide dispersion σ_q, so that intermodal dispersion dominates and $\sigma_\tau \approx \frac{1}{2}(\tau_{\max} - \tau_{\min})$. In multimode graded-index fibers, $\frac{1}{2}(\tau_{\max} - \tau_{\min})$ may be comparable to σ_q, so that the overall pulse width involves all dispersion effects. In single-mode fibers, there is obviously no modal dispersion and the transmission of pulses is limited by material and waveguide dispersion. The lowest overall dispersion is achieved in a single-mode fiber operating at the wavelength for which the combined material–waveguide dispersion vanishes.

READING LIST

Books

See also the books on optical waveguides in Chapter 7.

P. K. Cheo, *Fiber Optics and Optoelectronics*, Prentice Hall, Englewood Cliffs, NJ, 1985, 2nd ed. 1990.

F. C. Allard, *Fiber Optics Handbook for Engineers and Scientists*, McGraw-Hill, New York, 1990.

C. Yeh, *Handbook of Fiber Optics: Theory and Applications*, Academic Press, Orlando, FL, 1990.

L. B. Jeunhomme, *Single-Mode Fiber Optics*, Marcel Dekker, New York, 1983, 2nd ed. 1990.

P. W. France, ed., *Fluoride Glass Optical Fibers*, CRC Press, Boca Raton, FL, 1989.

P. Diament, *Wave Transmission and Fiber Optics*, Macmillan, New York, 1989.

W. B. Jones, Jr., *Introduction to Optical Fiber Communication Systems*, Holt, Rinehart and Winston, New York, 1988.

H. Murata, *Handbook of Optical Fibers and Cables*, Marcel Dekker, New York, 1988.

E. G. Neuman, *Single-Mode Fibers—Fundamentals*, Springer-Verlag, New York, 1988.

E. L. Safford, Jr. and J. A. McCann, *Fiberoptics and Laser Handbook*, Tab Books, Blue Ridge Summit, PA, 2nd ed. 1988.

S. E. Miller and I. Kaminow, *Optical Fiber Telecommunications II*, Academic Press, Boston, MA, 1988.

J. Gowar, *Optical Communication Systems*, Prentice Hall, Englewood Cliffs, NJ, 1984.

R. G. Seippel, *Fiber Optics*, Reston Publishing, Reston, VA, 1984.

ANSI/IEEE Standards 812-1984, *IEEE Standard Definitions of Terms Relating to Fiber Optics*, IEEE, New York, 1984.

A. H. Cherin, *An Introduction to Optical Fibers*, McGraw-Hill, New York, 1983.

G. E. Keiser, *Optical Fiber Communications*, McGraw-Hill, New York, 1983.

C. Hentschel, *Fiber Optics Handbook*, Hewlett-Packard, Palo Alto, CA, 1983.

Y. Suematsu and K. Iga, *Introduction to Optical Fiber Communications*, Wiley, New York, 1982.

T. Okoshi, *Optical Fibers*, Academic Press, New York, 1982.

C. K. Kao, *Optical Fiber Systems*, McGraw-Hill, New York, 1982.

E. A. Lacy, *Fiber Optics*, Prentice-Hall, Englewood Cliffs, NJ, 1982.

D. Marcuse, *Light Transmission Optics*, Van Nostrand Reinhold, New York, 1972, 2nd ed. 1982.

D. Marcuse, *Principles of Optical Fiber Measurements*, Academic Press, New York, 1981.

A. B. Sharma, S. J. Halme, and M. M. Butusov, *Optical Fiber Systems and Their Components*, Springer-Verlag, Berlin, 1981.

CSELT (Centro Studi e Laboratori Telecomunicazioni), *Optical Fibre Communications*, McGraw-Hill, New York, 1981.

M. K. Barnoski, ed., *Fundamentals of Optical Fiber Communications*, Academic Press, New York, 1976, 2nd ed. 1981.

C. P. Sandbank, ed., *Optical Fibre Communication Systems*, Wiley, New York, 1980.

M. J. Howes and D. V. Morgan, eds., *Optical Fibre Communications*, Wiley, New York, 1980.

H. F. Wolf, ed., *Handbook of Fiber Optics*, Garland STPM Press, New York, 1979.

D. B. Ostrowsky, ed., *Fiber and Integrated Optics*, Plenum Press, New York, 1979.

J. E. Midwinter, *Optical Fibers for Transmission*, Wiley, New York, 1979.

S. E. Miller and A. G. Chynoweth, *Optical Fiber Telecommunications*, Academic Press, New York, 1979.

G. R. Elion and H. A. Elion, *Fiber Optics in Communication Systems*, Marcel Dekker, New York, 1978.

H. G. Unger, *Planar Optical Waveguides and Fibers*, Clarendon Press, Oxford, 1977.

J. A. Arnaud, *Beam and Fiber Optics*, Academic Press, New York, 1976.

W. B. Allan, *Fibre Optics: Theory and Practice*, Plenum Press, New York, 1973.

N. S. Kapany, *Fiber Optics: Principles and Applications*, Academic Press, New York, 1967.

Special Journal Issues

Special issue on fiber-optic sensors, *Journal of Lightwave Technology*, vol. LT-5, no. 7, 1987.

Special issue on fiber, cable, and splicing technology, *Journal of Lightwave Technology*, vol. LT-4, no. 8, 1986.

Special issue on low-loss fibers, *Journal of Lightwave Technology*, vol. LT-2, no. 10, 1984.

Special issue on fiber optics, *IEEE Transactions on Communications*, vol. COM-26, no. 7, 1978.

Articles

M. G. Drexhage and C. T. Moynihan, Infrared Optical Fibers, *Scientific American*, vol. 259, no. 5, pp. 110–114, 1988.

S. R. Nagel, Optical Fiber—the Expanding Medium, *IEEE Communications Magazine*, vol. 25, no. 4, pp. 33–43, 1987.

R. H. Stolen and R. P. DePaula, Single-Mode Fiber Components, *Proceedings of the IEEE*, vol. 75, pp. 1498–1511, 1987.

P. S. Henry, Lightwave Primer, *IEEE Journal of Quantum Electronics*, vol. QE-21, pp. 1862–1879, 1985.

T. Li, Advances in Optical Fiber Communications: An Historical Perspective, *IEEE Journal on Selected Areas in Communications*, vol. SAC-1, pp. 356–372, 1983.

I. P. Kaminow, Polarization in Optical Fibers, *IEEE Journal of Quantum Electronics*, vol. QE-17, pp. 15–22, 1981.

P. J. B. Clarricoats, Optical Fibre Waveguides—A Review, in *Progress in Optics*, vol. 14, E. Wolf, ed., North-Holland, Amsterdam, 1977.

D. Gloge, Weakly Guiding Fibers, *Applied Optics*, vol. 10, pp. 2252–2258, 1971.

D. Gloge, Dispersion in Weakly Guiding Fibers, *Applied Optics*, vol. 10, pp. 2442–2445, 1971.

PROBLEMS

8.1-1 **Coupling Efficiency.** (a) A source emits light with optical power P_0 and a distribution $I(\theta) = (1/\pi)P_0 \cos\theta$, where $I(\theta)$ is the power per unit solid angle in the direction making an angle θ with the axis of a fiber. Show that the power collected

by the fiber is $P = (\mathrm{NA})^2 P_0$, i.e., the coupling efficiency is NA^2 where NA is the numerical aperture of the fiber.
(b) If the source is a planar light-emitting diode of refractive index n_s bonded to the fiber, and assuming that the fiber cross-sectional area is larger than the LED emitting area, calculate the numerical aperture of the fiber and the coupling efficiency when $n_1 = 1.46$, $n_2 = 1.455$, and $n_s = 3.5$.

8.1-2 **Modes.** A step-index fiber has radius $a = 5\ \mu$m, core refractive index $n_1 = 1.45$, and fractional refractive-index change $\Delta = 0.002$. Determine the shortest wavelength λ_c for which the fiber is a single-mode waveguide. If the wavelength is changed to $\lambda_c/2$, identify the indices (l, m) of all the guided modes.

8.1-3 **Modal Dispersion.** A step-index fiber of numerical aperture NA = 0.16, core radius $a = 45\ \mu$m and core refractive index $n_1 = 1.45$ is used at $\lambda_o = 1.3\ \mu$m, where material dispersion is negligible. If a pulse of light of very short duration enters the fiber at $t = 0$ and travels a distance of 1 km, sketch the shape of the received pulse:
(a) Using ray optics and assuming that only meridional rays are allowed.
(b) Using wave optics and assuming that only meridional ($l = 0$) modes are allowed.

8.1-4 **Propagation Constants and Group Velocities.** A step-index fiber with refractive indices $n_1 = 1.444$ and $n_2 = 1.443$ operates at $\lambda_o = 1.55\ \mu$m. Determine the core radius at which the fiber V parameter is 10. Use Fig. 8.1-6 to estimate the propagation constants of all the guided modes with $l = 0$. If the core radius is now changed so that $V = 4$, use Fig. 8.1-10(b) to determine the propagation constant and the group velocity of the LP_{01} mode. *Hint*: Derive an expression for the group velocity $v = (d\beta/d\omega)^{-1}$ in terms of $d\beta/dV$ and use Fig. 8.1-10(b) to estimate $d\beta/dV$. Ignore the effect of material dispersion.

8.2-1 **Numerical Aperture of a Graded-Index Fiber.** Compare the numerical apertures of a step-index fiber with $n_1 = 1.45$ and $\Delta = 0.01$ and a graded-index fiber with $n_1 = 1.45$, $\Delta = 0.01$, and a parabolic refractive-index profile ($p = 2$). (See Exercise 1.3-2 on page 24.)

8.2-2 **Propagation Constants and Wavevector (Step-Index Fiber).** A step-index fiber of radius $a = 20\ \mu$m and refractive indices $n_1 = 1.47$ and $n_2 = 1.46$ operates at $\lambda_o = 1.55\ \mu$m. Using the quasi-plane wave theory and considering only guided modes with azimuthal index $l = 1$:
(a) Determine the smallest and largest propagation constants.
(b) For the mode with the smallest propagation constant, determine the radii of the cylindrical shell within which the wave is confined, and the components of the wavevector **k** at $r = 5\ \mu$m.

8.2-3 **Propagation Constants and Wavevector (Graded-Index Fiber).** Repeat Problem 8.2-2 for a graded-index fiber with parabolic refractive-index profile with $p = 2$.

8.3-1 **Scattering Loss.** At $\lambda_o = 820$ nm the absorption loss of a fiber is 0.25 dB/km and the scattering loss is 2.25 dB/km. If the fiber is used instead at $\lambda_o = 600$ nm and calorimetric measurements of the heat generated by light absorption give a loss of 2 dB/km, estimate the total attenuation at $\lambda_o = 600$ nm.

8.3-2 **Modal Dispersion in Step-Index Fibers.** Determine the core radius of a multimode step-index fiber with a numerical aperture NA = 0.1 if the number of modes $M = 5000$ when the wavelength is 0.87 μm. If the core refractive index $n_1 = 1.445$, the group index $N_1 = 1.456$, and Δ is approximately independent of wavelength, determine the modal-dispersion response time σ_τ for a 2-km fiber.

8.3-3 **Modal Dispersion in Graded-Index Fibers.** Consider a graded-index fiber with $a/\lambda_o = 10$, $n_1 = 1.45$, $\Delta = 0.01$, and a power-law profile with index p. Determine the number of modes M, and the modal-dispersion pulse-broadening rate σ_τ/L for $p = 1.9$, 2, 2.1, and ∞.

8.3-4 **Pulse Propagation.** A pulse of initial width τ_0 is transmitted through a graded-index fiber of length L kilometers and power-law refractive-index profile with profile index p. The peak refractive index n_1 is wavelength-dependent with $D_\lambda = -(\lambda_o/c_o)d^2n_1/d\lambda_o^2$, Δ is approximately independent of wavelength, σ_λ is the source's spectral width, and λ_o is the operating wavelength. Discuss the effect of increasing each of the following parameters on the width of the received pulse: L, τ_0, $p, |D_\lambda|, \sigma_\lambda$, and λ_o.

CHAPTER

9

RESONATOR OPTICS

9.1 PLANAR-MIRROR RESONATORS
- A. Resonator Modes
- B. The Resonator as a Spectrum Analyzer
- C. Two- and Three-Dimensional Resonators

9.2 SPHERICAL-MIRROR RESONATORS
- A. Ray Confinement
- B. Gaussian Modes
- C. Resonance Frequencies
- D. Hermite - Gaussian Modes
- *E. Finite Apertures and Diffraction Loss

Charles Fabry (1867–1945)

Alfred Perot (1863–1925)

Fabry and Perot constructed an optical resonator for use as an interferometer. Now known as the Fabry–Perot etalon, it is used extensively in lasers.

An optical resonator, the optical counterpart of an electronic resonant circuit, confines and stores light at certain resonance frequencies. It may be viewed as an optical transmission system incorporating feedback; light circulates or is repeatedly reflected within the system, without escaping. The simplest resonator comprises two parallel planar mirrors between which light is repeatedly reflected with little loss. Typical optical resonator configurations are depicted in Fig. 9.0-1.

The frequency selectivity of an optical resonator makes it useful as an optical filter or spectrum analyzer. Its most important use, however, is as a "container" within which laser light is generated. The laser is an optical resonator containing a medium that amplifies light. The resonator determines the frequency and spatial distribution of the laser beam. Because resonators have the capability of storing energy, they can also be used to generate pulses of laser energy. Lasers are discussed in Chap. 14; the material in this chapter is essential to their understanding.

Several approaches are useful for describing the operation of an optical resonator:

- The simplest approach is based on *ray optics* (Chap. 1). Optical rays are traced as they reflect within the resonator; the geometrical conditions under which they remain confined are determined.
- *Wave optics* (Chap. 2) is used to determine the modes of the resonator, i.e., the resonance frequencies and wavefunctions of the optical waves that exist self-consistently within the resonator. This analysis is similar to that used in Sec. 7.1 to determine the modes of a planar-mirror waveguide.

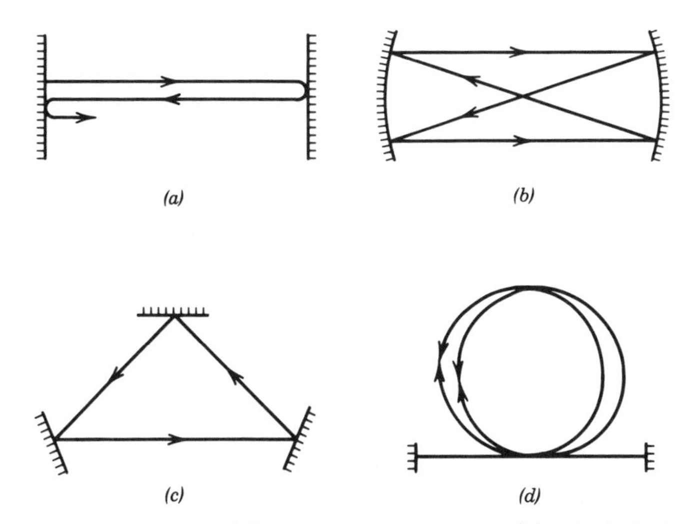

Figure 9.0-1 Optical resonators: (*a*) planar-mirror resonator; (*b*) spherical-mirror resonator; (*c*) ring resonator; (*d*) optical-fiber resonator.

- The modes of a resonator with spherical mirrors are Gaussian and Hermite–Gaussian optical beams. The study of *beam optics* (Chap. 3) is therefore useful for understanding the behavior of spherical-mirror resonators.
- *Fourier optics* and the theory of propagation and diffraction of light (Chap. 4) are necessary for understanding the effect of the finite size of the resonator's mirrors on its loss and on the spatial distribution of the modes.

The optical resonator evidently provides an excellent arena for applying the different theories of light presented in earlier chapters.

9.1 PLANAR-MIRROR RESONATORS

A. Resonator Modes

In this section we examine the modes of a resonator constructed of two parallel, highly reflective, flat mirrors separated by a distance d (Fig. 9.1-1). This simple one-dimensional resonator is known as a **Fabry–Perot etalon**. We first consider an ideal resonator whose mirrors are lossless; the effect of losses is included subsequently.

Resonator Modes as Standing Waves

A monochromatic wave of frequency ν has a wavefunction

$$u(\mathbf{r}, t) = \mathrm{Re}\{U(\mathbf{r}) \exp(j2\pi\nu t)\},$$

which represents the transverse component of the electric field. The complex amplitude $U(\mathbf{r})$ satisfies the Helmholtz equation, $\nabla^2 U + k^2 U = 0$, where $k = 2\pi\nu/c$ is the wavenumber and c is the speed of light in the medium (see Secs. 2.2, 5.3, and 5.4). The modes of a resonator are the basic solutions of the Helmholtz equation subject to the appropriate boundary conditions. For the planar-mirror resonator, the transverse components of the electric field vanish at the surfaces of the mirrors, so that $U(\mathbf{r}) = 0$ at the planes $z = 0$ and $z = d$ in Fig. 9.1-2. The standing wave

$$U(\mathbf{r}) = A \sin kz, \tag{9.1-1}$$

where A is a constant, satisfies the Helmholtz equation and vanishes at $z = 0$ and $z = d$ if k satisfies the condition $kd = q\pi$, where q is an integer. This restricts k to

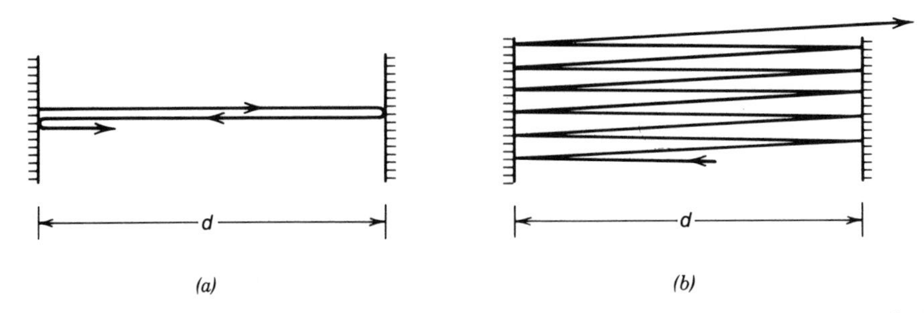

Figure 9.1-1 Two-mirror planar resonator (Fabry–Perot etalon). (*a*) Light rays perpendicular to the mirrors reflect back and forth without escaping. (*b*) Rays that are only slightly inclined eventually escape. Rays also escape if the mirrors are not perfectly parallel.

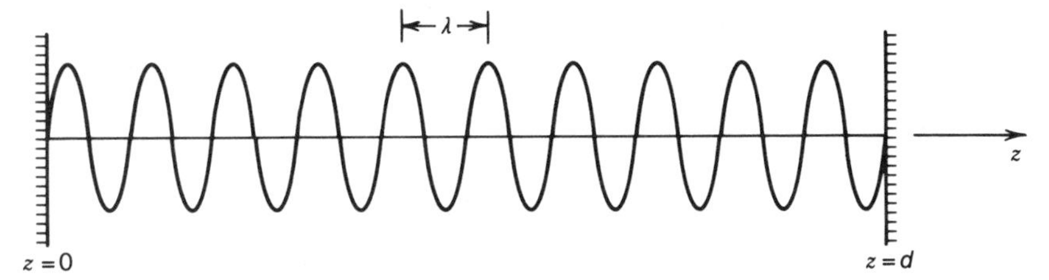

Figure 9.1-2 Complex amplitude of a resonator mode ($q = 20$).

the values

$$k_q = \frac{q\pi}{d}, \tag{9.1-2}$$

so that the modes have complex amplitudes $U(\mathbf{r}) = A_q \sin k_q z$, where the A_q are constants. Negative values of q do not constitute independent modes since $\sin k_{-q} z = -\sin k_q z$. The value $q = 0$ is associated with a mode that carries no energy since $k_0 = 0$ and $\sin k_0 z = 0$. The modes of the resonator are therefore the standing waves $A_q \sin k_q z$, where the positive integer $q = 1, 2, \ldots$ is called the mode number. An arbitrary wave inside the resonator can be written as a superposition of the resonator modes, $U(\mathbf{r}) = \Sigma_q A_q \sin k_q z$.

It follows from (9.1-2) that the frequency $\nu = ck/2\pi$ is restricted to the discrete values

$$\nu_q = q\frac{c}{2d}, \qquad q = 1, 2, \ldots, \tag{9.1-3}$$

which are the resonance frequencies of the resonator. As shown in Fig. 9.1-3 adjacent resonance frequencies are separated by a constant frequency difference

$$\boxed{\nu_F = \frac{c}{2d}.} \tag{9.1-4}$$

Frequency Spacing of Adjacent Resonator Modes

The resonance wavelengths are, of course, $\lambda_q = c/\nu_q = 2d/q$. At resonance, the length of the resonator, $d = q\lambda/2$, is an integer number of half wavelengths. Note that $c = c_o/n$ is the speed of light in the medium embedded between the two mirrors, and the λ_q represent wavelengths in the medium.

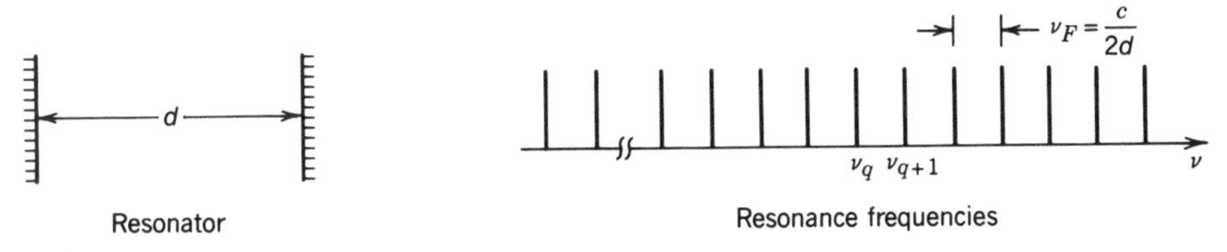

Figure 9.1-3 The resonance frequencies of a planar-mirror resonator are separated by $\nu_F = c/2d$. If the resonator is 15 cm long (d = 15 cm) and $n = 1$, for example, then $\nu_F = 1$ GHz.

Resonator Modes as Traveling Waves

The resonator modes can alternatively be determined by following a wave as it travels back and forth between the two mirrors [Fig. 9.1-4(*a*)]. A mode is a self-reproducing wave, i.e., a wave that reproduces itself after a single round trip (see Appendix C). The phase shift imparted by the two mirror reflections is 0 or 2π (π at each mirror). The phase shift imparted by a single round trip of propagation (a distance $2d$), $\varphi = k2d = 4\pi\nu d/c$, must therefore be a multiple of 2π,

$$\varphi = k2d = q2\pi, \qquad q = 1, 2, \ldots. \tag{9.1-5}$$

This leads to the relation $kd = q\pi$ and the resonance frequencies in (9.1-3). Equation (9.1-5) may be regarded as a condition of positive feedback in the system shown in Fig. 9.1-4(*b*); this requires that the output of the system be fed back *in phase* with the input.

We now show that only self-reproducing waves, or combinations thereof, can exist within the resonator in the steady state. Consider a monochromatic plane wave of

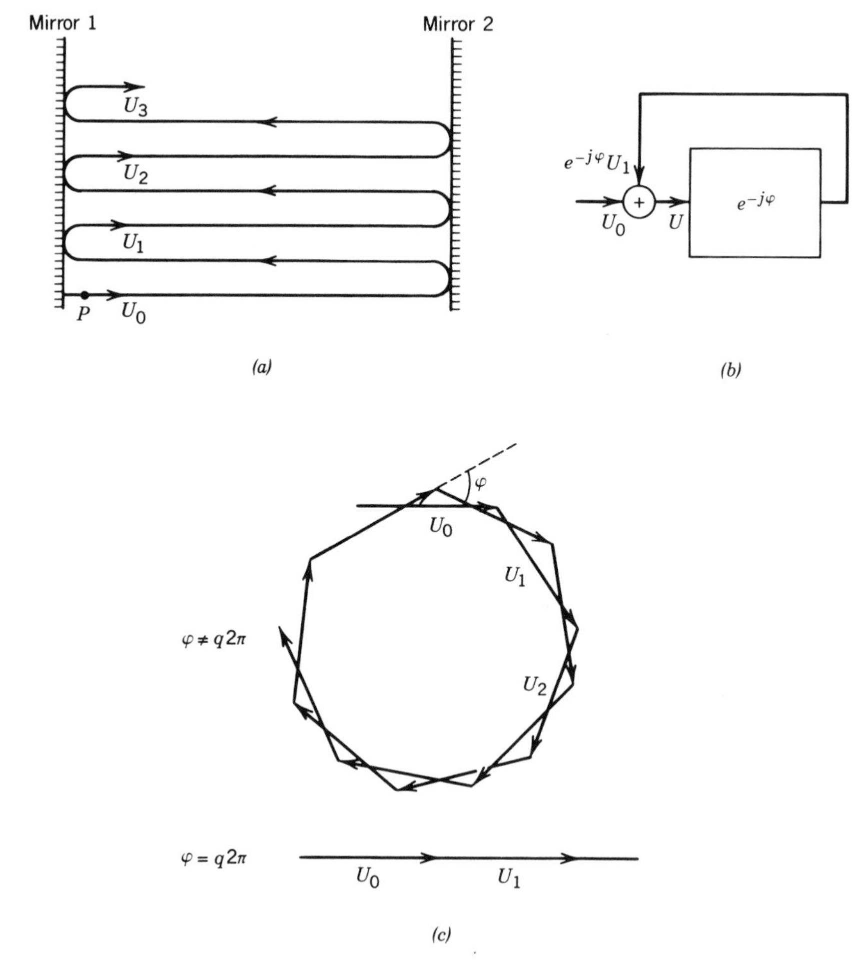

Figure 9.1-4 (*a*) A wave reflects back and forth between the resonator mirrors, suffering a phase shift φ each round trip. (*b*) Block diagram representing the optical feedback system. (*c*) Phasor diagram representing the sum $U = U_0 + U_1 + \cdots$ for $\varphi \neq q2\pi$ and $\varphi = q2\pi$.

complex amplitude U_0 at point P traveling to the right along the axis of the resonator [see Fig. 9.1-4(*a*)]. The wave is reflected from mirror 2 and propagates back to mirror 1 where it is again reflected. Its amplitude at P then becomes U_1. Yet another round trip results in a wave of complex amplitude U_2, and so on *ad infinitum*. Because the original wave U_0 is monochromatic, it is "eternal." Indeed, all of the partial waves $U_0, U_1, U_2, \dots$ are monochromatic and perpetually coexist. Furthermore, their magnitudes are identical because there is no loss associated with the reflection and propagation. The total wave U is therefore represented by the sum of an infinite number of phasors of equal magnitude,

$$U = U_0 + U_1 + U_2 + \cdots, \tag{9.1-6}$$

as shown in Fig. 9.1-4(*c*).

The phase difference of two consecutive phasors imparted by a single round trip of propagation is $\varphi = k2d$. If the magnitude of the initial phasor is infinitesimal, the magnitude of each of these phasors must be infinitesimal. The magnitude of the sum of this infinite number of infinitesimal phasors is itself infinitesimal unless they are aligned, i.e., unless $\varphi = q2\pi$. Thus, an infinitesimal initial wave can result in the buildup of finite power in the resonator, but only if $\varphi = q2\pi$.

EXERCISE 9.1-1

Resonance Frequencies of a Ring Resonator. Derive an expression for the resonance frequencies of the three-mirror ring resonator shown in Fig. 9.1-5. Assume that each mirror reflection introduces a phase shift of π.

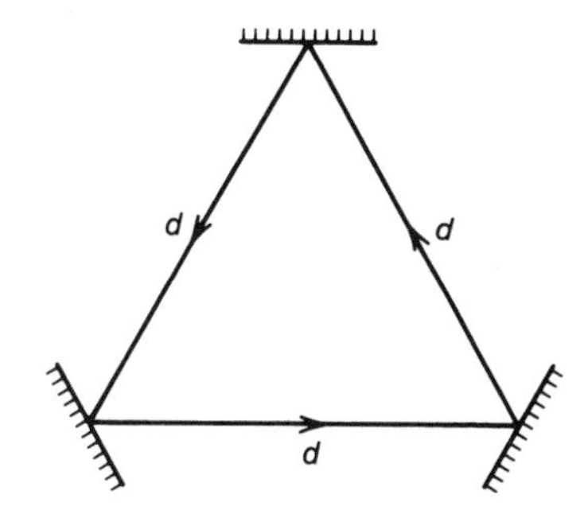

Figure 9.1-5 Three-mirror ring resonator.

Density of Modes

The number of modes per unit frequency is $1/\nu_F = 2d/c$ in each of the two orthogonal polarizations. Thus the density of modes $M(\nu)$, which is the number of modes per unit frequency per unit length of the resonator, is

$$\boxed{M(\nu) = \frac{4}{c}.} \tag{9.1-7}$$

Density of Modes
(One-Dimensional Resonator)

The number of modes in a resonator of length d within the frequency interval $\Delta\nu$ is therefore $(4/c)d\,\Delta\nu$. This represents the number of degrees of freedom for the optical waves existing in the resonator, i.e., the number of independent ways in which these waves may be arranged.

Losses and Resonance Spectral Width

The strict condition on the frequencies of optical waves that are permitted to exist inside a resonator is relaxed when the resonator has losses. Consider again Fig. 9.1-4(*a*) and follow a wave U_0 in its excursions between the two mirrors. The result is an infinite sum of phasors as in Fig. 9.1-4(*c*). As previously, the phase difference imparted by a single propagation round trip is

$$\varphi = 2kd = \frac{4\pi\nu d}{c}. \tag{9.1-8}$$

In the presence of loss, however, the phasors are not of equal magnitude. The magnitude ratio of two consecutive phasors is the round-trip amplitude attenuation factor $\mathscr{r}$ introduced by the two mirror reflections and by absorption in the medium. The intensity attenuation factor is therefore $\mathscr{r}^2$. Thus $U_1 = hU_0$, where $h = \mathscr{r}e^{-j\varphi}$. The phasor U_2 is related to U_1 by this same complex factor h, as are all consecutive phasors. The net result is the superposition of an infinite number of waves, separated by equal phase shifts, but with amplitudes that are geometrically reduced. It is readily seen that $U = U_0 + U_1 + U_2 + \cdots = U_0 + hU_0 + h^2U_0 + \cdots = U_0(1 + h + h^2 + \cdots) = U_0/(1 - h)$. The relation $U = U_0/(1 - h)$ may be easily verified from the simple feedback configuration provided in Fig. 9.1-4(*b*). The intensity in the resonator

$$I = |U|^2 = |U_0|^2/|1 - \mathscr{r}e^{-j\varphi}|^2 = I_0\big/\left[(1 - \mathscr{r}\cos\varphi)^2 + (\mathscr{r}\sin\varphi)^2\right]$$

$$= I_0\big/\left(1 + \mathscr{r}^2 - 2\mathscr{r}\cos\varphi\right) = I_0\big/\left[(1 - \mathscr{r})^2 + 4\mathscr{r}\sin^2(\varphi/2)\right]$$

is found to be

$$I = \frac{I_{\max}}{1 + (2\mathscr{F}/\pi)^2\sin^2(\varphi/2)}, \qquad I_{\max} = \frac{I_0}{(1 - \mathscr{r})^2}. \tag{9.1-9}$$

Here $I_0 = |U_0|^2$ is the intensity of the initial wave, and

$$\mathscr{F} = \frac{\pi\mathscr{r}^{1/2}}{1 - \mathscr{r}} \tag{9.1-10}$$

is a parameter known as the **finesse** of the resonator.

The intensity I is a periodic function of φ with period 2π. If $\mathscr{F}$ is large, then I has sharp peaks centered about the values $\varphi = q2\pi$ (when all the phasors are aligned). The peaks have a full width at half maximum (FWHM) given by $\Delta\varphi = 2\pi/\mathscr{F}$. The width of each peak is $\mathscr{F}$ times smaller than the period. The treatment given here is not unlike that provided in Sec. 2.5B on pages 70–72. One superficial difference (that has no bearing on the results) is the choice of $h = \mathscr{r}e^{-j\varphi}$, which is used since successive phasors arise from the delay of the wave as it bounces between the mirrors.

The dependence of I on ν, which is the spectral response of the resonator, has a similar periodic behavior since $\varphi = 4\pi\nu d/c$ is proportional to ν. This resonance profile,

$$I = \frac{I_{\max}}{1 + (2\mathscr{F}/\pi)^2 \sin^2(\pi\nu/\nu_F)}, \tag{9.1-11}$$

Spectral Response of the Fabry–Perot Resonator

is shown in Fig. 9.1-6, where $\nu_F = c/2d$. The maximum $I = I_{\max}$ is achieved at the resonance frequencies

$$\nu = \nu_q = q\nu_F, \qquad q = 1, 2, \ldots, \tag{9.1-12}$$

whereas the minimum value

$$I_{\min} = \frac{I_{\max}}{1 + (2\mathscr{F}/\pi)^2}, \tag{9.1-13}$$

occurs at the midpoints between the resonances. When the finesse is large ($\mathscr{F} \gg 1$), the resonator spectral response is sharply peaked about the resonance frequencies and $I_{\min}/I_{\max}$ is small. In that case, the FWHM of the resonance peak is $\delta\nu = (c/4\pi d)\,\Delta\varphi = \nu_F/\mathscr{F}$, as was shown in Sec. 2.5B.

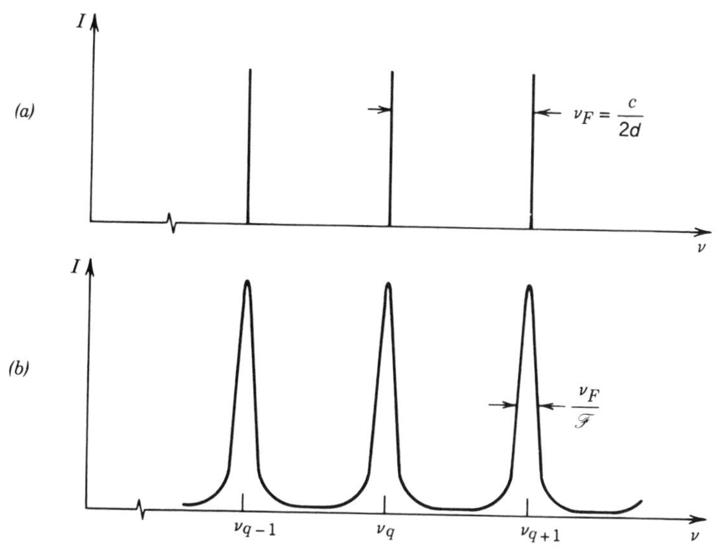

Figure 9.1-6 (*a*) A lossless resonator ($\mathscr{F} = \infty$) in the steady state can sustain light waves only at the precise resonance frequencies ν_q. (*b*) A lossy resonator sustains waves at all frequencies, but the attenuation resulting from destructive interference increases at frequencies away from the resonances.

In short, two parameters characterize the spectral response of the Fabry–Perot resonator:

- The spacing between adjacent resonance frequencies

$$\nu_F = \frac{c}{2d}; \tag{9.1-14}$$

Frequency Spacing of Adjacent Resonator Modes

- The width of the resonances $\delta\nu$. When $\mathscr{F} \gg 1$,

$$\delta\nu \approx \frac{\nu_F}{\mathscr{F}}. \tag{9.1-15}$$

Spectral Width of Resonator Modes

The resonance linewidth is inversely proportional to the finesse. Since the finesse decreases with increasing loss, the spectral width increases with increasing loss.

Sources of Resonator Loss

The two principal sources of loss in optical resonators are:

- Losses attributable to absorption and scattering in the medium between the mirrors. The round-trip power attenuation factor associated with these processes is $\exp(-2\alpha_s d)$, where α_s is the absorption coefficient of the medium.
- Losses arising from imperfect reflection at the mirrors. There are two underlying sources of reduced reflection: (1) A partially transmitting mirror is often used in a resonator to permit light to escape from it; and (2) the finite size of the mirrors causes a fraction of the light to leak around the mirrors and thereby to be lost. This also modifies the spatial distribution of the reflected wave by truncating it to the size of the mirror. The reflected light produces a diffraction pattern at the opposite mirror which is again truncated. Such diffraction loss may be regarded as an effective reduction of the mirror reflectance. Further details regarding diffraction loss are provided in Sec. 9.2E.

For mirrors of reflectances $\mathscr{R}_1 = r_1^2$ and $\mathscr{R}_2 = r_2^2$, the wave intensity decreases by the factor $\mathscr{R}_1\mathscr{R}_2$ in the course of the two reflections associated with a single round trip. The overall intensity attenuation factor is therefore

$$r^2 = \mathscr{R}_1\mathscr{R}_2 \exp(-2\alpha_s d), \tag{9.1-16}$$

which is usually written in the form

$$r^2 = \exp(-2\alpha_r d), \tag{9.1-17}$$

where α_r is an effective overall distributed-loss coefficient. Equations (9.1-16) and (9.1-17) provide

$$\alpha_r = \alpha_s + \frac{1}{2d} \ln \frac{1}{\mathscr{R}_1\mathscr{R}_2}. \tag{9.1-18}$$

Loss Coefficient

This can also be written as

$$\alpha_r = \alpha_s + \alpha_{m1} + \alpha_{m2},$$

where the quantities

$$\alpha_{m1} = \frac{1}{2d} \ln \frac{1}{\mathscr{R}_1}, \qquad \alpha_{m2} = \frac{1}{2d} \ln \frac{1}{\mathscr{R}_2}$$

represent the loss coefficients attributed to mirrors 1 and 2, respectively.

The loss coefficient can be cast in a simpler form for mirrors of high reflectance. If $\mathscr{R}_1 \approx 1$, then $\ln(1/\mathscr{R}_1) = -\ln(\mathscr{R}_1) = -\ln[1 - (1 - \mathscr{R}_1)] \approx 1 - \mathscr{R}_1$, where we have used the approximation $\ln(1 - \Delta) \approx -\Delta$, which is valid for $|\Delta| \ll 1$. This allows us to write

$$\alpha_{m1} \approx \frac{1 - \mathscr{R}_1}{2d}. \tag{9.1-19}$$

Similarly, if $\mathscr{R}_2 \approx 1$, we have $\alpha_{m2} \approx (1 - \mathscr{R}_2)/2d$. If, furthermore, $\mathscr{R}_1 = \mathscr{R}_2 = \mathscr{R} \approx 1$, then

$$\alpha_r \approx \alpha_s + \frac{1 - \mathscr{R}}{d}. \tag{9.1-20}$$

The finesse $\mathscr{F}$ can be expressed as a function of the effective loss coefficient α_r by substituting (9.1-17) in (9.1-10), which provides

$$\mathscr{F} = \frac{\pi \exp(-\alpha_r d/2)}{1 - \exp(-\alpha_r d)}. \tag{9.1-21}$$

The finesse decreases with increasing loss, as shown in Fig. 9.1-7. If the loss factor

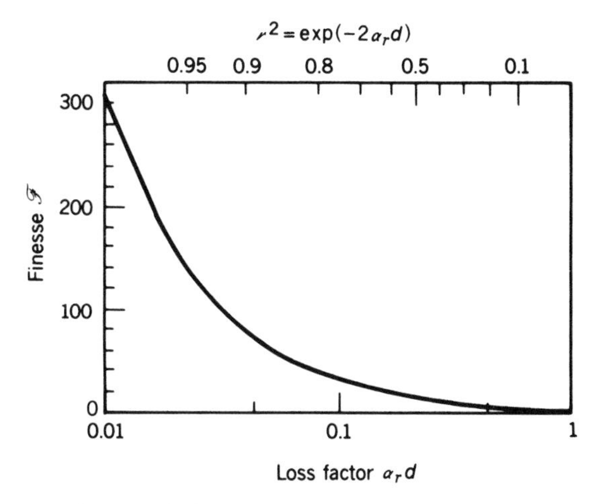

Figure 9.1-7 Finesse of an optical resonator versus the loss factor $\alpha_r d$. The round-trip attenuation factor $r^2 = \exp(-2\alpha_r d)$.

$\alpha_r d \ll 1$, then $\exp(-\alpha_r d) \approx 1 - \alpha_r d$, whereupon

$$\mathscr{F} \approx \frac{\pi}{\alpha_r d}. \tag{9.1-22}$$

Relation Between Finesse and Loss Factor

This demonstrates that the finesse is inversely proportional to the loss factor $\alpha_r d$ in this limit.

EXERCISE 9.1-2

Resonator Modes and Spectral Width. Determine the frequency spacing, and spectral width, of the modes of a Fabry–Perot resonator whose mirrors have reflectances 0.98 and 0.99 and are separated by a distance $d = 100$ cm. The medium has refractive index $n = 1$ and negligible losses. Is the approximation used to derive (9.1-22) appropriate in this case?

Photon Lifetime

The relationship between the resonance linewidth and the resonator loss may be viewed as a manifestation of the time–frequency uncertainty relation. Substituting (9.1-14) and (9.1-22) in (9.1-15), we obtain

$$\delta\nu \approx \frac{c/2d}{\pi/\alpha_r d} = \frac{c\alpha_r}{2\pi}. \tag{9.1-23}$$

Because α_r is the loss per unit length, $c\alpha_r$ is the loss per unit time. Defining the characteristic decay time

$$\tau_p = \frac{1}{c\alpha_r} \tag{9.1-24}$$

as the **resonator lifetime** or **photon lifetime**, we obtain

$$\delta\nu = \frac{1}{2\pi\tau_p}. \tag{9.1-25}$$

The time-frequency uncertainty product is therefore $\delta\nu \cdot \tau_p = 1/2\pi$. The resonance line broadening is seen to be governed by the decay of optical energy arising from resonator losses. An electric field that decays as $\exp(-t/2\tau_p)$, which corresponds to an energy that decays as $\exp(-t/\tau_p)$, has a Fourier transform that is proportional to $1/(1 + j4\pi\nu\tau_p)$ with a FWHM spectral width $\delta\nu = 1/2\pi\tau_p$.

In summary, three parameters are convenient for characterizing the losses in an optical resonator of length d: the finesse $\mathscr{F}$, the loss coefficient α_r (cm^{-1}), and the photon lifetime $\tau_p = 1/c\alpha_r$ (seconds). In addition, the quality factor Q can also be used for this purpose, as outlined below.

*The Quality Factor Q

The quality factor Q is often used to characterize electrical resonance circuits and microwave resonators. This parameter is defined as

$$Q = \frac{2\pi \text{ (stored energy)}}{\text{energy loss per cycle}}.$$

Large Q factors are associated with low-loss resonators. A series RLC circuit has resonance frequency $\nu_0 \approx 1/2\pi(LC)^{1/2}$ and quality factor $Q = 2\pi\nu_0 L/R$, where R, C, and L are the resistance, capacitance, and inductance of the resonance circuit, respectively.

The Q factor of an optical resonator may be determined by observing that stored energy is lost at the rate $c\alpha_r$ (per unit time), which is equivalent to the rate $c\alpha_r/\nu_0$ (per cycle), so that $Q = 2\pi[1/(c\alpha_r/\nu_0)]$. Since $\delta\nu = c\alpha_r/2\pi$,

$$Q = \frac{\nu_0}{\delta\nu}. \tag{9.1-26}$$

The quality factor is related to the resonator lifetime (photon lifetime) $\tau_p = 1/c\alpha_r$ by

$$Q = 2\pi\nu_0\tau_p. \tag{9.1-27}$$

By using (9.1-15), we find that Q is related to the finesse of the resonator by

$$Q = \frac{\nu_0}{\nu_F}\mathscr{F}. \tag{9.1-28}$$

Since optical resonator frequencies ν_0 are typically much greater than the mode spacing ν_F, $Q \gg \mathscr{F}$. The quality factor of an optical resonator is typically far greater than that of a resonator at microwave frequencies.

B. The Resonator as a Spectrum Analyzer

What fraction of the intensity of an optical wave of frequency ν incident on a Fabry–Perot etalon is transmitted through it? We proceed to demonstrate that the transmittance is high if the frequency of the optical wave coincides with one of the resonance frequencies ($\nu = \nu_q$). The attenuation at other frequencies depends on the lossiness of the resonator. A low-loss resonator can therefore be used as a spectrum analyzer.

A plane wave of complex amplitude U_i and intensity I_i entering a resonator undergoes multiple reflections and transmissions, as illustrated in Fig. 9.1-8. Defining the complex amplitude and intensity of the transmitted wave as U_t and I_t, respectively, we proceed to obtain an expression for the intensity transmittance $\mathscr{T}(\nu) = I_t/I_i$, as a function of the frequency of the wave ν.

Let r_1 and r_2 be the amplitude reflectances of the inner surfaces of mirrors 1 and 2, and t_1 and t_2 the amplitude transmittances of the mirrors, respectively. In accordance with our previous analysis, the intensity I of the sum U of the internal waves $U_0, U_1, \ldots$ is related to the intensity I_0 of the initial wave U_0 by (9.1-9), with $r = r_1 r_2$. The transmitted intensity I_t is, however, related to the total internal intensity by $I_t = |t_2|^2 I$, while the initial intensity I_0 is related to the incident intensity by $I_0 = |t_1|^2 I_i$. Thus $I_t/I_i = |t|^2(I/I_0)$, where $t = t_1 t_2$. Finally, using (9.1-9), we obtain an

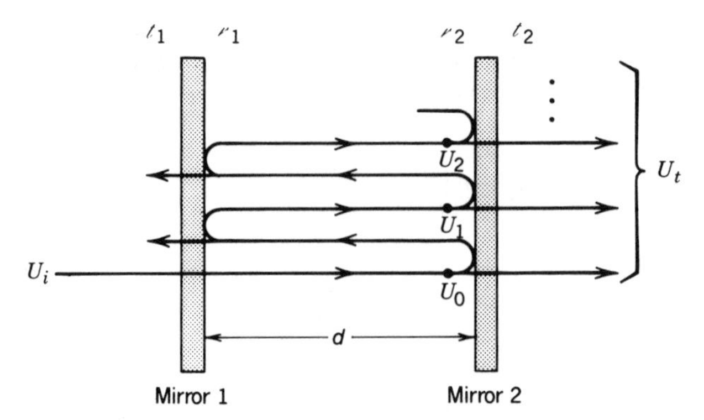

Figure 9.1-8 Transmission of a plane wave across a planar-mirror resonator (Fabry–Perot etalon).

expression for $\mathscr{T}(\nu) = I_t/I_i$:

$$\mathscr{T}(\nu) = \frac{\mathscr{T}_{\max}}{1 + (2\mathscr{F}/\pi)^2 \sin^2(\pi\nu/\nu_F)}, \tag{9.1-29}$$

Transmittance of a Fabry – Perot Resonator

where

$$\mathscr{T}_{\max} = \frac{|t|^2}{(1-r)^2}, \qquad t = t_1 t_2, \quad r = r_1 r_2, \tag{9.1-30}$$

and again

$$\mathscr{F} = \frac{\pi r^{1/2}}{1-r}. \tag{9.1-31}$$

We conclude that the resonator transmittance $\mathscr{T}(\nu)$ has the same dependence on ν as that of the internal wave—sharply peaked functions surrounding the resonance frequencies. The width of each of these resonance peaks is a factor $\mathscr{F}$ smaller than the spacing between them.

A Fabry–Perot etalon may therefore be used as a sharply tuned optical filter or spectrum analyzer. Because of the periodic nature of the spectral response, however, the spectral width of the measured light must be narrower than the frequency spacing $\nu_F = c/2d$ in order to avoid ambiguity. The quantity ν_F is therefore known as the **free spectral range**. The filter is tuned (i.e., the resonance frequencies are shifted) by adjusting the distance d between the mirrors. A slight change in mirror spacing Δd shifts the resonance frequency $\nu_q = qc/2d$ by a relatively large amount $\Delta\nu_q = -(qc/2d^2)\,\Delta d = -\nu_q\,\Delta d/d$. Although the frequency spacing ν_F also changes, it is by the far smaller amount $-\nu_F\,\Delta d/d$. Using an example with mirror separation $d = 1.5$ cm leads to a free spectral range $\nu_F = 10$ GHz when $n = 1$. For a typical optical frequency ($\nu = 10^{14}$ Hz), a change of d by a factor of 10^{-4} ($\Delta d = 1.5$ μm) translates the peak frequency by $\Delta\nu_q = 10$ GHz, whereas the free spectral range is altered by only 1 MHz becoming 9.999 GHz.

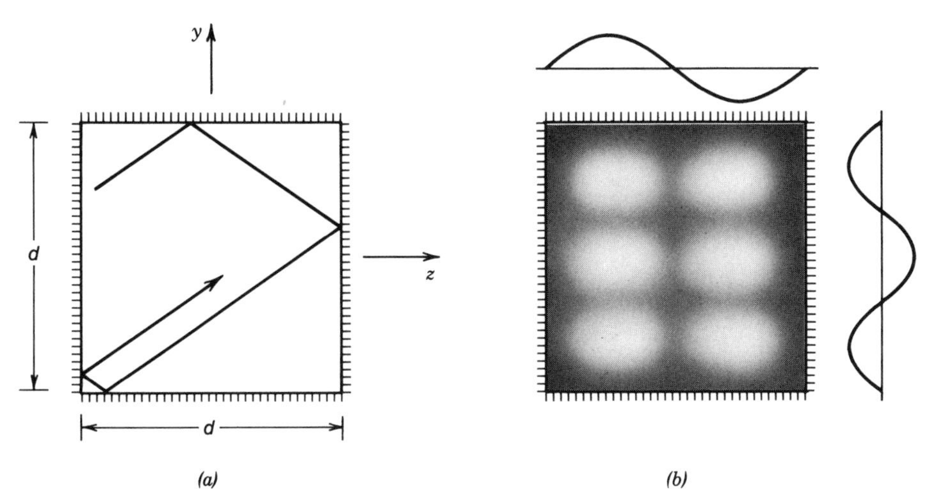

Figure 9.1-9 A two-dimensional planar-mirror resonator: (*a*) ray pattern; (*b*) standing-wave pattern with mode numbers $q_y = 3$ and $q_z = 2$.

C. Two- and Three-Dimensional Resonators†

Two-Dimensional Resonators

A two-dimensional planar-mirror resonator is constructed from two orthogonal pairs of parallel mirrors, e.g., a pair normal to the z axis and another pair normal to the y axis. Light is confined by a sequence of ray reflections in the z-y plane as illustrated in Fig. 9.1-9(*a*).

As for the one-dimensional Fabry–Perot resonator, the boundary conditions establish the resonator modes. If the mirror spacing is d, then the components of the wavevector $\mathbf{k} = (k_y, k_z)$ for standing waves are restricted to the values

$$k_y = \frac{q_y \pi}{d}, \quad k_z = \frac{q_z \pi}{d}, \qquad q_y = 1, 2, \ldots, \quad q_z = 1, 2, \ldots, \tag{9.1-32}$$

where q_y and q_z are mode numbers in the y and z directions, respectively. This condition is a generalization of (9.1-2). Each pair of integers (q_y, q_z) represents a resonator mode $U(\mathbf{r}) \propto \sin(q_y \pi y/d)\sin(q_z \pi z/d)$, as illustrated in Fig. 9.1-9(*b*). The lowest-order mode is the $(1, 1)$ mode since the modes $(q_y, 0)$ and $(0, q_z)$ have zero amplitude, viz., $U(\mathbf{r}) = 0$. Modes are conveniently represented by dots that mark their values of k_y and k_z on a periodic lattice of spacing π/d (Fig. 9.1-10).

The wavenumber k of a mode is the distance of the dot from the origin. Its frequency is $\nu = ck/2\pi$. The frequencies of the resonator modes are determined by using the relation

$$k^2 = k_y^2 + k_z^2 = \left(\frac{2\pi\nu}{c}\right)^2. \tag{9.1-33}$$

The number of modes in a given frequency band, $\nu_1 < \nu < \nu_2$, is determined by drawing two circles, of radii $k_1 = 2\pi\nu_1/c$ and $k_2 = 2\pi\nu_2/c$, and counting the number

†Although the material contained in this section is not used in the remainder of this chapter, it is required for Chap. 12 (Secs. 12.2B and 12.3B).

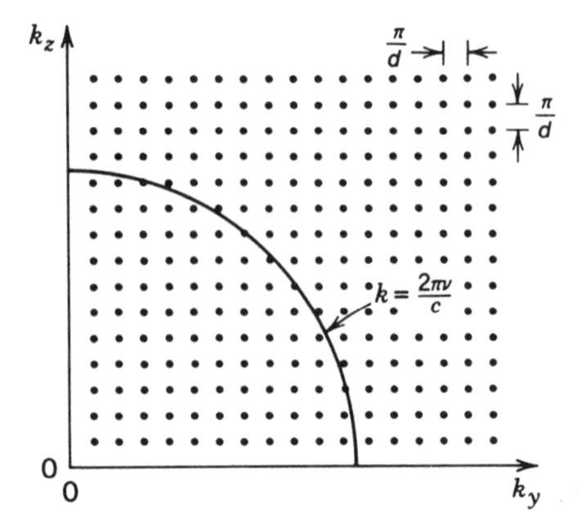

Figure 9.1-10 Dots denote the endpoints of the wavevectors $\mathbf{k} = (k_y, k_z)$ of modes in a two-dimensional resonator.

of dots that lie within that area. This procedure converts the allowed values of the vector $\mathbf{k}$ into allowed values of the frequency ν.

EXERCISE 9.1-3

Density of Modes in a Two-Dimensional Resonator

(a) Determine an approximate expression for the number of modes in a two-dimensional resonator with frequencies lying between 0 and ν, assuming that $2\pi\nu/c \gg \pi/d$, i.e., $d \gg \lambda/2$, and allowing for two orthogonal polarizations per mode number.

(b) Show that the number of modes per unit area lying within the frequency interval between ν and $\nu + d\nu$ is $M(\nu)\,d\nu$, where the density of modes $M(\nu)$ (modes per unit area per unit frequency) at frequency ν is given by

$$M(\nu) = \frac{4\pi\nu}{c^2}. \tag{9.1-34}$$

Density of Modes (Two-Dimensional Resonator)

Three-Dimensional Resonators

Consider now a resonator constructed of three pairs of parallel mirrors forming the walls of a closed box of size d. The structure is a three-dimensional resonator. Standing-wave solutions within the resonator require that components of the wavevector $\mathbf{k} = (k_x, k_y, k_z)$ are discretized to obey

$$k_x = \frac{q_x\pi}{d}, \quad k_y = \frac{q_y\pi}{d}, \quad k_z = \frac{q_z\pi}{d}, \qquad q_x, q_y, q_z = 1, 2, \ldots, \tag{9.1-35}$$

where q_x, q_y, and q_z are positive integers representing the respective mode numbers. Each mode, which is characterized by the three integers (q_x, q_y, q_z), is represented by a

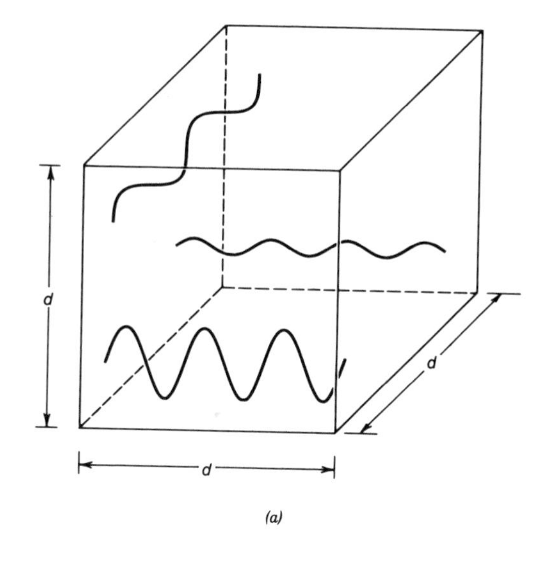

(a)

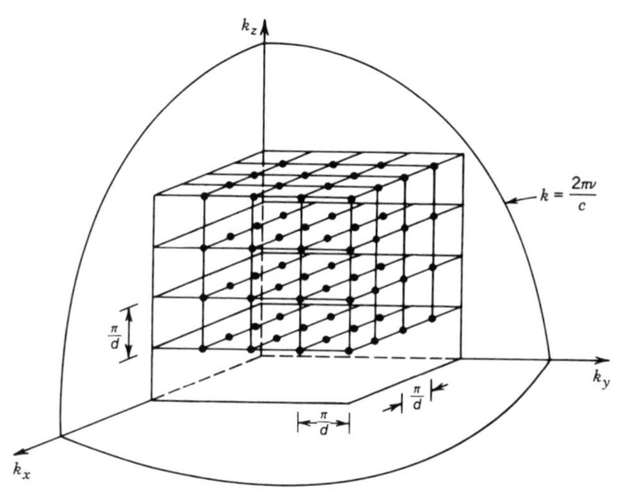

(b)

Figure 9.1-11 (*a*) Waves in a three-dimensional cubic resonator. (*b*) The endpoints of the wavevectors (k_x, k_y, k_z) of the modes in a three-dimensional resonator are marked by dots. The wavenumber k of a mode is the distance from the origin to the dot. All modes of frequency smaller than ν lie inside the positive octant of a sphere of radius $k = 2\pi\nu/c$.

dot in the (k_x, k_y, k_z)-space in Fig. 9.1-11. The values of the wavenumbers k and the corresponding resonance frequencies ν satisfy

$$k^2 = k_x^2 + k_y^2 + k_z^2 = \left(\frac{2\pi\nu}{c}\right)^2. \tag{9.1-36}$$

The surface of constant frequency ν is a sphere of radius $k = 2\pi\nu/c$.

Density of Modes

The number of modes lying in the frequency interval between 0 and ν corresponds to the number of points lying in the volume of the positive octant of a sphere of radius k in the k diagram [Fig. 9.1-11(b)]. Because it is analytically difficult to enumerate these modes, we resort to a continuous approximation, the validity of which depends on the relative values of the bandwidth of interest and the frequency interval between successive modes. The number of modes in the positive octant of a sphere of radius k is $2(\frac{1}{8})(4\pi k^3/3)/(\pi/d)^3 = (k^3/3\pi^2)d^3$. The initial factor of 2 accounts for the two possible polarizations of each mode, whereas the denominator $(\pi/d)^3$ represents the volume in k space per point. Since $k = 2\pi\nu/c$, the number of modes lying between 0 and ν is $[(2\pi\nu/c)^3/3\pi^2]d^3 = (8\pi\nu^3/3c^3)d^3$. The number of modes in the incremental frequency interval lying between ν and $\nu + \Delta\nu$ is therefore given by $(d/d\nu)[(8\pi\nu^3/3c^3)d^3]\,\Delta\nu = (8\pi\nu^2/c^3)d^3\,\Delta\nu$. The density of modes $M(\nu)$, i.e., the number of modes per unit volume of the resonator per unit bandwidth surrounding the frequency ν, is therefore

$$M(\nu) = \frac{8\pi\nu^2}{c^3}. \qquad (9.1\text{-}37)$$

Density of Modes
(Three-Dimensional Resonator)

The number of modes per unit volume within an arbitrary frequency interval $\nu_1 < \nu < \nu_2$ is the integral $\int_{\nu_1}^{\nu_2} M(\nu)\, d\nu$.

The density of modes $M(\nu)$ is a quadratically increasing function of frequency so that the number of modes within a fixed bandwidth $\Delta\nu$ increases with frequency ν in the manner indicated in Fig. 9.1-12. At $\nu = 3 \times 10^{14}$ ($\lambda_o = 1$ μm), $M(\nu) = 0.08$ modes/cm^3-Hz. Within a band of width 1 GHz, for example, there are $\approx 8 \times 10^7$ modes/cm^3.

The density of modes in two and three dimensions were derived on the basis of square and cubic geometry, respectively. Nevertheless, the results are applicable for arbitrary geometries, provided that the resonator dimensions are large in comparison with the wavelength.

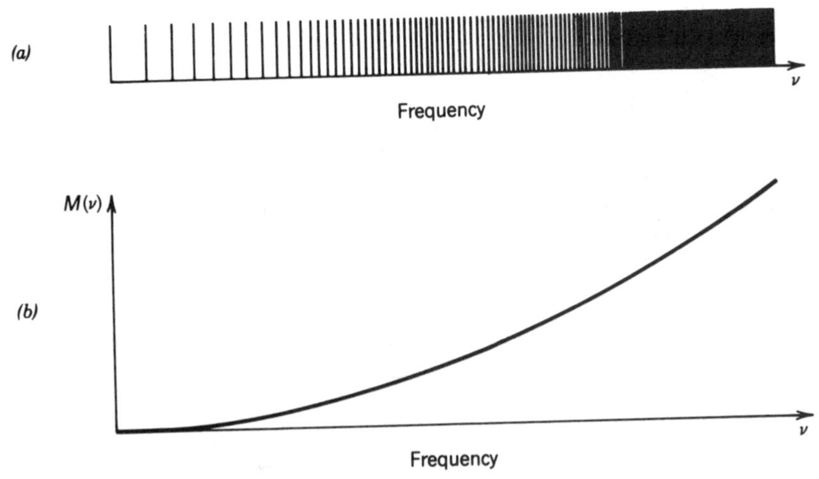

Figure 9.1-12 (a) The frequency spacing between adjacent modes decreases as the frequency increases. (b) The density of modes $M(\nu)$ for a three-dimensional optical resonator is a quadratically increasing function of frequency.

Finally, we point out that the enumeration of the electromagnetic modes presented here is mathematically identical to the calculation of the allowed quantum states of electrons confined within perfectly reflecting walls. The latter model is of importance in determining the density of allowed electron states as a function of energy in a semiconductor material (see Sec. 15.1C).

9.2 SPHERICAL-MIRROR RESONATORS

The planar-mirror resonator configuration discussed in the preceding section is highly sensitive to misalignment. If the mirrors are not perfectly parallel, or the rays are not perfectly normal to the mirror surfaces, they undergo a sequence of lateral displacements that eventually causes them to wander out of the resonator. Spherical-mirror resonators, in contrast, provide a more stable configuration for the confinement of light that renders them less sensitive to misalignment under certain geometrical conditions.

A spherical-mirror resonator is constructed of two spherical mirrors of radii R_1 and R_2 separated by a distance d (Fig. 9.2-1). The centers of the mirrors define the optical axis (z axis), about which the system exhibits circular symmetry. Each of the mirrors can be concave ($R < 0$) or convex ($R > 0$). The planar-mirror resonator is a special case for which $R_1 = R_2 = \infty$. We first examine the conditions for the confinement of optical rays. Then we determine the resonator modes. Finally, the effect of finite mirror size is discussed briefly.

A. Ray Confinement

Our initial approach is to use ray optics to determine the conditions of confinement for light rays in a spherical-mirror resonator. We consider only meridional rays (rays lying in a plane that passes through the optical axis) and limit ourselves to paraxial rays (rays that make small angles with the optical axis). The matrix-optics methods introduced in Sec. 1.4, which are valid only for paraxial rays, are used to study the trajectories of rays as they travel inside the resonator.

A resonator is a periodic optical system, since a ray travels through the same system after a round trip of two reflections. We may therefore make use of the analysis of periodic optical systems presented in Sec. 1.4D. Let y_m and θ_m be the position and inclination of an optical ray after m round trips, as illustrated in Fig. 9.2-2. Given y_m and θ_m, y_{m+1} and θ_{m+1} can be determined by tracing the ray through the system.

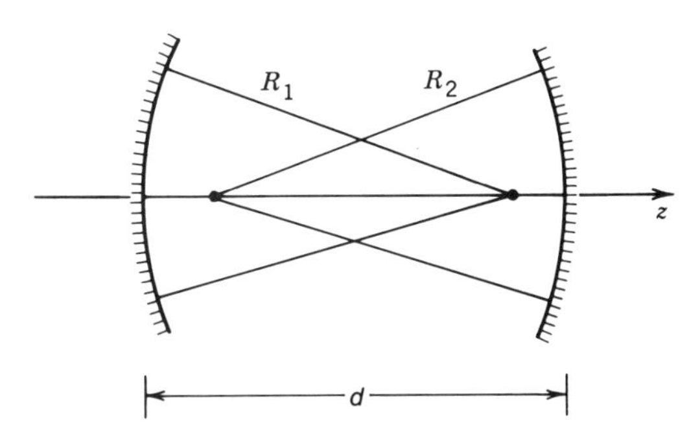

Figure 9.2-1 Geometry of a spherical-mirror resonator. In this case both mirrors are concave (their radii of curvature are negative).

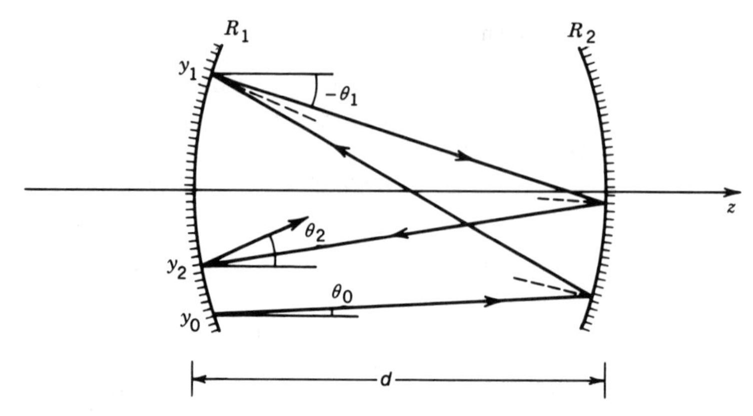

Figure 9.2-2 The position and inclination of a ray after m round trips are represented by y_m and θ_m, respectively, where $m = 0, 1, 2, \ldots$. In this diagram, $\theta_1 < 0$ since the ray is going downward.

For paraxial rays, where all angles are small, the relation between (y_{m+1}, θ_{m+1}) and (y_m, θ_m) is linear and can be written in the matrix form

$$\begin{bmatrix} y_{m+1} \\ \theta_{m+1} \end{bmatrix} = \begin{bmatrix} A & B \\ C & D \end{bmatrix} \begin{bmatrix} y_m \\ \theta_m \end{bmatrix}. \tag{9.2-1}$$

The round-trip ray-transfer matrix for Fig. 9.2-2:

$$\begin{bmatrix} A & B \\ C & D \end{bmatrix} = \begin{bmatrix} 1 & 0 \\ \frac{2}{R_1} & 1 \end{bmatrix} \begin{bmatrix} 1 & d \\ 0 & 1 \end{bmatrix} \begin{bmatrix} 1 & 0 \\ \frac{2}{R_2} & 1 \end{bmatrix} \begin{bmatrix} 1 & d \\ 0 & 1 \end{bmatrix}$$

is a product of ray-transfer matrices representing, from right to left [see (1.4-3) and (1.4-8)]:

propagation a distance d through free space,
reflection from a mirror of radius R_2,
propagation a distance d through free space,
reflection from a mirror of radius R_1.

As shown in Sec. 1.4D, the solution of the difference equation (9.2-1) is $y_m = y_{\max} F^m \sin(m\varphi + \varphi_0)$, where $F^2 = AD - BC$, $\varphi = \cos^{-1}(b/F)$, $b = (A + D)/2$, and $y_{\max}$ and φ_0 are constants to be determined from the initial position and inclination of the ray. For the case at hand $F = 1$, so that

$$y_m = y_{\max} \sin(m\varphi + \varphi_0), \tag{9.2-2}$$

$$\varphi = \cos^{-1} b, \qquad b = 2\left(1 + \frac{d}{R_1}\right)\left(1 + \frac{d}{R_2}\right) - 1.$$

The solution (9.2-2) is harmonic (and therefore bounded) provided that $\varphi = \cos^{-1} b$ is real. This is ensured if $|b| \le 1$, i.e., if $-1 \le b \le 1$ or $0 \le (1 + d/R_1)(1 + d/R_2) \le 1$. It is convenient to write this condition in terms of the parameters $g_1 = 1 + d/R_1$ and

$g_2 = 1 + d/R_2$, which are known as the **g parameters**,

$$0 \le g_1 g_2 \le 1. \tag{9.2-3}$$

Confinement Condition

When this condition is not satisfied, φ is imaginary so that y_m in (9.2-2) becomes a hyperbolic sine function of m which increases without bound. The resonator is then said to be **unstable**. At the boundary of the confinement condition (when the inequalities are equalities), the resonator is said to be **conditionally stable**; slight errors in alignment render it unstable.

A useful graphical representation of the confinement condition (Fig. 9.2-3) identifies each combination (g_1, g_2) of the two g parameters of a resonator as a point in a g_2 versus g_1 diagram. The left inequality in (9.2-3) is equivalent to $\{g_1 \ge 0$ and $g_2 \ge 0$; or $g_1 \le 0$ and $g_2 \le 0\}$; i.e., all stable points (g_1, g_2) must lie in the first or third quadrant. The right inequality in (9.2-3) signifies that stable points (g_1, g_2) must lie in a region bounded by the hyperbola $g_1 g_2 = 1$. The unshaded area in Fig. 9.2-3 represents the region for which both inequalities are satisfied, indicating that the resonator is stable.

Symmetrical resonators, by definition, have identical mirrors ($R_1 = R_2 = R$) so that $g_1 = g_2 = g$. The condition of stability is then $g^2 \le 1$, or $-1 \le g \le 1$, so that

$$0 \le \frac{d}{(-R)} \le 2. \tag{9.2-4}$$

Confinement Condition (Symmetrical Resonator)

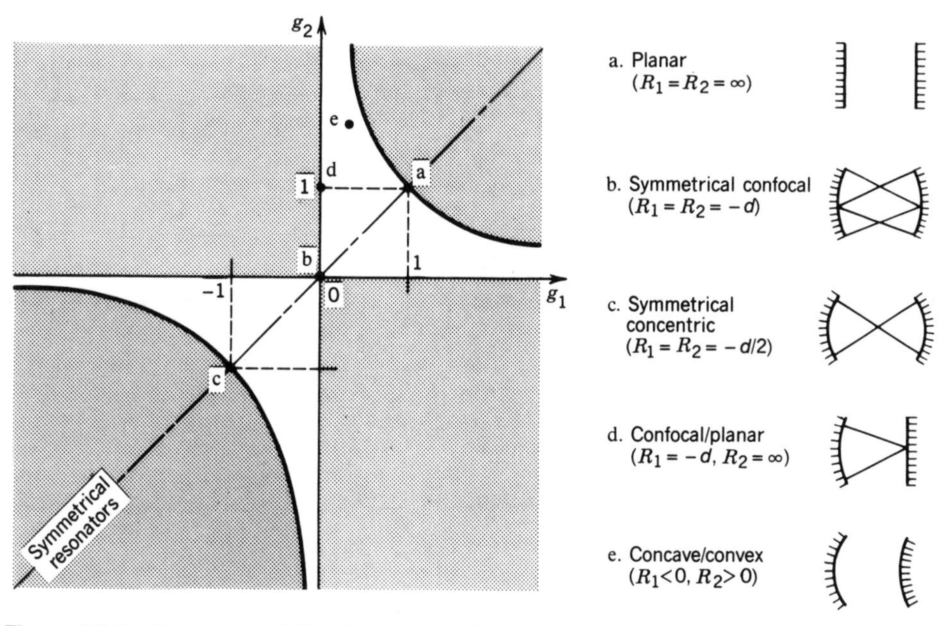

Figure 9.2-3 Resonator stability diagram. A spherical-mirror resonator is stable if the parameters $g_1 = 1 + d/R_1$ and $g_2 = 1 + d/R_2$ lie in the unshaded regions bounded by the lines $g_1 = 0$ and $g_2 = 0$, and the hyperbola $g_2 = 1/g_1$. R is negative for a concave mirror and positive for a convex mirror. Various special configurations are indicated by letters. All symmetrical resonators lie along the line $g_2 = g_1$.

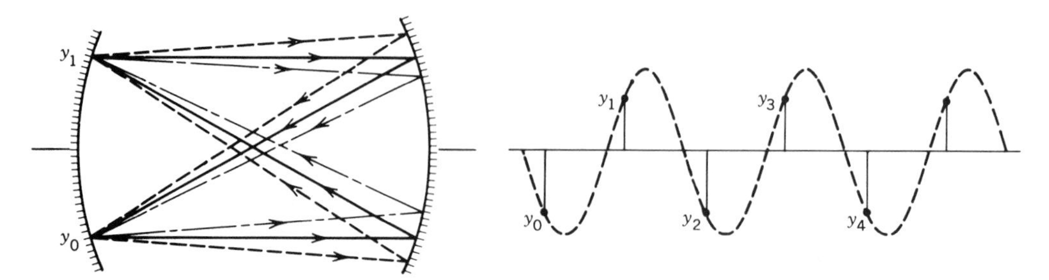

Figure 9.2-4 All paraxial rays in a symmetrical confocal resonator retrace themselves after two round trips, regardless of their original position and inclination. Angles are exaggerated in this drawing for the purpose of illustration.

These resonators are represented in Fig. 9.2-3 by points along the line $g_2 = g_1$. To satisfy (9.2-4) a stable symmetrical resonator must use concave mirrors ($R < 0$) whose radii are greater than half the resonator length. Three points within this interval are of special interest: $d/(-R) = 0$, 1, and 2, corresponding to **planar, confocal**, and **concentric resonators**, respectively.

In the symmetrical confocal resonator, $(-R) = d$, so that the center of curvature of each mirror lies on the other. Thus in (9.2-2), $b = -1$, $\varphi = \pi$, and the ray position is $y_m = y_{\max}\sin(m\pi + \varphi_0)$, i.e., $y_m = (-1)^m y_0$. Rays initiated at position y_0, at any inclination, are imaged to position $y_1 = -y_0$, then imaged again to position $y_2 = y_0$, and so on, repeatedly. Each ray retraces itself after two round trips (Fig. 9.2-4). All paraxial rays are therefore confined, no matter what their original position and inclination. This is to be compared with the planar-mirror resonator, for which only rays of zero inclination retrace themselves.

In summary, the confinement condition for paraxial rays in a spherical-mirror resonator, constructed of mirrors of radii R_1 and R_2 separated by a distance d, is $0 \le g_1 g_2 \le 1$, where $g_1 = 1 + d/R_1$ and $g_2 = 1 + d/R_2$.

EXERCISE 9.2-1

Maximum Resonator Length for Confined Rays. A resonator is constructed using concave mirrors of radii 50 cm and 100 cm. Determine the maximum resonator length for which rays satisfy the confinement condition.

B. Gaussian Modes

Although the ray-optics approach considered in the preceding section is useful for determining the geometrical conditions under which rays are confined, it cannot provide information about the spatial intensity distributions and resonance frequencies of the resonator modes. We now proceed to show that Gaussian beams are modes of

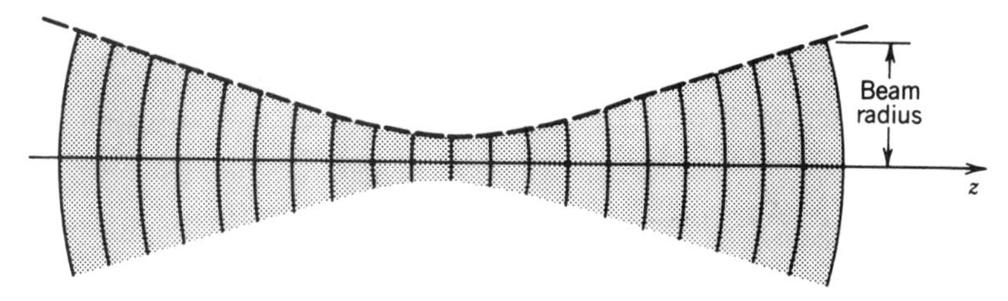

Figure 9.2-5 Gaussian beam wavefronts (solid curves) and beam radius (dashed curve).

the spherical-mirror resonator; Gaussian beams provide solutions of the Helmholtz equation under the boundary conditions imposed by the spherical-mirror resonator.

Gaussian Beams

As discussed in Chap. 3, a Gaussian beam is a circularly symmetric wave whose energy is confined about its axis (the z axis) and whose wavefront normals are paraxial rays (Fig. 9.2-5). In accordance with (3.1-12), at an axial distance z from the beam waist the beam intensity I varies in the transverse x–y plane as the Gaussian distribution $I = I_0[W_0/W(z)]^2 \exp[-2(x^2 + y^2)/W^2(z)]$. Its width is given by (3.1-8):

$$W(z) = W_0\left[1 + \left(\frac{z}{z_0}\right)^2\right]^{1/2}, \tag{9.2-5}$$

where z_0 is the distance, known as the Rayleigh range, at which the beam wavefronts are most curved. The beam width (radius) $W(z)$ increases in both directions from its minimum value W_0 at the beam waist ($z = 0$). The radius of curvature of the wavefronts, which is given by (3.1-9),

$$R(z) = z + \frac{z_0^2}{z} \tag{9.2-6}$$

decreases from ∞ at $z = 0$, to a minimum value at $z = z_0$, and thereafter grows linearly with z for large z. For $z > 0$, the wave diverges and $R(z) > 0$; for $z < 0$, the wave converges and $R(z) < 0$. The Rayleigh range z_0 is related to the beam waist radius W_0 by (3.1-11):

$$z_0 = \frac{\pi W_0^2}{\lambda}. \tag{9.2-7}$$

The Gaussian Beam Is a Mode of the Spherical-Mirror Resonator

A Gaussian beam reflected from a spherical mirror will retrace the incident beam if the radius of curvature of its wavefront is the same as the mirror radius (see Sec. 3.2C). Thus, if the radii of curvature of the wavefronts of a Gaussian beam at planes separated by a distance d match the radii of two mirrors separated by the same distance d, a beam incident on the first mirror will reflect and retrace itself to the second mirror, where it once again will reflect and retrace itself back to the first mirror, and so on. The beam can then exist self-consistently within the spherical-mirror resonator, satisfying the Helmholtz equation and the boundary conditions imposed by the mirrors. The Gaussian beam is then said to be a mode of the spherical-mirror resonator (provided that the phase also retraces itself, as discussed in Sec. 9.2C).

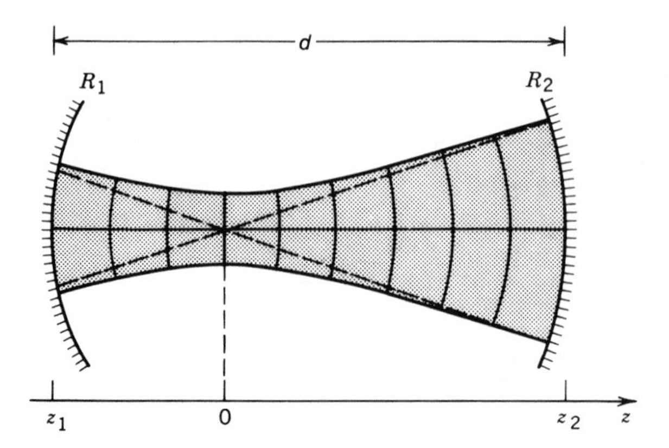

Figure 9.2-6 Fitting a Gaussian beam to two mirrors of radii R_1 and R_2 separated by a distance d. In this diagram both mirrors are concave (R_1, R_2, and z_1 are negative).

We now proceed to determine the Gaussian beam that matches a spherical-mirror resonator with mirrors of radii R_1 and R_2 separated by the distance d. This is illustrated in Fig. 9.2-6 for the special case when both mirrors are concave ($R_1 < 0$ and $R_2 < 0$).

The z axis is defined by the centers of the mirrors. The center of the beam, which is yet to be determined, is assumed to be located at the origin $z = 0$; mirrors R_1 and R_2 are located at positions z_1 and

$$z_2 = z_1 + d, \tag{9.2-8}$$

respectively. (A negative value for z_1 indicates that the center of the beam lies to the right of mirror 1; a positive value indicates that it lies to the left.) The values of z_1 and z_2 are determined by matching the radius of curvature of the beam, $R(z) = z + z_0^2/z$, to the radii R_1 at z_1 and R_2 at z_2. Careful attention must be paid to the signs. If both mirrors are concave, they have negative radii. But the beam radius of curvature was defined to be positive for $z > 0$ (at mirror 2) and negative for $z < 0$ (at mirror 1). We therefore equate $R_1 = R(z_1)$ but $-R_2 = R(z_2)$, i.e.,

$$R_1 = z_1 + \frac{z_0^2}{z_1} \tag{9.2-9}$$

$$-R_2 = z_2 + \frac{z_0^2}{z_2}. \tag{9.2-10}$$

Solving (9.2-8), (9.2-9), and (9.2-10) for z_1, z_2, and z_0 leads to

$$z_1 = \frac{-d(R_2 + d)}{R_2 + R_1 + 2d}, \qquad z_2 = z_1 + d \tag{9.2-11}$$

$$z_0^2 = \frac{-d(R_1 + d)(R_2 + d)(R_2 + R_1 + d)}{(R_2 + R_1 + 2d)^2}. \tag{9.2-12}$$

Having determined the location of the beam center and the depth of focus $2z_0$, everything about the beam is known (see Sec. 3.1B). The waist radius is $W_0 =$

$(\lambda z_0/\pi)^{1/2}$, and the beam radii at the mirrors are

$$W_i = W_0\left[1 + \left(\frac{z_i}{z_0}\right)^2\right]^{1/2}, \qquad i = 1, 2. \tag{9.2-13}$$

A similar problem has been addressed in Chap. 3 (Exercise 3.1-5).

In order that the solution (9.2-11)–(9.2-12) indeed represent a Gaussian beam, z_0 must be real. An imaginary value of z_0 signifies that the Gaussian beam is in fact a paraboloidal wave, which is an unconfined solution (see Sec. 3.1A). Using (9.2-12), it is not difficult to show that the condition $z_0^2 > 0$ is equivalent to

$$0 \leq \left(1 + \frac{d}{R_1}\right)\left(1 + \frac{d}{R_2}\right) \leq 1, \tag{9.2-14}$$

which is precisely the confinement condition required by ray optics, as set forth in (9.2-3).

EXERCISE 9.2-2

A Plano-Concave Resonator. When mirror 1 is planar ($R_1 = \infty$), determine the confinement condition, the depth of focus, and the beam radius at the waist and at each of the mirrors, as a function of $d/|R_2|$.

Gaussian Mode of a Symmetrical Spherical-Mirror Resonator

The results obtained in (9.2-11)–(9.2-13) simplify considerably for symmetrical resonators with concave mirrors. Substituting $R_1 = R_2 = -|R|$ into (9.2-11) provides $z_1 = -d/2$, $z_2 = d/2$. Thus the beam center lies at the center of the resonator, and

$$z_0 = \frac{d}{2}\left(2\frac{|R|}{d} - 1\right)^{1/2} \tag{9.2-15}$$

$$W_0^2 = \frac{\lambda d}{2\pi}\left(2\frac{|R|}{d} - 1\right)^{1/2} \tag{9.2-16}$$

$$W_1^2 = W_2^2 = \frac{\lambda d/\pi}{\{(d/|R|)[2 - (d/|R|)]\}^{1/2}}. \tag{9.2-17}$$

The confinement condition (9.2-14) becomes

$$0 \leq \frac{d}{|R|} \leq 2. \tag{9.2-18}$$

Given a resonator of fixed mirror separation d, we now examine the effect of increasing mirror curvature (increasing $d/|R|$) on the beam radius at the waist W_0, and at the mirrors $W_1 = W_2$. The results are illustrated in Fig. 9.2-7. For a planar-mirror resonator, $d/|R| = 0$, so that W_0 and W_1 are infinite, corresponding to a plane wave

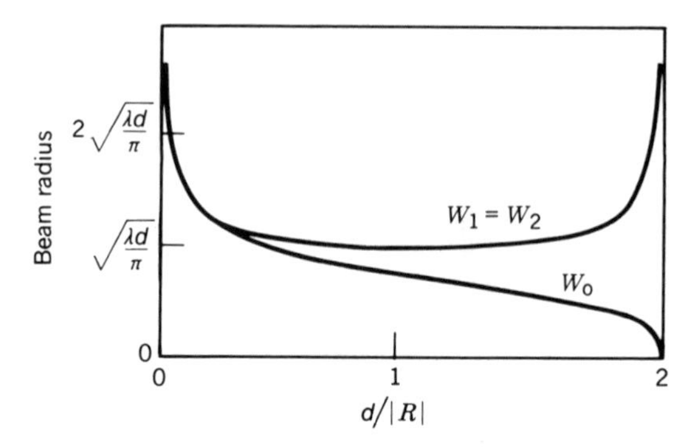

Figure 9.2-7 The beam radius at the waist, W_0, and at the mirrors, $W_1 = W_2$, for a symmetrical spherical-mirror resonator with concave mirrors as a function of the ratio $d/|R|$. Symmetrical confocal and concentric resonators correspond to $d/|R| = 1$ and $d/|R| = 2$, respectively.

rather than a Gaussian beam. As $d/|R|$ increases, W_0 decreases until it vanishes for the concentric resonator ($d/|R| = 2$); at this point $W_1 = W_2 = \infty$. This is not surprising inasmuch as a spherical wave fits within a symmetrical concentric resonator (see Fig. 9.2-3).

The radius of the beam at the mirrors has its minimum value, $W_1 = W_2 = (\lambda d/\pi)^{1/2}$, when $d/|R| = 1$, i.e., for the symmetrical confocal resonator. In this case

$$z_0 = \frac{d}{2} \tag{9.2-19}$$

$$W_0 = \left(\frac{\lambda d}{2\pi}\right)^{1/2} \tag{9.2-20}$$

$$W_1 = W_2 = \sqrt{2}\, W_0. \tag{9.2-21}$$

The depth of focus $2z_0$ is then equal to the length of the resonator d, as shown in Fig. 9.2-8. This explains why the parameter $2z_0$ is sometimes called the confocal parameter. A long resonator has a long depth of focus. The waist radius is proportional to the square root of the mirror spacing. A Gaussian beam at $\lambda_o = 633$ nm (one of the wavelengths of the helium–neon laser) in a resonator with $d = 100$ cm, for example,

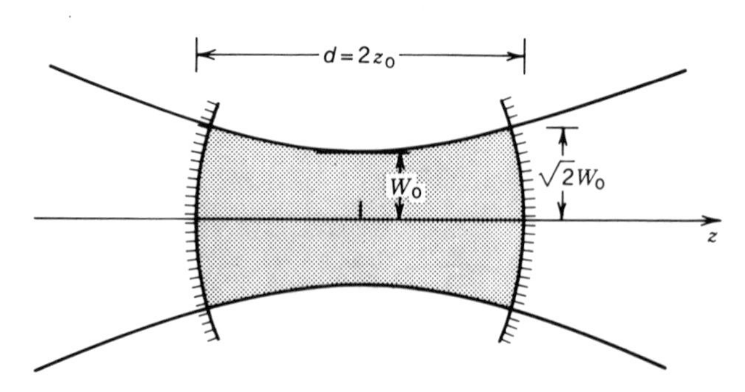

Figure 9.2-8 Gaussian beam in a symmetrical confocal resonator with concave mirrors. The depth of focus $2z_0$ equals the length of the resonator d. The beam radius at the mirrors is a factor of $\sqrt{2}$ greater than that at the waist.

has a waist radius $W_0 = (\lambda d/2\pi)^{1/2} = 0.32$ mm, whereas a 25-cm-long resonator supports a Gaussian beam of waist radius 0.16 mm at the same wavelength. The radius of the beam at each of the mirrors is greater than at the waist by a factor of $\sqrt{2}$.

C. Resonance Frequencies

As indicated in Sec. 9.2B, a Gaussian beam is a mode of the spherical-mirror resonator provided that the wavefront normals reflect onto themselves, always retracing the same path, and that the phase retraces itself as well.

The phase of a Gaussian beam, in accordance with (3.1-22), is

$$\varphi(\rho, z) = kz - \zeta(z) + \frac{k\rho^2}{2R(z)}$$

where $\zeta(z) = \tan^{-1}(z/z_0)$ and $\rho^2 = x^2 + y^2$. At points on the optical axis ($\rho = 0$), $\varphi(0, z) = kz - \zeta(z)$, so that the phase retardation relative to a plane wave is $\zeta(z)$. At the locations of the mirrors z_1 and z_2,

$$\varphi(0, z_1) = kz_1 - \zeta(z_1)$$

$$\varphi(0, z_2) = kz_2 - \zeta(z_2).$$

Because the mirror surface coincides with the wavefronts, all points on each mirror share the same phase. As the beam propagates from mirror 1 to mirror 2 its phase changes by

$$\varphi(0, z_2) - \varphi(0, z_1) = k(z_2 - z_1) - [\zeta(z_2) - \zeta(z_1)]$$

$$= kd - \Delta\zeta, \tag{9.2-22}$$

where

$$\Delta\zeta = \zeta(z_2) - \zeta(z_1). \tag{9.2-23}$$

As the traveling wave completes a round trip between the two mirrors, therefore, its phase changes by $2kd - 2\Delta\zeta$.

In order that the beam truly retrace itself, the round-trip phase change must be a multiple of 2π, i.e., $2kd - 2\Delta\zeta = 2\pi q$, $q = 0, \pm 1, \pm 2, \ldots$. Substituting $k = 2\pi\nu/c$ and $\nu_F = c/2d$, the frequencies ν_q that satisfy this condition are

$$\boxed{\nu_q = q\nu_F + \frac{\Delta\zeta}{\pi}\nu_F.} \tag{9.2-24}$$

Spherical-Mirror Resonator
Resonance Frequencies
(Gaussian Modes)

The frequency spacing of adjacent modes is $\nu_F = c/2d$, which is the same result as that obtained in Sec. 9.1A for the planar-mirror resonator. For spherical-mirror resonators, this frequency spacing is independent of the curvatures of the mirrors. The second term in (9.2-24), which does depend on the mirror curvatures, simply represents a displacement of all resonance frequencies.

EXERCISE 9.2-3

Resonance Frequencies of a Confocal Resonator. A symmetrical confocal resonator has a length $d = 30$ cm, and the medium has refractive index $n = 1$. Determine the frequency spacing ν_F and the displacement frequency $(\Delta\zeta/\pi)\nu_F$. Determine all the resonance frequencies that lie within the band $5 \times 10^{14} \pm 2 \times 10^9$ Hz.

D. Hermite–Gaussian Modes

In Sec. 3.3 it was shown that the Gaussian beam is not the only beam-like solution of the paraxial Helmholtz equation. An entire family of solutions, the Hermite–Gaussian family, exists. Although a Hermite–Gaussian beam of order (l, m) has the same wavefronts as a Gaussian beam, its amplitude distribution differs. The design of a resonator that "matches" a given beam (or the design of a beam that "fits" a given resonator) is therefore the same as in the Gaussian-beam case, regardless of (l, m). It follows that the entire family of Hermite–Gaussian beams represents modes of the spherical-mirror resonator.

The resonance frequencies of the (l, m) mode do, however, depend on the indices (l, m). This is because of the dependence of the axial phase delay on l and m. Using (3.3-9), the phase of the (l, m) mode on the beam axis is

$$\varphi(0, z) = kz - (l + m + 1)\zeta(z). \tag{9.2-25}$$

The phase shift encountered by a traveling wave undergoing a single round trip through a resonator of length d should be set equal to a multiple of 2π in order that the beam retrace itself. Thus

$$2kd - 2(l + m + 1)\Delta\zeta = 2\pi q, \qquad q = 0, \pm 1, \pm 2, \ldots, \tag{9.2-26}$$

where, as before, $\Delta\zeta = [\zeta(z_2) - \zeta(z_1)]$ and z_1, z_2 are the positions of the two mirrors. With $\nu_F = c/2d$, this yields the resonance frequencies

$$\boxed{\nu_{l,m,q} = q\nu_F + (l + m + 1)\frac{\Delta\zeta}{\pi}\nu_F.} \tag{9.2-27}$$

Spherical-Mirror Resonator Resonance Frequencies (Hermite–Gaussian Modes)

Modes of different q, but the same (l, m), have identical intensity distributions [see (3.3-11)]. They are known as **longitudinal** or **axial modes**. The indices (l, m) label different spatial dependences on the transverse coordinates x, y; these represent different **transverse modes**, as illustrated in Fig. 3.3-2.

Equation (9.2-27) indicates that the resonance frequencies of the Hermite–Gaussian modes satisfy the following properties:

- Longitudinal modes corresponding to a given transverse mode (l, m) have resonance frequencies spaced by $\nu_F = c/2d$, i.e., $\nu_{l,m,q+1} - \nu_{l,m,q} = \nu_F$.
- All transverse modes, for which the sum of the indices $l + m$ is the same, have the same resonance frequencies.

- Two transverse modes (l, m), (l', m') corresponding to the same longitudinal mode q have resonance frequencies spaced by

$$\nu_{l,m,q} - \nu_{l',m',q} = [(l+m) - (l'+m')]\frac{\Delta\zeta}{\pi}\nu_F. \qquad (9.2\text{-}28)$$

This expression determines the frequency shift between the sets of longitudinal modes of indices (l, m) and (l', m').

EXERCISE 9.2-4

Resonance Frequencies of the Symmetrical Confocal Resonator. Show that for a symmetrical confocal resonator the longitudinal modes associated with two transverse modes are either the same or are displaced by $\nu_F/2$, as illustrated in Fig. 9.2-9.

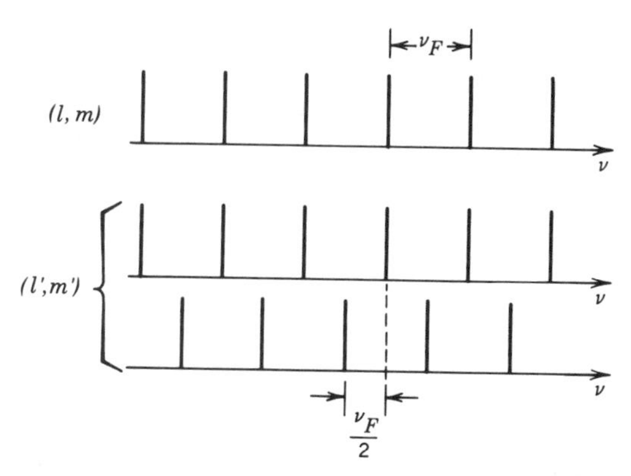

Figure 9.2-9 In a symmetrical confocal resonator, the longitudinal modes associated with two transverse modes of indices (l, m) and (l', m') are either aligned or displaced by half a longitudinal mode spacing.

---

*E. Finite Apertures and Diffraction Loss

Since Gaussian and Hermite–Gaussian beams have infinite transverse extent and since the resonator mirrors are of finite extent, a portion of the optical power escapes from the resonator on each pass. An estimate of the power loss may be determined by calculating the fractional power of the beam that is not intercepted by the mirror. If the beam is Gaussian with radius W and the mirror is circular with radius $a = 2W$, for example, a small fraction, $\exp(-2a^2/W^2) \approx 3.35 \times 10^{-4}$, of the beam power escapes on each pass [see (3.1-16)], the remainder being reflected (or absorbed in the mirror). Higher-order transverse modes suffer greater losses since they have greater spatial extent in the transverse plane.

When the mirror radius a is smaller than $2W$, the losses are greater. However, the Gaussian and Hermite–Gaussian beams no longer provide good approximations for the resonator modes. The problem of determining the modes of a spherical-mirror resonator with finite-size mirrors is difficult. A wave is a mode if it retraces its amplitude

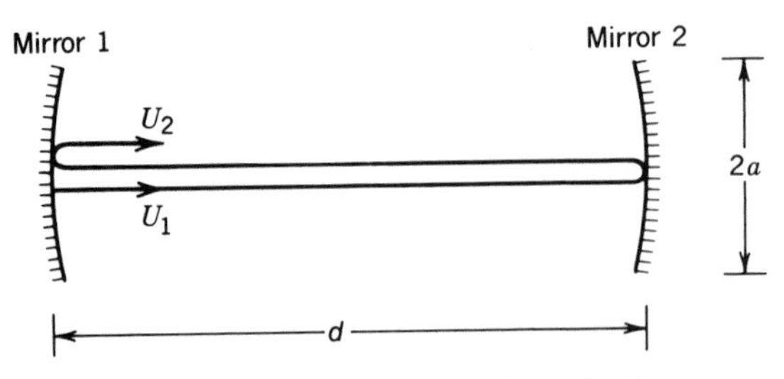

Figure 9.2-10 Propagation of a wave through a spherical-mirror resonator. The complex amplitude $U_1(x, y)$ corresponds to a mode if it reproduces itself after a round trip, i.e., if $U_2(x, y) = \mu U_1(x, y)$ and $\arg\{\mu\} = q2\pi$.

(to within a multiplicative constant) and reproduces its phase (to within an integer multiple of 2π) after completing a round trip through the resonator. One often-used method of determining the modes involves following a wave repeatedly as it bounces through the resonator, thereby determining its amplitude and phase, much as we determined the position and inclination of a ray bouncing within a resonator. After many round trips this process converges to one of the modes.

If $U_1(x, y)$ is the complex amplitude of a wave immediately to the right of mirror 1 in Fig. 9.2-10, and if $U_2(x, y)$ is the complex amplitude after one round trip of travel through the resonator, then $U_1(x, y)$ is a mode provided that $U_2(x, y) = \mu U_1(x, y)$ and $\arg\{\mu\}$ is an integer multiple of 2π (i.e., μ is real and positive). After a single round trip, the mode intensity is attenuated by the factor μ^2, and the phase is reproduced. The methods of Fourier optics (Chap. 4) may be used to determine $U_2(x, y)$ from $U_1(x,y)$. These quantities may be regarded as the output and input, respectively, of a linear system (see Appendix B) characterized by an impulse-response function $h(x, y; x', y')$,

$$U_2(x, y) = \int_{-\infty}^{\infty}\int_{-\infty}^{\infty} h(x, y; x', y')U_1(x', y')\, dx'\, dy'.$$

If the impulse-response function h is known, the modes can be determined by solving the eigenvalue problem described by the integral equation (see Appendix C)

$$\int_{-\infty}^{\infty}\int_{-\infty}^{\infty} h(x, y; x', y')U(x', y')\, dx'\, dy' = \mu U(x, y). \tag{9.2-29}$$

The solutions determine the eigenfunctions $U_{l,m}(x, y)$, and the eigenvalues $\mu_{l,m}$, labeled by the indices (l, m). The modes are the eigenfunctions and the round-trip multiplicative factor is the eigenvalue. The squared magnitude $|\mu_{l,m}|^2$ is the round-trip intensity reduction factor for the (l, m) mode. Clearly, when the mirrors are infinite in size and the paraxial approximation is satisfied, the modes reduce to the family of Hermite–Gaussian beams discussed earlier.

It remains to determine $h(x, y; x', y')$ and to solve the integral equation (9.2-29). A single pass inside the resonator involves traveling a distance d, truncation by the mirror aperture, and reflection by the mirror. The remaining pass, needed to comprise a single round trip, is similar. The impulse-response function $h(x, y; x', y')$ can then be determined by application of the theory of Fresnel diffraction (Sec. 4.3B). In general, however, the modes and their associated losses can be determined only by numerically solving the integral equation (9.2-29). An iterative numerical solution begins with an initial guess U_1, from which U_2 is computed and passed through the system one more round trip, and so on until the process converges.

This technique has been used to determine the losses associated with the various modes of a spherical-mirror resonator with circular apertures of radius a. The results

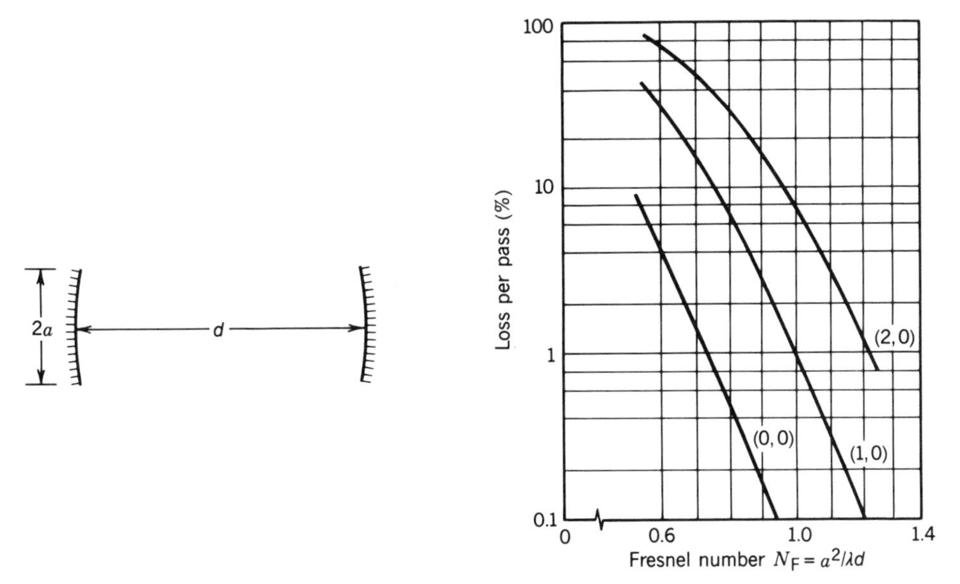

Figure 9.2-11 Percent diffraction loss per pass (half a round trip) in a symmetrical confocal resonator as a function of the Fresnel number $N_F = a^2/\lambda d$ for the (0, 0), (1, 0), and (2, 0) modes. (Adapted from A. E. Siegman, *Lasers*, University Science Books, Mill Valley, CA, 1986.)

are illustrated in Fig. 9.2-11. For a symmetrical confocal resonator the loss is governed by a single parameter, the Fresnel number $N_F = a^2/\lambda d$. This is because the Fresnel number governs Fresnel diffraction between the two mirrors, as discussed in Sec. 4.3B. For the symmetrical confocal resonator described by (9.2-20) and (9.2-21), the beam radius at the mirrors is $W = (\lambda d/\pi)^{1/2}$, so that $\lambda d = \pi W^2$, from which the Fresnel number is readily determined to be $N_F = a^2/\pi W^2$. N_F is therefore proportional to the ratio a^2/W^2; a higher Fresnel number corresponds to a smaller loss. From Fig. 9.2-11, for example, the loss per pass of the lowest-order mode $(l, m) = (0, 0)$ is about 0.1% when $N_F \approx 0.94$. This Fresnel number corresponds to $a/W = 1.72$. For a Gaussian beam of radius W, the percentage of power contained outside a circle of radius $a = 1.72W$ is $\exp(-2a^2/W^2) \approx 0.27\%$. Higher-order modes suffer from greater losses because of their larger spatial extent.

READING LIST

Books on Resonators

J. M. Vaughan, *The Fabry–Perot Interferometer*, Adam Hilger, Bristol, England, 1989.

Y. Anan'ev, *Résonateurs optiques et problème de divergence du rayonnement laser*, Mir, Moscow, Russian original 1979, French translation 1982.

L. A. Weinstein, *Open Resonators and Open Waveguides*, Golem, Boulder, CO, 1969.

Books on Lasers with Chapters on Resonators

See also the reading list in Chapter 13.

A. Yariv, *Optical Electronics*, Holt, Rinehart and Winston, New York, 4th ed. 1991.

J. T. Verdeyen, *Laser Electronics*, Prentice-Hall, Englewood Cliffs, NJ, 2nd ed. 1989.

O. Svelto, *Principles of Lasers*, Plenum Press, New York, 3rd ed. 1989.

P. W. Milonni and J. H. Eberly, *Lasers*, Wiley, New York, 1988.

J. Wilson and J. F. B. Hawkes, *Lasers: Principles and Applications*, Prentice-Hall, Englewood Cliffs, NJ, 1987.

W. Witteman, *The Laser*, Springer-Verlag, New York, 1987.

A. E. Siegman, *Lasers*, University Science Books, Mill Valley, CA, 1986.

K. Shimoda, *Introduction to Laser Physics*, Springer-Verlag, New York, 2nd ed. 1986.

A. E. Siegman, *An Introduction to Lasers and Masers*, McGraw-Hill, New York, 1971.

A. Maitland and M. H. Dunn, *Laser Physics*, North-Holland, Amsterdam, 1969.

Articles

A. E. Siegman, Unstable Optical Resonators, *Applied Optics*, vol. 13, pp. 353–367, 1974.

H. Kogelnik and T. Li, Laser Beams and Resonators, *Applied Optics*, vol. 5, pp. 1550–1567, 1966 (published simultaneously in *Proceedings of the IEEE*, vol. 54, pp. 1312–1329, 1966).

A. E. Siegman, Unstable Optical Resonators for Laser Applications, *Proceedings of the IEEE*, vol. 53, pp. 277–287, 1965.

A. G. Fox and T. Li, Resonant Modes in a Maser Interferometer, *Bell System Technical Journal*, vol. 40, pp. 453–488, 1961.

G. D. Boyd and J. P. Gordon, Confocal Multimode Resonator for Millimeter Through Optical Wavelength Masers, *Bell System Technical Journal*, vol. 40, pp. 489–508, 1961.

PROBLEMS

9.1-1 **Resonance Frequencies of a Resonator with an Etalon.** (a) Determine the spacing between adjacent resonance frequencies in a resonator constructed of two parallel planar mirrors separated by a distance $d = 15$ cm in air ($n = 1$).
(b) A transparent plate of thickness $d_1 = 2.5$ cm and refractive index $n = 1.5$ is placed inside the resonator and is tilted slightly to prevent light reflected from the plate from reaching the mirrors. Determine the spacing between the resonance frequencies of the resonator.

9.1-2 **Mirrorless Resonators.** Semiconductor lasers are often fabricated from crystals whose surfaces are cleaved along crystal planes. These surfaces act as reflectors and therefore serve as the resonator mirrors. The reflectance is given in (6.2-14). Consider a crystal with refractive index $n = 3.6$ placed in air ($n = 1$). The light reflects between two parallel surfaces separated by the distance $d = 0.2$ mm. Determine the spacing between resonance frequencies ν_F, the overall distributed loss coefficient α_r, the finesse $\mathcal{F}$, and the spectral width $\delta\nu$. Assume that the loss coefficient $\alpha_s = 1\ \text{cm}^{-1}$.

9.1-3 **Resonator Spectral Response.** The transmittance of a symmetrical Fabry–Perot resonator was measured by using light from a tunable monochromatic light source. The transmittance versus frequency exhibits periodic pulses of period 150 MHz, each of width (FWHM) 5 MHz. Assuming that the medium within the resonator mirrors is a gas with $n = 1$, determine the length and finesse of the resonator. Assuming that the only source of loss is associated with the mirrors, find their reflectances.

9.1-4 **Optical Decay Time.** What time does it take for the optical energy stored in a resonator of finesse $\mathcal{F} = 100$, length $d = 50$ cm, and refractive index $n = 1$, to decay to one-half of its initial value?

9.1-5 **Number of Modes.** Consider light of wavelength $\lambda_o = 1.06\ \mu$m and spectral width $\Delta\nu = 120$ GHz. How many modes have frequencies within this linewidth in the following resonators ($n = 1$):
(a) A one-dimensional resonator of length $d = 10$ cm?
(b) A $10 \times 10\ \text{cm}^2$ two-dimensional resonator?
(c) A $10 \times 10 \times 10\ \text{cm}^3$ three-dimensional resonator?

9.2-1 **Stability of Spherical-Mirror Resonators.** (a) Can a resonator with two convex mirrors ever be stable?
(b) Can a resonator with one convex and one concave mirror ever be stable?

9.2-2 **A Planar-Mirror Resonator Containing a Lens.** A lens of focal length f is placed inside a planar-mirror resonator constructed of two flat mirrors separated by a distance d. The lens is located at a distance $d/2$ from each of the mirrors.
(a) Determine the ray-transfer matrix for a ray that begins at one of the mirrors and travels a round trip inside the resonator.
(b) Determine the condition of stability of the resonator.
(c) Under stable conditions sketch the Gaussian beam that fits this resonator.

9.2-3 **Self-Reproducing Rays.** Consider a symmetrical resonator using two concave mirrors of radii R separated by a distance $d = 3\,|R|/2$. After how many round trips through the resonator will a ray retrace its path?

9.2-4 **Ray Position in Unstable Resonators.** Show that for an unstable resonator the ray position after m round trips is given by $y_m = \alpha_1 h_1^m + \alpha_2 h_2^m$, where α_1 and α_2 are constants, and where $h_1 = b + (b^2 - 1)^{1/2}$ and $h_2 = b - (b^2 - 1)^{1/2}$, and $b = 2(1 + d/R_1)(1 + d/R_2) - 1$. *Hint*: Use the results in Sec. 1.4D.

9.2-5 **Ray Position in Unstable Symmetrical Resonators.** Verify that a symmetrical resonator using two concave mirrors of radii $R = -30$ cm separated by a distance $d = 65$ cm is unstable. Find the position y_1 of a ray that begins at one of the mirrors at position $y_0 = 0$ with an angle $\theta_0 = 0.1°$ after one round trip. If the mirrors have 5-cm-diameter apertures, after how many round trips does the ray leave the resonator? Write a computer program to plot y_m, $m = 2, 3, \dots,$ for $d = 50$ cm and $d = 65$ cm. You may use the results of Problem 9.2-4.

9.2-6 **Gaussian-Beam Standing Waves.** Consider a wave formed by the sum of two identical Gaussian beams propagating in the $+z$ and $-z$ directions. Show that the result is a standing wave. Using the boundary conditions at two ideal mirrors placed such that they coincide with the wavefronts, derive the resonance frequencies (9.2-24).

9.2-7 **Gaussian Beam in a Symmetrical Confocal Resonator.** A symmetrical confocal resonator with mirror spacing $d = 16$ cm, mirror reflectances 0.995, and $n = 1$ is used in a laser operating at $\lambda_o = 1\ \mu$m.
(a) Find the radii of curvature of the mirrors.
(b) Find the waist of the (0, 0) (Gaussian) mode.
(c) Sketch the intensity distribution of the (1, 0) mode at one of the mirrors and determine the distance between its two peaks.
(d) Determine the resonance frequencies of the (0, 0) and (1, 0) modes.
(e) Assuming that the only losses result from imperfect mirror reflectances, determine the resonator loss coefficient α_r.

*9.2-8 **Diffraction Loss.** The percent diffraction loss per pass for the different low-order modes of a symmetrical confocal resonator is given in Fig. 9.2-11, as a function of the Fresnel number $N_F = a^2/\lambda d$ (where d is the mirror spacing and a is the radius of the mirror aperture). Using the parameters provided in Problem 9.2-7, determine the mirror radius for which the loss per pass of the (1, 0) mode is 1%.

CHAPTER

10

STATISTICAL OPTICS

10.1 STATISTICAL PROPERTIES OF RANDOM LIGHT
- A. Optical Intensity
- B. Temporal Coherence and Spectrum
- C. Spatial Coherence
- D. Longitudinal Coherence

10.2 INTERFERENCE OF PARTIALLY COHERENT LIGHT
- A. Interference of Two Partially Coherent Waves
- B. Interference and Temporal Coherence
- C. Interference and Spatial Coherence

*10.3 TRANSMISSION OF PARTIALLY COHERENT LIGHT THROUGH OPTICAL SYSTEMS
- A. Propagation of Partially Coherent Light
- B. Image Formation with Incoherent Light
- C. Gain of Spatial Coherence by Propagation

10.4 PARTIAL POLARIZATION

Max Born (1882–1970)

Emil Wolf (born 1922)

Principles of Optics, first published in 1959 by Max Born and Emil Wolf, brought attention to the importance of coherence in optics. Emil Wolf is responsible for many advances in the theory of optical coherence.

Statistical optics is the study of the properties of random light. Randomness in light arises because of unpredictable fluctuations of the light source or of the medium through which light propagates. Natural light, e.g., light radiated by a hot object, is random because it is a superposition of emissions from a very large number of atoms radiating independently and at different frequencies and phases. Randomness in light may also be a result of scattering from rough surfaces, diffused glass, or turbulent fluids, which impart random variations to the optical wavefront. The study of the random fluctuations of light is also known as the **theory of optical coherence**.

In the preceding chapters it was assumed that light is deterministic or "coherent." An example of coherent light is the monochromatic wave $u(\mathbf{r}, t) = \operatorname{Re}\{U(\mathbf{r}) \exp(j2\pi\nu t)\}$, for which the complex amplitude $U(\mathbf{r})$ is a deterministic complex function, e.g., $U(\mathbf{r}) = A \exp(-jkr)/r$ in the case of a spherical wave [Fig. 10.0-1(*a*)]. The dependence of the wavefunction on time and position is perfectly periodic and predictable. On the other hand, for random light, the dependence of the wavefunction on time and position [Fig. 10.0-1(*b*)] is not totally predictable and cannot generally be described without resorting to statistical methods.

How can we extract from the fluctuations of a random optical wave some meaningful measures that characterize it and distinguish it from other random waves? Examine, for instance, the three random optical waves whose wavefunctions at some position vary with time as in Fig. 10.0-2. It is apparent that wave (*b*) is more "intense" than wave (*a*) and that the envelope of wave (*c*) fluctuates "faster" than the envelopes of the other two waves. To translate these casual qualitative observations into quantitative measures, we use the concept of statistical averaging to define a number of nonrandom measures. Because the random function $u(\mathbf{r}, t)$ satisfies certain laws (the wave equation and boundary conditions) its statistical averages must also satisfy certain laws. The theory of optical coherence deals with the definitions of these statistical averages, with

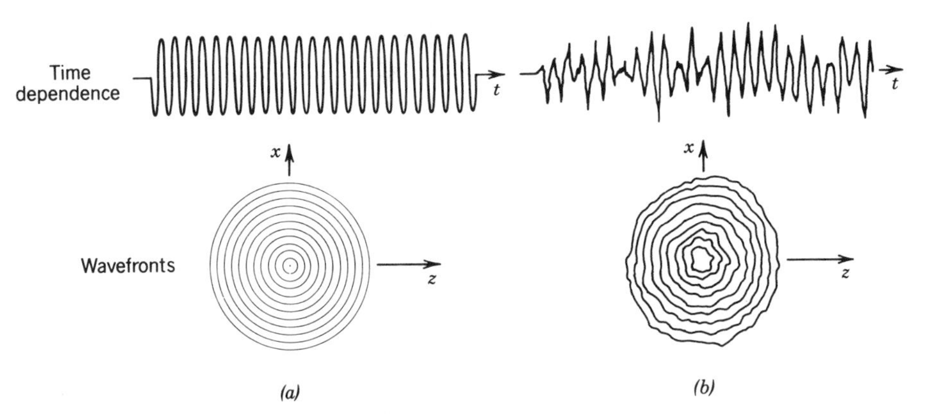

Figure 10.0-1 Time dependence and wavefronts of (*a*) a monochromatic spherical wave, which is an example of coherent light; (*b*) random light.

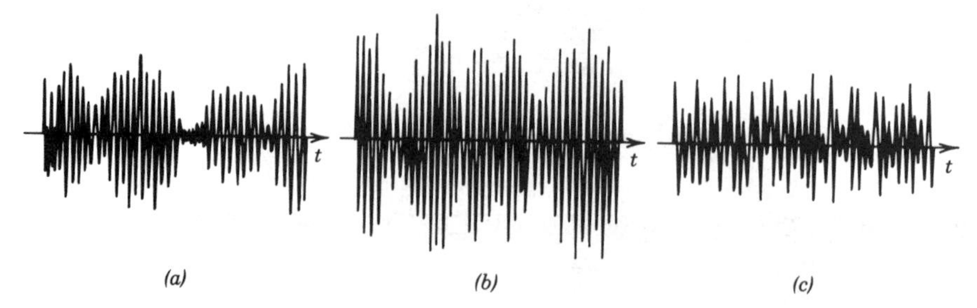

Figure 10.0-2 Time dependence of the wavefunctions of three random waves.

the laws that govern them, and with measures by which light is classified as **coherent**, **incoherent**, or, in general, **partially coherent**.

Familiarity with the theory of random fields (random functions of many variables—space and time) is necessary for a full understanding of the theory of optical coherence. However, the ideas presented in this chapter are limited in scope, so that knowledge of the concept of statistical averaging is sufficient.

In Sec. 10.1 we define two statistical averages used to describe random light: the optical intensity and the mutual coherence function. Temporal and spatial coherence are delineated, and the connection between temporal coherence and monochromaticity is established. The examples of partially coherent light provided in Sec. 10.1 demonstrate that spatially coherent light need not be temporally coherent, and that monochromatic light need not be spatially coherent. One of the basic manifestations of the coherence of light is its ability to produce visible interference fringes. Section 10.2 is devoted to the laws of interference of random light. The transmission of partially coherent light in free space and through different optical systems, including image-formation systems, is the subject of Sec. 10.3. A brief introduction to the theory of polarization of random light (partial polarization) is provided in Sec. 10.4.

10.1 STATISTICAL PROPERTIES OF RANDOM LIGHT

An arbitrary optical wave is described by a wavefunction $u(\mathbf{r}, t) = \mathrm{Re}\{U(\mathbf{r}, t)\}$, where $U(\mathbf{r}, t)$ is the complex wavefunction. For example, $U(\mathbf{r}, t)$ may take the form $U(\mathbf{r})\exp(j2\pi\nu t)$ for monochromatic light, or it may be a sum of many similar functions of different ν for polychromatic light (see Sec. 2.6A for a discussion of the complex wavefunction). For random light, both functions, $u(\mathbf{r}, t)$ and $U(\mathbf{r}, t)$, are random and are characterized by a number of statistical averages introduced in this section.

A. Optical Intensity

The intensity $I(\mathbf{r}, t)$ of coherent (deterministic) light is the absolute square of the complex wavefunction $U(\mathbf{r}, t)$,

$$I(\mathbf{r}, t) = |U(\mathbf{r}, t)|^2. \tag{10.1-1}$$

(see Sec. 2.2A, and 2.6A). For monochromatic deterministic light the intensity is independent of time, but for pulsed light it is time varying.

For random light, $U(\mathbf{r}, t)$ is a random function of time and position. The intensity $|U(\mathbf{r}, t)|^2$ is therefore also random. The **average intensity** is then defined as

$$I(\mathbf{r}, t) = \langle |U(\mathbf{r}, t)|^2 \rangle. \tag{10.1-2}$$

Average Intensity

where the symbol $\langle \cdot \rangle$ now denotes an ensemble average over many realizations of the random function. This means that the wave is produced repeatedly under the same conditions, with each trial yielding a different wavefunction, and the average intensity at each time and position is determined. When there is no ambiguity we shall simply call $I(\mathbf{r}, t)$ the *intensity* of light (with the word "average" implied). The quantity $|U(\mathbf{r}, t)|^2$ is called the **random** or **instantaneous intensity**. For deterministic light, the averaging operation is unnecessary since all trials produce the same wavefunction, so that (10.1-2) is equivalent to (10.1-1).

The average intensity may be time independent or may be a function of time, as illustrated in Figs. 10.1-1(*a*) and (*b*), respectively. The former case applies when the optical wave is statistically **stationary**; that is, its statistical averages are invariant to time. The instantaneous intensity $|U(\mathbf{r}, t)|^2$ fluctuates randomly with time, but its average is constant. We will denote it, in this case, by $I(\mathbf{r})$. Stationarity does not

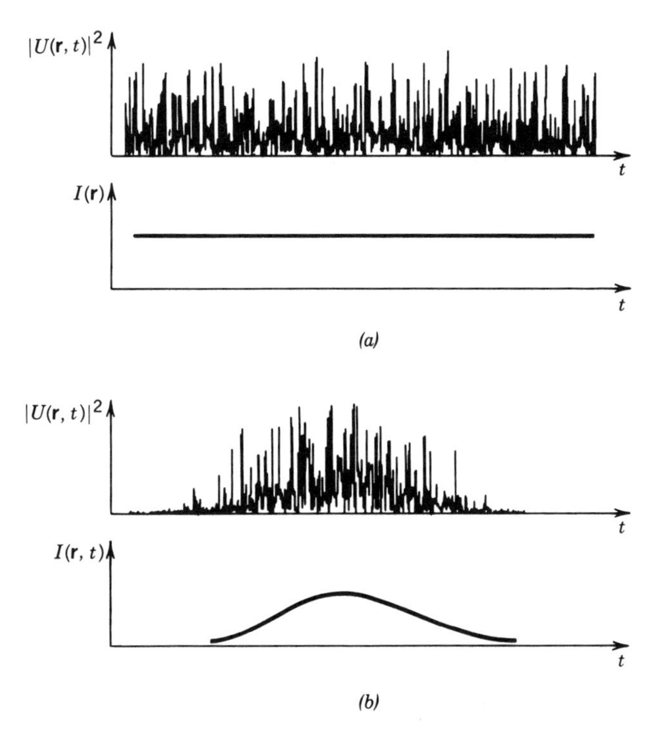

Figure 10.1-1 (*a*) A statistically stationary wave has an average intensity that does not vary with time. (*b*) A statistically nonstationary wave has a time-varying intensity. These plots represent, e.g., the intensity of light from an incandescent lamp driven by a constant electric current in (*a*) and a pulse of electric current in (*b*).

necessarily mean constancy. It means constancy of the average properties. An example of stationary random light is that from an ordinary incandescent lamp heated by a constant electric current. The average intensity $I(\mathbf{r})$ is a function of distance from the lamp, but it does not vary with time. However, the random intensity $|U(\mathbf{r}, t)|^2$ fluctuates with both position and time, as illustrated in Fig. 10.1-1(a).

When the light is stationary, the statistical averaging operation in (10.1-2) can usually be determined by time averaging over a long time duration (instead of averaging over many realizations of the wave), whereupon

$$I(\mathbf{r}) = \lim_{T\to\infty} \frac{1}{2T} \int_{-T}^{T} |U(\mathbf{r}, t)|^2 \, dt. \tag{10.1-3}$$

B. Temporal Coherence and Spectrum

Consider the fluctuations of *stationary* light at a fixed position $\mathbf{r}$ as a function of time. The stationary random function $U(\mathbf{r}, t)$ has a constant intensity $I(\mathbf{r}) = \langle |U(\mathbf{r}, t)|^2 \rangle$. For brevity, we drop the $\mathbf{r}$ dependence (since $\mathbf{r}$ is fixed), so that $U(\mathbf{r}, t) = U(t)$ and $I(\mathbf{r}) = I$.

The random fluctuations of $U(t)$ are characterized by a time scale representing the "memory" of the random function. Fluctuations at points separated by a time interval longer than the memory time are independent, so that the process "forgets" itself. The function appears to be smooth within its memory time, but "rough" and "erratic" when examined over longer time scales (see Fig. 10.0-2). A quantitative measure of this temporal behavior is established by defining a statistical *average* known as the autocorrelation function. This function describes the extent to which the wavefunction fluctuates in unison at two instants of time separated by a given time delay, so that it establishes the time scale of the process that underlies the generation of the wavefunction.

Temporal Coherence Function

The autocorrelation function of a stationary complex random function $U(t)$ is the average of the product of $U^*(t)$ and $U(t + \tau)$ as a function of the time delay τ

$$\boxed{G(\tau) = \langle U^*(t) U(t + \tau) \rangle} \tag{10.1-4}$$

Temporal Coherence Function

or

$$G(\tau) = \lim_{T\to\infty} \frac{1}{2T} \int_{-T}^{T} U^*(t) U(t + \tau) \, dt$$

(see Sec. A.1 in Appendix A).

To understand the significance of the definition in (10.1-4), consider the case in which the average value of the complex wavefunction $\langle U(t) \rangle = 0$. This is applicable when the phase of the phasor $U(t)$ is equally likely to have any value between 0 and 2π, as illustrated in Fig. 10.1-2. The phase of the product $U^*(t)U(t + \tau)$ is the angle between phasors $U(t)$ and $U(t + \tau)$. If $U(t)$ and $U(t + \tau)$ are uncorrelated, the angle between their phasors varies randomly between 0 and 2π. The phasor $U^*(t)U(t + \tau)$ then has a totally uncertain angle, so that it is equally likely to take any direction,

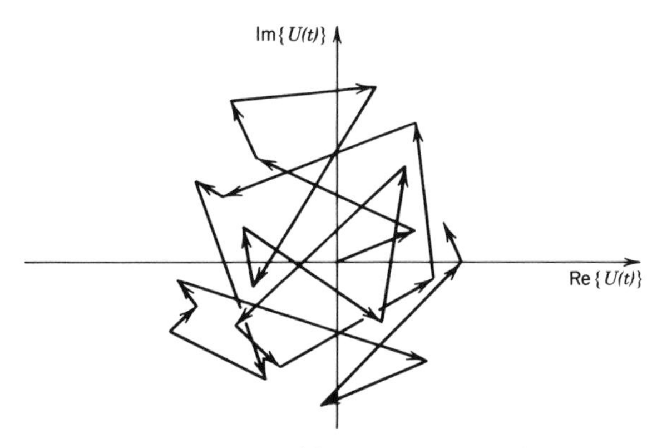

Figure 10.1-2 Variation of the phasor $U(t)$ with time when its argument is uniformly distributed between 0 and 2π. The average values of its real and imaginary parts are zero, so that $\langle U(t)\rangle = 0$.

making its average, the autocorrelation function $G(\tau)$, vanish. On the other hand if, for a given τ, $U(t)$ and $U(t+\tau)$ are correlated, their phasors will maintain some relationship. Their fluctuations are then linked together so that the product phasor $U^*(t)U(t+\tau)$ has a preferred direction and its average $G(\tau)$ will not vanish.

In the language of optical coherence theory, the autocorrelation function $G(\tau)$ is known as the **temporal coherence function**. It is easy to show that $G(\tau)$ is a function with Hermitian symmetry, $G(-\tau) = G^*(\tau)$, and that the intensity I, defined by (10.1-2), is equal to $G(\tau)$ when $\tau = 0$,

$$I = G(0). \tag{10.1-5}$$

Degree of Temporal Coherence

The temporal coherence function $G(\tau)$ carries information about both the intensity $I = G(0)$ and the degree of correlation (coherence) of stationary light. A measure of coherence that is insensitive to the intensity is provided by the normalized autocorrelation function,

$$g(\tau) = \frac{G(\tau)}{G(0)} = \frac{\langle U^*(t)U(t+\tau)\rangle}{\langle U^*(t)U(t)\rangle}, \tag{10.1-6}$$

Complex Degree of Temporal Coherence

which is called the **complex degree of temporal coherence**. Its absolute value cannot exceed unity,

$$0 \leq |g(\tau)| \leq 1. \tag{10.1-7}$$

The value of $|g(\tau)|$ is a measure of the degree of correlation between $U(t)$ and $U(t+\tau)$. When the light is deterministic and monochromatic, i.e., $U(t) = A\exp(j2\pi\nu_0 t)$, where A is a constant, (10.1-6) gives

$$g(\tau) = \exp(j2\pi\nu_0\tau), \tag{10.1-8}$$

so that $|g(\tau)| = 1$ for all τ. The variables $U(t)$ and $U(t+\tau)$ are then completely correlated for all time delays τ. Usually, $|g(\tau)|$ drops from its largest value $|g(0)| = 1$ as τ increases and the fluctuations become uncorrelated for sufficiently large time delay τ.

Coherence Time

If $|g(\tau)|$ decreases monotonically with time delay, the value τ_c at which it drops to a prescribed value ($\frac{1}{2}$ or $1/e$, for example) serves as a measure of the memory time of the fluctuations known as the **coherence time** (see Fig. 10.1-3). For $\tau < \tau_c$ the fluctuations are "strongly" correlated whereas for $\tau > \tau_c$ they are "weakly" correlated. In general, τ_c is the width of the function $|g(\tau)|$. Although the definition of the width of a function is rather arbitrary (see Sec. A.2 of Appendix A), the power-equivalent width

$$\tau_c = \int_{-\infty}^{\infty} |g(\tau)|^2 \, d\tau \qquad (10.1\text{-}9)$$

Coherence Time

is commonly used as the definition of coherence time [see (A.2-8) and note that $g(0) = 1$]. The coherence time of monochromatic light is infinite since $|g(\tau)| = 1$ everywhere.

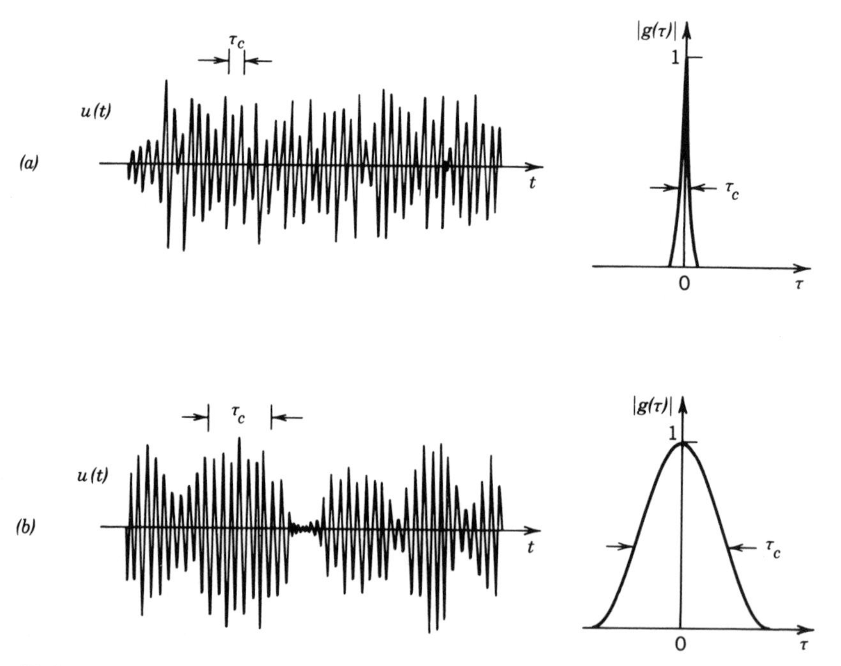

Figure 10.1-3 Illustrative examples of the wavefunction, the magnitude of the complex degree of temporal coherence $|g(\tau)|$, and the coherence time τ_c for an optical field with (*a*) short coherence time and (*b*) long coherence time. The amplitude and phase of the wavefunction vary randomly with time constants approximately equal to the coherence time. In both cases the coherence time τ_c is greater than the duration of an optical cycle. Within the coherence time, the wave is rather predictable and can be approximated as a sinusoid. However, given the amplitude and phase of the wave at a particular time, one cannot predict the amplitude and phase at times beyond the coherence time.

EXERCISE 10.1-1

Coherence Time. Verify that the following expressions for the complex degree of temporal coherence are consistent with the definition of τ_c given in (10.1-9):

$$g(\tau) = \begin{cases} \exp\left(-\dfrac{|\tau|}{\tau_c}\right) & \text{(exponential)} \\ \exp\left(-\dfrac{\pi\tau^2}{2\tau_c^2}\right) & \text{(Gaussian).} \end{cases}$$

By what factor does $|g(\tau)|$ drop as τ increases from 0 to τ_c in each case?

Light for which the coherence time τ_c is much longer than the differences of the time delays encountered in the optical system of interest is effectively completely coherent. Thus light is effectively coherent if the distance $c\tau_c$ is much greater than all optical path-length differences encountered. The distance

$$l_c = c\tau_c \tag{10.1-10}$$

Coherence Length

is known as the **coherence length**.

Power Spectral Density

To determine the *average* spectrum of random light, we carry out a Fourier decomposition of the random function $U(t)$. The amplitude of the component with frequency ν is the Fourier transform (see Appendix A)

$$V(\nu) = \int_{-\infty}^{\infty} U(t)\exp(-j2\pi\nu t)\,dt.$$

The average energy per unit area of those components with frequencies in the interval between ν and $\nu + d\nu$ is $\langle|V(\nu)|^2\rangle\,d\nu$, so that $\langle|V(\nu)|^2\rangle$ represents the energy spectral density of the light (energy per unit area per unit frequency). Note that the complex wavefunction $U(t)$ has been defined so that $V(\nu) = 0$ for negative ν (see Sec. 2.6A).

Since a truly stationary function $U(t)$ is eternal and carries infinite energy, we consider instead the *power* spectral density. We first determine the energy spectral density of the function $U(t)$ observed over a window of time width T by finding the truncated Fourier transform

$$V_T(\nu) = \int_{-T/2}^{T/2} U(t)\exp(-j2\pi\nu t)\,dt \tag{10.1-11}$$

and then determine the energy spectral density $\langle |V_T(\nu)|^2 \rangle$. The power spectral density is the energy per unit time $(1/T)\langle |V_T(\nu)|^2 \rangle$. We can now extend the time window to infinity by taking the limit $T \to \infty$. The result

$$S(\nu) = \lim_{T\to\infty} \frac{1}{T} \langle |V_T(\nu)|^2 \rangle, \tag{10.1-12}$$

is called the **power spectral density**. It is nonzero only for positive frequencies. Because $U(t)$ was defined such that $|U(t)|^2$ represents power per unit area, or intensity (W/cm^2), $S(\nu)\,d\nu$ represents the average power per unit area carried by frequencies between ν and $\nu + d\nu$, so that $S(\nu)$ actually represents the intensity spectral density ($\text{W/cm}^2\text{-Hz}$). It is often referred to simply as the **spectral density** or the **spectrum**. The total average intensity is the integral

$$I = \int_0^\infty S(\nu)\,d\nu. \tag{10.1-13}$$

The autocorrelation function $G(\tau)$, defined by (10.1-4), and the spectral density $S(\nu)$ defined by (10.1-12) can be shown to form a Fourier transform pair (see Problem 10.1-2),

$$S(\nu) = \int_{-\infty}^{\infty} G(\tau) \exp(-j2\pi\nu\tau)\,d\tau. \tag{10.1-14}$$

Power Spectral Density

This relation is known as the **Wiener–Khinchin theorem**.

An optical wave representing a color image, such as the illustration in Fig. 10.1-4, has a spectrum that varies with position **r**; each spectral profile shown corresponds to a perceived color.

Figure 10.1-4 Variation of the spectral density as a function of wavelength at three positions in a color image (Bouquet of Flowers in a White Vase, Henri Matisse, Pushkin Museum of Fine Arts, Moscow).

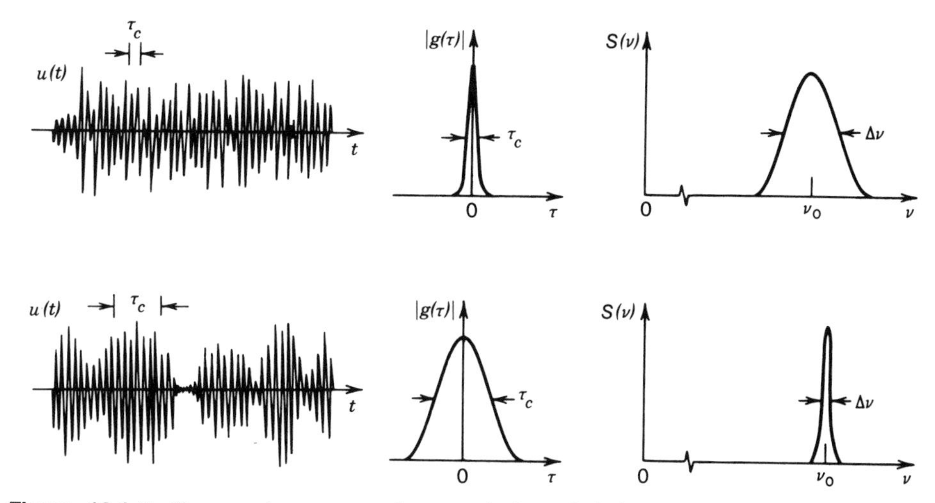

Figure 10.1-5 Two random waves, the magnitudes of their complex degree of temporal coherence, and their spectral densities.

Spectral Width

The spectrum of light is often confined to a narrow band centered about a central frequency ν_0. The **spectral width**, or **linewidth**, of light is the width $\Delta\nu$ of the spectral density $S(\nu)$. Because of the Fourier-transform relation between $S(\nu)$ and $G(\tau)$, their widths are inversely related. A light source of broad spectrum has a short coherence time, whereas a light source with narrow linewidth has a long coherence time, as illustrated in Fig. 10.1-5. In the limiting case of monochromatic light, $G(\tau) = I\exp(j2\pi\nu_0\tau)$, so that the corresponding intensity spectral density $S(\nu) = I\delta(\nu - \nu_0)$ contains only a single frequency component, ν_0. Thus $\tau_c = \infty$ and $\Delta\nu = 0$. The coherence time of a light source can be increased by using an optical filter to reduce its spectral width. The resultant gain of coherence comes at the expense of losing light energy.

There are several definitions for the spectral width. The most common is the full width of the function $S(\nu)$ at half its maximum value (FWHM). The relation between the coherence time and the spectral width depends on the spectral profile, as indicated in Table 10.1-1 (see also Appendix A, Sec. A.2).

TABLE 10.1-1 Relation Between Spectral Width and Coherence Time

Spectral Density	Spectral Width $\Delta\nu_{\text{FWHM}}$
Rectangular	$\dfrac{1}{\tau_c}$
Lorentzian	$\dfrac{1}{\pi\tau_c} \approx \dfrac{0.32}{\tau_c}$
Gaussian	$\dfrac{(2\ln 2/\pi)^{1/2}}{\tau_c} \approx \dfrac{0.66}{\tau_c}$

Another convenient definition of the spectral width is

$$\Delta\nu_c = \frac{\left(\int_0^\infty S(\nu)\,d\nu\right)^2}{\int_0^\infty S^2(\nu)\,d\nu}. \tag{10.1-15}$$

By this definition it can be shown that

$$\boxed{\Delta\nu_c = \frac{1}{\tau_c},} \tag{10.1-16}$$

Spectral Width

regardless of the spectral profile (see Exercise 10.1-2). If $S(\nu)$ is a rectangular function extending over a frequency interval from $\nu_0 - B/2$ to $\nu_0 + B/2$, for example, then (10.1-15) yields $\Delta\nu_c = B$. The two definitions of bandwidth, $\Delta\nu_c$ and $\Delta\nu_{\text{FWHM}} \equiv \Delta\nu$, differ by a factor that ranges from $1/\pi \approx 0.32$ to 1 for the profiles listed in Table 10.1-1.

EXERCISE 10.1-2

Relation Between Spectral Width and Coherence Time. Show that the coherence time τ_c defined by (10.1-9) is related to the spectral width $\Delta\nu_c$ defined in (10.1-15) by the simple inverse relation $\tau_c = 1/\Delta\nu_c$. *Hint*: Use the definitions of $\Delta\nu_c$ and τ_c, the Fourier transform relation between $S(\nu)$ and $G(\tau)$, and Parseval's theorem [see (A.1-7) in Appendix A].

Representative spectral bandwidths for different light sources, and their associated coherence times and coherence lengths $l_c = c\tau_c$, are provided in Table 10.1-2.

TABLE 10.1-2 Spectral Widths of a Number of Light Sources Together with Their Coherence Times and Coherence Lengths in Free Space

Source	$\Delta\nu_c$ (Hz)	$\tau_c = 1/\Delta\nu_c$	$l_c = c\tau_c$
Filtered sunlight (λ_o = 0.4–0.8 μm)	3.75×10^{14}	2.67 fs	800 nm
Light-emitting diode (λ_o = 1 μm, $\Delta\lambda_o$ = 50 nm)	1.5×10^{13}	67 fs	20 μm
Low-pressure sodium lamp	5×10^{11}	2 ps	600 μm
Multimode He–Ne laser (λ_o = 633 nm)	1.5×10^{9}	0.67 ns	20 cm
Single-mode He–Ne laser (λ_o = 633 nm)	1×10^{6}	1 μs	300 m

EXAMPLE 10.1-1. ***A Wave Comprising a Random Sequence of Wavepackets.*** Light emitted from an incoherent source may be modeled as a sequence of wavepackets emitted at random times (Fig. 10.1-6). Each wavepacket has a random phase since it is emitted by a different atom. The wavepackets may be sinusoidal with an exponentially decaying envelope, for example, so that a wavepacket emitted at $t = 0$ has a complex wavefunction (at a given position)

$$U_p(t) = \begin{cases} A_p \exp\left(-\dfrac{t}{\tau_c}\right) \exp(j2\pi\nu_0 t), & t \geq 0 \\ 0, & t < 0. \end{cases}$$

The emission times are totally random, and the random independent phases of the different emissions are included in A_p. The statistical properties of the total field may be determined by performing the necessary averaging operations using the rules of mathematical statistics. The result yields a complex degree of coherence given by $g(\tau) = \exp(-|\tau|/\tau_c)\exp(j2\pi\nu_0\tau)$ whose magnitude is a double-sided exponential function. The corresponding power spectral density is Lorentzian, $S(\nu) = (\Delta\nu/2\pi)/[(\nu - \nu_0)^2 + (\Delta\nu/2)^2]$, where $\Delta\nu = 1/\pi\tau_c$ (see Table A.1-1 in Appendix A). The coherence time τ_c in this case is exactly the width of a wavepacket. The statement that this light is correlated within the coherence time therefore means that it is correlated within the duration of an individual wavepacket.

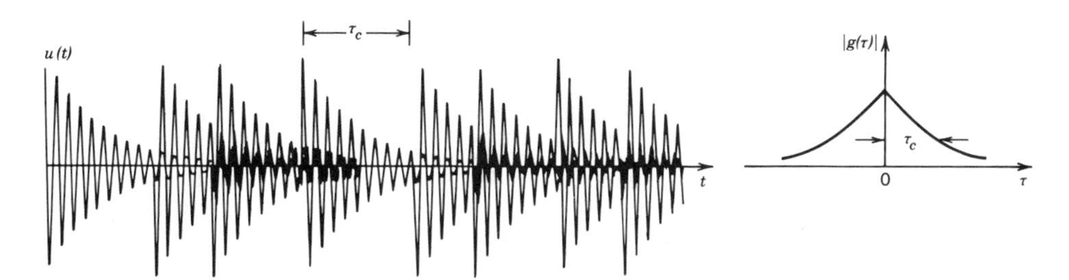

Figure 10.1-6 Light comprised of wavepackets emitted at random times has a coherence time equal to the duration of a wavepacket.

C. Spatial Coherence

Mutual Coherence Function

An important descriptor of the spatial and temporal fluctuations of the random function $U(\mathbf{r}, t)$ is the cross-correlation function of $U(\mathbf{r}_1, t)$ and $U(\mathbf{r}_2, t)$ at pairs of positions $\mathbf{r}_1$ and $\mathbf{r}_2$,

$$G(\mathbf{r}_1, \mathbf{r}_2, \tau) = \langle U^*(\mathbf{r}_1, t) U(\mathbf{r}_2, t + \tau) \rangle. \tag{10.1-17}$$

Mutual Coherence Function

This function of the time delay τ is known as the **mutual coherence function**. Its normalized form,

$$g(\mathbf{r}_1, \mathbf{r}_2, \tau) = \frac{G(\mathbf{r}_1, \mathbf{r}_2, \tau)}{\left[I(\mathbf{r}_1) I(\mathbf{r}_2)\right]^{1/2}}, \tag{10.1-18}$$

Complex Degree of Coherence

is called the **complex degree of coherence**. When the two points coincide so that $\mathbf{r}_1 = \mathbf{r}_2 = \mathbf{r}$, (10.1-17) and (10.1-18) reproduce the temporal coherence function and the complex degree of temporal coherence defined in (10.1-4) and (10.1-6) at the position $\mathbf{r}$. Ultimately, when $\tau = 0$, the intensity is $I(\mathbf{r}) = G(\mathbf{r}, \mathbf{r}, 0)$ at the position $\mathbf{r}$.

The complex degree of coherence $g(\mathbf{r}_1, \mathbf{r}_2, \tau)$ is the cross-correlation coefficient of the random variables $U^*(\mathbf{r}_1, t)$ and $U(\mathbf{r}_2, t + \tau)$. Its absolute value is bounded between zero and unity,

$$0 \le |g(\mathbf{r}_1, \mathbf{r}_2, \tau)| \le 1. \tag{10.1-19}$$

It is therefore considered a measure of the degree of correlation between the fluctuations at $\mathbf{r}_1$ and those at $\mathbf{r}_2$ at a time τ later.

When the two phasors $U(\mathbf{r}_1, t)$ and $U(\mathbf{r}_2, t)$ fluctuate independently and their phases are totally random (each having equally probable phase between 0 and 2π), $|g(\mathbf{r}_1, \mathbf{r}_2, \tau)| = 0$ since the average of the product $U^*(\mathbf{r}_1, t)U(\mathbf{r}_2, t + \tau)$ vanishes. The light fluctuations at the two points are then uncorrelated. The other limit, $|g(\mathbf{r}_1, \mathbf{r}_2, \tau)| = 1$, applies when the light fluctuations at $\mathbf{r}_1$, and at $\mathbf{r}_2$ a time τ later, are fully correlated. Note that $|g(\mathbf{r}_1, \mathbf{r}_2, 0)|$ is not necessarily unity, however by definition $|g(\mathbf{r}, \mathbf{r}, 0)| = 1$.

The dependence of $g(\mathbf{r}_1, \mathbf{r}_2, \tau)$ on time delay and on the positions characterizes the temporal and spatial coherence of light. Two examples of the dependence of $|g(\mathbf{r}_1, \mathbf{r}_2, \tau)|$ on the distance $|\mathbf{r}_1 - \mathbf{r}_2|$ and the time delay τ are illustrated in Fig. 10.1-7.

The temporal and spatial fluctuations of light are intimately related since light propagates in waves and the complex wavefunction $U(\mathbf{r}, t)$ must satisfy the wave

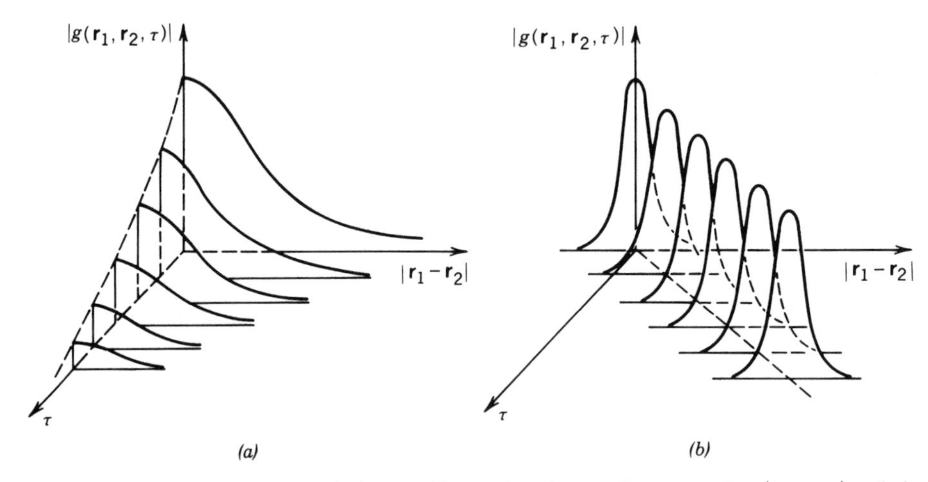

Figure 10.1-7 Two examples of $|g(\mathbf{r}_1, \mathbf{r}_2, \tau)|$ as a function of the separation $|\mathbf{r}_1 - \mathbf{r}_2|$ and the time delay τ. In (*a*) the maximum correlation for a given $|\mathbf{r}_1 - \mathbf{r}_2|$ occurs at $\tau = 0$. In (*b*) the maximum correlation occurs at $|\mathbf{r}_1 - \mathbf{r}_2| = c\tau$.

equation. This imposes certain conditions on the mutual coherence function (see Exercise 10.1-3). To illustrate this point, consider, for example, a plane wave of random light traveling in the z direction in a homogeneous and nondispersive medium with velocity c. Fluctuations at the points $\mathbf{r}_1 = (0, 0, z_1)$ and $\mathbf{r}_2 = (0, 0, z_2)$ are completely correlated when the time delay is $\tau = \tau_0 \equiv |z_2 - z_1|/c$, so that $|g(\mathbf{r}_1, \mathbf{r}_2, \tau_0)| = 1$. As a function of τ, $|g(\mathbf{r}_1, \mathbf{r}_2, \tau)|$ has a peak at $\tau = \tau_0$, as illustrated in Fig. 10.1-7(*b*). This example will be discussed again in Sec. 10.1D.

EXERCISE 10.1-3

Differential Equations Governing the Mutual Coherence Function. In free space, $U(\mathbf{r}, t)$ must satisfy the wave equation, $\nabla^2 U - (1/c^2)\partial^2 U/\partial t^2 = 0$. Use the definition (10.1-17) to show that the mutual coherence function $G(\mathbf{r}_1, \mathbf{r}_2, \tau)$ satisfies the two partial differential equations

$$\nabla_1^2 G - \frac{1}{c^2}\frac{\partial^2 G}{\partial \tau^2} = 0 \qquad (10.1\text{-}20a)$$

$$\nabla_2^2 G - \frac{1}{c^2}\frac{\partial^2 G}{\partial \tau^2} = 0, \qquad (10.1\text{-}20b)$$

where ∇_1^2 and ∇_2^2 are the Laplacian operators with respect to $\mathbf{r}_1$ and $\mathbf{r}_2$, respectively.

Mutual Intensity

The spatial correlation of light may be assessed by examining the dependence of the mutual coherence function on position for a fixed time delay τ. In many situations the point $\tau = 0$ is the most appropriate, as in the example in Fig. 10.1-7(*a*). However, this need not always be the case, as in the example in Fig. 10.1-7(*b*). The mutual coherence function at $\tau = 0$,

$$G(\mathbf{r}_1, \mathbf{r}_2, 0) = \langle U^*(\mathbf{r}_1, t) U(\mathbf{r}_2, t)\rangle,$$

is known as the **mutual intensity** and is denoted by $G(\mathbf{r}_1, \mathbf{r}_2)$ for simplicity. The diagonal values of the mutual intensity ($\mathbf{r}_1 = \mathbf{r}_2 = \mathbf{r}$) provide the intensity $I(\mathbf{r}) = G(\mathbf{r}, \mathbf{r})$.

When the optical path differences encountered in an optical system are much shorter than the coherence length $l_c = c\tau_c$, the light may be considered to effectively possess complete temporal coherence, so that the mutual coherence function is a harmonic function of time:

$$G(\mathbf{r}_1, \mathbf{r}_2, \tau) = G(\mathbf{r}_1, \mathbf{r}_2) \exp(j2\pi\nu_0\tau), \qquad (10.1\text{-}21)$$

where ν_0 is the central frequency. In this case the light is said to be **quasi-monochromatic** and the mutual intensity $G(\mathbf{r}_1, \mathbf{r}_2)$ describes the spatial coherence completely.

The complex degree of coherence $g(\mathbf{r}_1, \mathbf{r}_2, 0)$ is similarly denoted by $g(\mathbf{r}_1, \mathbf{r}_2)$. Thus

$$g(\mathbf{r}_1, \mathbf{r}_2) = \frac{G(\mathbf{r}_1, \mathbf{r}_2)}{[I(\mathbf{r}_1) I(\mathbf{r}_2)]^{1/2}} \qquad (10.1\text{-}22)$$

Normalized Mutual Intensity

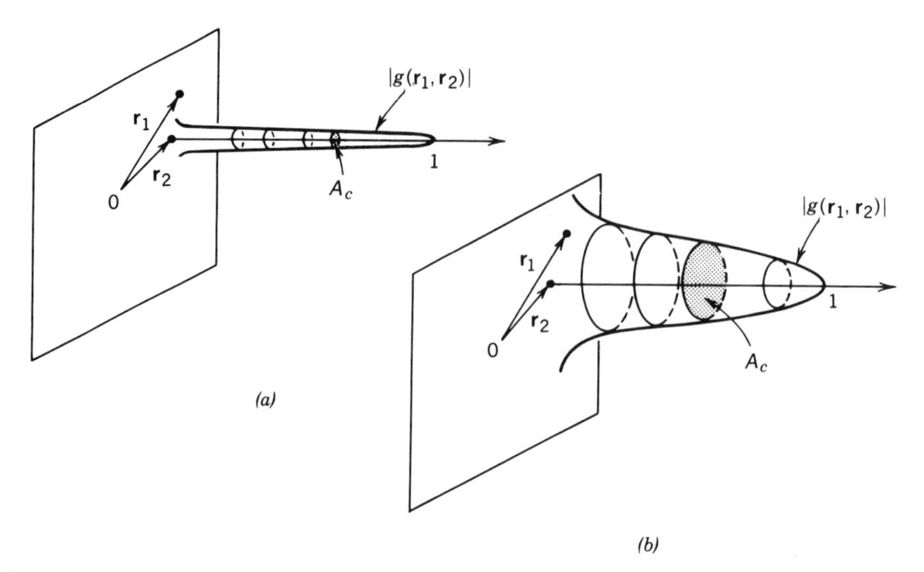

Figure 10.1-8 Two illustrative examples of the magnitude of the normalized mutual intensity as a function of $\mathbf{r}_1$ in the vicinity of a fixed point $\mathbf{r}_2$. The coherence area in (a) is smaller than that in (b).

is the normalized mutual intensity. The magnitude $|g(\mathbf{r}_1, \mathbf{r}_2)|$ is bounded between zero and unity and is regarded as a measure of the degree of spatial coherence (when the time delay τ is zero). If the complex wavefunction $U(\mathbf{r}, t)$ is deterministic, $|g(\mathbf{r}_1, \mathbf{r}_2)| = 1$ for all $\mathbf{r}_1$ and $\mathbf{r}_2$, so that the light is completely correlated everywhere.

Coherence Area

The spatial coherence of quasi-monochromatic light in a given plane in the vicinity of a given position $\mathbf{r}_2$ is described by $|g(\mathbf{r}_1, \mathbf{r}_2)|$ as a function of the distance $|\mathbf{r}_1 - \mathbf{r}_2|$. This function is unity when $\mathbf{r}_1 = \mathbf{r}_2$ and drops as $|\mathbf{r}_1 - \mathbf{r}_2|$ increases (but it need not be monotonic). The area scanned by the point $\mathbf{r}_1$ within which the function $|g(\mathbf{r}_1, \mathbf{r}_2)|$ is greater than some prescribed value ($\frac{1}{2}$ or $\frac{1}{e}$, for example) is called the **coherence area.** It represents the spatial extent of $|g(\mathbf{r}_1, \mathbf{r}_2)|$ as a function of $\mathbf{r}_1$ for fixed $\mathbf{r}_2$, as illustrated in Fig. 10.1-8. In the ideal limit of coherent light the coherence area is infinite.

The coherence area is an important parameter that characterizes random light. This parameter must be considered in relation to other pertinent dimensions of the optical system. For example, if the area of coherence is greater than the size of the aperture through which light is transmitted, so that $|g(\mathbf{r}_1, \mathbf{r}_2)| \approx 1$ at all points of interest, the light may be regarded as coherent, as if the coherence area were infinite. Similarly, if the coherence area is smaller than the resolution of the optical system, it can be regarded as infinitesimal, i.e., $g(\mathbf{r}_1, \mathbf{r}_2) = 0$ for practically all $\mathbf{r}_1 \neq \mathbf{r}_2$. In this limit the light is said to be **incoherent**.

Light radiated from an extended radiating hot surface has an area of coherence on the order of λ^2, where λ is the central wavelength, so that for most practical cases it may be regarded as incoherent. Thus complete coherence and incoherence are only idealizations representing the two limits of partial coherence.

Cross-Spectral Density

The mutual coherence function $G(\mathbf{r}_1, \mathbf{r}_2, \tau)$ describes the spatial correlation at each time delay τ. The time delay $\tau = 0$ is selected to define the mutual intensity $G(\mathbf{r}_1, \mathbf{r}_2)$

$= G(\mathbf{r}_1, \mathbf{r}_2, 0)$, which is suitable for describing the spatial coherence of quasi-monochromatic light. A useful alternative is to describe spatial coherence in the frequency domain by examining the spatial correlation at a fixed frequency. The **cross-spectral density** (or the cross-power spectrum) is defined as the Fourier transform of $G(\mathbf{r}_1, \mathbf{r}_2, \tau)$ with respect to τ:

$$\mathcal{S}(\mathbf{r}_1, \mathbf{r}_2, \nu) = \int_{-\infty}^{\infty} G(\mathbf{r}_1, \mathbf{r}_2, \tau) \exp(-j2\pi\nu\tau)\, d\tau. \tag{10.1-23}$$

Cross-Spectral Density

When $\mathbf{r}_1 = \mathbf{r}_2 = \mathbf{r}$, the cross-spectral density becomes the power-spectral density $\mathcal{S}(\nu)$ at position $\mathbf{r}$, as defined in (10.1-14).

The normalized cross-spectral density is defined by

$$s(\mathbf{r}_1, \mathbf{r}_2, \nu) = \frac{\mathcal{S}(\mathbf{r}_1, \mathbf{r}_2, \nu)}{[\mathcal{S}(\mathbf{r}_1, \mathbf{r}_1, \nu)\mathcal{S}(\mathbf{r}_2, \mathbf{r}_2, \nu)]^{1/2}}, \tag{10.1-24}$$

and its magnitude can be shown to be bounded between zero and unity, so that it serves as a measure of the degree of spatial coherence at the frequency ν. It represents the correlatedness of the fluctuation components of frequency ν at positions $\mathbf{r}_1$ and $\mathbf{r}_2$.

In certain cases, the cross-spectral density factors into a product of one function of position and another of frequency, $\mathcal{S}(\mathbf{r}_1, \mathbf{r}_2, \nu) = G(\mathbf{r}_1, \mathbf{r}_2)s(\nu)$, so that the spatial and spectral properties are separable. The light is then said to be **cross-spectrally pure**. The mutual coherence function must then also factor into a product of a function of position and another of time, $G(\mathbf{r}_1, \mathbf{r}_2, \tau) = G(\mathbf{r}_1, \mathbf{r}_2)g(\tau)$, where $g(\tau)$ is the inverse Fourier transform of $s(\nu)$. If the factorization parts are selected such that $\int s(\nu)\, d\nu = 1$, then $G(\mathbf{r}_1, \mathbf{r}_2) = G(\mathbf{r}_1, \mathbf{r}_2, 0)$, so that $G(\mathbf{r}_1, \mathbf{r}_2)$ is nothing but the mutual intensity. Cross-spectrally pure light has two important properties:

- At a single position $\mathbf{r}$, $\mathcal{S}(\mathbf{r}, \mathbf{r}, \nu) = G(\mathbf{r}, \mathbf{r})s(\nu) = I(\mathbf{r})s(\nu)$. The spectrum has the same profiles at all positions. If the light represents a visible image, it would appear to have the same color everywhere but with varying brightness.
- The normalized cross-spectral density

$$s(\mathbf{r}_1, \mathbf{r}_2, \nu) = G(\mathbf{r}_1, \mathbf{r}_2) / [G(\mathbf{r}_1, \mathbf{r}_1)G(\mathbf{r}_2, \mathbf{r}_2)]^{1/2} = g(\mathbf{r}_1, \mathbf{r}_2)$$

  is independent of frequency. In this case the normalized mutual intensity $g(\mathbf{r}_1, \mathbf{r}_2)$ describes spatial coherence at all frequencies.

D. Longitudinal Coherence

In this section the concept of longitudinal coherence is introduced by taking examples of random waves with fixed wavefronts, such as planar and spherical waves.

Partially Coherent Plane Wave

Consider a plane wave

$$U(\mathbf{r}, t) = a\left(t - \frac{z}{c}\right) \exp\left[j2\pi\nu_0\left(t - \frac{z}{c}\right)\right] \tag{10.1-25}$$

traveling in the z direction in a homogeneous medium with velocity c. As shown in Sec.

2.6A, $U(\mathbf{r}, t)$ satisfies the wave equation for an arbitrary function $a(t)$. If $a(t)$ is a random function, $U(\mathbf{r}, t)$ represents partially coherent light. The mutual coherence function defined in (10.1-17) is

$$G(\mathbf{r}_1, \mathbf{r}_2, \tau) = G_a\left(\tau - \frac{z_2 - z_1}{c}\right) \exp\left[j2\pi\nu_0\left(\tau - \frac{z_2 - z_1}{c}\right)\right], \quad (10.1\text{-}26)$$

where z_1 and z_2 are the z components of $\mathbf{r}_1$ and $\mathbf{r}_2$ and $G_a(\tau) = \langle a^*(t)a(t+\tau)\rangle$ is the autocorrelation function of $a(t)$, assumed to be independent of t.

The intensity $I(\mathbf{r}) = G(\mathbf{r}, \mathbf{r}, 0) = G_a(0)$ is constant everywhere in space. Temporal coherence is characterized by the time function $G(\mathbf{r}, \mathbf{r}, \tau) = G_a(\tau)\exp(j2\pi\nu_0\tau)$, which is independent of position. The complex degree of coherence is $g(\mathbf{r}, \mathbf{r}, \tau) = g_a(\tau)\exp(j2\pi\nu_0\tau)$, where $g_a(\tau) = G_a(\tau)/G_a(0)$. The width of $|g_a(\tau)| = |g(\mathbf{r}, \mathbf{r}, \tau)|$, defined by an expression similar to (10.1-9), is the coherence time τ_c. It is the same at all positions.

The power spectral density is the Fourier transform of $G(\mathbf{r}, \mathbf{r}, \tau)$ with respect to τ. From (10.1-26), $S(\nu)$ is seen to be equal to the Fourier transform of $G_a(\tau)$ shifted by a frequency ν_0 (in accordance with the frequency shift property of the Fourier transform defined in Appendix A, Sec. A.1). The wave therefore has the same power spectral density everywhere in space.

The spatial coherence properties are described by

$$G(\mathbf{r}_1, \mathbf{r}_2, 0) = G_a\left(\frac{z_1 - z_2}{c}\right) \exp\left[\frac{j2\pi\nu_0(z_1 - z_2)}{c}\right] \quad (10.1\text{-}27)$$

and its normalized version,

$$g(\mathbf{r}_1, \mathbf{r}_2, 0) = g_a\left(\frac{z_1 - z_2}{c}\right) \exp\left[\frac{j2\pi\nu_0(z_1 - z_2)}{c}\right]. \quad (10.1\text{-}28)$$

If the two points $\mathbf{r}_1$ and $\mathbf{r}_2$ lie in the same transverse plane, i.e., $z_1 = z_2$, then $|g(\mathbf{r}_1, \mathbf{r}_2, 0)| = |g_a(0)| = 1$. This means that fluctuations at points on a wavefront (a plane normal to the z axis) are completely correlated; the coherence area in any transverse plane is infinite (Fig. 10.1-9). On the other hand, fluctuations at two points separated by an axial distance $z_2 - z_1$ such that $|z_2 - z_1|/c > \tau_c$, or $|z_2 - z_1| > l_c$, where $l_c = c\tau_c$ is the coherence length, are approximately uncorrelated.

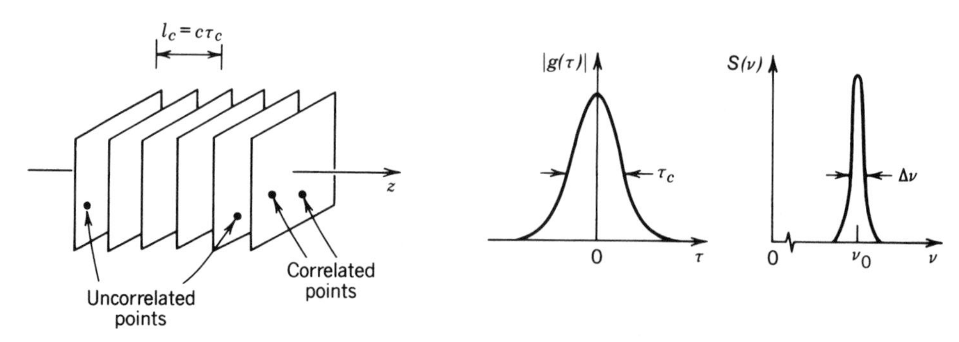

Figure 10.1-9 The fluctuations of a partially coherent plane wave at points on any wavefront (transverse plane) are completely correlated, whereas those at points on wavefronts separated by an axial distance greater than the coherence length $l_c = c\tau_c$ are approximately uncorrelated.

In summary: The partially coherent plane wave is spatially coherent within each transverse plane, but partially coherent in the axial direction. The axial (longitudinal) spatial coherence of the wave has a one-to-one correspondence with the temporal coherence. The ratio of the coherence length $l_c = c\tau_c$ to the maximum optical path difference l_{max} in the system governs the role played by coherence. If $l_c \gg l_{max}$, the wave is effectively completely coherent. The coherence lengths of a number of light sources are listed in Table 10.1-2.

Partially Coherent Spherical Wave

A partially coherent spherical wave is described by the complex wavefunction (see Secs. 2.2B and 2.6A)

$$U(\mathbf{r}, t) = \frac{1}{r} a\left(t - \frac{r}{c}\right) \exp\left[j2\pi\nu_0\left(t - \frac{r}{c}\right)\right], \tag{10.1-29}$$

where $a(t)$ is a random function. The corresponding mutual coherence function is

$$G(\mathbf{r}_1, \mathbf{r}_2, \tau) = \frac{1}{r_1 r_2} G_a\left(\tau - \frac{r_2 - r_1}{c}\right) \exp\left[j2\pi\nu_0\left(\tau - \frac{r_2 - r_1}{c}\right)\right], \tag{10.1-30}$$

with $G_a(\tau) = \langle a^*(t) a(t + \tau)\rangle$.

The intensity $I(\mathbf{r}) = G_a(0)/r^2$ varies in accordance with an inverse-square law. The coherence time τ_c is the width of the function $|g_a(\tau)| = |G_a(\tau)/G_a(0)|$. It is the same everywhere in space. So is the power spectral density. For $\tau = 0$, fluctuations at all points on a wavefront (a sphere) are completely correlated, whereas fluctuations at points on two wavefronts separated by the radial distance $|r_2 - r_1| \gg l_c = c\tau_c$ are uncorrelated (see Fig. 10.1-10).

An arbitrary partially coherent wave transmitted through a pinhole generates a partially coherent spherical wave. This process therefore imparts spatial coherence to the incoming wave (points on any sphere centered about the pinhole become completely correlated). However, the wave remains temporally partially coherent. Points at different distances from the pinhole are only partially correlated. *The pinhole imparts spatial coherence but not temporal coherence to the wave.*

Suppose now that an optical filter of very narrow spectral width is placed at the pinhole, so that the transmitted wave becomes approximately monochromatic. The wave will then have complete temporal, as well as spatial, coherence. Temporal coherence is introduced by the narrowband filter, whereas spatial coherence is imparted by the pinhole, which acts as a spatial filter. The price for obtaining this ideal wave is, of course, the loss of optical energy introduced by the temporal and spatial filtering processes.

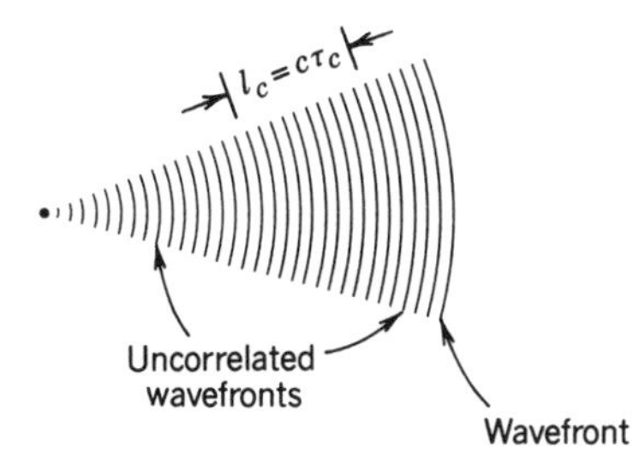

Figure 10.1-10 A partially coherent spherical wave has complete spatial coherence at all points on a wavefront, but not at points with different radial distances.

10.2 INTERFERENCE OF PARTIALLY COHERENT LIGHT

The interference of *coherent* light was discussed in Sec. 2.5. This section is devoted to the interference of *partially coherent* light.

A. Interference of Two Partially Coherent Waves

The statistical properties of two partially coherent waves U_1 and U_2 are described not only by their own mutual coherence functions but also by a measure of the degree to which their fluctuations are correlated. At a given position **r** and time t, the intensities of the two waves are $I_1 = \langle |U_1|^2 \rangle$ and $I_2 = \langle |U_2|^2 \rangle$, whereas their cross-correlation is described by the statistical average $G_{12} = \langle U_1^* U_2 \rangle$, and its normalized version

$$g_{12} = \frac{\langle U_1^* U_2 \rangle}{(I_1 I_2)^{1/2}}. \tag{10.2-1}$$

When the two waves are superposed, the average intensity of their sum is

$$\begin{aligned} I &= \langle |U_1 + U_2|^2 \rangle = \langle |U_1|^2 \rangle + \langle |U_2|^2 \rangle + \langle U_1^* U_2 \rangle + \langle U_1 U_2^* \rangle \\ &= I_1 + I_2 + G_{12} + G_{12}^* = I_1 + I_2 + 2\,\mathrm{Re}\{G_{12}\} \\ &= I_1 + I_2 + 2(I_1 I_2)^{1/2}\,\mathrm{Re}\{g_{12}\}, \end{aligned} \tag{10.2-2}$$

from which

$$I = I_1 + I_2 + 2(I_1 I_2)^{1/2} |g_{12}| \cos\varphi, \tag{10.2-3}$$

Interference Equation

where $\varphi = \arg\{g_{12}\}$ is the phase of g_{12}. The third term on the right-hand side of (10.2-3) represents optical interference.

There are two important limits:

- For two *completely correlated* waves with $g_{12} = \exp(j\varphi)$ and $|g_{12}| = 1$, we recover the interference formula (2.5-4) for two coherent waves of phase difference φ.
- For two *uncorrelated* waves with $g_{12} = 0$, $I = I_1 + I_2$, so that there is no interference.

In the general case, the normalized intensity I versus the phase φ assumes the form of a sinusoidal pattern, as shown in Fig. 10.2-1. The strength of the interference is measured by the visibility $\mathscr{V}$, also called the modulation depth or the contrast of the interference pattern

$$\mathscr{V} = \frac{I_{\max} - I_{\min}}{I_{\max} + I_{\min}},$$

where $I_{\max}$ and $I_{\min}$ are the maximum and minimum values that I takes as φ is varied. Since $\cos\varphi$ stretches between 1 and -1, (10.2-3) yields

$$\mathscr{V} = \frac{2(I_1 I_2)^{1/2}}{(I_1 + I_2)} |g_{12}|. \tag{10.2-4}$$

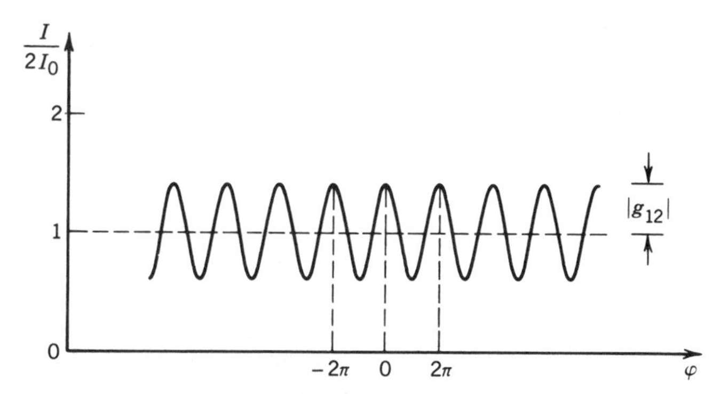

Figure 10.2-1 The normalized intensity $I/2I_0$ of the sum of two partially coherent waves of equal intensities $I_1 = I_2 = I_0$ as a function of the phase φ of their normalized cross-correlation g_{12}. This sinusoidal pattern has visibility $\mathscr{V} = |g_{12}|$.

The visibility is therefore proportional to the absolute value of the normalized cross-correlation $|g_{12}|$. If $I_1 = I_2$,

$$\mathscr{V} = |g_{12}|. \tag{10.2-5}$$

Visibility

The interference equation (10.2-3) will now be applied to a number of special cases to illustrate the effects of temporal and spatial coherence on the interference of partially coherent light.

B. Interference and Temporal Coherence

Consider a partially coherent wave $U(t)$ with intensity I_0 and complex degree of temporal coherence $g(\tau) = \langle U^*(t)U(t+\tau)\rangle / I_0$. If $U(t)$ is added to a replica of itself delayed by the time τ, $U(t+\tau)$, what is the intensity I of the superposition?

Using the interference formula (10.2-3) with $U_1 = U(t)$, $U_2 = U(t+\tau)$, $I_1 = I_2 = I_0$, and $g_{12} = \langle U_1^* U_2\rangle / I_0 = \langle U^*(t)U(t+\tau)\rangle / I_0 = g(\tau)$, we obtain

$$I = 2I_0[1 + \mathrm{Re}\{g(\tau)\}] = 2I_0[1 + |g(\tau)|\cos\varphi(\tau)], \tag{10.2-6}$$

where $\varphi(\tau) = \arg\{g(\tau)\}$. *The ability of a wave to interfere with a time delayed replica of itself is governed by its complex degree of temporal coherence at that time delay.*

A wave may be added to a time-delayed replica of itself by using a beamsplitter to generate two identical waves, one of which is made to travel a longer optical path before the two waves are recombined using another (or the same) beamsplitter. This may be achieved by using a Mach–Zehnder or a Michelson interferometer, for example (see Fig. 2.5-3).

Consider, as an example, the partially coherent plane wave introduced in Sec. 10.1D [equation (10.1-25)] whose complex degree of temporal coherence is $g(\tau) = g_a(\tau)\exp(j2\pi\nu_0\tau)$. The spectral width of the wave is $\Delta\nu_c = 1/\tau_c$, where τ_c, the width of $|g_a(\tau)|$, is the coherence time. Substituting into (10.2-6), we obtain

$$I = 2I_0\{1 + |g_a(\tau)|\cos[2\pi\nu_0\tau + \varphi_a(\tau)]\}, \tag{10.2-7}$$

where $\varphi_a(\tau) = \arg\{g_a(\tau)\}$.

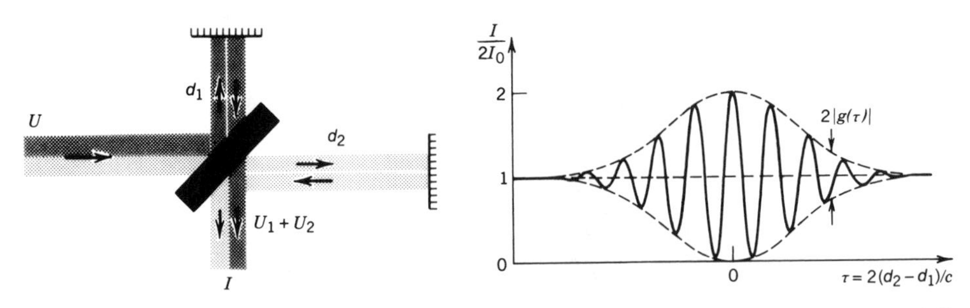

Figure 10.2-2 The normalized intensity $I/2I_0$ as a function of time delay τ when a partially coherent plane wave is introduced into a Michelson interferometer. The visibility equals the magnitude of the complex degree of temporal coherence.

The relation between I and τ is known as an **interferogram** (Fig. 10.2-2). Assuming that $\Delta\nu_c \ll \nu_0$, the functions $|g_a(\tau)|$ and $\varphi_a(\tau)$ vary slowly in comparison to the period $1/\nu_0$ since $\Delta\nu_c = 1/\tau_c \ll \nu_0$. The visibility of this interferogram in the vicinity of a particular time delay τ is $\mathscr{V} = |g(\tau)| = |g_a(\tau)|$. It has a peak value of unity near $\tau = 0$ and vanishes for $\tau \gg \tau_c$, i.e., when the optical path difference is much greater than the coherence length $l_c = c\tau_c$. For the Michelson interferometer shown in Fig. 10.2-2, $\tau = 2(d_2 - d_1)/c$. Interference occurs only when the optical path difference is smaller than the coherence length.

The magnitude of the complex degree of temporal coherence of a wave $|g(\tau)|$ may therefore be measured by monitoring the visibility of the interference pattern as a function of time delay. The phase of $g(\tau)$ may be measured by observing the locations of the peaks of the pattern.

It is revealing to write (10.2-6) in terms of the power spectral density. Using the Fourier transform relation between $G(\tau)$ and $S(\nu)$,

$$G(\tau) = I_0 g(\tau) = \int_0^\infty S(\nu)\exp(j2\pi\nu\tau)\,d\nu,$$

substituting into (10.2-6), and noting that $S(\nu)$ is real and $\int_0^\infty S(\nu)\,d\nu = I_0$, we obtain

$$I = 2\int_0^\infty S(\nu)[1 + \cos(2\pi\nu\tau)]\,d\nu. \tag{10.2-8}$$

This equation can be interpreted as a weighted superposition of interferograms produced by each of the monochromatic components of the wave. Each component ν produces an interferogram with period $1/\nu$ and unity visibility, but the composite interferogram has reduced visibility as a result of the different periods.

Equation (10.2-8) suggests a technique for determining the spectral density $S(\nu)$ of a light source by measuring the interferogram I versus τ and then inverting it by means of Fourier-transform methods. This technique is known as **Fourier-transform spectroscopy**.

C. Interference and Spatial Coherence

The effect of spatial coherence on interference is demonstrated by considering the Young's double-pinhole interference experiment, discussed in Exercise 2.5-2 for coherent light. A partially coherent optical wave $U(\mathbf{r}, t)$ illuminates an opaque screen with two pinholes located at positions $\mathbf{r}_1$ and $\mathbf{r}_2$. The wave has mutual coherence function

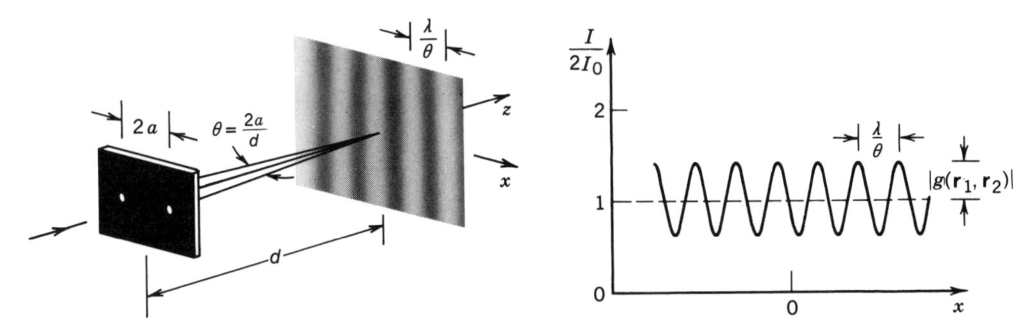

Figure 10.2-3 Young's double-pinhole interferometer. The incident wave is quasi-monochromatic and the normalized mutual intensity at the pinholes is $g(\mathbf{r}_1, \mathbf{r}_2)$. The normalized intensity $I/2I_0$ in the observation plane at a large distance is a sinusoidal function of x with period λ/θ and visibility $\mathscr{V} = |g(\mathbf{r}_1, \mathbf{r}_2)|$.

$G(\mathbf{r}_1, \mathbf{r}_2, \tau) = \langle U^*(\mathbf{r}_1, t) U(\mathbf{r}_2, t + \tau)\rangle$ and complex degree of coherence $g(\mathbf{r}_1, \mathbf{r}_2, \tau)$. The intensities at the pinholes are assumed to be equal.

Light is diffracted in the form of two spherical waves centered at the pinholes. The two waves interfere, and the intensity I of their sum is observed at a point $\mathbf{r}$ in the observation plane a distance d from the screen sufficiently large so that the paraboloidal approximation is applicable. In Cartesian coordinates (Fig. 10.2-3) $\mathbf{r}_1 = (-a, 0, 0)$, $\mathbf{r}_2 = (a, 0, 0)$, and $\mathbf{r} = (x, 0, d)$. The intensity is observed as a function of x. An important geometrical parameter is the angle $\theta \approx 2a/d$ subtended by the two pinholes.

In the paraboloidal (Fresnel) approximation [see (2.2-16)], the two diffracted spherical waves are approximately related to $U(\mathbf{r}, t)$ by

$$U_1(\mathbf{r}, t) \propto U\left(\mathbf{r}_1, t - \frac{|\mathbf{r} - \mathbf{r}_1|}{c}\right) \approx U\left(\mathbf{r}_1, t - \frac{d + (x + a)^2/2d}{c}\right) \tag{10.2-9a}$$

$$U_2(\mathbf{r}, t) \propto U\left(\mathbf{r}_2, t - \frac{|\mathbf{r} - \mathbf{r}_2|}{c}\right) \approx U\left(\mathbf{r}_2, t - \frac{d + (x - a)^2/2d}{c}\right), \tag{10.2-9b}$$

and have approximately equal intensities, $I_1 = I_2 = I_0$. The normalized cross-correlation between the two waves at $\mathbf{r}$ is

$$g_{12} = \frac{\langle U_1^*(\mathbf{r}, t) U_2(\mathbf{r}, t)\rangle}{I_0} = g(\mathbf{r}_1, \mathbf{r}_2, \tau_x), \tag{10.2-10}$$

where

$$\tau_x = \frac{|\mathbf{r} - \mathbf{r}_1| - |\mathbf{r} - \mathbf{r}_2|}{c} = \frac{(x + a)^2 - (x - a)^2}{2dc} = \frac{2ax}{dc} = \frac{\theta}{c} x \tag{10.2-11}$$

is the difference in the time delays encountered by the two waves.

Substituting (10.2-10) into the interference formula (10.2-3) gives rise to an observed intensity $I \equiv I(x)$:

$$I(x) = 2I_0[1 + |g(\mathbf{r}_1, \mathbf{r}_2, \tau_x)| \cos\varphi_x], \tag{10.2-12}$$

where $\varphi_x = \arg\{g(\mathbf{r}_1, \mathbf{r}_2, \tau_x)\}$. This equation describes the pattern of observed intensity as a function of position x in the observation plane, in terms of the magnitude and phase of the complex degree of coherence at the pinholes at time delay $\tau_x = \theta x/c$.

Quasi-Monochromatic Light

If the light is quasi-monochromatic with central frequency ν_0, i.e., if $g(\mathbf{r}_1, \mathbf{r}_2, \tau) \approx g(\mathbf{r}_1, \mathbf{r}_2) \exp(j2\pi\nu_0\tau)$, then (10.2-12) gives

$$I(x) = 2I_0\left[1 + \mathscr{V} \cos\left(\frac{2\pi\theta}{\lambda}x + \varphi\right)\right], \tag{10.2-13}$$

where $\lambda = c/\nu_0$, $\mathscr{V} = |g(\mathbf{r}_1, \mathbf{r}_2)|$, $\tau_x = \theta x/c$, and $\varphi = \arg\{g(\mathbf{r}_1, \mathbf{r}_2)\}$. The interference fringe pattern is therefore sinusoidal with spatial period λ/θ and visibility $\mathscr{V}$. In analogy with the temporal case, *the visibility of the interference pattern equals the magnitude of the complex degree of spatial coherence at the two pinholes* (Fig. 10.2-3). The locations of the peaks depend on the phase φ.

Interference with Light from an Extended Source

If the incident wave in Young's interferometer is a coherent plane wave traveling in the z direction, $U(\mathbf{r}, t) = \exp(-jkz)\exp(j2\pi\nu_0 t)$, then $g(\mathbf{r}_1, \mathbf{r}_2) = 1$, so that $|g(\mathbf{r}_1, \mathbf{r}_2)| = 1$, and $\arg\{g(\mathbf{r}_1, \mathbf{r}_2)\} = 0$. The interference pattern therefore has unity visibility and a peak at $x = 0$. But if the illumination is, instead, a tilted plane wave arriving from a direction in the x–z plane making a small angle θ_x with respect to the z axis, i.e., $U(\mathbf{r}, t) \approx \exp[-j(kz + k\theta_x x)]\exp(j2\pi\nu_0 t)$, then $g(\mathbf{r}_1, \mathbf{r}_2) = \exp(-jk\theta_x 2a)$. The visibility remains $\mathscr{V} = 1$, but the tilt results in a phase shift $\varphi = -k\theta_x 2a = -2\pi\theta_x 2a/\lambda$, so that the interference pattern is shifted laterally by a fraction $(2a\theta_x/\lambda)$ of a period. When $\varphi = 2\pi$, the pattern is shifted one period.

Suppose now that the incident light is a collection of independent plane waves arriving from a source that subtends an angle θ_s at the pinhole plane (Fig. 10.2-4). The phase shift φ then takes values in the range $\pm 2\pi(\theta_s/2)2a/\lambda = \pm 2\pi\theta_s a/\lambda$ and the fringe pattern is a superposition of displaced sinusoids. If $\theta_s = \lambda/2a$ then φ takes on values in the range $\pm\pi$, which is sufficient to wash out the interference pattern and reduce its visibility to zero.

We conclude that the degree of spatial coherence at the two pinholes is very small when the angle subtended by the source is $\theta_s = \lambda/2a$ (or greater). Consequently, the distance

$$\rho_c \approx \frac{\lambda}{\theta_s} \tag{10.2-14}$$

Coherence Distance

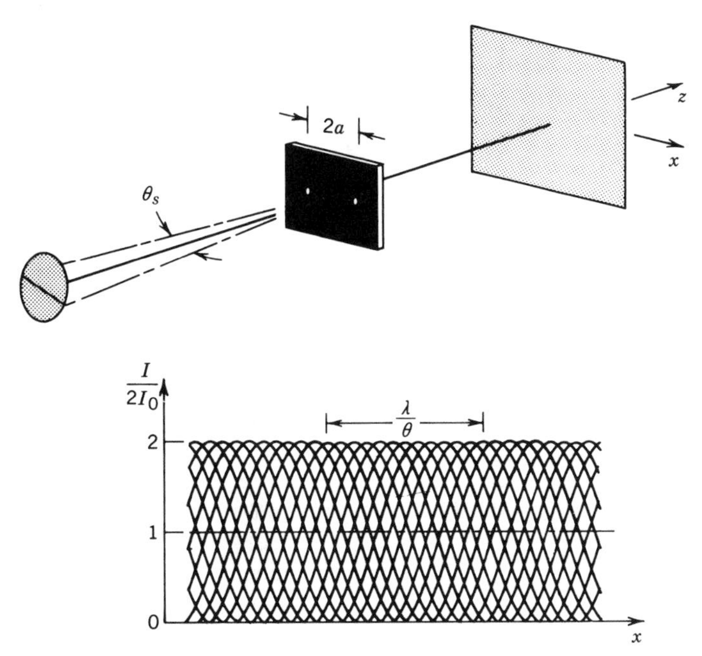

Figure 10.2-4 Young's interference fringes are washed out if the illumination emanates from a source of angular diameter $\theta_s > \lambda/2a$. If the distance $2a$ is smaller than λ/θ_s, the fringes become visible.

is a measure of the coherence distance in the plane of the screen and

$$A_c \approx \left(\frac{\lambda}{\theta_s}\right)^2 \tag{10.2-15}$$

is a measure of the coherence area of light emitted from a source subtending an angle θ_s. The angle subtended by the sun, for example, is 0.5°, so that the coherence distance for filtered sunlight of wavelength λ is $\rho_c \approx \lambda/\theta_s \approx 115\lambda$. At $\lambda = 0.5$ μm, $\rho_c \approx 57.5$ μm.

A more rigorous analysis (see Sec. 10.3C) shows that the transverse coherence distance ρ_c for a circular incoherent light source of uniform intensity is

$$\rho_c = 1.22\frac{\lambda}{\theta_s}. \tag{10.2-16}$$

Effect of Spectral Width on Interference

Finally, we examine the effect of the spectral width on interference in the Young's double-pinhole interferometer. The power spectral density of the incident wave is assumed to be a narrow function of width $\Delta\nu_c$ centered about ν_0, and $\Delta\nu_c \ll \nu_0$. The complex degree of coherence then has the form

$$g(\mathbf{r}_1, \mathbf{r}_2, \tau) = g_a(\mathbf{r}_1, \mathbf{r}_2, \tau)\exp(j2\pi\nu_0\tau), \tag{10.2-17}$$

where $g_a(\mathbf{r}_1, \mathbf{r}_2, \tau)$ is a slowly varying function of τ (in comparison with the period

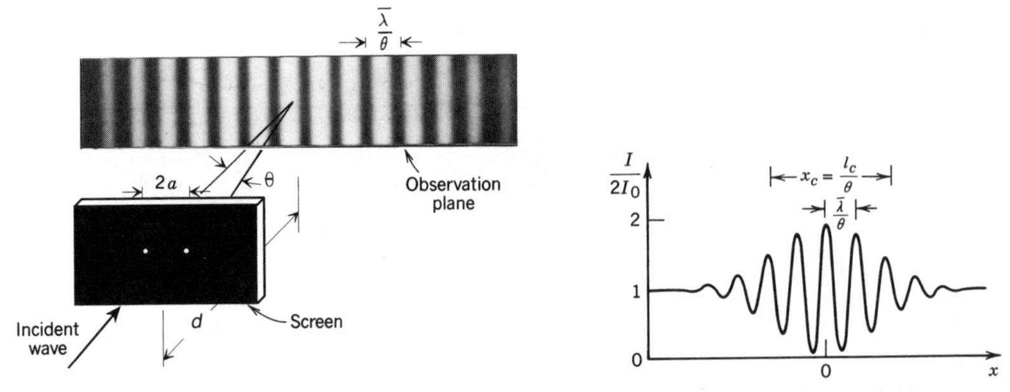

Figure 10.2-5 The visibility of Young's interference fringes at position x is the magnitude of the complex degree of coherence at the pinholes at a time delay $\tau_x = \theta x/c$. For spatially coherent light the number of observable fringes is the ratio of the coherence length to the central wavelength, or the ratio of the central frequency to the spectral linewidth.

$1/\nu_0$). Substituting (10.2-17) into (10.2-12), we obtain

$$I(x) = 2I_0\left[1 + \mathscr{V}_x \cos\left(\frac{2\pi\theta}{\bar{\lambda}}x + \varphi_x\right)\right], \tag{10.2-18}$$

where $\mathscr{V}_x = |g_a(\mathbf{r}_1, \mathbf{r}_2, \tau_x)|$, $\varphi_x = \arg\{g_a(\mathbf{r}_1, \mathbf{r}_2, \tau_x)\}$, $\tau_x = \theta x/c$, and $\bar{\lambda} = c/\nu_0$.

Thus the interference pattern is sinusoidal with period $\bar{\lambda}/\theta$ but with a varying visibility $\mathscr{V}_x$ and varying phase φ_x equal to the magnitude and phase of the complex degree of coherence at the two pinholes, respectively, evaluated at the time delay $\tau_x = \theta x/c$. If $|g_a(\mathbf{r}_1, \mathbf{r}_2, \tau)| = 1$ at $\tau = 0$, decreases with increasing τ, and vanishes for $\tau \gg \tau_c$, the visibility $\mathscr{V}_x = 1$ at $x = 0$, decreases with increasing x, and vanishes for $x \gg x_c = c\tau_c/\theta$. The interference pattern is then visible over a distance

$$x_c = \frac{l_c}{\theta}, \tag{10.2-19}$$

where $l_c = c\tau_c$ is the coherence length and θ is the angle subtended by the two pinholes (Fig. 10.2-5).

The number of observable fringes is thus $x_c/(\bar{\lambda}/\theta) = l_c/\bar{\lambda} = c\tau_c/\bar{\lambda} = \nu_0/\Delta\nu_c$. It equals the ratio $l_c/\bar{\lambda}$ of the coherence length to the central wavelength, or the ratio $\nu_0/\Delta\nu_c$ of the central frequency to the linewidth. Clearly, if $|g(\mathbf{r}_1, \mathbf{r}_2, 0)| < 1$, i.e., if the source is not spatially coherent, the visibility will be further reduced and even fewer fringes will be observable.

*10.3 TRANSMISSION OF PARTIALLY COHERENT LIGHT THROUGH OPTICAL SYSTEMS

The transmission of coherent light through thin optical components, through apertures, and through free space was discussed in Chaps. 2 and 4. In this section we pursue the same goal for quasi-monochromatic partially coherent light. We assume that the spectral width is sufficiently small so that the coherence length $l_c = c\tau_c = c/\Delta\nu_c$ is much greater than the differences of optical path lengths in the system. The mutual coherence function may then be approximated by $G(\mathbf{r}_1, \mathbf{r}_2, \tau) \approx G(\mathbf{r}_1, \mathbf{r}_2)\exp(j2\pi\nu_0\tau)$, where $G(\mathbf{r}_1, \mathbf{r}_2)$ is the mutual intensity and ν_0 is the central frequency.

It is noted at the outset that the transmission laws that apply to the deterministic function $U(\mathbf{r})$, which represents coherent light, apply also to the random function $U(\mathbf{r})$, which represents partially coherent light. However, for partially coherent light our interest is in the laws that govern statistical averages: the intensity $I(\mathbf{r})$ and the mutual intensity $G(\mathbf{r}_1, \mathbf{r}_2)$.

A. Propagation of Partially Coherent Light

Transmission Through Thin Optical Components

When a partially coherent wave is transmitted through a thin optical component characterized by an amplitude transmittance $t(x, y)$ the incident and transmitted waves are related by $U_2(\mathbf{r}) = t(\mathbf{r})U_1(\mathbf{r})$, where $\mathbf{r} = (x, y)$ is the position in the plane of the component (see Fig. 10.3-1). Using the definition of the mutual intensity, $G(\mathbf{r}_1, \mathbf{r}_2) = \langle U^*(\mathbf{r}_1)U(\mathbf{r}_2)\rangle$, we obtain

$$G_2(\mathbf{r}_1, \mathbf{r}_2) = t^*(\mathbf{r}_1)t(\mathbf{r}_2)G_1(\mathbf{r}_1, \mathbf{r}_2), \tag{10.3-1}$$

where $G_1(\mathbf{r}_1, \mathbf{r}_2)$ and $G_2(\mathbf{r}_1, \mathbf{r}_2)$ are the mutual intensities of the incident and transmitted light, respectively.

Since the intensity at position $\mathbf{r}$ equals the mutual intensity at $\mathbf{r}_1 = \mathbf{r}_2 = \mathbf{r}$,

$$I_2(\mathbf{r}) = |t(\mathbf{r})|^2 I_1(\mathbf{r}). \tag{10.3-2}$$

The normalized mutual intensities defined by (10.1-22) therefore satisfy

$$|g_2(\mathbf{r}_1, \mathbf{r}_2)| = |g_1(\mathbf{r}_1, \mathbf{r}_2)|. \tag{10.3-3}$$

Although transmission through a thin optical component may change the intensity of partially coherent light, it does not alter the magnitude of its degree of spatial coherence. Naturally, if the complex amplitude transmittance of the component itself were random, the coherence of the transmitted light would be altered.

Transmission Through an Arbitrary Optical System

We next consider an arbitrary optical system—one that includes propagation in free space or transmission through thick optical components. It was shown in Chap. 4 that the complex amplitude $U_2(\mathbf{r})$ at a point $\mathbf{r} = (x, y)$ in the output plane of such a system is generally a weighted superposition integral comprising contributions from the complex amplitudes $U_1(\mathbf{r})$ at points $\mathbf{r}' = (x', y')$ in the input plane (see Fig. 10.3-2),

$$U_2(\mathbf{r}) = \int h(\mathbf{r}; \mathbf{r}')U_1(\mathbf{r}')\, d\mathbf{r}', \tag{10.3-4}$$

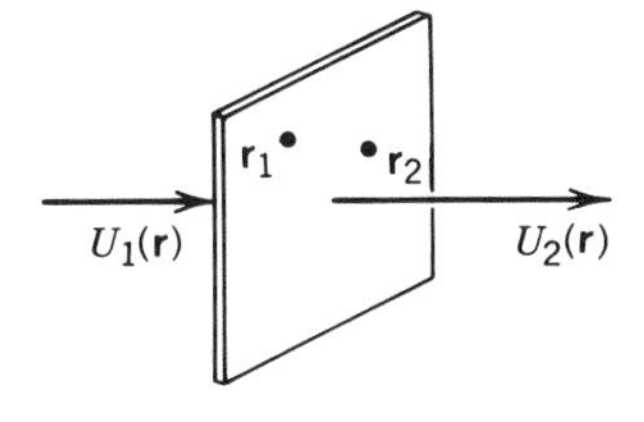

Figure 10.3-1 The absolute value of the degree of spatial coherence is not altered by transmission through a thin optical component.

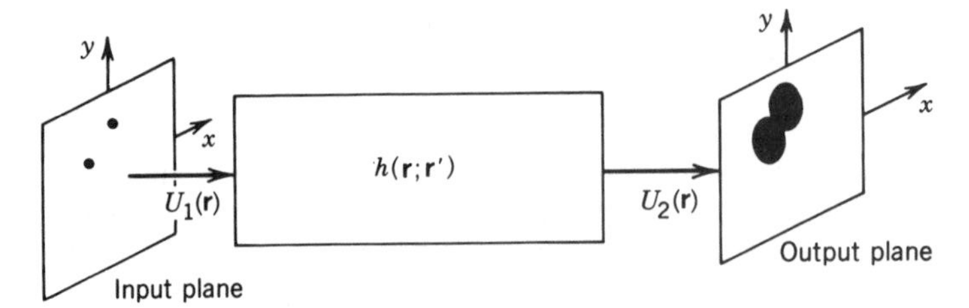

Figure 10.3-2 An optical system is characterized by its impulse-response function $h(\mathbf{r};\mathbf{r}')$.

where $h(\mathbf{r};\mathbf{r}')$ is the impulse-response function of the system. The integral in (10.3-4) is a double integral with respect to $\mathbf{r}' = (x', y')$ extending over the entire input plane.

To translate this relation between the random functions $U_2(\mathbf{r})$ and $U_1(\mathbf{r})$ into a relation between their mutual intensities, we substitute (10.3-4) into the definition $G_2(\mathbf{r}_1, \mathbf{r}_2) = \langle U_2^*(\mathbf{r}_1) U_2(\mathbf{r}_2) \rangle$ and use the definition $G_1(\mathbf{r}_1, \mathbf{r}_2) = \langle U_1^*(\mathbf{r}_1) U_1(\mathbf{r}_2) \rangle$ to obtain

$$G_2(\mathbf{r}_1, \mathbf{r}_2) = \iint h^*(\mathbf{r}_1; \mathbf{r}_1') h(\mathbf{r}_2; \mathbf{r}_2') G_1(\mathbf{r}_1', \mathbf{r}_2') \, d\mathbf{r}_1' \, d\mathbf{r}_2'. \tag{10.3-5}$$

Image Mutual Intensity

If the mutual intensity $G_1(\mathbf{r}_1, \mathbf{r}_2)$ of the input light and the impulse-response function $h(\mathbf{r};\mathbf{r}')$ of the system are known, the mutual intensity of the output light $G_2(\mathbf{r}_1, \mathbf{r}_2)$ can be determined by carrying out the integrals in (10.3-5).

The intensity of the output light is obtained by using the definition $I_2(\mathbf{r}) = G_2(\mathbf{r}, \mathbf{r})$, which reduces (10.3-5) to

$$I_2(\mathbf{r}) = \iint h^*(\mathbf{r}; \mathbf{r}_1') h(\mathbf{r}; \mathbf{r}_2') G_1(\mathbf{r}_1', \mathbf{r}_2') \, d\mathbf{r}_1' \, d\mathbf{r}_2'. \tag{10.3-6}$$

Image Intensity

To determine the intensity of the output light, we must know the mutual intensity of the input light. *Knowledge of the input intensity $I_1(\mathbf{r})$ by itself is generally not sufficient to determine the output intensity $I_2(\mathbf{r})$.*

B. Image Formation with Incoherent Light

We now consider the special case when the input light is incoherent. The mutual intensity $G_1(\mathbf{r}_1, \mathbf{r}_2)$ vanishes when $\mathbf{r}_2$ is only slightly separated from $\mathbf{r}_1$ so that the coherence distance is much smaller than other pertinent dimensions in the system (for example, the resolution distance of an imaging system). The mutual intensity may then be written in the form $G_1(\mathbf{r}_1, \mathbf{r}_2) = [I_1(\mathbf{r}_1) I_1(\mathbf{r}_2)]^{1/2} g(\mathbf{r}_1 - \mathbf{r}_2)$, where $g(\mathbf{r}_1 - \mathbf{r}_2)$ is a very narrow function. When $G_1(\mathbf{r}_1, \mathbf{r}_2)$ appears under the integral in (10.3-5) or (10.3-6) it is convenient to replace $g(\mathbf{r}_1 - \mathbf{r}_2)$ with a delta function, $g(\mathbf{r}_1 - \mathbf{r}_2) = \sigma\delta(\mathbf{r}_1 - \mathbf{r}_2)$, where $\sigma = \int g(\mathbf{r}) \, d\mathbf{r}$ is the area under $g(\mathbf{r})$, so that

$$G_1(\mathbf{r}_1, \mathbf{r}_2) \approx \sigma \left[I_1(\mathbf{r}_1) I_1(\mathbf{r}_2) \right]^{1/2} \delta(\mathbf{r}_1 - \mathbf{r}_2). \tag{10.3-7}$$

Since the mutual intensity must remain finite and $\delta(0) \to \infty$, this equation is clearly not generally accurate. It is valid only for the purpose of evaluating integrals such as in (10.3-6). Substituting (10.3-7) into (10.3-6), the delta function reduces the double integral into a single integral and we obtain

$$I_2(\mathbf{r}) = \int I_1(\mathbf{r}')h_i(\mathbf{r};\mathbf{r}')\,d\mathbf{r}', \tag{10.3-8}$$

Imaging Equation (Incoherent Illumination)

where

$$h_i(\mathbf{r};\mathbf{r}') = \sigma|h(\mathbf{r};\mathbf{r}')|^2. \tag{10.3-9}$$

Impulse-Response Function (Incoherent Illumination)

Under these conditions, the relation between the intensities at the input and output planes describes a linear system of impulse-response function $h_i(\mathbf{r};\mathbf{r}')$, also called the **point-spread function**. When the input light is completely incoherent, therefore, the intensity of the light at each point $\mathbf{r}$ in the output plane is a weighted superposition of contributions from intensities at many points $\mathbf{r}'$ of the input plane; intereference does not occur and the intensities simply add (Fig. 10.3-3). This is to be contrasted with the completely coherent system, for which the complex amplitudes rather than intensities are related by a superposition integral, as in (10.3-4).

In certain optical systems the impulse-response function $h(\mathbf{r};\mathbf{r}')$ is a function of $\mathbf{r} - \mathbf{r}'$, say $h(\mathbf{r} - \mathbf{r}')$. The system is then said to be **shift invariant** or **isoplanatic** (see Appendix B). In this case $h_i(\mathbf{r};\mathbf{r}') = h_i(\mathbf{r} - \mathbf{r}')$. The integrals in (10.3-4) and (10.3-8) are then two-dimensional convolutions and the systems can be described by transfer functions $\mathcal{H}(\nu_x, \nu_y)$ and $\mathcal{H}_i(\nu_x, \nu_y)$, which are the Fourier transforms of $h(\mathbf{r}) = h(x, y)$ and $h_i(\mathbf{r}) = h_i(x, y)$, respectively.

As an example, we apply the relations above to an imaging system. It was shown in Sec. 4.4C that with coherent illumination, the impulse-response function of the

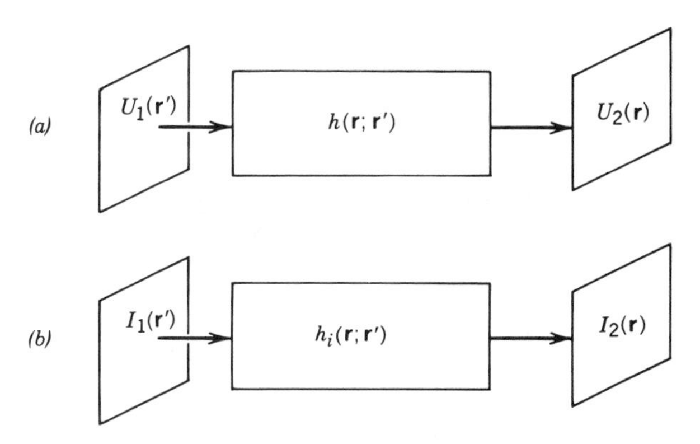

Figure 10.3-3 (*a*) The complex *amplitudes* of light at the input and output planes of an optical system illuminated by *coherent* light are related by a linear system with impulse-response function $h(\mathbf{r};\mathbf{r}')$. (*b*) The *intensities* of light at the input and output planes of an optical system illuminated by *incoherent* light are related by a linear system with impulse-response function $h_i(\mathbf{r};\mathbf{r}') = \sigma|h(\mathbf{r};\mathbf{r}')|^2$.

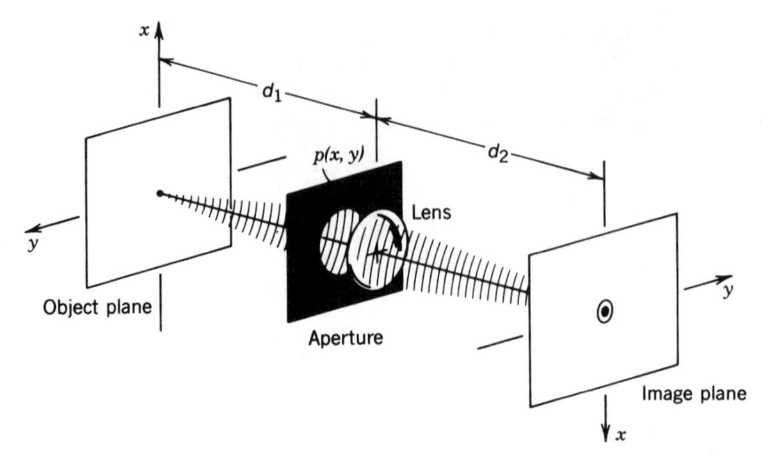

Figure 10.3-4 A single-lens imaging system.

single-lens focused imaging system illustrated in Fig. 10.3-4 in the Fresnel approximation is

$$h(\mathbf{r}) \propto P\left(\frac{x}{\lambda d_2}, \frac{y}{\lambda d_2}\right), \tag{10.3-10}$$

where $P(\nu_x, \nu_y)$ is the Fourier transform of the pupil function $p(x, y)$ and d_2 is the distance from the lens to the image plane. The pupil function is unity within the aperture and zero elsewhere.

When the illumination is quasi-monochromatic and spatially incoherent, the intensities of light at the object and image plane are linearly related by a system with impulse-response function

$$h_i(\mathbf{r}) = \sigma|h(\mathbf{r})|^2 \propto \left|P\left(\frac{x}{\lambda d_2}, \frac{y}{\lambda d_2}\right)\right|^2, \tag{10.3-11}$$

where λ is the wavelength corresponding to the central frequency ν_0.

EXAMPLE 10.3-1. ***Imaging System with a Circular Aperture.*** If the aperture is a circle of radius a, the pupil function $p(x, y) = 1$ for x, y inside the circle, and 0 elsewhere. Its Fourier transform is

$$P(\nu_x, \nu_y) = \frac{aJ_1(2\pi\nu_\rho a)}{\nu_\rho}, \qquad \nu_\rho = \left(\nu_x^2 + \nu_y^2\right)^{1/2},$$

where $J_1(\cdot)$ is the Bessel function (see Appendix A, Sec. A.3). The impulse-response function of the coherent system is obtained by substituting into (10.3-10),

$$h(x, y) \propto \left[\frac{J_1(2\pi\nu_s\rho)}{\pi\nu_s\rho}\right], \qquad \rho = \left(x^2 + y^2\right)^{1/2}, \tag{10.3-12}$$

where

$$\nu_s = \frac{\theta}{2\lambda}, \qquad \theta = \frac{2a}{d_2}. \tag{10.3-13}$$

For incoherent illumination, the impulse-response function is therefore

$$h_i(x, y) \propto \left[\frac{J_1(2\pi\nu_s\rho)}{\pi\nu_s\rho}\right]^2. \tag{10.3-14}$$

The response functions $h(x, y)$ and $h_i(x, y)$ are illustrated in Fig. 10.3-5. Both functions reach their first zero when $2\pi\nu_s\rho = 3.832$, or $\rho = \rho_s \approx 3.832/2\pi\nu_s = 3.832\lambda/\pi\theta$, from which

$$\boxed{\rho_s \approx 1.22\frac{\lambda}{\theta}.} \tag{10.3-15}$$

Two-Point Resolution

Thus the image of a point (impulse) in the input plane is a patch of intensity $h_i(x, y)$ and radius ρ_s. When the input distribution is composed of two points (impulses) separated by a distance ρ_s, the image of one point vanishes at the center of the image of the other point. The distance ρ_s is therefore a measure of the resolution of the imaging system.

The transfer functions of linear systems (see Appendix B) with the impulse-response functions $h(x, y)$ and $h_i(x, y)$ are the Fourier transforms (see Appendix A),

$$\mathcal{H}(\nu_x, \nu_y) = \begin{cases} 1, & \nu_\rho < \nu_s \\ 0, & \text{otherwise,} \end{cases} \tag{10.3-16}$$

and

$$\mathcal{H}_i(\nu_x, \nu_y) = \begin{cases} \dfrac{2}{\pi}\left\{\cos^{-1}\dfrac{\nu_\rho}{2\nu_s} - \dfrac{\nu_\rho}{2\nu_s}\left[1 - \left(\dfrac{\nu_\rho}{2\nu_s}\right)^2\right]^{1/2}\right\}, & \nu_\rho < 2\nu_s \\ 0, & \text{otherwise,} \end{cases} \tag{10.3-17}$$

where $\nu_\rho = (\nu_x^2 + \nu_y^2)^{1/2}$. Both functions have been normalized such that their values at $\nu_\rho = 0$ are 1. These functions are illustrated in Fig. 10.3-5. For coherent illumination, the transfer function is flat and has a cutoff frequency $\nu_s = \theta/2\lambda$ lines/mm. For incoherent illumination, the transfer function drops approximately linearly with the spatial frequency and has a cutoff frequency $2\nu_s = \theta/\lambda$ lines/mm.

If the object is placed at infinity, i.e., $d_1 = \infty$, then $d_2 = f$, the focal length of the lens. The angle $\theta = 2a/f$ is then the inverse of the lens F-number, $F_\# = f/2a$. The cutoff frequencies ν_s and $2\nu_s$ are related to the lens F-number by

$$\boxed{\begin{array}{c}\text{Cutoff frequency} \\ \text{(lines/mm)}\end{array} = \begin{cases} \dfrac{1}{2\lambda F_\#} & \text{(coherent illumination)} \\ \dfrac{1}{\lambda F_\#} & \text{(incoherent illumination).} \end{cases}} \tag{10.3-18}$$

One should not draw the false conclusion that incoherent illumination is superior to coherent illumination since it has twice the spatial bandwidth. The transfer functions of the two systems should not be compared directly since one describes imaging of the complex amplitude, whereas the other describes imaging of the intensity.

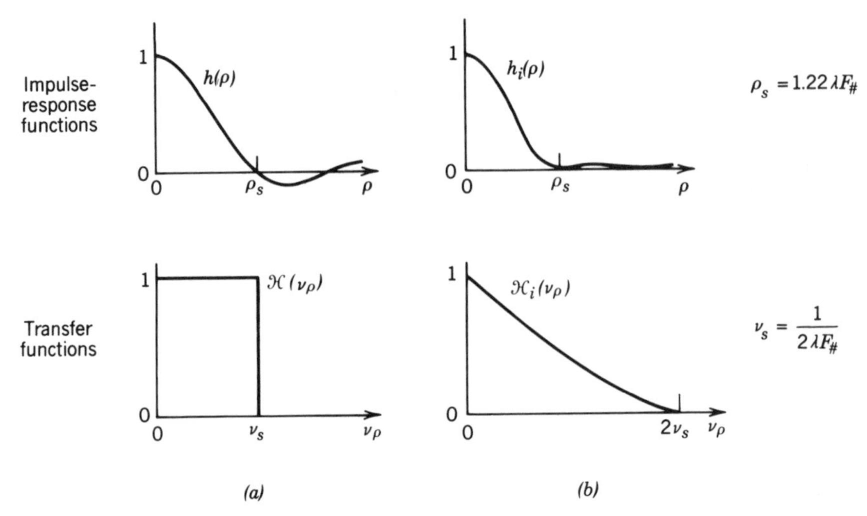

Figure 10.3-5 Impulse-response functions and transfer functions of a single-lens focused diffraction-limited imaging system with a circular aperture and F-number $F_{\#}$ under (a) coherent and (b) incoherent illumination.

C. Gain of Spatial Coherence by Propagation

Equation (10.3-5) describes the change of the mutual intensity when the light propagates through an optical system of impulse-response function $h(\mathbf{r}; \mathbf{r}')$. When the input light is incoherent, the mutual intensity $G_1(\mathbf{r}_1, \mathbf{r}_2)$ may be replaced by $\sigma[I_1(\mathbf{r}_1)I_1(\mathbf{r}_2)]^{1/2}\delta(\mathbf{r}_1 - \mathbf{r}_2)$ and substituted in the double integral in (10.3-5) to obtain the single integral,

$$G_2(\mathbf{r}_1, \mathbf{r}_2) = \sigma \int h^*(\mathbf{r}_1; \mathbf{r}) h(\mathbf{r}_2; \mathbf{r}) I_1(\mathbf{r})\, d\mathbf{r}. \tag{10.3-19}$$

Image Mutual Intensity

It is evident that the received light is no longer incoherent. In general, *light gains spatial coherence by the mere act of propagation*. This is not surprising. Although light fluctuations at different points of the input plane are uncorrelated, the radiation from each point spreads and overlaps with that from the neighboring points. The light reaching two points in the output plane comes from many points of the input plane, some of which are common (see Fig. 10.3-6). These common contributions create partial correlation between fluctuations at the output points.

This is not unlike the transmission of an uncorrelated time signal (white noise) through a low-pass filter. The filter smooths the function and reduces its spectral bandwidth, so that its coherence time increases and it is no longer uncorrelated. The propagation of light through an optical system is a form of spatial filtering that cuts the spatial bandwidth and therefore increases the coherence area.

Van Cittert–Zernike Theorem

There is a mathematical similarity between the gain of coherence of initially *incoherent* light propagating through an optical system, and the change of the amplitude of

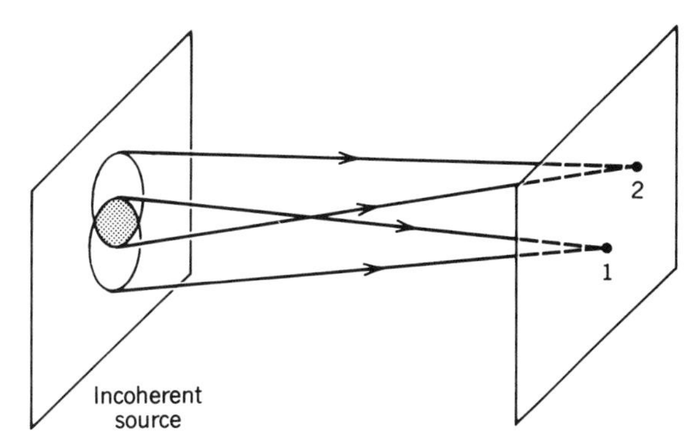

Figure 10.3-6 Gain of coherence by propagation is a result of the spreading of light. Although the light is completely uncorrelated at the source, the light fluctuations at points 1 and 2 share a common origin, the shaded area, and are therefore partially correlated.

coherent light traveling through the same system. In reference to (10.3-19), if the observation point $\mathbf{r}_1$ is fixed, for example at the origin $\mathbf{0}$, and the mutual intensity $G_2(\mathbf{0}, \mathbf{r}_2)$ is examined as a function of $\mathbf{r}_2$, then

$$G_2(\mathbf{0}, \mathbf{r}_2) = \sigma \int h^*(\mathbf{0}; \mathbf{r}) h(\mathbf{r}_2; \mathbf{r}) I_1(\mathbf{r})\, d\mathbf{r}. \tag{10.3-20}$$

Defining $U_2(\mathbf{r}_2) = G_2(\mathbf{0}, \mathbf{r}_2)$ and $U_1(\mathbf{r}) = \sigma h^*(\mathbf{0}; \mathbf{r}) I_1(\mathbf{r})$, (10.3-20) may be written in the familiar form

$$U_2(\mathbf{r}_2) = \int h(\mathbf{r}_2; \mathbf{r}) U_1(\mathbf{r})\, d\mathbf{r}, \tag{10.3-21}$$

which is exactly the integral (10.3-4) that governs the propagation of coherent light. Thus the observed mutual intensity $G(\mathbf{0}, \mathbf{r}_2)$ at the output of an optical system whose input is incoherent is mathematically identical to the observed complex amplitude if a coherent wave of complex amplitude $U_1(\mathbf{r}) = \sigma h^*(\mathbf{0}; \mathbf{r}) I_1(\mathbf{r})$ were the input to the same system.

As an example, suppose that the incoherent input wave has uniform intensity and extends over an aperture $p(\mathbf{r})$ [$p(\mathbf{r}) = 1$ within the aperture, and zero elsewhere], i.e., $I_1(\mathbf{r}) = p(\mathbf{r})$; and assume that the optical system is free space, i.e., $h(\mathbf{r}'; \mathbf{r}) = \exp(-jk|\mathbf{r}' - \mathbf{r}|)/|\mathbf{r}' - \mathbf{r}|$. The mutual intensity $G_2(\mathbf{0}, \mathbf{r}_2)$ is then identical to the amplitude $U_2(\mathbf{r}_2)$ obtained when a coherent wave with input amplitude $U_1(\mathbf{r}) = \sigma h^*(\mathbf{0}; \mathbf{r}) p(\mathbf{r}) = \sigma p(\mathbf{r}) \exp(jkr)/r$ is transmitted through the same system. This is a spherical wave converging to the point $\mathbf{0}$ in the output plane and transmitted through the aperture.

This similarity between the diffraction of coherent light and the gain of spatial coherence of incoherent light traveling through the same system is known as the **Van Cittert–Zernike theorem**.

Gain of Coherence in Free Space

Consider the optical system of free-space propagation between two parallel planes separated by a distance d (Fig. 10.3-7). Light in the input plane is quasi-monochromatic, spatially incoherent, and has intensity $I(x, y)$ extending over a finite area. The distance d is sufficiently large so that for points of interest in the output plane the Fraunhofer approximation is valid. Under these conditions the impulse-response func-

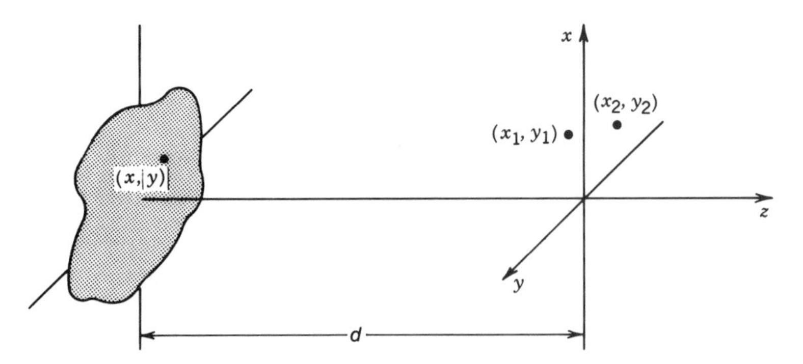

Figure 10.3-7 Radiation from an incoherent source in free space.

tion of the optical system is described by the Fraunhofer diffraction formula [see (4.2-3)]

$$h(\mathbf{r};\mathbf{r}') = h_0 \exp\left(-j\pi\frac{x^2+y^2}{\lambda d}\right)\exp\left(j2\pi\frac{xx'+yy'}{\lambda d}\right), \qquad (10.3\text{-}22)$$

where $\mathbf{r} = (x, y, d)$ and $\mathbf{r}' = (x', y', 0)$ are the coordinates of points in the output and input planes, respectively, and $h_0 = (j/\lambda d)\exp(-j2\pi d/\lambda)$ is a constant.

To determine the mutual coherence function $G(x_1, y_1, x_2, y_2)$ at two points (x_1, y_1) and (x_2, y_2) in the output plane, we substitute (10.3-22) into (10.3-19) and obtain

$$|G(x_1, y_1, x_2, y_2)| = \sigma_1\left|\iint_{-\infty}^{\infty} \exp\left\{j\frac{2\pi}{\lambda d}[(x_2-x_1)x + (y_2-y_1)y]\right\} I(x,y)\,dx\,dy\right|, \qquad (10.3\text{-}23)$$

where $\sigma_1 = \sigma|h_0|^2 = \sigma/\lambda^2 d^2$ is another constant. Given $I(x, y)$, one can easily determine $|G(x_1, y_1, x_2, y_2)|$ in terms of the two-dimensional Fourier transform of $I(x, y)$,

$$\mathcal{I}(\nu_x, \nu_y) = \int_{-\infty}^{\infty}\int_{-\infty}^{\infty} \exp\left[j2\pi(\nu_x x + \nu_y y)\right] I(x,y)\,dx\,dy \qquad (10.3\text{-}24)$$

evaluated at $\nu_x = (x_2 - x_1)/\lambda d$ and $\nu_y = (y_2 - y_1)/\lambda d$. The magnitude of the corresponding normalized mutual intensity is

$$\boxed{|g(x_1, y_1, x_2, y_2)| = \left|\mathcal{I}\left(\frac{x_2-x_1}{\lambda d}, \frac{y_2-y_1}{\lambda d}\right)\right| \Big/ \mathcal{I}(0,0).} \qquad (10.3\text{-}25)$$

This Fourier transform relation between the intensity profile of an incoherent source and the degree of spatial coherence of its far field is similar to the Fourier transform relation between the amplitude of coherent light at the input and output planes (see Sec. 4.2A). The similarity is expected in view of the Van Cittert–Zernike theorem.

The implications of (10.3-25) are profound. If the area of the source, i.e., the spatial extent of $I(x, y)$, is small, its Fourier transform $\mathcal{I}(\nu_x, \nu_y)$ is wide, so that the mutual intensity in the output plane extends over a wide area and the area of coherence in the output plane is large. In the extreme limit in which light in the input plane originates from a point, the area of coherence is infinite and the radiated field is spatially completely coherent. This confirms our earlier discussions in Sec. 10.1D regarding the coherence of spherical waves. On the other hand, if the input incoherent light originates from a large extended source, the propagated light has a small area of coherence.

EXAMPLE 10.3-2. ***Radiation from an Incoherent Circular Source.*** For input light with uniform intensity $I(x, y) = I_0$ confined to a circular aperture of radius a, (10.3-25) yields

$$|g(x_1, y_1, x_2, y_2)| = \left| \frac{2J_1(\pi \rho \theta_s / \lambda)}{\pi \rho \theta_s / \lambda} \right|, \tag{10.3-26}$$

where $\rho = [(x_2 - x_1)^2 + (y_2 - y_1)^2]^{1/2}$ is the distance between the two points, $\theta_s = 2a/d$ is the angle subtended by the source, and $J_1(\cdot)$ is the Bessel function. This relation is plotted in Fig. 10.3-8. The Bessel function reaches its first zero when its argument is 3.832. We can therefore define the area of coherence as a circle of radius $\rho_c = 3.832(\lambda/\pi\theta_s)$, so that

$$\boxed{\rho_c = 1.22 \frac{\lambda}{\theta_s}.} \tag{10.3-27}$$

Coherence Distance

A similar result, (10.2-14), was obtained using a less rigorous analysis. The area of coherence is inversely proportional to θ_s^2. An incoherent light source of wavelength $\lambda = 0.6$ μm and radius 1 cm observed at a distance $d = 100$ m, for example, has a coherence distance $\rho_c \approx 3.7$ mm.

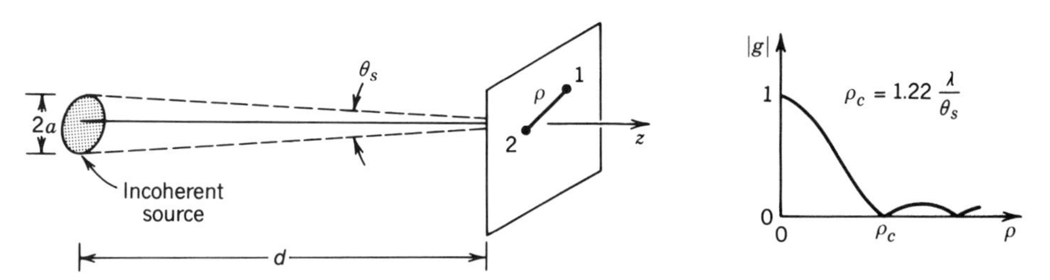

Figure 10.3-8 The magnitude of the degree of spatial coherence of light radiated from an incoherent circular light source subtending an angle θ_s, as a function of the separation ρ.

Measurement of the Angular Diameter of Stars; The Michelson Stellar Interferometer

Equation (10.3-27) is the basis of a method for measuring the angular diameters of stars. If the star is regarded as an incoherent disk of diameter $2a$ with uniform brilliance, then at an observation plane a distance d away from the star, the coherence function drops to 0 when the separation between the two observation points reaches

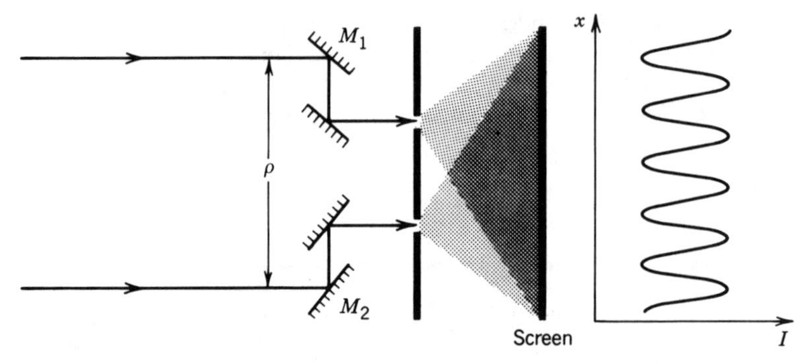

Figure 10.3-9 Michelson stellar interferometer. The angular diameter of a star is estimated by measuring the mutual intensity at two points with variable separation ρ using Young's double-slit interferometer. The distance ρ between mirrors M_1 and M_2 is varied and the visibility of the interference fringes is measured. When $\rho = \rho_c = 1.22\lambda/\theta_s$, the visibility = 0.

$\rho_c = 1.22\lambda/\theta_s$. Measuring ρ_c for a given λ permits us to determine the angular diameter $\theta_s = 2a/d$.

As an example, taking the angular diameter of the sun to be 0.5°, $\theta_s = 8.7 \times 10^{-3}$ radians, and assuming that the intensity is uniform, we obtain $\rho_c \approx 140\lambda$. For $\lambda =$ 0.5 μm, $\rho_c = 70$ μm. To observe interference fringes in a Young's double-slit apparatus, the holes would have to be separated by a distance smaller than 70 μm. Stars of smaller angular diameter have correspondingly larger areas of coherence. For example, the first star whose angular diameter was measured using this technique (α-Orion) has an angular diameter $\theta_s = 22.6 \times 10^{-8}$, so that for $\lambda = 0.57$ μm, $\rho_c = 3.1$ m. A Young's interferometer can be modified to accommodate such large slit separations by using movable mirrors, as shown in Fig. 10.3-9.

10.4 PARTIAL POLARIZATION

As we have seen in Chap. 6, the scalar theory of light is often inadequate and a vector theory including the polarization of light is necessary. This section provides a brief discussion of the statistical theory of *random* light, including the effects of polarization. The **theory of partial polarization** is based on characterizing the components of the optical field vector by correlations and cross-correlations similar to those defined earlier in this chapter.

To simplify the presentation, we shall not be concerned with spatial effects. We therefore limit ourselves to light described by a transverse electromagnetic (TEM) plane wave traveling in the z direction. The electric-field vector has two components in the x and y directions with complex wavefunctions $U_x(t)$ and $U_y(t)$ that are generally random. Each function is characterized by its autocorrelation function (the temporal coherence function),

$$G_{xx}(\tau) = \langle U_x^*(t)U_x(t+\tau)\rangle \qquad (10.4\text{-}1)$$

$$G_{yy}(\tau) = \langle U_y^*(t)U_y(t+\tau)\rangle. \qquad (10.4\text{-}2)$$

An additional descriptor of the wave is the cross-correlation function of $U_x(t)$ and $U_y(t)$,

$$G_{xy}(\tau) = \langle U_x^*(t)U_y(t+\tau)\rangle. \qquad (10.4\text{-}3)$$

The normalized function

$$g_{xy}(\tau) = \frac{G_{xy}(\tau)}{\left[G_{xx}(0)G_{yy}(0)\right]^{1/2}} \tag{10.4-4}$$

is the cross-correlation coefficient of $U_x^*(t)$ and $U_y(t+\tau)$. It satisfies the inequality $0 \le |g_{xy}(\tau)| \le 1$. When the two components are uncorrelated at all times, $|g_{xy}(\tau)| = 0$; and when they are completely correlated at all times, $|g_{xy}(\tau)| = 1$.

The spectral properties are, in general, tied to the polarization properties, so that the autocorrelation and cross-correlation functions have different dependences on τ. However, for quasimonochromatic light all dependences on τ in (10.4-1) to (10.4-4) are approximately of the form $\exp(j2\pi\nu_0\tau)$, so that the polarization properties are described by the values at $\tau = 0$. The three numbers $G_{xx}(0)$, $G_{yy}(0)$, and $G_{xy}(0)$, hereafter denoted G_{xx}, G_{yy}, and G_{xy}, are then used to describe the polarization of the wave. Note that $G_{xx} = I_x$ and $G_{yy} = I_y$ are real numbers that represent the intensities of the x and y components, but G_{xy} is complex and $G_{yx} = G_{xy}^*$, as can easily be verified from the definition.

Coherency Matrix

It is convenient to write the four variables G_{xx}, G_{xy}, G_{yx}, and G_{yy} in the form of a 2×2 Hermitian matrix

$$\mathbf{G} = \begin{bmatrix} G_{xx} & G_{xy} \\ G_{yx} & G_{yy} \end{bmatrix} \tag{10.4-5}$$

called the **coherency matrix**. The diagonal elements are the intensities I_x and I_y, and the off-diagonal elements are the cross-correlations. The trace of the matrix $\mathrm{Tr}\,\mathbf{G} = I_x + I_y \equiv \bar{I}$ is the total intensity.

The coherency matrix may also be written in terms of the Jones vector, $\mathbf{J} = \begin{bmatrix} U_x \\ U_y \end{bmatrix}$, defined in terms of the complex wavefunctions and complex amplitudes (instead of in terms of the complex envelopes as in Sec. 6.1),

$$\langle \mathbf{J}^*\mathbf{J}^\dagger \rangle = \left\langle \begin{bmatrix} U_x^* \\ U_y^* \end{bmatrix} \begin{bmatrix} U_x & U_y \end{bmatrix} \right\rangle = \begin{bmatrix} \langle U_x^* U_x \rangle & \langle U_x^* U_y \rangle \\ \langle U_y^* U_x \rangle & \langle U_y^* U_y \rangle \end{bmatrix} = \mathbf{G}, \tag{10.4-6}$$

where † denotes the transpose of a matrix, and U_x and U_y denote $U_x(t)$ and $U_y(t)$, respectively.

The Jones vector is transformed by polarization devices, such as polarizers and retarders, in accordance with the rule $\mathbf{J}' = \mathbf{TJ}$ [see (6.1-10)], where $\mathbf{T}$ is the Jones matrix representing the device [see (6.1-11) to (6.1-18)]. The coherency matrix is therefore transformed in accordance with $\mathbf{G}' = \langle \mathbf{T}^*\mathbf{J}^*(\mathbf{TJ})^\dagger \rangle = \langle \mathbf{T}^*\mathbf{J}^*\mathbf{J}^\dagger\mathbf{T}^\dagger \rangle = \mathbf{T}^*\langle \mathbf{J}^*\mathbf{J}^\dagger \rangle \mathbf{T}^\dagger$, so that

$$\boxed{\mathbf{G}' = \mathbf{T}^*\mathbf{G}\mathbf{T}^\dagger.} \tag{10.4-7}$$

We thus have a formalism for determining the effect of polarization devices on the coherency matrix of partially polarized light.

To understand the significance of the coherency matrix, we examine next two limiting cases.

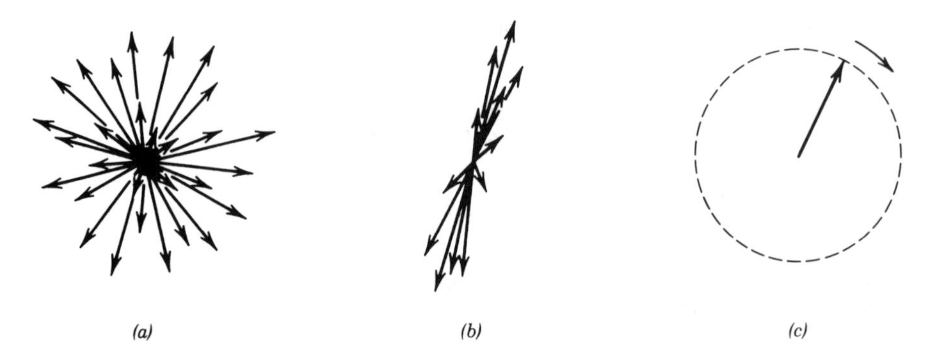

Figure 10.4-1 Fluctuations of the electric field vector for (*a*) unpolarized light; (*b*) partially polarized light; (*c*) polarized light with circular polarization.

Unpolarized Light

Light of intensity $\bar{I}$ is said to be **unpolarized** if its two components have the same intensity and are uncorrelated, $I_x = I_y \equiv \frac{1}{2}\bar{I}$ and $G_{xy} = 0$. The coherency matrix is then

$$\mathbf{G} = \tfrac{1}{2}\bar{I}\begin{bmatrix} 1 & 0 \\ 0 & 1 \end{bmatrix}. \tag{10.4-8}$$

By use of (10.4-7) and (6.1-15), it can be shown that (10.4-8) is invariant to rotation of the coordinate system, so that the two components always have equal intensities and are uncorrelated. Unpolarized light therefore has an electric field vector that is statistically isotropic; it is equally likely to have any direction in the x–y plane, as illustrated in Fig. 10.4-1(a).

When passed through a polarizer, unpolarized light becomes linearly polarized, but it remains random with an average intensity $\frac{1}{2}\bar{I}$. A wave retarder has no effect on unpolarized light since it only introduces a phase shift between two components that have a totally random phase to begin with. Similarly, unpolarized light transmitted through a polarization rotator remains unpolarized. These effects may be shown formally by use of (10.4-7) and (10.4-8) together with (6.1-11), (6.1-12), and (6.1-13).

Polarized Light

If the cross-correlation coefficient $g_{xy} = G_{xy}/[I_x I_y]^{1/2}$ has unit magnitude, $|g_{xy}| = 1$, the two components of the optical field are perfectly correlated and the light is said to be completely polarized (or simply **polarized**). Since $g_{xy} = G_{xy}/[I_x I_y]^{1/2}$, the coherency matrix takes the form

$$\mathbf{G} = \begin{bmatrix} I_x & (I_x I_y)^{1/2} e^{j\varphi} \\ (I_x I_y)^{1/2} e^{-j\varphi} & I_y \end{bmatrix}, \tag{10.4-9}$$

where φ is the argument of g_{xy}. Defining $U_x = I_x^{1/2}$ and $U_y = I_y^{1/2} e^{j\varphi}$,

$$\mathbf{G} = \begin{bmatrix} U_x^* U_x & U_x^* U_y \\ U_y^* U_x & U_y^* U_y \end{bmatrix} = \mathbf{J}^* \mathbf{J}^{\dagger}, \tag{10.4-10}$$

where $\mathbf{J}$ is a Jones matrix with components U_x and U_y. Thus $\mathbf{G}$ has the same form as the coherency matrix of a coherent wave.

Using the Jones vectors listed in Table 6.1-1 on page 198, we can determine the coherency matrices for different states of polarization. Two examples are:

Linearly polarized in the x direction	$\mathbf{G} = \bar{I}\begin{bmatrix} 1 & 0 \\ 0 & 0 \end{bmatrix}$	Right-circularly polarized	$\mathbf{G} = \frac{1}{2}\bar{I}\begin{bmatrix} 1 & j \\ -j & 1 \end{bmatrix}$

It is instructive to examine the distinction between unpolarized light and circularly polarized light. In both cases the intensities of the x and y components are equal ($I_x = I_y$). For circularly polarized light the two components are completely correlated, but for unpolarized light they are uncorrelated. Circularly polarized light may be transformed into linearly polarized light by the use of a wave retarder, but unpolarized light remains unpolarized upon passage through such a device.

Degree of Polarization

Partial polarization is a general state of random polarization that lies between the two ideal limits of unpolarized and polarized light. One measure of the **degree of polarization** is defined in terms of the determinant and the trace of the coherency matrix:

$$\mathcal{P} = \left\{1 - \frac{4 \det \mathbf{G}}{(\operatorname{Tr} \mathbf{G})^2}\right\}^{1/2} \tag{10.4-11}$$

$$= \left\{1 - 4\left[\frac{I_x I_y}{(I_x + I_y)^2}\right]\left(1 - |g_{xy}|^2\right)\right\}^{1/2}. \tag{10.4-12}$$

This measure is meaningful because of the following considerations:

- It satisfies the inequality $0 \le \mathcal{P} \le 1$.
- For polarized light, $\mathcal{P}$ has its highest value of 1, as can easily be seen by substituting $|g_{xy}| = 1$ into (10.4-12). For unpolarized light it has its lowest value $\mathcal{P} = 0$, since $I_x = I_y$ and $g_{xy} = 0$.
- It is invariant to rotation of the coordinate system (since the determinant and the trace of a matrix are invariants to unitary transformations).
- It can be shown (Exercise 10.4-1) that a partially polarized wave can always be regarded as a mixture of two uncorrelated waves: a completely polarized wave and an unpolarized wave, with the ratio of the intensity of the polarized component to the total intensity equal to the degree of polarization $\mathcal{P}$.

EXERCISE 10.4-1

Partially Polarized Light. Show that the superposition of unpolarized light of intensity $(I_x + I_y)(1 - \mathcal{P})$, and linearly polarized light with intensity $(I_x + I_y)\mathcal{P}$, where $\mathcal{P}$ is given by (10.4-12), yields light whose x and y components have intensities I_x and I_y and normalized cross-correlation $|g_{xy}|$.

READING LIST

General

G. Reynolds, J. B. DeVelis, G. Parrent, and B. J. Thompson, *The New Physical Optics Notebook: Tutorials in Fourier Optics*, SPIE—The International Society for Optical Engineering, Bellingham, WA, and American Institute of Physics, New York, 1989.

J. W. Goodman, *Statistical Optics*, Wiley, New York, 1985.

J. Peřina, *Coherence of Light*, Van Nostrand-Reinhold, London, 1971, 2nd ed. 1985.

B. R. Frieden, *Probability, Statistical Optics, and Data Testing*, Springer-Verlag, Berlin, 1983.

A. S. Marathay, *Elements of Optical Coherence Theory*, Wiley, New York, 1982.

M. Born and E. Wolf, *Principles of Optics*, Pergamon Press, New York, 1959, 6th ed. 1980, Chap. 10.

B. Saleh, *Photoelectron Statistics with Applications to Spectroscopy and Optical Communication*, Springer-Verlag, Berlin, 1978.

B. Crosignani, P. Di Porto, and M. Bertolotti, *Statistical Properties of Scattered Light*, Academic Press, New York, 1975.

J. C. Dainty, ed., *Laser Speckle and Related Phenomena*, Springer-Verlag, Berlin, 1975.

R. Hanbury-Brown, *The Intensity Interferometer*, Taylor and Francis, London, 1974.

G. J. Troup, *Optical Coherence Theory*, Methuen, London, 1967.

M. J. Beran and G. B. Parrent, Jr., *Theory of Partial Coherence*, Prentice-Hall, Englewood Cliffs, NJ, 1964.

E. L. O'Neil, *Introduction to Statistical Optics*, Addison-Wesley, Reading, MA, 1963.

Books on Random Functions

C. W. Helstrom, *Probability and Stochastic Processes for Engineers and Scientists*, Macmillan, New York, 2nd ed. 1991.

A. Papoulis, *Probability, Random Variables, and Stochastic Processes*, McGraw-Hill, New York, 1965, 2nd ed. 1984.

E. Vanmarcke, *Random Fields*, MIT Press, Cambridge, MA, 1983.

E. Parzen, *Modern Probability Theory and Its Applications*, Wiley, New York, 1960.

Articles

Journal of the Optical Society of America, Feature issues on applications of coherence and statistical optics, no. 7, 1986 and no. 8, 1986.

F. T. S. Yu, Principles of Optical Processing with Partially Coherent Light, in *Progress in Optics*, vol. 23, E. Wolf, ed., North-Holland, Amsterdam, 1986.

W. J. Tango and R. Q. Twiss, Michelson Stellar Interferometry, in *Progress in Optics*, vol. 17, E. Wolf, ed., North-Holland, Amsterdam, 1980.

G. O. Reynolds and J. B. DeVelis, Review of Optical Coherence Effects in Instrument Design, *SPIE Proceedings*, vol. 194, p. 2, 1979.

E. Wolf, Coherence and Radiometry, *Journal of the Optical Society of America*, vol. 68, pp. 6–17, 1978.

H. P. Baltes, J. Geist, and A. Walther, Radiometry and Coherence, in *Inverse Source Problems in Optics*, H. P. Baltes, ed., Springer-Verlag, New York, 1978.

L. Mandel and E. Wolf, eds., *Selected Papers on Coherence and Fluctuations of Light*, vols. 1 and 2, Dover, New York, 1970.

B. J. Thompson, Image Formation with Partially Coherent Light, in *Progress in Optics*, vol. 7, E. Wolf, ed., North-Holland, Amsterdam, 1969.

L. Mandel and E. Wolf, Coherence Properties of Optical Fields, *Reviews of Modern Physics*, vol. 37, pp. 231–287, 1965.

PROBLEMS

10.1-1 **Lorentzian Spectrum.** A light-emitting diode (LED) emits light of Lorentzian spectrum with a linewidth $\Delta\nu$ (FWHM) $= 10^{13}$ Hz centered about a frequency corresponding to a wavelength $\lambda_o = 0.7\ \mu$m. Determine the linewidth $\Delta\lambda_o$ (in units of nm), the coherence time τ_c, and the coherence length l_c. What is the maximum time delay within which the magnitude of the complex degree of temporal coherence $|g(\tau)|$ is greater than 0.5?

10.1-2 **Proof of the Wiener–Khinchin Theorem.** Use the definitions in (10.1-4), (10.1-11), and (10.1-12) to prove that the spectral density $S(\nu)$ is the Fourier transform of the autocorrelation function $G(\tau)$. Prove that the intensity I is the integral of the power spectral density $S(\nu)$.

10.1-3 **Mutual Intensity.** The mutual intensity of an optical wave at points on the x axis is given by

$$G(x_1, x_2) = I_0 \exp\left[-\frac{(x_1^2 + x_2^2)}{W_0^2}\right] \exp\left[-\frac{(x_1 - x_2)^2}{\rho_c^2}\right],$$

where I_0, W_0, and ρ_c are constants. Sketch the intensity distribution as a function of x. Derive an expression for the normalized mutual intensity $g(x_1, x_2)$ and sketch it as a function of $x_1 - x_2$. What is the physical meaning of the parameters I_0, W_0, and ρ_c?

10.1-4 **Mutual Coherence Function.** An optical wave has a mutual coherence function at points on the x axis,

$$G(x_1, x_2, \tau) = \exp\left(-\frac{\pi\tau^2}{2\tau_c^2}\right) \exp[j2\pi u(x_1, x_2)\tau] \exp\left[-\frac{(x_1 - x_2)^2}{\rho_c^2}\right],$$

where $u(x_1, x_2) = 5 \times 10^{14}\ \mathrm{s}^{-1}$ for $x_1 + x_2 > 0$, and $6 \times 10^{14}\ \mathrm{s}^{-1}$ for $x_1 + x_2 < 0$, $\rho_c = 1$ mm, and $\tau_c = 1\ \mu$s. Determine the intensity, the power spectral density, the coherence length, and the coherence distance in the transverse plane. Which of these quantities is position dependent? If this wave is recorded on color film, what would the recorded image look like?

10.1-5 **Coherence Length.** Show that light of narrow spectral width has a coherence length $l_c \approx \lambda^2/\Delta\lambda$, where $\Delta\lambda$ is the linewidth in wavelength units. Show that for light of broad uniform spectrum extending between the wavelengths $\lambda_{\min}$ and $\lambda_{\max} = 2\lambda_{\min}$, the coherence length $l_c = \lambda_{\max}$.

10.1-6 **Effect of Spectral Width on Spatial Coherence.** A point source at the origin $(0, 0, 0)$ of a Cartesian coordinate system emits light with a Lorentzian spectrum and coherence time $\tau_c = 10$ ps. Determine an expression for the normalized mutual intensity of the light at the points $(0, 0, d)$ and $(x, 0, d)$, where $d = 10$ cm. Sketch the magnitude of the normalized mutual intensity as a function of x.

10.1-7 **Gaussian Mutual Intensity.** An optical wave in free space has a mutual coherence function $G(\mathbf{r}_1, \mathbf{r}_2, \tau) = J(\mathbf{r}_1 - \mathbf{r}_2)\exp(j2\pi\nu_0\tau)$. (a) Show that the function $J(\mathbf{r})$ must satisfy the Helmholtz equation $\nabla^2 J + k_o^2 J = 0$, where $k_o = 2\pi\nu_0/c$. (b) An approximate solution of the Helmholtz equation is the Gaussian-beam solution

$$J(\mathbf{r}) = \frac{1}{q(z)} \exp\left[-\frac{jk_o(x^2+y^2)}{2q(z)}\right] \exp(-jk_o z),$$

where $q(z) = z + jz_0$ and z_0 is a constant. This solution has been studied extensively in Chap. 3 in connection with Gaussian beams. Determine an expression for the coherence area near the z axis and show that it increases with $|z|$, so that the wave gains coherence with propagation away from the origin.

10.2-1 **Effect of Spectral Width on Fringe Visibility.** Light from a sodium lamp of Lorentzian spectral linewidth $\Delta\nu = 5 \times 10^{11}$ Hz is used in a Michelson interferometer. Determine the maximum path-length difference for which the visibility of the interferogram $\mathscr{V} > \frac{1}{2}$.

10.2-2 **Number of Observable Fringes in Young's Interferometer.** Determine the number of observable fringes in Young's interferometer if each of the sources in Table 10.1-2 on page 352 is used. Assume full spatial coherence in all cases.

10.2-3 **Spectrum of a Superposition of Two Waves.** An optical wave is a superposition of two waves $U_1(t)$ and $U_2(t)$ with identical spectra $S_1(\nu) = S_2(\nu)$, which are Gaussian with spectral width $\Delta\nu$ and central frequency ν_0. The waves are not necessarily uncorrelated. Determine an expression for the power spectral density $S(\nu)$ of the superposition $U(t) = U_1(t) + U_2(t)$. Explore the possibility that $S(\nu)$ is also Gaussian, with a shifted central frequency $\nu_1 \neq \nu_0$. If this were possible, our faith in using the Doppler shift as a method to determine the velocity of stars would be shaken, since frequency shifts could originate from something other than the Doppler effect.

*10.3-1 **Partially Coherent Gaussian Beam.** A quasi-monochromatic light wave of wavelength λ travels in free space in the z direction. Its intensity in the $z = 0$ plane is a Gaussian function $I(x) = I_0 \exp(-2x^2/W_0^2)$ and its normalized mutual intensity is also a Gaussian function $g(x_1, x_2) = \exp[-(x_1 - x_2)^2/\rho_c^2]$. Show that the intensity at a distance z satisfying conditions of the Fraunhofer approximation is also a Gaussian function $I_z(x) \propto \exp[-2x^2/W^2(z)]$ and derive an expression for the beam radius $W(z)$ as a function of z and the parameters W_0, ρ_c, and λ. Discuss the effect of spatial coherence on beam divergence.

*10.3-2 **Fourier-Transform Lens.** Quasi-monochromatic spatially incoherent light of uniform intensity illuminates a transparency of intensity transmittance $f(x, y)$ and the emerging light is transmitted between the front and back focal planes of a lens. Determine an expression for the intensity of the observed light. Compare your results with the case of coherent light in which the lens performs the Fourier transform (see Sec. 4.2).

*10.3-3 **Light from a Two-Point Incoherent Source.** A spatially incoherent quasi-monochromatic source of light emits only at two points separated by a distance $2a$. Determine an expression for the normalized mutual intensity at a distance d from the source (use the Fraunhofer approximation).

*10.3-4 **Coherence of Light Transmitted Through a Fourier-Transform Optical System.** Light from a quasi-monochromatic spatially incoherent source with uniform intensity is transmitted through a thin slit of width $2a$ and travels between the front and back focal planes of a lens. Determine an expression for the normalized mutual intensity in the back focal plane.

10.4-1 **Partially Polarized Light.** The intensities of the two components of a partially polarized wave are $I_x = I_y = \frac{1}{2}$, and the argument of the cross-correlation coefficient g_{xy} is $\pi/2$.
(a) Plot the degree of polarization $\mathscr{P}$ versus the magnitude of the cross-correlation coefficient $|g_{xy}|$.
(b) Determine the coherency matrix if $\mathscr{P} = 0, 0.5$, and 1, and describe the nature of the light in each case.
(c) If the light is transmitted through a polarizer with its axis in the x direction, what is the intensity of the light transmitted?

CHAPTER

11

PHOTON OPTICS

11.1 THE PHOTON
- A. Photon Energy
- B. Photon Position
- C. Photon Momentum
- D. Photon Polarization
- E. Photon Interference
- F. Photon Time

11.2 PHOTON STREAMS
- A. Mean Photon Flux
- B. Randomness of Photon Flux
- C. Photon-Number Statistics
- D. Random Partitioning of Photon Streams

*11.3 QUANTUM STATES OF LIGHT
- A. Coherent-State Light
- B. Squeezed-State Light

Max Planck (1858–1947) suggested that the emission and absorption of light by matter occur in quanta of energy.

Albert Einstein (1879–1955) advanced the hypothesis that light itself consists of quanta of energy.

Electromagnetic optics (Chap. 5) provides the most complete treatment of light within the confines of **classical optics**. It encompasses wave optics, which in turn encompasses ray optics (Fig. 11.0-1). Although classical electromagnetic theory is capable of providing explanations for a great many effects in optics, as attested to by the earlier chapters in this book, it nevertheless fails to account for certain optical phenomena. This failure, which became evident about the turn of this century, ultimately led to the formulation of a quantum electromagnetic theory known as **quantum electrodynamics**. For optical phenomena, this theory is also referred to as **quantum optics**. Quantum electrodynamics (QED) is more general than classical electrodynamics and it is today accepted as a theory that is useful for explaining virtually all known optical phenomena.

In the framework of QED, the electric and magnetic fields **E** and **H** are mathematically treated as operators in a vector space. They are assumed to satisfy certain operator equations and commutation relations that govern their time dynamics and their interdependence. The equations of QED are required to accurately describe the interactions of electromagnetic fields with matter in the same way that Maxwell's equations are used in classical electrodynamics. The use of QED can lead to results that are characteristically quantum in nature and cannot be explained classically.

The formal treatment of QED is beyond the scope of this book. Nevertheless, it is possible to derive many of the quantum-mechanical properties of light and its interaction with matter by supplementing electromagnetic optics with a few simple relationships drawn from QED that represent the corpuscularity, localization, and fluctuations of electromagnetic fields and energy. This set of rules, which we call **photon optics**, permits us to deal with optical phenomena that are beyond the reach of classical theory, while retaining classical optics as a limiting case. However, photon optics is not intended to be a theory that is capable of providing an explanation for all optical effects.

In Sec. 11.1 we introduce the concept of the photon and its properties in the form of a number of rules that govern the behavior of photon energy, momentum, polarization, position, time, and interference. These rules take the form of deceptively simple relationships with far-reaching consequences. This is followed, in Sec. 11.2, by a

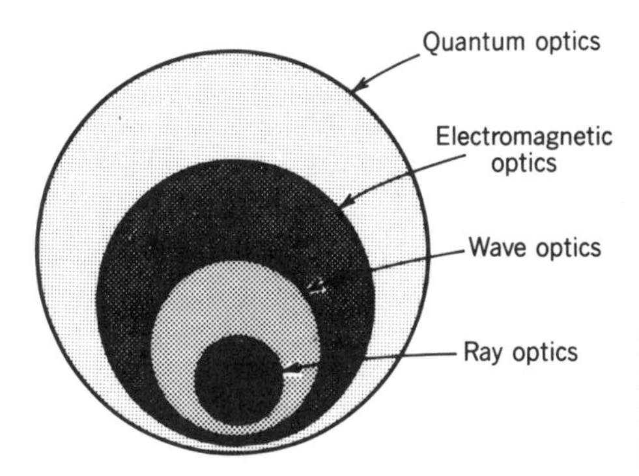

Figure 11.0-1 The theory of quantum optics provides an explanation for virtually all optical phenomena. It is more general than electromagnetic optics, which was shown earlier to encompass wave optics and ray optics.

discussion of the properties of photon streams. The number of photons emitted by a light source in a given time is almost always random, with statistical properties that depend on the nature of the source. The photon-number statistics for several important optical sources, including the laser and thermal radiators, are discussed. The effects of simple optical components (such as a beamsplitter and a filter) on the randomness of a photon stream are also examined. In Sec. 11.3 we use quantum optics to discuss the random fluctuations of the magnitude and phase of the electromagnetic field and to provide a brief introduction to coherent and squeezed states of light. The interaction of photons with atoms is discussed in Chap. 12.

11.1 THE PHOTON

Light consists of particles called **photons**. A photon has zero rest mass and carries electromagnetic energy and momentum. It also carries an intrinsic angular momentum (or spin) that governs its polarization properties. The photon travels at the speed of light in vacuum (c_o); its speed is retarded in matter. Photons also have a wavelike character that determines their localization properties in space and the rules by which they interfere and diffract.

The notion of the photon initially grew out of an attempt by Planck to resolve a long-standing riddle concerning the spectrum of blackbody radiation. He finally achieved this goal by quantizing the allowed energy values of each of the electromagnetic modes in a cavity from which radiation was emanating (this subject is discussed in Chap. 12). The concept of the photon and the rules of photon optics are introduced in this section by considering light inside an optical resonator (a cavity). This is a convenient choice because it restricts the space under consideration to a simple geometry. The presence of the resonator turns out not to be an important restriction in the argument; the results can be shown to be independent of its presence.

Electromagnetic-Optics Theory of Light in a Resonator

In accordance with electromagnetic optics, light inside a lossless resonator of volume V is completely characterized by an electromagnetic field that takes the form of a sum of discrete orthogonal modes of different frequencies, different spatial distributions, and different polarizations. The electric field vector is $\mathcal{E}(\mathbf{r}, t) = \text{Re}\{\mathbf{E}(\mathbf{r}, t)\}$, where

$$\mathbf{E}(\mathbf{r}, t) = \sum_{\mathbf{q}} A_{\mathbf{q}} U_{\mathbf{q}}(\mathbf{r}) \exp(j2\pi\nu_{\mathbf{q}} t)\hat{\mathbf{e}}_{\mathbf{q}}. \tag{11.1-1}$$

The qth mode has complex amplitude $A_{\mathbf{q}}$, frequency $\nu_{\mathbf{q}}$, polarization along the direction of the unit vector $\hat{\mathbf{e}}_{\mathbf{q}}$, and a spatial distribution characterized by the complex function $U_{\mathbf{q}}(\mathbf{r})$, which is normalized such that $\int_V |U_{\mathbf{q}}(\mathbf{r})|^2 \, d\mathbf{r} = 1$. The choice of the expansion functions $U_{\mathbf{q}}(\mathbf{r})$ and $\hat{\mathbf{e}}_{\mathbf{q}}$ is not unique.

In a cubic resonator of dimension d, one convenient choice of the spatial expansion functions is the set of standing waves

$$U_{\mathbf{q}}(\mathbf{r}) = \left(\frac{2}{d}\right)^{3/2} \sin\frac{q_x \pi x}{d} \sin\frac{q_y \pi y}{d} \sin\frac{q_z \pi z}{d}, \tag{11.1-2}$$

where q_x, q_y, and q_z are integers denoted collectively by the index $\mathbf{q} = (q_x, q_y, q_z)$ [see Sec. 9.1 and Fig. 11.1-1(a)]. The energy contained in the mode is

$$E_{\mathbf{q}} = \tfrac{1}{2}\epsilon \int_V \mathbf{E}(\mathbf{r}, t) \cdot \mathbf{E}^*(\mathbf{r}, t) \, d\mathbf{r} = \tfrac{1}{2}\epsilon |A_{\mathbf{q}}|^2.$$

In classical electromagnetic theory, the energy E_q can assume an arbitrary nonnegative value, no matter how small. The total energy is the sum of the energies in all the modes.

Photon-Optics Theory of Light in a Resonator

The electromagnetic-optics theory described above is maintained in photon optics, but a restriction is placed on the energy that is allowed to be carried by each mode. Rather than assuming a continuous range, the energy of a mode is restricted to a discrete set of values equally separated by a fixed energy. The energy of a mode is said to be quantized, with only integral units of this fixed energy allowed. Each unit of energy is carried by a photon.

> Light in a resonator is comprised of a set of modes, each containing an integral number of identical photons. Characteristics of the mode, such as its frequency, spatial distribution, direction of propagation, and polarization, are assigned to the photon.

A. Photon Energy

Photon optics provides that the energy of an electromagnetic mode is quantized to discrete levels separated by the energy of a photon (Fig. 11.1-1). The energy of a photon in a mode of frequency ν is

$$E = h\nu = \hbar\omega, \tag{11.1-3}$$

Photon Energy

where $h = 6.63 \times 10^{-34}$ J-s is **Planck's constant** and $\hbar \equiv h/2\pi$. Energy may be added to, or taken from, this mode only in units of $h\nu$.

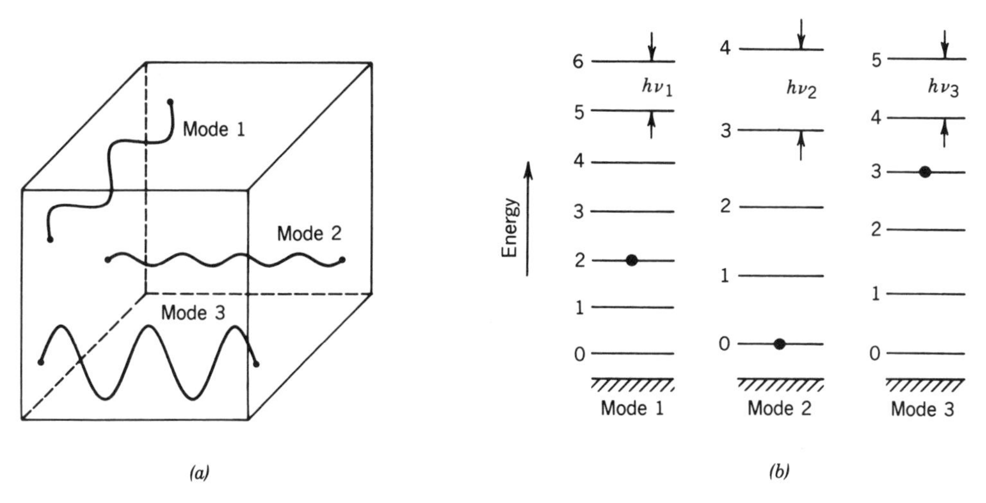

Figure 11.1-1 (*a*) Three modes of different frequencies and directions in a cubic resonator. (*b*) Allowed energies of three modes of frequencies ν_1, ν_2, and ν_3. The solid circles indicate the number of photons in each mode; modes 1, 2, and 3 contain 2, 0, and 3 photons, respectively.

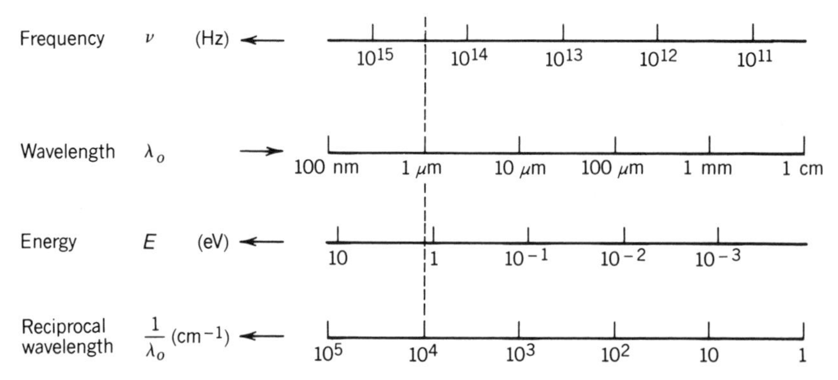

Figure 11.1-2 Relationships between photon frequency ν (Hz), wavelength λ_o, energy E (eV), and reciprocal wavelength $1/\lambda_o$ (cm^{-1}). A photon of wavelength 1 cm has reciprocal wavelength 1 cm^{-1}. A photon of frequency $\nu = 3 \times 10^{14}$ Hz has wavelength $\lambda_o = 1$ μm, energy 1.24 eV, and reciprocal wavelength 10,000 cm^{-1}.

A mode containing zero photons nevertheless carries an energy $E_0 = \frac{1}{2}h\nu$, which is called the zero-point energy. When it carries n photons, therefore, the mode has total energy

$$E_n = \left(n + \tfrac{1}{2}\right)h\nu, \qquad n = 0, 1, 2, \ldots . \tag{11.1-4}$$

In most experiments the zero-point energy is not directly observable because only energy differences [such as $E_{n2} - E_{n1}$ in (11.1-4)] are measured. The presence of the zero-point energy can, however, be manifested in subtle ways when matter is exposed to static fields. It plays a crucial role in the process of spontaneous emission from an atom, as discussed in Chap. 12.

The order of magnitude of photon energy is easily estimated. An infrared photon of wavelength $\lambda_o = 1$ μm has frequency 3×10^{14} Hz since $\lambda_o \nu = c_o$ in vacuum. Its energy is thus $h\nu = 1.99 \times 10^{-19}$ J $= 1.24$ eV (electron volts), which is the same as the kinetic energy of an electron that has been accelerated through a potential difference of 1.24 V. The conversion formula between wavelength (μm) and photon energy (eV) is therefore simply $\lambda_o(\mu\text{m}) = 1.24/E(\text{eV})$.

As another example, a microwave photon with a wavelength of 1 cm has an energy that is 10^4 times smaller, $h\nu = 1.24 \times 10^{-4}$ eV. The reciprocal wavelength is often also used as a unit of energy. It is specified in cm^{-1}, also called wavenumbers (1 cm^{-1} corresponds to 1.24×10^{-4} eV and 1 eV corresponds to 8068.1 cm^{-1}). The relationship between photon frequency, wavelength, energy, and reciprocal wavelength is illustrated in Fig. 11.1-2.

Because photons of higher frequency carry larger energy, the particle nature of light becomes increasingly important as the frequency of the radiation increases. Furthermore, wavelike effects such as diffraction and interference become more difficult to discern as the wavelength becomes shorter. X-rays and gamma-rays almost always behave like collections of particles, in contrast to radio waves, which almost always behave like waves. The frequency of light in the optical region is such that both particle-like and wavelike behavior occur, thus spurring the need for photon optics.

B. Photon Position

Associated with each photon is a wave described by the complex wavefunction $AU(\mathbf{r})\exp(j2\pi\nu t)\hat{\mathbf{e}}$ of the mode. However, when a photon impinges on a detector of small area dA located normal to the direction of propagation at the position $\mathbf{r}$, its indivisibility causes it to be either wholly detected or not detected at all. The location at which the photon is registered is not precisely determined. It is governed by the optical intensity $I(\mathbf{r}) \propto |U(\mathbf{r})|^2$, in accordance with the following probabilistic law:

The probability $p(\mathbf{r})\,dA$ of observing a photon at a point $\mathbf{r}$ within an incremental area dA, at any time, is proportional to the local optical intensity $I(\mathbf{r}) \propto |U(\mathbf{r})|^2$, i.e.,

$$p(\mathbf{r})\,dA \propto I(\mathbf{r})\,dA. \tag{11.1-5}$$

Photon Position

The photon is more likely to be found at those locations where the intensity is high. A photon in a mode described by a standing wave with the intensity distribution $I(x, y, z) \propto \sin^2(\pi z/d)$, where $0 \le z \le d$, for example, is most likely to be detected at $z = d/2$, but will never be detected at $z = 0$ or $z = d$. In contrast to waves, which are extended in space, and particles, which are localized, optical photons behave as extended *and* localized entities. This behavior is called **wave–particle duality**. The localized nature of photons becomes evident when they are detected.

EXERCISE 11.1-1

Photons in a Gaussian Beam

(a) Consider a single photon described by a Gaussian beam (i.e., a $\text{TEM}_{0,0}$ mode of a spherical-mirror resonator; see Secs. 3.1B, 5.4A, and 9.2B). What is the probability of detecting the photon at a point within a circle whose radius is the waist radius of the beam W_0? Recall that at the waist ($z = 0$), $I(\rho, z = 0) \propto \exp(-2\rho^2/W_0^2)$, where ρ is the radial coordinate.

(b) If the beam carries a large number N of independent photons, estimate the average number of photons that lie within this circle.

Transmission of a Single Photon Through a Beamsplitter

An ideal beamsplitter is an optical device that losslessly splits a beam of light into two beams emerging at right angles. It is characterized by a transmittance $\mathscr{T}$ and a reflectance $\mathscr{R} = 1 - \mathscr{T}$. The intensity of the transmitted wave I_t and the intensity of the reflected wave I_r can be calculated from the intensity of the incident wave I using the electromagnetic relations $I_r = (1 - \mathscr{T})I$ and $I_t = \mathscr{T}I$.

Because a photon is indivisible, it must choose between the two possible directions permitted by the beamsplitter. A single photon incident on it follows one of the two possible paths in accordance with the probabilistic photon-position rule (11.1-5). The probability that the photon is transmitted is proportional to I_t and is therefore equal to the transmittance $\mathscr{T}$. The probability that it is reflected is $1 - \mathscr{T}$. From a probability point of view, the problem is identical to that of flipping a coin. Figure 11.1-3 illustrates the process.

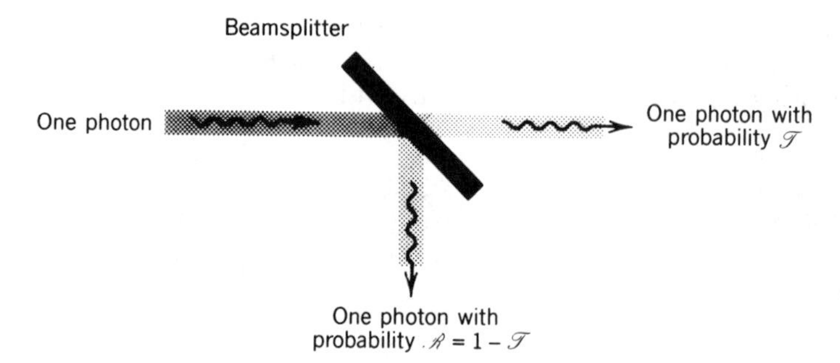

Figure 11.1-3 Probabilistic reflection or transmission of a photon at a beamsplitter.

C. Photon Momentum

The momentum of a photon is related to the wavevector of its associated wavefunction by the following rule:

A photon in a mode described by the plane wave

$$\mathbf{E}(\mathbf{r}, t) = A \exp(-j\mathbf{k} \cdot \mathbf{r}) \exp(j2\pi\nu t)\hat{\mathbf{e}}$$

has a momentum vector

$$\mathbf{p} = \hbar\mathbf{k}. \tag{11.1-6}$$

The photon travels in the direction of the wavevector and the magnitude of the momentum is $p = \hbar k = \hbar 2\pi/\lambda$, i.e.,

$$p = \frac{h}{\lambda}. \tag{11.1-7}$$

Electromagnetic optics leads to the same energy–momentum relationship $\mathbf{p} = (E/c)\hat{\mathbf{k}}$ for a plane wave, where p is the momentum content per unit volume of the wave, E is the energy content per unit volume, and $\hat{\mathbf{k}}$ is a unit vector in the direction of $\mathbf{k}$. Of course, the concept of the photon does not exist in electromagnetic optics, so that the expressions in (11.1-6) and (11.1-7) containing $\hbar$ are unique to photon optics.

*Momentum of a Localized Wave

A wave more general than a plane wave, with a complex wavefunction of the form $AU(\mathbf{r})\exp(j2\pi\nu t)\hat{\mathbf{e}}$, can be expanded as a sum of plane waves of different wavevectors by using the techniques of Fourier optics (see Chap. 4). The component with wavevector $\mathbf{k}$ may be written in the form $A(\mathbf{k})\exp(-j\mathbf{k} \cdot \mathbf{r})\exp(j2\pi\nu t)\hat{\mathbf{e}}$, where $A(\mathbf{k})$ is its amplitude.

The momentum of a photon described by an arbitrary complex wavefunction $AU(\mathbf{r})\exp(j2\pi\nu t)\hat{\mathbf{e}}$ is uncertain. It has the value

$$\mathbf{p} = \hbar\mathbf{k},$$

with probability proportional to $|A(\mathbf{k})|^2$, where $A(\mathbf{k})$ is the amplitude of the plane-wave Fourier component of $U(\mathbf{r})$ with wavevector $\mathbf{k}$.

If $f(x, y) = U(x, y, 0)$ is the complex amplitude at the $z = 0$ plane, the plane-wave Fourier component of wavevector $\mathbf{k} = (k_x, k_y, k_z)$ has an amplitude $A(\mathbf{k}) = F(k_x/2\pi, k_y/2\pi)$, where $F(\nu_x, \nu_y)$ is the two-dimensional Fourier transform of $f(x, y)$ (see Chap. 4). Because the functions $f(x, y)$ and $F(\nu_x, \nu_y)$ are a Fourier transform pair, their widths are inversely related and satisfy the duration–bandwidth relation (see Appendix A, (A.2-6)). The uncertainty relation between the position of the photon and the direction of its momentum is established because the position of the photon at the $z = 0$ plane is probabilistically determined by $|U(\mathbf{r})|^2 = |f(x, y)|^2$, and the direction of its momentum is probabilistically determined by $|A(\mathbf{k})|^2 = |F(k_x/2\pi, k_y/2\pi)|^2$. Thus if, at the plane $z = 0$, σ_x is the position uncertainty in the x direction, and $\sigma_\theta = \sin^{-1}(\sigma_{kx}/k) \approx (\lambda/2\pi)\sigma_{kx}$ is the angular uncertainty about the z axis (assumed $\ll 1$), then the uncertainty relation $\sigma_x \sigma_{kx} \geq \frac{1}{2}$ is equivalent to $\sigma_x \sigma_\theta \geq \lambda/4\pi$.

A plane-wave photon has a known momentum (fixed direction and magnitude), so that $\sigma_\theta = 0$, but its position is totally uncertain ($\sigma_x = \infty$); it is equally likely to be detected anywhere in the $z = 0$ plane. When a plane-wave photon passes through an aperture, its position is localized, at the expense of a spread in the direction of its momentum. The position–momentum uncertainty therefore parallels the theory of diffraction described in Chap. 4. At the other extreme from the plane wave is the spherical-wave photon. It is well localized in position (at the center of the wave), but its momentum has a direction that is totally uncertain.

Radiation Pressure

Because momentum is conserved, its association with a photon means that the emitting atom experiences a recoil of magnitude $h\nu/c$. Furthermore, the momentum associated with a photon can be transferred to objects of finite mass, giving rise to a force and causing mechanical motion. As an example, light beams can be used to deflect atomic beams traveling perpendicular to the photons. The term **radiation pressure** is often used to describe this phenomenon (pressure is force/area).

EXERCISE 11.1-2

Photon-Momentum Recoil. Calculate the recoil velocity imparted to a ^{198}Hg atom that has emitted a photon of energy 4.88 eV. Compare this with the root-mean-square thermal velocity v of the atom at $T = 300$ K (obtained by setting the average kinetic energy equal to the average thermal energy, $\frac{1}{2}mv^2 = \frac{3}{2}k_B T$).

D. Photon Polarization

As indicated earlier, light is characterized as a sum of modes of different frequencies, directions, and polarizations.

The polarization of a photon is that of its mode.

The choice of a particular set of modes is not unique, however. This important concept is best explained by examining the polarization properties of light from the perspective of photon optics.

Linearly Polarized Photons

Consider light described by a superposition of two plane-wave modes propagating in the z direction, one linearly polarized in the x direction and the other linearly

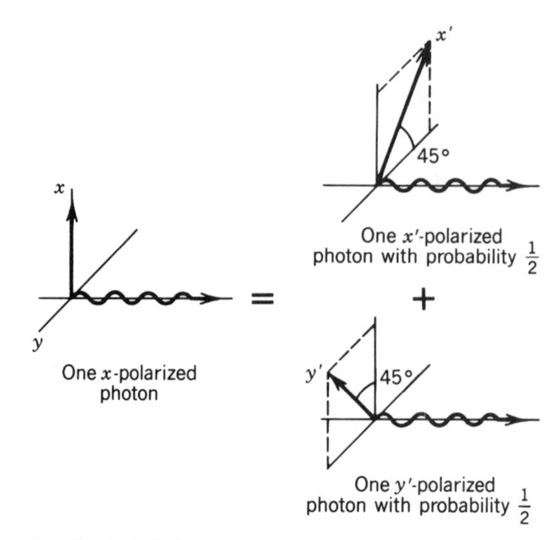

Figure 11.1-4 Probabilistic outcomes for a linearly polarized photon.

polarized in the y direction:

$$\mathbf{E}(\mathbf{r}, t) = (A_x \hat{\mathbf{x}} + A_y \hat{\mathbf{y}}) \exp(-jkz) \exp(j2\pi\nu t).$$

However, the very same electromagnetic field may also be represented in a different coordinate system (x', y') (e.g., one that makes a 45° angle with the initial coordinate system). Thus we can equally well view the field in terms of two modes carrying photons polarized along the x' and y' directions, i.e.,

$$\mathbf{E}(\mathbf{r}, t) = (A_{x'} \hat{\mathbf{x}}' + A_{y'} \hat{\mathbf{y}}') \exp(-jkz) \exp(j2\pi\nu t),$$

where

$$A_{x'} = \frac{1}{\sqrt{2}}(A_x - A_y), \qquad A_{y'} = \frac{1}{\sqrt{2}}(A_x + A_y).$$

If we know that the x-polarized mode is occupied by a photon, and the y-polarized mode is empty, what can be said about the possibility of finding a photon polarized along the x' direction? This question is addressed in photon optics by invoking the usual probabilistic approach. The probabilities of finding a photon with x, y, x', or y' polarization are proportional to the intensities $|A_x|^2$, $|A_y|^2$, $|A_{x'}|^2$, and $|A_{y'}|^2$, respectively. In our example $|A_x|^2 = 1$, $|A_y|^2 = 0$, so that $|A_{x'}|^2 = |A_{y'}|^2 = \frac{1}{2}$. Therefore, given that there is one photon polarized along the x direction and no photon polarized along the y direction, the probabilities of finding a photon polarized along the x' or y' directions are both $\frac{1}{2}$. This is illustrated schematically in Fig. 11.1-4.

EXAMPLE 11.1-1. ***Transmission of a Linearly Polarized Photon Through a Polarizer.*** Consider a plane wave, linearly polarized at an angle θ with respect to the x axis, directed onto a polarizer which has its transmission axis along the x direction (see Fig. 11.1-5). The polarizer transmits light that is linearly polarized in the x direction but blocks light that is linearly polarized in the y direction. It is known from classical polarization optics that the intensity of the transmitted light $I_t = I_i \cos^2 \theta$, where I_i is the intensity of the incident light (see Sec. 6.1B). What happens if only a single photon impinges on the polarizer? If the photon is polarized along the x axis, it always passes through. If it is

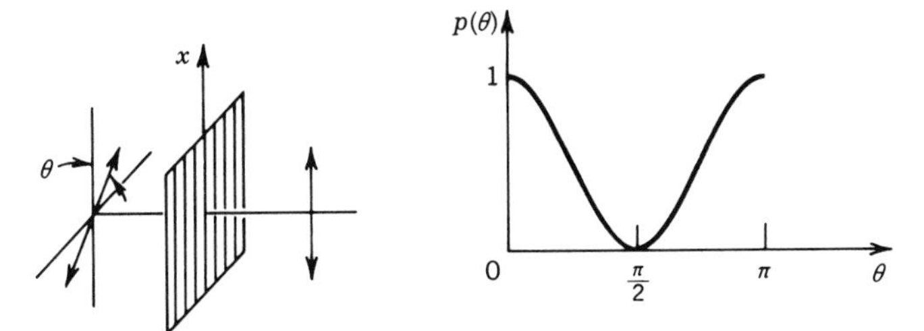

Figure 11.1-5 Probability of observing a linearly polarized photon after transmission through a polarizer at an angle θ.

polarized along the y axis, it is always blocked. The probability for the passage of the photon is determined by the classical intensity I_t. Thus the probability of passage of a photon polarized at an angle θ with the polarizer is $p(\theta) = \cos^2\theta$. The probability that the photon is blocked is therefore $1 - p(\theta) = \sin^2\theta$.

Circularly Polarized Photons

A modal expansion in terms of two circularly polarized plane-wave modes, one right-handed and one left-handed, can also be used, i.e.,

$$\mathbf{E}(\mathbf{r}, t) = [A_R\hat{\mathbf{e}}_R + A_L\hat{\mathbf{e}}_L]\exp(-jkz)\exp(j2\pi\nu t),$$

where $\hat{\mathbf{e}}_R = (1/\sqrt{2})(\hat{\mathbf{x}} + j\hat{\mathbf{y}})$ and $\hat{\mathbf{e}}_L = (1/\sqrt{2})(\hat{\mathbf{x}} - j\hat{\mathbf{y}})$ (see Sec. 6.1B). These modes carry right-handed and left-handed circularly polarized photons, respectively. Again, the probabilities of finding a photon with these polarizations are proportional to the intensities $|A_R|^2$ and $|A_L|^2$. As illustrated in Fig. 11.1-6, a linearly polarized photon is equivalent to the superposition of a right-handed and a left-handed circularly polarized photon, each with probability $\frac{1}{2}$. Conversely, when a circularly polarized photon is passed through a linear polarizer, the probability of detecting it is $\frac{1}{2}$.

Photon Spin

Photons possess intrinsic angular momentum (spin). The magnitude of the **photon spin** is quantized to the two values

$$\mathcal{S} = \pm\hbar. \qquad (11.1\text{-}8)$$

Photon Spin

Right-handed (left-handed) circularly polarized photons have their spin vector parallel (antiparallel) to their momentum vector. Linearly polarized photons have an equal

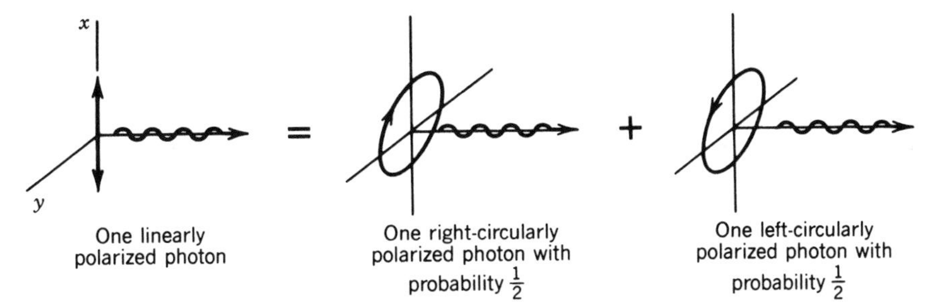

Figure 11.1-6 A linearly polarized photon is equivalent to the superposition of a right- and left-circularly polarized photon, each with probability $\frac{1}{2}$.

probability of exhibiting parallel and antiparallel spin. In the same way that photons can transfer linear momentum to an object, circularly polarized photons can exert a torque on an object. For example, a circularly polarized photon will exert a torque on a half-wave plate of quartz.

E. Photon Interference

Young's two-pinhole interference experiment is generally invoked to demonstrate the wave nature of light (see Exercise 2.5-2 on page 67). However, Young's experiment can be carried out even when there is only a single photon in the apparatus at a given time. The outcome of this experiment can be understood in the context of photon optics by using the photon-position rule. The intensity at the observation plane is calculated using electromagnetic (wave) optics and the result is converted to a probability density function that specifies the random position of the detected photon. The interference arises from phase differences in the two paths.

Consider a plane wave illuminating a screen with two pinholes, as shown in Fig. 11.1-7. This generates two spherical waves that interfere at the observation plane. In the Fresnel approximation these produce a sinusoidal intensity given by (see Exercise 2.5-2)

$$I(x) = 2I_0\left(1 + \cos\frac{2\pi\theta x}{\lambda}\right), \tag{11.1-9}$$

where I_0 is the intensity of each of the waves at the observation plane, λ is the wavelength, and θ is the angle subtended by the two pinholes at the observation plane (Fig. 11.1-7). The line that joins the holes defines the x axis. The result in (11.1-9) describes the intensity pattern that is experimentally observed when the incident light is strong.

Now if only a single photon is present in the apparatus, the probability of detecting it at position x is proportional to $I(x)$, in accordance with (11.1-5). It is most likely to be detected at those values of x for which $I(x)$ is maximum. It will never be detected at values for which $I(x) = 0$. If a histogram of the locations of the detected photon is constructed by repeating the experiment many times, as Taylor did in 1909, the classical interference pattern obtained by carrying out the experiment once with a strong beam of light emerges. The interference pattern represents the probability distribution of the position at which the photon is observed.

The occurrence of interference results from the extended nature of the photon, which permits it to pass through *both* holes of the apparatus. This gives it knowledge of the entire geometry of the experiment when it reaches the observation plane, where it

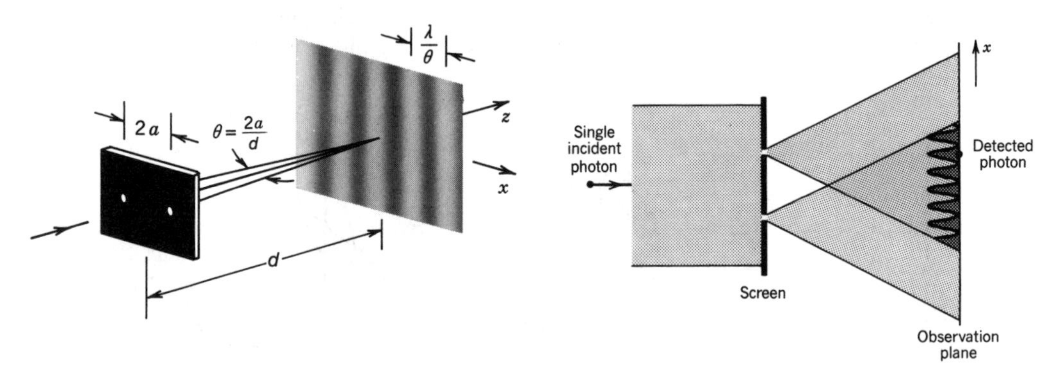

Figure 11.1-7 Young's two-pinhole experiment with a single photon. The interference pattern $I(x)$ is proportional to the probability density of detecting the photon at position x.

is detected as a single entity. If one of the holes were to be covered, the interference pattern would disappear because the photon was forced to pass through the other hole, depriving it of knowledge of the whole apparatus.

EXERCISE 11.1-3

Photon in a Mach–Zehnder Interferometer. Consider a plane wave of light of wavelength λ that is split into two parts at a beamsplitter (see Sec. 11.1B) and recombined in a Mach–Zehnder interferometer, as shown in Fig. 11.1-8 [see also Fig. 2.5-3(*a*)].

If the wave contains only a single photon, plot the probability of finding the photon at the detector as a function of d/λ (for $0 \le d/\lambda \le 1$), where d is the difference between the two optical paths of the light. Assume that the mirrors and beamsplitters are perfectly flat and lossless, and that the beamsplitters have a 50% reflectance. Where might the photon be located when the probability of finding it at the detector is not unity?

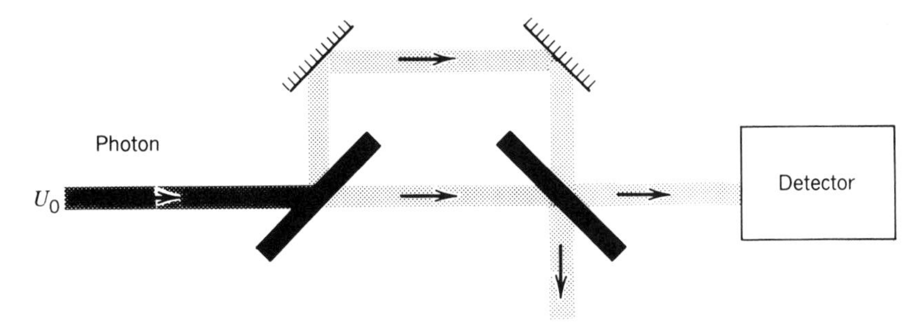

Figure 11.1-8 Mach–Zehnder interferometer.

F. Photon Time

The modal expansion provided in (11.1-1) represents monochromatic (single-frequency) modes which are "eternal" harmonic functions of time. A photon in a monochromatic mode is equally likely to be detected at any time. However, as indicated previously, a modal expansion of the radiation inside (or outside) a resonator is not unique. A more general expansion may be made in terms of polychromatic modes (time-localized wavepackets, for example). The probability of detecting the photon described by the complex wavefunction $U(\mathbf{r}, t)$ (see Sec. 2.6A) at any position, in the incremental time interval between t and $t + dt$, is proportional to $I(\mathbf{r}, t)\,dt \propto |U(\mathbf{r}, t)|^2\,dt$.

The photon-position rule presented in (11.1-5) may therefore be generalized to include photon time localization:

The probability of observing a photon at a point $\mathbf{r}$ within the incremental area dA, and during the incremental time interval dt following time t, is proportional to the intensity of the mode at $\mathbf{r}$ and t, i.e.,

$$p(\mathbf{r}, t)\,dA\,dt \propto I(\mathbf{r}, t)\,dA\,dt \propto |U(\mathbf{r}, t)|^2\,dA\,dt. \qquad (11.1\text{-}10)$$

Photon Position and Time

Time–Energy Uncertainty

The time during which a photon in a monochromatic mode of frequency ν may be detected is totally uncertain, whereas the value of its frequency ν (and its energy $h\nu$) is absolutely certain. On the other hand, a photon in a wavepacket mode with an intensity function $I(t)$ of duration σ_t must be localized within this time. Bounding the photon time in this way engenders an uncertainty in the photon's frequency (and energy) as a result of the properties of the Fourier transform. The result is a "polychromatic" photon. The frequency uncertainty is readily determined by Fourier expanding $U(t)$ in terms of its harmonic components,

$$U(t) = \int_{-\infty}^{\infty} V(\nu) \exp(j2\pi\nu t)\, d\nu \qquad (11.1\text{-}11)$$

where $V(\nu)$ is the Fourier transform of $U(t)$ (see Sec. A.1, Appendix A). The $\mathbf{r}$ dependence has been suppressed for simplicity. The width σ_ν of $|V(\nu)|^2$ represents the spectral width. If σ_t is the rms width of the function $|U(t)|^2$ (i.e., the power-rms width), then σ_t and σ_ν must satisfy the duration–bandwidth reciprocity relation $\sigma_\nu \sigma_t \geq 1/4\pi$, or $\sigma_\omega \sigma_t \geq \frac{1}{2}$ (see Sec. A.2, Appendix A for the definitions of σ_t and σ_ν that lead to this uncertainty relation).

The energy of the photon $h\nu$ then cannot be specified to an accuracy better than $\sigma_E = h\sigma_\nu$. It follows that the energy uncertainty of a photon, and the time during which it may be detected, must satisfy

$$\boxed{\sigma_E \sigma_t \geq \frac{\hbar}{2},} \qquad (11.1\text{-}12)$$

Time–Energy Uncertainty

known as the **time–energy uncertainty relation**. This relation is analogous to that between position and wavenumber (momentum), which sets a limit on the precision with which the position and momentum of a photon can be simultaneously specified. The average energy $\bar{E}$ of this polychromatic photon is $\bar{E} = h\bar{\nu} = \hbar\bar{\omega}$.

To summarize: A monochromatic photon ($\sigma_\nu \to 0$) has an eternal duration within which it can be observed ($\sigma_t \to \infty$). In contrast, a photon associated with an optical wavepacket is localized in time and is therefore polychromatic with a corresponding energy uncertainty. Thus a wavepacket photon can be viewed as a confined traveling packet of energy.

EXERCISE 11.1-4

Single Photon in a Gaussian Wavepacket. Consider a plane-wave wavepacket (see Sec. 2.6A) containing a single photon traveling in the z direction, with complex wavefunction

$$U(\mathbf{r}, t) = a\left(t - \frac{z}{c}\right)$$

where

$$a(t) = \exp\left(-\frac{t^2}{4\tau^2}\right) \exp(j2\pi\nu_0 t).$$

(a) Show that the uncertainties in its time and z position are $\sigma_t = \tau$ and $\sigma_z = c\sigma_t$, respectively.

(b) Show that the uncertainties in its energy and momentum satisfy the minimum uncertainty relations

$$\sigma_E \sigma_t = \frac{\hbar}{2} \tag{11.1-13}$$

$$\sigma_z \sigma_p = \frac{\hbar}{2}. \tag{11.1-14}$$

Equation (11.1-14) is the minimum-uncertainty limit of the **Heisenberg position-momentum uncertainty relation** [see (A.2-7) in Appendix A].

Summary

Electromagnetic radiation may be described as a sum of modes, e.g., monochromatic uniform plane waves of the form

$$\mathbf{E}(\mathbf{r}, t) = \sum_{\mathbf{q}} A_{\mathbf{q}} \exp(-j\mathbf{k}_{\mathbf{q}} \cdot \mathbf{r}) \exp(j2\pi\nu_{\mathbf{q}} t)\hat{\mathbf{e}}_{\mathbf{q}}.$$

Each plane wave has two orthogonal polarization states (e.g., vertical/horizontal-linearly polarized, right/left-circularly polarized, etc.) represented by the vectors $\hat{\mathbf{e}}_{\mathbf{q}}$. When the energy of a mode is measured, the result is an integer (in general, random) number of energy quanta (photons). Each of the photons associated with the mode **q** has the following properties:

- Energy $E = h\nu_{\mathbf{q}}$.
- Momentum $\mathbf{p} = \hbar\mathbf{k}$.
- Spin $\mathcal{S} = \pm\hbar$, if it is circularly polarized.
- The photon is equally likely to be found anywhere in space, and at any time, since the wavefunction of the mode is a monochromatic plane wave.

The choice of modes is not unique. A modal expansion in terms of nonmonochromatic (quasimonochromatic), nonplanar waves,

$$\mathbf{E}(\mathbf{r}, t) = \sum_{\mathbf{q}} A_{\mathbf{q}} U_{\mathbf{q}}(\mathbf{r}, t)\hat{\mathbf{e}}_{\mathbf{q}},$$

is also possible. The photons associated with the mode **q** then have the following properties:

- Photon position and time are governed by the complex wavefunction $U_{\mathbf{q}}(\mathbf{r}, t)$. The probability of detecting a photon in the incremental time between t and $t + dt$, in an incremental area dA at position $\mathbf{r}$, is proportional to $|U_{\mathbf{q}}(\mathbf{r}, t)|^2 \, dA \, dt$.
- If $U_{\mathbf{q}}(\mathbf{r}, t)$ has a finite time duration σ_t, i.e., if the photon is localized in time, then the photon energy $h\nu_{\mathbf{q}}$ has an uncertainty $h\sigma_\nu \geq h/4\pi\sigma_t$.
- If $U_{\mathbf{q}}(\mathbf{r}, t)$ has a finite spatial extent in the transverse ($z = 0$) plane, i.e., if the photon is localized in the x direction, for example, then the direction of photon momentum is uncertain. The spread in photon momentum can be determined by analyzing $U_{\mathbf{q}}(\mathbf{r}, t)$ as a sum of plane waves, the wave with wavevector $\mathbf{k}$ corresponding to photon momentum $\hbar\mathbf{k}$. Localization of the photon in the transverse plane results in a spread of the uncertainty of the photon-momentum direction.

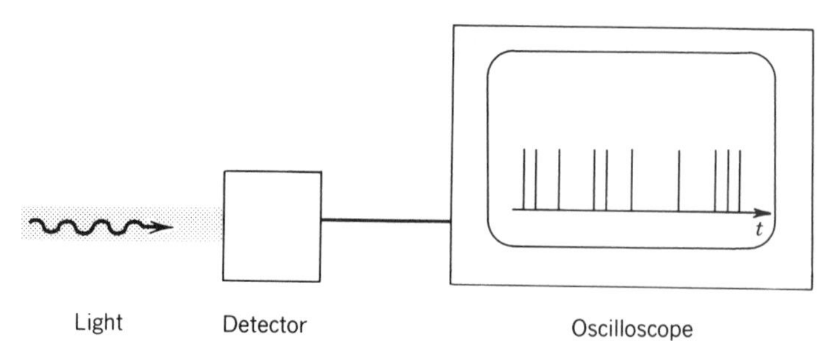

Figure 11.2-1 Photon registrations at random localized instants of time.

11.2 PHOTON STREAMS

In Sec. 11.1 we concentrated on the properties and behavior of single photons. We now consider the properties of collections of photons. As a result of the processes by which photons are created (e.g., emissions from atoms; see Chap. 12), the number of photons occupying a mode is generally random. The probability distribution obeyed by the photon number is governed by the quantum state of the mode, which is determined by the nature of the light source (see Sec. 11.3). Real photon streams often contain numerous propagating modes, each carrying a random number of photons.

If an experiment is carried out in which a weak stream of photons falls on a light-sensitive surface, the photons are registered (detected) at random localized instants of time and at random points in space, in accordance with (11.1-10). This space–time process can be discerned by viewing an object with the naked eye in a dimly lit room.

The time course of such photon registrations can be highlighted by looking at the temporal and spatial behavior separately. Consider the use of a detector that integrates light over a finite area, as illustrated in Fig. 11.2-1. The probability of detecting a photon in the incremental time interval between t and $t + dt$ is proportional to the optical power $P(t)$ at the time t. The photons will be registered at random times.

On the other hand, the spatial pattern of photon registrations is readily manifested by using a detector that integrates over a fixed exposure time T (e.g., photographic film). In accordance with (11.1-10), the probability of observing a photon in an incremental area dA surrounding the point $\mathbf{r}$ is proportional to the integrated local intensity $\int_0^T I(\mathbf{r}, t)\, dt$. This is illustrated by the "grainy" photographic image of Max Planck provided in Fig. 11.2-2. This image was obtained by rephotographing, under

Figure 11.2-2 Random photon registrations with a spatial density that follows the local optical intensity. This image of Max Planck taken with a weak stream of photons should be compared with the photograph on page 384 taken with intense light.

very low light conditions, the picture of Max Planck shown on page 384. Each of the white dots represents a random photon registration; the density of registrations follows the local intensity.

A. Mean Photon Flux

We begin by introducing a number of definitions that relate the mean photon flux to classical electromagnetic intensity, power, and energy. These definitions are related to the probability law (11.1-10) governing the position and time at which a single photon is observed. We then discuss the randomness of the photon flux and the photon-number statistics for different sources of light. Finally, we consider the random partitioning of a photon stream.

Mean Photon-Flux Density

Monochromatic light of frequency ν and classical intensity $I(\mathbf{r})$ (watts/cm^2) carries a mean **photon-flux density**

$$\phi(\mathbf{r}) = \frac{I(\mathbf{r})}{h\nu}, \tag{11.2-1}$$

Mean Photon-Flux Density

where $h\nu$ is the energy of each photon. This equation converts a classical measure (with units of energy/s-cm^2) into a quantum measure (with units of photons/s-cm^2). For quasimonochromatic light of central frequency $\bar{\nu}$, all photons have approximately the same energy $h\bar{\nu}$, so that the mean photon-flux density is approximately

$$\phi(\mathbf{r}) = \frac{I(\mathbf{r})}{h\bar{\nu}}. \tag{11.2-2}$$

Typical values of $\phi(\mathbf{r})$ for some common sources of light are provided in Table 11.2-1. It is clear from these numbers that trillions of photons rain down on each square centimeter of us each second.

TABLE 11.2-1 Mean Photon-Flux Density for Several Light Sources

Source	Mean Photon-Flux Density (photons/s-cm^2)
Starlight	10^6
Moonlight	10^8
Twilight	10^{10}
Indoor light	10^{12}
Sunlight	10^{14}
Laser light (10-mW He–Ne laser beam at $\lambda_o = 633$ nm focused to a 20-μm-diameter spot)	10^{22}

Mean Photon Flux

The mean photon flux Φ (with units of photons/s) is obtained by integrating the mean photon-flux density over a specified area,

$$\Phi = \int_A \phi(\mathbf{r})\, dA = \frac{P}{h\bar{\nu}}, \qquad (11.2\text{-}3)$$

Mean Photon Flux

where again $h\bar{\nu}$ is the average energy of a photon, and

$$P = \int_A I(\mathbf{r})\, dA \qquad (11.2\text{-}4)$$

is the optical power (watts). As an example, 1 nW of optical power, at a wavelength $\lambda_o = 0.2\ \mu\text{m}$, delivers to an object an average photon flux $\Phi \approx 10^9$ photons per second. Roughly speaking, one photon will therefore strike the object every nanosecond, i.e.,

$$1\ \text{nW at}\ \lambda_o = 0.2\ \mu\text{m} \rightarrow 1\ \text{photon/ns}. \qquad (11.2\text{-}5)$$

A $\lambda_o = 1\ \mu\text{m}$ photon carries one-fifth of the energy, so that 1 nW corresponds to an average of 5 photons/ns.

Mean Number of Photons

The mean number of photons $\bar{n}$ detected in the area A and the time interval T is obtained by multiplying the photon flux Φ by the time duration,

$$\bar{n} = \Phi T = \frac{E}{h\bar{\nu}}, \qquad (11.2\text{-}6)$$

Mean Photon Number

where $E = PT$ is the optical energy (joules).

To summarize, the relations between the classical and quantum measures are:

Classical		Quantum	
Optical intensity	$I(\mathbf{r})$	Photon-flux density	$\phi(\mathbf{r}) = \dfrac{I(\mathbf{r})}{h\bar{\nu}}$
Optical power	P	Photon flux	$\Phi = \dfrac{P}{h\bar{\nu}}$
Optical energy	E	Photon number	$\bar{n} = \dfrac{E}{h\bar{\nu}}$

Spectral Densities of Photon Flux

For polychromatic light of broad bandwidth, it is useful to define spectral densities of the classical intensity, power, and energy, and their quantum counterparts: spectral photon-flux density, spectral photon flux, and spectral photon number:

Classical	Quantum
I_ν (W/cm^2-Hz)	$\phi_\nu = \dfrac{I_\nu}{h\nu}$ (photons/s-cm^2-Hz)
P_ν (W/Hz)	$\Phi_\nu = \dfrac{P_\nu}{h\nu}$ (photons/s-Hz)
E_ν (J/Hz)	$\bar{n}_\nu = \dfrac{E_\nu}{h\nu}$ (photons/Hz)

For example, $P_\nu \, d\nu$ represents the optical power in the frequency range ν to $\nu + d\nu$; and $\Phi_\nu \, d\nu$ represents the flux of photons whose frequency lies between ν and $\nu + d\nu$.

Time-Varying Light

If the light intensity is time varying, the photon-flux density is a function of time,

$$\phi(\mathbf{r}, t) = \frac{I(\mathbf{r}, t)}{h\bar{\nu}}. \tag{11.2-7}$$

The optical power and the photon flux are also, then, functions of time:

$$\Phi(t) = \int_A \phi(\mathbf{r}, t) \, dA = \frac{P(t)}{h\bar{\nu}}, \tag{11.2-8}$$

Mean Photon Flux

where

$$P(t) = \int_A I(\mathbf{r}, t) \, dA. \tag{11.2-9}$$

The mean number of photons registered in a time interval between $t = 0$ and $t = T$ also varies with time. It is obtained by integrating the photon flux,

$$\bar{n} = \int_0^T \Phi(t) \, dt = \frac{E}{h\bar{\nu}}, \tag{11.2-10}$$

Mean Photon Number

where

$$E = \int_0^T P(t) \, dt = \int_0^T \int_A I(\mathbf{r}, t) \, dA \, dt. \tag{11.2-11}$$

is the optical energy (the intensity integrated over time and area).

B. Randomness of Photon Flux

Even if the classical intensity $I(\mathbf{r}, t)$ is constant, the time of arrival and position of registration of a single photon are governed by probabilistic laws, as we have seen in Sec. 11.1 (see Fig. 11.2-1). If a source provides exactly one photon, the probability density of detecting that photon at the space–time point $(\mathbf{r}, t)$ is proportional to $I(\mathbf{r}, t)$, in accordance with (11.1-10). We shall see in this section that the classical electromag-

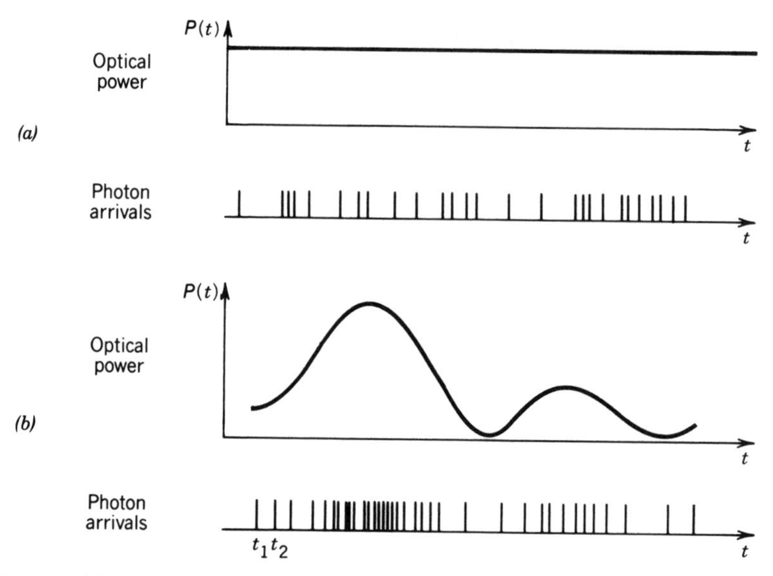

Figure 11.2-3 (*a*) Constant optical power and the corresponding random photon arrival times. (*b*) Time-varying optical power and the corresponding random photon arrival times.

netic intensity $I(\mathbf{r}, t)$ governs the behavior of photon streams as well as single photons. The interpretation ascribed to $I(\mathbf{r}, t)$ differs, however. *For photon streams, the classical intensity $I(\mathbf{r}, t)$ determines the mean photon-flux density $\phi(\mathbf{r}, t)$. The properties of the light source determine the fluctuations in $\phi(\mathbf{r}, t)$.*

If the optical power $P(t)$ varies with time, the density of random times at which the associated photons are detected generally follows the function $P(t)$, as schematically illustrated in Fig. 11.2-3. The mean flux $\Phi(t)$ is $P(t)/h\bar{\nu}$, but the actual times at which the photons are detected are random. Where the power is large, there are, on the average, more photons; where the power is small, the photons are sparse. Even when P is constant, the times at which the photons are detected is random, with behavior determined by the source (Figs. 11.2-3(*a*) and 11.2-4). For example, at $\lambda_o = 1.24\ \mu\text{m}$, 1 nW carries an *average* of 6.25 photons/ns, or 0.00625 photons every picosecond. Of course, only integral numbers of photons may be detected. An average of 0.00625 photons/ps means that if 10^5 time intervals (each of duration $T = 1$ ps) were examined, most of the time intervals would be empty (no photons), about 625 intervals would contain one photon, and very few intervals would contain two or more photons.

The image of Max Planck in Fig. 11.2-2 shows the same behavior in the spatial domain. The locations of the detected photons generally follow the classical intensity distribution, with a high density of photons where the intensity is large and low photon density where the intensity is small. But there is considerable graininess (noise) in the image. Fluctuations in the photon-flux density are most discernible when its mean value is small, as in the case of Fig. 11.2-2. When the mean photon-flux density becomes large everywhere in the image, the graininess disappears and the classical intensity distribution is recovered, as seen in the picture of Max Planck on page 384.

The study of the randomness of photon numbers is important for applications such as noise in weak images and optical information transmission. In a fiber-optic communication system, for example, information is carried on a photon stream (see Sec. 22.3). Only the mean number of photons emitted by the source is controlled at the transmitter. The actual number of emitted photons is unpredictable, the nature of the source determining the form of its randomness. The unpredictability of the photon number results in errors in the transmission of information.

C. Photon-Number Statistics

The statistical distribution of the number of photons depends on the nature of the light source and must generally be treated by use of the quantum theory of light, as described briefly in Sec. 11.3. However, under certain conditions, the arrival of photons may be regarded as the independent occurrences of a sequence of random events at a rate equal to the photon flux, which is proportional to the optical power. The optical power may be deterministic (as in coherent light) or random (as in partially coherent light). For partially coherent light, the power fluctuations are correlated, so that the arrival of photons is no longer a sequence of independent events; the photon statistics are then significantly different.

Coherent Light

Consider light of constant optical power P. The corresponding mean photon flux $\Phi = P/h\bar{\nu}$ (photons/s) is also constant, but the actual times of registration of the photons are random as shown in Fig. 11.2-4. Given a time interval of duration T, let the number of detected photons be n. We already know that the mean value of n is $\bar{n} = \Phi T = PT/h\bar{\nu}$. We wish to obtain an expression for the probability distribution $p(n)$, i.e., the probability $p(0)$ of detecting no photons, the probability $p(1)$ of detecting one photon, and so on.

An expression for the probability distribution $p(n)$ can be derived under the assumption that the registrations of photons are statistically independent. The result is the Poisson distribution

$$p(n) = \frac{\bar{n}^n \exp(-\bar{n})}{n!}, \qquad n = 0, 1, 2, \ldots . \tag{11.2-12}$$

Poisson Distribution

This distribution, known as the **Poisson distribution**, is displayed on a semilogarithmic plot in Fig. 11.2-5 for several values of the mean $\bar{n}$. The curves become progressively broader as $\bar{n}$ increases.

Derivation of the Poisson Distribution

Divide the time interval T into a large number N of subintervals of sufficiently small width T/N each, such that each interval carries one photon with probability $p = \bar{n}/N$ and no photons with probability $1 - p$. The probability of finding n independent

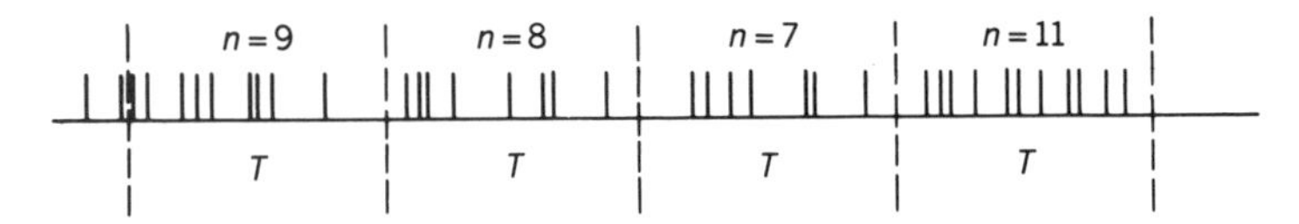

Figure 11.2-4 Random arrival of photons in a light beam of power P within intervals of duration T. Although the optical power is constant, the number n of photons arriving within each interval is random.

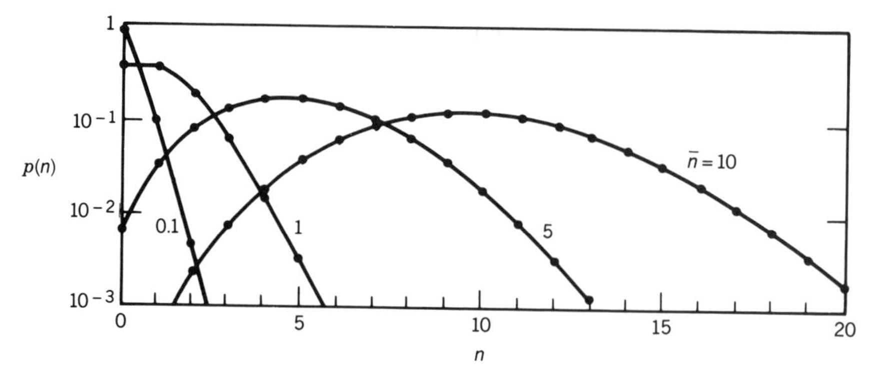

Figure 11.2-5 Poisson distribution $p(n)$ of the photon number n.

photons in the N intervals, like the flips of a biased coin, then follows the binomial distribution

$$p(n) = \frac{N!}{n!\,(N-n)!} p^n (1-p)^{N-n},$$

$$= \frac{N!}{n!\,(N-n)!} \left(\frac{\bar{n}}{N}\right)^n \left(1 - \frac{\bar{n}}{N}\right)^{N-n}.$$

T/N
0
N t
T

In the limit as $N \to \infty$, $N!/(N-n)!\,N^n \to 1$, and $[1 - (\bar{n}/N)]^{N-n} \to \exp(-\bar{n})$, so that (11.2-12) is obtained.

Mean and Variance

Two important parameters characterize any random number n—its mean value,

$$\bar{n} = \sum_{n=0}^{\infty} n p(n), \tag{11.2-13}$$

and its variance

$$\sigma_n^2 = \sum_{n=0}^{\infty} (n - \bar{n})^2 p(n), \tag{11.2-14}$$

which is the average of the squared deviation from the mean. The standard deviation σ_n (the square root of the variance) is a measure of the width of the distribution. The quantities $p(n)$, $\bar{n}$, and σ_n are collectively called the photon-number statistics. Although the function $p(n)$ contains more information than just its mean and variance, these are useful measures.

It is not difficult to show [by use of (11.2-12) in (11.2-13) and (11.2-14)] that the mean of the Poisson distribution is indeed $\bar{n}$ and its variance is equal to its mean,

$$\boxed{\sigma_n^2 = \bar{n}.} \tag{11.2-15}$$

Variance of the Poisson Distribution

For example, when $\bar{n} = 100$, $\sigma_n = 10$; i.e., the generation of 100 photons is accompanied by an inaccuracy of about ± 10 photons.

The Poisson photon-number distribution applies for many light sources, including an ideal laser emitting a beam of monochromatic coherent light in a single mode (see Chap. 14). This distribution corresponds to a quantum state of light known as the coherent state (see Sec. 11.3A).

Signal-to-Noise Ratio

The randomness of the number of photons constitutes a fundamental source of noise that we have to contend with when using light to transmit a signal. Representing the mean of the signal as $\bar{n}$ and its noise by the root mean square value σ_n, a useful measure of the performance of light as an information-carrying medium is the signal-to-noise ratio (SNR). The SNR of the random number n is defined as

$$\text{SNR} \equiv \frac{(\text{mean})^2}{\text{variance}} = \frac{\bar{n}^2}{\sigma_n^2}. \tag{11.2-16}$$

For the Poisson distribution

$$\boxed{\text{SNR} = \bar{n},} \tag{11.2-17}$$

Poisson Photon-Number Signal-to-Noise Ratio

i.e., the signal-to-noise ratio increases without limit as the mean photon number increases.

Although the SNR is a useful measure of the randomness of a signal, in some applications it is necessary to know the probability distribution itself. For example, if one communicates by sending a mean number of photons $\bar{n} = 20$, according to (11.2-12) the probability that no photons are received is $p(0) \approx 2 \times 10^{-9}$. This represents a probability of error in the transmission of information. This topic is addressed in Chap. 22.

Thermal Light

When the photon arrival times are correlated, the photon number statistics obey distributions other than the Poisson. This is the case for thermal light. Consider an optical resonator whose walls are maintained at temperature T kelvins (K), so that photons are emitted into the modes of the resonator. In accordance with the laws of statistical mechanics, under conditions of thermal equilibrium the probability distribution for the electromagnetic energy E_n in one of its modes satisfies the **Boltzmann probability distribution**

$$\boxed{P(E_n) \propto \exp\left(-\frac{E_n}{k_B T}\right).} \tag{11.2-18}$$

Boltzmann Distribution

Here k_B is **Boltzmann's constant** ($k_B = 1.38 \times 10^{-23}$ J/K). The energy associated with each mode is random. Higher energies are relatively less probable than lower energies, in accordance with a simple exponential law governed by the quantity $k_B T$. The smaller the value of $k_B T$, the less likely are higher energies. At room temperature ($T = 300$ K), $k_B T = 0.026$ eV, which is equivalent to 208 cm^{-1}. The Boltzmann

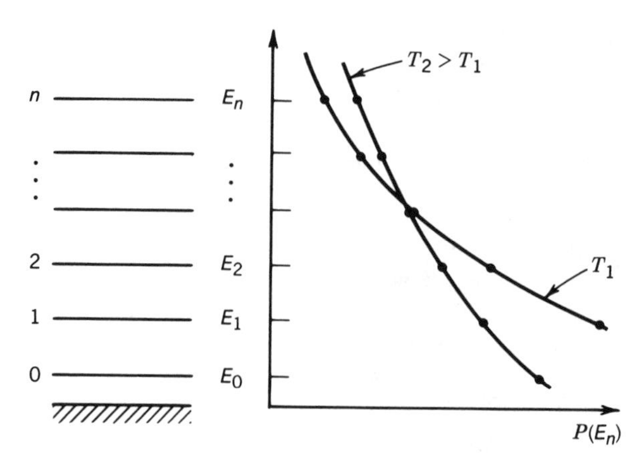

Figure 11.2-6 Boltzmann probability distribution $P(E_n)$ versus energy E_n.

distribution for this single mode is sketched in Fig. 11.2-6 with temperature as a parameter.

It follows from (11.2-18) and the photon-energy quantization relation given by $E_n = (n + \frac{1}{2})h\nu$ that the probability of finding n photons in a single mode of a resonator in thermal equilibrium is given by

$$p(n) \propto \exp\left(-\frac{nh\nu}{k_B T}\right)$$

$$= \left[\exp\left(-\frac{h\nu}{k_B T}\right)\right]^n, \qquad n = 0, 1, 2, \ldots . \tag{11.2-19}$$

Using the condition that the probability distribution must have a sum equal to unity, i.e., $\sum_{n=0}^{\infty} p(n) = 1$, the normalization constant is determined to be $[1 - \exp(-h\nu/k_B T)]$. The zero-point energy $E_0 = \frac{1}{2}h\nu$ disappears into the normalization and does not affect the results, in accordance with the discussion in Sec. 11.1A.

The result is most simply written in terms of its mean $\bar{n}$ as

$$\boxed{p(n) = \frac{1}{\bar{n}+1}\left(\frac{\bar{n}}{\bar{n}+1}\right)^n,} \tag{11.2-20}$$

Bose–Einstein Distribution

where

$$\bar{n} = \frac{1}{\exp(h\nu/k_B T) - 1}, \tag{11.2-21}$$

as determined from (11.2-13). In the parlance of probability theory, this distribution is called the **geometric distribution** since $p(n)$ is a geometrically decreasing function of n. In physics it is referred to as the **Bose–Einstein probability distribution.**

The Bose–Einstein distribution is displayed on a semilogarithmic plot in Fig. 11.2-7, for several values of $\bar{n}$ (or equivalently, for several values of the temperature T). Its exponential character is evident in the straight-line behavior in the plot. Comparing

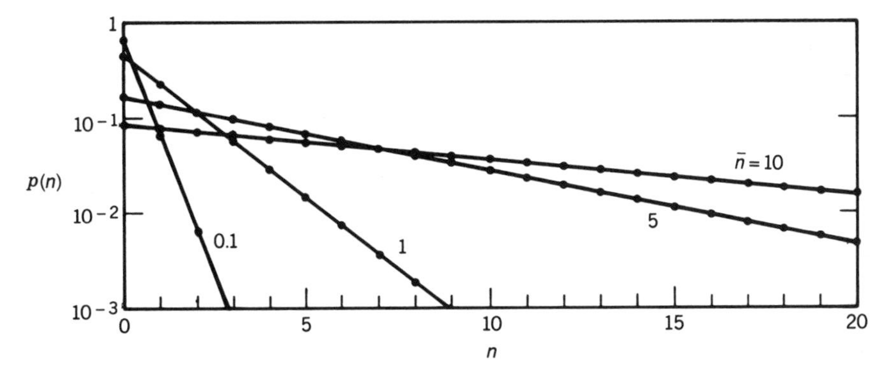

Figure 11.2-7 Bose–Einstein distribution $p(n)$ of the photon number n.

Figs. 11.2-7 with 11.2-5 demonstrates that the photon-number distribution for thermal light is far broader than that for coherent light.

Using (11.2-14), the photon-number variance turns out to be

$$\sigma_n^2 = \bar{n} + \bar{n}^2. \tag{11.2-22}$$

Bose–Einstein Variance

Comparing this expression to the variance for the Poisson distribution, which is simply $\bar{n}$, we see that thermal light has a larger variance corresponding to more uncertainty and a greater range of fluctuations of the photon number. The signal-to-noise ratio of the Bose–Einstein distribution is

$$\mathrm{SNR} = \frac{\bar{n}}{\bar{n}+1};$$

it is always smaller than unity no matter how large the optical power. The amplitude and phase of thermal light behave like random quantities, as described in Chapter 10. This randomness results in a broadening of the photon-number distribution. Indeed, this form of light is too noisy to be used in high-data-rate information transmission.

EXERCISE 11.2-1

Average Energy in a Resonator Mode. Show that the average energy of a resonator mode of frequency ν, under conditions of thermal equilibrium at temperature T, is given by

$$\bar{E} = k_{\mathrm{B}}T \frac{h\nu/k_{\mathrm{B}}T}{\exp(h\nu/k_{\mathrm{B}}T) - 1}. \tag{11.2-23}$$

Sketch the dependence of $\bar{E}$ on ν for several values of $k_{\mathrm{B}}T/h$. Use a Taylor series expansion of the denominator to obtain a simplified approximate expression for $\bar{E}$ in the limit $h\nu/k_{\mathrm{B}}T \ll 1$. Explain the result on a physical basis.

---

*Other Sources of Light

As mentioned earlier, for a certain class of light sources the photon arrivals can be regarded as a sequence of independent events, arriving at a rate proportional to the optical power. For coherent light, the power is deterministic, and the photon number obeys the Poisson distribution $p(n) = \mathscr{W}^n e^{-\mathscr{W}}/n!$, where

$$\mathscr{W} = \frac{1}{h\nu}\int_0^T P(t)\,dt = \frac{1}{h\nu}\int_0^T \int_A I(\mathbf{r}, t)\,dA\,dt. \qquad (11.2\text{-}24)$$

The integrated optical power normalized to units of photon number, $\mathscr{W}$, is a constant representing the mean photon number $\bar{n}$.

When the intensity $I(\mathbf{r}, t)$ itself fluctuates randomly in time and/or space, the optical power $P(t)$ also undergoes random fluctuations [see Fig. 11.2-3(*b*)], and its integral $\mathscr{W}$ is therefore also random. As a result, not only is the photon number random but so is its mean $\mathscr{W}$. Because of this added source of randomness, the photon-number statistics for partially coherent light will differ from the Poisson distribution. If the fluctuations in the mean photon number $\mathscr{W}$ are described by a probability density function $p(\mathscr{W})$, the unconditional probability distribution for partially coherent light may be obtained by averaging the conditional Poisson distribution $p(n|\mathscr{W}) = \mathscr{W}^n e^{-\mathscr{W}}/n!$ over all permitted values of $\mathscr{W}$, each weighted by its probability density $p(\mathscr{W})$. The resultant photon-number distribution is then

$$p(n) = \int_0^\infty \frac{\mathscr{W}^n e^{-\mathscr{W}}}{n!} p(\mathscr{W})\,d\mathscr{W}, \qquad (11.2\text{-}25)$$

Mandel's Formula

which is known as Mandel's formula. Equation (11.2-25) is also referred to as the doubly stochastic Poisson counting distribution because of the two sources of randomness that contribute to it: the photons themselves (which behave in Poisson fashion) and the intensity fluctuations arising from the noncoherent nature of the light (which must be specified).

Note that this theory of photon statistics is applicable only to a certain class of light (called classical light); a more general theory based on a quantum description of the state of light is described briefly in Sec. 11.3.

The photon-number mean and variance for partially coherent light, which can be derived by using (11.2-13) and (11.2-14) in conjunction with (11.2-25), are

$$\bar{n} = \overline{\mathscr{W}} \qquad (11.2\text{-}26)$$

and

$$\sigma_n^2 = \bar{n} + \sigma_{\mathscr{W}}^2, \qquad (11.2\text{-}27)$$

respectively. Here $\sigma_{\mathscr{W}}^2$ signifies the variance of $\mathscr{W}$. Note that the variance of the photon number is the sum of two contributions—the first term is the basic contribution of the Poisson distribution, and the second is an additional contribution due to the classical fluctuations of the optical power.

In one important example of statistical fluctuations, the normalized integrated optical power $\mathscr{W}$ obeys the exponential probability density function

$$p(\mathscr{W}) = \begin{cases} \dfrac{1}{\overline{\mathscr{W}}} \exp\left(-\dfrac{\mathscr{W}}{\overline{\mathscr{W}}}\right), & \mathscr{W} \geq 0 \\ 0, & \mathscr{W} < 0. \end{cases} \qquad (11.2\text{-}28)$$

This distribution is applicable to quasi-monochromatic spatially coherent light, when the real and imaginary components of the complex amplitude are independent and have normal (Gaussian) probability distributions. The spectral width must be sufficiently small so that the coherence time τ_c is much greater than the counting time T, and the coherence area A_c must be much larger than the area of the detector A. The photon-number distribution $p(n)$ corresponding to (11.2-28) can be obtained by substitution into (11.2-25) and evaluation of the integral. The result turns out to be the Bose–Einstein distribution given in (11.2-20). The Gaussian-distributed optical field therefore has photon statistics identical to those of single-mode thermal light. When the area A and the time T are not small, the statistics are modified; they describe multimode thermal light (see Probs. 11.2-5 to 11.2-7).

D. Random Partitioning of Photon Streams

A photon stream is said to be partitioned when it is subjected to the removal of some of its photons. The photons removed may be either diverted or destroyed. The process is called random partitioning when they are diverted and random deletion when they are destroyed. There are numerous ways in which this can occur. Perhaps the simplest example of random partitioning is provided by an ideal lossless beamsplitter. Photons are randomly selected to join either of the two emerging streams (see Fig. 11.2-8). An example of random deletion is provided by the action of an optical absorption filter on a light beam. Photons are randomly selected either to pass through the filter or to be destroyed (and converted into heat).

We restrict our treatment to situations in which the possibility of each photon being removed behaves in accordance with an independent random (Bernoulli) trial. In terms of the beamsplitter, this is satisfied if a photon stream impinges on only one of the input ports (Fig. 11.2-8). This eliminates the possibility of interference, which, in general, invalidates the independent-trial assumption. Although the results derived below are couched in terms of random partitioning, they apply equally well to random deletion.

Consider a lossless beamsplitter with transmittance $\mathscr{T}$ and reflectance $\mathscr{R} = 1 - \mathscr{T}$. In electromagnetic optics, the intensity of the transmitted wave I_t is related to the intensity of the incident wave I by $I_t = \mathscr{T} I$. The result of a single photon impinging on a beamsplitter was examined in Sec. 11.1B; it was shown that the probability of transmission is equal to the transmittance $\mathscr{T}$. We now proceed to calculate the

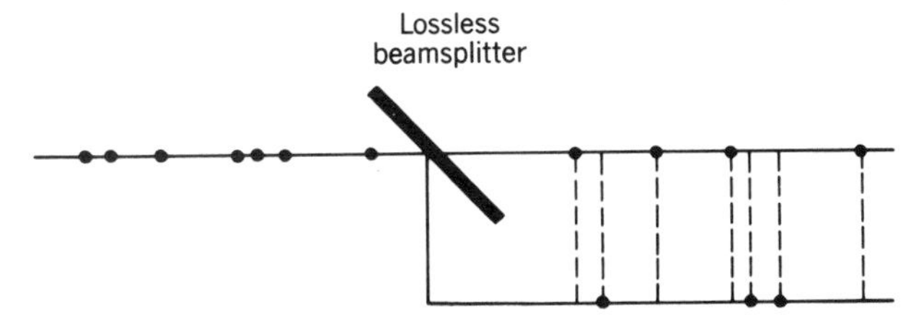

Figure 11.2-8 Random partitioning of photons by a beamsplitter.

outcome when a photon stream of mean flux Φ is incident, so that a mean number of photons $\bar{n} = \Phi T$ strikes the beamsplitter in the time interval T.

In accordance with (11.2-6), the mean number of photons in a beam is proportional to the optical energy. The mean number of transmitted and reflected photons in this time must therefore be $\mathscr{T}\bar{n}$ and $(1 - \mathscr{T})\bar{n}$, respectively. We now consider a more general question: what happens to the photon-number statistics $p(n)$ of the photon stream on partitioning by a beamsplitter?

A single photon falling on the beamsplitter is transmitted with probability $\mathscr{T}$ and reflected with probability $1 - \mathscr{T}$ (see Fig. 11.1-3). If the incident beam contains precisely n photons, the probability $p(m)$ that m photons are transmitted is the same as that of flipping a coin n times, where the probability of achieving a head (being transmitted) is $\mathscr{T}$. From elementary probability theory we know that the outcome is the binomial distribution

$$p(m) = \binom{n}{m}\mathscr{T}^m(1 - \mathscr{T})^{n-m}, \qquad m = 0, 1, \ldots, n, \tag{11.2-29}$$

where $\binom{n}{m} = n!/m!(n - m)!$. The mean number of transmitted photons is easily shown to be

$$\bar{m} = \mathscr{T}n. \tag{11.2-30}$$

The variance for the binomial distribution is given by

$$\sigma_m^2 = \mathscr{T}(1 - \mathscr{T})n = (1 - \mathscr{T})\bar{m}. \tag{11.2-31}$$

Because of the symmetry of the problem, the results for the reflected beam are obtained immediately. As the average number of transmitted photons $\bar{m}$ increases, the signal-to-noise ratio, represented by $\bar{m}^2/\sigma_m^2 = \bar{m}/(1 - \mathscr{T})$ increases. Therefore, for large intensities, the photons will be partitioned between the two streams in good accord with $\mathscr{T}$ and $(1 - \mathscr{T})$, indicating that the laws of classical optics are recovered.

The expressions provided above are useful because they permit us to calculate the effect of a beamsplitter on photons obeying various photon-number statistics. The solution is obtained by recognizing that in these cases the number of photons n at the input to the beamsplitter is random rather than fixed. Let the probability that there are exactly n photons present be $p_0(n)$. If we treat the photons as independent events, the photon-number probability distribution in the transmitted stream will be a weighted sum of binomial distributions, with n taking on the random value n. The weighting is in accordance with the probability that n photons were present. The probability of finding m photons transmitted through the beamsplitter, when the input photon-number distribution is $p_0(n)$, is therefore given by $p(m) = \sum_n p(m|n)p_0(n)$, where $p(m|n) = \binom{n}{m}\mathscr{T}^m(1 - \mathscr{T})^{n-m}$ is the binomial distribution. Explicitly, then,

$$p(m) = \sum_{n=m}^{\infty} \binom{n}{m}\mathscr{T}^m(1 - \mathscr{T})^{n-m} p_0(n), \qquad m = 0, 1, 2, \ldots. \tag{11.2-32}$$

Photon-Number Statistics under Random Partitioning

When $p_0(n)$ is the Poisson distribution (coherent light) or the Bose–Einstein distribution (thermal light), the results turn out to be quite simple: $p(m)$ has exactly the same form of photon-number distribution as $p_0(n)$. These distributions retain their form under random partitioning. Thus single-mode laser light transmitted through a beamsplitter remains Poisson and thermal light remains Bose–Einstein, but of course with a reduced photon-number mean. Light with a deterministic number of photons (see Sec. 11.3B), on the other hand, does not retain its form under random partitioning, and this unfortunate property accounts for its lack of robustness.

The signal-to-noise ratio of m is easily calculated for photon streams that have undergone partitioning or deletion. For coherent light and single-mode thermal light, the results are

$$\text{SNR} = \begin{cases} \mathcal{T}\bar{n} & \text{coherent light} \quad (11.2\text{-}33) \\ \dfrac{\mathcal{T}\bar{n}}{\mathcal{T}\bar{n}+1} & \text{thermal light.} \quad (11.2\text{-}34) \end{cases}$$

Since $\mathcal{T} \leq 1$ it is clear that random partitioning decreases the signal-to-noise ratio. Another way of stating this is that random partitioning introduces noise. The effect is most severe for deterministic photon-number light.

The same results are also applicable to the *detection* of photons. If every photon has an independent chance of being detected, then out of n incident photons, m photons would be detected where $p(m)$ is related to $p_0(n)$ by (11.2-32). This result will be useful in the theory of photon detection (Chap. 17).

*11.3 QUANTUM STATES OF LIGHT

The position, momentum, and number of photons in an electromagnetic mode are generally random quantities. In this section it will be shown that the electric field itself is also generally random. Consider a plane-wave monochromatic electromagnetic mode in a volume V, described by the electric field $\text{Re}\{\mathbf{E}(\mathbf{r}, t)\}$, where

$$\mathbf{E}(\mathbf{r}, t) = A \exp(-j\mathbf{k}\cdot\mathbf{r}) \exp(j2\pi\nu t)\hat{\mathbf{e}}.$$

According to classical electromagnetic optics, the energy of the mode is fixed at $\frac{1}{2}\epsilon|A|^2V$. We define a complex variable a, such that $\frac{1}{2}\epsilon|A|^2V = h\nu|a|^2$, which allows $|a|^2$ to be interpreted as the energy of the mode in units of photon number. The electric field may then be written as

$$\mathbf{E}(\mathbf{r}, t) = \left(\frac{2h\nu}{\epsilon V}\right)^{1/2} a \exp(-j\mathbf{k}\cdot\mathbf{r}) \exp(j2\pi\nu t)\hat{\mathbf{e}}, \tag{11.3-1}$$

where the complex variable a determines the complex amplitude of the field.

In classical electromagnetic optics, $a \exp(j2\pi\nu t)$ is a rotating phasor whose projection on the real axis determines the sinusoidal electric field (see Fig. 11.3-1). The real and imaginary parts $x = \text{Re}\{a\}$ and $p = \text{Im}\{a\}$ are called the quadrature components of the phasor a because they are a quarter cycle (90°) out of phase with each other. They determine the amplitude and phase of the sine wave that represents the temporal variation of the electric field. The rotating phasor $a \exp(j2\pi\nu t)$ also describes the motion of a harmonic oscillator; the real component x is proportional to position and the imaginary component p to momentum. From a mathematical point of view, a classical monochromatic mode of the electromagnetic field and a classical harmonic oscillator behave identically.

Similarly, a quantum monochromatic electromagnetic mode and a one-dimensional quantum-mechanical harmonic oscillator have identical behavior. We therefore review the quantum theory of a simple harmonic oscillator before proceeding.

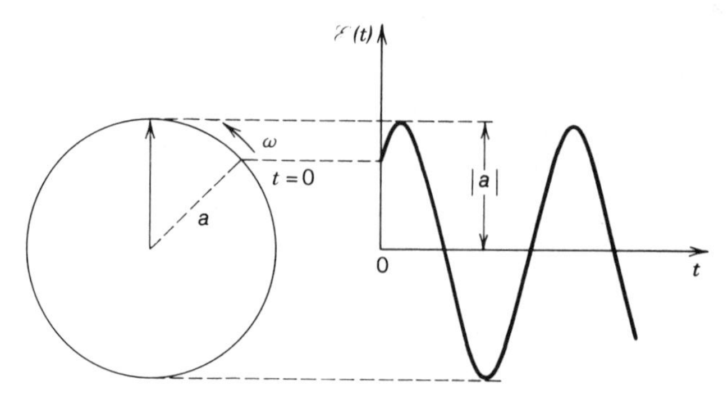

Figure 11.3-1 The real and imaginary parts of the variable $a \exp(j2\pi\nu t)$, which governs the complex amplitude of a classical electromagnetic field of frequency ν. The time dynamics are identical to those of a harmonic oscillator of angular frequency $\omega = 2\pi\nu$.

Quantum Theory of the Harmonic Oscillator

A particle of mass m, position x, momentum p, and potential energy $V(x) = \frac{1}{2}\kappa x^2$, where κ is the elastic constant, is a harmonic oscillator of total energy $\frac{1}{2}p^2/m + \frac{1}{2}\kappa x^2$ and oscillation frequency $\omega = (\kappa/m)^{1/2}$. In accordance with quantum mechanics its behavior may be described by a complex wavefunction $\psi(x)$ satisfying the time-independent Schrödinger equation

$$-\frac{\hbar^2}{2m}\frac{d^2\psi}{dx^2} + V(x)\psi(x) = E\psi(x), \tag{11.3-2}$$

where E is the particle energy. For the harmonic oscillator the solutions of the Schrödinger equation give rise to discrete values of energy given by

$$E_n = \left(n + \tfrac{1}{2}\right)h\nu, \qquad n = 0, 1, 2, \ldots; \tag{11.3-3}$$

adjacent energy levels are separated by a quantum of energy $h\nu = \hbar\omega$. The corresponding eigenfunctions $\psi_n(x)$ are normalized Hermite–Gaussian functions,

$$\psi_n(x) = (2^n n!)^{-1/2}\left(\frac{2m\omega}{h}\right)^{1/4} H_n\left[\left(\frac{m\omega}{\hbar}\right)^{1/2} x\right]\exp\left(-\frac{m\omega x^2}{2\hbar}\right),$$

where $H_n(x)$ is the Hermite polynomial of order n [see (3.3-5) to (3.3-7) and (3.3-10)].

An arbitrary wavefunction $\psi(x)$ may be expanded in terms of the orthonormal eigenfunctions $\{\psi_n(x)\}$ as the superposition $\psi(x) = \sum_n c_n\psi_n(x)$. Given the wavefunction $\psi(x)$, which determines the state of the system, the behavior of the particle may be determined as follows:

- The probability $p(n)$ that the harmonic oscillator carries n quanta of energy is given by the coefficient $|c_n|^2$.
- The probability density of finding the particle at the position x is given by $|\psi(x)|^2$.
- The probability density that the momentum of the particle is p is given by $|\phi(p)|^2$, where $\phi(p)$ is proportional to the inverse Fourier transform of $\psi(x)$ evaluated at the frequency p/h,

$$\phi(p) = \frac{1}{\sqrt{h}}\int_{-\infty}^{\infty}\psi(x)\exp\left(j2\pi\frac{p}{h}x\right)dx. \tag{11.3-4}$$

The Fourier transform relation between the variables x and p/h implies a Heisenberg position–momentum uncertainty relation

$$\frac{\sigma_x \sigma_p}{h} \geq \frac{1}{4\pi} \qquad \text{or} \qquad \sigma_x \sigma_p \geq \frac{\hbar}{2}.$$

Analogy Between an Optical Mode and a Harmonic Oscillator

The energy of an electromagnetic mode is $h\nu|a|^2 = h\nu(x^2 + p^2)$. The analogy with a harmonic oscillator of energy $\frac{1}{2}(p^2/m + \kappa x^2)$ is established by effecting the substitutions

$$x = (2h\nu)^{-1/2}\omega x \qquad \text{and} \qquad p = (2h\nu)^{-1/2} p.$$

The mode energy then becomes $\frac{1}{2}(p^2 + \omega^2 x^2)$, which is the same as the energy of a harmonic oscillator of mass $m = 1$ (for which $\omega = \sqrt{\kappa}$). Because the analogy is complete, we conclude that the energy of a quantum electromagnetic mode, like that of a quantum-mechanical harmonic oscillator, is quantized to the values $(n + \frac{1}{2})h\nu$, as suggested earlier. With the use of proper scaling factors, the behavior of the position x and momentum p of the harmonic oscillator also describe the quadrature components of the electromagnetic field x and p.

Summary

An electromagnetic mode of frequency ν is described by a complex wavefunction $\psi(x)$ that governs the uncertainties of the quadrature components x and p and the statistics of the number of photons in the mode.

- The probability $p(n)$ that the mode contains n photons is given by $|c_n|^2$, where the c_n are coefficients of the expansion of $\psi(x)$ in terms of the eigenfunctions $\psi_n(x)$,

$$\psi(x) = \sum_n c_n \psi_n(x).$$

- The probability densities of the quadrature components x and p are given by the functions $|\psi(x)|^2$ and $|\phi(p)|^2$, where $\psi(\cdot)$ and $\phi(\cdot)$ are related by

$$\phi(p) = \frac{1}{\sqrt{\pi}} \int_{-\infty}^{\infty} \psi(x) \exp(j2px)\, dx. \tag{11.3-5}$$

- If $\psi(x)$ is known, then $\phi(p)$ may be calculated and the probability densities of x and p determined. The complex wavefunction $\psi(x)$ therefore determines the uncertainties of the quadrature components of the complex amplitude.

As shown in Sec. A.2 of Appendix A, the Fourier transform relation between $\psi(x)$ and $\phi(p)$ indicates that there is an uncertainty relation between the power-rms widths of the quadrature components given by

$$\boxed{\sigma_x \sigma_p \geq \tfrac{1}{4}.} \tag{11.3-6}$$

Quadrature Uncertainty

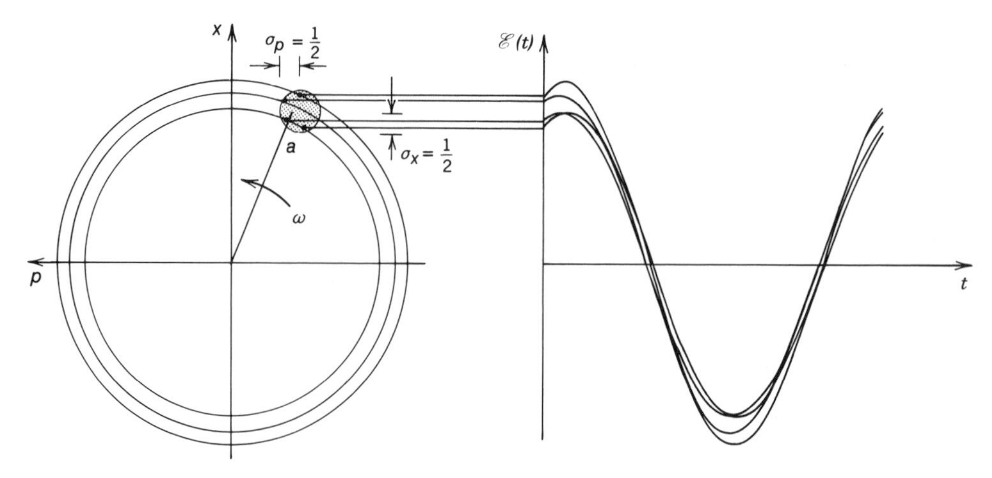

Figure 11.3-2 Uncertainties for the coherent state. Representative values of $\mathscr{E}(t) \propto a \exp(j2\pi\nu t)$ are drawn by choosing arbitrary points within the uncertainty circle. The coefficient of proportionality is chosen to be unity.

The real and imaginary components of the electric field cannot both be determined simultaneously with arbitrary precision.

A. Coherent-State Light

The uncertainty product $\sigma_x \sigma_p$ attains its minimum value of $\frac{1}{4}$ when the function $\psi(x)$ is Gaussian (see Sec. A.2 of Appendix A). In that case

$$\psi(x) \propto \exp\left[-(x - \alpha_x)^2\right], \tag{11.3-7}$$

whereupon its Fourier transform is also Gaussian, so that

$$\phi(p) \propto \exp\left[-(p - \alpha_p)^2\right]. \tag{11.3-8}$$

Here, α_x and α_p are arbitrary values that represent the means of x and p. The quadrature uncertainties, determined from $|\psi(x)|^2$ and $|\phi(p)|^2$, are then given by

$$\sigma_x = \sigma_p = \tfrac{1}{2}. \tag{11.3-9}$$

Under these conditions the electromagnetic field is said to be in a coherent state. The one-standard-deviation range of uncertainty in the quadrature components x and p, as well as in the complex amplitude a and in the electric field $\mathscr{E}(t)$, are illustrated in Fig. 11.3-2 for coherent-state light. The squared-magnitude $|c_n|^2$ of the coefficient of the expansion of $\psi(x)$ in the Hermite-Gaussian basis equals $\bar{n}^n \exp(-\bar{n})/n!$, where $\bar{n} = \alpha_x^2 + \alpha_p^2$; thus the photon-counting probability $p(n)$ is Poisson. Unlike its status in electromagnetic optics, in the context of photon optics coherent light is *not* deterministic.

The uncertainty of the coherent state is most pronounced when α_x and α_p are small. The time behavior of the electric field is illustrated in Fig. 11.3-3 in the limit when $\alpha_x = \alpha_p = 0$. This corresponds to the case when the mode contains zero photons and has only the residual zero-point energy $\frac{1}{2}h\nu$; this is called the vacuum state.

B. Squeezed-State Light

Quadrature-Squeezed Light

Although the uncertainty product $\sigma_x \sigma_p$ cannot be reduced below its minimum value of $\frac{1}{4}$, the uncertainty of one of the quadrature components may be reduced (squeezed) below $\frac{1}{2}$, of course at the expense of an increased uncertainty in the other component.

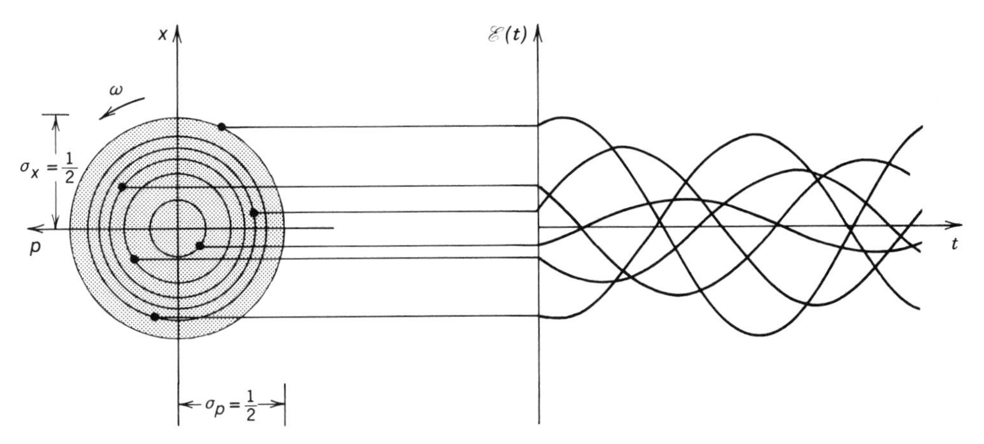

Figure 11.3-3 Representative uncertainties for the vacuum state.

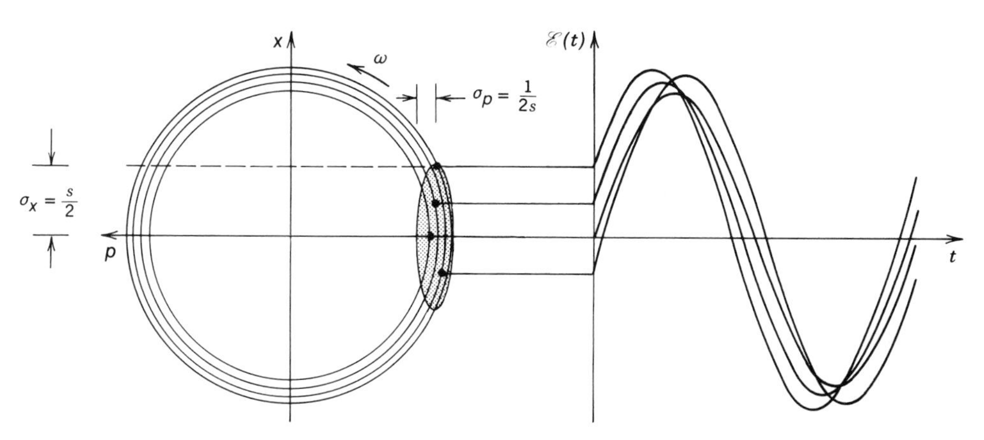

Figure 11.3-4 Representative uncertainties for a quadrature-squeezed state.

The light is then said to be **quadrature squeezed**. For example, a state for which $\psi(x)$ is a Gaussian function with a (stretched) width $\sigma_x = s/2$ $(s > 1)$ corresponds to a Gaussian $\phi(p)$ with a (squeezed) width $\sigma_p = 1/2s$. The product $\sigma_x \sigma_p$ maintains its minimum value of $\frac{1}{4}$, but the uncertainty circle of the phasor a is squeezed into an elliptical form, as shown in Fig. 11.3-4. The asymmetry in the uncertainties of the two quadrature components is manifested in the time course of the electric field by periodic occurrences of increased uncertainty followed, each quarter cycle later, by occurrences of decreased uncertainty. If the field were to be measured only at those times when its uncertainty is minimal, its noise would be reduced below that of the coherent state. The selection of those times may be achieved by heterodyning the squeezed field with a coherent optical field of appropriate phase (see Sec. 22.5). Because of its reduced noisiness, squeezed light is useful in precision measurements and in information transmission.

Photon-Number-Squeezed Light

Quadrature-squeezed light exhibits an uncertainty in one of its quadrature components that is reduced relative to that of the coherent state. Another form of nonclassical light is **photon-number-squeezed** or **sub-Poisson** light. It has a photon-number variance that is squeezed below the coherent-state (Poisson) value, i.e., $\sigma_n^2 < \bar{n}$. Photon-number fluctuations obeying this relation are nonclassical since (11.2-27) cannot be satisfied.

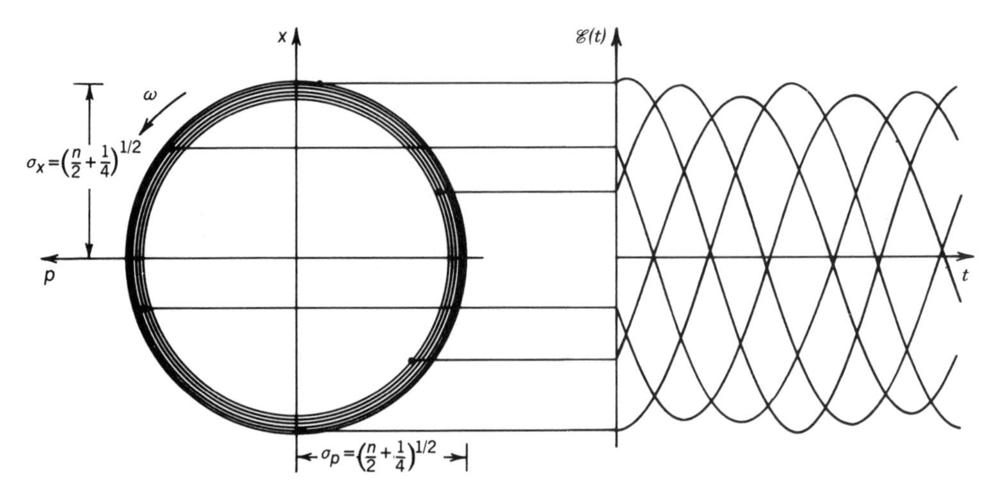

Figure 11.3-5 Representative uncertainties for the number state. This state is photon-number squeezed but not quadrature squeezed.

Photon-number squeezed light, like quadrature-squeezed light, has applications in precision measurements and information transmission. It can be generated by specially designed semiconductor injection lasers.

As an example of photon-number-squeezed light, consider an electromagnetic mode described by the harmonic oscillator eigenstate $\psi(x) = \psi_{n0}(x)$. This is called a number state because $p(n) = |c_n|^2 = 1$ for $n = n_0$, while all other coefficients (c_n for $n \neq n_0$) vanish, so that the number of photons carried by the mode is precisely n_0. Number-state light therefore has a deterministic photon number. The mean photon number is obviously $\bar{n} = n_0$ and the variance is zero (since there are no photon-number fluctuations). The case $n_0 = 1$ corresponds to the presence of precisely one photon.

The uncertainties of number-state light are shown in Fig. 11.3-5. Although the quadrature components, as well as the phasor magnitude and phase, are all uncertain, the photon number is absolutely certain. The question arises as to whether it is possible to carry out experiments requiring a fixed number of photons by using coherent-state light in a selective manner. Could this, for example, be achieved by monitoring the photons from a coherent source in successive time intervals, and then using the photons only in those time intervals where the desired photon number is observed? The problem with this approach is that it is difficult to observe the photons without annihilating them. One way of circumventing the problem is to generate photons in correlated pairs by means of a process such as parametric downconversion (see Secs. 19.2C and 19.4C). With two "copies" of a photon stream, one can be observed and used to indicate or control the photon number in the other.

READING LIST

Books on Quantum Mechanics

L. E. Ballentine, *Quantum Mechanics*, Prentice-Hall, Englewood Cliffs, NJ, 1990.

W. Greiner, *Quantum Mechanics*, Springer-Verlag, New York, 1989.

A. Yariv, *Introduction to the Theory and Applications of Quantum Mechanics*, Wiley, New York, 1982.

L. I. Schiff, *Quantum Mechanics*, McGraw-Hill, New York, 3rd ed. 1968.

R. P. Feynman, R. B. Leighton, and M. Sands, *The Feynman Lectures on Physics*, vol. 3, *Quantum Mechanics*, Addison-Wesley, Reading, MA, 1965.

R. P. Feynman, *Quantum Electrodynamics*, W. A. Benjamin, New York, 1962.

A. Messiah, *Quantum Mechanics*, vols. 1 and 2, North-Holland/Wiley, Amsterdam/New York, 1961/1962.

E. Merzbacher, *Quantum Mechanics*, Wiley, New York, 1961.

R. M. Eisberg, *Fundamentals of Modern Physics*, Wiley, New York, 1961.

P. A. M. Dirac, *The Principles of Quantum Mechanics*, Oxford University Press, New York, 4th ed. 1958.

L. D. Landau and E. M. Lifshitz, *Quantum Mechanics*, Addison-Wesley, Reading, MA, 1958.

Books on Quantum Optics

J. Peřina, *Quantum Statistics of Linear and Nonlinear Optical Phenomena*, Reidel, Dordrecht, The Netherlands, 2nd ed. 1991.

P. Meystre and M. Sargent III, *Elements of Quantum Optics*, Springer-Verlag, New York, 1990.

W. H. Louisell, *Quantum Statistical Properties of Radiation*, Wiley, New York, 1973, 1990.

C. Cohen-Tannoudji, J. Dupont-Roc, and G. Grynberg, *Photons and Atoms*, Wiley, New York, 1989.

E. R. Pike and H. Walther, eds., *Photons and Quantum Fluctuations*, Adam Hilger, Bristol, England, 1988.

F. Haake, L. M. Narducci, and D. F. Walls, eds., *Coherence, Cooperation, and Fluctuations*, Cambridge University Press, New York, 1986.

E. R. Pike and S. Sarkar, eds., *Frontiers in Quantum Optics*, Adam Hilger, Bristol, England, 1986.

R. P. Feynman, *QED*, Princeton University Press, Princeton, NJ, 1985.

J. Peřina, *Coherence of Light*, Reidel, Dordrecht, The Netherlands, 2nd ed. 1985.

R. Loudon, *The Quantum Theory of Light*, Oxford University Press, New York, 2nd ed. 1983.

R. L. Knight and L. Allen, *Concepts of Quantum Optics*, Pergamon Press, Oxford, 1983.

E. Goldin, *Waves and Photons: An Introduction to Quantum Optics*, Wiley, New York, 1982.

H. Haken, *Light: Waves, Photons, Atoms*, vol. 1, North-Holland, Amsterdam, 1981.

D. Marcuse, *Principles of Quantum Electronics*, Academic Press, New York, 1980.

B. Saleh, *Photoelectron Statistics*, Springer-Verlag, New York, 1978.

H. M. Nussenzveig, *Introduction to Quantum Optics*, Gordon and Breach, New York, 1973.

D. Marcuse, *Engineering Quantum Electrodynamics*, Harcourt, Brace & World, New York, 1970.

J. R. Klauder and E. C. G. Sudarshan, *Fundamentals of Quantum Optics*, W. A. Benjamin, New York, 1968.

C. DeWitt, A. Blandin, and C. Cohen-Tannoudji, eds., *Quantum Optics and Electronics*, Gordon and Breach, New York, 1965.

W. H. Louisell, *Radiation and Noise in Quantum Electronics*, McGraw-Hill, New York, 1964.

W. Heitler, *The Quantum Theory of Radiation*, Clarendon Press, Oxford, England, 3rd ed. 1954.

Special Journal Issues

Special issue on the statistical efficiency of natural and artificial vision, Part II, *Journal of the Optical Society of America A*, vol. 5, no. 4, 1988.

Special issue on the statistical efficiency of natural and artificial vision, *Journal of the Optical Society of America A*, vol. 4, no. 12, 1987.

Special issue on squeezed states of the electromagnetic field, *Journal of the Optical Society of America B*, vol. 4, no. 10, 1987.

Special issue on squeezed light, *Journal of Modern Optics*, vol. 34, no. 6/7, 1987.

Special issue on quantum-limited imaging and image processing, *Journal of the Optical Society of America A*, vol. 3, no. 12, 1986.

Special issue on the mechanical effects of light, *Journal of the Optical Society of America B*, vol. 2, no. 11, 1985.

Articles

M. C. Teich and B. E. A. Saleh, Squeezed and Antibunched Light, *Physics Today*, vol. 43, no. 6, pp. 26–34, 1990.

M. C. Teich and B. E. A. Saleh, Squeezed States of Light, *Quantum Optics*, vol. 1, pp. 153–191, 1989.

M. C. Teich and B. E. A. Saleh, Photon Bunching and Antibunching, in *Progress in Optics*, vol. 26, E. Wolf, ed., North-Holland, Amsterdam, 1988, pp. 1–104.

R. E. Slusher and B. Yurke, Squeezed Light, *Scientific American*, vol. 258, no. 5, pp. 50–56, 1988.

M. D. Levenson and R. M. Shelby, Deamplification of Quantum Noise and Quantum Nondemolition Detection in Optical Fibers, *Optics News*, vol. 14, no. 1, pp. 7–12, 1988.

R. W. Henry and S. C. Glotzer, A Squeezed-State Primer, *American Journal of Physics*, vol. 56, pp. 318–328, 1988.

G. Leuchs, Squeezing the Quantum Fluctuations of Light, *Contemporary Physics*, vol. 29, pp. 299–314, 1988.

R. Loudon and P. L. Knight, Squeezed Light, *Journal of Modern Optics*, vol. 34, pp. 709–759, 1987.

M.-A. Bouchiat and L. Pottier, Optical Experiments and Weak Interactions, *Science*, vol. 234, pp. 1203–1210, 1986.

D. F. Walls, Squeezed States of Light, *Nature*, vol. 306, pp. 141–146, 1983.

H. Paul, Photon Antibunching, *Reviews of Modern Physics*, vol. 54, pp. 1061–1102, 1982.

E. Wolf, Einstein's Researches on the Nature of Light, *Optics News*, vol. 5, no. 1, pp. 24–39, 1979.

S. Weinberg, Light as a Fundamental Particle, *Physics Today*, vol. 28, no. 6, pp. 32–37, 1975.

M. O. Scully and M. Sargent III, The Concept of the Photon, *Physics Today*, vol. 25, no. 3, pp. 38–47, 1972.

H. Risken, Statistical Properties of Laser Light, in *Progress in Optics*, vol. 8, E. Wolf, ed., North-Holland, Amsterdam, 1970.

L. Mandel and E. Wolf, eds., *Selected Papers on Coherence and Fluctuations of Light*, vols. 1 and 2, Dover, New York, 1970.

L. Mandel and E. Wolf, Coherence Properties of Optical Fields, *Reviews of Modern Physics*, vol. 37, pp. 231–287, 1965.

L. Mandel, Fluctuations of Light Beams, in *Progress in Optics*, vol. 2, E. Wolf, ed., North-Holland, Amsterdam, 1963.

PROBLEMS

11.1-1 **Photon Energy.** (a) What voltage should be applied to accelerate an electron from zero velocity in order that it acquire the same energy as a photon of wavelength $\lambda_o = 0.87\ \mu\text{m}$?
(b) A photon of wavelength 1.06 μm is combined with a photon of wavelength 10.6 μm to create a photon whose energy is the sum of the energies of the two photons. What is the wavelength of the resultant photon? Photon interactions of this type are discussed in Chap. 19.

11.1-2 **Position of a Single Photon at a Screen.** Consider a monochromatic light beam of wavelength λ_o falling on an infinite screen in the plane $z = 0$, with an intensity $I(\rho) = I_0 \exp(-\rho/\rho_0)$, where $\rho = (x^2 + y^2)^{1/2}$. Assume that the intensity of the source is reduced to a level at which only a single photon strikes the screen.
(a) Find the probability that the photon strikes the screen within a radius ρ_0 of the origin.

(b) If the beam contains exactly 10^6 photons, on the average how many photons strike within a circle of radius ρ_0?

11.1-3 **Momentum of a Free Photon.** Compare the total momentum of the photons in a 10-J laser pulse with that of a 1-g mass moving at a velocity of 1 cm/s and with an electron moving at a velocity $c_o/10$.

*11.1-4 **Momentum of a Photon in a Gaussian Beam.** (a) What is the probability that the momentum vector of a photon associated with a Gaussian beam of waist radius W_0 lies within the beam divergence angle θ_0? Refer to Sec. 3.1 for definitions. (b) Does the relation $p = E/c_o$ hold in this case?

11.1-5 **Levitation by Light Pressure.** Consider an isolated hydrogen atom of mass 1.66×10^{-27} kg.
(a) Find the gravitational force on this hydrogen atom near the surface of the earth (assume that at sea level the gravitational acceleration constant $g = 9.8$ m/s^2).
(b) Let an upwardly directed laser beam emitting 1-eV photons be focused in such a way that the full momentum of each of its photons is transferred to the atom. Find the average upward force on the atom provided by one photon striking it each second.
(c) Find the number of photons that must strike the atom per second and the corresponding optical power for it not to fall under the effect of gravity, given idealized conditions in vacuum.
(d) How many photons per second would be required to keep the atom from falling if it were perfectly reflecting?

*11.1-6 **Single Photon in a Fabry–Perot Resonator.** Consider a Fabry–Perot resonator of length $d = 1$ cm containing nonabsorbing material of refractive index $n = 1.5$ and perfectly reflecting mirrors. Assume that there is exactly one photon in the mode described by the standing wave $\sin(10^5 \pi x/d)$.
(a) Determine the photon wavelength and energy (in eV).
(b) Estimate the uncertainty in the photon's position and momentum (magnitude and direction). Compare with the value obtained from the relation $\sigma_p \sigma_x \approx \hbar/2$.

11.1-7 **Single-Photon Beating (Time Interference).** Consider a detector illuminated by a polychromatic plane wave consisting of two plane-parallel superposed monochromatic waves represented by

$$U_1(t) = \sqrt{I_1}\, \exp(j2\pi\nu_1 t) \qquad \text{and} \qquad U_2(t) = \sqrt{I_2}\, \exp(j2\pi\nu_2 t),$$

with frequencies ν_1 and ν_2 and intensities I_1 and I_2, respectively. According to wave optics (see Sec. 2.6B), the intensity of this wave is given by $I(t) = I_1 + I_2 + 2(I_1 I_2)^{1/2} \cos[2\pi(\nu_2 - \nu_1)t]$. Assume that the two constituent plane waves have equal intensities ($I_1 = I_2$). Assume also that the wave is sufficiently weak that only a single polychromatic photon reaches the detector during the time interval $T = 1/|\nu_2 - \nu_1|$.
(a) Plot the probability density $p(t)$ for the detection time of the photon for $0 \le t \le 1/|\nu_2 - \nu_1|$. At what time instant during T is the probability zero that the photon will be detected?
(b) An attempt to discover from which of the two constituent waves the photon came would entail an energy measurement to a precision better than

$$\sigma_E < h|\nu_2 - \nu_1|.$$

Use the time–energy uncertainty relation to show that the time required for such a measurement would be of the order of the beat-frequency period so that the very process of measurement would wash out the interference.

11.1-8 **Photon Momentum Exchange at a Beamsplitter.** Consider a single photon, in a mode described by a plane wave, impinging on a lossless beamsplitter. What is the momentum vector of the photon before it impinges on the mirror? What are the possible values of the photon's momentum vector, and the probabilities of observing these values, after the beamsplitter?

11.2-1 **Photon Flux.** Show that the power of a monochromatic optical beam that carries an average of one photon per optical cycle is inversely proportional to the squared wavelength.

11.2-2 **The Poisson Distribution.** Verify that the Poisson probability distribution given by (11.2-12) is normalized to unity and has mean $\bar{n}$ and variance $\sigma_n^2 = \bar{n}$.

11.2-3 **Photon Statistics of a Coherent Gaussian Beam.** Assume that a 100-pW He–Ne single-mode laser emits light at 633 nm in a $\text{TEM}_{0,0}$ Gaussian beam (see Chap. 3).
(a) What is the mean number of photons crossing a circle of radius equal to the waist radius of the beam W_0 in a time $T = 100$ ns?
(b) What is the root-mean-square value of the number of photon counts in (a)?
(c) What is the probability that no photons are counted in (a)?

11.2-4 **The Bose–Einstein Distribution.** (a) Verify that the Bose–Einstein probability distribution given by (11.2-20) is normalized and has a mean $\bar{n}$ and variance $\sigma_n^2 = \bar{n} + \bar{n}^2$.
(b) If a beam of photons obeying Bose–Einstein statistics contains an average of $\Phi = 1$ photon per nanosecond, what is the probability that zero photons will be detected in a 20-ns time interval?

*11.2-5 **The Negative-Binomial Distribution.** It is well known in the literature of probability theory that the sum of $\mathscr{M}$ identically distributed random variables, each with a geometric (Bose–Einstein) distribution, obeys the negative binomial distribution

$$p(n) = \binom{n + \mathscr{M} - 1}{n} \frac{(\bar{n}/\mathscr{M})^n}{(1 + \bar{n}/\mathscr{M})^{n+\mathscr{M}}}.$$

Verify that the negative-binomial distribution reduces to the Bose–Einstein distribution for $\mathscr{M} = 1$ and to the Poisson distribution as $\mathscr{M} \to \infty$.

*11.2-6 **Photon Statistics for Multimode Thermal Light in a Cavity.** Consider $\mathscr{M}$ modes of thermal radiation sufficiently close to each other in frequency that each can be considered to be occupied in accordance with a Bose–Einstein distribution of the same mean photon number $1/[\exp(h\nu/k_B T) - 1]$. Show that the variance of the *total* number of photons n is related to its mean by

$$\sigma_n^2 = \bar{n} + \frac{\bar{n}^2}{\mathscr{M}},$$

indicating that multimode thermal light has less variance than does single-mode thermal light. The presence of the multiple modes provides averaging, thereby reducing the noisiness of the light.

*11.2-7 **Photon Statistics for a Beam of Multimode Thermal Light.** A multimode thermal light source that carries $\mathscr{M}$ identical modes, each with exponentially distributed (random) integrated rate, has a probability density $p(\mathscr{W})$ describable by the gamma distribution

$$p(\mathscr{W}) = \frac{1}{(\mathscr{M}-1)!}\left(\frac{\mathscr{M}}{\langle\mathscr{W}\rangle}\right)^{\mathscr{M}} \mathscr{W}^{\mathscr{M}-1} \exp\left(-\frac{\mathscr{M}\mathscr{W}}{\langle\mathscr{W}\rangle}\right), \qquad \mathscr{W} \geq 0.$$

Use Mandel's formula (11.2-25) to show that the resulting photon-number distribution assumes the form of the negative-binomial distribution defined in Problem 11.2-5.

*11.2-8 **Mean and Variance of the Doubly Stochastic Poisson Distribution.** Prove (11.2-26) and (11.2-27).

11.2-9 **Random Partitioning of Coherent Light.** (a) Use (11.2-32) to show that the photon-number distribution of randomly partitioned coherent light retains its Poisson form.
(b) Show explicitly that the mean photon number for light reflected from a lossless beamsplitter is $(1-\mathscr{T})\bar{n}$.
(c) Prove (11.2-33) for coherent light.

11.2-10 **Random Partitioning of Single-Mode Thermal Light.** (a) Use (11.2-32) to show that the photon-number distribution of randomly partitioned single-mode thermal light retains its Bose–Einstein form.
(b) Show explicitly that the mean photon number for light reflected from a lossless beamsplitter is $(1-\mathscr{T})\bar{n}$.
(c) Prove (11.2-34) for single-mode thermal light.

*11.2-11 **Exponential Decay of Mean Photon Number in an Absorber.** (a) Consider an absorptive material of thickness d and absorption coefficient α (cm^{-1}). If the average number of photons that enters the material is $\bar{n}_0$, write a differential equation to find the average number of photons $\bar{n}(x)$ at position x, where x is the depth into the filter $(0 \leq x \leq d)$.
(b) Solve the differential equation. State the reason that your result is the exponential intensity decay law obtained from electromagnetic optics (Sec. 5.5A).
(c) Write an expression for the photon-number distribution at an arbitrary position x in the absorber, $p(n)$, when coherent light is incident on it.
(d) What is the probability of survival of a single photon incident on the absorber?

*11.3-1 **Statistics of the Binomial Photon-Number Distribution.** The binomial probability distribution may be written $p(n) = [M!/(M-n)!\,n!]p^n(1-p)^{M-n}$. It describes certain photon-number-squeezed sources of light.
(a) Indicate a possible mechanism for converting number-state light into light described by binomial photon statistics.
(b) Prove that the binomial probability distribution is normalized to unity.
(c) Find the count mean $\bar{n}$ and the count variance σ_n^2 of the binomial probability distribution in terms of its two parameters, p and M.
(d) Find an expression for the SNR in terms of $\bar{n}$ and p. Evaluate it for the limiting cases $p \to 0$ and $p \to 1$. To what kinds of light do these two limits correspond?

*11.3-2 **Noisiness of a Hypothetical Photon Source.** Consider a hypothetical light source that produces a photon stream with a photon-number distribution that is discrete-uniform, given by

$$p(n) = \begin{cases} \dfrac{1}{2\bar{n} + 1}, & 0 \le n \le 2\bar{n} \\ 0, & \text{otherwise.} \end{cases}$$

(a) Verify that the distribution is normalized to unity and has mean $\bar{n}$. Calculate the photon-number variance σ_n^2 and the signal-to-noise ratio (SNR) and compare them to those of the Bose–Einstein and Poisson distributions of the same mean.
(b) In terms of SNR, would this source be quieter or noisier than an ideal single-mode laser when $\bar{n} < 2$? When $\bar{n} = 2$? When $\bar{n} > 2$?
(c) By what factor is the SNR for this light larger than that for single-mode thermal light?
[*Useful formulas*:

$$1 + 2 + 3 + \cdots + j = \frac{j(j+1)}{2}$$

$$1^2 + 2^2 + 3^2 + \cdots + j^2 = \frac{j(j+1)(2j+1)}{6}.]$$

CHAPTER

12

PHOTONS AND ATOMS

12.1 ATOMS, MOLECULES, AND SOLIDS
- A. Energy Levels
- B. Occupation of Energy Levels in Thermal Equilibrium

12.2 INTERACTIONS OF PHOTONS WITH ATOMS
- A. Interaction of Single-Mode Light with an Atom
- B. Spontaneous Emission
- C. Stimulated Emission and Absorption
- D. Line Broadening
- *E. Laser Cooling and Trapping of Atoms

12.3 THERMAL LIGHT
- A. Thermal Equilibrium Between Photons and Atoms
- B. Blackbody Radiation Spectrum

12.4 LUMINESCENCE LIGHT

Niels Bohr (1885–1962)

Albert Einstein (1879–1955)

Bohr and Einstein laid the theoretical foundations for describing the interaction of light with matter.

Photons interact with matter because matter contains electric charges. The electric field of light exerts forces on the electric charges and dipoles in atoms, molecules, and solids, causing them to vibrate or accelerate. Conversely, vibrating electric charges emit light.

Atoms, molecules, and solids have specific allowed energy levels determined by the rules of quantum mechanics. Light interacts with an atom through changes in the potential energy arising from forces on the electric charges induced by the time-varying electric field of the light. A photon may interact with an atom if its energy matches the difference between two energy levels. The photon may impart its energy to the atom, raising it to a higher energy level. The photon is then said to be absorbed (or annihilated). An alternative process can also occur. The atom can undergo a transition to a lower energy level, resulting in the emission (or creation) of a photon of energy equal to the difference between the energy levels.

Matter constantly undergoes upward and downward transitions among its allowed energy levels. Some of these transitions are caused by thermal excitations and lead to photon emission and absorption. The result is the generation of electromagnetic radiation from all objects with temperatures above absolute zero. As the temperature of the object increases, higher energy levels become increasingly accessible, resulting in a radiation spectrum that moves toward higher frequencies (shorter wavelengths). Thermal equilibrium between a collection of photons and atoms is reached as a result of these random processes of photon emission and absorption, together with thermal transitions among the allowed energy levels. The radiation emitted has a spectrum that is ultimately determined by this equilibrium condition. Light emitted from atoms, molecules, and solids, under conditions of thermal equilibrium and in the absence of other external energy sources, is known as **thermal light**. Photon emission may also be induced by the presence of other external sources of energy, such as an external source of light, an electron current or a chemical reaction. The excited atoms can then emit nonthermal light called **luminescence light**.

The purpose of this chapter is to introduce the laws that govern the interaction of light with matter and lead to the emission of thermal and luminescence light. The chapter begins with a brief review (Sec. 12.1) of the energy levels of different types of atoms, molecules, and solids. In Sec. 12.2 the laws governing the interaction of a photon with an atom, i.e., photon emission and absorption, are introduced. The interaction of many photons with many atoms, under conditions of thermal equilibrium, is then discussed in Sec. 12.3. A brief description of luminescence light is provided in Sec. 12.4.

12.1 ATOMS, MOLECULES, AND SOLIDS

Matter consists of atoms. These may exist in relative isolation, as in the case of a dilute atomic gas, or they may interact with neighboring atoms to form molecules and matter in the liquid or solid state. The motion of the constituents of matter follow the laws of quantum mechanics.

The behavior of a single nonrelativistic particle of mass m (e.g., an electron), with a potential energy $V(\mathbf{r}, t)$, is governed by a complex wavefunction $\Psi(\mathbf{r}, t)$ satisfying the **Schrödinger equation**

$$-\frac{\hbar^2}{2m}\nabla^2\Psi(\mathbf{r}, t) + V(\mathbf{r}, t)\Psi(\mathbf{r}, t) = j\hbar\frac{\partial\Psi(\mathbf{r}, t)}{\partial t}. \tag{12.1-1}$$

The potential energy is determined by the environment surrounding the particle and is responsible for the great variety of solutions to the equation. Systems with multiple particles, such as atoms, molecules, liquids, and solids, obey a more complex but similar equation; the potential energy then contains terms permitting interactions among the particles and with externally applied fields. Equation (12.1-1) is not unlike the paraxial Helmholtz equation [see (2.2-22) and (5.6-18)].

The Born postulate of quantum mechanics specifies that the probability of finding the particle within an incremental volume dV surrounding the position $\mathbf{r}$, within the time interval between t and $t + dt$, is

$$p(\mathbf{r}, t)\, dV dt = |\Psi(\mathbf{r}, t)|^2\, dV dt. \tag{12.1-2}$$

Equation (12.1-2) is similar to (11.1-10), which gives the photon position and time.

If we wish simply to determine the allowed energy levels E of the particle in the absence of time-varying interactions, the technique of separation of variables may be used in (12.1-1) to obtain $\Psi(\mathbf{r}, t) = \psi(\mathbf{r})\exp[j(E/\hbar)t]$, where $\psi(\mathbf{r})$ satisfies the **time-independent Schrödinger equation**

$$-\frac{\hbar^2}{2m}\nabla^2\psi(\mathbf{r}) + V(\mathbf{r})\psi(\mathbf{r}) = E\psi(\mathbf{r}). \tag{12.1-3}$$

Systems of multiple particles obey a generalized form of (12.1-3). The solutions provide the allowed values of the energy of the system E. These values are sometimes discrete (as for an atom), sometimes continuous (as for a free particle), and sometimes take the form of densely packed discrete levels called bands (as for a semiconductor). The presence of thermal excitation or an external field, such as light shining on the material, can induce the system to move from one of its energy levels to another. It is by these means that the system exchanges energy with the outside world.

A. Energy Levels

The energy levels of a molecular system arise from the potential energy of the electrons in the presence of the atomic nuclei and other electrons, as well as from molecular vibrations and rotations. In this section we illustrate various kinds of energy levels for a number of specific atoms, molecules, and solids.

Vibrational and Rotational Energy Levels of Molecules

Vibrations of a Diatomic Molecule. The vibrations of a diatomic molecule, such as N_2, CO, and HCl, may be modeled by two masses m_1 and m_2 connected by a spring. The intermolecular attraction provides a restoring force that is approximately proportional to the change x in the distance separating the atoms. A molecular spring constant κ can be defined so that the potential energy is $V(x) = \frac{1}{2}\kappa x^2$. The molecular vibrations then take on the set of allowed energy levels appropriate for the quantum-mechanical

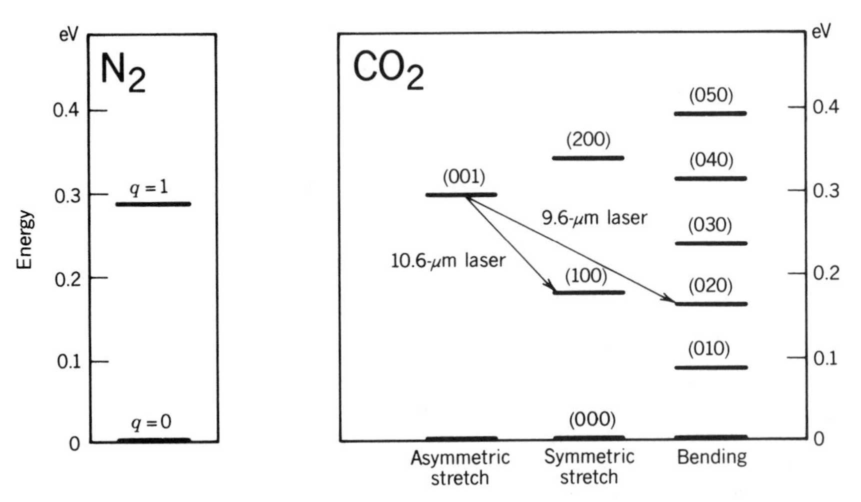

Figure 12.1-1 Lowest vibrational energy levels of the N_2 and CO_2 molecules (the zero of energy is chosen at $q = 0$). The transitions marked by arrows represent energy exchanges corresponding to photons of wavelengths 10.6 μm and 9.6 μm, as indicated. These transitions are used in CO_2 lasers, as discussed in Chaps. 13 and 14.

harmonic oscillator. These are

$$E_q = \left(q + \tfrac{1}{2}\right)\hbar\omega, \qquad q = 0, 1, 2, \ldots, \tag{12.1-4}$$

where $\omega = (\kappa/m_r)^{1/2}$ is the oscillation frequency and $m_r = m_1 m_2/(m_1 + m_2)$ is the reduced mass of the system. The energy levels are equally spaced. Typical values of $\hbar\omega$ lie between 0.05 and 0.5 eV, which corresponds to the energy of a photon in the infrared spectral region (the relations between the different units of energy are provided in Fig. 11.1-2 and inside the back cover of the book). The two lowest-lying vibrational energy levels of N_2 are shown in Fig. 12.1-1. Equation (12.1-4) is identical to the expression for the allowed energies of a mode of the electromagnetic field [see (11.1-4)].

Vibrations of the CO_2 Molecule. A CO_2 molecule may undergo independent vibrations of three kinds: asymmetric stretching (AS), symmetric stretching (SS), and bending (B). Each of these vibrational modes behaves like a harmonic oscillator, with its own spring constant and therefore its own value of $\hbar\omega$. The allowed energy levels are specified by (12.1-4) in terms of the three modal quantum numbers (q_1, q_2, q_3) corresponding to the SS, B, and AS modes, as illustrated in Fig. 12.1-1.

Rotations of a Diatomic Molecule. The rotations of a diatomic molecule about its axes are similar to those of a rigid rotor with moment of inertia $\mathscr{I}$. The rotational energy is quantized to the values

$$E_q = q(q + 1)\frac{\hbar^2}{2\mathscr{I}}, \qquad q = 0, 1, 2, \ldots. \tag{12.1-5}$$

These levels are not evenly spaced. Typical rotational energy levels are separated by values in the range 0.001 to 0.01 eV, so that the energy differences correspond to photons in the far infrared region of the spectrum. Each of the vibrational levels shown

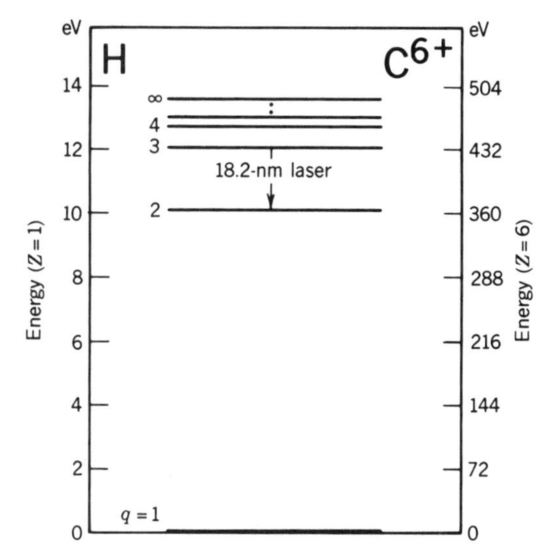

Figure 12.1-2 Energy levels of H ($Z = 1$) and C^{6+} (an H-like atom with $Z = 6$). The $q = 3$ to $q = 2$ transition marked by an arrow corresponds to the C^{6+} x-ray laser transition at 18.2 nm, as discussed in Chap. 14. The arbitrary zero of energy is taken at $q = 1$.

in Fig. 12.1-1 is actually split into many closely spaced rotational levels, with energies given approximately by (12.1-5).

Electron Energy Levels of Atoms and Molecules

Isolated Atoms. An isolated hydrogen atom has a potential energy that derives from the Coulomb law of attraction between the proton and the electron. The solution of the Schrödinger equation leads to an infinite number of discrete levels with energies

$$E_q = -\frac{m_r Z^2 e^4}{2\hbar^2 q^2}, \qquad q = 1, 2, 3, \ldots, \tag{12.1-6}$$

where m_r is the reduced mass of the atom, e is the electron charge, and Z is the number of protons in the nucleus ($Z = 1$ for hydrogen). These levels are shown in Fig. 12.1-2 for $Z = 1$ and $Z = 6$.

The computation of the energy levels of more complex atoms is difficult, however, because of the interactions among the electrons and the effects of electron spin. All atoms have discrete energy levels with energy differences that typically lie in the optical region (up to several eV). Some of the energy levels of He and Ne atoms are illustrated in Fig. 12.1-3.

Dye Molecules. Organic dye molecules are large and complex. They may undergo electronic, vibrational, and rotational transitions so that they typically have many energy levels. Levels exist in both singlet (S) and triplet (T) states. Singlet states have an excited electron whose spin is antiparallel to the spin of the remainder of the dye molecule; triplet states have parallel spins. The energy differences correspond to photons covering broad regions of the optical spectrum, as illustrated schematically in Fig. 12.1-4.

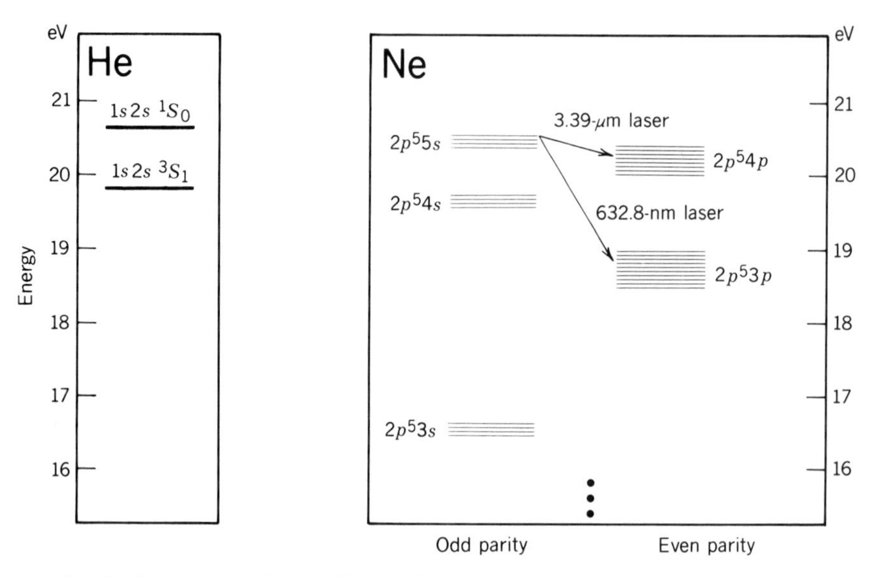

Figure 12.1-3 Some energy levels of He and Ne atoms. The Ne transitions marked by arrows correspond to photons of wavelengths 3.39 μm and 632.8 nm, as indicated. These transitions are used in He–Ne lasers, as discussed in Chaps. 13 and 14.

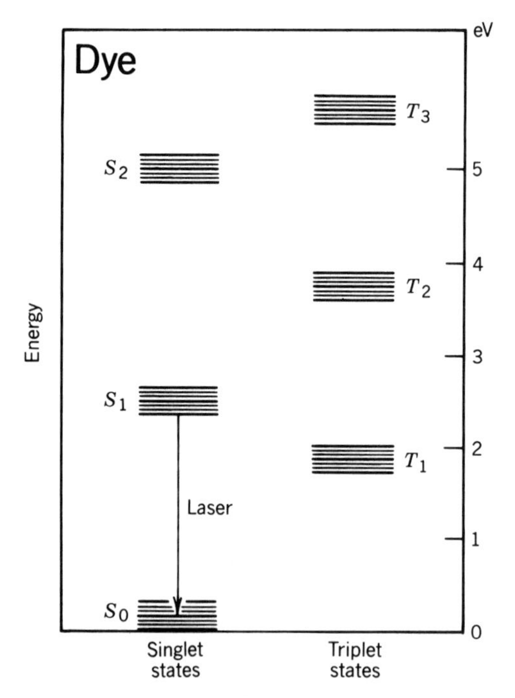

Figure 12.1-4 Schematic illustration of rotational (thinner lines), vibrational (thicker lines), and electronic energy bands of a typical dye molecule. A representative dye laser transition is indicated; the organic dye laser is discussed in Chaps. 13 and 14.

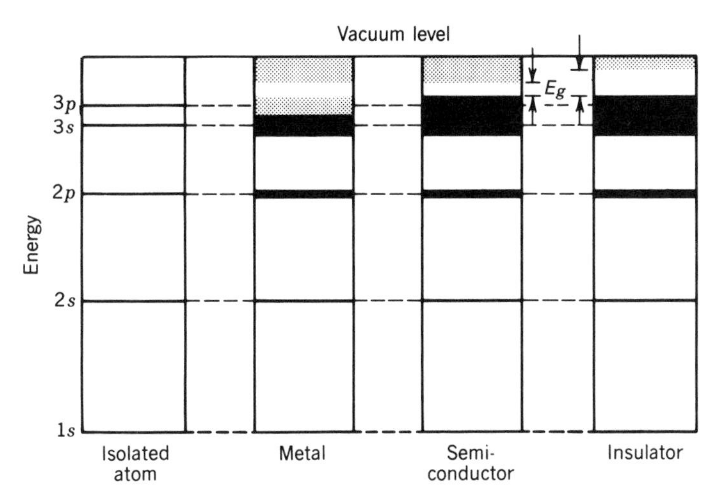

Figure 12.1-5 Broadening of the discrete energy levels of an isolated atom into bands for solid-state materials.

Electron Energy Levels in Solids

Isolated atoms and molecules exhibit discrete energy levels, as shown in Figs. 12.1-1 to 12.1-4. For solids, however, the atoms, ions, or molecules lie in close proximity to each other and cannot therefore be considered as simple collections of isolated atoms; rather, they must be treated as a many-body system.

The energy levels of an isolated atom, and three generic solids with different electrical properties (metal, semiconductor, insulator) are illustrated in Fig. 12.1-5. The lower energy levels in the solids (denoted $1s$, $2s$, and $2p$ levels in this example) are similar to those of the isolated atom. They are not broadened because they are filled by core atomic electrons that are well shielded from the external fields produced by neighboring atoms. In contrast, the energies of the higher-lying discrete atomic levels split into closely spaced discrete levels and form bands. The highest partially occupied band is called the conduction band; the valence band lies below it. They are separated by an energy E_g called the energy bandgap. The lowest-energy bands are filled first.

Conducting solids such as metals have a partially filled conduction band at all temperatures. The availability of many unoccupied states in this band (lightly shaded region in Fig. 12.1-5) means that the electrons can move about easily; this gives rise to the large conductivity in these materials. Intrinsic semiconductors (at $T = 0$ K) have a filled valence band (solid region) and an empty conduction band. Since there are no available free states in the valence band and no electrons in the conduction band, the conductivity is theoretically zero. As the temperature is raised above absolute zero, however, the increasing numbers of electrons from the valence band that are thermally excited into the conduction band contribute to the conductivity. Insulators, which also have a filled valence band, have a larger energy gap (typically > 3 eV) than do semiconductors, so that fewer electrons can attain sufficient thermal energy to contribute to the conductivity. Typical values of the conductivity for metals, semiconductors, and insulators at room temperature are 10^6 $(\Omega\text{-cm})^{-1}$, 10^{-6} to 10^3 $(\Omega\text{-cm})^{-1}$, and 10^{-12} $(\Omega\text{-cm})^{-1}$, respectively. The energy levels of some representative solid-state materials are considered below.

Ruby Crystal. Ruby is an insulator. It is alumina (also known as sapphire, with the chemical formula Al_2O_3) in which a small fraction of the Al^{3+} ions are replaced by

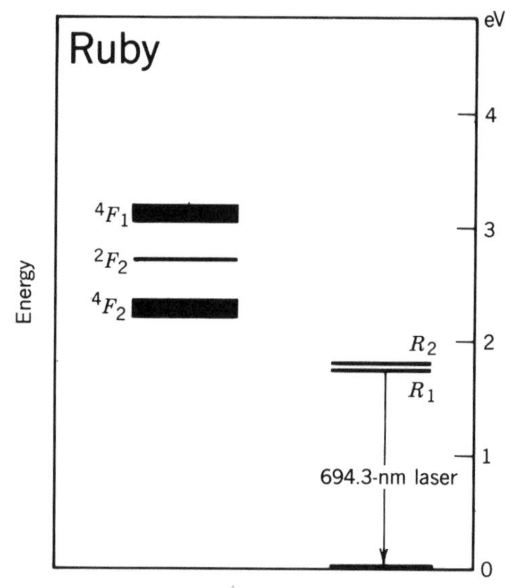

Figure 12.1-6 Discrete energy levels and bands in ruby (Cr^{3+}:Al_2O_3) crystal. The transition indicated by an arrow corresponds to the ruby-laser wavelength of 694.3 nm, as described in Chaps. 13 and 14.

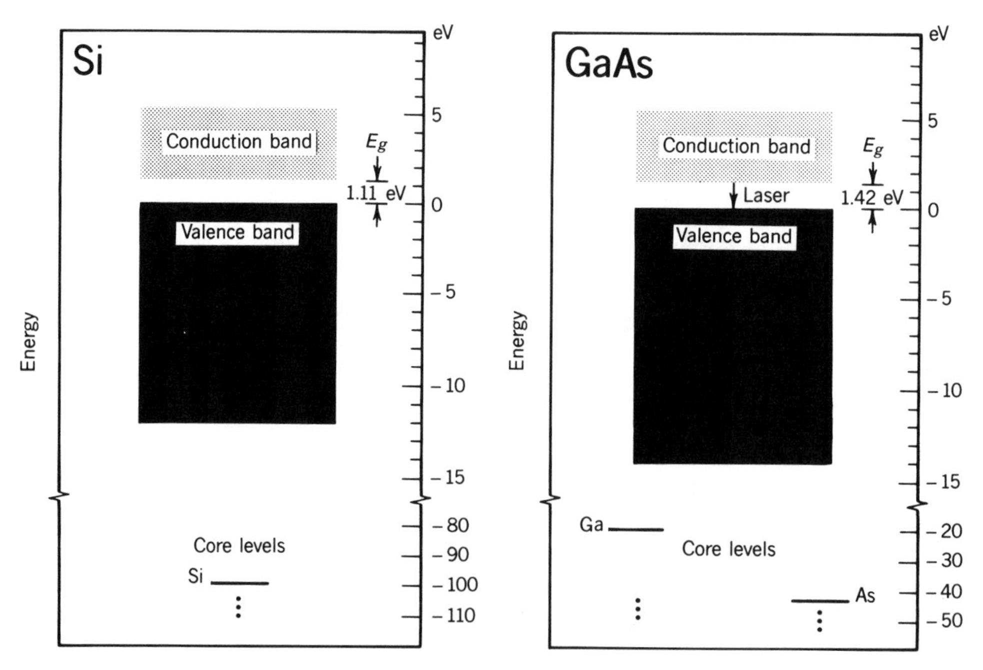

Figure 12.1-7 Energy bands of Si and GaAs semiconductor crystals. The zero of energy is (arbitrarily) defined at the top of the valence band. The GaAs semiconductor injection laser operates on the electron transition between the conduction and valence bands, in the near-infrared region of the spectrum (see Chap. 16).

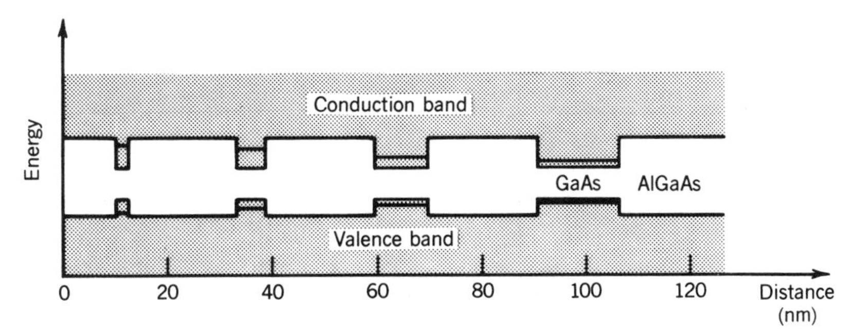

Figure 12.1-8 Quantized energies in a single-crystal AlGaAs/GaAs multiquantum-well structure. The well widths can be arbitrary (as shown) or periodic.

Cr^{3+} ions. The interaction of the constituent ions in this crystal is such that some energy levels are discrete, whereas others form bands, as shown in Fig. 12.1-6. The green and violet absorption bands (indicated by the group-theory notations ${}^4F_2$ and ${}^4F_1$, respectively) give the material its characteristic pink color.

Semiconductors. Semiconductors have closely spaced allowed electron energy levels that take the form of bands as shown in Fig. 12.1-7. The bandgap energy E_g, which separates the valence and conduction bands, is 1.11 eV for Si and 1.42 eV for GaAs at room temperature. The Ga and As $(3d)$ core levels, and the Si $(2p)$ core level are quite narrow, as seen in Fig. 12.1-7. The valence band of Si is formed from the $3s$ and $3p$ levels (as illustrated schematically in Fig. 12.1-5), whereas in GaAs it is formed from the $4s$ and $4p$ levels. The properties of semiconductors are examined in more detail in Chap. 15.

Quantum Wells and Superlattices. Crystal-growth techniques, such as molecular-beam epitaxy and vapor-phase epitaxy, can be used to grow materials with specially designed band structures. In semiconductor quantum-well structures, the energy bandgap is engineered to vary with position in a specified manner, leading to materials with unique electronic and optical properties. An example is the **multiquantum-well** structure illustrated in Fig. 12.1-8. It consists of ultrathin (2 to 15 nm) layers of GaAs alternating with thin (20 nm) layers of AlGaAs. The bandgap of the GaAs is smaller than that of the AlGaAs. For motion perpendicular to the layer, the allowed energy levels for electrons in the conduction band, and for holes in the valence band, are discrete and well separated, like those of the square-well potential in quantum mechanics; the lowest energies are shown schematically in each of the quantum wells. When the AlGaAs barrier regions are also made ultrathin, so that electrons in adjacent wells can readily couple to each other via quantum-mechanical tunneling, these discrete energy levels broaden into miniature bands. The material is then called a **superlattice** structure because these minibands arise from a lattice that is super to (i.e., greater than) the spacing of the natural atomic lattice structure.

EXERCISE 12.1-1

Energy Levels of an Infinite Quantum Well. Solve the Schrödinger equation (12.1-3) to show that the allowed energies of an electron of mass m, in an infinitely deep one-dimensional rectangular potential well [$V(x) = 0$ for $0 < x < d$ and $= \infty$ otherwise], are $E_q =$

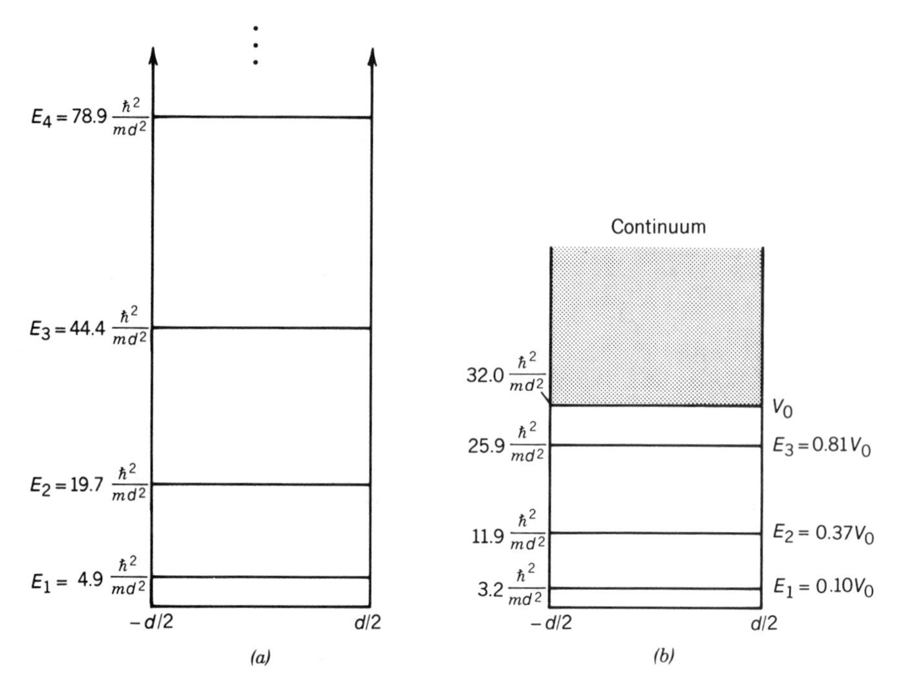

Figure 12.1-9 Energy levels of (*a*) a one-dimensional infinite rectangular potential well and (*b*) a finite square quantum well with an energy depth $V_0 = 32\hbar^2/md^2$. Quantum wells may be made by using modern semiconductor-material growth techniques.

$\hbar^2(q\pi/d)^2/2m$, $q = 1, 2, 3, \ldots$, as shown in Fig. 12.1-9(*a*). Compare these energies with those for the particular finite square quantum well shown in Fig. 12.1-9(*b*).

B. Occupation of Energy Levels in Thermal Equilibrium

As indicated earlier, each atom or molecule in a collection continuously undergoes random transitions among its different energy levels. Such random transitions are described by the rules of statistical physics, in which temperature plays the key role in determining both the average behavior and the fluctuations.

Boltzmann Distribution

Consider a collection of identical atoms (or molecules) in a medium such as a dilute gas. Each atom is in one of its allowed energy levels $E_1, E_2, \ldots$. If the system is in thermal equilibrium at temperature T (i.e., the atoms are kept in contact with a large heat bath maintained at temperature T and their motion reaches a steady state in which the fluctuations are, on the average, invariant to time), the probability $P(E_m)$ that an arbitrary atom is in energy level E_m is given by the Boltzmann distribution

$$P(E_m) \propto \exp(-E_m/k_B T), \qquad m = 1, 2, \ldots, \tag{12.1-7}$$

where k_B is the Boltzmann constant and the coefficient of proportionality is such that $\sum_m P(E_m) = 1$. The occupation probability $P(E_m)$ is an exponentially decreasing function of E_m (see Fig. 12.1-10).

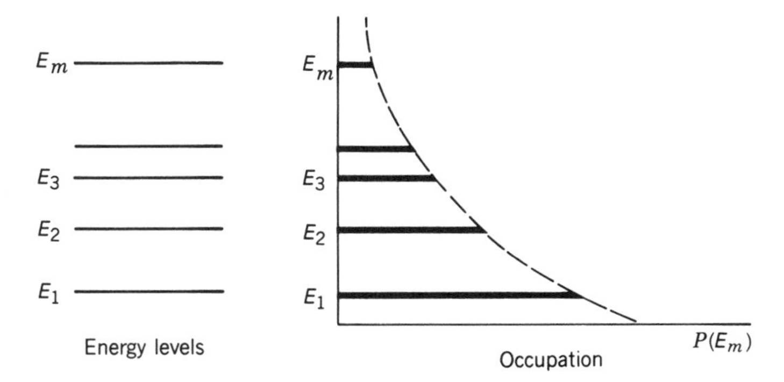

Figure 12.1-10 The Boltzmann distribution gives the probability that energy level E_m of an arbitrary atom is occupied; it is an exponentially decreasing function of E_m.

Thus, for a large number N of atoms, if N_m is the number of atoms occupying energy level E_m, the fraction $N_m/N \approx P(E_m)$. If N_1 atoms occupy level 1 and N_2 atoms occupy a higher level 2, the population ratio is, on the average,

$$\frac{N_2}{N_1} = \exp\left(-\frac{E_2 - E_1}{k_B T}\right). \tag{12.1-8}$$

This is the same probability distribution that governs the occupation of energy levels of an electromagnetic mode by photons in thermal equilibrium, as discussed in Sec. 11.2C (see Fig. 11.2-6). In this case, however, the electronic energy levels E_m are not generally equally spaced.

The Boltzmann distribution depends on the temperature T. At $T = 0$ K, all atoms are in the lowest energy level (ground state). As the temperature increases the populations of the higher energy levels increase. Under equilibrium conditions, the population of a given energy level is always greater than that of a higher-lying level. This does not necessarily hold under nonequilibrium conditions, however. A higher energy level can have a greater population than a lower energy level. This condition, which is called a **population inversion**, provides the basis for laser action (see Chaps. 13 and 14).

It was assumed above that there is a unique way in which an atom can find itself in one of its energy levels. It is often the case, however, that several different quantum states can correspond to the same energy (e.g., different states of angular momentum). To account for these degeneracies, (12.1-8) should be written in the more general form

$$\frac{N_2}{N_1} = \frac{g_2}{g_1} \exp\left(-\frac{E_2 - E_1}{k_B T}\right). \tag{12.1-9}$$

The degeneracy parameters g_2 and g_1 represent the number of states corresponding to the energy levels E_2 and E_1, respectively.

Fermi–Dirac Distribution

Electrons in a semiconductor obey a different occupation law. Since the atoms are located in close proximity to each other, the material must be treated as a single system within which the electrons are shared. A very large number of energy levels exist, forming bands. Because of the **Pauli exclusion principle**, each state can be occupied by at most one electron. A state is therefore either occupied or empty, so that the number of electrons N_m in state m is either 0 or 1.

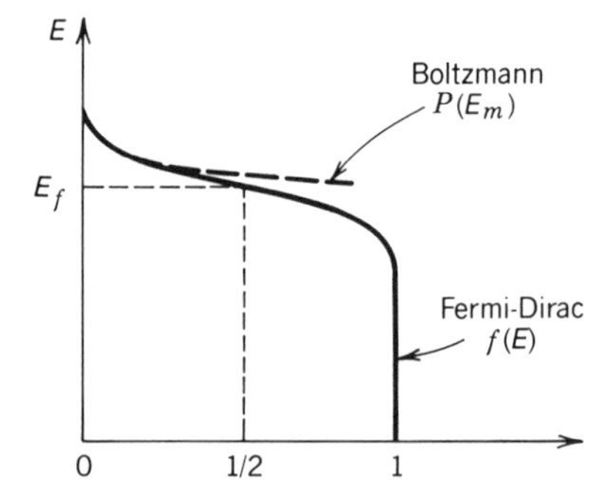

Figure 12.1-11 The Fermi–Dirac distribution $f(E)$ is well approximated by the Boltzmann distribution $P(E_m)$ when $E \gg E_f$.

The probability that energy level E is occupied is given by the **Fermi–Dirac distribution**

$$f(E) = \frac{1}{\exp\left[(E - E_f)/k_B T\right] + 1}, \tag{12.1-10}$$

where E_f is a constant known as the Fermi energy. This distribution has a maximum value of unity, which indicates that the energy level E is definitely occupied. $f(E)$ decreases monotonically as E increases, assuming the value $\frac{1}{2}$ at $E = E_f$. Although $f(E)$ is a distribution (sequence) of probabilities rather than a probability density function, when $E \gg E_f$ it behaves like the Boltzmann distribution

$$P(E) \propto \exp\left[-\frac{E - E_f}{k_B T}\right],$$

as is evident from (12.1-10). The Fermi–Dirac and Boltzmann distributions are compared in Fig. 12.1-11. The Fermi–Dirac distribution is discussed in further detail in Chap. 15.

12.2 INTERACTIONS OF PHOTONS WITH ATOMS

A. Interaction of Single-Mode Light with an Atom

As is known from atomic theory, an atom may emit (create) or absorb (annihilate) a photon by undergoing downward or upward transitions between its energy levels, conserving energy in the process. The laws that govern these processes are described in this section.

Interaction Between an Atom and an Electromagnetic Mode

Consider the energy levels E_1 and E_2 of an atom placed in an optical resonator of volume V that can sustain a number of electromagnetic modes. We are particularly interested in the interaction between the atom and the photons of a *prescribed* radiation mode of frequency $\nu \approx \nu_0$, where $h\nu_0 = E_2 - E_1$, since photons of this energy match the atomic energy-level difference. Such interactions are formally studied by the use of quantum electrodynamics. The key results are presented below, without proof. Three forms of interaction are possible—spontaneous emission, absorption, and stimulated emission.

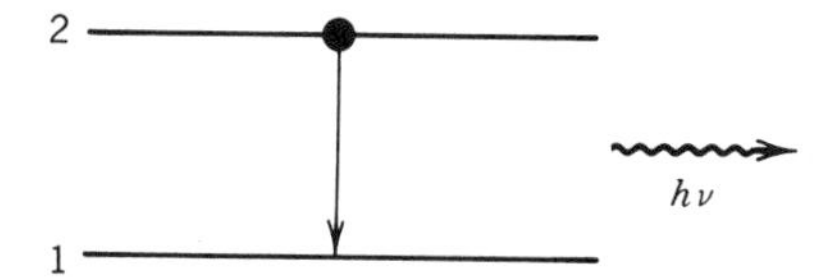

Figure 12.2-1 Spontaneous emission of a photon into the mode of frequency ν by an atomic transition from energy level 2 to energy level 1. The photon energy $h\nu \approx E_2 - E_1$.

Spontaneous Emission

If the atom is initially in the upper energy level, it may drop spontaneously to the lower energy level and release its energy in the form of a photon (Fig. 12.2-1). The photon energy $h\nu$ is added to the energy of the electromagnetic mode. The process is called **spontaneous emission** because the transition is independent of the number of photons that may already be in the mode.

In a cavity of volume V, the probability density (per second), or rate, of this spontaneous transition depends on ν in a way that characterizes the atomic transition.

$$p_{sp} = \frac{c}{V}\sigma(\nu). \qquad (12.2\text{-}1)$$

Probability Density of Spontaneous Emission into a Single Prescribed Mode

The function $\sigma(\nu)$ is a narrow function of ν centered about the atomic resonance frequency ν_0; it is known as the **transition cross section**. The significance of this name will become apparent subsequently, but it is clear that its dimensions are area (since p_{sp} has dimensions of second^{-1}). In principle, $\sigma(\nu)$ can be calculated from the Schrödinger equation; the calculations are usually so complex, however, that $\sigma(\nu)$ is usually determined experimentally rather than calculated. Equation (12.2-1) applies separately to every mode. Because they can have different directions or polarizations, more than one mode can have the same frequency ν.

The term "probability density" signifies that the probability of an emission taking place in an incremental time interval between t and $t + \Delta t$ is simply $p_{sp}\,\Delta t$. Because it is a probability density, p_{sp} can be greater than 1 (s^{-1}), although of course $p_{sp}\,\Delta t$ must always be smaller than 1. Thus, if there are a large number N of such atoms, a fraction of approximately $\Delta N = (p_{sp}\,\Delta t)N$ atoms will undergo the transition within the time interval Δt. We can therefore write $dN/dt = -p_{sp}N$, so that the number of atoms $N(t) = N(0)\exp(-p_{sp}t)$ decays exponentially with time constant $1/p_{sp}$, as illustrated in Fig. 12.2-2.

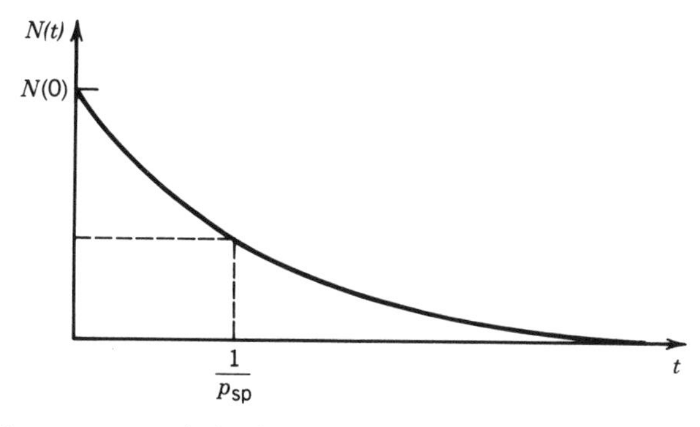

Figure 12.2-2 Spontaneous emission into a single mode causes the number of excited atoms to decrease exponentially with time constant $1/p_{sp}$.

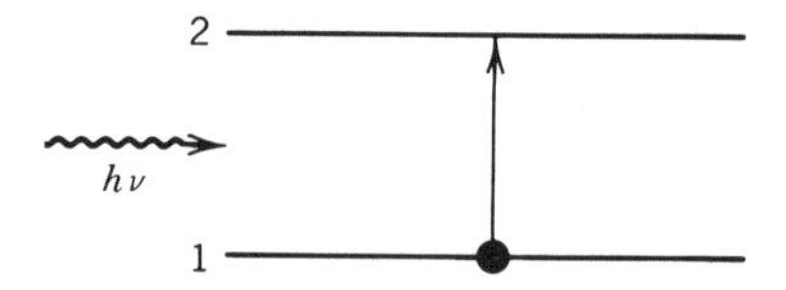

Figure 12.2-3 Absorption of a photon $h\nu$ leads to an upward transition of the atom from energy level 1 to energy level 2.

Absorption

If the atom is initially in the lower energy level and the radiation mode contains a photon, the photon may be absorbed, thereby raising the atom to the upper energy level (Fig. 12.2-3). The process is called **absorption**. Absorption is a transition **induced** by the photon. It can occur only when the mode contains a photon.

The probability density for the absorption of a photon from a given mode of frequency ν in a cavity of volume V is governed by the *same* law that governs spontaneous emission into that mode,

$$p_{\mathrm{ab}} = \frac{c}{V}\sigma(\nu). \tag{12.2-2}$$

However, if there are n photons in the mode, the probability density that the atom absorbs *one* photon is n times greater (since the events are mutually exclusive), i.e.,

$$\boxed{P_{\mathrm{ab}} = n\frac{c}{V}\sigma(\nu).} \tag{12.2-3}$$

Probability Density of Absorbing One Photon from a Mode Containing n Photons

Stimulated Emission

Finally, if the atom is in the upper energy level and the mode contains a photon, the atom may be stimulated to emit another photon into the same mode. The process is known as **stimulated emission**. It is the inverse of absorption. The presence of a photon in a mode of specified frequency, direction of propagation, and polarization stimulates the emission of a duplicate ("clone") photon with precisely the same characteristics as the original photon (Fig. 12.2-4). This photon amplification process is the phenomenon underlying the operation of laser amplifiers and lasers, as will be shown in later chapters. Again, the probability density p_{st} that this process occurs in a cavity of volume V is governed by the *same* transition cross section,

$$p_{\mathrm{st}} = \frac{c}{V}\sigma(\nu). \tag{12.2-4}$$

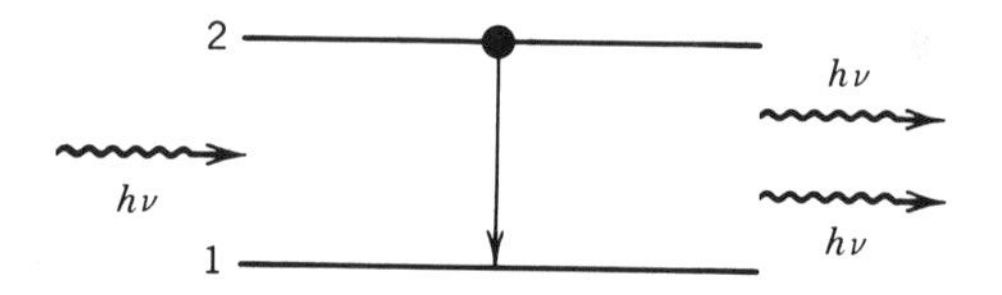

Figure 12.2-4 Stimulated emission is a process whereby a photon $h\nu$ stimulates the atom to emit a clone photon as it undergoes a downward transition.

As in the case of absorption, if the mode originally carries n photons, the probability density that the atom is stimulated to emit an additional photon is

$$P_{st} = n\frac{c}{V}\sigma(\nu). \tag{12.2-5}$$

Probability Density of Stimulated Emission of One Photon into a Mode in Which n Photons Are Present

After the emission, the radiation mode carries $n + 1$ photons. Since $P_{st} = P_{ab}$, we use the notation W_i for the probability density of both stimulated emission and absorption.

Since spontaneous emission occurs in addition to the stimulated emission, the total probability density of the atom emitting a photon into the mode is $p_{sp} + P_{st} = (n + 1)(c/V)\sigma(\nu)$. In fact, from a quantum electrodynamic point of view, spontaneous emission may be regarded as stimulated emission induced by the zero-point fluctuations of the mode. Because the zero-point energy is inaccessible for absorption, P_{ab} is proportional to n rather than to $(n + 1)$.

The three possible interactions between an atom and a cavity radiation mode (spontaneous emission, absorption, and stimulated emission) obey the fundamental relations provided above. These should be regarded as the laws governing photon–atom interactions, supplementing the rules of photon optics provided in Chap. 11. We now proceed to discuss the character and consequences of these rather simple relations in some detail.

The Lineshape Function

The transition cross section $\sigma(\nu)$ specifies the character of the interaction of the atom with the radiation. Its area,

$$S = \int_0^\infty \sigma(\nu)\, d\nu,$$

which has units of cm^2-Hz, is called the **transition strength** or **oscillator strength**, and represents the strength of the interaction. Its shape governs the relative magnitude of the interaction with photons of different frequencies. The shape (profile) of $\sigma(\nu)$ is readily separated from its overall strength by defining a normalized function with units of Hz^{-1} and unity area, $g(\nu) = \sigma(\nu)/S$, known as the **lineshape function**, so that $\int_0^\infty g(\nu)\, d\nu = 1$. The transition cross section can therefore be written in terms of its strength and its profile as

$$\sigma(\nu) = Sg(\nu). \tag{12.2-6}$$

The lineshape function $g(\nu)$ is centered about the frequency where $\sigma(\nu)$ is largest (viz., the transition resonance frequency ν_0) and drops sharply for ν different from ν_0. Transitions are therefore most likely for photons of frequency $\nu \approx \nu_0$. The width of the function $g(\nu)$ is known as the transition **linewidth**. The linewidth $\Delta\nu$ is defined as the full width of the function $g(\nu)$ at half its maximum value (FWHM). In general, the width of $g(\nu)$ is inversely proportional to its central value (since its area is unity),

$$\Delta\nu \propto \frac{1}{g(\nu_0)}. \tag{12.2-7}$$

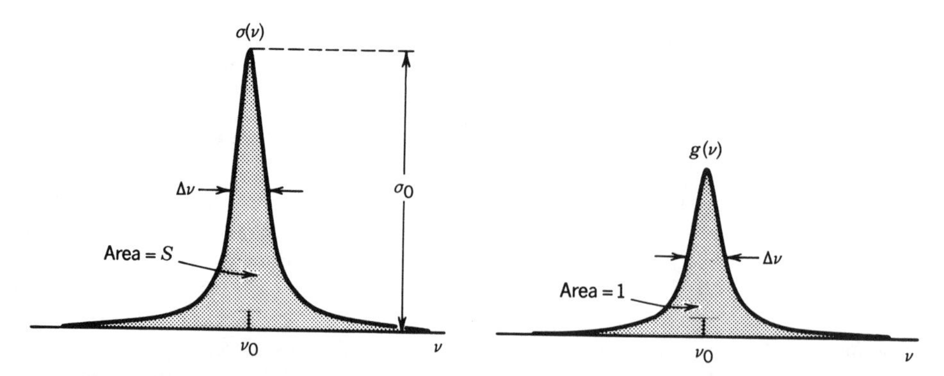

Figure 12.2-5 The transition cross section $\sigma(\nu)$ and the lineshape function $g(\nu)$.

It is also useful to define the peak transition cross section, which occurs at the resonance frequency, $\sigma_0 = \sigma(\nu_0)$. The function $\sigma(\nu)$ is therefore characterized by its height σ_0, width $\Delta\nu$, area S, and profile $g(\nu)$, as Fig. 12.2-5 illustrates.

B. Spontaneous Emission

Total Spontaneous Emission into All Modes

Equation (12.2-1) provides the probability density p_{sp} for spontaneous emission into a *specific* mode of frequency ν (regardless of whether the mode contains photons). As shown in Sec. 9.1C, the density of modes for a three-dimensional cavity is $M(\nu) = 8\pi\nu^2/c^3$. This quantity approximates the number of modes (per unit volume of the cavity per unit bandwidth) that have the frequency ν; it increases in quadratic fashion. An atom may spontaneously emit *one* photon of frequency ν into *any* of these modes, as shown schematically in Fig. 12.2-6.

The probability density of spontaneous emission into a single prescribed mode must therefore be weighted by the modal density. The overall spontaneous emission probability density is thus

$$P_{sp} = \int_0^\infty \left[\frac{c}{V}\sigma(\nu)\right][VM(\nu)]\, d\nu = c\int_0^\infty \sigma(\nu)M(\nu)\, d\nu.$$

For simplicity, this expression assumes that spontaneous emission into modes of the same frequency ν, but with different directions or polarizations, is equally likely.

Because the function $\sigma(\nu)$ is sharply peaked, it is narrow in comparison with the function $M(\nu)$. Since $\sigma(\nu)$ is centered about ν_0, $M(\nu)$ is essentially constant at $M(\nu_0)$,

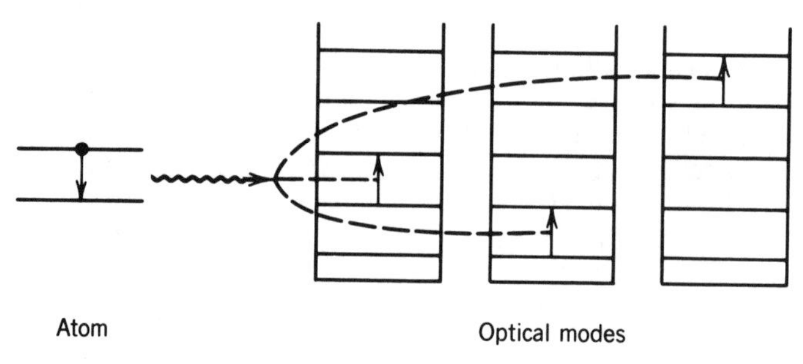

Figure 12.2-6 An atom may spontaneously emit a photon into any one (but only one) of the many modes with frequencies $\nu \approx \nu_0$.

so that it can be removed from the integral. The probability density of spontaneous emission of one photon into any mode therefore becomes

$$P_{sp} = M(\nu_0)cS = \frac{8\pi S}{\lambda^2}, \qquad (12.2\text{-}8)$$

where $\lambda = c/\nu_0$ is the wavelength in the medium. We define a time constant t_{sp}, known as the spontaneous lifetime of the $2 \to 1$ transition, such that $1/t_{sp} \equiv P_{sp} = M(\nu_0)cS$. Thus

$$P_{sp} = \frac{1}{t_{sp}}, \qquad (12.2\text{-}9)$$

Probability Density of Spontaneous Emission of One Photon into Any Mode

which, it is important to note, is independent of the cavity volume V. We can therefore express S as

$$S = \frac{\lambda^2}{8\pi t_{sp}}; \qquad (12.2\text{-}10)$$

consequently, the transition strength is determined from an experimental measurement of the spontaneous lifetime t_{sp}. This is useful because an analytical calculation of S would require knowledge about the quantum-mechanical behavior of the system and is usually too difficult to carry out.

Typical values of t_{sp} are $\approx 10^{-8}$ s for atomic transitions (e.g., the first excited state of atomic hydrogen); however, t_{sp} can vary over a large range (from subpicoseconds to minutes).

EXERCISE 12.2-1

Frequency of Spontaneously Emitted Photons. Show that the probability density of an excited atom spontaneously emitting a photon of frequency between ν and $\nu + d\nu$ is $P_{sp}(\nu)\,d\nu = (1/t_{sp})g(\nu)\,d\nu$. Explain why the spectrum of spontaneous emission from an atom is proportional to its lineshape function $g(\nu)$ after a large number of photons have been emitted.

Relation Between the Transition Cross Section and the Spontaneous Lifetime

The substitution of (12.2-10) into (12.2-6) shows that the transition cross section is related to the spontaneous lifetime and the lineshape function by

$$\sigma(\nu) = \frac{\lambda^2}{8\pi t_{sp}} g(\nu). \qquad (12.2\text{-}11)$$

Transition Cross Section

Furthermore, the transition cross section at the central frequency ν_0 is

$$\sigma_0 \equiv \sigma(\nu_0) = \frac{\lambda^2}{8\pi t_{sp}} g(\nu_0). \qquad (12.2\text{-}12)$$

Because $g(\nu_0)$ is inversely proportional to $\Delta\nu$, according to (12.2-7), the peak transition cross section σ_0 is inversely proportional to the linewidth $\Delta\nu$ for a given t_{sp}.

C. Stimulated Emission and Absorption

Transitions Induced by Monochromatic Light

We now consider the interaction of single-mode light with an atom when a stream of photons impinges on it, rather than when it is in a resonator of volume V as considered above. Let monochromatic light of frequency ν, intensity I, and mean photon-flux density (photons/cm^2-s)

$$\phi = \frac{I}{h\nu} \tag{12.2-13}$$

interact with an atom having a resonance frequency ν_0. We wish to determine the probability densities for stimulated emission and absorption $W_i = P_{ab} = P_{st}$ in this configuration.

The number of photons n involved in the interaction process is determined by constructing a volume in the form of a cylinder of area A and height c whose axis is parallel to the direction of propagation of the light (its **k** vector). The cylinder has a volume $V = cA$. The photon flux across the cylinder base is ϕA (photons per second). Because photons travel at the speed of light c, within one second all of the photons within the cylinder cross the cylinder base. It follows that at any time the cylinder contains $n = \phi A$, or

$$n = \phi \frac{V}{c}, \tag{12.2-14}$$

photons so that $\phi = (c/V)n$. To determine W_i, we substitute (12.2-14) into (12.2-3) to obtain

$$\boxed{W_i = \phi\sigma(\nu).} \tag{12.2-15}$$

It is apparent that $\sigma(\nu)$ is the coefficient of proportionality between the probability density of an induced transition and the photon-flux density. Hence the name "transition cross section": ϕ is the photon flux per cm^2, $\sigma(\nu)$ is the effective cross-sectional area of the atom (cm^2), and $\phi\sigma(\nu)$ is the photon flux "captured" by the atom for the purpose of absorption or stimulated emission.

Whereas the spontaneous emission rate is enhanced by the many modes into which an atom can decay, stimulated emission involves decay only into modes that contain photons. Its rate is enhanced by the possible presence of a large number of photons in few modes.

Transitions in the Presence of Broadband Light

Consider now an atom in a cavity of volume V containing multimode polychromatic light of spectral energy density $\varrho(\nu)$ (energy per unit bandwidth per unit volume) that is broadband in comparison with the atomic linewidth. The average number of photons in the ν to $\nu + d\nu$ band is $\varrho(\nu)V\,d\nu/h\nu$, each with a probability density $(c/V)\sigma(\nu)$ of initiating an atomic transition, so that the overall probability of absorption or stimulated emission is

$$W_i = \int_0^\infty \frac{\varrho(\nu)V}{h\nu}\left[\frac{c}{V}\sigma(\nu)\right] d\nu. \tag{12.2-16}$$

Since the radiation is broadband, the function $\varrho(\nu)$ varies slowly in comparison with the sharply peaked function $\sigma(\nu)$. We can therefore replace $\varrho(\nu)/\nu$ under the integral with $\varrho(\nu_0)/\nu_0$ to obtain

$$W_i = \frac{\varrho(\nu_0)}{h\nu_0} c \int_0^\infty \sigma(\nu)\, d\nu = \frac{\varrho(\nu_0)}{h\nu_0} cS.$$

Using (12.2-10), we have

$$W_i = \frac{\lambda^3}{8\pi h t_{sp}} \varrho(\nu_0), \tag{12.2-17}$$

where $\lambda = c/\nu_0$ is the wavelength (in the medium) at the central frequency ν_0.

The approach followed here is similar to that used for calculating the probability density of spontaneous emission into multiple modes, which gives rise to $P_{sp} = M(\nu_0)cS$. Defining

$$\bar{n} = \frac{\lambda^3}{8\pi h} \varrho(\nu_0),$$

which represents the mean number of photons per mode, we write (12.2-17) in the convenient form

$$W_i = \frac{\bar{n}}{t_{sp}}. \tag{12.2-18}$$

The interpretation of $\bar{n}$ follows from the ratio $W_i/P_{sp} = \varrho(\nu_0)/h\nu_0 M(\nu_0)$. The probability density W_i is a factor of $\bar{n}$ greater than that for spontaneous emission since each of the modes contains an average of $\bar{n}$ photons.

Einstein's $\mathbb{A}$ and $\mathbb{B}$ Coefficients

Einstein did not have knowledge of (12.2-17). However, based on an analysis of the exchange of energy between atoms and radiation under conditions of thermal equilibrium, he was able to postulate certain expressions for the probability densities of the different kinds of transitions an atom may undergo when it interacts with broadband radiation of spectral energy density $\varrho(\nu)$. The expressions he obtained were as follows:

$$P_{sp} = \mathbb{A} \tag{12.2-19}$$

$$W_i = \mathbb{B}\varrho(\nu_0). \tag{12.2-20}$$

Einstein's Postulates

The constants $\mathbb{A}$ and $\mathbb{B}$ are known as **Einstein's A and B coefficients**. By a simple comparison with our expressions (12.2-9) and (12.2-17), the $\mathbb{A}$ and $\mathbb{B}$ coefficients are identified as

$$\mathbb{A} = \frac{1}{t_{sp}} \tag{12.2-21}$$

$$\mathbb{B} = \frac{\lambda^3}{8\pi h t_{sp}}, \tag{12.2-22}$$

so that

$$\frac{\mathbb{B}}{\mathbb{A}} = \frac{\lambda^3}{8\pi h}. \qquad (12.2\text{-}23)$$

It is important to note that the relation between the $\mathbb{A}$ and $\mathbb{B}$ coefficients is a result of the microscopic (rather than macroscopic) probability laws of interaction between an atom and the photons of each mode. We shall present an analysis similar to that of Einstein in Sec. 12.3.

EXAMPLE 12.2-1. ***Comparison Between Rates of Spontaneous and Stimulated Emission.*** Whereas the rate of spontaneous emission for an atom in the upper state is constant (at $\mathbb{A} = 1/t_{sp}$), the rate of stimulated emission in the presence of broadband light $\mathbb{B}\varrho(\nu_0)$ is proportional to the spectral energy density of the light $\varrho(\nu_0)$. The two rates are equal when $\varrho(\nu_0) = \mathbb{A}/\mathbb{B} = 8\pi h/\lambda^3$; for greater spectral energy densities, the rate of stimulated emission exceeds that of spontaneous emission. If $\lambda = 1\ \mu\text{m}$, for example, $\mathbb{A}/\mathbb{B} = 1.66 \times 10^{-14}$ J/m^3-Hz. This corresponds to an optical intensity spectral density $c\varrho(\nu_0) \approx 5 \times 10^{-6}$ W/m^2-Hz in free space. Thus for a linewidth $\Delta\nu = 10^7$ Hz, the optical intensity at which the stimulated emission rate equals the spontaneous emission rate is 50 W/m^2 or 5 mW/cm^2.

Summary

An atomic transition is characterized by its resonance frequency ν_0, its spontaneous lifetime t_{sp}, and its lineshape function $g(\nu)$, whose width is the linewidth $\Delta\nu$. The transition cross section is

$$\sigma(\nu) = \frac{\lambda^2}{8\pi t_{sp}} g(\nu). \qquad (12.2\text{-}11)$$

Spontaneous Emission

- If the atom is in a cavity of volume V in the upper level, the probability density (per second) of emitting spontaneously into one *prescribed* mode of frequency ν is

$$p_{sp} = \frac{c}{V}\sigma(\nu). \qquad (12.2\text{-}1)$$

- The probability density of spontaneous emission into *any* of the available modes is

$$P_{sp} = \frac{1}{t_{sp}}. \tag{12.2-9}$$

- The probability density of emitting into modes lying only in the frequency band between ν and $\nu + d\nu$ is $P_{sp}(\nu)\, d\nu = (1/t_{sp}) g(\nu)\, d\nu$. The spectrum of spontaneously emitted light is therefore proportional to the lineshape function $g(\nu)$.

Stimulated Emission and Absorption

- If the atom in the cavity is in the upper level and a radiation mode contains n photons, the probability density of emitting a photon into that mode is

$$W_i = n \frac{c}{V} \sigma(\nu). \tag{12.2-5}$$

  If the atom is instead in the lower level, and a mode contains n photons, the probability of absorption of a photon from that mode is also given by (12.2-5).
- If instead of being in a cavity the atom is illuminated by a monochromatic beam of light of frequency ν, with mean photon-flux density ϕ (photons per second per unit area), the probability density of stimulated emission (if the atom is in the upper level) or absorption (if the atom is in the lower level) is

$$W_i = \phi \sigma(\nu). \tag{12.2-15}$$

- If the light illuminating the atom is polychromatic but narrowband in comparison with the atomic linewidth, and has a mean photon-flux spectral density ϕ_ν (photons per second per unit area per unit frequency), the probability density of stimulated emission/absorption is

$$W_i = \int \phi_\nu \sigma(\nu)\, d\nu. \tag{12.2-24}$$

- If the light illuminating the atom has a spectral energy density $\varrho(\nu)$ that is broadband in comparison with the atomic linewidth, the probability density of stimulated emission/absorption is

$$W_i = \mathbb{B} \varrho(\nu_0), \tag{12.2-20}$$

  where $\mathbb{B} = (\lambda^3 / 8\pi h t_{sp})$ is the Einstein $\mathbb{B}$ coefficient.

In all of these formulas, $c = c_o/n$ is the velocity of light, $\lambda = \lambda_o/n$ is the wavelength of light in the atomic medium, and n is the refractive index.

D. Line Broadening

Because the lineshape function $g(\nu)$ plays an important role in atom–photon interactions, we devote this subsection to a brief discussion of its origins. The same lineshape function is applicable for spontaneous emission, absorption, and stimulated emission.

Lifetime Broadening

Atoms can undergo transitions between energy levels by both radiative and nonradiative means. Radiative transitions result in photon absorption and emission. Nonradiative transitions permit energy transfer by mechanisms such as lattice vibrations, inelastic collisions among the constituent atoms, and inelastic collisions with the walls of the vessel. Each atomic energy level has a lifetime τ, which is the inverse of the rate at which its population decays, radiatively or nonradiatively, to all lower levels.

The lifetime τ_2 of energy level 2 shown in Fig. 12.2-1 represents the inverse of the rate at which the population of that level decays to level 1 and to all other lower energy levels (none of which are shown in the figure), by either radiative or nonradiative means. Since $1/t_{sp}$ is the radiative decay rate from level 2 to level 1, the overall decay rate $1/\tau_2$ must be more rapid, i.e., $1/\tau_2 \geq 1/t_{sp}$, so that $\tau_2 \leq t_{sp}$. The lifetime τ_1 of level 1 is defined similarly. Clearly, if level 1 is the lowest allowed energy level (the ground state), $\tau_1 = \infty$.

Lifetime broadening is, in essence, a Fourier transform effect. The lifetime τ of an energy level is related to the time uncertainty of the occupation of that level. As shown in Appendix A, the Fourier transform of an exponentially decaying harmonic function of time $e^{-t/2\tau} e^{j2\pi\nu_0 t}$, which has an energy that decays as $e^{-t/\tau}$ (with time constant τ), is proportional to $1/[1 + j4\pi(\nu - \nu_0)\tau]$. The full width at half-maximum (FWHM) of the square magnitude of this Lorentzian function of frequency is $\Delta\nu = 1/2\pi\tau$. This spectral uncertainty corresponds to an energy uncertainty $\Delta E = h\,\Delta\nu = h/2\pi\tau$. An energy level with lifetime τ therefore has an energy spread $\Delta E = h/2\pi\tau$, provided that we can model the decay process as a simple exponential. In this picture, spontaneous emission can be viewed in terms of a damped harmonic oscillator which generates an exponentially decaying harmonic function.

Thus, if the energy spreads of levels 1 and 2 are $\Delta E_1 = h/2\pi\tau_1$ and $\Delta E_2 = h/2\pi\tau_2$, respectively, the spread in the energy difference, which corresponds to the transition between the two levels, is

$$\Delta E = \Delta E_1 + \Delta E_2 = \frac{h}{2\pi}\left(\frac{1}{\tau_1} + \frac{1}{\tau_2}\right) = \frac{h}{2\pi}\frac{1}{\tau}, \tag{12.2-25}$$

where $\tau^{-1} = (\tau_1^{-1} + \tau_2^{-1})$ and τ is the transition lifetime. The corresponding spread of the transition frequency, which is called the lifetime-broadening linewidth, is therefore

$$\Delta\nu = \frac{1}{2\pi}\left(\frac{1}{\tau_1} + \frac{1}{\tau_2}\right). \tag{12.2-26}$$

Lifetime-Broadening Linewidth

This spread is centered about the frequency $\nu_0 = (E_2 - E_1)/h$, and the lineshape function has a Lorentzian profile,

$$g(\nu) = \frac{\Delta\nu/2\pi}{(\nu - \nu_0)^2 + (\Delta\nu/2)^2}. \tag{12.2-27}$$

Lorentzian Lineshape Function

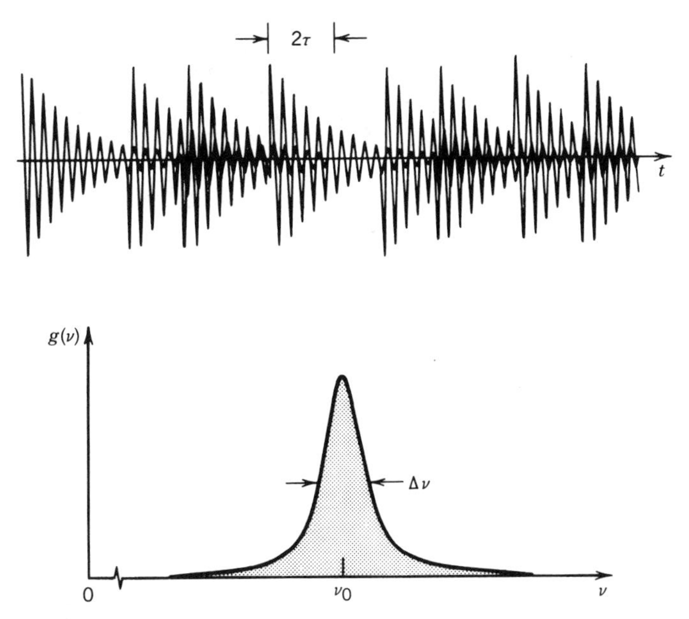

Figure 12.2-7 Wavepacket emissions at random times from a lifetime broadened atomic system with transition lifetime τ. The light emitted has a Lorentzian power spectral density of width $\Delta\nu = 1/2\pi\tau$.

The lifetime broadening from an atom or a collection of atoms may be more generally modeled as follows. Each of the photons emitted from the transition represents a wavepacket of central frequency ν_0 (the transition resonance frequency), with an exponentially decaying envelope of decay time 2τ (i.e., with energy decay time equal to the transition lifetime τ), as shown in Fig. 12.2-7. The radiated light is taken to be a sequence of such wavepackets emitted at random times. As discussed in Example 10.1-1, this corresponds to random (partially coherent) light whose power spectral density is precisely the Lorentzian function given in (12.2-27), with $\Delta\nu = 1/2\pi\tau$.

The value of the Lorentzian lineshape function at the central frequency ν_0 is $g(\nu_0) = 2/\pi\Delta\nu$, so that the peak transition cross section, given by (12.2-12), becomes

$$\sigma_0 = \frac{\lambda^2}{2\pi}\frac{1}{2\pi t_{sp}\Delta\nu}. \tag{12.2-28}$$

The largest transition cross section occurs under ideal conditions when the decay is entirely radiative so that $\tau_2 = t_{sp}$ and $1/\tau_1 = 0$ (which is the case when level 1 is the ground state from which no decay is possible). Then $\Delta\nu = 1/2\pi t_{sp}$ and

$$\sigma_0 = \frac{\lambda^2}{2\pi}, \tag{12.2-29}$$

indicating that the peak cross-sectional area is of the order of one square wavelength. When level 1 is not the ground state or when nonradiative transitions are significant,

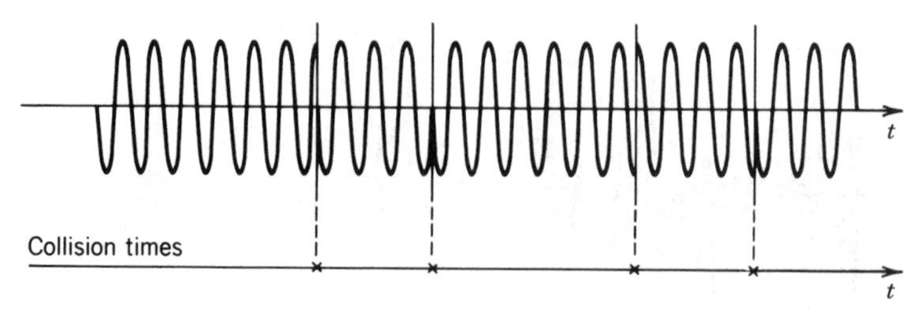

Figure 12.2-8 A sinewave interrupted at the rate f_{col} by random phase jumps has a Lorentzian spectrum of width $\Delta\nu = f_{col}/\pi$.

$\Delta\nu$ can be $\gg 1/t_{sp}$ in which case σ_0 can be significantly smaller than $\lambda^2/2\pi$. For example, for optical transitions in the range $\lambda = 0.1$ to $10\ \mu$m, $\lambda^2/2\pi \approx 10^{-11}$ to 10^{-7} cm^2, whereas typical values of σ_0 for optical transitions fall in the range 10^{-20} to 10^{-11} cm^2 (see, e.g., Table 13.2-1 on page 480).

Collision Broadening

Inelastic collisions, in which energy is exchanged, result in atomic transitions between energy levels. This contribution to the decay rates affects the lifetimes of all levels involved and hence the linewidth of the radiated field, as indicated above.

Elastic collisions, on the other hand, do not involve energy exchange. Rather, they cause random phase shifts of the wavefunction associated with the energy level, which in turn results in a random phase shift of the radiated field at each collision time. Collisions between atoms provide a source of such line broadening. A sinewave whose phase is modified by a random shift at random times (collision times), as illustrated in Fig. 12.2-8, exhibits spectral broadening. The determination of the spectrum of such a randomly dephased function is a problem that can be solved using the theory of random processes. The spectrum turns out to be Lorentzian, with width $\Delta\nu = f_{col}/\pi$, where f_{col} is the collision rate (mean number of collisions per second).†

Adding the linewidths arising from lifetime and collision broadening therefore results in an overall Lorentzian lineshape of linewidth

$$\Delta\nu = \frac{1}{2\pi}\left(\frac{1}{\tau_1} + \frac{1}{\tau_2} + 2f_{col}\right). \qquad (12.2\text{-}30)$$

Inhomogeneous Broadening

Lifetime broadening and collision broadening are forms of **homogeneous broadening** that are exhibited by the atoms of a medium. All of the atoms are assumed to be identical and to have identical lineshape functions. In many situations, however, the different atoms constituting a medium have different lineshape functions or different center frequencies. In this case we can define an average lineshape function

$$\bar{g}(\nu) = \langle g_\beta(\nu) \rangle, \qquad (12.2\text{-}31)$$

where $\langle \cdot \rangle$ represents an average with respect to the variable β, which is used to label

†See, e.g., A. E. Siegman, *Lasers*, University Science Books, Mill Valley, CA, 1986, Sec. 3.2.

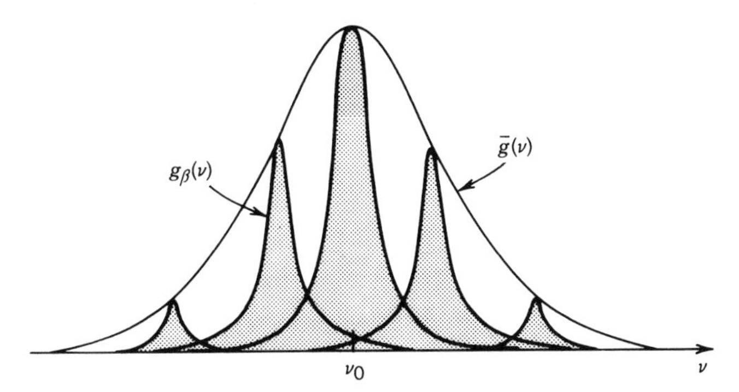

Figure 12.2-9 The average lineshape function of an inhomogeneously broadened collection of atoms.

those atoms with lineshape function $g_\beta(\nu)$. Thus $g_\beta(\nu)$ is weighted with the fraction of the atomic population having the property β, as shown in Fig. 12.2-9.

One inhomogeneous broadening mechanism is Doppler broadening. As a result of the **Doppler effect**, an atom moving with velocity v along a given direction exhibits a spectrum that is shifted by the frequency $\pm(v/c)\nu_0$, where ν_0 is its central frequency, when viewed along that direction. The shift is in the direction of higher frequency (+ sign) if the atom is moving toward the observer, and in the direction of lower frequency (− sign) if it is moving away. For an arbitrary direction of observation, the frequency shift is $\pm(v_\parallel/c)\nu_0$, where $v_\parallel$ is the component of velocity parallel to the direction of observation. Since a collection of atoms in a gas exhibits a distribution of velocities, the light they emit exhibits a range of frequencies, resulting in **Doppler broadening**, as illustrated in Fig. 12.2-10.

In the case of Doppler broadening, the velocity v therefore plays the role of the parameter β; $\bar{g}(\nu) = \langle g_v(\nu) \rangle$. Thus if $p(v)\,dv$ is the probability that the velocity of a given atom lies between v and $v + dv$, the overall inhomogeneous Doppler-broadened lineshape is (see Fig. 12.2-11)

$$\bar{g}(\nu) = \int_{-\infty}^{\infty} g\left(\nu - \nu_0 \frac{v}{c}\right) p(v)\, dv. \tag{12.2-32}$$

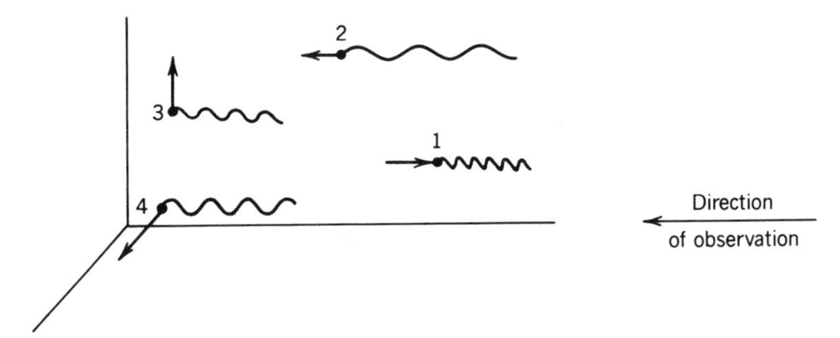

Figure 12.2-10 The radiated frequency is dependent on the direction of atomic motion relative to the direction of observation. Radiation from atom 1 has higher frequency than that from atoms 3 and 4. Radiation from atom 2 has lower frequency.

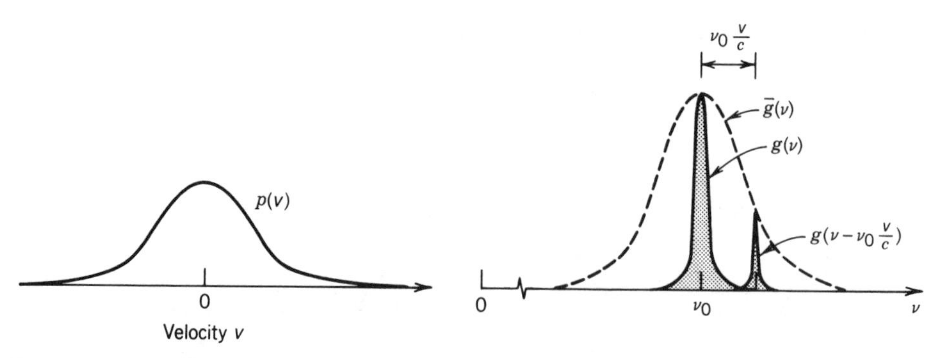

Figure 12.2-11 The velocity distribution and average lineshape function of a Doppler-broadened atomic system.

EXERCISE 12.2-2

Doppler-Broadened Lineshape Function

(a) The component of velocity v of atoms of a gas along a particular direction is known to have a Gaussian probability density function

$$p(\mathsf{v}) = \frac{1}{\sqrt{2\pi}\,\sigma_{\mathsf{v}}} \exp\left(-\frac{\mathsf{v}^2}{2\sigma_{\mathsf{v}}^2}\right),$$

where $\sigma_{\mathsf{v}}^2 = k_B T/M$ and M is the atomic mass. If each atom has a Lorentzian natural lineshape function of width $\Delta\nu$ and central frequency ν_0, derive an expression for the average lineshape function $\bar{g}(\nu)$.

(b) Show that if $\Delta\nu \ll \nu_0\sigma_{\mathsf{v}}/c$, $\bar{g}(\nu)$ may be approximated by the Gaussian lineshape function

$$\bar{g}(\nu) = \frac{1}{\sqrt{2\pi}\,\sigma_D} \exp\left[-\frac{(\nu-\nu_0)^2}{2\sigma_D^2}\right], \tag{12.2-33}$$

where

$$\sigma_D = \nu_0 \frac{\sigma_{\mathsf{v}}}{c} = \frac{1}{\lambda}\left(\frac{k_B T}{M}\right)^{1/2}. \tag{12.2-34}$$

The full-width half-maximum (FWHM) Doppler linewidth $\Delta\nu_D$ is then

$$\Delta\nu_D = (8\ln 2)^{1/2}\sigma_D \approx 2.35\sigma_D. \tag{12.2-35}$$

(c) Compute the Doppler linewidth for the $\lambda_o = 632.8$ nm transition in Ne, and for the $\lambda_o = 10.6\ \mu$m transition in CO_2 at room temperature, assuming that $\Delta\nu \ll \nu_0\sigma_{\mathsf{v}}/c$. These transitions are used in the He–Ne and CO_2 lasers, respectively.

(d) Show that the maximum value of the transition cross section for the Gaussian lineshape in (12.2-33) is

$$\sigma_0 = \frac{\lambda^2}{8\pi}\left(\frac{4\ln 2}{\pi}\right)^{1/2}\frac{1}{t_{sp}\,\Delta\nu_D}$$

$$\approx 0.94\frac{\lambda^2}{8\pi}\frac{1}{t_{sp}\,\Delta\nu_D}. \qquad (12.2\text{-}36)$$

Compare with (12.2-28) for the Lorentzian lineshape function.

Many atom–photon interactions exhibit broadening that is intermediate between purely homogeneous and purely inhomogeneous. Such mixed broadening can be modeled by an intermediate lineshape function known as the Voight profile.

*E. Laser Cooling and Trapping of Atoms

The broadening associated with the Doppler effect often masks the natural lineshape function; the magnitude of the latter is often of interest. One way to minimize Doppler broadening is to use a carefully controlled atomic beam in which the velocities of the atoms are well regulated. However, the motion of atoms can also be controlled by means of radiation pressure (see Sec. 11.1C).

Photons from a laser beam of narrow linewidth, tuned above the atomic line center, can be absorbed by a beam of atoms moving toward the laser beam. After absorption, the atom can return to the ground state by either stimulated or spontaneous emission. If it returns by stimulated emission, the momentum of the emitted photon is the same as that of the absorbed photon, leaving the atom with no net change of momentum. If it returns by spontaneous emission, on the other hand, the direction of photon emission is random so that repeated absorptions result in a net decrease of the atomic momentum in the direction pointing toward the laser beam. The result is a decrease in the velocity of those atoms, as shown schematically in Fig. 12.2-12. Ultimately, the

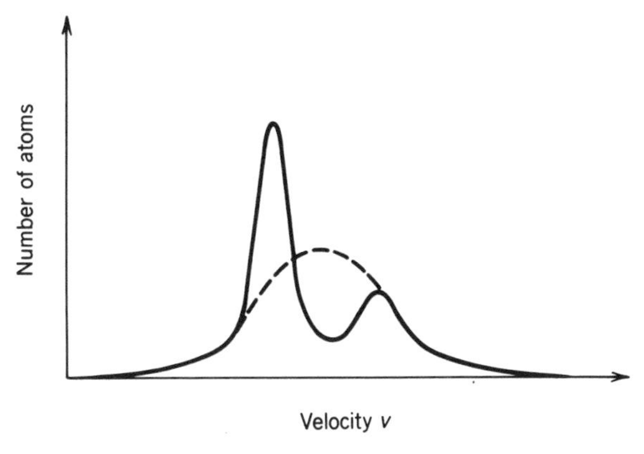

Figure 12.2-12 The thermal velocity distribution (dashed curve) and the laser-cooled distribution (solid curve).

change of atomic momentum (and therefore velocity) results in the atoms moving out of resonance with the laser beam, so that they no longer absorb light.

Once the atoms have been cooled in this manner, photon beams can be used to construct an optical trap in which large numbers of atoms can be confined to a limited region of space for long periods of time (many seconds). Although this is relatively easy to achieve for ionized atoms because of their electric charge, it can also be achieved when the atoms are electrically neutral. The trapped atoms can then be rapidly moved about simply by redirecting the laser beam. For trapping to occur, however, the collection of atoms must be very cold (their kinetic energy must be sufficiently low so that they cannot jump out of the trap). A set of mutually orthogonal laser beams can be directed at the atoms in such a way that they experience a viscous retarding force in any direction in which they move.

By the use of such cooling and trapping processes, temperatures as low as 1 μK have been achieved with neutral atoms. Furthermore, it has been found that crystal-like structures can be formed when even a few ions are confined to a trap. Phase transitions between an ordered "crystalline" state and a disordered cloud can be induced by changes in the degree of laser cooling.

12.3 THERMAL LIGHT

Light emitted from atoms, molecules, and solids, under conditions of thermal equilibrium and in the absence of other external energy sources, is known as thermal light. In this section we determine the properties of thermal light by examining the interaction between photons and atoms in equilibrium.

A. Thermal Equilibrium Between Photons and Atoms

We make use of (12.2-9) and (12.2-18), which govern the interaction between photons and an atom, to develop the macroscopic laws of interaction of many photons with many atoms in thermal equilibrium. Consider a cavity of unit volume whose walls have a large number of atoms that have two energy levels, denoted 1 and 2, that are separated by an energy difference $h\nu$. The cavity supports broadband radiation. Let $N_2(t)$ and $N_1(t)$ represent the number of atoms per unit volume occupying energy levels 2 and 1, at time t, respectively. Spontaneous emission creates radiation in the cavity, assuming that some atoms are initially in level 2 (this is ensured by the external finite temperature). This radiation induces absorption or stimulated emission. The three processes coexist and a steady state (equilibrium) is reached. We assume that an average of $\bar{n}$ photons occupies *each* of the radiation modes whose frequencies lie within the atomic linewidth, as shown in (12.2-18).

We first consider spontaneous emission. The probability that a single atom in the upper level undergoes spontaneous emission into any of the modes within the time increment from t to $t + \Delta t$ is $P_{sp}\,\Delta t = \Delta t / t_{sp}$. There are $N_2(t)$ such atoms. The average number of emitted photons within Δt is therefore $N_2(t)\,\Delta t / t_{sp}$. This is also the number of atoms that depart level 2 during the time interval Δt. Therefore, the rate of increase of $N_2(t)$ due to spontaneous emission is negative and is given by the differential equation

$$\frac{dN_2}{dt} = -\frac{N_2}{t_{sp}}, \tag{12.3-1}$$

whose solution $N_2(t) = N_2(0)\exp(-t/t_{sp})$ is an exponentially decaying function of time, as shown in Fig. 12.3-1. Given a sufficient time, the number of atoms in the upper

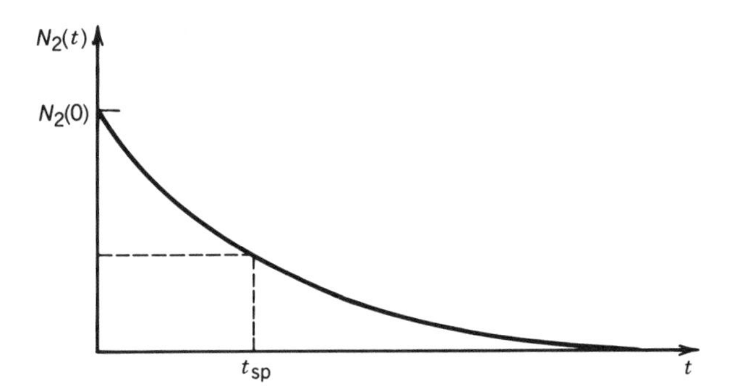

Figure 12.3-1 Decay of the upper-level population caused by spontaneous emission alone.

level N_2 decays to zero with time constant t_{sp}. The energy is carried off by the spontaneously emitted photons.

Spontaneous emission is not the only form of interaction, however. In the presence of radiation, absorption and stimulated emission also contribute to changes in the populations $N_1(t)$ and $N_2(t)$. Let us consider absorption first. Since there are N_1 atoms capable of absorbing, the rate of increase of the population of atoms in the upper energy level due to absorption is, using (12.2-18),

$$\frac{dN_2}{dt} = N_1 W_i = \frac{N_1 \bar{n}}{t_{sp}}. \tag{12.3-2}$$

Similarly, stimulated emission gives rise to a rate of increase of atoms in the upper state (which is negative), given by

$$\frac{dN_2}{dt} = -\frac{N_2 \bar{n}}{t_{sp}}. \tag{12.3-3}$$

It is apparent that the rates of atomic absorption and stimulated emission are proportional to $\bar{n}$, the average number of photons in each mode.

We can now combine (12.3-1), (12.3-2), and (12.3-3) to write an equation for the rate of change of the population density $N_2(t)$ arising from spontaneous emission, absorption, and stimulated emission,

$$\boxed{\frac{dN_2}{dt} = -\frac{N_2}{t_{sp}} + \frac{\bar{n}N_1}{t_{sp}} - \frac{\bar{n}N_2}{t_{sp}}.} \tag{12.3-4}$$

Rate Equation

This equation does not include transitions into or out of level 2 arising from other effects, such as interactions with other energy levels, nonradiative transitions, and external sources of excitation. In the steady state $dN_2/dt = 0$, and we have

$$\frac{N_2}{N_1} = \frac{\bar{n}}{1 + \bar{n}}. \tag{12.3-5}$$

where $\bar{n}$ is the average number of photons per mode. Clearly, $N_2/N_1 \leq 1$.

If we now use the fact that the atoms are in thermal equilibrium, (12.1-8) dictates that their populations obey the Boltzmann distribution, i.e.,

$$\frac{N_2}{N_1} = \exp\left(-\frac{E_2 - E_1}{k_B T}\right) = \exp\left(-\frac{h\nu}{k_B T}\right). \tag{12.3-6}$$

Substituting (12.3-6) into (12.3-5) and solving for $\bar{n}$ leads to

$$\bar{n} = \frac{1}{\exp(h\nu/k_B T) - 1} \tag{12.3-7}$$

for the average number of photons in a mode of frequency ν.

The foregoing derivation is predicated on the interaction of two energy levels, coupled by absorption as well as stimulated and spontaneous emission at a frequency near ν. The applicability of (12.3-7) is, however, far broader. Consider a cavity whose walls are made of solid materials and possess a continuum of energy levels at all energy separations, and therefore all values of ν. Atoms of the walls spontaneously emit into the cavity. The emitted light subsequently interacts with the atoms, giving rise to absorption and stimulated emission. If the walls are maintained at a temperature T, the combined system of atoms and radiation reaches thermal equilibrium.

Equation (12.3-7) is identical to (11.2-21)—the expression for the mean photon number in a mode of thermal light [for which the occupation of the mode energy levels follows a Boltzmann, or Bose–Einstein, distribution, $p(n) \propto \exp(-nh\nu/k_B T)$]. This result indicates a self-consistency in our analysis. Photons interacting with atoms in thermal equilibrium at temperature T are themselves in thermal equilibrium at the same temperature T (see Sec. 11.2C).

B. Blackbody Radiation Spectrum

The average energy $\bar{E}$ of a radiation mode in the situation described in Sec. 12.3A is simply $\bar{n}h\nu$, so that

$$\bar{E} = \frac{h\nu}{\exp(h\nu/k_B T) - 1}. \tag{12.3-8}$$

Average Energy of a Mode in Thermal Equilibrium

The dependence of $\bar{E}$ on ν is shown in Fig. 12.3-2. Note that for $h\nu \ll k_B T$ (i.e., when the energy of a photon is sufficiently small), $\exp(h\nu/k_B T) \approx 1 + h\nu/k_B T$ and $\bar{E} \approx k_B T$. This is the classical value for a harmonic oscillator with two degrees of freedom, as expected from statistical mechanics.

Multiplying this expression for the average energy per mode $\bar{E}$, by the modal density $M(\nu) = 8\pi\nu^2/c^3$, gives rise to a spectral energy density (energy per unit bandwidth per unit cavity volume) $\varrho(\nu) = M(\nu)\bar{E}$, i.e.,

$$\varrho(\nu) = \frac{8\pi h\nu^3}{c^3} \frac{1}{\exp(h\nu/k_B T) - 1}. \tag{12.3-9}$$

Spectral Energy Density of Blackbody Radiation

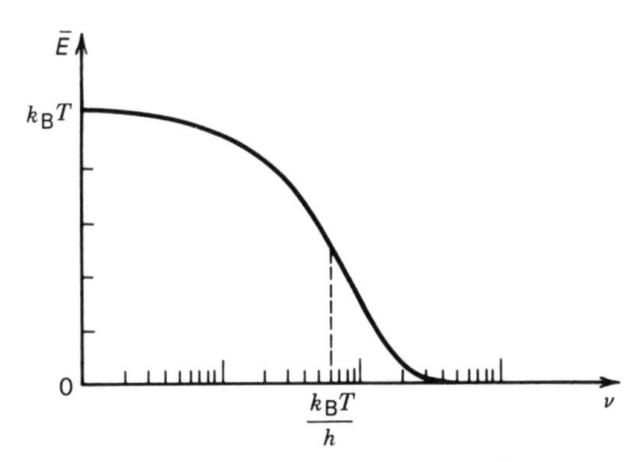

Figure 12.3-2 Semilogarithmic plot of the average energy $\bar{E}$ of an electromagnetic mode in thermal equilibrium at temperature T as a function of the mode frequency ν. At $T = 300$ K, $k_B T/h = 6.25$ THz, which corresponds to a wavelength of 48 μm.

This formula, known as the blackbody radiation law, is plotted in Fig. 12.3-3. The dependence of the radiation density on temperature is illustrated in Fig. 12.3-4.

The spectrum of blackbody radiation played an important role in the discovery of the quantum (photon) nature of light (Sec. 11.1). Based on classical electromagnetic theory, it was known that the modal density should be $M(\nu)$ as given above. However, based on classical statistical mechanics (in which electromagnetic energy is not quantized) the average energy per mode was known to be $\bar{E} = k_B T$. This gives an incorrect result for $\varrho(\nu)$ (its integral diverges). It was Max Planck who, in 1900, saw that a way to

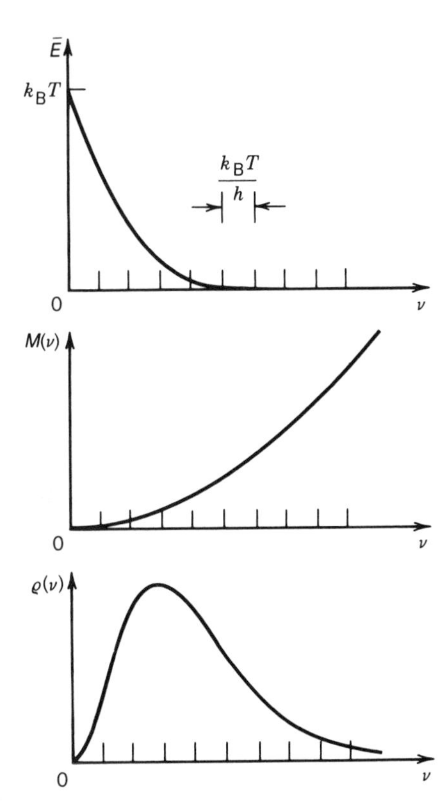

Figure 12.3-3 Frequency dependence of the energy per mode $\bar{E}$, the density of modes $M(\nu)$, and the spectral energy density $\varrho(\nu) = M(\nu)\bar{E}$ on a linear-linear scale.

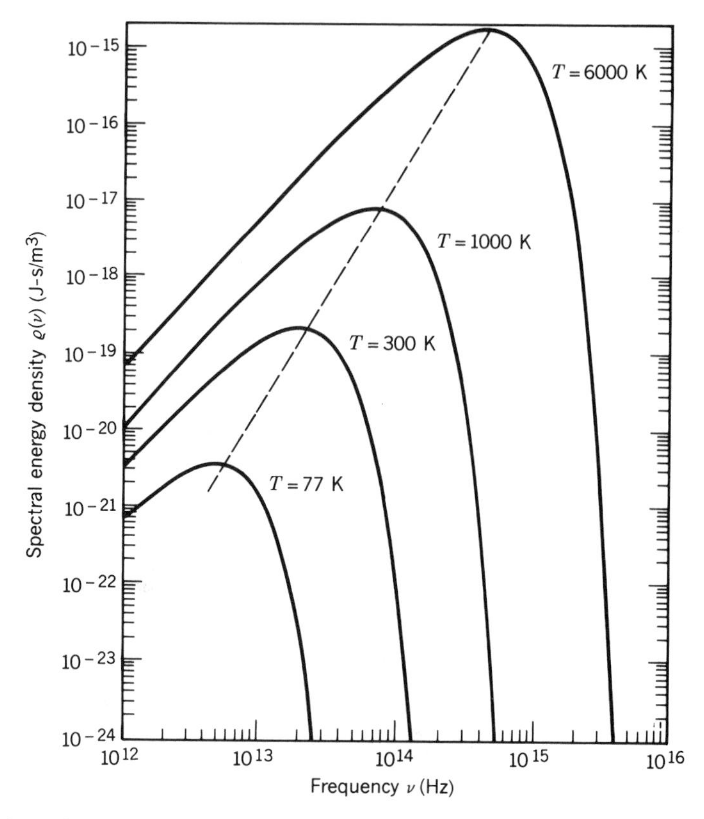

Figure 12.3-4 Dependence of the spectral energy density $\varrho(\nu)$ on frequency for different temperatures, on a double-logarithmic scale.

obtain the correct blackbody spectrum was to quantize the energy of each mode and suggested using the correct quantum expression for $\bar{E}$ given in (12.3-8).

EXERCISE 12.3-1

Frequency of Maximum Blackbody Energy Density. Using the blackbody radiation law $\varrho(\nu)$, show that the frequency ν_p at which the spectral energy density is maximum satisfies the equation $3(1 - e^{-x}) = x$, where $x = h\nu_p/k_B T$. Find x approximately and determine ν_p at $T = 300$ K.

12.4 LUMINESCENCE LIGHT

An applied external source of energy may cause an atomic or molecular system to undergo transitions to higher energy levels. In the course of decaying to a lower energy, the system may subsequently emit optical radiation. Such "nonthermal" radiators are generally called **luminescent radiators** and the radiation process is called **luminescence**. Luminescent radiators are classified according to the source of excitation energy, as indicated by the following examples.

- **Cathodoluminescence** is caused by accelerated electrons that collide with the atoms of a target. An example is the cathode ray tube where electrons deliver their energy to a phosphor. The term **betaluminescence** is used when the fast electrons are the product of nuclear beta decay rather than an electron gun, as in the cathode-ray tube.
- **Photoluminescence** is caused by energetic optical photons. An example is the glow emitted by some crystals after irradiation by ultraviolet light. The term **radioluminescence** is applied when the energy source is x-ray or gamma-ray photons, or other ionizating radiation. Indeed, such high-energy radiation is often detected by the use of luminescent (scintillation) materials such as NaI, special plastics, or $PbCO_3$ in conjunction with optical detectors.
- **Chemiluminescence** provides energy through a chemical reaction. An example is the glow of phosphorus as it oxidizes in air. **Bioluminescence**, which characterizes the light given off by living organisms (e.g., fireflies and glowworms), provides another example of chemiluminescence.
- **Electroluminescence** results from energy provided by an applied electric field. An important example is **injection electroluminescence**, which occurs when electric current is injected into a forward-biased semiconductor junction diode. As injected electrons drop from the conduction band to the valence band, they emit photons. An example is the light-emitting diode (LED).
- **Sonoluminescence** is caused by energy acquired from a sound wave. The light emitted by water under irradiation by a strong ultrasonic beam is an example.

Injection electroluminescence is discussed in the context of semiconductor photon sources in Chap. 16. The following section provides a brief introduction to photoluminescence.

Photoluminescence

Photoluminescence occurs when a system is excited to a higher energy level by absorbing a photon, and then spontaneously decays to a lower energy level, emitting a photon in the process. To conserve energy, the emitted photon cannot have more energy than the exciting photon, unless two or more excitation photons act in tandem. Several examples of transitions that lead to photoluminescence are depicted schematically in Fig. 12.4-1. Intermediate nonradiative downward transitions are possible, as

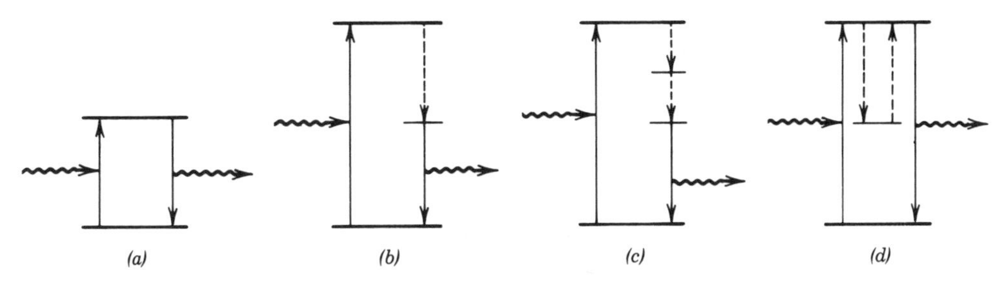

Figure 12.4-1 Various forms of photoluminescence.

shown by the dashed lines in (b) and (c). The electron can be stored in an intermediate state (e.g., a trap) for a long time, resulting in delayed luminescence. Ultraviolet light can be converted to visible light by this mechanism. Intermediate downward nonradiative transitions, followed by upward nonradiative transitions, can also occur, as shown in the example provided in (d).

If the radiative transitions are spin-allowed, i.e., if they take place between two states with equal multiplicity (singlet–singlet or triplet–triplet transitions; see Fig. 12.1-4, for example), the luminescence process is called **fluorescence**. In contrast, luminescence from spin-forbidden transitions (e.g., triplet–singlet) is called **phosphorescence**. Fluorescence lifetimes are usually short (0.1 to 10 ns), so that the luminescence photon is promptly emitted after excitation. This is in contrast to phosphorescence, which because the transitions are "forbidden," involves longer lifetimes (1 ms to 10 s) and therefore substantial delay between excitation and emission.

Photoluminescence occurs in many materials, including simple inorganic molecules (e.g., N_2, CO_2, Hg), noble gases, inorganic crystals (e.g., diamond, ruby, zinc sulfide), and aromatic molecules. A semiconductor can also act as a photoluminescent material. The process, which is of the form depicted in Fig. 12.4-1(c), involves electron–hole generation induced by photon absorption, followed by fast nonradiative relaxation to lower energy levels of the conduction band, and finally, by photon emission accompanying band-to-band electron–hole recombination. Intraband relaxation is very fast in comparison with band-to-band recombination.

Frequency Upconversion

The successive absorption of two or more photons may result in the emission of one photon of shorter wavelength, as illustrated in Fig. 12.4-2. The process readily occurs when there are traps in the material that can store the electron elevated by one photon for a time that is long enough for another photon to come along to excite it further. Materials that behave in this manner can be used for the detection of infrared radiation. The effect occurs in various phosphors doped with rare-earth ions such as Yb^{3+} and Er^{3+}. In certain materials, the traps can be charged up in minutes by daylight or fluorescent light (which provides $h\nu_2$ in Fig. 12.4-2); an infrared signal photon ($h\nu_1$ in Fig. 12.4-2) then releases the electron from the trap, causing a visible luminescence photon to be emitted [$h(\nu_1 + \nu_2)$ in Fig. 12.4-2].

Useful devices often take the form of a small (50 mm × 50 mm) card consisting of fine upconverting powder laminated between plastic sheets. The upconverting powder can also be dispersed in a three-dimensional polymer for three-dimensional viewing. The spatial distribution of an infrared beam, such as that produced by an infrared laser, can be visibly displayed by this means. The conversion efficiency is, however, usually substantially less than 1%. The relative spectral sensitivity and emission spectrum of a particular commercially available card is shown in Fig. 12.4-3.

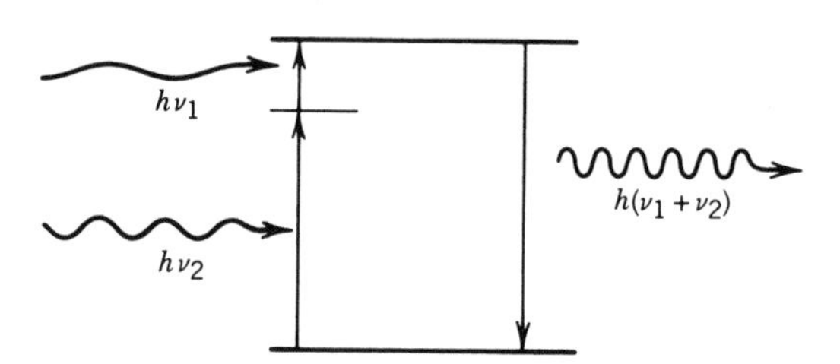

Figure 12.4-2 Detection of a long-wavelength photon $h\nu_1$ by upconversion to a short-wavelength photon $h\nu_3 = h(\nu_1 + \nu_2)$. An auxiliary photon $h\nu_2$ provides the additional energy.

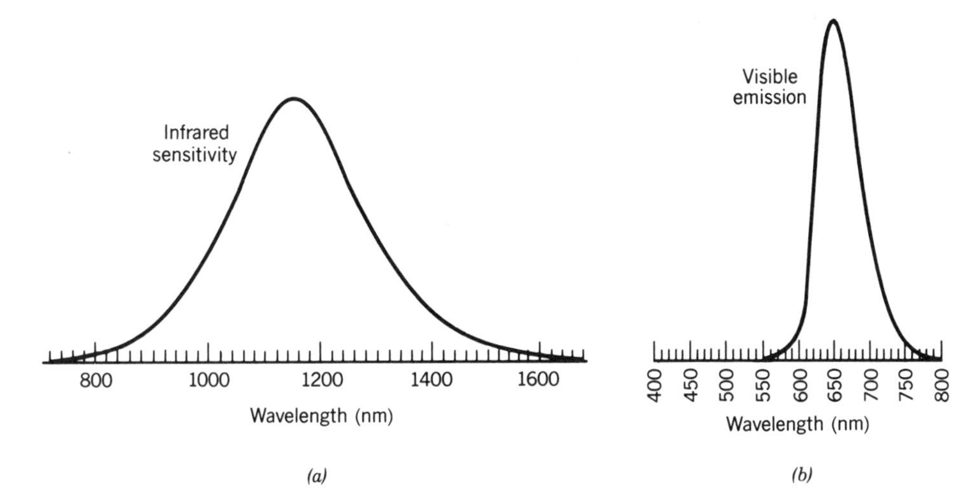

Figure 12.4-3 (*a*) Infrared spectral sensitivity of an upconversion phosphor card. (*b*) Spectrum of visible emission.

READING LIST

Books

See also the books on lasers in Chapter 13.

V. S. Letokhov, ed., *Laser Spectroscopy of Highly Vibrationally Excited Molecules*, Adam Hilger, Bristol, England, 1989.

R. M. Eisberg, R. Resnick, D. O. Caldwell, and J. R. Christman, *Quantum Physics of Atoms, Molecules, Solids, Nuclei, and Particles*, Wiley, New York, 2nd ed. 1985.

R. Loudon, *The Quantum Theory of Light*, Oxford University Press, New York, 2nd ed. 1983.

R. G. Breene, Jr., *Theories of Spectral Line Shape*, Wiley, New York, 1981.

C. Kittel, *Thermal Physics*, W. H. Freeman, San Francisco, 2nd ed. 1980.

L. Allen and J. H. Eberly, *Optical Resonance and Two-Level Atoms*, Wiley, New York, 1975.

H. G. Kuhn, *Atomic Spectra*, Academic Press, New York 1969.

G. Herzberg, *Electronic Spectra and Electronic Structure of Polyatomic Molecules*, Van Nostrand Reinhold, Princeton, NJ, 1966.

D. L. Livesey, *Atomic and Nuclear Physics*, Blaisdell, Waltham, MA, 1966.

R. P. Feynman, R. B. Leighton, and M. Sands, *The Feynman Lectures on Physics*, vol. 3, *Quantum Mechanics*, Addison-Wesley, Reading, MA, 1965.

M. Garbuny, *Optical Physics*, Academic Press, New York, 1965.

F. Reif, *Fundamentals of Statistical and Thermal Physics*, McGraw-Hill, New York, 1965.

R. P. Feynman, R. B. Leighton, and M. Sands, *The Feynman Lectures on Physics*, vol. 1, *Mainly Mechanics, Radiation, and Heat*, Addison-Wesley, Reading, MA, 1963.

A. C. G. Mitchell and M. W. Zemansky, *Resonance Radiation and Excited Atoms*, Cambridge University Press, New York, 1961.

J. C. Slater, *Quantum Theory of Atomic Structure*, vol. 1, McGraw-Hill, New York, 1960.

E. U. Condon and G. H. Shortley, *The Theory of Atomic Spectra*, Cambridge University Press, New York, 1959.

M. Born, *Atomic Physics*, Hafner Press, New York, 1959.

C. Kittel, *Elementary Statistical Physics*, Wiley, New York, 1958.

G. Herzberg, *Molecular Spectra and Molecular Structure*, vol. 1, *Spectra of Diatomic Molecules*, Van Nostrand, New York, 2nd ed. 1950.

G. Herzberg, *Atomic Spectra and Atomic Structure*, Dover, New York, 2nd ed. 1944.

Books on Luminescence

M. Pazzagli, E. Cadenas, L. J. Kricka, A. Roda, and P. E. Stanley, eds., *Bioluminescence and Chemiluminescence*, Wiley, New York, 1989.

J. Scholmerich, R. Andreesen, R. Kapp, M. Ernst, and W. G. Woods, eds., *Bioluminescence and Chemiluminescence: New Perspectives*, Wiley, New York, 1987.

W. Elenbaas, *Light Sources*, Macmillan, London, 1972.

H. K. Henisch, *Electroluminescence*, Pergamon Press, New York, 1962.

Special Journal Issue

Special issue on laser cooling and trapping of atoms, *Journal of the Optical Society of America B*, vol. 6, no. 11, 1989.

Articles

C. Foot and A. Steane, The Coolest Atoms Yet, *Physics World*, vol. 3, no. 10, pp. 25–27, 1990.

R. Pool, Making Atoms Jump Through Hoops, *Science*, vol. 248, pp. 1076–1078, 1990.

S. Haroche and D. Kleppner, Cavity Quantum Electrodynamics, *Physics Today*, vol. 42, no. 1, pp. 24–30, 1989.

R. Blümel, J. M. Chen, E. Peik, W. Quint, W. Schleich, Y. R. Shen, and H. Walther, Phase Transitions of Stored Laser-Cooled Ions, *Nature*, vol. 334, pp. 309–313, 1988.

W. D. Phillips and H. J. Metcalf, Cooling and Trapping of Atoms, *Scientific American*, vol. 256, no. 3, pp. 50–56, 1987.

H. J. Metcalf, Laser Cooling and Electromagnetic Trapping of Atoms, *Optics News*, vol. 13, no. 3, pp. 6–10, 1987.

E. Wolf, Einstein's Researches on the Nature of Light, *Optics News*, vol. 5, no. 1, pp. 24–39, 1979.

J. H. van Vleck and D. L. Huber, Absorption, Emission, and Linebreadths: A Semihistorical Perspective, *Reviews of Modern Physics*, vol. 49, pp. 939–959, 1977.

V. F. Weisskopf, How Light Interacts with Matter, *Scientific American*, vol. 219, no. 3, pp. 60–71, 1968.

A. Javan, The Optical Properties of Materials, *Scientific American*, vol. 217, no. 3, pp. 239–248, 1967.

G. R. Fowles, Quantum Dynamical Description of Atoms and Radiative Processes, *American Journal of Physics*, vol. 31, pp. 407–409, 1963.

A. Einstein, Zur Quantentheorie der Strahlung (On the Quantum Theory of Radiation), *Physikalische Zeitschrift*, vol. 18, pp. 121–128, 1917.

PROBLEMS

12.2-1 **Comparison of Stimulated and Spontaneous Emission.** An atom with two energy levels corresponding to the transition ($\lambda_o = 0.7\ \mu$m, $t_{sp} = 3$ ms, $\Delta\nu = 50$ GHz, Lorentzian lineshape) is placed in a resonator of volume $V = 100$ cm^3 and refractive index $n = 1$. Two radiation modes (one at the center frequency ν_0 and the other at $\nu_0 + \Delta\nu$) are excited with 1000 photons each. Determine the probability density for stimulated emission (or absorption). If N_2 such atoms are excited to energy level 2, determine the time constant for the decay of N_2 due to stimulated *and* spontaneous emission. How many photons (rather than 1000) should be present so that the decay rate due to stimulated emission equals that due to spontaneous emission?

12.2-2 **Spontaneous Emission into Prescribed Modes.** (a) Given a 1-μm^3 cubic cavity, with a medium of refractive index $n = 1$, what are the mode numbers (q_1, q_2, q_3) of the lowest- and next-higher-frequency modes? (See Sec. 9.1C.) Show that these frequencies are 260 and 367 THz.
(b) Consider a single excited atom in the cavity in the absence of photons. Let p_{sp1} be the probability density (s^{-1}) that the atom spontaneously emits a photon into the (2, 1, 1) mode, and let p_{sp2} be the probability density that the atom spontaneously emits a photon with frequency 367 THz. Determine the ratio p_{sp2}/p_{sp1}.

12.3-1 **Rate Equations for Broadband Radiation.** A resonator of unit volume contains atoms having two energy levels, labeled 1 and 2, corresponding to a transition of resonance frequency ν_0 and linewidth $\Delta\nu$. There are N_1 and N_2 atoms in the lower and upper levels, 1 and 2, respectively, and a total of $\bar{n}$ photons in each of the modes within a broad band surrounding ν_0. Photons are lost from the resonator at a rate $1/\tau_p$ as a result of imperfect reflection at the cavity walls. Assuming that there are no nonradiative transitions between levels 2 and 1, write rate equations for N_2 and $\bar{n}$.

12.3-2 **Inhibited Spontaneous Emission.** Consider a hypothetical two-dimensional blackbody radiator (e.g., a square plate of area A) in thermal equilibrium at temperature T.
(a) Determine the density of modes $M(\nu)$ and the spectral energy density (i.e., the energy in the frequency range between ν and $\nu + d\nu$ per unit area) of the emitted radiation $\varrho(\nu)$ (see Sec. 9.1C).
(b) Find the probability density of spontaneous emission P_{sp} for an atom located in a cavity that permits radiation only in two dimensions.

12.3-3 **Comparison of Stimulated and Spontaneous Emission in Blackbody Radiation.** Find the temperature of a thermal-equilibrium blackbody cavity emitting a spectral energy density $\varrho(\nu)$, when the rates of stimulated and spontaneous emission from the atoms in the cavity walls are equal at $\lambda_o = 1\ \mu m$.

12.3-4 **Wien's Law.** Derive an expression for the spectral energy density $\varrho_\lambda(\lambda)$ [the energy per unit volume in the wavelength region between λ and $\lambda + d\lambda$ is $\varrho_\lambda(\lambda)\, d\lambda$]. Show that the wavelength λ_p at which the spectral energy density is maximum satisfies the equation $5(1 - e^{-y}) = y$, where $y = hc/\lambda_p k_B T$, demonstrating that the relationship $\lambda_p T =$ constant (Wien's law) is satisfied. Find $\lambda_p T$ approximately. Show that $\lambda_p \neq c/\nu_p$, where ν_p is the frequency at which the blackbody energy density $\varrho(\nu)$ is maximum (see Exercise 12.3-1 on page 454). Explain.

12.3-5 **Spectral Energy Density of One-Dimensional Blackbody Radiation.** Consider a hypothetical one-dimensional blackbody radiator of length L in thermal equilibrium at temperature T.
(a) What is the density of modes $M(\nu)$ (number of modes per unit frequency per unit length) in one dimension.
(b) Using the average energy $\bar{E}$ of a mode of frequency ν, determine the spectral energy density (i.e., the energy in the frequency range between ν and $\nu + d\nu$ per unit length) of the blackbody radiation $\varrho(\nu)$. Sketch $\varrho(\nu)$ versus ν.

*12.4-1 **Statistics of Cathodoluminescence Light.** Consider a beam of electrons impinging on the phosphor of a cathode-ray tube. Let $\bar{m}$ be the mean number of electrons striking a unit area of the phosphor in unit time. If the number m of electrons arriving in a fixed time is random with a Poisson distribution and the number of photons emitted per electron is also Poisson distributed, but with mean G, find the overall distribution $p(n)$ of the emitted cathodoluminescence photons. The result is called the **Neyman type-A distribution**. Determine expressions for the mean $\bar{n}$ and the variance σ_n^2. *Hint*: Use conditional probability.

CHAPTER

13

LASER AMPLIFIERS

13.1 THE LASER AMPLIFIER
- A. Amplifier Gain
- B. Amplifier Phase Shift

13.2 AMPLIFIER POWER SOURCE
- A. Rate Equations
- B. Four- and Three-Level Pumping Schemes
- C. Examples of Laser Amplifiers

13.3 AMPLIFIER NONLINEARITY AND GAIN SATURATION
- A. Gain Coefficient
- B. Gain
- *C. Gain of Inhomogeneously Broadened Amplifiers

*13.4 AMPLIFIER NOISE

Charles H. Townes (born 1915)

Nikolai G. Basov (born 1922)

Aleksandr M. Prokhorov (born 1916)

Townes, Basov, and Prokhorov developed the principle of *l*ight *a*mplification by the *s*timulated *e*mission of *r*adiation (*laser*). They received the Nobel Prize in 1964.

A **coherent optical amplifier** is a device that increases the amplitude of an optical field while maintaining its phase. If the optical field at the input to such an amplifier is monochromatic, the output will also be monochromatic, with the same frequency. The output amplitude is increased relative to the input while the phase is unchanged or shifted by a fixed amount. In contrast, an amplifier that increases the intensity of an optical wave without preserving the phase is called an **incoherent optical amplifier**.

This chapter is concerned with coherent optical amplifiers. Such amplifiers are important for various applications; examples include the amplification of weak optical pulses such as those that have traveled through a long length of optical fiber, and the production of highly intense optical pulses such as those required for laser-fusion applications. Furthermore, it is important to understand the principles underlying the operation of optical amplifiers as a prelude to the discussion of optical oscillators (lasers) in Chap. 14.

The underlying principle for achieving the coherent amplification of light is light amplification by the stimulated emission of radiation, known by its acronym as the LASER process. Stimulated emission (see Sec. 12.2) allows a photon in a given mode to induce an atom in an upper energy level to undergo a transition to a lower energy level and, in the process, to emit a clone photon into the same mode as the initial photon (viz., a photon with the same frequency, direction, and polarization). These two photons, in turn, can serve to stimulate the emission of two additional photons, and so on, while preserving these properties. The result is coherent light amplification. Because stimulated emission occurs when the photon energy is nearly equal to the atomic-transition energy difference, the process is restricted to a band of frequencies determined by the atomic linewidth.

Laser amplification differs in a number of respects from electronic amplification. Electronic amplifiers rely on devices in which small changes in an injected electric current or applied voltage result in large changes in the rate of flow of charge carriers, such as electrons and holes in a semiconductor field-effect transistor (FET) or bipolar junction transistor. Tuned electronic amplifiers make use of resonant circuits (e.g., a capacitor and an inductor) or resonators (metal cavities) to limit the amplifier's gain to the band of frequencies of interest. In contrast, atomic, molecular, and solid-state laser amplifiers rely on their energy-level differences to provide the primary frequency selection. These act as natural resonators that select the amplifier's bandwidth and frequencies of operation. Optical cavities (resonant circuits) are often used to provide auxiliary frequency tuning.

Light transmitted through matter in thermal equilibrium is attenuated rather than amplified. This is because absorption by the large population of atoms in the lower energy level is more prevalent than stimulated emission by the smaller population of atoms in the upper level. An essential ingredient for achieving laser amplification is the presence of a greater number of atoms in the upper energy level than in the lower level, which is clearly a nonequilibrium situation. Achieving such a **population inversion** requires a source of power to excite (pump) the atoms into the higher energy level, as illustrated in Fig. 13.0-1. Although the presentation throughout this chapter is couched in terms of "atoms" and "atomic levels," these appelations are to be more broadly understood as "active medium" and "laser energy levels," respectively.

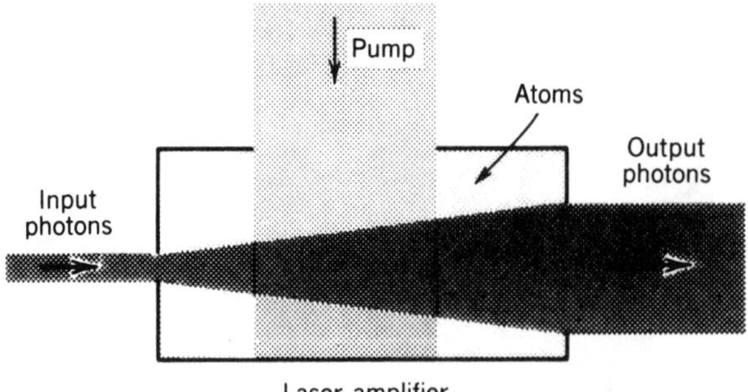

Figure 13.0-1 The laser amplifier. An external power source (called a pump) excites the active medium (represented by a collection of atoms), producing a population inversion. Photons interact with the atoms; when stimulated emission is more prevalent than absorption, the medium acts as a coherent amplifier.

The properties of an ideal (optical or electronic) coherent amplifier are displayed schematically in Fig. 13.0-2(*a*). It is a linear system that increases the amplitude of the input signal by a fixed factor, called the amplifier gain. A sinusoidal input leads to a sinusoidal output at the same frequency, but with larger amplitude. The gain of the ideal amplifier is constant for all frequencies within the amplifier spectral bandwidth. The amplifier may impart to the input signal a phase shift that varies linearly with frequency, corresponding to a time delay of the output with respect to the input (see Appendix B).

Real coherent amplifiers deliver a gain and phase shift that are frequency dependent, typically in the manner illustrated in Fig. 13.0-2(*b*). The gain and phase shift constitute the amplifier's transfer function. For a sufficiently high input amplitude, furthermore, real amplifiers may exhibit saturation, a form of nonlinear behavior in which the output amplitude fails to increase in proportion to the input amplitude. Saturation introduces harmonic components into the output, provided that the ampli-

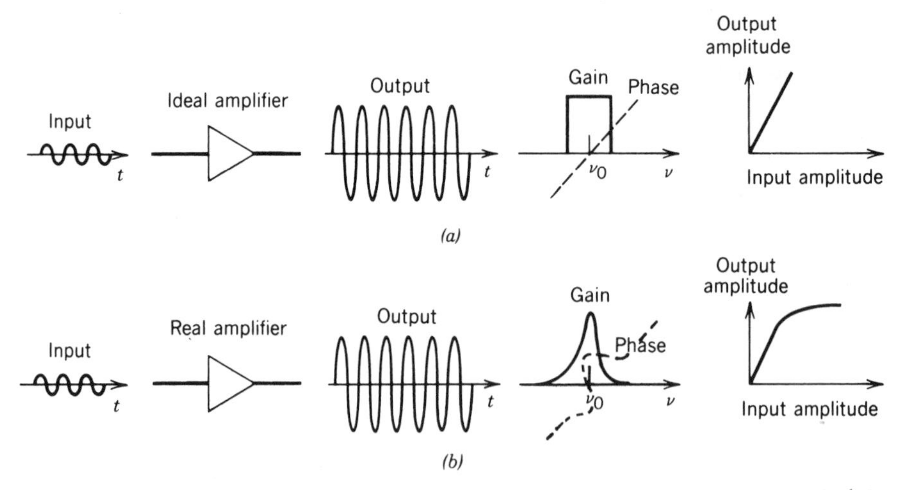

Figure 13.0-2 (*a*) An ideal amplifier is linear. It increases the amplitude of signals (whose frequencies lie within its bandwidth) by a constant gain factor, possibly introducing a linear phase shift. (*b*) A real amplifier typically has a gain and phase shift that are functions of frequency, as shown. For large inputs the output signal saturates; the amplifier exhibits nonlinearity.

fier bandwidth is sufficiently broad to pass them. Real amplifiers also introduce noise, so that a randomly fluctuating component is always present at the output, regardless of the input.

An amplifier may therefore be characterized by the following features:

- Gain
- Bandwidth
- Phase shift
- Power source
- Nonlinearity and gain saturation
- Noise

We proceed to discuss these characteristics in turn. In Sec. 13.1 the theory of laser amplification is developed, leading to expressions for the amplifier gain, spectral bandwidth, and phase shift. The mechanisms by which an amplifier power source can achieve a population inversion are examined in Sec. 13.2. Sections 13.3 and 13.4 are devoted to gain saturation and noise in the amplification process, respectively. This chapter relies on material presented in Chap. 12, especially in Sec. 12.2.

13.1 THE LASER AMPLIFIER

A monochromatic optical plane wave traveling in the z direction with frequency ν, electric field $\mathrm{Re}\{E(z)\exp(j2\pi\nu t)\}$, intensity $I(z) = |E(z)|^2/2\eta$, and photon-flux density $\phi(z) = I(z)/h\nu$ (photons per second per unit area) will interact with an atomic medium, provided that the atoms of the medium have two relevant energy levels whose energy difference nearly matches the photon energy $h\nu$. The numbers of atoms per unit volume in the lower and upper energy levels are N_1 and N_2, respectively. The wave is amplified with a gain coefficient $\gamma(z)$ (per unit length) and undergoes a phase shift $\varphi(z)$ (per unit length). We proceed to determine expressions for $\gamma(\nu)$ and $\varphi(\nu)$. Positive $\gamma(\nu)$ corresponds to amplification; negative $\gamma(\nu)$, to attenuation.

A. Amplifier Gain

Three forms of photon–atom interaction are possible (see Sec. 12.2). If the atom is in the lower energy level, the photon may be absorbed, whereas if it is in the upper energy level, a clone photon may be emitted by the process of stimulated emission. These two processes lead to attenuation and amplification, respectively. The third form of interaction, spontaneous emission, in which an atom in the upper energy level emits a photon independently of the presence of other photons, is responsible for amplifier noise as discussed in Sec. 13.4.

The probability density (s^{-1}) that an unexcited atom absorbs a single photon is, according to (12.2-15) and (12.2-11),

$$W_i = \phi\sigma(\nu), \qquad (13.1\text{-}1)$$

where $\sigma(\nu) = (\lambda^2/8\pi t_{\mathrm{sp}})g(\nu)$ is the transition cross section at the frequency ν, $g(\nu)$ is the normalized lineshape function, t_{sp} is the spontaneous lifetime, and λ is the wavelength of light in the medium. The probability density for stimulated emission is also given by (13.1-1).

The average density of absorbed photons (number of photons per unit time per unit volume) is $N_1 W_i$. Similarly, the average density of clone photons generated as a result

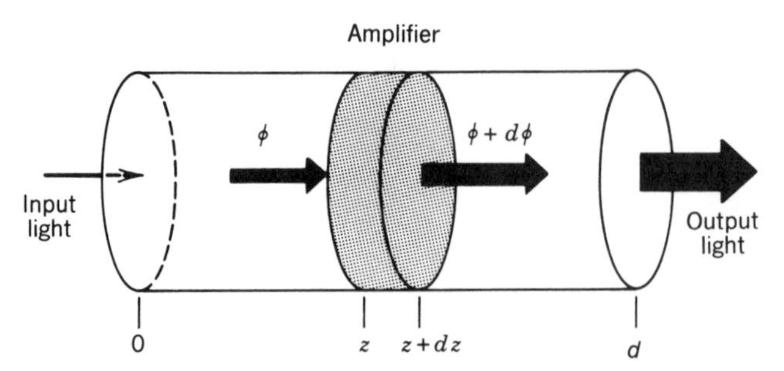

Figure 13.1-1 The photon-flux density ϕ (photons/cm^2-s) entering an incremental cylinder containing excited atoms grows to $\phi + d\phi$ after length dz.

of stimulated emission is $N_2 W_i$. The net number of photons gained per second per unit volume is therefore NW_i, where $N = N_2 - N_1$ is the population density difference. For convenience, N is simply referred to as the population difference. If N is positive, a population inversion exists, in which case the medium can act as an amplifier and the photon-flux density can increase. If it is negative, the medium acts as an attenuator and the photon-flux density decreases. If $N = 0$, the medium is transparent.

Since the incident photons travel in the z direction, the stimulated-emission photons will also travel in this direction, as illustrated in Fig. 13.1-1. An external pump providing a population inversion ($N > 0$) will then cause the photon-flux density $\phi(z)$ to increase with z. Because emitted photons stimulate further emissions, the growth at any position z is proportional to the population at that position; $\phi(z)$ will thus increase exponentially.

To demonstrate this process explicitly, consider an incremental cylinder of length dz and unit area as shown in Fig. 13.1-1. If $\phi(z)$ and $\phi(z) + d\phi(z)$ are the photon-flux densities entering and exiting the cylinder, respectively, then $d\phi(z)$ must be the photon-flux density emitted from within the cylinder. This incremental number of photons per unit area per unit time $d\phi(z)$ is simply the number of photons gained per unit time per unit volume, NW_i, multiplied by the thickness of the cylinder dz, i.e.,

$$d\phi = NW_i \, dz. \tag{13.1-2}$$

With the help of (13.1-1), (13.1-2) can be written in the form of a differential equation,

$$\frac{d\phi(z)}{dz} = \gamma(\nu)\phi(z), \tag{13.1-3}$$

where

$$\gamma(\nu) = N\sigma(\nu) = N\frac{\lambda^2}{8\pi t_{sp}} g(\nu). \tag{13.1-4}$$

Gain Coefficient of a Laser Medium

The coefficient $\gamma(\nu)$ represents the net gain in the photon-flux density per unit length of the medium. The solution of (13.1-3) is the exponentially increasing function

$$\phi(z) = \phi(0) \exp[\gamma(\nu) z]. \tag{13.1-5}$$

Since the optical intensity $I(z) = h\nu\phi(z)$, (13.1-5) can also be written in terms of I as

$$I(z) = I(0)\exp[\gamma(\nu)z]. \tag{13.1-6}$$

Thus $\gamma(\nu)$ also represents the gain in the intensity per unit length of the medium.

The amplifier gain coefficient $\gamma(\nu)$ is seen to be proportional to the population difference $N = N_2 - N_1$. Although N was considered to be positive in the example provided above, the derivation is valid whatever the sign of N. In the absence of a population inversion, N is negative ($N_2 < N_1$) and so is the gain coefficient. The medium will then attenuate (rather than amplify) light traveling in the z direction, in accordance with the exponentially decreasing function $\phi(z) = \phi(0)\exp[-\alpha(\nu)z]$, where the attenuation coefficient $\alpha(\nu) = -\gamma(\nu) = -N\sigma(\nu)$. A medium in thermal equilibrium therefore cannot provide laser amplification.

For an interaction region of total length d (see Fig. 13.1-1), the overall gain of the laser amplifier $G(\nu)$ is defined as the ratio of the photon-flux density at the output to the photon-flux density at the input, $G(\nu) = \phi(d)/\phi(0)$, so that

$$G(\nu) = \exp[\gamma(\nu)d]. \tag{13.1-7}$$

Amplifier Gain

Amplifier Bandwidth

The dependence of the gain coefficient $\gamma(\nu)$ on the frequency of the incident light ν is contained in its proportionality to the lineshape function $g(\nu)$, as given in (13.1-4). The latter is a function of width $\Delta\nu$ centered about the atomic resonance frequency $\nu_0 = (E_2 - E_1)/h$, where E_2 and E_1 are the atomic energies. The laser amplifier is therefore a resonant device, with a resonance frequency and bandwidth determined by the lineshape function of the atomic transition. This is because stimulated emission and absorption are governed by the atomic transition. The linewidth $\Delta\nu$ is measured either in units of frequency (Hz) or in units of wavelength (nm). These linewidths are related by $\Delta\lambda = |\Delta(c_o/\nu)| = +(c_o/\nu^2)\,\Delta\nu = (\lambda_o^2/c_o)\,\Delta\nu$. Thus a linewidth $\Delta\nu = 10^{12}$ Hz at $\lambda_o = 0.6\ \mu$m corresponds to $\Delta\lambda = 1.2$ nm.

For example, if the lineshape function is Lorentzian, (12.2-27) provides

$$g(\nu) = \frac{\Delta\nu/2\pi}{(\nu - \nu_0)^2 + (\Delta\nu/2)^2}. \tag{13.1-8}$$

The gain coefficient is then also Lorentzian with the same width, i.e.,

$$\gamma(\nu) = \gamma(\nu_0)\frac{(\Delta\nu/2)^2}{(\nu - \nu_0)^2 + (\Delta\nu/2)^2}, \tag{13.1-9}$$

as illustrated in Fig. 13.1-2, where $\gamma(\nu_0) = N(\lambda^2/4\pi^2 t_{sp}\,\Delta\nu)$ is the gain coefficient at the central frequency ν_0.

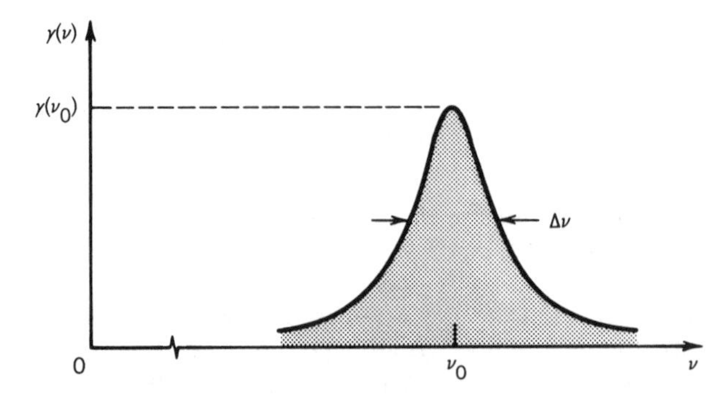

Figure 13.1-2 Gain coefficient $\gamma(\nu)$ of a Lorentzian-lineshape laser amplifier.

EXERCISE 13.1-1

Attenuation and Gain in a Ruby Laser Amplifier

(a) Consider a ruby crystal with two energy levels separated by an energy difference corresponding to a free-space wavelength $\lambda_o = 694.3$ nm, with a Lorentzian lineshape of width $\Delta\nu = 60$ GHz. The spontaneous lifetime is $t_{sp} = 3$ ms and the refractive index of ruby is $n = 1.76$. If $N_1 + N_2 = N_a = 10^{22}$ cm^{-3}, determine the population difference $N = N_2 - N_1$ and the attenuation coefficient at the line center $\alpha(\nu_0)$ under conditions of thermal equilibrium (so that the Boltzmann distribution is obeyed) at $T = 300$ K.

(b) What value should the population difference N assume to achieve a gain coefficient $\gamma(\nu_0) = 0.5$ cm^{-1} at the central frequency?

(c) How long should the crystal be to provide an overall gain of 4 at the central frequency when $\gamma(\nu_0) = 0.5$ cm^{-1}?

B. Amplifier Phase Shift

Because the gain of the resonant medium is frequency dependent, the medium is dispersive (see Sec. 5.5) and a frequency-dependent phase shift must be associated with its gain. The phase shift imparted by the laser amplifier can be determined by considering the interaction of light with matter in terms of the electric field rather than the photon-flux density or the intensity.

We proceed with an alternative approach, in which the mathematical properties of a causal system are used to determine the phase shift. For homogeneously broadened media, the phase-shift coefficient $\varphi(\nu)$ (phase shift per unit length of the amplifier medium) is related to the gain coefficient $\gamma(\nu)$ by the Kramers–Kronig (Hilbert transform) relations (see Sec. B.1 of Appendix B and Sec. 5.5), so that knowledge of $\gamma(\nu)$ at all frequencies uniquely determines $\varphi(\nu)$.

The optical intensity and field are related by $I(z) = |E(z)|^2/2\eta$. Since $I(z) = I(0)\exp[\gamma(\nu)z]$ in accordance with (13.1-6), the optical field obeys the relation

$$E(z) = E(0)\exp\left[\tfrac{1}{2}\gamma(\nu)z\right]\exp[-j\varphi(\nu)z], \qquad (13.1\text{-}10)$$

where $\varphi(\nu)$ is the phase-shift coefficient. The field evaluated at $z + \Delta z$ is therefore

$$E(z + \Delta z) = E(z) \exp\left[\tfrac{1}{2}\gamma(\nu)\,\Delta z\right] \exp\left[-j\varphi(\nu)\,\Delta z\right]$$

$$\approx E(z)\left[1 + \tfrac{1}{2}\gamma(\nu)\,\Delta z - j\varphi(\nu)\,\Delta z\right], \tag{13.1-11}$$

where we have made use of a Taylor-series approximation for the exponential functions. The incremental change in the electric field $\Delta E(z) = E(z + \Delta z) - E(z)$ therefore satisfies the equation

$$\frac{\Delta E(z)}{\Delta z} = E(z)\left[\tfrac{1}{2}\gamma(\nu) - j\varphi(\nu)\right]. \tag{13.1-12}$$

This incremental amplifier may be regarded as a linear system whose input and output are $E(z)$ and $\Delta E(z)/\Delta z$, respectively, and whose transfer function is

$$\mathcal{H}(\nu) = \tfrac{1}{2}\gamma(\nu) - j\varphi(\nu). \tag{13.1-13}$$

Because this incremental amplifier represents a physical system, it must be causal. But the real and imaginary parts of the transfer function of a linear causal system are related by the Hilbert transform (see Appendix B). It follows that $-\varphi(\nu)$ is the Hilbert transform of $\frac{1}{2}\gamma(\nu)$ [see (5.5-11)] so that the amplifier phase shift function is determined by its gain coefficient.

A simple example is provided by the Lorentzian atomic lineshape function with narrow width $\Delta\nu \ll \nu_0$, for which the gain coefficient $\gamma(\nu)$ is given by (13.1-9). The corresponding phase shift coefficient $\varphi(\nu)$ is provided in (B.1-13) of Appendix B,

$$\varphi(\nu) = \frac{\nu - \nu_0}{\Delta\nu}\gamma(\nu). \tag{13.1-14}$$

Phase-Shift Coefficient (Lorentzian Lineshape)

The Lorentzian gain and phase-shift coefficients are plotted in Fig. 13.1-3 as functions of frequency. At resonance, the gain coefficient is maximum and the phase-shift

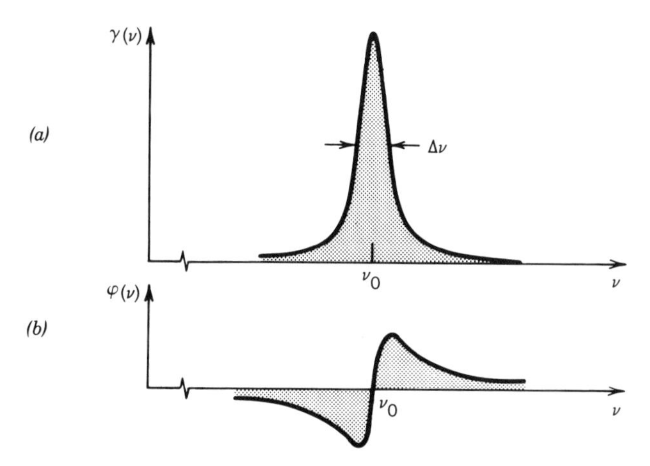

Figure 13.1-3 (*a*) Gain coefficient $\gamma(\nu)$ and (*b*) phase-shift coefficient $\varphi(\nu)$ for a laser amplifier with a Lorentzian lineshape function.

coefficient is zero. The phase-shift coefficient is negative for frequencies below resonance and positive for frequencies above resonance.

13.2 AMPLIFIER POWER SOURCE

Laser amplifiers, like other amplifiers, require an external source of power to provide the energy to be added to the input signal. The pump supplies this power through mechanisms that excite the electrons in the atoms, causing them to move from lower to higher atomic energy levels. To achieve amplification, the pump must provide a population inversion on the transition of interest ($N = N_2 - N_1 > 0$). The mechanics of pumping often involves the use of ancillary energy levels other than those directly involved in the amplification process, however. The pumping of atoms from level 1 into level 2 might be most readily achieved, for example, by pumping them from level 1 into level 3 and then by relying on the natural processes of decay from level 3 to populate level 2.

The pumping may be achieved optically (e.g., with a flashlamp or laser), electrically (e.g., through a gas discharge, an electron or ion beam, or by means of injected electron and holes as in semiconductor laser amplifiers), chemically (e.g., through a flame), or even by means of a nuclear explosion to achieve x-ray laser action. For continuous-wave (CW) operation, the rates of excitation and decay of all of the different energy levels participating in the process must be balanced to maintain a steady-state inverted population for the 1–2 transition. The equations that describe the rates of change of the population densities N_1 and N_2 as a result of pumping, radiative, and nonradiative transitions are called the **rate equations**. They are not unlike the equations presented in Sec. 12.3, but selective external pumping is now permitted so that thermal equilibrium conditions no longer prevail.

A. Rate Equations

Consider the schematic energy-level diagram of Fig. 13.2-1. We focus on levels 1 and 2, which have overall lifetimes τ_1 and τ_2, respectively, permitting transitions to lower levels. The lifetime of level 2 has two contributions—one associated with decay from 2 to 1 (τ_{21}), and the other (τ_{20}) associated with decay from 2 to all other lower levels. When several modes of decay are possible, the overall transition rate is a sum of the component transition rates. Since the rates are inversely proportional to the decay times, the reciprocals of the decay times must be added,

$$\tau_2^{-1} = \tau_{21}^{-1} + \tau_{20}^{-1}. \tag{13.2-1}$$

Multiple modes of decay therefore shorten the overall lifetime (i.e., they render the decay more rapid). Aside from the radiative spontaneous emission component (of time

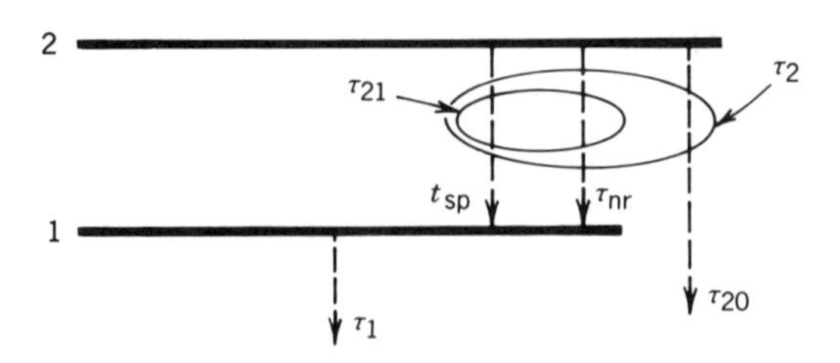

Figure 13.2-1 Energy levels 1 and 2 and their decay times.

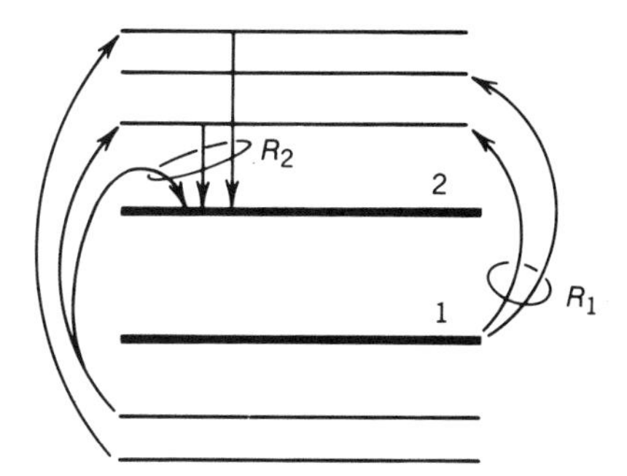

Figure 13.2-2 Energy levels 1 and 2, together with surrounding higher and lower energy levels.

constant t_{sp}) in τ_{21}, a nonradiative contribution τ_{nr} may also be present (arising, for example, from a collision of the atom with the wall of the container thereby resulting in a depopulation), so that

$$\tau_{21}^{-1} = t_{sp}^{-1} + \tau_{nr}^{-1}.$$

If a system like that illustrated in Fig. 13.2-1 is allowed to reach steady state, the population densities N_1 and N_2 will vanish by virtue of all the electrons ultimately decaying to lower energy levels.

Steady-state populations of levels 1 and 2 can be maintained, however, if energy levels above level 2 are continuously excited and leak downward into level 2, as shown in the more realistic energy level diagram of Fig. 13.2-2. Pumping can bring atoms from levels other than 1 and 2 out of level 1 and into level 2, at rates R_1 and R_2 (per unit volume per second), respectively, as shown in simplified form in Fig. 13.2-3. Consequently, levels 1 and 2 can achieve nonzero steady-state populations.

We now proceed to write the rate equations for this system both in the absence and in the presence of amplifier radiation (which is the radiation resonant with the 2–1 transition).

Rate Equations in the Absence of Amplifier Radiation

The rates of increase of the population densities of levels 2 and 1 arising from pumping and decay are

$$\frac{dN_2}{dt} = R_2 - \frac{N_2}{\tau_2} \tag{13.2-2}$$

$$\frac{dN_1}{dt} = -R_1 - \frac{N_1}{\tau_1} + \frac{N_2}{\tau_{21}}. \tag{13.2-3}$$

Under steady-state conditions ($dN_1/dt = dN_2/dt = 0$), (13.2-2) and (13.2-3) can be solved for N_1 and N_2, and the population difference $N = N_2 - N_1$ can be found. The

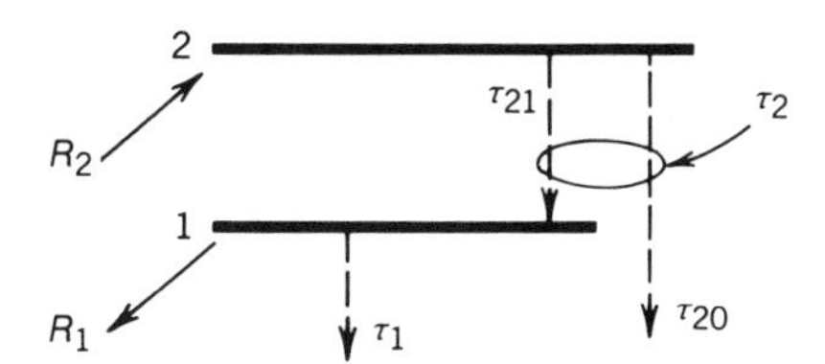

Figure 13.2-3 Energy levels 1 and 2 and their decay times. By means of pumping, the population density of level 2 is increased at the rate R_2 while that of level 1 is decreased at the rate R_1.

result is

$$N_0 = R_2\tau_2\left(1 - \frac{\tau_1}{\tau_{21}}\right) + R_1\tau_1, \tag{13.2-4}$$

Steady-State Population Difference (in Absence of Amplifier Radiation)

where the symbol N_0 represents the steady-state population difference N in the absence of amplifier radiation.

A large gain coefficient clearly requires a large population difference, i.e., a large positive value of N_0. Equation (13.2-4) shows that this may be achieved by:

- Large R_1 and R_2.
- Long τ_2 (but t_{sp}, which contributes to τ_2 through τ_{21}, must be sufficiently short so as to make the radiative transition rate large, as will be seen subsequently).
- Short τ_1 if $R_1 < (\tau_2/\tau_{21})R_2$.

The physical reasons underlying these conditions make good sense. The upper level should be pumped strongly and decay slowly so that it retains its population. The lower level should depump strongly so that it quickly disposes of its population. Ideally, it is desirable to have $\tau_{21} \approx t_{sp} \ll \tau_{20}$ so that $\tau_2 \approx t_{sp}$, and $\tau_1 \ll t_{sp}$. Under these conditions we obtain a simplified result:

$$N_0 \approx R_2 t_{sp} + R_1\tau_1. \tag{13.2-4a}$$

In the absence of depumping ($R_1 = 0$), or when $R_1 \ll (t_{sp}/\tau_1)R_2$, this result further simplifies to

$$N_0 \approx R_2 t_{sp}. \tag{13.2-4b}$$

EXERCISE 13.2-1

Optical Pumping. Assume that $R_1 = 0$ and that R_2 is realized by exciting atoms from the ground state $E = 0$ to level 2 using photons of frequency E_2/h absorbed with a transition probability W. Assume that $\tau_2 \approx t_{sp}$ and $\tau_1 \ll t_{sp}$ so that in steady state $N_1 \approx 0$ and $N_0 \approx R_2 t_{sp}$. If N_a is the total population of levels 0, 1, and 2, show that $R_2 \approx (N_a - 2N_0)W$, so that the population difference is $N_0 \approx N_a t_{sp} W/(1 + 2t_{sp}W)$.

Rate Equations in the Presence of Amplifier Radiation

The presence of radiation near the resonance frequency ν_0 enables transitions between levels 1 and 2 to take place by the processes of stimulated emission and absorption as well. These are characterized by the probability density $W_i = \phi\sigma(\nu)$, as provided in (13.1-1) and illustrated in Fig. 13.2-4. The rate equations (13.2-2) and (13.2-3) must

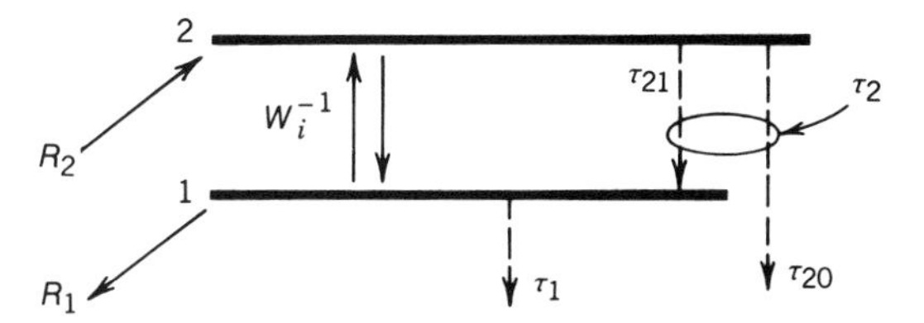

Figure 13.2-4 The population densities N_1 and N_2 (cm^{-3}-s^{-1}) of atoms in energy levels 1 and 2 are determined by three processes: decay (at the rates $1/\tau_1$ and $1/\tau_2$, respectively, which includes the effects of spontaneous emission), pumping (at the rates $-R_1$ and R_2, respectively), and absorption and stimulated emission (at the rate W_i).

then be extended to include this source of population loss and gain in each of the levels:

$$\frac{dN_2}{dt} = R_2 - \frac{N_2}{\tau_2} - N_2 W_i + N_1 W_i \tag{13.2-5}$$

$$\frac{dN_1}{dt} = -R_1 - \frac{N_1}{\tau_1} + \frac{N_2}{\tau_{21}} + N_2 W_i - N_1 W_i . \tag{13.2-6}$$

The population density of level 2 is decreased by stimulated emission from level 2 to level 1 and increased by absorption from level 1 to level 2. The spontaneous emission contribution is contained in τ_{21}.

Under steady-state conditions ($dN_1/dt = dN_2/dt = 0$), (13.2-5) and (13.2-6) are readily solved for N_1 and N_2, and for the population difference $N = N_2 - N_1$. The result is

$$N = \frac{N_0}{1 + \tau_s W_i} \tag{13.2-7}$$

Steady-State Population Difference (in Presence of Amplifier Radiation)

$$\tau_s = \tau_2 + \tau_1\left(1 - \frac{\tau_2}{\tau_{21}}\right), \tag{13.2-8}$$

Saturation Time Constant

where N_0 is the steady-state population difference in the absence of amplifier radiation, given by (13.2-4). The characteristic time τ_s is always positive since $\tau_2 \leq \tau_{21}$.

In the absence of amplifier radiation, $W_i = 0$ so that (13.2-7) provides $N = N_0$, as expected. Because τ_s is positive, the steady-state population difference in the presence of radiation always has a smaller absolute value than in the absence of radiation, i.e., $|N| \leq |N_0|$. If the radiation is sufficiently weak so that $\tau_s W_i \ll 1$ (the **small-signal approximation**), we may take $N \approx N_0$. As the radiation becomes stronger, W_i increases and N approaches zero regardless of the initial sign of N_0, as shown in Fig. 13.2-5. This arises because stimulated emission and absorption dominate the interaction when W_i is very large and they have equal probability densities. It is apparent that even very strong radiation cannot convert a negative population difference into a positive population difference, nor vice versa. The quantity τ_s plays the role of a **saturation time constant**, as is evident from Fig. 13.2-5.

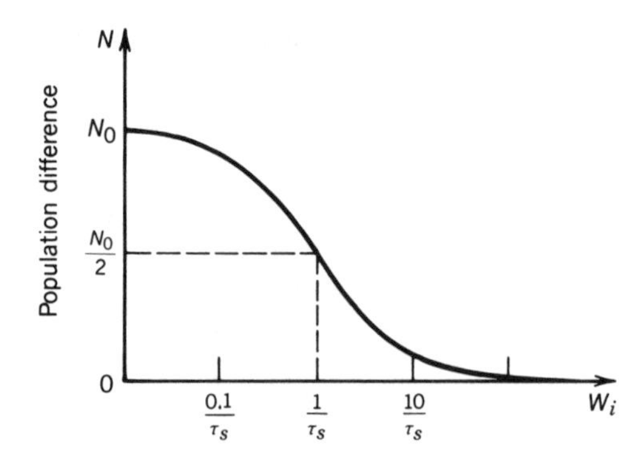

Figure 13.2-5 Depletion of the steady-state population difference $N = N_2 - N_1$ as the rate of absorption and stimulated emission W_i increases. When $W_i = 1/\tau_s$, N is reduced by a factor of 2 from its value when $W_i = 0$.

EXERCISE 13.2-2

Saturation Time Constant. Show that if $t_{sp} \ll \tau_{nr}$ (the nonradiative part of the lifetime τ_{21} of the 2–1 transition), $\ll \tau_{20}$, and $\gg \tau_1$, then $\tau_s \approx t_{sp}$.

We now proceed to examine specific (four- and three-level) schemes that are used in practice to achieve a population inversion. The object of these arrangements is to make use of an excitation process to increase the number of atoms in level 2 while decreasing the number in level 1.

B. Four- and Three-Level Pumping Schemes

Four-Level Pumping Schemes

In this arrangement, shown in Fig. 13.2-6, level 1 lies above the ground state (which is designated as the lowest energy level 0). In thermal equilibrium, level 1 will be virtually unpopulated, provided that $E_1 \gg k_B T$, which is of course highly desirable. Pumping is accomplished by making use of the energy level (or collection of energy levels) lying

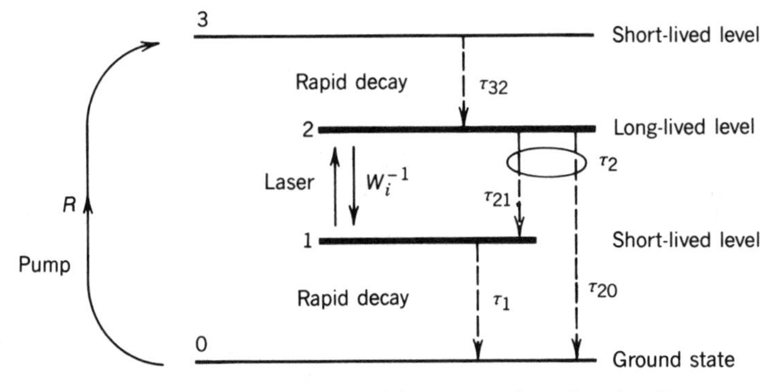

Figure 13.2-6 Energy levels and decay rates for a four-level system.

above level 2 and designated level 3. The 3–2 transition has a short lifetime (decay occurs rapidly) so that there is little accumulation in level 3. For reasons that are made clear in Problem 13.2-1, level 2 is pumped through level 3 rather than directly. Level 2 is long-lived, so that it accumulates population, whereas level 1 is short-lived so that it sustains little accumulation. All told, four energy levels are involved in the process but the optical interaction of interest is restricted to only two of them (levels 1 and 2).

An external source of energy (e.g., photons with frequency E_3/h) pumps atoms from level 0 to level 3 at a rate R. If the decay from level 3 to 2 is sufficiently rapid, it may be considered to be instantaneous, in which case pumping to level 3 is equivalent to pumping level 2 at the rate $R_2 = R$. In this configuration, atoms are neither pumped into nor out of level 1, so that $R_1 = 0$. The situation is then the same as that shown in Fig. 13.2-4. Thus the expressions in (13.2-7) and (13.2-8) apply. In the absence of amplifier radiation ($W_i = \phi = 0$), the steady-state population difference is given by (13.2-4) with $R_1 = 0$, i.e.,

$$N_0 = R\tau_2\left(1 - \frac{\tau_1}{\tau_{21}}\right). \tag{13.2-9}$$

In most four-level systems, the nonradiative decay component in the transition between 2 and 1 is typically negligible ($t_{sp} \ll \tau_{nr}$) and $\tau_{20} \gg t_{sp} \gg \tau_1$ (see Exercise 13.2-2), so that

$$N_0 \approx Rt_{sp}, \tag{13.2-10}$$

$$\tau_s \approx t_{sp}, \tag{13.2-11}$$

and therefore

$$N \approx Rt_{sp}/(1 + t_{sp}W_i). \tag{13.2-12}$$

Implicit in the preceding derivation is the assumption that the pumping rate R is independent of the population difference $N = N_2 - N_1$. This is not always the case, however, because the population densities of the ground state and level 3, N_g and N_3, are related to N_1 and N_2 by

$$N_g + N_1 + N_2 + N_3 = N_a, \tag{13.2-13}$$

where the total atomic density in the system N_a is a constant. If the pumping involves a transition between the ground state and level 3 with transition probability W, then $R = (N_g - N_3)W$. If levels 1 and 3 are short-lived ($N_1 \approx N_3 \approx 0$), then $N_g + N_2 \approx N_a$ so that $N_g \approx N_a - N_2 \approx N_a - N$.

Under these conditions, the pumping rate can be approximated as

$$R \approx (N_a - N)W. \tag{13.2-14}$$

It is seen to be a linearly decreasing function of the population difference N and is therefore clearly not independent of it. Substituting $R = (N_a - N)W$ into (13.2-12) and reorganizing terms, we obtain

$$N \approx \frac{t_{sp}N_aW}{1 + t_{sp}W_i + t_{sp}W}. \tag{13.2-15}$$

Finally, the population difference can be written in the generic form of (13.2-7),

$$N = \frac{N_0}{1 + \tau_s W_i},$$

but where now N_0 and τ_s are given by

$$\boxed{N_0 \approx \frac{t_{sp} N_a W}{1 + t_{sp} W},} \tag{13.2-16}$$

and

$$\boxed{\tau_s \approx \frac{t_{sp}}{1 + t_{sp} W},} \tag{13.2-17}$$

rather than by (13.2-10) and (13.2-11). Under conditions of weak pumping ($W \ll 1/t_{sp}$), $N_0 \approx t_{sp} N_a W$ is proportional to W (the pumping transition probability density), and $\tau_s \approx t_{sp}$, giving rise to the results obtained previously. However, as the pumping increases, N_0 saturates and τ_s decreases.

Three-Level Pumping Schemes

A three-level pumping arrangement, in contrast, makes use of the ground state ($E_1 = 0$) as the lower laser level 1, as shown in Fig. 13.2-7. Again, an auxiliary third level (designated 3) is involved. The 3–2 decay is rapid so that there is no buildup of population in level 3. The 3–1 decay is slow (i.e., $\tau_{32} \ll \tau_{31}$) so that the pumping ends up populating the upper laser level. Level 2 is long-lived so that it accumulates population. Atoms are pumped from level 1 to level 3 (e.g., by absorbing radiation at the frequency E_3/h) at a rate R; their fast (nonradiative) decay to level 2 provides the pumping rate $R_2 = R$.

It is not difficult to see that under rapid 3–2 decay, the three-level system displayed in Fig. 13.2-7 is a special case of the system shown in Fig. 13.2-4 (provided that R is independent of N) with the parameters

$$R_1 = R_2 = R, \qquad \tau_1 = \infty, \qquad \tau_2 = \tau_{21}.$$

To avoid algebraic problems in connection with the value $\tau_1 = \infty$, rather than substituting these special values into (13.2-7) and (13.2-8), we return to the original rate equations (13.2-5) and (13.2-6). In the steady state, both (13.2-5) and (13.2-6) result in

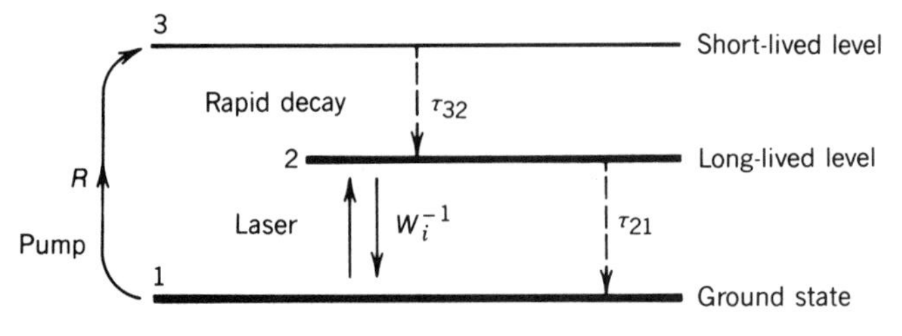

Figure 13.2-7 Energy levels and decay rates for a three-level system.

the same equation,

$$0 = R - \frac{N_2}{\tau_{21}} - N_2 W_i + N_1 W_i. \tag{13.2-18}$$

It is not possible to determine both N_1 and N_2 from a single equation relating them. However, knowledge of the total atomic density N_a in the system (in levels 1, 2, and 3) provides an auxiliary condition that does permit N_1 and N_2 to be determined. Since τ_{32} is very short, level 3 retains a negligible steady-state population; all of the atoms that are raised to it immediately decay to level 2. Thus

$$N_1 + N_2 = N_a, \tag{13.2-19}$$

which enables us to solve (13.2-18) for N_1 and N_2 and thereby to determine the population difference $N = N_2 - N_1$ and the saturation time τ_s. The result may be cast in the usual form of (13.2-7), $N = N_0/(1 + \tau_s W_i)$, where now

$$N_0 = 2R\tau_{21} - N_a \tag{13.2-20}$$

$$\tau_s = 2\tau_{21}. \tag{13.2-21}$$

When nonradiative decay from levels 2 to 1 is negligible ($t_{sp} \ll \tau_{nr}$), τ_{21} may be replaced by t_{sp}, whereupon

$$N_0 \approx 2Rt_{sp} - N_a \tag{13.2-22}$$

$$\tau_s \approx 2t_{sp}. \tag{13.2-23}$$

Note that $\tau_s \approx t_{sp}$ for four-level pumping schemes [see (13.2-11)].

It is of interest to compare these equations with the analogous results (13.2-10) and (13.2-11) for a four-level pumping scheme. Attaining a population inversion ($N > 0$ and therefore $N_0 > 0$) in the three-level system requires a pumping rate $R > N_a/2t_{sp}$. Thus, just to make the population density N_2 equal to N_1 (i.e., $N_0 = 0$) requires a substantial pump power density, given by $E_3 N_a/2t_{sp}$. The large population in the ground state (which is the lowest laser level) provides an inherent obstacle to achieving a population inversion in a three-level system that is avoided in four-level systems (in which level 1 is normally empty).

The dependence of the pumping rate R on the population difference N can be included in the analysis of the three-level system by writing $R = (N_1 - N_3)W$, $N_3 \approx 0$, and $N_1 = \frac{1}{2}(N_a - N)$, from which $R \approx \frac{1}{2}(N_a - N)W$. Substituting in the principal equation $N = (2Rt_{sp} - N_a)/(1 + 2t_{sp}W_i)$, and reorganizing terms, we again obtain

$$N = \frac{N_0}{1 + \tau_s W_i},$$

but now with

$$\boxed{N_0 = \frac{N_a(t_{sp}W - 1)}{1 + t_{sp}W},} \tag{13.2-24}$$

and

$$\tau_s = \frac{2t_{sp}}{1 + t_{sp}W}. \tag{13.2-25}$$

Thus, as in the four-level scheme, N_0 and τ_s are in general nonlinear functions of the pumping transition probability W.

EXERCISE 13.2-3

Pumping Powers in Three- and Four-Level Systems

(a) Determine the pumping transition probability W required to achieve a zero population difference in a three- and a four-level laser amplifier.

(b) If the pumping transition probability $W = 2/t_{sp}$ in the three-level system and $= 1/2t_{sp}$ in the four-level system, show that $N_0 = N_a/3$. Compare the pumping powers required to achieve this population difference.

Examples of Pumping Methods

As indicated earlier, pumping may be achieved by many methods, including the use of electrical, optical, and chemical means. A number of common methods of electrical and optical pumping are illustrated schematically in Fig. 13.2-8.

It is important to note that R_1 and R_2 represent the numbers of atoms/cm^3-s that are pumped successfully. The pumping process is generally quite inefficient. In optical pumping, for example, many of the photons supplied by the pump fail to raise the atoms to the upper laser level and are therefore wasted.

C. Examples of Laser Amplifiers

Laser amplification can take place in a great variety of materials. The energy-level diagrams for several atoms, molecules, and solids that exhibit laser action were shown in Sec. 12.1A. Practical laser systems usually involve many interacting energy levels that influence N_1 and N_2, the populations of the transition of interest, as illustrated in Fig. 13.2-2. Nevertheless, the essential principles of laser amplifier operation may be understood by classifying lasers as either three- or four-level systems.

This is illustrated by three **solid-state laser** amplifiers which are discussed in turn below: the three-level ruby laser amplifier, the four-level neodymium-doped yttrium–aluminum garnet laser amplifier, and the three-level erbium-doped silica fiber laser amplifier. Although most laser amplifiers and oscillators operate on the basis of a four-level pumping scheme, two notable exceptions are ruby and Er^{3+}-doped silica fiber. Laser amplification can also be achieved with **gas lasers** and **liquid lasers**, as indicated briefly near the end of this section. All of the laser amplifiers discussed here also operate as laser oscillators (see Sec. 14.2E).

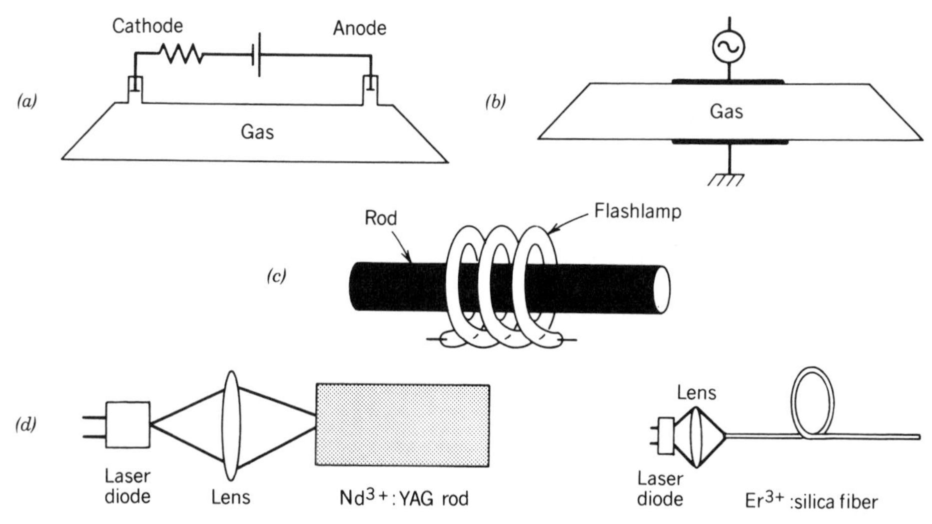

Figure 13.2-8 Examples of electrical and optical pumping. (*a*) Direct current (dc) is often used to pump gas lasers. The current may be passed either along the laser axis, to give a longitudinal discharge, or transverse to it. The latter configuration is often used for high-pressure pulsed lasers, such as the transversely excited atmospheric (TEA) CO_2 laser. (*b*) Radio-frequency (RF) discharge currents are also used for pumping gas lasers. (*c*) Flashlamps are effective for optically pumping ruby and rare-earth solid-state lasers. (*d*) A semiconductor injection laser diode (or array of laser diodes) can be used to optically pump Nd^{3+}:YAG or Er^{3+}:silica fiber lasers.

Ruby

Ruby ($Cr^{3+}:Al_2O_3$) is sapphire (Al_2O_3), in which chromium ions (Cr^{3+}) replace a small percentage of the aluminum ions (see Sec. 12.1A). As with most materials, laser action can take place on a variety of transitions. The energy levels pertinent to the well-known red ruby laser transition are shown in Fig. 13.2-9 (these are labeled in group-theory notation). Ruby is the first material in which laser action was observed. In

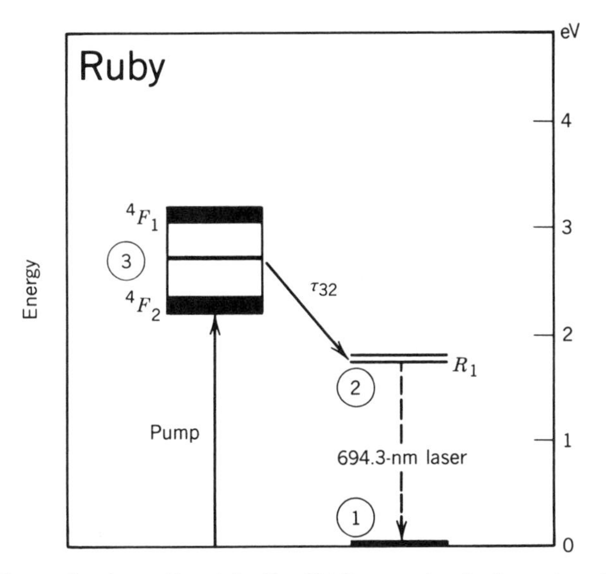

Figure 13.2-9 Energy levels pertinent to the 694.3-nm red ruby laser transition. The three interacting levels are indicated in circles.

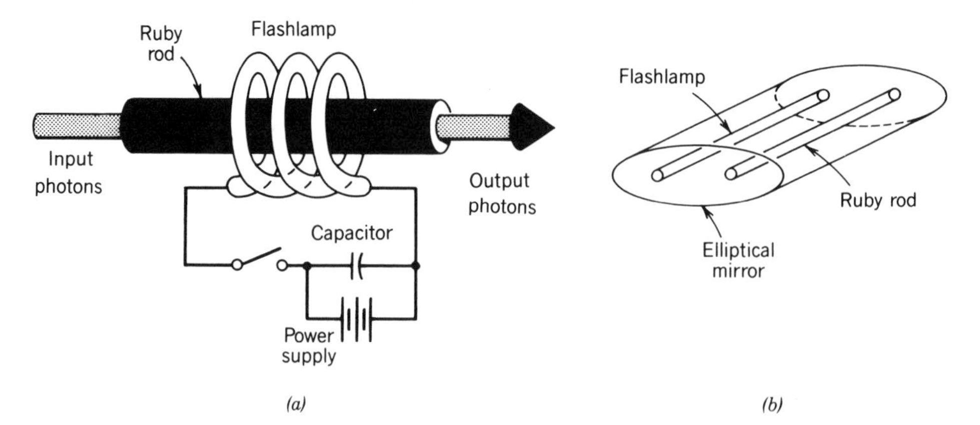

Figure 13.2-10 The ruby laser amplifier. (*a*) Geometry used in the first laser oscillator built by Maiman in 1960 (see Chap. 14). (*b*) Cross section of a high-efficiency geometry using a rod-shaped flashlamp and a reflecting elliptical cylinder.

essence, ruby is a three-level system in which level 1 is the ground state, level 2 consists of a pair of closely spaced discrete levels (the lower of which corresponds to the red laser transition at $\lambda_o = 694.3$ nm), and level 3 comprises two bands of energies centered at about 550 nm (green) and 400 nm (violet). These absorption bands are responsible for the pink color of the material.

The material may be optically pumped from level 1 to level 3 by surrounding the ruby rod by a flashlamp or enclosing it with a rod-shaped flashlamp within a reflecting cylinder of elliptical cross section, as shown in Fig. 13.2-10. The flashlamp emits broad-spectrum radiation, some fraction of which is absorbed and results in the excitation of the Cr^{3+} ions to level 3. The broad nature of level 3 is useful in maximizing the percentage of pump light absorbed. Excited Cr^{3+} ions rapidly decay from level 3 to level 2 (τ_{32} is of the order of picoseconds), whereas the spontaneous lifetime for the 2–1 transition is relatively long ($t_{sp} \approx 3$ ms), in agreement with the scheme shown in Fig. 13.2-7. Nonradiative decay is negligible ($\tau_{21} \approx t_{sp}$). The transition has a homogeneously broadened linewidth $\Delta\nu = 60$ GHz, arising principally from elastic collisions with lattice phonons.

Commercially available ruby laser amplifiers use rods that are typically 5 to 20 cm in length. They can deliver a small-signal gain of about 20 in the pulsed mode. The properties of a typical ruby laser oscillator are provided in Table 14.2-1.

Nd^{3+}:YAG and Nd^{3+}:Glass

A useful near-infrared four-level laser amplifier makes use of neodymium in the form of impurity ions in a crystal of yttrium–aluminum garnet ($Nd_xY_{3-x}Al_5O_{12}$, usually written as Nd^{3+}:YAG). The crystal is pale purple in color. The energy levels pertinent to the $\lambda_o = 1.064$-μm transition are shown in Fig. 13.2-11; spectroscopic notation is used. Level 1 has an energy ≈ 0.2 eV above the ground state. This energy is substantially larger than $k_B T \approx 0.026$ eV at room temperature, so that the thermal population of the lower laser level is negligible. Level 3 is a collection of four ≈ 30-nm-wide absorption bands centered at about 810, 750, 585, and 525 nm. The 2–1 transition is homogeneously broadened (as a result of collisions with lattice phonons), with a room-temperature linewidth $\Delta\nu \approx 120$ GHz. The excited ions rapidly decay from level 3 to level 2 ($\tau_{32} \approx 100$ ns), the spontaneous lifetime t_{sp} is 1.2 ms, and τ_1 is short (≈ 30 ns), in agreement with the four-level scheme shown in Fig. 13.2-6. The gain is substantially greater than that of ruby by virtue of it being a four-level system.

Nd^{3+}:YAG can also be optically pumped directly to the upper laser level; an efficient laser system making use of this scheme has recently been developed in which the pump is a semiconductor injection laser [see Fig. 13.2-8(*d*)].

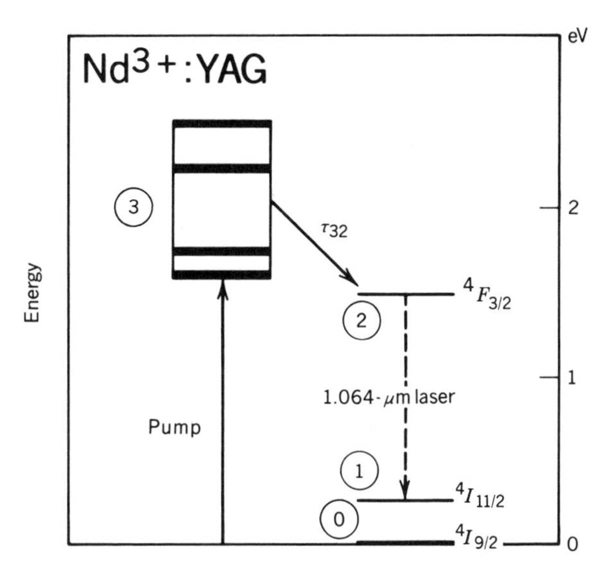

Figure 13.2-11 Energy levels pertinent to the 1.064-μm Nd^{3+}:YAG laser transition. The energy levels for Nd^{3+}:glass are similar but the absorption bands are broader.

The neodymium in glass laser amplifier (Nd^{3+}:glass) has characteristics that are quite similar to those of Nd^{3+}:YAG, with the notable exception that it is inhomogeneously broadened; this is a result of the amorphous nature of glass, which presents a different environment at each ionic location. Nd^{3+}:glass therefore has a far larger room-temperature linewidth, $\Delta\nu \approx 3000$ GHz, which turns out to be desirable for mode-locked pulsed lasers (see Chap. 14). Nd^{3+}:glass amplifiers can be made in very large sizes and have been used extensively in laser fusion experiments (particularly in the 10-beam NOVA laser system at the Lawrence Livermore National Laboratory in California, which is capable of delivering 10^5 J in a 1-ns pulse and in the GEKKO system at Osaka University in Japan). The characteristics of typical Nd^{3+}:YAG and Nd^{3+}:glass laser oscillators are provided in Table 14.2-1.

Er^{3+}:Silica Fiber

Rare-earth-doped silica fibers can serve as useful laser amplifying media while offering the advantages of single-mode guided-wave optics (see Chaps. 7 and 8). In particular they offer polarization-independent gain and low insertion loss. The core of the silica fiber may be doped with any of a number of rare-earth ions (e.g., Nd, Er, Yb, Pr, Sm). Pumping is achieved by transmitting laser light (e.g., light from a semiconductor injection laser, dye laser, color-center laser, Ti^{3+}:Al_2O_3 laser, or Ar^+ ion laser) through the fiber [see Fig. 13.2-8(d)]. Fiber laser amplifiers can be made to operate over a broad range of wavelengths (e.g., 1.3 μm, 1.55 μm, 2 to 3 μm).

Er^{3+}:silica fibers, in particular, have a broad laser transition ($\Delta\nu \approx 4000$ GHz) near $\lambda = 1.55$ μm, which coincides with the wavelength of maximum transmission for silica fibers (see Fig. 8.3-2). Because of their high gain, erbium-doped silica fibers offer substantial promise for use as optical amplifiers and repeaters in fiber-optic communication systems. In one configuration, an 807-nm semiconductor laser pump is used to drive a 1-m-long SiO_2:GeO_2 fiber (typical fiber lengths lie in the range between 0.5 and 10 m) doped with ≈ 500 parts per million (ppm) erbium. This wavelength, as well as 980 nm, are convenient because of the presence of strong pumping bands in Er^{3+}. However, pumping at 807 nm gives rise to undesirable excited-state absorption. The laser transition can instead be directly pumped at 1.48 μm by light from an InGaAsP semiconductor laser in which case excited-state absorption does not occur. Efficient

TABLE 13.2-1 Characteristics of a Number of Important Laser Transitions

Laser Medium	Transition Wavelength λ_o (μm)	Transition Cross Section σ_0 (cm^2)	Spontaneous Lifetime t_{sp}	Transition Linewidth[a] $\Delta\nu$		Refractive Index n
He–Ne	0.6328	1×10^{-13}	0.7 μs	1.5 GHz	I	≈ 1
Ruby	0.6943	2×10^{-20}	3.0 ms	60 GHz	H	1.76
Nd^{3+}:YAG	1.064	4×10^{-19}	1.2 ms	120 GHz	H	1.82
Nd^{3+}:glass	1.06	3×10^{-20}	0.3 ms	3 THz	I	1.5
Er^{3+}:silica fiber	1.55	6×10^{-21}	10.0 ms	4 THz	H/I	1.46
Rhodamine-6G dye	0.56–0.64	2×10^{-16}	3.3 ns	5 THz	H/I	1.33
Ti^{3+}:Al_2O_3	0.66–1.18	3×10^{-19}	3.2 μs	100 THz	H	1.76
CO_2	10.6	3×10^{-18}	2.9 s	60 MHz	I	≈ 1
Ar^+	0.515	3×10^{-12}	10.0 ns	3.5 GHz	I	≈ 1

[a] H and I indicate line broadening dominated by homogeneous and inhomogeneous mechanisms, respectively.

light amplification is possible because of the frequency shift that exists between the fluorescence and absorption bands of this transition. Currently, gains $\approx$ 30 dB are available by launching $\approx$ 5 mW of pump power (from a diode laser pump operated at either 980 nm or 1.48 μm) into a roughly 50-m length of fiber containing $\approx$ 300 ppm Er_2O_3. Optical bandwidths $\approx$ 30 nm can be obtained, although larger bandwidths are possible with reduced gain.

The Er^{3+}:silica fiber system behaves as a three-level laser at $T = 300$ K and as a four-level laser when cooled to $T = 77$ K. The broadening is a mixture of homogeneous (phonon mediated) and inhomogeneous (arising from local field variations in the glass).

Other Laser Amplifiers

The transition cross section, spontaneous lifetime, transition linewidths, and refractive indices of several important laser transitions are provided in Table 13.2-1. The free-space wavelength λ_o shown in the table represents the most commonly used transition in each laser medium. The He–Ne gas laser system, for example, is most often used on its red-orange line at 0.633 μm, but it is also extensively used at 0.543, 1.15, and 3.39 μm (it also has laser transitions at hundreds of other wavelengths). CO_2 is a commonly used laser amplifying medium in the middle-infrared region of the spectrum. The values reported in the table are typical for low-pressure operation (the atomic linewidth in a gas depends on its pressure because of the role of collision broadening, which is a homogeneous broadening mechanism).

The tunable rhodamine-6G dye laser, which is usually pumped by an Ar^+ laser, provides gain over a continuous band of wavelengths stretching from 560 to 640 nm. Other dyes cover different wavelength regions. Dye laser amplifiers enjoy broad application and are effective for the amplification of femtosecond optical pulses. The Ti^{3+}:Al_2O_3 laser enjoys even broader tunability than the rhodamine-6G dye laser and at the same time is far easier to operate. Free-electron laser systems are also often used for amplification. The semiconductor laser amplifier is discussed in Chap. 16.

13.3 AMPLIFIER NONLINEARITY AND GAIN SATURATION

A. Gain Coefficient

It has been established that the gain coefficient $\gamma(\nu)$ of a laser medium depends on the population difference N [see (13.1-4)]; that N depends on the transition rate W_i [see (13.2-7)]; and that W_i, in turn, depends on the radiation photon-flux density ϕ [see

(13.1-1)]. It follows that the gain coefficient of a laser medium is dependent on the photon-flux density that is to be amplified. This is the origin of gain saturation and laser amplifier nonlinearity, as we now show.

Substituting (13.1-1) into (13.2-7) provides

$$N = \frac{N_0}{1 + \phi/\phi_s(\nu)} \tag{13.3-1}$$

where

$$\frac{1}{\phi_s(\nu)} = \tau_s \sigma(\nu) = \frac{\lambda^2}{8\pi} \frac{\tau_s}{t_{sp}} g(\nu). \tag{13.3-2}$$

Saturation Photon-Flux Density

This represents the dependence of the population difference N on the photon-flux density ϕ. Now, substituting (13.3-1) into the expression for the gain coefficient (13.1-4) leads directly to the saturated gain coefficient for homogeneously broadened media:

$$\gamma(\nu) = \frac{\gamma_0(\nu)}{1 + \phi/\phi_s(\nu)}, \tag{13.3-3}$$

Saturated Gain Coefficient

where

$$\gamma_0(\nu) = N_0 \sigma(\nu) = N_0 \frac{\lambda^2}{8\pi t_{sp}} g(\nu). \tag{13.3-4}$$

Small-Signal Gain Coefficient

The gain coefficient is a decreasing function of the photon-flux density ϕ, as illustrated in Fig. 13.3-1. The quantity $\phi_s(\nu) = 1/\tau_s \sigma(\nu)$ represents the photon-flux density at which the gain coefficient decreases to half its maximum value; it is therefore called the **saturation photon-flux density**. When $\tau_s \approx t_{sp}$ the interpretation of $\phi_s(\nu)$ is straightforward: Roughly one photon can be emitted during each spontaneous emission time into each transition cross-sectional area $[\sigma(\nu)\phi_s(\nu)t_{sp} = 1]$.

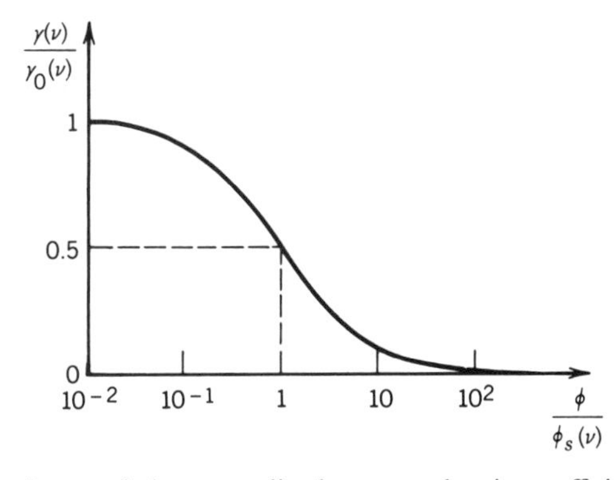

Figure 13.3-1 Dependence of the normalized saturated gain coefficient $\gamma(\nu)/\gamma_0(\nu)$ on the normalized photon-flux density $\phi/\phi_s(\nu)$. When ϕ equals its saturation value $\phi_s(\nu)$, the gain coefficient is reduced by a factor of 2.

EXERCISE 13.3-1

Saturation Photon-Flux Density for Ruby. Determine the saturation photon-flux density, and the corresponding saturation intensity, for the $\lambda_o = 694.3$-nm ruby laser transition at $\nu = \nu_0$. Use the parameters provided in Table 13.2-1 on page 480. Assume that $\tau_s \approx 2t_{sp}$, in accordance with (13.2-23)

EXERCISE 13.3-2

Spectral Broadening of a Saturated Amplifier. Consider a homogeneously broadened amplifying medium with a Lorentzian lineshape of width $\Delta\nu$ [see (13.1-8)]. Show that when the photon-flux density is ϕ, the amplifier gain coefficient $\gamma(\nu)$ assumes a Lorentzian lineshape with width:

$$\Delta\nu_s = \Delta\nu\left(1 + \frac{\phi}{\phi_s(\nu_0)}\right)^{1/2}. \tag{13.3-5}$$

Linewidth of Saturated Amplifier

This demonstrates that gain saturation is accompanied by an increase in bandwidth (i.e., reduced frequency selectivity), as shown in Fig. 13.3-2.

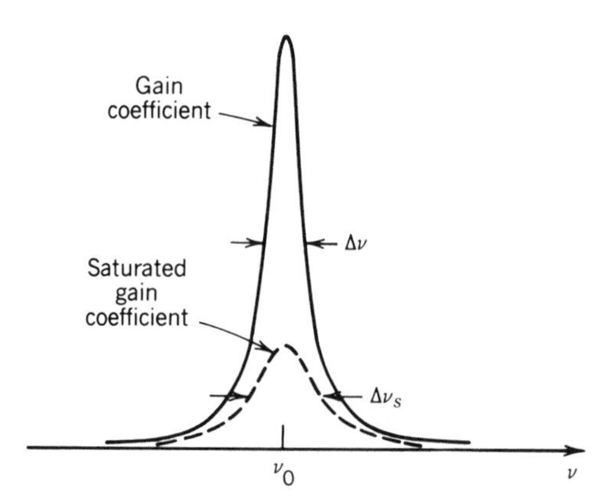

Figure 13.3-2 Gain coefficient reduction and bandwidth increase resulting from saturation when $\phi = 2\phi_s(\nu_0)$.

B. Gain

Having determined the effect of saturation on the gain coefficient (gain per unit length), we embark on determining the behavior of the overall gain for a homogeneously broadened laser amplifier of length d. For simplicity, we suppress the frequency dependencies of $\gamma(\nu)$ and $\phi_s(\nu)$, using the symbols γ and ϕ_s instead.

If the photon-flux density at position z is $\phi(z)$, then in accordance with (13.3-3) the gain coefficient at that position is also a function of z. We know from (13.1-3) that the incremental increase of photon-flux density at the position z is $d\phi = \gamma\phi\, dz$, which leads to the differential equation

$$\frac{d\phi}{dz} = \frac{\gamma_0 \phi}{1 + \phi/\phi_s}. \tag{13.3-6}$$

Rewriting this equation as $(1/\phi + 1/\phi_s)\, d\phi = \gamma_0\, dz$, and integrating, we obtain

$$\ln \frac{\phi(z)}{\phi(0)} + \frac{\phi(z) - \phi(0)}{\phi_s} = \gamma_0 z. \tag{13.3-7}$$

The relation between the input photon-flux density to the amplifier $\phi(0)$ and the output $\phi(d)$ is therefore

$$[\ln(Y) + Y] = [\ln(X) + X] + \gamma_0 d, \tag{13.3-8}$$

where $X = \phi(0)/\phi_s$ and $Y = \phi(d)/\phi_s$ are the input and output photon-flux densities normalized to the saturation photon-flux density, respectively.

The solution for the gain $G = \phi(d)/\phi(0) = Y/X$ can be examined in two limiting cases:

- If both X and Y are much smaller than unity (i.e., the photon-flux densities are much smaller than the saturation photon-flux density), then X and Y are negligible in comparison with $\ln(X)$ and $\ln(Y)$, whereupon we obtain the approximate relation $\ln(Y) \approx \ln(X) + \gamma_0 d$, from which

$$Y \approx X \exp(\gamma_0 d). \tag{13.3-9}$$

  In this case the relation between Y and X is linear, with a gain $G = Y/X \approx \exp(\gamma_0 d)$. This accords with (13.1-7) which was obtained under the small-signal approximation, valid when the gain coefficient is independent of the photon-flux density, i.e., $\gamma \approx \gamma_0$.
- When $X \gg 1$, we can neglect $\ln(X)$ in comparison with X, and $\ln(Y)$ in comparison with Y, whereupon

$$Y \approx X + \gamma_0 d$$

  or

$$\phi(d) \approx \phi(0) + \gamma_0 \phi_s d$$

$$\approx \phi(0) + \frac{N_0 d}{\tau_s}. \tag{13.3-10}$$

  Under these heavily saturated conditions, the atoms of the medium are "busy" emitting a constant photon-flux density $N_0 d/\tau_s$. Incoming input photons therefore simply leak through to the output, augmented by a constant photon-flux density that is independent of the amplifier input.

For intermediate values of X and Y, (13.3-8) must be solved numerically. A plot of the solution is shown as the solid curve in Fig. 13.3-3(*b*). The linear input–output

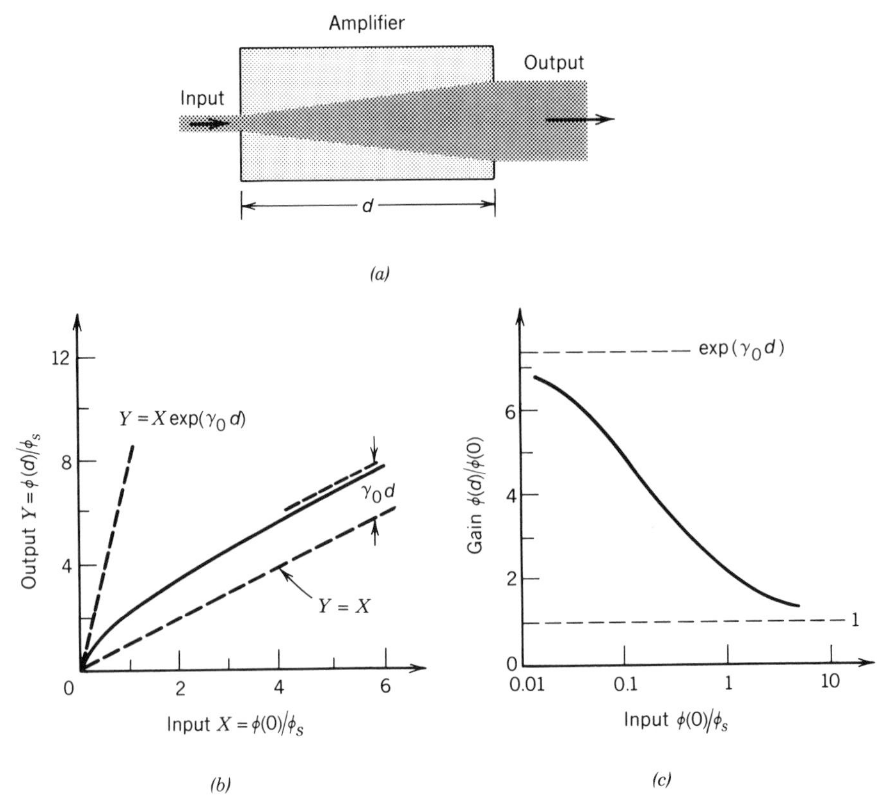

Figure 13.3-3 (*a*) A nonlinear (saturated) amplifier. (*b*) Relation between the normalized output photon-flux density $Y = \phi(d)/\phi_s$ and the normalized input photon-flux density $X = \phi(0)/\phi_s$. For $X \ll 1$, the gain $Y/X \approx \exp(\gamma_0 d)$. For $X \gg 1$, $Y \approx X + \gamma_0 d$. (*c*) Gain as a function of the input normalized photon-flux density X in an amplifier of length d when $\gamma_0 d = 2$.

relationship obtained for $X \ll 1$, and the saturated relationship for $X \gg 1$, are evident as limiting cases of the numerical solution. The gain $G = Y/X$ is plotted in Fig. 13.3-3(*c*). It achieves its maximum value $\exp(\gamma_0 d)$ for small values of the input photon-flux density ($X \ll 1$), and decreases toward unity as $X \to \infty$.

Saturable Absorbers

If the gain coefficient γ_0 is negative, i.e., if the population is normal rather than inverted ($N_0 < 0$), the medium provides attenuation rather than amplification. The attenuation coefficient $\alpha(\nu) = -\gamma(\nu)$ also suffers from saturation, in accordance with the relation $\alpha(\nu) = \alpha_0(\nu)/[1 + \phi/\phi_s(\nu)]$. This indicates that there is less absorption for large values of the photon-flux density. A material exhibiting this property is called a **saturable absorber**.

The relation between the output and input photon-flux densities, $\phi(d)$ and $\phi(0)$, for an absorber of length d is governed by (13.3-8) with negative γ_0. The overall transmittance of the absorber $Y/X = \phi(d)/\phi(0)$ is presented as a function of $X = \phi(0)/\phi_s$ in the solid curve of Fig. 13.3-4. The transmittance increases as $\phi(0)$ increases, ultimately reaching a limiting value of unity. This effect occurs because the population difference $N \to 0$, so that there is no net absorption.

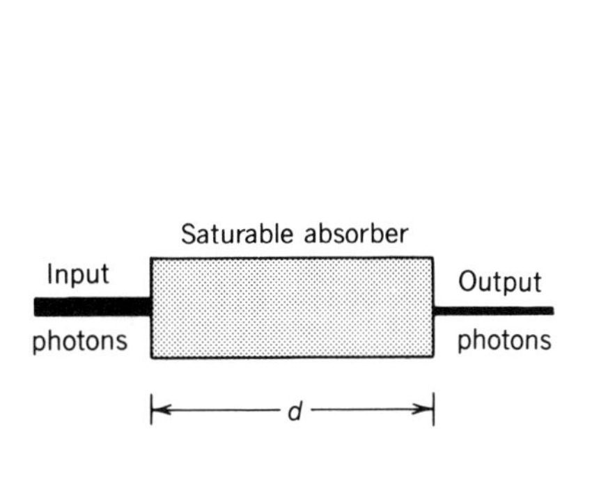

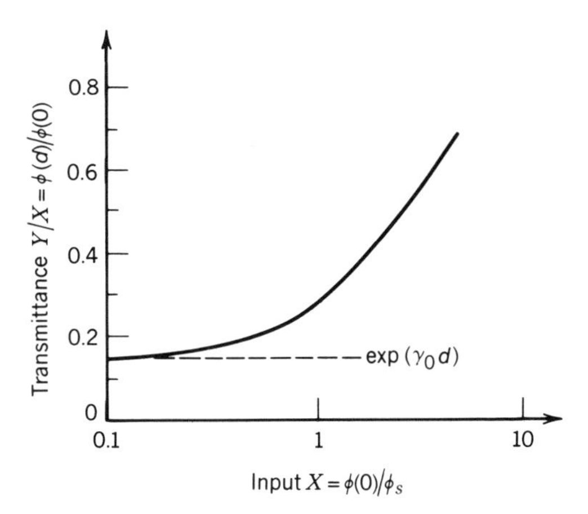

Figure 13.3-4 The transmittance of a saturable absorber $Y/X = \phi(d)/\phi(0)$ versus the normalized input photon-flux density $X = \phi(0)/\phi_s$, for $\gamma_0 d = -2$. The transmittance increases with increasing input photon-flux density.

*C. Gain of Inhomogeneously Broadened Amplifiers

An inhomogeneously broadened medium comprises a collection of atoms with different properties. As discussed in Sec. 12.2D, the subset of atoms labeled β has a homogeneously broadened lineshape function $g_\beta(\nu)$. The overall inhomogeneous average lineshape function of the medium is described by $\bar{g}(\nu) = \langle g_\beta(\nu) \rangle$, where $\langle \cdot \rangle$ represents an average with respect to β.

Because the small-signal gain coefficient $\gamma_0(\nu)$ is proportional to $g(\nu)$, as provided in (13.3-4), different subsets β of atoms have different gain coefficients $\gamma_{0\beta}(\nu)$. The average small-signal gain coefficient is therefore

$$\bar{\gamma}_0(\nu) = N_0 \frac{\lambda^2}{8\pi t_{sp}} \bar{g}(\nu). \tag{13.3-11}$$

Obtaining the saturated gain coefficient is more subtle, however, because the saturation photon-flux density $\phi_s(\nu)$, being inversely proportional to $g(\nu)$ as provided in (13.3-2), is itself dependent on the subset of atoms β. An average gain coefficient may be defined by using (13.3-3) and (13.3-2),

$$\bar{\gamma}(\nu) = \langle \gamma_\beta(\nu) \rangle, \tag{13.3-12}$$

where

$$\gamma_\beta(\nu) = \frac{\gamma_{0\beta}(\nu)}{1 + \phi/\phi_{s\beta}(\nu)}$$

$$= b \frac{g_\beta(\nu)}{1 + \phi a^2 g_\beta(\nu)}, \tag{13.3-13}$$

with $b = N_0(\lambda^2/8\pi t_{sp})$ and $a^2 = (\lambda^2/8\pi)(\tau_s/t_{sp})$. Evaluating the average of (13.3-13) requires care because the average of a ratio is not equal to the ratio of the averages.

Doppler-Broadened Medium

Although all of the atoms in a Doppler-broadened medium share a $g(\nu)$ of identical shape, the center frequency of the subset β is shifted by an amount ν_β proportional to the velocity v_β of the subset. If $g(\nu)$ is Lorentzian with width $\Delta\nu$, (13.1-8) provides $g(\nu) = (\Delta\nu/2\pi)/[(\nu - \nu_0)^2 + (\Delta\nu/2)^2]$ and $g_\beta(\nu) = g(\nu - \nu_\beta)$. Substituting $g_\beta(\nu)$ into (13.3-13) provides

$$\gamma_\beta(\nu) = \frac{b(\Delta\nu/2\pi)}{(\nu - \nu_\beta - \nu_0)^2 + (\Delta\nu_s/2)^2}, \tag{13.3-14}$$

where

$$\Delta\nu_s = \Delta\nu\left(1 + \frac{\phi}{\phi_s(\nu_0)}\right)^{1/2} \tag{13.3-15}$$

and

$$\phi_s^{-1}(\nu_0) = \frac{2a^2}{\pi\Delta\nu} = \frac{\lambda^2}{8\pi}\frac{\tau_s}{t_{sp}}\frac{2}{\pi\Delta\nu}$$

$$= \frac{\lambda^2}{8\pi}\frac{\tau_s}{t_{sp}}g(\nu_0). \tag{13.3-16}$$

Equation (13.3-15) was obtained for the homogeneously broadened saturated amplifier considered in Exercise 13.3-2 [see (13.3-5)]. It is evident that the subset of atoms with velocity v_β has a saturated gain coefficient $\gamma_\beta(\nu)$ with a Lorentzian shape of width $\Delta\nu_s$ that increases as the photon-flux density becomes larger.

The average of $\gamma_\beta(\nu)$ specified in (13.3-12) is obtained by recalling that the shifts ν_β follow a zero-mean Gaussian probability density function $p(\nu_\beta) = (2\pi\sigma_D^2)^{-1/2}\exp(-\nu_\beta^2/2\sigma_D^2)$ with standard deviation σ_D (see Exercise 12.2-2). Thus $\bar{\gamma}(\nu) = \langle\gamma_\beta(\nu)\rangle$ is given by

$$\bar{\gamma}(\nu) = \int_{-\infty}^{\infty}\gamma_\beta(\nu)p(\nu_\beta)\,d\nu_\beta. \tag{13.3-17}$$

If $p(\nu_\beta)$ is much broader than $\gamma_\beta(\nu)$ (i.e., the Doppler broadening is much wider than $\Delta\nu_s$), we may regard the broad function $p(\nu_\beta)$ as constant and remove it from the integral when evaluating $\bar{\gamma}(\nu_0)$. Setting $\nu = \nu_0$ and $\nu_\beta = 0$ in the exponential provides

$$\bar{\gamma}(\nu_0) = \frac{bp(0)}{(1 + 2\phi a^2/\pi\Delta\nu)^{1/2}} = \frac{\bar{\gamma}_0}{[1 + \phi/\phi_s(\nu_0)]^{1/2}}, \tag{13.3-18}$$

where the average small-signal gain coefficient $\bar{\gamma}_0$ is

$$\bar{\gamma}_0 = N_0\frac{\lambda^2}{8\pi t_{sp}}(2\pi\sigma_D^2)^{-1/2}. \tag{13.3-19}$$

Equation (13.3-18) provides an expression for the average saturated gain coefficient of a Doppler broadened medium at the central frequency ν_0, as a function of the photon-flux density ϕ at $\nu = \nu_0$. The gain coefficient saturates as ϕ increases in

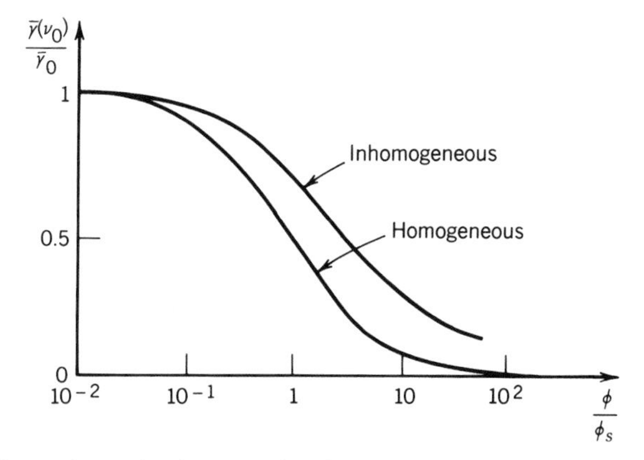

Figure 13.3-5 Comparison of gain saturation in homogeneously and inhomogeneously broadened media.

accordance with a square-root law. The gain coefficient in an inhomogeneously broadened medium therefore saturates more slowly than the gain coefficient in a homogeneously broadened medium [see (13.3-3)], as illustrated in Fig. 13.3-5.

Hole Burning

When a large flux density of monochromatic photons at frequency ν_1 is applied to an inhomogeneously broadened medium, the gain saturates only for those atoms whose lineshape function overlaps ν_1. Other atoms simply do not interact with the photons and remain unsaturated. When the saturated medium is probed by a weak monochromatic light source of varying frequency ν, the profile of the gain coefficient therefore exhibits a hole centered around ν_1, as illustrated in Fig. 13.3-6. This phenomenon is known as **hole burning**. Since the gain coefficient $\gamma_\beta(\nu)$ of the subset of atoms with velocity v_β has a Lorentzian shape with width $\Delta\nu_s$, given by (13.3-15), it follows that the width of the hole is $\Delta\nu_s$. As the flux density of saturating photons at ν_1 increases, both the depth and the width of the hole increase.

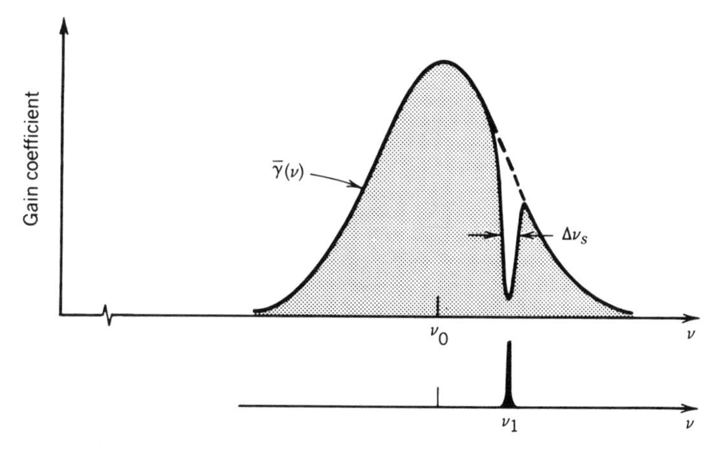

Figure 13.3-6 The gain coefficient of an inhomogeneously broadened medium is locally saturated by a large flux density of monochromatic photons at frequency ν_1.

*13.4 AMPLIFIER NOISE

The resonant medium that provides amplification by the process of stimulated emission also generates spontaneous emission. The light arising from the latter process, which is independent of the input to the amplifier, represents a fundamental source of laser amplifier noise. Whereas the amplified signal has a specific frequency, direction, and polarization, the amplified spontaneous emission (ASE) noise is broadband, multidirectional, and unpolarized. As a consequence it is possible to filter out some of this noise by following the amplifier with a narrow bandpass optical filter, a collection aperture, and a polarizer.

The probability density (per second) that an atom in the upper laser level spontaneously emits a photon of frequency between ν and $\nu + d\nu$ is (see Exercise 12.2-1).

$$P_{sp}(\nu)\, d\nu = \frac{1}{t_{sp}} g(\nu)\, d\nu. \tag{13.4-1}$$

The probability density of spontaneously emitting a photon of any frequency is, of course, $P_{sp} = 1/t_{sp}$. If N_2 is the atomic density in the upper energy level, the average spontaneously emitted photon density is $N_2 P_{sp}(\nu)$. The average spontaneously emitted power per unit volume per unit frequency is therefore $h\nu N_2 P_{sp}(\nu)$. This power density is emitted uniformly in all directions and is equally divided between the two polarizations. If the amplifier output is collected from a solid angle $d\Omega$ (as illustrated in Fig. 13.4-1), and from only one of the polarizations, it contains only a fraction $\frac{1}{2}\, d\Omega/4\pi$ of the spontaneously emitted power. Furthermore, if the receiver is sensitive only to photons within a narrow frequency band B centered about the amplified signal frequency ν, the number of photons added by spontaneous emission from an incremental volume of unit area and length dz is $\xi_{sp}(\nu)\, dz$, where

$$\xi_{sp}(\nu) = N_2 \frac{1}{t_{sp}} g(\nu) B \frac{d\Omega}{8\pi} \tag{13.4-2}$$

is the noise photon-flux density per unit length.

In determining the noise photon-flux density contributed by the amplifier, it is incorrect to simply multiply the photon-flux density per unit length by the length of the amplifier. The spontaneous-emission noise itself is amplified by the medium; noise generated near the input end of the amplifier provides a greater contribution than noise generated near the output end. One way in which spontaneous-emission noise

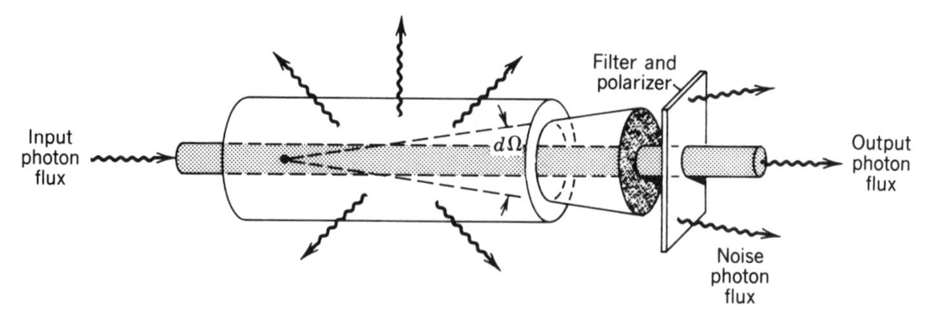

Figure 13.4-1 Spontaneous emission is a source of amplifier noise. It is broadband, radiated in all directions, and unpolarized. Only light within a narrow optical band, solid angle $d\Omega$, and a single polarization is collected by the optics at the output of the amplifier.

may be accounted for is to replace the differential equation governing the growth of photon-flux density (13.1-3) by

$$\frac{d\phi}{dz} = \gamma(\nu)\phi + \xi_{sp}(\nu). \tag{13.4-3}$$

Equation (13.4-3) permits the calculation of the photon-flux density arising from the amplified signal and spontaneous-emission photons.

EXERCISE 13.4-1

Amplified Spontaneous Emission (ASE)

(a) Use (13.4-3) to show that in the absence of any input signal, spontaneous emission produces a photon-flux density at the output of an unsaturated amplifier $[\gamma(\nu) \approx \gamma_0(\nu)]$ of length d given by $\phi(d) = \phi_{sp}\{\exp[\gamma_0(\nu)d] - 1\}$, where $\phi_{sp} = \xi_{sp}(\nu)/\gamma_0(\nu)$.

(b) Since both $\xi_{sp}(\nu)$ and $\gamma_0(\nu)$ are proportional to $g(\nu)$, ϕ_{sp} is independent of $g(\nu)$ so that the frequency dependence of $\phi(d)$ is governed by the factor $\{\exp[\gamma_0(\nu)d] - 1\}$. If $\gamma_0(\nu)$ is Lorentzian with width $\Delta\nu$, i.e., $\gamma_0(\nu) = \gamma_0(\nu_0)(\Delta\nu/2)^2/[(\nu - \nu_0)^2 + (\Delta\nu/2)^2]$, show that the bandwidth of the factor $\{\exp[\gamma_0(\nu)d] - 1\}$ is smaller than $\Delta\nu$, i.e., that the amplification of spontaneous emission is accompanied by spectral narrowing.

In the process of amplification, the photon-number statistics of the incoming light are altered (see Sec. 11.2C). A coherent signal presented to the input of the amplifier has a number of photons counted in time T that obeys Poisson statistics, with a variance σ_S^2 equal to the mean signal photon number $\bar{n}_S$. The ASE photons, on the other hand, obey Bose–Einstein statistics exhibiting $\sigma_{\mathrm{ASE}}^2 = \bar{n}_{\mathrm{ASE}} + \bar{n}_{\mathrm{ASE}}^2$ and are therefore considerably noisier than Poisson statistics. The photon-number statistics of the light after amplification, comprising both signal and spontaneous-emission contributions, obey a probability law intermediate between the two. If the counting time is short and the emerging light is linearly polarized, these statistics can be well approximated by the Laguerre-polynomial photon-number distribution (see Problem 13.4-2), which has a variance given by

$$\sigma_n^2 = \bar{n}_S + \left(\bar{n}_{\mathrm{ASE}} + \bar{n}_{\mathrm{ASE}}^2\right) + 2\bar{n}_S\bar{n}_{\mathrm{ASE}}. \tag{13.4-4}$$

The photon-number fluctuations are seen to contain contributions from the signal alone and from the spontaneous emission alone, as well as added fluctuations from the interference of the two components.

READING LIST

Books on Laser Theory

A. Yariv, *Quantum Electronics*, Wiley, New York, 3rd ed. 1989.

J. T. Verdeyen, *Laser Electronics*, Prentice-Hall, Englewood Cliffs, NJ, 2nd ed. 1989.

O. Svelto, *Principles of Lasers*, Plenum Press, New York, 3rd ed. 1989.

J. Wilson and J. F. B. Hawkes, *Optoelectronics*, Prentice-Hall, Englewood Cliffs, NJ, 2nd ed. 1989.

P. W. Milonni and J. H. Eberly, *Lasers*, Wiley, New York, 1988.

W. Witteman, *The Laser*, Springer-Verlag, New York, 1987.

K. A. Jones, *Introduction to Optical Electronics*, Harper & Row, New York, 1987.

J. Wilson and J. F. B. Hawkes, *Lasers: Principles and Applications*, Prentice-Hall, Englewood Cliffs, NJ, 1987.

A. E. Siegman, *Lasers*, University Science Books, Mill Valley, CA, 1986.

K. Shimoda, *Introduction to Laser Physics*, Springer-Verlag, Berlin, 2nd ed. 1986.

B. B. Laud, *Lasers and Nonlinear Optics*, Wiley, New York, 1986.

A. Yariv, *Optical Electronics*, Holt, Rinehart and Winston, New York, 3rd ed. 1985.

H. Haken, *Light: Laser Light Dynamics*, vol. 2, North-Holland, Amsterdam, 1985.

H. Haken, *Laser Theory*, Springer-Verlag, Berlin, 1984.

R. Loudon, *The Quantum Theory of Light*, Oxford University Press, New York, 2nd ed. 1983.

B. E. A. Saleh, *Photoelectron Statistics*, Springer-Verlag, New York, 1978.

D. C. O'Shea, W. R. Callen, and W. T. Rhodes, *Introduction to Lasers and Their Applications*, Addison-Wesley, Reading, MA, 1977.

M. Sargent III, M. O. Scully, and W. E. Lamb, Jr., *Laser Physics*, Addison-Wesley, Reading, MA, 1974.

F. T. Arecchi and E. O. Schulz-Dubois, eds., *Laser Handbook*, vol. 1, North-Holland/Elsevier, Amsterdam/New York, 1972.

A. E. Siegman, *An Introduction to Lasers and Masers*, McGraw-Hill, New York, 1971.

B. A. Lengyel, *Lasers*, Wiley, New York, 2nd ed. 1971.

A. Maitland and M. H. Dunn, *Laser Physics*, North-Holland, Amsterdam, 1969.

W. S. C. Chang, *Principles of Quantum Electronics*, Addison-Wesley, Reading, MA, 1969.

R. H. Pantell and H. E. Puthoff, *Fundamentals of Quantum Electronics*, Wiley, New York, 1969.

D. Ross, *Lasers, Light Amplifiers, and Oscillators*, Academic Press, New York, 1969.

E. L. Steele, *Optical Lasers in Electronics*, Wiley, New York, 1968.

A. K. Levine, ed., *Lasers*, vols. 1–4, Marcel Dekker, New York, 1966–1976.

G. Birnbaum, *Optical Masers*, Academic Press, New York, 1964.

G. Troup, *Masers and Lasers*, Methuen, London, 2nd ed. 1963.

Articles

R. Baker, Optical Amplification, *Physics World*, vol. 3, no. 3, pp. 41–44, 1990.

D. O'Shea and D. C. Peckham, *Lasers: Selected Reprints*, American Association of Physics Teachers, Stony Brook, NY, 1982.

M. J. Mumma, D. Buhl, G. Chin, D. Deming, F. Espenak, and T. Kostiuk, Discovery of Natural Gain Amplification in the 10 μm CO_2 Laser Bands on Mars: A Natural Laser, *Science*, vol. 212, pp. 45–49, 1981.

F. S. Barnes, ed., *Laser Theory*, IEEE Press Reprint Series, IEEE Press, New York, 1972.

A. L. Schawlow, ed., *Lasers and Light—Readings from Scientific American*, W. H. Freeman, San Francisco, 1969.

J. H. Shirley, Dynamics of a Simple Maser Model, *American Journal of Physics*, vol. 36, pp. 949–963, 1968.

J. Weber, ed., *Lasers: Selected Reprints with Editorial Comment*, Gordon and Breach, New York, 1967.

C. Cohen-Tannoudji and A. Kastler, Optical Pumping, in *Progress in Optics*, vol. 5, E. Wolf, ed., North-Holland, Amsterdam, 1966.

W. E. Lamb, Jr., Theory of an Optical Maser, *Physical Review*, vol. 134, pp. A1429–A1450, 1964.

A. Yariv and J. P. Gordon, The Laser, *Proceedings of the IEEE*, vol. 51, pp. 4–29, 1963.

T. H. Maiman, Stimulated Optical Radiation in Ruby, *Nature*, vol. 187, pp. 493–494, 1960.

A. L. Schawlow and C. H. Townes, Infrared and Optical Masers, *Physical Review*, vol. 112, pp. 1940–1949, 1958.

Historical

J. Hecht, ed., *Laser Pioneer Interviews*, High Tech Publications, Torrance, CA, 1985.

A. Kastler, Birth of the Maser and Laser, *Nature*, vol. 316, pp. 307–309, 1985.

M. Bertolotti, *Masers and Lasers: An Historical Approach*, Adam Hilger, Bristol, England, 1983.

C. H. Townes, Science, Technology, and Invention: Their Progress and Interactions, *Proceedings of the National Academy of Sciences (USA)*, vol. 80, pp. 7679–7683, 1983.

D. C. O'Shea and D. C. Peckham, Resource Letter L-1: Lasers, *American Journal of Physics*, vol. 49, pp. 915–925, 1981.

C. H. Townes, The Laser's Roots: Townes Recalls the Early Days, *Laser Focus Magazine*, vol. 14, no. 8, pp. 52–58, 1978.

A. L. Schawlow, Masers and Lasers, *IEEE Transactions on Electron Devices*, vol. ED-23, pp. 773–779, 1976.

A. L. Schawlow, From Maser to Laser, in *Impact of Basic Research on Technology*, B. Kursunoglu and A. Perlmutter, eds., Plenum Press, New York, 1973.

W. E. Lamb, Jr., Physical Concepts in the Development of the Maser and Laser, in *Impact of Basic Research on Technology*, B. Kursunoglu and A. Perlmutter, eds., Plenum Press, New York, 1973.

A. Kastler, Optical Methods for Studying Hertzian Resonances, in *Nobel Lectures in Physics, 1963–1970*, Elsevier, Amsterdam, 1972.

C. H. Townes, Production of Coherent Radiation by Atoms and Molecules, in *Nobel Lectures in Physics, 1963–1970*, Elsevier, Amsterdam, 1972.

N. G. Basov, Semiconductor Lasers, in *Nobel Lectures in Physics, 1963–1970*, Elsevier, Amsterdam, 1972.

A. M. Prokhorov, Quantum Electronics, in *Nobel Lectures in Physics, 1963–1970*, Elsevier, Amsterdam, 1972.

C. H. Townes, Quantum Electronics and Surprise in the Development of Technology, *Science*, vol. 159, pp. 699–703, 1968.

B. A. Lengyel, Evolution of Masers and Lasers, *American Journal of Physics*, vol. 34, pp. 903–913, 1966.

R. H. Dicke, Molecular Amplification and Generation Systems and Methods, U.S. Patent 2,851,652, Sept. 9, 1958.

J. P. Gordon, H. J. Zeiger, and C. H. Townes, The Maser—New Type of Microwave Amplifier, Frequency Standard, and Spectrometer, *Physical Review*, vol. 99, pp. 1264–1274, 1955.

N. G. Basov and A. M. Prokhorov, Possible Methods of Obtaining Active Molecules for a Molecular Oscillator, *Soviet Physics—JETP*, vol. 1, pp. 184–185, 1955 [*Zhurnal Eksperimental'noi i Teoreticheskoi Fiziki (USSR)*, vol. 28, pp. 249–250, 1955].

V. A. Fabrikant, The Emission Mechanism of Gas Discharges, *Trudi Vsyesoyuznogo Elektrotekhnicheskogo Instituta* (Reports of the All-Union Electrotechnical Institute, Moscow), vol. 41, *Elektronnie i Ionnie Pribori* (Electron and Ion Devices), pp. 236–296, 1940.

PROBLEMS

13.1-1 **Amplifier Gain and Rod Length.** A commercially available ruby laser amplifier using a 15-cm-long rod has a small-signal gain of 12. What is the small-signal gain of a 20-cm-long rod? Neglect gain saturation effects.

13.1-2 **Laser Amplifier Gain and Population Difference.** A 15-cm-long rod of Nd^{3+}:glass used as a laser amplifier has a total small-signal gain of 10 at $\lambda_o = 1.06\ \mu m$. Use the data in Table 13.2-1 on page 480 to determine the population difference N required to achieve this gain (Nd^{3+} ions per cm^3).

13.1-3 **Amplification of a Broadband Signal.** The transition between two energy levels exhibits a Lorentzian lineshape of central frequency $\nu_0 = 5 \times 10^{14}$ with a linewidth

$\Delta\nu = 10^{12}$ Hz. The population is inverted so that the maximum gain coefficient $\gamma(\nu_0) = 0.1\ \text{cm}^{-1}$. The medium has an additional loss coefficient $\alpha_s = 0.05\ \text{cm}^{-1}$, which is independent of ν. Approximately how much loss or gain is encountered by a light wave in 1 cm if it has a uniform power spectral density centered about ν_0 with a bandwidth $2\Delta\nu$?

13.2-1 **The Two-Level Pumping System.** Write the rate equations for a two-level system, showing that a steady-state population inversion cannot be achieved by using direct optical pumping between levels 1 and 2.

13.2-2 **Two Laser Lines.** Consider an atomic system with four levels: 0 (ground state), 1, 2, and 3. Two pumps are applied: between the ground state and level 3 at a rate R_3, and between ground state and level 2 at a rate R_2. Population inversion can occur between levels 3 and 1 and/or between levels 2 and 1 (as in a four-level laser). Assuming that decay from level 3 to 2 is not possible and that decay from levels 3 and 2 to the ground state are negligible, write the rate equations for levels 1, 2, and 3 in terms of the lifetimes τ_1, τ_{31}, and τ_{21}. Determine the steady state populations N_1, N_2, and N_3 and examine the possibility of simultaneous population inversions between 3 and 1, and between 2 and 1. Show that the presence of radiation at the 2–1 transition reduces the population difference for the 3–1 transition.

13.3-1 **Significance of the Saturation Photon-Flux Density.** In the general two-level atomic system of Fig. 13.2-3, τ_2 represents the lifetime of level 2 in the absence of stimulated emission. In the presence of stimulated emission, the rate of decay from level 2 increases and the effective lifetime decreases. Find the photon-flux density ϕ at which the lifetime decreases to half its value. How is that flux density related to the saturation photon-flux density ϕ_s?

13.3-2 **Saturation Optical Intensity.** Determine the saturation photon-flux density $\phi_s(\nu_0)$ and the corresponding saturation optical intensity $I_s(\nu_0)$, for the homogeneously broadened ruby and Nd^{3+}:YAG laser transitions provided in Table 13.2-1.

13.3-3 **Growth of the Photon-Flux Density in a Saturated Amplifier.** The growth of the photon-flux density $\phi(z)$ in a laser amplifier is described by (13.3-7). Use a computer to plot $\phi(z)/\phi_s$ versus $\gamma_0 z$ for $\phi(0)/\phi_s = 0.05$. Identify the onset of saturation in this amplifier.

13.3-4 **Resonant Absorption of a Medium in Thermal Equilibrium.** A unity refractive index medium of volume 1 cm^3 contains $N_a = 10^{23}$ atoms in thermal equilibrium. The ground state is energy level 1; level 2 has energy 2.48 eV above the ground state ($\lambda_o = 0.5\ \mu\text{m}$). The transition between these two levels is characterized by a spontaneous lifetime $t_{sp} = 1$ ms, and a Lorentzian lineshape of width $\Delta\nu = 1$ GHz. Consider two temperatures, T_1 and T_2, such that $k_B T_1 = 0.026$ eV and $k_B T_2 = 0.26$ eV.

(a) Determine the populations N_1 and N_2.
(b) Determine the number of photons emitted spontaneously every second.
(c) Determine the attenuation coefficient of this medium at $\lambda_o = 0.5\ \mu\text{m}$ assuming that the incident photon flux is small.
(d) Sketch the dependence of the attenuation coefficient on frequency, indicating on the sketch the important parameters.
(e) Find the value of photon-flux density at which the attenuation coefficient decreases by a factor of 2 (i.e., the saturation photon-flux density).
(f) Sketch the dependence of the transmitted photon-flux density $\phi(d)$ on the incident photon-flux density $\phi(0)$ for $\nu = \nu_0$ and $\nu = \nu_0 + \Delta\nu$ when $\phi(0)/\phi_s \ll 1$.

13.3-5 **Gain in a Saturated Amplifying Medium.** Consider a homogeneously broadened laser amplifying medium of length $d = 10$ cm and saturation photon flux density $\phi_s = 4 \times 10^{18}$ photons/cm^2-s. It is known that a photon-flux density at the input $\phi(0) = 4 \times 10^{15}$ photons/cm^2-s produces a photon-flux density at the output $\phi(d) = 4 \times 10^{16}$ photons/cm^2-s.
(a) Determine the small-signal gain of the system G_0.
(b) Determine the small-signal gain coefficient γ_0.
(c) What is the photon-flux density at which the gain coefficient decreases by a factor of 5?
(d) Determine the gain coefficient when the input photon-flux density is $\phi(0) = 4 \times 10^{19}$ photons/cm^2-s. Under these conditions, is the gain of the system greater than, less than, or the same as the small-signal gain determined in part (a)?

*13.4-1 **Ratio of Signal Power to ASE Power.** An unsaturated laser amplifier of length d and gain coefficient $\gamma_0(\nu)$ amplifies an input signal $\phi_S(0)$ of frequency ν and introduces amplified spontaneous emission (ASE) at a rate ξ_{sp} (per unit length). The amplified signal photon-flux density is $\phi_S(d)$ and the ASE at the output is ϕ_{ASE}. Sketch the dependence of the ratio $\phi_S(d)/\phi_{\text{ASE}}$ on the amplifier gain coefficient-length product $\gamma_0(\nu)d$.

*13.4-2 **Photon-Number Distribution for Amplified Coherent Light.** A linearly polarized superposition of interfering thermal and coherent light serves as a suitable model for the light emerging from a laser amplifier. This superposition is known to have random energy fluctuations $\mathscr{W}$ that obey the noncentral-chi-square probability distribution

$$p(\mathscr{W}) = \frac{1}{\overline{\mathscr{W}}_{\text{ASE}}} \exp\left(-\frac{\mathscr{W} + \mathscr{W}_S}{\overline{\mathscr{W}}_{\text{ASE}}}\right) I_0\left[\frac{2(\mathscr{W}_S \mathscr{W})^{1/2}}{\overline{\mathscr{W}}_{\text{ASE}}}\right],$$

provided that the measurement time is sufficiently short.† Here I_0 denotes the modified Bessel function, $\overline{\mathscr{W}}_{\text{ASE}}$ is the mean energy of the ASE, and $\mathscr{W}_S$ is the (constant) energy of the amplified coherent signal.
(a) Calculate the mean and variance of $\mathscr{W}$.
(b) Use (11.2-26) and (11.2-27) to determine the photon-number mean $\bar{n}$ and variance σ_n^2, confirming the validity of (13.4-4).
(c) Use (11.2-25) to show that the photon-number distribution is given by

$$p(n) = \frac{\bar{n}_{\text{ASE}}^n}{(1 + \bar{n}_{\text{ASE}})^{n+1}} \exp\left(-\frac{\bar{n}_S}{1 + \bar{n}_{\text{ASE}}}\right) L_n\left(-\frac{\bar{n}_S/\bar{n}_{\text{ASE}}}{1 + \bar{n}_{\text{ASE}}}\right),$$

where L_n represents the Laguerre polynomial

$$L_n(-x) = \sum_{k=0}^{n} \binom{n}{k} \frac{x^k}{k!},$$

and $\bar{n}_S$ and $\bar{n}_{\text{ASE}}$ are the mean signal and amplified-spontaneous-emission photon numbers, respectively.
(d) Use a computer to plot $p(n)$ for $\bar{n}_S/\bar{n} = 0$, 0.5, 0.8, and 1, when $\bar{n} = 5$, demonstrating that it reduces to the Bose–Einstein distribution for $\bar{n}_S/\bar{n} = 0$ and to the Poisson distribution for $\bar{n}_S/\bar{n} = 1$.

†See, for example, B. E. A. Saleh, *Photoelectron Statistics*, Springer-Verlag, New York, 1978.

CHAPTER

14

LASERS

14.1 THEORY OF LASER OSCILLATION
   A. Optical Amplification and Feedback
   B. Conditions for Laser Oscillation
14.2 CHARACTERISTICS OF THE LASER OUTPUT
   A. Power
   B. Spectral Distribution
   C. Spatial Distribution and Polarization
   D. Mode Selection
   E. Characteristics of Common Lasers
14.3 PULSED LASERS
   A. Methods of Pulsing Lasers
   *B. Analysis of Transient Effects
   *C. Q-Switching
   D. Mode Locking

Arthur L. Schawlow (born 1921)

Theodore H. Maiman (born 1927)

In 1958 Schawlow, together with Charles Townes, showed how to extend the principle of the maser to the optical region. He shared the 1981 Nobel Prize with Nicolaas Bloembergen. Maiman demonstrated the first successful operation of the ruby laser in 1960.

The laser is an optical oscillator. It comprises a resonant optical amplifier whose output is fed back into its input with matching phase (Fig. 14.0-1). In the absence of such an input there is no output, so that the fed-back signal is also zero. However, this is an unstable situation. The presence at the input of even a small amount of noise (containing frequency components lying within the amplifier bandwidth) is unavoidable and may initiate the oscillation process. The input is amplified and the output is fed back to the input, where it undergoes further amplification. The process continues indefinitely until a large output is produced. Saturation of the amplifier gain limits further growth of the signal, and the system reaches a steady state in which an output signal is created at the frequency of the resonant amplifier.

Two conditions must be satisfied for oscillation to occur:

- The amplifier gain must be greater than the loss in the feedback system so that net gain is incurred in a round trip through the feedback loop.
- The total phase shift in a single round trip must be a multiple of 2π so that the fedback input phase matches the phase of the original input.

If these conditions are satisfied, the system becomes unstable and oscillation begins. As the oscillation power grows, however, the amplifier saturates and the gain diminishes below its initial value. A stable condition is reached when the reduced gain is equal to the loss (Fig. 14.0-2). The gain then just compensates the loss so that the cycle of amplification and feedback is repeated without change and steady-state oscillation ensues.

Because the gain and phase shift are functions of frequency, the two oscillation conditions are satisfied only at one (or several) frequencies, called the resonance frequencies of the oscillator. The useful output is extracted by coupling a portion of the power out of the oscillator. In summary, an oscillator comprises:

- An amplifier with a gain-saturation mechanism
- A feedback system
- A frequency-selection mechanism
- An output coupling scheme

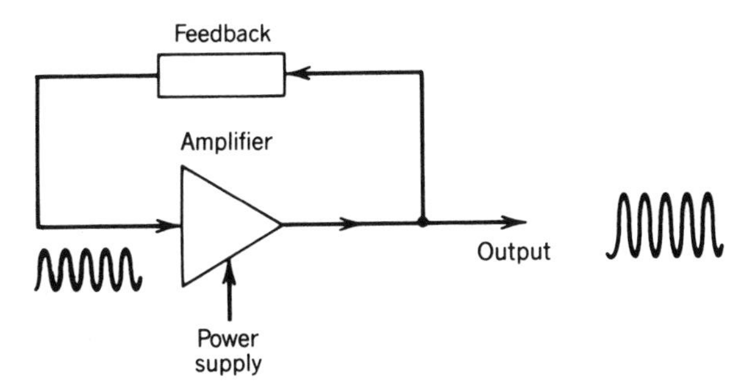

Figure 14.0-1 An oscillator is an amplifier with positive feedback.

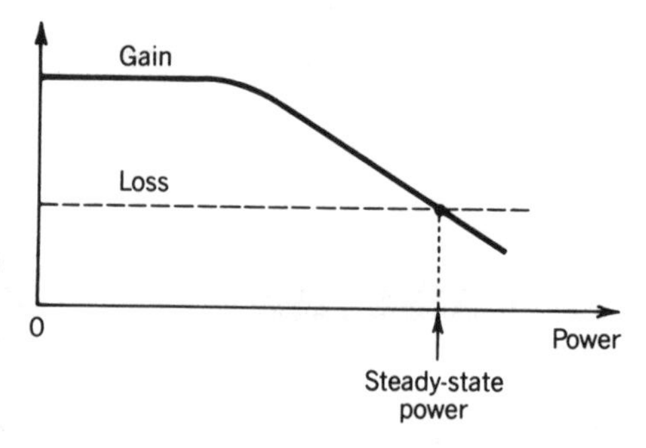

Figure 14.0-2 If the initial amplifier gain is greater than the loss, oscillation may initiate. The amplifier then saturates whereupon its gain decreases. A steady-state condition is reached when the gain just equals the loss.

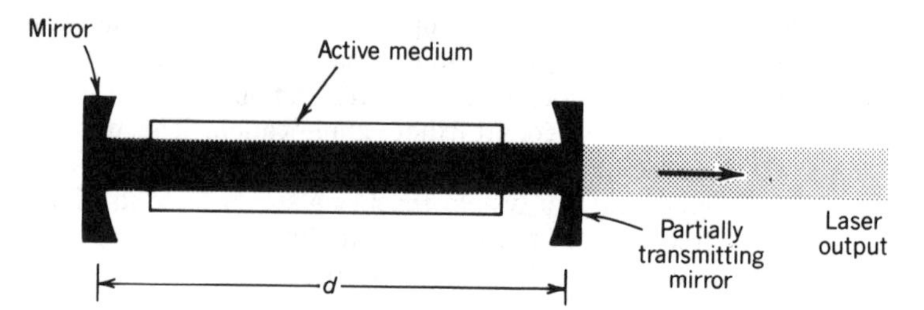

Figure 14.0-3 A laser consists of an optical amplifier (employing an active medium) placed within an optical resonator. The output is extracted through a partially transmitting mirror.

The laser is an optical oscillator (see Fig. 14.0-3) in which the amplifier is the pumped active medium discussed in Chap. 13. Gain saturation is a basic property of laser amplifiers. Feedback is obtained by placing the active medium in an optical resonator, which reflects the light back and forth between its mirrors, as discussed in Chap. 9. Frequency selection is achieved by the resonant amplifier *and* by the resonator, which admits only certain modes. Output coupling is accomplished by making one of the resonator mirrors partially transmitting.

Lasers are used in a great variety of scientific and technical applications including communications, computing, image processing, information storage, holography, lithography, materials processing, geology, metrology, rangefinding, biology, and clinical medicine.

This chapter provides an introduction to the operation of lasers. In Sec. 14.1 the behavior of the laser amplifier and the laser resonator are summarized, and the oscillation conditions of the laser are derived. The characteristics of the laser output (power, spectral distribution, spatial distribution, and polarization) are discussed in Sec. 14.2, and typical parameters for various kinds of lasers are provided. Whereas Secs. 14.1 and 14.2 are concerned with continuous-wave (CW) laser oscillation, Sec. 14.3 is devoted to the operation of pulsed lasers.

14.1 THEORY OF LASER OSCILLATION

We begin this section with a summary of the properties of the two basic components of the laser—the amplifier and the resonator. Although these topics have been discussed in detail in Chaps. 13 and 9, they are reviewed here for convenience.

A. Optical Amplification and Feedback

Laser Amplification

The laser amplifier is a narrowband coherent amplifier of light. Amplification is achieved by stimulated emission from an atomic or molecular system with a transition whose population is inverted (i.e., the upper energy level is more populated than the lower). The amplifier bandwidth is determined by the linewidth of the atomic transition, or by an inhomogeneous broadening mechanism such as the Doppler effect in gas lasers.

The laser amplifier is a distributed-gain device characterized by its gain coefficient (gain per unit length) $\gamma(\nu)$, which governs the rate at which the photon-flux density ϕ (or the optical intensity $I = h\nu\phi$) increases. When the photon-flux density ϕ is small, the gain coefficient is

$$\gamma_0(\nu) = N_0\sigma(\nu) = N_0\frac{\lambda^2}{8\pi t_{sp}}g(\nu), \tag{14.1-1}$$

Small-Signal Gain Coefficient

where

N_0 = equilibrium population density difference (density of atoms in the upper energy state minus that in the lower state); N_0 increases with increasing pumping rate
$\sigma(\nu) = (\lambda^2/8\pi t_{sp})g(\nu)$ = transition cross section
t_{sp} = spontaneous lifetime
$g(\nu)$ = transition lineshape
λ = wavelength in the medium = λ_o/n, where n = refractive index.

As the photon-flux density increases, the amplifier enters a region of nonlinear operation. It saturates and its gain decreases. The amplification process then depletes the initial population difference N_0, reducing it to $N = N_0/[1 + \phi/\phi_s(\nu)]$ for a homogeneously broadened medium, where

$\phi_s(\nu) = [\tau_s\sigma(\nu)]^{-1}$ = saturation photon-flux density
τ_s = saturation time constant, which depends on the decay times of the energy levels involved; in an ideal four-level pumping scheme, $\tau_s \approx t_{sp}$, whereas in an ideal three-level pumping scheme, $\tau_s \approx 2t_{sp}$.

The gain coefficient of the saturated amplifier is therefore reduced to $\gamma(\nu) = N\sigma(\nu)$, so that for homogeneous broadening

$$\gamma(\nu) = \frac{\gamma_0(\nu)}{1 + \phi/\phi_s(\nu)}. \tag{14.1-2}$$

Saturated Gain Coefficient

The laser amplification process also introduces a phase shift. When the lineshape is Lorentzian with linewidth $\Delta\nu$, $g(\nu) = (\Delta\nu/2\pi)/[(\nu - \nu_0)^2 + (\Delta\nu/2)^2]$, the amplifier

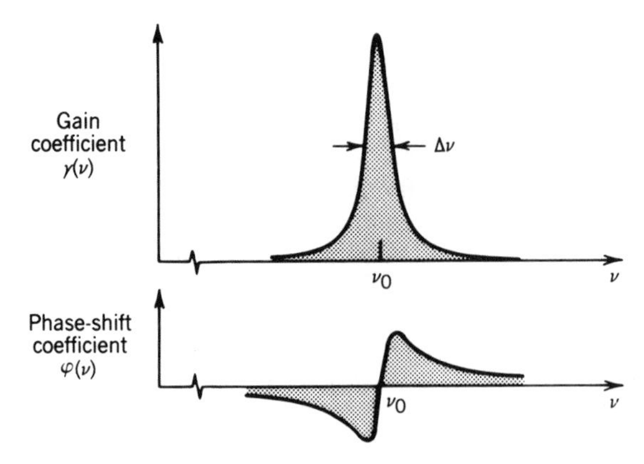

Figure 14.1-1 Spectral dependence of the gain and phase-shift coefficients for an optical amplifier with Lorentzian lineshape function.

phase shift per unit length is

$$\varphi(\nu) = \frac{\nu - \nu_0}{\Delta\nu}\gamma(\nu). \tag{14.1-3}$$

Phase-Shift Coefficient (Lorentzian Lineshape)

This phase shift is in addition to that introduced by the medium hosting the laser atoms. The gain and phase-shift coefficients for an amplifier with Lorentzian lineshape function are illustrated in Fig. 14.1-1.

Feedback and Loss: The Optical Resonator

Optical feedback is achieved by placing the active medium in an optical resonator. A Fabry–Perot resonator, comprising two mirrors separated by a distance d, contains the medium (refractive index n) in which the active atoms of the amplifier reside. Travel through the medium introduces a phase shift per unit length equal to the wavenumber

$$k = \frac{2\pi\nu}{c}. \tag{14.1-4}$$

Phase-Shift Coefficient

The resonator also contributes to losses in the system. Absorption and scattering of light in the medium introduces a distributed loss characterized by the attenuation coefficient α_s (loss per unit length). In traveling a round trip through a resonator of length d, the photon-flux density is reduced by the factor $\mathscr{R}_1\mathscr{R}_2\exp(-2\alpha_s d)$, where $\mathscr{R}_1$ and $\mathscr{R}_2$ are the reflectances of the two mirrors. The overall loss in one round trip can therefore be described by a total effective distributed loss coefficient α_r, where

$$\exp(-2\alpha_r d) = \mathscr{R}_1\mathscr{R}_2\exp(-2\alpha_s d),$$

so that

$$\alpha_r = \alpha_s + \alpha_{m1} + \alpha_{m2}$$
$$\alpha_{m1} = \frac{1}{2d} \ln \frac{1}{\mathscr{R}_1} \qquad (14.1\text{-}5)$$
$$\alpha_{m2} = \frac{1}{2d} \ln \frac{1}{\mathscr{R}_2},$$

Loss Coefficient

where α_{m1} and α_{m2} represent the contributions of mirrors 1 and 2, respectively. The contribution from both mirrors is

$$\alpha_m = \alpha_{m1} + \alpha_{m2} = \frac{1}{2d} \ln \frac{1}{\mathscr{R}_1 \mathscr{R}_2}.$$

Since α_r represents the total loss of energy (or number of photons) per unit length, $\alpha_r c$ represents the loss of photons per second. Thus

$$\tau_p = \frac{1}{\alpha_r c} \qquad (14.1\text{-}6)$$

represents the **photon lifetime**.

The resonator sustains only frequencies that correspond to a round-trip phase shift that is a multiple of 2π. For a resonator devoid of active atoms (i.e., a "cold" resonator), the round-trip phase shift is simply $k2d = 4\pi\nu d/c = q2\pi$, corresponding to modes of frequencies

$$\nu_q = q\nu_F, \qquad q = 1, 2, \ldots, \qquad (14.1\text{-}7)$$

where $\nu_F = c/2d$ is the resonator mode spacing and $c = c_o/n$ is the speed of light in the medium (Fig. 14.1-2). The (full width at half maximum) spectral width of these resonator modes is

$$\delta\nu \approx \frac{\nu_F}{\mathscr{F}} \qquad (14.1\text{-}8)$$

where $\mathscr{F}$ is the finesse of the resonator (see Sec. 9.1A). When the resonator losses are

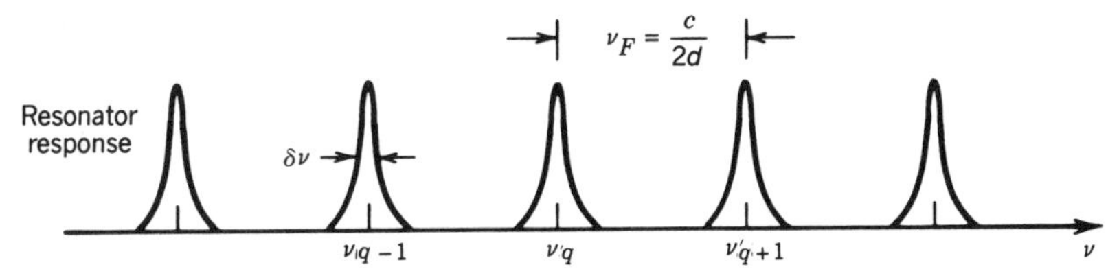

Figure 14.1-2 Resonator modes are separated by the frequency $\nu_F = c/2d$ and have linewidths $\delta\nu = \nu_F/\mathscr{F} = 1/2\pi\tau_p$.

small and the finesse is large,

$$\mathscr{F} \approx \frac{\pi}{\alpha_r d} = 2\pi \tau_p \nu_F. \tag{14.1-9}$$

B. Conditions for Laser Oscillation

Two conditions must be satisfied for the laser to oscillate (lase). The *gain condition* determines the minimum population difference, and therefore the pumping threshold, required for lasing. The *phase condition* determines the frequency (or frequencies) at which oscillation takes place.

Gain Condition: Laser Threshold

The initiation of laser oscillation requires that the small-signal gain coefficient be greater than the loss coefficient, i.e.,

$$\boxed{\gamma_0(\nu) > \alpha_r.} \tag{14.1-10}$$

Threshold Gain Condition

In accordance with (14.1-1), the small-signal gain coefficient $\gamma_0(\nu)$ is proportional to the equilibrium population density difference N_0, which in turn is known from Chap. 13 to increase with the pumping rate R. Indeed, (14.1-1) may be used to translate (14.1-10) into a condition on the population difference, i.e., $N_0 = \gamma_0(\nu)/\sigma(\nu) > \alpha_r/\sigma(\nu)$. Thus

$$N_0 > N_t, \tag{14.1-11}$$

where the quantity

$$N_t = \frac{\alpha_r}{\sigma(\nu)} \tag{14.1-12}$$

is called the threshold population difference. N_t, which is proportional to α_r, determines the minimum pumping rate R_t for the initiation of laser oscillation.

Using (14.1-6), α_r may alternatively be written in terms of the photon lifetime, $\alpha_r = 1/c\tau_p$, whereupon (14.1-12) takes the form

$$N_t = \frac{1}{c\tau_p \sigma(\nu)}. \tag{14.1-13}$$

The threshold population density difference is therefore directly proportional to α_r and inversely proportional to τ_p. Higher loss (shorter photon lifetime) requires more vigorous pumping to achieve lasing.

Finally, use of the standard formula for the transition cross section, $\sigma(\nu) = (\lambda^2/8\pi t_{sp})g(\nu)$, leads to yet another expression for the threshold population difference,

$$\boxed{N_t = \frac{8\pi}{\lambda^2 c} \frac{t_{sp}}{\tau_p} \frac{1}{g(\nu)},} \tag{14.1-14}$$

Threshold Population Difference

from which it is clear that N_t is a function of the frequency ν. The threshold is lowest, and therefore lasing is most readily achieved, at the frequency where the lineshape function is greatest, i.e., at its central frequency $\nu = \nu_0$. For a Lorentzian lineshape function, $g(\nu_0) = 2/\pi\,\Delta\nu$, so that the minimum population difference for oscillation at the central frequency ν_0 turns out to be

$$N_t = \frac{2\pi}{\lambda^2 c}\,\frac{2\pi\,\Delta\nu\,t_{sp}}{\tau_p}. \qquad (14.1\text{-}15)$$

It is directly proportional to the linewidth $\Delta\nu$.

If, furthermore, the transition is limited by lifetime broadening with a decay time t_{sp}, $\Delta\nu$ assumes the value $1/2\pi t_{sp}$ (see Sec. 12.2D), whereupon (14.1-15) simplifies to

$$N_t = \frac{2\pi}{\lambda^2 c\tau_p} = \frac{2\pi\alpha_r}{\lambda^2}. \qquad (14.1\text{-}16)$$

This formula shows that the minimum threshold population difference required to achieve oscillation is a simple function of the wavelength λ and the photon lifetime τ_p. It is clear that laser oscillation becomes more difficult to achieve as the wavelength decreases. As a numerical example, if $\lambda_o = 1\ \mu\text{m}$, $\tau_p = 1$ ns, and the refractive index $n = 1$, we obtain $N_t \approx 2.1 \times 10^7\ \text{cm}^{-3}$.

EXERCISE 14.1-1

Threshold of a Ruby Laser

(a) At the line center of the $\lambda_o = 694.3$-nm transition, the absorption coefficient of ruby in thermal equilibrium (i.e., without pumping) at $T = 300$ K is $\alpha(\nu_0) \equiv -\gamma(\nu_0) \approx 0.2\ \text{cm}^{-1}$. If the concentration of Cr^{3+} ions responsible for the transition is $N_a = 1.58 \times 10^{19}\ \text{cm}^{-3}$, determine the transition cross section $\sigma_0 = \sigma(\nu_0)$.

(b) A ruby laser makes use of a 10-cm-long ruby rod (refractive index $n = 1.76$) of cross-sectional area 1 cm^2 and operates on this transition at $\lambda_o = 694.3$ nm. Both of its ends are polished and coated so that each has a reflectance of 80%. Assuming that there are no scattering or other extraneous losses, determine the resonator loss coefficient α_r and the resonator photon lifetime τ_p.

(c) As the laser is pumped, $\gamma(\nu_0)$ increases from its initial thermal equilibrium value of $-0.2\ \text{cm}^{-1}$ and changes sign, thereby providing gain. Determine the threshold population difference N_t for laser oscillation.

Phase Condition: Laser Frequencies

The second condition of oscillation requires that the phase shift imparted to a light wave completing a round trip within the resonator must be a multiple of 2π, i.e.,

$$2kd + 2\varphi(\nu)d = 2\pi q, \qquad q = 1, 2, \ldots. \qquad (14.1\text{-}17)$$

If the contribution arising from the active laser atoms $[2\varphi(\nu)d]$ is small, dividing (14.1-17) by $2d$ gives the cold-resonator result obtained earlier, $\nu = \nu_q = q(c/2d)$.

In the presence of the active medium, when $2\varphi(\nu)d$ contributes, the solution of (14.1-17) gives rise to a set of oscillation frequencies ν_q' that are slightly displaced from the cold-resonator frequencies ν_q. It turns out that the cold-resonator modal frequencies are all pulled slightly toward the central frequency of the atomic transition, as shown below.

*Frequency Pulling

Using the relation $k = 2\pi\nu/c$, and the phase-shift coefficient for the Lorentzian lineshape function provided in (14.1-3), the phase-shift condition (14.1-17) provides

$$\nu + \frac{c}{2\pi}\frac{\nu - \nu_0}{\Delta\nu}\gamma(\nu) = \nu_q. \tag{14.1-18}$$

This equation can be solved for the oscillation frequency $\nu = \nu_q'$ corresponding to each cold-resonator mode ν_q. Because the equation is nonlinear, a graphical solution is useful. The left-hand side of (14.1-18) is designated $\psi(\nu)$ and plotted in Fig. 14.1-3 (it is the sum of a straight line representing ν plus the Lorentzian phase-shift coefficient shown schematically in Fig. 14.1-1). The value of $\nu = \nu_q'$ that makes $\psi(\nu) = \nu_q$ is graphically determined. It is apparent from the figure that the cold-resonator modes ν_q are always frequency pulled toward the central frequency of the resonant medium ν_0.

An approximate analytic solution of (14.1-18) can also be obtained. We write (14.1-18) in the form

$$\nu = \nu_q - \frac{c}{2\pi}\frac{\nu - \nu_0}{\Delta\nu}\gamma(\nu). \tag{14.1-19}$$

When $\nu = \nu_q' \approx \nu_q$, the second term of (14.1-19) is small, whereupon ν may be replaced with ν_q without much loss of accuracy. Thus

$$\nu_q' \approx \nu_q - \frac{c}{2\pi}\frac{\nu_q - \nu_0}{\Delta\nu}\gamma(\nu_q), \tag{14.1-20}$$

which is an explicit expression for the oscillation frequency ν_q' as a function of the

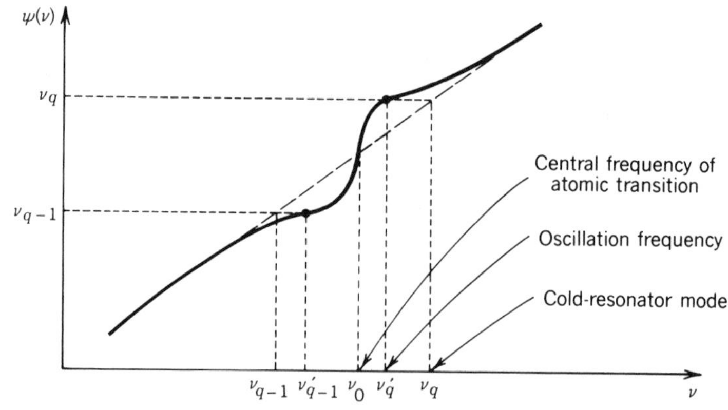

Figure 14.1-3 The left-hand side of (14.1-18), $\psi(\nu)$, plotted as a function of ν. The frequency ν for which $\psi(\nu) = \nu_q$ is the solution of (14.1-18). Each "cold" resonator frequency ν_q corresponds to a "hot" resonator frequency ν_q', which is shifted in the direction of the atomic resonance central frequency ν_0.

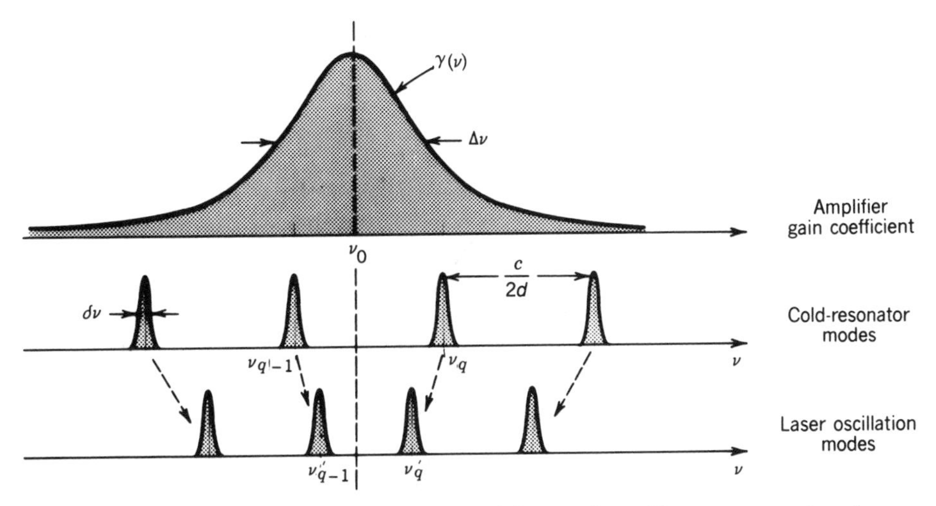

Figure 14.1-4 The laser oscillation frequencies fall near the cold-resonator modes; they are pulled slightly toward the atomic resonance central frequency ν_0.

cold-resonator frequency ν_q. Furthermore, under steady-state conditions, the gain equals the loss so that $\gamma(\nu_q) = \alpha_r \approx \pi/\mathcal{F}d = (2\pi/c)\delta\nu$, where $\delta\nu$ is the spectral width of the cold resonator modes. Substituting this relation into (14.1-20) leads to

$$\nu_q' \approx \nu_q - (\nu_q - \nu_0)\frac{\delta\nu}{\Delta\nu}. \tag{14.1-21}$$

Laser Frequencies

The cold-resonator frequency ν_q is therefore pulled toward the atomic resonance frequency ν_0 by a fraction $\delta\nu/\Delta\nu$ of its original distance from the central frequency $(\nu_q - \nu_0)$, as shown in Fig. 14.1-4. The sharper the resonator mode (the smaller the value of $\delta\nu$), the less significant the pulling effect. By contrast, the narrower the atomic resonance linewidth (the smaller the value of $\Delta\nu$), the more effective the pulling.

14.2 CHARACTERISTICS OF THE LASER OUTPUT

A. Power

Internal Photon-Flux Density

A laser pumped above threshold ($N_0 > N_t$) exhibits a small-signal gain coefficient $\gamma_0(\nu)$ that is greater than the loss coefficient α_r, as shown in (14.1-10). Laser oscillation may then begin, provided that the phase condition (14.1-17) is satisfied. As the photon-flux density ϕ inside the resonator increases (Fig. 14.2-1), the gain coefficient $\gamma(\nu)$ begins to decrease in accordance with (14.1-2) for homogeneously broadened media. As long as the gain coefficient remains larger than the loss coefficient, the photon flux continues to grow.

Finally, when the saturated gain coefficient becomes equal to the loss coefficient (or equivalently $N = N_t$), the photon flux ceases its growth and the oscillation reaches steady-state conditions. The result is **gain clamping** at the value of the loss. The steady-state laser internal photon-flux density is therefore determined by equating the

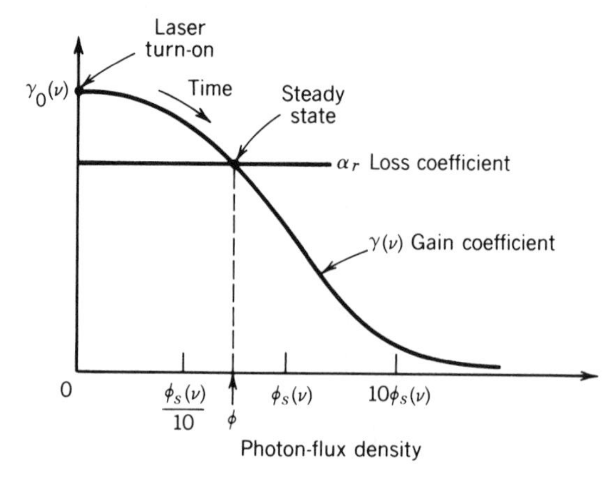

Figure 14.2-1 Determination of the steady-state laser photon-flux density ϕ. At the time of laser turn on, $\phi = 0$ so that $\gamma(\nu) = \gamma_0(\nu)$. As the oscillation builds up in time, the increase in ϕ causes $\gamma(\nu)$ to decrease through gain saturation. When γ reaches α_r, the photon-flux density ceases its growth and steady-state conditions are achieved. The smaller the loss, the greater the value of ϕ.

large-signal (saturated) gain coefficient to the loss coefficient $\gamma_0(\nu)/[1 + \phi/\phi_s(\nu)] = \alpha_r$, which provides

$$\phi = \begin{cases} \phi_s(\nu)\left(\dfrac{\gamma_0(\nu)}{\alpha_r} - 1\right), & \gamma_0(\nu) > \alpha_r \\ 0, & \gamma_0(\nu) \leq \alpha_r. \end{cases} \qquad (14.2\text{-}1)$$

Equation (14.2-1) represents the steady-state photon-flux density arising from laser action. This is the mean number of photons per second crossing a unit area in both directions, since photons traveling in both directions contribute to the saturation process. The photon-flux density for photons traveling in a single direction is therefore $\phi/2$. Spontaneous emission has been neglected in this simplified treatment. Of course, (14.2-1) represents the mean photon-flux density; there are random fluctuations about this mean as discussed in Sec. 11.2.

Since $\gamma_0(\nu) = N_0\sigma(\nu)$ and $\alpha_r = N_t\sigma(\nu)$, (14.2-1) may be written in the form

$$\phi = \begin{cases} \phi_s(\nu)\left(\dfrac{N_0}{N_t} - 1\right), & N_0 > N_t \\ 0, & N_0 \leq N_t. \end{cases} \qquad (14.2\text{-}2)$$

Steady-State
Laser Internal
Photon-Flux Density

Below threshold, the laser photon-flux density is zero; any increase in the pumping rate is manifested as an increase in the spontaneous-emission photon flux, but there is no sustained oscillation. Above threshold, the steady-state internal laser photon-flux density is directly proportional to the initial population difference N_0, and therefore increases with the pumping rate R [see (13.2-10) and (13.2-22)]. If N_0 is twice the threshold value N_t, the photon-flux density is precisely equal to the saturation value $\phi_s(\nu)$, which is the photon-flux density at which the gain coefficient decreases to half its

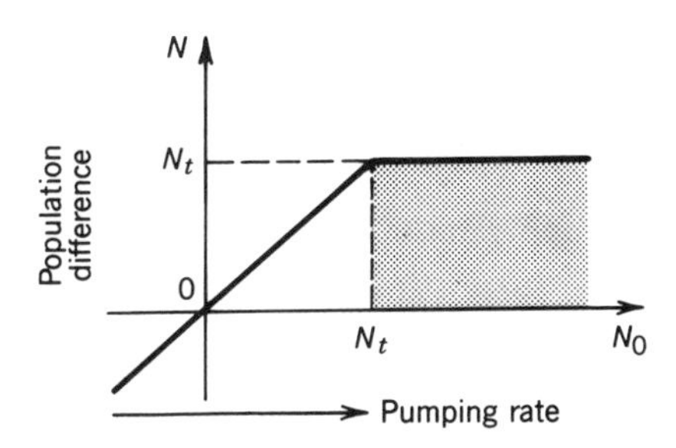

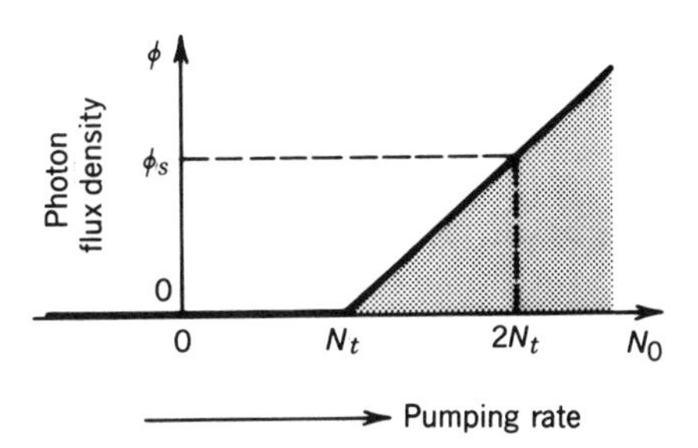

Figure 14.2-2 Steady-state values of the population difference N, and the laser internal photon-flux density ϕ, as functions of N_0 (the population difference in the absence of radiation; N_0 increases with the pumping rate R). Laser oscillation occurs when N_0 exceeds N_t; the steady-state value of N then saturates, clamping at the value N_t [just as $\gamma_0(\nu)$ is clamped at α_r]. Above threshold, ϕ is proportional to $N_0 - N_t$.

maximum value. Both the population difference N and the photon-flux density ϕ are shown as functions of N_0 in Fig. 14.2-2.

Output Photon-Flux Density

Only a portion of the steady-state internal photon-flux density determined by (14.2-2) leaves the resonator in the form of useful light. The output photon-flux density ϕ_o is that part of the internal photon-flux density that propagates toward mirror 1 ($\phi/2$) and is transmitted by it. If the transmittance of mirror 1 is $\mathcal{T}$, the output photon-flux density is

$$\phi_o = \frac{\mathcal{T}\phi}{2}. \tag{14.2-3}$$

The corresponding optical intensity of the laser output I_o is

$$I_o = \frac{h\nu\mathcal{T}\phi}{2}, \tag{14.2-4}$$

and the laser output power is $P_o = I_o A$, where A is the cross-sectional area of the laser beam. These equations, together with (14.2-2), permit the output power of the laser to be explicitly calculated in terms of $\phi_s(\nu)$, N_0, N_t, $\mathcal{T}$, and A.

Optimization of the Output Photon-Flux Density

The useful photon-flux density at the laser output diminishes the internal photon-flux density and therefore contributes to the losses of the laser oscillator. Any attempt to increase the fraction of photons allowed to escape from the resonator (in the expectation of increasing the useful light output) results in increased losses so that the steady-state photon-flux density inside the resonator decreases. The net result may therefore be a decrease, rather than an increase, in the useful light output.

We proceed to show that there is an optimal transmittance $\mathcal{T}$ $(0 < \mathcal{T} < 1)$ that maximizes the laser output intensity. The output photon-flux density $\phi_o = \mathcal{T}\phi/2$ is a product of the mirror's transmittance $\mathcal{T}$ and the internal photon-flux density $\phi/2$. As $\mathcal{T}$ is increased, ϕ decreases as a result of the greater losses. At one extreme, when $\mathcal{T} = 0$, the oscillator has the least loss (ϕ is maximum), but there is no laser output whatever ($\phi_o = 0$). At the other extreme, when the mirror is removed so that $\mathcal{T} = 1$, the increased losses make $\alpha_r > \gamma_0(\nu)$ $(N_t > N_0)$, thereby preventing laser oscillation.

In this case $\phi = 0$, so that again $\phi_o = 0$. The optimal value of $\mathscr{T}$ lies somewhere between these two extremes.

To determine it, we must obtain an explicit relation between ϕ_o and $\mathscr{T}$. We assume that mirror 1, with a reflectance $\mathscr{R}_1$ and a transmittance $\mathscr{T} = 1 - \mathscr{R}_1$, transmits the useful light. The loss coefficient α_r is written as a function of $\mathscr{T}$ by substituting in (14.1-5) the loss coefficient due to mirror 1,

$$\alpha_{m1} = \frac{1}{2d} \ln \frac{1}{\mathscr{R}_1} = -\frac{1}{2d} \ln(1 - \mathscr{T}), \qquad (14.2\text{-}5)$$

to obtain

$$\alpha_r = \alpha_s + \alpha_{m2} - \frac{1}{2d} \ln(1 - \mathscr{T}), \qquad (14.2\text{-}6)$$

where the loss coefficient due to mirror 2 is

$$\alpha_{m2} = \frac{1}{2d} \ln \frac{1}{\mathscr{R}_2}. \qquad (14.2\text{-}7)$$

We now use (14.2-1), (14.2-3), and (14.2-6) to obtain an equation for the transmitted photon-flux density ϕ_o as a function of the mirror transmittance

$$\phi_o = \frac{1}{2}\phi_s \mathscr{T}\left[\frac{g_0}{L - \ln(1 - \mathscr{T})} - 1\right], \qquad g_0 = 2\gamma_0(\nu)d, \quad L = 2(\alpha_s + \alpha_{m2})d, \qquad (14.2\text{-}8)$$

which is plotted in Fig. 14.2-3. Note that the transmitted photon-flux density is directly related to the small-signal gain coefficient. The optimal transmittance $\mathscr{T}_{\text{op}}$ is found by setting the derivative of ϕ_o with respect to $\mathscr{T}$ equal to zero. When $\mathscr{T} \ll 1$ we can

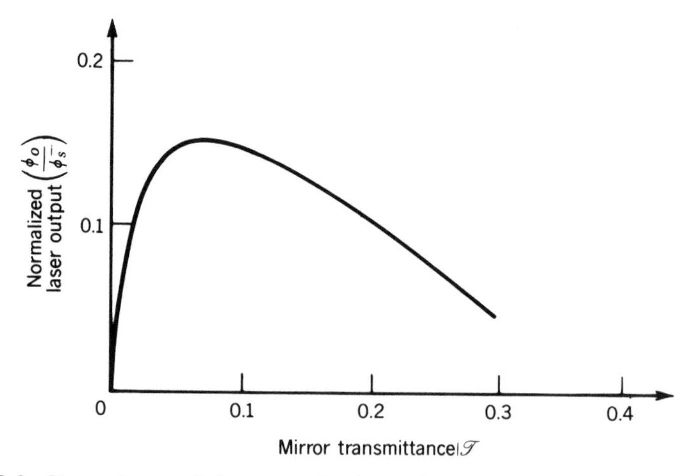

Figure 14.2-3 Dependence of the transmitted steady-state photon-flux density ϕ_o on the mirror transmittance $\mathscr{T}$. For the purposes of this illustration, the gain factor $g_0 = 2\gamma_0 d$ has been chosen to be 0.5 and the loss factor $L = 2(\alpha_s + \alpha_{m2})d$ is 0.02 (2%). The optimal transmittance $\mathscr{T}_{\text{op}}$ turns out to be 0.08.

make use of the approximation $\ln(1 - \mathscr{T}) \approx -\mathscr{T}$ to obtain

$$\mathscr{T}_{\mathrm{op}} \approx (g_0 L)^{1/2} - L. \tag{14.2-9}$$

Internal Photon-Number Density

The steady-state number of photons per unit volume inside the resonator n is related to the steady-state internal photon-flux density ϕ (for photons traveling in both directions) by the simple relation

$$n = \frac{\phi}{c}. \tag{14.2-10}$$

This is readily visualized by considering a cylinder of area A, length c, and volume cA (c is the velocity of light in the medium), whose axis lies parallel to the axis of the resonator. For a resonator containing n photons per unit volume, the cylinder contains cAn photons. These photons travel in both directions, parallel to the axis of the resonator, half of them crossing the base of the cylinder in each second. Since the base of the cylinder also receives an equal number of photons from the other side, however, the photon-flux density (photons per second per unit area in both directions) is $\phi = 2(\frac{1}{2}cAn)/A = cn$, from which (14.2-10) follows.

The photon-number density corresponding to the steady-state internal photon-flux density in (14.2-2) is

$$n = n_s\left(\frac{N_0}{N_t} - 1\right), \qquad N_0 > N_t, \tag{14.2-11}$$

Steady-State Photon-Number Density

where $n_s = \phi_s(\nu)/c$ is the photon-number density saturation value. Using the relations $\phi_s(\nu) = [\tau_s \sigma(\nu)]^{-1}$, $\alpha_r = \gamma(\nu)$, $\alpha_r = 1/c\tau_p$, and $\gamma(\nu) = N\sigma(\nu) = N_t\sigma(\nu)$, (14.2-11) may be written in the form

$$n = (N_0 - N_t)\frac{\tau_p}{\tau_s}, \qquad N_0 > N_t. \tag{14.2-12}$$

Steady-State Photon-Number Density

This relation admits a simple and direct interpretation: $(N_0 - N_t)$ is the population difference (per unit volume) in excess of threshold, and $(N_0 - N_t)/\tau_s$ represents the rate at which photons are generated which, by virtue of steady-state operation, is equal to the rate at which photons are lost, n/τ_p. The fraction τ_p/τ_s is the ratio of the rate at which photons are emitted to the rate at which they are lost.

Under ideal pumping conditions in a four-level laser system, (13.2-10) and (13.2-11) provide that $\tau_s \approx t_{\mathrm{sp}}$ and $N_0 \approx Rt_{\mathrm{sp}}$, where R is the rate (s^{-1}-cm^{-3}) at which atoms are pumped. Equation (14.2-12) can thus be rewritten as

$$\frac{n}{\tau_p} = R - R_t, \qquad R > R_t, \tag{14.2-13}$$

where $R_t = N_t/t_{sp}$ is the threshold value of the pumping rate. Under steady-state conditions, therefore, the overall photon-density loss rate n/τ_p is precisely equal to the excess pumping rate $R - R_t$.

Output Photon Flux and Efficiency

If transmission through the laser output mirror is the only source of resonator loss (which is accounted for in τ_p), and V is the volume of the active medium, (14.2-13) provides that the total output photon flux Φ_o (photons per second) is

$$\Phi_o = (R - R_t)V, \qquad R > R_t. \tag{14.2-14}$$

If there are loss mechanisms other than through the output laser mirror, the output photon flux can be written as

$$\boxed{\Phi_o = \eta_e(R - R_t)V,} \tag{14.2-15}$$

Laser Output Photon Flux

where the emission efficiency η_e is the ratio of the loss arising from the extracted useful light to all of the total losses in the resonator α_r.

If the useful light exits only through mirror 1, (14.1-6) and (14.2-5) for α_r and α_{m1} may be used to write η_e as

$$\eta_e = \frac{\alpha_{m1}}{\alpha_r} = \frac{c}{2d}\tau_p \ln\frac{1}{\mathcal{R}_1}. \tag{14.2-16}$$

If, furthermore, $\mathcal{T} = 1 - \mathcal{R}_1 \ll 1$, (14.2-16) provides

$$\boxed{\eta_e \approx \frac{\tau_p}{T_F}\mathcal{T},} \tag{14.2-17}$$

Emission Efficiency

where we have defined $1/T_F = c/2d$, indicating that the emission efficiency η_e can be understood in terms of the ratio of the photon lifetime to its round-trip travel time, multiplied by the mirror transmittance. The output laser power is then $P_o = h\nu\Phi_o = \eta_e h\nu(R - R_t)V$. With the help of a few algebraic manipulations it can be confirmed that this expression accords with that obtained from (12.2-4).

Losses also result from other sources such as inefficiency in the pumping process. The overall efficiency η of the laser (also called the power conversion efficiency or wall-plug efficiency) is given in Table 14.2-1 for various types of lasers.

B. Spectral Distribution

The spectral distribution of the generated laser light is determined both by the atomic lineshape of the active medium (including whether it is homogeneously or inhomogeneously broadened) and by the resonator modes. This is illustrated in the two conditions for laser oscillation:

- The gain condition requiring that the initial gain coefficient of the amplifier be greater than the loss coefficient [$\gamma_0(\nu) > \alpha_r$] is satisfied for all oscillation fre-

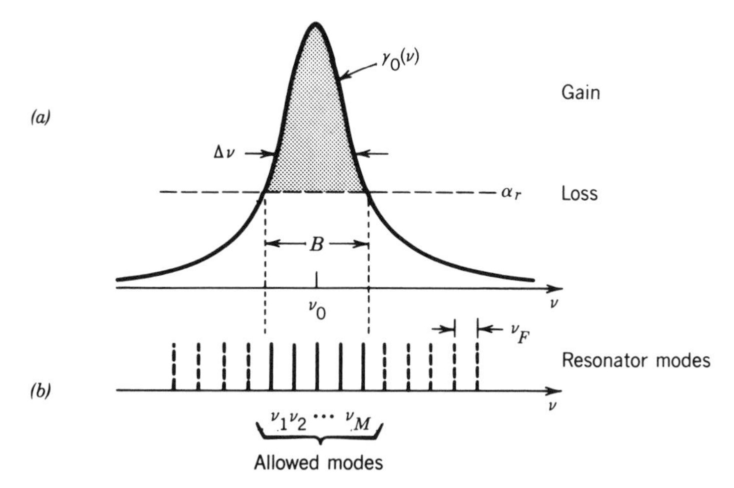

Figure 14.2-4 (*a*) Laser oscillation can occur only at frequencies for which the gain coefficient is greater than the loss coefficient (stippled region). (*b*) Oscillation can occur only within $\delta\nu$ of the resonator modal frequencies (which are represented as lines for simplicity of illustration).

quencies lying within a continuous spectral band of width B centered about the atomic resonance frequency ν_0, as illustrated in Fig. 14.2-4(*a*). The width B increases with the atomic linewidth $\Delta\nu$ and the ratio $\gamma_0(\nu_0)/\alpha_r$; the precise relation depends on the shape of the function $\gamma_0(\nu)$.

- The phase condition requires that the oscillation frequency be one of the resonator modal frequencies ν_q (assuming, for simplicity, that mode pulling is negligible). The FWHM linewidth of each mode is $\delta\nu \approx \nu_F/\mathscr{F}$ [Fig. 14.2-4(*b*)].

It follows that only a finite number of oscillation frequencies $(\nu_1, \nu_2, \ldots, \nu_M)$ are *possible*. The number of possible laser oscillation modes is therefore

$$\boxed{M \approx \frac{B}{\nu_F},} \tag{14.2-18}$$

Number of Possible Laser Modes

where $\nu_F = c/2d$ is the approximate spacing between adjacent modes. However, of these M possible modes, the number of modes that actually carry optical power depends on the nature of the atomic line broadening mechanism. It will be shown below that for an inhomogeneously broadened medium all M modes oscillate (albeit at different powers), whereas for a homogeneously broadened medium these modes engage in some degree of competition, making it more difficult for as many modes to oscillate simultaneously.

The approximate FWHM linewidth of each laser mode might be expected to be $\approx \delta\nu$, but it turns out to be far smaller than this. It is limited by the so-called Schawlow–Townes linewidth, which decreases inversely as the optical power. Almost all lasers have linewidths far greater than the Schawlow–Townes limit as a result of extraneous effects such as acoustic and thermal fluctuations of the resonator mirrors, but the limit can be approached in carefully controlled experiments.

EXERCISE 14.2-1

Number of Modes in a Gas Laser. A Doppler-broadened gas laser has a gain coefficient with a Gaussian spectral profile (see Sec. 12.2D and Exercise 12.2-2) given by $\gamma_0(\nu) = \gamma_0(\nu_0)\exp[-(\nu - \nu_0)^2/2\sigma_D^2]$, where $\Delta\nu_D = (8\ln 2)^{1/2}\sigma_D$ is the FWHM linewidth.

(a) Derive an expression for the allowed oscillation band B as a function of $\Delta\nu_D$ and the ratio $\gamma_0(\nu_0)/\alpha_r$, where α_r is the resonator loss coefficient.

(b) A He–Ne laser has a Doppler linewidth $\Delta\nu_D = 1.5$ GHz and a midband gain coefficient $\gamma_0(\nu_0) = 2 \times 10^{-3}$ cm^{-1}. The length of the laser resonator is $d = 100$ cm, and the reflectances of the mirrors are 100% and 97% (all other resonator losses are negligible). Assuming that the refractive index $n = 1$, determine the number of laser modes M.

Homogeneously Broadened Medium

Immediately after being turned on, all laser modes for which the initial gain is greater than the loss begin to grow [Fig. 14.2-5(*a*)]. Photon-flux densities $\phi_1, \phi_2, \dots, \phi_M$ are created in the M modes. Modes whose frequencies lie closest to the transition central frequency ν_0 grow most quickly and acquire the highest photon-flux densities. These photons interact with the medium and reduce the gain by depleting the population difference. The saturated gain is

$$\gamma(\nu) = \frac{\gamma_0(\nu)}{1 + \sum_{j=1}^{M} \phi_j/\phi_s(\nu_j)}, \qquad (14.2\text{-}19)$$

where $\phi_s(\nu_j)$ is the saturation photon-flux density associated with mode j. The validity of (14.2-19) may be verified by carrying out an analysis similar to that which led to (13.3-3). The saturated gain is shown in Fig. 14.2-5(*b*).

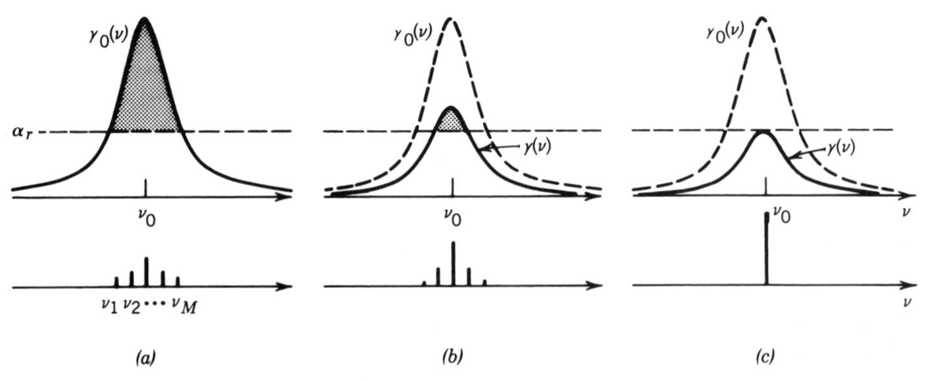

Figure 14.2-5 Growth of oscillation in an ideal homogeneously broadened medium. (*a*) Immediately following laser turn-on, all modal frequencies $\nu_1, \nu_2, \dots, \nu_M$, for which the gain coefficient exceeds the loss coefficient, begin to grow, with the central modes growing at the highest rate. (*b*) After a short time the gain saturates so that the central modes continue to grow while the peripheral modes, for which the loss has become greater than the gain, are attenuated and eventually vanish. (*c*) In the absence of spatial hole burning, only a single mode survives.

Because the gain coefficient is reduced uniformly, for modes sufficiently distant from the line center the loss becomes greater than the gain; these modes lose power while the more central modes continue to grow, albeit at a slower rate. Ultimately, only a single surviving mode (or two modes in the symmetrical case) maintains a gain equal to the loss, with the loss exceeding the gain for all other modes. Under ideal steady-state conditions, the power in this preferred mode remains stable, while laser oscillation at all other modes vanishes [Fig. 14.2-5(*c*)]. The surviving mode has the frequency lying closest to ν_0; values of the gain for its competitors lie below the loss line. Given the frequency of the surviving mode, its photon-flux density may be determined by means of (14.2-2).

In practice, however, homogeneously broadened lasers do indeed oscillate on multiple modes because the different modes occupy different spatial portions of the active medium. When oscillation on the most central mode in Fig. 14.2-5 is established, the gain coefficient can still exceed the loss coefficient at those locations where the standing-wave electric field of the most central mode vanishes. This phenomenon is called **spatial hole burning**. It allows another mode, whose peak fields are located near the energy nulls of the central mode, the opportunity to lase as well.

Inhomogeneously Broadened Medium

In an inhomogeneously broadened medium, the gain $\bar{\gamma}_0(\nu)$ represents the composite envelope of gains of different species of atoms (see Sec. 12.2D), as shown in Fig. 14.2-6.

The situation immediately after laser turn-on is the same as in the homogeneously broadened medium. Modes for which the gain is larger than the loss begin to grow and the gain decreases. If the spacing between the modes is larger than the width $\Delta\nu$ of the constituent atomic lineshape functions, different modes interact with different atoms. Atoms whose lineshapes fail to coincide with any of the modes are ignorant of the presence of photons in the resonator. Their population difference is therefore not affected and the gain they provide remains the small-signal (unsaturated) gain. Atoms whose frequencies coincide with modes deplete their inverted population and their gain saturates, creating "holes" in the gain spectral profile [Fig. 14.2-7(*a*)]. This process is known as **spectral hole burning**. The width of a spectral hole increases with the photon-flux density in accordance with the square-root law $\Delta\nu_s = \Delta\nu(1 + \phi/\phi_s)^{1/2}$ obtained in (13.3-15).

This process of saturation by hole burning progresses independently for the different modes until the gain is equal to the loss for each mode in steady state. Modes do not compete because they draw power from different, rather than shared, atoms. Many modes oscillate independently, with the central modes burning deeper holes and

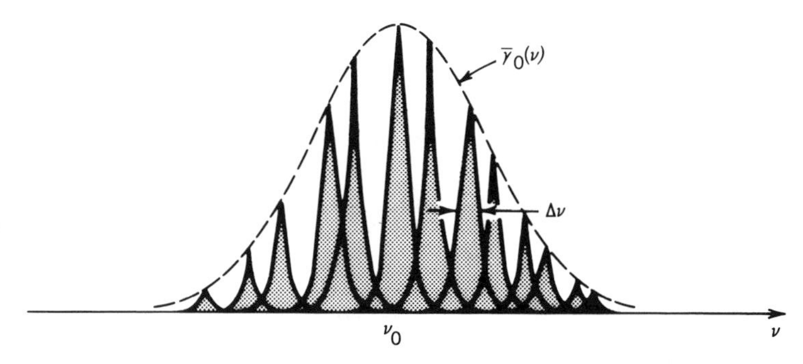

Figure 14.2-6 The lineshape of an inhomogeneously broadened medium is a composite of numerous constituent atomic lineshapes, associated with different properties or different environments.

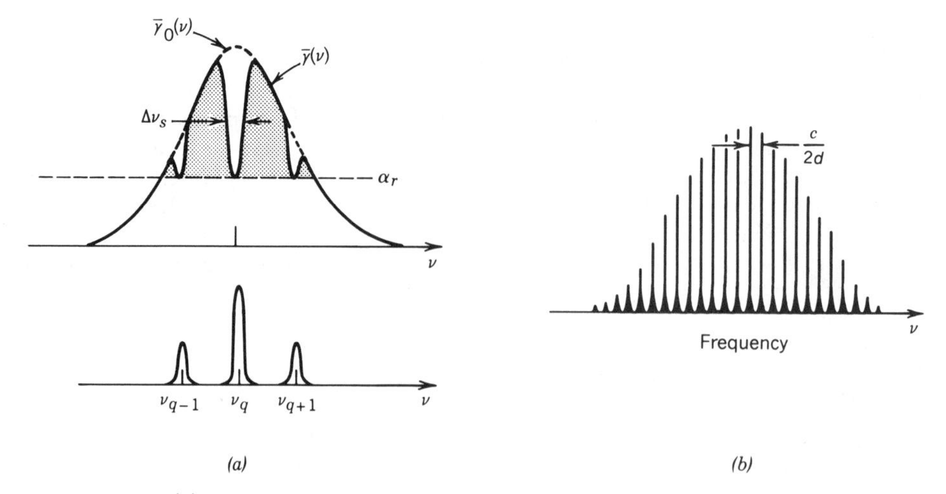

Figure 14.2-7 (*a*) Laser oscillation occurs in an inhomogeneously broadened medium by each mode independently burning a hole in the overall spectral gain profile. The gain provided by the medium to one mode does not influence the gain it provides to other modes. The central modes garner contributions from more atoms, and therefore carry more photons than do the peripheral modes. (*b*) Spectrum of a typical inhomogeneously broadened multimode gas laser.

growing larger, as illustrated in Fig. 14.2-7(*a*). The spectrum of a typical multimode inhomogeneously broadened gas laser is shown in Fig. 14.2-7(*b*). The number of modes is typically larger than that in homogeneously broadened media since spatial hole burning generally sustains fewer modes than spectral hole burning.

*Spectral Hole Burning in a Doppler-Broadened Medium

The lineshape of a gas at temperature T arises from the collection of Doppler-shifted emissions from the individual atoms, which move at different velocities (see Sec. 12.2D and Exercise 12.2-2). A stationary atom interacts with radiation of frequency ν_0. An atom moving with velocity v toward the direction of propagation of the radiation interacts with radiation of frequency $\nu_0(1 + v/c)$, whereas an atom moving away from the direction of propagation of the radiation interacts with radiation of frequency

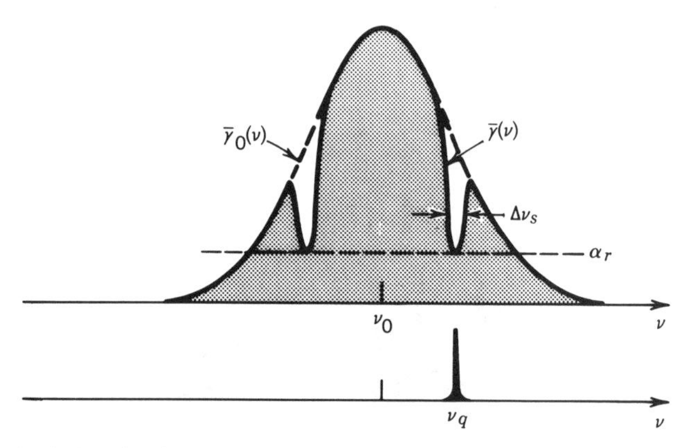

Figure 14.2-8 Hole burning in a Doppler-broadened medium. A probe wave at frequency ν_q saturates those atomic populations with velocities $v = \pm c(\nu_q/\nu_0 - 1)$ on both sides of the central frequency, burning two holes in the gain profile.

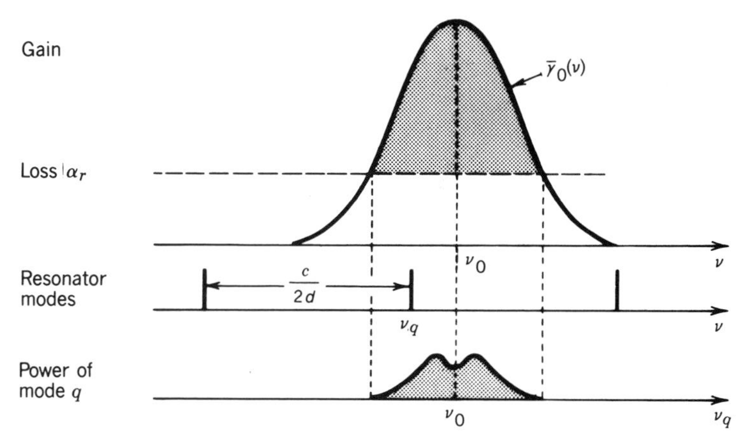

Figure 14.2-9 Power in a single laser mode of frequency ν_q in a Doppler-broadened medium whose gain coefficient is centered about ν_0. Rather than providing maximum power at $\nu_q = \nu_0$, it exhibits the Lamb dip.

$\nu_0(1 - \mathsf{v}/c)$. Because a radiation mode of frequency ν_q travels in both directions as it bounces back and forth between the mirrors of the resonator, it interacts with atoms of two velocity classes: those traveling with velocity $+\mathsf{v}$ and those traveling with velocity $-\mathsf{v}$, such that $\nu_q - \nu_0 = \pm\nu_0 \mathsf{v}/c$. It follows that the mode ν_q saturates the populations of atoms on both sides of the central frequency and burns two holes in the gain profile, as shown in Fig. 14.2-8. If $\nu_q = \nu_0$, of course, only a single hole is burned in the center of the profile.

The steady-state power of a mode increases with the depth of the hole(s) in the gain profile. As the frequency ν_q moves toward ν_0 from either side, the depth of the holes increases, as does the power in the mode. As the modal frequency ν_q begins to approach ν_0, however, the mode begins to interact with only a single group of atoms instead of two, so that the two holes collapse into one. This decrease in the number of available active atoms when $\nu_q = \nu_0$ causes the power of the mode to decrease slightly. Thus the power in a mode, plotted as a function of its frequency ν_q, takes the form of a bell-shaped curve with a central depression, known as the **Lamb dip**, at its center (Fig. 14.2-9).

C. Spatial Distribution and Polarization

Spatial Distribution

The spatial distribution of the emitted laser light depends on the geometry of the resonator and on the shape of the active medium. In the laser theory developed to this point we have ignored transverse spatial effects by assuming that the resonator is constructed of two parallel planar mirrors of infinite extent and that the space between them is filled with the active medium. In this idealized geometry the laser output is a plane wave propagating along the axis of the resonator. But as is evident from Chap. 9, this planar-mirror resonator is highly sensitive to misalignment.

Laser resonators usually have spherical mirrors. As indicated in Sec. 9.2, the spherical-mirror resonator supports a Gaussian beam (which was studied in detail in Chap. 3). A laser using a spherical-mirror resonator may therefore give rise to an output that takes the form of a Gaussian beam.

It was also shown (in Sec. 9.2D) that the spherical-mirror resonator supports a hierarchy of transverse electric and magnetic modes denoted $\mathrm{TEM}_{l,m,q}$. Each pair of indices (l, m) defines a transverse mode with an associated spatial distribution. The

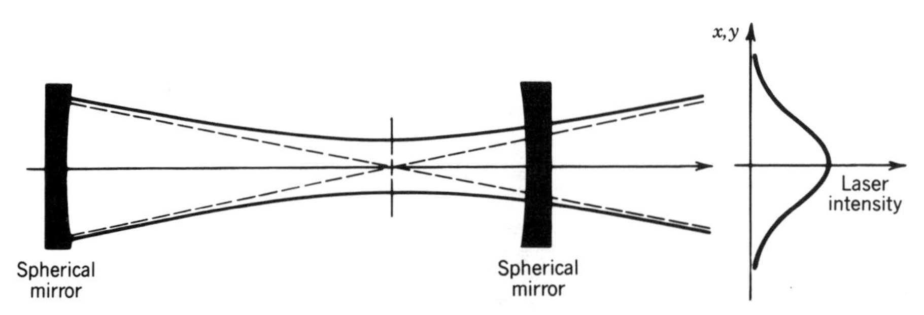

Figure 14.2-10 The laser output for the (0, 0) transverse mode of a spherical-mirror resonator takes the form of a Gaussian beam.

(0, 0) transverse mode is the Gaussian beam (Fig. 14.2-10). Modes of higher l and m form Hermite–Gaussian beams (see Sec. 3.3 and Fig. 3.3-2). For a given (l, m), the index q defines a number of longitudinal (axial) modes of the same spatial distribution but of different frequencies ν_q (which are always separated by the longitudinal-mode spacing $\nu_F = c/2d$, regardless of l and m). The resonance frequencies of two sets of longitudinal modes belonging to two different transverse modes are, in general, displaced with respect to each other by some fraction of the mode spacing ν_F [see (9.2-28)].

Because of their different spatial distributions, different transverse modes undergo different gains and losses. The (0, 0) Gaussian mode, for example, is the most confined about the optical axis and therefore suffers the least diffraction loss at the boundaries of the mirrors. The (1, 1) mode vanishes at points on the optical axis (see Fig. 3.3-2); thus if the laser mirror were blocked by a small central obstruction, the (1, 1) mode would be completely unaffected, whereas the (0, 0) mode would suffer significant loss. Higher-order modes occupy a larger volume and therefore can have larger gain. This disparity between the losses and/or gains of different transverse modes in different geometries determines their competitive edge in contributing to the laser oscillation, as Fig. 14.2-11 illustrates.

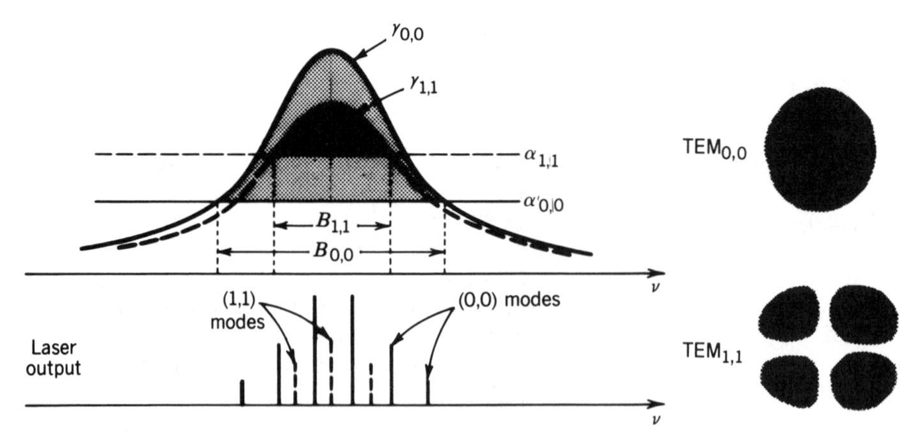

Figure 14.2-11 The gains and losses for two transverse modes, say (0, 0) and (1, 1), usually differ because of their different spatial distributions. A mode can contribute to the output if it lies in the spectral band (of width B) within which the gain coefficient exceeds the loss coefficient. The allowed longitudinal modes associated with each transverse mode are shown.

In a homogeneously broadened laser, the strongest mode tends to suppress the gain for the other modes, but spatial hole burning can permit a few longitudinal modes to oscillate. Transverse modes can have substantially different spatial distributions so that they can readily oscillate simultaneously. A mode whose energy is concentrated in a given transverse spatial region saturates the atomic gain in that region, thereby burning a spatial hole there. Two transverse modes that do not spatially overlap can coexist without competition because they draw their energy from different atoms. Partial spatial overlap between different transverse modes and atomic migrations (as in gases) allow for mode competition.

Lasers are often designed to operate on a single transverse mode; this is usually the $(0,0)$ Gaussian mode because it has the smallest beam diameter and can be focused to the smallest spot size (see Chap. 3). Oscillation on higher-order modes can be desirable, on the other hand, for purposes such as generating large optical power.

Polarization

Each (l, m, q) mode has two degrees of freedom, corresponding to two independent orthogonal polarizations. These two polarizations are regarded as two independent modes. Because of the circular symmetry of the spherical-mirror resonator, the two polarization modes of the same l and m have the same spatial distributions. If the resonator and the active medium provide equal gains and losses for both polarizations, the laser will oscillate on the two modes simultaneously, independently, and with the same intensity. The laser output is then unpolarized (see Sec. 10.4).

Unstable Resonators

Although our discussion has focused on laser configurations that make use of stable resonators (see Fig. 9.2-3), the use of **unstable resonators** offers a number of advantages in the operation of high-power lasers. These include (1) a greater portion of the gain medium contributing to the laser output power as a result of the availability of a larger modal volume; (2) higher output powers attained from operation on the lowest-order transverse mode, rather than on higher-order transverse modes as in the case of stable resonators; and (3) high output power with minimal optical damage to the resonator mirrors, as a result of the use of purely reflective optics that permits the laser light to spill out around the mirror edges (this configuration also permits the optics to be water-cooled and thereby to tolerate high optical powers without damage).

D. Mode Selection

A multimode laser may be operated on a single mode by making use of an element inside the resonator to provide loss sufficient to prevent oscillation of the undesired modes.

Selection of a Laser Line

An active medium with multiple transitions (atomic lines) whose populations are inverted by the pumping mechanism will produce a multiline laser output. A particular line may be selected for oscillation by placing a prism inside the resonator, as shown schematically in Fig. 14.2-12. The prism is adjusted such that only light of the desired wavelength strikes the highly reflecting mirror at normal incidence and can therefore be reflected back to complete the feedback process. By rotating the prism, one wavelength at a time may be selected. Argon-ion lasers, as an example, often contain a rotatable prism in the resonator to allow the choice of one of six common laser lines, stretching from 488 nm in the blue to 514.5 nm in the blue-green. A prism can only be used to select a line if the other lines are well separated from it. It cannot be used, for example, to select one longitudinal mode; adjacent modes are so closely spaced that the dispersive refraction provided by the prism cannot distinguish them.

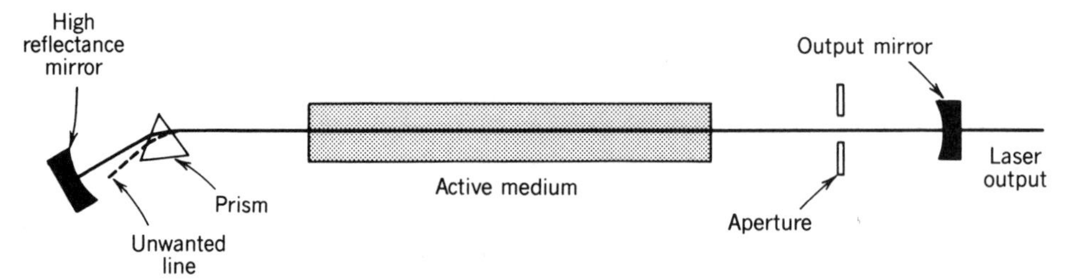

Figure 14.2-12 A particular atomic line may be selected by the use of a prism placed inside the resonator. A transverse mode may be selected by means of a spatial aperture of carefully chosen shape and size.

Selection of a Transverse Mode

Different transverse modes have different spatial distributions, so that an aperture of controllable shape placed inside the resonator may be used to selectively attenuate undesired modes (Fig. 14.2-12). The laser mirrors may also be designed to favor a particular transverse mode.

Selection of a Polarization

A polarizer may be used to convert unpolarized light into polarized light. It is advantageous, however, to place the polarizer inside the resonator rather than outside it. An external polarizer wastes half the output power generated by the laser. The light transmitted by the external polarizer can also suffer from noise arising from the fluctuation of power between the two polarization modes (mode hopping). An internal polarizer creates high losses for one polarization so that oscillation in its corresponding mode never begins. The atomic gain is therefore provided totally to the surviving polarization. An internal polarizer is usually implemented with the help of Brewster windows (see Sec. 6.2 and Exercise 6.2-1), as illustrated in Fig. 14.2-13.

Selection of a Longitudinal Mode

The selection of a single longitudinal mode is also possible. The number of longitudinal modes in an inhomogeneously broadened laser (e.g., a Doppler broadened gas laser) is the number of resonator modes contained in a frequency band B within which the atomic gain is greater than the loss (see Fig. 14.2-4). There are two alternatives for

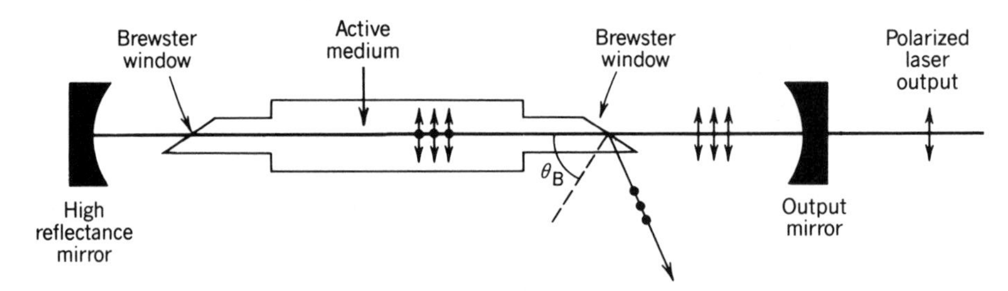

Figure 14.2-13 The use of Brewster windows in a gas laser provides a linearly polarized laser beam. Light polarized in the plane of incidence (the TM wave) is transmitted without reflection loss through a window placed at the Brewster angle. The orthogonally polarized (TE) mode suffers reflection loss and therefore does not oscillate.

operating a laser in a single longitudinal mode:

- Increase the loss sufficiently so that only the mode with the largest gain oscillates. This means, however, that the surviving mode would itself be weak.
- Increase the longitudinal-mode spacing, $\nu_F = c/2d$ by reducing the resonator length. This means, however, that the length of the active medium is reduced, so that the volume of the active medium, and therefore the available laser power, is diminished. In some cases, this approach is impractical. In an argon-ion laser, for example, $\Delta\nu_D = 3.5$ GHz. Thus if $B = \Delta\nu_D$ and $n = 1$, $M = \Delta\nu_D/(c/2d)$, so that the resonator must be shorter than about 4.3 cm to obtain single longitudinal-mode operation.

A number of techniques making use of intracavity frequency-selective elements have been devised for altering the frequency spacing of the resonator modes:

- An *intracavity tilted etalon* (Fabry–Perot resonator) whose mirror separation d_1 is much shorter (thinner) than the laser resonator may be used for mode selection (Fig. 14.2-14). Modes of the etalon have a large spacing $c/2d_1 > B$, so that only one etalon mode can fit within the laser amplifier bandwidth. The etalon is designed so that one of its modes coincides with the resonator longitudinal mode exhibiting the highest gain (or any other desired mode). The etalon may be fine-tuned by means of a slight rotation, by changing its temperature, or by slightly changing its width d_1 with the help of a piezoelectric (or other) transducer. The etalon is slightly tilted with respect to the resonator axis to prevent

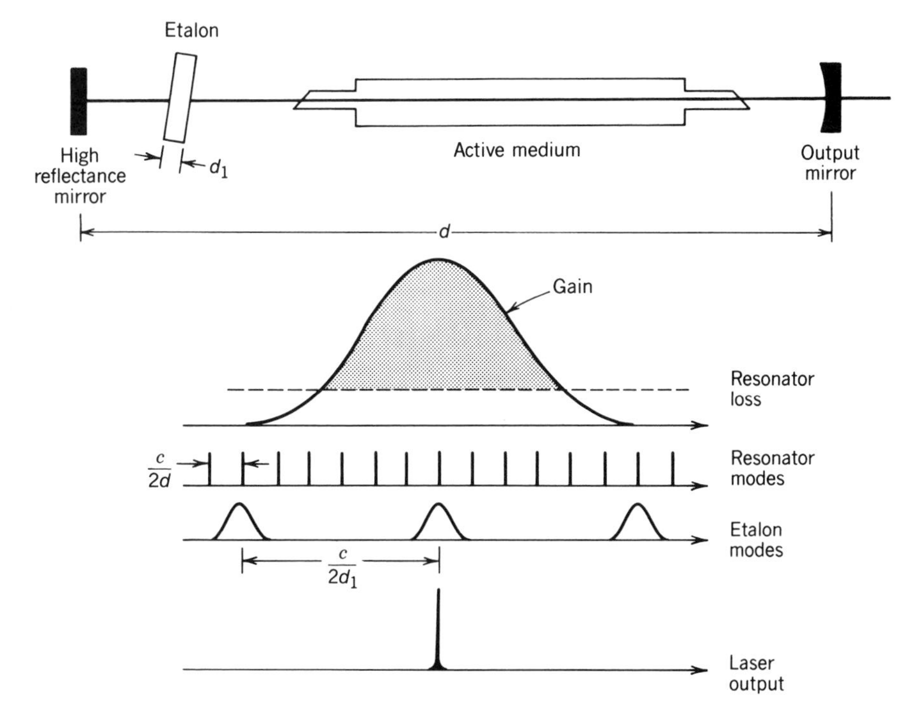

Figure 14.2-14 Longitudinal mode selection by the use of an intracavity etalon. Oscillation occurs at frequencies where a mode of the resonator coincides with an etalon mode; both must, of course, lie within the spectral window where the gain of the medium exceeds the loss.

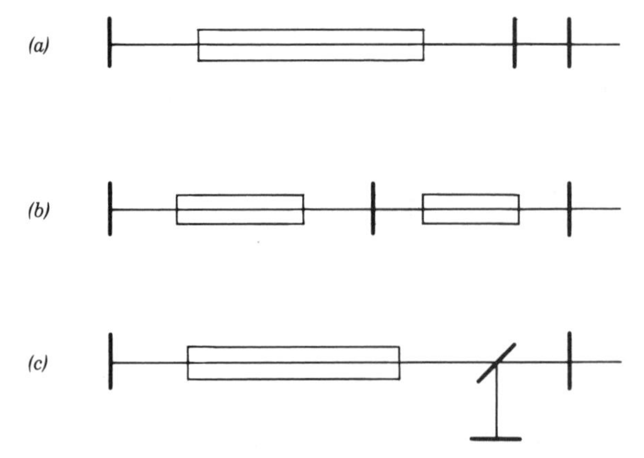

Figure 14.2-15 Longitudinal mode selection by use of (*a*) two coupled resonators (one passive and one active); (*b*) two coupled active resonators; (*c*) a coupled resonator-interferometer.

reflections from its surfaces from reaching the resonator mirrors and thereby creating undesired additional resonances. The etalon is usually temperature stabilized to assure frequency stability.

- *Multiple-mirror resonators* can also be used for mode selection. Several configurations are illustrated in Fig. 14.2-15. Mode selection may be achieved by means of two coupled resonators of different lengths [Fig. 14.2-15(*a*)]. The resonator in Fig. 14.2-15(*b*) consists of two coupled cavities, each with its own gain—in essence, two coupled lasers. This is the configuration used for the C^3 (cleaved-coupled-cavity) semiconductor laser discussed in Chap. 16. Another technique makes use of a resonator coupled with an interferometer [Fig. 14.2-15(*c*)]. The theory of coupled resonators and coupled resonator/interferometers is not addressed here.

E. Characteristics of Common Lasers

Laser amplification and oscillation is ubiquitous and can take place in a great variety of media, including solids (crystals, glasses, and fibers), gases (atomic, ionic, molecular, and excimeric), liquids (organic and inorganic solutes), and plasmas (in which x-ray laser action occurs). The active medium can also be provided by the energy levels of an electron in a magnetic field, as in the case of the free electron laser.

Solid-State Lasers

In Sec. 13.2C we discussed several **solid-state laser** amplifiers in some detail: ruby, Nd^{3+}:YAG, Nd^{3+}:glass, and Er^{3+}:silica fiber. When placed in an optical resonator that provides feedback, all of these materials behave as laser oscillators.

Nd^{3+}:YAG, in particular, enjoys widespread use (see Fig. 13.2-11 for the energy levels of Nd^{3+}:YAG). Its threshold is about an order of magnitude lower than that of ruby by virtue of it being a four-level system. Because it can be optically pumped to its upper laser level by light from a semiconductor laser diode, as shown schematically in Fig. 13.2-8, Nd^{3+}:YAG serves as an efficient compact source of 1.064-μm laser radiation powered by a battery. Nd^{3+}:YAG crystals with lengths as small as a fraction of a millimeter operate as single-frequency (microchip) lasers. Furthermore, neodymium laser light can be passed through a second-harmonic generating crystal (see Sec. 19.2A) which doubles its frequency, thereby providing a strong source of radiation at 532 nm in the green.

Because the transitions in Nd^{3+} arise from inner electrons, which are well shielded from their surroundings, this ionic impurity can, in fact, be made to lase near 1.06 μm in a broad variety of hosts, including glasses of various types, yttrium lithium fluoride (YLF) and yttrium scandium gallium garnet (YSGG). The use of scandium in place of the aluminum in YAG serves to increase the efficiency by about a factor of 2, whereas the gallium aids in the crystal growth. Nd^{3+} ions may even be dissolved in selenium oxychloride which operates as a liquid Nd^{3+} laser. Transitions in other rare-earth ions exhibit similar robustness.

Rare-earth-doped silica fibers can, with proper resonator design, be operated as single-longitudinal-mode lasers (see Fig. 13.2-8). An example is provided by a 5-m-long Er^{3+}:silica fiber laser operated in a Fabry–Perot configuration. A mirror reflectance of 99% at one end and 4% at the other end (simple Fresnel reflection) provides an output of about 8 mW with a semiconductor pump power of 90 mW at 1.46 μm. Alternatively, cavities can be constructed in the form of fiber ring resonators or fiber loop reflectors. Doped silica fiber lasers can also function in pulsed Q-switched and mode-locked configurations (see Sec. 14.3). At 300 K this system behaves as a three-level laser, while at 77 K it behaves as a four-level laser. The distinction is important because there is an optimal fiber length for achieving minimum threshold in a three-level system, whereas in a four-level system the threshold power decreases inversely with the active fiber length.

Aside from ruby, Nd^{3+}, and Er^{3+}, other commonly encountered optically pumped solid-state laser amplifiers and oscillators include alexandrite ($Cr^{3+}:Al_2BeO_4$), which offers a tunable output in the wavelength range between 700 and 800 nm; $Ti^{3+}:Al_2O_3$ (Ti:sapphire), which is tunable over an even broader range, from 660 to 1180 nm; and Er^{3+}:YAG, which is often operated at 1.66 μm.

Gas Lasers

The **gas laser** is probably the most frequently encountered type of laser oscillator. The red-orange, green, and blue beams of the He–Ne, Ar^+, and He–Cd gas lasers, respectively, are by now familiar to many (see Fig. 12.1-3 for the energy levels of He and Ne). The Kr^+ laser readily produces hundreds of milliwatts of optical power at wavelengths ranging from 350 nm in the ultraviolet to 647 nm in the red. It can be operated simultaneously on a number of lines to produce "white laser light." These lasers can all be operated on innumerable other lines. Small He–Ne lasers are so commonplace and inexpensive that they are used by lecturers as pointers and in supermarkets as bar-code readers.

Molecular gas lasers such as CO_2 (see Fig. 12.1-1 for the energy levels of CO_2) and CO, which operate in the middle-infrared region of the spectrum, are highly efficient and can produce copious amounts of power. Indeed, most molecular transitions in the infrared region can be made to lase; even simple water vapor (H_2O) lases at many wavelengths in the far infrared.

A gas laser of high current importance in the ultraviolet region is the **excimer laser**. Excimers (e.g., KrF) exist only in the form of excited electronic states since the constituents are repulsive in the ground state. The lower laser level is therefore always empty, providing a built-in population inversion. Rare-gas halides readily form complexes in the excited state because the chemical behavior of an excited rare gas atom is similar to that of an alkali atom, which readily reacts with a halogen.

Liquid Lasers

The importance of liquid **dye lasers** stems principally from their tunability. The active medium of a dye laser is a solution of an organic dye compound in alcohol or water (see Fig. 12.1-4 for a schematic illustration of the energy levels of a dye molecule). Polymethine dyes provide oscillation in the red or near infrared ($\approx$ 0.7 to 1.5 μm), xanthene dyes lase in the visible (500 to 700 nm), coumarin dyes oscillate in the blue-green (400 to 500 nm), and scintillator dyes lase in the ultraviolet region of the spectrum (< 400 nm). Rhodamine-6G dye, for example, can be tuned over the range from 560 to 640 nm.

Plasma X-Ray Lasers

A number of different types of **x-ray lasers** have been operated during the past decade. The difficulty in achieving x-ray laser action stems from several factors. The threshold population difference N_t, according to (14.1-14), is proportional to $1/\lambda^2\tau_p$. It is therefore increasingly difficult to attain threshold as λ decreases. Furthermore, it is technically difficult to fabricate high-quality mirrors in the x-ray region because the refractive index does not vary appreciably from material to material. Dielectric mirrors therefore require a very large number of layers, rendering the resonator loss coefficient α_r large and the photon lifetime τ_p small. Improved x-ray optical components are in the offing, however.

X-ray laser action was apparently first achieved in a dramatic experiment carried out by researchers at the Lawrence Livermore National Laboratory (LLNL) in 1980. An underground nuclear detonation was used to create x-rays, which, in turn, served to pump the atoms in an assembly of metal rods. The x-ray laser pulse was generated before the detonation vaporized the apparatus.

In a series of more controlled experiments at the Princeton Plasma Physics Laboratory (PPPL) in New Jersey, a solid carbon disk was used as the x-ray laser medium. A 10.6-μm-wavelength CO_2 laser pulse, of 50 ns duration and 300 J energy, was focused onto the carbon. The infrared laser pulse generated sufficient heat to strip all the electrons away from some of the carbon atoms, thereby creating a plasma of ionized carbon (C^{6+}) and serving as the pump. The plasma was radially confined by the use of a magnetic field. The cooling of the plasma at the termination of the laser pulse led to the capture of electrons in the $q = 3$ orbits of the hydrogen-like C^{6+} ions, and simultaneously to a dearth of electrons in the $q = 2$ orbits, resulting in a population inversion (see Fig. 12.1-2).

As expected from (12.1-6), the decay of electrons from $q = 3$ to $q = 2$ was accompanied by the emission of x-ray photons of energy

$$E = \frac{m_r Z^2 e^4}{2\hbar^2}\left(\frac{1}{2^2} - \frac{1}{3^2}\right).$$

With $Z = 6$ this corresponds to a photon of energy 68 eV and wavelength $\lambda_o = (1.24/68)$ μm = 18.2 nm. These spontaneously emitted photons caused the stimulated emission of x-ray photons from other atoms, resulting in amplified spontaneous emission (ASE). These experiments exhibited a single-pass gain coefficient-length product $\gamma d \approx 6$, so that in accordance with (13.1-7) the gain was $G \approx e^6$. The result was the generation of a 20-ns pulse of soft x-ray ASE with a power of 100 kW, an energy of 2 mJ, and a divergence of 5 mrad.

More recently, the gigantic NOVA 1.06-μm Nd^{3+}:glass laser system at LLNL was used to vaporize thin foils of tantalum and tungsten metal, creating nickel-like Ta^{45+} and W^{46+} ions respectively, and producing 250-ps x-ray laser pulses at wavelengths as short as $\lambda_o = 4.3$ nm.

Potential x-ray laser applications include x-ray microlithography for producing the next generation of densely packed semiconductor chips, and the dynamic imaging and holography of individual cellular structures in biological specimens.

Free Electron Lasers

The **free electron laser** (FEL) makes use of a magnetic "wiggler" field, which is produced by a periodic assembly of magnets of alternating polarity. The active medium is a relativistic electron beam moving in the wiggler field. The electrons are not bound to atoms, but they are nevertheless not truly free since their motion is governed by the wiggler field. The emission wavelength can be tuned over a broad range by changing the electron-beam energy and the magnet period. Depending on their design, FELs can emit at wavelengths that range from the vacuum ultraviolet to the far infrared. Several examples of operating FELs are: the ultraviolet FEL at the University of Paris, which operates near 0.2 μm; the visible FEL at Stanford University (California), which operates in the region from 0.5 to 10 μm; the middle-infrared FEL at the Los Alamos National Laboratory (LANL) in New Mexico, which operates in the region from 9 to 40 μm; and the far-infrared FEL at the University of California at Santa Barbara, which operates in the wavelength band from 400 to 1000 μm.

Tabulation of Selected Laser Transitions

In Table 14.2-1 we provide a list, in order of increasing wavelength, of the representative parameters and characteristics of some well-known laser transitions. The broad range of transition wavelengths, overall efficiencies, and power outputs for the different lasers is noteworthy.

The transition cross section, spontaneous lifetime, and atomic linewidth for a number of these laser transitions are listed in Table 13.2-1. The linewidth of the laser

TABLE 14.2-1 Typical Characteristics and Parameters for a Number of Well-Known Gas (g), Solid (s), Liquid (l), and Plasma (p) Laser Transitions

Laser Medium	Transition Wavelength λ_o	Single Mode (S) or Multimode (M)	CW or Pulsed[a]	Approximate Overall Efficiency $\eta(\%)$[b]	Output Power or Energy[c]	Energy-Level Diagram
C^{6+} (p)	18.2 nm	M	Pulsed	10^{-5}	2 mJ	Fig. 12.1-2
ArF excimer (g)	193 nm	M	Pulsed	1.	500 mJ	
KrF excimer (g)	248 nm	M	Pulsed	1.	500 mJ	
He–Cd (g)	442 nm	S/M	CW	0.1	10 mW	
Ar^+ (g)	515 nm	S/M	CW	0.05	10 W	
Rhodamine-6G dye (l)	560–640 nm	S/M	CW	0.005	100 mW	Fig. 12.1-4
He–Ne (g)	633 nm	S/M	CW	0.05	1 mW	Fig. 12.1-3
Kr^+ (g)	647 nm	S/M	CW	0.01	500 mW	
Ruby (s)	694 nm	M	Pulsed	0.1	5 J	Fig. 13.2-9
Ti^{3+}:Al_2O_3 (s)	0.66–1.18 μm	S/M	CW	0.01	10 W	
Nd^{3+}:glass (s)	1.06 μm	M	Pulsed	1.	50 J	Fig. 13.2-11
Nd^{3+}:YAG (s)	1.064 μm	S/M	CW	0.5	10 W	Fig. 13.2-11
KF color center (s)	1.25–1.45 μm	S/M	CW	0.005	500 mW	
He–Ne (g)	3.39 μm	S/M	CW	0.05	1 mW	Fig. 12.1-3
FEL (LANL)	9–40 μm	M	Pulsed	0.5	1 mJ	
CO_2 (g)	10.6 μm	S/M	CW	10.	100 W	Fig. 12.1-1
H_2O (g)	118.7 μm	S/M	CW	0.001	10 μW	
HCN (g)	336.8 μm	S/M	CW	0.001	1 mW	

[a]Lasers designated "CW" can alternatively be operated in a pulsed mode; lasers designated "pulsed" are usually operated in that mode.
[b]The overall efficiency (also called the power conversion efficiency or wall-plug efficiency) is the ratio of light power output to electrical power input (or, for pulsed lasers, light energy output to electrical energy input). The record for high overall efficiency ($\approx$ 65%) belongs to the semiconductor injection laser, which is discussed in Chap. 16.
[c]The output power (for CW systems) and output energy per pulse (for pulsed systems) vary over a substantial range, in part because of the wide range of pulse durations; representative values are provided.

output is generally many orders of magnitude smaller than the atomic linewidths given in Table 13.2-1; this is because of the additional frequency selectivity imposed by the optical resonator. Some laser systems cannot sustain a continuous population inversion and therefore operate only in a pulsed mode.

14.3 PULSED LASERS

A. Methods of Pulsing Lasers

The most direct method of obtaining pulsed light from a laser is to use a continuous-wave (CW) laser in conjunction with an external switch or modulator that transmits the light only during selected short time intervals. This simple method has two distinct disadvantages, however. First, the scheme is inefficient since it blocks (and therefore wastes the light) energy during the off-time of the pulse train. Second, the peak power of the pulses cannot exceed the steady power of the CW source, as illustrated in Fig. 14.3-1(*a*).

More efficient pulsing schemes are based on turning the laser itself on and off by means of an internal modulation process, designed so that energy is stored during the off-time and released during the on-time. Energy may be stored either in the resonator, in the form of light that is periodically permitted to escape, or in the atomic system, in the form of a population inversion that is released periodically by allowing the system to oscillate. These schemes permit short laser pulses to be generated with peak powers far in excess of the constant power deliverable by CW lasers, as illustrated in Fig. 14.3-1(*b*).

Four common methods used for the internal modulation of laser light are: gain switching, Q-switching, cavity dumping, and mode locking. These are considered in turn.

Gain Switching

In this rather direct approach, the gain is controlled by turning the laser pump on and off (Fig. 14.3-2). In the flashlamp-pumped pulsed ruby laser, for example, the pump (flashlamp) is switched on periodically for brief periods of time by a sequence of electrical pulses. During the on-times, the gain coefficient exceeds the loss coefficient and laser light is produced. Most pulsed semiconductor lasers are gain switched because it is easy to modulate the electric current used for pumping, as discussed in Chap. 16. The laser-pulse rise and fall times achievable with gain switching are determined in Sec. 14.3B.

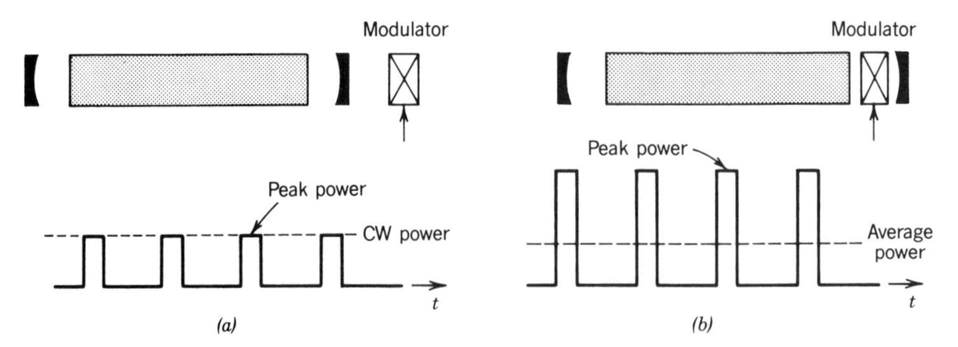

Figure 14.3-1 Comparison of pulsed laser outputs achievable with (*a*) an external modulator, and (*b*) an internal modulator.

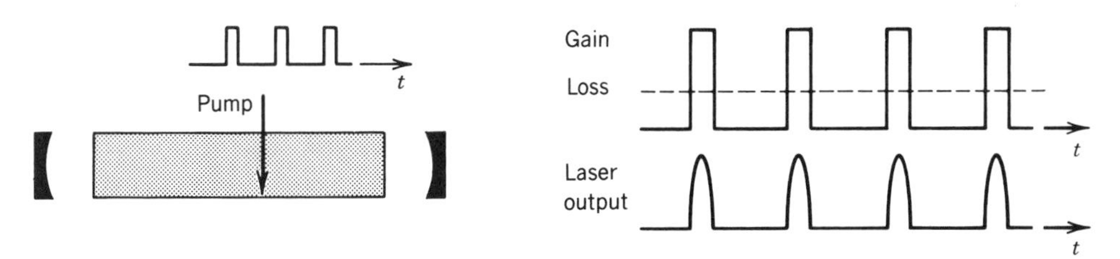

Figure 14.3-2 Gain switching.

Q-Switching

In this scheme, the laser output is turned off by increasing the resonator loss (spoiling the resonator quality factor Q) periodically with the help of a modulated absorber insider the resonator (Fig. 14.3-3). Thus Q-switching is loss switching. Because the pump continues to deliver constant power at all time, energy is stored in the atoms in the form of an accumulated population difference during the off (high-loss)-times. When the losses are reduced during the on-times, the large accumulated population difference is released, generating intense (usually short) pulses of light. An analysis of this method is provided in Sec. 14.3C.

Cavity Dumping

This technique is based on storing photons (rather than a population difference) in the resonator during the off-times, and releasing them during the on-times. It differs from Q-switching in that the resonator loss is modulated by altering the mirror transmittance (see Fig. 14.3-4). The system operates like a bucket into which water is poured from a hose at a constant rate. After a period of time of accumulating water, the bottom of the bucket is suddenly removed so that the water is "dumped." The bucket bottom is subsequently returned and the process repeated. A constant flow of water is therefore converted into a pulsed flow. For the cavity-dumped laser, of course, the bucket represents the resonator, the water hose represents the constant pump, and the bucket bottom represents the laser output mirror. The leakage of light from the resonator, *including useful light*, is not permitted during the off-times. This results in negligible resonator losses, thereby increasing the optical power inside the laser resonator. Photons are stored in the resonator and cannot escape. The mirror is suddenly removed altogether (e.g., by rotating it out of alignment), increasing its transmittance to 100% during the on-times. As the accumulated photons leave the resonator, the sudden increase in the loss arrests the oscillation. The result is a strong pulse of laser light. The analysis for cavity dumping is not provided here inasmuch as it is closely related to that of Q-switching. This is because the variation of the gain and loss with time are similar, as may be seen by comparing Fig. 14.3-4 with Fig. 14.3-3.

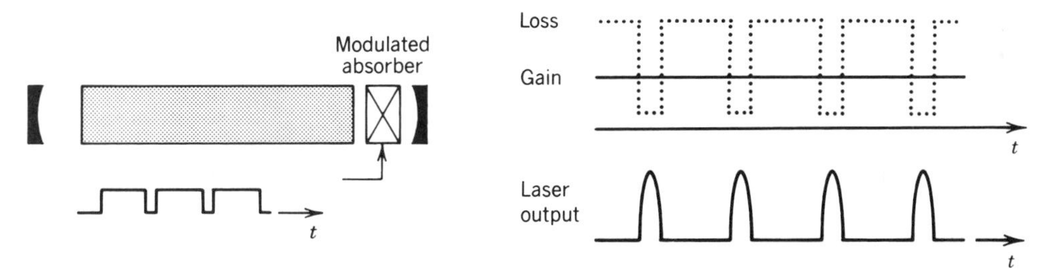

Figure 14.3-3 Q-switching.

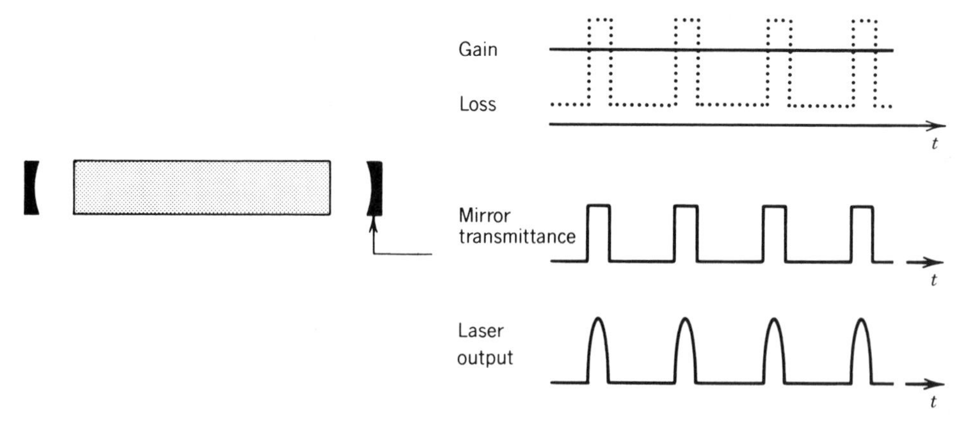

Figure 14.3-4 Cavity dumping. One of the mirrors is removed altogether to dump the stored photons as useful light.

Mode Locking

Mode locking is distinct from the previous three techniques. Pulsed laser action is attained by coupling together the modes of a laser and locking their phases to each other. For example, the longitudinal modes of a multimode laser, which oscillate at frequencies that are equally separated by the intermodal frequency $c/2d$, may be made to behave in this fashion. When the phases of these components are locked together, they behave like the Fourier components of a periodic function, and therefore form a periodic pulse train. The coupling of the modes is achieved by periodically modulating the losses inside the resonator. Mode locking is examined in Sec. 14.3D.

*B. Analysis of Transient Effects

An analytical description of the operation of pulsed lasers requires an understanding of the dynamics of the laser oscillation process, i.e., the time course of laser oscillation onset and termination. The steady-state solutions presented earlier in the chapter are inadequate for this purpose. The lasing process is governed by two variables: the number of photons per unit volume in the resonator, $n(t)$, and the atomic population difference per unit volume, $N(t) = N_2(t) - N_1(t)$; both are functions of the time t.

Rate Equation for the Photon-Number Density

The photon-number density n is governed by the rate equation

$$\frac{dn}{dt} = -\frac{n}{\tau_p} + NW_i. \tag{14.3-1}$$

The first term represents photon loss arising from leakage from the resonator, at a rate given by the inverse photon lifetime $1/\tau_p$. The second term represents net photon gain, at a rate NW_i, arising from stimulated emission and absorption. $W_i = \phi\sigma(\nu) = cn\sigma(\nu)$ is the probability density for induced absorption/emission. Spontaneous emission is assumed to be small. With the help of the relation $N_t = \alpha_r/\sigma(\nu) = 1/c\tau_p\sigma(\nu)$, where N_t is the threshold population difference [see (14.1-13)], we write $\sigma(\nu) = 1/c\tau_p N_t$, from which

$$W_i = \frac{n}{N_t \tau_p}.$$

Substituting this into (14.3-1) provides a simple differential equation for the photon number density n,

$$\boxed{\frac{dn}{dt} = -\frac{n}{\tau_p} + \frac{N}{N_t}\frac{n}{\tau_p}.} \tag{14.3-2}$$

Photon-Number Rate Equation

As long as $N > N_t$, dn/dt will be positive and n will increase. When steady state $(dn/dt = 0)$ is reached, $N = N_t$.

Rate Equation for the Population Difference

The dynamics of the population difference $N(t)$ depends on the pumping configuration. A three-level pumping scheme (see Sec. 13.2B) is analyzed here. The rate equation for the population of the upper energy level of the transition is, according to (13.2-5),

$$\frac{dN_2}{dt} = R - \frac{N_2}{t_{sp}} - W_i(N_2 - N_1), \tag{14.3-3}$$

where it is assumed that $\tau_2 = t_{sp}$. R is the pumping rate, which is assumed to be independent of the population difference N. Denoting the total atomic number density $N_2 + N_1$ by N_a, so that $N_1 = (N_a - N)/2$ and $N_2 = (N_a + N)/2$, we obtain a differential equation for the population difference $N = N_2 - N_1$,

$$\frac{dN}{dt} = \frac{N_0}{t_{sp}} - \frac{N}{t_{sp}} - 2W_iN, \tag{14.3-4}$$

where the small-signal population difference $N_0 = 2Rt_{sp} - N_a$ [see (13.2-22)]. Substituting the relation $W_i = n/N_t\tau_p$ obtained above into (14.3-4) then yields

$$\boxed{\frac{dN}{dt} = \frac{N_0}{t_{sp}} - \frac{N}{t_{sp}} - 2\frac{N}{N_t}\frac{n}{\tau_p}.} \tag{14.3-5}$$

Population-Difference Rate Equation (Three-Level System)

The third term on the right-hand side of (14.3-5) is twice the second term on the right-hand side of (14.3-2), and of opposite sign. This reflects the fact that the generation of one photon by an induced transition reduces the population of level 2 by one atom while increasing the population of level 1 by one atom, thereby decreasing the population difference by two atoms.

Equations (14.3-2) and (14.3-5) are coupled nonlinear differential equations whose solution determines the transient behavior of the photon number density $n(t)$ and the population difference $N(t)$. Setting $dN/dt = 0$ and $dn/dt = 0$ leads to $N = N_t$

and $n = (N_0 - N_t)(\tau_p/2t_{sp})$. These are indeed the steady-state values of N and n obtained previously, as is evident from (14.2-12) with $\tau_s = 2t_{sp}$, as provided by (13.2-23) for a three-level pumping scheme.

EXERCISE 14.3-1

Population-Difference Rate Equation for a Four-Level System. Obtain the population-difference rate equation for a four-level system for which $\tau_1 \ll t_{sp}$. Explain the absence of the factor of 2 that appears in (14.3-5).

Gain Switching

Gain switching is accomplished by turning the pumping rate R on and off; this, in turn, is equivalent to modulating the small-signal population difference $N_0 = 2Rt_{sp} - N_a$. A schematic illustration of the typical time evolution of the population difference $N(t)$ and the photon-number density $n(t)$, as the laser is pulsed by varying N_0 is provided in Fig. 14.3-5. The following regimes are evident in the process:

- For $t < 0$, the population difference $N(t) = N_{0a}$ lies below the threshold N_t and oscillation cannot occur.
- The pump is turned on at $t = 0$, which increases N_0 from a value N_{0a} below threshold to a value N_{0b} above threshold in step-function fashion. The population difference $N(t)$ begins to increase as a result. As long as $N(t) < N_t$, however, the photon-number density $n = 0$. In this region (14.3-5) therefore becomes $dN/dt = (N_0 - N)/t_{sp}$, indicating that $N(t)$ grows exponentially toward its equilibrium value N_{0b} with time constant t_{sp}.
- Once $N(t)$ crosses the threshold N_t, at $t = t_1$, laser oscillation begins and $n(t)$ increases. The population inversion then begins to deplete so that the rate of

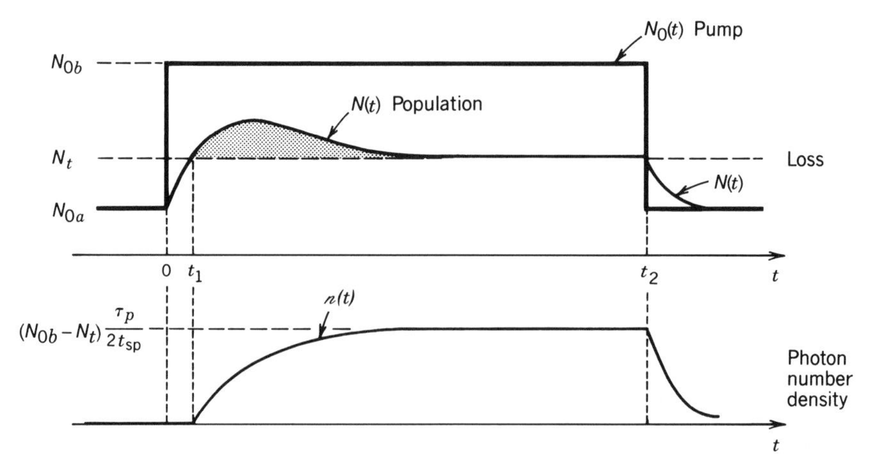

Figure 14.3-5 Variation of the population difference $N(t)$ and the photon-number density $n(t)$ with time, as a square pump pulse results in N_0 suddenly increasing from a low value N_{0a} to a high value N_{0b}, and then decreasing back to a low value N_{0a}.

increase of $N(t)$ slows. As $n(t)$ becomes larger, the depletion becomes more effective so that $N(t)$ begins to decay toward N_t. $N(t)$ finally reaches N_t, at which time $n(t)$ reaches its steady-state value.

- The pump is turned off at time $t = t_2$, which reduces N_0 to its initial value N_{0a}. $N(t)$ and $n(t)$ decay to the values N_{0a} and 0, respectively.

The actual profile of the buildup and decay of $n(t)$ is obtained by numerically solving (14.3-2) and (14.3-5). The precise shape of the solution depends on t_{sp}, τ_p, N_t, as well as on N_{0a} and N_{0b} (see Problem 14.3-1).

*C. *Q*-Switching

Q-switched laser pulsing is achieved by switching the resonator loss coefficient α_r from a large value during the off-time to a small value during the on-time. This may be accomplished in any number of ways, such as by placing a modulator that periodically introduces large losses in the resonator. Since the lasing threshold population difference N_t is proportional to the resonator loss coefficient α_r [see (14.1-12) and (14.1-5)], the result of switching α_r is to decrease N_t from a high value N_{ta} to a low value N_{tb}, as illustrated in Fig. 14.3-6. In *Q*-switching, therefore, N_t is modulated while N_0 remains fixed, whereas in gain switching N_0 is modulated while N_t remains fixed (see Fig. 14.3-5). The population and photon-number densities behave as follows:

- At $t = 0$, the pump is turned on so that N_0 follows a step function. The loss is maintained at a level that is sufficiently high ($N_t = N_{ta} > N_0$) so that laser oscillation cannot begin. The population difference $N(t)$ therefore builds up (with time constant t_{sp}). Although the medium is now a high-gain amplifier, the loss is sufficiently large so that oscillation is prevented.
- At $t = t_1$, the loss is suddenly decreased so that N_t diminishes to a value $N_{tb} < N_0$. Oscillation therefore begins and the photon-number density rises sharply. The presence of the radiation causes a depletion of the population inversion (gain saturation) so that $N(t)$ begins to decrease. When $N(t)$ falls below N_{tb}, the loss again exceeds the gain, resulting in a rapid decrease of the photon-number density (with a time constant of the order of the photon lifetime τ_p).

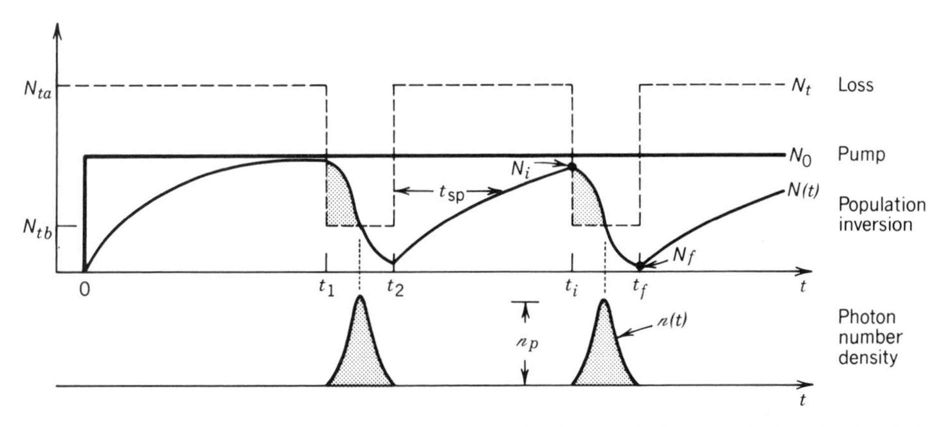

Figure 14.3-6 Operation of a *Q*-switched laser. Variation of the population threshold N_t (which is proportional to the resonator loss), the pump parameter N_0, the population difference $N(t)$, and the photon number $n(t)$.

- At $t = t_2$, the loss is reinstated, insuring the availability of a long period of population-inversion buildup to prepare for the next pulse. The process is repeated periodically so that a periodic optical pulse train is generated.

We now undertake an analysis to determine the peak power, energy, width, and shape of the optical pulse generated by a Q-switched laser in the steady pulsed state. We rely on the two basic rate equations (14.3-2) and (14.3-5) for $n(t)$ and $N(t)$, respectively, which we solve during the on-time t_i to t_f indicated in Fig. 14.3-6. The problem can, of course, be solved numerically. However, it simplifies sufficiently to permit an analytic solution if we assume that the first two terms of (14.3-5) are negligible. This assumption is suitable if both the pumping and the spontaneous emission are negligible in comparison with the effects of induced transitions during the short time interval from t_i to t_f. This approximation turns out to be reasonable if the width of the generated optical pulse is much shorter than t_{sp}. When this is the case, (14.3-2) and (14.3-5) become

$$\frac{dn}{dt} = \left(\frac{N}{N_t} - 1\right)\frac{n}{\tau_p} \tag{14.3-6}$$

$$\frac{dN}{dt} = -2\frac{N}{N_t}\frac{n}{\tau_p}. \tag{14.3-7}$$

These are two coupled differential equations in $n(t)$ and $N(t)$ with initial conditions $n = 0$ and $N = N_i$ at $t = t_i$. Throughout the time interval from t_i to t_f, N_t is fixed at its low value N_{th}.

Dividing (14.3-6) by (14.3-7), we obtain a single differential equation relating n and N,

$$\frac{dn}{dN} \approx \frac{1}{2}\left(\frac{N_t}{N} - 1\right), \tag{14.3-8}$$

which we integrate to obtain

$$n \approx \tfrac{1}{2}N_t \ln(N) - \tfrac{1}{2}N + \text{constant}. \tag{14.3-9}$$

Using the initial condition $n = 0$ when $N = N_i$ finally leads to

$$n \approx \frac{1}{2}N_t \ln\frac{N}{N_i} - \frac{1}{2}(N - N_i). \tag{14.3-10}$$

Pulse Power

According to (14.2-10) and (14.2-3), the internal photon-flux density (comprising both directions) is given by $\phi = nc$, whereas the external photon-flux density emerging from mirror 1 (which has transmittance $\mathcal{T}$) is $\phi_o = \frac{1}{2}\mathcal{T}nc$. Assuming that the photon-flux density is uniform over the cross-sectional area A of the emerging beam, the corresponding optical output power is

$$P_o = h\nu A\phi_o = \frac{1}{2}h\nu c\mathscr{T}A n = h\nu\mathscr{T}\frac{c}{2d}Vn, \tag{14.3-11}$$

where $V = Ad$ is the volume of the resonator. According to (14.2-17), if $\mathscr{T} \ll 1$, the fraction of the resonator loss that contributes to useful light at the output is $\eta_e \approx \mathscr{T}(c/2d)\tau_p$, so that we obtain

$$P_o = \eta_e h\nu \frac{nV}{\tau_p}. \tag{14.3-12}$$

Equation (14.3-12) is easily interpreted since the factor nV/τ_p is the number of photons lost from the resonator per unit time.

Peak Pulse Power

As discussed earlier and illustrated in Fig. 14.3-6, n reaches its peak value n_p when $N = N_t = N_{th}$. This is corroborated by setting $dn/dt = 0$ in (14.3-6), which leads immediately to $N = N_t$. Substituting this into (14.3-10) therefore provides

$$n_p = \frac{1}{2}N_i\left(1 + \frac{N_t}{N_i}\ln\frac{N_t}{N_i} - \frac{N_t}{N_i}\right). \tag{14.3-13}$$

Using this result in conjunction with (14.3-11) gives the peak power

$$P_p = h\nu\mathscr{T}\frac{c}{2d}Vn_p. \tag{14.3-14}$$

When $N_i \gg N_t$, as must be the case for pulses of large peak power, $N_t/N_i \ll 1$, whereupon (14.3-13) gives

$$n_p \approx \tfrac{1}{2}N_i, \tag{14.3-15}$$

The peak photon-number density is then equal to one-half the initial population density difference. In this case, the peak power assumes the particularly simple form

$$P_p \approx \frac{1}{2}h\nu\mathscr{T}\frac{c}{2d}VN_i. \tag{14.3-16}$$

Peak Pulse Power

Pulse Energy

The pulse energy is given by

$$E = \int_{t_i}^{t_f} P_o\, dt,$$

which, in accordance with Eq. (14.3-11), can be written as

$$E = h\nu\mathscr{T}\frac{c}{2d}V\int_{t_i}^{t_f} n(t)\, dt = h\nu\mathscr{T}\frac{c}{2d}V\int_{N_i}^{N_f} n(t)\frac{dt}{dN}\, dN. \tag{14.3-17}$$

Using (14.3-7) in (14.3-17), we obtain

$$E = \frac{1}{2} h\nu \mathscr{T} \frac{c}{2d} V N_t \tau_p \int_{N_f}^{N_i} \frac{dN}{N}, \tag{14.3-18}$$

which integrates to

$$E = \frac{1}{2} h\nu \mathscr{T} \frac{c}{2d} V N_t \tau_p \ln \frac{N_i}{N_f}. \tag{14.3-19}$$

The final population difference N_f is determined by setting $n = 0$ and $N = N_f$ in (14.3-10) which provides

$$\ln \frac{N_i}{N_f} = \frac{N_i - N_f}{N_t}. \tag{14.3-20}$$

Substituting this into (14.3-19) gives

$$E = \frac{1}{2} h\nu \mathscr{T} \frac{c}{2d} V \tau_p (N_i - N_f). \tag{14.3-21}$$

Q-Switched Pulse Energy

When $N_i \gg N_f$, $E \approx \frac{1}{2} h\nu \mathscr{T}(c/2d) V \tau_p N_i$, as expected. It remains to solve (14.3-20) for N_f. One approach is to rewrite it in the form $Y \exp(-Y) = X \exp(-X)$, where $X = N_i/N_t$ and $Y = N_f/N_t$. Given $X = N_i/N_t$, we can easily solve for Y numerically or by using the graph provided in Fig. 14.3-7.

Pulse Width

A rough estimate of the pulse width is the ratio of the pulse energy to the peak pulse power. Using (14.3-13), (14.3-14), and (14.3-21), we obtain

$$\tau_{\text{pulse}} = \tau_p \frac{N_i/N_t - N_f/N_t}{N_i/N_t - \ln(N_i/N_t) - 1}. \tag{14.3-22}$$

Pulse Width

When $N_i \gg N_t$ and $N_i \gg N_f$, $\tau_{\text{pulse}} \approx \tau_p$.

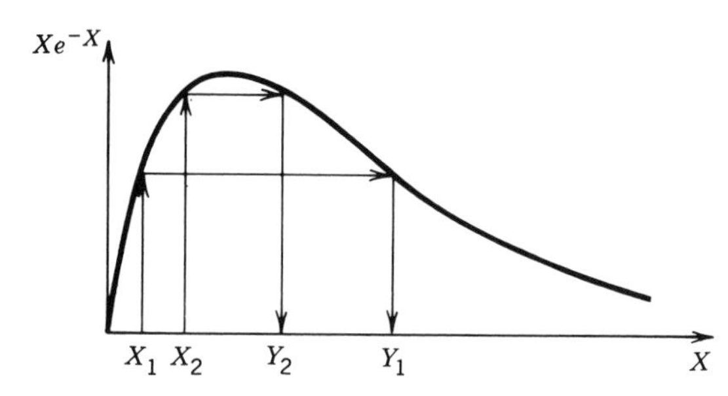

Figure 14.3-7 Graphical construction for determining N_f from N_i, where $X = N_i/N_t$ and $Y = N_f/N_t$. For $X = X_1$ the ordinate represents the value $X_1 \exp(-X_1)$. Since the corresponding solution Y_1 obeys $Y_1 \exp(-Y_1) = X_1 \exp(-X_1)$, it must have the same value of the ordinate.

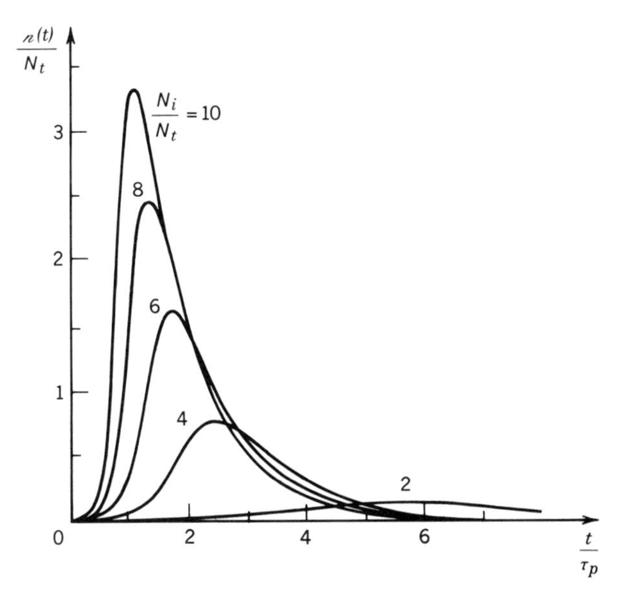

Figure 14.3-8 Typical Q-switched pulse shapes obtained from numerical integration of the approximate rate equations. The photon-number density $n(t)$ is normalized to the threshold population difference $N_t = N_{tb}$ and the time t is normalized to the photon lifetime τ_p. The pulse narrows and achieves a higher peak value as the ratio N_i/N_t increases. In the limit $N_i/N_t \gg 1$, the peak value of $n(t)$ approaches $\frac{1}{2}N_i$.

Pulse Shape

The optical pulse shape, along with all of the pulse characteristics described above, can be determined by numerically integrating (14.3-6) and (14.3-7). Examples of the resulting pulse shapes are shown in Fig. 14.3-8.

EXERCISE 14.3-2

Pulsed Ruby Laser. Consider the ruby laser discussed in Exercise 14.1-1 on page 501. If the laser is now Q-switched so that at the end of the pumping cycle (at $t = t_i$ in Fig. 14.3-6) the population difference $N_i = 6N_t$, use Fig. 14.3-8 to estimate the shape of the laser pulse, its width, peak power, and total energy.

D. Mode Locking

A laser can oscillate on many longitudinal modes, with frequencies that are equally separated by the intermodal spacing $\nu_F = c/2d$. Although these modes normally oscillate independently (they are then called free-running modes), external means can be used to couple them and lock their phases together. The modes can then be regarded as the components of a Fourier-series expansion of a periodic function of time of period $T_F = 1/\nu_F = 2d/c$, in which case they constitute a periodic pulse train. After examining the properties of a mode-locked laser pulse train, we discuss methods of locking the phases of the modes together.

Properties of a Mode-Locked Pulse Train

If each of the laser modes is approximated by a uniform plane wave propagating in the z direction with a velocity $c = c_o/n$, we may write the total complex wavefunction of the field in the form of a sum:

$$U(z,t) = \sum_q A_q \exp\left[j2\pi\nu_q\left(t - \frac{z}{c}\right)\right], \tag{14.3-23}$$

where

$$\nu_q = \nu_0 + q\nu_F, \qquad q = 0, \pm 1, \pm 2, \ldots \tag{14.3-24}$$

is the frequency of mode q, and A_q is its complex envelope. For convenience we assume that the $q = 0$ mode coincides with the central frequency ν_0 of the atomic lineshape. The magnitudes $|A_q|$ may be determined from knowledge of the spectral profile of the gain and the resonator loss (see Sec. 14.2B). Since the modes interact with different groups of atoms in an inhomogeneously broadened medium, their phases $\arg\{A_q\}$ are random and statistically independent.

Substituting (14.3-24) into (14.3-23) provides

$$U(z,t) = \mathscr{A}\left(t - \frac{z}{c}\right)\exp\left[j2\pi\nu_0\left(t - \frac{z}{c}\right)\right], \tag{14.3-25}$$

where the complex envelope $\mathscr{A}(t)$ is the function

$$\mathscr{A}(t) = \sum_q A_q \exp\left(\frac{jq2\pi t}{T_F}\right) \tag{14.3-26}$$

and

$$T_F = \frac{1}{\nu_F} = \frac{2d}{c}. \tag{14.3-27}$$

The complex envelope $\mathscr{A}(t)$ in (14.3-26) is a periodic function of the period T_F, and $\mathscr{A}(t - z/c)$ is a periodic function of z of period $cT_F = 2d$. If the magnitudes and phases of the complex coefficients A_q are properly chosen, $\mathscr{A}(t)$ may be made to take the form of periodic narrow pulses.

Consider, for example, M modes ($q = 0, \pm 1, \ldots, \pm S$, so that $M = 2S + 1$), whose complex coefficients are all equal, $A_q = A$, $q = 0, \pm 1, \ldots, \pm S$. Then

$$\mathscr{A}(t) = A \sum_{q=-S}^{S} \exp\left(\frac{jq2\pi t}{T_F}\right) = A \sum_{q=-S}^{S} x^q = A\frac{x^{S+1} - x^{-S}}{x - 1} = A\frac{x^{S+\frac{1}{2}} - x^{-S-\frac{1}{2}}}{x^{\frac{1}{2}} - x^{-\frac{1}{2}}},$$

where $x = \exp(j2\pi t/T_F)$ (see Sec. 2.6B for more details). After a few algebraic manipulations, $\mathscr{A}(t)$ can be cast in the form

$$\mathscr{A}(t) = A\frac{\sin(M\pi t/T_F)}{\sin(\pi t/T_F)}.$$

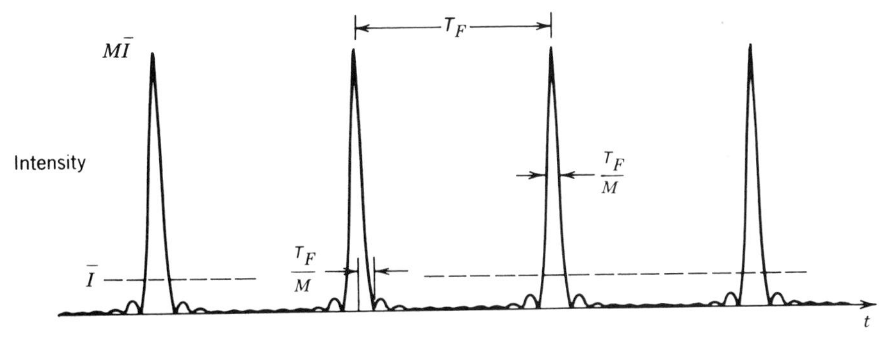

Figure 14.3-9 Intensity of the periodic pulse train resulting from the sum of M laser modes of equal magnitudes and phases. Each pulse has a width that is M times smaller than the period T_F and a peak intensity that is M times greater than the mean intensity.

The optical intensity is then given by $I(t, z) = |\mathscr{A}(t - z/c)|^2$ or

$$I(t, z) = |A|^2 \frac{\sin^2 [M\pi(t - z/c)/T_F]}{\sin^2 [\pi(t - z/c)/T_F]}. \tag{14.3-28}$$

As illustrated in Fig. 14.3-9, this is a periodic function of time.

The shape of the mode-locked laser pulse train is therefore dependent on the number of modes M, which is proportional to the atomic linewidth $\Delta\nu$. The pulse width τ_{pulse} is therefore inversely proportional to the atomic linewidth $\Delta\nu$. If $M \approx \Delta\nu/\nu_F$, then $\tau_{\text{pulse}} = T_F/M \approx 1/\Delta\nu$. Because $\Delta\nu$ can be quite large, very narrow mode-locked laser pulses can be generated. The ratio between the peak and mean intensities is equal to the number of modes M, which can also be quite large.

The period of the pulse train is $T_F = 2d/c$. This is just the time for a single round trip of reflection within the resonator. Indeed, the light in a mode-locked laser can be regarded as a single narrow pulse of photons reflecting back and forth between the mirrors of the resonator (see Fig. 14.3-10). At each reflection from the output mirror, a fraction of the photons is transmitted in the form of a pulse of light. The transmitted

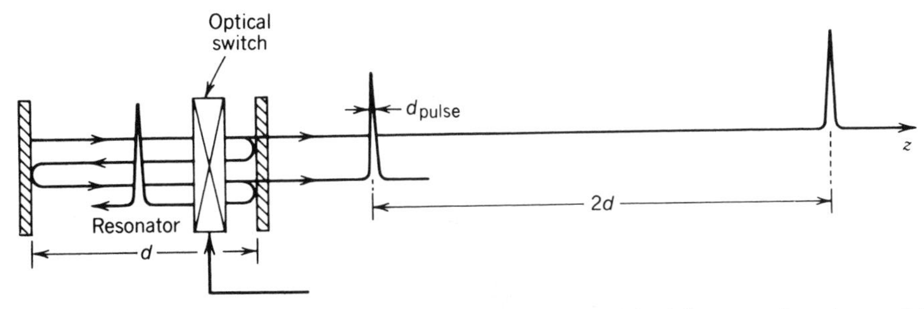

Figure 14.3-10 The mode-locked laser pulse reflects back and forth between the mirrors of the resonator. Each time it reaches the output mirror it transmits a short optical pulse. The transmitted pulses are separated by the distance $2d$ and travel with velocity c. The switch opens only when the pulse reaches it and only for the duration of the pulse. The periodic pulse train is therefore unaffected by the presence of the switch. Other wave patterns, however, suffer losses and are not permitted to oscillate.

TABLE 14.3-1 Characteristic Properties of a Mode-Locked Pulse Train

Temporal period	$T_F = \dfrac{2d}{c}$
Pulse width	$\tau_{\text{pulse}} = \dfrac{T_F}{M} = \dfrac{1}{M\nu_F}$
Spatial period	$2d$
Pulse length	$d_{\text{pulse}} = c\tau_{\text{pulse}} = \dfrac{2d}{M}$
Mean intensity	$\bar{I} = M\|A\|^2$
Peak intensity	$I_p = M^2\|A\|^2 = M\bar{I}$

pulses are separated by the distance $c(2d/c) = 2d$ and have a spatial width $d_{\text{pulse}} = 2d/M$.

A summary of the properties of a mode-locked laser pulse train is provided in Table 14.3-1.

As a particular example, we consider a Nd^{3+}:glass laser operating at $\lambda_o = 1.06\ \mu m$. It has a refractive index $n = 1.5$ and a linewidth $\Delta\nu = 3 \times 10^{12}$ Hz. Thus the pulse width $\tau_{\text{pulse}} = 1/\Delta\nu \approx 0.33$ ps and the pulse length $d_{\text{pulse}} \approx 67\ \mu m$. If the resonator has a length $d = 10$ cm, the mode separation is $\nu_F = c/2d = 1$ GHz, which means that $M = \Delta\nu/\nu_F = 3000$ modes. The peak intensity is therefore 3000 times greater than the average intensity. In media with broad linewidths, mode locking is generally more advantageous than Q-switching for obtaining short pulses. Gas lasers generally have narrow atomic linewidths, on the other hand, so that ultrashort pulses cannot be obtained by mode locking.

Although the formulas provided above were derived for the special case in which the modes have equal amplitudes and phases, calculations based on more realistic behavior provide similar results.

EXERCISE 14.3-3

Demonstration of Pulsing by Mode Locking. Write a computer program to plot the intensity $I(t) = |\mathscr{A}(t)|^2$ of a wave whose envelope $\mathscr{A}(t)$ is given by the sum in (14.3-26). Assume that the number of modes $M = 11$ and use the following choices for the complex coefficients A_q:

(a) Equal magnitudes and equal phases (this should reproduce the results of the foregoing example).

(b) Magnitudes that obey the Gaussian spectral profile $|A_q| = \exp[-\frac{1}{2}(q/5)^2]$ and equal phases.

(c) Equal magnitudes and random phases (obtain the phases by using a random number generator to produce a random variable uniformly distributed between 0 and 2π).

Methods of Mode Locking

We have found so far that if a large number M of modes are locked in phase, they form a giant narrow pulse of photons that reflects back and forth between the mirrors of the resonator. The spatial length of the pulse is a factor of M smaller than twice the resonator length. The question that remains is how the modes can be locked together so that they have the same phase. This can be accomplished with the help of a modulator or switch placed inside the resonator, as we now show.

Suppose that an optical switch (e.g., an electro-optic or acousto-optic switch, as discussed in Chaps. 18, 20, and 21) is placed inside the resonator, which blocks the light at all times, except when the pulse is about to cross it, whereupon it opens for the duration of the pulse (Fig. 14.3-10). Since the pulse itself is permitted to pass, it is not affected by the presence of the switch and the pulse train continues uninterrupted. In the absence of phase locking, the individual modes have different phases that are determined by the random conditions at the onset of their oscillation. If the phases happen, by accident, to take on equal values, the sum of the modes will form a giant pulse that would not be affected by the presence of the switch. Any other combination of phases would form a field distribution that is totally or partially blocked by the switch, which adds to the losses of the system. Therefore, in the presence of the switch, only the case where the modes have equal phases can lase. The laser waits for the lucky accident of such phases, but once the oscillations start, they continue to be locked.

The problem can also be examined mathematically. An optical field must satisfy the wave equation with the boundary conditions imposed by the presence of the switch. The multimode optical field of (14.3-23) does indeed satisfy the wave equation for any combination of phases. The case of equal phases also satisfies the boundary conditions imposed by the switch; therefore, it must be a unique solution.

A passive switch such as a saturable absorber may also be used for mode locking. A saturable absorber (see Sec. 13.3B) is a medium whose absorption coefficient decreases as the intensity of the light passing through it increases; thus it transmits intense pulses with relatively little absorption and absorbs weak ones. Oscillation can therefore occur only when the phases of the different modes are related to each other in such a way that they form an intense pulse which can then pass through the switch. Active and passive switches are also used for the mode locking of homogeneously broadened media.

Examples of Mode-Locked Lasers

Table 14.3-2 is a list, in order of increasing observed pulse width, of some mode-locked laser media. A broad range of observed pulse widths is represented. The observed pulse widths, which for a given medium can vary greatly, depend on the method used to achieve mode locking. Rhodamine-6G dye lasers, for example, can be constructed in a **colliding pulse mode** (CPM) ring-resonator configuration. The oppositely traveling ultrashort laser pulses collide at a very thin jet of dye serving as a saturable absorber.

TABLE 14.3-2 Typical Observed Pulse Widths for a Number of Homogeneously (H) and Inhomogeneously (I) Broadened, Mode-Locked Lasers

Laser Medium		Transition Linewidth[a] $\Delta\nu$	Calculated Pulse Width $\tau_{\text{pulse}} = 1/\Delta\nu$	Observed Pulse Width
$Ti^{3+}:Al_2O_3$	H	100 THz	10 fs	30 fs
Rhodamine-6G dye	H/I	5 THz	200 fs	500 fs
Nd^{3+}:glass	I	3 THz	333 fs	500 fs
Er^{3+}:silica fiber	H/I	4 THz	250 fs	7 ps
Ruby	H	60 GHz	16 ps	10 ps
Nd^{3+}:YAG	H	120 GHz	8 ps	50 ps
Ar^+	I	3.5 GHz	286 ps	150 ps
He–Ne	I	1.5 GHz	667 ps	600 ps
CO_2	I	60 MHz	16 ns	20 ns

[a]The transition linewidths $\Delta\nu$ are obtained from Table 13.2-1.

Only during the brief time that the optical pulses pass each other in the thin absorber is the intensity increased and the loss minimized. Proper positioning of the active medium relative to the saturable absorber can give rise to pulse widths as low as 25 fs. In a conventional configuration, the pulse width is far greater ($\approx$ 500 fs).

READING LIST

Books and Articles on Laser Theory

See also the reading list in Chapter 13.

Books on Lasers

C. A. Brau, *Free-Electron Lasers*, Academic Press, Orlando, FL, 1990.

F. P. Schäfer, ed., *Dye Lasers*, Springer-Verlag, New York, 3rd ed. 1990.

R. C. Elton, *X-Ray Lasers*, Academic Press, Orlando, FL, 1990.

N. G. Basov, A. S. Bashkin, V. I. Igoshin, A. N. Oraevsky, and A. A. Shcheglov, *Chemical Lasers*, Springer-Verlag, New York, 1990.

P. K. Das, *Lasers and Optical Engineering*, Springer-Verlag, New York, 1990.

A. A. Kaminskii, *Laser Crystals*, Springer-Verlag, New York, 2nd ed. 1990.

N. G. Douglas, *Millimetre and Submillimetre Lasers*, Springer-Verlag, New York, 1989.

P. K. Cheo, ed., *Handbook of Solid-State Lasers*, Marcel Dekker, New York, 1988.

P. K. Cheo, ed., *Handbook of Molecular Lasers*, Marcel Dekker, New York, 1987.

L. F. Mollenauer and J. C. White, eds., *Tunable Lasers*, Springer-Verlag, Berlin, 1987.

T. C. Marshall, *Free Electron Lasers*, Macmillan, New York, 1985.

P. Hammerling, A. B. Budgor, and A. Pinto, eds., *Tunable Solid State Lasers*, Springer-Verlag, New York, 1985.

C. K. Rhodes, ed., *Excimer Lasers*, Springer-Verlag, Berlin, 2nd ed. 1984.

G. Brederlow, E. Fill, and K. J. Witte, *The High-Power Iodine Laser*, Springer-Verlag, Berlin, 1983.

D. C. Brown, *High Peak Power Nd:Glass Laser Systems*, Springer-Verlag, Berlin, 1981.

S. A. Losev, *Gasdynamic Laser*, Springer-Verlag, Berlin, 1981.

A. L. Bloom, *Gas Lasers*, R. E. Krieger, Huntington, NY, 1978.

E. R. Pike, ed., *High-Power Gas Lasers*, Institute of Physics, Bristol, England, 1975.

C. S. Willett, *Introduction to Gas Lasers: Population Inversion Mechanisms*, Pergamon Press, New York, 1974.

R. J. Pressley, *Handbook of Lasers*, Chemical Rubber Company, Cleveland, OH, 1971.

D. C. Sinclair and W. E. Bell, *Gas Laser Technology*, Holt, Rinehart and Winston, New York, 1969.

L. Allen and D. G. C. Jones, *Principles of Gas Lasers*, Plenum Press, New York, 1967.

C. G. B. Garrett, *Gas Lasers*, McGraw-Hill, New York, 1967.

W. V. Smith and P. P. Sorokin, *The Laser*, McGraw-Hill, New York, 1966.

Books on Laser Applications

F. J. Duarte and L. W. Hillman, *Dye Laser Principles with Applications*, Academic Press, Orlando, FL, 1990.

P. G. Cielo, *Optical Techniques for Industrial Inspection*, Academic Press, New York, 1988.

W. Guimaraes, C. T. Lin, and A. Mooradian, *Lasers and Applications*, Springer-Verlag, Berlin, 1987.

H. Koebner, *Industrial Applications of Lasers*, Wiley, New York, 1984.

W. W. Duley, *Laser Processing and Analysis of Materials*, Plenum Press, New York, 1983.

H. M. Muncheryan, *Principles and Practice of Laser Technology*, Tab Books, Blue Summit, PA, 1983.

F. Durst, A. Mellino, and J. H. Whitelaw, *Principles and Practice of Laser-Doppler Anemometry*, Academic Press, New York, 1981.

L. E. Drain, *The Laser Doppler Technique*, Wiley, New York, 1980.

M. J. Beesley, *Lasers and Their Applications*, Halsted Press, New York, 1978.

J. F. Ready, *Industrial Applications of Lasers*, Academic Press, New York, 1978.

W. E. Kock, *Engineering Applications of Lasers and Holography*, Plenum Press, New York, 1975.

F. T. Arecchi and E. O. Schulz-Dubois, eds., *Laser Handbook*, vol. 2, North-Holland/Elsevier, Amsterdam/New York, 1972.

S. S. Charschan, ed., *Lasers in Industry*, Van Nostrand Reinhold, New York, 1972.

J. W. Goodman and M. Ross, eds., *Laser Applications*, vols. 1–5, Academic Press, New York, 1971–1984.

S. L. Marshall, ed., *Laser Technology and Applications*, McGraw-Hill, New York, 1968.

D. Fishlock, ed., *A Guide to the Laser*, Elsevier, New York, 1967.

Special Journal Issues

Special issue on laser technology, *Lincoln Laboratory Journal*, vol. 3, no. 3, 1990.

Special issue on novel laser system optics, *Journal of the Optical Society of America B*, vol. 5, no. 9, 1988.

Special issue on solid-state lasers, *IEEE Journal of Quantum Electronics*, vol. QE-24, no. 6, 1988.

Special issue on nonlinear dynamics of lasers, *Journal of the Optical Society of America B*, vol. 5, no. 5, 1988.

Special issue on lasers in biology and medicine, *IEEE Journal of Quantum Electronics*, vol. QE-23, no. 10, 1987.

Special issue on free electron lasers, *IEEE Journal of Quantum Electronics*, vol. QE-23, no. 9, 1987.

Special issue on the generation of coherent XUV and soft-X-ray radiation, *Journal of the Optical Society of America B*, vol. 4, no. 4, 1987.

Special issue on solid-state laser materials, *Journal of the Optical Society of America B*, vol. 3, no. 1, 1986.

Special issue: "Twenty-five years of the laser," *Optica Acta* (*Journal of Modern Optics*), vol. 32, no. 9/10, 1985.

Special issue on ultrasensitive laser spectroscopy, *Journal of the Optical Society of America B*, vol. 2, no. 9, 1985.

Third special issue on free electron lasers, *IEEE Journal of Quantum Electronics*, vol. QE-21, no. 7, 1985.

Special issue on infrared spectroscopy with tunable lasers, *Journal of the Optical Society of America B*, vol. 2, no. 5, 1985.

Special issue on lasers in biology and medicine, *IEEE Journal of Quantum Electronics*, vol. QE-20, no. 12, 1984.

Centennial issue, *IEEE Journal of Quantum Electronics*, vol. QE-20, no. 6, 1984.

Special issue on laser materials interactions, *IEEE Journal of Quantum Electronics*, vol. QE-17, no. 10, 1981.

Special issue on free electron lasers, *IEEE Journal of Quantum Electronics*, vol. QE-17, no. 8, 1981.

Special issue on laser photochemistry, *IEEE Journal of Quantum Electronics*, vol. QE-16, no. 11, 1980.

Special issue on excimer lasers, *IEEE Journal of Quantum Electronics*, vol. QE-15, no. 5, 1979.

Special issue on quantum electronics, *Proceedings of the IEEE*, vol. 51, no. 1, 1963.

Articles

E. Desurvire, Erbium-Doped Fiber Amplifiers for New Generations of Optical Communication Systems, *Optics & Photonics News*, vol. 2, no. 1, pp. 6–11, 1991.

K.-J. Kim and A. Sessler, Free-Electron Lasers: Present Status and Future Prospects, *Science*, vol. 250, pp. 88–93, 1990.

G. New, Femtofascination, *Physics World*, vol. 3, no. 7, pp. 33–37, 1990.

P. F. Moulton, Ti: Sapphire Lasers: Out of the Lab and Back In Again, *Optics & Photonics News*, vol. 1, no. 8, pp. 20–23, 1990.

R. D. Petrasso, Plasmas Everywhere, *Nature*, vol. 343, pp. 21–22, 1990.

S. Suckewer and A. R. DeMeo, Jr., X-Ray Laser Microscope Developed at Princeton, *Princeton Plasma Physics Laboratory Digest*, May 1989.

H. P. Freund and R. K. Parker, Free-Electron Lasers, *Scientific American*, vol. 260, no. 4, pp. 84–89, 1989.

P. Urquhart, Review of Rare Earth Doped Fibre Lasers and Amplifiers, *Institution of Electrical Engineers Proceedings—Part J*, vol. 135, pp. 385–407, 1988.

D. L. Matthews and M. D. Rosen, Soft X-Ray Lasers, *Scientific American*, vol. 259, no. 6, pp. 86–91, 1988.

C. A. Brau, Free-Electron Lasers, *Science*, vol. 239, pp. 1115–1121, 1988.

R. L. Byer, Diode Laser-Pumped Solid-State Lasers, *Science*, vol. 239, pp. 742–747, 1988.

J. A. Pasour, Free-Electron Lasers, *IEEE Circuits and Devices Magazine*, vol. 3, no. 2, pp. 55–64, 1987.

J. G. Eden, Photochemical Processing of Semiconductors: New Applications for Visible and Ultraviolet Lasers, *IEEE Circuits and Devices Magazine*, vol. 2, no. 1, pp. 18–24, 1986.

J. F. Holzricher, High-Power Solid-State Lasers, *Nature*, vol. 316, pp. 309–314, 1985.

W. L. Wilson, Jr., F. K. Tittel, and W. Nighan, Broadband Tunable Excimer Lasers, *IEEE Circuits and Devices Magazine*, vol. 1, no. 1, pp. 55–62, 1985.

P. Sprangle and T. Coffey, New Sources of High-Power Coherent Radiation, *Physics Today*, vol. 37, no. 3, pp. 44–51, 1984.

A. L. Schawlow, Spectroscopy in a New Light, (Nobel lecture), *Reviews of Modern Physics*, vol. 54, pp. 697–707, 1982.

P. W. Smith, Mode Selection in Lasers, *Proceedings of the IEEE*, vol. 60, pp. 422–440, 1972.

L. Allen and D. G. C. Jones, Mode Locking in Gas Lasers, in *Progress in Optics*, vol. 9, E. Wolf, ed., North-Holland, Amsterdam, 1971.

P. W. Smith, Mode-Locking of Lasers, *Proceedings of the IEEE*, vol. 58, pp. 1342–1359, 1970.

D. R. Herriott, Applications of Laser Light, *Scientific American*, vol. 219, no. 3, pp. 141–156, 1968.

C. K. N. Patel, High-Power Carbon Dioxide Lasers, *Scientific American*, vol. 219, no. 2, pp. 22–33, 1968.

A. Lempicki and H. Samelson, Liquid Lasers, *Scientific American*, vol. 216, no. 6, pp. 80–90, 1967.

PROBLEMS

14.2-1 **Number of Longitudinal Modes.** An Ar^+-ion laser has a resonator of length 100 cm. The refractive index $n = 1$.
(a) Determine the frequency spacing ν_F between the resonator modes.
(b) Determine the number of longitudinal modes that the laser can sustain if the FWHM Doppler-broadened linewidth is $\Delta\nu_D = 3.5$ GHz and the loss coefficient is half the peak small-signal gain coefficient.

(c) What would the resonator length d have to be to achieve operation on a single longitudinal mode? What would that length be for a CO_2 laser that has a much smaller Doppler linewidth $\Delta\nu_D = 60$ MHz under the same conditions?

14.2-2 **Frequency Drift of the Laser Modes.** A He–Ne laser has the following characteristics: (1) A resonator with 97% and 100% mirror reflectances and negligible internal losses; (2) a Doppler-broadened atomic transition with Doppler linewidth $\Delta\nu_D = 1.5$ GHz; and (3) a small-signal peak gain coefficient $\gamma_0(\nu_0) = 2.5 \times 10^{-3}$ cm^{-1}. While the laser is running, the frequencies of its longitudinal modes drift with time as a result of small thermally induced changes in the length of the resonator. Find the allowable range of resonator lengths such that the laser will always oscillate in one or two (but not more) longitudinal modes. The refractive index $n = 1$.

14.2-3 **Mode Control Using an Etalon.** A Doppler-broadened gas laser operates at 515 nm in a resonator with two mirrors separated by a distance of 50 cm. The photon lifetime is 0.33 ns. The spectral window within which oscillation can occur is of width $B = 1.5$ GHz. The refractive index $n = 1$. To select a single mode, the light is passed into an etalon (a passive Fabry–Perot resonator) whose mirrors are separated by the distance d and its finesse is $\mathcal{F}$. The etalon acts as a filter. Suggest suitable values of d and $\mathcal{F}$. Is it better to place the etalon inside or outside the laser resonator?

14.2-4 **Modal Powers in a Multimode Laser.** A He–Ne laser operating at $\lambda_o = 632.8$ nm produces 50 mW of multimode power at its output. It has an inhomogeneously broadened gain profile with a Doppler linewidth $\Delta\nu_D = 1.5$ GHz and the refractive index $n = 1$. The resonator is 30 cm long.
(a) If the maximum small-signal gain coefficient is twice the loss coefficient, determine the number of longitudinal modes of the laser.
(b) If the mirrors are adjusted to maximize the intensity of the strongest mode, estimate its power.

14.2-5 **Output of a Single-Mode Gas Laser.** Consider a 10-cm-long gas laser operating at the center of the 600-nm line in a single longitudinal and single transverse mode. The mirror reflectances are $\mathcal{R}_1 = 99\%$ and $\mathcal{R}_2 = 100\%$. The refractive index $n = 1$ and the effective area of the output beam is 1 mm^2. The small-signal gain coefficient $\gamma_0(\nu_0) = 0.1$ cm^{-1} and the saturation photon-flux density $\phi_s = 1.43 \times 10^{19}$ photons/cm^2-s.
(a) Determine the distributed loss coefficients, α_{m1} and α_{m2}, associated with each of the mirrors separately. Assuming that $\alpha_s = 0$, find the resonator loss coefficient α_r.
(b) Find the photon lifetime τ_p.
(c) Determine the output photon flux density ϕ_o and the output power P_o.

14.2-6 **Threshold Population Difference for an Ar^+-Ion Laser.** An Ar^+-ion laser has a 1-m-long resonator with 98% and 100% mirror reflectances. Other loss mechanisms are negligible. The atomic transition has a central wavelength $\lambda_o = 515$ nm, spontaneous lifetime $t_{sp} = 10$ ns, and linewidth $\Delta\lambda = 0.003$ nm. The lower energy level has a very short lifetime and hence zero population. The diameter of the oscillating mode is 1 mm. Determine (a) the photon lifetime and (b) the threshold population difference for laser action.

14.2-7 **Transmittance of a Laser Resonator.** Monochromatic light from a tunable optical source is transmitted through the optical resonator of an unpumped gas laser. The observed transmittance, as a function of frequency, is shown in Fig. P14.2-7.

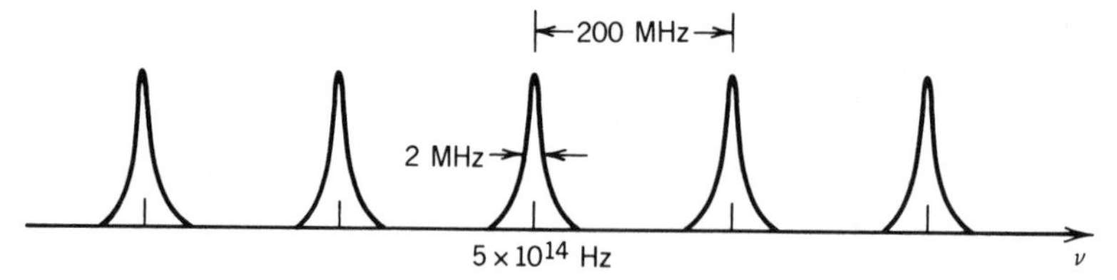

Figure P14.2-7 Transmittance of a laser resonator.

(a) Determine the resonator length, the photon lifetime, and the threshold gain coefficient of the laser. Assume that the refractive index $n = 1$.
(b) Assuming that the central frequency of the laser transition is 5×10^{14} Hz, sketch the transmittance versus frequency if the laser is now pumped but the pumping is not sufficient for laser oscillation to occur.

14.2-8 **Rate Equations in a Four-Level Laser.** Consider a four-level laser with an active volume $V = 1\ \text{cm}^3$. The population densities of the upper and lower laser levels are N_2 and N_1 and $N = N_2 - N_1$. The pumping rate is such that the steady-state population difference N in the absence of stimulated emission and absorption is N_0. The photon-number density is n and the photon lifetime is τ_p. Write the rate equations for N_2, N_1, N, and n in terms of N_0, the transition cross section $\sigma(\nu)$, and the times t_{sp}, τ_1, τ_2, τ_{21}, and τ_p. Determine the steady state values of N and n.

*14.3-1 **Transients in a Gain-Switched Laser**
(a) Introduce the new variables $X = n/\tau_p$, $Y = N/N_t$, and the normalized time $s = t/\tau_p$, to demonstrate that the rate equations (14.3-2) and (14.3-5) take the form

$$\frac{dX}{ds} = -X + XY$$

$$\frac{dY}{ds} = a(Y_0 - Y) - 2XY,$$

where $a = \tau_p/t_{sp}$ and $Y_0 = N_0/N_t$.
(b) Write a computer program to solve these two equations for both switching on and switching off. Assume that Y_0 is switched from 0 to 2 to turn the laser on, and from 2 to 0 to turn it off. Assume further that an initially very small photon flux corresponding to $X = 10^{-5}$ starts the oscillation at $t = 0$. Speculate on the possible origin of this flux. Determine the switching transient times for $a = 10^{-3}$, 1, and 10^3. Comment on the significance of your results.

*14.3-2 ***Q*-Switched Ruby Laser Power.** A Q-switched ruby laser makes use of a 15-cm-long rod of cross-sectional area 1 cm^2 placed in a resonator of length 20 cm. The mirrors have reflectances $\mathscr{R}_1 = 0.95$ and $\mathscr{R}_2 = 0.7$. The Cr^{3+} density is 1.58×10^{19} atoms/cm^3, and the transition cross section $\sigma(\nu_0) = 2 \times 10^{-20}$ cm^2. The laser is pumped to an initial population of 10^{19} atoms/cm^3 in the upper state with negligible population in the lower state. The pump band (level 3) is centered at ≈ 450 nm and the decay from level 3 to level 2 is fast. The lifetime of level 2 is ≈ 3 ms.

(a) How much pump power is required to maintain the population in level 2 at 10^{19} cm^{-3}?

(b) How much power is spontaneously radiated before the Q-switch is operated?
(c) Determine the peak power, energy, and width of the Q-switched pulse.

*14.3-3 **Operation of a Cavity-Dumped Laser.** Sketch the variation of the threshold population difference N_t (which is proportional to the loss), the population difference $N(t)$, the internal photon number density $n(t)$, and the external photon flux density $\phi_o(t)$, during two cycles of operation of a pulsed cavity-dumped laser.

14.3-4 **Mode Locking with Lorentzian Amplitudes.** Assume that the envelopes of the modes of a mode-locked laser are

$$A_q = \sqrt{P}\,\frac{(\Delta\nu/2)^2}{(q\nu_F)^2 + (\Delta\nu/2)^2}, \qquad q = -\infty, \dots, \infty,$$

and the phases are equal. Determine expressions for the following parameters of the generated pulse train:
(a) Mean power
(b) Peak power
(c) Pulse width (FWHM).

14.3-5 **Second-Harmonic Generation.** Crystals with nonlinear optical properties are often used for second-harmonic generation, as explained in Chap. 19. In this process, two photons of frequency ν are converted into a single photon of frequency 2ν. Assume that such a crystal is placed inside a laser resonator with an active medium providing gain at frequency ν. The frequencies ν and 2ν correspond to two modes of the resonator. If the rate of second-harmonic conversion is ζn (s^{-1}-m^{-3}) and the rate of photon production by the laser process (net effect of stimulated emission and absorption) is ξn (s^{-1}-m^{-3}), where ζ and ξ are constants, write the rate equations for the photon number densities n and n_2 at the frequencies ν and 2ν. Assume that the photon lifetimes at ν and 2ν are τ_p and τ_{p2}, respectively. Determine the steady-state values of n and n_2.

CHAPTER

15

PHOTONS IN SEMICONDUCTORS

15.1 SEMICONDUCTORS
- A. Energy Bands and Charge Carriers
- B. Semiconducting Materials
- C. Electron and Hole Concentrations
- D. Generation, Recombination, and Injection
- E. Junctions
- F. Heterojunctions
- *G. Quantum Wells and Superlattices

15.2 INTERACTIONS OF PHOTONS WITH ELECTRONS AND HOLES
- A. Band-to-Band Absorption and Emission
- B. Rates of Absorption and Emission
- C. Refractive Index

William P. Shockley (1910–1989), left, **Walter H. Brattain (1902–1987)**, center, and **John Bardeen (1908–1991)**, right, shared the Nobel Prize in 1956 for showing that semiconductor devices could be used to achieve amplification.

Electronics is the technology of controlling the flow of electrons whereas photonics is the technology of controlling the flow of photons. Electronics and photonics have been joined together in semiconductor optoelectronic devices where photons generate mobile electrons, and electrons generate and control the flow of photons. The compatibility of semiconductor optoelectronic devices and electronic devices has, in recent years, led to substantive advances in both technologies. Semiconductors are used as optical detectors, sources (light-emitting diodes and lasers), amplifiers, waveguides, modulators, sensors, and nonlinear optical elements.

Semiconductors absorb and emit photons by undergoing transitions between different allowed energy levels, in accordance with the general theory of photon–atom interactions described in Chap. 12. However, as we indicated briefly there, semiconductors have properties that are unique in certain respects:

- A semiconductor material cannot be viewed as a collection of noninteracting atoms, each with its own individual energy levels. The proximity of the atoms in a solid results in one set of energy levels representing the entire system.
- The energy levels of semiconductors take the form of groups of closely spaced levels that form bands. In the absence of thermal excitations (at $T = 0$ K), these are either completely occupied by electrons or completely empty. The highest filled band is called the valence band, and the empty band above it is called the conduction band. The two bands are separated by an energy gap.
- Thermal and optical interactions can impart energy to an electron, causing it to jump across the gap from the valence band into the conduction band (leaving behind an empty state called a hole). The inverse process can also occur. An electron can decay from the conduction band into the valence band to fill an empty state (provided that one is accessible) by means of a process called electron–hole recombination. We therefore have two types of particles that carry electric current and can interact with photons: electrons and holes.

Two processes are fundamental to the operation of almost all semiconductor optoelectronic devices:

- *The absorption of a photon can create an electron–hole pair*. The mobile charge carriers resulting from absorption can alter the electrical properties of the material. One such effect, photoconductivity, is responsible for the operation of certain semiconductor photodetectors.
- *The recombination of an electron and a hole can result in the emission of a photon*. This process is responsible for the operation of semiconductor light sources. Spontaneous radiative electron–hole recombination is the underlying process of light generation in the light-emitting diode. Stimulated electron–hole recombination is the source of photons in the semiconductor laser.

In Sec. 15.1 we begin with a review of the properties of semiconductors that are important in semiconductor photonics; the reader is expected to be familiar with the basic principles of semiconductor physics. Section 15.2 provides an introduction to the optical properties of semiconductors. A simplified theory of absorption, spontaneous emission, and stimulated emission is developed using the theory of radiative atomic transitions developed in Chap. 12.

This, and the following two chapters, are to be regarded as a single unit. Chapter 16 deals with semiconductor optical sources such as the light-emitting diode and the injection laser diode. Chapter 17 is devoted to semiconductor photon detectors.

15.1 SEMICONDUCTORS

A semiconductor is a crystalline or amorphous solid whose electrical conductivity is typically intermediate between that of a metal and an insulator and can be changed significantly by altering the temperature or the impurity content of the material, or by illumination with light. The unique energy-level structure of semiconductor materials leads to special electrical and optical properties, as described later in this chapter. Electronic devices principally make use of silicon (Si) as a semiconductor material, but compounds such as gallium arsenide (GaAs) are of utmost importance to photonics (see Sec. 15.1B for a selected tabulation of other semiconductor materials).

A. Energy Bands and Charge Carriers

Energy Bands in Semiconductors

Atoms of solid-state materials have a sufficiently strong interaction that they cannot be treated as individual entities. Valence electrons are not attached (bound) to individual atoms; rather, they belong to the system of atoms as a whole. The solution of the Schrödinger equation for the electron energy, in the periodic potential created by the collection of atoms in a crystal lattice, results in a splitting of the atomic energy levels and the formation of energy bands (see Sec. 12.1). Each band contains a large number of finely separated discrete energy levels that can be approximated as a continuum. The valence and conduction bands are separated by a "forbidden" energy gap of width E_g (see Fig. 15.1-1), called the **bandgap energy**, which plays an important role in determining the electrical and optical properties of the material. Materials with a filled valence band and a large energy gap (> 3 eV) are electrical insulators; those for which the gap is small or nonexistent are conductors (see Fig. 12.1-5). Semiconductors have energy gaps that lie roughly in the range 0.1 to 3 eV.

Electrons and Holes

In accordance with the **Pauli exclusion principle**, no two electrons can occupy the same quantum state. Lower energy levels are filled first. In elemental semiconductors, such as Si and Ge, there are four valence electrons per atom; the valence band has a number of quantum states such that in the absence of thermal excitations the valence band is completely filled and the conduction band is completely empty. Consequently, the material cannot conduct electricity.

As the temperature increases, however, some electrons will be thermally excited into the empty conduction band where there is an abundance of unoccupied states (see Fig. 15.1-2). There, the electrons can act as mobile carriers; they can drift in the crystal lattice under the effect of an applied electric field and thereby contribute to the electric current. Furthermore, the departure of an electron from the valence band provides an empty quantum state, allowing the remaining electrons in the valence band to exchange

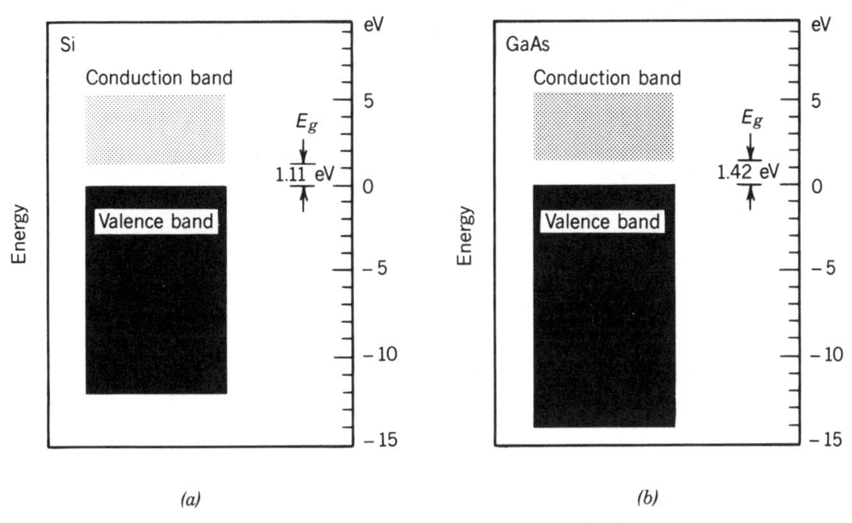

Figure 15.1-1 Energy bands: (*a*) in Si, and (*b*) in GaAs.

places with each other under the influence of an electric field. A motion of the "collection" of remaining electrons in the valence band occurs. This can equivalently be regarded as the motion, in the opposite direction, of the hole left behind by the departed electron. The hole therefore behaves as if it has a positive charge $+e$. The result of each electron excitation is, then, the creation of a free electron in the conduction band and a free hole in the valence band. The two charge carriers are free to drift under the effect of the applied electric field and thereby to generate an electric current. The material behaves as a *semi*conductor whose conductivity increases sharply with temperature as an increasing number of mobile carriers are thermally generated.

Energy-Momentum Relations

The energy E and momentum $\mathbf{p}$ of an electron in free space are related by $E = p^2/2m_0 = \hbar^2 k^2/2m_0$, where p is the magnitude of the momentum and k is the magnitude of the wavevector $\mathbf{k} = \mathbf{p}/\hbar$ associated with the electron's wavefunction, and m_0 is the electron mass (9.1×10^{-31} kg). The E–k relation is a simple parabola.

The motion of electrons in the conduction band, and holes in the valence band, of a semiconductor are subject to different dynamics. They are governed by the Schrödinger

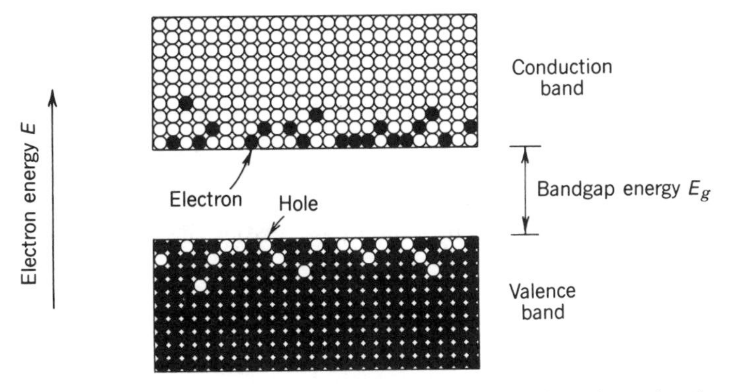

Figure 15.1-2 Electrons in the conduction band and holes in the valence band at $T > 0$ K.

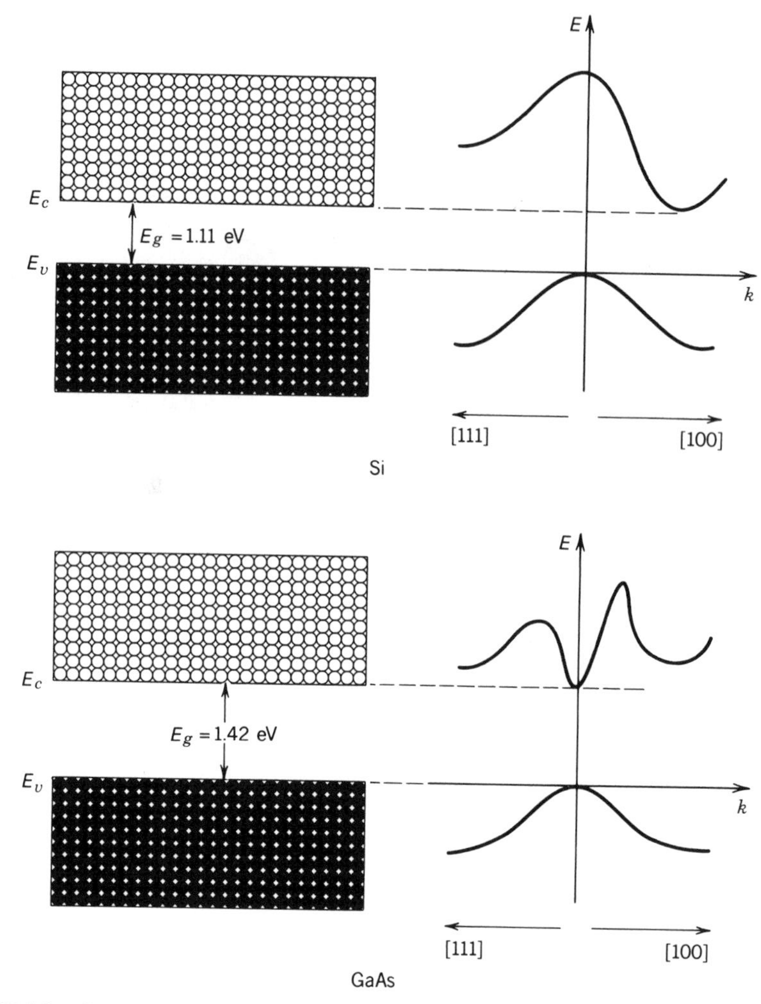

Figure 15.1-3 Cross section of the E–k function for Si and GaAs along the crystal directions [111] and [100].

equation and the periodic lattice of the material. The E–k relations are illustrated in Fig. 15.1-3 for Si and GaAs. The energy E is a periodic function of the components (k_1, k_2, k_3) of the vector **k**, with periodicities $(\pi/a_1, \pi/a_2, \pi/a_3)$, where a_1, a_2, a_3 are the crystal lattice constants. Figure 15.1-3 shows cross sections of this relation along two different directions of **k**. The energy of an electron in the conduction band depends not only on the magnitude of its momentum, but also on the direction in which it is traveling in the crystal.

Effective Mass

Near the bottom of the conduction band, the E–k relation may be approximated by the parabola

$$E = E_c + \frac{\hbar^2 k^2}{2m_c}, \qquad (15.1\text{-}1)$$

where E_c is the energy at the bottom of the conduction band and m_c is a constant representing the **effective mass** of the electron in the conduction band (see Fig. 15.1-4).

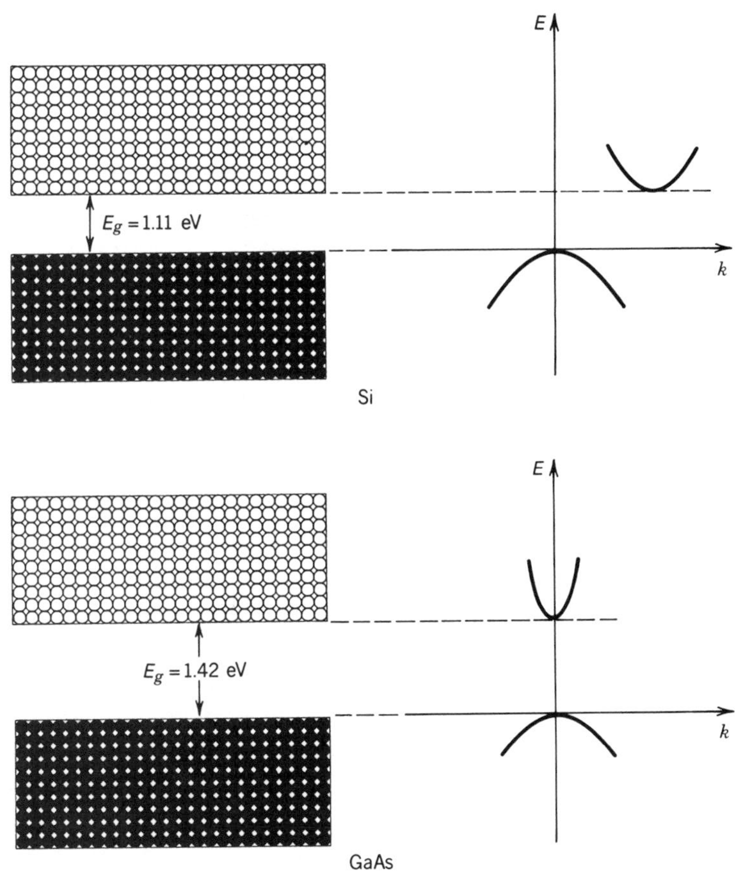

Figure 15.1-4 Approximating the E–k diagram at the bottom of the conduction band and at the top of the valence band of Si and GaAs by parabolas.

Similarly, near the top of the valence band,

$$E = E_v - \frac{\hbar^2 k^2}{2m_v}, \tag{15.1-2}$$

where $E_v = E_c - E_g$ is the energy at the top of the valence band and m_v is the effective mass of a hole in the valence band. In general, the effective mass depends on the crystal orientation and the particular band under consideration. Typical ratios of the averaged effective masses to the mass of the free electron m_0 are provided in Table 15.1-1 for Si and GaAs.

Direct- and Indirect-Gap Semiconductors

Semiconductors for which the valence-band maximum and the conduction-band minimum correspond to the same momentum (same k) are called **direct-gap** materials.

TABLE 15.1-1 Average Effective Masses of Electrons and Holes in Si and GaAs

	m_c/m_0	m_v/m_0
Si	0.33	0.5
GaAs	0.07	0.5

TABLE 15.1-2 A Section of the Periodic Table

II	III	IV	V	VI
	Aluminum (Al)	Silicon (Si)	Phosphorus (P)	Sulfur (S)
Zinc (Zn)	Gallium (Ga)	Germanium (Ge)	Arsenic (As)	Selenium (Se)
Cadmium (Cd)	Indium (In)		Antimony (Sb)	Tellurium (Te)
Mercury (Hg)				

Semiconductors for which this is not the case are known as **indirect-gap** materials. The distinction is important; a transition between the top of the valence band and the bottom of the conduction band in an indirect-gap semiconductor requires a substantial change in the electron's momentum. As is evident in Fig. 15.1-4, Si is an indirect-gap semiconductor, whereas GaAs is a direct-gap semiconductor. It will be shown subsequently that direct-gap semiconductors such as GaAs are efficient photon emitters, whereas indirect-gap semiconductors such as Si cannot be efficiently used as light emitters.

B. Semiconducting Materials

Table 15.1-2 reproduces a section of the periodic table of the elements, containing some of the important elements involved in semiconductor electronics and optoelectronics technology. Both elemental and compound semiconductors are of importance.

Elemental Semiconductors

Several elements in group IV of the periodic table are semiconductors. Most important are **silicon** (Si) and **germanium** (Ge). At present most commercial electronic integrated circuits and devices are fabricated from Si. However, these materials are not useful for fabricating photon emitters because of their indirect bandgap. Nevertheless, both are widely used for making photon detectors.

Binary Semiconductors

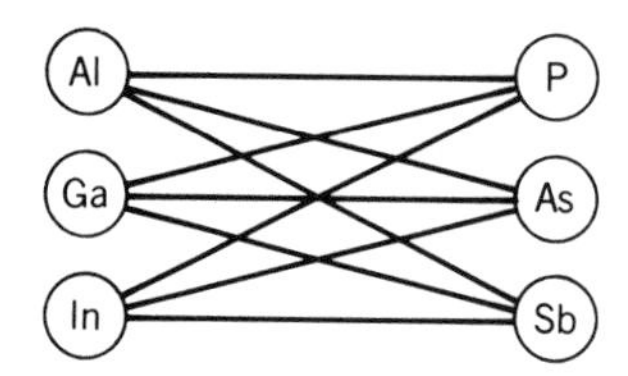

Compounds formed by combining an element in group III, such as aluminum (Al), gallium (Ga), or indium (In), with an element in group V, such as phosphorus (P), arsenic (As), or antimony (Sb), are important semiconductors. There are nine such III–V compounds. These are listed in Table 15.1-3, along with their bandgap energy E_g, bandgap wavelength $\lambda_g = hc_o/E_g$ (which is the free-space wavelength of a photon of energy E_g), and gap type (direct or indirect). The bandgap energies and the lattice constants of these compounds are also provided in Fig. 15.1-5. Various of these compounds are used for making photon detectors and sources (light-emitting diodes and lasers). The most important binary semiconductor for optoelectronic devices is gallium arsenide (GaAs). Furthermore,

GaAs is becoming increasingly important (relative to Si) as the basis of fast electronic devices and circuits.

Ternary Semiconductors

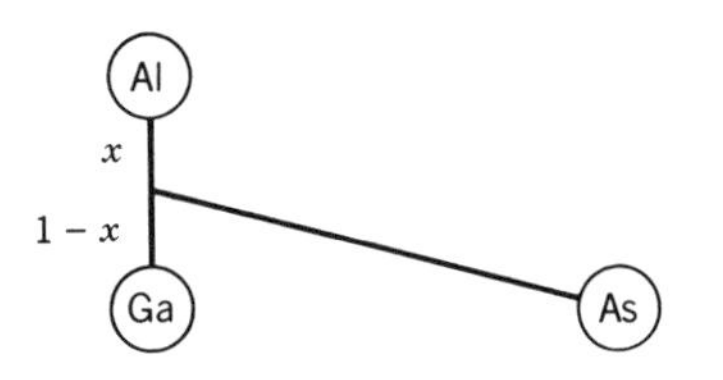

Compounds formed from two elements of group III with one element of group V (or one from group III with two from Group V) are important ternary semiconductors. **(Al_xGa_{1-x})As**, for example, is a ternary compound with properties intermediate between those of AlAs and GaAs, depending on the compositional mixing ratio x (where x denotes the fraction of Ga atoms in GaAs replaced by Al atoms). The bandgap energy E_g for this material varies between 1.42 eV for GaAs and 2.16 eV for AlAs, as x is varied between 0 and 1. The material is represented by the line connecting GaAs and AlAs in Fig. 15.1-5. Because this line is nearly horizontal, $Al_xGa_{1-x}As$ is lattice matched to GaAs (i.e., they have the same lattice constant). This means that a layer of a given composition can be grown on a layer of different composition without introducing strain in the material. The combination $Al_xGa_{1-x}As/GaAs$ is highly important in current LED and semiconductor laser technology. Other III–V compound semiconductors of various compositions and bandgap types (direct/indirect) are indicated in the lattice-constant versus bandgap-energy diagram in Fig. 15.1-5.

Quaternary Semiconductors

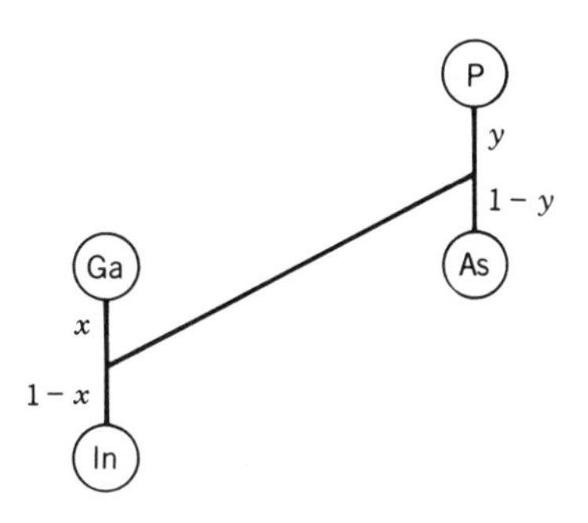

These compounds are formed from a mixture of two elements from Group III with two elements from group V. Quaternary semiconductors offer more flexibility for the synthesis of materials with desired properties than do ternary semiconductors, since they provide an extra degree of freedom. An example is provided by the quaternary **($In_{1-x}Ga_x$)($As_{1-y}P_y$)**, whose bandgap energy E_g varies between 0.36 eV (InAs) and 2.26 eV (GaP) as the compositional mixing ratios x and y vary between 0 and 1. The shaded area in Fig. 15.1-5 indicates the range of energy gaps and lattice constants spanned by this compound. For mixing ratios x and y that satisfy $y = 2.16(1 - x)$, ($In_{1-x}Ga_x$)($As_{1-y}P_y$) can be very well lattice matched to InP and therefore conveniently grown on it. These compounds are used in making semiconductor lasers and detectors.

TABLE 15.1-3 Selected Elemental and III–V Binary Semiconductors and Their Bandgap Energies E_g at $T = 300$ K, Bandgap Wavelengths $\lambda_g = hc_o / E_g$, and Type of Gap (I = Indirect, D = Direct)

Material	Bandgap Energy E_g (eV)	Bandgap Wavelength λ_g (μm)	Type
Ge	0.66	1.88	I
Si	1.11	1.15	I
AlP	2.45	0.52	I
AlAs	2.16	0.57	I
AlSb	1.58	0.75	I
GaP	2.26	0.55	I
GaAs	1.42	0.87	D
GaSb	0.73	1.70	D
InP	1.35	0.92	D
InAs	0.36	3.5	D
InSb	0.17	7.3	D

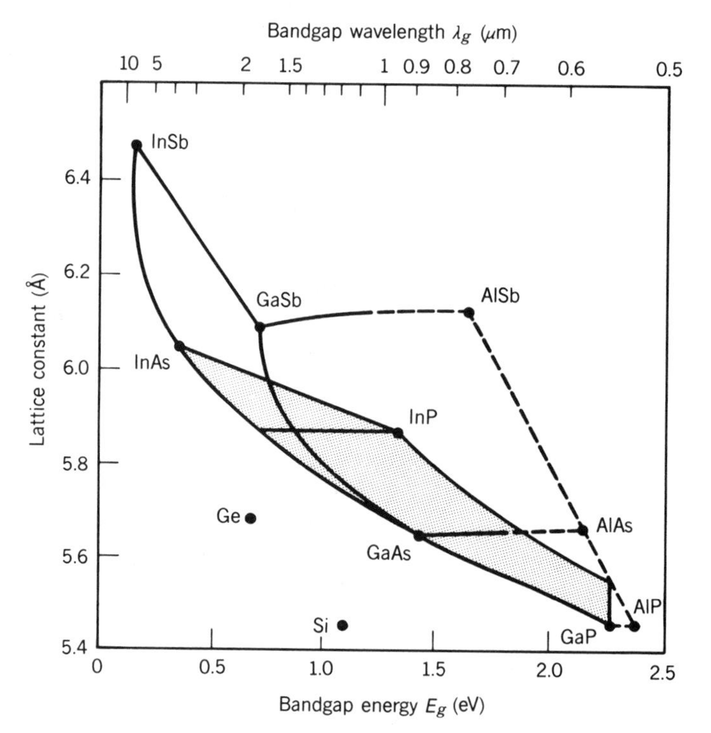

Figure 15.1-5 Lattice constants, bandgap energies, and bandgap wavelengths for Si, Ge, and nine III–V binary compounds. Ternary compounds can be formed from binary materials by motion along the line joining the two points that represent the binary materials. For example, $Al_xGa_{1-x}As$ is represented by points on the line connecting GaAs and AlAs. As x varies from 0 to 1, the point moves along the line from GaAs to AlAs. Since this line is nearly horizontal, $Al_xGa_{1-x}As$ is lattice matched to GaAs. Solid and dashed curves represent direct-gap and indirect-gap compositions, respectively. A material may have direct bandgap for one mixing ratio x and an indirect bandgap for a different x. A quaternary compound is represented by a point in the area formed by its four binary components. For example, $(In_{1-x}Ga_x)(As_{1-y}P_y)$ is represented by the shaded area with vertices at InAs, InP, GaP, and GaAs; the upper horizontal line represents compounds that are lattice matched to InP.

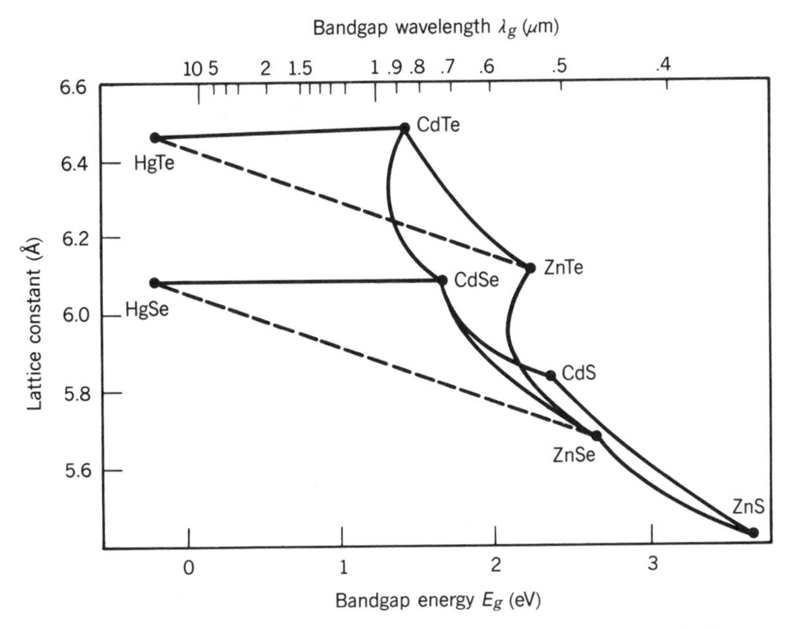

Figure 15.1-6 Lattice constants, bandgap energies, and bandgap wavelengths for some important II–VI binary compounds.

Compounds using elements from group II (e.g., Zn, Cd, Hg) and group VI (e.g., S, Se, Te) of the periodic table also form useful semiconductors, particularly at wavelengths shorter than 0.5 μm and longer than 5.0 μm, as shown in Fig. 15.1-6. HgTe and CdTe, for example, are nearly lattice matched, so that the ternary semiconductor $Hg_xCd_{1-x}Te$ is a useful material for fabricating photon detectors in the middle-infrared region of the spectrum. Also used in this range are IV–VI compounds such as $Pb_xSn_{1-x}Te$ and $Pb_xSn_{1-x}Se$. Applications include night vision, thermal imaging, and long-wavelength lightwave communications.

Doped Semiconductors

The electrical and optical properties of semiconductors can be substantially altered by adding small controlled amounts of specially chosen impurities, or **dopants**, which alter the concentration of mobile charge carriers by many orders of magnitude. Dopants with excess valence electrons (called **donors**) can be used to replace a small proportion of the normal atoms in the crystal lattice and thereby to create a predominance of mobile electrons; the material is then said to be an ***n*-type** semiconductor. Thus atoms from group V (e.g., P or As) replacing some of the group IV atoms in an elemental semiconductor, or atoms from group VI (e.g., Se or Te) replacing some of the group V atoms in a III–V binary semiconductor, produce an *n*-type material. Similarly, a ***p*-type** material can be made by using dopants with a deficiency of valence electrons, called **acceptors**. The result is a predominance of holes. Group-IV atoms in an elemental semiconductor replaced with some group-III atoms (e.g., B or In), or group-III atoms in a III–V binary semiconductor replaced with some group-II atoms (e.g., Zn or Cd), produce a *p*-type material. Group IV atoms act as donors in group III and as acceptors in group V, and therefore can be used to produce an excess of both electrons and holes in III–V materials.

Undoped semiconductors (i.e., semiconductors with no intentional doping) are referred to as **intrinsic** materials, whereas doped semiconductors are called **extrinsic** materials. The concentrations of mobile electrons and holes are equal in an intrinsic semiconductor, $n = p \equiv n_i$, where n_i increases with temperature at an exponential rate. The concentration of mobile electrons in an n-type semiconductor (called **majority carriers**) is far greater than the concentration of holes (called **minority carriers**), i.e., $n \gg p$. The opposite is true in p-type semiconductors, for which holes are majority carriers and $p \gg n$. Doped semiconductors at room temperature typically have a majority carrier concentration that is approximately equal to the impurity concentration.

C. Electron and Hole Concentrations

Determining the concentration of carriers (electrons and holes) as a function of energy requires knowledge of:

- The density of allowed energy levels (density of states).
- The probability that each of these levels is occupied.

Density of States

The quantum state of an electron in a semiconductor material is characterized by its energy E, its wavevector $\mathbf{k}$ [the magnitude of which is approximately related to E by (15.1-1) or (15.1-2)], and its spin. The state is described by a wavefunction satisfying certain boundary conditions.

An electron near the conduction band edge may be approximately described as a particle of mass m_c confined to a three-dimensional cubic box (of dimension d) with perfectly reflecting walls, i.e., a three-dimensional infinite rectangular potential well. The standing-wave solutions require that the components of the wavevector $\mathbf{k} = (k_x, k_y, k_z)$ assume the discrete values $\mathbf{k} = (q_1\pi/d, q_2\pi/d, q_3\pi/d)$, where the respective mode numbers, q_1, q_2, q_3, are positive integers. This result is a three-dimensional generalization of the one-dimensional case discussed in Exercise 12.1-1. The tip of the vector $\mathbf{k}$ must lie on the points of a lattice whose cubic unit cell has dimension π/d. There are therefore $(d/\pi)^3$ points per unit volume in $\mathbf{k}$-space. The number of states whose wavevectors $\mathbf{k}$ have magnitudes between 0 and k is determined by counting the number of points lying within the positive octant of a sphere of radius k [with volume $\approx (\frac{1}{8})4\pi k^3/3 = \pi k^3/6$]. Because of the two possible values of the electron spin, each point in $\mathbf{k}$-space corresponds to two states. There are therefore approximately $2(\pi k^3/6)/(\pi/d)^3 = (k^3/3\pi^2)d^3$ such points in the volume d^3 and $(k^3/3\pi^2)$ points per unit volume. It follows that the number of states with electron wavenumbers between k and $k + \Delta k$, per unit volume, is $\varrho(k)\,\Delta k = [(d/dk)(k^3/3\pi^2)]\,\Delta k = (k^2/\pi^2)\,\Delta k$, so that the density of states is

$$\varrho(k) = \frac{k^2}{\pi^2}. \tag{15.1-3}$$

Density of States

This derivation is identical to that used for counting the number of modes that can be supported in a three-dimensional electromagnetic resonator (see Sec. 9.1C). In the case of electromagnetic modes there are two degrees of freedom associated with the field polarization (i.e., two photon spin values), whereas in the semiconductor case there are two spin values associated with the electron state. In resonator optics the allowed electromagnetic solutions for $\mathbf{k}$ were converted into allowed frequencies through the linear frequency–wavenumber relation $\nu = ck/2\pi$. In semiconductor physics, on the other hand, the allowed solutions for $\mathbf{k}$ are converted into allowed energies through the quadratic energy–wavenumber relations given in (15.1-1) and (15.1-2).

If $\varrho_c(E)\,\Delta E$ represents the number of conduction-band energy levels (per unit volume) lying between E and $E + \Delta E$, then, because of the one-to-one correspondence between E and k governed by (15.1-1), the densities $\varrho_c(E)$ and $\varrho(k)$ must be related by $\varrho_c(E)\,dE = \varrho(k)\,dk$. Thus the density of allowed energies in the conduction band is $\varrho_c(E) = \varrho(k)/(dE/dk)$. Similarly, the density of allowed energies in the valence band is $\varrho_v(E) = \varrho(k)/(dE/dk)$, where E is given by (15.1-2). The approximate quadratic E–k relations (15.1-1) and (15.1-2), which are valid near the edges of the conduction band and valence band, respectively, are used to evaluate the derivative dE/dk for each band. The result that obtains is

$$\varrho_c(E) = \frac{(2m_c)^{3/2}}{2\pi^2\hbar^3}(E - E_c)^{1/2}, \quad E \geq E_c \tag{15.1-4}$$

$$\varrho_v(E) = \frac{(2m_v)^{3/2}}{2\pi^2\hbar^3}(E_v - E)^{1/2}, \quad E \leq E_v. \tag{15.1-5}$$

Density of States Near Band Edges

The square-root relation is a result of the quadratic energy–wavenumber formulas for electrons and holes near the band edges. The dependence of the density of states on energy is illustrated in Fig. 15.1-7. It is zero at the band edge, increasing away from it at a rate that depends on the effective masses of the electrons and holes. The values of m_c and m_v for Si and GaAs that were provided in Table 15.1-1 are actually averaged values suitable for calculating the density of states.

Probability of Occupancy

In the absence of thermal excitation (at $T = 0$ K), all electrons occupy the lowest possible energy levels, subject to the Pauli exclusion principle. The valence band is then completely filled (there are no holes) and the conduction band is completely empty (it

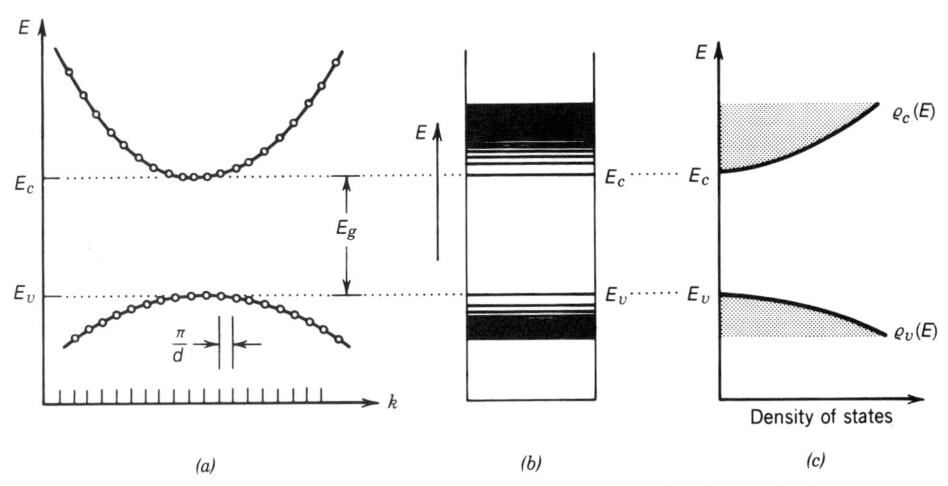

Figure 15.1-7 (*a*) Cross section of the E–k diagram (e.g., in the direction of the k_1 component with k_2 and k_3 fixed). (*b*) Allowed energy levels (at all **k**). (*c*) Density of states near the edges of the conduction and valence bands. $\varrho_c(E)\,dE$ is the number of quantum states of energy between E and $E + dE$, per unit volume, in the conduction band. $\varrho_v(E)$ has an analogous interpretation for the valence band.

contains no electrons). When the temperature is raised, thermal excitations raise some electrons from the valence band to the conduction band, leaving behind empty states in the valence band (holes). The laws of statistical mechanics dictate that under conditions of thermal equilibrium at temperature T, the probability that a given state of energy E is occupied by an electron is determined by the **Fermi function**

$$f(E) = \frac{1}{\exp\left[(E - E_f)/k_B T\right] + 1}, \tag{15.1-6}$$

Fermi Function

where k_B is Boltzmann's constant (at $T = 300$ K, $k_B T = 0.026$ eV) and E_f is a constant known as the **Fermi energy** or **Fermi level**. This function is also known as the **Fermi–Dirac distribution**. The energy level E is either occupied [with probability $f(E)$], or it is empty [with probability $1 - f(E)$]. The probabilities $f(E)$ and $1 - f(E)$ depend on the energy E in accordance with (15.1-6). The function $f(E)$ is not itself a probability distribution, and it does not integrate to unity; rather, it is a sequence of occupation probabilities of successive energy levels.

Because $f(E_f) = \frac{1}{2}$ whatever the temperature T, the Fermi level is that energy level for which the probability of occupancy (if there were an allowed state there) would be $\frac{1}{2}$. The Fermi function is a monotonically decreasing function of E (Fig. 15.1-8). At $T = 0$ K, $f(E)$ is 0 for $E > E_f$ and 1 for $E \leq E_f$. This establishes the significance of E_f; it is the division between the occupied and unoccupied energy levels at $T = 0$ K. Since $f(E)$ is the probability that the energy level E is occupied, $1 - f(E)$ is the probability that it is empty, i.e., that it is occupied by a hole (if E lies in the valence band). Thus for energy level E:

$$f(E) = \text{probability of occupancy by an electron}$$

$$1 - f(E) = \text{probability of occupancy by a hole (valence band)}.$$

These functions are symmetric about the Fermi level.

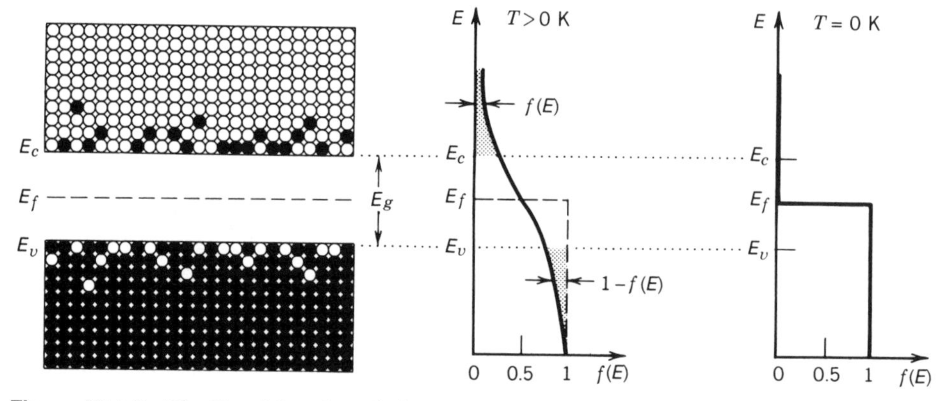

Figure 15.1-8 The Fermi function $f(E)$ is the probability that an energy level E is filled with an electron; $1 - f(E)$ is the probability that it is empty. In the valence band, $1 - f(E)$ is the probability that energy level E is occupied by a hole. At $T = 0$ K, $f(E) = 1$ for $E < E_f$, and $f(E) = 0$ for $E > E_f$; i.e., there are no electrons in the conduction band and no holes in the valence band.

When $E - E_f \gg k_B T$, $f(E) \approx \exp[-(E - E_f)/k_B T]$, so that the high-energy tail of the Fermi function in the conduction band decreases exponentially with increasing energy. The Fermi function is then proportional to the Boltzmann distribution, which describes the exponential energy dependence of the fraction of a population of atoms excited to a given energy level (see Sec. 12.1B). By symmetry, when $E < E_f$ and $E_f - E \gg k_B T$, $1 - f(E) \approx \exp[-(E_f - E)/k_B T]$; i.e., the probability of occupancy by holes in the valence band decreases exponentially as the energy decreases well below the Fermi level.

Thermal-Equilibrium Carrier Concentrations

Let $\mathfrak{n}(E)\,\Delta E$ and $\mathfrak{p}(E)\,\Delta E$ be the number of electrons and holes per unit volume, respectively, with energy lying between E and $E + \Delta E$. The densities $\mathfrak{n}(E)$ and $\mathfrak{p}(E)$ can be obtained by multiplying the densities of states at energy level E by the probabilities of occupancy of the level by electrons or holes, so that

$$\mathfrak{n}(E) = \varrho_c(E) f(E), \qquad \mathfrak{p}(E) = \varrho_v(E)[1 - f(E)]. \tag{15.1-7}$$

The concentrations (populations per unit volume) of electrons and holes $\mathfrak{n}$ and $\mathfrak{p}$ are then obtained from the integrals

$$\mathfrak{n} = \int_{E_c}^{\infty} \mathfrak{n}(E)\, dE, \qquad \mathfrak{p} = \int_{-\infty}^{E_v} \mathfrak{p}(E)\, dE. \tag{15.1-8}$$

In an intrinsic (pure) semiconductor at any temperature, $\mathfrak{n} = \mathfrak{p}$ because thermal excitations always create electrons and holes in pairs. The Fermi level must therefore be placed at an energy level such that $\mathfrak{n} = \mathfrak{p}$. If $m_v = m_c$, the functions $\mathfrak{n}(E)$ and $\mathfrak{p}(E)$ are symmetric, so that E_f must lie precisely in the middle of the bandgap (Fig. 15.1-9). In most intrinsic semiconductors the Fermi level does indeed lie near the middle of the bandgap.

The energy-band diagrams, Fermi functions, and equilibrium concentrations of electrons and holes for *n*-type and *p*-type doped semiconductors are illustrated in Figs. 15.1-10 and 15.1-11, respectively. Donor electrons occupy an energy E_D slightly below the conduction-band edge so that they are easily raised to it. If $E_D = 0.01$ eV, for example, at room temperature ($k_B T = 0.026$ eV) most donor electrons will be ther-

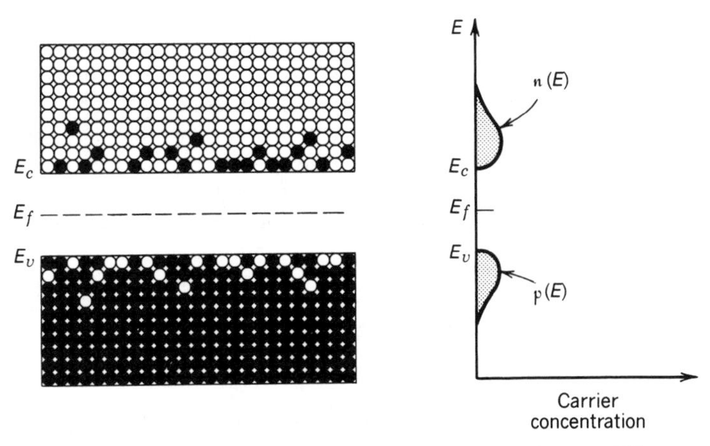

Figure 15.1-9 The concentrations of electrons and holes, $\mathfrak{n}(E)$ and $\mathfrak{p}(E)$, as a function of energy E in an intrinsic semiconductor. The total concentrations of electrons and holes are $\mathfrak{n}$ and $\mathfrak{p}$, respectively.

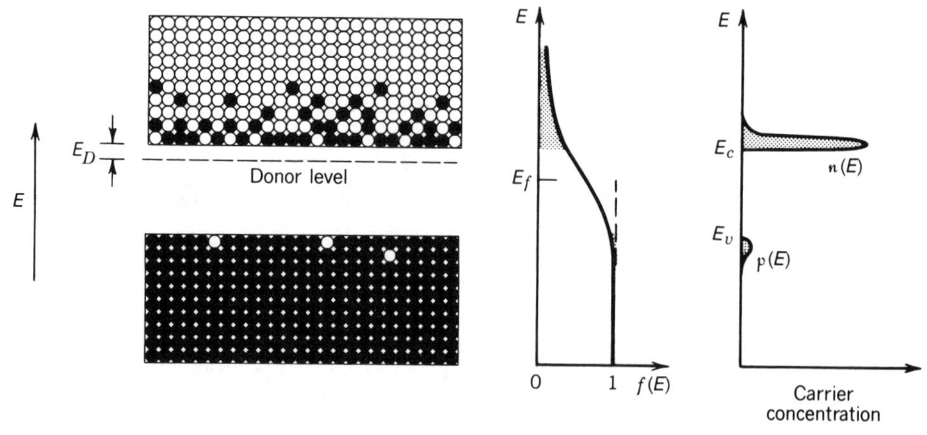

Figure 15.1-10 Energy-band diagram, Fermi function $f(E)$, and concentrations of mobile electrons and holes $n(E)$ and $p(E)$ in an n-type semiconductor.

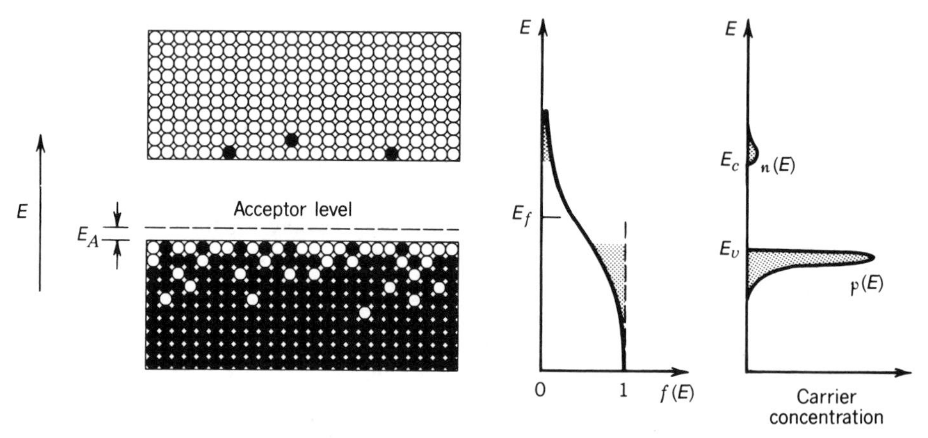

Figure 15.1-11 Energy-band diagram, Fermi function $f(E)$, and concentrations of mobile electrons and holes $n(E)$ and $p(E)$ in a p-type semiconductor.

mally excited into the conduction band. As a result, the Fermi level [where $f(E_f) = \frac{1}{2}$] lies above the middle of the bandgap. For a p-type semiconductor, the acceptor energy level lies at an energy E_A just above the valence-band edge so that the Fermi level is below the middle of the bandgap. Our attention has been directed to the mobile carriers in doped semiconductors. These materials are, of course, electrically neutral as assured by the fixed donor and acceptor ions, so that $n + N_A = p + N_D$ where N_A and N_D are, respectively, the number of ionized acceptors and donors per unit volume.

EXERCISE 15.1-1

Exponential Approximation of the Fermi Function. When $E - E_f \gg k_B T$, the Fermi function $f(E)$ may be approximated by an exponential function. Similarly, when $E_f - E \gg k_B T$, $1 - f(E)$ may be approximated by an exponential function. These conditions apply when the Fermi level lies within the bandgap, but away from its edges by an energy of at least several times $k_B T$ (at room temperature $k_B T \approx 0.026$ eV whereas $E_g = 1.11$ eV in Si and 1.42 eV in GaAs). Using these approximations, which apply for both intrinsic and doped semiconductors, show that (15.1-8) gives

$$n = N_c \exp\left(-\frac{E_c - E_f}{k_B T}\right) \qquad (15.1\text{-}9a)$$

$$p = N_v \exp\left(-\frac{E_f - E_v}{k_B T}\right) \qquad (15.1\text{-}9b)$$

$$np = N_c N_v \exp\left(-\frac{E_g}{k_B T}\right), \qquad (15.1\text{-}10a)$$

where $N_c = 2(2\pi m_c k_B T/h^2)^{3/2}$ and $N_v = 2(2\pi m_v k_B T/h^2)^{3/2}$. Verify that if E_f is closer to the conduction band and $m_v = m_c$, then $n > p$ whereas if it is closer to the valence band, then $p > n$.

Law of Mass Action

Equation (15.1-10a) reveals that the product

$$np = 4\left(\frac{2\pi k_B T}{h^2}\right)^3 (m_c m_v)^{3/2} \exp\left(-\frac{E_g}{k_B T}\right) \qquad (15.1\text{-}10b)$$

is independent of the location of the Fermi level E_f within the bandgap and the semiconductor doping level, provided that the exponential approximation to the Fermi function is valid. The constancy of the concentration product is called the **law of mass action**. For an intrinsic semiconductor, $n = p \equiv n_i$. Combining this relation with (15.1-10a) then leads to

$$n_i \approx (N_c N_v)^{1/2} \exp\left(-\frac{E_g}{2k_B T}\right), \qquad (15.1\text{-}11)$$

Intrinsic Carrier Concentration

revealing that the intrinsic concentration of electrons and holes increases with temperature T at an exponential rate. The law of mass action may therefore be written in the form

$$np = n_i^2. \qquad (15.1\text{-}12)$$

Law of Mass Action

The values of n_i for different materials vary because of differences in the bandgap energies and effective masses. For Si and GaAs, the room temperature values of intrinsic carrier concentrations are provided in Table 15.1-4.

The law of mass action is useful for determining the concentrations of electrons and holes in doped semiconductors. A moderately doped n-type material, for example, has

TABLE 15.1-4 Intrinsic Concentrations in Si and GaAs at $T = 300$ K[a]

	n_i (cm^{-3})
Si	1.5×10^{10}
GaAs	1.8×10^{6}

[a]Substitution of the values of m_c and m_v given in Table 15.1-1, and E_g given in Table 15.1-3, into (15.1-11) will not yield the precise values of n_i given here because of the sensitivity of the formula to the precise values of the parameters.

a concentration of electrons n that is essentially equal to the donor concentration N_D. Using the law of mass action, the hole concentration can be determined from $p = n_i^2/N_D$. Knowledge of n and p allows the Fermi level to be determined by the use of (15.1-8). As long as the Fermi level lies within the bandgap, at an energy greater than several times $k_B T$ from its edges, the approximate relations in (15.1-9) can be used to determine it directly.

If the Fermi level lies inside the conduction (or valence) band, the material is referred to as a **degenerate semiconductor**. In that case, the exponential approximation to the Fermi function cannot be used, so that $np \neq n_i^2$. The carrier concentrations must then be obtained by numerical solution. Under conditions of very heavy doping, the donor (acceptor) impurity band actually merges with the conduction (valence) band to become what is called the **band tail**. This results in an effective decrease of the bandgap.

Quasi-Equilibrium Carrier Concentrations

The occupancy probabilities and carrier concentrations provided above are applicable only for a semiconductor in thermal equilibrium. They are not valid when thermal equilibrium is disturbed. There are, nevertheless, situations in which the conduction-band electrons are in thermal equilibrium among themselves, as are the valence-band holes, but the electrons and holes are not in mutual thermal equilibrium. This can occur, for example, when an external electric current or photon flux induces band-to-band transitions at too high a rate for interband equilibrium to be achieved. This situation, which is known as quasi-equilibrium, arises when the relaxation (decay) times for transitions within each of the bands are much shorter than the relaxation time between the two bands. Typically, the intraband relaxation time $< 10^{-12}$ s, whereas the radiative electron–hole recombination time $\approx 10^{-9}$ s.

Under these circumstances, it is appropriate to use a separate Fermi function for each band; the two Fermi levels are then denoted E_{fc} and E_{fv} and are known as **quasi-Fermi levels** (Fig. 15.1-12). When E_{fc} and E_{fv} lie well inside the conduction and valence bands, respectively, the concentrations of *both* electrons and holes can be quite large.

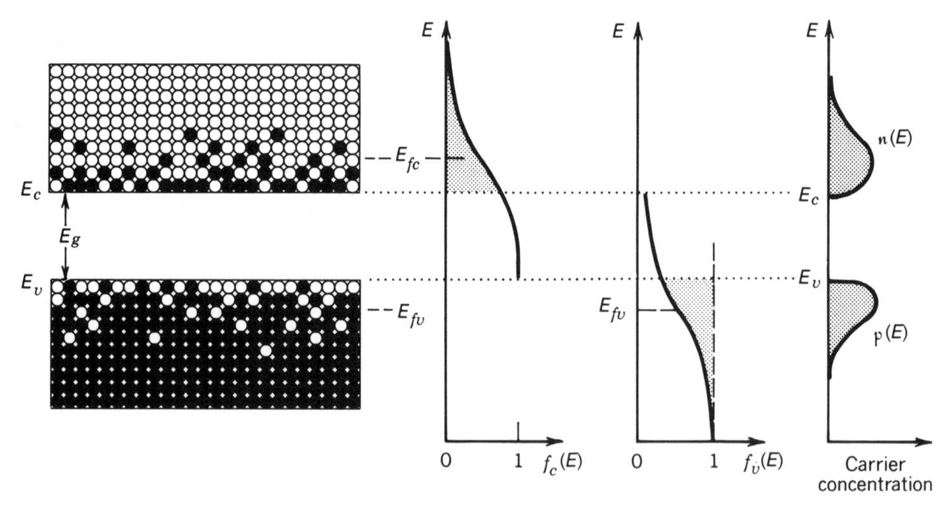

Figure 15.1-12 A semiconductor in quasi-equilibrium. The probability that a particular conduction-band energy level E is occupied by an electron is $f_c(E)$, the Fermi function with Fermi level E_{fc}. The probability that a valence-band energy level E is occupied by a hole is $1 - f_v(E)$, where $f_v(E)$ is the Fermi function with Fermi level E_{fv}. The concentrations of electrons and holes are $n(E)$ and $p(E)$, respectively. Both can be large.

EXERCISE 15.1-2

Determination of the Quasi-Fermi Levels Given the Electron and Hole Concentrations

(a) Given the concentrations of electrons n and holes p in a semiconductor at $T = 0$ K, use (15.1-7) and (15.1-8) to show that the quasi-Fermi levels are

$$E_{fc} = E_c + (3\pi^2)^{2/3} \frac{\hbar^2}{2m_c} n^{2/3} \tag{15.1-13a}$$

$$E_{fv} = E_v - (3\pi^2)^{2/3} \frac{\hbar^2}{2m_v} p^{2/3}. \tag{15.1-13b}$$

(b) Show that these equations are approximately applicable at an arbitrary temperature T if n and p are sufficiently large so that $E_{fc} - E_c \gg k_B T$ and $E_v - E_{fv} \gg k_B T$, i.e., if the quasi-Fermi levels lie deeply within the conduction and valence bands.

D. Generation, Recombination, and Injection

Generation and Recombination in Thermal Equilibrium

The thermal excitation of electrons from the valence band into the conduction band results in the *generation* of electron–hole pairs (Fig. 15.1-13). Thermal equilibrium requires that this generation process be accompanied by a simultaneous reverse process of deexcitation. This process, called **electron–hole recombination**, occurs when an electron decays from the conduction band to fill a hole in the valence band (Fig. 15.1-13). The energy released by the electron may take the form of an emitted photon, in which case the process is called **radiative recombination. Nonradiative recombination** can occur via a number of independent competing processes, including the transfer of energy to lattice vibrations (creating one or more phonons) or to another free electron (Auger process).

Recombination may also occur indirectly via traps or defect centers. These are energy levels associated with impurities or defects due to grain boundaries, dislocations, or other lattice imperfections, that lie within the energy bandgap. An impurity or defect state can act as a recombination center if it is capable of trapping both the

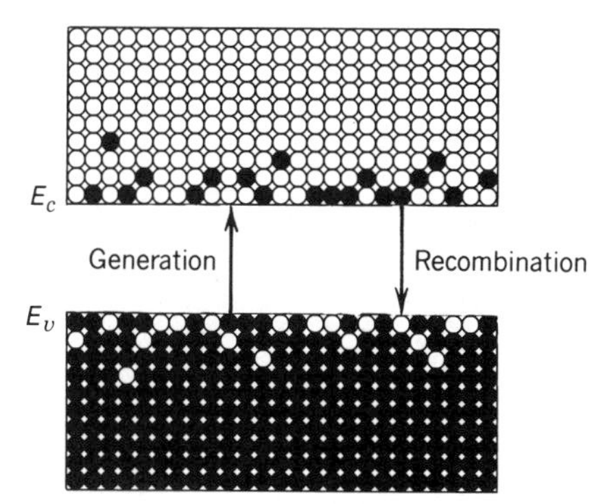

Figure 15.1-13 Electron–hole generation and recombination.

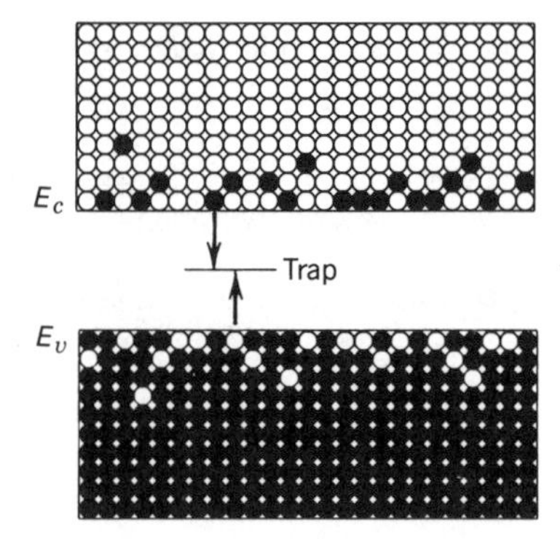

Figure 15.1-14 Electron–hole recombination via a trap.

electron and the hole, thereby increasing their probability of recombining (Fig. 15.1-14). Impurity-assisted recombination may be radiative or nonradiative.

Because it takes both an electron *and* a hole for a recombination to occur, the rate of recombination is proportional to the product of the concentrations of electrons and holes, i.e.,

$$\textit{rate of recombination} = \mathfrak{r} np, \tag{15.1-14}$$

where $\mathfrak{r}$ (cm^3/s) is a parameter that depends on the characteristics of the material, including its composition and defects, and on temperature; it also depends relatively weakly on the doping.

The equilibrium concentrations of electrons and holes n_0 and p_0 are established when the generation and recombination rates are in balance. In the steady state, the rate of recombination must equal the rate of generation. If G_0 is the rate of thermal electron–hole generation at a given temperature, then, in thermal equilibrium,

$$G_0 = \mathfrak{r} n_0 p_0.$$

The product of the electron and hole concentrations $n_0 p_0 = G_0/\mathfrak{r}$ is approximately the same whether the material is n-type, p-type, or intrinsic. Thus $n_i^2 = G_0/\mathfrak{r}$, which leads directly to the law of mass action $n_0 p_0 = n_i^2$. This law is therefore seen to be a consequence of the balance between generation and recombination in thermal equilibrium.

Electron–Hole Injection

A semiconductor in thermal equilibrium with carrier concentrations n_0 and p_0 has equal rates of generation and recombination, $G_0 = \mathfrak{r} n_0 p_0$. Now let additional electron–hole pairs be generated at a steady rate R (pairs per unit volume per unit time) by means of an external (nonthermal) injection mechanism. A new steady state will be reached in which the concentrations are $n = n_0 + \Delta n$ and $p = p_0 + \Delta p$. It is clear, however, that $\Delta n = \Delta p$ since the electrons and holes are created in pairs. Equating the new rates of generation and recombination, we obtain

$$G_0 + R = \mathfrak{r} np. \tag{15.1-15}$$

Substituting $G_0 = \mathfrak{r} n_0 p_0$ into (15.1-15) leads to

$$R = \mathfrak{r}(np - n_0 p_0) = \mathfrak{r}(n_0 \,\Delta n + p_0 \,\Delta n + \Delta n^2) = \mathfrak{r}\,\Delta n(n_0 + p_0 + \Delta n),$$

which we write in the form

$$R = \frac{\Delta n}{\tau}, \tag{15.1-16}$$

with

$$\tau = \frac{1}{\mathfrak{r}[(n_0 + p_0) + \Delta n]}. \tag{15.1-17}$$

For an injection rate such that $\Delta n \ll n_0 + p_0$,

$$\boxed{\tau \approx \frac{1}{\mathfrak{r}(n_0 + p_0)}.} \tag{15.1-18}$$

Excess-Carrier Recombination Lifetime

In an n-type material, where $n_0 \gg p_0$, the recombination lifetime $\tau \approx 1/\mathfrak{r} n_0$ is inversely proportional to the electron concentration. Similarly, for a p-type material where $p_0 \gg n_0$, we obtain $\tau \approx 1/\mathfrak{r} p_0$. This simple formulation is not applicable when traps play an important role in the process.

The parameter τ may be regarded as the **electron–hole recombination lifetime** of the injected excess electron–hole pairs. This is readily understood by noting that the injected carrier concentration is governed by the rate equation

$$\frac{d(\Delta n)}{dt} = R - \frac{\Delta n}{\tau},$$

which is similar to (13.2-2). In the steady state $d(\Delta n)/dt = 0$ whereupon (15.1-16), which is like (13.2-10), is recovered. If the source of injection is suddenly removed (R becomes 0) at the time t_0, then Δn decays exponentially with time constant τ, i.e., $\Delta n(t) = \Delta n(t_0) \exp[-(t - t_0)/\tau]$. In the presence of strong injection, on the other hand, τ is itself a function of Δn, as evident from (15.1-17), so that the rate equation is nonlinear and the decay is no longer exponential.

If the injection rate R is known, the steady-state injected concentration may be determined from

$$\boxed{\Delta n = R\tau,} \tag{15.1-19}$$

permitting the total concentrations $n = n_0 + \Delta n$ and $p = p_0 + \Delta n$ to be determined. Furthermore, if quasi-equilibrium is assumed, (15.1-8) may be used to determine the quasi-Fermi levels. Quasi-equilibrium is not inconsistent with the balance of generation and recombination assumed in the analysis above; it simply requires that the intraband equilibrium time be short in comparison with the recombination time τ.

This type of analysis will prove useful in developing theories of the semiconductor light-emitting diode and the semiconductor diode laser, which are based on enhancing light emission by means of carrier injection (see Chap. 16).

EXERCISE 15.1-3

Electron–Hole Pair Injection in GaAs. Assume that electron–hole pairs are injected into n-type GaAs ($E_g = 1.42$ eV, $m_c \approx 0.07m_0$, $m_v \approx 0.5m_0$) at a rate $R = 10^{23}$ per cm^3 per second. The thermal equilibrium concentration of electrons is $n_0 = 10^{16}$ cm^{-3}. If the recombination parameter $\mathfrak{r} = 10^{-11}$ cm^3/s and $T = 300$ K, determine:

(a) The equilibrium concentration of holes p_0.
(b) The recombination lifetime τ.
(c) The steady-state excess concentration Δn.
(d) The separation between the quasi-Fermi levels $E_{fc} - E_{fv}$, assuming that $T = 0$ K.

Internal Quantum Efficiency

The **internal quantum efficiency** η_i of a semiconductor material is defined as the ratio of the radiative electron–hole recombination rate to the total (radiative and nonradiative) recombination rate. This parameter is important because it determines the efficiency of light generation in a semiconductor material. The total rate of recombination is given by (15.1-14). If the parameter $\mathfrak{r}$ is split into a sum of radiative and nonradiative parts, $\mathfrak{r} = \mathfrak{r}_r + \mathfrak{r}_{nr}$, the internal quantum efficiency is

$$\eta_i = \frac{\mathfrak{r}_r}{\mathfrak{r}} = \frac{\mathfrak{r}_r}{\mathfrak{r}_r + \mathfrak{r}_{nr}}. \tag{15.1-20}$$

The internal quantum efficiency may also be written in terms of the recombination lifetimes since τ is inversely proportional to $\mathfrak{r}$ [see (15.1-18)]. Defining the radiative and nonradiative lifetimes τ_r and τ_{nr}, respectively, leads to

$$\frac{1}{\tau} = \frac{1}{\tau_r} + \frac{1}{\tau_{nr}}. \tag{15.1-21}$$

The internal quantum efficiency is then $\mathfrak{r}_r/\mathfrak{r} = (1/\tau_r)/(1/\tau)$, or

$$\boxed{\eta_i = \frac{\tau}{\tau_r} = \frac{\tau_{nr}}{\tau_r + \tau_{nr}}.} \tag{15.1-22}$$

Internal Quantum Efficiency

The radiative recombination lifetime τ_r governs the rates of photon absorption and emission, as explained in Sec. 15.2B. Its value depends on the carrier concentrations and the material parameter $\mathfrak{r}_r$. For low to moderate injection rates,

$$\tau_r \approx \frac{1}{\mathfrak{r}_r(n_0 + p_0)}, \tag{15.1-23}$$

in accordance with (15.1-18). The nonradiative recombination lifetime is governed by a similar equation. However, if nonradiative recombination takes place via defect centers in the bandgap, τ_{nr} is more sensitive to the concentration of these centers than to the electron and hole concentrations.

TABLE 15.1-5 Approximate Values for Radiative Recombination Rates $\mathfrak{r}_r$, Recombination Lifetimes, and Internal Quantum Efficiency η_i in Si and GaAs[a]

	$\mathfrak{r}_r$ (cm^3/s)	τ_r	τ_{nr}	τ	η_i
Si	10^{-15}	10 ms	100 ns	100 ns	10^{-5}
GaAs	10^{-10}	100 ns	100 ns	50 ns	0.5

[a]Under conditions of doping, temperature, and defect concentration specified in the text.

Approximate values for recombination rates and lifetimes in Si and GaAs are provided in Table 15.1-5. Order-of-magnitude values are given for $\mathfrak{r}_r$ and τ_r (assuming n-type material with a carrier concentration $n_0 = 10^{17}$ cm^{-3} at $T = 300$ K), τ_{nr} (assuming defect centers with a concentration of 10^{15} cm^{-3}), τ, and the internal quantum efficiency η_i.

The radiative lifetime for Si is orders of magnitude larger than its overall lifetime, principally because it has an indirect bandgap. This results in a small internal quantum efficiency. For GaAs, on the other hand, the decay is largely via radiative transitions (it has a direct bandgap), and consequently the internal quantum efficiency is large. GaAs and other direct-gap materials are therefore useful for fabricating light-emitting structures, whereas Si and other indirect-gap materials are not.

E. Junctions

Junctions between differently doped regions of a semiconductor material are called **homojunctions**. An important example is the p–n junction, which is discussed in this subsection. Junctions between different semiconductor materials are called **heterojunctions**. These are discussed subsequently.

The p–n Junction

The p–n junction is a homojunction between a p-type and an n-type semiconductor. It acts as a diode which can serve in electronics as a rectifier, logic gate, voltage regulator (Zener diode), and tuner (varactor diode); and in optoelectronics as a light-emitting diode (LED), laser diode, photodetector, and solar cell.

A p–n junction consists of a p-type and an n-type section of the same semiconducting materials in metallurgical contact with each other. The p-type region has an abundance of holes (majority carriers) and few mobile electrons (minority carriers); the n-type region has an abundance of mobile electrons and few holes (Fig. 15.1-15). Both charge carriers are in continuous random thermal motion in all directions.

When the two regions are brought into contact (Fig. 15.1-16), the following sequence of events takes place:

- Electrons and holes diffuse from areas of high concentration toward areas of low concentration. Thus electrons diffuse away from the n-region into the p-region, leaving behind positively charged ionized donor atoms. In the p-region the electrons recombine with the abundant holes. Similarly, holes diffuse away from the p-region, leaving behind negatively charged ionized acceptor atoms. In the n-region the holes recombine with the abundant mobile electrons. This diffusion process cannot continue indefinitely, however, because it causes a disruption of the charge balance in the two regions.
- As a result, a narrow region on both sides of the junction becomes almost totally depleted of *mobile* charge carriers. This region is called the **depletion layer**. It

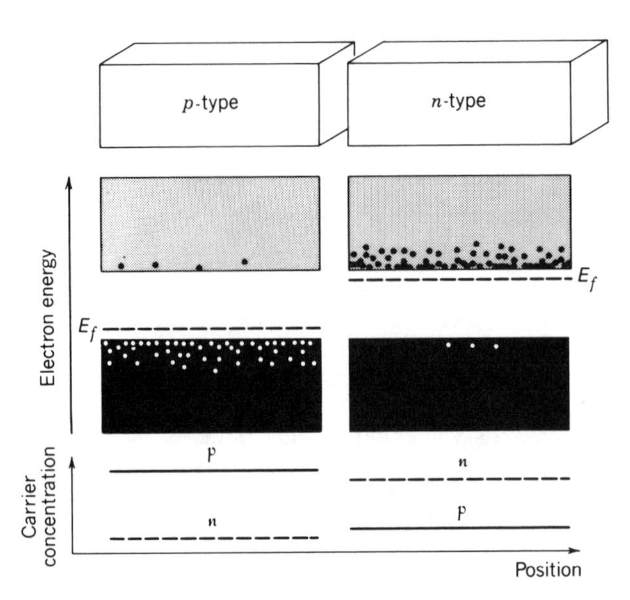

Figure 15.1-15 Energy levels and carrier concentrations of a p-type and an n-type semiconductor before contact.

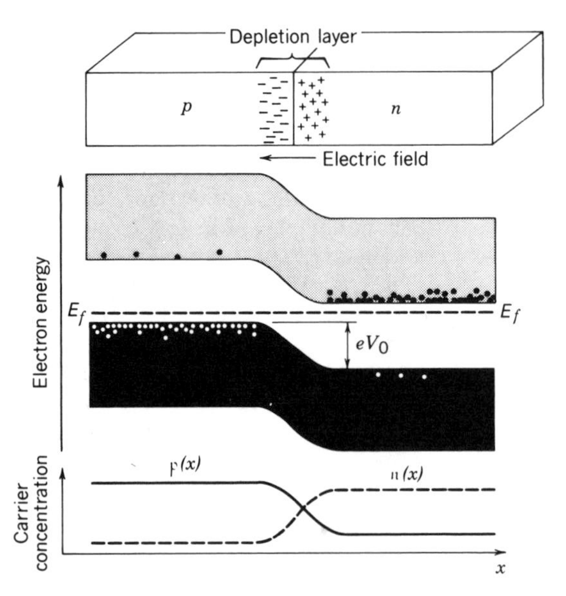

Figure 15.1-16 A p–n junction in thermal equilibrium at $T > 0$ K. The depletion-layer, energy-band diagram, and concentrations (on a logarithmic scale) of mobile electrons $n(x)$ and holes $p(x)$ are shown as functions of position x. The built-in potential difference V_0 corresponds to an energy eV_0, where e is the magnitude of the electron charge.

contains only the *fixed* charges (positive ions on the n-side and negative ions on the p-side). The thickness of the depletion layer in each region is inversely proportional to the concentration of dopants in the region.

- The fixed charges create an electric field in the depletion layer which points from the n-side toward the p-side of the junction. This **built-in field** obstructs the diffusion of further mobile carriers through the junction region.
- An equilibrium condition is established that results in a net built-in potential difference V_0 between the two sides of the depletion layer, with the n-side exhibiting a higher potential than the p-side.
- The built-in potential provides a lower potential energy for an electron on the n-side relative to the p-side. As a result, the energy bands bend as shown in Fig. 15.1-16. In thermal equilibrium there is only a single Fermi function for the entire structure so that the Fermi levels in the p- and n-regions must align.
- No *net* current flows across the junction. The diffusion and drift currents cancel for the electrons and holes independently.

The Biased Junction

An externally applied potential will alter the potential difference between the p- and n-regions. This, in turn, will modify the flow of majority carriers, so that the junction can be used as a "gate." If the junction is **forward biased** by applying a positive voltage V to the p-region (Fig. 15.1-17), its potential is increased with respect to the n-region, so that an electric field is produced in a direction opposite to that of the built-in field. The presence of the external bias voltage causes a departure from equilibrium and a misalignment of the Fermi levels in the p- and n-regions, as well as in the depletion layer. The presence of two Fermi levels in the depletion layer, E_{fc} and E_{fv}, represents a state of quasi-equilibrium.

The net effect of the forward bias is a reduction in the height of the potential-energy hill by an amount eV. The majority carrier current turns out to increase by an exponential factor $\exp(eV/k_BT)$ so that the net current becomes $i = i_s \exp(eV/k_BT) - i_s$, where i_s is a constant. The excess majority carrier holes and electrons that enter

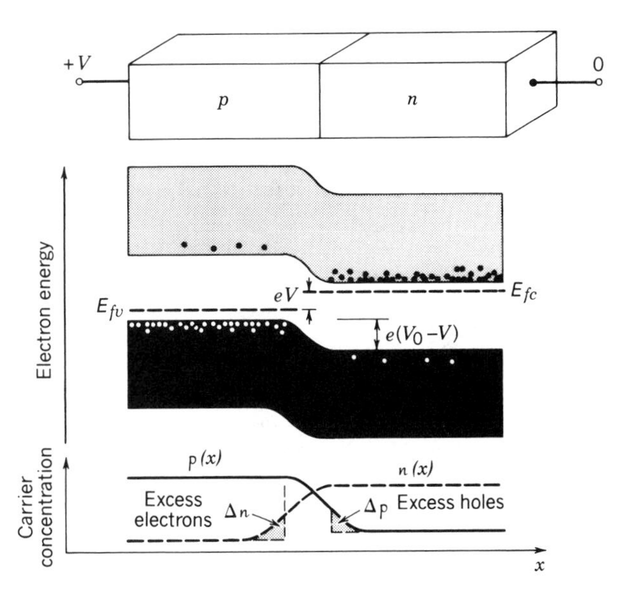

Figure 15.1-17 Energy-band diagram and carrier concentrations in a forward-biased p–n junction.

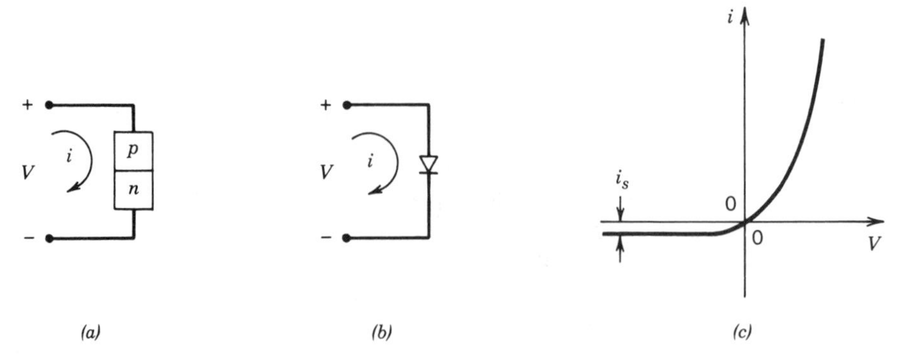

Figure 15.1-18 (*a*) Voltage and current in a *p–n* junction. (*b*) Circuit representation of the *p–n* junction diode. (*c*) Current–voltage characteristic of the ideal *p–n* junction diode.

the *n*- and *p*-regions, respectively, become minority carriers and recombine with the local majority carriers. Their concentration therefore decreases with distance from the junction as shown in Fig. 15.1-17. This process is known as **minority carrier injection**.

If the junction is **reverse biased** by applying a negative voltage V to the *p*-region, the height of the potential-energy hill is augmented by eV. This impedes the flow of majority carriers. The corresponding current is multiplied by the exponential factor $\exp(eV/k_BT)$, where V is negative; i.e., it is reduced. The net result for the current is $i = i_s \exp(eV/k_BT) - i_s$, so that a small current of magnitude $\approx i_s$ flows in the reverse direction when $|V| \gg k_BT/e$.

A *p–n* junction therefore acts as a diode with a current–voltage (i–V) characteristic

$$i = i_s\left[\exp\left(\frac{eV}{k_BT}\right) - 1\right], \tag{15.1-24}$$

Ideal Diode Characteristic

as illustrated in Fig. 15.1-18.

The response of a *p–n* junction to a dynamic (ac) applied voltage is determined by solving the set of differential equations governing the processes of electron and hole diffusion, drift (under the influence of the built-in and external electric fields), and recombination. These effects are important for determining the speed at which the diode can be operated. They may be conveniently modeled by two capacitances, a junction capacitance and a diffusion capacitance, in parallel with an ideal diode. The **junction capacitance** accounts for the time necessary to change the fixed positive and negative charges stored in the depletion layer when the applied voltage changes. The thickness l of the depletion layer turns out to be proportional to $(V_0 - V)^{1/2}$; it therefore increases under reverse-bias conditions (negative V) and decreases under forward-bias conditions (positive V). The junction capacitance $C = \epsilon A/l$ (where A is the area of the junction) is therefore inversely proportional to $(V_0 - V)^{1/2}$. The junction capacitance of a reverse-biased diode is smaller (and the RC response time is therefore shorter) than that of a forward-biased diode. The dependence of C on V is used to make voltage-variable capacitors (varactors).

Minority carrier injection in a forward-biased diode is described by the **diffusion capacitance**, which depends on the minority carrier lifetime and the operating current.

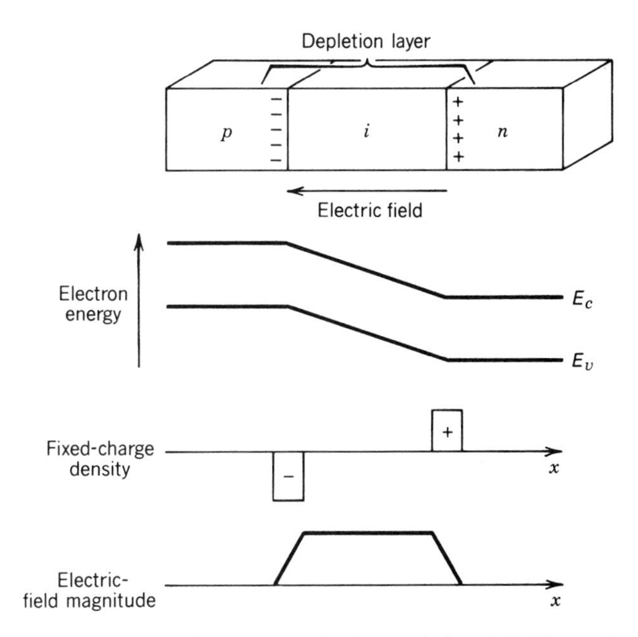

Figure 15.1-19 Electron energy, fixed-charge density, and electric field magnitude for a p–i–n diode in thermal equilibrium.

The p–i–n Junction Diode

A p–i–n diode is made by inserting a layer of intrinsic (or lightly doped) semiconductor material between a p-type region and an n-type region (Fig. 15.1-19). Because the depletion layer extends into each side of a junction by a distance inversely proportional to the doping concentration, the depletion layer of the p–i junction penetrates deeply into the i-region. Similarly, the depletion layer of the i–n junction extends well into the i-region. As a result, the p–i–n diode can behave like a p–n junction with a depletion layer that encompasses the entire intrinsic region. The electron energy, density of fixed charges, and the electric field in a p–i–n diode in thermal equilibrium are illustrated in Fig. 15.1-19. One advantage of using a diode with a large depletion layer is its small junction capacitance and its consequent fast response. For this reason, p–i–n diodes are favored over p–n diodes for use as semiconductor photodiodes. The large depletion layer also permits an increased fraction of the incident light to be captured, thereby increasing the photodetection efficiency (see Sec. 17.3B).

F. Heterojunctions

Junctions between different semiconductor materials are called heterojunctions. Their development has been made possible by modern material growth techniques. Heterojunctions are used in novel bipolar and field-effect transistors, and in optical sources and detectors. They can provide substantial improvement in the performance of electronic and optoelectronic devices. In particular, in photonics the juxtaposition of different semiconductors can be advantageous in several respects:

- Junctions between materials of different bandgap create localized jumps in the energy-band diagram. A potential energy discontinuity provides a barrier that can be useful in preventing selected charge carriers from entering regions where they are undesired. This property may be used in a p–n junction, for example, to

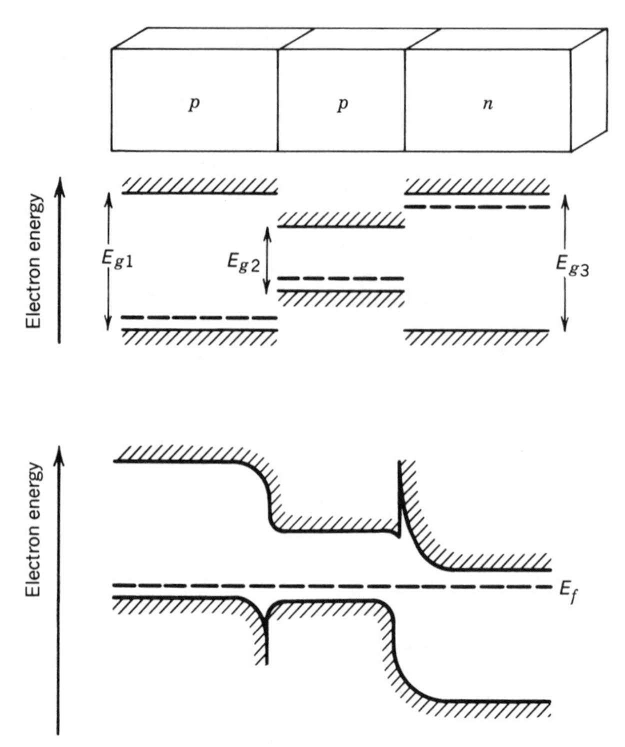

Figure 15.1-20 The p–p–n double heterojunction structure. The middle layer is of narrower bandgap than the outer layers. In equilibrium, the Fermi levels align so that the edge of the conduction band drops sharply at the p–p junction and the edge of the valence band drops sharply at the p–n junction. The ratio of the difference in conduction-band energies to the difference in valence-band energies is known as the **band offset**. When the device is forward biased, these jumps act as barriers that confine the injected minority carriers. Electrons injected from the n-region, for example, are prevented from diffusing beyond the barrier at the p–p junction. Similarly, holes injected from the p-region are not permitted to diffuse beyond the energy barrier at the p–n junction. This double heterostructure therefore forces electrons and holes to occupy a narrow common region. This is essential for the efficient operation of an injection laser diode (see Secs. 16.2 and 16.3).

reduce the proportion of current carried by minority carriers, and thus to increase injection efficiency (see Fig. 15.1-20).

- Discontinuities in the energy-band diagram created by two heterojunctions can be useful for confining charge carriers to a desired region of space. For example, a layer of narrow bandgap material can be sandwiched between two layers of a wider bandgap material, as shown in the p–p–n structure illustrated in Fig. 15.1-20 (which consists of a p–p heterojunction and a p–n heterojunction). This double heterostructure is effectively used in the fabrication of diode lasers, as explained in Sec. 16.3.
- Heterojunctions are useful for creating energy-band discontinuities that accelerate carriers at specific locations. The additional kinetic energy suddenly imparted to a carrier can be useful for selectively enhancing the probability of impact ionization in a multilayer avalanche photodiode (see Sec. 17.4A).
- Semiconductors of different bandgap type (direct and indirect) can be used in the same device to select regions of the structure where light is emitted. Only semiconductors of the direct-gap type can efficiently emit light (see Sec. 15.2).

- Semiconductors of different bandgap can be used in the same device to select regions of the structure where light is absorbed. Semiconductor materials whose bandgap energy is larger than the incident photon energy will be transparent, acting as a "window layer."
- Heterojunctions of materials with different refractive indices can be used to create optical waveguides that confine and direct photons.

*G. Quantum Wells and Superlattices

Heterostructures of thin layers of semiconductor materials can be grown epitaxially, i.e., as lattice-matched layers of one semiconductor material over another, by using techniques such as molecular-beam epitaxy (MBE), liquid-phase epitaxy (LPE), and vapor-phase epitaxy (VPE), of which a common variant is metal-organic chemical vapor deposition (MOCVD). MBE makes use of molecular beams of the constituent elements that are caused to impinge on an appropriately prepared substrate in a high-vacuum environment, LPE uses the cooling of a saturated solution containing the constituents in contact with the substrate, and MOCVD uses gases in a reactor. The compositions and dopings of the individual layers are determined by manipulating the arrival rates of the molecules and the temperature of the substrate surface and can be made as thin as monolayers.

When the layer thickness is comparable to, or smaller than, the de Broglie wavelength of thermalized electrons (e.g., in GaAs the de Broglie wavelength ≈ 50 nm), the energy–momentum relation for a bulk semiconductor material no longer applies. Three structures offer substantial advantages for use in photonics: quantum wells, quantum wires, and quantum dots. The appropriate energy–momentum relations for these structures are derived below. Applications are deferred to subsequent chapters (see Secs. 16.3B and 17.4A).

Quantum Wells

A quantum well is a double heterojunction structure consisting of an ultrathin ($\lesssim 50$ nm) layer of semiconductor material whose bandgap is smaller than that of the surrounding material (Fig. 15.1-21). An example is provided by a thin layer of GaAs surrounded by AlGaAs (see Fig. 12.1-8). The sandwich forms conduction- and valence-band rectangular potential wells within which electrons and holes are confined: electrons in the conduction-band well and holes in the valence-band well. A sufficiently deep potential well can be approximated as an infinite potential well (see Fig. 12.1-9).

The energy levels E_q of a particle of mass m (m_c for electrons and m_v for holes) confined to a *one-dimensional* infinite rectangular well of full width d are determined by solving the time-independent Schrödinger equation. From Exercise 12.1-1,

$$E_q = \frac{\hbar^2(q\pi/d)^2}{2m}, \qquad q = 1, 2, \ldots . \tag{15.1-25}$$

As an example, the allowed energy levels of electrons in an infinitely deep GaAs well ($m_c = 0.07m_0$) of width $d = 10$ nm are $E_q = 54, 216, 486, \ldots$ meV (recall that at $T = 300$ K, $k_B T = 26$ meV). The smaller the width of the well, the larger the separation between adjacent energy levels.

In the quantum-well structure shown in Fig. 15.1-21, electrons (and holes) are confined in the x direction to within a distance d_1 (the well thickness). However, they extend over much larger dimensions ($d_2, d_3 \gg d_1$) in the plane of the confining layer. Thus in the y–z plane, they behave as if they were in bulk semiconductor. The

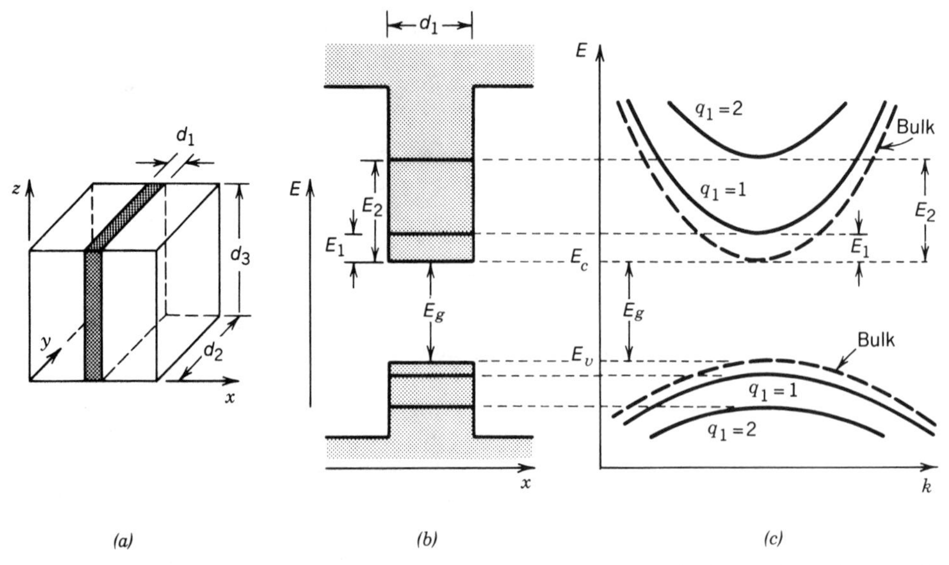

Figure 15.1-21 (a) Geometry of the quantum-well structure. (b) Energy-level diagram for electrons and holes in a quantum well. (c) Cross section of the E–k relation in the direction of k_2 or k_3. The energy subbands are labeled by their quantum number $q_1 = 1, 2, \ldots$. The E–k relation for bulk semiconductor is indicated by the dashed curves.

energy–momentum relation is

$$E = E_c + \frac{\hbar^2 k_1^2}{2m_c} + \frac{\hbar^2 k_2^2}{2m_c} + \frac{\hbar^2 k_3^2}{2m_c},$$

where $k_1 = q_1\pi/d_1$, $k_2 = q_2\pi/d_2$, $k_3 = q_3\pi/d_3$, and $q_1, q_2, q_3 = 1, 2, \ldots$. Since $d_1 \ll d_2, d_3$, k_1 takes on well-separated discrete values, whereas k_2 and k_3 have finely spaced discrete values which may be approximated as a continuum. It follows that the energy–momentum relation for electrons in the conduction band of a quantum well is given by

$$E = E_c + E_{q1} + \frac{\hbar^2 k^2}{2m_c}, \qquad q_1 = 1, 2, 3, \ldots, \tag{15.1-26}$$

where k is the magnitude of a two-dimensional $\mathbf{k} = (k_2, k_3)$ vector in the y–z plane. Each quantum number q_1 corresponds to a subband whose lowest energy is $E_c + E_{q1}$. Similar relations apply for the valence band.

The energy–momentum relation for a bulk semiconductor is given by (15.1-1), where k is the magnitude of a three-dimensional wavevector $\mathbf{k} = (k_1, k_2, k_3)$. The sole distinction is that for the quantum well, k_1 takes on well-separated discrete values. As a result, the density of states associated with a quantum-well structure differs from that associated with bulk material, where the density of states is determined from the magnitude of the three-dimensional wavevector with components $k_1 = q_1\pi/d$, $k_2 = q_2\pi/d$, and $k_3 = q_3\pi/d$ for $d_1 = d_2 = d_3 = d$. The result is [see (15.1-3)] $\varrho(k) = k^2/\pi^2$ per unit volume, which yields the density of conduction-band states [see (15.1-4) and Fig. 15.1-7]

$$\varrho_c(E) = \frac{\sqrt{2}\,m_c^{3/2}}{\pi^2\hbar^3}(E - E_c)^{1/2}, \quad E > 0. \tag{15.1-27}$$

In a quantum-well structure the density of states is obtained from the magnitude of the *two*-dimensional wavevector (k_2, k_3). For each quantum number q_1 the density of states is therefore $\varrho(k) = k/\pi$ states per unit area in the y–z plane, and therefore $k/\pi d_1$ per unit volume. The densities $\varrho_c(E)$ and $\varrho(k)$ are related by $\varrho_c(E)\,dE = \varrho(k)\,dk = (k/\pi d_1)\,dk$. Finally, using the E–k relation (15.1-26) we obtain $dE/dk = \hbar^2 k/m_c$, from which

$$\varrho_c(E) = \begin{cases} \dfrac{m_c}{\pi\hbar^2 d_1}, & E > E_c + E_{q1} \\ 0, & E < E_c + E_{q1}, \end{cases}$$

$$q_1 = 1, 2, \ldots . \tag{15.1-28}$$

Thus for each quantum number q_1, the density of states per unit volume is constant when $E > E_c + E_{q1}$. The overall density of states is the sum of the densities for all values of q_1, so that it exhibits the staircase distribution shown in Fig. 15.1-22. Each step of the staircase corresponds to a different quantum number q_1 and may be regarded as a subband within the conduction band (Fig. 15.1-21). The bottoms of these subbands move progressively higher for higher quantum numbers. It can be shown by substituting $E = E_c + E_{q1}$ in (15.1-27), and by using (15.1-25), that at $E = E_c + E_{q1}$ the quantum-well density of states is the same as that for the bulk. The density of states in the valence band has a similar staircase distribution.

In contrast with bulk semiconductor, the quantum-well structure exhibits a substantial density of states at its lowest allowed conduction-band energy level and at its highest allowed valence-band energy level. This property has a dramatic effect on the optical properties of the material, as discussed in Sec. 16.3G.

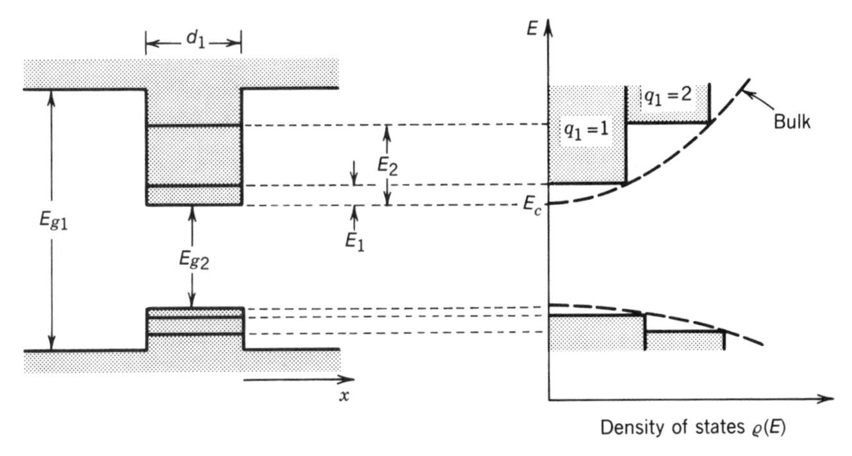

Figure 15.1-22 Density of states for a quantum-well structure (solid) and for a bulk semiconductor (dashed).

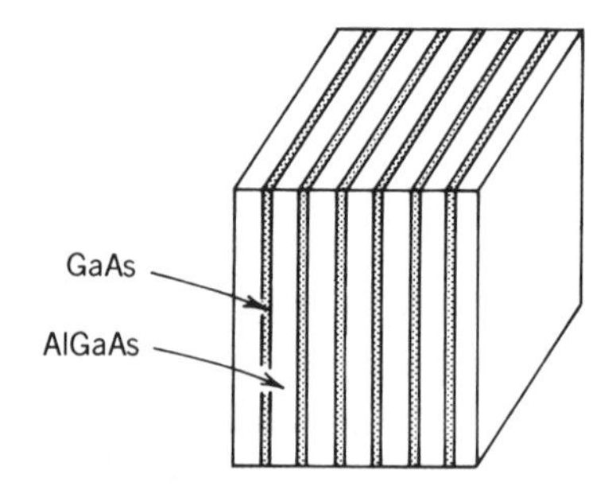

Figure 15.1-23 A multiquantum-well structure fabricated from alternating layers of AlGaAs and GaAs.

Multiquantum Wells and Superlattices

Multiple-layered structures of different semiconductor materials that alternate with each other are called **multiquantum-well** (MQW) structures (see Fig. 15.1-23). They can be fabricated such that the energy bandgap varies with position in any number of ways (see, e.g., Fig. 12.1-8). If the energy barriers between the adjacent wells are sufficiently thin so that electrons can readily tunnel through (quantum mechanically penetrate) the barriers between them, the discrete energy levels broaden into miniature bands in which case the multiquantum-well structure is also referred to as a **superlattice structure**. Multiquantum-well structures are used in lasers and photodetectors, and as nonlinear optical elements. A typical MQW structure might consist of 100 layers, each of which has thickness $\approx$ 10 nm and contains some 40 atomic planes, so that the total thickness of the structure is $\approx 1\ \mu$m. Such a structure would take about 1 hour to grow in an MBE machine.

Quantum Wires and Quantum Dots

A semiconductor material that takes the form of a thin wire of rectangular cross section, surrounded by a material of wider bandgap, is called a **quantum-wire** structure (Fig. 15.1-24). The wire acts as a potential well that narrowly confines electrons (and holes) in two directions (x, y). Assuming that the cross-sectional area is $d_1 d_2$, the energy–momentum relation in the conduction band is

$$E = E_c + E_{q1} + E_{q2} + \frac{\hbar^2 k^2}{2m_c}, \tag{15.1-29}$$

where

$$E_{q1} = \frac{\hbar^2 (q_1 \pi / d_1)^2}{2m_c}, \qquad E_{q2} = \frac{\hbar^2 (q_2 \pi / d_2)^2}{2m_c}, \qquad q_1, q_2 = 1, 2, \ldots. \tag{15.1-30}$$

and k is the wavevector component in the z direction (along the axis of the wire).

Each pair of quantum numbers (q_1, q_2) is associated with an energy subband with a density of states $\varrho(k) = 1/\pi$ per unit length of the wire and therefore $1/\pi d_1 d_2$ per unit volume. The corresponding density of states (per unit volume), as a function of energy, is

$$\varrho_c(E) = \begin{cases} \dfrac{(1/d_1 d_2)\left(m_c^{1/2}/\sqrt{2}\,\pi\hbar\right)}{\left(E - E_c - E_{q1} - E_{q2}\right)^{1/2}}, & E > E_c + E_{q1} + E_{q2} \\ 0, & \text{otherwise}, \end{cases}$$
$$q_1, q_2 = 1, 2, \ldots. \tag{15.1-31}$$

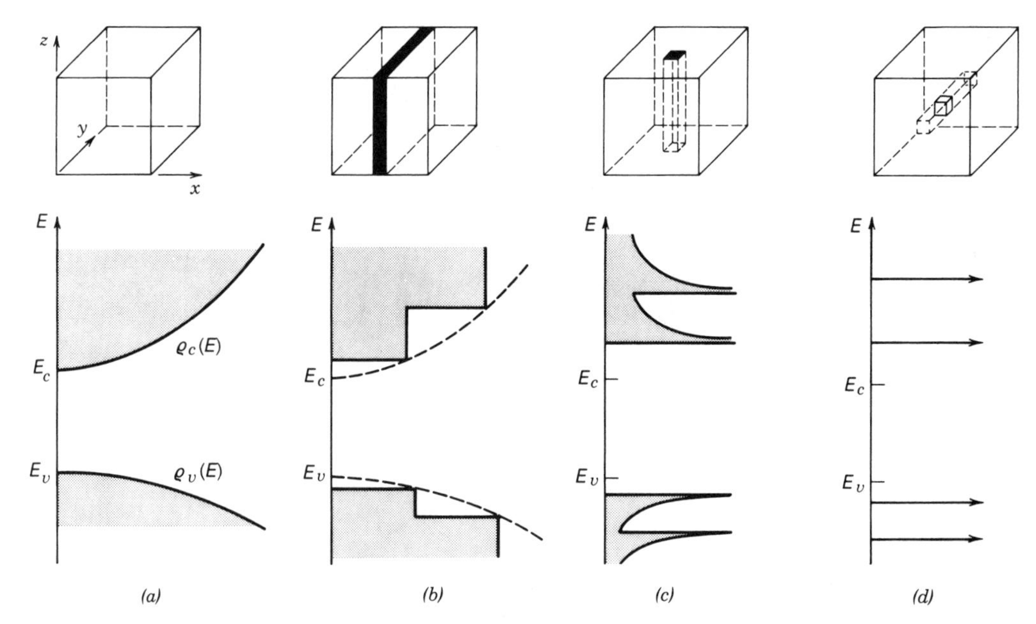

Figure 15.1-24 The density of states in different confinement configurations: (*a*) bulk; (*b*) quantum well; (*c*) quantum wire; (*d*) quantum dot. The conduction and valence bands split into overlapping subbands that become successively narrower as the electron motion is restricted in more dimensions.

These are decreasing functions of energy, as illustrated in Fig. 15.1-24(*c*). The energy subbands in a quantum wire are narrower than those in a quantum well.

In a **quantum-dot** structure, the electrons are narrowly confined in all three directions within a box of volume $d_1 d_2 d_3$. The energy is therefore quantized to

$$E = E_c + E_{q1} + E_{q2} + E_{q3},$$

where

$$E_{q1} = \frac{\hbar^2 (q_1 \pi / d_1)^2}{2m_c}, \qquad E_{q2} = \frac{\hbar^2 (q_2 \pi / d_2)^2}{2m_c}, \qquad E_{q3} = \frac{\hbar^2 (q_3 \pi / d_3)^2}{2m_c},$$

$$q_1, q_2, q_3 = 1, 2, \ldots . \qquad (15.1\text{-}32)$$

The allowed energy levels are discrete and well separated so that the density of states is represented by a sequence of impulse functions (delta functions) at the allowed energies, as illustrated in Fig. 15.1-24(*d*). Quantum dots are often called artificial atoms. Even though they consist of perhaps tens of thousands of strongly interacting natural atoms, the discrete energy levels of the quantum dot can, in principle, be chosen at will by selecting a proper design.

15.2 INTERACTIONS OF PHOTONS WITH ELECTRONS AND HOLES

We now consider the basic optical properties of semiconductors, with an emphasis on the processes of absorption and emission that are important in the operation of photon sources and detectors.

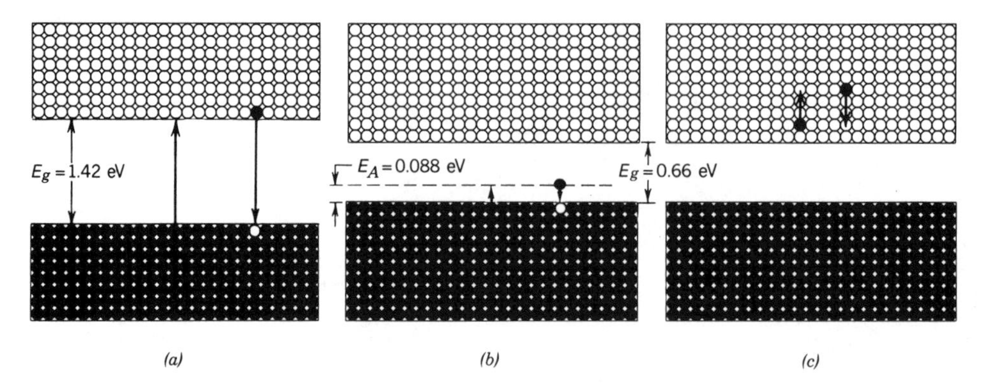

Figure 15.2-1 Examples of absorption and emission of photons in a semiconductor. (*a*) Band-to-band transitions in GaAs can result in the absorption or emission of photons of wavelength $< \lambda_g = hc_o/E_g = 0.87$ μm. (*b*) The absorption of a photon of wavelength $\lambda_A = hc_o/E_A = 14$ μm results in a valence-band to acceptor-level transition in Hg-doped Ge (Ge:Hg). (*c*) A free-carrier transition within the conduction band.

Several mechanisms can lead to the absorption and emission of photons in a semiconductor. The most important of these are:

- *Band-to-Band (Interband) Transitions.* An absorbed photon can result in an electron in the valence band making an upward transition to the conduction band, thereby creating an electron–hole pair [Fig. 15.2-1(*a*)]. Electron–hole recombination can result in the emission of a photon. Band-to-band transitions may be assisted by one or more phonons. A phonon is a quantum of the lattice vibrations that results from the thermal vibrations of the atoms in the material.
- *Impurity-to-Band Transitions.* An absorbed photon can result in a transition between a donor (or acceptor) level and a band in a doped semiconductor. In a *p*-type material, for example, a low-energy photon can lift an electron from the valence band to the acceptor level, where it becomes trapped by an acceptor atom [Fig. 15.2-1(*b*)]. A hole is created in the valence band and the acceptor atom is ionized. Or a hole may be trapped by an ionized acceptor atom; the result is that the electron decays from its acceptor level to recombine with the hole. The energy may be released radiatively (in the form of an emitted photon) or nonradiatively (in the form of phonons). The transition may also be assisted by traps in defect states, as illustrated in Fig. 15.1-14.
- *Free-Carrier (Intraband) Transitions.* An absorbed photon can impart its energy to an electron in a given band, causing it to move higher within that band. An electron in the conduction band, for example, can absorb a photon and move to a higher energy level within the conduction band [Fig. 15.2-1(*c*)]. This is followed by thermalization, a process whereby the electron relaxes down to the bottom of the conduction band while releasing its energy in the form of lattice vibrations.
- *Phonon Transitions.* Long-wavelength photons can release their energy by directly exciting lattice vibrations, i.e., by creating phonons.
- *Excitonic Transitions.* The absorption of a photon can result in the formation of an electron and a hole at some distance from each other but which are nevertheless bound together by their mutual Coulomb interaction. This entity, which is much like a hydrogen atom but with a hole rather than a proton, is called an **exciton**. A photon may be emitted as a result of the electron and hole recombining, thereby annihilating the exciton.

These transitions all contribute to the overall absorption coefficient, which is shown in Fig. 15.2-2 for Si and GaAs, and at greater magnification in Fig. 15.2-3 for a number

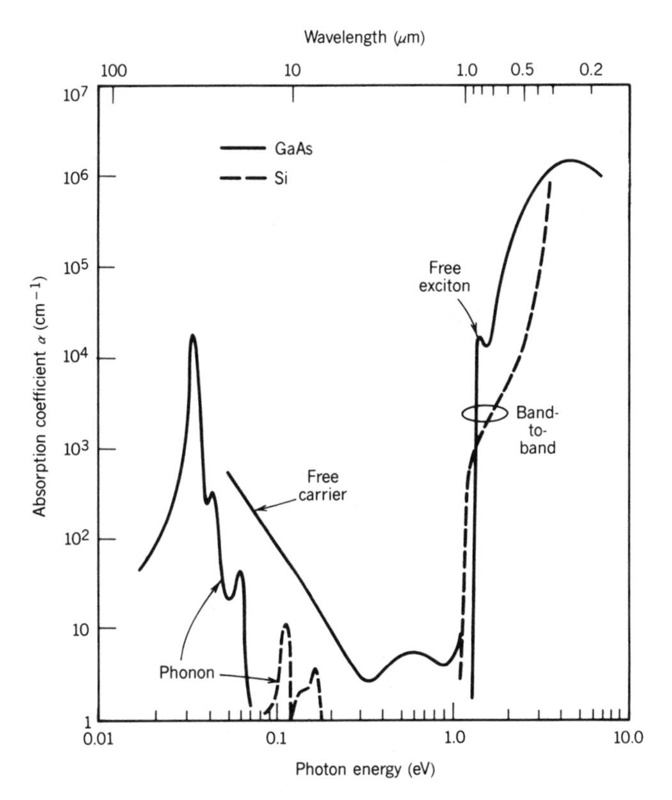

Figure 15.2-2 Observed optical absorption coefficient α versus photon energy for Si and GaAs in thermal equilibrium at $T = 300$ K. The bandgap energy E_g is 1.11 eV for Si and 1.42 eV for GaAs. Si is relatively transparent in the band $\lambda_o \approx 1.1$ to 12 μm, whereas intrinsic GaAs is relatively transparent in the band $\lambda_o \approx 0.87$ to 12 μm (see Fig. 5.5-1).

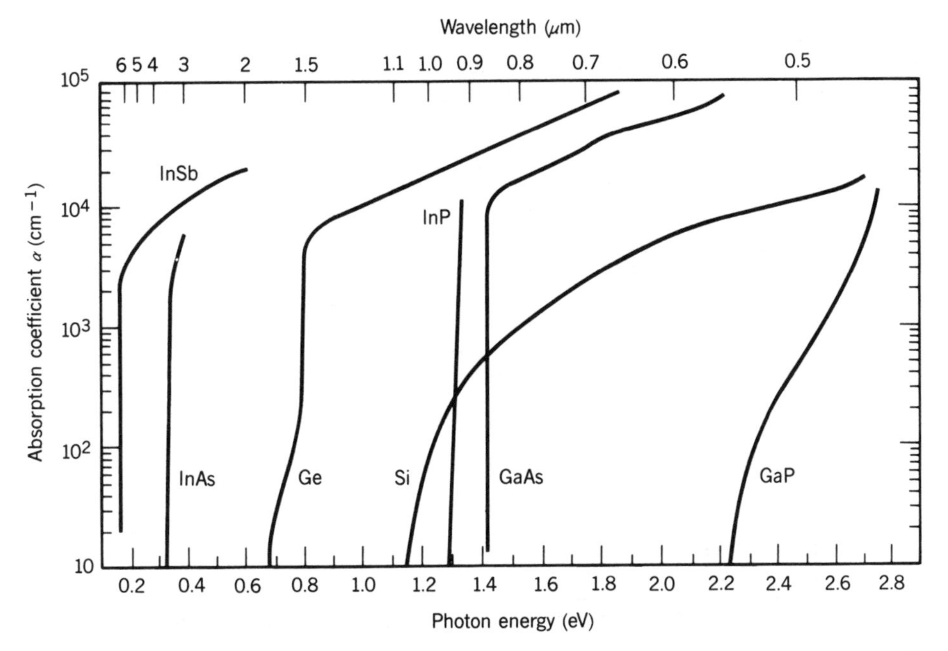

Figure 15.2-3 Absorption coefficient versus photon energy for Ge, Si, GaAs, and selected other III–V binary semiconductors at $T = 300$ K, on an expanded scale (Adapted from G. E. Stillman, V. M. Robbins, and N. Tabatabaie, III–V Compound Semiconductor Devices: Optical Detectors, *IEEE Transactions on Electron Devices*, vol. ED-31, pp. 1643–1655, © 1984 IEEE.)

of semiconductor materials. For photon energies greater than the bandgap energy E_g, the absorption is dominated by band-to-band transitions which form the basis of most photonic devices. The spectral region where the material changes from being relatively transparent ($h\nu < E_g$) to strongly absorbing ($h\nu > E_g$) is known as the **absorption edge**. Direct-gap semiconductors have a more abrupt absorption edge than indirect-gap materials, as is apparent from Figs. 15.2-2 and 15.2-3.

A. Band-to-Band Absorption and Emission

We now proceed to develop a simple theory of direct band-to-band photon absorption and emission, ignoring the other types of transitions.

Bandgap Wavelength

Direct band-to-band absorption and emission can take place only at frequencies for which the photon energy $h\nu > E_g$. The minimum frequency ν necessary for this to occur is $\nu_g = E_g/h$, so that the corresponding maximum wavelength is $\lambda_g = c_o/\nu_g = hc_o/E_g$. If the bandgap energy is given in eV (rather than joules), the bandgap wavelength $\lambda_g = hc_o/eE_g$ in μm is given by

$$\lambda_g = \frac{1.24}{E_g}. \tag{15.2-1}$$

Bandgap Wavelength
λ_g (μm) and E_g (eV)

The quantity λ_g is called the **bandgap wavelength** (or the **cutoff wavelength**); it is provided in Table 15.1-3 and in Figs. 15.1-5 and 15.1-6 for a number of semiconductor materials. The bandgap wavelength λ_g can be adjusted over a substantial range (from the infrared to the visible) by using III–V ternary and quaternary semiconductors of different composition, as is evident in Fig. 15.2-4.

Absorption and Emission

Electron excitation from the valence to the conduction band may be induced by the absorption of a photon of appropriate energy ($h\nu > E_g$). An electron–hole pair is generated [Fig. 15.2-5(*a*)]. This adds to the concentration of mobile charge carriers and

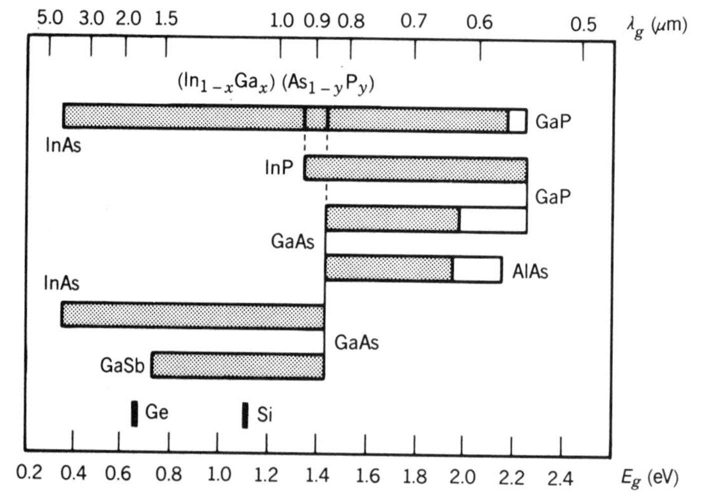

Figure 15.2-4 Bandgap energy E_g and corresponding bandgap wavelength λ_g for selected elemental and III–V binary, ternary, and quaternary semiconductor materials. The shaded regions represent compositions for which the materials are direct-gap semiconductors.

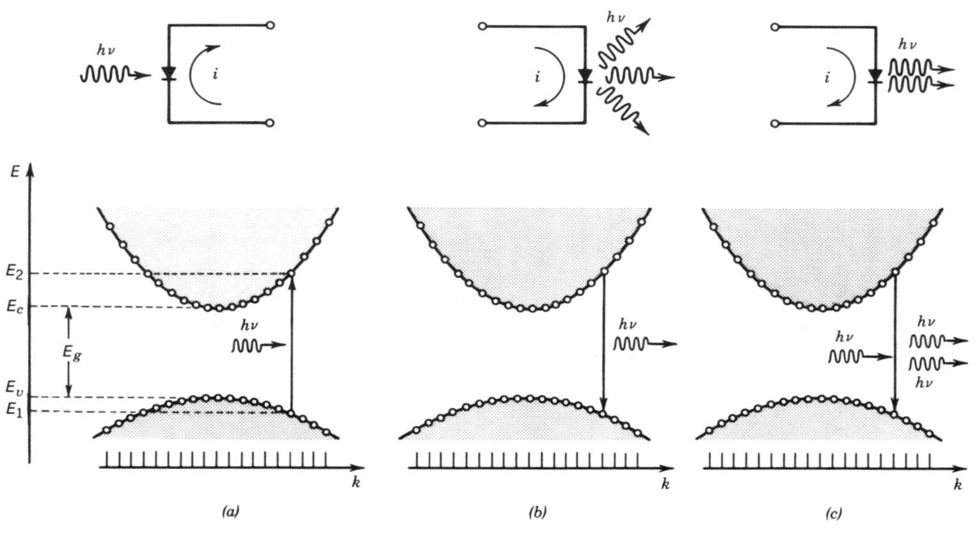

Figure 15.2-5 (*a*) The absorption of a photon results in the generation of an electron–hole pair. This process is used in the photodetection of light. (*b*) The recombination of an electron–hole pair results in the spontaneous emission of a photon. Light-emitting diodes (LEDs) operate on this basis. (*c*) Electron–hole recombination can be stimulated by a photon. The result is the induced emission of an identical photon. This is the underlying process responsible for the operation of semiconductor injection lasers.

increases the conductivity of the material. The material behaves as a photoconductor with a conductivity proportional to the photon flux. This effect is used to detect light, as discussed in Chap. 17.

Electron deexcitation from the conduction to the valence band (electron–hole recombination) may result in the spontaneous emission of a photon of energy $h\nu > E_g$ [Fig. 15.2-5(b)], or in the stimulated emission of a photon (see Sec. 12.2), provided that a photon of energy $h\nu > E_g$ is present [Fig. 15.2-5(c)]. Spontaneous emission is the underlying phenomenon on which the light-emitting diode is based, as will be seen in Sec. 16.1. Stimulated emission is responsible for the operation of semiconductor amplifiers and lasers, as will be seen in Secs. 16.2 and 16.3.

Conditions for Absorption and Emission

- *Conservation of Energy.* The absorption or emission of a photon of energy $h\nu$ requires that the energies of the two states involved in the interaction (E_1 and E_2 in the valence band and conduction band, respectively) be separated by $h\nu$. Thus, for photon emission to occur by electron–hole recombination, for example, an electron occupying an energy level E_2 must interact with a hole occupying an energy level E_1, such that energy is conserved, i.e.,

$$E_2 - E_1 = h\nu. \tag{15.2-2}$$

- *Conservation of Momentum.* Momentum must also be conserved in the process of photon emission/absorption, so that $p_2 - p_1 = h\nu/c = h/\lambda$, or $k_2 - k_1 = 2\pi/\lambda$. The photon-momentum magnitude h/λ is, however, very small in comparison with the range of values that electrons and holes can assume. The semiconductor E–k diagram extends to values of k of the order $2\pi/a$, where the lattice constant a is much smaller than the wavelength λ, so that $2\pi/\lambda \ll 2\pi/a$. The momenta of the electron and the hole involved in interaction with the photon are therefore roughly equal. This condition, $k_2 \approx k_1$, is called the **k-selection rule**. Transitions that obey this rule are represented in the E–k diagram (Fig. 15.2-5) by vertical lines, indicating that the change in k is negligible on the scale of the diagram.
- *Energies and Momenta of the Electron and Hole with Which a Photon Interacts.* As is apparent from Fig. 15.2-5, conservation of energy and momentum require that a photon of frequency ν interact with electrons and holes of specific energies and momentum determined by the semiconductor E–k relation. Using (15.1-1) and (15.1-2) to approximate this relation for a direct-gap semiconductor by two parabolas, and writing $E_c - E_v = E_g$, (15.2-2) may be written in the form

$$E_2 - E_1 = \frac{\hbar^2 k^2}{2m_v} + E_g + \frac{\hbar^2 k^2}{2m_c} = h\nu, \tag{15.2-3}$$

from which

$$k^2 = \frac{2m_r}{\hbar^2}(h\nu - E_g), \tag{15.2-4}$$

where

$$\frac{1}{m_r} = \frac{1}{m_v} + \frac{1}{m_c}. \tag{15.2-5}$$

Substituting (15.2-4) into (15.1-1), the energy levels E_1 and E_2 with which the photon interacts are therefore

$$E_2 = E_c + \frac{m_r}{m_c}(h\nu - E_g) \tag{15.2-6}$$

$$E_1 = E_v - \frac{m_r}{m_v}(h\nu - E_g) = E_2 - h\nu. \tag{15.2-7}$$

Energies of Electron and Hole Interacting with a Photon $h\nu$

In the special case where $m_c = m_v$, we obtain $E_2 = E_c + \frac{1}{2}(h\nu - E_g)$, as required by symmetry.

- *Optical Joint Density of States.* We now determine the density of states $\varrho(\nu)$ with which a photon of energy $h\nu$ interacts under conditions of energy and momentum conservation in a direct-gap semiconductor. This quantity incorporates the density of states in both the conduction and valence bands and is called the optical joint density of states. The one-to-one correspondence between E_2 and ν, embodied in (15.2-6), permits us to readily relate $\varrho(\nu)$ to the density of states $\varrho_c(E_2)$ in the conduction band by use of the incremental relation $\varrho_c(E_2)\,dE_2 = \varrho(\nu)\,d\nu$, from which $\varrho(\nu) = (dE_2/d\nu)\varrho_c(E_2)$, so that

$$\varrho(\nu) = \frac{hm_r}{m_c}\varrho_c(E_2). \tag{15.2-8}$$

Using (15.1-4) and (15.2-6), we finally obtain the number of states per unit volume per unit frequency:

$$\varrho(\nu) = \frac{(2m_r)^{3/2}}{\pi\hbar^2}(h\nu - E_g)^{1/2}, \qquad h\nu \geq E_g, \tag{15.2-9}$$

Optical Joint Density of States

which is illustrated in Fig. 15.2-6. The one-to-one correspondence between E_1 and ν in (15.2-7), together with $\varrho_v(E_1)$ from (15.1-5), results in an expression for $\varrho(\nu)$ identical to (15.2-9).

- *Photon Emission Is Unlikely in an Indirect-Gap Semiconductor.* Radiative electron–hole recombination is unlikely in an indirect-gap semiconductor. This is because transitions from near the bottom of the conduction band to near the top of the valence band (where electrons and holes, respectively, are most likely to reside) requires an exchange of momentum that cannot be accommodated by the emitted photon. Momentum may be conserved, however, by the participation of

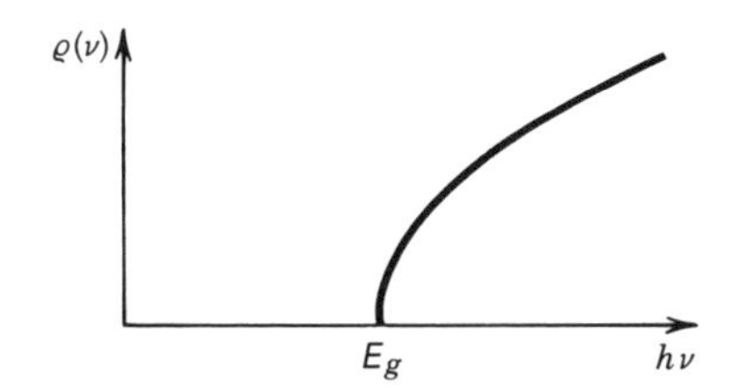

Figure 15.2-6 The density of states with which a photon of energy $h\nu$ interacts increases with $h\nu - E_g$ in accordance with a square-root law.

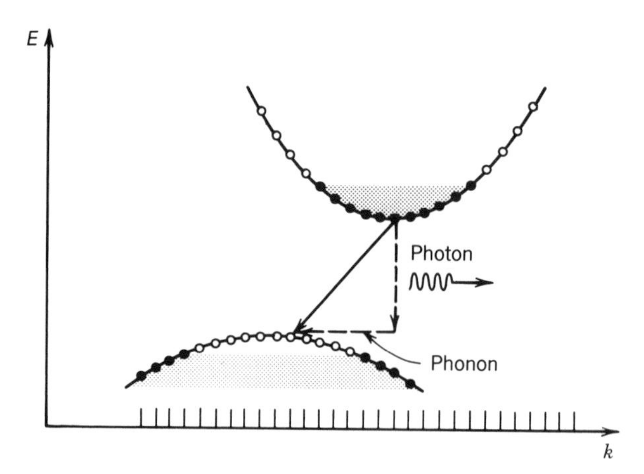

Figure 15.2-7 Photon emission in an indirect-gap semiconductor. The recombination of an electron near the bottom of the conduction band with a hole near the top of the valence band requires the exchange of energy *and* momentum. The energy may be carried off by a photon, but one or more phonons are required to conserve momentum. This type of multiparticle interaction is unlikely.

phonons in the interaction. Phonons can carry relatively large momenta but typically have small energies (≈ 0.01 to 0.1 eV; see Fig. 15.2-2), so their transitions appear horizontal on the E–k diagram (see Fig. 15.2-7). The net result is that momentum is conserved, but the k-selection rule is violated. Because phonon-assisted emission involves the participation of three bodies (electron, photon, and phonon), the probability of their occurrence is quite low. Thus Si, which is an indirect-gap semiconductor, has a substantially lower radiative recombination rate than does GaAs, which is a direct-gap semiconductor (see Table 15.1-5). Si is therefore not an efficient light emitter, whereas GaAs is.

- *Photon Absorption is Not Unlikely in an Indirect-Gap Semiconductor.* Although photon absorption also requires energy and momentum conservation in an indirect-gap semiconductor, this is readily achieved by means of a two-step process (Fig. 15.2-8). The electron is first excited to a high energy level within the

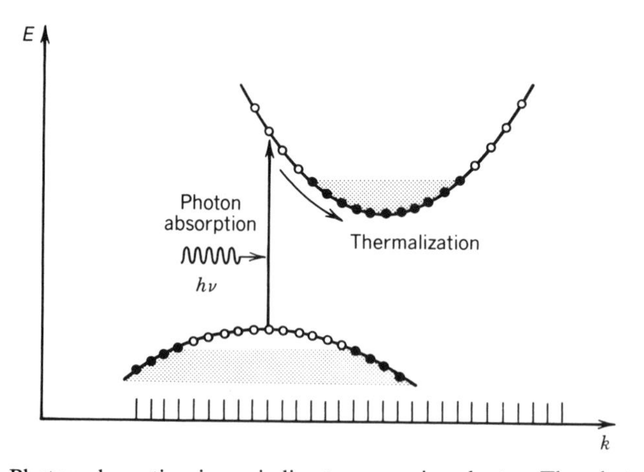

Figure 15.2-8 Photon absorption in an indirect-gap semiconductor. The photon generates an excited electron and a hole by a vertical transition; the carriers then undergo fast transitions to the bottom of the conduction band and top of the valence band, respectively, releasing their energy in the form of phonons. Since the process is sequential it is not unlikely.

conduction band by a vertical transition. It then quickly relaxes to the bottom of the conduction band by a process called thermalization in which its momentum is transferred to phonons. The generated hole behaves similarly. Since the process occurs sequentially, it does not require the simultaneous presence of three bodies and is thus not unlikely. Si is therefore an efficient photon detector, as is GaAs.

B. Rates of Absorption and Emission

We now proceed to determine the probability densities of a photon of energy $h\nu$ being emitted or absorbed by a semiconductor material in a direct band-to-band transition. Conservation of energy and momentum, in the form of (15.2-6), (15.2-7), and (15.2-4), determine the energies E_1 and E_2, and the momentum $\hbar k$, of the electrons and holes with which the photon may interact.

Three factors determine these probability densities: the occupancy probabilities, the transition probabilities, and the density of states. We consider these in turn.

Occupancy Probabilities

The occupancy conditions for photon emission and absorption by means of transitions between the discrete energy levels E_1 and E_2 are the following:

Emission condition: A conduction-band state of energy E_2 is filled (with an electron) *and* a valence-band state of energy E_1 is empty (i.e., filled with a hole).

Absorption condition: A conduction-band state of energy E_2 is empty *and* a valence-band state of energy E_1 is filled.

The probabilities that these occupancy conditions are satisfied for various values of E_1 and E_2 are determined from the appropriate Fermi functions $f_c(E)$ and $f_v(E)$ associated with the conduction and valence bands of a semiconductor in quasi-equilibrium. Thus the probability $f_e(\nu)$ that the emission condition is satisfied for a photon of energy $h\nu$ is the product of the probabilities that the upper state is filled and that the lower state is empty (these are independent events), i.e.,

$$f_e(\nu) = f_c(E_2)[1 - f_v(E_1)]. \tag{15.2-10}$$

E_1 and E_2 are related to ν by (15.2-6) and (15.2-7). Similarly, the probability $f_a(\nu)$ that the absorption condition is satisfied is

$$f_a(\nu) = [1 - f_c(E_2)] f_v(E_1). \tag{15.2-11}$$

EXERCISE 15.2-1

Requirement for the Photon Emission Rate to Exceed the Absorption Rate

(a) For a semiconductor in thermal equilibrium, show that $f_e(\nu)$ is always smaller than $f_a(\nu)$ so that the rate of photon emission cannot exceed the rate of photon absorption.

(b) For a semiconductor in quasi-equilibrium ($E_{fc} \neq E_{fv}$), with radiative transitions occurring between a conduction-band state of energy E_2 and a valence-band state of energy

E_1 with the same k, show that emission is more likely than absorption if the separation between the quasi-Fermi levels is larger than the photon energy, i.e., if

$$E_{fc} - E_{fv} > h\nu. \tag{15.2-12}$$

Condition for Net Emission

What does this condition imply about the locations of E_{fc} relative to E_c and E_{fv} relative to E_v?

Transition Probabilities

Satisfying the emission/absorption occupancy condition does not assure that the emission/absorption actually takes place. These processes are governed by the probabilistic laws of interaction between photons and atomic systems examined at length in Secs. 12.2A to C (see also Exercise 12.2-1). As they relate to semiconductors, these laws are generally expressed in terms of emission into (or absorption from) a narrow band of frequencies between ν and $\nu + d\nu$:

A radiative transition between two discrete energy levels E_1 and E_2 is characterized by a transition cross section $\sigma(\nu) = (\lambda^2/8\pi t_{sp})g(\nu)$, where ν is the frequency, t_{sp} is the spontaneous lifetime, and $g(\nu)$ is the lineshape function (which has linewidth $\Delta\nu$ centered about the transition frequency $\nu_0 = (E_2 - E_1)/h$ and has unity area). In semiconductors the radiative electron–hole recombination lifetime τ_r, which was discussed in Sec. 15.1D, plays the role of t_{sp} so that

$$\sigma(\nu) = \frac{\lambda^2}{8\pi\tau_r} g(\nu). \tag{15.2-13}$$

- If the occupancy condition for emission is satisfied, the probability density (per unit time) for the spontaneous emission of a photon into any of the available radiation modes in the narrow frequency band between ν and $\nu + d\nu$ is

$$P_{sp}(\nu)\, d\nu = \frac{1}{\tau_r} g(\nu)\, d\nu. \tag{15.2-14}$$

- If the occupancy condition for emission is satisfied *and* a mean photon-flux spectral density ϕ_ν (photons per unit time per unit area per unit frequency) at frequency ν is present, the probability density (per unit time) for the stimulated emission of one photon into the narrow frequency band between ν and $\nu + d\nu$ is

$$W_i(\nu)\, d\nu = \phi_\nu \sigma(\nu)\, d\nu = \phi_\nu \frac{\lambda^2}{8\pi\tau_r} g(\nu)\, d\nu. \tag{15.2-15}$$

- If the occupancy condition for absorption is satisfied *and* a mean photon-flux spectral density ϕ_ν at frequency ν is present, the probability density for the absorption of one photon from the narrow frequency band between ν and $\nu + d\nu$ is also given by (15.2-15).

Since each transition has a different central frequency ν_0, and since we are considering a collection of such transitions, we explicitly label the central frequency of the transition by writing $g(\nu)$ as $g_{\nu 0}(\nu)$. In semiconductors the homogeneously broadened lineshape function $g_{\nu 0}(\nu)$ associated with a pair of energy levels generally has its origin in electron–phonon collision broadening. It therefore typically exhibits a Lorentzian lineshape [see (12.2-27) and (12.2-30)] with width $\Delta\nu \approx 1/\pi T_2$, where the electron–phonon collision time T_2 is of the order of picoseconds. If $T_2 = 1$ ps, for example, then $\Delta\nu = 318$ GHz, corresponding to an energy width $h\,\Delta\nu \approx 1.3$ meV. The radiative lifetime broadening of the levels is negligible in comparison with collisional broadening.

Overall Emission and Absorption Transition Rates

For a pair of energy levels separated by $E_2 - E_1 = h\nu_0$, the rates of spontaneous emission, stimulated emission, and absorption of photons of energy $h\nu$ (photons per second per hertz per cm^3 of the semiconductor) at the frequency ν are obtained as follows. The appropriate transition probability density $P_{sp}(\nu)$ or $W_i(\nu)$ [as given in (15.2-14) or (15.2-15)] is multiplied by the appropriate occupation probability $f_e(\nu_0)$ or $f_a(\nu_0)$ [as given in (15.2-10) or (15.2-11)], and by the density of states that can interact with the photon $\varrho(\nu_0)$ [as given in (15.2-9)]. The overall transition rate for all allowed frequencies ν_0 is then calculated by integrating over ν_0.

The rate of spontaneous emission at frequency ν, for example, is therefore given by

$$r_{sp}(\nu) = \int [(1/\tau_r) g_{\nu 0}(\nu)] f_e(\nu_0) \varrho(\nu_0)\, d\nu_0.$$

When the collision-broadened width $\Delta\nu$ is substantially less than the width of the function $f_e(\nu_0)\varrho(\nu_0)$, which is the usual situation, $g_{\nu 0}(\nu)$ may be approximated by $\delta(\nu - \nu_0)$, whereupon the transition rate simplifies to $r_{sp}(\nu) = (1/\tau_r)\varrho(\nu) f_e(\nu)$. The rates of stimulated emission and absorption are obtained in similar fashion, so that the following formulas emerge:

$$r_{sp}(\nu) = \frac{1}{\tau_r} \varrho(\nu) f_e(\nu) \tag{15.2-16}$$

$$r_{st}(\nu) = \phi_\nu \frac{\lambda^2}{8\pi\tau_r} \varrho(\nu) f_e(\nu) \tag{15.2-17}$$

$$r_{ab}(\nu) = \phi_\nu \frac{\lambda^2}{8\pi\tau_r} \varrho(\nu) f_a(\nu). \tag{15.2-18}$$

Rates of Spontaneous Emission Stimulated Emission and Absorption

These equations, together with (15.2-9) to (15.2-11), permit the rates of spontaneous emission, stimulated emission, and absorption arising from direct band-to-band transitions (photons per second per hertz per cm^3) to be calculated in the presence of a mean photon-flux spectral density ϕ_ν (photons per second per cm^2 per hertz). The products $\varrho(\nu) f_e(\nu)$ and $\varrho(\nu) f_a(\nu)$ are similar to the products of the lineshape function and the atomic number densities in the upper and lower levels, $g(\nu)N_2$ and $g(\nu)N_1$, respectively, used in Chaps. 12 to 14 to study emission and absorption in atomic systems.

The determination of the occupancy probabilities $f_e(\nu)$ and $f_a(\nu)$ requires knowledge of the quasi-Fermi levels E_{fc} and E_{fv}. It is through the control of these two parameters (by the application of an external bias to a p–n junction, for example) that the emission and absorption rates are modified to produce semiconductor photonic devices that carry out different functions. Equation (15.2-16) is the basic result that describes the operation of the light-emitting diode (LED), a semiconductor photon source based on spontaneous emission (see Sec. 16.1). Equation (15.2-17) is applicable to semiconductor optical amplifiers and injection lasers, which operate on the basis of stimulated emission (see Secs. 16.2 and 16.3). Equation (15.2-18) is appropriate for semiconductor photon detectors which function by means of photon absorption (see Chap. 17).

Spontaneous Emission Spectral Density in Thermal Equilibrium

A semiconductor in thermal equilibrium has only a single Fermi function so that (15.2-10) becomes $f_e(\nu) = f(E_2)\,[1 - f(E_1)]$. If the Fermi level lies within the bandgap, away from the band edges by at least several times $k_B T$, use may be made of the exponential approximations to the Fermi functions, $f(E_2) \approx \exp[-(E_2 - E_f)/k_B T]$ and $1 - f(E_1) \approx \exp[-(E_f - E_1)/k_B T]$, whereupon $f_e(\nu) \approx \exp[-(E_2 - E_1)/k_B T]$, i.e.,

$$f_e(\nu) \approx \exp\left(-\frac{h\nu}{k_B T}\right). \tag{15.2-19}$$

Substituting (15.2-9) for $\varrho(\nu)$ and (15.2-19) for $f_e(\nu)$ into (15.2-16) therefore provides

$$r_{sp}(\nu) \approx D_0 (h\nu - E_g)^{1/2} \exp\left(-\frac{h\nu - E_g}{k_B T}\right), \qquad h\nu \geq E_g, \tag{15.2-20}$$

where

$$D_0 = \frac{(2m_r)^{3/2}}{\pi \hbar^2 \tau_r} \exp\left(-\frac{E_g}{k_B T}\right) \tag{15.2-21}$$

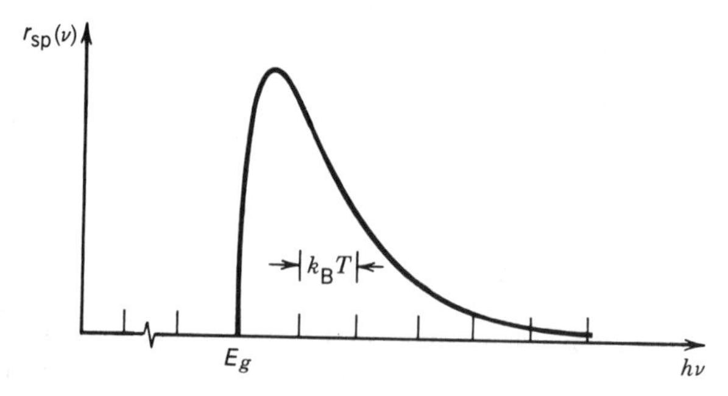

Figure 15.2-9 Spectral density of the direct band-to-band spontaneous emission rate $r_{sp}(\nu)$ (photons per second per hertz per cm^3) from a semiconductor in thermal equilibrium as a function of $h\nu$. The spectrum has a low-frequency cutoff at $\nu = E_g/h$ and extends over a width of approximately $2k_B T/h$.

is a parameter that increases with temperature at an exponential rate. The spontaneous emission rate, which is plotted versus $h\nu$ in Fig. 15.2-9, takes the form of two factors: a power-law increasing function of $h\nu - E_g$ arising from the density of states and an exponentially decreasing function of $h\nu - E_g$ arising from the Fermi function.

The spontaneous emission rate can be increased by increasing $f_e(\nu)$. In accordance with (15.2-10), this can be achieved by purposely causing the material to depart from thermal equilibrium in such a way that $f_c(E_2)$ is made large and $f_v(E_1)$ is made small. This assures an abundance of *both* electrons and holes, which is the desired condition for the operation of an LED, as discussed in Sec. 16.1.

Gain Coefficient in Quasi-Equilibrium

The net gain coefficient $\gamma_0(\nu)$ corresponding to the rates of stimulated emission and absorption in (15.2-17) and (15.2-18) is determined by taking a cylinder of unit area and incremental length dz and assuming that a mean photon-flux spectral density is directed along its axis (as shown in Fig. 13.1-1). If $\phi_\nu(z)$ and $\phi_\nu(z) + d\phi_\nu(z)$ are the mean photon-flux spectral densities entering and leaving the cylinder, respectively, $d\phi_\nu(z)$ must be the mean photon-flux spectral density emitted from within the cylinder. The incremental number of photons, per unit time per unit frequency per unit area, is simply the number of photons gained, per unit time per unit frequency per unit volume $[r_{\text{st}}(\nu) - r_{\text{ab}}(\nu)]$ multiplied by the thickness of the cylinder dz, i.e., $d\phi_\nu(z) = [r_{\text{st}}(\nu) - r_{\text{ab}}(\nu)]\, dz$. Substituting from (15.2-17) and (15.2-18), we obtain

$$\frac{d\phi_\nu(z)}{dz} = \frac{\lambda^2}{8\pi\tau_r}\varrho(\nu)[f_e(\nu) - f_a(\nu)]\phi_\nu(z) = \gamma_0(\nu)\phi_\nu(z). \quad (15.2\text{-}22)$$

The net gain coefficient is therefore

$$\gamma_0(\nu) = \frac{\lambda^2}{8\pi\tau_r}\varrho(\nu) f_g(\nu), \quad (15.2\text{-}23)$$

Gain Coefficient

where the Fermi inversion factor is given by

$$f_g(\nu) \equiv f_e(\nu) - f_a(\nu) = f_c(E_2) - f_v(E_1), \quad (15.2\text{-}24)$$

as may be seen from (15.2-10) and (15.2-11), with E_1 and E_2 related to ν by (15.2-6) and (15.2-7). Using (15.2-9), the gain coefficient may be cast in the form

$$\gamma_0(\nu) = D_1(h\nu - E_g)^{1/2} f_g(\nu), \qquad h\nu > E_g, \quad (15.2\text{-}25a)$$

with

$$D_1 = \frac{\sqrt{2}\, m_r^{3/2}\lambda^2}{h^2\tau_r}. \quad (15.2\text{-}25b)$$

The sign and spectral form of the Fermi inversion factor $f_g(\nu)$ are governed by the quasi-Fermi levels E_{fc} and E_{fv}, which, in turn, depend on the state of excitation of the carriers in the semiconductor. As shown in Exercise 15.2-1, this factor is positive (corresponding to a population inversion and net gain) only when $E_{fc} - E_{fv} > h\nu$. When the semiconductor is pumped to a sufficiently high level by means of an external energy source, this condition may be satisfied and net gain achieved, as we shall see in

Sec. 16.2. This is the physics underlying the operation of semiconductor optical amplifiers and injection lasers.

Absorption Coefficient in Thermal Equilibrium

A semiconductor in thermal equilibrium has only a single Fermi level $E_f = E_{fc} = E_{fv}$, so that

$$f_c(E) = f_v(E) = f(E) = \frac{1}{\exp[(E - E_f)/k_B T] + 1}. \qquad (15.2\text{-}26)$$

The factor $f_g(\nu) = f_c(E_2) - f_v(E_1) = f(E_2) - f(E_1) < 0$, and therefore the gain coefficient $\gamma_0(\nu)$ is always negative [since $E_2 > E_1$ and $f(E)$ decreases monotonically with E]. This is true whatever the location of the Fermi level E_f. Thus a semiconductor in thermal equilibrium, whether it be intrinsic or doped, always attenuates light. The attenuation (or absorption) coefficient, $\alpha(\nu) = -\gamma_0(\nu)$, is therefore

$$\boxed{\alpha(\nu) = D_1(h\nu - E_g)^{1/2}[f(E_1) - f(E_2)],} \qquad (15.2\text{-}27)$$

Absorption Coefficient

where E_1 and E_2 are given by (15.2-7) and (15.2-6), respectively, and D_1 is given by (15.2-25b).

If E_f lies within the bandgap but away from the band edges by an energy of at least several times $k_B T$, then $f(E_1) \approx 1$ and $f(E_2) \approx 0$ so that $[f(E_1) - f(E_2)] \approx 1$. In that case, the direct band-to-band contribution to the absorption coefficient is

$$\alpha(\nu) \approx \frac{\sqrt{2}\, c^2 m_r^{3/2}}{\tau_r} \frac{1}{(h\nu)^2} (h\nu - E_g)^{1/2}. \qquad (15.2\text{-}28)$$

As the temperature increases, $f(E_1) - f(E_2)$ decreases below unity and the absorption coefficient is reduced. Equation (15.2-28) is plotted in Fig. 15-2.10 for GaAs, using the following parameters: $n = 3.6$, $m_c = 0.07 m_0$, $m_v = 0.5 m_0$, $m_0 = 9.1 \times 10^{-31}$ kg, a

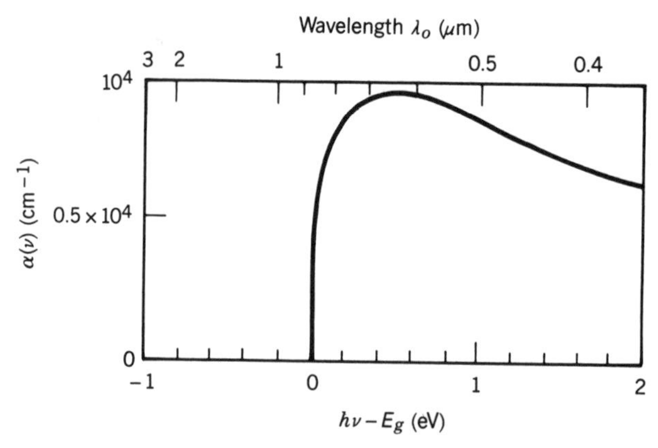

Figure 15.2-10 Calculated absorption coefficient $\alpha(\nu)$ (cm^{-1}) resulting from direct band-to-band transitions as a function of the photon energy $h\nu$ (eV) and wavelength λ_o (μm) for GaAs. This should be compared with the experimental result shown in Fig. 15.2-3, which includes all absorption mechanisms.

doping level such that $\tau_r = 0.4$ ns (this differs from that given in Table 15.1-5 because of the difference in doping level), $E_g = 1.42$ eV, and a temperature such that $[f(E_1) - f(E_2)] \approx 1$.

EXERCISE 15.2-2

Wavelength of Maximum Band-to-Band Absorption. Use (15.2-28) to determine the (free-space) wavelength λ_p at which the absorption coefficient of a semiconductor in thermal equilibrium is maximum. Calculate the value of λ_p for GaAs. Note that this result applies only to absorption by direct band-to-band transitions.

C. Refractive Index

The ability to control the refractive index of a semiconductor is important in the design of many photonic devices, particularly those that make use of optical waveguides, integrated optics, and injection laser diodes. Semiconductor materials are dispersive, so that the refractive index is dependent on the wavelength. Indeed, it is related to the absorption coefficient $\alpha(\nu)$ inasmuch as the real and imaginary parts of the susceptibility must satisfy the Kramers–Kronig relations (see Sec. 5.5B and Sec. B.1 of Appendix B). The refractive index also depends on temperature and on doping level, as is clear from the curves in Fig. 15.2-11 for GaAs.

The refractive indices of selected elemental and binary semiconductors, under specific conditions and near the bandgap wavelength, are provided in Table 15.2-1.

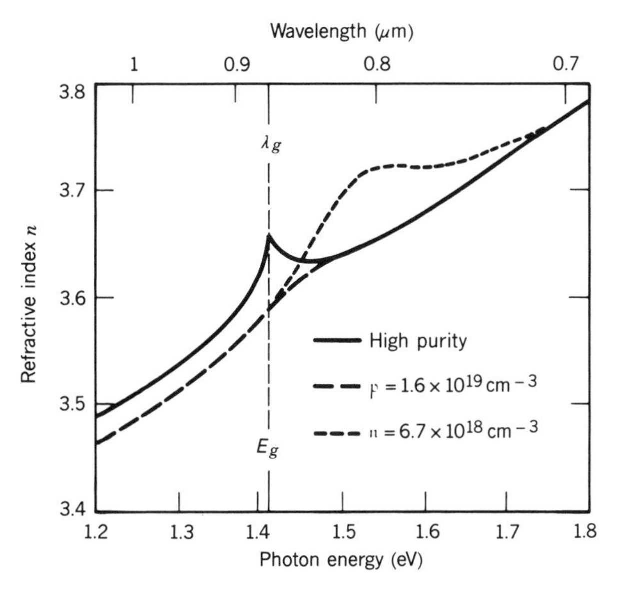

Figure 15.2-11 Refractive index for high-purity, *p*-type, and *n*-type GaAs at 300 K, as a function of photon energy (wavelength). The peak in the high-purity curve at the bandgap wavelength is associated with free excitons. (Adapted from H. C. Casey, Jr., and M. B. Panish, *Heterostructure Lasers*, part A, *Fundamental Principles*, Academic Press, New York, 1978.)

TABLE 15.2-1 Refractive Indices of Selected Semiconductor Materials at $T = 300$ K for Photon Energies Near the Bandgap Energy of the Material ($h\nu \approx E_g$)[a]

Material	Refractive Index
Elemental semiconductors	
Ge	4.0
Si	3.5
III–V binary semiconductors	
AlP	3.0
AlAs	3.2
AlSb	3.8
GaP	3.3
GaAs	3.6
GaSb	4.0
InP	3.5
InAs	3.8
InSb	4.2

[a]The refractive indices of ternary and quaternary semiconductors can be approximated by linear interpolation between the refractive indices of their components.

READING LIST

Books on Semiconductor Physics and Devices

B. G. Streetman, *Solid State Electronic Devices*, Prentice-Hall, Englewood Cliffs, NJ, 3rd ed. 1990.

S. Wang, *Fundamentals of Semiconductor Theory and Device Physics*, Prentice-Hall, Englewood Cliffs, NJ, 1989.

E. S. Yang, *Microelectronic Devices*, McGraw-Hill, New York, 1988.

K. Hess, *Advanced Theory of Semiconductor Devices*, Prentice-Hall, Englewood Cliffs, NJ, 1988.

C. Kittel, *Introduction to Solid State Physics*, Wiley, New York, 6th ed. 1986.

D. A. Fraser, *The Physics of Semiconductor Devices*, Clarendon Press, Oxford, 4th ed. 1986.

S. M. Sze, *Semiconductor Devices: Physics and Technology*, Wiley, New York, 1985.

K. Seeger, *Semiconductor Physics*, Springer-Verlag, Berlin, 2nd ed. 1982.

S. M. Sze, *Physics of Semiconductor Devices*, Wiley, New York, 2nd ed. 1981.

O. Madelung, *Introduction to Solid State Theory*, Springer-Verlag, Berlin, 1978.

R. A. Smith, *Semiconductors*, Cambridge University Press, New York, 2nd ed. 1978.

N. W. Ashcroft and N. D. Mermin, *Solid State Physics*, Holt, Rinehart and Winston, New York, 1976.

A. van der Ziel, *Solid State Physical Electronics*, Prentice-Hall, Englewood Cliffs, NJ, 3rd ed. 1976.

D. H. Navon, *Electronic Materials and Devices*, Houghton Mifflin, Boston, 1975.

W. A. Harrison, *Solid State Theory*, McGraw-Hill, New York, 1970.

C. A. Wert and R. M. Thomson, *Physics of Solids*, McGraw-Hill, New York, 1970.

J. M. Ziman, *Principles of the Theory of Solids*, Wiley, New York, 1968.

A. S. Grove, *Physics and Technology of Semiconductor Devices*, Wiley, New York, 1967.

Books on Optoelectronics

J. Wilson and J. F. B. Hawkes, *Optoelectronics*, Prentice-Hall, Englewood Cliffs, NJ, 2nd ed. 1989.

M. L. Cohen and J. R. Chelikowsky, *Electronic Structure and Optical Properties of Semiconductors*, Springer-Verlag, New York, 2nd ed. 1989.

J. Gowar, *Optical Communication Systems*, Prentice-Hall, Englewood Cliffs, NJ, 1984.

H. Kressel, ed., *Semiconductor Devices for Optical Communications*, Springer-Verlag, New York, 2nd ed. 1982.

T. S. Moss, G. J. Burrell, and B. Ellis, *Semiconductor Opto-electronics*, Wiley, New York, 1973.

J. I. Pankove, *Optical Processes in Semiconductors*, Prentice-Hall, Englewood Cliffs, NJ, 1971; Dover, New York, 1975.

Books on Heterostructures and Quantum-Well Structures

C. Weisbuch and B. Vinter, *Quantum Semiconductor Structures*, Academic Press, Orlando, FL, 1991.

F. Capasso, ed., *Physics of Quantum Electron Devices*, Springer-Verlag, New York, 1990.

R. Dingle, *Applications of Multiquantum Wells, Selective Doping, and Super-Lattices*, Academic Press, New York, 1987.

F. Capasso and G. Margaritondo, eds., *Heterojunction Band Discontinuities*, North-Holland, Amsterdam, 1987.

H. C. Casey, Jr., and M. B. Panish, *Heterostructure Lasers*, part A, *Fundamental Principles*, Academic Press, New York, 1978.

H. C. Casey, Jr., and M. B. Panish, *Heterostructure Lasers*, part B, *Materials and Operating Characteristics*, Academic Press, New York, 1978.

H. Kressel and J. K. Butler, *Semiconductor Lasers and Heterojunction LEDs*, Academic Press, New York, 1977.

A. G. Milnes and D. L. Feucht, *Heterojunctions and Metal–Semiconductor Junctions*, Academic Press, New York, 1972.

Special Journal Issues

Special issue on quantum-well heterostructures and superlattices, *IEEE Journal of Quantum Electronics*, vol. QE-24, no. 8, 1988.

Special issue on semiconductor quantum wells and superlattices: physics and applications, *IEEE Journal of Quantum Electronics*, vol. QE-22, no. 9, 1986.

Articles

E. Corcoran, Diminishing Dimensions, *Scientific American*, vol. 263, no. 5, pp. 122–131, 1990.

D. A. B. Miller, Optoelectronic Applications of Quantum Wells, *Optics and Photonics News*, vol. 1, no. 2, pp. 7–15, 1990.

S. Schmitt-Rink, D. S. Chemla, and D. A. B. Miller, Linear and Nonlinear Optical Properties of Semiconductor Quantum Wells, *Advances in Physics*, vol. 38, pp. 89–188, 1989.

W. D. Goodhue, Using Molecular-Beam Epitaxy to Fabricate Quantum-Well Devices, *Lincoln Laboratory Journal*, vol. 2, no. 2, pp. 183–206, 1989.

S. R. Forrest, Organic-on-Inorganic Semiconductor Heterojunctions: Building Block for the Next Generation of Optoelectronic Devices?, *IEEE Circuits and Devices Magazine*, vol. 5, no. 3, pp. 33–37, 41, 1989.

A. M. Glass, Optical Materials, *Science*, vol. 235, pp. 1003–1009, 1987.

L. Esaki, A Bird's-Eye View on the Evolution of Semiconductor Superlattices and Quantum Wells," *IEEE Journal of Quantum Electronics*, vol. QE-22, pp. 1611–1624, 1986.

D. S. Chemla, Quantum Wells for Photonics, *Physics Today*, vol. 38, no. 5, pp. 56–64, 1985.

PROBLEMS

15.1-1 **Fermi Level of an Intrinsic Semiconductor.** Given the expressions for the thermal equilibrium carrier concentrations in the conduction and valence bands [(15.1-9a) and (15.1-9b)]:
(a) Determine an expression for the Fermi level E_f of an intrinsic semiconductor and show that it falls exactly in the middle of the bandgap only when the effective mass of the electrons m_c is precisely equal to the effective mass of the holes m_v.
(b) Determine an expression for the Fermi level of a doped semiconductor as a function of the doping level and the Fermi level determined in part (a).

15.1-2 **Electron–Hole Recombination Under Strong Injection.** Consider electron–hole recombination under conditions of strong carrier-pair injection such that the recombination lifetime can be approximated by $\tau = 1/\mathfrak{r}\,\Delta n$, where $\mathfrak{r}$ is the recombination parameter of the material and Δn is the injection-generated excess carrier concentration. Assuming that the source of injection R is set to zero at $t = t_0$, find an analytic expression for $\Delta n(t)$, demonstrating that it exhibits power-law rather than exponential behavior.

*15.1-3 **Energy Levels in a GaAs/AlGaAs Quantum Well.** (a) Draw the energy-band diagram of a single-crystal multiquantum-well structure of GaAs/AlGaAs to scale on the energy axis when the AlGaAs has the composition $Al_{0.3}Ga_{0.7}As$. The bandgap of GaAs, E_g(GaAs), is 1.42 eV; the bandgap of AlGaAs increases above that of GaAs by $\approx$ 12.47 meV for each 1% Al increase in the composition. Because of the inherent characteristics of these two materials, the depth of the GaAs conduction-band quantum well is about 60% of the total conduction-plus-valence band quantum-well depths.
(b) Assume that a GaAs conduction-band well has depth as determined in part (a) above and precisely the same energy levels as the finite square well shown in Fig. 12.1-9(*b*), for which $(mV_0 d^2/2\hbar^2)^{1/2} = 4$, where V_0 is the depth of the well. Find the total width d of the GaAs conduction-band well. The effective mass of an electron in the conduction band of GaAs is $m_c \approx 0.07 m_0 = 0.64 \times 10^{-31}$ kg.

15.2-1 **Validity of the Approximation for Absorption/Emission Rates.** The derivation of the rate of spontaneous emission made use of the approximation $g_{\nu 0}(\nu) \approx \delta(\nu - \nu_0)$ in the course of evaluating the integral

$$r_{sp}(\nu) = \int \left[\frac{1}{\tau_r} g_{\nu 0}(\nu)\right] f_e(\nu_0) \varrho(\nu_0)\, d\nu_0.$$

(a) Demonstrate that this approximation is satisfactory for GaAs by plotting the functions $g_{\nu 0}(\nu)$, $f_e(\nu_0)$, and $\varrho(\nu_0)$ at $T = 300$ K and comparing their widths. GaAs is collisionally lifetime broadened with $T_2 \approx 1$ ps.
(b) Repeat part (a) for the rate of absorption in thermal equilibrium.

15.2-2 **Peak Spontaneous Emission Rate in Thermal Equilibrium.** (a) Determine the photon energy $h\nu_p$ at which the direct band-to-band spontaneous emission rate from a semiconductor material in thermal equilibrium achieves its maximum value when the Fermi level lies within the bandgap and away from the band edges by at least several times $k_B T$.
(b) Show that this peak rate (photons per second per hertz per cm^3) is given by

$$r_{sp}(\nu_p) = \frac{D_0}{\sqrt{2e}} (k_B T)^{1/2} = \frac{2(m_r)^{3/2}}{\sqrt{e}\,\pi \hbar^2 \tau_r} (k_B T)^{1/2} \exp\left(-\frac{E_g}{k_B T}\right).$$

(c) What is the effect of doping on this result?
(d) Assuming that $\tau_r = 0.4$ ns, $m_c = 0.07m_0$, $m_v = 0.5m_0$, and $E_g = 1.42$ eV, find the peak rate in GaAs at $T = 300$ K.

15.2-3 **Radiative Recombination Rate in Thermal Equilibrium.** (a) Show that the direct band-to-band spontaneous emission rate integrated over all emission frequencies (photons per second per cm^3) is given by

$$\int_0^{\infty} r_{sp}(\nu)\, d\nu = D_0 \frac{\sqrt{\pi}}{2h}(k_B T)^{3/2} = \frac{(m_r)^{3/2}}{\sqrt{2}\,\pi^{3/2}\hbar^3 \tau_r}(k_B T)^{3/2} \exp\left(-\frac{E_g}{k_B T}\right),$$

provided that the Fermi level is within the semiconductor energy gap and away from the band edges. [*Note*: $\int_0^{\infty} x^{1/2} e^{-\mu x}\, dx = (\sqrt{\pi}/2)\mu^{-3/2}$.]
(b) Compare this with the approximate integrated rate obtained by multiplying the peak rate obtained in Problem 15.2-2 by the approximate frequency width $2k_B T/h$ shown in Fig. 15.2-9.
(c) Using (15.1-10b), set the phenomenological equilibrium radiative recombination rate $\mathfrak{r}_r np = \mathfrak{r}_r n_i^2$ (photons per second per cm^3) introduced in Sec. 15.1D equal to the direct band-to-band result derived in (a) to obtain the expression for the radiative recombination rate

$$\mathfrak{r}_r = \frac{\sqrt{2}\,\pi^{3/2}\hbar^3}{(m_c + m_v)^{3/2}} \frac{1}{(k_B T)^{3/2}\tau_r}.$$

(d) Use the result in (c) to find the value of $\mathfrak{r}_r$ for GaAs at $T = 300$ K using $m_c = 0.07m_0$, $m_v = 0.5m_0$, and $\tau_r = 0.4$ ns. Compare this with the value provided in Table 15.1-5 on page 563 ($\mathfrak{r}_r \approx 10^{-10}$ cm^3/s).

CHAPTER

16

SEMICONDUCTOR PHOTON SOURCES

16.1 LIGHT-EMITTING DIODES
- A. Injection Electroluminescence
- B. LED Characteristics

16.2 SEMICONDUCTOR LASER AMPLIFIERS
- A. Gain
- B. Pumping
- C. Heterostructures

16.3 SEMICONDUCTOR INJECTION LASERS
- A. Amplification, Feedback, and Oscillation
- B. Power
- C. Spectral Distribution
- D. Spatial Distribution
- E. Mode Selection
- F. Characteristics of Typical Lasers
- *G. Quantum-Well Lasers

The operation of semiconductor injection lasers was reported nearly simultaneously in 1962 by independent research teams from **General Electric** Corporation, **IBM** Corporation, and **Lincoln Laboratory** of the Massachusetts Institute of Technology.

Light can be emitted from a semiconductor material as a result of electron–hole recombination. However, materials capable of emitting such light do not glow at room temperature because the concentrations of thermally excited electrons and holes are too low to produce discernible radiation. On the other hand, an external source of energy can be used to excite electron–hole pairs in sufficient numbers such that they produce large amounts of spontaneous recombination radiation, causing the material to glow or luminesce. A convenient way of achieving this is to forward bias a p–n junction, which has the effect of injecting electrons and holes into the same region of space; the resulting recombination radiation is then called **injection electroluminescence**.

A light-emitting diode (LED) is a forward-biased p–n junction fabricated from a direct-gap semiconductor material that emits light via injection electroluminescence [Fig. 16.0-1(a)]. If the forward voltage is increased beyond a certain value, the number of electrons and holes in the junction region can become sufficiently large so that a population inversion is achieved, whereupon stimulated emission (viz., emission induced by the presence of photons) becomes more prevalent than absorption. The junction may then be used as a diode laser amplifier [Fig. 16.0-1(b)] or, with appropriate feedback, as an injection laser diode [Fig. 16.0-1(c)].

Semiconductor photon sources, in the form of both LEDs and injection lasers, serve as highly efficient electronic-to-photonic transducers. They are convenient because they are readily modulated by controlling the injected current. Their small size, high efficiency, high reliability, and compatibility with electronic systems are important factors in their successful use in many applications. These include lamp indicators;

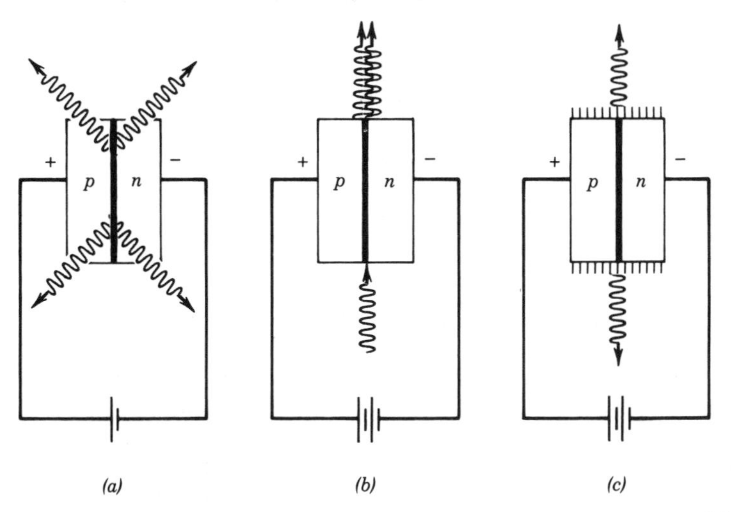

Figure 16.0-1 A forward-biased semiconductor p–n junction diode operated as (a) an LED, (b) a semiconductor optical amplifier, and (c) a semiconductor injection laser.

display devices; scanning, reading, and printing systems; fiber-optic communication systems; and optical data storage systems such as compact-disc players.

This chapter is devoted to the study of the light-emitting diode (Sec. 16.1), the semiconductor laser amplifier (Sec. 16.2), and the semiconductor injection laser (Sec. 16.3). Our treatment draws on the material contained in Chap. 15. The analysis of semiconductor laser amplification and oscillation is closely related to that developed in Chaps. 13 and 14.

16.1 LIGHT-EMITTING DIODES

A. Injection Electroluminescence

Electroluminescence in Thermal Equilibrium

Electron–hole radiative recombination results in the emission of light from a semiconductor material. At room temperature the concentration of thermally excited electrons and holes is so small, however, that the generated photon flux is very small.

EXAMPLE 16.1-1. ***Photon Emission from GaAs in Thermal Equilibrium.*** At room temperature, the intrinsic concentration of electrons and holes in GaAs is $n_i \approx 1.8 \times 10^6$ cm^{-3} (see Table 15.1-4). Since the radiative electron–hole recombination parameter $\mathfrak{r}_r \approx 10^{-10}$ cm^3/s (as specified in Table 15.1-5 for certain conditions), the electroluminescence rate $\mathfrak{r}_r np = \mathfrak{r}_r n_i^2 \approx 324$ photons/cm^3-s, as discussed in Sec. 15.1D. Using the bandgap energy for GaAs, $E_g = 1.42$ eV $= 1.42 \times 1.6 \times 10^{-19}$ J, this emission rate corresponds to an optical power density $= 324 \times 1.42 \times 1.6 \times 10^{-19} \approx 7.4 \times 10^{-17}$ W/cm^3. A 2-μm layer of GaAs therefore produces an intensity $I \approx 1.5 \times 10^{-20}$ W/cm^2, which is negligible. Light emitted from a layer of GaAs thicker than about 2 μm suffers reabsorption.

If thermal equilibrium conditions are maintained, this intensity cannot be appreciably increased (or decreased) by doping the material. In accordance with the law of mass action provided in (15.1-12), the product np is fixed at n_i^2 if the material is not too heavily doped so that the recombination rate $\mathfrak{r}_r np = \mathfrak{r}_r n_i^2$ depends on the doping level only through $\mathfrak{r}_r$. An abundance of electrons *and* holes is required for a large recombination rate; in an *n*-type semiconductor n is large but p is small, whereas the converse is true in a *p*-type semiconductor.

Electroluminescence in the Presence of Carrier Injection

The photon emission rate can be appreciably increased by using external means to produce excess electron–hole pairs in the material. This may be accomplished, for example, by illuminating the material with light, but it is typically achieved by forward biasing a *p–n* junction diode, which serves to inject carrier pairs into the junction region. This process is illustrated in Fig. 15.1-17 and will be explained further in Sec. 16.1B. The photon emission rate may be calculated from the electron–hole pair injection rate R (pairs/cm^3-s), where R plays the role of the laser pumping rate (see Sec. 13.2). The photon flux Φ (photons per second), generated within a volume V of

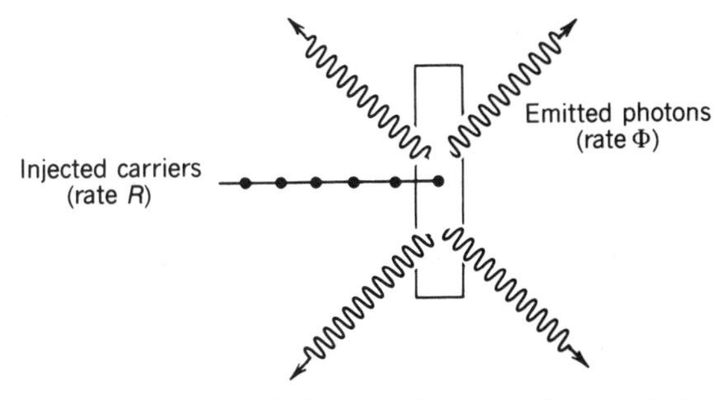

Figure 16.1-1 Spontaneous photon emission resulting from electron–hole radiative recombination, as might occur in a forward-biased p–n junction.

the semiconductor material, is directly proportional to the carrier-pair injection rate (see Fig. 16.1-1).

Denoting the equilibrium concentrations of electrons and holes in the absence of pumping as n_0 and p_0, respectively, we use $n = n_0 + \Delta n$ and $p = p_0 + \Delta p$ to represent the steady-state carrier concentrations in the presence of pumping (see Sec. 15.1D). The excess electron concentration Δn is precisely equal to the excess hole concentration Δp because electrons and holes are produced in pairs. It is assumed that the excess electron–hole pairs recombine at the rate $1/\tau$, where τ is the overall (radiative and nonradiative) electron–hole recombination time. Under steady-state conditions, the generation (pumping) rate must precisely balance the recombination (decay) rate, so that $R = \Delta n/\tau$. Thus the steady-state excess-carrier concentration is proportional to the pumping rate, i.e.,

$$\Delta n = R\tau. \tag{16.1-1}$$

For carrier injection rates that are sufficiently low, as explained in Sec. 15.1D, we have $\tau \approx 1/\mathfrak{r}(n_0 + p_0)$, where $\mathfrak{r}$ is the (radiative and nonradiative) recombination parameter, so that $R \approx \mathfrak{r}\Delta n(n_0 + p_0)$.

Only radiative recombinations generate photons, however, and the internal quantum efficiency $\eta_i = \mathfrak{r}_r/\mathfrak{r} = \tau/\tau_r$, defined in (15.1-20) and (15.1-22), accounts for the fact that only a fraction of the recombinations are radiative in nature. The injection of RV carrier pairs per second therefore leads to the generation of a photon flux $\Phi = \eta_i RV$ photons/s, i.e.,

$$\boxed{\Phi = \eta_i RV = \eta_i \frac{V\Delta n}{\tau} = \frac{V\Delta n}{\tau_r}.} \tag{16.1-2}$$

The internal photon flux Φ is proportional to the carrier-pair injection rate R and therefore to the steady-state concentration of excess electron–hole pairs Δn.

The internal quantum efficiency η_i plays a crucial role in determining the performance of this electron-to-photon transducer. Direct-gap semiconductors are usually used to make LEDs (and injection lasers) because η_i is substantially larger than for indirect-gap semiconductors (e.g., $\eta_i \approx 0.5$ for GaAs, whereas $\eta_i \approx 10^{-5}$ for Si, as shown in Table 15.1-5). The internal quantum efficiency η_i depends on the doping, temperature, and defect concentration of the material.

EXAMPLE 16.1-2. ***Injection Electroluminescence Emission from GaAs.*** Under certain conditions, $\tau = 50$ ns and $\eta_i = 0.5$ for GaAs (see Table 15.1-5), so that a steady-state excess concentration of injected electron–hole pairs $\Delta n = 10^{17}\ \text{cm}^{-3}$ will give rise to a photon flux concentration $\eta_i \Delta n/\tau \approx 10^{24}$ photons/cm^3-s. This corresponds to an optical power density $\approx 2.3 \times 10^5\ \text{W/cm}^3$ for photons at the bandgap energy $E_g = 1.42$ eV. A 2-μm-thick slab of GaAs therefore produces an optical intensity of $\approx 46\ \text{W/cm}^2$, which is a factor of 10^{21} greater than the thermal equilibrium value calculated in Example 16.1-1. Under these conditions the power emitted from a device of area 200 μm $\times$ 10 μm is ≈ 0.9 mW.

Spectral Density of Electroluminescence Photons

The spectral density of injection electroluminescence light may be determined by using the direct band-to-band emission theory developed in Sec. 15.2. The rate of spontaneous emission $r_{sp}(\nu)$ (number of photons per second per hertz per unit volume), as provided in (15.2-16), is

$$r_{sp}(\nu) = \frac{1}{\tau_r}\varrho(\nu) f_e(\nu), \tag{16.1-3}$$

where τ_r is the radiative electron–hole recombination lifetime. The optical joint density of states for interaction with photons of frequency ν, as given in (15.2-9), is

$$\varrho(\nu) = \frac{(2m_r)^{3/2}}{\pi\hbar^2}(h\nu - E_g)^{1/2},$$

where m_r is related to the effective masses of the holes and electrons by $1/m_r = 1/m_v + 1/m_c$ [as given in (15.2-5)], and E_g is the bandgap energy. The emission condition [as given in (15.2-10)] provides

$$f_e(\nu) = f_c(E_2)[1 - f_v(E_1)], \tag{16.1-4}$$

which is the probability that a conduction-band state of energy

$$E_2 = E_c + \frac{m_r}{m_c}(h\nu - E_g) \tag{16.1-5}$$

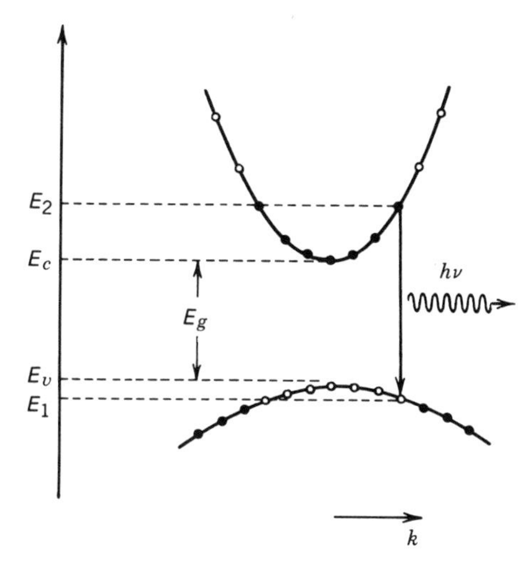

Figure 16.1-2 The spontaneous emission of a photon resulting from the recombination of an electron of energy E_2 with a hole of energy $E_1 = E_2 - h\nu$. The transition is represented by a vertical arrow because the momentum carried away by the photon, $h\nu/c$, is negligible on the scale of the figure.

is filled *and* a valence-band state of energy

$$E_1 = E_2 - h\nu \tag{16.1-6}$$

is empty, as provided in (15.2-6) and (15.2-7) and illustrated in Fig. 16.1-2. Equations (16.1-5) and (16.1-6) guarantee that energy and momentum are conserved. The Fermi functions $f_c(E) = 1/\{\exp[(E - E_{fc})/k_B T] + 1\}$ and $f_v(E) = 1/\{\exp[(E - E_{fv})/k_B T] + 1\}$ that appear in (16.1-4), with quasi-Fermi levels E_{fc} and E_{fv}, apply to the conduction and valence bands, respectively, under conditions of quasi-equilibrium.

The semiconductor parameters E_g, τ_r, m_v and m_c, and the temperature T determine the spectral distribution $r_{sp}(\nu)$, given the quasi-Fermi levels E_{fc} and E_{fv}. These, in turn, are determined from the concentrations of electrons and holes given in (15.1-7) and (15.1-8),

$$\int_{E_c}^{\infty} \varrho_c(E) f_c(E)\, dE = n = n_0 + \Delta n; \qquad \int_{-\infty}^{E_v} \varrho_v(E)[1 - f_v(E)]\, dE = p = p_0 + \Delta n. \tag{16.1-7}$$

The densities of states near the conduction- and valence-band edges are, respectively, as per (15.1-4) and (15.1-5),

$$\varrho_c(E) = \frac{(2m_c)^{3/2}}{2\pi^2\hbar^3}(E - E_c)^{1/2}; \qquad \varrho_v(E) = \frac{(2m_v)^{3/2}}{2\pi^2\hbar^3}(E_v - E)^{1/2},$$

where n_0 and p_0 are the concentrations of electrons and holes in thermal equilibrium (in the absence of injection), and $\Delta n = R\tau$ is the steady-state injected-carrier concentration. For sufficiently weak injection, such that the Fermi levels lie within the bandgap and away from the band edges by several $k_B T$, the Fermi functions may be approximated by their exponential tails. The spontaneous photon flux (integrated over all frequencies) is then obtained from the spectral density $r_{sp}(\nu)$ by

$$\Phi = V\int_0^\infty r_{sp}(\nu)\,d\nu = \frac{V(m_r)^{3/2}}{\sqrt{2}\,\pi^{3/2}\hbar^3\tau_r}(k_B T)^{3/2}\exp\left(\frac{E_{fc} - E_{fv} - E_g}{k_B T}\right),$$

as is readily extrapolated from Problem 15.2-3.

Increasing the pumping level R causes Δn to increase, which, in turn, moves E_{fc} toward (or further into) the conduction band, and E_{fv} toward (or further into) the valence band. This results in an increase in the probability $f_c(E_2)$ of finding the conduction-band state of energy E_2 filled with an electron, and the probability $1 - f_v(E_1)$ of finding the valence-band state of energy E_1 empty (filled with a hole). The net result is that the emission-condition probability $f_e(\nu) = f_c(E_2)[1 - f_v(E_1)]$ increases with R, thereby enhancing the spontaneous emission rate given in (16.1-3) and the spontaneous photon flux Φ given above.

EXERCISE 16.1-1

Quasi-Fermi Levels of a Pumped Semiconductor

(a) Under ideal conditions at $T = 0$ K, when there is no thermal electron–hole pair generation [see Fig. 16.1-3(*a*)], show that the quasi-Fermi levels are related to the concentrations of injected electron–hole pairs Δn by

$$E_{fc} = E_c + (3\pi^2)^{2/3}\frac{\hbar^2}{2m_c}(\Delta n)^{2/3} \tag{16.1-8a}$$

$$E_{fv} = E_v - (3\pi^2)^{2/3}\frac{\hbar^2}{2m_v}(\Delta n)^{2/3}, \tag{16.1-8b}$$

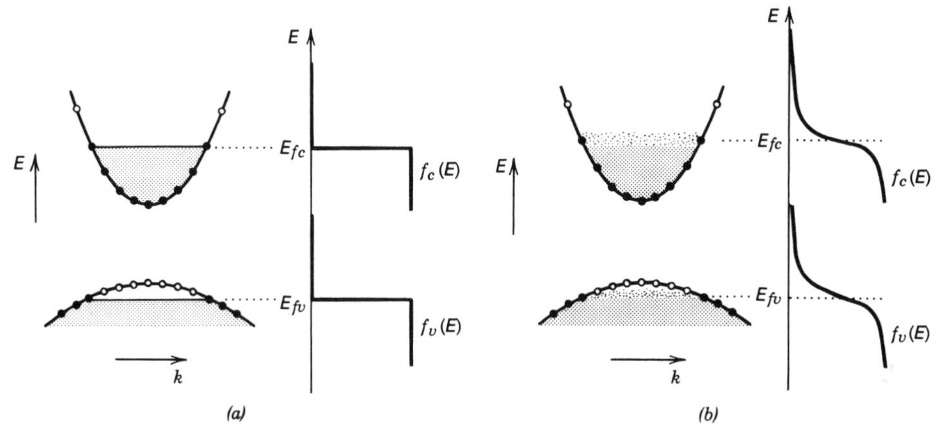

Figure 16.1-3 Energy bands and Fermi functions for a semiconductor in quasi-equilibrium (*a*) at $T = 0$ K, and (*b*) at $T > 0$ K.

so that

$$E_{fc} - E_{fv} = E_g + (3\pi^2)^{2/3} \frac{\hbar^2}{2m_r} (\Delta n)^{2/3}, \qquad (16.1\text{-}8c)$$

where $\Delta n \gg n_0, p_0$. Under these conditions all Δn electrons occupy the lowest allowed energy levels in the conduction band, and all Δp holes occupy the highest allowed levels in the valence band. Compare with the results of Exercise 15.1-2.

(b) Sketch the functions $f_e(\nu)$ and $r_{sp}(\nu)$ for two values of Δn. Given the effect of temperature on the Fermi functions, as illustrated in Fig. 16.1-3(*b*), determine the effect of increasing the temperature on $r_{sp}(\nu)$.

EXERCISE 16.1-2

Spectral Density of Injection Electroluminescence Under Weak Injection. For sufficiently weak injection, such that $E_c - E_{fc} \gg k_B T$ and $E_{fv} - E_v \gg k_B T$, the Fermi functions may be approximated by their exponential tails. Show that the luminescence rate can then be expressed as

$$r_{sp}(\nu) = D(h\nu - E_g)^{1/2} \exp\left(-\frac{h\nu - E_g}{k_B T}\right), \qquad h\nu \geq E_g, \qquad (16.1\text{-}9a)$$

where

$$D = \frac{(2m_r)^{3/2}}{\pi \hbar^2 \tau_r} \exp\left(\frac{E_{fc} - E_{fv} - E_g}{k_B T}\right) \qquad (16.1\text{-}9b)$$

is an exponentially increasing function of the separation between the quasi-Fermi levels $E_{fc} - E_{fv}$. The spectral density of the spontaneous emission rate is shown in Fig. 16.1-4; it has precisely the same shape as the thermal-equilibrium spectral density shown in Fig. 15.2-9, but its magnitude is increased by the factor $D/D_0 = \exp[(E_{fc} - E_{fv})/k_B T]$, which can be very large in the presence of injection. In thermal equilibrium $E_{fc} = E_{fv}$, so that (15.2-20) and (15.2-21) are recovered.

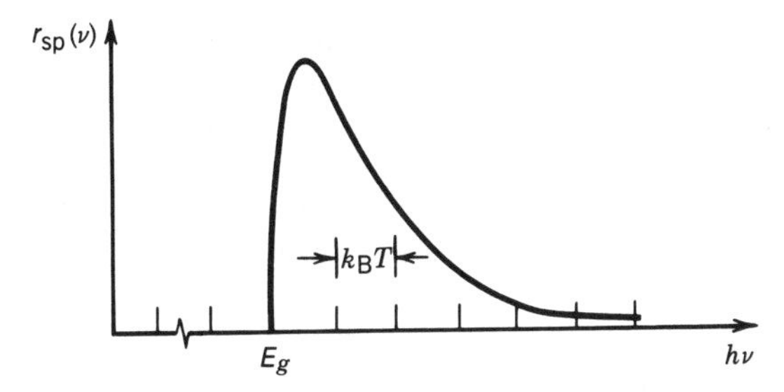

Figure 16.1-4 Spectral density of the direct band-to-band injection-electroluminescence rate $r_{sp}(\nu)$ (photons per second per hertz per cm^3), versus $h\nu$, from (16.1-9), under conditions of weak injection.

EXERCISE 16.1-3

Electroluminescence Spectral Linewidth

(a) Show that the spectral density of the emitted light described by (16.1-9) attains its peak value at a frequency ν_p determined by

$$h\nu_p = E_g + \frac{k_B T}{2}. \qquad (16.1\text{-}10)$$

Peak Frequency

(b) Show that the full width at half-maximum (FWHM) of the spectral density is

$$\Delta\nu \approx \frac{1.8 k_B T}{h}, \qquad (16.1\text{-}11)$$

Spectral Width (Hz)

(c) Show that this width corresponds to a wavelength spread $\Delta\lambda \approx 1.8\lambda_p^2 k_B T/hc$, where $\lambda_p = c/\nu_p$. For $k_B T$ expressed in eV and the wavelength expressed in μm, show that

$$\Delta\lambda \approx 1.45\lambda_p^2 k_B T. \qquad (16.1\text{-}12)$$

(d) Calculate $\Delta\nu$ and $\Delta\lambda$ at $T = 300$ K, for $\lambda_p = 0.8\ \mu$m and $\lambda_p = 1.6\ \mu$m.

B. LED Characteristics

As is clear from the foregoing discussion, the simultaneous availability of electrons and holes substantially enhances the flux of spontaneously emitted photons from a semiconductor. Electrons are abundant in n-type material, and holes are abundant in p-type material, but the generation of copious amounts of light requires that both electrons and holes be plentiful in the same region of space. This condition may be readily achieved in the junction region of a forward-biased p–n diode (see Sec. 15.1E). As shown in Fig. 16.1-5, forward biasing causes holes from the p side and electrons from the n side to be forced into the common junction region by the process of minority carrier injection, where they recombine and emit photons.

The light-emitting diode (LED) is a *forward-biased p–n junction* with a large radiative recombination rate arising from injected minority carriers. The semiconductor material is usually *direct-gap* to ensure high quantum efficiency. In this section we determine the output power, and spectral and spatial distributions of the light emitted from an LED and derive expressions for the efficiency, responsivity, and response time.

Internal Photon Flux

A schematic representation of a simple p–n junction diode is provided in Fig. 16.1-6. An injected dc current i leads to an increase in the steady-state carrier concentrations Δn, which, in turn, result in radiative recombination in the active-region volume V.

Since the total number of carriers per second passing through the junction region is i/e, where e is the magnitude of the electronic charge, the carrier injection (pumping)

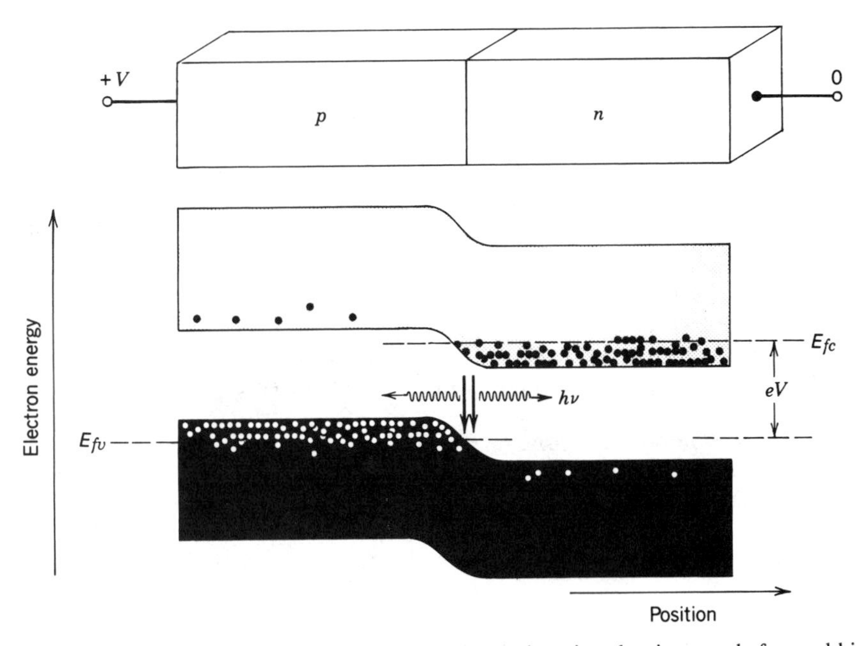

Figure 16.1-5 Energy diagram of a heavily doped p–n junction that is strongly forward biased by an applied voltage V. The dashed lines represent the quasi-Fermi levels, which are separated as a result of the bias. The simultaneous abundance of electrons and holes within the junction region results in strong electron–hole radiative recombination (injection electroluminescence).

rate (carriers per second per cm^3) is simply

$$R = \frac{i/e}{V}. \tag{16.1-13}$$

Equation (16.1-1) provides that $\Delta n = R\tau$, which results in a steady-state carrier concentration

$$\Delta n = \frac{(i/e)\tau}{V}. \tag{16.1-14}$$

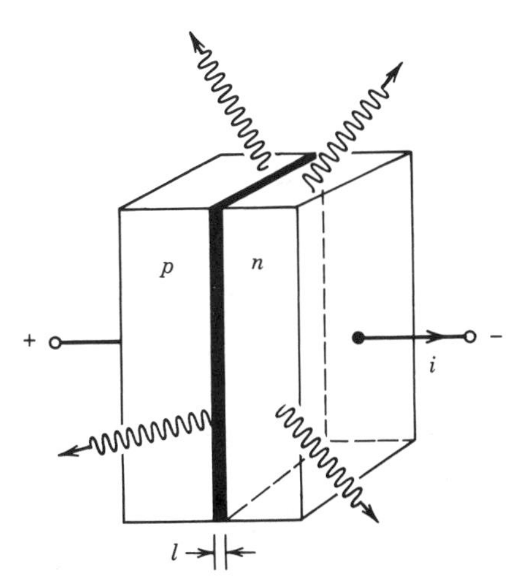

Figure 16.1-6 A simple forward-biased LED. The photons are emitted spontaneously from the junction region.

In accordance with (16.1-2), the generated photon flux Φ is then $\eta_i RV$, which, using (16.1-13), gives

$$\Phi = \eta_i \frac{i}{e}. \qquad (16.1\text{-}15)$$

Internal Photon Flux

This simple and intuitively appealing formula governs the production of photons by electrons in an LED: a fraction η_i of the injected electron flux i/e (electrons per second) is converted into photon flux. The **internal quantum efficiency** η_i is therefore simply the ratio of the generated photon flux to the injected electron flux.

Output Photon Flux and Efficiency

The photon flux generated in the junction is radiated uniformly in all directions; however, the flux that emerges from the device depends on the direction of emission. This is readily illustrated by considering the photon flux transmitted through the material along three possible ray directions, denoted A, B, and C in the geometry of Fig. 16.1-7:

- The photon flux traveling in the direction of ray A is attenuated by the factor

$$\eta_1 = \exp(-\alpha l_1), \qquad (16.1\text{-}16)$$

where α is the absorption coefficient of the n-type material and l_1 is the distance from the junction to the surface of the device. Furthermore, for normal incidence, reflection at the semiconductor–air boundary permits only a fraction of the light,

$$\eta_2 = 1 - \frac{(n-1)^2}{(n+1)^2} = \frac{4n}{(n+1)^2}, \qquad (16.1\text{-}17)$$

to be transmitted, where n is the refractive index of the semiconductor material [see Fresnel's equations (6.2-14)]. For GaAs, $n = 3.6$, so that $\eta_2 = 0.68$. The overall transmittance for the photon flux traveling in the direction of ray A is therefore $\eta_A = \eta_1 \eta_2$.

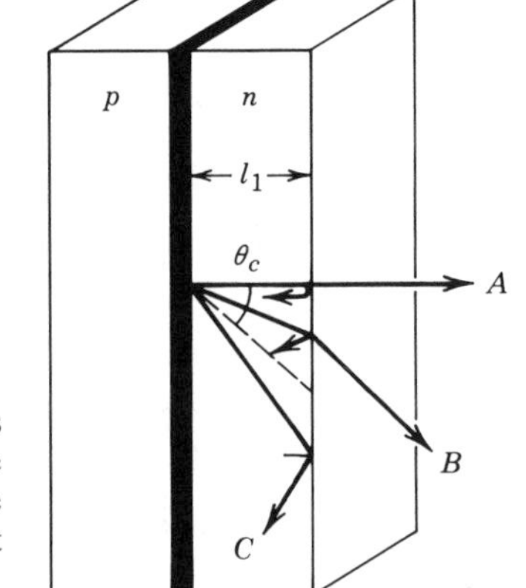

Figure 16.1-7 Not all light generated in an LED emerges from it. Ray A is partly reflected. Ray B suffers more reflection. Ray C lies outside the critical angle and therefore undergoes total internal reflection, so that, ideally, it cannot escape from the structure.

- The photon flux traveling in the direction of ray B has farther to travel and therefore suffers a larger absorption; it also has greater reflection losses. Thus $\eta_B < \eta_A$.
- The photon flux emitted along directions lying outside a cone of (critical) angle $\theta_c = \sin^{-1}(1/n)$, such as illustrated by ray C, suffer total internal reflection in an ideal material and are not transmitted at all [see (1.2-5)]. The fraction of emitted light lying within this cone is

$$\eta_3 = 1 - \cos\theta_c = 1 - \left(1 - \frac{1}{n^2}\right)^{1/2} \approx \frac{1}{2n^2}. \tag{16.1-18}$$

Thus, for $n = 3.6$, only 3.9% of the total generated photon flux can be transmitted. For a parallelepiped of refractive index $n > \sqrt{2}$, the ratio of isotropically generated light energy that can emerge, to the total generated light energy, is $3[1 - (1 - 1/n^2)^{1/2}]$, as shown in Exercise 1.2-6. However, in real LEDs, photons emitted outside the critical angle can be absorbed and re-emitted within this angle, so that in practice, η_3 may assume a value larger than that indicated in (16.1-18).

The output photon flux Φ_o is related to the internal photon flux by

$$\Phi_o = \eta_e \Phi = \eta_e \eta_i \frac{i}{e}, \tag{16.1-19}$$

where η_e is the overall transmission efficiency with which the internal photons can be extracted from the LED structure, and η_i relates the internal photon flux to the injected electron flux. A single quantum efficiency that accommodates both kinds of losses is the **external quantum efficiency** η_{ex},

$$\boxed{\eta_{ex} \equiv \eta_e \eta_i.} \tag{16.1-20}$$

External Quantum Efficiency

The output photon flux in (16.1-19) can therefore be written as

$$\boxed{\Phi_o = \eta_{ex} \frac{i}{e};} \tag{16.1-21}$$

External Photon Flux

η_{ex} is simply the ratio of the externally produced photon flux Φ_o to the injected electron flux. Because the pumping rate generally varies locally within the junction region, so does the generated photon flux.

The LED output optical power P_o is related to the output photon flux. Each photon has energy $h\nu$, so that

$$\boxed{P_o = h\nu\Phi_o = \eta_{ex} h\nu \frac{i}{e}.} \tag{16.1-22}$$

Output Power

Although η_i can be near unity for certain LEDs, η_{ex} generally falls well below unity, principally because of reabsorption of the light in the device and internal reflection at its boundaries. As a consequence, the external quantum efficiency of commonly encountered LEDs, such as those used in pocket calculators, is typically less than 1%.

Another measure of performance is the **overall quantum efficiency** η (also called the **power-conversion efficiency** or **wall-plug efficiency**), which is defined at the ratio of the emitted optical power P_o to the applied electrical power,

$$\eta \equiv \frac{P_o}{iV} = \eta_{ex}\frac{h\nu}{eV}, \tag{16.1-23}$$

where V is the voltage drop across the device. For $h\nu \approx eV$, as is the case for commonly encountered LEDs, it follows that $\eta \approx \eta_{ex}$.

Responsivity

The responsivity $\Re$ of an LED is defined as the ratio of the emitted optical power P_o to the injected current i, i.e., $\Re = P_o/i$. Using (16.1-22), we obtain

$$\Re = \frac{P_o}{i} = \frac{h\nu\Phi_o}{i} = \eta_{ex}\frac{h\nu}{e}. \tag{16.1-24}$$

The responsivity in W/A, when λ_o is expressed in μm, is then

$$\boxed{\Re = \eta_{ex}\frac{1.24}{\lambda_o}.} \tag{16.1-25}$$

LED Responsivity (W/A)
λ_o in μm

For example, if $\lambda_o = 1.24\ \mu$m, then $\Re = \eta_{ex}$ W/A; if η_{ex} were unity, the maximum optical power that could be produced by an injection current of 1 mA would be 1 mW. However, as indicated above, typical values of η_{ex} for LEDs are in the range of 1 to 5%, so that LED responsivities are in the vicinity of 10 to 50 μW/mA.

In accordance with (16.1-22), the LED output power P_o should be proportional to the injected current i. In practice, however, this relationship is valid only over a restricted range. For the particular device whose light–current characteristic is shown in Fig. 16.1-8, the emitted optical power is proportional to the injection (drive) current only when the latter is less than about 75 mA. In this range, the responsivity has a

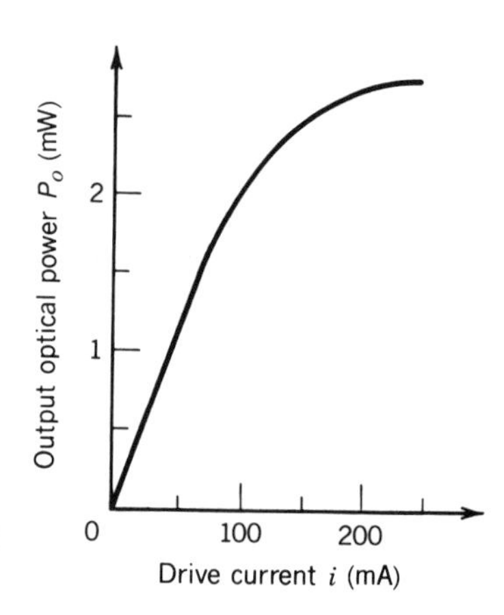

Figure 16.1-8 Optical power at the output of an actual LED versus injection (drive) current.

constant value of about 25 μW/mA, as determined from the slope of the curve. For larger drive currents, saturation causes the proportionality to fail; the responsivity is then no longer constant but rather declines with increasing drive current.

Spectral Distribution

The spectral density $r_{sp}(\nu)$ of light spontaneously emitted from a semiconductor in quasi-equilibrium has been determined, as a function of the concentration of injected carriers Δn, in Exercises 16.1-2 and 16.1-3. This theory is applicable to the electroluminescence light emitted from an LED in which quasi-equilibrium conditions are established by injecting current into a p–n junction.

Under conditions of weak pumping, such that the quasi-Fermi levels lie within the bandgap and are at least a few $k_B T$ away from the band edges, the spectral density achieves its peak value at the frequency $\nu_p = (E_g + k_B T/2)/h$ (see Exercise 16.1-3). In accordance with (16.1-11) and (16.1-12), the FWHM of the spectral density is $\Delta\nu \approx 1.8 k_B T/h$ ($\Delta\nu = 10$ THz for $T = 300$ K), which is independent of ν. The width expressed in terms of the wavelength does depend on λ,

$$\Delta\lambda \approx 1.45 \lambda_p^2 k_B T, \tag{16.1-26}$$

Spectral Width (μm)

where $k_B T$ is expressed in eV, the wavelength is expressed in μm, and $\lambda_p = c/\nu_p$.

The proportionality of $\Delta\lambda$ to λ_p^2 is apparent in Fig. 16.1-9, which illustrates the observed wavelength spectral densities for a number of LEDs that operate in the visible and near-infrared regions. If $\lambda_p = 1$ μm at $T = 300$ K, for example, (16.1-26) provides $\Delta\lambda \approx 36$ nm.

Materials

LEDs have been operated from the near ultraviolet to the infrared, as illustrated in Fig. 16.1-9. In the near infrared, many binary semiconductor materials serve as highly efficient LED materials because of their direct-band gap nature. Examples of III–V

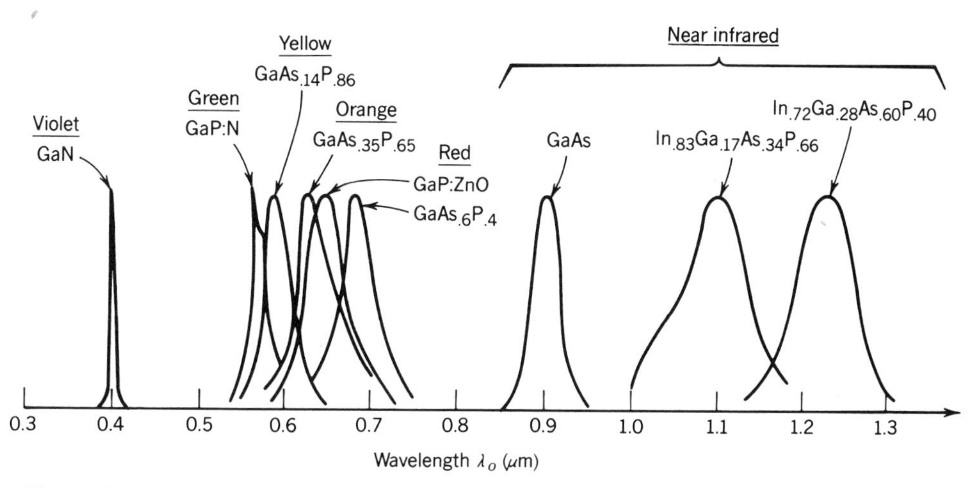

Figure 16.1-9 Spectral densities versus wavelength for semiconductor LEDs with different bandgaps. The peak intensities are normalized to the same value. The increasing spectral linewidth is a result of its proportionality to λ_p^2. (Adapted from S. M. Sze, *Physics of Semiconductor Devices*, Wiley, New York, 2nd ed. 1981.)

binary materials include (as shown in Table 15.1-3 and Fig. 15.1-5) GaAs ($\lambda_g = 0.87$ μm), GaSb (1.7 μm), InP (0.92 μm), InAs (3.5 μm), and InSb (7.3 μm). Ternary and quaternary compounds are also direct-gap over a wide range of compositions (see Fig. 15.1-5). These materials have the advantage that their emission wavelength can be compositionally tuned. Particularly important among the III–V compounds is ternary $Al_xGa_{1-x}As$ (0.75 to 0.87 μm) and quaternary $In_{1-x}Ga_xAs_{1-y}P_y$ (1.1 to 1.6 μm).

At short wavelengths (in the ultraviolet and most of the visible spectrum) materials such as GaN, GaP, and $GaAs_{1-x}P_x$ are typically used despite their low internal quantum efficiencies. These materials are often doped with elements that serve to enhance radiative recombination by acting as recombination centers. LEDs that emit blue light can also be made by using a phosphor to up-convert near-infrared photons from a GaAs LED (see Fig. 12.4-2).

Response Time

The response time of an LED is limited principally by the lifetime τ of the injected minority carriers that are responsible for radiative recombination. For a sufficiently small injection rate R, the injection/recombination process can be described by a first-order linear differential equation (see Sec. 15.1D), and therefore by the response to sinusoidal signals. An experimental determination of the highest frequency at which an LED can be effectively modulated is easily obtained by measuring the output light power in response to sinusoidal electric currents of different frequencies. If the injected current assumes the form $i = i_0 + i_1 \cos(\Omega t)$, where i_1 is sufficiently small so that the emitted optical power P varies linearly with the injected current, the emitted optical power behaves as $P = P_0 + P_1 \cos(\Omega t + \varphi)$.

The associated transfer function, which is defined as $\mathcal{H}(\Omega) = (P_1/i_1)\exp(j\varphi)$, assumes the form

$$\mathcal{H}(\Omega) = \frac{\Re}{1 + j\Omega\tau}, \tag{16.1-27}$$

which is characteristic of a resistor–capacitor circuit. The rise time of the LED is τ (seconds) and its 3-dB bandwidth is $B = 1/2\pi\tau$ (Hz). A larger bandwidth B is therefore attained by decreasing the rise time τ, which comprises contributions from both the radiative lifetime τ_r and the nonradiative lifetime τ_{nr} through the relation $1/\tau = 1/\tau_r + 1/\tau_{nr}$. However, reducing τ_{nr} results in an undesirable reduction of the internal quantum efficiency $\eta_i = \tau/\tau_r$. It may therefore be desirable to maximize the internal quantum efficiency–bandwidth product $\eta_i B = 1/2\pi\tau_r$ rather than maximizing the bandwidth alone. This requires a reduction of only the radiative lifetime τ_r, without a reduction of τ_{nr}, which may be achieved by careful choice of semiconductor material and doping level. Typical rise times of LEDs fall in the range 1 to 50 ns, corresponding to bandwidths as large as hundreds of MHz.

Device Structures

LEDs may be constructed either in surface-emitting or edge-emitting configurations (Fig. 16.1-10). The surface-emitting LED emits light from a face of the device that is parallel to the junction plane. Light emitted from the opposite face is absorbed by the substrate and lost or, preferably, reflected from a metallic contact (which is possible if a transparent substrate is used). The edge-emitting LED emits light from the edge of the junction region. The latter structure has usually been used for diode lasers as well, although surface-emitting laser diodes (SELDs) are being increasingly used. Surface-emitting LEDs are generally more efficient than edge-emitting LEDs. Heterostructure LEDs, with configurations such as those described in Sec. 16.2C, provide superior performance.

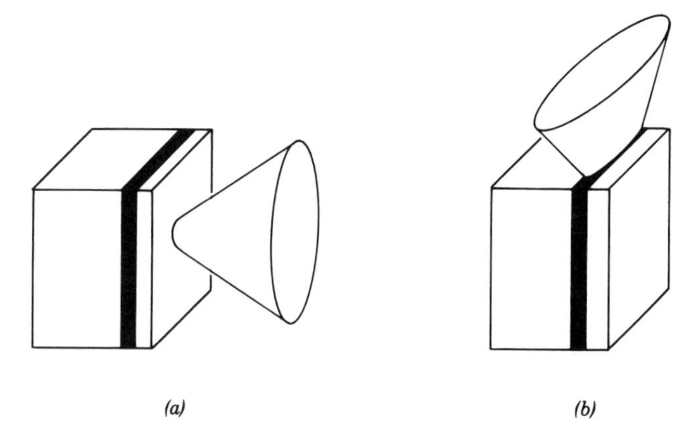

Figure 16.1-10 (*a*) Surface-emitting LED. (*b*) Edge-emitting LED.

Examples of surface-emitting LED structures are illustrated in Fig. 16.1-11. A flat-diode-configuration $GaAs_{1-x}P_x$ LED on a GaAs substrate is shown in Fig. 16.1-11(*a*). A layer of graded $GaAs_{1-y}P_y$, placed between the substrate and the *n*-type layer, reduces the lattice mismatch. The bandgap of GaAs is smaller than the photon energy of the emitted red light so that the radiation emitted toward the substrate is absorbed. Alternatively, transparent substrates such as GaP can be used in conjunction with a reflective contact to increase the external quantum efficiency. The Burrus-type LED, shown in Fig. 16.1-11(*b*), makes use of an etched well to permit the light to be collected directly from the junction region. This structure is particularly suitable for efficient coupling of the emitted light into an optical fiber, which may be brought into close proximity with the active region (see Fig. 22.1-5).

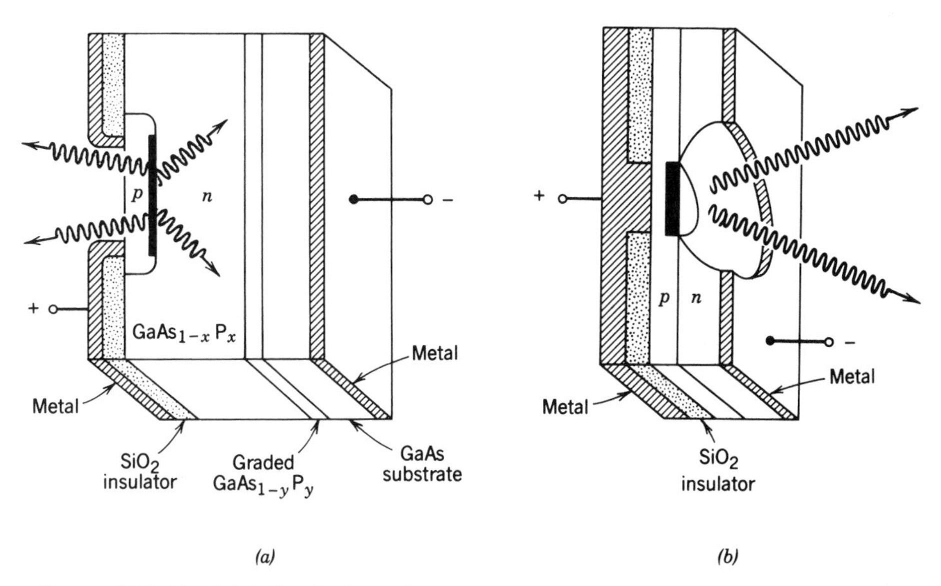

Figure 16.1-11 (*a*) A flat-diode-configuration $GaAs_{1-x}P_x$ LED. (*b*) A Burrus-type LED.

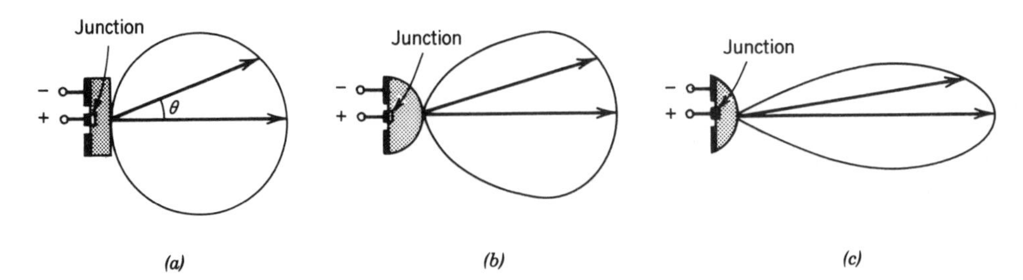

Figure 16.1-12 Radiation patterns of surface-emitting LEDs: (*a*) Lambertian pattern of a surface-emitting LED in the absence of a lens; (*b*) pattern of an LED with a hemispherical lens; (*c*) pattern of an LED with a parabolic lens.

Spatial Pattern of Emitted Light

The far-field radiation pattern from a surface-emitting LED is similar to that from a Lambertian radiator; the intensity varies as $\cos\theta$, where θ is the angle from the emission-plane normal. The intensity decreases to half its value at $\theta = 60°$. Epoxy lenses are often placed on the LED to reduce this angular spread. Differently shaped lenses alter the angular dependence of the emission pattern in specified ways as shown schematically in Fig. 16.1-12.

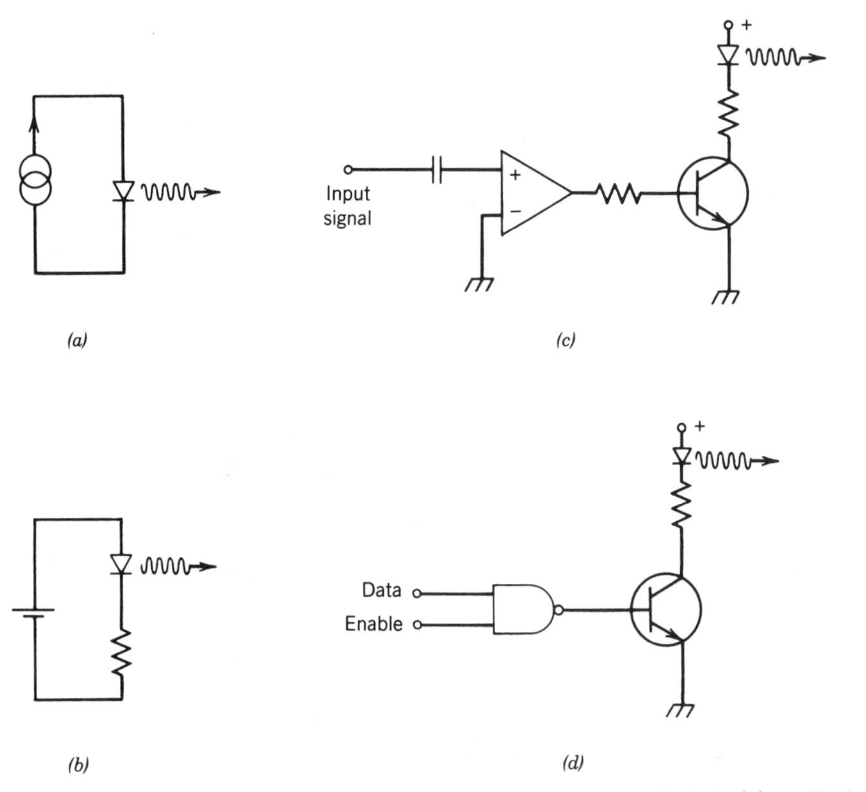

Figure 16.1-13 Various circuits can be used to drive an LED. These include (*a*) an ideal dc current source; (*b*) a dc current source provided by a constant-voltage source in series with a resistor; (*c*) transistor control of the current injected into the LED to provide analog modulation of the emitted light; (*d*) transistor switching of the current injected into the LED to provide digital modulation of the emitted light.

The radiation emitted from edge-emitting LEDs (and laser diodes) usually has a narrower radiation pattern. This pattern can often be well modeled by the function $\cos^s(\theta)$, where $s > 1$. If $s = 10$, for example, the intensity drops to half its value at $\theta \approx 21°$.

Electronic Circuitry

An LED is usually driven by a current source, as shown schematically in Fig. 16.1-13(*a*), for example by use of a constant-voltage source in series with a resistor, as illustrated in Fig. 16.1-13(*b*). The emitted light may be readily modulated (in either analog or digital format) simply by modulating the injected current. Two examples of such circuitry are the analog circuit shown in Fig. 16.1-13(*c*) and the digital circuit shown in Fig. 16.1-13(*d*). The performance of these circuits may be improved by adding bias current regulators, impedance matching circuitry, and nonlinear compensation circuitry. Furthermore, fluctuations in the intensity of the emitted light may be stabilized by the use of optical feedback in which the emitted light is monitored and used to control the injected current.

16.2 SEMICONDUCTOR LASER AMPLIFIERS

The principle underlying the operation of a semiconductor laser amplifier is the same as that for other laser amplifiers: the creation of a population inversion that renders stimulated emission more prevalent than absorption. The population inversion is usually achieved by electric current injection in a p–n junction diode; a forward bias voltage causes carrier pairs to be injected into the junction region, where they recombine by means of stimulated emission.

The theory of the semiconductor laser amplifier is somewhat more complex than that presented in Chap. 13 for other laser amplifiers, inasmuch as the transitions take place between bands of closely spaced energy levels rather than well-separated discrete levels. For purposes of comparison, nevertheless, the semiconductor laser amplifier may be viewed as a four-level laser system (see Fig. 13.2-6) in which the upper two levels lie in the conduction band and the lower two levels lie in the valence band.

The extension of the laser amplifier theory given in Chap. 13 to semiconductor structures has been provided in Chap. 15. In this section we use the results derived in Sec. 15.2 to obtain expressions for the gain and bandwidth of semiconductor laser amplifiers. We also review pumping schemes used for attaining a population inversion and briefly discuss semiconductor amplifier structures of current interest. The theoretical underpinnings of semiconductor laser amplifiers form the basis of injection laser operation, considered in Sec. 16.3.

Most semiconductor laser amplifiers fabricated to date are designed to operate in 1.3- to 1.55-μm lightwave communication systems as nonregenerative repeaters, optical preamplifiers, or narrowband electrically tunable amplifiers. In comparison with Er^{3+}:silica fiber amplifiers, semiconductor amplifiers have both advantages and disadvantages. They are smaller in size and are readily incorporated into optoelectronic integrated circuits. Their bandwidths can be as large as 10 THz, which is greater than that of fiber amplifiers. On the negative side, semiconductor amplifiers currently have greater insertion losses (typically 3 to 5 dB per facet) than fiber amplifiers. Furthermore, temperature instability, as well as polarization sensitivity, are difficult to overcome.

If a semiconductor laser amplifier is to be operated as a broadband single-pass device (i.e., as a traveling-wave amplifier), care must be taken to reduce the facet reflectances to very low values. Failure to do so would result in multiple reflections and a gain profile modulated by the resonator modes; this could also lead to oscillation, which, of course, obviates the possibility of controllable amplification. The response time is determined by complex carrier dynamics; the shortest value to date is $\approx$ 100 ps.

A. Gain

Light of frequency ν can interact with the carriers of a semiconductor material of bandgap energy E_g via band-to-band transitions, provided that $\nu > E_g/h$. The incident photons may be absorbed resulting in the generation of electron–hole pairs, or they may produce additional photons through stimulated electron–hole recombination radiation (see Fig. 16.2-1). When emission is more likely than absorption, net optical gain ensues and the material can serve as a coherent optical amplifier.

Expressions for the rate of photon absorption $r_{ab}(\nu)$ and the rate of stimulated emission $r_{st}(\nu)$ were provided in (15.2-18) and (15.2-17). These quantities depend on the photon-flux spectral density ϕ_ν, the quantum-mechanical strength of the transition for the particular material under consideration (which is implicit in the value of the electron–hole radiative recombination lifetime τ_r), the optical joint density of states $\varrho(\nu)$, and the occupancy probabilities for emission and absorption, $f_e(\nu)$ and $f_a(\nu)$.

The optical joint density of states $\varrho(\nu)$ is determined by the E–k relations for electrons and holes and by the conservation of energy and momentum. With the help of the parabolic approximation for the E–k relations near the conduction- and valence-band edges, it was shown in (15.2-6) and (15.2-7) that the energies of the electron and hole that interact with a photon of energy $h\nu$ are

$$E_2 = E_c + \frac{m_r}{m_c}(h\nu - E_g), \qquad E_1 = E_2 - h\nu, \tag{16.2-1}$$

respectively, where m_c and m_v are their effective masses and $1/m_r = 1/m_c + 1/m_v$. The resulting optical joint density of states that interacts with a photon of energy $h\nu$ was determined to be [see (15.2-9)]

$$\varrho(\nu) = \frac{(2m_r)^{3/2}}{\pi\hbar^2}(h\nu - E_g)^{1/2}, \qquad h\nu \geq E_g. \tag{16.2-2}$$

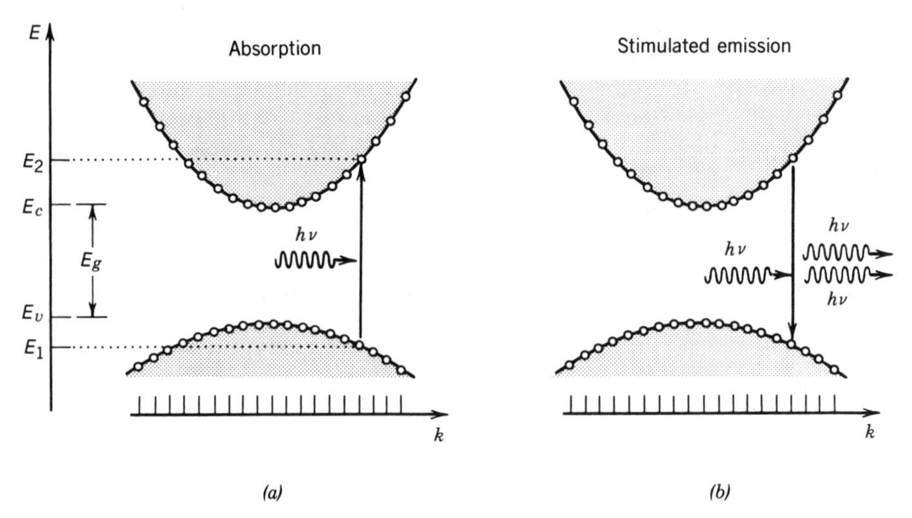

Figure 16.2-1 (*a*) The absorption of a photon results in the generation of an electron–hole pair. (*b*) Electron–hole recombination can be induced by a photon; the result is the stimulated emission of an identical photon.

It is apparent that $\varrho(\nu)$ increases as the square root of photon energy above the bandgap.

The occupancy probabilities $f_e(\nu)$ and $f_a(\nu)$ are determined by the pumping rate through the quasi-Fermi levels E_{fc} and E_{fv}. $f_e(\nu)$ is the probability that a conduction-band state of energy E_2 is filled with an electron and a valence-band state of energy E_1 is filled with a hole. $f_a(\nu)$, on the other hand, is the probability that a conduction-band state of energy E_2 is empty and a valence-band state of energy E_1 is filled with an electron. The Fermi inversion factor [see (15.2-24)]

$$f_g(\nu) = f_e(\nu) - f_a(\nu) = f_c(E_2) - f_v(E_1) \tag{16.2-3}$$

represents the degree of population inversion. $f_g(\nu)$ depends on both the Fermi function for the conduction band, $f_c(E) = 1/\{\exp[(E - E_{fc})/k_B T] + 1\}$, and the Fermi function for the valence band, $f_v(E) = 1/\{\exp[(E - E_{fv})/k_B T] + 1\}$. It is a function of temperature and of the quasi-Fermi levels E_{fc} and E_{fv}, which, in turn, are determined by the pumping rate. Because a complete population inversion can in principle be achieved in a semiconductor laser amplifier $[f_g(\nu) = 1]$, it behaves like a four-level system.

The results provided above were combined in (15.2-23) to give an expression for the net gain coefficient, $\gamma_0(\nu) = [r_{st}(\nu) - r_{ab}(\nu)]/\phi_\nu$,

$$\gamma_0(\nu) = \frac{\lambda^2}{8\pi\tau_r}\varrho(\nu) f_g(\nu). \tag{16.2-4}$$

Gain Coefficient

Comparing (16.2-4) with (13.1-4), it is apparent that the quantity $\varrho(\nu)f_g(\nu)$ in the semiconductor laser amplifier plays the role of $Ng(\nu)$ in other laser amplifiers.

Amplifier Bandwidth

In accordance with (16.2-3) and (16.2-4), a semiconductor medium provides net optical gain at the frequency ν when $f_c(E_2) > f_v(E_1)$. Conversely, net attenuation ensues when $f_c(E_2) < f_v(E_1)$. Thus a semiconductor material in thermal equilibrium (undoped or doped) cannot provide net gain whatever its temperature; this is because the conduction- and valence-band Fermi levels coincide ($E_{fc} = E_{fv} = E_f$). External pumping is required to separate the Fermi levels of the two bands in order to achieve amplification.

The condition $f_c(E_2) > f_v(E_1)$ is equivalent to the requirement that the photon energy be smaller than the separation between the quasi-Fermi levels, i.e., $h\nu < E_{fc} - E_{fv}$, as demonstrated in Exercise 15.2-1. Of course, the photon energy must be larger than the bandgap energy ($h\nu > E_g$) in order that laser amplification occur by means of band-to-band transitions. Thus if the pumping rate is sufficiently large that the separation between the two quasi-Fermi levels exceeds the bandgap energy E_g, the medium can act as an amplifier for optical frequencies in the band

$$\frac{E_g}{h} < \nu < \frac{E_{fc} - E_{fv}}{h}. \tag{16.2-5}$$

Amplifier Bandwidth

For $h\nu < E_g$ the medium is transparent, whereas for $h\nu > E_{fc} - E_{fv}$ it is an attenuator instead of an amplifier. Equation (16.2-5) demonstrates that the amplifier bandwidth increases with $E_{fc} - E_{fv}$, and therefore with pumping level. In this respect it is unlike the atomic laser amplifier, which has an unsaturated bandwidth $\Delta\nu$ that is independent of pumping level (see Fig. 13.1-2).

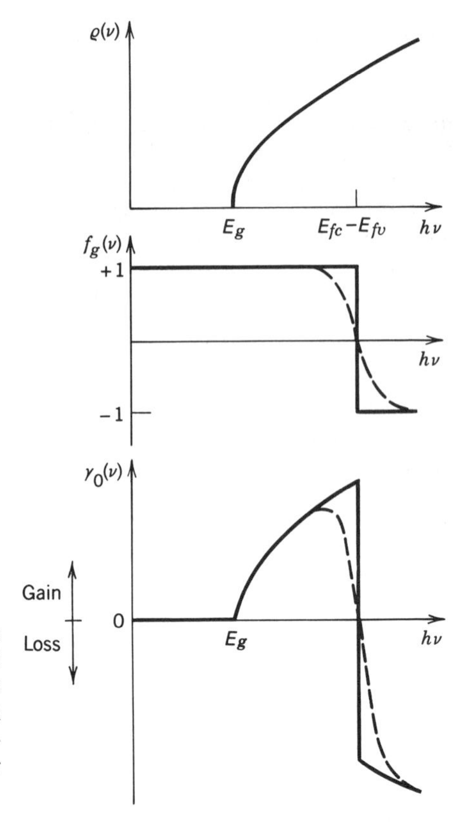

Figure 16.2-2 Dependence on energy of the joint optical density of states $\varrho(\nu)$, the Fermi inversion factor $f_g(\nu)$, and the gain coefficient $\gamma_0(\nu)$ at $T = 0$ K (solid curves) and at room temperature (dashed curves). Photons whose energy lies between E_g and $E_{fc} - E_{fv}$ undergo laser amplification.

Computation of the gain properties is simplified considerably if thermal excitations can be ignored (viz., $T = 0$ K). The Fermi functions are then simply $f_c(E_2) = 1$ for $E_2 < E_{fc}$ and 0 otherwise; $f_v(E_1) = 1$ for $E_1 < E_{fv}$ and 0 otherwise. In that case the Fermi inversion factor is

$$f_g(\nu) = \begin{cases} +1, & h\nu < E_{fc} - E_{fv} \\ -1, & \text{otherwise}. \end{cases} \qquad (16.2\text{-}6)$$

Schematic plots of the functions $\varrho(\nu)$, $f_g(\nu)$, and the gain coefficient $\gamma_0(\nu)$ are presented in Fig. 16.2-2, illustrating how $\gamma_0(\nu)$ changes sign and turns into a loss coefficient when $h\nu > E_{fc} - E_{fv}$. The ν^{-2} dependence of $\gamma_0(\nu)$, arising from the λ^2 factor in the numerator of (16.2-4), is sufficiently slow that it may be ignored. Finite temperature smoothes the functions $f_g(\nu)$ and $\gamma_0(\nu)$, as shown by the dashed curves in Fig. 16.2-2.

Dependence of the Gain Coefficient on Pumping Level

The gain coefficient $\gamma_0(\nu)$ increases both in its width and in its magnitude as the pumping rate R is elevated. As provided in (16.1-1), a constant pumping rate R (number of injected excess electron–hole pairs per cm^3 per second) establishes a steady-state concentration of injected electron–hole pairs in accordance with $\Delta n = \Delta p = R\tau$, where τ is the electron–hole recombination lifetime (which includes both radiative and nonradiative contributions). Knowledge of the steady–steady total concentrations of electrons and holes, $n = n_0 + \Delta n$ and $p = p_0 + \Delta n$, respectively, permits the Fermi levels E_{fc} and E_{fv} to be determined via (16.1-7). Once the Fermi levels are known, the computation of the gain coefficient can proceed using (16.2-4). The dependence of $\gamma_0(\nu)$ on Δn and thereby on R, is illustrated in Example 16.2-1. The onset of gain saturation and the noise performance of semiconductor laser amplifiers is similar to that of other amplifiers, as considered in Secs. 13.3 and 13.4.

EXAMPLE 16.2-1. ***InGaAsP Laser Amplifier.*** A room-temperature ($T = 300$ K) sample of $In_{0.72}Ga_{0.28}As_{0.6}P_{0.4}$ with $E_g = 0.95$ eV is operated as a semiconductor laser amplifier at $\lambda_o = 1.3$ μm. The sample is undoped but has residual concentrations of $\approx 2 \times 10^{17}$ cm^{-3} donors and acceptors, and a radiative electron–hole recombination lifetime $\tau_r \approx 2.5$ ns. The effective masses of the electrons and holes are $m_c \approx 0.06 m_0$ and $m_v \approx 0.4 m_0$, respectively, and the refractive index $n \approx 3.5$. Given the steady-state injected-carrier concentration Δn (which is controlled by the injection rate R and the overall recombination time τ), the gain coefficient $\gamma_0(\nu)$ may be computed from (16.2-4) in conjunction with (16.1-7). As illustrated in Fig. 16.2-3, both the amplifier bandwidth and the peak value of the gain coefficient γ_p increase with Δn. The energy at which the peak

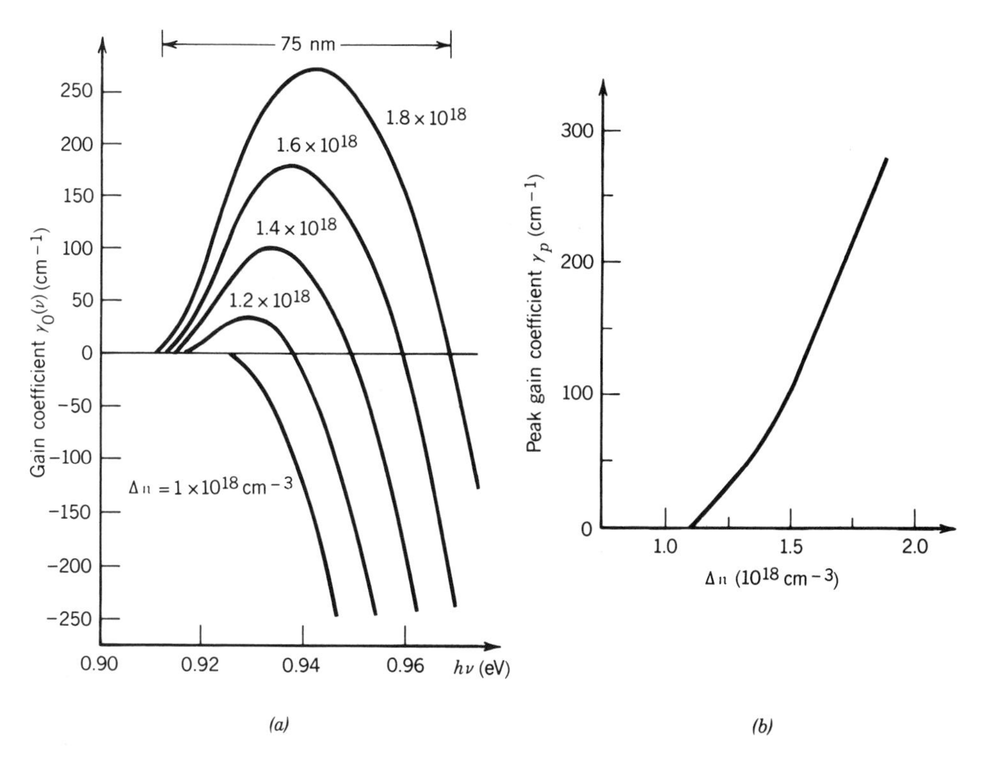

Figure 16.2-3 (*a*) Calculated gain coefficient $\gamma_0(\nu)$ for an InGaAsP laser amplifier versus photon energy $h\nu$, with the injected-carrier concentration Δn as a parameter ($T = 300$ K). The band of frequencies over which amplification occurs (centered near 1.3 μm) increases with increasing Δn. At the largest value of Δn shown, the full amplifier bandwidth is 15 THz, corresponding to 0.06 eV in energy, and 75 nm in wavelength. (Adapted from N. K. Dutta, Calculated Absorption, Emission, and Gain in $In_{0.72}Ga_{0.28}As_{0.6}P_{0.4}$, *Journal of Applied Physics*, vol. 51, pp. 6095–6100, 1980.) (*b*) Calculated peak gain coefficient γ_p as a function of Δn. At the largest value of Δn, the peak gain coefficient ≈ 270 cm^{-1}. (Adapted from N. K. Dutta and R. J. Nelson, The Case for Auger Recombination in $In_{1-x}Ga_xAs_yP_{1-y}$, *Journal of Applied Physics*, vol. 53, pp. 74–92, 1982.)

occurs also increases with Δn, as expected from the behavior shown in Fig. 16.2-2. Furthermore, the minimum energy at which amplification occurs decreases slightly with increasing Δn as a result of band-tail states, which reduce the bandgap energy. At the largest value of Δn shown ($\Delta n = 1.8 \times 10^{18}$ cm^{-3}), photons with energies falling between 0.91 and 0.97 eV undergo amplification. This corresponds to a full amplifier bandwidth of 15 THz, and a wavelength range of 75 nm, which is large in comparison with most atomic linewidths (see Table 13.2-1). The calculated peak gain coefficient $\gamma_p = 270$ cm^{-1} at this value of Δn is also large in comparison with most atomic laser amplifiers.

Approximate Peak Gain Coefficient

The complex dependence of the gain coefficient on the injected-carrier concentration makes the analysis of the semiconductor amplifier (and laser) somewhat difficult. Because of this, it is customary to adopt an empirical approach in which the peak gain coefficient γ_p is assumed to be linearly related to Δn for values of Δn near the operating point. As the example in Fig. 16.2-3(*b*) illustrates, this approximation is reasonable when γ_p is large. The dependence of the peak gain coefficient γ_p on Δn may then be modeled by the linear equation

$$\gamma_p \approx \alpha\left(\frac{\Delta n}{\Delta n_T} - 1\right), \tag{16.2-7}$$

Peak Gain Coefficient (Linear Approximation)

which is illustrated in Fig. 16.2-4. The parameters α and Δn_T are chosen to satisfy the following limits:

- When $\Delta n = 0$, $\gamma_p = -\alpha$, where α represents the absorption coefficient of the semiconductor in the absence of current injection.
- When $\Delta n = \Delta n_T$, $\gamma_p = 0$. Thus Δn_T is the injected-carrier concentration at which emission and absorption just balance so that the medium is transparent.

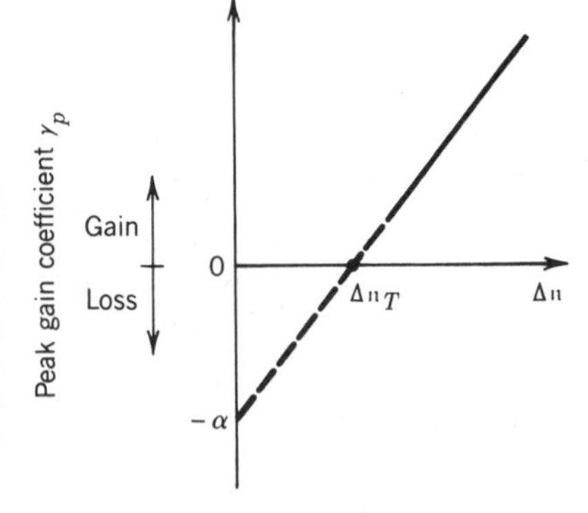

Figure 16.2-4 Peak value of the gain coefficient γ_p as a function of injected carrier concentration Δn for the approximate linear model. α represents the attenuation coefficient in the absence of injection, whereas Δn_T represents the injected carrier concentration at which emission and absorption just balance each other. The solid portion of the straight line matches the more realistic calculation considered in the preceding subsection.

EXAMPLE 16.2-2. ***InGaAsP Laser Amplifier.*** The peak gain coefficient γ_p versus Δn for InGaAsP presented in Fig. 16.2-3(*b*) may be approximately fit by a linear relation in the form of (16.2-7) with the parameters $\Delta n_T \approx 1.25 \times 10^{18}$ cm^{-3} and $\alpha = 600$ cm^{-1}. For $\Delta n = 1.4\,\Delta n_T = 1.75 \times 10^{18}$ cm^{-3}, the linear model yields a peak gain $\gamma_p = 240$ cm^{-1}. For an InGaAsP crystal of length $d = 350$ μm, this corresponds to a total gain of $\exp(\gamma_p d) \approx 4447$ or 36.5 dB. It must be kept in mind, however, that coupling losses are typically 3 to 5 dB per facet.

Increasing the injected-carrier concentration from below to above the transparency value Δn_T results in the semiconductor changing from a strong absorber of light $[f_g(\nu) < 0]$ into a high-gain amplifier of light $[f_g(\nu) > 0]$. The very same large transition probability that makes the semiconductor a good absorber also makes it a good amplifier, as may be understood by comparing (15.2-17) and (15.2-18).

B. Pumping

Optical Pumping

Pumping may be achieved by the use of external light, as depicted in Fig. 16.2-5, provided that its photon energy is sufficiently large ($> E_g$). Pump photons are absorbed by the semiconductor, resulting in the generation of carrier pairs. The generated electrons and holes decay to the bottom of the conduction band and the top of the valence band, respectively. If the intraband relaxation time is much shorter than the interband relaxation time, as is usually the case, a steady-state population inversion between the bands may be established as discussed in Sec. 13.2.

Electric-Current Pumping

A more practical scheme for pumping a semiconductor is by means of electron–hole injection in a heavily doped *p–n* junction—a diode. As with the LED (see Sec. 16.1) the junction is forward biased so that minority carriers are injected into the junction region (electrons into the *p*-region and holes into the *n*-region). Figure 16.1-5 shows the energy-band diagram of a forward-biased heavily doped *p–n* junction. The conduction-band and valence-band quasi-Fermi levels E_{fc} and E_{fv} lie within the conduction and valence bands, respectively, and a state of quasi-equilibrium exists within the junction region. The quasi-Fermi levels are sufficiently well separated so that a population inversion is achieved and net gain may be obtained over the bandwidth

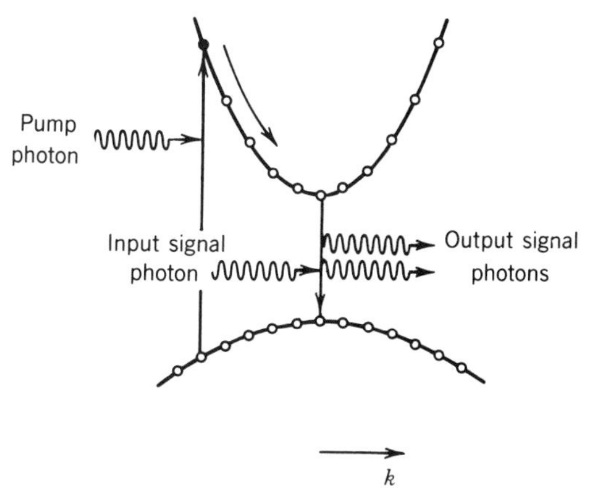

Figure 16.2-5 Optical pumping of a semiconductor laser amplifier.

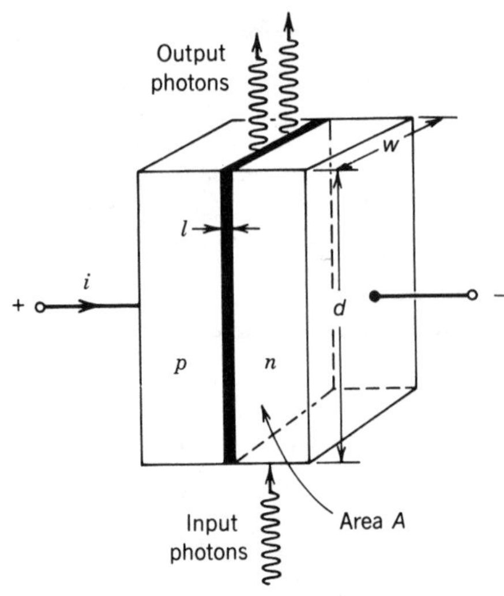

Figure 16.2-6 Geometry of a simple laser amplifier. Charge carriers travel perpendicularly to the p–n junction, whereas photons travel in the plane of the junction.

$E_g \leq h\nu \leq E_{fc} - E_{fv}$ within the active region. The thickness l of the active region is an important parameter of the diode that is determined principally by the diffusion lengths of the minority carriers at both sides of the junction. Typical values of l for InGaAsP are 1 to 3 μm.

If an electric current i is injected through an area $A = wd$, where w and d are the width and height of the device, respectively, into a volume lA (as shown in Fig. 16.2-6), then the steady-state carrier injection rate is $R = i/elA = J/el$ per second per unit volume, where $J = i/A$ is the injected current density. The resulting injected carrier concentration is then

$$\Delta n = \tau R = \frac{\tau}{elA} i = \frac{\tau}{el} J. \qquad (16.2\text{-}8)$$

The injected carrier concentration is therefore directly proportional to the injected current density and the results shown in Figs. 16.2-3 and 16.2-4 with Δn as a parameter may just as well have J as a parameter. In particular, it follows from (16.2-7) and (16.2-8) that within the linear approximation implicit in (16.2-7), the peak gain coefficient is linearly related to the injected current density J, i.e.,

$$\gamma_p \approx \alpha \left(\frac{J}{J_T} - 1 \right). \qquad (16.2\text{-}9)$$

Peak Gain Coefficient

The transparency current density J_T is given by

$$J_T = \frac{el}{\eta_i \tau_r} \Delta n_T, \qquad (16.2\text{-}10)$$

Transparency Current Density

where $\eta_i = \tau/\tau_r$ again represents the internal quantum efficiency.

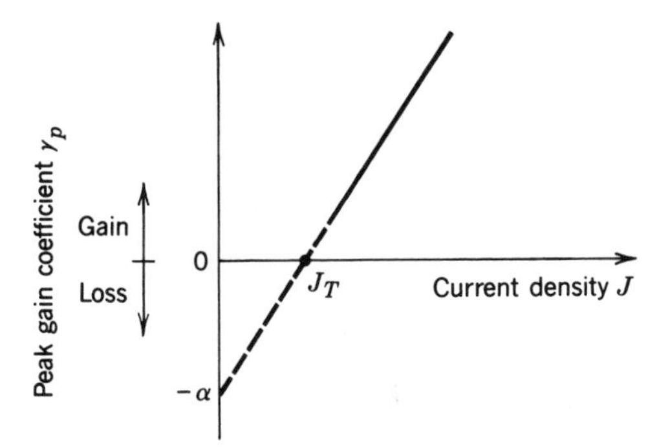

Figure 16.2-7 Peak optical gain coefficient γ_p as a function of current density J for the approximate linear model. When $J = J_T$ the material is transparent and exhibits neither gain nor loss.

When $J = 0$, the peak gain coefficient $\gamma_p = -\alpha$ becomes the attenuation coefficient, as is apparent in Fig. 16.2-7. When $J = J_T$, $\gamma_p = 0$ and the material is transparent and neither amplifies nor attenuates. Net gain can only be achieved when the injected current density J exceeds its transparency value J_T. Note that J_T is directly proportional to the junction thickness l so that a lower transparency current density J_T is achieved by using a narrower active-region thickness. This is an important consideration in the design of semiconductor amplifiers (and lasers).

EXAMPLE 16.2-3. *InGaAsP Laser Amplifier.* An InGaAsP diode laser amplifier operates at 300 K and has the following parameters: $\tau_r = 2.5$ ns, $\eta_i = 0.5$, $\Delta n_T = 1.25 \times 10^{18}$ cm^{-3}, and $\alpha = 600$ cm^{-1}. The junction has thickness $l = 2$ μm, length $d = 200$ μm, and width $w = 10$ μm. Using (16.2-10), the current density that just makes the semiconductor transparent is $J_T = 3.2 \times 10^4$ A/cm^2. A slightly larger current density $J = 3.5 \times 10^4$ A/cm^2 provides a peak gain coefficient $\gamma_p \approx 56$ cm^{-1} as is clear from (16.2-9). This gives rise to an amplifier gain $G = \exp(\gamma_p d) = \exp(1.12) \approx 3$. However, since the junction area $A = wd = 2 \times 10^{-5}$ cm^2, a rather large injection current $i = JA = 700$ mA is required to produce this current density.

Motivation for Heterostructures

If the thickness l of the active region in Example 16.2-3 were able to be reduced from 2 μm to, say, 0.1 μm, the current density J_T would be reduced by a factor of 20, to the more reasonable value 1600 A/cm^2. Because proportionately less volume would have to be pumped, the amplifier could then provide the same gain with a far lower injected current density. Reducing the thickness of the active region poses a problem, however, because the diffusion lengths of the electrons and holes in InGaAsP are several μm; the carriers would therefore tend to diffuse out of this smaller region. Is there a way in which these carriers can be confined to an active region whose thickness is smaller than their diffusion lengths? The answer is yes, by using a heterostructure device. These devices also make it possible to confine a light beam to an active region smaller than its wavelength, which provides further substantial advantage.

C. Heterostructures

As is apparent from (16.2-9) and (16.2-10), the diode-laser peak amplifier gain coefficient γ_p varies inversely with the thickness l of the active region. It is therefore advantageous to use the narrowest thickness possible. The active region is defined by the diffusion distances of minority carriers on both sides of the junction. The concept of the double heterostructure is to form heterojunction potential barriers on both sides of the p–n junction to provide a potential well that limits the distance over which minority carriers may diffuse. The junction barriers define a region of space within which minority carriers are confined, so that active regions of thickness l as small as 0.1 μm can be achieved. (Even thinner confinement regions, ≈ 0.01 μm, can be achieved with quantum-well lasers, as will be discussed in Sec. 16.3G.)

Electromagnetic confinement of the amplified optical beam can simultaneously be achieved if the material of the active layer is selected such that its refractive index is slightly greater than that of the two surrounding layers, so that the structure acts as an optical waveguide (see Chap. 7).

The double-heterostructure design therefore calls for three layers of different lattice-matched materials (see Fig. 16.2-8):

Layer 1: p-type, energy gap E_{g1}, refractive index n_1.
Layer 2: p-type, energy gap E_{g2}, refractive index n_2.
Layer 3: n-type, energy gap E_{g3}, refractive index n_3.

The materials are selected such that E_{g1} and E_{g3} are greater than E_{g2} to achieve carrier confinement, while n_2 is greater than n_1 and n_3 to achieve light confinement. The active layer (layer 2) is made quite thin (0.1 to 0.2 μm) to minimize the

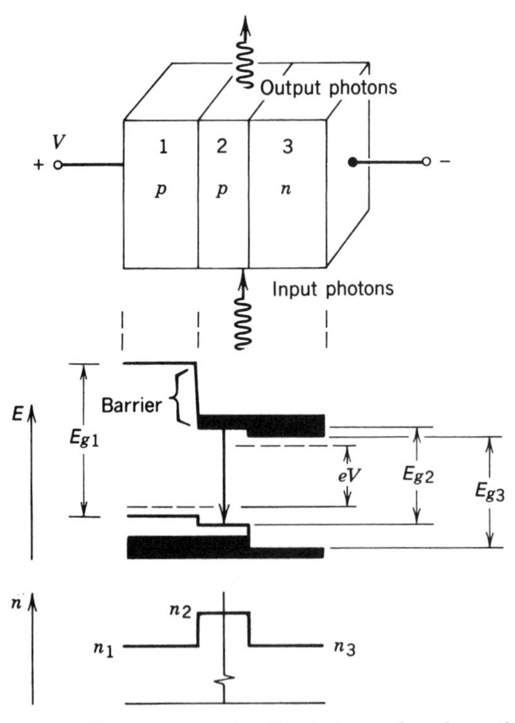

Figure 16.2-8 Energy-band diagram and refractive index as functions of position for a double-heterostructure semiconductor laser amplifier.

transparency current density J_T and maximize the peak gain coefficient γ_p. Stimulated emission takes place in the p–n junction region between layers 2 and 3.

In summary, the double-heterostructure design offers the following advantages:

- Increased amplifier gain, for a given injected current density, resulting from a decreased active-layer thickness [see (16.2-9) and (16.2-10)]. Injected minority carriers are confined within the thin active layer between the two heterojunction barriers and are prevented from diffusing to the surrounding layers.
- Increased amplifier gain resulting from the confinement of light within the active layer caused by its larger refractive index. The active medium acts as an optical waveguide.
- Reduced loss, resulting from the inability of layers 1 and 3 to absorb the guided photons because their bandgaps E_{g1} and E_{g3} are larger than the photon energy (i.e., $h\nu = E_{g2} < E_{g1}, E_{g2}$).

Two examples of double-heterostructure laser amplifiers are:

- *InGaAsP / InP Double-Heterostructure Laser Diode Amplifier.* The active layer is $In_{1-x}Ga_xAs_{1-y}P_y$, while the surrounding layers are InP. The ratios x and y are selected so that the materials are lattice matched. Operation is thereby restricted to a range of values of x and y for which E_{g2} corresponds to the band 1.1 to 1.7 μm.
- *GaAs / AlGaAs Double-Heterostructure Laser Diode Amplifier.* The active layer (layer 2) is fabricated from GaAs ($E_{g2} = 1.42$ eV, $n_2 = 3.6$). The surrounding layers (1 and 3) are fabricated from $Al_xGa_{1-x}As$ with $E_g > 1.43$ eV and $n < 3.6$ (by 5 to 10%). This amplifier typically operates within the 0.82- to 0.88-μm wavelength band using AlGaAs with $x = 0.35$ to 0.5.

16.3 SEMICONDUCTOR INJECTION LASERS

A. Amplification, Feedback, and Oscillation

A semiconductor injection laser is a semiconductor laser amplifier that is provided with a path for optical feedback. As discussed in the preceding section, a semiconductor laser amplifier is a forward-biased heavily doped p–n junction fabricated from a direct-gap semiconductor material. The injected current is sufficiently large to provide optical gain. The optical feedback is provided by mirrors, which are usually obtained by cleaving the semiconductor material along its crystal planes. The sharp refractive index difference between the crystal and the surrounding air causes the cleaved surfaces to act as reflectors. Thus the semiconductor crystal acts both as a gain medium and as an optical resonator, as illustrated in Fig. 16.3-1. Provided that the gain coefficient is sufficiently large, the feedback converts the optical amplifier into an optical oscillator (a laser). The device is called a semiconductor injection laser, or a laser diode.

The laser diode (LD) is similar to the light-emitting diode (LED) discussed in Sec. 16.1. In both devices, the source of energy is an electric current injected into a p–n junction. However, the light emitted from an LED is generated by spontaneous emission, whereas the light from an LD arises from stimulated emission.

In comparison with other types of lasers, injection lasers have a number of advantages: small size, high efficiency, integrability with electronic components, and ease of pumping and modulation by electric current injection. However, the spectral linewidth of semiconductor lasers is typically larger than that of other lasers.

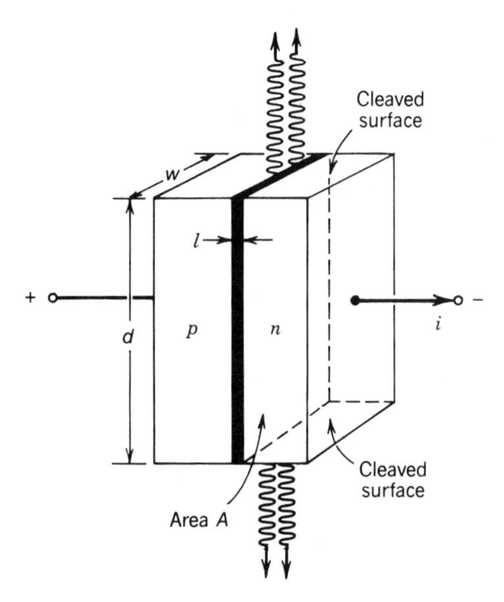

Figure 16.3-1 An injection laser is a forward-biased p–n junction with two parallel surfaces that act as reflectors.

We begin our study of the conditions required for laser oscillation, and the properties of the emitted light, with a brief summary of the basic results that describe the semiconductor laser amplifier and the optical resonator.

Laser Amplification

The gain coefficient $\gamma_0(\nu)$ of a semiconductor laser amplifier has a peak value γ_p that is approximately proportional to the injected carrier concentration, which, in turn, is proportional to the injected current density J. Thus, as provided in (16.2-9) and (16.2-10) and illustrated in Fig. 16.2-7,

$$\gamma_p \approx \alpha\left(\frac{J}{J_T} - 1\right), \qquad J_T = \frac{el}{\eta_i \tau_r}\,\Delta n_T, \tag{16.3-1}$$

where τ_r is the radiative electron–hole recombination lifetime, $\eta_i = \tau/\tau_r$ is the internal quantum efficiency, l is the thickness of the active region, α is the thermal-equilibrium absorption coefficient, and Δn_T and J_T are the injected-carrier concentration and current density required to just make the semiconductor transparent.

Feedback

The feedback is usually obtained by cleaving the crystal planes normal to the plane of the junction, or by polishing two parallel surfaces of the crystal. The active region of the p–n junction illustrated in Fig. 16.3-1 then also serves as a planar-mirror optical resonator of length d and cross-sectional area lw. Semiconductor materials typically have large refractive indices, so that the power reflectance at the semiconductor–air interface

$$\mathscr{R} = \left(\frac{n-1}{n+1}\right)^2 \tag{16.3-2}$$

is substantial (see (6.2-14) and Table 15.2-1). Thus if the gain of the medium is sufficiently large, the refractive index discontinuity itself can serve as an adequate reflective surface and no external mirrors are necessary. For GaAs, for example, $n = 3.6$, so that (16.3-2) yields $\mathscr{R} = 0.32$.

Resonator Losses

The principal source of resonator loss arises from the partial reflection at the surfaces of the crystal. This loss constitutes the transmitted useful laser light. For a resonator of length d the reflection loss coefficient is [see (9.1-18)]

$$\alpha_m = \alpha_{m1} + \alpha_{m2} = \frac{1}{2d} \ln \frac{1}{\mathscr{R}_1 \mathscr{R}_2}; \tag{16.3-3}$$

if the two surfaces have the same reflectance $\mathscr{R}_1 = \mathscr{R}_2 = \mathscr{R}$, then $\alpha_m = (1/d)\ln(1/\mathscr{R})$. The total loss coefficient is

$$\alpha_r = \alpha_s + \alpha_m, \tag{16.3-4}$$

where α_s represents other sources of loss, including free carrier absorption in the semiconductor material (see Fig. 15.2-2) and scattering from optical inhomogeneities. α_s increases as the concentration of impurities and interfacial imperfections in heterostructures increase. It can attain values in the range 10 to 100 cm^{-1}.

Of course, the term $-\alpha$ in the expression for the gain coefficient (16.3-1), corresponding to absorption in the material, also contributes substantially to the losses. This contribution is accounted for, however, in the *net* peak gain coefficient γ_p given by (16.3-1). This is apparent from the expression for $\gamma_0(\nu)$ given in (15.2-23), which is proportional to $f_g(\nu) = f_e(\nu) - f_a(\nu)$ (i.e., to stimulated emission less absorption).

Another important contribution to the loss results from the spread of optical energy outside the active layer of the amplifier (in the direction *perpendicular* to the junction plane). This can be especially detrimental if the thickness of the active layer l is small. The light then propagates through a thin amplifying layer (the active region) surrounded by a lossy medium so that large losses are likely. This problem may be alleviated by the use of a double heterostructure (see Sec. 16.2C and Fig. 16.2-8), in which the middle layer is fabricated from a material of elevated refractive index that acts as a waveguide confining the optical energy.

Losses caused by optical spread may be phenomenologically accounted for by defining a **confinement factor** Γ to represent the fraction of the optical energy lying within the active region (Fig. 16.3-2). Assuming that the energy outside the active region is totally wasted, Γ is therefore the factor by which the gain coefficient is reduced, or equivalently, the factor by which the loss coefficient is increased. Equation (16.3-4) must therefore be modified to reflect this increase, so that

$$\alpha_r = \frac{1}{\Gamma}(\alpha_s + \alpha_m). \tag{16.3-5}$$

There are basically three types of laser-diode structures based on the mechanism used for confining the carriers or light in the lateral direction (viz., *in* the junction plane): **broad-area** (in which there is no mechanism for lateral confinement), **gain-guided** (in which lateral variations of the gain are used for confinement), and **index-guided** (in which lateral refractive index variations are used for confinement). Index-guided lasers are generally preferred because of their superior properties.

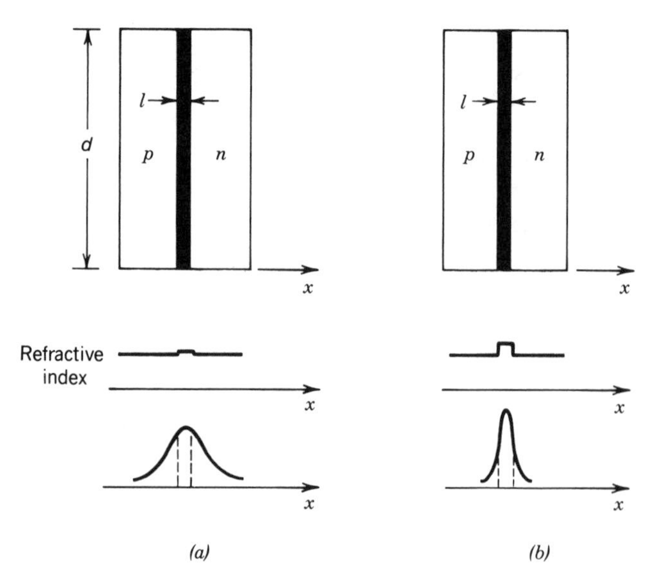

Figure 16.3-2 Spatial spread of the laser light in the direction perpendicular to the plane of the junction for (a) homostructure, and (b) heterostructure lasers.

Gain Condition: Laser Threshold

The laser oscillation condition is that the gain exceed the loss, $\gamma_p > \alpha_r$, as indicated in (14.1-10). The threshold gain coefficient is therefore α_r. Setting $\gamma_p = \alpha_r$ and $J = J_t$ in (16.3-1) corresponds to a threshold injected current density J_t given by

$$J_t = \frac{\alpha_r + \alpha}{\alpha} J_T, \qquad (16.3\text{-}6)$$

Threshold Current Density

where the transparency current density,

$$J_T = \frac{el}{\eta_i \tau_r} \Delta n_T, \qquad (16.3\text{-}7)$$

Transparency Current Density

is the current density that just makes the medium transparent. The threshold current density is larger than the transparency current density by the factor $(\alpha_r + \alpha)/\alpha$, which is ≈ 1 when $\alpha \gg \alpha_r$. Since the current $i = JA$, where $A = wd$ is the cross-sectional area of the active region, we can define $i_T = J_T A$ and $i_t = J_t A$, corresponding to the currents required to achieve transparency of the medium and laser oscillation threshold, respectively.

The threshold current density J_t is a key parameter in characterizing the diode-laser performance; smaller values of J_t indicate superior performance. In accordance with (16.3-6) and (16.3-7), J_t is minimized by maximizing the internal quantum efficiency η_i, and by minimizing the resonator loss coefficient α_r, the transparency injected-carrier concentration Δn_T, and the active-region thickness l. As l is reduced beyond a certain point, however, the loss coefficient α_r becomes larger because the confinement factor Γ decreases [see (16.3-5)]. Consequently, J_t decreases with decreasing l until it reaches

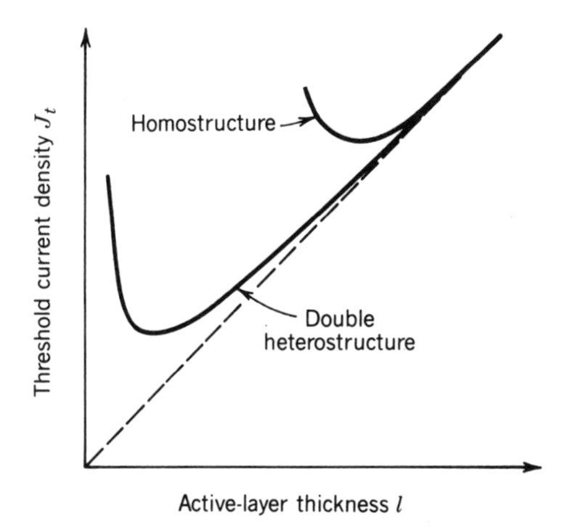

Figure 16.3-3 Dependence of the threshold current density J_t on the thickness of the active layer l. The double-heterostructure laser exhibits a lower value of J_t than the homostructure laser, and therefore superior performance.

a minimum value, beyond which any further reduction causes J_t to increase (see Fig. 16.3-3). In double-heterostructure lasers, however, the confinement factor remains near unity for lower values of l because the active layer behaves as an optical waveguide (see Fig. 16.3-2). The result is a lower minimum value of J_t, as shown in Fig. 16.3-3, and therefore superior performance. The reduction in J_t is illustrated in the following examples.

Because the parameters Δn_T and α in (16.3-1) strongly depend on temperature, so does the threshold current density J_t and the frequency at which the peak gain occurs. As a result, temperature control is required to stabilize the laser output. Indeed, frequency tuning is often achieved by deliberate modification of the temperature of operation.

EXAMPLE 16.3-1. ***Threshold Current for an InGaAsP Homostructure Laser Diode.*** Consider an InGaAsP homostructure semiconductor injection laser with the same material parameters as in Examples 16.2-1 and 16.2-2: $\Delta n_T = 1.25 \times 10^{18}$ cm^{-3}, $\alpha = 600$ cm^{-1}, $\tau_r = 2.5$ ns, $n = 3.5$, and $\eta_i = 0.5$ at $T = 300$ K. Assume that the dimensions of the junction are $d = 200$ μm, $w = 10$ μm, and $l = 2$ μm. The current density necessary for transparency is then calculated to be $J_T = 3.2 \times 10^4$ A/cm^2. We now determine the threshold current density for laser oscillation. Using (16.3-2), the surface reflectance is $\mathscr{R} = 0.31$. The corresponding mirror loss coefficient is $\alpha_m = (1/d)\ln(1/\mathscr{R}) = 59$ cm^{-1}. Assuming that the loss coefficient due to other effects is also $\alpha_s = 59$ cm^{-1} and that the confinement factor $\Gamma \approx 1$, the total loss coefficient is then $\alpha_r = 118$ cm^{-1}. The threshold current density is therefore $J_t = [(\alpha_r + \alpha)/\alpha]J_T = [(118 + 600)/600][3.2 \times 10^4] = 3.8 \times 10^4$ A/cm^2. The corresponding threshold current $i_t = J_t wd \approx 760$ mA, which is rather high. Homostructure lasers are no longer used because continuous-wave (CW) operation of devices with such large currents is not possible unless they are cooled substantially below $T = 300$ K to dissipate the heat.

EXAMPLE 16.3-2. ***Threshold Current for an InGaAsP Heterostructure Laser Diode.*** We turn now to an InGaAsP/InP double-heterostructure semiconductor injection laser (see Fig. 16.2-8) with the same parameters and dimensions as in Example 16.3-1 except for the active-layer thickness, which is now $l = 0.1$ μm instead of 2 μm. If the confinement of light is assumed to be perfect ($\Gamma = 1$), we may use the same values for the resonator loss coefficient α_r. The transparency current density is then reduced by a factor of 20 to $J_T = 1600$ A/cm^2, and the threshold current density assumes a more reasonable value of $J_t = 1915$ A/cm^2. The corresponding threshold current is $i_t = 38$ mA. It is this significant reduction in threshold current that makes CW operation of the double-heterostructure laser diode at room temperature feasible.

B. Power

Internal Photon Flux

When the laser current density is increased above its threshold value (i.e., $J > J_t$), the amplifier peak gain coefficient γ_p exceeds the loss coefficient α_r. Stimulated emission then outweighs absorption and other resonator losses so that oscillation can begin and the photon flux Φ in the resonator can increase. As with other homogeneously broadened lasers, saturation sets in as the photon flux becomes larger and the population difference becomes depleted [see (14.1-2)]. As shown in Fig. 14.2-1, the gain coefficient then decreases until it becomes equal to the loss coefficient, whereupon steady state is reached.

As with the internal photon-flux density and the internal photon-number density considered for other types of lasers [see (14.2-2) and (14.2-13)], the steady-state internal photon flux Φ is proportional to the difference between the pumping rate R and the threshold pumping rate R_t. Since $R \propto i$ and $R_t \propto i_t$, in accordance with (16.2-8), Φ may be written as

$$\Phi = \begin{cases} \eta_i \dfrac{i - i_t}{e}, & i > i_t \\ 0, & i \le i_t. \end{cases} \tag{16.3-8}$$

Steady-State Laser Internal Photon Flux

Thus the steady-state laser internal photon flux (photons per second generated within the active region) is equal to the electron flux (which is the number of injected electrons per second) in excess of that required for threshold, multiplied by the internal quantum efficiency η_i.

The internal laser power above threshold is simply related to the internal photon flux Φ by the relation $P = h\nu\Phi$, so that we obtain

$$P = \eta_i (i - i_t) \frac{1.24}{\lambda_o}, \tag{16.3-9}$$

Internal Laser Power λ_o (μm), P (W), i (A)

provided that λ_o is expressed in μm, i in amperes, and P in Watts.

Output Photon Flux and Efficiency

The laser output photon flux Φ_o is the product of the internal photon flux Φ and the **emission efficiency** η_e [see (14.2-16)], which is the ratio of the loss associated with the useful light transmitted through the mirrors to the total resonator loss α_r. If only the light transmitted through mirror 1 is used, then $\eta_e = \alpha_{m1}/\alpha_r$; on the other hand, if the light transmitted through both mirrors is used, then $\eta_e = \alpha_m/\alpha_r$. In the latter case, if both mirrors have the same reflectance $\mathscr{R}$, we obtain $\eta_e = [(1/d)\ln(1/\mathscr{R})]/\alpha_r$. The laser output photon flux is therefore given by

$$\Phi_o = \eta_e \eta_i \frac{i - i_t}{e}. \tag{16.3-10}$$

Laser Output Photon Flux

It is clear from (16.3-10) that the proportionality between the laser output photon flux and the injected electron flux above threshold is governed by the **external differential quantum efficiency**

$$\eta_d = \eta_e \eta_i. \tag{16.3-11}$$

External Differential Quantum Efficiency

η_d therefore represents the rate of change of the output photon flux with respect to the injected electron flux above threshold, i.e.,

$$\eta_d = \frac{d\Phi_o}{d(i/e)}. \tag{16.3-12}$$

The laser output power above threshold is $P_o = h\nu\Phi_o = \eta_d(i - i_t)(h\nu/e)$, which may therefore be written as

$$P_o = \eta_d(i - i_t)\frac{1.24}{\lambda_o}, \tag{16.3-13}$$

Laser Output Power λ_o (μm), P_o (W), i (A)

provided that λ_o is expressed in μm. The output power is plotted against the injected (drive) current i as the straight line in Fig. 16.3-4 with the parameters $i_t \approx 21$ mA and $\eta_d = 0.4$. This is called the **light-current curve**. The solid curve in Fig. 16.3-4 represents data obtained from both output faces of a 1.3-μm InGaAsP semiconductor

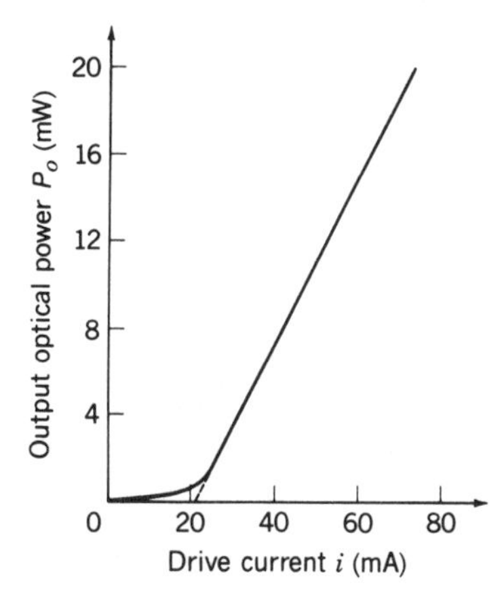

Figure 16.3-4 Ideal (straight line) and actual (solid curve) laser light-current curve for a strongly index-guided buried-heterostructure (see Fig. 16.3-7) InGaAsP injection laser operated at a wavelength of 1.3 μm. Nonlinearities, which are not accounted for by the simple theory presented here, cause the optical output power to saturate for currents greater than about 75 mA (not shown).

injection laser. The agreement between the simple theory presented here and the data is very good and shows clearly that the emitted optical power does indeed increase linearly with the drive current (over the range 23 to 73 mA in this example).

From (16.3-13) it is clear that the slope of the light-current curve above threshold is given by

$$\Re_d = \frac{dP_o}{di} = \eta_d \frac{1.24}{\lambda_o} \qquad [\lambda_o\,(\mu\text{m}),\, P_o\,(\text{W}),\, i\,(\text{A})]. \tag{16.3-14}$$

$\Re_d$ is called the **differential responsivity** of the laser (W/A); it represents the ratio of the optical power increase to the electric current increase above threshold. For the data shown in Fig. 16.3-4, $dP_o/di \approx 0.38$ W/A.

The **overall efficiency** (power-conversion efficiency) η is defined as the ratio of the emitted laser light power to the electrical input power iV, where V is the forward-bias voltage applied to the diode. Since $P_o = \eta_d(i - i_t)(h\nu/e)$, we obtain

$$\eta = \eta_d\left(1 - \frac{i_t}{i}\right)\frac{h\nu}{eV}. \tag{16.3-15}$$

Overall Efficiency

For operation well above threshold, which provides $i \gg i_t$, and for $eV \approx h\nu$, as is usually the case, we obtain $\eta \approx \eta_d$. The data illustrated in Fig. 16.3-4 therefore exhibit an overall efficiency $\eta \approx 40\%$, which is greater than that of any other type of laser (see Table 14.2-1). Indeed, this value is somewhat below the record high value reported to date, which is $\approx 65\%$. The electrical power that fails to be transformed into light becomes heat. Because laser diodes do, in fact, generate substantial amounts of heat they are usually mounted on heat sinks which help to dissipate the heat and stabilize the temperature.

Summary

There are four efficiencies associated with the semiconductor injection laser: the internal quantum efficiency $\eta_i = \tau_r/\tau = \tau/\tau_r$, which accounts for the fact that only a fraction of the electron–hole recombinations are radiative in nature; the emission efficiency η_e, which accounts for the fact that only a portion of the light lost from the cavity is useful; the external differential quantum efficiency $\eta_d = \eta_e\eta_i$, which accounts for both of these effects; and the power-conversion efficiency η, which is the overall efficiency. The differential responsivity $\Re_d$ (W/A) is also used as a measure of performance.

EXAMPLE 16.3-3. ***InGaAsP Double-Heterostructure Laser Diode.*** Consider again Example 16.3-2 for the InGaAsP/InP double-heterostructure semiconductor injection laser with $\eta_i = 0.5$, $\alpha_m = 59\ \text{cm}^{-1}$, $\alpha_r = 118\ \text{cm}^{-1}$, and $i_t = 38$ mA. If the light from both output faces is used, the emission efficiency is $\eta_e = \alpha_m/\alpha_r = 0.5$, while the external differential quantum efficiency is $\eta_d = \eta_e\eta_i = 0.25$. At $\lambda_o = 1.3\ \mu$m, the differential responsivity of this laser is $dP_o/di = 0.24$ W/A. If, for example, $i = 50$ mA, we obtain $i - i_t = 12$ mA and $P_o = 12 \times 0.24 = 2.9$ mW. Comparison of these numbers with those obtained from the data in Fig. 16.3-4 shows that the double-heterostructure laser has a higher threshold, and a lower efficiency and differential responsivity, than the buried-heterostructure laser. This illustrates the superiority of the strongly index-guided buried-heterostructure device over the strongly gain-guided double-heterostructure device.

Comparison of Laser Diode and LED Operation

Laser diodes produce light even below threshold, as is apparent from Fig. 16.3-4. This light arises from spontaneous emission, which was examined in Sec. 16.1 in connection with the LED, but which has been ignored in the present laser theory. When operated below threshold, the semiconductor laser diode acts as an edge-emitting LED. In fact, most LEDs are simply edge-emitting double-heterostructure devices. Laser diodes with sufficiently strong injection so that stimulated emission is much greater than spontaneous emission, but with little feedback so that the lasing threshold is high, are called **superluminescent LEDs**.

As discussed in Sec. 16.1, there are four efficiencies associated with the LED: the internal quantum efficiency η_i, which accounts for the fact that only a fraction of the electron–hole recombinations are radiative in nature; the transmittance efficiency η_e, which accounts for the fact that only a small fraction of the light generated in the junction region can escape from the high-index medium; the external quantum efficiency $\eta_{ex} = \eta_i \eta_e$, which accounts for both of these effects; and the power-conversion efficiency η, which is the overall efficiency. The responsivity $\mathfrak{R}$ is also used as a measure of LED performance.

There is a one-to-one correspondence between the quantities η_i, η_e, and η for the LED and the laser diode. Furthermore, there is a correspondence between η_{ex} and η_d, $\mathfrak{R}$ and $\mathfrak{R}_d$, and i and $(i - i_t)$. The superior performance of the laser results from the fact that η_e can be much greater than in the LED. This stems from the fact that the laser operates on the basis of stimulated (rather than spontaneous) emission, which has several important consequences. The stimulated emission in an above-threshold device causes the laser light rays to travel perpendicularly to the facets of the material where the loss is minimal. This provides three advantages: a net gain in place of absorption, the prevention of light rays from becoming trapped because they impinge on the inner surfaces of the material perpendicularly (and therefore at an angle less than the critical angle), and multiple opportunities for the rays to emerge as useful light from the facet as they execute multiple round trips within the cavity. LED light, by contrast, is subject to absorption and trapping and has only a single opportunity to escape; if it is not successful, it is lost. The net result is that a laser diode operated above threshold has a value of η_d (typically $\approx 40\%$) that far exceeds the value of η_{ex} (typically $\approx 2\%$) for an LED, as is evident in the comparison of Figs. 16.3-4 and 16.1-8.

C. Spectral Distribution

The spectral distribution of the laser light generated is governed by three factors, as described in Sec. 14.2B:

- The spectral width B within which the active medium small-signal gain coefficient $\gamma_0(\nu)$ is greater than the loss coefficient α_r.
- The homogeneous or inhomogeneous nature of the line-broadening mechanism (see Sec. 12.2D).
- The resonator modes, in particular the approximate frequency spacing between the longitudinal modes $\nu_F = c/2d$, where d is the resonator length.

Semiconductor lasers are characterized by the following features:

- The spectral width of the gain coefficient is relatively large because transitions occur between two energy bands rather than between two discrete energy levels.
- Because intraband processes are very fast, semiconductors tend to be homogeneously broadened. Nevertheless, spatial hole burning permits the simultaneous oscillation of many longitudinal modes (see Sec. 14.2B). Spatial hole burning is particularly prevalent in short cavities in which there are few standing-wave cycles. This permits the fields of different longitudinal modes, which are distributed along the resonator axis, to overlap less, thereby allowing partial spatial hole burning to occur.
- The semiconductor resonator length d is significantly smaller than that of most other lasers. The frequency spacing of adjacent resonator modes $\nu_F = c/2d$ is therefore relatively large. Nevertheless, many of these can generally fit within the broad band B over which the small-signal gain exceeds the loss (the number of possible laser modes is $M = B/\nu_F$).

EXAMPLE 16.3-4. *Number of Longitudinal Modes in an InGaAsP Laser.* An InGaAsP crystal ($n = 3.5$) of length $d = 400$ μm has resonator modes spaced by $\nu_F = c/2d = c_o/2nd \approx 107$ GHz. Near the central wavelength $\lambda_o = 1.3$ μm, this frequency spacing corresponds to a free-space wavelength spacing λ_F, where $\lambda_F/\lambda_o = \nu_F/\nu$, so that $\lambda_F = \lambda_o \nu_F/\nu = \lambda_o^2/2nd \approx 0.6$ nm. If the spectral width $B = 1.2$ THz (corresponding to a wavelength width of 7 nm), then approximately 11 longitudinal modes may oscillate. A typical spectral distribution consisting of a single transverse mode and about 11 longitudinal modes is illustrated in Fig. 16.3-5. The overall spectral width of semiconductor injection lasers is greater than that of most other lasers, particularly gas lasers (see Table 13.2-1). To reduce the number of modes to one, the resonator length d would have to be reduced so that $B = c/2d$, requiring a cavity of length $d \approx 36$ μm.

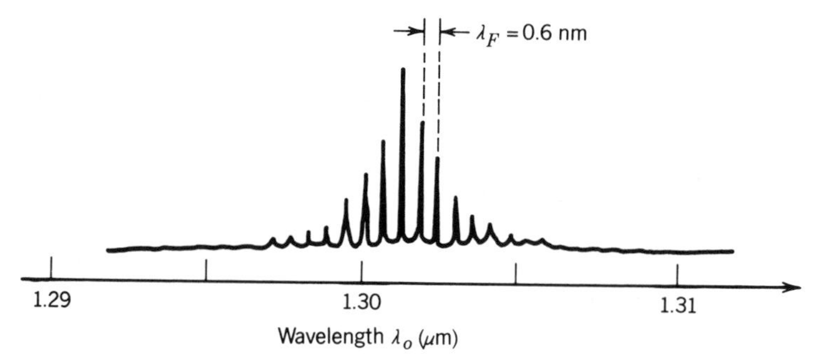

Figure 16.3-5 Spectral distribution of a 1.3-μm InGaAsP index-guided buried-heterostructure laser. This distribution is considerably narrower, and differs in shape, from that of the $\lambda_o \approx 1.3$-μm InGaAsP LED shown in Fig. 16.1-9. The number of modes decreases as the injection current increases; the mode closest to the gain maximum increases in power while the side peaks saturate. (Adapted from R. J. Nelson, R. B. Wilson, P. D. Wright, P. A. Barnes, and N. K. Dutta, CW Electrooptical Properties of InGaAsP ($\lambda = 1.3$ μm) Buried-Heterostructure Lasers, *IEEE Journal of Quantum Electronics*, vol. QE-17, pp. 202–207, © 1981 IEEE.)

The approximate linewidth of each longitudinal mode is typically ≈ 0.01 nm (corresponding to a few GHz) for gain-guided lasers, but generally far smaller (≈ 30 MHz) for index-guided lasers.

D. Spatial Distribution

Like in other lasers, oscillation in semiconductor injection lasers takes the form of transverse and longitudinal modes. In Sec. 14.2C the indices (l, m) were used to characterize the spatial distributions in the transverse direction, while the index q was used to represent variation along the direction of wave propagation or temporal behavior. In most other lasers, the laser beam lies totally within the active medium so that the spatial distributions of the different modes are determined by the shapes of the mirrors and their separations. In circularly symmetric systems, the transverse modes can be represented in terms of Laguerre–Gaussian or, more conveniently, Hermite–Gaussian beams (see Sec. 9.2D). The situation is different in semiconductor lasers since the laser beam extends outside the active layer. The transverse modes are modes of the dielectric waveguide created by the different layers of the semiconductor diode.

The transverse modes can be determined by using the theory presented in Sec. 7.3 for an optical waveguide with rectangular cross section of dimensions l and w. If l/λ_o is sufficiently small (as it usually is in double-heterostructure lasers), the waveguide will admit only a single mode in the transverse direction perpendicular to the junction plane. However, w is usually larger than λ_o, so that the waveguide will support several modes in the direction parallel to the plane of the junction, as illustrated in Fig. 16.3-6. Modes in the direction parallel to the junction plane are called **lateral modes**. The larger the ratio w/λ_o, the greater the number of lateral modes possible.

Because higher-order lateral modes have a wider spatial spread, they are less confined; their loss coefficient α_r is therefore greater than that for lower-order modes. Consequently, some of the highest-order modes will fail to satisfy the oscillation conditions; others will oscillate at a lower power than the fundamental (lowest-order) mode. To achieve high-power single-spatial-mode operation, the number of waveguide modes must be reduced by decreasing the width w of the active layer. The attendant reduction of the junction area also has the effect of reducing the threshold current.

An example of a design using a laterally confined active layer is the buried-heterostructure laser illustrated in Fig. 16.3-7. The lower-index material on either side of the active region produces lateral confinement in this (and other laterally confined) index-guided lasers.

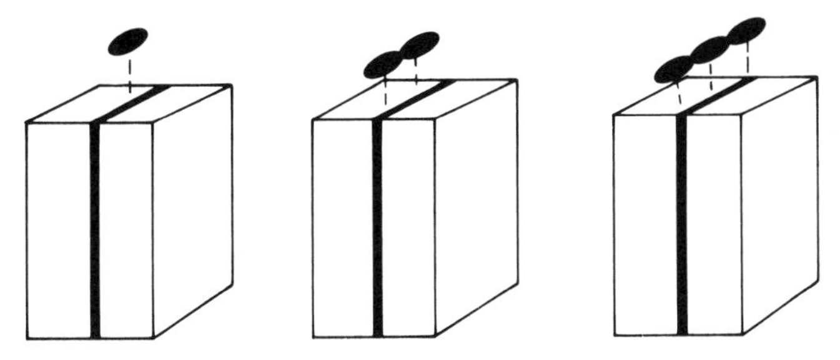

Figure 16.3-6 Schematic illustration of spatial distributions of the optical intensity for the laser waveguide modes $(l, m) = (1, 1)$, $(1, 2)$, and $(1, 3)$.

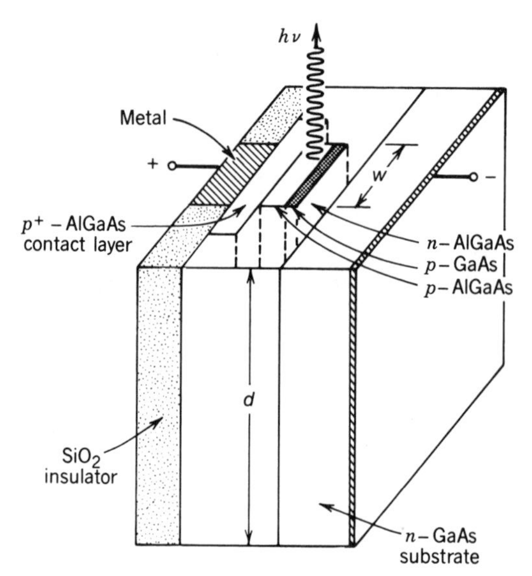

Figure 16.3-7 Schematic diagram of an AlGaAs/GaAs buried-heterostructure semiconductor injection laser. The junction width w is typically 1 to 3 μm, so that the device is strongly index guided.

Far-Field Radiation Pattern

A laser diode with an active layer of dimensions l and w emits light with far-field angular divergence $\approx \lambda_o/l$ (radians) in the plane perpendicular to the junction and $\approx \lambda_o/w$ in the plane parallel to the junction (see Fig. 16.3-8). (Recall from Sec. 3.1B, for example, that for a Gaussian beam of diameter $2W_0$ the divergence angle is $\theta \approx (2/\pi)(\lambda_o/2W_0) = \lambda_o/\pi W_0$ when $\theta \ll 1$). The angular divergence determines the far-field radiation pattern (see Sec. 4.3). Because of its small active layer, the semicon-

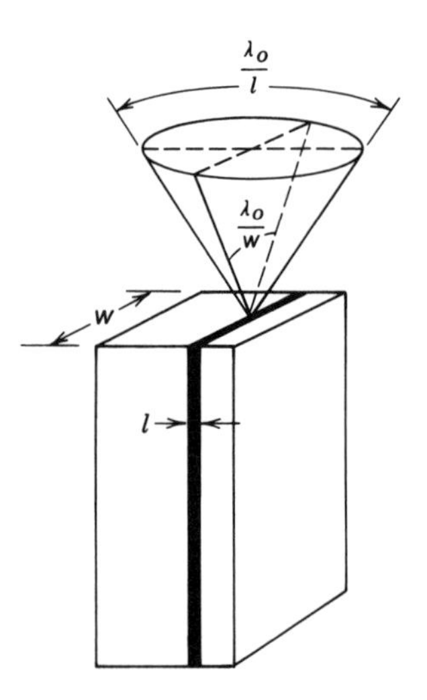

Figure 16.3-8 Angular distribution of the optical beam emitted from a laser diode.

ductor injection laser is characterized by an angular divergence larger than that of most other lasers. As an example, if $l = 2$ μm, $w = 10$ μm, and $\lambda_o = 0.8$ μm, the divergence angles are calculated to be $\approx 23°$ and 5°. Light from a single-transverse-mode laser diode, for which w is smaller, has an even larger angular divergence. The spatial distribution of the far-field light within the radiation cone depends on the number of transverse modes and on their optical powers. The highly asymmetric elliptical distribution of laser-diode light can make its collimation tricky.

E. Mode Selection

Single-Frequency Operation

As indicated above, a semiconductor injection laser may be operated on a single-transverse mode by reducing the dimensions of the active-layer cross section (l and w), so that it acts as a single-mode waveguide. Single-frequency operation may be achieved by reducing the length d of the resonator so that the frequency spacing between adjacent longitudinal modes exceeds the spectral width of the amplifying medium.

Other methods of single-mode operation include the use of multiple-mirror resonators, as discussed in Sec. 14.2D and illustrated in Fig. 14.2-15. A double-resonator diode laser (coupled-cavity laser) can be implemented by cleaving a groove perpendicular to the active layer, as shown in Fig. 16.3-9. This creates two coupled cavities so that the structure is known as a **cleaved-coupled-cavity** (C^3) **laser**. The standing wave in the laser must satisfy boundary conditions at the surfaces of both cavities, thereby providing a more stringent restriction that can be satisfied only at a single frequency. In practice, the usefulness of this approach is limited by thermal drift.

An alternative approach is to replace the cleaved surfaces usually used as mirrors with frequency-selective reflectors such as gratings parallel to the junction plane [Fig. 16.3-10(*a*)]. The grating is a periodic structure that reflects light only when the grating period Λ satisfies $\Lambda = q\lambda/2$, where q is an integer (see Sec. 2.4B). These are called **distributed Bragg reflectors** and the device is known as a DBR laser.

Yet another approach places the grating itself directly adjacent to the active layer by using a spatially corrugated waveguide as shown in Fig. 16.3-10(*b*). The grating then acts as a distributed reflector, substituting for the lumped reflections provided by the mirrors of a Fabry–Perot laser. The surfaces of the crystal are antireflection coated to minimize surface reflections. This structure is known as a **distributed-feedback** (DFB) **laser**. DFB lasers operate with spectral widths as small as 10 MHz (without modulation) and offer modulation bandwidths well into the GHz range. They are used in many

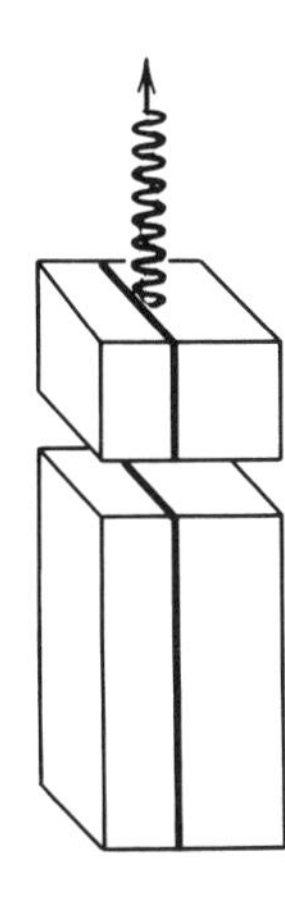

Figure 16.3-9 Cleaved-coupled-cavity (C^3) laser.

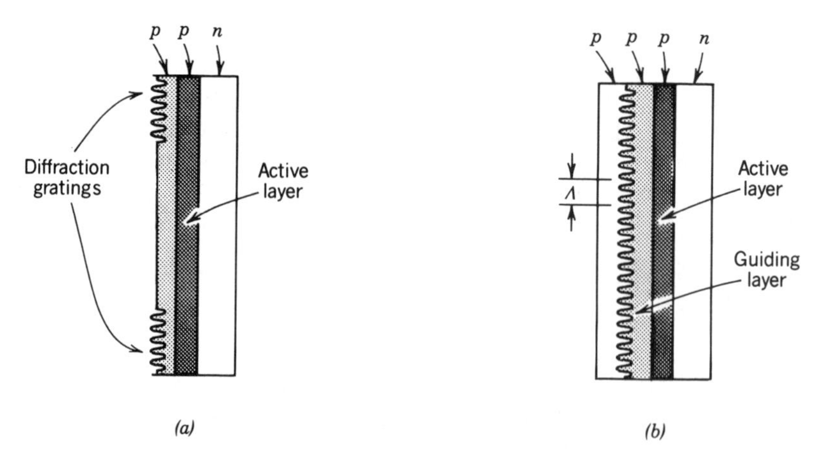

Figure 16.3-10 (*a*) External diffraction gratings serve as mirrors in a DBR laser. (*b*) The distributed feedback (DFB) laser has a periodic layer that acts as a distributed reflector.

applications including fiber-optic communications in the wavelength range 1.3 to 1.55 μm.

F. Characteristics of Typical Lasers

Semiconductor lasers have been operated at wavelengths stretching from the near ultraviolet to the far infrared, as illustrated in Fig. 16.3-11. They have been operated with power outputs reaching 100 mW, but laser-diode arrays (with closely spaced active regions) offer narrow coherent beams with powers in excess of 10 W. Surface-emitting laser diodes (SELDs) are becoming increasingly common.

Laser diodes operating in the visible band are usually fabricated from $Ga_{0.5}In_{0.5}P$ and generate light at $\lambda_o \approx 670$ nm. They use either gain-guided or index-guided structures. CW output powers are typically ≈ 5 mW at $T = 300$ K; an off-the-shelf device might operate at a voltage of 2.1 V and a current of 85 mA. Powers as high as 50 mW have been achieved using index-guided lateral confinement. The efficiency of a GaInP laser is substantially greater, and the size substantially smaller, than a 5-mW He–Ne laser operating at 633 nm. Room-temperature CW lasers operating at 584 nm (in the yellow) can be fabricated by using AlInP instead of GaInP.

In the near infrared, direct-bandgap ternary and quaternary materials are often used because their wavelengths can be compositionally tuned and CW operation at room temperature is possible. Temperature tuning can be used to adjust the output wavelength on a fine scale. As with LEDs, $Al_xGa_{1-x}As$ ($\lambda_o = 0.75$ to 0.87 μm) and $In_{1-x}Ga_xAs_{1-y}P_y$ ($\lambda_o = 1.1$ to 1.6 μm) are particularly important.

Laser diodes can also be operated throughout the middle-infrared region, although cooling is then required for efficient operation. II–VI direct-gap compounds such as $Hg_xCd_{1-x}Te$, and IV–VI materials such as $Pb_xSn_{1-x}Te$, are used over a broad range of this region from about 3 to 35 μm. When operated at very low temperatures, $Bi_{1-x}Sb_x$ lases out to wavelengths as long as ≈ 100 μm.

*G. Quantum-Well Lasers

As emphasized earlier, the laser threshold current density may be reduced by decreasing the thickness of the active layer. We have already discussed the way that heterostructures are used to confine electrons and photons within the active layer. When

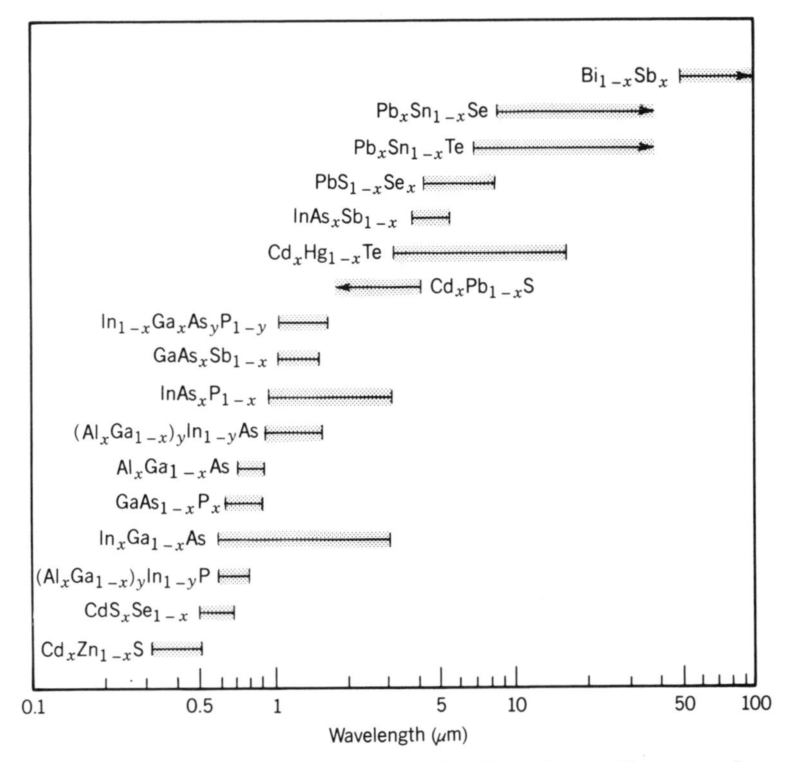

Figure 16.3-11 Compound materials used for semiconductor lasers. The range of wavelengths reaches from the near ultraviolet to the far infrared. Semiconductor lasers operating at $\lambda_o > 3$ μm usually require cooling below $T = 300$ K. Some of these materials require optical or electron-beam pumping to lase.

the thickness of the active layer is made sufficiently narrow (i.e., smaller than the de Broglie wavelength of a thermalized electron), quantum effects begin to play a dramatic role. Since the active layer in a double heterostructure has a bandgap energy smaller than the surrounding layers, the structure then acts as a quantum well (see Sec. 15.1G) and the laser is called a single-quantum well (SQW) laser or simply a quantum-well laser.

The band structure and energy–momentum (E–k) relations of a quantum well are different from bulk material. The conduction band is split into a number of subbands, labeled by the quantum number $q = 1, 2, \ldots,$ each with its own energy–momentum relation and density of states. The bottoms of these subbands have energies $E_c + E_q$, where $E_q = \hbar^2(q\pi/l)^2/2m_c$, $q = 1, 2, \ldots,$ are the energies of an electron of effective mass m_c in a one-dimensional quantum well of thickness l (see Figs. 15.1-21 and 15.1-22; q_1 and d_1 in Chap. 15 correspond to q and l here). Each subband has a parabolic E–k relation and a constant density of states that is independent of energy. The overall density of states in the conduction band, $\varrho_c(E)$, therefore assumes a staircase distribution [see (15.1-28)] with steps at energies $E_c + E_q$, $q = 1, 2, \ldots$. The valence band has similar subbands at energies $E_v - E'_q$, where $E'_q = \hbar^2(q\pi/l)^2/2m_v$ are the energies of a hole of effective mass m_v in a quantum well of thickness l.

The interactions of photons with electrons and holes in a quantum well take the form of energy- and momentum-conserving transitions between the conduction and valence bands. The transitions must also conserve the quantum number q, as illustrated in Fig. 16.3-12; they obey rules similar to those that govern transitions between the conduction and valence bands in bulk semiconductors. The expressions for the transition probabilities and gain coefficient in the bulk material (see Sec. 15.2) apply to the quantum-well structure if we simply replace the bandgap energy E_g with the energy gap between the subbands, $E_{gq} = E_g + E_q + E'_q$, and use a constant density of states rather than one that varies as the square root of energy. The total gain coefficient is the sum of the gain coefficients provided by all of the subbands ($q = 1, 2, \ldots$).

Density of States

Consider transitions between the two subbands of quantum number q. To satisfy the conservation of energy and momentum, a photon of energy $h\nu$ interacts with states of energies $E = E_c + E_q + (m_r/m_c)(h\nu - E_{gq})$ in the upper subband and $E - h\nu$ in the lower. The optical joint density of states $\varrho(\nu)$ is related to $\varrho_c(E)$ by $\varrho(\nu) = (dE/d\nu)\varrho_c(E) = (hm_r/m_c)\varrho_c(E)$. It follows from (15.1-28) that

$$\varrho(\nu) = \begin{cases} \dfrac{hm_r}{m_c}\dfrac{m_c}{\pi\hbar^2 l} = \dfrac{2m_r}{\hbar l}, & h\nu > E_g + E_q + E'_q \\ 0, & \text{otherwise.} \end{cases} \tag{16.3-16}$$

Including transitions between all subbands $q = 1, 2, \ldots$, we arrive at a $\varrho(\nu)$ that has a staircase distribution with steps at the energy gaps between subbands of the same quantum number (Fig. 16.3-12).

Gain Coefficient

The gain coefficient of the laser is given by the usual expression [see (15.2-23)]

$$\gamma_0(\nu) = \frac{\lambda^2}{8\pi\tau_r}\varrho(\nu) f_g(\nu), \tag{16.3-17}$$

where the Fermi inversion factor $f_g(\nu)$ depends on the quasi-Fermi levels and tempera-

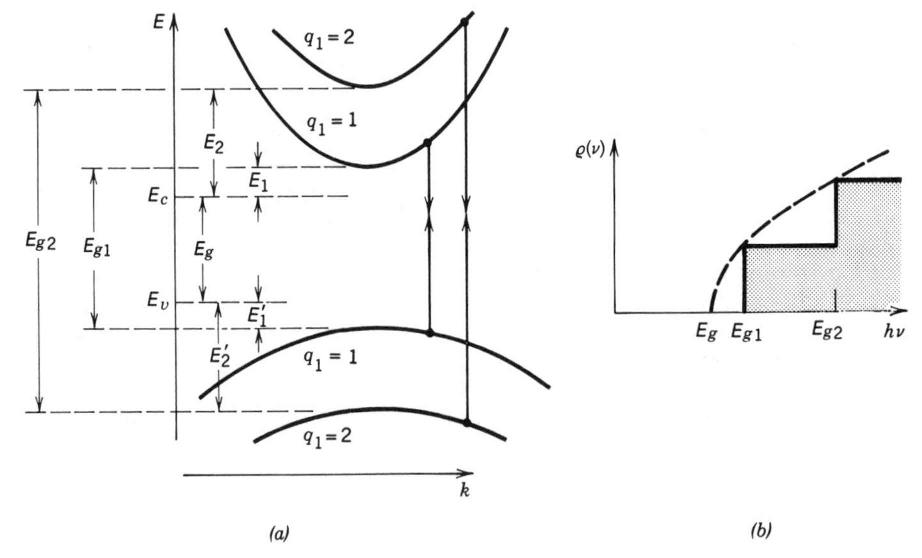

Figure 16.3-12 (*a*) E–k relations of different subbands. (*b*) Optical joint density of states for a quantum-well structure (staircase curve) and for a bulk semiconductor (dashed curve). The first jump occurs at energy $E_{g1} = E_g + E_1 + E'_1$ (where E_1 and E'_1 are, respectively, the lowest energies of an electron and a hole in the quantum well).

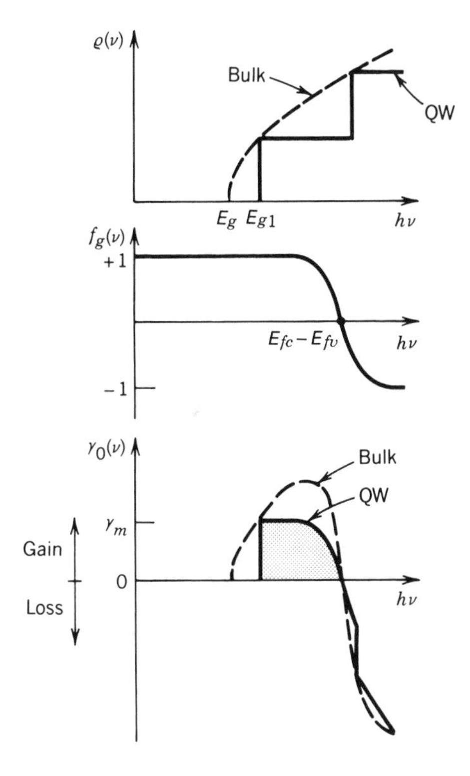

Figure 16.3-13 Density of states $\varrho(\nu)$, Fermi inversion factor $f_g(\nu)$, and gain coefficient $\gamma_0(\nu)$ in quantum-well (solid-curves) and bulk (dashed curves) structures.

ture and is the same for bulk and quantum-well lasers. The density of states $\varrho(\nu)$, however, differs in the two cases as we have shown. The frequency dependences of $\varrho(\nu)$, $f_g(\nu)$, and their product are illustrated in Fig. 16.3-13 for quantum-well and bulk double-heterostructure configurations. The quantum-well laser has a smaller peak gain and a narrower gain profile.

It is assumed in Fig. 16.3-13 that only a single step of the staircase function $\varrho(\nu)$ occurs at an energy smaller than $E_{fc} - E_{fv}$. This is the case under usual injection conditions. The maximum gain γ_m may then be determined by substituting $f_g(\nu) = 1$ and $\varrho(\nu) = 2m_r/\hbar l$ in (16.3-17), yielding

$$\gamma_m = \frac{\lambda^2 m_r}{2\tau_r h l}. \tag{16.3-18}$$

Relation Between Gain Coefficient and Current Density

By increasing the injected current density J, the concentration of excess electrons and holes Δn is increased and, therefore, so is the separation between the quasi-Fermi levels $E_{fc} - E_{fv}$. The effect of this increase on the gain coefficient $\gamma_0(\nu)$ may be assessed by examining the diagrams in Fig. 16.3-13. For sufficiently small J there is no gain. When J is such that $E_{fc} - E_{fv}$ just exceeds the gap E_{g1} between the $q = 1$ subbands, the medium provides gain. The peak gain coefficient increases sharply and saturates at the value γ_m. A further increase of J increases the gain spectral width but not its peak value. If J is increased yet further, to the point where $E_{fc} - E_{fv}$ exceeds the gap E_{g2} between the $q = 2$ subbands, the peak gain coefficient undergoes another jump, and so on. The gain profile can therefore be quite broad, providing the possibility of a wide tuning range for such lasers. The dependences of γ_p on J for quantum-well and bulk double-heterostructure semiconductor lasers are illustrated schematically in

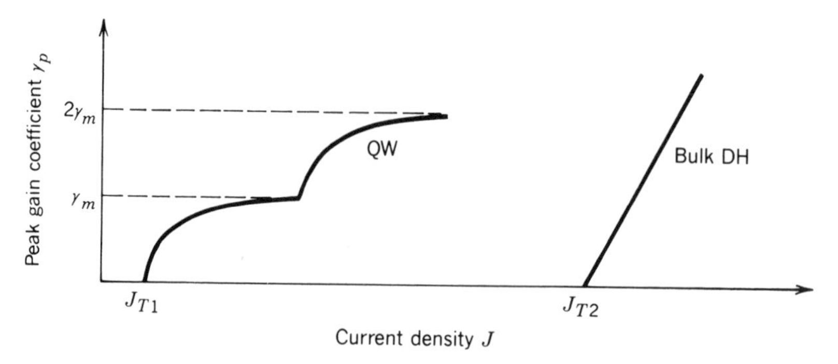

Figure 16.3-14 Schematic relations between peak gain coefficient γ_p and current density J in quantum-well (QW) and bulk double-heterostructure (DH) lasers.

Fig. 16.3-14. The quantum-well laser has a far smaller value of J_T (current density required for transparency), but its gain saturates at a lower value.

The threshold current density for QW laser oscillation is considerably smaller than that for bulk (DH) laser oscillation because of the reduction in active-layer thickness. Additional factors that make quantum-well lasers attractive include the narrower spectrum of the gain coefficient, the smaller linewidth of the laser modes, the possibility of achieving higher modulation frequencies, and the reduced temperature dependence.

The active-layer thickness of a SQW laser is typically < 10 nm, which is to be compared with 100 nm for a DH laser and 2 μm for an old-fashioned homojunction semiconductor laser. SQW threshold currents are roughly ≈ 0.5 mA, as compared with 20 mA for DH lasers (see Fig. 16.3-4). The spectral width of the light emitted from a SQW laser is usually < 10 MHz, which is substantially narrower than that from DH lasers. The output power of single quantum-well lasers is limited to about 100 mW to avoid facet damage. However, arrays of AlGaAs/GaAs quantum-well lasers can emit as much as 50 W of incoherent CW optical power in a line of dimensions 1 μm $\times$ 1 cm, making them excellent candidates for the side-pumping of solid-state lasers such as Nd^{3+}:YAG (see Sec. 13.2). Remarkably, the overall quantum efficiency η of such arrays is $> 50\%$ and the differential quantum efficiency η_d can exceed 80%.

Semiconductor lasers have also been fabricated in quantum-wire configurations (see Sec. 15.1G). Threshold currents < 0.1 mA are expected for devices in which l and w are both ≈ 10 nm. Arrays of quantum-dot lasers would offer yet lower threshold currents.

Multiquantum-Well Lasers

The gain coefficient may be increased by using a parallel stack of quantum wells. This structure, illustrated in Fig. 16.3-15, is known as a multiquantum-well (MQW) laser. The gain of an N-well MQW laser is N times the gain of each of its wells. However, a fair comparison of the performance of single quantum-well (SQW) and MQW lasers requires that both be injected by the same current. Assume that a single quantum well is injected with an excess carrier density Δn and has a peak gain coefficient γ_p. Each of the N wells in the MQW structure would then be injected with only $\Delta n/N$ carriers. Because of the nonlinear dependence of the gain on Δn, the gain coefficient of each well is $\xi\gamma_p/N$, where ξ may be smaller or greater than 1, depending on the operating conditions. The total gain provided by the MQW laser is $N(\xi\gamma_p/N) = \xi\gamma_p$. It is not clear which of the two structures produces higher gain. It turns out that at low current densities, the SQW is superior, while at high current densities, the MQW is superior (but by a factor of less than N).

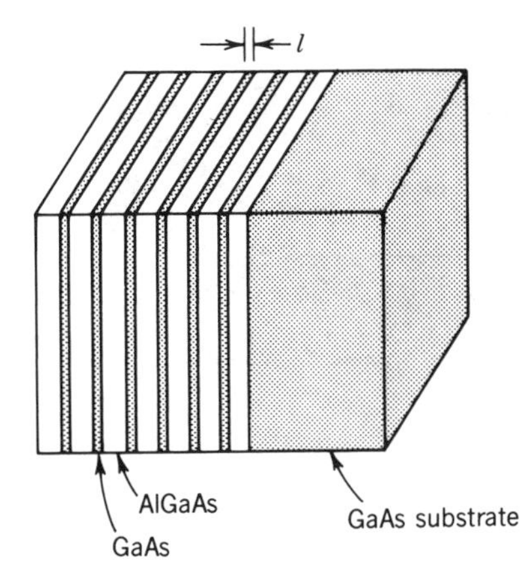

Figure 16.3-15 An AlGaAs/GaAs multiquantum-well laser with $l = 10$ nm.

Strained-Layer Lasers

Surprising as it may seem, the introduction of strain can provide a salutatory effect on the performance of semiconductor injection lasers. **Strained-layer lasers** can have superior properties, and can operate at wavelengths other than those accessible by means of compositional tuning. These lasers have been fabricated from III–V semiconductor materials, using both single-quantum-well and multiquantum-well configurations. Rather than being lattice-matched to the confining layers, the active layer is purposely chosen to have a different lattice constant. If sufficiently thin, it can accommodate its atomic spacings to those of the surrounding layers, and in the process become strained (if the layer is too thick it will not properly accommodate and the material will contain dislocations). The InGaAs active layer in an InGaAs/AlGaAs strained-layer laser, for example, has a lattice constant that is significantly greater than that of its AlGaAs confining layers. The thin InGaAs layer therefore experiences a biaxial compression in the plane of the layer, while its atomic spacings are increased above their usual values in the direction perpendicular to the layer.

The compressive strain alters the band structure in three significant ways: (1) it increases the bandgap E_g; (2) it removes the degeneracy at $k = 0$ between the heavy and light hole bands; and (3) it makes the valence bands anisotropic so that in the direction parallel to the plane of the layer the highest band has a light effective mass, whereas in the perpendicular direction the highest band has a heavy effective mass.

This behavior can significantly improve the performance of lasers. First, the laser wavelength is altered by virtue of the dependence of E_g on the strain. Second, the laser threshold current density can be reduced by the presence of the strain. Achieving a population inversion requires that the separation of the quasi-Fermi levels be greater than the bandgap energy, i.e., $E_{fc} - E_{fv} > E_g$. The reduced hole mass more readily allows E_{fv} to descend into the valence band, thereby permitting this condition to be satisfied at a lower injection current.

Strained-layer InGaAs lasers have been fabricated in many different configurations using a variety of confining materials, including AlGaAs and InGaAsP. They have been operated over a broad range of wavelengths from 0.9 to 1.55 μm. In one particular example that uses a MQW configuration, a device constructed of several 2-nm-thick $In_{0.78}Ga_{0.22}As$ quantum-well layers, separated by 20-nm barriers and 40-nm confining layers of InGaAsP, operates at $\lambda_o = 1.55$ μm with a sub-milliampere threshold current. As another example, GaInP/InGaAlP strained-layer quantum-well lasers emit more than $\frac{1}{2}$ W at 634 nm.

Surface-Emitting Quantum-Well Laser-Diode Arrays

Surface-emitting quantum-well laser diodes (SELDs) are of increasing interest, and offer the advantages of high packing densities on a wafer scale. An array of about 1

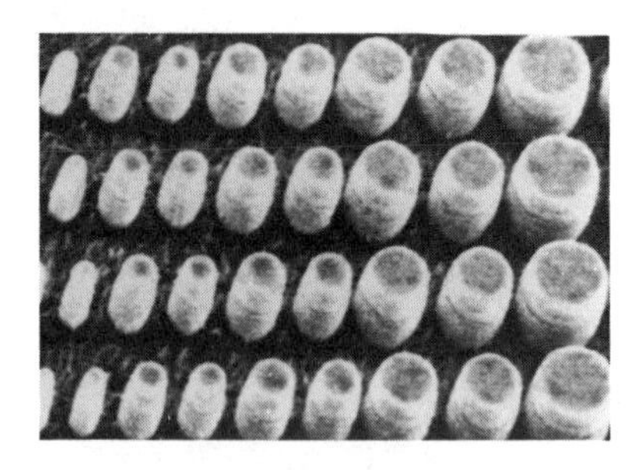

Figure 16.3-16 Scanning electron micrograph of a small portion of an array of vertical-cavity quantum-well lasers with diameters between 1 and 5 μm. (After J. L. Jewell *et al.*, Low Threshold Electrically-Pumped Vertical-Cavity Surface-Emitting Micro-Lasers, *Optics News*, vol. 15, no. 12, pp. 10–11, 1989.)

million electrically pumped tiny vertical-cavity cylindrical $In_{0.2}Ga_{0.8}As$ quantum-well SELDs (diameter ≈ 2 μm, height ≈ 5.5 μm), with lasing wavelengths in the vicinity of 970 nm, has been fabricated on a single 1-cm^2 chip of GaAs. These particular devices have thresholds $i_t \approx 1.5$ mA, for $T = 300$ K CW operation, and single-facet external differential quantum efficiencies $\eta_d \approx 16\%$. A scanning electron micrograph of a small portion of such an array is shown in Fig. 16.3-16. The circular output beams have the advantage of providing easy coupling to optical fibers. More recently, the lasing thresholds of devices of this type have been reduced to ≈ 0.2 mA. Their very small active-material volume (≈ 0.05 μm^3) can, in principle, permit thresholds as low as 10 μA.

READING LIST

Books and Articles on Laser Theory

See the reading list in Chapter 13.

Books and Articles on Semiconductor Physics

See the reading list in Chapter 15.

Books on LEDs and Semiconductor Injection Lasers

Y. Yamamoto, ed., *Coherence, Amplification, and Quantum Effects in Semiconductor Lasers*, Wiley, New York, 1991.

P. K. Cheo, ed., *Handbook of Solid-State Lasers*, Marcel Dekker, New York, 1988.

G. P. Agrawal and N. K. Dutta, *Long-Wavelength Semiconductor Lasers*, Van Nostrand Reinhold, New York, 1986.

R. K. Willardson and A. C. Beer, eds., *Semiconductors and Semimetals*, vol. 22, *Lightwave Communications Technology*, W. T. Tsang, ed., part B, *Semiconductor Injection Lasers, I*, Academic Press, New York, 1985.

R. K. Willardson and A. C. Beer, eds., *Semiconductors and Semimetals*, vol. 22, *Lightwave Communications Technology*, W. T. Tsang, ed., part C, *Semiconductor Injection Lasers, II and Light Emitting Diodes*, Academic Press, New York, 1985.

H. Kressel, ed., *Semiconductor Devices for Optical Communication*, Springer-Verlag, Berlin, 2nd ed. 1982.

G. H. B. Thomson, *Physics of Semiconductor Lasers*, Wiley, New York, 1981.

H. C. Casey, Jr., and M. B. Panish, *Heterostructure Lasers*, part A, *Fundamental Principles*, Academic Press, New York, 1978.

H. C. Casey, Jr., and M. B. Panish, *Heterostructure Lasers*, part B, *Materials and Operating Characteristics*, Academic Press, New York, 1978.

H. Kressel and J. K. Butler, *Semiconductor Lasers and Heterojunction LEDs*, Academic Press, New York, 1977.

E. W. Williams and R. Hall, *Luminescence and the Light Emitting Diode*, Pergamon Press, New York, 1977.

A. A. Bergh and P. J. Dean, *Light Emitting Diodes*, Clarendon Press, Oxford, 1976.

C. H. Gooch, *Injection Electroluminescent Devices*, Wiley, New York, 1973.

R. W. Campbell and F. M. Mims III, *Semi-Conductor Diode Lasers*, Howard Sams, Indianapolis, IN, 1972.

Special Journal Issues

Special issue on laser technology, *Lincoln Laboratory Journal*, vol. 3, no. 3, 1990.

Special issue on semiconductor diode lasers, *IEEE Journal of Quantum Electronics*, vol. QE-25, no. 6, 1989.

Special issue on semiconductor lasers, *IEEE Journal of Quantum Electronics*, vol. QE-23, no. 6, 1987.

Special issue on semiconductor quantum wells and superlattices: physics and applications, *IEEE Journal of Quantum Electronics*, vol. QE-22, no. 9, 1986.

Special issue on semiconductor lasers, *IEEE Journal of Quantum Electronics*, vol. QE-21, no. 6, 1985.

Special issue on optoelectronics, *Physics Today*, vol. 38, no. 5, 1985.

Special issue on light emitting diodes and long-wavelength photodetectors, *IEEE Transactions on Electron Devices*, vol. ED-30, no. 4, 1983.

Special issue on optoelectronic devices, *IEEE Transactions on Electron Devices*, vol. ED-29, no. 9, 1982.

Special issue on light sources and detectors, *IEEE Transactions on Electron Devices*, vol. ED-28, no. 4, 1981.

Special issue on quaternary compound semiconductor materials and devices—sources and detectors, *IEEE Journal of Quantum Electronics*, vol. QE-17, no. 2, 1981.

Special joint issue on optoelectronic devices and circuits, *IEEE Transactions on Electron Devices*, vol. ED-25, no. 2, 1978.

Special issue on semiconductor lasers, *IEEE Journal of Quantum Electronics*, vol. QE-6, no. 6, 1970.

Articles

J. Jewell, Surface-Emitting Lasers: A New Breed, *Physics World*, vol. 3, no. 7, pp. 28–30, 1990.

R. Baker, Optical Amplification, *Physics World*, vol. 3, no. 3, pp. 41–44, 1990.

D. A. B. Miller, Optoelectronic Applications of Quantum Wells, *Optics and Photonics News*, vol. 1, no. 2, pp. 7–15, 1990.

J. L. Jewell, A. Scherer, S. L. McCall, Y. H. Lee, S. J. Walker, J. P. Harbison, and L. T. Florez, Low Threshold Electrically-Pumped Vertical-Cavity Surface-Emitting Micro-Lasers, *Optics News*, vol. 15, no. 12, pp. 10–11, 1989.

A. Yariv, Quantum Well Semiconductor Lasers Are Taking Over, *IEEE Circuits and Devices Magazine*, vol. 5, no. 6, pp. 25–28, 1989.

G. Eisenstein, Semiconductor Optical Amplifiers, *IEEE Circuits and Devices Magazine*, vol. 5, no. 4, pp. 25–30, 1989.

D. Welch, W. Streifer, and D. Scifres, High Power, Coherent Laser Diodes, *Optics News*, vol. 15, no. 3, pp. 7–10, 1989.

G. P. Agrawal, Single-Longitudinal-Mode Semiconductor Lasers, in *Progress in Optics*, E. Wolf, ed., vol. 26, North-Holland, Amsterdam, 1988.

M. Ohtsu and T. Tako, Coherence in Semiconductor Lasers, in *Progress in Optics*, E. Wolf, ed., vol. 25, North-Holland, Amsterdam, 1988.

I. Hayashi, Future Prospects of the Semiconductor Laser, *Optics News*, vol. 14, no. 10, pp. 7–12, 1988.

M. Ettenberg, Laser Diode Systems and Devices, *IEEE Circuits and Devices Magazine*, vol. 3, no. 5, pp. 22–26, 1987.

G. L. Harnagel, W. Streifer, D. R. Scifres, and D. F. Welch, Ultrahigh-Power Semiconductor Diode Laser Arrays, *Science*, vol. 237, pp. 1305–1309, 1987.

Y. Suematsu, Advances in Semiconductor Lasers, *Physics Today*, vol. 38, no. 5, pp. 32–39, 1985.

D. Botez, Laser Diodes are Power-Packed, *IEEE Spectrum*, vol. 22, no. 6, pp. 43–53, 1985.

A. Mooradian, Laser Linewidth, *Physics Today*, vol. 38, no. 5, pp. 42–48, 1985.

W. T. Tsang, The C^3 Laser, *Scientific American*, vol. 251, no. 5, pp. 149–161, 1984.

S. Kobayashi and T. Kimura, Semiconductor Optical Amplifiers, *IEEE Spectrum*, vol. 21, no. 5, pp. 26–33, 1984.

F. Stern, Semiconductor Lasers: Theory, in *Laser Handbook*, F. T. Arecchi and E. O. Schultz-Du Bois, eds., North-Holland, Amsterdam, 1972.

Historical

R. D. Dupuis, An Introduction to the Development of the Semiconductor Laser, *IEEE Journal of Quantum Electronics*, vol. QE-23, pp. 651–657, 1987.

N. G. Basov, Quantum Electronics at the P. N. Lebedev Physics Institute of the Academy of Sciences of the USSR (FIAN), *Soviet Physics–Uspekhi*, vol. 29, pp. 179–185, 1986 [*Uspekhi Fizicheskikh Nauk*, vol. 148, pp. 313–324, 1986].

J. K. Butler, ed., *Semiconductor Injection Lasers*, IEEE Press, New York, 1980.

N. G. Basov, Semiconductor Lasers, in *Nobel Lectures in Physics, 1963–1970*, Elsevier, Amsterdam, 1972.

T. M. Quist, R. H. Rediker, R. J. Keyes, W. E. Krag, B. Lax, A. L. McWhorter, and H. J. Zeiger, Semiconductor Maser of GaAs, *Applied Physics Letters*, vol. 1, pp. 91–92, 1962.

N. Holonyak, Jr., and S. F. Bevacqua, Coherent (Visible) Light Emission from $Ga(As_{1-x}P_x)$ Junctions, *Applied Physics Letters*, vol. 1, pp. 82–83, 1962.

M. I. Nathan, W. P. Dumke, G. Burns, F. H. Dill, Jr., and G. Lasher, Stimulated Emission of Radiation from GaAs *p–n* Junctions, *Applied Physics Letters*, vol. 1, pp. 62–64, 1962.

R. N. Hall, G. E. Fenner, J. D. Kingsley, T. J. Soltys, and R. O. Carlson, Coherent Light Emission from GaAs Junctions, *Physical Review Letters*, vol. 9, pp. 366–368, 1962.

R. J. Keyes and T. M. Quist, Recombination Radiation Emitted by Gallium Arsenide, *Proceedings of the IRE*, vol. 50, pp. 1822–1823, 1962.

N. G. Basov, O. N. Krokhin, and Yu. M. Popov, Production of Negative-Temperature States in *p–n* Junctions of Degenerate Semiconductors, *Soviet Physics–JETP*, vol. 13, pp. 1320–1321, 1961 [*Zhurnal Eksperimental'noi i Teoreticheskoi Fiziki* (*USSR*), vol. 40, pp. 1879–1880, 1961].

M. G. A. Bernard and G. Duraffourg, Laser Conditions in Semiconductors, *Physica Status Solidi*, vol. 1, pp. 699–703, 1961.

J. von Neumann, in unpublished calculations sent to E. Teller in September 1953, showed that it was in principle possible to upset the equilibrium concentration of carriers in a semiconductor and thereby obtain light amplification by stimulated emission, e.g., from the recombination of electrons and holes injected into a *p–n* junction [see J. von Neumann, Notes on the Photon-Disequilibrium-Amplification Scheme (JvN), Sept. 16, 1953, *IEEE Journal of Quantum Electronics*, vol. QE-23, pp. 658–673, 1987].

PROBLEMS

16.1-1 **LED Spectral Widths.** Estimate the spectral widths of the $In_{0.72}Ga_{0.28}As_{0.6}P_{0.4}$, GaAs, and $GaAs_{0.6}P_{0.4}$ LEDs from the spectra provided in Fig. 16.1-9, in units of nm, Hz, and eV. Compare these estimates with the results calculated from the formulas given in Exercise 16.1-3.

16.1-2 **External Quantum Efficiency of an LED.** Derive an expression for η_e, the efficiency for the extraction of internal unpolarized light from an LED, that includes the angular dependence of Fresnel reflection at the semiconductor–air boundary (see Sec. 6.2).

16.1-3 **Coupling Light from an LED into an Optical Fiber.** Calculate the fraction of optical power emitted from an LED that is accepted by a step-index optical fiber of numerical aperture NA = 0.1 in air and core refractive index 1.46 (see Sec. 8.1). Assume that the LED has a planar surface, a refractive index $n = 3.6$, and an angular dependence of optical power that is proportional to $\cos^4(\theta)$. Assume further that the LED is bonded to the core of the fiber and that the emission area is smaller than the fiber core.

16.2-1 **Bandwidth of Semiconductor Laser Amplifier.** Use the data in Fig. 16.2-3(*a*) to plot the full bandwidth of the InGaAsP amplifier against the injected carrier concentration Δn. Find an approximate linear formula for this bandwidth as a function of Δn and plot the amplifier gain coefficient versus bandwidth.

16.2-2 **Peak Gain Coefficient at $T = 0$ K.** (a) Show that the peak value γ_p of the gain coefficient $\gamma_0(\nu)$ at $T = 0$ K is located at $\nu = (E_{fc} - E_{fv})/h$.
(b) Obtain an analytic expression for the peak gain coefficient γ_p as a function of the injected carrier concentration Δn at $T = 0$ K.
(c) Plot γ_p versus Δn for an InGaAsP amplifier ($\lambda_o = 1.3$ μm, $n = 3.5$, $\tau_r = 2.5$ ns, $m_c = 0.06 m_0$, $m_v = 0.4 m_0$) for values of Δn in the range 1×10^{18} to 2×10^{18} cm^{-3}.
(d) Compare the results with the data provided in Fig. 16.2-3b.

*16.2-3 **Gain Coefficient of a GaAs Amplifier.** A room-temperature ($T = 300$ K) *p*-type GaAs laser amplifier ($E_g \approx 1.40$ eV, $m_c = 0.07 m_0$, $m_v = 0.5 m_0$), with refractive index $n = 3.6$, is doped ($p_0 = 1.2 \times 10^{18}$) such that the radiative recombination lifetime $\tau_r \approx 2$ ns.
(a) Given the steady-state injected-carrier concentration Δn (which is controlled by the injection rate R and the overall recombination time τ), use (16.2-2)–(16.2-4) to compute the gain coefficient $\gamma_0(\nu)$ versus the photon energy $h\nu$, assuming that $T = 0$ K.
(b) Carry out the same calculation using a computer, assuming that $T = 300$ K.
(c) Plot the peak gain coefficient as a function of Δn for both cases.
(d) Determine the loss coefficient α and the transparency concentration Δn_T using the linear approximation model.
(e) Plot the full amplifier bandwidth (in Hz, nm, and eV) as a function of Δn for both cases.
(f) Compare your results with the gain coefficient and peak gain coefficient curves calculated by Panish and shown in Fig. P16.2-3.

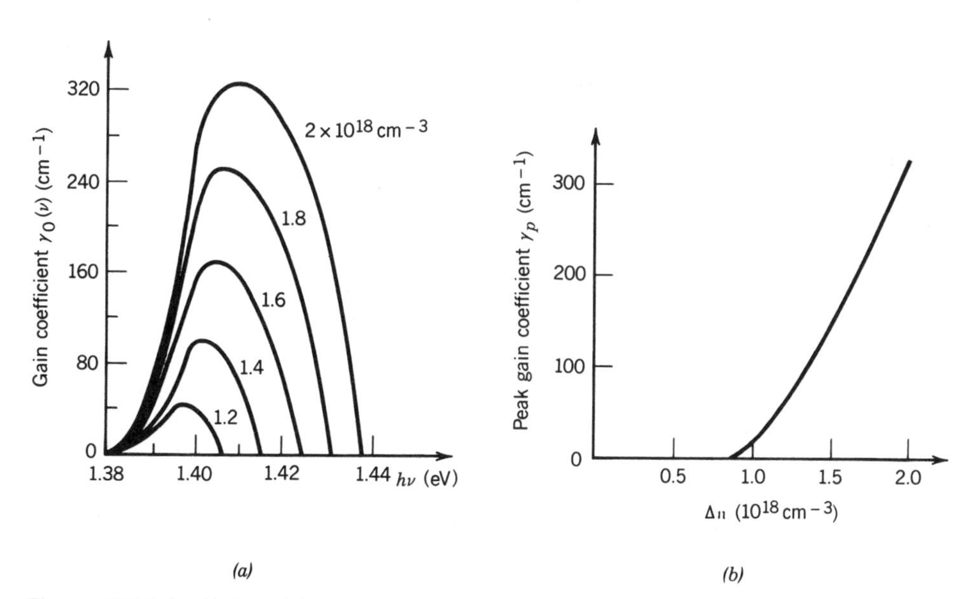

Figure P16.2-3 (Adapted from M. B. Panish, Heterostructure Injection Lasers, *Proceedings of the IEEE*, vol. 64, pp. 1512–1540, © 1976 IEEE).

16.2-4 **Bandgap Reduction Arising from Band-Tail States.** The bandgap reduction ΔE_g arising from band-tail states in InGaAsP and GaAs can be empirically expressed as

$$\Delta E_g(\text{eV}) \approx (-1.6 \times 10^{-8})(p^{1/3} + n^{1/3}),$$

where n and p are the carrier concentrations (cm^{-3}) provided by doping, carrier injection, or both.
(a) For *p*-type InGaAsP and GaAs, determine the concentration p that reduces the bandgap by approximately 0.02 eV.
(b) For undoped InGaAsP and GaAs, determine the injected carrier density Δn that reduces the bandgap by approximately 0.02 eV. Assume that n_i is negligible.
(c) Compute $E_g + \Delta E_g$ and compare the result with the energy at which the gain coefficient in Fig. 16.2-3(a) is zero on the low-frequency side.

16.2-5 **Amplifier Gain and Bandwidth.** GaAs has an intrinsic carrier concentration $n_i = 1.8 \times 10^6\ \text{cm}^{-3}$, a recombination lifetime $\tau = 50$ ns, a bandgap energy $E_g = 1.42$ eV, an effective electron mass $m_c = 0.07m_0$, and an effective hole mass $m_v = 0.5m_0$. Assume that $T = 0$ K.
(a) Determine the center frequency, bandwidth, and peak net gain within the bandwidth for a GaAs amplifier of length $d = 200\ \mu\text{m}$, width $w = 10\ \mu\text{m}$, and thickness $l = 2\ \mu\text{m}$, when 1 mA of current is passed through the device.
(b) Determine the number of voice messages that can be supported by the bandwidth determined above, given that each message occupies a bandwidth of 4 kHz.
(c) Determine the bit rate that can be passed through the amplifier given that each voice channel requires 64 kbits/s.

16.2-6 **Transition Cross Section.** Determine the transition cross section $\sigma(\nu)$ for GaAs as a function of Δn at $T = 0$ K. The transition probability is $\phi\sigma(\nu)$, where ϕ is the photon-flux density. Why is the transition cross section less useful for semiconductor laser amplifiers than for other laser amplifiers?

*16.2-7 **Gain Profile.** Consider a 1.55-μm InGaAsP amplifier ($n = 3.5$) of the configuration shown in Fig. 16.2-6, with identical antireflection coatings on its input and output facets. Calculate the maximum reflectivity of each of the facets that can be tolerated if it is desired to maintain the variations in the gain profile arising from the frequency dependence of the Fabry–Perot transmittance to less than 10% [see (9.1-29)].

16.3-1 **Dependence of Output Power on Refractive Index.** Identify the terms in the output photon flux Φ_o given in (16.3-10) that depend on the refractive index of the crystal.

16.3-2 **Longitudinal Modes.** A current is injected into an InGaAsP diode of bandgap energy $E_g = 0.91$ eV and refractive index $n = 3.5$ such that the difference in Fermi levels is $E_{fc} - E_{fv} = 0.96$ eV. If the resonator is of length $d = 250\ \mu\text{m}$ and has no losses, determine the maximum number of longitudinal modes that can oscillate.

16.3-3 **Minimum Gain Required for Lasing.** A 500-μm-long InGaAsP crystal operates at a wavelength where its refractive index $n = 3.5$. Neglecting scattering and other losses, determine the gain coefficient required to barely compensate for reflection losses at the crystal boundaries.

*16.3-4 **Modal Spacings with a Wavelength-Dependent Refractive Index.** The frequency separation of the modes of a laser diode is complicated by the fact that the refractive index is wavelength dependent [i.e., $n = n(\lambda_o)$]. A laser diode of length 430 μm oscillates at a central wavelength $\lambda_c = 650$ nm. Within the emission bandwidth, $n(\lambda_o)$ may be assumed to be linearly dependent on λ_o [i.e., $n(\lambda_o) = n_0 - a(\lambda_o - \lambda_c)$, where $n_0 = n(\lambda_c) = 3.4$ and $a = dn/d\lambda_o$].

(a) The separation between the laser modes with wavelength near λ_c was observed to be $\Delta\lambda \approx 0.12$ nm. Explain why this does not correspond to the usual modal spacing $\nu_F = c/2d$.

(b) Find an estimate of a.

(c) Explain the phenomenon of mode pulling in a gas laser and compare it with the effect described above in semiconductor lasers.

CHAPTER

17

SEMICONDUCTOR PHOTON DETECTORS

17.1 PROPERTIES OF SEMICONDUCTOR PHOTODETECTORS
- A. Quantum Efficiency
- B. Responsivity
- C. Response Time

17.2 PHOTOCONDUCTORS

17.3 PHOTODIODES
- A. The *p-n* Photodiode
- B. The *p-i-n* Photodiode
- C. Heterostructure Photodiodes
- D. Array Detectors

17.4 AVALANCHE PHOTODIODES
- A. Principles of Operation
- B. Gain and Responsivity
- C. Response Time

17.5 NOISE IN PHOTODETECTORS
- A. Photoelectron Noise
- B. Gain Noise
- C. Circuit Noise
- D. Signal-to-Noise Ratio and Receiver Sensitivity

Heinrich Hertz (1857–1894) discovered photoemission in 1887.

Siméon Poisson (1781–1840) developed the probability distribution that describes photodetector noise.

A photodetector is a device that measures photon flux or optical power by converting the energy of the absorbed photons into a measurable form. Photographic film is probably the most ubiquitous of photodetectors. Two principal classes of photodetectors are in common use: **thermal detectors** and **photoelectric detectors**:

- Thermal detectors operate by converting photon energy into heat. However, most thermal detectors are rather inefficient and relatively slow as a result of the time required to change their temperature. Consequently, they are not suitable for most applications in photonics.
- The operation of photoelectric detectors is based on the **photoeffect**, in which the absorption of photons by some materials results directly in an electronic transition to a higher energy level and the generation of mobile charge carriers. Under the effect of an electric field these carriers move and produce a measurable electric current.

We consider only photoelectric detectors in this chapter.

The photoeffect takes two forms: external and internal. The former process involves **photoelectric emission**, in which the photogenerated electrons escape from the material as free electrons. In the latter process, **photoconductivity**, the excited carriers remain within the material, usually a semiconductor, and serve to increase its conductivity.

The External Photoeffect: Photoelectron Emission

If the energy of a photon illuminating the surface of a material in vacuum is sufficiently large, the excited electron can escape over the potential barrier of the material surface and be liberated into the vacuum as a free electron. The process, called **photoelectron emission**, is illustrated in Fig. 17.0-1(a). A photon of energy $h\nu$ incident on a metal releases a free electron from within the partially filled conduction band. Energy conservation requires that electrons emitted from below the Fermi level, where they

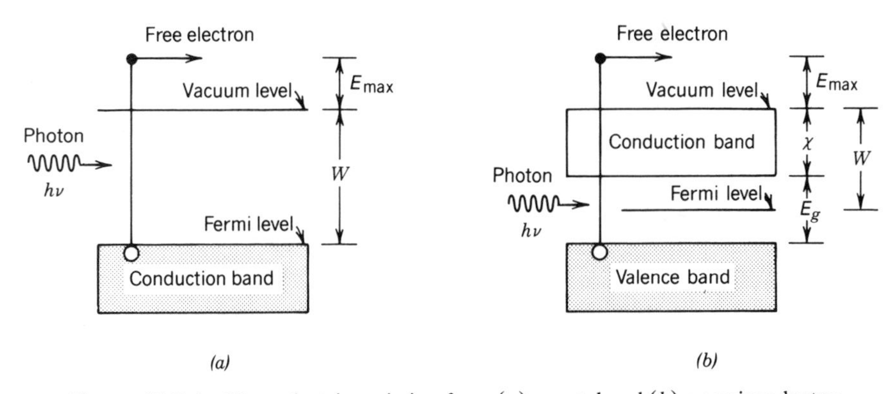

Figure 17.0-1 Photoelectric emission from (a) a metal and (b) a semiconductor.

are plentiful, have a maximum kinetic energy

$$E_{\max} = h\nu - W, \tag{17.0-1}$$

where the work function W is the energy difference between the vacuum level and the Fermi level of the material. Equation (17.0-1) is known as Einstein's photoemission equation. Only if the electron initially lies at the Fermi level can it receive the maximum kinetic energy specified in (17.0-1); the removal of a deeper-lying electron requires additional energy to transport it to the Fermi level, thereby reducing the kinetic energy of the liberated electron. The lowest work function for a metal (Cs) is about 2 eV, so that optical detectors based on the external photoeffect from pure metals are useful only in the visible and ultraviolet regions of the spectrum.

Photoelectric emission from a semiconductor is shown schematically in Fig. 17.0-1(*b*). Photoelectrons are usually released from the valence band, where electrons are plentiful. The formula analogous to Eq. (17.0-1) is

$$E_{\max} = h\nu - (E_g + \chi), \tag{17.0-2}$$

where E_g is the energy gap and χ is the electron affinity of the material (the energy difference between the vacuum level and the bottom of the conduction band). The energy $E_g + \chi$ can be as low as 1.4 eV for certain materials (e.g., NaKCsSb, which forms the basis for the S–20 photocathode), so that semiconductor photoemissive detectors can operate in the near infrared, as well as in the visible and ultraviolet. Furthermore, negative-electron-affinity semiconductors have been developed in which the bottom of the conduction band lies above the vacuum level in the bulk of the material, so that $h\nu$ need only exceed E_g for photoemission to occur (at the surface of the material the bands bend so that the conduction band does indeed lie below the vacuum level). These detectors are therefore responsive to slightly longer wavelengths in the near infrared, and exhibit improved quantum efficiency and reduced dark current. Photocathodes constructed of multiple layers or inhomogeneous materials, such as the S–1 photocathode, can also be used in the near infrared.

Photodetectors based on photoelectric emission usually take the form of vacuum tubes called **phototubes**. Electrons are emitted from the surface of a photoemissive material (cathode) and travel to an electrode (anode), which is maintained at a higher electric potential [Fig. 17.0-2(*a*)]. As a result of the electron transport between the cathode and anode, an electric current proportional to the photon flux is created in the circuit. The photoemitted electrons may also impact other specially placed metal or semiconductor surfaces in the tube, called dynodes, from which a cascade of electrons is emitted by the process of **secondary emission**. The result is an amplification of the generated electric current by a factor as high as 10^7. This device, illustrated in Fig. 10.0-2(*b*), is known as a **photomultiplier tube**.

A modern imaging device that makes use of this principle is the **microchannel plate**. It consists of an array of millions of capillaries (of internal diameter $\approx 10\ \mu$m) in a glass plate of thickness ≈ 1 mm. Both faces of the plate are coated with thin metal films that act as electrodes and a voltage is applied across them [Fig. 17.0-2(*c*)]. The interior walls of each capillary are coated with a secondary-electron-emissive material and behave as a continuous dynode, multiplying the photoelectron current emitted at that position [Fig. 17.0-2(*d*)]. The local photon flux in an image can therefore be rapidly converted into a substantial electron flux that can be measured directly. Furthermore, the electron flux can be reconverted into an (amplified) optical image by using a phosphor coating as the rear electrode to provide electroluminescence; this combination provides an **image intensifier**.

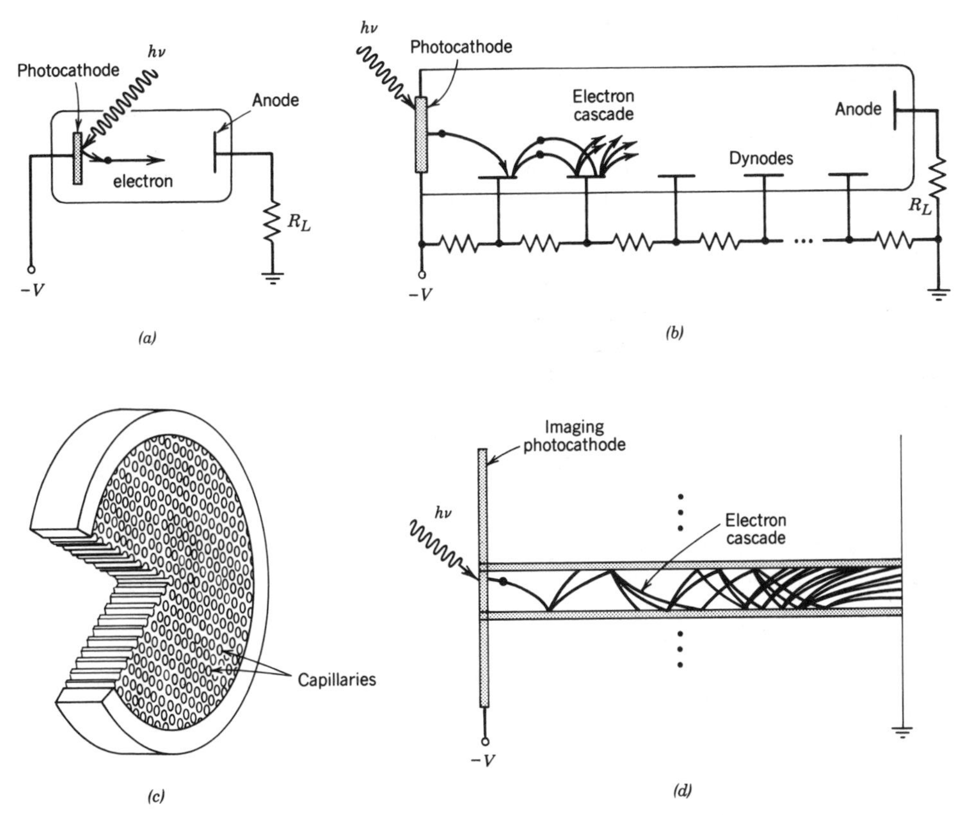

Figure 17.0-2 (*a*) Phototube. (*b*) Photomultiplier tube with semitransparent photocathode. (*c*) Cutaway view of microchannel plate. (*d*) Single capillary in a microchannel plate.

The Internal Photoeffect

Many modern photodetectors operate on the basis of the internal photoeffect, in which the photoexcited carriers (electrons and holes) remain within the sample. The most important of the internal photoeffects is **photoconductivity**. Photoconductor detectors rely directly on the light-induced increase in the electrical conductivity, which is exhibited by almost all semiconductor materials. The absorption of a photon by an intrinsic photoconductor results in the *generation* of a free electron excited from the valence band to the conduction band (Fig. 17.0-3). Concurrently, a hole is generated in the valence band. The application of an electric field to the material results in the *transport* of both electrons and holes through the material and the consequent production of an electric current in the electrical circuit of the detector.

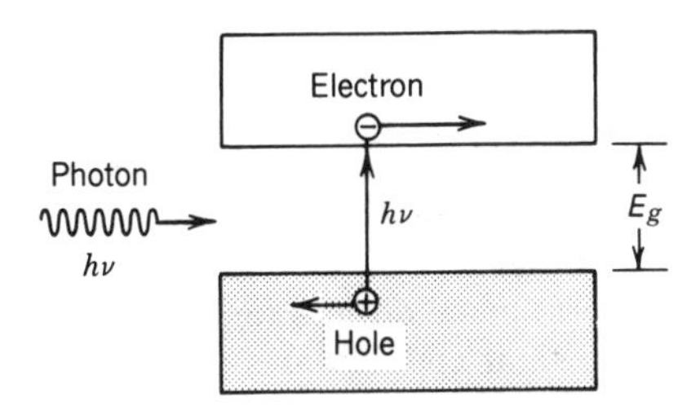

Figure 17.0-3 Electron–hole photogeneration in a semiconductor.

The semiconductor **photodiode** detector is a *p-n* junction structure that is also based on the internal photoeffect. Photons absorbed in the depletion layer *generate* electrons and holes which are subjected to the local electric field within that layer. The two carriers drift in opposite directions. Such a *transport* process induces an electric current in the external circuit.

Some photodetectors incorporate internal gain mechanisms so that the photoelectron current can be physically amplified within the detector and thus make the signal more easily detectable. If the depletion-layer electric field in a photodiode is increased by applying a sufficiently large reverse bias across the junction, the electrons and holes generated may acquire sufficient energy to liberate more electrons and holes within this layer by a process of impact ionization. Devices in which this internal *amplification* process occurs are known as **avalanche photodiodes** (APDs). Such detectors can be used as an alternative to (or in conjunction with) a laser amplifier (see Chaps. 13 and 16), in which the optical signal is amplified before detection. Each of these amplification mechanisms introduces its own form of noise, however.

In brief, semiconductor photoelectric detectors with gain involve the following three basic processes:

- *Generation:* Absorbed photons generate free carriers.
- *Transport:* An applied electric field induces these carriers to move, which results in a circuit current.
- *Amplification:* In APDs, large electric fields impart sufficient energy to the carriers so that they, in turn, free additional carriers by impact ionization. This internal amplification process enhances the responsivity of the detector.

This chapter is devoted to three types of semiconductor photodetectors: photoconductors, photodiodes, and avalanche photodiodes. All of these rely on the internal photoeffect as the generation mechanism. In Sec. 17.1 several important general properties of these detectors are discussed, including quantum efficiency, responsivity, and response time. The properties of photoconductor detectors are addressed in Sec. 17.2. The operation of photodiodes and avalanche photodiodes are considered in Secs. 17.3 and 17.4, respectively.

To assess the performance of semiconductor photodetectors in various applications, it is important to understand their noise properties, and these are set forth in Sec. 17.5. Noise in the output circuit of a photoelectric detector arises from several sources: the photon character of the light itself (photon noise), the conversion of photons to photocarriers (photoelectron noise), the generation of secondary carriers by internal amplification (gain noise), as well as receiver circuit noise. A brief discussion of the performance of an optical receiver is provided; we return to this topic in Sec. 22.4 in connection with the performance of fiber-optic communication systems.

17.1 PROPERTIES OF SEMICONDUCTOR PHOTODETECTORS

Certain fundamental rules govern all semiconductor photodetectors. Before studying details of the particular detectors of interest, we examine the quantum efficiency, responsivity, and response time of photoelectric detectors from a general point of view.

Semiconductor photodetectors and semiconductor photon sources are inverse devices. Detectors convert an input photon flux to an output electric current; sources achieve the opposite. The same materials are often used to make devices for both. The performance measures discussed in this section all have their counterparts in sources, as has been discussed in Chap. 16.

A. Quantum Efficiency

The **quantum efficiency** η $(0 \le \eta \le 1)$ of a photodetector is defined as the probability that a single photon incident on the device generates a photocarrier pair that contributes to the detector current. When many photons are incident, as is almost always the case, η is the ratio of the flux of generated electron–hole pairs that contribute to the detector current to the flux of incident photons. Not all incident photons produce electron–hole pairs because not all incident photons are absorbed. This is illustrated in Fig. 17.1-1. Some photons simply fail to be absorbed because of the probabilistic nature of the absorption process (the rate of photon absorption in a semiconductor material was derived in Sec. 15.2B). Others may be reflected at the surface of the detector, thereby reducing the quantum efficiency further. Furthermore, some electron–hole pairs produced near the surface of the detector quickly recombine because of the abundance of recombination centers there and are therefore unable to contribute to the detector current. Finally, if the light is not properly focused onto the active area of the detector, some photons will be lost. This effect is not included in the definition of the quantum efficiency, however, because it is associated with the use of the device rather than with its intrinsic properties.

The quantum efficiency can therefore be written as

$$\eta = (1 - \mathscr{R})\zeta[1 - \exp(-\alpha d)], \tag{17.1-1}$$

Quantum Efficiency

where $\mathscr{R}$ is the optical power reflectance at the surface, ζ the fraction of electron–hole pairs that contribute successfully to the detector current, α the absorption coefficient of the material (cm^{-1}) discussed in Sec. 15.2B, and d the photodetector depth. Equation (17.1-1) is a product of three factors:

- The first factor $(1 - \mathscr{R})$ represents the effect of reflection at the surface of the device. Reflection can be reduced by the use of antireflection coatings.
- The second factor ζ is the fraction of electron–hole pairs that successfully avoid recombination at the material surface and contribute to the useful photocurrent. Surface recombination can be reduced by careful material growth.
- The third factor, $\int_0^d e^{-\alpha x}\,dx / \int_0^\infty e^{-\alpha x}\,dx = [1 - \exp(-\alpha d)]$, represents the fraction of the photon flux absorbed in the bulk of the material. The device should have a sufficiently large value of d to maximize this factor.

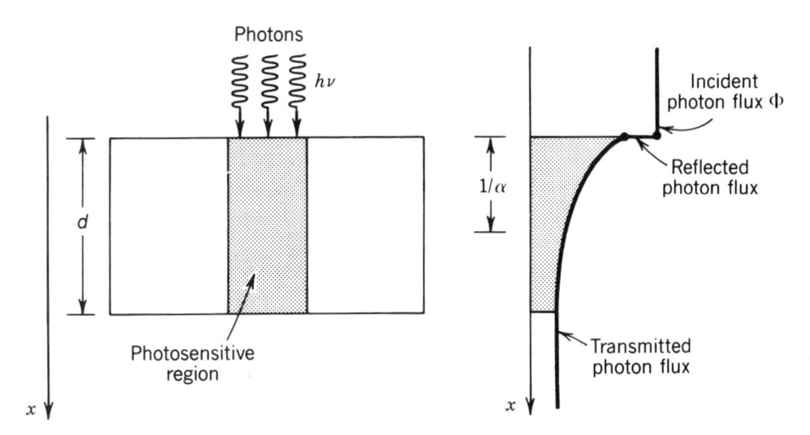

Figure 17.1-1 Effect of absorption on the quantum efficiency η.

It should be noted that some definitions of the quantum efficiency η exclude reflection at the surface, which must then be considered separately.

Dependence of η on Wavelength

The quantum efficiency η is a function of wavelength, principally because the absorption coefficient α depends on wavelength (see Fig. 15.2-2). For photodetector materials of interest, η is large within a spectral window that is determined by the characteristics of the material. For sufficiently large λ_o, η becomes small because absorption cannot occur when $\lambda_o \geq \lambda_g = hc_o/E_g$ (the photon energy is then insufficient to overcome the bandgap). The bandgap wavelength λ_g is the **long-wavelength limit** of the semiconductor material (see Chap. 15). Representative values of E_g and λ_g are shown in Figs. 15.1-5 and 15.1-6 (see also Table 15.1-3) for selected intrinsic semiconductor materials. For sufficiently small values of λ_o, η also decreases, because most photons are then absorbed near the surface of the device (e.g., for $\alpha = 10^4$ cm^{-1}, most of the light is absorbed within a distance $1/\alpha = 1$ μm). The recombination lifetime is quite short near the surface, so that the photocarriers recombine before being collected.

B. Responsivity

The **responsivity** relates the electric current flowing in the device to the incident optical power. If every photon were to generate a single photoelectron, a photon flux Φ (photons per second) would produce an electron flux Φ, corresponding to a short-circuit electric current $i_p = e\Phi$. An optical power $P = h\nu\Phi$ (watts) at frequency ν would then give rise to an electric current $i_p = eP/h\nu$. Since the fraction of photons producing detected photoelectrons is η rather than unity, the electric current is

$$i_p = \eta e\Phi = \frac{\eta eP}{h\nu} = \Re P. \tag{17.1-2}$$

The proportionality factor $\Re$, between the electric current and the optical power, is defined as the responsivity $\Re$ of the device. $\Re = i_p/P$ has units of A/W and is given by

$$\boxed{\Re = \frac{\eta e}{h\nu} = \eta \frac{\lambda_o}{1.24}.} \tag{17.1-3}$$

Photodetector Responsivity (A/W) (λ_o in μm)

$\Re$ increases with λ_o because photoelectric detectors are responsive to the photon flux rather than to the optical power. As λ_o increases, a given optical power is carried by more photons, which, in turn, produce more electrons. The region over which $\Re$ increases with λ_o is limited, however, since the wavelength dependence of η comes into play for both long and short wavelengths. It is important to distinguish the detector responsivity defined here (A/W) from the light-emitting-diode responsivity (W/A) defined in (16.1-25).

The responsivity can be degraded if the detector is presented with an excessively large optical power. This condition, which is called detector saturation, limits the detector's **linear dynamic range**, which is the range over which it responds linearly with the incident optical power.

An appreciation for the order of magnitude of the responsivity is gained by setting $\eta = 1$ in (17.1-3), whereupon $\Re = 1$ A/W, i.e., 1 nW $\rightarrow$ 1 nA, at $\lambda_o = 1.24$ μm. The linear increase of the responsivity with wavelength, for a given fixed value of η, is

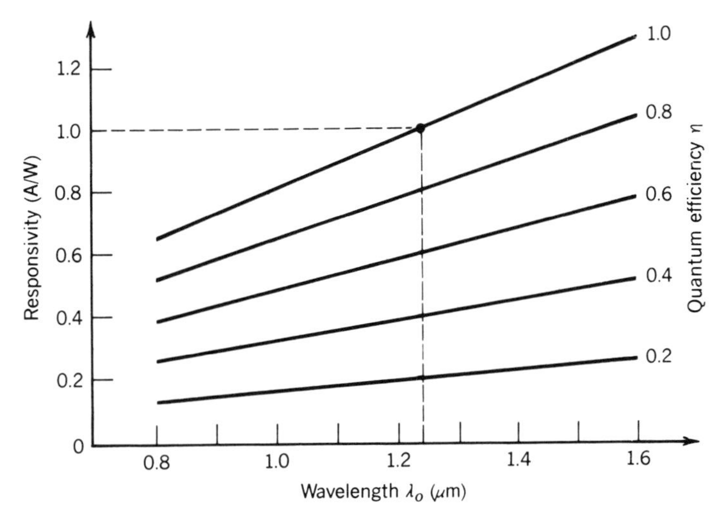

Figure 17.1-2 Responsivity $\mathfrak{R}$ (A/W) versus wavelength λ_o with the quantum efficiency η as a parameter. $\mathfrak{R} = 1$ A/W at $\lambda_o = 1.24\ \mu$m when $\eta = 1$.

illustrated in Fig. 17.1-2. $\mathfrak{R}$ is also seen to increase linearly with η if λ_o is fixed. For thermal detectors $\mathfrak{R}$ is independent of λ_o because they respond directly to optical power rather than to the photon flux.

Devices with Gain

The formulas presented above are predicated on the assumption that each carrier produces a charge e in the detector circuit. However, many devices produce a charge q in the circuit that differs from e. Such devices are said to exhibit gain. The gain G is the average number of circuit electrons generated per photocarrier pair. G should be distinguished from η, which is the probability that an incident photon produces a detectable photocarrier pair. The gain, which is defined as

$$G = \frac{q}{e}, \tag{17.1-4}$$

can be either greater than or less than unity, as will be seen subsequently. Therefore, more general expressions for the photocurrent and responsivity are

$$\boxed{i_p = \eta q \Phi = G \eta e \Phi = \frac{G \eta e P}{h\nu}} \tag{17.1-5}$$

Photocurrent

and

$$\boxed{\mathfrak{R} = \frac{G \eta e}{h\nu} = G\eta \frac{\lambda_o}{1.24},} \tag{17.1-6}$$

Responsivity in the Presence of Gain (A/W) (λ_o in μm)

respectively.

Other useful measures of photodetector behavior, such as signal-to-noise ratio and receiver sensitivity, must await a discussion of the detector noise properties presented in Sec. 17.5.

C. Response Time

One might be inclined to argue that the charge generated in an external circuit should be $2e$ when a photon generates an electron–hole pair in a photodetector material, since there are two charge carriers. In fact, the charge generated is e, as we will show below. Furthermore, the charge delivered to the external circuit by carrier motion in the photodetector material is not provided instantaneously but rather occupies an extended time. It is as if the motion of the charged carriers in the material draws charge slowly from the wire on one side of the device and pushes it slowly into the wire at the other side so that each charge passing through the external circuit is spread out in time. This phenomenon is known as **transit-time spread**. It is an important limiting factor for the speed of operation of all semiconductor photodetectors.

Consider an electron–hole pair generated (by photon absorption, for example) at an arbitrary position x in a semiconductor material of width w to which a voltage V is applied, as shown in Fig. 17.1-3(a). We restrict our attention to motion in the x direction. A carrier of charge Q (a hole of charge $Q = e$ or an electron of charge $Q = -e$) moving with a velocity $v(t)$ in the x direction creates a current in the external circuit given by

$$i(t) = -\frac{Q}{w}v(t). \tag{17.1-7}$$

Ramo's Theorem

This important formula, known as **Ramo's theorem**, can be proved with the help of an

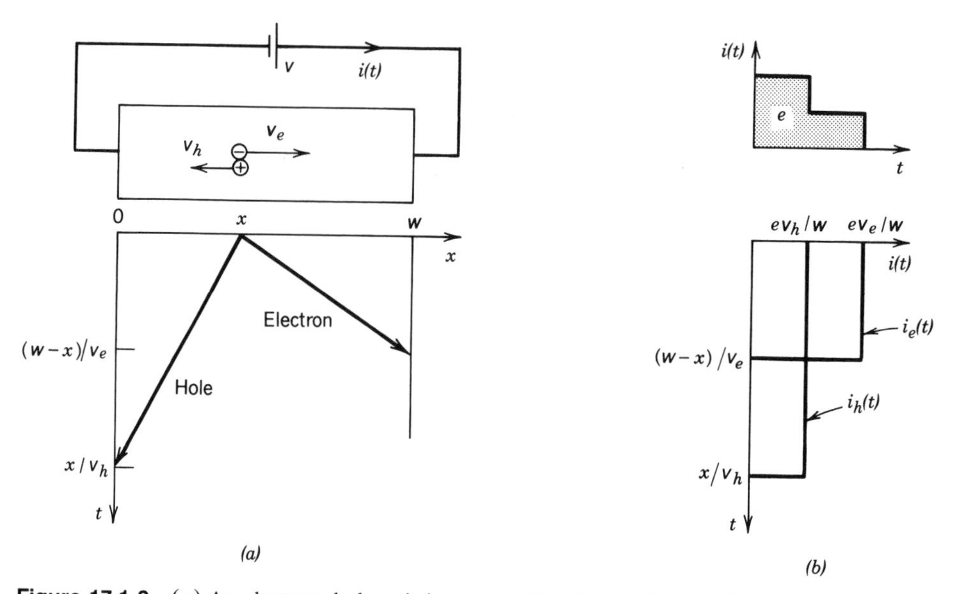

Figure 17.1-3 (a) An electron–hole pair is generated at the position x. The hole moves to the left with velocity v_h and the electron moves to the right with velocity v_e. The process terminates when the carriers reach the edge of the material. (b) Hole current $i_h(t)$, electron current $i_e(t)$, and total current $i(t)$ induced in the circuit. The total charge induced in the circuit is e.

energy argument. If the charge moves a distance dx in the time dt, under the influence of an electric field of magnitude $E = V/w$, the work done is $-QE\,dx = -Q(V/w)\,dx$. This work must equal the energy provided by the external circuit, $i(t)V\,dt$. Thus $i(t)V\,dt = -Q(V/w)\,dx$ from which $i(t) = -(Q/w)(dx/dt) = -(Q/w)v(t)$, as promised.

In the presence of a uniform charge density ϱ, instead of a single point charge Q, the total charge is ϱAw, where A is the cross-sectional area, so that (17.1-7) gives $i(t) = -(\varrho Aw/w)v(t) = -\varrho Av(t)$ from which the current density in the x direction $J(t) = -i(t)/A = \varrho v(t)$.

In the presence of an electric field E, a charge carrier in a semiconductor will drift at a mean velocity

$$v = \mu E, \tag{17.1-8}$$

where μ is the carrier mobility. Thus, $J = \sigma E$, where $\sigma = \mu\varrho$ is the conductivity.

Assuming that the hole moves with constant velocity v_h to the left, and the electron moves with constant velocity v_e to the right, (17.1-7) tells us that the hole current $i_h = -e(-v_h)/w$ and the electron current $i_e = -(-e)v_e/w$, as illustrated in Fig. 17.1-3(b). Each carrier contributes to the current as long as it is moving. If the carriers continue their motion until they reach the edge of the material, the hole moves for a time x/v_h and the electron moves for a time $(w - x)/v_e$ [see Fig. 17.1-3(a)]. In semiconductors, v_e is generally larger than v_h so that the full width of the transit-time spread is x/v_h.

The total charge q induced in the external circuit is the sum of the areas under i_e and i_h

$$q = e\frac{v_h}{w}\frac{x}{v_h} + e\frac{v_e}{w}\frac{w - x}{v_e} = e\left(\frac{x}{w} + \frac{w - x}{w}\right) = e,$$

as promised. The result is independent of the position x at which the electron–hole pair was created.

The transit-time spread is even more severe if the electron–hole pairs are generated uniformly throughout the material, as shown in Fig. 17.1-4. For $v_h < v_e$, the full width of the transit-time spread is then w/v_h rather than x/v_h. This occurs because uniform illumination produces carrier pairs everywhere, including at $x = w$, which is the point at which the holes have the farthest to travel before being able to recombine at $x = 0$.

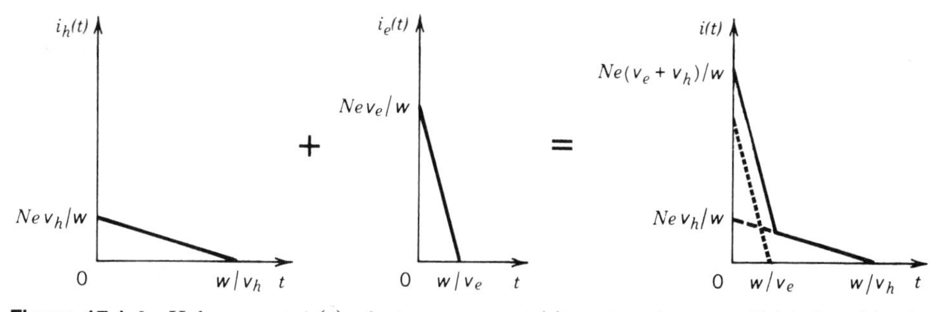

Figure 17.1-4 Hole current $i_h(t)$, electron current $i_e(t)$, and total current $i(t)$ induced in the circuit for electron–hole generation by N photons uniformly distributed between 0 and w (see Problem 17.1-4). The tail in the total current results from the motion of the holes. $i(t)$ can be viewed as the impulse-response function (see Appendix B) for a uniformly illuminated detector subject to transit-time spread.

Another response-time limit of semiconductor detectors is the RC time constant formed by the resistance R and capacitance C of the photodetector and its circuitry. The combination of resistance and capacitance serves to integrate the current at the output of the detector, and thereby to lengthen the impulse-response function. The impulse-response function in the presence of transit-time *and* simple RC time-constant spread is determined by convolving $i(t)$ in Fig. 17.1-4 with the exponential function $(1/RC)\exp(-t/RC)$ (see Appendix B, Sec. B.1). Photodetectors of different types have other specific limitations on their speed of response; these are considered at the appropriate point.

As a final point, we mention that photodetectors of a given material and structure often exhibit a fixed gain–bandwidth product. Increasing the gain results in a decrease of the bandwidth, and vice versa. This trade-off between sensitivity and frequency response is associated with the time required for the gain process to take place.

17.2 PHOTOCONDUCTORS

When photons are absorbed by a semiconductor material, mobile charge carriers are *generated* (an electron–hole pair for every absorbed photon). The electrical conductivity of the material increases in proportion to the photon flux. An electric field applied to the material by an external voltage source causes the electrons and holes to be *transported*. This results in a measurable electric current in the circuit, as shown in Fig. 17.2-1. **Photoconductor detectors** operate by registering either the photocurrent i_p, which is proportional to the photon flux Φ, or the voltage drop across a load resistor R placed in series with the circuit.

The semiconducting material may take the form of a slab or a thin film. The anode and cathode contacts are often placed on the same surface of the material, interdigitating with each other to maximize the light transmission while minimizing the transit time (see Fig. 17.2-1). Light can also be admitted from the bottom of the device if the substrate has a sufficiently large bandgap (so that it is not absorptive).

The increase in conductivity arising from a photon flux Φ (photons per second) illuminating a semiconductor volume wA (see Fig. 17.2-1) may be calculated as follows. A fraction η of the incident photon flux is absorbed and gives rise to excess

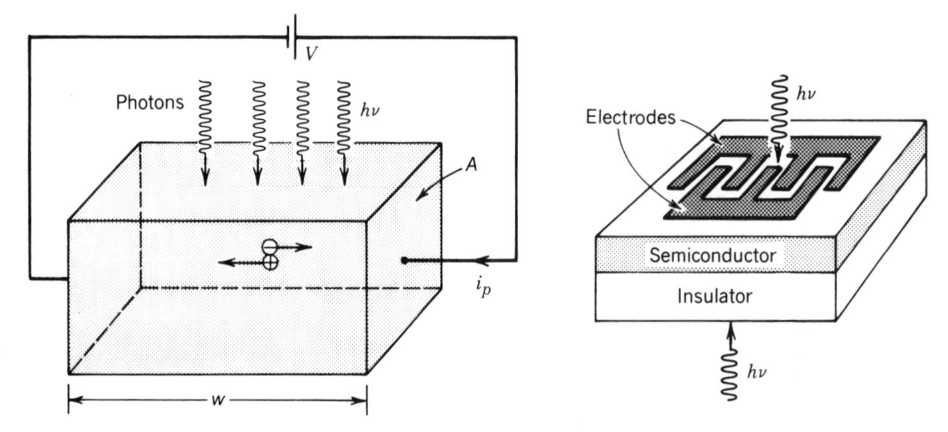

Figure 17.2-1 The photoconductor detector. Photogenerated carrier pairs move in response to the applied voltage V, generating a photocurrent i_p proportional to the incident photon flux. The interdigitated electrode structure shown is designed to maximize both the light reaching the semiconductor and the device bandwidth (by minimizing the carrier transit time).

electron–hole pairs. The pair-production rate R (per unit volume) is therefore $R = \eta\Phi/wA$. If τ is the excess-carrier recombination lifetime, electrons are lost at the rate $\Delta n/\tau$ where Δn is the photoelectron concentration (see Chap. 15). Under steady-state conditions both rates are equal ($R = \Delta n/\tau$) so that $\Delta n = \eta\tau\Phi/wA$. The increase in the charge carrier concentration therefore results in an increase in the conductivity given by

$$\Delta\sigma = e\Delta n(\mu_e + \mu_h) = \frac{e\eta\tau(\mu_e + \mu_h)}{wA}\Phi, \tag{17.2-1}$$

where μ_e and μ_h are the electron and hole mobilities. Thus the increase in conductivity is proportional to the photon flux.

Since the current density $J_p = \Delta\sigma E$ and $v_e = \mu_e E$ and $v_h = \mu_h E$ where E is the electric field, (17.2-1) gives $J_p = [e\eta\tau(v_e + v_h)/wA]\Phi$ corresponding to an electric current $i_p = AJ_p = [e\eta\tau(v_e + v_h)/w]\Phi$. If $v_h \ll v_e$ and $\tau_e = w/v_e$,

$$i_p \approx e\eta\frac{\tau}{\tau_e}\Phi. \tag{17.2-2}$$

In accordance with (17.1-5), the ratio τ/τ_e in (17.2-2) corresponds to the detector gain $G = \tau/\tau_e$, as explained subsequently.

Gain

The responsivity of a photoconductor is given by (17.1-6). The device exhibits an internal gain which, simply viewed, comes about because the recombination lifetime and transit time generally differ. Suppose that electrons travel faster than holes (see Fig. 17.2-1) and that the recombination lifetime is very long. As the electron and hole are transported to opposite sides of the photoconductor, the electron completes its trip sooner than the hole. The requirement of current continuity forces the external circuit to provide another electron immediately, which enters the device from the wire at the left. This new electron moves quickly toward the right, again completing its trip before the hole reaches the left edge. This process continues until the electron recombines with the hole. A single photon absorption can therefore result in an electron passing through the external circuit many times. The expected number of trips that the electron makes before the process terminates is

$$G = \frac{\tau}{\tau_e}, \tag{17.2-3}$$

where τ is the excess-carrier recombination lifetime and $\tau_e = w/v_e$ is the electron transit time across the sample. The charge delivered to the circuit by a single electron–hole pair in this case is $q = Ge > e$ so that the device exhibits gain.

However, the recombination lifetime may be sufficiently short such that the carriers recombine before reaching the edge of the material. This can occur provided that there is a ready availability of carriers of the opposite type for recombination. In that case $\tau < \tau_e$ and the gain is less than unity so that, on average, the carriers contribute only a fraction of the electronic charge e to the circuit. Charge is, of course, conserved and the many carrier pairs present deliver an integral number of electronic charges to the circuit.

The photoconductor gain $G = \tau/\tau_e$ can be interpreted as the fraction of the sample length traversed by the average excited carrier before it undergoes recombination. The transit time τ_e depends on the dimensions of the device and the applied voltage via (17.1-8); typical values of $w = 1$ mm and $v_e = 10^7$ cm/s give $\tau_e \approx 10^{-8}$ s. The

TABLE 17.2-1 Selected Extrinsic Semiconductor Materials with Their Activation Energy and Long-Wavelength Limit

Semiconductor:Dopant	E_A (eV)	λ_A (μm)
Ge:Hg	0.088	14
Ge:Cu	0.041	30
Ge:Zn	0.033	38
Ge:B	0.010	124
Si:B	0.044	28

recombination lifetime τ can range from 10^{-13} s to many seconds, depending on the photoconductor material and doping [see (15.1-17)]. Thus G can assume a broad range of values, both below unity and above unity, depending on the parameters of the material, the size of the device, and the applied voltage. The gain of a photoconductor cannot generally exceed 10^6, however, because of the restrictions imposed by space-charge-limited current flow, impact ionization, and dielectric breakdown.

Spectral Response

The spectral sensitivity of photoconductors is governed principally by the wavelength dependence of η, as discussed in Sec. 17.1A. Different intrinsic semiconductors have different long-wavelength limits, as indicated in Chap. 15. Ternary and quaternary compound semiconductors are also used. Photoconductor detectors (unlike photoemissive detectors) can operate well into the infrared region on band-to-band transitions. However, operation at wavelengths beyond about 2 μm requires that the devices be cooled to minimize the thermal excitation of electrons into the conduction band in these low-gap materials.

At even longer wavelengths extrinsic photoconductors can be used as detectors. **Extrinsic photoconductivity** operates on transitions involving forbidden-gap energy levels. It takes place when the photon interacts with a bound electron at a donor site, producing a free electron and a bound hole [or conversely, when it interacts with a bound hole at an acceptor site, producing a free hole and a bound electron as shown in Fig. 15.2-1(*b*)]. Donor and acceptor levels in the bandgap of doped semiconductor materials can have very low activation energies E_A. In this case the long-wavelength limit is $\lambda_A = hc_o/E_A$. These detectors must be cooled to avoid thermal excitation; liquid He at 4 K is often used. Representative values of E_A and λ_A are provided in Table 17.2-1 for selected extrinsic semiconductor materials.

The spectral responses of several extrinsic photoconductor detectors are shown in Fig. 17.2-2. The responsivity increases approximately linearly with λ_o, in accordance

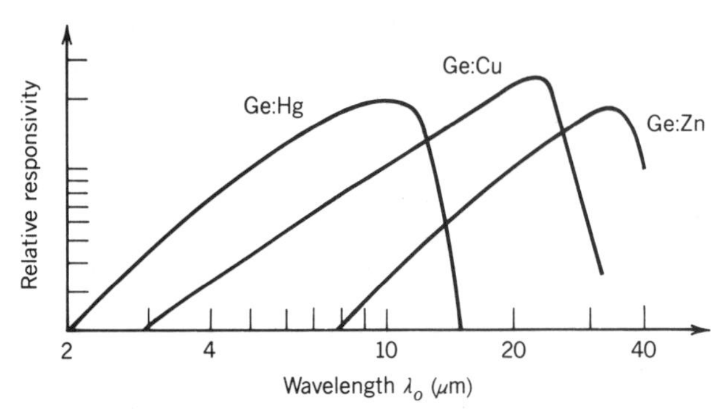

Figure 17.2-2 Relative responsivity versus wavelength λ_o (μm) for three doped-Ge extrinsic infrared photoconductor detectors.

with (17.1-6), peaks slightly below the long-wavelength limit λ_A and falls off beyond it. The quantum efficiency for these detectors can be quite high (e.g., $\eta \approx 0.5$ for Ge:Cu), although the gain may be low under usual operating conditions (e.g., $G \approx 0.03$ for Ge:Hg).

Response Time

The response time of photoconductor detectors is, of course, constrained by the transit-time and *RC* time-constant considerations presented in Sec. 17.1C. The carrier-transport response time is approximately equal to the recombination time τ, so that the carrier-transport bandwidth B is inversely proportional to τ. Since the gain G is proportional to τ in accordance with (17.2-3), increasing τ increases the gain, which is desirable, but it also decreases the bandwidth, which is undesirable. Thus the gain–bandwidth product GB is roughly independent of τ. Typical values of GB extend up to $\approx 10^9$.

17.3 PHOTODIODES

A. The *p-n* Photodiode

As with photoconductors, photodiode detectors rely on photogenerated charge carriers for their operation. A photodiode is a *p-n* junction (see Sec. 15.1E) whose reverse current increases when it absorbs photons. Although *p-n* and *p-i-n* photodiodes are generally faster than photoconductors, they do not exhibit gain.

Consider a reverse-biased *p-n* junction under illumination, as depicted in Fig. 17.3-1. Photons are absorbed everywhere with absorption coefficient α. Whenever a photon is absorbed, an electron–hole pair is *generated*. But only where an electric field is present can the charge carriers be *transported* in a particular direction. Since a *p-n* junction can support an electric field only in the depletion layer, this is the region in which it is desirable to generate photocarriers.

There are, however, three possible locations where electron–hole pairs can be generated:

- Electrons and holes generated in the depletion layer (region 1) quickly drift in opposite directions under the influence of the strong electric field. Since the

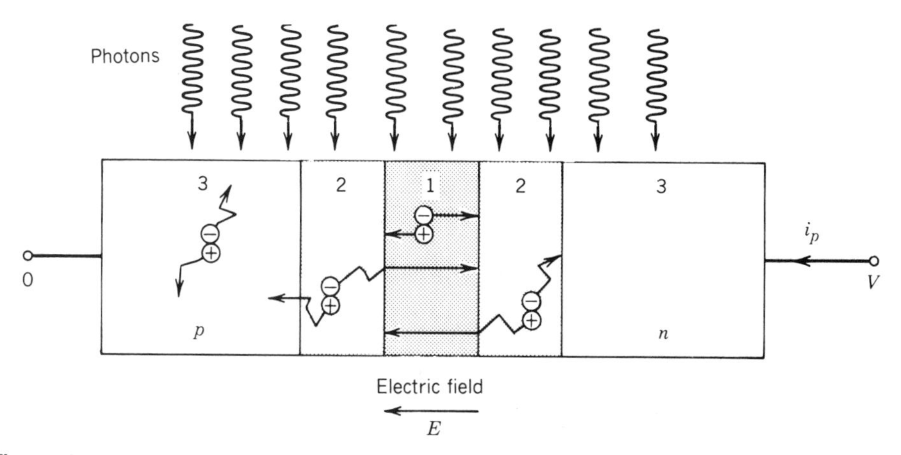

Figure 17.3-1 Photons illuminating an idealized reverse-biased *p-n* photodiode detector. The drift and diffusion regions are indicated by 1 and 2, respectively.

electric field always points in the n-p direction, electrons move to the n side and holes to the p side. As a result, the photocurrent created in the external circuit is always in the reverse direction (from the n to the p region). Each carrier pair generates in the external circuit an electric current pulse of area e ($G = 1$) since recombination does not take place in the depleted region.

- Electrons and holes generated away from the depletion layer (region 3) cannot be transported because of the absence of an electric field. They wander randomly until they are annihilated by recombination. They do not contribute a signal to the external electric current.
- Electron–hole pairs generated outside the depletion layer, but in its vicinity (region 2), have a chance of entering the depletion layer by random diffusion. An electron coming from the p side is quickly transported across the junction and therefore contributes a charge e to the external circuit. A hole coming from the n side has a similar effect.

Photodiodes have been fabricated from many of the semiconductor materials listed in Table 15.1-3, as well as from ternary and quaternary compound semiconductors such as InGaAs and InGaAsP. Devices are often constructed in such a way that the light impinges normally on the p-n junction instead of parallel to it. In that case the additional carrier diffusion current in the depletion region acts to enhance η, but this is counterbalanced by the decreased thickness of the material which acts to reduce η.

Response Time

The transit time of carriers drifting across the depletion layer (w_d/v_e for electrons and w_d/v_h for holes) and the *RC* time response play a role in the response time of photodiode detectors, as discussed in Sec. 17.1C. The resulting circuit current is shown in Fig. 17.1-3(*b*) for an electron–hole pair generated at the position x, and in Fig. 17.1-4 for uniform electron–hole pair generation.

In photodiodes there is an additional contribution to the response time arising from diffusion. Carriers generated outside the depletion layer, but sufficiently close to it, take time to diffuse into it. This is a relatively slow process in comparison with drift. The maximum times allowed for this process are, of course, the carrier lifetimes (τ_p for electrons in the p region and τ_n for holes in the n region). The effect of diffusion time can be decreased by using a p-i-n diode, as will be seen subsequently.

Nevertheless, photodiodes are generally faster than photoconductors because the strong field in the depletion region imparts a large velocity to the photogenerated carriers. Furthermore, photodiodes are not affected by many of the trapping effects associated with photoconductors.

Bias

As an electronic device, the photodiode has an i–V relation given by

$$i = i_s\left[\exp\left(\frac{eV}{k_B T}\right) - 1\right] - i_p,$$

illustrated in Fig. 17.3-2. This is the usual i–V relation of a p-n junction [see (15.1-24)] with an added photocurrent $-i_p$ proportional to the photon flux.

There are three classical modes of photodiode operation: open circuit (photovoltaic), short-circuit, and reverse biased (photoconductive). In the open-circuit mode (Fig. 17.3-3), the light generates electron–hole pairs in the depletion region. The additional electrons freed on the n side of the layer recombine with holes on the p side, and vice versa. The net result is an increase in the electric field, which produces a photovoltage

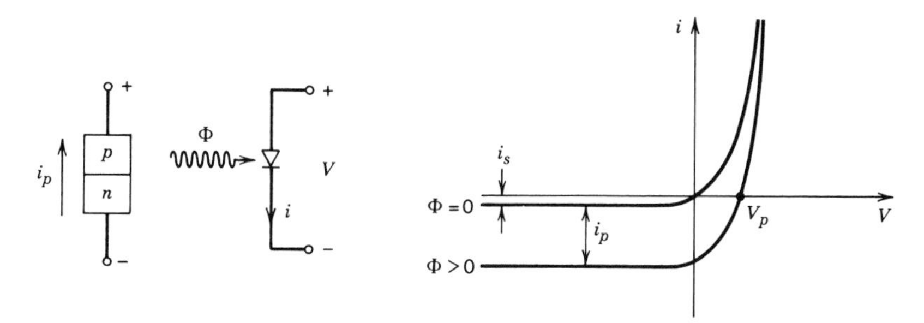

Figure 17.3-2 Generic photodiode and its i–V relation.

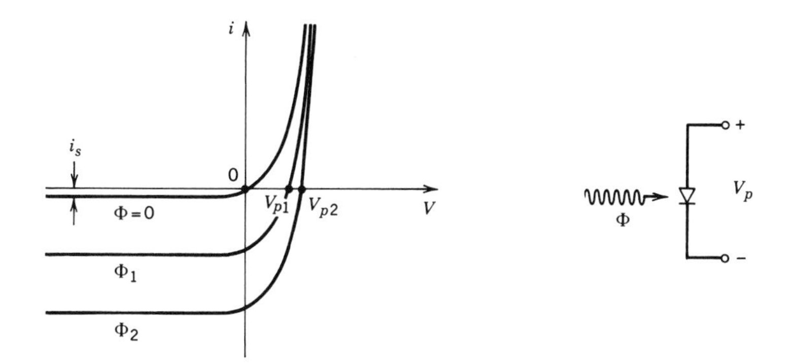

Figure 17.3-3 Photovoltaic operation of a photodiode.

V_p across the device that increases with increasing photon flux. This mode of operation is used, for example, in solar cells. The responsivity of a photovoltaic photodiode is measured in V/W rather than in A/W. The short-circuit ($V = 0$) mode is illustrated in Fig. 17.3-4. The short-circuit current is then simply the photocurrent i_p. Finally, a photodiode may be operated in its reverse-biased or "photoconductive" mode, as shown in Fig. 17.3-5(a). If a series-load resistor is inserted in the circuit, the operating conditions are those illustrated in Fig. 17.3-5(b).

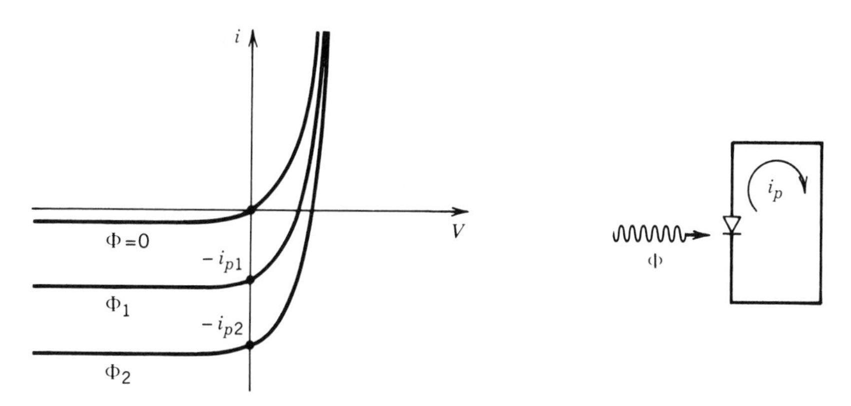

Figure 17.3-4 Short-circuit operation of a photodiode.

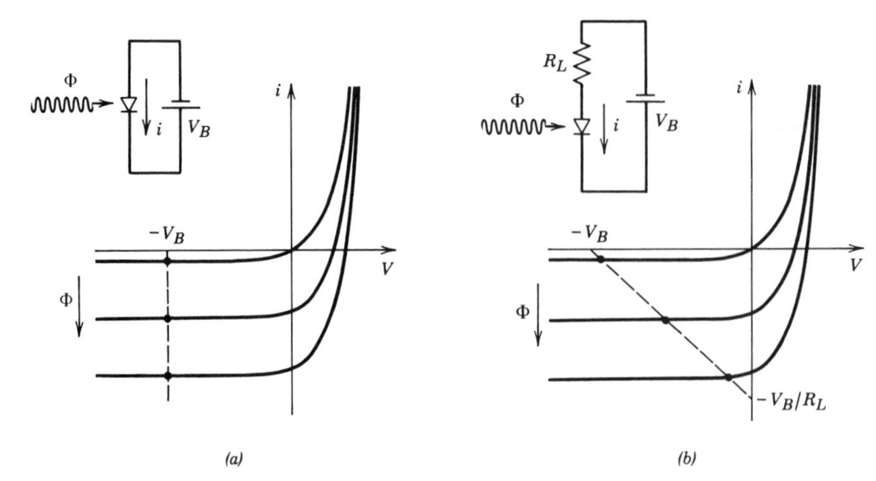

Figure 17.3-5 (*a*) Reverse-biased operation of a photodiode without a load resistor and (*b*) with a load resistor. The operating point lies on the dashed line.

Photodiodes are usually operated in the strongly reverse-biased mode for the following reasons:

- A strong reverse bias creates a strong electric field in the junction which increases the drift velocity of the carriers, thereby reducing transit time.
- A strong reverse bias increases the width of the depletion layer, thereby reducing the junction capacitance and improving the response time.
- The increased width of the depletion layer leads to a larger photosensitive area, making it easier to collect more light.

B. The *p-i-n* Photodiode

As a detector, the *p-i-n* photodiode has a number of advantages over the *p-n* photodiode. A *p-i-n* diode is a *p-n* junction with an intrinsic (usually lightly doped) layer sandwiched between the *p* and *n* layers (see Sec. 15.1E). It may be operated under the variety of bias conditions discussed in the preceding section. The energy-band diagram, charge distribution, and electric field distribution for a reverse-biased *p-i-n* diode are illustrated in Fig. 17.3-6. This structure serves to extend the width of the region supporting an electric field, in effect widening the depletion layer.

Photodiodes with the *p-i-n* structure offer the following advantages:

- Increasing the width of the depletion layer of the device (where the generated carriers can be transported by drift) increases the area available for capturing light.
- Increasing the width of the depletion layer reduces the junction capacitance and thereby the *RC* time constant. On the other hand, the transit time increases with the width of the depletion layer.
- Reducing the ratio between the diffusion length and the drift length of the device results in a greater proportion of the generated current being carried by the faster drift process.

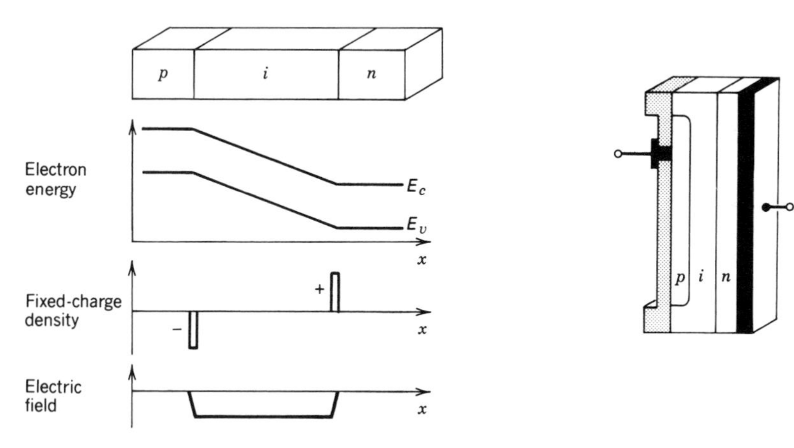

Figure 17.3-6 The *p-i-n* photodiode structure, energy diagram, charge distribution, and electric field distribution. The device can be illuminated either perpendicularly or parallel to the junction.

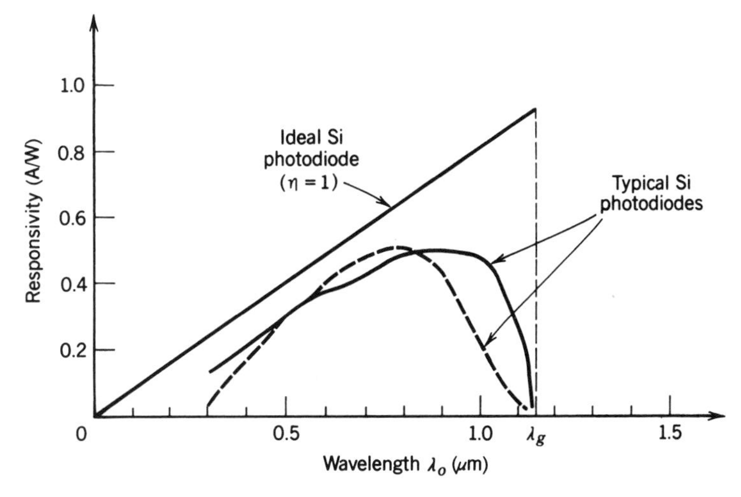

Figure 17.3-7 Responsivity versus wavelength (μm) for ideal and commercially available silicon *p-i-n* photodiodes.

Response times in the tens of ps, corresponding to bandwidths $\approx$ 50 GHz, are achievable. The responsivity of two commercially available silicon *p-i-n* photodiodes is compared with that of an ideal device in Fig. 17.3-7. It is interesting to note that the responsivity maximum occurs for wavelengths substantially shorter than the bandgap wavelength. This is because Si is an indirect-gap material. The photon-absorption transitions therefore typically take place from the valence-band to conduction-band states that typically lie well above the conduction-band edge (see Fig. 15.2-8).

C. Heterostructure Photodiodes

Heterostructure photodiodes, formed from two semiconductors of different bandgaps, can exhibit advantages over *p-n* junctions fabricated from a single material. A heterojunction comprising a large-bandgap material ($E_g > h\nu$), for example, can make use of its transparency to minimize optical absorption outside the depletion region. The large-bandgap material is then called a **window layer**. The use of different materials can also provide devices with a great deal of flexibility. Several material systems are of particular interest (see Figs. 15.1-5 and 15.1-6):

- $Al_xGa_{1-x}As/GaAs$ (AlGaAs lattice matched to a GaAs substrate) is useful in the wavelength range 0.7 to 0.87 μm.
- $In_{0.53}Ga_{0.47}As/InP$ operates at 1.65 μm in the near infrared ($E_g = 0.75$ eV). Typical values for the responsivity and quantum efficiency of detectors fabricated from these materials are $\mathfrak{R} \approx 0.7$ A/W and $\eta \approx 0.75$. The gap wavelength can be compositionally tuned over the range of interest for fiber-optic communication, 1.3–1.6 μm.
- $Hg_xCd_{1-x}Te/CdTe$ is a material that is highly useful in the middle-infrared region of the spectrum. This is because HgTe and CdTe have nearly the same lattice parameter and can therefore be lattice matched at nearly all compositions. This material provides a compositionally tunable bandgap that operates in the wavelength range between 3 and 17 μm.
- Quaternary materials, such as $In_{1-x}Ga_xAs_{1-y}P_y/InP$ and $Ga_{1-x}Al_xAs_ySb_{1-y}/GaSb$, which are useful over the range 0.92 to 1.7 μm, are of particular interest because the fourth element provides an additional degree of freedom that allows lattice matching to be achieved for different compositionally determined values of E_g.

Schottky-Barrier Photodiodes

Metal–semiconductor photodiodes (also called **Schottky-barrier photodiodes**) are formed from metal–semiconductor heterojunctions. A thin semitransparent metallic film is used in place of the *p*-type (or *n*-type) layer in the *p*-*n* junction photodiode. The thin film is sometimes made of a metal–semiconductor alloy that behaves like a metal. The Schottky-barrier structure and its energy-band diagram are shown schematically in Fig. 17.3-8.

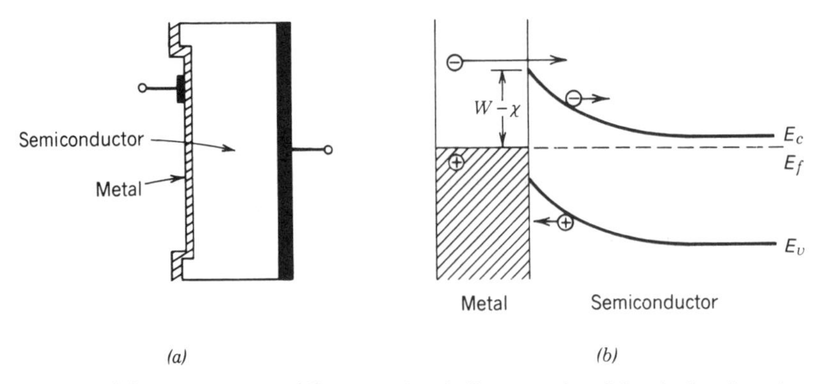

Figure 17.3-8 (*a*) Structure and (*b*) energy-band diagram of a Schottky-barrier photodiode formed by depositing a metal on an *n*-type semiconductor. These photodetectors are responsive to photon energies greater than the Schottky barrier height, $h\nu > W - \chi$. Schottky photodiodes can be fabricated from many materials, such as Au on *n*-type Si (which operates in the visible) and platinum silicide (PtSi) on *p*-type Si (which operates over a range of wavelengths stretching from the near ultraviolet to the infrared).

There are a number of reasons why Schottky-barrier photodiodes are useful:

- Not all semiconductors can be prepared in both *p*-type and *n*-type forms; Schottky devices are of particular interest in these materials.
- Semiconductors used for the detection of visible and ultraviolet light with photon energies well above the bandgap energies have a large absorption coefficient. This gives rise to substantial surface recombination and a reduction of the quantum efficiency. The metal–semiconductor junction has a depletion layer present immediately at the surface, thus eliminating surface recombination.
- The response speed of *p-n* and *p-i-n* junction photodiodes is in part limited by the slow diffusion current associated with photocarriers generated close to, but outside of, the depletion layer. One way of decreasing this unwanted absorption is to decrease the thickness of one of the junction layers. However, this should be achieved without substantially increasing the series resistance of the device because such an increase has the undesired effect of reducing the speed by increasing the *RC* time constant. The Schottky-barrier structure achieves this because of the low resistance of the metal. Furthermore Schottky barrier structures are majority-carrier devices and therefore have inherently fast responses and large operating bandwidths. Response times in the picosecond regime, corresponding to bandwidths $\approx$ 100 GHz, are readily available.

Representative quantum efficiencies for Schottky-barrier and *p-i-n* photodiode detectors are shown in Fig. 17.3-9; η can approach unity for carefully constructed Si devices that include antireflection coatings.

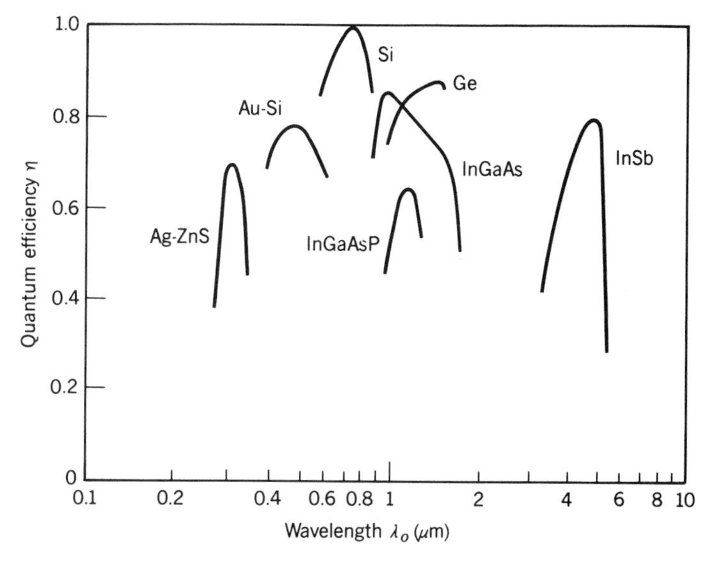

Figure 17.3-9 Quantum efficiency η versus wavelength λ_o (μm) for various photodiodes. Si *p-i-n* photodiodes can be fabricated with nearly unity quantum efficiency if an antireflection coating is applied to the surface of the device. The optimal response wavelength of ternary and quaternary *p-i-n* photodetectors is compositionally tunable (the quantum efficiency for a range of wavelengths is shown for InGaAs). Long-wavelength photodetectors (e.g., InSb) must be cooled to minimize thermal excitation. (Adapted from S. M. Sze, *Physics of Semiconductor Devices*, Wiley, New York, 2nd ed. 1981.)

D. Array Detectors

An individual photodetector registers the photon flux striking it as a function of time. In contrast, an array containing a large number of photodetectors can simultaneously register the photon fluxes (as functions of time) from many spatial points. Such detectors therefore permit electronic versions of optical images to be formed. One type of array detector, the microchannel plate [see Fig. 17.0-2(*c*)], has already been discussed.

Modern microelectronics technology permits other types of arrays containing large numbers of individual semiconductor photodetectors (called pixels) to be fabricated. One example of current interest, illustrated in Fig. 17.3-10, makes use of an array of nearly 40,000 tiny Schottky-barrier photodiodes of PtSi on p-type Si. The device is sensitive to a broad band of wavelengths stretching from the near ultraviolet to about 6 μm in the infrared, which corresponds to the Schottky barrier height of about 0.2 eV.

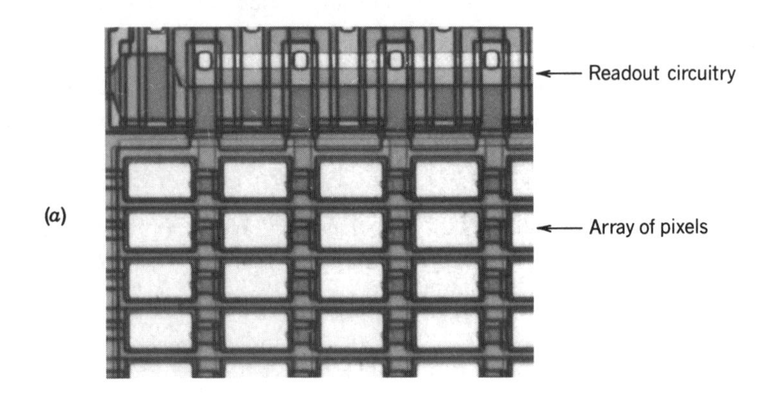

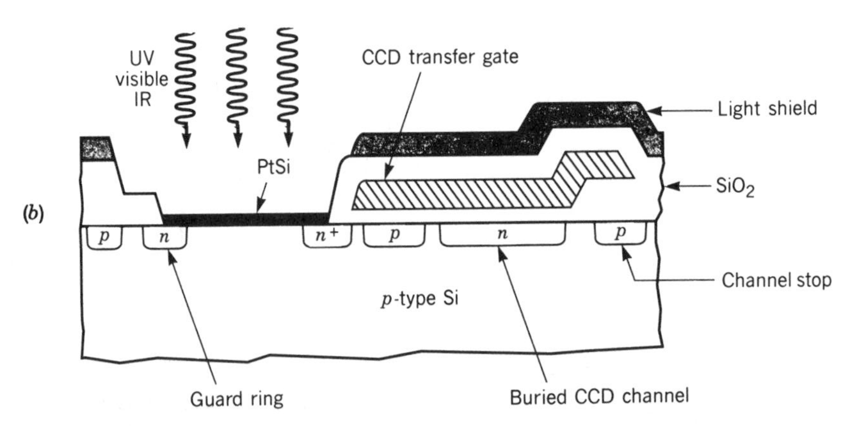

Figure 17.3-10 (*a*) Corner of an array of 160 × 244 PtSi/Si Schottky-barrier photodiodes. Each pixel is 40 μm × 80 μm in size. Portions of the readout circuitry are visible. (Courtesy of W. F. Kosonocky.) (*b*) Cross section of a single pixel in the CCD array. The light shield prevents the generation of photocarriers in the CCD transfer gate and buried channel. The guard ring minimizes dark-current spikes and the channel stop confines the signal charge in the lateral direction. (Adapted from B.-Y. Tsaur, C. K. Chen, and J. P. Mattia, PtSi Schottky-Barrier Focal Plane Arrays for Multispectral Imaging in Ultraviolet, Visible and Infrared Spectral Bands, *IEEE Electron Device Letters*, vol. 11, pp. 162–164, 1990, copyright © IEEE.)

The quantum efficiency η ranges between 35% and 60% in the ultraviolet and visible regions (from $\lambda_o = 290$ nm to about 900 nm) where the photon energy exceeds the bandgap of Si. At these wavelengths, the light transmitted through the PtSi film generates copious numbers of electron–hole pairs in the Si substrate [this is illustrated in Fig. 17.3-8(*b*) for a Schottky barrier with an *n*-type semiconductor]. At longer wavelengths, corresponding to photon energies below the bandgap of Si, the photogenerated carriers are produced by absorption in the PtSi film and η slowly decreases from about 3% at 1.5 μm to about 0.02% at 6 μm. At all wavelengths, the array must be cooled to 77 K because of the low Schottky barrier height. However, similar devices have recently been fabricated from PtSi on *n*-type Si; these have a higher barrier height and can therefore be operated without cooling but they are only sensitive in the ultraviolet and visible. IrSi devices are also regularly used.

When illuminated, carriers with sufficient energy (holes in the *p*-type case) climb the Schottky barrier and enter the Si. This leaves a residue of negative charge (proportional to the number of photons absorbed by the pixel) to accumulate on the PtSi electrode. The electronic portion of the detection process is accomplished by transferring the negative charge from the PtSi electrode into a **charge-coupled device** (CCD) readout structure. The CCD transfer gate [Fig. 17.3-10(*b*)] permits the charge to be transferred to a buried CCD channel at a specified time. Many different kinds of electrode structures and clocking schemes have been developed for periodically reading out the charge accumulated by each pixel and thereby generating an electronic data stream representing the image.

The multispectral imaging capabilities of a Schottky-barrier-diode CCD array detector such as that described above is clearly illustrated in Fig. 17.3-11. The images are the radiation from a coffee mug that was partially filled with warm water and focused on the array by a lens. The left portion of the figure represents the infrared image (in the wavelength range from 3 to 5 μm) whereas the right portion is the visible image obtained in room light using wavelengths shorter than ≈ 2 μm. The infrared image clearly shows that the top of the mug and its handle are cooler than the rest. Of course, the photosensitive elements in a CCD array need not be Schottky-barrier diodes; *p-n* photodiodes are used as well.

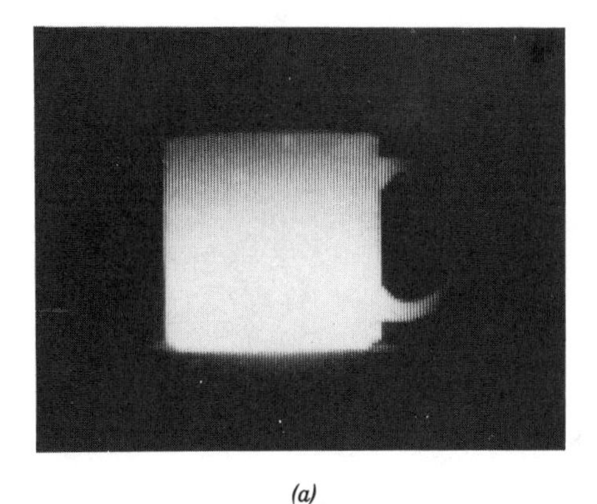

(*a*)

(*b*)

Figure 17.3-11 (*a*) Infrared and (*b*) visible images of a coffee mug partially filled with warm water obtained from a 160 × 244-element PtSi/Si Schottky-barrier-diode focal-plane CCD array detector operated at 77 K. (After B.-Y. Tsaur, C. K. Chen, and J. P. Mattia, PtSi Schottky-Barrier Focal Plane Arrays for Multispectral Imaging in Ultraviolet, Visible and Infrared Spectral Bands, *IEEE Electron Device Letters*, vol. 11, pp. 162–164, 1990, copyright © IEEE.)

17.4 AVALANCHE PHOTODIODES

An **avalanche photodiode** (APD) operates by converting each detected photon into a *cascade* of moving carrier pairs. Weak light can then produce a current that is sufficient to be readily detected by the electronics following the APD. The device is a strongly reverse-biased photodiode in which the junction electric field is large; the charge carriers therefore accelerate, acquiring enough energy to excite new carriers by the process of **impact ionization**.

A. Principles of Operation

The history of a typical electron–hole pair in the depletion region of an APD is depicted in Fig. 17.4-1. A photon is absorbed at point 1, creating an electron–hole pair (an electron in the conduction band and a hole in the valence band). The electron accelerates under the effect of the strong electric field, thereby increasing its energy with respect to the bottom of the conduction band. The acceleration process is constantly interrupted by random collisions with the lattice in which the electron loses some of its acquired energy. These competing processes cause the electron to reach an average saturation velocity. Should the electron be lucky and acquire an energy larger than E_g at any time during the process, it has an opportunity to generate a second electron–hole pair by impact ionization (say at point 2). The two electrons then accelerate under the effect of the field, and each of them may be the source for a further impact ionization. The holes generated at points 1 and 2 also accelerate, moving toward the left. Each of these also has a chance of impact ionizing should they acquire sufficient energy, thereby generating a hole-initiated electron–hole pair (e.g., at point 3).

Ionization Coefficients

The abilities of electrons and holes to impact ionize are characterized by the **ionization coefficients** α_e and α_h. These quantities represent ionization probabilities per unit length (rates of ionization, cm^{-1}); the inverse coefficients, $1/\alpha_e$ and $1/\alpha_h$, represent

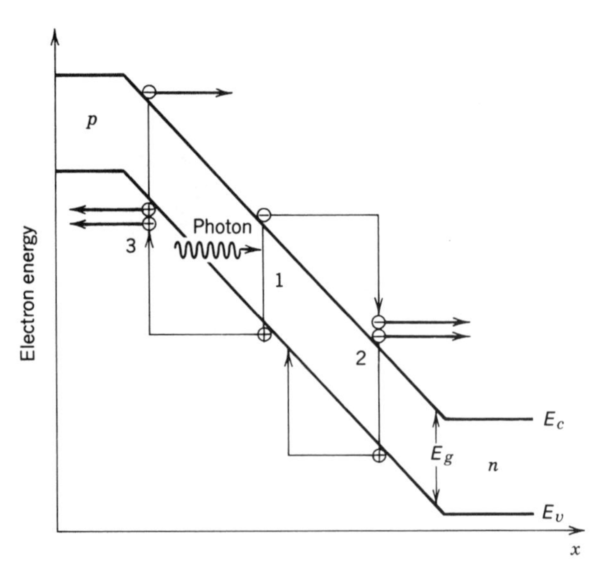

Figure 17.4-1 Schematic representation of the multiplication process in an APD.

the average distances between consecutive ionizations. The ionization coefficients increase with the depletion-layer electric field (since it provides the acceleration) and decrease with increasing device temperature. The latter occurs because increasing temperature causes an increase in the frequency of collisions, diminishing the opportunity a carrier has of gaining sufficient energy to ionize. The simple theory considered here assumes that α_e and α_h are constants that are independent of position and carrier history.

An important parameter for characterizing the performance of an APD is the **ionization ratio**

$$\mathscr{k} = \frac{\alpha_h}{\alpha_e}.$$

When holes do not ionize appreciably [i.e., when $\alpha_h \ll \alpha_e$ ($\mathscr{k} \ll 1$)], most of the ionization is achieved by electrons. The avalanching process then proceeds principally from left to right (i.e., from the p side to the n side) in Fig. 17.4-1. It terminates some time later when all the electrons arrive at the n side of the depletion layer. If electrons and holes both ionize appreciably ($\mathscr{k} \approx 1$), on the other hand, those holes moving to the left create electrons that move to the right, which, in turn, generate further holes moving to the left, in a possibly unending circulation. Although this feedback process increases the gain of the device (i.e., the total generated charge in the circuit per photocarrier pair q/e), it is nevertheless undesirable for several reasons:

- It is time consuming and therefore reduces the device bandwidth.
- It is random and therefore increases the device noise.
- It can be unstable, thereby causing avalanche breakdown.

It is therefore desirable to fabricate APDs from materials that permit only one type of carrier (either electrons or holes) to impact ionize. If electrons have the higher ionization coefficient, for example, optimal behavior is achieved by injecting the electron of a photocarrier pair at the p edge of the depletion layer and by using a material whose value of $\mathscr{k}$ is as low as possible. If holes are injected, the hole of a photocarrier pair should be injected at the n edge of the depletion layer and $\mathscr{k}$ should be as large as possible. The ideal case of single-carrier multiplication is achieved when $\mathscr{k} = 0$ or ∞.

Design

As with any photodiode, the geometry of the APD should maximize photon absorption, for example by assuming the form of a *p-i-n* structure. On the other hand, the multiplication region should be thin to minimize the possibility of localized uncontrolled avalanches (instabilities or microplasmas) being produced by the strong electric field. Greater electric-field uniformity can be achieved in a thin region.

These two conflicting requirements call for an APD design in which the absorption and multiplication regions are separate [**separate-absorption-multiplication** (SAM) **APD**]. Its operation is most readily understood by considering a device with $\mathscr{k} \approx 0$ (e.g., Si). Photons are absorbed in a large intrinsic or lightly doped region. The photoelectrons drift across it under the influence of a moderate electric field, and finally enter a thin multiplication layer with a strong electric field where avalanching occurs. The reach-through p^+-π-p-n^+ APD structure illustrated in Fig. 17.4-2 accomplishes this. Photon absorption occurs in the wide π region (very lightly doped p region). Electrons drift through the π region into a thin p-n^+ junction, where they experience a sufficiently strong electric field to cause avalanching. The reverse bias applied across

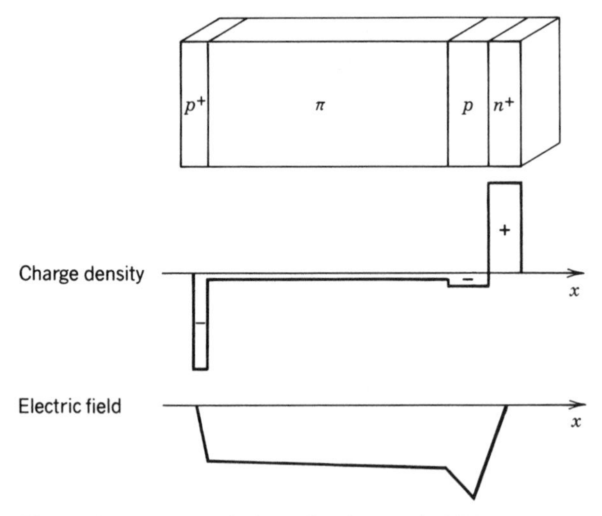

Figure 17.4-2 Reach-through p^+-π-p-n^+ APD structure.

the device is large enough for the depletion layer to reach through the p and π regions into the p^+ contact layer.

*Multilayer Devices

The noise inherent in the APD multiplication process can be reduced, at least in principle, by use of a multilayer avalanche photodiode. One such structure, called the staircase APD, has an energy-band diagram as shown in Fig. 17.4-3. A three-stage device is illustrated in both unbiased and reverse-biased conditions. The bandgap is

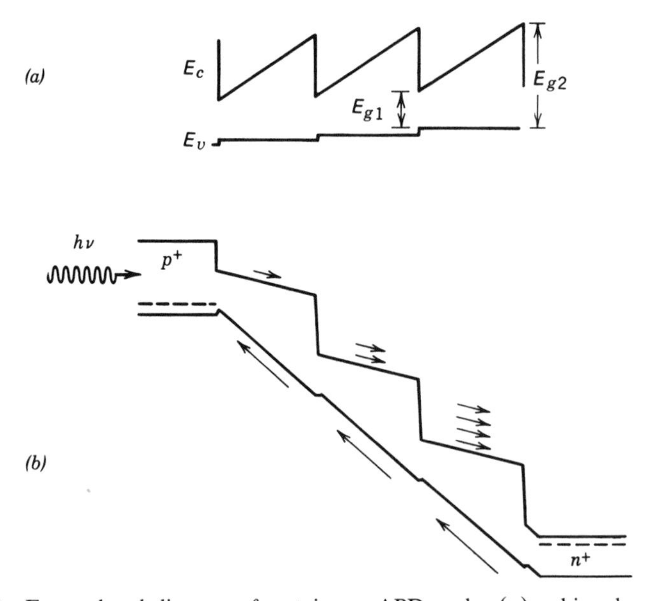

Figure 17.4-3 Energy-band diagram of a staircase APD under (a) unbiased and (b) reverse-biased conditions. The conduction-band steps encourage electron ionizations at discrete locations. (After F. Capasso, W. T. Tsang, and G. F. Williams, Staircase Solid-State Photomultipliers and Avalanche Photodiodes with Enhanced Ionization Rates Ratio, *IEEE Transactions on Electron Devices*, vol. ED-30, pp. 381–390, 1983, copyright © IEEE.)

compositionally graded (over a distance ≈ 10 nm), from a low value E_{g1} (e.g., GaAs) to a high value E_{g2} (e.g., AlGaAs). Because of the material properties, hole-induced ionizations are discouraged, thereby reducing the value of the ionization ratio k. Other potential advantages of such devices include the discrete locations of the multiplications (at the jumps in the conduction band edge), the low operating voltage, which minimizes tunneling, and the fast time response resulting from the reduced avalanche buildup time. Graded-gap devices of this kind are, however, difficult to fabricate.

B. Gain and Responsivity

As a prelude to determining the gain of an APD in which both kinds of carriers cause multiplication, the simpler problem of single-carrier (electron) multiplication ($\alpha_h = 0$, $k = 0$) is addressed first. Let $J_e(x)$ be the electric current density carried by electrons at location x, as shown in Fig. 17.4-4. Within a distance dx, on the average, the current is incremented by the factor

$$dJ_e(x) = \alpha_e J_e(x)\,dx,$$

from which we obtain the differential equation

$$\frac{dJ_e}{dx} = \alpha_e J_e(x),$$

whose solution is the exponential function $J_e(x) = J_e(0)\exp(\alpha_e x)$. The gain $G = J_e(w)/J_e(0)$ is therefore

$$G = \exp(\alpha_e w). \tag{17.4-1}$$

The electric current density increases exponentially with the product of the ionization coefficient α_e and the multiplication layer width w.

The double-carrier multiplication problem requires knowledge of both the electron current density $J_e(x)$ and the hole current density $J_h(x)$. It is assumed that only electrons are injected into the multiplication region. Since hole ionizations also produce electrons, however, the growth of $J_e(x)$ is governed by the differential equation

$$\frac{dJ_e}{dx} = \alpha_e J_e(x) + \alpha_h J_h(x). \tag{17.4-2}$$

As a result of charge neutrality, $dJ_e/dx = -dJ_h/dx$, so that the sum $J_e(x) + J_h(x)$ must remain constant for all x under steady-state conditions. This is clear from Fig.

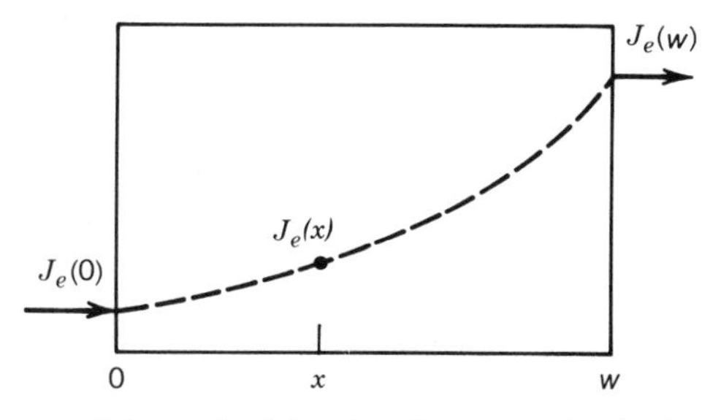

Figure 17.4-4 Exponential growth of the electric current density in a single-carrier APD.

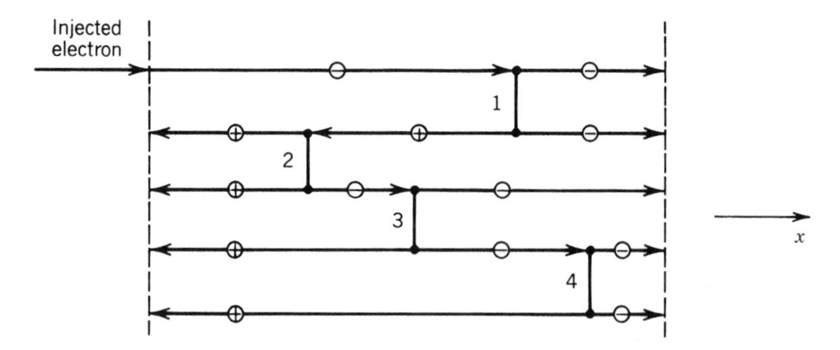

Figure 17.4-5 Constancy of the sum of the electron and hole current densities across a plane at any x.

17.4-5; the total number of charge carriers crossing any plane is the same regardless of position (four impact ionizations and five electrons-plus-holes crossing every plane are shown by way of illustration).

Since it is assumed that no holes are injected at $x = w$, $J_h(w) = 0$, so that

$$J_e(x) + J_h(x) = J_e(w), \tag{17.4-3}$$

as shown in Fig. 17.4-6. $J_h(x)$ can therefore be eliminated in (17.4-2) to obtain

$$\frac{dJ_e}{dx} = (\alpha_e - \alpha_h)J_e(x) + \alpha_h J_e(w). \tag{17.4-4}$$

This first-order differential equation is readily solved for the gain $G = J_e(w)/J_e(0)$. For $\alpha_e \neq \alpha_h$, the result is $G = (\alpha_e - \alpha_h)/\{\alpha_e \exp[-(\alpha_e - \alpha_h)w] - \alpha_h\}$, from which

$$G = \frac{1 - k}{\exp[-(1 - k)\alpha_e w] - k}. \tag{17.4-5}$$

APD Gain

The single-carrier multiplication result (17.4-1), with its exponential growth, is recovered when $k = 0$. When $k = \infty$, the gain remains unity since only electrons are injected and electrons do not multiply. For $k = 1$, (17.4-5) is indeterminate and the gain must be obtained directly from (17.4-4); the result is then $G = 1/(1 - \alpha_e w)$. An instability is reached when $\alpha_e w = 1$. The dependence of the gain on $\alpha_e w$ for several values of the

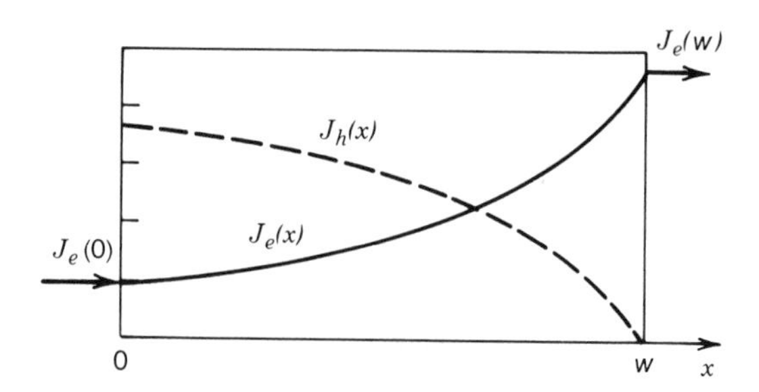

Figure 17.4-6 Growth of the electron and hole currents as a result of avalanche multiplication.

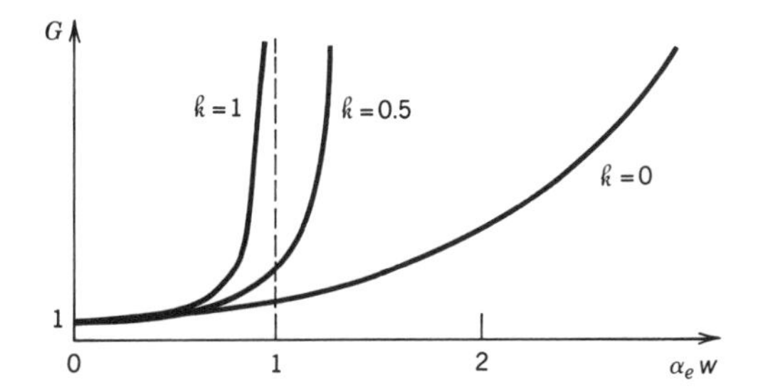

Figure 17.4-7 Growth of the gain G with multiplication-layer width for several values of the ionization ratio k, assuming pure electron injection.

ionization ratio k is illustrated in Fig. 17.4-7. The responsivity $\mathfrak{R}$ is obtained by using (17.4-5) in the general relation (17.1-6).

The materials of interest are the same as those for photodiodes, with the additional proviso that they should have the lowest (or highest) possible value of ionization ratio k. APDs with values of k as low as 0.006 have been fabricated from silicon, providing excellent performance in the wavelength region 0.7 to 0.9 μm.

C. Response Time

Aside from the usual transit, diffusion, and *RC* effects that govern the response time of photodiodes, APDs suffer from an additional multiplication time called the **avalanche buildup time**. The response time of a two-carrier-multiplication APD is illustrated in Fig. 17.4-8 by following the history of a photoelectron generated at the edge of the absorption region (point 1). The electron drifts with a saturation velocity v_e, reaching the multiplication region (point 2) after a transit time w_d/v_e. Within the multiplication region the electron also travels with a velocity v_e. Through impact ionization it creates electron–hole pairs, say at points 3 and 4, generating two additional electron–hole pairs. The holes travel in the opposite direction with their saturation velocity v_h. The holes can also cause impact ionizations resulting in electron–hole pairs as shown, for example, at points 5 and 6. The resulting carriers can themselves cause impact ionizations, sustaining the feedback loop. The process is terminated when the last hole leaves the multiplication region (at point 7) and crosses the drift region to point 8. The total time τ required for the entire process (between points 1 and 8) is the sum of the transit times (from 1 to 2 and from 7 to 8) and the multiplication time denoted τ_m,

$$\tau = \frac{w_d}{v_e} + \frac{w_d}{v_h} + \tau_m. \tag{17.4-6}$$

Because of the randomness of the multiplication process, the multiplication time τ_m is random. In the special case $k = 0$ (no hole multiplication) the maximum value of τ_m is readily seen from Fig. 17.4-8 to be

$$\tau_m = \frac{w_m}{v_e} + \frac{w_m}{v_h}. \tag{17.4-7}$$

For a large gain G, and for electron injection with $0 < k < 1$, an order of magnitude of

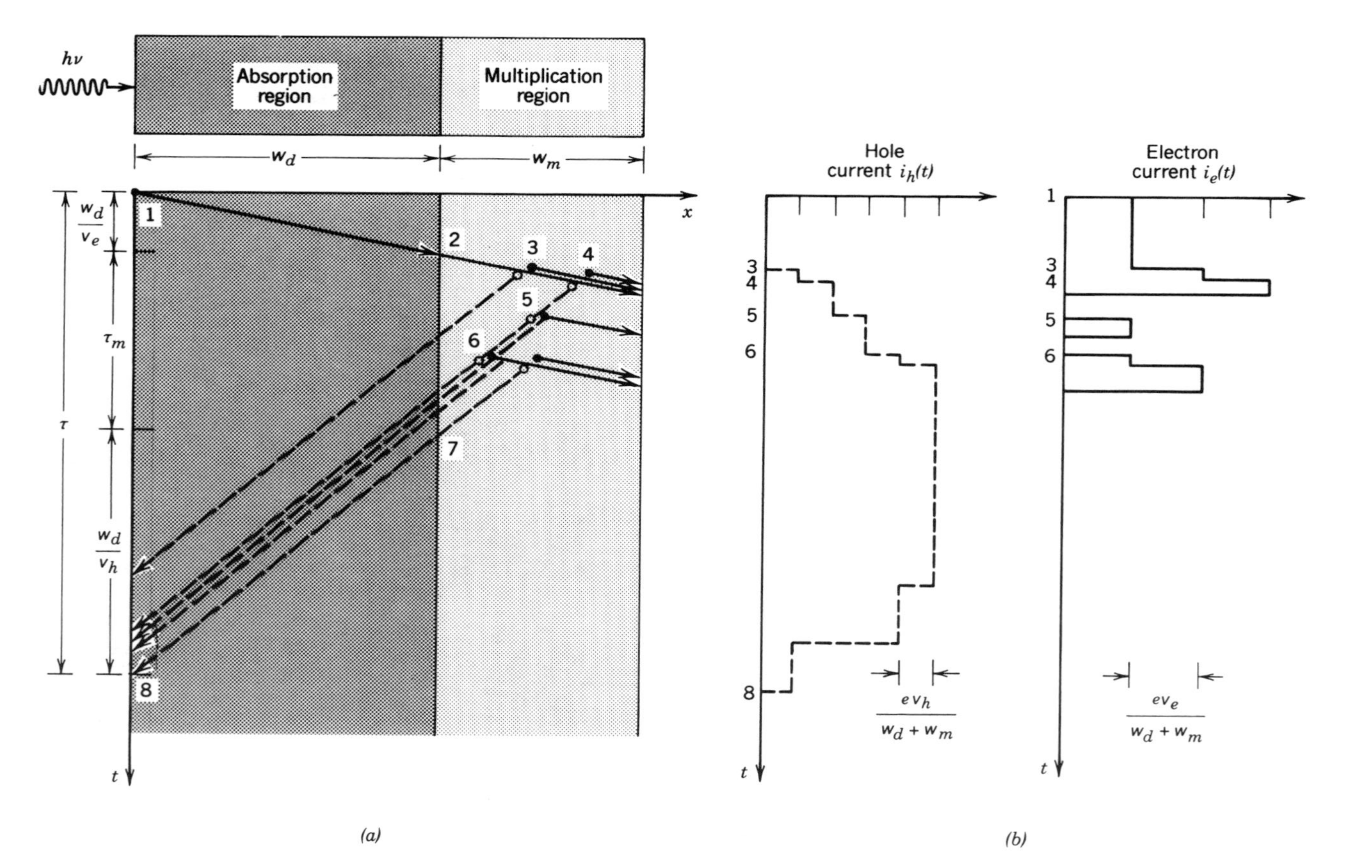

Figure 17.4-8 (*a*) Tracing the course of the avalanche buildup time in an APD with the help of a position–time graph. The solid lines represent electrons, and the dashed lines represent holes. Electrons move to the right with velocity v_e and holes move to the left with velocity v_h. Electron–hole pairs are produced in the multiplication region. The carriers cease moving when they reach the edge of the material. (*b*) Hole current $i_h(t)$ and electron current $i_e(t)$ induced in the circuit. Each carrier pair induces a charge e in the circuit. The total induced charge q, which is the area under the $i_e(t) + i_h(t)$ versus t curve, is Ge. This figure is a generalization of Fig. 17.1-3, which applies for a single electron–hole pair.

the average value of τ_m is obtained by increasing the first term of (17.4-7) by the factor $G\mathscr{k}$,

$$\tau_m \approx \frac{G\mathscr{k} w_m}{v_e} + \frac{w_m}{v_h}. \tag{17.4-8}$$

A more accurate theory is rather complex.

EXAMPLE 17.4-1. ***Avalanche Buildup Time in a Si APD.*** Consider a Si APD with $w_d = 50$ μm, $w_m = 0.5$ μm, $v_e = 10^7$ cm/s, $v_h = 5 \times 10^6$ cm/s, $G = 100$, and $\mathscr{k} = 0.1$. Equation (17.4-7) yields $\tau_m = 5 + 10 = 15$ ps, whereupon (17.4-6) gives $\tau = 1020$ ps $= 1.02$ ns. On the other hand, (17.4-8) yields $\tau_m = 60$ ps, so that (17.4-6) provides $\tau = 1065$ ps $= 1.07$ ns. For a *p-i-n* photodiode with the same values of w_d, v_e, and v_h, the transit time is $w_d/v_e + w_d/v_h \approx 1$ ns. These results do not differ greatly because τ_m is quite low in a silicon device.

17.5 NOISE IN PHOTODETECTORS

The photodetector is a device that measures photon flux (or optical power). Ideally, it responds to a photon flux Φ (optical power $P = h\nu\Phi$) by generating a proportional electric current $i_p = \eta e \Phi = \Re P$ [see (17.1-2)]. In actuality, the device generates a *random* electric current i whose value fluctuates above and below its average, $\bar{i} \equiv i_p = \eta e \Phi = \Re P$. These random fluctuations, which are regarded as noise, are characterized by the standard deviation σ_i, where $\sigma_i^2 = \langle (i - \bar{i})^2 \rangle$. For a current of zero mean ($\bar{i} = 0$), the standard deviation is the same as the root-mean-square (rms) value of the current, i.e., $\sigma_i = \langle i^2 \rangle^{1/2}$.

Several sources of noise are inherent in the process of photon detection:

- *Photon Noise.* The most fundamental source of noise is associated with the random arrivals of the photons themselves (which are usually described by Poisson statistics), as discussed in Sec. 11.2.
- *Photoelectron Noise.* For a photon detector with quantum efficiency $\eta < 1$, a single photon generates a photoelectron–hole pair with probability η but fails to do so with probability $1 - \eta$. Because of the inherent randomness in this process of carrier generation, it serves as a source of noise.
- *Gain Noise.* The amplification process that provides internal gain in some photodetectors (such as APDs) is random. Each detected photon generates a random number G of carriers with an average value $\overline{G}$ but with an uncertainty that is dependent on the nature of the amplification mechanism.
- *Receiver Circuit Noise.* The various components in the electrical circuitry of an optical receiver, such as resistors and transistors, contribute to the receiver circuit noise.

These four sources of noise are illustrated schematically in Fig. 17.5-1. The signal entering the detector (input signal) has an intrinsic photon noise. The photoeffect converts the photons into photoelectrons. In the process, the mean signal decreases by the factor η. The noise also decreases but by a lesser amount than the signal; thus the

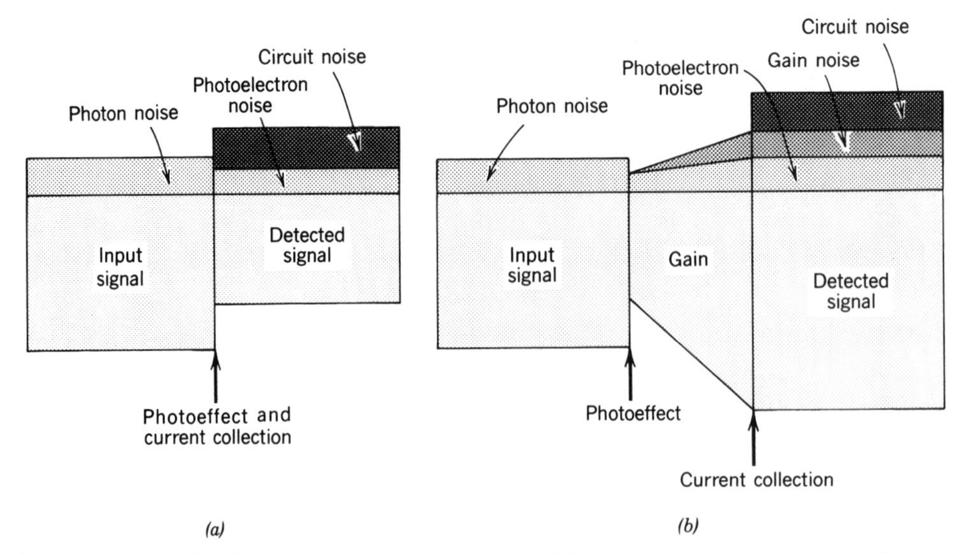

Figure 17.5-1 Signal and various noise sources for (*a*) a photodetector without gain (e.g., a *p-i-n* photodiode) and (*b*) a photodetector with gain (e.g., an APD).

signal-to-noise ratio of the photoelectrons is lower than that of the incident photons. If a photodetector gain mechanism is present, it amplifies both the signal and the photoelectron noise, and introduces its own gain noise as well. Finally, circuit noise enters at the point of current collection.

An optical receiver as a component in an information transmission system can be characterized by the following performance measures:

- The **signal-to-noise ratio** (SNR). The SNR of a random variable is defined as SNR = (mean)^2/variance; thus the SNR of the current i is SNR $= \bar{i}^2/\sigma_i^2$, whereas the SNR of the photon number is SNR $= \bar{n}^2/\sigma_n^2$.
- The **minimum-detectable signal**, which is defined as the mean signal that yields SNR = 1.
- The **receiver sensitivity**, which, like the minimum-detectable signal, is defined as the signal corresponding to a prescribed SNR = SNR$_0$. Rather than selecting SNR$_0$ = 1, however, a higher value is usually chosen to ensure a good level of accuracy (e.g., SNR$_0$ = 10 to 10^3, corresponding to 10 to 30 dB).

We proceed to derive expressions for the signal-to-noise ratio (SNR) for optical detectors with these sources of noise. Other sources of noise that are not explicitly considered here include background noise and dark-current noise. **Background noise** is the photon noise associated with light reaching the detector from extraneous sources (i.e., from sources other than the signal of interest, such as sunlight and starlight). Background noise is particularly deleterious in middle- and far-infrared detection systems because objects at room temperature emit copious thermal radiation in this region (see Fig. 12.3-4). Photodetection devices also generate **dark-current noise**, which, as the name implies, is present even in the absence of light. Dark-current noise results from random electron–hole pairs generated thermally or by tunneling. Also ignored are leakage current and $1/f$ noise.

A. Photoelectron Noise

Photon Noise

As described in Sec. 11.2, the photon flux associated with a fixed optical power P is inherently uncertain. The mean photon flux $\Phi = P/h\nu$ (photons/s), but it fluctuates randomly in accordance with a probability law dependent on the nature of the light source. The number of photons n counted in a time interval T is random with mean $\bar{n} = \Phi T$. The photon number for light from an ideal laser, or from a thermal source of spectral width much greater than $1/T$, obeys the Poisson probability distribution, for which $\sigma_n^2 = \bar{n}$. Thus the fluctuations associated with an average of 100 photons cause the actual number of photons to lie approximately within the range 100 ± 10.

The photon-number signal-to-noise ratio $\bar{n}^2/\sigma_n^2$ is therefore

$$\boxed{\text{SNR} = \bar{n}} \tag{17.5-1}$$

Photon-Number Signal-to-Noise Ratio

and the minimum detectable photon number is $\bar{n} = 1$ photon. If the observation time $T = 1\ \mu\text{s}$ and the wavelength $\lambda_o = 1.24\ \mu\text{m}$, this is equivalent to a minimum detectable power of 0.16 pW. The receiver sensitivity, the signal required to attain SNR $= 10^3$ (30 dB), is 1000 photons. If the time interval $T = 10$ ns, this is equivalent to a sensitivity of 10^{11} photons/s or an optical power sensitivity of 16 nW (at $\lambda_o = 1.24\ \mu\text{m}$).

Photoelectron Noise

A photon incident on a photodetector of quantum efficiency η either generates a photoevent (i.e., liberates a photoelectron or creates a photoelectron–hole pair) with probability η, or fails to do so with probability $1 - \eta$. Photoevents are assumed to be selected at random from the photon stream so that an incident mean photon flux Φ (photons/s) results in a mean photoelectron flux $\eta\Phi$ (photoelectrons/s). The number of photoelectrons detected in the time interval T is a random number m with mean

$$\bar{m} = \eta\bar{n}, \tag{17.5-2}$$

where $\bar{n} = \Phi T$ is the mean number of incident photons in the same time interval T. If the photon number is distributed in Poisson fashion, so is the photoelectron number, as can be ascertained by using an argument similar to that developed in Sec. 11.2D. It follows that the photoelectron-number variance is then precisely equal to $\bar{m}$, so that

$$\sigma_m^2 = \bar{m} = \eta\bar{n}. \tag{17.5-3}$$

It is apparent from this relationship that the photoelectron noise and the photon noise are nonadditive.

The underlying randomness inherent in the photon number, which constitutes a fundamental source of noise with which we must contend when using light to transmit a signal, therefore results in a photoelectron-number signal-to-noise ratio

$$\boxed{\text{SNR} = \bar{m} = \eta\bar{n}.} \tag{17.5-4}$$

Photoelectron-Number Signal-to-Noise Ratio

The minimum-detectable photoelectron number for SNR $= 1$ corresponds to $\bar{m} = \eta\bar{n} = 1$ (i.e., one photoelectron or $1/\eta$ photons). The receiver sensitivity for SNR $= 10^3$ is 1000 photoelectrons or $1000/\eta$ photons.

Photocurrent Noise

We now examine the properties of the electric current $i(t)$ induced in a circuit by a random photoelectron flux with mean $\eta\Phi$. The treatment we provide includes the effects of photon noise, photoelectron noise, and the characteristic time response of the detector and circuitry (filtering). Every photoelectron–hole pair generates a pulse of electric current of charge (area) e and time duration τ_p in the external circuit of the photodetector (Fig. 17.5-2). A photon stream incident on a photodetector therefore results in a stream of electrical pulses which add together to constitute an electric current $i(t)$. The randomness of the photon stream is transformed into a fluctuating electric current. If the incident photons are Poisson distributed, these fluctuations are known as shot noise. More generally, for detectors with gain G, the generated charge in each pulse is $q = Ge$.

Before providing an analytical derivation, we first show in a simplified way that the photocurrent i in a circuit of bandwidth B, generated by a photon flux Φ, can be determined by considering a characteristic time interval $T = 1/2B$ (the resolution time of the circuit) and by relating the random number m of photoelectrons counted within that interval to the photocurrent $i(t)$, where t is the instant of time immediately at the end of the interval T. For rectangular current pulses of duration T, the current and photoelectron-number random variables are then related by $i = (e/T)m$, so that the mean and variance are given by

$$\bar{i} = \frac{e}{T}\overline{m}$$

$$\sigma_i^2 = \left(\frac{e}{T}\right)^2 \sigma_m^2,$$

where $\overline{m} = \eta\Phi T = \eta\Phi/2B$ is the number of photoelectrons collected in the time interval $T = 1/2B$. Substituting $\sigma_m^2 = \overline{m}$ for the Poisson law yields the photocurrent mean and variance:

$$\boxed{\bar{i} = e\eta\Phi} \qquad (17.5\text{-}5)$$

Photocurrent Mean

$$\boxed{\sigma_i^2 = 2e\bar{i}B.} \qquad (17.5\text{-}6)$$

Photocurrent Variance

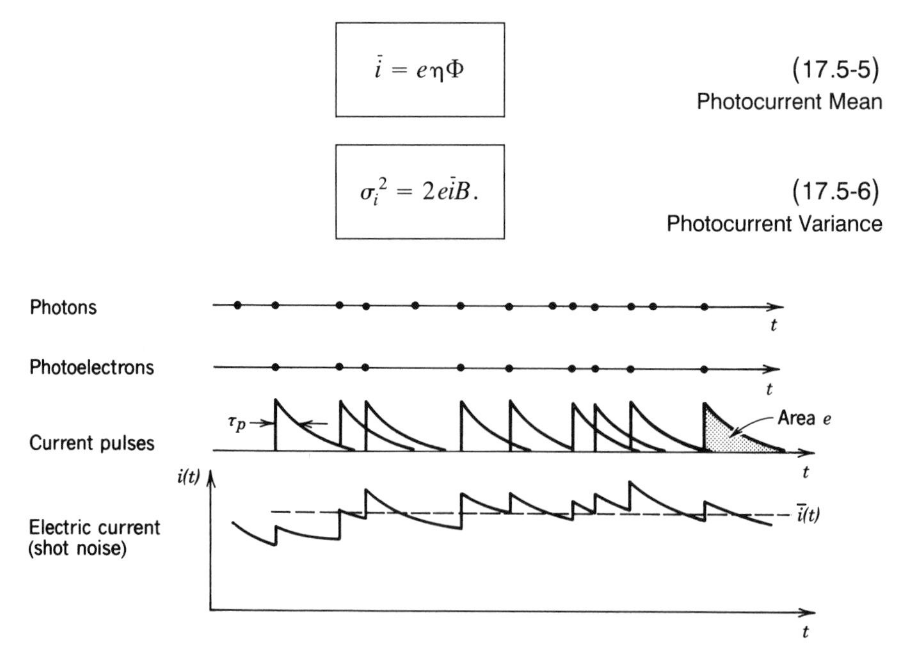

Figure 17.5-2 The electric current in a photodetector circuit comprises a superposition of electrical pulses, each associated with a detected photon. The individual pulses illustrated are exponential but they can assume an arbitrary shape (see, e.g., Figs. 17.1-3(*b*) and 17.1-4).

It follows that the signal-to-noise ratio of the photoelectric current, SNR $= \bar{i}^2/\sigma_i^2$, is

$$\mathrm{SNR} = \frac{\bar{i}}{2eB} = \frac{\eta\Phi}{2B} = \bar{m}. \tag{17.5-7}$$

Photocurrent Signal-to-Noise Ratio

The SNR is directly proportional to the photon flux Φ and inversely proportional to the electrical bandwidth of the circuit B.

EXAMPLE 17.5-1. ***SNR and Receiver Sensitivity.*** For $\bar{i} = 10$ nA and $B = 100$ MHz, $\sigma_i \approx 0.57$ nA, corresponding to a signal-to-noise ratio SNR $= 310$ or 25 dB. An average of 310 photoelectrons are detected in every time interval $1/2B = 5$ ns. The minimum-detectable photon flux is $\Phi = 2B/\eta$, and the receiver sensitivity for SNR $= 10^3$ is $\Phi = 1000(2B/\eta) = 2 \times 10^{11}/\eta$ photons/s.

Equations (17.5-5) and (17.5-6) will now be proved for current pulses of arbitrary shape.

Derivation of the Photocurrent Mean and Variance

Assume that a photoevent generated at $t = 0$ produces an electric pulse $h(t)$, of area e, in the external circuit. A photoelectron generated at time t_1 then produces a displaced pulse $h(t - t_1)$. If the time axis is divided into incremental time intervals Δt, the probability p that a photoevent occurs within an interval is $p = \eta\Phi\,\Delta t$. The electric current i at time t is written as

$$i(t) = \sum_l X_l h(t - l\,\Delta t), \tag{17.5-8}$$

where X_l has either the value 1 with probability p, or 0 with probability $1 - p$. The variables $\{X_l\}$ are independent. The mean value of X_l is $0 \times (1 - p) + 1 \times p = p$. Its mean-square value is $\langle X_l^2 \rangle = 0^2 \times (1 - p) + 1^2 \times p = p$. The mean of the product $X_l X_k$ is p^2 if $l \neq k$, and p if $l = k$. The mean and mean-square values of $i(t)$ are now determined as follows:

$$\bar{i} = \langle i \rangle = \sum_l p h(t - l\,\Delta t) \tag{17.5-9}$$

$$\langle i^2 \rangle = \sum_l \sum_k \langle X_l X_k \rangle h(t - l\,\Delta t) h(t - k\,\Delta t)$$

$$= \sum_{l \neq k}\sum p^2 h(t - l\,\Delta t) h(t - k\,\Delta t) + \sum_l p h^2(t - l\,\Delta t). \tag{17.5-10}$$

Substituting $p = \eta\Phi\,\Delta t$, and taking the limit $\Delta t \to 0$, so that the summations become integrals, (17.5-9) and (17.5-10) yield

$$\bar{i} = \eta\Phi\int_0^\infty h(t)\,dt = e\eta\Phi \tag{17.5-11}$$

$$\langle i^2 \rangle = (e\eta\Phi)^2 + \eta\Phi\int_0^\infty h^2(t)\,dt. \tag{17.5-12}$$

It follows that the variance of i is $\sigma_i^2 = \langle i^2 \rangle - \langle i \rangle^2$, or

$$\sigma_i^2 = \eta\Phi\int_0^\infty h^2(t)\,dt. \tag{17.5-13}$$

Defining

$$B = \frac{1}{2e^2}\int_0^\infty h^2(t)\,dt = \frac{1}{2}\frac{\int_0^\infty h^2(t)\,dt}{\left[\int_0^\infty h(t)\,dt\right]^2}, \tag{17.5-14}$$

we finally obtain (17.5-5) and (17.5-6).

The parameter B defined by (17.5-14) represents the device–circuit bandwidth. This is readily verified by noting that the Fourier transform of $h(t)$ is its transfer function $\mathcal{H}(\nu)$. The area under $h(t)$ is simply $\mathcal{H}(0) = e$. In accordance with Parseval's theorem [see (A.1-7) in Appendix A], the area under $h^2(t)$ is equal to the area under the symmetric function $|\mathcal{H}(\nu)|^2$, so that

$$B = \int_0^\infty \left| \frac{\mathcal{H}(\nu)}{\mathcal{H}(0)} \right|^2 d\nu. \tag{17.5-15}$$

The quantity B is therefore the power-equivalent spectral width of the function $|\mathcal{H}(\nu)|$ (i.e., the bandwidth of the device–circuit combination), in accordance with (A.2-10) of Appendix A. As an example, if $\mathcal{H}(\nu) = 1$ for $-\nu_c < \nu < \nu_c$ and 0 elsewhere, (17.5-15) yields $B = \nu_c$.

These relations are applicable for all photoelectric detection devices without gain (e.g., phototubes and junction photodiodes). Use of the formulas requires knowledge of the bandwidth of the device, biasing circuit, and amplifier; B is determined by inserting the transfer function of the overall system into (17.5-15).

B. Gain Noise

The photocurrent mean and variance for a device with fixed (deterministic) gain G is determined by replacing e with $q = Ge$ in (17.5-5) and (17.5-6):

$$\bar{i} = eG\eta\Phi = \frac{eG\eta P}{h\nu} \tag{17.5-16}$$

$$\sigma_i^2 = 2eG\bar{i}B = 2e^2G^2\eta B\Phi. \tag{17.5-17}$$

The signal-to-noise ratio, which is given by

$$\text{SNR} = \frac{\bar{i}}{2eGB} = \frac{\eta\Phi}{2B} = \bar{m}, \tag{17.5-18}$$

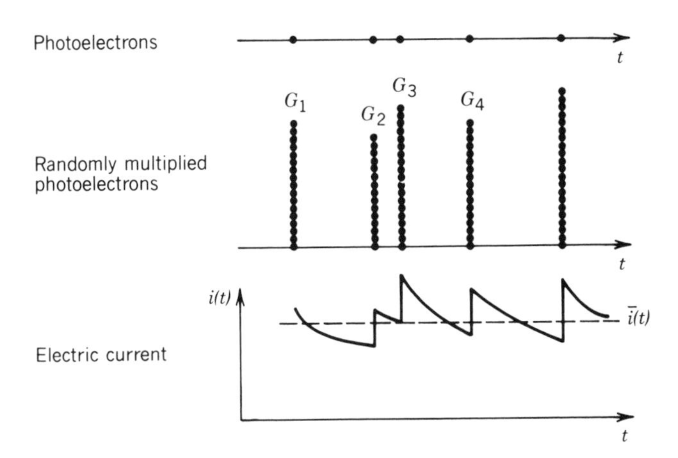

Figure 17.5-3 Each photoelectron–hole pair in a photodetector with gain generates a random number G_l of electron–hole pairs, each of which produces an electrical current pulse of area eG_l in the detector circuit. The total electric current $i(t)$ is the superposition of these pulses.

then turns out to be independent of G. This is because the mean current $\bar{i}$ and its rms value σ_i are both increased by the same factor G as a result of the gain.

This simple result does not apply when the gain itself is random, as is the case in a photomultiplier tube, a photoconductor, and an avalanche photodiode. The derivation of the photocurrent mean and variance given in the previous section must be generalized to account for the randomness in G. The electric current (17.5-8) should then be written in the form

$$i(t) = \sum_l X_l G_l h(t - l\,\Delta t),$$

where, as before, X_l takes the value 1 with probability $p = \eta\Phi\,\Delta t$, and 0 with probability $1 - p$. Included now are the G_l, which are independent random numbers representing the gain imparted to a photoelectron–hole pair generated in the lth time slot. The process is illustrated in Fig. 17.5-3. If the random variable G_l has mean value $\langle G \rangle = \overline{G}$ and mean-square value $\langle G^2 \rangle$, an analysis similar to that provided in (17.5-8) through (17.5-14) leads to

$$\bar{i} = e\overline{G}\eta\Phi \tag{17.5-19}$$

Photocurrent Mean
(Detector with Random Gain)

$$\sigma_i^2 = 2e\overline{G}\bar{i}BF, \tag{17.5-20}$$

Photocurrent Variance
(Detector with Random Gain)

where

$$F = \frac{\langle G^2 \rangle}{\langle G \rangle^2} \tag{17.5-21}$$

Excess Noise Factor

is called the **excess noise factor**.

The excess noise factor is related to the variance of the gain σ_G^2 by the relation $F = 1 + \sigma_G^2/\langle G \rangle^2$. When the gain is deterministic $\sigma_G^2 = 0$ and $F = 1$ so that (17.5-20) properly reduces to (17.5-17). When the gain is random, $\sigma_G^2 > 0$ and $F > 1$; both increase with the severity of the gain fluctuations. The resulting electric current i is then more noisy than shot noise.

In the presence of random gain, the signal-to-noise ratio $\bar{i}^2/\sigma_i^2$ becomes

$$\text{SNR} = \frac{\bar{i}}{2e\overline{G}BF} = \frac{\eta\Phi/2B}{F} = \frac{\overline{m}}{F}, \tag{17.5-22}$$

Signal-to-Noise Ratio (Detector with Random Gain)

where $\overline{m}$ is the mean number of photoelectrons collected in the time $T = 1/2B$. This is smaller than the deterministic-gain expression by the factor F; the reduction in the SNR arises from the randomness of the gain.

Excess Noise Factor for an APD

When photoelectrons are injected at the edge of a uniformly multiplying APD, the gain G of the device is given by (17.4-5). It depends on the electron ionization coefficient α_e and the ionization ratio $k = \alpha_h/\alpha_e$, as well as on the width of the multiplication region w. The use of a similar (but more complex) analysis, incorporating the randomness associated with the gain process, leads to an expression for the mean-square gain $\langle G^2 \rangle$, and therefore for the excess noise factor F. This more general derivation gives rise to an expression for the mean gain $\overline{G}$ which is identical to that given by (17.4-5). The excess noise factor F turns out to be related to the mean gain and the ionization ratio by

$$F = k\overline{G} + (1 - k)\left(2 - \frac{1}{\overline{G}}\right). \tag{17.5-23}$$

Excess Noise Factor for an APD

A plot of this result is presented in Fig. 17.5-4.

Equation (17.5-23) is valid when electrons are injected at the edge of the depletion layer, but both electrons and holes have the capability of initiating impact ionizations. If only holes are injected, the same expression applies, provided that k is replaced by $1/k$. Gain noise is minimized by injecting the carrier with the higher ionization coefficient, and by fabricating a structure with the lowest possible value of k if electrons are injected, or the highest possible value of k if holes are injected. Thus the ionization coefficients for the two carriers should be as different as possible.

Equation (17.5-23) is said to be valid under conditions of single-carrier-initiated double-carrier multiplication since both types of carrier have the capacity to impact ionize, even when only one type is injected. If electrons and holes are injected simultaneously, the overall result is the sum of the two partial results.

The gain noise introduced in a conventional APD arises from two sources: the randomness in the locations at which ionizations may occur and the feedback process associated with the fact that both kinds of carrier can produce impact ionizations. The first of these sources of noise is present even when only one kind of carrier can multiply; it gives rise to a minimum excess noise factor $F = 2$ at large values of the mean gain $\overline{G}$, as is apparent by setting $k = 0$ and letting $\overline{G}$ become large in (17.5-23). The second source of noise (the feedback process) is potentially more detrimental since it can result in a far larger increase in F. In a photomultiplier tube, there is only one

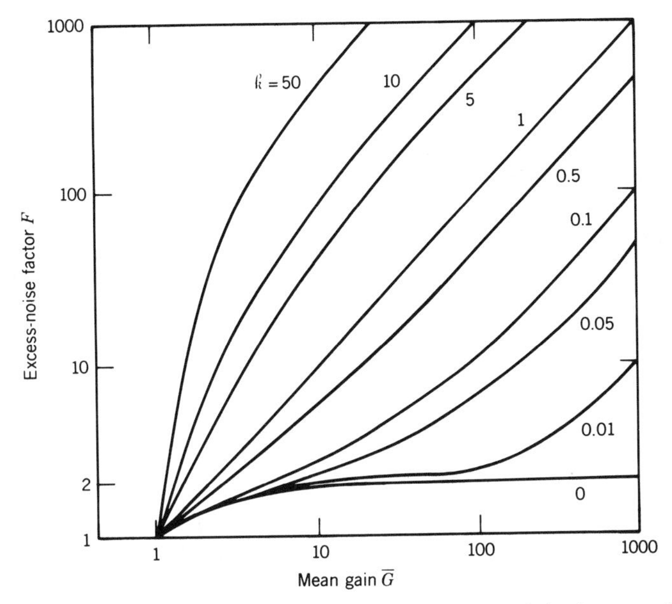

Figure 17.5-4 Excess noise factor F for an APD, under electron injection, as a function of the mean gain $\overline{G}$ for different values of the ionization ratio k. For hole injection, $1/k$ replaces k.

kind of carrier (electrons) and ionizations always occur at the dynodes, so there is no location randomness. In general, therefore, $1 \lesssim F < 2$ for photomultiplier tubes.

EXAMPLE 17.5-2. ***Excess Noise Factor of a Si APD.*** A Si APD (for which $k \approx 0.1$) with $\overline{G} = 100$ (and electron injection) has an excess noise factor $F = 11.8$. Use of the APD therefore increases the mean value of the detected current by a factor of 100, while reducing the signal-to-noise ratio by a factor of 11.8. We show in the next section, however, that in the presence of circuit noise the use of an APD can increase the overall SNR.

---

*Excess Noise Factor for a Multilayer APD

In principle, both sources of gain noise (location randomness and feedback) can be eliminated by the use of a multilayer avalanche photodiode structure. One example of such a structure, the graded-gap staircase device, is illustrated in Fig. 17.4-3. An electron in the conduction band can gain sufficient energy to promote an impact ionization only at discrete locations (at the risers in the staircase), which eliminates the location randomness. The feedback noise is reduced by the valence-band-edge discontinuity, which is in the wrong direction for fostering impact ionizations and therefore should result in a small value of k. In theory, totally noise-free multiplication ($F = 1$) can be achieved. As indicated earlier, however, devices of this type are very difficult to fabricate.

C. Circuit Noise

Yet additional noise is introduced by the electronic circuitry associated with an optical receiver. Circuit noise results from the thermal motion of charged carriers in resistors and other dissipative elements (thermal noise), and from fluctuations of charge carriers in transistors used in the receiver amplifier.

Thermal Noise

Thermal noise (also called **Johnson noise** or **Nyquist noise**) arises from the random motions of mobile carriers in resistive electrical materials at finite temperatures; these motions give rise to a random electric current $i(t)$ even in the absence of an external electrical power source. The thermal electric current in a resistance R is therefore a random function $i(t)$ whose mean value $\langle i(t) \rangle = 0$, i.e., it is equally likely to be in either direction. The variance of the current σ_i^2 (which is the same as the mean-square value since the mean vanishes) increases with the temperature T.

Using an argument based on statistical mechanics, presented subsequently, it can be shown that a resistance R at temperature T exhibits a random electric current $i(t)$ characterized by a power spectral density (defined in Sec. 10.1B)

$$S_i(f) = \frac{4}{R} \frac{hf}{\exp(hf/k_B T) - 1}, \tag{17.5-24}$$

where f is the frequency. In the region $f \ll k_B T/h$, which is of principal interest since $k_B T/h = 6.25$ THz at room temperature, $\exp(hf/k_B T) \approx 1 + hf/k_B T$ so that

$$S_i(f) \approx 4k_B T/R. \tag{17.5-25}$$

The variance of the electric current is the integral of the power spectral density over all frequencies within the bandwidth B of the circuit, i.e.,

$$\sigma_i^2 = \int_0^B S_i(f)\, df.$$

When $B \ll k_B T/h$, we obtain

$$\boxed{\sigma_i^2 \approx 4k_B TB/R.} \tag{17.5-26}$$

(17.5-26) Thermal Noise Current Variance in a Resistance R

Thus, as shown in Fig. 17.5-5, a resistor R at temperature T in a circuit of bandwidth B behaves as a noiseless resistor in parallel with a source of noise current with zero mean and an rms value σ_i determined by (17.5-26).

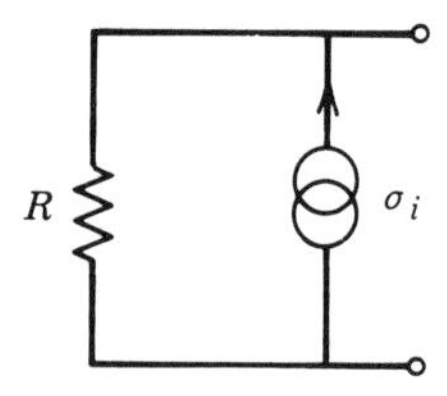

Figure 17.5-5 A resistance R at temperature T is equivalent to a noiseless resistor in parallel with a noise current source of variance $\sigma_i^2 = \langle i^2 \rangle \approx 4k_B TB/R$, where B is the circuit bandwidth.

EXAMPLE 17.5-3. ***Thermal Noise in a Resistor.*** A 1-kΩ resistor at $T = 300$ K, in a circuit of bandwidth $B = 100$ MHz, exhibits an rms thermal noise current $\sigma_i \approx 41$ nA.

---

*Derivation of the Power Spectral Density of Thermal Noise

We now derive (17.5-24) by showing that the electrical power associated with the thermal noise in a resistance is identical to the electromagnetic power radiated by a one-dimensional blackbody. The factor $hf/[\exp(hf/k_BT) - 1]$ in (17.5-24) is recognized as the mean energy $\bar{E}$ of an electromagnetic mode of frequency f (the symbol ν is reserved for optical frequencies) in thermal equilibrium at temperature T [see (12.3-8)]. This equation may therefore be written as $S_i(f)R = 4\bar{E}$. The electrical power dissipated by a noise current i passing through a resistance R is $\langle i^2 \rangle R = \sigma_i^2 R$, so that the term $S_i(f)R$ represents the electrical power density (per Hz) dissipated by the noise current $i(t)$ through R. We proceed to demonstrate that $4\bar{E}$ is the power density radiated by a one-dimensional blackbody.

As discussed in Sec. 12.3B, an atomic system in thermal equilibrium with the electromagnetic modes in a cavity radiates a spectral energy density $\varrho(\nu) = M(\nu)\bar{E}$, where $M(\nu) = 8\pi\nu^2/c^3$ is the three-dimensional density of modes, and the spectral intensity density is $c\varrho(\nu)$. Although the charge carriers in a resistor move in all directions, only motion in the direction of the circuit current flow contributes. The density of modes in a single dimension is $M(f) = 4/c$ modes/m-Hz [see (9.1-7)] so that the corresponding energy density is $\varrho(f) = M(f)\bar{E} = 4\bar{E}/c$ and the radiated power density is $c\varrho(f) = 4\bar{E}$ as promised.

Circuit-Noise Parameter: Resistance-Limited and Amplifier-Limited Optical Receivers

It is convenient to lump the various sources of circuit noise (thermal noise in resistors as well as noise in transistors and other circuit devices) into a single random current source i_r at the receiver input that produces the same total noise at the receiver output (Fig. 17.5-6). The mean value of i_r is zero and the variance σ_r^2 depends on temperature, receiver bandwidth, circuit parameters, and type of devices.

Furthermore, it is convenient to define a dimensionless circuit-noise parameter

$$\sigma_q = \frac{\sigma_r T}{e} = \frac{\sigma_r}{2Be}, \tag{17.5-27}$$

where B is the receiver bandwidth and $T = 1/2B$ is the receiver resolution time. Since

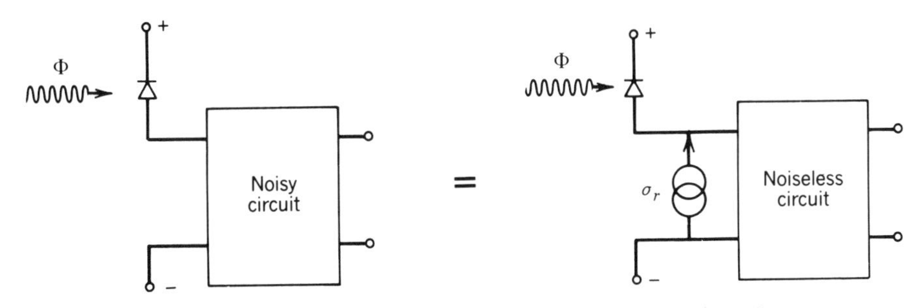

Figure 17.5-6 Noise in the receiver circuit can be replaced with a single random current source with rms value σ_r.

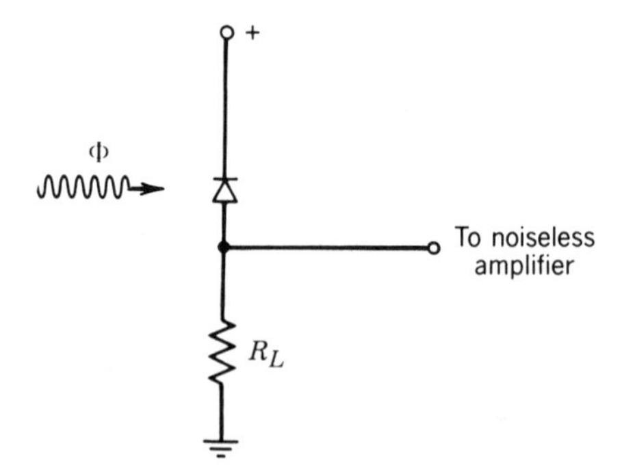

Figure 17.5-7 The resistance-limited optical receiver.

σ_r is the rms value of the electric noise current, σ_r/e is the rms electron flux (electrons/s) arising from circuit noise, and $\sigma_q = (\sigma_r/e)T$ therefore represents the rms number of circuit-noise electrons collected in the time T. It will become apparent in Sec. 17.5D that the circuit-noise parameter σ_q is a figure of merit that characterizes the quality of the optical receiver circuit.

An optical receiver comprising a photodiode, in series with a load resistor R_L followed by an amplifier, is illustrated in Fig. 17.5-7. This simple receiver is said to be **resistance limited** if the circuit-noise current arising from thermal noise in the load resistor substantially exceeds contributions from other sources. The amplifier may then be regarded as noiseless and the circuit-noise mean-square current is simply $\sigma_r^2 = 4k_B TB/R_L$. The circuit-noise parameter defined by (17.5-27) is therefore

$$\sigma_q = \left(\frac{k_B T}{e^2 R_L B} \right)^{1/2}, \tag{17.5-28}$$

which is inversely proportional to the square-root of the bandwidth B.

EXAMPLE 17.5-4. *Circuit-Noise Parameter.* At room temperature, a resistance $R_L = 50\ \Omega$ in a circuit of bandwidth $B = 100$ MHz generates a random current of rms value $\sigma_r = 0.18\ \mu$A. This corresponds to a circuit-noise parameter $\sigma_q \approx 5700$.

Receivers using well-designed low-noise amplifiers can yield lower circuit-noise parameters than does the resistance-limited receiver. Consider a receiver using an *FET amplifier*. If the noise arising from the high input resistance of the amplifier can be neglected, the receiver is limited by thermal noise in the channel between the FET source and drain. With the use of an equalizer to boost the high frequencies attenuated by the capacitive input impedance of the circuit, the circuit-noise parameter at room temperature for typical circuit component values turns out to be

$$\sigma_q \approx \frac{B^{1/2}}{100} \quad (B \text{ in Hz}). \tag{17.5-29}$$

Circuit-Noise Parameter (FET Amplifier Receiver)

For example, if $B = 100$ MHz, then $\sigma_q = 100$. This is significantly smaller than the circuit-noise parameter associated with a 50-Ω resistance-limited amplifier of the same bandwidth. The circuit-noise parameter σ_q increases with B because of the effect of the equalizer.†

Receivers using *bipolar transistor amplifiers*, in contrast, have a circuit-noise parameter σ_q that is independent of the bandwidth B over a wide range of frequencies.† For bandwidths between 100 MHz and 2 GHz, σ_q is typically ≈ 500, provided that appropriate transistors are used and that they are optimally biased.

D. Signal-to-Noise Ratio and Receiver Sensitivity

The simplest measure of the quality of reception is the signal-to-noise ratio (SNR). The SNR of the current at the input to the noiseless circuit represented in Fig. 17.5-6 is the ratio of the square of the mean current to the sum of the variances of the constituent sources of noise, i.e.,

$$\text{SNR} = \frac{\bar{i}^2}{2e\bar{G}\bar{i}BF + \sigma_r^2} = \frac{(e\bar{G}\eta\Phi)^2}{2e^2\bar{G}^2\eta B\Phi F + \sigma_r^2}. \tag{17.5-30}$$

Signal-to-Noise Ratio of an Optical Receiver

The first term in each of the denominators represents photoelectron and gain noise [see (17.5-20)], whereas the second term represents circuit noise. For a detector without gain, $\bar{G} = 1$ and $F = 1$. Even if it provides amplification, the noiseless circuit does not alter the signal-to-noise ratio.

EXERCISE 17.5-1

Signal-to-Noise Ratio of the Resistance-Limited Optical Receiver. Assume that the optical receiver shown in Fig. 17.5-7 uses an ideal *p-i-n* photodiode ($\eta = 1$) and the resistance R_L is 50 Ω at room temperature ($T = 300$ K). The bandwidth is $B = 100$ MHz. At what value of photon flux Φ is the photoelectron-noise current variance equal to the resistor thermal-noise current variance? What is the corresponding optical power at $\lambda_o = 1.55$ μm?

It is useful to write the SNR in (17.5-30) in terms of the mean number of detected photons $\bar{m}$ in the resolution time of the receiver $T = 1/2B$,

$$\bar{m} = \eta\Phi T = \frac{\eta\Phi}{2B}, \tag{17.5-31}$$

and the circuit noise-parameter $\sigma_q = \sigma_r/2Be$. The resulting expression is

$$\text{SNR} = \frac{\bar{G}^2\bar{m}^2}{\bar{G}^2F\bar{m} + \sigma_q^2}. \tag{17.5-32}$$

Signal-to-Noise Ratio of an Optical Receiver

†For further details, see S. D. Personick, *Optical Fiber Transmission Systems*, Plenum Press, New York, 1981, Sec. 3.4; note that the parameter σ_q is equivalent to $Z/2$ in this reference.

Equation (17.5-32) has a simple interpretation. The numerator is the square of the mean number of multiplied photoelectrons detected in the receiver resolution time $T = 1/2B$. The denominator is the sum of the variances of the number of photoelectrons and the number of circuit-noise electrons collected in T. For a photodiode without gain $\overline{G} = F = 1$, so that (17.5-32) reduces to

$$\text{SNR} = \frac{\overline{m}^2}{\overline{m} + \sigma_q^2}. \tag{17.5-33}$$

Signal-to-Noise Ratio of an Optical Receiver in the Absence of Gain

The relative magnitudes of $\overline{m}$ and σ_q^2 determine the relative importance of photoelectron noise and circuit noise. The manner in which the parameter σ_q characterizes the circuit's performance as an optical receiver is now apparent. For example, if $\sigma_q = 100$, then circuit noise dominates photoelectron noise as long as the mean number of photoelectrons recorded per resolution time lies below 10,000.

We proceed now to examine the dependence of the SNR on photon flux Φ, circuit bandwidth B, receiver circuit-noise parameter σ_q, and gain $\overline{G}$. This will allow us to determine when the use of an APD is beneficial and will permit us to select an appropriate preamplifier for a given photon flux. In undertaking this parametric study, we rely on the expressions for the SNR provided in (17.5-30), (17.5-32), and (17.5-33).

Dependence of the SNR on Photon Flux

The dependence of the SNR on $\overline{m} = \eta\Phi/2B$ provides an indication of how the SNR varies with the photon flux Φ. Consider first a photodiode without gain, in which case (17.5-33) applies. Two limiting cases are of interest:

- *Circuit-Noise Limit:* If Φ is sufficiently small, such that $\overline{m} \ll \sigma_q^2$ ($\Phi \ll 2B\sigma_q^2/\eta$), the photon noise is negligible and circuit noise dominates, yielding

$$\text{SNR} \approx \frac{\overline{m}^2}{\sigma_q^2}. \tag{17.5-34}$$

- *Photon-Noise Limit:* If the photon flux Φ is sufficiently large, such that $\overline{m} \gg \sigma_q^2$ ($\Phi \gg 2B\sigma_q^2/\eta$), the circuit-noise term can be neglected, whereupon

$$\text{SNR} \approx \overline{m}. \tag{17.5-35}$$

For small $\overline{m}$, therefore, the SNR is proportional to $\overline{m}^2$ and thereby to Φ^2, whereas for large $\overline{m}$, it is proportional to $\overline{m}$ and thereby to Φ, as illustrated in Fig. 17.5-8. For all levels of light the SNR increases with increasing incident photon flux Φ; the presence of more light improves receiver performance.

When the Use of an APD Provides an Advantage

We now compare two receivers that are identical in all respects except that one exhibits no gain, whereas the other exhibits gain $\overline{G}$ and excess noise factor F (e.g., an APD). For sufficiently small $\overline{m}$ (or photon flux Φ), circuit noise dominates. Amplifying the photocurrent above the level of the circuit noise should then improve the SNR. The APD receiver would then be superior. For sufficiently large $\overline{m}$ (or photon flux),

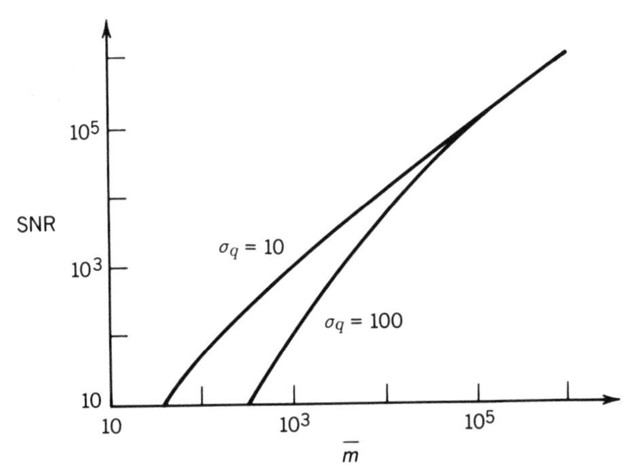

Figure 17.5-8 Signal-to-noise ratio (SNR) as a function of the mean number of photoelectrons per receiver resolution time, $\overline{m} = \eta\Phi/2B$, for a photodiode at two values of the circuit-noise parameter σ_q.

circuit noise is negligible. Amplifying the photocurrent then introduces gain noise, thereby reducing the SNR. The photodiode receiver would then be superior. Comparing (17.5-32) and (17.5-33) shows that the SNR of the APD receiver is greater than that of the photodiode receiver when $\overline{m} < \sigma_q^2(1 - 1/\overline{G}^2)/(F - 1)$. For $\overline{G} \gg 1$, the APD provides an advantage when $\overline{m} < \sigma_q^2/(F - 1)$. If this condition is not satisfied, the use of an APD compromises rather than enhances receiver performance. When σ_q is very small, for example, it is evident from (17.5-32) that the APD SNR $= \overline{m}/F$ is inferior to the photodiode SNR $= \overline{m}$. The SNR is plotted as a function of $\overline{m}$ for the two receivers in Fig. 17.5-9.

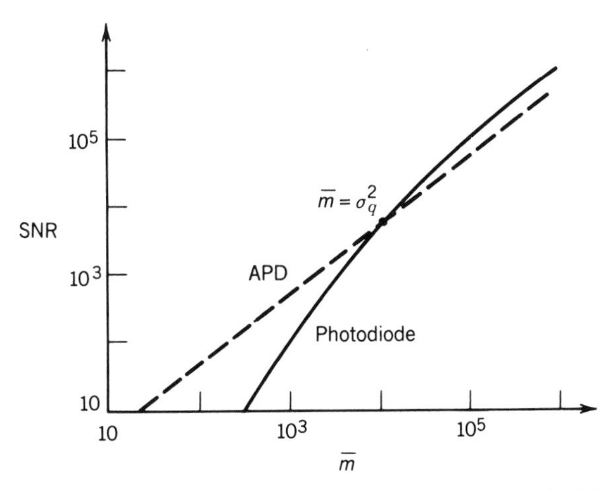

Figure 17.5-9 SNR versus $\overline{m} = \eta\Phi/2B$ for a photodiode receiver (solid curve) and for an APD receiver with mean gain $\overline{G} = 100$ and excess noise factor $F = 2$ (dashed curve) obtained from (17.5-32). The circuit noise parameter $\sigma_q = 100$ in both cases. For small photon flux (circuit-noise-limited case), the APD yields a higher SNR than the photodiode. For large photon flux (photon-noise limit), the photodiode receiver is superior to the APD receiver. The transition between the two regions occurs at $\overline{m} \approx \sigma_q^2/(F - 1) = 10^4$.

Dependence of the SNR on APD Gain

The use of an APD is beneficial for a sufficiently small photon flux such that $\overline{m} < \sigma_q^2/(F-1)$. The optimal gain of an APD is now determined by making use of (17.5-32),

$$\text{SNR} = \frac{\overline{G}^2\overline{m}}{\overline{G}^2 F + \sigma_q^2/\overline{m}}. \qquad (17.5\text{-}36)$$

For an APD, the excess noise factor F is itself a function of $\overline{G}$, in accordance with (17.5-23). Substitution yields

$$\text{SNR} = \frac{\overline{G}^2\overline{m}}{k\overline{G}^3 + (1-k)(2\overline{G}^2 - \overline{G}) + \sigma_q^2/\overline{m}}, \qquad (17.5\text{-}37)$$

where k is the APD carrier ionization ratio. This expression is plotted in Fig. 17.5-10 for $\overline{m} = 1000$ and $\sigma_q = 500$. For the single-carrier multiplication ($k = 0$) APD, the SNR increases with gain and eventually saturates. For the double-carrier multiplication ($k > 0$) APD, the SNR also increases with increasing gain, but it reaches a maximum at an optimal value of the gain, beyond which it decreases as a result of the sharp increase in gain noise. In general, there is therefore an optimal choice of APD gain.

Dependence of the SNR on Receiver Bandwidth

The relation between the SNR and the bandwidth B is implicit in (17.5-30). It is governed by the dependence of the circuit-noise current variance σ_r^2 on B. Consider three receivers:

- The *resistance-limited receiver* exhibits $\sigma_r^2 \propto B$ [see (17.5-26)] so that

$$\text{SNR} \propto B^{-1}. \qquad (17.5\text{-}38)$$

- The *FET amplifier* receiver obeys $\sigma_q \propto B^{1/2}$ [see (17.5-29)] so that $\sigma_r = 2eB\sigma_q \propto B^{3/2}$. This indicates that the dependence of the SNR on B in (17.5-30) assumes

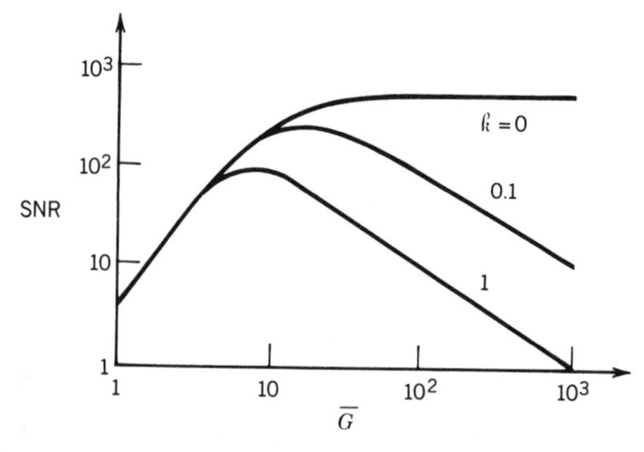

Figure 17.5-10 Dependence of the SNR on the APD mean gain $\overline{G}$ for different ionization ratios k when $\overline{m} = 1000$ and $\sigma_q = 500$.

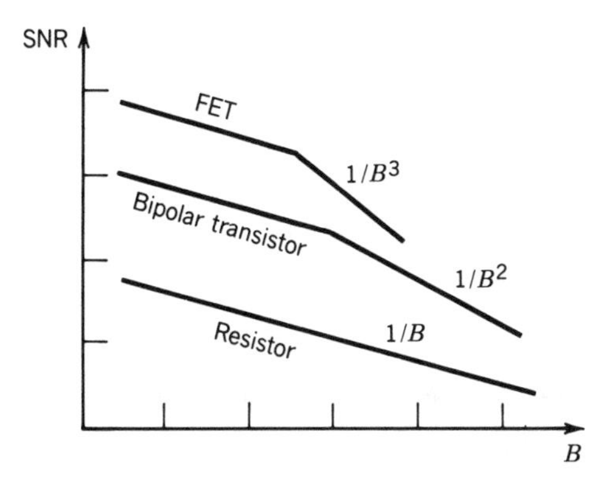

Figure 17.5-11 Double-logarithmic plot of the dependence of the SNR on the bandwidth B for three types of receivers.

the form

$$\text{SNR} \propto (B + sB^3)^{-1}, \tag{17.5-39}$$

where s is a constant.

- The *bipolar-transistor amplifier* has a circuit-noise parameter σ_q that is approximately independent of B. Thus $\sigma_r \propto B$, so that (17.5-30) take the form

$$\text{SNR} \propto (B + s'B^2)^{-1}, \tag{17.5-40}$$

where s' is a constant.

These relations are illustrated schematically in Fig. 17.5-11. The SNR always decreases with increasing B. For sufficiently small bandwidths, all of the receivers exhibit an SNR that varies as B^{-1}. For large bandwidths, the SNR of the FET and bipolar transistor-amplifier receivers declines more sharply with bandwidth.

Receiver Sensitivity

The receiver sensitivity is the minimum photon flux Φ_0, with its corresponding optical power $P_0 = h\nu\Phi_0$ and corresponding mean number of photoelectrons $\overline{m}_0 = \eta\Phi_0/2B$, required to achieve a prescribed value of signal-to-noise ratio SNR_0. The quantity $\overline{m}_0$ can be determined by solving (17.5-32) for $\text{SNR} = \text{SNR}_0$. We shall consider only the case of the unity-gain receiver, leaving the more general solution as an exercise.

Solving the quadratic equation (17.5-33) for $\overline{m}_0$, we obtain

$$\overline{m}_0 = \tfrac{1}{2}\left[\text{SNR}_0 + \left(\text{SNR}_0^2 + 4\sigma_q^2 \text{SNR}_0\right)^{1/2}\right]. \tag{17.5-41}$$

Two limiting cases emerge:

$$\text{Photon-noise limit } \left(\sigma_q^2 \ll \frac{\text{SNR}_0}{4}\right): \quad \overline{m}_0 = \text{SNR}_0 \tag{17.5-42}$$

$$\text{Circuit-noise limit } \left(\sigma_q^2 \gg \frac{\text{SNR}_0}{4}\right): \quad \overline{m}_0 = (\text{SNR}_0)^{1/2}\sigma_q. \tag{17.5-43}$$

Receiver Sensitivity

EXAMPLE 17.5-5. ***Receiver Sensitivity.*** We assume that $\mathrm{SNR}_0 = 10^4$, which corresponds to an acceptable signal-to-noise ratio of 40 dB. If the receiver circuit-noise parameter $\sigma_q \ll 50$, the receiver is photon-noise limited and its sensitivity is $\overline{m}_0 = 10{,}000$ photoelectrons per receiver resolution time. In the more likely situation for which $\sigma_q \gg 50$, the receiver sensitivity $\approx 100\sigma_q$. If $\sigma_q = 500$, for example, the sensitivity is $\overline{m}_0 = 50{,}000$, which corresponds to $2B\overline{m}_0 = 10^5 B$ photoelectrons/s. The optical power sensitivity $P_0 = 2B\overline{m}_0 h\nu/\eta = 10^5 Bh\nu/\eta$ is directly proportional to the bandwidth. If $B = 100$ MHz and $\eta = 0.8$, then at $\lambda_o = 1.55$ μm the receiver sensitivity is $P_0 \approx 1.6$ μW.

When using (17.5-41) to determine the receiver sensitivity, it should be kept in mind that the circuit-noise parameter σ_q is, in general, a function of the bandwidth B, in accordance with:

Resistance-limited receiver:	$\sigma_q \propto B^{-1/2}$
FET amplifier:	$\sigma_q \propto B^{1/2}$
Bipolar-transistor amplifier:	σ_q independent of B

For these receivers, the sensitivity $\overline{m}_0$ depends on bandwidth B as illustrated in Fig. 17.5-12. The optimal choice of receiver therefore depends in part on the bandwidth B.

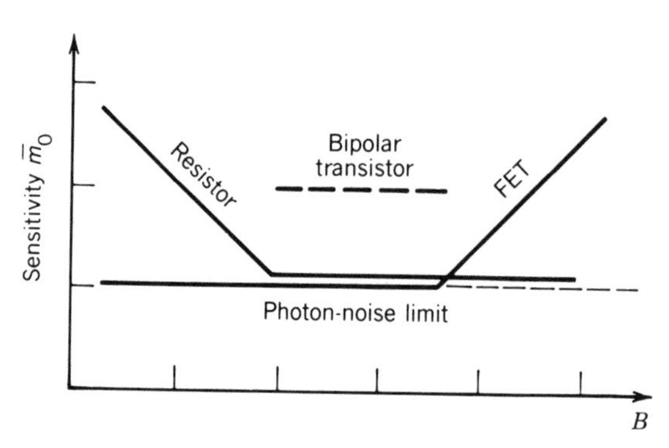

Figure 17.5-12 Double-logarithmic plot of receiver sensitivity $\overline{m}_0$ (the minimum mean number of photoelectrons per resolution time $T = 1/2B$ guaranteeing a minimum signal-to-noise ratio SNR_0) as a function of bandwidth B for three types of receivers. The curves approach the photon-noise limit at values of B for which $\sigma_q^2 \ll \mathrm{SNR}_0/4$. In the photon-noise limit (i.e., when circuit noise is negligible), $\overline{m}_0 = \mathrm{SNR}_0$ in all cases.

EXERCISE 17.5-2

Sensitivity of the APD Receiver. Derive an expression analogous to (17.5-41) for the sensitivity of a receiver incorporating an APD of gain $\overline{G}$ and excess noise factor F. Show that in the limit of negligible circuit noise, the receiver sensitivity reduces to

$$\overline{m}_0 = F \cdot \mathrm{SNR}_0.$$

READING LIST

Books

See also the reading list in Chapter 15.

J. D. Vincent, *Fundamentals of Infrared Detector Operation and Testing*, Wiley, New York, 1990.

N. V. Joshi, *Photoconductivity*, Marcel Dekker, New York, 1990.

P. N. J. Dennis, *Photodetectors*, Plenum Press, New York, 1986.

A. van der Ziel, *Noise in Solid State Devices and Circuits*, Wiley-Interscience, New York, 1986.

R. K. Willardson and A. C. Beer, eds., *Semiconductors and Semimetals*, vol. 22, *Lightwave Communications Technology*, W. T. Tsang, ed., part D, *Photodetectors*, Academic Press, New York, 1985.

E. L. Dereniak and D. G. Crowe, *Optical Radiation Detectors*, Wiley, New York, 1984.

R. W. Boyd, *Radiometry and the Detection of Optical Radiation*, Wiley, New York, 1983.

W. Budde, ed., *Physical Detectors of Optical Radiation*, Academic Press, New York, 1983.

M. J. Buckingham, *Noise in Electron Devices and Systems*, Wiley, New York, 1983.

R. J. Keyes, ed., *Optical and Infrared Detectors*, vol. 19, *Topics in Applied Physics*, Springer-Verlag, Berlin, 2nd ed. 1980.

D. F. Barbe, ed., *Charge Coupled Devices*, vol. 39, *Topics in Applied Physics*, Springer-Verlag, Berlin, 1980.

R. W. Engstrom, *RCA Photomultiplier Handbook* (PMT-62), RCA Electro Optics and Devices, Lancaster, PA, 1980.

B. O. Seraphin, ed., *Solar Energy Conversion: Solid-State Physics Aspects*, vol. 31, *Topics in Applied Physics*, Springer-Verlag, Berlin, 1979.

R. H. Kingston, *Detection of Optical and Infrared Radiation*, Springer-Verlag, New York, 1978.

B. Saleh, *Photoelectron Statistics*, Springer-Verlag, New York, 1978.

A. Rose, *Concepts in Photoconductivity and Allied Problems*, Wiley-Interscience, New York, 1963; R. E. Krieger, Huntington, NY, 2nd ed. 1978.

M. Cardona and L. Ley, eds., *Photoemission in Solids*, vol. 26, *Topics in Applied Physics*, Springer-Verlag, Berlin, 1978.

R. K. Willardson and A. C. Beer, eds., *Semiconductors and Semimetals*, vol. 12, *Infrared Detectors II*, Academic Press, New York, 1977.

J. Mort and D. M. Pai, eds., *Photoconductivity and Related Phenomena*, Elsevier, New York, 1976.

R. K. Willardson and A. C. Beer, eds., *Semiconductors and Semimetals*, vol. 5, *Infrared Detectors*, R. J. Keyes, ed., Academic Press, New York, 1970.

A. H. Sommer, *Photoemissive Materials*, Wiley, New York, 1968.

Special Journal Issues

Special issue on quantum well heterostructures and superlattices, *IEEE Journal of Quantum Electronics*, vol. QE-24, no. 8, 1988.

Special issue on semiconductor quantum wells and superlattices: physics and applications, *IEEE Journal of Quantum Electronics*, vol. QE-22, no. 9, 1986.

Special issue on light emitting diodes and long-wavelength photodetectors, *IEEE Transactions on Electron Devices*, vol. ED-30, no. 4, 1983.

Special issue on optoelectronic devices, *IEEE Transactions on Electron Devices*, vol. ED-29, no. 9, 1982.

Special issue on light sources and detectors, *IEEE Transactions on Electron Devices*, vol. ED-28, no. 4, 1981.

Special issue on quaternary compound semiconductor materials and devices—sources and detectors, *IEEE Journal of Quantum Electronics*, vol. QE-17, no. 2, 1981.

Special joint issue on optoelectronic devices and circuits, *IEEE Transactions on Electron Devices*, vol. ED-25, no. 2, 1978.

Special joint issue on optical electronics, *Proceedings of the IEEE*, vol. 54, no. 10, 1966.

Articles

D. Parker, Optical Detectors: Research to Reality, *Physics World*, vol. 3, no. 3, pp. 52–54, 1990.

S. R. Forrest, Optical Detectors for Lightwave Communication, in *Optical Fiber Telecommunications II*, S. E. Miller and I. P. Kaminow, eds., Academic Press, New York, 1988.

G. Margaritondo, 100 Years of Photoemission, *Physics Today*, vol. 44, no. 4, pp. 66–72, 1988.

F. Capasso, Band-Gap Engineering: From Physics and Materials to New Semiconductor Devices, *Science*, vol. 235, pp. 172–176, 1987.

S. R. Forrest, Optical Detectors: Three Contenders, *IEEE Spectrum*, vol. 23, no. 5, pp. 76–84, 1986.

M. C. Teich, K. Matsuo, and B. E. A. Saleh, Excess Noise Factors for Conventional and Superlattice Avalanche Photodiodes and Photomultiplier Tubes, *IEEE Journal of Quantum Electronics*, vol. QE-22, pp. 1184–1193, 1986.

D. S. Chemla, Quantum Wells for Photonics, *Physics Today*, vol. 38, no. 5, pp. 56–64, 1985.

F. Capasso, Multilayer Avalanche Photodiodes and Solid-State Photomultipliers, *Laser Focus/Electro-Optics*, vol. 20, no. 7, pp. 84–101, 1984.

P. P. Webb and R. J. McIntyre, Recent Developments in Silicon Avalanche Photodiodes, *RCA Engineer*, vol. 27, pp. 96–102, 1982.

H. Melchior, Detectors for Lightwave Communication, *Physics Today*, vol. 30, no. 11, pp. 32–39, 1977.

P. P. Webb, R. J. McIntyre, and J. Conradi, Properties of Avalanche Photodiodes, *RCA Review*, vol. 35, pp. 234–278, 1974.

R. J. Keyes and R. H. Kingston, A Look at Photon Detectors, *Physics Today*, vol. 25, no. 3, pp. 48–54, 1972.

H. Melchior, Demodulation and Photodetection Techniques, in F. T. Arecchi and E. O. Schulz-Dubois, eds., *Laser Handbook*, vol. 1, North-Holland, Amsterdam, 1972, pp. 725–835.

W. E. Spicer and F. Wooten, Photoemission and Photomultipliers, *Proceedings of the IEEE*, vol. 51, pp. 1119–1126, 1963.

PROBLEMS

17.1-1 **Effect of Reflectance on Quantum Efficiency.** Determine the factor $1 - \mathscr{R}$ in the expression for the quantum efficiency, under normal and 45° incidence, for an unpolarized light beam incident from air onto Si, GaAs, and InSb (see Sec. 6.2 and Table 15.2-1 on page 588).

17.1-2 **Responsivity.** Find the maximum responsivity of an ideal (unity quantum efficiency and unity gain) semiconductor photodetector made of (a) Si; (b) GaAs; (c) InSb.

17.1-3 **Transit Time.** Referring to Fig. 17.1-3, assume that a photon generates an electron–hole pair at the position $x = w/3$, that $v_e = 3v_h$ (in semiconductors v_e is generally larger than v_h), and that the carriers recombine at the contacts. For each carrier, find the magnitudes of the currents, i_h and i_e, and the durations of the currents, τ_h and τ_e. Express your results in terms of e, w, and v_e. Verify that the total charge induced in the circuit is e. For $v_e = 6 \times 10^7$ cm/s and $w = 10$ μm, sketch the time course of the currents.

17.1-4 **Current Response with Uniform Illumination.** Consider a semiconductor material (as in Fig. 17.1-3) exposed to an impulse of light at $t = 0$ that generates N electron–hole pairs uniformly distributed between 0 and w. Let the electron and hole velocities in the material be v_e and v_h, respectively. Show that the hole current can be written as

$$i_h(t) = \begin{cases} -\dfrac{Nev_h^2}{w^2}t + \dfrac{Nev_h}{w} & 0 \le t \le \dfrac{w}{v_h} \\ 0, & \text{elsewhere,} \end{cases}$$

while the electron current is

$$i_e(t) = \begin{cases} -\dfrac{Nev_e^2}{w^2}t + \dfrac{Nev_e}{w}, & 0 \le t \le \dfrac{w}{v_e} \\ 0, & \text{elsewhere,} \end{cases}$$

and that the total current is therefore

$$i(t) = \begin{cases} \dfrac{Ne}{w}\left[(v_h + v_e) - \dfrac{1}{w}(v_h^2 + v_e^2)t\right], & 0 \le t \le \dfrac{w}{v_e} \\ \dfrac{Nev_h}{w}\left[1 - \dfrac{v_h}{w}t\right], & \dfrac{w}{v_e} \le t \le \dfrac{w}{v_h}. \end{cases}$$

The various currents are illustrated in Fig. 17.1-4. Verify that the electrons and holes each contribute a charge $Ne/2$ to the external circuit so that the total charge generated is Ne.

*17.1-5 **Two-Photon Detectors.** Consider a beam of photons of energy $h\nu$ and photon flux density ϕ (photons/cm^2-s) incident on a semiconductor detector with bandgap $h\nu < E_g < 2h\nu$, such that one photon cannot provide sufficient energy to raise an electron from the valence band to the conduction band. Nevertheless, two photons can occasionally conspire to jointly give up their energy to the electron. Assume that the current *density* induced in such a detector is given by $J_p = \zeta\phi^2$, where ζ is a constant. Show that the responsivity (A/W) is given by $\Re = [\zeta/(hc_o)^2]\lambda_o^2 P/A$ for the two-photon detector, where P is the optical power and A is the detector area illuminated. Explain physically the proportionality to λ_o^2 and P/A.

17.2-1 **Photoconductor Circuit.** A photoconductor detector is often connected in series with a load resistor R and a dc voltage source V, and the voltage V_p across the load resistor is measured. If the conductance of the detector is proportional to the optical power P, sketch the dependence of V_p on P. Under what conditions is this dependence linear?

17.2-2 **Photoconductivity.** The concentration of charge carriers in a sample of intrinsic Si is $n_i = 1.5 \times 10^{10}$ cm^{-3} and the recombination lifetime $\tau = 10$ μs. If the material is illuminated with light, and an optical power density of 1 mW/cm^3 at $\lambda_o = 1$ μm is absorbed by the material, determine the percentage increase in its conductivity. The quantum efficiency $\eta = \frac{1}{2}$.

17.3-1 **Quantum Efficiency of a Photodiode Detector.** For a particular *p-i-n* photodiode, a pulse of light containing 6×10^{12} incident photons at wavelength $\lambda_o = 1.55$ μm gives rise to, on average, 2×10^{12} electrons collected at the terminals of the device. Determine the quantum efficiency η and the responsivity $\Re$ of the photodiode at this wavelength.

17.4-1 **Quantum Efficiency of an APD.** A conventional APD with gain $\overline{G} = 20$ operates at a wavelength $\lambda_o = 1.55$ μm. If its responsivity at this wavelength is $\Re = 12$ A/W, calculate its quantum efficiency η. What is the photocurrent at the output of the device if a photon flux $\Phi = 10^{10}$ photons/s, at this same wavelength, is incident on it?

17.4-2 **Gain of an APD.** Show that an APD with ionization ratio $k = 1$, such as germanium, has a gain given by $\overline{G} = 1/(1 - \alpha_e w)$, where α_e is the electron ionization coefficient and w is the width of the multiplication layer. [*Note*: Equation (17.4-5) does not give a proper answer for the gain when $k = 1$.]

17.5-1 **Excess Noise Factor for a Single-Carrier APD.** Show that an APD with pure electron injection and no hole multiplication ($k = 0$) has an excess noise factor $F \approx 2$ for all appreciable values of the gain. Use (17.4-5) to show that the mean gain is then $G = \exp(\alpha_e w)$. Calculate the responsivity of a Si APD for photons with energy equal to the bandgap energy E_g, assuming that the quantum efficiency $\eta = 0.8$ and the gain $\overline{G} = 70$. Find the excess noise factor for a double-carrier-multiplication Si APD when $k = 0.01$. Compare it with the value $F \approx 2$ obtained in the single-carrier-multiplication limit.

*17.5-2 **Gain of a Multilayer APD.** Use the Bernoulli probability law to show that the mean gain of a single-carrier-multiplication multilayer APD is $\overline{G} = (1 + P)^l$, where P is the probability of impact ionization at each stage and l is the number of stages. Show that the result reduces to that of the conventional APD when $P \to 0$ and $l \to \infty$.

*17.5-3 **Excess Noise Factor for a One-Stage Photomultiplier Tube.** Derive an expression for the excess noise factor F of a one-stage photomultiplier tube assuming that the number of secondary emission electrons per incident primary electron is Poisson distributed with mean δ.

*17.5-4 **Excess Noise Factor for a Photoconductor Detector.** The gain of a photoconductor detector was shown in Sec. 17.2 to be $G = \tau/\tau_e$, where τ is the electron–hole recombination lifetime and τ_e is the electron transit time across the sample. Actually, G is random because τ can be thought of as random. Show that an exponential probability density function for the random recombination lifetime, $P(\tau) = (1/\bar{\tau})\exp(-\tau/\bar{\tau})$, results in an excess noise factor $F = 2$, confirming that photoconductor **generation-recombination** (GR) **noise** degrades the SNR by a factor of 2.

17.5-5 **Bandwidth of an RC Circuit.** Using the definition of bandwidth provided in (17.5-14), show that a circuit of impulse response function $h(t) = (e/\tau)\exp(-t/\tau)$ has a bandwidth $B = 1/4\tau$. What is the bandwidth of an RC circuit? Determine the thermal noise current for a resistance $R = 1$ kΩ at $T = 300$ K connected to a capacitance $C = 5$ pF.

17.5-6 **Signal-to-Noise Ratio of an APD Receiver.** By what factor does the signal-to-noise ratio of a receiver using an APD of mean gain $\overline{G} = 100$ change if the ionization ratio k is increased from $k = 0.1$ to 0.2. Assume that circuit noise is negligible. Show that if the mean gain $\overline{G} \gg 1$ and $\gg 2(1 - k)/k$, the SNR is approximately inversely proportional to $\overline{G}$.

17.5-7 **Noise in an APD Receiver.** An optical receiver using an APD has the following parameters: quantum efficiency $\eta = 0.8$; mean gain $\overline{G} = 100$; ionization ratio $k = 0.5$; load resistance $R_L = 1$ kΩ; bandwidth $B = 100$ kHz; dark and leakage current = 1 nA. An optical signal of power 10 nW at $\lambda_o = 0.87$ μm is received. Determine the rms values of the different noise currents, and the SNR. Assume that the dark and leakage current has a noise variance that obeys the same law as photocurrent noise and that the receiver is resistance limited.

17.5-8 **Optimal Gain in an APD.** A receiver using a *p-i-n* photodiode has a ratio of circuit noise variance to photoelectron noise variance of 1000. If an APD with ionization ratio $k = 0.2$ is used instead, determine the optimal mean gain for maximizing the signal-to-noise ratio and the corresponding improvement in signal-to-noise ratio.

17.5-9 **Receiver Sensitivity.** Determine the receiver sensitivity (i.e., optical power required to achieve a SNR $= 10^3$) for a photodetector of quantum efficiency $\eta = 0.8$ at $\lambda_o = 1.3$ μm in a circuit of bandwidth $B = 100$ MHz when there is no circuit noise. The receiver measures the electric current i.

17.5-10 **Noise Comparison of Three Photodetectors.** Consider three photodetectors in series with a 50-Ω load resistor at 77 K (liquid nitrogen temperature) that are to be used with a 1-μm wavelength optical system with a bandwidth of 1 GHz: (1) a *p-i-n* photodiode with quantum efficiency $\eta = 0.9$; (2) an APD with quantum efficiency $\eta = 0.6$, gain $\overline{G} = 100$, and ionization ratio $k = 0$; (3) a 10-stage photomultiplier tube (PMT) with quantum efficiency $\eta = 0.3$, overall mean gain $\overline{G} = 4^{10}$, and overall gain variance $\sigma_G^2 = \overline{G}^2/4$.
(a) For each detector, find the photocurrent SNR when the detector is illuminated by a photon flux of 10^{10} s^{-1}.
(b) By which devices is the signal detectable?

*17.5-11 **A Single-Dynode Photomultiplier Tube.** Consider a photomultiplier tube with quantum efficiency $\eta = 1$ and only one dynode. Incident on the cathode is light from a hypothetical photon source that gives rise to a probability of observing n photons in the counting time $T = 1.3$ ns, which is given by

$$p(n) = \begin{cases} \frac{1}{2}, & n = 0, 1 \\ 0, & \text{otherwise.} \end{cases}$$

When one electron strikes the dynode, either two or three secondary electrons are emitted and these proceed to the anode. The gain distribution $P(G)$ is given by

$$P(G) = \begin{cases} \frac{1}{3}, & G = 2 \\ \frac{2}{3}, & G = 3 \\ 0, & \text{otherwise.} \end{cases}$$

Thus it is twice as likely that three electrons are emitted as two.
(a) Calculate the SNR of the input photon number and compare the result with that of a Poisson photon number of the same mean.
(b) Find the probability distribution for the photoelectron number $p(m)$ and the SNR of the photoelectron number.
(c) Find the mean gain $\langle G \rangle$ and the mean-square gain $\langle G^2 \rangle$.
(d) Find the excess noise factor F.
(e) Find the mean anode current $\bar{i}$ in a circuit of bandwidth $B = 1/2T$.
(f) Find the responsivity of this photomultiplier tube if the wavelength of the light is $\lambda_o = 1.5$ μm.
(g) Explain why (17.5-20) for σ_i^2 is not applicable.

CHAPTER

18

ELECTRO-OPTICS

18.1 PRINCIPLES OF ELECTRO-OPTICS
- A. Pockels and Kerr Effects
- B. Electro-Optic Modulators and Switches
- C. Scanners
- D. Directional Couplers
- E. Spatial Light Modulators

*18.2 ELECTRO-OPTICS OF ANISOTROPIC MEDIA
- A. Pockels and Kerr Effects
- B. Modulators

18.3 ELECTRO-OPTICS OF LIQUID CRYSTALS
- A. Wave Retarders and Modulators
- B. Spatial Light Modulators

*18.4 PHOTOREFRACTIVE MATERIALS

Friedrich Pockels (1865–1913) was first to describe the linear electro-optic effect in 1893.

John Kerr (1824–1907) discovered the quadratic electro-optic effect in 1875.

Certain materials change their optical properties when subjected to an electric field. This is caused by forces that distort the positions, orientations, or shapes of the molecules constituting the material. The **electro-optic effect** is the change in the refractive index resulting from the application of a dc or low-frequency electric field (Fig. 18.0-1). A field applied to an *anisotropic* electro-optic material modifies its refractive indices and thereby its effect on polarized light.

The dependence of the refractive index on the applied electric field takes one of two forms:

- The refractive index changes in proportion to the applied electric field, in which case the effect is known as the **linear electro-optic effect** or the **Pockels effect.**
- The refractive index changes in proportion to the square of the applied electric field, in which case the effect is known as the **quadratic electro-optic effect** or the **Kerr effect.**

The change in the refractive index is typically very small. Nevertheless, its effect on an optical wave propagating a distance much greater than a wavelength of light in the medium can be significant. If the refractive index increases by 10^{-5}, for example, an optical wave propagating a distance of 10^5 wavelengths will experience an additional phase shift of 2π.

Materials whose refractive index can be modified by means of an applied electric field are useful for producing electrically controllable optical devices, as indicated by the following examples:

- A lens made of a material whose refractive index can be varied is a lens of controllable focal length.
- A prism whose beam bending ability is controllable can be used as an optical scanning device.
- Light transmitted through a transparent plate of controllable refractive index undergoes a controllable phase shift. The plate can be used as an optical phase modulator.

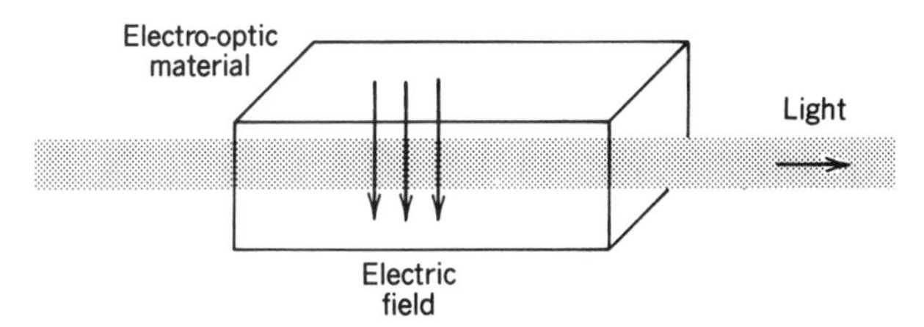

Figure 18.0-1 A steady electric field applied to an electro-optic material changes its refractive index. This, in turn, changes the effect of the material on light traveling through it. The electric field therefore controls the light.

- An anisotropic crystal whose refractive indices can be varied serves as a wave retarder of controllable retardation; it may be used to change the polarization properties of light.
- A wave retarder placed between two crossed polarizers gives rise to transmitted light whose intensity is dependent on the phase retardation (see Sec. 6.6B). The transmittance of the device is therefore electrically controllable, so that it can be used as an optical intensity modulator or an optical switch.

These are useful components for optical communication and optical signal-processing applications.

We begin with a simple description of the electro-optic effect and the principles of electro-optic modulation and scanning (Sec. 18.1). The initial presentation is simplified by deferring the detailed consideration of anisotropic effects to Sec. 18.2.

Section 18.3 is devoted to the electro-optic properties of liquid crystals. An electric field applied to the molecules of a liquid crystal causes them to alter their orientations. This leads to changes in the optical properties of the medium, i.e., it exhibits an electro-optic effect. The molecules of a twisted nematic liquid crystal are organized in a helical pattern so that they normally act as polarization rotators. An applied electric field removes the helical pattern, thereby deactivating the polarization rotatory power of the material. Removal of the electric field results in the material regaining its helical structure and therefore its rotatory power. The device therefore acts as a dynamic polarization rotator. The use of additional fixed polarizers permits such a polarization rotator to serve as an intensity modulator or a switch. This behavior is the basis of most liquid-crystal display devices.

The electro-optic properties of photorefractive media are considered in Sec. 18.4. These are materials in which the absorption of light creates an internal electric field which, in turn, initiates an electro-optic effect that alters the optical properties of the medium. Thus the optical properties of the medium are indirectly controlled by the light incident on it. Photorefractive devices therefore permit *light to control light*.

18.1 PRINCIPLES OF ELECTRO-OPTICS

A. Pockels and Kerr Effects

The refractive index of an electro-optic medium is a function $n(E)$ of the applied electric field E. This function varies only slightly with E so that it can be expanded in a Taylor's series about $E = 0$,

$$n(E) = n + a_1 E + \tfrac{1}{2} a_2 E^2 + \cdots, \qquad (18.1\text{-}1)$$

where the coefficients of expansion are $n = n(0)$, $a_1 = (dn/dE)|_{E=0}$, and $a_2 = (d^2n/dE^2)|_{E=0}$. For reasons that will become apparent subsequently, it is conventional to write (18.1-1) in terms of two new coefficients $\mathfrak{r} = -2a_1/n^3$ and $\mathfrak{s} = -a_2/n^3$, known as the electro-optic coefficients, so that

$$n(E) = n - \tfrac{1}{2}\mathfrak{r} n^3 E - \tfrac{1}{2}\mathfrak{s} n^3 E^2 + \cdots. \qquad (18.1\text{-}2)$$

The second- and higher-order terms of this series are typically many orders of magnitude smaller than n. Terms higher than the third can safely be neglected.

For future use it is convenient to derive an expression for the electric impermeability, $\eta = \epsilon_o/\epsilon = 1/n^2$, of the electro-optic medium as a function of E. The parameter η is useful in describing the optical properties of anisotropic media (see Sec. 6.3A). The incremental change $\Delta\eta = (d\eta/dn)\,\Delta n = (-2/n^3)(-\tfrac{1}{2}\mathfrak{r} n^3 E - \tfrac{1}{2}\mathfrak{s} n^3 E^2) = \mathfrak{r}E +$

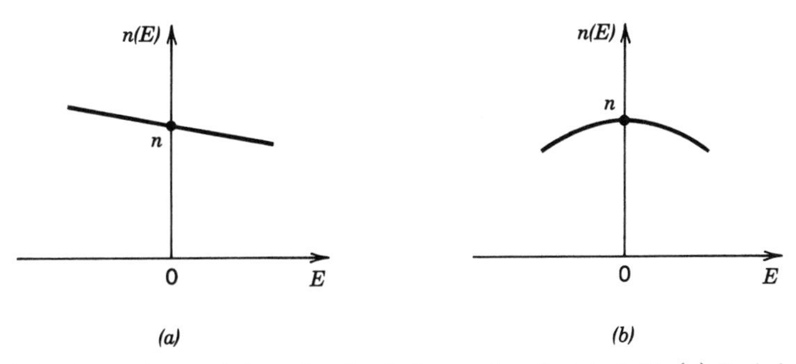

Figure 18.1-1 Dependence of the refractive index on the electric field: (a) Pockels medium; (b) Kerr medium.

$\mathfrak{s}E^2$, so that

$$\eta(E) \approx \eta + \mathfrak{r}E + \mathfrak{s}E^2, \tag{18.1-3}$$

where $\eta = \eta(0)$. The electro-optic coefficients $\mathfrak{r}$ and $\mathfrak{s}$ are therefore simply the coefficients of proportionality of the two terms of $\Delta\eta$ with E and E^2, respectively. This explains the seemingly odd definitions of $\mathfrak{r}$ and $\mathfrak{s}$ in (18.1-2).

The values of the coefficients $\mathfrak{r}$ and $\mathfrak{s}$ depend on the direction of the applied electric field and the polarization of the light, as will be discussed in Sec. 18.2.

Pockels Effect

In many materials the third term of (18.1-2) is negligible in comparison with the second, whereupon

$$n(E) \approx n - \tfrac{1}{2}\mathfrak{r}n^3E, \tag{18.1-4}$$

Pockels Effect

as illustrated in Fig. 18.1-1(a). The medium is then known as a Pockels medium (or a Pockels cell). The coefficient $\mathfrak{r}$ is called the **Pockels coefficient** or the linear electro-optic coefficient. Typical values of $\mathfrak{r}$ lie in the range 10^{-12} to 10^{-10} m/V (1 to 100 pm/V). For $E = 10^6$ V/m (10 kV applied across a cell of thickness 1 cm), for example, the term $\frac{1}{2}\mathfrak{r}n^3E$ in (18.1-4) is on the order of 10^{-6} to 10^{-4}. Changes in the refractive index induced by electric fields are indeed very small. The most common crystals used as Pockels cells include $NH_4H_2PO_4$ (ADP), KH_2PO_4 (KDP), $LiNbO_3$, $LiTaO_3$, and CdTe.

Kerr Effect

If the material is centrosymmetric, as is the case for gases, liquids, and certain crystals, $n(E)$ must be an even symmetric function [see Fig. 18.1-1(b)] since it must be invariant to the reversal of E. Its first derivative then vanishes, so that the coefficient $\mathfrak{r}$ must be zero, whereupon

$$n(E) \approx n - \tfrac{1}{2}\mathfrak{s}n^3E^2. \tag{18.1-5}$$

Kerr Effect

The material is then known as a Kerr medium (or a Kerr cell). The parameter $\mathfrak{s}$ is called the **Kerr coefficient** or the quadratic electro-optic coefficient. Typical values of $\mathfrak{s}$ are 10^{-18} to 10^{-14} m^2/V^2 in crystals and 10^{-22} to 10^{-19} m^2/V^2 in liquids. For $E = 10^6$ V/m the term $\frac{1}{2}\mathfrak{s}n^3E^2$ in (18.1-5) is on the order of 10^{-6} to 10^{-2} in crystals and 10^{-10} to 10^{-7} in liquids.

B. Electro-Optic Modulators and Switches

Phase Modulators

When a beam of light traverses a Pockels cell of length L to which an electric field E is applied, it undergoes a phase shift $\varphi = n(E)k_oL = 2\pi n(E)L/\lambda_o$, where λ_o is the free-space wavelength. Using (18.1-4), we have

$$\varphi \approx \varphi_0 - \pi \frac{\mathfrak{r}n^3EL}{\lambda_o}, \tag{18.1-6}$$

where $\varphi_0 = 2\pi nL/\lambda_o$. If the electric field is obtained by applying a voltage V across two faces of the cell separated by distance d, then $E = V/d$, and (18.1-6) gives

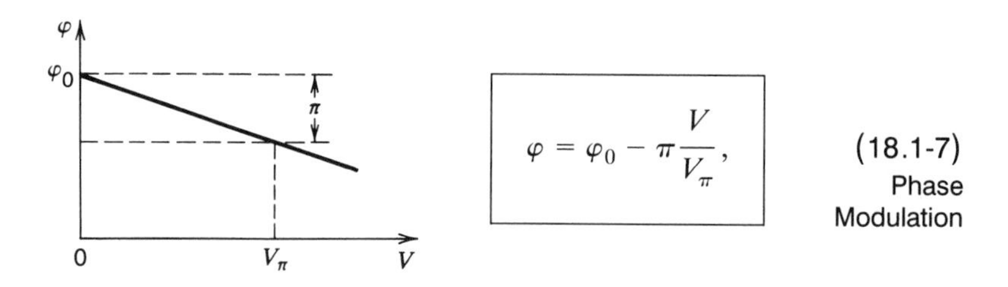

$$\varphi = \varphi_0 - \pi \frac{V}{V_\pi}, \tag{18.1-7}$$

Phase Modulation

where

$$V_\pi = \frac{d}{L} \frac{\lambda_o}{\mathfrak{r}n^3}. \tag{18.1-8}$$

Half-Wave Voltage

The parameter V_π, known as the **half-wave voltage**, is the applied voltage at which the phase shift changes by π. Equation (18.1-7) expresses a linear relation between the optical phase shift and the voltage. One can therefore modulate the phase of an optical wave by varying the voltage V that is applied across a material through which the light passes. The parameter V_π is an important characteristic of the modulator. It depends on the material properties (n and $\mathfrak{r}$), on the wavelength λ_o, and on the aspect ratio d/L.

The electric field may be applied in a direction perpendicular to the direction of light propagation (transverse modulators) or parallel to that direction (longitudinal modulators), in which case $d = L$ (Fig. 18.1-2). The value of the electro-optic coefficient $\mathfrak{r}$ depends on the directions of propagation and the applied field since the crystal is anisotropic (as explained in Sec. 18.2). Typical values of the half-wave voltage are in the vicinity of 1 to a few kilovolts for longitudinal modulators, and hundreds of volts for transverse modulators.

The speed at which an electro-optic modulator operates is limited by electrical capacitive effects and by the transit time of the light through the material. If the

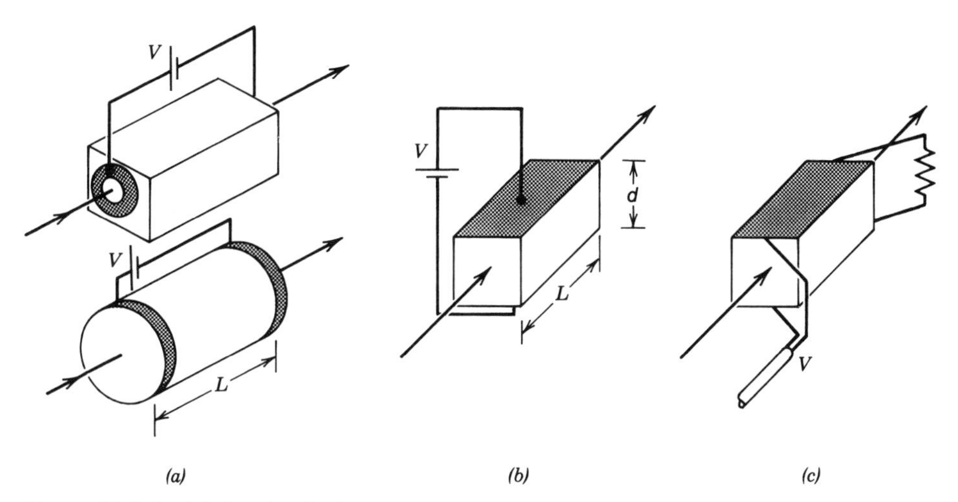

Figure 18.1-2 (*a*) Longitudinal modulator. The electrodes may take the shape of washers or bands, or may be transparent conductors. (*b*) Transverse modulator. (*c*) Traveling-wave transverse modulator.

electric field $E(t)$ varies significantly within the light transit time T, the traveling optical wave will be subjected to different electric fields as it traverses the crystal. The modulated phase at a given time t will then be proportional to the average electric field $E(t)$ at times from $t - T$ to t. As a result, the transit-time-limited modulation bandwidth is $\approx 1/T$. One method of reducing this time is to apply the voltage V at one end of the crystal while the electrodes serve as a transmission line, as illustrated in Fig. 18.1-2(c). If the velocity of the traveling electrical wave matches that of the optical wave, transit time effects can, in principle, be eliminated. Commercial modulators in the forms shown in Fig. 18.1-2 generally operate at several hundred MHz, but modulation speeds of several GHz are possible.

Electro-optic modulators can also be constructed as integrated-optical devices. These devices operate at higher speeds and lower voltages than do bulk devices. An optical waveguide is fabricated in an electro-optic substrate (often $LiNbO_3$) by indiffusing a material such as titanium to increase the refractive index. The electric field is applied to the waveguide using electrodes, as shown in Fig. 18.1-3. Because the configuration is transverse and the width of the waveguide is much smaller than its length ($d \ll L$), the half-wave voltage can be as small as a few volts. These modulators have been operated at speeds in excess of 100 GHz. Light can be conveniently coupled into, and out of, the modulator by the use of optical fibers.

Dynamic Wave Retarders

An anisotropic medium has two linearly polarized normal modes that propagate with different velocities, say c_o/n_1 and c_o/n_2 (see Sec. 6.3B). If the medium exhibits the Pockels effect, then in the presence of a steady electric field E the two refractive indices are modified in accordance with (18.1-4), i.e.,

$$n_1(E) \approx n_1 - \tfrac{1}{2}\mathfrak{r}_1 n_1^3 E$$

$$n_2(E) \approx n_2 - \tfrac{1}{2}\mathfrak{r}_2 n_2^3 E,$$

where $\mathfrak{r}_1$ and $\mathfrak{r}_2$ are the appropriate Pockels coefficients (anisotropic effects are examined in detail in Sec. 18.2). After propagation a distance L, the two modes

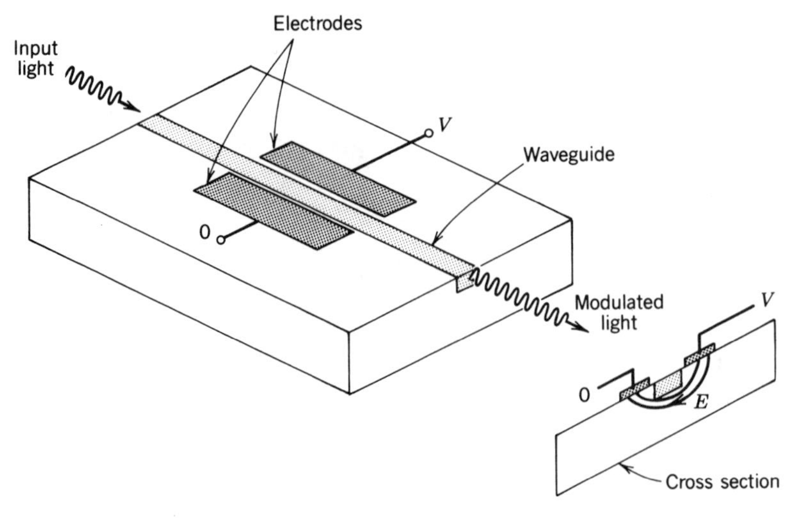

Figure 18.1-3 An integrated-optical phase modulator using the electro-optic effect.

undergo a phase retardation (with respect to each other) given by

$$\Gamma = k_o[n_1(E) - n_2(E)]L = k_o(n_1 - n_2)L - \tfrac{1}{2}k_o(\mathfrak{r}_1 n_1^3 - \mathfrak{r}_2 n_2^3)EL. \quad (18.1\text{-}9)$$

If E is obtained by applying a voltage V between two surfaces of the medium separated by a distance d, (18.1-9) can be written in the compact form

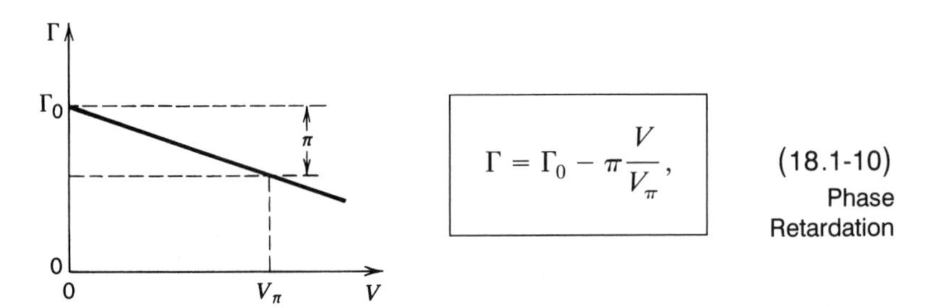

$$\Gamma = \Gamma_0 - \pi \frac{V}{V_\pi}, \quad (18.1\text{-}10)$$

Phase Retardation

where $\Gamma_0 = k_o(n_1 - n_2)L$ is the phase retardation in the absence of the electric field and

$$V_\pi = \frac{d}{L} \frac{\lambda_o}{\mathfrak{r}_1 n_1^3 - \mathfrak{r}_2 n_2^3} \quad (18.1\text{-}11)$$

Retardation Half-Wave Voltage

is the applied voltage necessary to obtain a phase retardation π. Equation (18.1-10) indicates that the phase retardation is linearly related to the applied voltage. The medium serves as an electrically controllable dynamic wave retarder.

Intensity Modulators: Use of a Phase Modulator in an Interferometer

Phase delay (or retardation) alone does not affect the intensity of a light beam. However, a phase modulator placed in one branch of an interferometer can function as an intensity modulator. Consider, for example, the Mach–Zehnder interferometer

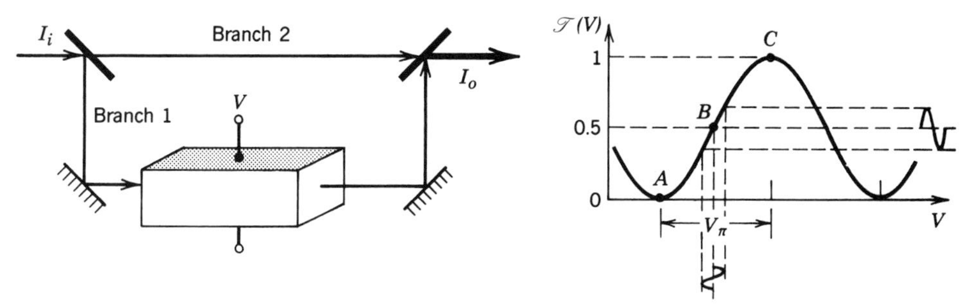

Figure 18.1-4 A phase modulator placed in one branch of a Mach–Zehnder interferometer can serve as an intensity modulator. The transmittance of the interferometer $\mathcal{T}(V) = I_o/I_i$ varies periodically with the applied voltage V. By operating in a limited region near point B, the device acts as a linear intensity modulator. If V is switched between points A and C, the device serves as an optical switch.

illustrated in Fig. 18.1-4. If the beamsplitters divide the optical power equally, the transmitted intensity I_o is related to the incident intensity I_i by

$$I_o = \tfrac{1}{2}I_i + \tfrac{1}{2}I_i \cos\varphi = I_i \cos^2\frac{\varphi}{2}$$

where $\varphi = \varphi_1 - \varphi_2$ is the difference between the phase shifts encountered by light as it travels through the two branches (see Sec. 2.5A). The transmittance of the interferometer is $\mathcal{T} = I_o/I_i = \cos^2(\varphi/2)$.

Because of the presence of the phase modulator in branch 1, according to (18.1-7) we have $\varphi_1 = \varphi_{10} - \pi V/V_\pi$, so that φ is controlled by the applied voltage V in accordance with the linear relation $\varphi = \varphi_1 - \varphi_2 = \varphi_0 - \pi V/V_\pi$, where the constant $\varphi_0 = \varphi_{10} - \varphi_2$ depends on the optical path difference. The transmittance of the device is therefore a function of the applied voltage V,

$$\mathcal{T}(V) = \cos^2\left(\frac{\varphi_0}{2} - \frac{\pi}{2}\frac{V}{V_\pi}\right). \qquad (18.1\text{-}12)$$

Transmittance

This function is plotted in Fig. 18.1-4 for an arbitrary value of φ_0. The device may be operated as a linear intensity modulator by adjusting the optical path difference so that $\varphi_0 = \pi/2$ and operating in the nearly linear region around $\mathcal{T} = 0.5$. Alternatively, the optical path difference may be adjusted so that φ_0 is a multiple of 2π. In this case $\mathcal{T}(0) = 1$ and $\mathcal{T}(V_\pi) = 0$, so that the modulator switches the light on and off as V is switched between 0 and V_π.

A Mach–Zehnder intensity modulator may also be constructed in the form of an integrated-optical device. Waveguides are placed on a substrate in the geometry shown in Fig. 18.1-5. The beamsplitters are implemented by the use of waveguide Y's. The optical input and output may be carried by optical fibers. Commercially available integrated-optical modulators generally operate at speeds of a few GHz but modulation speeds exceeding 25 GHz have been achieved.

Intensity Modulators: Use of a Retarder Between Crossed Polarizers

As described in Sec. 6.6B, a wave retarder (retardation Γ) sandwiched between two crossed polarizers, placed at 45° with respect to the retarder's axes (see Fig. 6.6-4), has

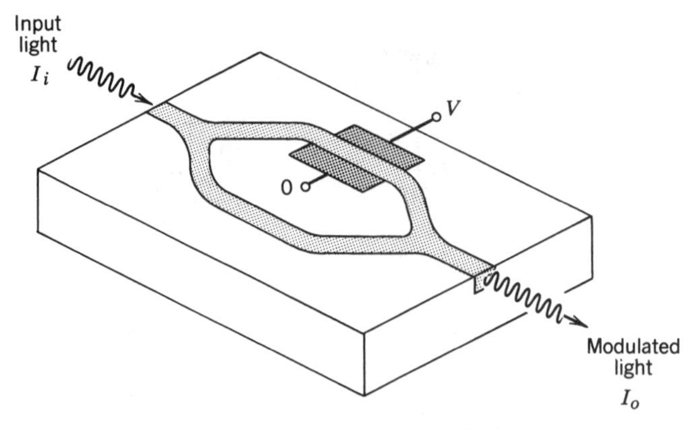

Figure 18.1-5 An integrated-optical intensity modulator (or optical switch). A Mach–Zehnder interferometer and an electro-optic phase modulator are implemented using optical waveguides fabricated from a material such as $LiNbO_3$.

an intensity transmittance $\mathscr{T} = \sin^2(\Gamma/2)$. If the retarder is a Pockels cell, then Γ is linearly dependent on the applied voltage V as provided in (18.1-10). The transmittance of the device is then a periodic function of V,

$$\mathscr{T}(V) = \sin^2\left(\frac{\Gamma_0}{2} - \frac{\pi}{2}\frac{V}{V_\pi}\right), \tag{18.1-13}$$

Transmittance

as shown in Fig. 18.1-6. By changing V, the transmittance can be varied between 0 (shutter closed) and 1 (shutter open). The device can also be used as a linear modulator if the system is operated in the region near $\mathscr{T}(V) = 0.5$. By selecting

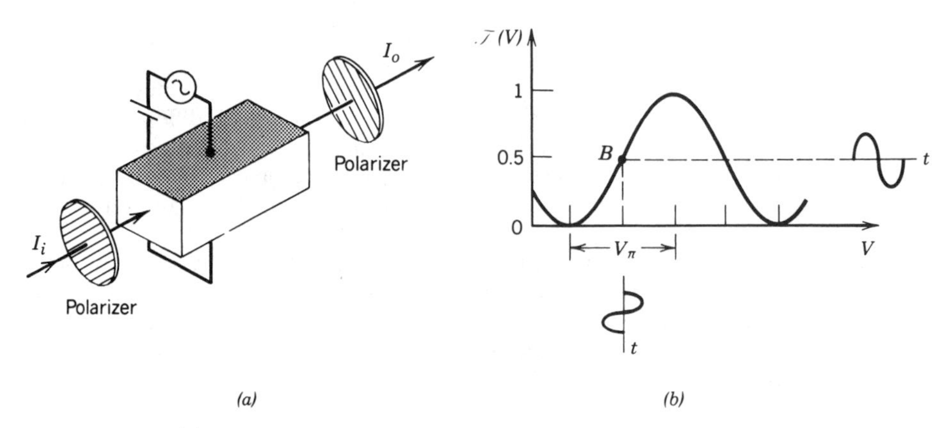

Figure 18.1-6 (a) An optical intensity modulator using a Pockels cell placed between two crossed polarizers. (b) Optical transmittance versus applied voltage for an arbitrary value of Γ_0; for linear operation the cell is biased near the point B.

$\Gamma_0 = \pi/2$ and $V \ll V_\pi$,

$$\mathscr{T}(V) = \sin^2\left(\frac{\pi}{4} - \frac{\pi}{2}\frac{V}{V_\pi}\right)$$

$$\approx \mathscr{T}(0) + \left.\frac{d\mathscr{T}}{dV}\right|_{V=0} V = \frac{1}{2} - \frac{\pi}{2}\frac{V}{V_\pi}, \tag{18.1-14}$$

so that $\mathscr{T}(V)$ is a linear function with slope $\pi/2V_\pi$ representing the sensitivity of the modulator. The phase retardation Γ_0 can be adjusted either optically (by assisting the modulator with an additional phase retarder, a compensator) or electrically by adding a constant bias voltage to V.

In practice, the maximum transmittance of the modulator is smaller than unity because of losses caused by reflection, absorption, and scattering. Furthermore, the minimum transmittance is greater than 0 because of misalignments of the direction of propagation and the directions of polarizations relative to the crystal axes and the polarizers. The ratio between the maximum and minimum transmittances is called the extinction ratio. Ratios higher than 1000 : 1 are possible.

C. Scanners

An optical beam can be deflected dynamically by using a prism with an electrically controlled refractive index. The angle of deflection introduced by a prism of small apex angle α and refractive index n is $\theta \approx (n - 1)\alpha$ [see (1.2-7)]. An incremental change of the refractive index Δn caused by an applied electric field E corresponds to an incremental change of the deflection angle,

$$\Delta\theta = \alpha\,\Delta n = -\tfrac{1}{2}\alpha\mathfrak{r}n^3 E = -\tfrac{1}{2}\alpha\mathfrak{r}n^3\frac{V}{d}, \tag{18.1-15}$$

where V is the applied voltage and d is the prism width [Fig. 18.1-7(a)]. By varying the applied voltage V, the angle $\Delta\theta$ varies proportionally, so that the incident light is scanned.

It is often more convenient to place triangularly shaped electrodes defining a prism on a rectangular crystal. Two, or several, prisms can be cascaded by alternating the direction of the electric field, as illustrated in Fig. 18.1-7(b).

An important parameter that characterizes a scanner is its resolution, i.e., the number of independent spots it can scan. An optical beam of width D and wavelength λ_o has an angular divergence $\delta\theta \approx \lambda_o/D$ [see (4.3-6)]. To minimize that angle, the beam should be as wide as possible, ideally covering the entire width of the prism itself. For a given maximum voltage V corresponding to a scanned angle $\Delta\theta$, the number of independent spots is given by

$$N \approx \frac{|\Delta\theta|}{\delta\theta} = \frac{\tfrac{1}{2}\alpha\mathfrak{r}n^3 V/d}{(\lambda_o/D)}. \tag{18.1-16}$$

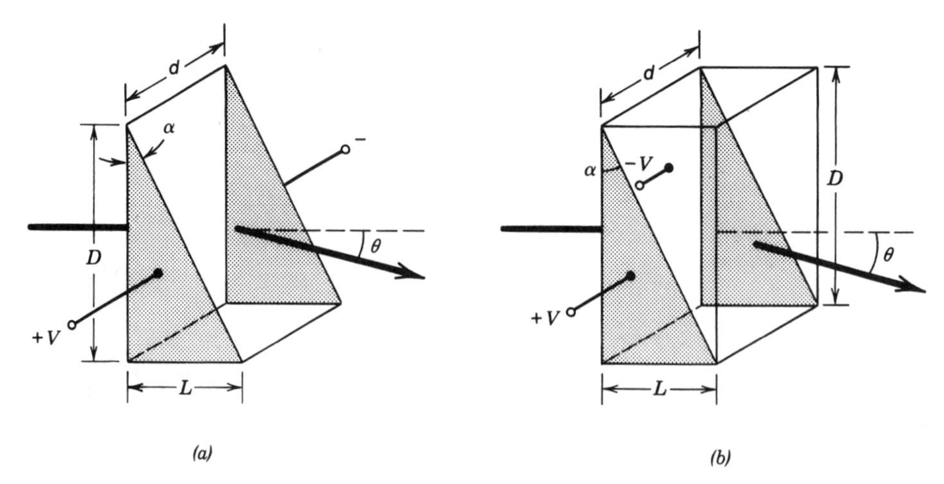

Figure 18.1-7 (*a*) An electro-optic prism. The deflection angle θ is controlled by the applied voltage. (*b*) An electro-optic double prism.

Substituting $\alpha \approx L/D$ and $V_\pi = (d/L)(\lambda_o/\mathfrak{r}n^3)$, we obtain

$$N \approx \frac{V}{2V_\pi}, \tag{18.1-17}$$

from which $V \approx 2NV_\pi$. This is a discouraging result. To scan N independent spots, a voltage $2N$ times greater than the half-wave voltage is necessary. Since V_π is usually large, making a useful scanner with $N \gg 1$ requires unacceptably high voltages. More popular scanners therefore include mechanical and acousto-optic scanners (see Secs. 20.2B and 21.1B).

The process of double refraction in anisotropic crystals (see Sec. 6.3E) introduces a lateral shift of an incident beam parallel to itself for one polarization and no shift for the other polarization. This effect can be used for switching a beam between two parallel positions by switching the polarization. A linearly polarized optical beam is

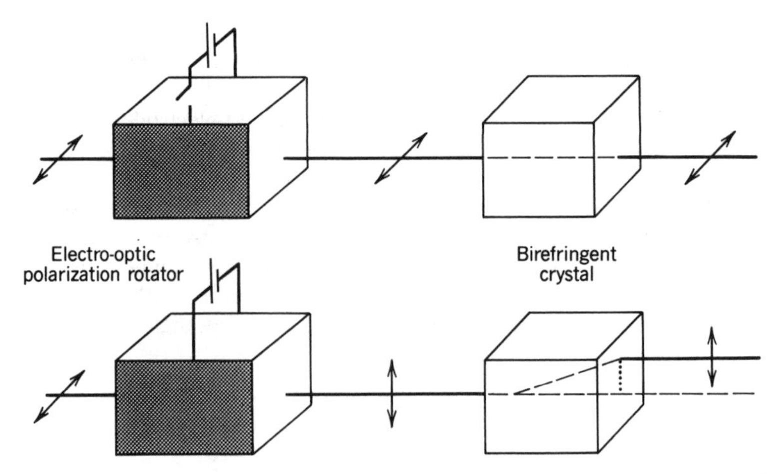

Figure 18.1-8 A position switch based on electro-optic phase retardation and double refraction.

transmitted first through an electro-optic wave retarder acting as a polarization rotator and then through the crystal. The rotator controls the polarization electrically, which determines whether the beam is shifted laterally or not, as illustrated in Fig. 18.1-8.

D. Directional Couplers

An important application of the electro-optic effect is in controlling the coupling between two parallel waveguides in an integrated-optical device. This can be used to transfer the light from one waveguide to the other, so that the device serves as an electrically controlled directional coupler.

The coupling of light between two parallel single-mode planar waveguides [Fig. 18.1-9(*a*)] was examined in Sec. 7.4B. It was shown that the optical powers carried by the two waveguides, $P_1(z)$ and $P_2(z)$, are exchanged periodically along the direction of propagation z. Two parameters govern the strength of this coupling process: the coupling coefficient $\mathcal{C}$ (which depends on the dimensions, wavelength, and refractive indices), and the mismatch of the propagation constants $\Delta\beta = \beta_1 - \beta_2 = 2\pi\,\Delta n/\lambda_o$, where Δn is the difference between the refractive indices of the waveguides. If the waveguides are identical, with $\Delta\beta = 0$ and $P_2(0) = 0$, then at a distance $z = L_0 = \pi/2\mathcal{C}$, called the transfer distance or coupling length, the power is transferred completely from waveguide 1 into waveguide 2, i.e., $P_1(L_0) = 0$ and $P_2(L_0) = P_1(0)$, as illustrated in Fig. 18.1-9(*a*).

For a waveguide of length L_0 and $\Delta\beta \neq 0$, the power-transfer ratio $\mathcal{T} = P_2(L_0)/P_1(0)$ is a function of the phase mismatch [see (7.4-12)],

$$\mathcal{T} = \left(\frac{\pi}{2}\right)^2 \operatorname{sinc}^2\left\{\frac{1}{2}\left[1 + \left(\frac{\Delta\beta\, L_0}{\pi}\right)^2\right]^{1/2}\right\}, \qquad (18.1\text{-}18)$$

where $\operatorname{sinc}(x) = \sin(\pi x)/(\pi x)$. Figure 18.1-9(*b*) illustrates this dependence. The ratio has its maximum value of unity at $\Delta\beta\, L_0 = 0$, decreases with increasing $\Delta\beta\, L_0$, and vanishes when $\Delta\beta\, L_0 = \sqrt{3}\,\pi$, at which point the optical power is *not* transferred to waveguide 2.

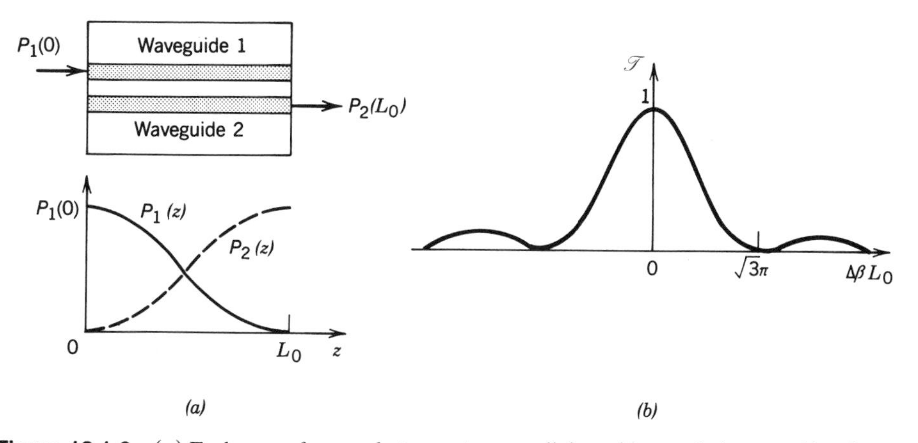

Figure 18.1-9 (*a*) Exchange of power between two parallel weakly coupled waveguides that are identical, with the same propagation constant β. At $z = 0$ all of the power is in waveguide 1. At $z = L_0$ all of the power is transferred into waveguide 2. (*b*) Dependence of the power-transfer ratio $\mathcal{T} = P_2(L_0)/P_1(0)$ on the phase mismatch parameter $\Delta\beta\, L_0$.

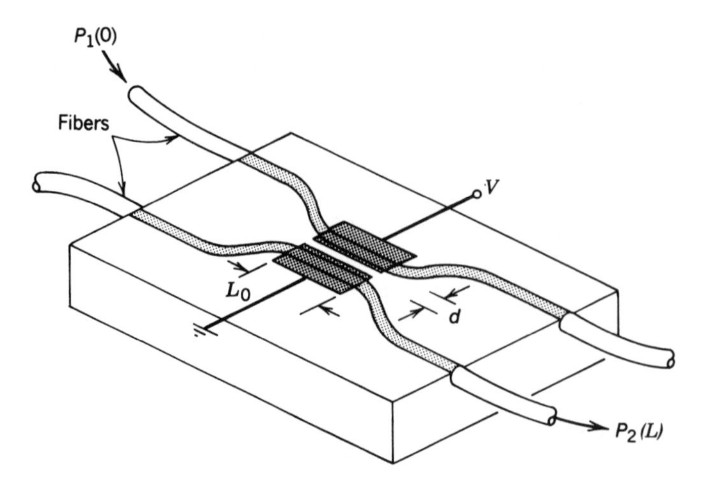

Figure 18.1-10 An integrated electro-optic directional coupler.

A dependence of the coupled power on the phase mismatch is the key to making electrically activated directional couplers. If the mismatch $\Delta\beta\, L_0$ is switched from 0 to $\sqrt{3}\,\pi$, the light remains in waveguide 1. Electrical control of $\Delta\beta$ is achieved by use of the electro-optic effect. An electric field E applied to one of two, otherwise identical, waveguides alters the refractive index by $\Delta n = -\frac{1}{2}n^3 \mathfrak{r} E$, where $\mathfrak{r}$ is the Pockels coefficient. This results in a phase shift $\Delta\beta\, L_0 = \Delta n(2\pi L_0/\lambda_o) = -(\pi/\lambda_o)n^3 \mathfrak{r} L_0 E$.

A typical electro-optic directional coupler has the geometry shown in Fig. 18.1-10. The electrodes are laid over two waveguides separated by a distance d. An applied voltage V creates an electric field $E \approx V/d$ in one waveguide and $-V/d$ in the other, where d is an effective distance determined by solving the electrostatics problem (the electric-field lines go downward at one waveguide and upward at the other). The refractive index is incremented in one guide and decremented in the other. The result is a net refractive index difference $2\Delta n = -n^3 \mathfrak{r}(V/d)$, corresponding to a phase mismatch factor $\Delta\beta\, L_0 = -(2\pi/\lambda_o)n^3 \mathfrak{r}(L_0/d)V$, which is proportional to the applied voltage V.

The voltage V_0 necessary to switch the optical power is that for which $|\Delta\beta\, L_0| = \sqrt{3}\,\pi$, i.e.,

$$V_0 = \sqrt{3}\,\frac{d}{L_0}\,\frac{\lambda_o}{2n^3\mathfrak{r}} = \frac{\sqrt{3}}{\pi}\,\frac{\mathcal{C}\lambda_o d}{n^3\mathfrak{r}}, \tag{18.1-19}$$

where $L_0 = \pi/2\mathcal{C}$ and $\mathcal{C}$ is the coupling coefficient. This is called the **switching voltage**.

Since $|\Delta\beta\, L_0| = \sqrt{3}\,\pi V/V_0$, (18.1-18) gives

$$\mathcal{T} = \left(\frac{\pi}{2}\right)^2 \operatorname{sinc}^2\left\{\frac{1}{2}\left[1 + 3\left(\frac{V}{V_0}\right)^2\right]^{1/2}\right\}. \tag{18.1-20}$$

Coupling Efficiency

This equation (plotted in Fig. 18.1-11) governs the coupling of power as a function of the applied voltage V.

An electro-optic directional coupler is characterized by its coupling length L_0, which is inversely proportional to the coupling coefficient $\mathcal{C}$, and its switching voltage V_0, which is directly proportional to $\mathcal{C}$. The key parameter is therefore $\mathcal{C}$, which is governed by the geometry and the refractive indices.

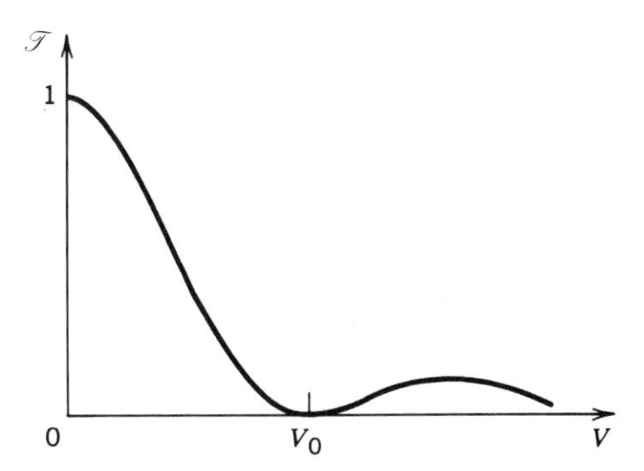

Figure 18.1-11 Dependence of the coupling efficiency on the applied voltage V. When $V = 0$, all of the optical power is coupled from waveguide 1 into waveguide 2; when $V = V_0$, all of the optical power remains in waveguide 1.

Integrated-optic directional couplers may be fabricated by diffusing titanium into high-purity $LiNbO_3$ substrates. The switching voltage V_0 is typically under 10 V, and the operating speeds can exceed 10 GHz. The light beams are focused to spot sizes of a few μm. The ends of the waveguide may be permanently attached to single-mode polarization-maintaining optical fibers (see Sec. 8.1C). Increased bandwidths can be obtained by using a traveling-wave version of this device.

EXERCISE 18.1-1

Spectral Response. Equation (18.1-19) indicates that the switching voltage V_0 is proportional to the wavelength. Assume that the applied voltage $V = V_0$ for a wavelength $\bar{\lambda}_o$; i.e., the coupling efficiency $\mathscr{T} = 0$ at $\bar{\lambda}_o$. If, instead, the incident wave has wavelength λ_o, plot the coupling efficiency $\mathscr{T}$ as a function of $\lambda_o - \bar{\lambda}_o$. Assume that the coupling coefficient $\mathcal{C}$ and the material parameters n and $\mathfrak{r}$ are approximately independent of wavelength.

E. Spatial Light Modulators

A spatial light modulator is a device that modulates the intensity of light at different positions by prescribed factors (Fig. 18.1-12). It is a planar optical element of control-

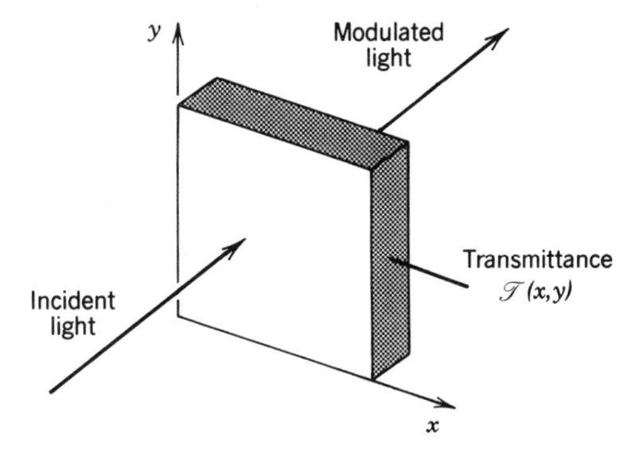

Figure 18.1-12 The spatial light modulator.

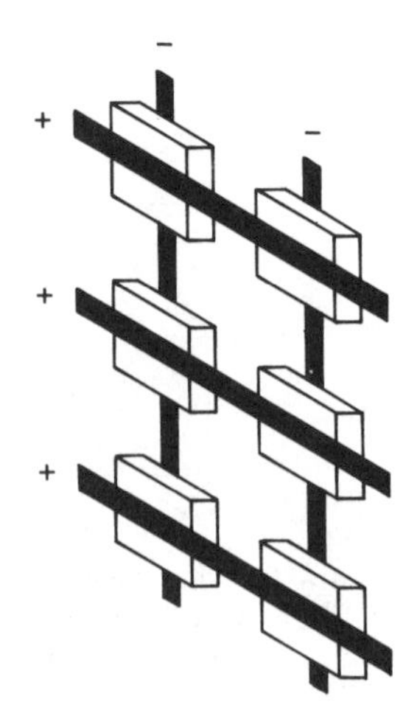

Figure 18.1-13 An electrically addressable array of longitudinal electro-optic modulators.

lable intensity transmittance $\mathscr{T}(x, y)$. The transmitted light intensity $I_o(x, y)$ is related to the incident light intensity $I_i(x, y)$ by the product $I_o(x, y) = I_i(x, y)\mathscr{T}(x, y)$. If the incident light is uniform [i.e., $I_i(x, y)$ is constant], the transmitted light intensity is proportional to $\mathscr{T}(x, y)$. The "image" $\mathscr{T}(x, y)$ is then imparted to the transmitted light, much like "reading" the image stored in a transparency by uniformly illuminating it in a slide projector. In a spatial light modulator, however, $\mathscr{T}(x, y)$ is controllable. In an electro-optic modulator the control is electrical.

To construct a spatial light modulator using the electro-optic effect, some mechanism must be devised for creating an electric field $E(x, y)$ proportional to the desired transmittance $\mathscr{T}(x, y)$ at each position. This is not easy. One approach is to place an array of transparent electrodes on small plates of electro-optic material placed between crossed polarizers and to apply on each electrode an appropriate voltage (Fig. 18.1-13). The voltage applied to the electrode centered at the position (x_i, y_i), $i = 1, 2, \ldots$ is made proportional to the desired value of $\mathscr{T}(x_i, y_i)$ (see, e.g., Fig. 18.1-6). If the number of electrodes is sufficiently large, the transmittance approximates $\mathscr{T}(x, y)$. The system is in effect a parallel array of longitudinal electro-optic modulators operated as intensity modulators. However, it is not practical to address a large number of these electrodes independently; nevertheless we will see that this scheme is practical in the liquid-crystal spatial light modulators used for display, since the required voltages are low (see Sec. 18.3B).

Optically Addressed Electro-Optic Spatial Light Modulators

One method of optically addressing an electro-optic spatial light modulator is based on the use of a thin layer of photoconductive material to create the electric field required to operate the modulator (Fig. 18.1-14). The conductivity of a photoconductive material is proportional to the intensity of light to which it is exposed (see Sec. 17.2). When illuminated by light of intensity distribution $I_W(x, y)$, a spatial pattern of conductance $G(x, y) \propto I_W(x, y)$ is created. The photoconductive layer is placed between two electrodes that act as a capacitor. The capacitor is initially charged and the electrical charge leakage at the position (x, y) is proportional to the local conductance $G(x, y)$. As a result, the charge on the capacitor is reduced in those regions where the conductance is high. The local voltage is therefore proportional to $1/G(x, y)$ and the corresponding electric field $E(x, y) \propto 1/G(x, y) \propto 1/I_W(x, y)$. If the transmittance $\mathscr{T}(x, y)$ [or the reflectance $\mathscr{R}(x, y)$] of the modulator is proportional to the applied field, it must be inversely proportional to the initial light intensity $I_W(x, y)$.

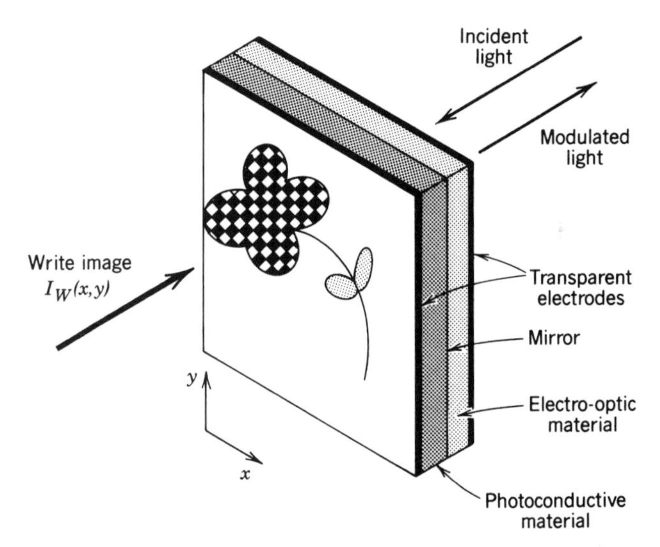

Figure 18.1-14 The electro-optic spatial light modulator uses a photoconductive material to create a spatial distribution of electric field which is used to control an electro-optic material.

The Pockels Readout Optical Modulator

An ingenious implementation of this principle is the Pockels readout optical modulator (PROM). The device uses a crystal of bismuth silicon oxide, $Bi_{12}SiO_{20}$ (BSO), which has an unusual combination of optical and electrical properties: (1) it exhibits the electro-optic (Pockels) effect; (2) it is photoconductive for blue light, but not for red light; and (3) it is a good insulator in the dark. The PROM (Fig. 18.1-15) is made of a thin wafer of BSO sandwiched between two transparent electrodes. The light that is to be modulated (read light) is transmitted through a polarizer, enters the BSO layer, and is reflected by a dichroic reflector, whereupon it crosses a second polarizer. The reflector reflects red light but is transparent to blue light. The PROM is operated as

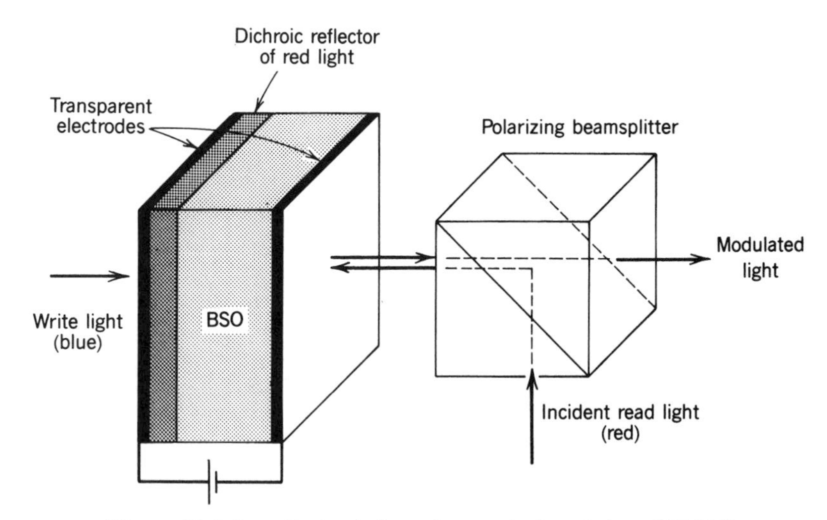

Figure 18.1-15 The Pockels readout optical modulator (PROM).

follows:

- *Priming:* A large potential difference (≈ 4 kV) is applied to the electrodes and the capacitor is charged (with no leakage since the crystal is a good insulator in the dark).
- *Writing:* Intense blue light of intensity distribution $I_W(x, y)$ illuminates the crystal. As a result, a spatial pattern of conductance $G(x, y) \propto I_W(x, y)$ is created, the voltage across the crystal is selectively lowered, and the electric field decreases proportionally at each position, so that $E(x, y) \propto 1/G(x, y) \propto 1/I_W(x, y)$. As a result of the electro-optic effect, the refractive indices of the BSO are altered, and a spatial pattern of refractive-index change $\Delta n(x, y) \propto 1/I_W(x, y)$ is created and stored in the crystal.
- *Reading:* Uniform red light is used to read $\Delta n(x, y)$ as with usual electro-optic intensity modulators [see Fig. 18.1-6(a)] with the polarizing beamsplitter playing the role of the crossed polarizers.
- *Erasing:* The refractive-index pattern is erased by the use of a uniform flash of blue light. The crystal is again primed by applying 4 kV, and a new cycle is started.

Incoherent-to-Coherent Optical Converters

In an optically addressed spatial light modulator, such as the PROM, the light used to write a spatial pattern into the modulator need not be coherent since photoconductive materials are sensitive to optical intensity. A spatial optical pattern (an image) may be written using incoherent light, and read using coherent light. This process of real-time conversion of a spatial distribution of natural incoherent light into a proportional spatial distribution of coherent light is useful in many optical data- and image-processing applications (see Sec. 21.5B).

*18.2 ELECTRO-OPTICS OF ANISOTROPIC MEDIA

The basic principles and applications of electro-optics have been presented in Sec. 18.1. For simplicity, polarization and anisotropic effects have been either ignored or introduced only generically. In this section a more complete analysis of the electro-optics of anisotropic media is presented. The following is a brief reminder of the important properties of anisotropic media, but the reader is expected to be familiar with the material in Sec. 6.3 on propagation of light in anisotropic media.

Crystal Optics: A Brief Refresher

The optical properties of an anisotropic medium are characterized by a geometric construction called the index ellipsoid,

$$\sum_{ij} \eta_{ij} x_i x_j = 1, \qquad i, j = 1, 2, 3, \tag{18.2-1}$$

where $\eta_{ij} = \eta_{ji}$ are elements of the impermeability tensor $\boldsymbol{\eta} = \epsilon_o \boldsymbol{\epsilon}^{-1}$. The ellipsoid's principal axes are the optical principal axes of the medium; its principal dimensions along these axes are the principal refractive indices n_1, n_2, and n_3 (Fig. 18.2-1). The index ellipsoid may be used to determine the polarizations and refractive indices n_a and n_b of the two normal modes of a wave traveling in an arbitrary direction. This is accomplished by drawing a plane

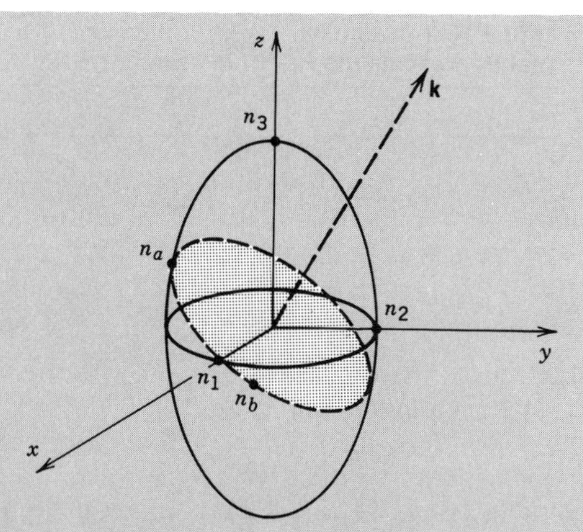

Figure 18.2-1 The index ellipsoid. The coordinates (x, y, z) are the principal axes and n_1, n_2, n_3 are the principal refractive indices. The refractive indices of the normal modes of a wave traveling in the direction **k** are n_a and n_b.

perpendicular to the direction of propagation and passing through the center of the ellipsoid. Its intersection with the ellipsoid is an ellipse whose major and minor axes have half-lengths equal to n_a and n_b (see Sec. 6.3C).

A. Pockels and Kerr Effects

When a steady electric field **E** with components (E_1, E_2, E_3) is applied to a crystal, elements of the tensor $\boldsymbol{\eta}$ are altered, so that each of the nine elements η_{ij} becomes a function of E_1, E_2, and E_3, i.e., $\eta_{ij} = \eta_{ij}(\mathbf{E})$. As a result, the index ellipsoid is modified (Fig. 18.2-2). Once we know the function $\eta_{ij}(\mathbf{E})$, we can determine the index ellipsoid and the optical properties at any applied electric field **E**. The problem is simple in principle, but the implementation is often lengthy.

Each of the elements $\eta_{ij}(\mathbf{E})$ is a function of the three variables $\mathbf{E} = (E_1, E_2, E_3)$, which may be expanded in a Taylor's series about $\mathbf{E} = \mathbf{0}$,

$$\eta_{ij}(\mathbf{E}) = \eta_{ij} + \sum_k \mathfrak{r}_{ijk} E_k + \sum_{kl} \mathfrak{s}_{ijkl} E_k E_l, \qquad i, j, k, l = 1, 2, 3, \qquad (18.2\text{-}2)$$

where $\eta_{ij} = \eta_{ij}(\mathbf{0})$, $\mathfrak{r}_{ijk} = \partial \eta_{ij} / \partial E_k$, $\mathfrak{s}_{ijkl} = \frac{1}{2} \partial^2 \eta_{ij} / \partial E_k \, \partial E_l$, and the derivatives are evaluated at $\mathbf{E} = \mathbf{0}$. Equation (18.2-2) is a generalization of (18.1-3), in which $\mathfrak{r}$ is replaced by $3^3 = 27$ coefficients $\{\mathfrak{r}_{ijk}\}$, and $\mathfrak{s}$ is replaced by $3^4 = 81$ coefficients $\{\mathfrak{s}_{ijkl}\}$. The coefficients $\{\mathfrak{r}_{ijk}\}$ are known as the linear electro-optic (Pockels) coefficients. They

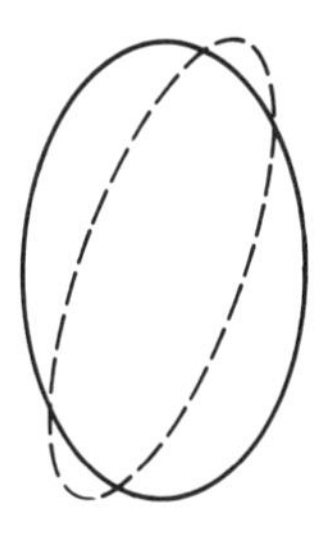

Figure 18.2-2 The index ellipsoid is modified as a result of applying a steady electric field.

TABLE 18.2-1 Lookup Table for the Index I That Represents the Pair of Indices (i, j)[a]

j	$i:1$	2	3
1	1	6	5
2	6	2	4
3	5	4	3

[a]The pair $(i, j) = (3, 2)$, for example, is labeled $I = 4$.

form a tensor of third rank. The coefficients $\{\mathfrak{s}_{ijkl}\}$ are the quadratic electro-optic (Kerr) coefficients. They form a fourth-rank tensor.

Symmetry

Because $\boldsymbol{\eta}$ is symmetric ($\eta_{ij} = \eta_{ji}$), $\mathfrak{r}$ and $\mathfrak{s}$ are invariant under permutations of the indices i and j, i.e., $\mathfrak{r}_{ijk} = \mathfrak{r}_{jik}$ and $\mathfrak{s}_{ijkl} = \mathfrak{s}_{jikl}$. Also, the coefficients $\mathfrak{s}_{ijkl} = \frac{1}{2}\partial^2\eta_{ij}/\partial E_k\, \partial E_l$ are invariant to permutations of k and l (because of the invariance to the order of differentiation), so that $\mathfrak{s}_{ijkl} = \mathfrak{s}_{ijlk}$. Because of this permutation symmetry the nine combinations of the indices i, j generate six instead of nine independent elements. The same reduction applies to the indices k, l. Consequently, $\mathfrak{r}_{ijk}$ has 6×3 independent elements, whereas $\mathfrak{s}_{ijkl}$ has 6×6 independent elements.

It is conventional to rename the pair of indices (i, j), $i, j = 1, 2, 3$ as a single index $I = 1, 2, \ldots, 6$ in accordance with Table 18.2-1. The pair (k, l) is similarly replaced by an index $K = 1, 2, \ldots, 6$, in accordance with the same rule. Thus the coefficients $\mathfrak{r}_{ijk}$ and $\mathfrak{s}_{ijkl}$ are replaced by $\mathfrak{r}_{Ik}$ and $\mathfrak{s}_{IK}$, respectively. For example, $\mathfrak{r}_{12k}$ is denoted as $\mathfrak{r}_{6k}$, $\mathfrak{s}_{1231}$ is renamed $\mathfrak{s}_{65}$, and so on. The third-rank tensor $\mathfrak{r}$ is therefore replaced by a 6×3 matrix and the fourth-rank tensor $\mathfrak{s}$ by a 6×6 matrix.

Crystal Symmetry

The symmetry of the crystal adds more constraints to the entries of the $\mathfrak{r}$ and $\mathfrak{s}$ matrices. Some entries must be zero and others must be equal, or equal in magnitude and opposite in sign, or related by some other rule. For centrosymmetric crystals $\mathfrak{r}$ vanishes and only the Kerr effect is exhibited. Lists of the coefficients of $\mathfrak{r}$ and $\mathfrak{s}$ and their symmetry relations for the 32 crystal point groups may be found in several of the books referenced in the reading list. Representative examples are provided in Tables 18.2-2 and 18.2-3.

Pockels Effect

To determine the optical properties of an anisotropic material exhibiting the Pockels effect in the presence of an electric field $\mathbf{E} = (E_1, E_2, E_3)$, the following sequence is

TABLE 18.2-2 Pockels Coefficients $\mathfrak{r}_{Ik}$ for Some Representative Crystal Groups

$\begin{bmatrix} 0 & 0 & 0 \\ 0 & 0 & 0 \\ 0 & 0 & 0 \\ \mathfrak{r}_{41} & 0 & 0 \\ 0 & \mathfrak{r}_{41} & 0 \\ 0 & 0 & \mathfrak{r}_{41} \end{bmatrix}$	$\begin{bmatrix} 0 & 0 & 0 \\ 0 & 0 & 0 \\ 0 & 0 & 0 \\ \mathfrak{r}_{41} & 0 & 0 \\ 0 & \mathfrak{r}_{41} & 0 \\ 0 & 0 & \mathfrak{r}_{63} \end{bmatrix}$	$\begin{bmatrix} 0 & -\mathfrak{r}_{22} & \mathfrak{r}_{13} \\ 0 & \mathfrak{r}_{22} & \mathfrak{r}_{13} \\ 0 & 0 & \mathfrak{r}_{33} \\ 0 & \mathfrak{r}_{51} & 0 \\ \mathfrak{r}_{51} & 0 & 0 \\ -\mathfrak{r}_{22} & 0 & 0 \end{bmatrix}$
Cubic $\bar{4}3m$ [e.g., GaAs, CdTe, InAs]	Tetragonal $\bar{4}2m$ [e.g., KDP, ADP]	Trigonal $3m$ [e.g., $LiNbO_3$, $LiTaO_3$]

TABLE 18.2-3 Kerr Coefficients $\mathfrak{s}_{IK}$ for an Isotropic Medium

$$\begin{bmatrix} \mathfrak{s}_{11} & \mathfrak{s}_{12} & \mathfrak{s}_{12} & 0 & 0 & 0 \\ \mathfrak{s}_{12} & \mathfrak{s}_{11} & \mathfrak{s}_{12} & 0 & 0 & 0 \\ \mathfrak{s}_{12} & \mathfrak{s}_{12} & \mathfrak{s}_{11} & 0 & 0 & 0 \\ 0 & 0 & 0 & \mathfrak{s}_{44} & 0 & 0 \\ 0 & 0 & 0 & 0 & \mathfrak{s}_{44} & 0 \\ 0 & 0 & 0 & 0 & 0 & \mathfrak{s}_{44} \end{bmatrix}, \qquad \mathfrak{s}_{44} = \frac{\mathfrak{s}_{11} - \mathfrak{s}_{12}}{2}$$

followed:

- Find the principal axes and principal refractive indices n_1, n_2, and n_3 in the absence of $\mathbf{E}$.
- Find the coefficients $\{\mathfrak{r}_{ijk}\}$ by using the appropriate matrix for $\mathfrak{r}_{Ik}$ (e.g., Table 18.2-2) together with the contraction rule relating i, j to I (Table 18.2-1);
- Determine the elements of the impermeability tensor using

  $$\eta_{ij}(\mathbf{E}) = \eta_{ij}(\mathbf{0}) + \sum_k \mathfrak{r}_{ijk} E_k,$$

  where $\eta_{ij}(\mathbf{0})$ is a diagonal matrix with elements $1/n_1^2$, $1/n_2^2$, and $1/n_3^2$.
- Write the equation for the modified index ellipsoid

  $$\sum_{ij} \eta_{ij}(\mathbf{E}) x_i x_j = 1.$$

- Determine the principal axes of the modified index ellipsoid by diagonalizing the matrix $\eta_{ij}(\mathbf{E})$, and find the corresponding principal refractive indices $n_1(\mathbf{E})$, $n_2(\mathbf{E})$, and $n_3(\mathbf{E})$.
- Given the direction of light propagation, find the normal modes and their associated refractive indices by using the index ellipsoid.

EXAMPLE 18.2-1. ***Trigonal 3m Crystals (e.g., LiNbO$_3$ and LiTaO$_3$).*** Trigonal $3m$ crystals are uniaxial ($n_1 = n_2 = n_o$, $n_3 = n_e$) with the matrix $\mathfrak{r}$ provided in Table 18.2-2. Assuming that $\mathbf{E} = (0, 0, E)$, i.e., that the electric field points along the optic axis (see Fig. 18.2-3), we find that the modified index ellipsoid is

$$\left(\frac{1}{n_o^2} + \mathfrak{r}_{13} E\right)\left(x_1^2 + x_2^2\right) + \left(\frac{1}{n_e^2} + \mathfrak{r}_{33} E\right) x_3^2 = 1. \tag{18.2-3}$$

This is an ellipsoid of revolution whose principal axes are independent of E. The ordinary and extraordinary indices $n_o(E)$ and $n_e(E)$ are given by

$$\frac{1}{n_o^2(E)} = \frac{1}{n_o^2} + \mathfrak{r}_{13} E \tag{18.2-4}$$

$$\frac{1}{n_e^2(E)} = \frac{1}{n_e^2} + \mathfrak{r}_{33} E. \tag{18.2-5}$$

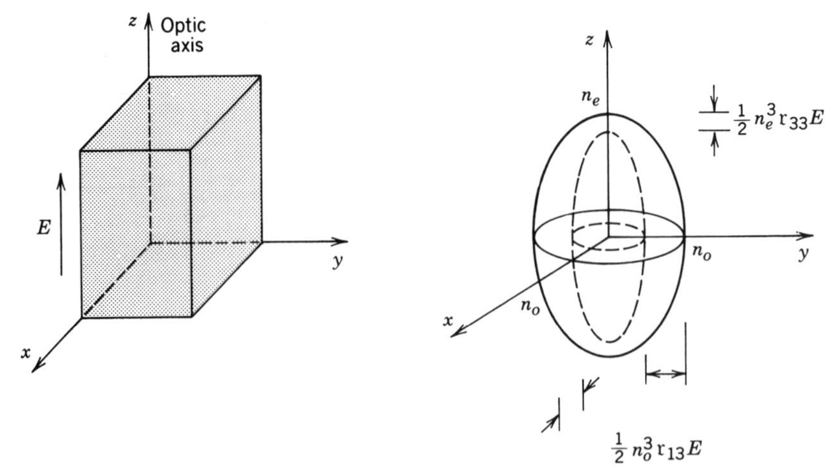

Figure 18.2-3 Modification of the index ellipsoid of a trigonal $3m$ crystal caused by an electric field in the direction of the optic axis.

Because the terms $r_{13}E$ and $r_{33}E$ in (18.2-4) and (18.2-5) are usually small, we use the approximation $(1 + \Delta)^{-1/2} \approx 1 - \frac{1}{2}\Delta$, when $|\Delta|$ is small, to obtain

$$n_o(E) \approx n_o - \tfrac{1}{2} n_o^3 r_{13} E \tag{18.2-6}$$

$$n_e(E) \approx n_e - \tfrac{1}{2} n_e^3 r_{33} E. \tag{18.2-7}$$

We conclude that when an electric field is applied along the optic axis of this uniaxial crystal it remains uniaxial with the same principal axes, but its refractive indices are modified in accordance with (18.2-6) and (18.2-7) (Fig. 18.2-3). Note the similarity between these equations and the generic equation (18.1-4).

EXAMPLE 18.2-2. ***Tetragonal $\bar{4}2m$ Crystals (e.g., KDP and ADP).*** Repeating the same steps for these uniaxial crystals and assuming that the electric field also points along the optic axis (Fig. 18.2-4), we obtain the equation of the index ellipsoid

$$\frac{x_1^2 + x_2^2}{n_o^2} + \frac{x_3^2}{n_e^2} + 2 r_{63} E x_1 x_2 = 1. \tag{18.2-8}$$

The modified principal axes are obtained by rotating the coordinate system 45° about the z axis. Substituting $u_1 = (x_1 - x_2)/\sqrt{2}$, $u_2 = (x_1 + x_2)/\sqrt{2}$, $u_3 = x_3$ in (18.2-8), we obtain

$$\frac{u_1^2}{n_1^2(E)} + \frac{u_2^2}{n_2^2(E)} + \frac{u_3^2}{n_3^2(E)} = 1,$$

where

$$\frac{1}{n_1^2(E)} = \frac{1}{n_o^2} + r_{63} E$$

$$\frac{1}{n_2^2(E)} = \frac{1}{n_o^2} - r_{63} E$$

$$n_3(E) = n_e.$$

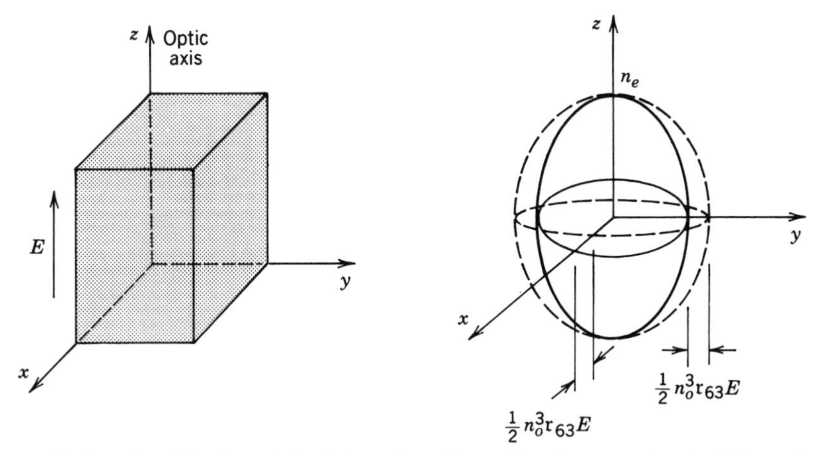

Figure 18.2-4 Modification of the index ellipsoid resulting from an electric field E along the direction of the optic axis of a uniaxial tetragonal $\bar{4}2m$ crystal.

Using the approximation $(1 + \Delta)^{-1/2} \approx 1 - \frac{1}{2}\Delta$ yields

$$n_1(E) \approx n_o - \tfrac{1}{2}n_o^3 r_{63} E \tag{18.2-9}$$

$$n_2(E) \approx n_o + \tfrac{1}{2}n_o^3 r_{63} E \tag{18.2-10}$$

$$n_3(E) = n_e. \tag{18.2-11}$$

Thus the originally uniaxial crystal becomes biaxial when subjected to an electric field in the direction of its optic axis (Fig. 18.2-4).

EXAMPLE 18.2-3. ***Cubic $\bar{4}3m$ Crystals (e.g., GaAs, CdTe, and InAs).*** These crystals are isotropic ($n_1 = n_2 = n_3 = n$). Without loss of generality, the coordinate system may be selected such that the applied electric field points in the z direction (Fig. 18.2-5). Following the same steps, an equation for the index ellipsoid is obtained,

$$\frac{x_1^2 + x_2^2 + x_3^2}{n^2} + 2r_{41}Ex_1x_2 = 1. \tag{18.2-12}$$

As in Example 18.2-2, the new principal axes are rotated 45° about the z axis and the principal refractive indices are

$$n_1(E) \approx n - \tfrac{1}{2}n^3 r_{41} E \tag{18.2-13}$$

$$n_2(E) \approx n + \tfrac{1}{2}n^3 r_{41} E \tag{18.2-14}$$

$$n_3(E) = n. \tag{18.2-15}$$

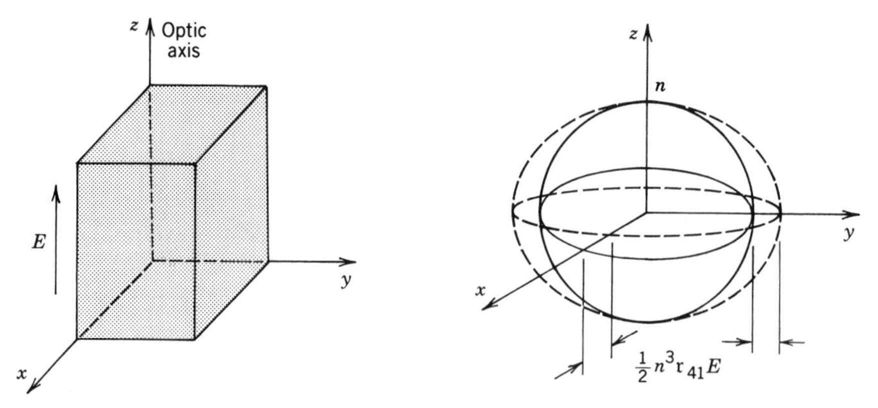

Figure 18.2-5 Modification of the index ellipsoid as a result of application of an electric field E to a cubic $\bar{4}3m$ crystal.

Thus the applied field changes the originally isotropic crystal into a biaxial crystal (Fig. 18.2-5).

The three examples above share the property that the principal axes do not change as the applied steady electric field E increases. The directions of polarization of the normal modes therefore remain the same, but their associated refractive indices are functions of E. The medium can then be used as a phase modulator, wave retarder, or intensity modulator in accordance with the generic theory provided in Sec. 18.1B. This principle is described further in Sec. 18.2B.

Kerr Effect

The optical properties of a Kerr medium can be determined by using the same procedure used for the Pockels medium, except that the coefficients $\eta_{ij}(\mathbf{E})$ are now given by $\eta_{ij}(\mathbf{E}) = \eta_{ij}(\mathbf{0}) + \sum_{kl} \mathfrak{s}_{ijkl} E_k E_l$.

EXAMPLE 18.2-4. ***Isotropic Medium.*** Using the Kerr coefficients $\mathfrak{s}_{IK}$ in Table 18.2-3 for an isotropic medium, and taking the z axis to point along the applied electric field E, we easily find the equation of the index ellipsoid,

$$\left(\frac{1}{n^2} + \mathfrak{s}_{12} E^2\right)\left(x_1^2 + x_2^2\right) + \left(\frac{1}{n^2} + \mathfrak{s}_{11} E^2\right) x_3^2 = 1. \tag{18.2-16}$$

This is the equation of an ellipsoid of revolution whose axis is the z axis. The principal refractive indices $n_o(E)$ and $n_e(E)$ are determined from

$$\frac{1}{n_o^2(E)} = \frac{1}{n^2} + \mathfrak{s}_{12} E^2 \tag{18.2-17}$$

$$\frac{1}{n_e^2(E)} = \frac{1}{n^2} + \mathfrak{s}_{11} E^2. \tag{18.2-18}$$

Noting that the second terms in (18.2-17) and (18.2-18) are small, and using the approximation $(1 + \Delta)^{-1/2} \approx 1 - \frac{1}{2}\Delta$ when $|\Delta| \ll 1$, we obtain

$$n_o(E) \approx n - \tfrac{1}{2}n^3\mathfrak{s}_{12}E^2 \qquad (18.2\text{-}19)$$

$$n_e(E) \approx n - \tfrac{1}{2}n^3\mathfrak{s}_{11}E^2. \qquad (18.2\text{-}20)$$

Thus a steady electric field E applied to an originally isotropic medium converts it into a uniaxial crystal with the optic axis pointing in the direction of the electric field. The ordinary and extraordinary indices are *quadratic* decreasing functions of E.

B. Modulators

The principles of phase and intensity modulation using the electro-optic effect were outlined in Sec. 18.1B. Anisotropic effects were introduced only generically. Using the anisotropic theory presented in this section, the generic parameters $\mathfrak{r}$ and $\mathfrak{s}$, which were used in Sec. 18.1, can now be determined for any given crystal and directions of the applied electric field and light propagation. Only Pockels modulators will be discussed, but the same approach can be applied to Kerr modulators. For simplicity, we assume that the direction of the electric field is such that the principal axes of the crystal are not altered as a result of modulation. We shall also assume that the direction of the wave relative to these axes is such that the planes of polarization of the normal modes are also not altered by the electric field.

Phase Modulators

A normal mode is characterized by a refractive index $n(E) \approx n - \frac{1}{2}\mathfrak{r}n^3E$, where n and $\mathfrak{r}$ are the appropriate refractive index and Pockels coefficient, respectively, and $E = V/d$ is the electric field obtained by applying a voltage V across a distance d. A wave traveling a distance L undergoes a phase shift

$$\varphi = \varphi_o - \pi\frac{V}{V_\pi} \qquad (18.2\text{-}21)$$

where $\varphi_o = 2\pi nL/\lambda_o$ and

$$V_\pi = \frac{d}{L}\frac{\lambda_o}{\mathfrak{r}n^3} \qquad (18.2\text{-}22)$$

is the half-wave voltage. The appropriate coefficients generically called n and $\mathfrak{r}$ can be easily determined as demonstrated in the following example.

EXAMPLE 18.2-5. ***Trigonal 3m Crystal ($LiNbO_3$ and $LiTaO_3$).*** When an electric field is directed along the optic axis of this type of uniaxial crystal, the crystal remains uniaxial with the same principal axes (see Fig. 18.2-3). The principal refractive indices are given by (18.2-6) and (18.2-7). The crystal can be used as a phase modulator in either of two configurations:

Longitudinal Modulator: If a linearly polarized optical wave travels along the direction of the optic axis (parallel to the electric field), the appropriate parameters for the phase modulator are $n = n_o$, $\mathfrak{r} = \mathfrak{r}_{13}$, and $d = L$. For $LiNbO_3$, $\mathfrak{r}_{13} = 9.6$ pm/V, and $n_o = 2.3$ at $\lambda_o = 633$ nm. Equation (18.2-22) then gives $V_\pi = 5.41$ kV, so that 5.41 kV is required to change the phase by π.

Transverse Modulator: If the wave travels in the x direction and is polarized in the z direction, the appropriate parameters are $n = n_e$ and $\mathfrak{r} = \mathfrak{r}_{33}$. The width d is generally not equal to the length L. For $LiNbO_3$ at $\lambda_o = 633$ nm, $r_{33} = 30.9$ pm/V, and $n_e = 2.2$, giving a half-wave voltage $V_\pi = 1.9(d/L)$ kV. If $d/L = 0.1$, we obtain $V_\pi \approx 190$ V, which is significantly lower than the half-wave voltage for the longitudinal modulator.

Intensity Modulators

The difference in the dependence on the applied field of the refractive indices of the two normal modes of a Pockels cell provides a voltage-dependent retardation,

$$\Gamma = \Gamma_0 - \pi \frac{V}{V_\pi}, \tag{18.2-23}$$

where

$$\Gamma_0 = \frac{2\pi(n_1 - n_2)L}{\lambda_o} \tag{18.2-24}$$

$$V_\pi = \frac{(d/L)\lambda_o}{\mathfrak{r}_1 n_1^3 - \mathfrak{r}_2 n_2^3}. \tag{18.2-25}$$

If the cell is placed between crossed polarizers, the system serves as an intensity modulator (see Sec. 18.1B). It is not difficult to determine the appropriate indices n_1 and n_2, and coefficients $\mathfrak{r}_1$ and $\mathfrak{r}_2$, as illustrated by the following example.

EXAMPLE 18.2-6. *Tetragonal $\bar{4}2m$ Crystal (e.g., KDP and ADP).* As described in Example 18.2-2, when an electric field is applied along the optic axis of this uniaxial crystal, it changes into a biaxial crystal. The new principal axes are the original axes rotated by 45° about the optic axis. Assume a longitudinal modulator configuration ($d/L = 1$) in which the wave travels along the optic axis. The two normal modes have refractive indices given by (18.2-9) and (18.2-10). The appropriate coefficients to be used in (18.2-25) are therefore $n_1 = n_2 = n_o$, $\mathfrak{r}_1 = \mathfrak{r}_{63}$, $\mathfrak{r}_2 = -\mathfrak{r}_{63}$, and $d = L$, so that $\Gamma_0 = 0$ and

$$V_\pi = \frac{\lambda_o}{2\mathfrak{r}_{63} n_o^3}. \tag{18.2-26}$$

For KDP at $\lambda_o = 633$ nm, $V_\pi = 8.4$ kV.

EXERCISE 18.2-1

Intensity Modulation Using the Kerr Effect. Use (18.2-19) and (18.2-20) to determine an expression for the phase shift φ, and the phase retardation Γ, in a longitudinal Kerr modulator made of an isotropic material, as functions of the applied voltage V. Derive expressions for the half-wave voltages V_π in each case.

18.3 ELECTRO-OPTICS OF LIQUID CRYSTALS

As described in Sec. 6.5, the elongated molecules of nematic liquid crystals tend to have ordered orientations that are altered when the material is subjected to mechanical or electric forces. Because of their anisotropic nature, liquid crystals can be arranged to serve as wave retarders or polarization rotators. In the presence of an electric field, their molecular orientation is modified, so that their effect on polarized light is altered. Liquid crystals can therefore be used as electrically controlled optical wave retarders, modulators, and switches. These devices are particularly useful in display technology.

A. Wave Retarders and Modulators

Electrical Properties of Nematic Liquid Crystals

The liquid crystals used to make electro-optic devices are usually of sufficiently high resistivity that they can be regarded as ideal dielectric materials. Because of the elongated shape of the constituent molecules, and their ordered orientation, liquid crystals have anisotropic dielectric properties with uniaxial symmetry (see Sec. 6.3A). The electric permittivity is $\epsilon_\parallel$ for electric fields pointing in the direction of the molecules and $\epsilon_\perp$ in the perpendicular direction. Liquid crystals for which $\epsilon_\parallel > \epsilon_\perp$, (positive uniaxial) are usually selected for electro-optic applications.

When a steady (or low frequency) electric field is applied, electric dipoles are induced and the resultant electric forces exert torques on the molecules. The molecules rotate in a direction such that the free electrostatic energy, $-\frac{1}{2}\mathbf{E}\cdot\mathbf{D} = -\frac{1}{2}[\epsilon_\perp E_1^2 + \epsilon_\perp E_2^2 + \epsilon_\parallel E_3^2]$, is minimized (here, E_1, E_2, and E_3 are components of $\mathbf{E}$ in the directions of the principal axes). Since $\epsilon_\parallel > \epsilon_\perp$, for a given direction of the electric field, minimum energy is achieved when the molecules are aligned with the field, so that $E_1 = E_2 = 0$, $\mathbf{E} = (0, 0, E)$, and the energy is then $-\frac{1}{2}\epsilon_\parallel E^2$. When the alignment is complete the molecular axis points in the direction of the electric field (Fig. 18.3-1). Evidently, a reversal of the electric field effects the same molecular rotation. An alternating field generated by an ac voltage also has the same effect.

Nematic Liquid-Crystal Retarders and Modulators

A nematic liquid-crystal cell is a thin layer of nematic liquid crystal placed between two parallel glass plates and rubbed so that the molecules are parallel to each other. The

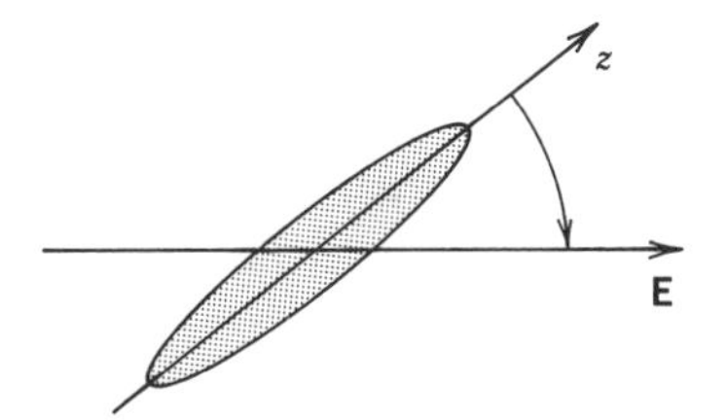

Figure 18.3-1 The molecules of a positive uniaxial liquid crystal rotate and align with the applied electric field.

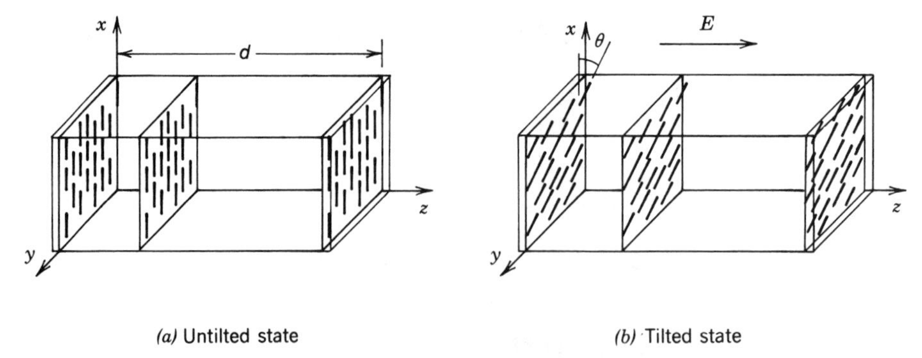

(a) Untilted state *(b)* Tilted state

Figure 18.3-2 Molecular orientation of a liquid-crystal cell (*a*) in the absence of a steady electric field and (*b*) when a steady electric field is applied. The optic axis lies along the direction of the molecules.

material then acts as a uniaxial crystal with the optic axis parallel to the molecular orientation. For waves traveling in the z direction (perpendicular to the glass plates), the normal modes are linearly polarized in the x and y directions, (parallel and perpendicular to the molecular directions), as illustrated in Fig. 18.3-2(*a*). The refractive indices are the extraordinary and ordinary indices n_e and n_o. A cell of thickness d provides a wave retardation $\Gamma = 2\pi(n_e - n_o)d/\lambda_o$.

If an electric field is applied in the z direction (by applying a voltage V across transparent conductive electrodes coated on the inside of the glass plates), the resultant electric forces tend to tilt the molecules toward alignment with the field, but the elastic forces at the surfaces of the glass plates resist this motion. When the applied electric field is sufficiently large, most of the molecules tilt, except those adjacent to the glass surfaces. The equilibrium tilt angle θ for most molecules is a monotonically increasing function of V, which can be described by[†]

$$\theta = \begin{cases} 0, & V \le V_c \\ \dfrac{\pi}{2} - 2\tan^{-1}\exp\left(-\dfrac{V - V_c}{V_0}\right), & V > V_c, \end{cases} \tag{18.3-1}$$

where V is the applied rms voltage, V_c a critical voltage at which the tilting process begins, and V_0 a constant. When $V - V_c = V_0$, $\theta \approx 50°$; as $V - V_c$ increases beyond V_0, θ approaches 90°, as indicated in Fig. 18.3-3(*a*).

When the electric field is removed, the orientations of the molecules near the glass surfaces are reasserted and all of the molecules tilt back to their original orientation (in planes parallel to the plates). In a sense, the liquid-crystal material may be viewed as a *liquid with memory*.

For a tilt angle θ, the normal modes of an optical wave traveling in the z direction are polarized in the x and y directions and have refractive indices $n(\theta)$ and n_o, where

$$\frac{1}{n^2(\theta)} = \frac{\cos^2\theta}{n_e^2} + \frac{\sin^2\theta}{n_o^2}, \tag{18.3-2}$$

so that the retardation becomes $\Gamma = 2\pi[n(\theta) - n_o]d/\lambda_o$ (see Sec. 6.3C). The retardation achieves its maximum value $\Gamma_{\max} = 2\pi(n_e - n_o)d/\lambda_o$ when the molecules are not

[†]See, e.g., P.-G. de Gennes, *The Physics of Liquid Crystals*, Clarendon Press, Oxford, 1974, Chap. 3.

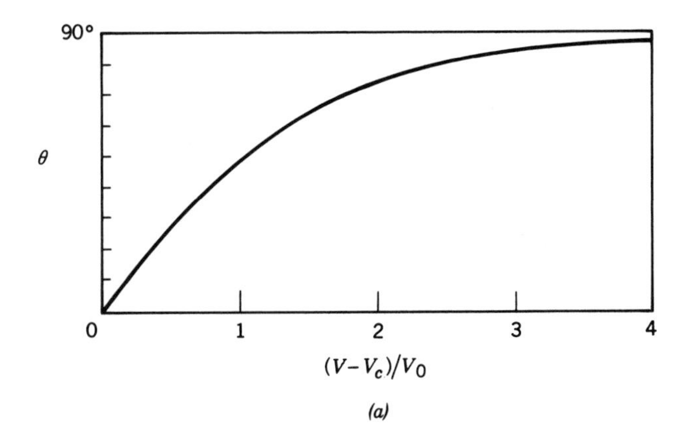

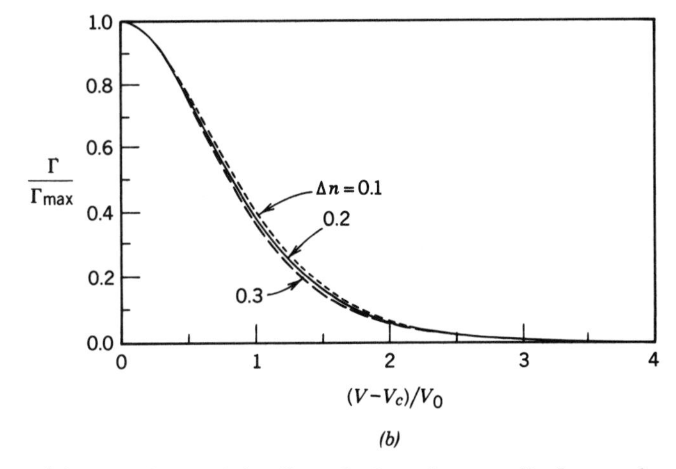

Figure 18.3-3 (*a*) Dependence of the tilt angle θ on the normalized rms voltage. (*b*) Dependence of the normalized retardation $\Gamma/\Gamma_{max} = [n(\theta) - n_o]/(n_e - n_o)$ on the normalized rms voltage when $n_o = 1.5$, for the values of $\Delta n = n_e - n_o$ indicated. This plot is obtained from (18.3-1) and (18.3-2).

tilted ($\theta = 0$), and decreases monotonically toward 0 when the tilt angle reaches 90°, as illustrated in Fig. 18.3-3(*b*).

The cell can readily be used as a voltage-controlled *phase modulator*. For an optical wave traveling in the z direction and linearly polarized in the x direction (parallel to the untilted molecular orientation), the phase shift is $\varphi = 2\pi n(\theta)d/\lambda_o$. For waves polarized at 45° to the x axis in the x-y plane, the cell serves as a voltage-controlled *wave retarder*. When placed between two crossed polarizers (at $\pm 45°$), a half-wave retarder ($\Gamma = \pi$) becomes a voltage-controlled *intensity modulator*. Similarly, a quarter-wave retarder ($\Gamma = \pi/2$) placed between a mirror and a polarizer at 45° with the x axis serves as an intensity modulator, as illustrated in Fig. 18.3-4.

The liquid-crystal cell is sealed between optically flat glass windows with antireflection coatings. A typical thickness of the liquid crystal layer is $d = 10\ \mu$m and typical values of $\Delta n = n_e - n_o = 0.1$ to 0.3. The retardation Γ is typically given in terms of the retardance $\varrho = (n_e - n_o)d$, so that the retardation $\Gamma = 2\pi\varrho/\lambda_o$. Retardances of several hundred nanometers are typical (e.g., a retardance of 300 nm corresponds to a retardation of π at $\lambda_o = 600$ nm).

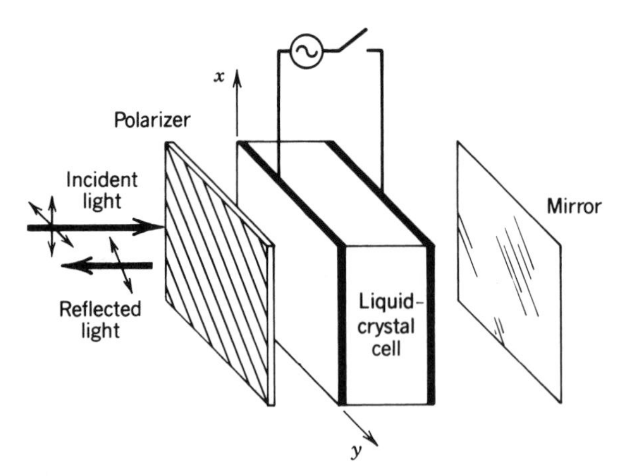

Figure 18.3-4 A liquid-crystal cell provides a retardation $\Gamma = \pi/2$ in the absence of the field ("off" state), and $\Gamma = 0$ in the presence of the field ("on" state). After reflection from the mirror and a round trip through the crystal, the plane of polarization rotates 90° in the "off" state, so that the light is blocked. In the "on" state, there is no rotation, and the reflected light is not blocked.

The applied voltage usually has a square waveform with a frequency in the range between tens of Hz and a few kHz. Operation at lower frequencies tends to cause electromechanical effects that disrupt the molecular alignment and reduce the lifetime of the device. Frequencies higher than 100 Hz result in greater power consumption because of the increased conductivity. The critical voltage V_c is typically a few volts rms.

Liquid crystals are slow. Their response time depends on the thickness of the liquid-crystal layer, the viscosity of the material, temperature, and the nature of the applied drive voltage. The rise time is of the order of tens of milliseconds if the operating voltage is near the critical voltage V_c, but decreases to a few milliseconds at higher voltages. The decay time is insensitive to the operating voltage but can be reduced by using cells of smaller thickness.

Twisted Nematic Liquid-Crystal Modulators

A *twisted* nematic liquid-crystal cell is a thin layer of nematic liquid crystal placed between two parallel glass plates and rubbed so that the molecular orientation rotates helically about an axis normal to the plates (the axis of twist). If the angle of twist is 90°, for example, the molecules point in the x direction at one plate and in the y direction at the other [Fig. 18.3-5(a)]. Transverse layers of the material act as uniaxial crystals, with the optic axes rotating helically about the axis of twist. It was shown in Sec. 6.5 that the polarization plane of linearly polarized light traveling in the direction of the axis of twist rotates with the molecules, so that the cell acts as a polarization rotator.

When an electric field is applied in the direction of the axis of twist (the z direction) the molecules tilt toward the field [Fig. 18.3-5(b)]. When the tilt is 90°, the molecules lose their twisted character (except for those adjacent to the glass surfaces), so that the polarization rotatory power is deactivated. If the electric field is removed, the orientations of the layers near the glass surfaces dominate, thereby causing the molecules to return to their original twisted state, and the polarization rotatory power to be regained.

Since the polarization rotatory power may be turned off and on by switching the electric field on and off, a shutter can be designed by placing a cell with 90° twist

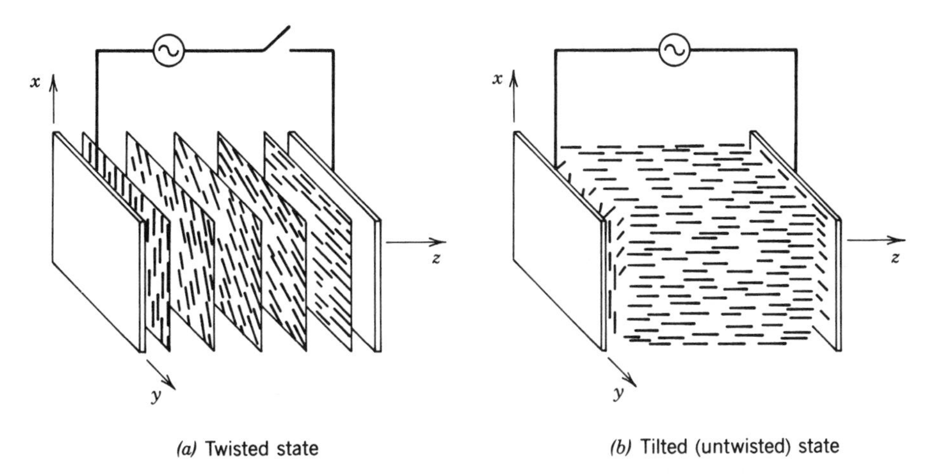

Figure 18.3-5 In the presence of a sufficiently large electric field, the molecules of a twisted nematic liquid crystal tilt and lose their twisted character.

between two crossed polarizers. The system transmits the light in the absence of an electric field and blocks it when the electric field is applied, as illustrated in Fig. 18.3-6.

Operation in the reflective mode is also possible, as illustrated in Fig. 18.3-7. Here, the twist angle is 45°; a mirror is placed on one side of the cell and a polarizer on the other side. When the electric field is absent the polarization plane rotates a total of 90° upon propagation a round trip through the cell; the reflected light is therefore blocked by the polarizer. When the electric field is present, the polarization rotatory power is suspended and the reflected light is transmitted through the polarizer. Other reflective and transmissive modes of operation with different angles of twist are also possible.

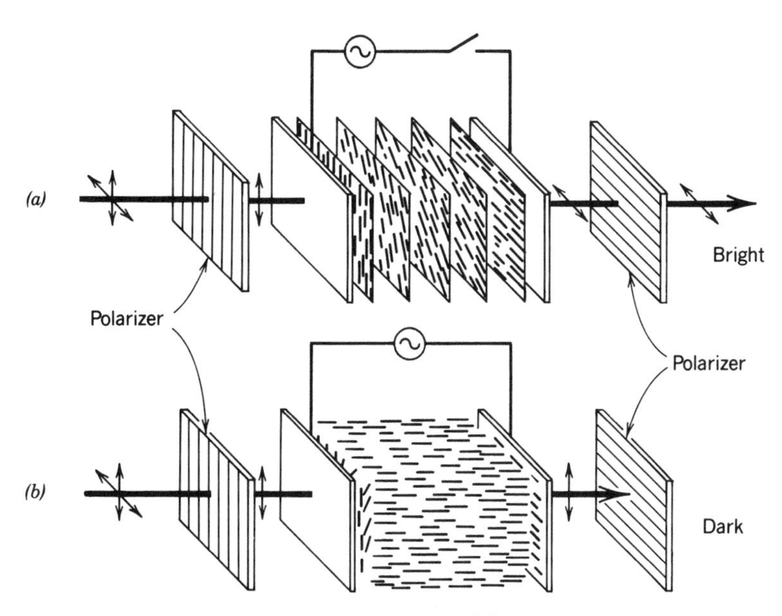

Figure 18.3-6 A twisted nematic liquid-crystal switch. (*a*) When the electric field is absent, the liquid-crystal cell acts as a polarization rotator; the light is transmitted. (*b*) When the electric field is present, the cell's rotatory power is suspended and the light is blocked.

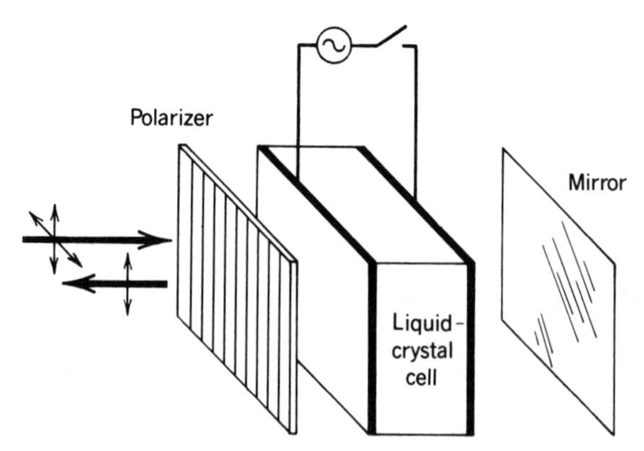

Figure 18.3-7 A twisted nematic liquid-crystal cell with 45° twist angle provides a round-trip polarization rotation of 90° in the absence of the electric field (blocked state) and no rotation when the field is applied (unblocked state). The device serves as a switch.

The twisted liquid-crystal cell placed between crossed polarizers may also be operated as an analog modulator. At intermediate tilt angles, there is a combination of polarization rotation and wave retardation. Analysis of the transmission of polarized light through tilted and twisted molecules is rather complex, but the overall effect is a partial intensity transmittance. There is an approximately linear range of transition between the total transmission of the fully twisted (untilted) state and zero transmission in the fully tilted (untwisted) state. However, the dynamic range is rather limited.

Ferroelectric Liquid Crystals

Smectic liquid crystals are organized in layers, as illustrated in Fig. 6.5-1(b). In the smectic-C phase, the molecular orientation is tilted by an angle θ with respect to the normal to the layers (the x axis), as illustrated in Fig. 18.3-8. The material has ferroelectric properties. When placed between two close glass plates the surface interactions permit only two stable states of molecular orientation at the angles $\pm\theta$, as shown in Fig. 18.3-8. When an electric field $+E$ is applied in the z direction, a torque is produced that switches the molecular orientation into the stable state $+\theta$ [Fig. 18.3-8(a)]. The molecules can be switched into the state $-\theta$ by use of an electric field of opposite polarity $-E$ [Fig. 18.3-8(b)]. Thus the cell acts as a uniaxial crystal whose optic axis may be switched between two orientations.

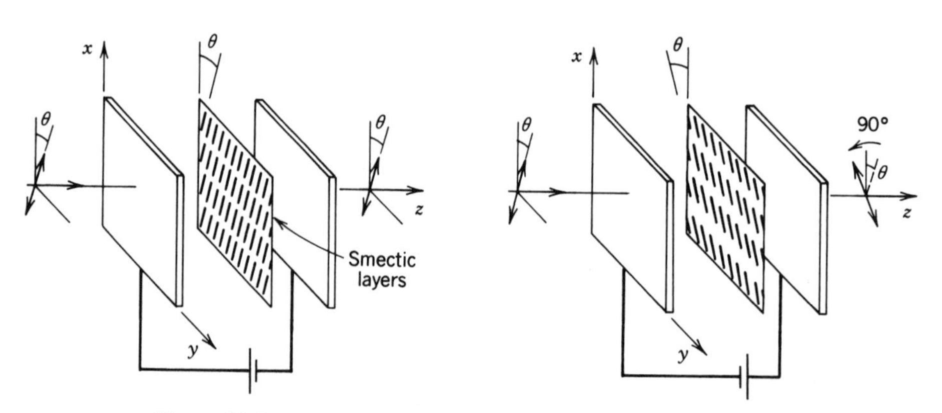

Figure 18.3-8 The two states of a ferroelectric liquid-crystal cell.

In the geometry of Fig. 18.3-8, the incident light is linearly polarized at an angle θ with the x axis in the x-y plane. In the $+\theta$ state, the polarization is parallel to the optic axis and the wave travels with the extraordinary refractive index n_e without retardation. In the $-\theta$ state, the polarization plane makes an angle 2θ with the optic axis. If $2\theta = 45°$, the wave undergoes a retardation $\Gamma = 2\pi(n_e - n_o)d/\lambda_o$, where d is the thickness of the cell and n_o is the ordinary refractive index. If d is selected such that $\Gamma = \pi$, the plane of polarization rotates 90°. Thus, reversing the applied electric field has the effect of rotating the plane of polarization by 90°.

An intensity modulator can be made by placing the cell between two crossed polarizers. The response time of ferroelectric liquid-crystal switches is typically < 20 μs at room temperature, which is far faster than that of nematic liquid crystals. The switching voltage is typically ± 10 V.

B. Spatial Light Modulators

Liquid-Crystal Displays

A liquid-crystal display (LCD) is constructed by placing transparent electrodes of different patterns on the glass plates of a reflective liquid-crystal (nematic, twisted-nematic, or ferroelectric) cell. By applying voltages to selected electrodes, patterns of reflective and nonreflective areas are created. Figure 18.3-9 illustrates a pattern for a seven-bar display of the numbers 0 to 9. Larger numbers of electrodes may be addressed sequentially. Indeed, charge-coupled devices (CCDs) can be used for addressing liquid-crystal displays. The resolution of the device depends on the number of segments per unit area. LCDs are used in consumer items such as digital watches, pocket calculators, computer monitors, and televisions.

Compared to light-emitting diode (LED) displays, the principal advantage of LCDs is their low electrical power consumption. However, LCDs have a number of disadvantages:

- They are passive devices that modulate light that is already present, rather than emitting their own light; thus they are not useful in the dark.
- Nematic liquid crystals are relatively slow.
- The optical efficiency is limited as a result of the use of polarizers that absorb at least 50% of unpolarized incident light.
- The angle of view is limited; the contrast of the modulated light is reduced as the angle of incidence/reflectance increases.

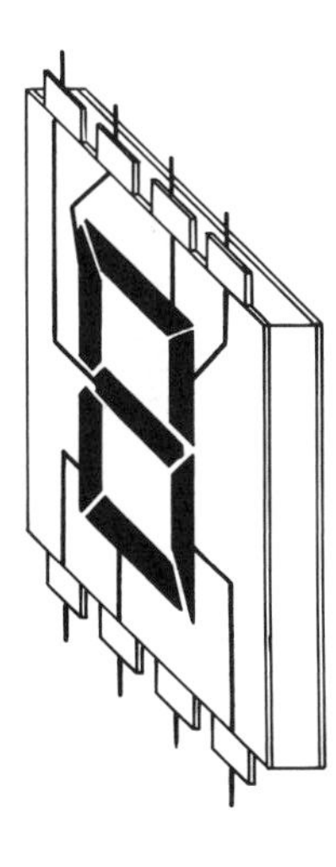

Figure 18.3-9 Electrodes of a seven-bar-segment LCD.

Optically Addressed Spatial Light Modulators

Most LCDs are addressed electrically. However, optically addressed spatial light modulators are attractive for applications involving image and optical data processing. Light with an intensity distribution $I_W(x, y)$, the "write" image, is converted by an optoelectronic sensor into a distribution of electric field $E(x, y)$, which controls the reflectance $\mathscr{R}(x, y)$ of a liquid-crystal cell operated in the reflective mode. Another optical wave of uniform intensity is reflected from the device and creates the "read" image $I(x, y) \propto \mathscr{R}(x, y)$. Thus the "read" image is controlled by the "write" image (see Fig. 18.1-14).

If the "write" image is carried by incoherent light, and the "read" image is formed by coherent light, the device serves as a spatial incoherent-to-coherent light converter, much like the PROM device discussed in Sec. 18.1E. Furthermore, the wavelengths of the "write" and "read" beams need not be the same. The "read" light may also be more intense than the "write" light, so that the device may serve as an image intensifier.

There are several means for converting the "write" image $I_W(x, y)$ into a pattern of electric field $E(x, y)$ for application to the liquid-crystal cell. A layer of photoconductive material placed between the electrodes of a capacitor may be used. When illuminated by the distribution $I_W(x, y)$, the conductance $G(x, y)$ is altered proportionally. The capacitor is discharged at each position in accordance with the local conductance, so that the resultant electric field $E(x, y) \propto 1/I_W(x, y)$ is a negative of the original image (much as in Fig. 18.1-14). An alternative is the use of a sheet photodiode [a *p-i-n* photodiode of hydrogenated amorphous silicon (a-Si : H), for example]. The reverse-biased photodiode conducts in the presence of light, thereby creating a potential difference proportional to the local light intensity.

An example of a liquid-crystal spatial light modulator is the **Hughes liquid-crystal light valve**. This device is essentially a capacitor with two low-reflectance transparent electrodes (indium–tin oxide) with a number of thin layers of materials between (Fig. 18.3-10). There are two principal layers: the liquid crystal, which is responsible for the modulation of the "read" light; and the photoconductor layer [cadmium sulfide (CdS)], which is responsible for sensing the "write" light distribution and converting it into an electric-field distribution. These two layers are separated by a dielectric mirror, which reflects the "read" light, and a light blocking dielectric material [cadmium telluride (CdTe)], which prevents the "write" light from reaching the "read" side of the device. The polarizers are placed externally (by use of a polarizing beamsplitter, for example).

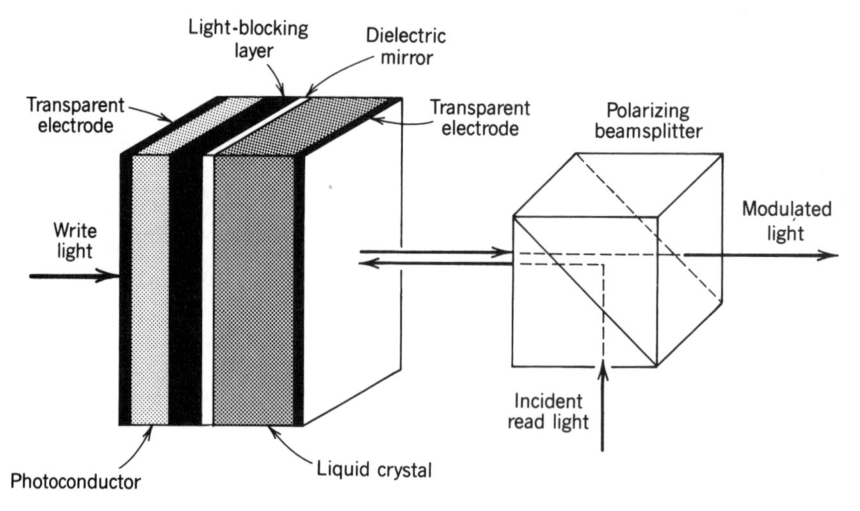

Figure 18.3-10 A liquid-crystal light valve is an optically addressed spatial light modulator.

*18.4 PHOTOREFRACTIVE MATERIALS

Photorefractive materials exhibit photoconductive and electro-optic behavior, and have the ability to detect and store spatial distributions of optical intensity in the form of spatial patterns of altered refractive index. Photoinduced charges create a space-charge distribution that produces an internal electric field, which, in turn, alters the refractive index by means of the electro-optic effect.

Ordinary *photoconductive* materials are often good insulators in the dark. Upon illumination, photons are absorbed, free charge carriers (electron–hole pairs) are generated, and the conductivity of the material increases. When the light is removed, the process of charge photogeneration ceases, and the conductivity returns to its dark value as the excess electrons and holes recombine. Photoconductors are used as photon *detectors* (see Sec. 17.2).

When a *photorefractive* material is exposed to light, free charge carriers (electrons or holes) are generated by excitation from impurity energy levels to an energy band, at a rate proportional to the optical power. This process is much like that in an extrinsic photoconductor (see Sec. 17.2). These carriers then diffuse away from the positions of high intensity where they were generated, leaving behind fixed charges of the opposite sign (associated with the impurity ions). The free carriers can be trapped by ionized impurities at other locations, depositing their charge there as they recombine. The result is the creation of an inhomogeneous space-charge distribution that can remain in place for a period of time after the light is removed. This charge distribution creates an *internal* electric field pattern that modulates the local refractive index of the material by virtue of the (Pockels) electro-optic effect. The image may be accessed optically by monitoring the spatial pattern of the refractive index using a probe optical wave. The material can be brought back to its original state (erased) by illumination with uniform light, or by heating. Thus the material can be used to record and store images, much like a photographic emulsion stores an image. The process is illustrated in Fig. 18.4-1 for doped lithium niobate ($LiNbO_3$).

Important photorefractive materials include barium titanate ($BaTiO_3$), bismuth silicon oxide ($Bi_{12}SiO_{20}$), lithium niobate ($LiNbO_3$), potassium niobate ($KNbO_3$), gallium arsenide (GaAs), and strontium barium niobate (SBN).

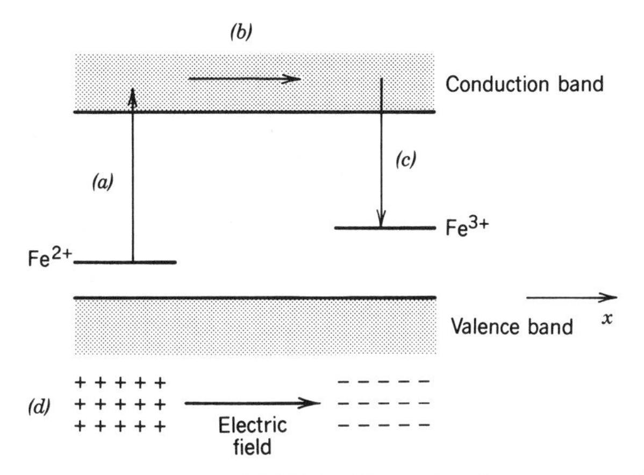

Figure 18.4-1 Energy-level diagram of $LiNbO_3$ illustrating the processes of (*a*) photoionization, (*b*) diffusion, (*c*) recombination, and (*d*) space-charge formation and electric-field generation. Fe^{2+} impurity centers act as donors, becoming Fe^{3+} when ionized, while Fe^{3+} centers act as traps, becoming Fe^{2+} after recombination.

Simplified Theory of Photorefractivity

When a photorefractive material is illuminated by light of intensity $I(x)$ that varies in the x direction, the refractive index changes by $\Delta n(x)$. The following is a step-by-step description of the processes that mediate this effect (illustrated in Fig. 18.4-1) and a simplified set of equations that govern them:

- *Photogeneration.* The absorption of a photon at position x raises an electron from the donor level to the conduction band. The rate of photoionization $G(x)$ is proportional both to the optical intensity and to the number density of nonionized donors. Thus

$$G(x) = s(N_D - N_D^+)I(x), \tag{18.4-1}$$

where N_D is the number density of donors, N_D^+ is the number density of ionized donors, and s is a constant known as the photoionization cross section.

- *Diffusion.* Since $I(x)$ is nonuniform, the number density of excited electrons $n(x)$ is also nonuniform. As a result, electrons diffuse from locations of high concentration to locations of low concentration.
- *Recombination.* The electrons recombine at a rate $R(x)$ proportional to their number density $n(x)$, and to the number density of ionized donors (traps) N_D^+, so that

$$R(x) = \gamma_R n(x) N_D^+, \tag{18.4-2}$$

where γ_R is a constant. In equilibrium, the rate of recombination equals the rate of photoionization, $R(x) = G(x)$, so that

$$sI(x)(N_D - N_D^+) = \gamma_R n(x) N_D^+, \tag{18.4-3}$$

from which

$$n(x) = \frac{s}{\gamma_R} \frac{N_D - N_D^+}{N_D^+} I(x). \tag{18.4-4}$$

- *Space Charge.* Each photogenerated electron leaves behind a positive ionic charge. When the electron is trapped (recombines), its negative charge is deposited at a different site. As a result, a nonuniform space-charge distribution is formed.
- *Electric Field.* This nonuniform space charge generates a position-dependent electric field $E(x)$, which may be determined by observing that in steady state the drift and diffusion electric-current densities must be of equal magnitude and opposite sign, so that the total current density vanishes, i.e.,

$$J = e\mu_e n(x) E(x) - k_B T \mu_e \frac{dn}{dx} = 0, \tag{18.4-5}$$

where μ_e is the electron mobility, k_B is Boltzmann's constant, and T is the temperature. Thus

$$E(x) = \frac{k_B T}{e} \frac{1}{n(x)} \frac{dn}{dx}. \tag{18.4-6}$$

- *Refractive Index.* Since the material is electro-optic, the internal electric field $E(x)$ locally modifies the refractive index in accordance with

$$\Delta n(x) = -\tfrac{1}{2}n^3 \mathfrak{r} E(x), \tag{18.4-7}$$

where n and $\mathfrak{r}$ are the appropriate values of refractive index and electro-optic coefficient for the material [see (18.1-4)].

The relation between the incident light intensity $I(x)$ and the resultant refractive index change $\Delta n(x)$ may readily be obtained if we assume that the ratio $(N_D/N_D^+ - 1)$ in (18.4-4) is approximately constant, independent of x. In that, case $n(x)$ is proportional to $I(x)$, so that (18.4-6) gives

$$E(x) = \frac{k_B T}{e} \frac{1}{I(x)} \frac{dI}{dx}. \tag{18.4-8}$$

Finally, substituting this into (18.4-7), provides an expression for the position-dependent refractive-index change as a function of intensity,

$$\Delta n(x) = -\frac{1}{2} n^3 \mathfrak{r} \frac{k_B T}{e} \frac{1}{I(x)} \frac{dI}{dx}. \tag{18.4-9}$$

Refractive-Index Change

This equation is readily generalized to two dimensions, whereupon it governs the operation of a photorefractive material as an image storage device.

Many assumptions have been made to keep the foregoing theory simple: In deriving (18.4-8) from (18.4-6) it was assumed that the ratio of number densities of unionized to ionized donors is approximately uniform, despite the spatial variation of the photoionization process. This assumption is approximately applicable when the ionization is caused by other more effective processes that are position independent in addition to the light pattern $I(x)$. Dark conductivity and volume photovoltaic effects were neglected. Holes were ignored. It was assumed that no external electric field was applied, when in fact this can be useful in certain applications. The theory is valid only in the steady state although the time dynamics of the photorefractive process are clearly important since they determine the speed with which the photorefractive material responds to the applied light. Yet in spite of all these assumptions, the simplified theory carries the essence of the behavior of photorefractive materials.

EXAMPLE 18.4-1. ***Detection of a Sinusoidal Spatial Intensity Pattern.*** Consider an intensity distribution in the form of a sinusoidal grating of period Λ, contrast m, and mean intensity I_0

$$I(x) = I_0\left(1 + m\cos\frac{2\pi x}{\Lambda}\right), \tag{18.4-10}$$

as shown in Fig. 18.4-2. Substituting this into (18.4-8) and (18.4-9), we obtain the internal

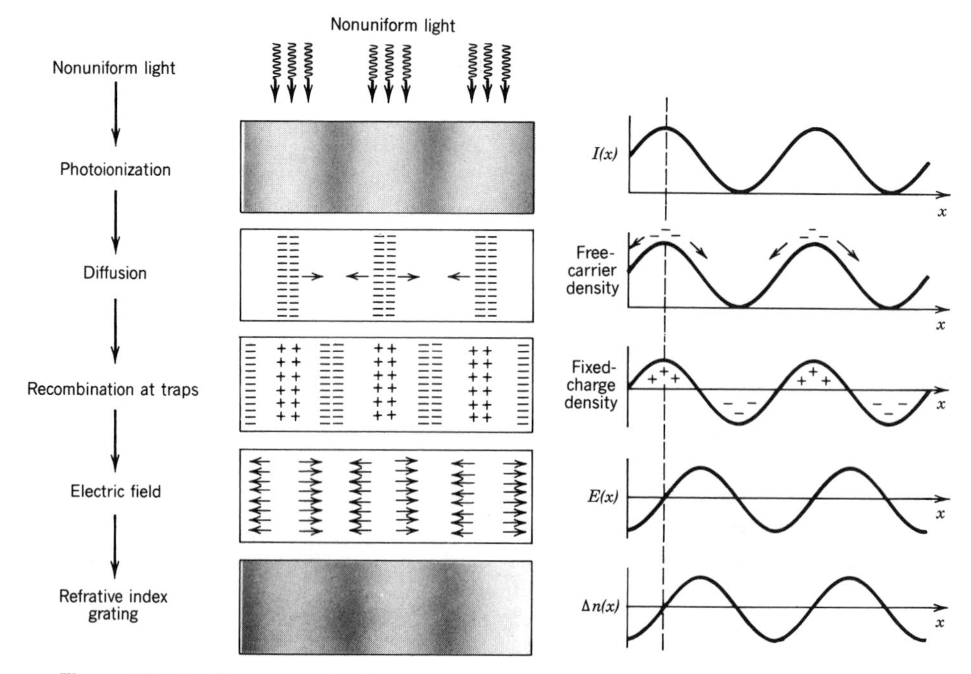

Figure 18.4-2 Response of a photorefractive material to a sinusoidal spatial light pattern.

electric field and refractive index distributions

$$E(x) = E_{\max} \frac{-\sin(2\pi x/\Lambda)}{1 + m\cos(2\pi x/\Lambda)}, \qquad \Delta n(x) = \Delta n_{\max} \frac{\sin(2\pi x/\Lambda)}{1 + m\cos(2\pi x/\Lambda)}, \quad (18.4\text{-}11)$$

where $E_{\max} = 2\pi(k_B T/e\Lambda)m$ and $\Delta n_{\max} = \frac{1}{2}n^3 \mathfrak{r} E_{\max}$ are the maximum values of $E(x)$ and $\Delta n(x)$, respectively.

If $\Lambda = 1$ μm, $m = 1$, and $T = 300$ K, for example, $E_{\max} = 1.6 \times 10^5$ V/m. This internal field is equivalent to applying 1.6 kV across a crystal of 1-cm width. The maximum refractive index change $\Delta n_{\max}$ is directly proportional to the contrast m and the electro-optic coefficient $\mathfrak{r}$, and inversely proportional to the spatial period Λ. The grating pattern $\Delta n(x)$ is totally insensitive to the uniform level of the illumination I_0.

When the image contrast m is small, the second term of the denominators in (18.4-11) may be neglected. The internal electric field and refractive index change are then sinusoidal patterns shifted by 90° relative to the incident light pattern,

$$\Delta n(x) \approx \Delta n_{\max} \sin \frac{2\pi x}{\Lambda}. \quad (18.4\text{-}12)$$

These patterns are illustrated in Fig. 18.4-2.

Applications of the Photorefractive Effect

An image $I(x, y)$ may be stored in a photorefractive crystal in the form of a refractive-index distribution $\Delta n(x, y)$. The image can be read by using the crystal as a spatial-phase modulator to encode the information on a uniform optical plane wave

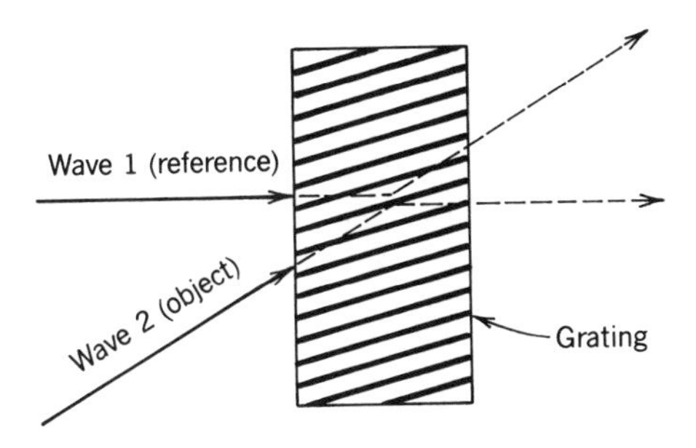

Figure 18.4-3 Two-wave mixing is a form of dynamic holography.

acting as a probe. Phase modulation may be converted into intensity modulation by placing the cell in an interferometer, for example.

Because of their capability to record images, photorefractive materials are attractive for use in real-time holography (see Sec. 4.5 for a discussion of holography). An object wave is holographically recorded by mixing it with a reference wave, as illustrated for two plane waves in Fig. 18.4-3. The intensity of the sum of two such waves forms a sinusoidal interference pattern, which is recorded in the photorefractive crystal in the form of a refractive-index variation. The crystal then serves as a volume phase hologram (see Sec. 4.5, Fig. 4.5-10). To reconstruct the stored object wave, the crystal is illuminated with the reference wave. Acting as a volume diffraction grating, the crystal reflects the reference wave and reproduces the object wave.

Since the recording process is relatively fast, the processes of recording and reconstruction can be carried out simultaneously. The object and reference waves travel together in the medium and exchange energy via reflection from the created grating. This process is called **two-wave mixing**. As shown in Fig. 18.4-3 (see also Fig. 4.5-8), waves 1 and 2 interfere and form a volume grating. Wave 1 reflects from the grating and adds to wave 2; wave 2 reflects from the grating and adds to wave 1. Thus the two waves are coupled together by the grating they create in the medium. Consequently, the transmission of wave 1 through the medium is controlled by the presence of wave 2, and *vice versa*. For example, wave 1 may be amplified at the expense of wave 2.

The mixing of two (or more) waves also occurs in other nonlinear optical materials with light-dependent optical properties, as discussed in Chap. 19. Wave mixing has numerous applications in optical data processing (see Chaps. 19 and 21), including image amplification, the removal of image aberrations, cross correlation of images, and optical interconnections.

READING LIST

General

M. A. Karim, *Electro-Optical Devices and Systems*, PWS-Kent, Boston, 1990.

A. Yariv, *Quantum Electronics*, Wiley, New York, 1967, 3rd ed. 1989.

M. J. Weber, ed., *Optical Materials*, in *Handbook of Laser Science and Technology*, vol. IV, Part 2, CRC Press, Boca Raton, FL, 1986–1987.

L. J. Pinson, *Electro-Optics*, Wiley, New York, 1985.

A. Yariv, *Optical Electronics*, Holt, Rinehart and Winston, New York, 1971, 3rd ed. 1985.

A. Yariv and P. Yeh, *Optical Waves in Crystals*, Wiley, New York, 1984.

H. A. Haus, *Waves and Fields in Optoelectronics*, Prentice-Hall, Englewood Cliffs, NJ, 1984.

J. F. Nye, *Physical Properties of Crystals: Their Representation by Tensors and Matrices*, Clarendon Press, Oxford, 1957; Oxford University Press, New York, 1984.

M. Gottlieb, C. L. M. Ireland, and J. M. Ley, *Electro-Optic and Acousto-Optic Scanning and Deflection*, Marcel Dekker, New York, 1983.

T. S. Narasimhamurty, *Photoelastic and Electro-Optic Properties of Crystals*, Plenum Press, New York, 1981.

J. I. Pankove, ed., *Display Devices*, vol. 40, *Topics in Applied Physics*, Springer-Verlag, Berlin, 1980.

D. F. Nelson, *Electric, Optic, and Acoustic Interactions in Dielectrics*, Wiley, New York, 1979.

G. R. Elion and H. A. Elion, *Electro-Optics Handbook*, Marcel Dekker, New York, 1979.

N. Bloembergen, *Nonlinear Optics*, W. A. Benjamin, Reading, MA, 1965, 1977.

I. P. Kaminow, *An Introduction to Electrooptic Devices*, Academic Press, New York, 1974.

F. Zernike and J. E. Midwinter, *Applied Nonlinear Optics*, Wiley, New York, 1973.

Liquid Crystals

L. M. Blinov, *Electro-Optical and Magneto-Optical Properties of Liquid Crystals*, Wiley, New York, 1983.

P.-G. de Gennes, *The Physics of Liquid Crystals*, Clarendon Press, Oxford, 1974, 1979.

M. E. Lines and A. M. Glass, *Principles and Applications of Ferroelectrics and Related Materials*, Clarendon Press, Oxford, 1977.

G. Meier, E. Sackmann, and J. G. Grabmaier, *Applications of Liquid Crystals*, Springer-Verlag, Berlin, 1975.

Photorefractive Materials

P. Günter and J.-P. Huignard, eds., *Photorefractive Materials and Their Applications II*, Springer-Verlag, New York, 1989.

P. Günter and J.-P. Huignard, eds., *Photorefractive Materials and Their Applications I*, Springer-Verlag, New York, 1988.

H. J. Eichler, P. Günter, and D. W. Pohl, *Laser-Induced Dynamic Gratings*, Springer-Verlag, New York, 1986.

B. Ya. Zel'dovich, N. F. Pilipetsky, and V. V. Shkunov, *Principles of Phase Conjugation*, Springer-Verlag, New York, 1985.

R. A. Fisher, ed., *Optical Phase Conjugation*, Academic Press, New York, 1983.

Special Journal Issues

Special issue on photorefractive materials, effects, and devices, *Journal of the Optical Society of America B*, vol. 7, no. 12, 1990.

Special issue on electrooptic materials and devices, *IEEE Journal of Quantum Electronics*, vol. QE-23, no. 12, 1987.

Articles

D. M. Pepper, J. Feinberg, and N. V. Kukhtarev, The Photorefractive Effect, *Scientific American*, vol. 263, no. 4, pp. 62–74, 1990.

I. Bennion and R. Walker, Guided-Wave Devices and Circuits, *Physics World*, vol. 3, no. 3, pp. 47–50, 1990.

J. Feinberg, Photorefractive Nonlinear Optics, *Physics Today*, vol. 41, no. 10, pp. 46–52, 1988.

R. C. Alferness, Titanium-Diffused Lithium Niobate Waveguide Devices, in *Guided-Wave Optoelectronics*, T. Tamir, ed., Springer-Verlag, Berlin, 1988.

F. J. Leonberger and J. F. Donnelly, Semiconductor Integrated Optic Devices, in *Guided-Wave Optoelectronics*, T. Tamir, ed., Springer-Verlag, Berlin, 1988.

S. K. Korotky and R. C. Alferness, Ti : $LiNbO_3$ Integrated Optic Technology: Fundamentals, Design Considerations, and Capabilities, in *Integrated Optical Circuits and Components*, L. D. Hutcheson, ed., Marcel Dekker, New York, 1987.

B. U. Chen, Integrated Optical Logic Devices, in *Integrated Optical Circuits and Components*, L. D. Hutcheson, ed., Marcel Dekker, New York, 1987.

A. D. Fisher and J. N. Lee, The Current Status of Two-Dimensional Spatial Light Modulator Technology, *SPIE Proceedings*, vol. 634, *Optical and Hybrid Computing*, 1987.

J. M. Hammer, Modulation and Switching of Light in Dielectric Waveguides, in *Integrated Optics*, T. Tamir, ed., Springer-Verlag, New York, 1979, Chap. 4.

W. R. Cook, Jr. and H. Jaffe, Electro-Optic Coefficients, in *Landolt-Bornstein*, new series, K. H. Hellwege, ed., vol. 11, Springer-Verlag, pp. 552–651, 1979.

S. H. Wemple and M. DiDomenico, Jr., Electro-Optical and Nonlinear Optical Properties of Crystals, in *Applied Solid State Science: Advances in Materials and Device Research*, vol. 3, R. Wolfe, ed., Academic Press, New York, 1972, pp. 263–383.

PROBLEMS

18.1-1 **Response Time of a Phase Modulator.** A GaAs crystal with refractive index $n = 3.6$ and electro-optic coefficient $\mathfrak{r} = 1.6$ pm/V is used as an electro-optic phase modulator operating at $\lambda_o = 1.3\ \mu$m in the longitudinal configuration. The crystal is 3 cm long and has a 1-cm^2 cross-sectional area. Determine the half-wave voltage V_π, the transit time of light through the crystal, and the electric capacitance of the device (the dielectric constant of GaAs is $\epsilon/\epsilon_o = 13.5$). The voltage is applied using a source with 50-Ω resistance. Which factor limits the speed of the device, the transit time of the light through the crystal or the response time of the electric circuit?

18.1-2 **Sensitivity of an Interferometric Electro-Optic Intensity Modulator.** An integrated-optic intensity modulator using the Mach–Zehnder configuration, illustrated in Fig. 18.1-5, is used as a linear analog modulator. If the half-wave voltage is $V_\pi = 10$ V, what is the sensitivity of the device (the incremental change of the intensity transmittance per unit incremental change of the applied voltage)?

18.1-3 **An Elasto-Optic Strain Sensor.** An elasto-optic material exhibits a change of the refractive index proportional to the strain. Design a strain sensor based on this effect. Consider an integrated-optical implementation. If the material is also electro-optic, consider a design based on compensating the elasto-optic and the electro-optic refractive index change, and measuring the electric field that nulls the reading of the photodetector in a Mach–Zehnder interferometer.

18.1-4 **Magneto-Optic Modulators.** Describe how a Faraday rotator (see Sec. 6.4B) may be used as an optical intensity modulator.

*18.2-1 **Cascaded Phase Modulators.** (a) A KDP crystal ($\mathfrak{r}_{41} = 8$ pm/V, $\mathfrak{r}_{63} = 11$ pm/V; $n_o = 1.507$, $n_e = 1.467$ at $\lambda_o = 633$ nm) is used as a longitudinal phase modulator. The orientation of the crystal axes and the applied electric field are as shown in Examples 18.2-2 and 18.2-6. Determine the half-wave voltage V_π at $\lambda_o = 633$ nm. (b) An electro-optic phase modulator consists of 9 KDP crystals separated by electrodes that are biased as shown in Fig. P18.2-1. How should the plates be oriented relative to each other so that the total phase modulation is maximized? Calculate V_π for the composite modulator.

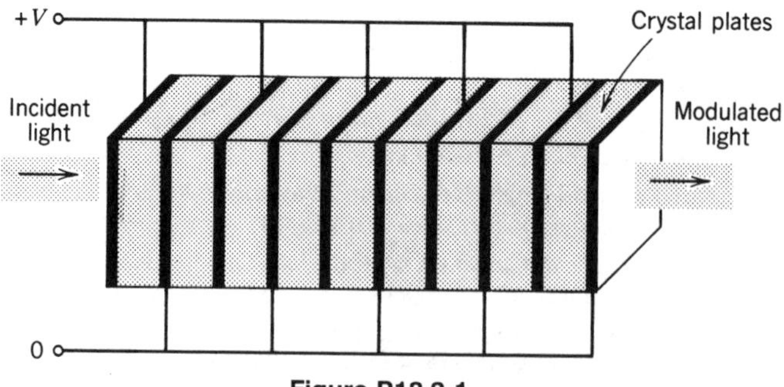

Figure P18.2-1

*18.2-2 **The "Push-Pull" Intensity Modulator.** An optical intensity modulator uses two integrated electro-optic phase modulators and a 3-dB directional coupler, as shown in Fig. P18.2-2. The input wave is split into two waves of equal amplitudes, each of which is phase modulated, reflected from a mirror, phase modulated once more, and the two returning waves are added by the directional coupler to form the output wave. Derive an expression for the intensity transmittance of the device in terms of the applied voltage, the wavelength, the dimensions, and the physical parameters of the phase modulator.

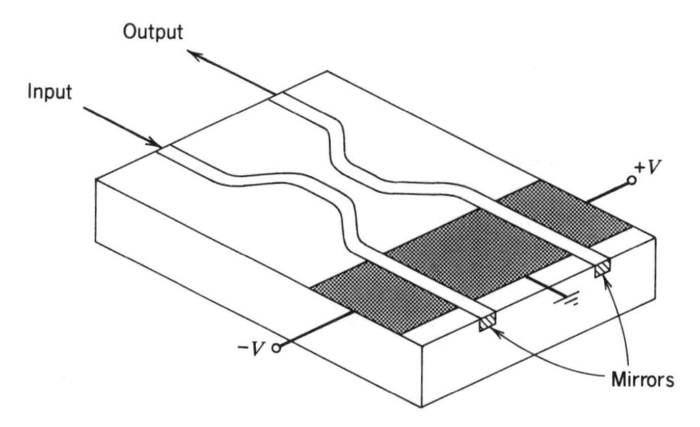

Figure P18.2-2

*18.2-3 **A $LiNbO_3$ Integrated-Optic Intensity Modulator.** Design a $LiNbO_3$ integrated-optic intensity modulator using the Mach–Zehnder interferometer shown in Fig. 18.1-5. Select the orientation of the crystal and the polarization of the guided wave for the smallest half-wave voltage V_π. Assume that the active region has length $L = 1$ mm and width $d = 5$ μm, the wavelength is $\lambda_o = 0.85$ μm, the refractive indices are $n_o = 2.29$, $n_e = 2.17$; and the electro-optic coefficients are $\mathfrak{r}_{33} = 30.9$, $\mathfrak{r}_{13} = 8.6$, $\mathfrak{r}_{22} = 3.4$, and $\mathfrak{r}_{42} = 28$ pm/V.

*18.2-4 **Double Refraction in an Electro-Optic Crystal.** (a) An unpolarized He–Ne laser beam ($\lambda_o = 633$ nm) is transmitted through a 1-cm-thick $LiNbO_3$ plate ($n_e = 2.17$, $n_o = 2.29$, $\mathfrak{r}_{33} = 30.9$ pm/V, $\mathfrak{r}_{13} = 8.6$ pm/V). The beam is orthogonal to the plate and the optic axis lies in the plane of incidence of the light at 45° with the beam. The beam is double refracted (see Sec. 6.3E). Determine the lateral displacement and the retardation between the ordinary and extraordinary beams. (b) If an electric field $E = 30$ V/m is applied in a direction parallel to the optic axis, what is the effect on the transmitted beams? What are possible applications of this device?

CHAPTER

19

NONLINEAR OPTICS

19.1 NONLINEAR OPTICAL MEDIA

19.2 SECOND-ORDER NONLINEAR OPTICS
- A. Second-Harmonic Generation and Rectification
- B. The Electro-Optic Effect
- C. Three-Wave Mixing

19.3 THIRD-ORDER NONLINEAR OPTICS
- A. Third-Harmonic Generation and Self-Phase Modulation
- B. Four-Wave Mixing
- C. Optical Phase Conjugation

*19.4 COUPLED-WAVE THEORY OF THREE-WAVE MIXING
- A. Second-Harmonic Generation
- B. Frequency Conversion
- C. Parametric Amplification and Oscillation

*19.5 COUPLED-WAVE THEORY OF FOUR-WAVE MIXING

*19.6 ANISOTROPIC NONLINEAR MEDIA

*19.7 DISPERSIVE NONLINEAR MEDIA

19.8 OPTICAL SOLITONS

Nicolaas Bloembergen (born 1920) has carried out pioneering studies in nonlinear optics since the early 1960s. He shared the 1981 Nobel Prize with Arthur Schawlow.

Throughout the long history of optics, and indeed until relatively recently, it was thought that all optical media were linear. The assumption of linearity of the optical medium has far-reaching consequences:

- The optical properties, such as the refractive index and the absorption coefficient, are independent of light intensity.
- The principle of superposition, a fundamental tenet of classical optics (as described in Sec. 2.1), holds.
- The frequency of light cannot be altered by its passage through the medium.
- Light cannot interact with light; two beams of light in the same region of a linear optical medium can have no effect on each other. Thus light cannot control light.

The invention of the laser in 1960 enabled us to examine the behavior of light in optical materials at higher intensities than previously possible. Many of the experiments carried out made it clear that optical media do in fact exhibit nonlinear behavior, as exemplified by the following observations:

- The refractive index, and consequently the speed of light in an optical medium, does change with the light intensity.
- The principle of superposition is violated.
- Light can alter its frequency as it passes through a nonlinear optical material (e.g., from red to blue!).
- Light can control light; photons do interact.

The field of nonlinear optics comprises many fascinating phenomena.

Linearity or nonlinearity is a property of the medium through which light travels, rather than a property of the light itself. Nonlinear behavior is not exhibited when light travels in free space. *Light interacts with light via the medium*. The presence of an optical field modifies the properties of the medium which, in turn, modify another optical field or even the original field itself.

It was pointed out in Sec. 5.2 that the properties of a dielectric medium through which an electromagnetic (optical) wave propagates are completely described by the relation between the polarization density vector $\mathscr{P}(\mathbf{r}, t)$ and the electric-field vector $\mathscr{E}(\mathbf{r}, t)$. It was suggested that $\mathscr{P}(\mathbf{r}, t)$ could be regarded as the output of a system whose input was $\mathscr{E}(\mathbf{r}, t)$. The mathematical relation between the vector functions $\mathscr{P}(\mathbf{r}, t)$ and $\mathscr{E}(\mathbf{r}, t)$ defines the system and is governed by the characteristics of the medium. The medium is said to be nonlinear if this relation is nonlinear.

In Sec. 5.2, dielectric media were further classified with respect to their homogeneity, isotropy, and dispersiveness. To focus on the principal effect of interest in this chapter—nonlinearity—the medium is initially assumed to be homogeneous, isotropic, and nondispersive. Sections 19.6 and 19.7 provide brief discussions of anisotropic and dispersive nonlinear optical media.

The theory of nonlinear optics and its applications is presented at two levels. A simplified approach is provided in Secs. 19.1 to 19.3. This is followed by a more detailed analysis of the same phenomena in Secs. 19.4 and 19.5.

Light propagation in media characterized by a second-order (quadratic) nonlinear relation between $\mathscr{P}$ and $\mathscr{E}$ is described in Secs. 19.2 and 19.4. Applications include the frequency doubling of a monochromatic wave (*second-harmonic generation*), the mixing of two monochromatic waves to generate a third wave whose frequency is the sum or difference of the frequencies of the original waves (*frequency conversion*), the use of two monochromatic waves to amplify a third wave (*parametric amplification*), and the addition of feedback to a parametric amplifier to create an oscillator (*parametric oscillation*). Wave propagation in a medium with a third-order $\mathscr{P}$–$\mathscr{E}$ relation is discussed in Secs. 19.3 and 19.5. Applications include *third-harmonic generation*, *self-phase modulation*, *self-focusing*, *four-wave mixing*, *optical amplification*, and *optical phase conjugation*.

Optical solitons are discussed in Sec. 19.8. These are optical pulses that propagate in a nonlinear dispersive medium without changing their shape. Changes in the pulse profile caused by the dispersive and nonlinear effects just compensate each other, so that the pulse shape is maintained for long propagation distances. *Optical bistability* is yet another nonlinear optical effect that has applications in photonic switching; its discussion is relegated to Chap. 21.

19.1 NONLINEAR OPTICAL MEDIA

A linear dielectric medium is characterized by a linear relation between the polarization density and the electric field, $\mathscr{P} = \epsilon_o \chi \mathscr{E}$, where ϵ_o is the permittivity of free space and χ is the electric susceptibility of the medium (see Sec. 5.2A). A nonlinear dielectric medium, on the other hand, is characterized by a nonlinear relation between $\mathscr{P}$ and $\mathscr{E}$, as illustrated in Fig. 19.1-1.

The nonlinearity may be of microscopic or macroscopic origin. The polarization density $\mathscr{P} = N\wp$ is a product of the individual dipole moment $\wp$, which is induced by the applied electric field $\mathscr{E}$, and the number density of dipole moments N. The nonlinear behavior may have its origin in either $\wp$ or in N.

The relation between $\wp$ and $\mathscr{E}$ is linear when $\mathscr{E}$ is small, but becomes nonlinear as $\mathscr{E}$ acquires values comparable with interatomic electric fields (typically, 10^5 to 10^8 V/m). This may be explained in terms of the simple Lorentz model in which the dipole moment is $\wp = -ex$, where x is the displacement of a mass with charge $-e$ to which an electric force $-e\mathscr{E}$ is applied (see Sec. 5.5C). If the restraining elastic force is

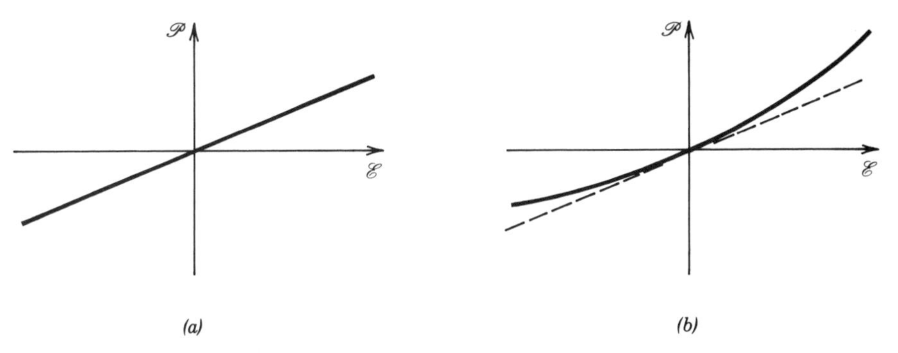

Figure 19.1-1 The $\mathscr{P}$–$\mathscr{E}$ relation for (*a*) a linear dielectric medium, and (*b*) a nonlinear medium.

proportional to the displacement (i.e., if Hooke's law is satisfied), the equilibrium displacement x is proportional to $\mathscr{E}$; $\mathscr{P}$ is then proportional to $\mathscr{E}$, and the medium is linear. However, if the restraining force is a nonlinear function of the displacement, the equilibrium displacement x and the polarization density $\mathscr{P}$ are nonlinear functions of $\mathscr{E}$ and, consequently, the medium is nonlinear. The time dynamics of an anharmonic oscillator model describing a dielectric medium with these features is discussed in Sec. 19.7.

Another possible origin of the nonlinear response of an optical material to light is the dependence of the number density N on the optical field. An example is a laser medium for which the number of atoms occupying the energy levels involved in the absorption and emission of light are dependent on the intensity of the light itself (see Sec. 13.3).

Since externally applied optical electric fields are typically small in comparison with characteristic interatomic or crystalline fields, even when focused laser light is used, the nonlinearity is usually weak. The relation between $\mathscr{P}$ and $\mathscr{E}$ is then approximately linear for small $\mathscr{E}$, deviating only slightly from linearity as $\mathscr{E}$ increases (see Fig. 19.1-1). Under these circumstances, it is possible to expand the function that relates $\mathscr{P}$ to $\mathscr{E}$ in a Taylor's series about $\mathscr{E} = 0$,

$$\mathscr{P} = a_1 \mathscr{E} + \tfrac{1}{2} a_2 \mathscr{E}^2 + \tfrac{1}{6} a_3 \mathscr{E}^3 + \cdots, \qquad (19.1\text{-}1)$$

and to use only few terms. The coefficients a_1, a_2, and a_3 are the first, second, and third derivatives of $\mathscr{P}$ with respect to $\mathscr{E}$ at $\mathscr{E} = 0$. These coefficients are characteristic constants of the medium. The first term, which is linear, dominates at small $\mathscr{E}$. Clearly, $a_1 = \epsilon_o \chi$, where χ is the linear susceptibility, which is related to the dielectric constant and the refractive index by $n^2 = \epsilon/\epsilon_o = 1 + \chi$. The second term represents a quadratic or second-order nonlinearity, the third term represents a third-order nonlinearity, and so on.

It is customary to write (19.1-1) in the form†

$$\boxed{\mathscr{P} = \epsilon_o \chi \mathscr{E} + 2 \mathscr{d} \mathscr{E}^2 + 4 \chi^{(3)} \mathscr{E}^3 + \cdots,} \qquad (19.1\text{-}2)$$

where $\mathscr{d} = \frac{1}{4} a_2$ and $\chi^{(3)} = \frac{1}{24} a_3$ are coefficients describing the second- and third-order nonlinear effects, respectively.

Equation (19.1-2) provides the basic description for a nonlinear optical medium. Anisotropy, dispersion, and inhomogeneity have been ignored both for simplicity and to enable us to focus on the basic nonlinear effect without the added algebraic complications brought about by these auxiliary effects. Sections 19.6 and 19.7 are devoted to anisotropic and dispersive nonlinear media.

In centrosymmetric media (these are media with inversion symmetry, so that the properties of the medium are not altered by the transformation $\mathbf{r} \to -\mathbf{r}$), the $\mathscr{P}$–$\mathscr{E}$ function must have odd symmetry, so that the reversal of $\mathscr{E}$ results in the reversal of $\mathscr{P}$ without any other change. The second-order nonlinear coefficient $\mathscr{d}$ must then vanish, and the lowest order nonlinearity is of third order.

Typical values of the second-order nonlinear coefficient $\mathscr{d}$ for dielectric crystals, semiconductors, and organic materials used in photonics applications lie in the range

†This nomenclature is used in a number of books, such as A. Yariv, *Quantum Electronics*, Wiley, New York, 3rd ed. 1989. An alternative relation, $\mathscr{P} = \epsilon_o(\chi \mathscr{E} + \chi^{(2)} \mathscr{E}^2 + \chi^{(3)} \mathscr{E}^3)$, is used in other books, e.g., Y. R. Shen, *The Principles of Nonlinear Optics*, Wiley, New York, 1984.

$\mathscr{d} = 10^{-24}$ to 10^{-21} (MKS units, A-s/V^2). Typical values of the third-order nonlinear coefficient $\chi^{(3)}$ for glasses, crystals, semiconductors, semiconductor-doped glasses, and organic materials of interest in photonics are $\chi^{(3)} = 10^{-34}$ to 10^{-29} (MKS units).

EXERCISE 19.1-1

Intensity of Light Necessary to Exhibit Nonlinear Effects

(a) Determine the intensity of light (in W/cm^2) at which the ratio of the second term to the first term in (19.1-2) is 1% in an ADP ($NH_4H_2PO_4$) crystal for which $n = 1.5$ and $\mathscr{d} = 6.8 \times 10^{-24}$ (MKS units) at $\lambda_o = 1.06$ μm.

(b) Determine the intensity of light at which the third term in (19.1-2) is 1% of the first term in carbon disulfide (CS_2) for which $n = 1.6$, $\mathscr{d} = 0$, and $\chi^{(3)} = 4.4 \times 10^{-32}$ (MKS units) at $\lambda_o = 694$ nm.

Note: The intensity of light is $I = \langle \mathscr{E}^2 \rangle / \eta$, where $\eta = \eta_o/n$ is the impedance of the medium and $\eta_o = (\mu_o/\epsilon_o)^{1/2} \approx 377$ Ω is the impedance of free space.

The Nonlinear Wave Equation

The propagation of light in a nonlinear medium is governed by the wave equation (5.2-19), which was derived from Maxwell's equations for an arbitrary homogeneous dielectric medium,

$$\nabla^2 \mathscr{E} - \frac{1}{c_o^2}\frac{\partial^2 \mathscr{E}}{\partial t^2} = \mu_o \frac{\partial^2 \mathscr{P}}{\partial t^2}. \qquad (19.1\text{-}3)$$

It is convenient to write $\mathscr{P}$ as a sum of linear and nonlinear parts,

$$\mathscr{P} = \epsilon_o \chi \mathscr{E} + \mathscr{P}_{NL}, \qquad (19.1\text{-}4)$$

$$\mathscr{P}_{NL} = 2\mathscr{d}\mathscr{E}^2 + 4\chi^{(3)}\mathscr{E}^3 + \cdots . \qquad (19.1\text{-}5)$$

Using (19.1-4) and the relations $n^2 = 1 + \chi$, $c_o = 1/(\mu_o \epsilon_o)^{1/2}$, and $c = c_o/n$, (19.1-3) may be written as

$$\nabla^2 \mathscr{E} - \frac{1}{c^2}\frac{\partial^2 \mathscr{E}}{\partial t^2} = -\mathscr{S} \qquad (19.1\text{-}6)$$

$$\mathscr{S} = -\mu_o \frac{\partial^2 \mathscr{P}_{NL}}{\partial t^2}. \qquad (19.1\text{-}7)$$

Wave Equation in a Nonlinear Medium

It is useful to regard (19.1-6) as a wave equation in which the term $\mathscr{S} = -\mu_o \partial^2 \mathscr{P}_{NL}/\partial t^2$ acts as a source radiating in a linear medium of refractive index n. Because $\mathscr{P}_{NL}$ (and therefore $\mathscr{S}$) is a nonlinear function of $\mathscr{E}$, (19.1-6) is a nonlinear partial differential equation in $\mathscr{E}$. This is the basic equation that underlies the theory of nonlinear optics.

There are two approximate approaches to solving the nonlinear wave equation. The first is an iterative approach known as the Born approximation. This approximation underlies the simplified introduction to nonlinear optics presented in Secs. 19.2 and 19.3. The second approach is a coupled-wave theory in which the nonlinear wave equation is used to derive linear coupled partial differential equations that govern the interacting waves. This is the basis of the more advanced study of wave interactions in nonlinear media, which is presented in Secs. 19.4 and 19.5.

Scattering Theory of Nonlinear Optics: The Born Approximation

The radiation source $\mathscr{S}$ in (19.1-6) is a function of the field $\mathscr{E}$ that it, itself, radiates. To emphasize this point we write $\mathscr{S} = \mathscr{S}(\mathscr{E})$ and illustrate the process by a simple diagram:

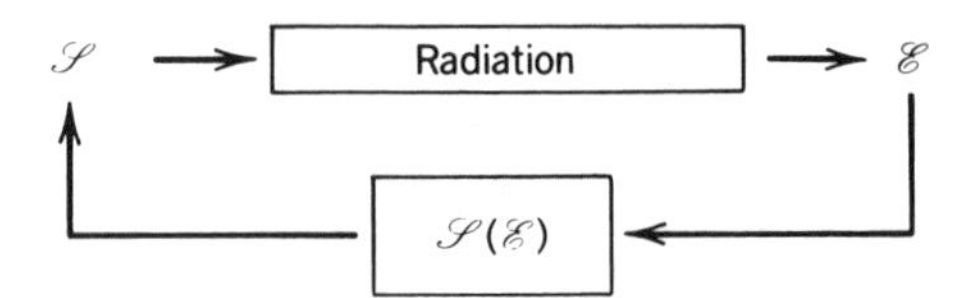

Suppose that an optical field $\mathscr{E}_0$ is incident on a nonlinear medium confined to some volume (see Fig. 19.1-2). This field creates a radiation source $\mathscr{S}(\mathscr{E}_0)$ that radiates an optical field $\mathscr{E}_1$. The corresponding radiation source $\mathscr{S}(\mathscr{E}_1)$ radiates a field $\mathscr{E}_2$, and so on. This process suggests an iterative solution, the first step of which is known as the **first Born approximation**. The second Born approximation carries the process an additional iteration, and so on.

The first Born approximation is adequate when the light intensity is sufficiently weak so that the nonlinearity is small. In this approximation, light propagation through the nonlinear medium is regarded as a scattering process in which the incident field is scattered by the medium. The scattered light is determined from the incident light in two steps:

- The incident field $\mathscr{E}_0$ is used to determine the nonlinear polarization density $\mathscr{P}_{\mathrm{NL}}$, from which the radiation source $\mathscr{S}(\mathscr{E}_0)$ is determined.
- The radiated (scattered) field $\mathscr{E}_1$ is determined from the radiation source by adding the spherical waves associated with the different source points (as in the theory of diffraction discussed in Sec. 4.3).

In many cases the amount of scattered light is very small, so that the depletion of the incident light is indeed negligible and the first Born approximation is adequate. Sections 19.2 and 19.3 are based on the first Born approximation. An initial field $\mathscr{E}_0$ containing one or several monochromatic waves of different frequencies is assumed.

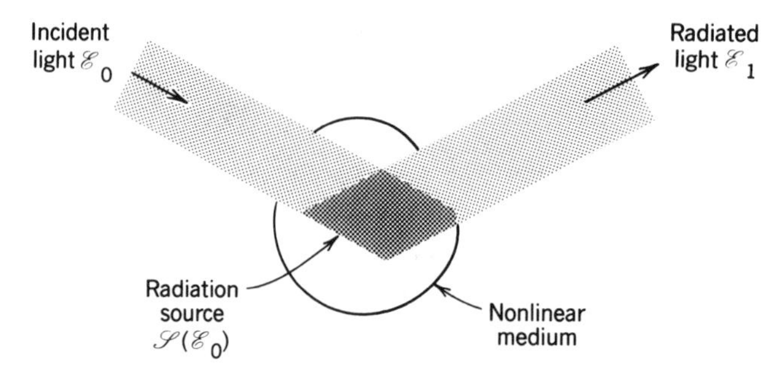

Figure 19.1-2 The first Born approximation. An incident optical field $\mathscr{E}_0$ creates a source $\mathscr{S}(\mathscr{E}_0)$, which radiates an optical field $\mathscr{E}_1$.

The corresponding nonlinear polarization $\mathscr{P}_{NL}$ is then determined using (19.1-5) and the source function $\mathscr{S}(\mathscr{E}_0)$ is evaluated using (19.1-7). Since $\mathscr{S}(\mathscr{E}_0)$ is a nonlinear function, new frequencies are created. The source therefore emits an optical field $\mathscr{E}_1$ with frequencies not present in the original wave $\mathscr{E}_0$. This leads to numerous interesting phenomena that have been utilized to make useful nonlinear-optics devices.

19.2 SECOND-ORDER NONLINEAR OPTICS

In this section we examine the optical properties of a nonlinear medium in which nonlinearities of order higher than the second are negligible, so that

$$\boxed{\mathscr{P}_{NL} = 2\,d\mathscr{E}^2.} \tag{19.2-1}$$

We consider an electric field $\mathscr{E}$ comprising one or two harmonic components and determine the spectral components of $\mathscr{P}_{NL}$. In accordance with the first Born approximation, the radiation source $\mathscr{S}$ contains the same spectral components as $\mathscr{P}_{NL}$, and so, therefore, does the emitted (scattered) field.

A. Second-Harmonic Generation and Rectification

Consider the response of this nonlinear medium to a harmonic electric field of angular frequency ω (wavelength $\lambda_o = 2\pi c_o/\omega$) and complex amplitude $E(\omega)$,

$$\mathscr{E}(t) = \mathrm{Re}\{E(\omega)\exp(j\omega t)\}. \tag{19.2-2}$$

The corresponding nonlinear polarization density $\mathscr{P}_{NL}$ is obtained by substituting (19.2-2) into (19.2-1),

$$\mathscr{P}_{NL}(t) = P_{NL}(0) + \mathrm{Re}\{P_{NL}(2\omega)\exp(j2\omega t)\}, \tag{19.2-3}$$

where

$$P_{NL}(0) = dE(\omega)E^*(\omega) \tag{19.2-4}$$

$$P_{NL}(2\omega) = dE(\omega)E(\omega). \tag{19.2-5}$$

This process is illustrated graphically in Fig. 19.2-1.

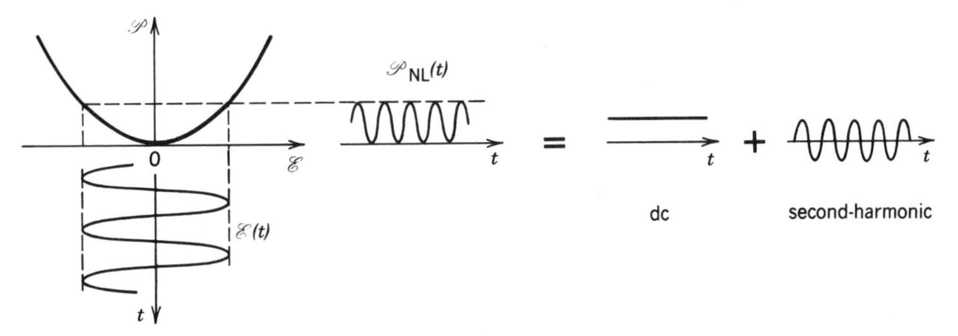

Figure 19.2-1 A sinusoidal electric field of angular frequency ω in a second-order nonlinear optical medium creates a polarization with a component at 2ω (second-harmonic) and a steady (dc) component.

Second-Harmonic Generation

The source $\mathcal{S}(t) = -\mu_o \partial^2 \mathcal{P}_{NL}/\partial t^2$ corresponding to (19.2-3) has a component at frequency 2ω and complex amplitude $S(2\omega) = 4\mu_o \omega^2 \mathscr{d} E(\omega) E(\omega)$, which radiates an optical field at frequency 2ω (wavelength $\lambda_o/2$). Thus the scattered optical field has a component at the second harmonic of the incident optical field. Since the amplitude of the emitted second-harmonic light is proportional to $S(2\omega)$, its intensity is proportional to $|S(2\omega)|^2 \propto \omega^4 \mathscr{d}^2 I^2$, where $I = |E(\omega)|^2/2\eta$ is the intensity of the incident wave. The intensity of the second-harmonic wave is therefore proportional to $\mathscr{d}^2$, to $1/\lambda_o^4$, and to I^2. Consequently, the efficiency of second-harmonic generation is proportional to $I = P/A$, where P is the incident power and A is the cross-sectional area. It is therefore essential that the incident wave have the largest possible power and be focused to the smallest possible area to produce strong second-harmonic radiation. Pulsed lasers are convenient in this respect since they deliver large peak powers.

To enhance the efficiency of second-harmonic generation, the interaction region should also be as long as possible. Since diffraction effects limit the distances within which light remains confined, guided wave structures that confine light for relatively long distances (see Chaps. 7 and 8) offer a clear advantage. Although glass fibers were initially ruled out for second-harmonic generation since glass is centrosymmetric (and therefore has $\mathscr{d} = 0$), efficient second-harmonic generation is, in fact, observed in silica glass fibers doped with germanium and phosphorus. It appears that defects can produce a non-centrosymmetric core with a value of $\mathscr{d}$ that is sufficiently large to achieve efficient second-harmonic generation.

Figure 19.2-2 illustrates several optical second-harmonic-generation configurations in bulk crystals and in waveguides, in which infrared light is converted to visible light and visible light is converted to the ultraviolet.

Optical Rectification

The component $P_{NL}(0)$ in (19.2-3) corresponds to a steady (non-time-varying) polarization density that creates a dc potential difference across the plates of a capacitor within which the nonlinear material is placed (Fig. 19.2-3). The generation of a dc voltage as a

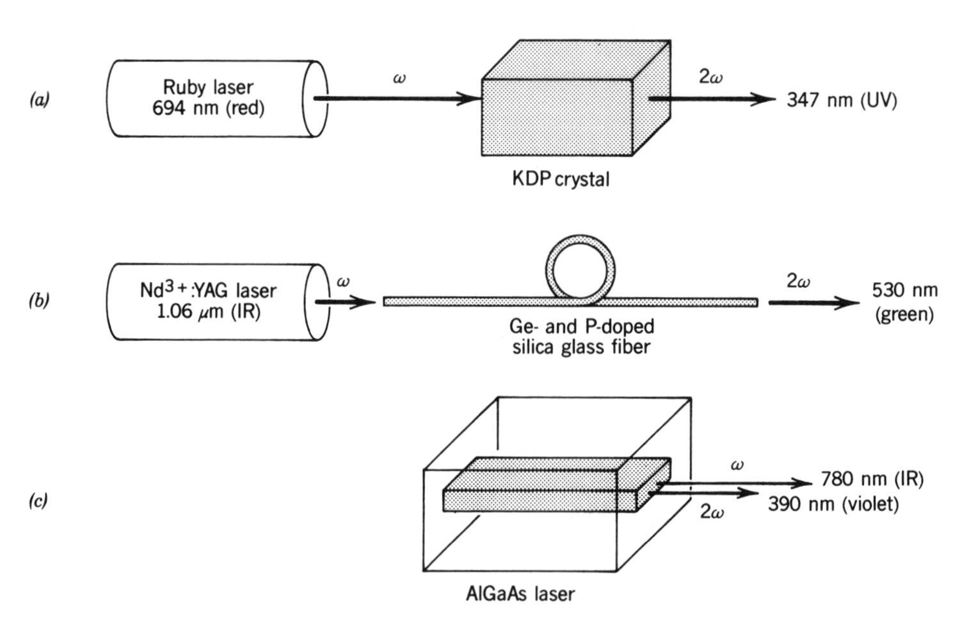

Figure 19.2-2 Optical second-harmonic generation in (*a*) a bulk crystal; (*b*) a glass fiber; (*c*) within the cavity of a semiconductor laser.

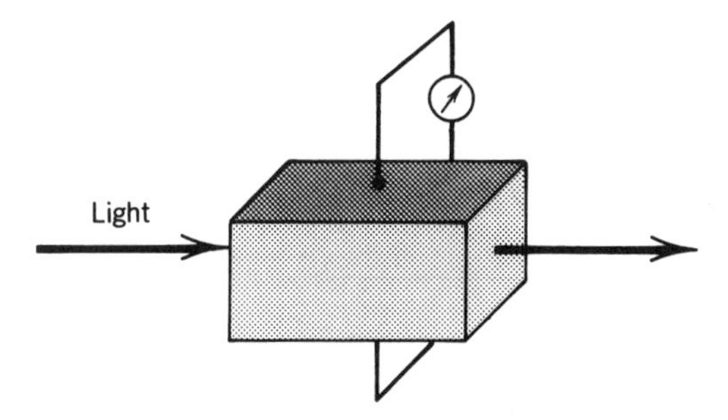

Figure 19.2-3 The transmission of an intense beam of light through a nonlinear crystal generates a dc voltage across it.

result of an intense optical field represents optical rectification (in analogy with the conversion of a sinusoidal ac voltage into a dc voltage in an ordinary electronic rectifier). An optical pulse of several MW peak power, for example, may generate a voltage of several hundred μV.

B. The Electro-Optic Effect

We now consider an electric field $\mathscr{E}(t)$ comprising a harmonic component at an optical frequency ω together with a steady component (at $\omega = 0$),

$$\mathscr{E}(t) = E(0) + \operatorname{Re}\{E(\omega)\exp(j\omega t)\}. \tag{19.2-6}$$

We distinguish between these two components by calling $E(0)$ the electric field and $E(\omega)$ the optical field. In fact, both components are electric fields.

Substituting (19.2-6) into (19.2-1), we obtain

$$\mathscr{P}_{\mathrm{NL}}(t) = P_{\mathrm{NL}}(0) + \operatorname{Re}\{P_{\mathrm{NL}}(\omega)\exp(j\omega t)\} + \operatorname{Re}\{P_{\mathrm{NL}}(2\omega)\exp(j2\omega t)\}, \tag{19.2-7}$$

where

$$P_{\mathrm{NL}}(0) = d\left[2E^2(0) + |E(\omega)|^2\right] \tag{19.2-8a}$$

$$P_{\mathrm{NL}}(\omega) = 4\,dE(0)E(\omega) \tag{19.2-8b}$$

$$P_{\mathrm{NL}}(2\omega) = dE(\omega)E(\omega), \tag{19.2-8c}$$

so that the polarization density contains components at the angular frequencies 0, ω, and 2ω.

If the optical field is substantially smaller in magnitude than the electric field, i.e., $|E(\omega)|^2 \ll |E(0)|^2$, the second-harmonic polarization component $P_{\mathrm{NL}}(2\omega)$ may be neglected in comparison with the components $P_{\mathrm{NL}}(0)$ and $P_{\mathrm{NL}}(\omega)$. This is equivalent to the linearization of $\mathscr{P}_{\mathrm{NL}}$ as a function of $\mathscr{E}$, i.e., approximating it by a straight line with a slope equal to the derivative at $\mathscr{E} = E(0)$, as illustrated in Fig. 19.2-4.

Equation (19.2-8b) provides a linear relation between $P_{\mathrm{NL}}(\omega)$ and $E(\omega)$ which we write in the form $P_{\mathrm{NL}}(\omega) = \epsilon_o\,\Delta\chi E(\omega)$, where $\Delta\chi = (4d/\epsilon_o)E(0)$ represents an increase in the susceptibility proportional to the electric field $E(0)$. The corresponding incremental change of the refractive index is obtained by differentiating the relation

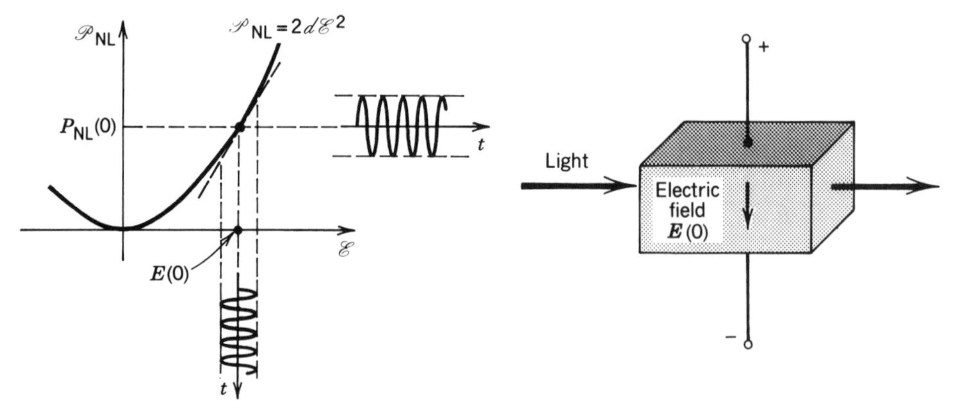

Figure 19.2-4 Linearization of the second-order nonlinear relation $\mathscr{P}_{\mathrm{NL}} = 2d\mathscr{E}^2$ in the presence of a strong electric field $E(0)$ and a weak optical field $E(\omega)$.

$n^2 = 1 + \chi$, to obtain $2n\,\Delta n = \Delta\chi$, from which

$$\Delta n = \frac{2d}{n\epsilon_o} E(0). \tag{19.2-9}$$

The medium is then effectively linear with a refractive index $n + \Delta n$ that is linearly controlled by the electric field $E(0)$.

The nonlinear nature of the medium creates a coupling between the electric field $E(0)$ and the optical field $E(\omega)$, causing one to control the other, so that the nonlinear medium exhibits the linear electro-optic effect (Pockels effect) discussed in Chapter 18. This effect is characterized by the relation $\Delta n = -\frac{1}{2}n^3 \mathfrak{r} E(0)$, where $\mathfrak{r}$ is the Pockels coefficient. Comparing this formula with (19.2-9), we conclude that the Pockels coefficient $\mathfrak{r}$ is related to the second-order nonlinear coefficient d by

$$\mathfrak{r} \approx -\frac{4}{\epsilon_o n^4} d. \tag{19.2-10}$$

Although this expression reveals the common underlying origin of the Pockels effect and the medium nonlinearity, it is not consistent with experimentally observed values of $\mathfrak{r}$ and d. This is because we have made the implicit assumption that the medium is nondispersive (i.e., that its response is insensitive to frequency). This assumption is clearly not satisfied when one of the components of the field is at the optical frequency ω and the other is a steady field with zero frequency. The role of dispersion is discussed in Sec. 19.7.

C. Three-Wave Mixing

Frequency Conversion

We now consider the case of a field $\mathscr{E}(t)$ comprising two harmonic components at optical frequencies ω_1 and ω_2,

$$\mathscr{E}(t) = \mathrm{Re}\{E(\omega_1)\exp(j\omega_1 t) + E(\omega_2)\exp(j\omega_2 t)\}.$$

The nonlinear component of the polarization $\mathscr{P}_{\mathrm{NL}} = 2\,d\mathscr{E}^2$ then contains components

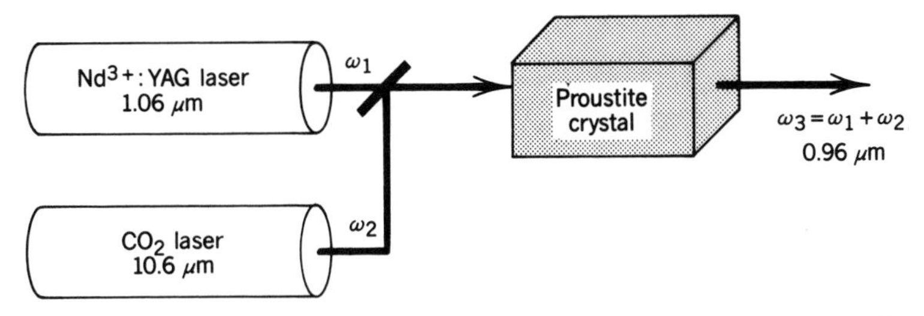

Figure 19.2-5 An example of frequency conversion in a nonlinear crystal.

at five frequencies, 0, $2\omega_1$, $2\omega_2$, $\omega_+ = \omega_1 + \omega_2$, and $\omega_- = \omega_1 - \omega_2$, with amplitudes

$$P_{\mathrm{NL}}(0) = d\left[|E(\omega_1)|^2 + |E(\omega_2)|^2\right] \tag{19.2-11a}$$

$$P_{\mathrm{NL}}(2\omega_1) = dE(\omega_1)E(\omega_1) \tag{19.2-11b}$$

$$P_{\mathrm{NL}}(2\omega_2) = dE(\omega_2)E(\omega_2) \tag{19.2-11c}$$

$$P_{\mathrm{NL}}(\omega_+) = 2\, dE(\omega_1)E(\omega_2) \tag{19.2-11d}$$

$$P_{\mathrm{NL}}(\omega_-) = 2\, dE(\omega_1)E^*(\omega_2). \tag{19.2-11e}$$

Thus the second-order nonlinear medium can be used to mix two optical waves of different frequencies and generate (among other things) a third wave at the difference frequency (down-conversion) or at the sum frequency (up-conversion). An example of frequency up-conversion using a proustite crystal, and two lasers with free-space wavelengths $\lambda_{o1} = 1.06\ \mu\mathrm{m}$ and $\lambda_{o2} = 10.6\ \mu\mathrm{m}$, to generate a wave with wavelength $\lambda_{o3} = 0.96\ \mu\mathrm{m}$ (where $\lambda_{o3}^{-1} = \lambda_{o1}^{-1} + \lambda_{o2}^{-1}$) is illustrated in Fig. 19.2-5.

Although the incident pair of waves at frequencies ω_1 and ω_2 produce polarization densities at frequencies 0, $2\omega_1$, $2\omega_2$, $\omega_1 + \omega_2$, and $\omega_1 - \omega_2$, all of these waves are not necessarily generated, since certain additional conditions (phase matching) must be satisfied, as explained presently.

Phase Matching

If waves 1 and 2 are plane waves with wavevectors $\mathbf{k}_1$ and $\mathbf{k}_2$, so that $E(\omega_1) = A_1 \exp(-j\mathbf{k}_1 \cdot \mathbf{r})$ and $E(\omega_2) = A_2 \exp(-j\mathbf{k}_2 \cdot \mathbf{r})$, then in accordance with (19.2-11d), $P_{\mathrm{NL}}(\omega_3) = 2\, dE(\omega_1)E(\omega_2) = 2\, dA_1A_2 \exp(-j\mathbf{k}_3 \cdot \mathbf{r})$, where

$$\boxed{\omega_3 = \omega_1 + \omega_2} \tag{19.2-12}$$

Frequency-Matching Condition

and

$$\boxed{\mathbf{k}_3 = \mathbf{k}_1 + \mathbf{k}_2.} \tag{19.2-13}$$

Phase-Matching Condition

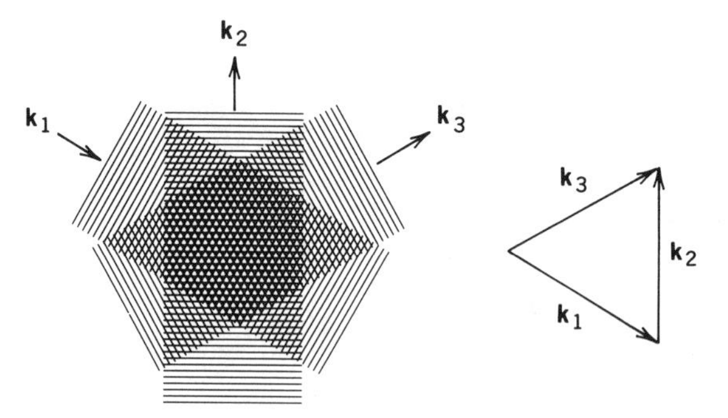

Figure 19.2-6 The phase-matching condition.

The medium therefore acts as a light source of frequency $\omega_3 = \omega_1 + \omega_2$, with a complex amplitude proportional to $\exp(-j\mathbf{k}_3 \cdot \mathbf{r})$, so that it radiates a wave of wavevector $\mathbf{k}_3 = \mathbf{k}_1 + \mathbf{k}_2$, as illustrated in Fig. 19.2-6. Equation (19.2-13) can be regarded as a condition of phase matching among the wavefronts of the three waves that is analogous to the frequency-matching condition $\omega_3 = \omega_1 + \omega_2$. Since the argument of the complex wavefunction is $\omega t - \mathbf{k} \cdot \mathbf{r}$, these two conditions ensure both the temporal and spatial phase matching of the three waves, which is necessary for their sustained mutual interaction over extended durations of time and regions of space.

If the three waves travel in the same direction, for example, the phase-matching condition is replaced by the scalar equation $n\omega_3/c_o = n\omega_1/c_o + n\omega_2/c_o$, which is automatically satisfied since $\omega_3 = \omega_1 + \omega_2$. In this case, frequency matching ensures phase matching. However, since all materials are in reality dispersive, the three waves actually travel at different velocities corresponding to their different refractive indices, n_1, n_2, and n_3. The phase-matching condition is then $n_3\omega_3/c_o = n_1\omega_1/c_o + n_2\omega_2/c_o$, from which we obtain $n_3\omega_3 = n_1\omega_1 + n_2\omega_2$. The phase-matching condition is then independent of the frequency-matching condition $\omega_3 = \omega_1 + \omega_2$; both conditions must be simultaneously satisfied. Precise control of the refractive indices at the three frequencies is often achieved by appropriate selection of the polarization (see Sec. 19.6) and in some cases by control of the temperature.

Three-Wave Mixing

Consider now the case of two optical waves of angular frequencies ω_1 and ω_2 traveling through a second-order nonlinear optical medium. These waves mix and produce a polarization density with components at a number of frequencies. We assume that only the component at the sum frequency $\omega_3 = \omega_1 + \omega_2$ satisfies the phase-matching condition. Other frequencies cannot be sustained by the medium since they are assumed not to satisfy the phase-matching condition.

Once wave 3 is generated, it interacts with wave 1 and generates a wave at the difference frequency $\omega_2 = \omega_3 - \omega_1$. Clearly, the phase-matching condition for this interaction is also satisfied. Waves 3 and 2 similarly combine and radiate at ω_1. The three waves therefore undergo mutual coupling in which each pair of waves interacts and contributes to the third wave. The process is called **three-wave mixing.**

Two-wave mixing is not, in general, possible. Two waves of arbitrary frequencies ω_1 and ω_2 cannot be coupled by the medium without the help of a third wave. Two-wave mixing can occur only in the degenerate case, $\omega_2 = 2\omega_1$, in which the second-harmonic of wave 1 contributes to wave 2; and the subharmonic $\omega_2/2$ of wave 2, which is at the frequency difference $\omega_2 - \omega_1$, contributes to wave 1.

Three-wave mixing is known as a **parametric interaction**. It takes a variety of forms, depending on which of the three waves is provided to the medium externally, and

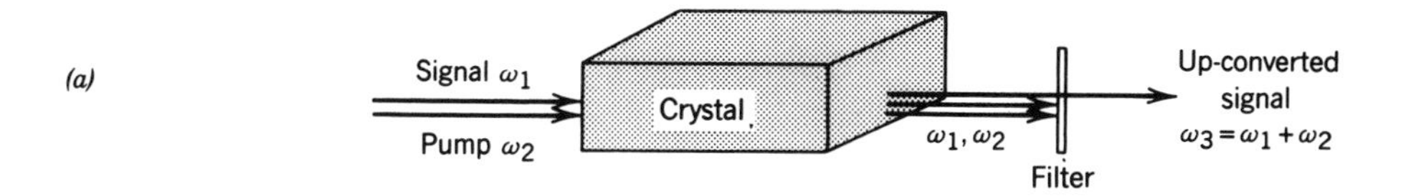

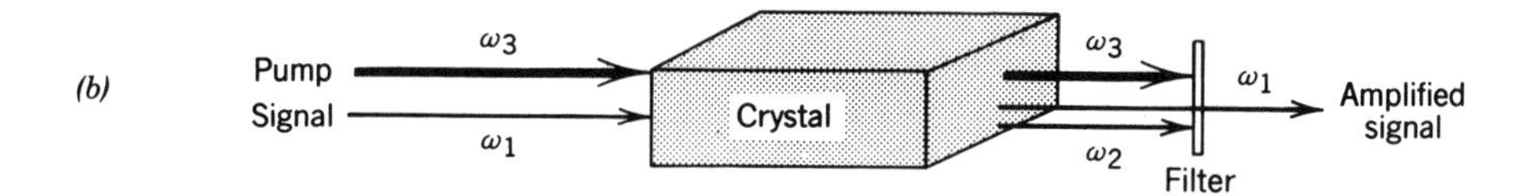

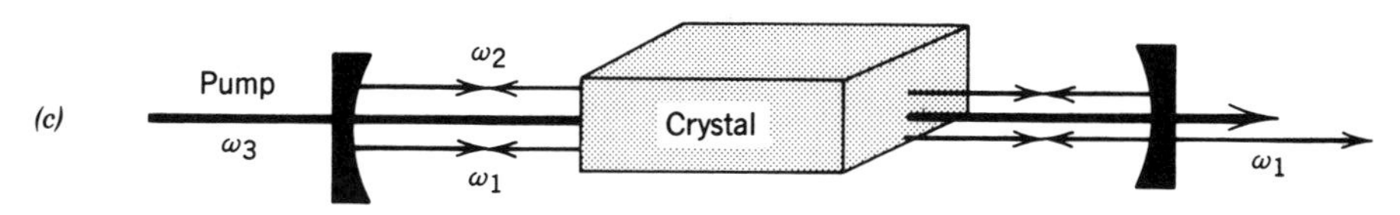

Figure 19.2-7 Optical parametric devices: (*a*) frequency up-converter; (*b*) parametric amplifier; (*c*) parametric oscillator.

which are extracted as outputs. The following examples are illustrated in Fig. 19.2-7:

- Waves 1 and 2 are mixed in an **up-converter**, generating a wave at a higher frequency $\omega_3 = \omega_1 + \omega_2$. This has already been illustrated in Fig. 19.2-5. A **down-converter** is realized by an interaction between waves 3 and 1 to generate wave 2, at the difference frequency $\omega_2 = \omega_3 - \omega_1$.
- Waves 1, 2, and 3 interact so that wave 1 grows. The device operates as an amplifier at frequency ω_1 and is known as a **parametric amplifier**. Wave 3, called the **pump**, provides the required energy, whereas wave 2 is an auxiliary wave known as the **idler** wave. The amplified wave is called the **signal**. Clearly, the gain of the amplifier depends on the power of the pump.
- With proper feedback, the parametric amplifier can operate as a **parametric oscillator**, in which only a pump wave is supplied.

Parametric devices are used for coherent light amplification, for the generation of coherent light at frequencies where no lasers are available (e.g., in the UV band), and for the detection of weak light at wavelengths for which sensitive detectors do not exist. Further details pertaining to the operation of parametric devices are provided in Sec. 19.4.

Wave Mixing as a Photon Interaction Process

The three-wave mixing process can be viewed from a photon optics perspective as a process of three-photon interaction. A photon of frequency ω_1 and wavevector $\mathbf{k}_1$ combines with a photon of frequency ω_2 and wavevector $\mathbf{k}_2$ to form a photon of frequency ω_3 and wavevector $\mathbf{k}_3$, as illustrated in Fig. 19.2-8(a). Since $\hbar\omega$ and $\hbar\mathbf{k}$ are

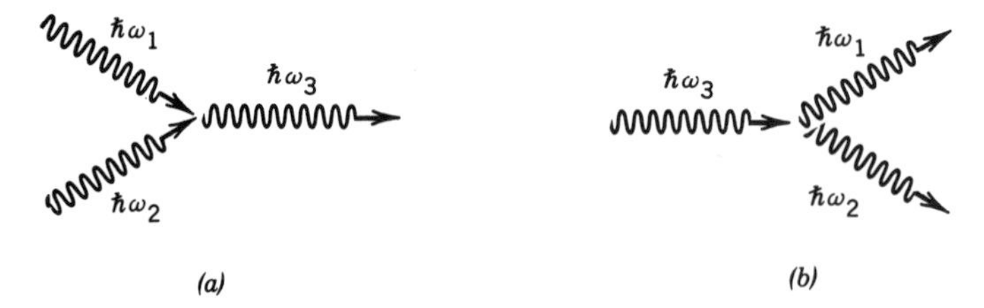

Figure 19.2-8 Mixing of three photons in a second-order nonlinear medium: (*a*) photon combining; (*b*) photon splitting.

the energy and momentum of a photon of frequency ω and wavevector $\mathbf{k}$ (see Sec. 11.1), conservation of energy and momentum require that

$$\hbar\omega_3 = \hbar\omega_1 + \hbar\omega_2 \tag{19.2-14}$$

$$\hbar\mathbf{k}_3 = \hbar\mathbf{k}_1 + \hbar\mathbf{k}_2, \tag{19.2-15}$$

so that the frequency- and phase-matching conditions presented in (19.2-12) and (19.2-13) are reproduced. The process of three-photon mixing may also take the form of a photon of frequency ω_3 splitting into two photons, one of frequency ω_1 and the other of frequency ω_2, as illustrated in Fig. 19.2-8(*b*). The same conditions of conservation of energy and momentum must also be satisfied.

The process of wave mixing involves an energy exchange among the interacting waves. Clearly, energy must be conserved, as is assured by the frequency-matching condition, $\omega_3 = \omega_1 + \omega_2$. Photon numbers must also be conserved, consistent with the photon interaction. Consider the photon-splitting process represented in Fig. 19.2-8(*b*). If $\Delta\Phi_1$, $\Delta\Phi_2$, and $\Delta\Phi_3$ are the net changes in the photon fluxes (photons per second) in the course of the interaction (the flux of photons leaving minus the flux of photons entering) at frequencies ω_1, ω_2, and ω_3, then $\Delta\Phi_1 = \Delta\Phi_2 = -\Delta\Phi_3$, so that for each of the ω_3 photons lost, one each of the ω_1 and ω_2 photons is gained.

If the three waves travel in the same direction, the z direction for example, then by taking a cylinder of unit area and incremental length $\Delta z \to 0$ as the interaction volume, we conclude that the photon flux densities ϕ_1, ϕ_2, ϕ_3 (photons/s-m^2) of the three waves must satisfy

$$\frac{d\phi_1}{dz} = \frac{d\phi_2}{dz} = -\frac{d\phi_3}{dz}. \tag{19.2-16}$$

Photon-Number Conservation

Since the wave intensities (W/m^2) are $I_1 = \hbar\omega_1\phi_1$, $I_2 = \hbar\omega_2\phi_2$, and $I_3 = \hbar\omega_3\phi_3$, (19.2-16) gives

$$\frac{d}{dz}\left(\frac{I_1}{\omega_1}\right) = \frac{d}{dz}\left(\frac{I_2}{\omega_2}\right) = -\frac{d}{dz}\left(\frac{I_3}{\omega_3}\right). \tag{19.2-17}$$

Manley – Rowe Relation

Equation (19.2-17) is known as the Manley–Rowe relation. It was derived in the context of wave interactions in nonlinear electronic systems. The Manley–Rowe relation can be derived using wave optics, without invoking the concept of the photon (see Exercise 19.4-3).

19.3 THIRD-ORDER NONLINEAR OPTICS

In media possessing centrosymmetry, the second-order nonlinear term is absent since the polarization must reverse exactly when the electric field is reversed. The dominant nonlinearity is then of third order,

$$\mathscr{P}_{\text{NL}} = 4\chi^{(3)}\mathscr{E}^3 \tag{19.3-1}$$

(see Fig. 19.3-1) and the material is called a **Kerr medium**. Kerr media respond to optical fields by generating third harmonics and sums and differences of triplets of frequencies.

EXERCISE 19.3-1

Third-Order Nonlinear Optical Media Exhibit the Kerr Electro-Optic Effect. A monochromatic optical field $E(\omega)$ is incident on a third-order nonlinear medium in the presence of a steady electric field $E(0)$. The optical field is much smaller than the electric field, so that $|E(\omega)|^2 \ll |E(0)|^2$. Use (19.3-1) to show that the component of $\mathscr{P}_{\text{NL}}$ of frequency ω is approximately given by $P_{\text{NL}}(\omega) \approx 12\chi^{(3)}E^2(0)E(\omega)$, when terms proportional to $E^2(\omega)$ and $E^3(\omega)$ are neglected. Show that this component of the polarization is equivalent to a refractive-index change $\Delta n = -\frac{1}{2}\mathfrak{s}n^3E^2(0)$, where

$$\mathfrak{s} = -\frac{12}{\epsilon_o n^4}\chi^{(3)}. \tag{19.3-2}$$

The proportionality between the refractive-index change and the squared electric field is the Kerr (quadratic) electro-optic effect described in Sec. 18.1A, where $\mathfrak{s}$ is the Kerr coefficient.

A. Third-Harmonic Generation and Self-Phase Modulation

Third-Harmonic Generation

In accordance with (19.3-1), the response of a third-order nonlinear medium to a monochromatic optical field $\mathscr{E}(t) = \text{Re}\{E(\omega)\exp(j\omega t)\}$ is a nonlinear polarization

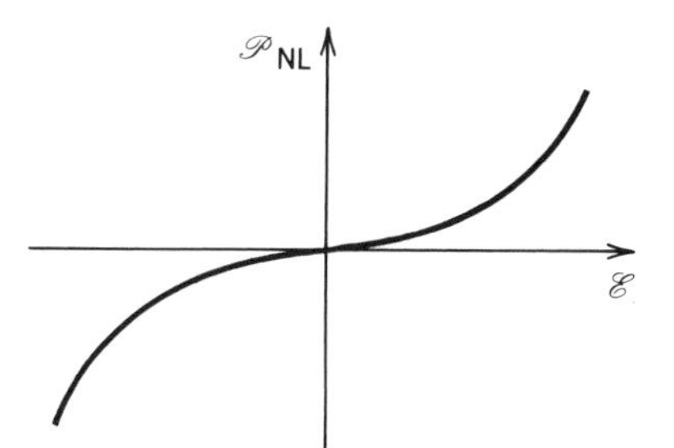

Figure 19.3-1 Third-order nonlinearity.

$\mathscr{P}_{NL}(t)$ containing a component at frequency ω and another at frequency 3ω,

$$P_{NL}(\omega) = 3\chi^{(3)}|E(\omega)|^2 E(\omega) \qquad (19.3\text{-}3a)$$

$$P_{NL}(3\omega) = \chi^{(3)} E^3(\omega). \qquad (19.3\text{-}3b)$$

The presence of a component of polarization at the frequency 3ω indicates that third-harmonic light is generated. However, in most cases the energy conversion efficiency is very low.

Optical Kerr Effect

The polarization component at frequency ω in (19.3-3a) corresponds to an incremental change of the susceptibility $\Delta\chi$ at frequency ω given by

$$\epsilon_o \Delta\chi = \frac{P_{NL}(\omega)}{E(\omega)} = 3\chi^{(3)}|E(\omega)|^2 = 6\chi^{(3)}\eta I,$$

where $I = |E(\omega)|^2/2\eta$ is the optical intensity of the initial wave. Since $n^2 = 1 + \chi$, this is equivalent to an incremental refractive index $\Delta n = (\partial n/\partial\chi)\Delta\chi = \Delta\chi/2n$, so that

$$\Delta n = \frac{3\eta}{\epsilon_o n}\chi^{(3)} I = n_2 I. \qquad (19.3\text{-}4)$$

Thus the change in the refractive index is proportional to the optical intensity. The overall refractive index is therefore a linear function of the optical intensity I,

$$\boxed{n(I) = n + n_2 I,} \qquad (19.3\text{-}5)$$

Optical Kerr Effect

where†

$$n_2 = \frac{3\eta_o}{n^2\epsilon_o}\chi^{(3)}. \qquad (19.3\text{-}6)$$

This effect is known as the **optical Kerr effect** because of its similarity to the electro-optic Kerr effect (for which Δn is proportional to the square of the steady electric field). The optical Kerr effect is a self-induced effect in which the phase velocity of the wave depends on the wave's own intensity.

The order of magnitude of the coefficient n_2 (in units of cm^2/W) is 10^{-16} to 10^{-14} in glasses, 10^{-14} to 10^{-7} in doped glasses, 10^{-10} to 10^{-8} in organic materials, and 10^{-10} to 10^{-2} in semiconductors. It is sensitive to the operating wavelength (see Sec. 19.7) and depends on the polarization.

†Equation (19.3-5) is also written in the alternative form, $n(I) = n + n_2|E|^2/2$ with n_2 differing from (19.3-6) by the factor η.

Self-Phase Modulation

As a result of the optical Kerr effect, an optical wave traveling in a third-order nonlinear medium undergoes **self-phase modulation**. The phase shift incurred by an optical beam of power P and cross-sectional area A, traveling a distance L in the medium, is $\varphi = 2\pi n(I)L/\lambda_o = 2\pi(n + n_2 P/A)L/\lambda_o$, so that it is altered by

$$\Delta\varphi = 2\pi n_2 \frac{L}{\lambda_o A} P, \tag{19.3-7}$$

which is proportional to the optical power P. Self-phase modulation is useful in applications in which light controls light.

To maximize the effect, L should be large and A small. These requirements are well served by the use of optical waveguides. The optical power at which $\Delta\varphi = \pi$ is achieved is $P_\pi = \lambda_o A/2Ln_2$. A doped-glass fiber of length $L = 1$ m, cross-sectional area $A = 10^{-2}$ mm^2, and $n_2 = 10^{-10}$ cm^2/W, operating at $\lambda_o = 1$ μm, for example, switches the phase by a factor of π at an optical power $P_\pi = 0.5$ W. Materials with larger values of n_2 can be used in centimeter-long channel waveguides to achieve a phase shift of π at powers of a few mW.

Phase modulation may be converted into intensity modulation by employing one of the schemes used in electro-optic modulators (see Sec. 18.1B): (1) using an interferometer (Mach–Zehnder, for example); (2) using the difference between the modulated phases of the two polarization components (birefringence) as a wave retarder placed between crossed polarizers; or (3) using an integrated-optic directional coupler (Sec. 7.4B). The result is an all-optical modulator in which a weak optical beam may be controlled by an intense optical beam. All-optical switches are discussed in Sec. 21.2.

Self-Focusing

Another interesting effect associated with self-phase modulation is **self-focusing**. If an intense optical beam is transmitted through a thin sheet of nonlinear material exhibiting the optical Kerr effect, as illustrated in Fig. 19.3-2, the refractive-index change maps the intensity pattern in the transverse plane. If the beam has its highest intensity at the center, for example, the maximum change of the refractive index is also at the center. The sheet then acts as a graded-index medium that imparts to the wave a nonuniform phase shift, thereby causing wavefront curvature. Under certain conditions the medium can act as a lens with a power-dependent focal length, as shown in Exercise 19.3-2.

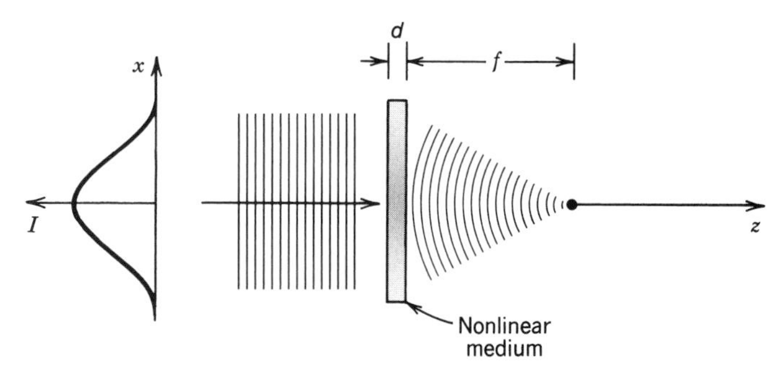

Figure 19.3-2 A third-order nonlinear medium acts as a lens whose focusing power depends on the intensity of the incident beam.

EXERCISE 19.3-2

Optical Kerr Lens. An optical beam traveling in the z direction is transmitted through a thin sheet of nonlinear optical material exhibiting the optical Kerr effect, $n(I) = n + n_2 I$. The sheet lies in the x-y plane and has a small thickness d so that its complex amplitude transmittance is $\exp(-jnk_o d)$. The beam has an approximately planar wavefront and an intensity distribution $I \approx I_0[1 - (x^2 + y^2)/W^2]$ at points near the beam axis ($x, y \ll W$), where I_0 is the peak intensity and W is the beam width. Show that the medium acts as a thin lens with a focal length inversely proportional to I_0. *Hint:* A lens of focal length f has a complex amplitude transmittance proportional to $\exp[jk_o(x^2 + y^2)/2f]$, as shown in (2.4-6); see also Exercise 2.4-6 on page 63.

Spatial Solitons

When an intense optical beam travels through a substantial thickness of nonlinear homogeneous medium, instead of a thin sheet, the refractive index is altered nonuniformly so that the medium can act as a graded-index waveguide. Thus the beam can create its own waveguide. If the intensity of the beam has the same spatial distribution in the transverse plane as one of the modes of the waveguide that the beam itself creates, the beam propagates self-consistently without changing its spatial distribution. Under such conditions, diffraction is compensated by the nonlinear effect, and the beam is confined to its self-created waveguide. Such self-guided beams are called **spatial solitons**.

The self-guiding of light in an optical Kerr medium is described mathematically by the Helmholtz equation, $[\nabla^2 + n^2(I)k_o^2]E = 0$, where $n(I) = n + n_2 I$, $k_o = \omega/c_o$, and $I = |E|^2/2\eta$. This is a nonlinear differential equation in E, which is simplified by writing $E = A\exp(-jkz)$, where $k = nk_o$, and assuming that the envelope $A = A(x, z)$ varies slowly in the z direction (in comparison with the wavelength $\lambda = 2\pi/k$) and does not vary in the y direction (see Sec. 2.2C). Using the approximation $(\partial^2/\partial z^2)[A\exp(-jkz)] \approx (-2jk\,\partial A/\partial z - k^2 A)\exp(-jkz)$, the Helmholtz equation becomes

$$\frac{\partial^2 A}{\partial x^2} - 2jk\frac{\partial A}{\partial z} + k_o^2\left[n^2(I) - n^2\right]A = 0. \tag{19.3-8}$$

Since the nonlinear effect is small ($n_2 I \ll n$), we write

$$\left[n^2(I) - n^2\right] = \left[n(I) - n\right]\left[n(I) + n\right] \approx \left[n_2 I\right]\left[2n\right] = \frac{2n_2 n|A|^2}{2\eta} = \frac{n^2 n_2}{\eta_o}|A|^2,$$

so that (19.3-8) becomes

$$\frac{\partial^2 A}{\partial x^2} + \frac{n_2}{\eta_o}k^2|A|^2 A = 2jk\frac{\partial A}{\partial z}. \tag{19.3-9}$$

Equation (19.3-9) is the nonlinear Schrödinger equation. One of its solutions is

$$A(x, z) = A_0 \operatorname{sech}\left(\frac{x}{W_0}\right)\exp\left(-j\frac{z}{4z_0}\right), \tag{19.3-10}$$

Spatial Soliton

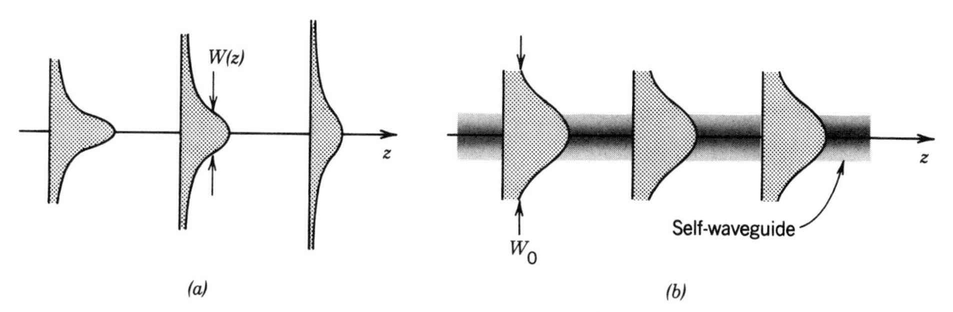

Figure 19.3-3 Comparison between (*a*) a Gaussian beam traveling in a linear medium, and (*b*) a spatial soliton (self-guided optical beam) traveling in a nonlinear medium.

where W_0 is a constant, sech$(\cdot)$ is the hyperbolic-secant function, A_0 satisfies $n_2(A_0^2/2\eta_o) = 1/k^2W_0^2$ and $z_0 = \frac{1}{2}kW_0^2 = \pi W_0^2/\lambda$ is the Rayleigh range [see (3.1-21)]. The intensity distribution

$$I(x,z) = \frac{|A(x,z)|^2}{2\eta} = \frac{A_0^2}{2\eta}\,\text{sech}^2\!\left(\frac{x}{W_0}\right)$$

is independent of z and has a width W_0, as illustrated in Fig. 19.3-3. The distribution in (19.3-10) is the mode of a graded-index waveguide with a refractive index $n + n_2 I = n[1 + (1/k^2W_0^2)\,\text{sech}^2(x/W_0)]$, so that self-consistency is assured. Since $E = A\exp(-jkz)$, the wave travels with a propagation constant $k + 1/4z_0 = k(1 + \lambda^2/8\pi^2W_0^2)$ and phase velocity $c/(1 + \lambda^2/8\pi^2W_0^2)$. The velocity is smaller than c for localized beams (small W_0) but approaches c for large W_0.

Raman Gain

The nonlinear coefficient $\chi^{(3)}$ is in general complex-valued, $\chi^{(3)} = \chi_R^{(3)} + j\chi_I^{(3)}$. The self-phase modulation in (19.3-7),

$$\Delta\varphi = 2\pi n_2 \frac{L}{\lambda_o A} P = \frac{6\pi\eta_o}{\epsilon_o}\frac{\chi^{(3)}}{n^2}\frac{L}{\lambda_o A}P,$$

is therefore also complex, so that the propagation phase factor $\exp(-j\varphi)$ is a combination of phase shift, $\Delta\varphi = (6\pi\eta_o/\epsilon_o)(\chi_R^{(3)}/n^2)(L/\lambda_o A)P$, and gain $\exp(\frac{1}{2}\gamma L)$, with a gain coefficient

$$\gamma = \frac{12\pi\eta_o}{\epsilon_o}\frac{\chi_I^{(3)}}{n^2}\frac{1}{\lambda_o A}P \qquad (19.3\text{-}11)$$

Raman Gain Coefficient

that is proportional to the optical power P. This effect, called **Raman gain**, has its origin in the coupling of light to the high-frequency vibrational modes of the medium, which act as an energy source providing the gain. For low-loss media, the Raman gain may exceed the loss at reasonable levels of power, so that the medium can act as an optical amplifier. With proper feedback, the amplifier can be made into a laser. This effect is exhibited in low-loss optical fibers. Fiber Raman lasers have been demonstrated.

B. Four-Wave Mixing

We have so far examined the response of a third-order nonlinear medium to a single monochromatic wave. In Exercise 19.3-3, the response to a superposition of two waves is explored, and in the remainder of this section the process of four-wave mixing is discussed.

EXERCISE 19.3-3

Two-Wave Mixing. Examine the response of a third-order nonlinear medium to an optical field comprising two monochromatic waves of angular frequencies ω_1 and ω_2, $\mathscr{E}(t) = \mathrm{Re}\{E(\omega_1)\exp(j\omega_1 t)\} + \mathrm{Re}\{E(\omega_2)\exp(j\omega_2 t)\}$. Determine the components $P_{NL}(\omega_1)$ and $P_{NL}(\omega_2)$ of the polarization density, showing that the two waves can be mutually coupled in a two-wave mixing process without the aid of other auxiliary waves. As we have seen in Sec. 19.2C, two-wave mixing is not possible in a second-order nonlinear medium (except in the degenerate case). The process of two-wave mixing in photorefractive media is illustrated in Fig. 18.4-3.

Three-wave mixing is generally not possible in a third-order nonlinear medium. Three waves of distinct frequencies ω_1, ω_2, and ω_3 cannot be coupled by the system without the help of an auxiliary fourth wave. For example, there is generally no contribution to the component $P_{NL}(\omega_1)$ by waves 2 and 3, except in degenerate cases (e.g., when $\omega_1 = 2\omega_3 - \omega_2$).

We now examine the case of **four-wave mixing** in a third-order nonlinear medium. We begin by determining the response of the medium to a superposition of three waves of angular frequencies ω_1, ω_2, and ω_3, with field

$$\mathscr{E}(t) = \mathrm{Re}\{E(\omega_1)\exp(j\omega_1 t)\} + \mathrm{Re}\{E(\omega_2)\exp(j\omega_2 t)\} + \mathrm{Re}\{E(\omega_3)\exp(j\omega_3 t)\}.$$

It is convenient to write $\mathscr{E}(t)$ as a sum of six terms

$$\mathscr{E}(t) = \sum_{q=\pm 1, \pm 2, \pm 3} \tfrac{1}{2} E(\omega_q)\exp(j\omega_q t), \tag{19.3-12}$$

where $\omega_{-q} = -\omega_q$ and $E(-\omega_q) = E^*(\omega_q)$. Substituting (19.3-12) into (19.3-1), we write $\mathscr{P}_{NL}$ as a sum of $6^3 = 216$ terms,

$$\mathscr{P}_{NL}(t) = \tfrac{1}{2}\chi^{(3)} \sum_{q,r,l=\pm 1, \pm 2, \pm 3} E(\omega_q)E(\omega_r)E(\omega_l)\exp\left[j(\omega_q + \omega_r + \omega_l)t\right]. \tag{19.3-13}$$

Thus $\mathscr{P}_{NL}$ is the sum of harmonic components of frequencies $\omega_1, \ldots, 3\omega_1, \ldots, 2\omega_1 \pm \omega_2, \ldots, \pm\omega_1 \pm \omega_2 \pm \omega_3$. The amplitude $P_{NL}(\omega_q + \omega_r + \omega_l)$ of the component of frequency $\omega_q + \omega_r + \omega_l$ can be determined by adding appropriate permutations of q, r, and l in (19.3-13). For example, $P_{NL}(\omega_3 + \omega_4 - \omega_1)$ involves six permutations,

$$P_{NL}(\omega_3 + \omega_4 - \omega_1) = 6\chi^{(3)}E(\omega_3)E(\omega_4)E^*(\omega_1). \tag{19.3-14}$$

EXERCISE 19.3-4

Optical Kerr Effect in the Presence of Three Waves. Three monochromatic waves with frequencies ω_1, ω_2, and ω_3 travel in a third-order nonlinear medium. Determine the complex amplitude of the component of $\mathscr{P}_{\mathrm{NL}}(t)$ in (19.3-13) at frequency ω_1. Show that this wave travels with a velocity $c_o/(n + n_2 I)$, where

$$n_2 = \frac{3\eta_o}{\epsilon_o n^2}\chi^{(3)}, \qquad (19.3\text{-}15)$$

and $I = I_1 + 2I_2 + 2I_3$, with $I_l = |E(\omega_l)|^2/2\eta$, $l = 1, 2, 3$. This effect is similar to the optical Kerr effect discussed earlier.

Equation (19.3-14) indicates that four waves of frequencies ω_1, ω_2, ω_3, and ω_4 are mixed by the medium if $\omega_2 = \omega_3 + \omega_4 - \omega_1$, or

$$\boxed{\omega_3 + \omega_4 = \omega_1 + \omega_2.} \qquad (19.3\text{-}16)$$

Frequency-Matching Condition

This equation constitutes the frequency-matching condition for four-wave mixing.

Assuming that waves 1, 3, and 4 are plane waves of wavevectors $\mathbf{k}_1$, $\mathbf{k}_3$, and $\mathbf{k}_4$, so that $E(\omega_q) \propto \exp(-j\mathbf{k}_q \cdot \mathbf{r})$, $q = 1, 3, 4$, then (19.3-14) gives

$$P_{\mathrm{NL}}(\omega_2) \propto \exp(-j\mathbf{k}_3 \cdot \mathbf{r}) \exp(-j\mathbf{k}_4 \cdot \mathbf{r}) \exp(j\mathbf{k}_1 \cdot \mathbf{r}) = \exp[-j(\mathbf{k}_3 + \mathbf{k}_4 - \mathbf{k}_1) \cdot \mathbf{r}],$$

so that wave 2 is also a plane wave with wavevector $\mathbf{k}_2 = \mathbf{k}_3 + \mathbf{k}_4 - \mathbf{k}_1$, from which

$$\boxed{\mathbf{k}_3 + \mathbf{k}_4 = \mathbf{k}_1 + \mathbf{k}_2.} \qquad (19.3\text{-}17)$$

Phase-Matching Condition

Equation (19.3-17) is the phase-matching condition for four-wave mixing.

The four-wave mixing process may also be interpreted as an interaction between four photons. A photon of frequency ω_3 combines with a photon of frequency ω_4 to produce a photon of frequency ω_1 and another of frequency ω_2, as illustrated in Fig. 19.3-4. Equations (19.3-16) and (19.3-17) represent conservation of energy and momentum, respectively.

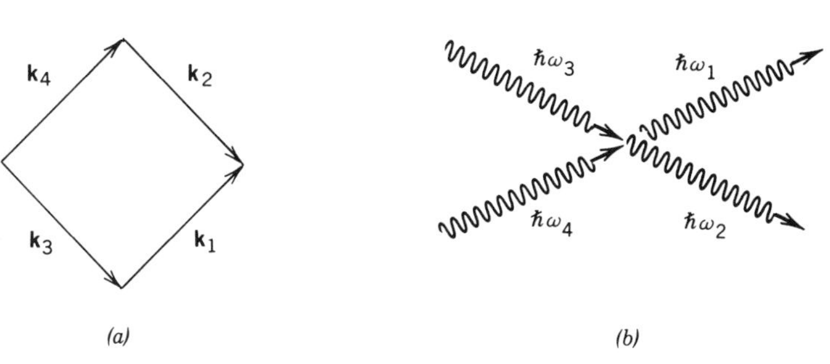

Figure 19.3-4 Four-wave mixing: (*a*) the phase-matching condition; (*b*) interaction of four photons.

C. Optical Phase Conjugation

The frequency-matching condition (19.3-16) is satisfied when all four waves are of the same frequency.

$$\omega_1 = \omega_2 = \omega_3 = \omega_4 = \omega. \qquad (19.3\text{-}18)$$

The process is then called **degenerate four-wave mixing**. Assuming further that two of the waves (waves 3 and 4) are uniform plane waves traveling in opposite directions,

$$E_3(\mathbf{r}) = A_3 \exp(-j\mathbf{k}_3 \cdot \mathbf{r}), \qquad E_4(\mathbf{r}) = A_4 \exp(-j\mathbf{k}_4 \cdot \mathbf{r}), \qquad (19.3\text{-}19)$$

with

$$\mathbf{k}_4 = -\mathbf{k}_3, \qquad (19.3\text{-}20)$$

and substituting (19.3-19) and (19.3-20) into (19.3-14), we see that the polarization density of wave 2 is $6\chi^{(3)} A_3 A_4 E_1^*(\mathbf{r})$. This term corresponds to a source emitting an optical wave (wave 2) of complex amplitude

$$E_2(\mathbf{r}) \propto A_3 A_4 E_1^*(\mathbf{r}). \qquad (19.3\text{-}21)$$

Phase Conjugation

Since A_3 and A_4 are constants, wave 2 is proportional to a conjugated version of wave 1. The device serves as a **phase conjugator**. Waves 3 and 4 are called the **pump** waves and waves 1 and 2 are called the **probe** and **conjugate** waves, respectively. As will be demonstrated shortly, the conjugate wave is identical to the probe wave except that it travels in the opposite direction. The phase conjugator is a special mirror that reflects the wave back onto itself without altering its wavefronts.

To understand the phase conjugation process consider two simple examples:

EXAMPLE 19.3-1. ***Conjugate of a Plane Wave.*** If wave 1 is a uniform plane wave, $E_1(\mathbf{r}) = A_1 \exp(-j\mathbf{k}_1 \cdot \mathbf{r})$, traveling in the direction $\mathbf{k}_1$, then $E_2(\mathbf{r}) = A_1^* \exp(j\mathbf{k}_1 \cdot \mathbf{r})$ is a uniform plane wave traveling in the opposite direction $\mathbf{k}_2 = -\mathbf{k}_1$, as illustrated in Fig. 19.3-5(*b*). Thus the phase-matching condition (19.3-17) is satisfied. The medium acts as a special "mirror" that reflects the incident plane wave back onto itself, no matter what the angle of incidence is.

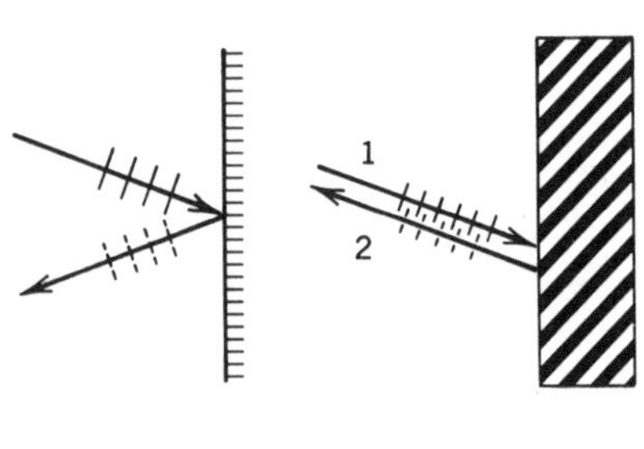

Figure 19.3-5 Reflection of a plane wave from (*a*) an ordinary mirror and (*b*) a phase conjugate mirror.

EXAMPLE 19.3-2. ***Conjugate of a Spherical Wave.*** If wave 1 is a spherical wave centered about the origin $\mathbf{r} = \mathbf{0}$, $E_1(\mathbf{r}) \propto (1/r)\exp(-jkr)$, then wave 2 has complex amplitude $E_2(\mathbf{r}) \propto (1/r)\exp(+jkr)$. This is a spherical wave traveling backward and converging toward the origin, as illustrated in Fig. 19.3-6(*b*).

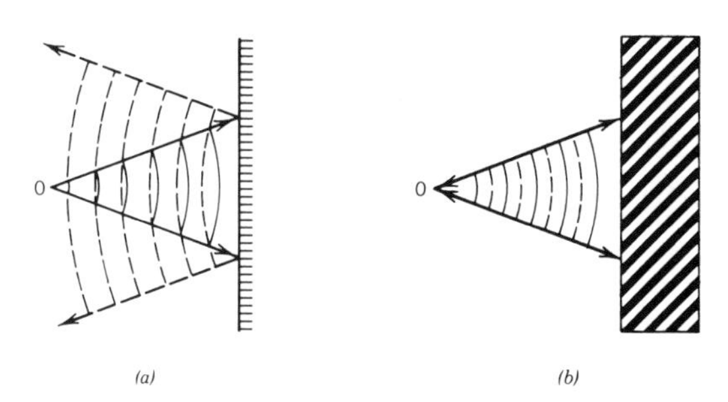

Figure 19.3-6 Reflection of a spherical wave from (*a*) an ordinary mirror and (*b*) a phase conjugate mirror.

Since an arbitrary probe wave may be regarded as a superposition of plane waves (see Chap. 4), each of which is reflected onto itself by the conjugator, the conjugate wave is identical to the incident wave everywhere, except for a reversed direction of propagation. The conjugate wave retraces the original wave by propagating backward, maintaining the same wavefronts.

Phase conjugation is analogous to time reversal. This may be understood by examining the field of the conjugate wave $\mathscr{E}_2(\mathbf{r}, t) = \mathrm{Re}\{E_2(\mathbf{r})\exp(j\omega t)\} \propto \mathrm{Re}\{E_1^*(\mathbf{r})\exp(j\omega t)\}$. Since the real part of a complex number equals the real part of its complex conjugate, $\mathscr{E}_2(\mathbf{r}, t) \propto \mathrm{Re}\{E_1(\mathbf{r})\exp(-j\omega t)\}$. Comparing this to the field of the probe wave $\mathscr{E}_1(\mathbf{r}, t) = \mathrm{Re}\{E_1(\mathbf{r})\exp(j\omega t)\}$, we readily see that one is obtained from the other by the transformation $t \to -t$, so that the conjugate wave appears as a time-reversed version of the probe wave.

The conjugate wave may carry more power than the probe wave. This can be seen by observing that the intensity of the conjugate wave (wave 2) is proportional to the product of the intensities of the pump waves 3 and 4 [see (19.3-21)]. When the powers of the pump waves are increased so that the conjugate wave (wave 2) carries more power than the probe wave (wave 1), the medium acts as an "amplifying mirror." An example of an optical setup for demonstrating phase conjugation is shown in Fig. 19.3-7.

Degenerate Four-Wave Mixing as a Form of Real-Time Holography

The degenerate four-wave mixing process is analogous to volume holography (see Sec. 4.5). Holography is a two-step process in which the interference pattern formed by the superposition of an object wave E_1 and a reference wave E_3 is recorded in a photographic emulsion. Another reference wave E_4 is subsequently transmitted through or reflected from the emulsion, creating the conjugate of the object wave $E_2 \propto E_4 E_3 E_1^*$, or its replica $E_2 \propto E_4 E_1 E_3^*$, depending on the geometry [see Fig. 4.5-10(*a*) and (*b*)]. The nonlinear medium permits a real-time simultaneous holographic recording and reconstruction process. This process occurs in both the Kerr medium and the photorefractive medium (see Sec. 18.4).

When four waves are mixed in a nonlinear medium, each pair of waves interferes and creates a grating, from which a third wave is reflected to produce the fourth wave. The roles of reference and object are exchanged among the four waves, so that there are two types of gratings as illustrated in Fig. 19.3-8. Consider first the process

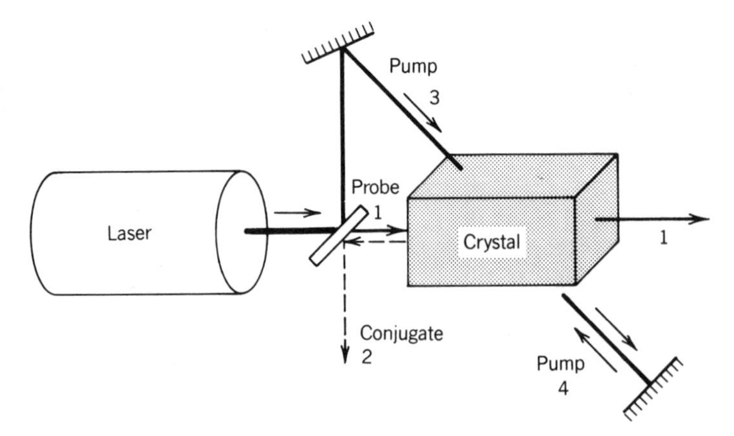

Figure 19.3-7 An optical system for degenerate four-wave mixing using a nonlinear crystal. The pump waves 3 and 4, and the probe wave 1 are obtained from a laser using a beamsplitter and two mirrors. The conjugate wave 2 is created within the crystal.

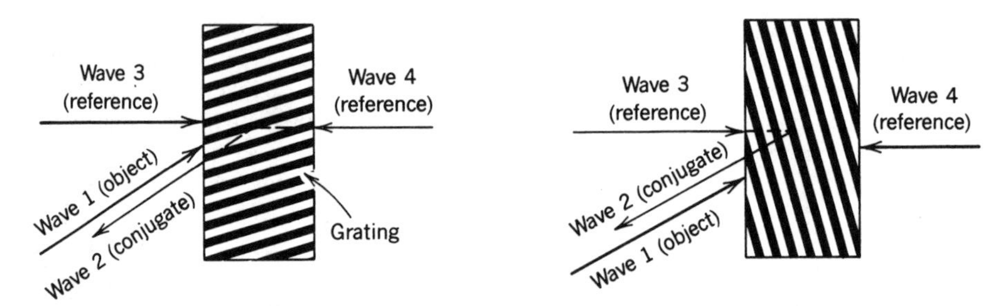

Figure 19.3-8 Four-wave mixing in a nonlinear medium. A reference and object wave interfere and create a grating from which the second reference wave reflects and produces a conjugate wave. There are two possibilities corresponding to (*a*) transmission and (*b*) reflection gratings.

illustrated in Fig. 19.3-8(*a*) [see also Fig. 4.5-10(*a*)]. Assume that the two reference waves (denoted as waves 3 and 4) are counter-propagating plane waves. The two steps of holography are:

Step 1. The object wave 1 is added to the reference wave 3 and the intensity of their sum is recorded in the medium in the form of a volume grating (hologram).

Step 2. The reconstruction reference wave 4 is Bragg reflected from the grating to create the conjugate wave (wave 2).

This grating is called the transmission grating.

The second possibility, illustrated in Fig. 19.3-8(*b*) is for the reference wave 4 to interfere with the object wave 1 and create a grating, called the reflection grating, from which the second reference wave 3 is reflected to create the conjugate wave 2. These two gratings can exist together but they usually have different efficiencies.

In summary, four-wave mixing can provide a means for real-time holography and phase conjugation, which have a number of applications in optical signal processing.

Use of Phase Conjugators in Wave Restoration

The ability to reflect a wave onto itself so that it retraces its path in the opposite direction suggests a number of useful applications, including the removal of wavefront aberrations. The idea is based on the principle of reciprocity, illustrated in Fig. 19.3-9. Rays traveling through a linear optical medium from left to right follow the same path if they reverse and travel back in the opposite direction. The same principle applies to waves.

If the wavefront of an optical beam is distorted by an aberrating medium, the original wave can be restored by use of a conjugator which reflects the beam onto itself and transmits it once more through the same medium, as illustrated in Fig. 19.3-10.

One important application is in optical resonators (see Chap. 9). If the resonator contains an aberrating medium, replacing one of the mirrors with a conjugate mirror ensures that the distortion is removed in each round trip, so that the resonator modes have undistorted wavefronts transmitted through the ordinary mirror, as illustrated in Fig. 19.3-11.

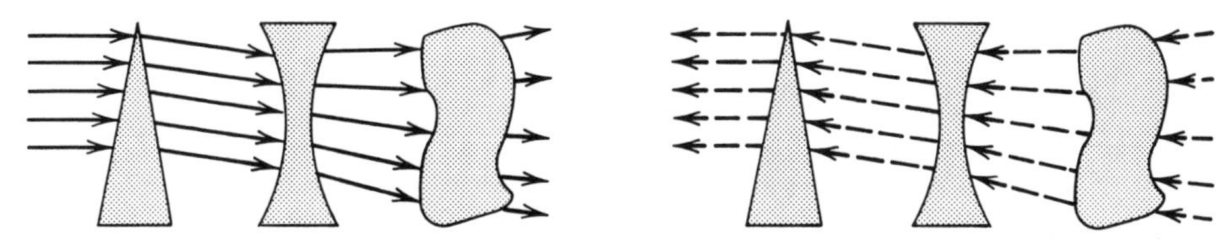

Figure 19.3-9 Optical reciprocity.

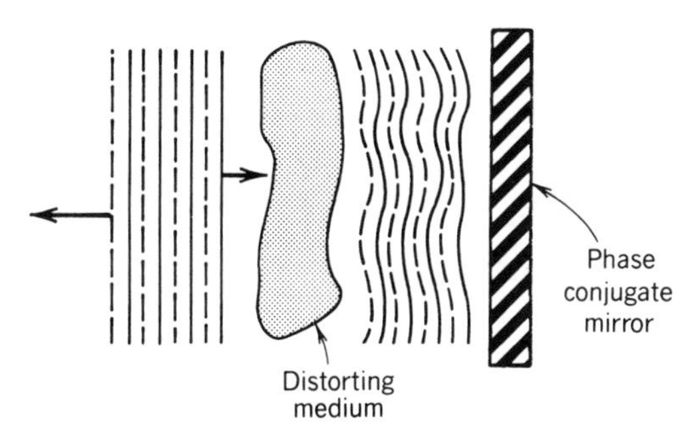

Figure 19.3-10 A phase conjugate mirror reflects a distorted wave onto itself, so that when it retraces its path, the distortion is compensated.

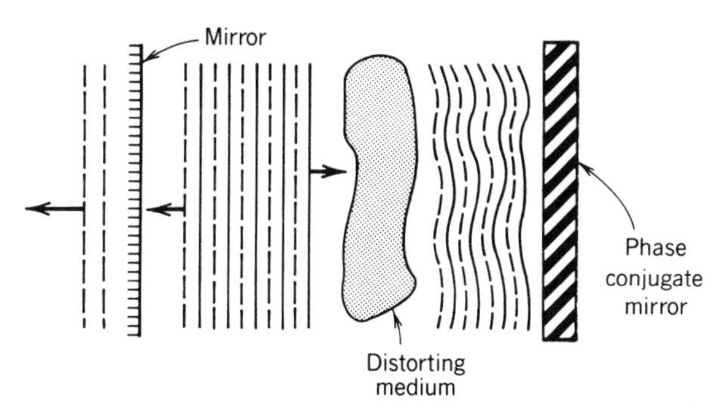

Figure 19.3-11 An optical resonator with an ordinary mirror and a phase conjugate mirror.

*19.4 COUPLED-WAVE THEORY OF THREE-WAVE MIXING

A quantitative analysis of the process of three-wave mixing in a second-order nonlinear optical medium is provided in this section using a coupled-wave theory. To simplify the analysis, the dispersive and anisotropic effects are not fully accounted for.

Coupled-Wave Equations

Wave propagation in a second-order nonlinear medium is governed by the basic wave equation

$$\nabla^2 \mathscr{E} - \frac{1}{c^2}\frac{\partial^2 \mathscr{E}}{\partial t^2} = -\mathscr{S}, \tag{19.4-1}$$

where

$$\mathscr{S} = -\mu_o \frac{\partial^2 \mathscr{P}_{\mathrm{NL}}}{\partial t^2}, \tag{19.4-2}$$

is regarded as a radiation source, and

$$\mathscr{P}_{\mathrm{NL}} = 2\, d \mathscr{E}^2 \tag{19.4-3}$$

is the nonlinear component of the polarization density.

The field $\mathscr{E}(t)$ is a superposition of three waves of angular frequencies ω_1, ω_2, and ω_3 and complex amplitudes E_1, E_2, and E_3, respectively,

$$\begin{aligned}\mathscr{E}(t) &= \sum_{q=1,2,3} \mathrm{Re}\left[E_q \exp(j\omega_q t)\right] \\ &= \sum_{q=1,2,3} \tfrac{1}{2}\left[E_q \exp(j\omega_q t) + E_q^* \exp(-j\omega_q t)\right].\end{aligned} \tag{19.4-4}$$

It is convenient to rewrite (19.4-4) in the compact form

$$\mathscr{E}(t) = \sum_{q=\pm 1, \pm 2, \pm 3} \tfrac{1}{2} E_q \exp(j\omega_q t), \tag{19.4-5}$$

where $\omega_{-q} = -\omega_q$ and $E_{-q} = E_q^*$. The corresponding polarization density obtained by substituting into (19.4-3) is a sum of 36 terms

$$\mathscr{P}_{\mathrm{NL}}(t) = \tfrac{1}{2} d \sum_{q,r=\pm 1, \pm 2, \pm 3} E_q E_r \exp\left[j(\omega_q + \omega_r)t\right], \tag{19.4-6}$$

and the corresponding radiation source,

$$\mathscr{S} = \tfrac{1}{2}\mu_o d \sum_{q,r=\pm 1, \pm 2, \pm 3} (\omega_q + \omega_r)^2 E_q E_r \exp\left[j(\omega_q + \omega_r)t\right], \tag{19.4-7}$$

is the sum of harmonic components of frequencies that are sums and differences of the original frequencies ω_1, ω_2, and ω_3.

Substituting (19.4-5) and (19.4-7) into the wave equation (19.4-1), we obtain a single differential equation with many terms, each of which is a harmonic function of some frequency. If the frequencies ω_1, ω_2, and ω_3 are distinct, we can separate this equation into three differential equations by equating terms on both sides of (19.4-1) at each of the frequencies ω_1, ω_2, and ω_3, separately. The result is cast in the form of three Helmholtz equations with sources,

$$(\nabla^2 + k_1^2)E_1 = -S_1 \tag{19.4-8a}$$

$$(\nabla^2 + k_2^2)E_2 = -S_2 \tag{19.4-8b}$$

$$(\nabla^2 + k_3^2)E_3 = -S_3, \tag{19.4-8c}$$

where S_q is the amplitude of the component of $\mathscr{S}$ with frequency ω_q and $k_q = n\omega_q/c_o$, $q = 1, 2, 3$. Each of the complex amplitudes of the three waves satisfies the Helmholtz equation with a source equal to the component of $\mathscr{S}$ at its frequency. Under certain conditions, the source for one wave depends on the electric fields of the other two waves, so that the three waves are coupled.

In the absence of nonlinearity, $\mathscr{d} = 0$ and the source term $\mathscr{S}$ vanishes so that each of the three waves satisfies the Helmholtz equation independently of the other two, as is expected in linear optics.

If the frequencies ω_1, ω_2, and ω_3 are not commensurate (one frequency is not the sum or difference of the other two, and one frequency is not twice another), then the source term $\mathscr{S}$ does not contain any components of frequencies ω_1, ω_2, or ω_3. The components S_1, S_2, and S_3 then vanish and the three waves do not interact.

For the three waves to be coupled by the medium, their frequencies must be commensurate. Assume, for example, that one frequency is the sum of the other two,

$$\boxed{\omega_3 = \omega_1 + \omega_2.} \tag{19.4-9}$$

Frequency-Matching Condition

The source $\mathscr{S}$ then contains components at the frequencies ω_1, ω_2, and ω_3. Examining the 36 terms of (19.4-7) we obtain

$$S_1 = 2\mu_o\omega_1^2\,\mathscr{d}E_3E_2^*$$

$$S_2 = 2\mu_o\omega_2^2\,\mathscr{d}E_3E_1^*$$

$$S_3 = 2\mu_o\omega_3^2\,\mathscr{d}E_1E_2.$$

The source for wave 1 is proportional to $E_3E_2^*$ (since $\omega_1 = \omega_3 - \omega_2$), so that waves 2 and 3 together contribute to the growth of wave 1. Similarly, the source for wave 3 is proportional to E_1E_2 (since $\omega_3 = \omega_1 + \omega_2$), so that waves 1 and 2 combine to amplify wave 3, and so on. The three waves are thus coupled or "mixed" by the medium in a process described by three coupled differential equations in E_1, E_2, and E_3,

$$(\nabla^2 + k_1^2)E_1 = -2\mu_o\omega_1^2\,\mathscr{d}E_3E_2^* \tag{19.4-10a}$$

$$(\nabla^2 + k_2^2)E_2 = -2\mu_o\omega_2^2\,\mathscr{d}E_3E_1^* \tag{19.4-10b}$$

$$(\nabla^2 + k_3^2)E_3 = -2\mu_o\omega_3^2\,\mathscr{d}E_1E_2. \tag{10.4-10c}$$

Three-Wave Mixing Coupled Equations

EXERCISE 19.4-1

Degenerate Three-Wave Mixing. Equations (19.4-10) are valid only when the frequencies ω_1, ω_2, and ω_3 are distinct. Consider now the degenerate case for which $\omega_1 = \omega_2 = \omega$ and $\omega_3 = 2\omega$, so that there are two, instead of three, waves with amplitudes E_1 and E_3. Show that these waves satisfy the Helmholtz equation with sources

$$S_1 = 2\mu_o\omega_1^2\, \mathscr{d} E_3 E_1^*$$

$$S_3 = \mu_o\omega_3^2\, \mathscr{d} E_1 E_1,$$

so that the coupled wave equations are

$$\left(\nabla^2 + k_1^2\right) E_1 = -2\mu_o\omega_1^2\, \mathscr{d} E_3 E_1^* \qquad \text{(19.4-11a)}$$

$$\left(\nabla^2 + k_3^2\right) E_3 = -\mu_o\omega_3^2\, \mathscr{d} E_1 E_1. \qquad \text{(19.4-11b)}$$

Note that these equations are not obtained from the three-wave-mixing equations (19.4-10) by substituting $E_1 = E_2$ [the factor of 2 is absent in (19.4-11b)].

Mixing of Three Collinear Uniform Plane Waves

Assume that the three waves are plane waves traveling in the z direction with complex amplitudes $E_q = A_q \exp(-jk_q z)$, complex envelopes A_q, and wavenumbers $k_q = \omega_q/c$, $q = 1, 2, 3$. It is convenient to normalize the complex envelopes by defining the variables $a_q = A_q/(2\eta\hbar\omega_q)^{1/2}$, where $\eta = \eta_o/n$ is the impedance of the medium, $\eta_o = (\mu_o/\epsilon_o)^{1/2}$ is the impedance of free space, and $\hbar\omega_q$ is the energy of a photon of angular frequency ω_q. Thus

$$E_q = \left(2\eta\hbar\omega_q\right)^{1/2} a_q \exp\left(-jk_q z\right), \qquad q = 1, 2, 3, \qquad \text{(19.4-12)}$$

and the intensities of the three waves are $I_q = |E_q|^2/2\eta = \hbar\omega_q |a_q|^2$. The photon flux densities (photons/s-m^2) associated with these waves are

$$\phi_q = \frac{I_q}{\hbar\omega_q} = |a_q|^2. \qquad \text{(19.4-13)}$$

The variable a_q therefore represents the complex envelope of wave q, scaled such that $|a_q|^2$ is the photon flux density. This scaling is convenient since the process of wave mixing must be governed by photon-number conservation (see Sec. 19.2C).

As a result of the interaction between the three waves, the complex envelopes a_q vary with z so that $a_q = a_q(z)$. If the interaction is weak, the $a_q(z)$ vary slowly with z, so that they can be assumed approximately constant within a distance of a wavelength. This makes it possible to use the slowly varying envelope approximation wherein $d^2 a_q/dz^2$ is neglected relative to $k_q\, da_q/dz = (2\pi/\lambda_q)\, da_q/dz$ and

$$\left(\nabla^2 + k_q^2\right)\left[a_q \exp\left(-jk_q z\right)\right] \approx -j2k_q \frac{da_q}{dz} \exp\left(-jk_q z\right) \qquad \text{(19.4-14)}$$

(see Sec. 2.2C). With this approximation (19.4-10) reduce to the simpler form

$$\frac{d\mathscr{a}_1}{dz} = -j\mathscr{g}\mathscr{a}_3\mathscr{a}_2^* \exp(-j\,\Delta k z) \tag{19.4-15a}$$

$$\frac{d\mathscr{a}_2}{dz} = -j\mathscr{g}\mathscr{a}_3\mathscr{a}_1^* \exp(-j\,\Delta k z) \tag{19.4-15b}$$

$$\frac{d\mathscr{a}_3}{dz} = -j\mathscr{g}\mathscr{a}_1\mathscr{a}_2 \exp(j\,\Delta k z), \tag{19.4-15c}$$

Three-Wave Mixing Coupled Equations

where

$$\mathscr{g}^2 = 2\hbar\omega_1\omega_2\omega_3\eta^3\mathscr{d}^2 \tag{19.4-16}$$

and

$$\Delta k = k_3 - k_2 - k_1 \tag{19.4-17}$$

represents the error in the phase-matching condition. The variations of $\mathscr{a}_1$, $\mathscr{a}_2$, and $\mathscr{a}_3$ with z are therefore governed by three coupled first-order differential equations (19.4-15), which we proceed to solve under the different boundary conditions corresponding to various applications. It is useful, however, first to derive some invariants of the wave-mixing process. These are functions of $\mathscr{a}_1$, $\mathscr{a}_2$, and $\mathscr{a}_3$ that are independent of z. Invariants are useful since they can be used to reduce the number of independent variables. Exercises 19.4-2 and 19.4-3 develop invariants based on conservation of energy and conservation of photons.

EXERCISE 19.4-2

Energy Conservation. Show that the sum of the intensities $I_q = \hbar\omega_q|\mathscr{a}_q|^2$, $q = 1, 2, 3$, of the three waves governed by (19.4-15) is invariant to z, so that

$$\frac{d}{dz}(I_1 + I_2 + I_3) = 0. \tag{19.4-18}$$

EXERCISE 19.4-3

Photon-Number Conservation: The Manley–Rowe Relation. Using (19.4-15), show that

$$\frac{d}{dz}|\mathscr{a}_1|^2 = \frac{d}{dz}|\mathscr{a}_2|^2 = -\frac{d}{dz}|\mathscr{a}_3|^2, \tag{19.4-19}$$

from which the Manley–Rowe relation (19.2-17), which was derived using photon-number conservation, follows. Equation (19.4-19) implies that $|\mathscr{a}_1|^2 + |\mathscr{a}_3|^2$ and $|\mathscr{a}_2|^2 + |\mathscr{a}_3|^2$ are also invariants of the wave-mixing process.

A. Second-Harmonic Generation

Second-harmonic generation is a degenerate case of three-wave mixing in which

$$\omega_1 = \omega_2 = \omega \quad \text{and} \quad \omega_3 = 2\omega. \tag{19.4-20}$$

Two forms of interaction occur:

- Two photons of frequency ω combine to form a photon of frequency 2ω (second harmonic).
- One photon of frequency 2ω splits into two photons, each of frequency ω.

The interaction of the two waves is described by the Helmholtz equations with sources. Conservation of momentum requires that

$$\mathbf{k}_3 = 2\mathbf{k}_1. \tag{19.4-21}$$

EXERCISE 19.4-4

Coupled-Wave Equations for Second-Harmonic Generation. Apply the slowly varying envelope approximation (19.4-14) to the Helmholtz equations (19.4-11), which describe two collinear waves in the degenerate case, to show that

$$\frac{da_1}{dz} = -j\mathscr{g}a_3a_1^* \exp(-j\Delta k z) \tag{19.4-22a}$$

$$\frac{da_3}{dz} = -j\frac{\mathscr{g}}{2}a_1a_1 \exp(j\Delta k z), \tag{19.4-22b}$$

where $\Delta k = k_3 - 2k_1$ and

$$\mathscr{g}^2 = 4\hbar\omega^3\eta^3\mathscr{d}^2. \tag{19.4-23}$$

Assuming two collinear waves with perfect phase matching ($\Delta k = 0$), equations (19.4-22) reduce to

$$\frac{da_1}{dz} = -j\mathscr{g}a_3a_1^* \tag{19.4-24a}$$

$$\frac{da_3}{dz} = -j\frac{\mathscr{g}}{2}a_1a_1. \tag{19.4-24b}$$

Coupled Equations (Second-Harmonic Generation)

At the input to the device ($z = 0$) the amplitude of the second-harmonic wave is assumed to be zero, $a_3(0) = 0$, and that of the fundamental wave, $a_1(0)$, is assumed to be real. With these boundary conditions, and using the photon-number conservation relation $|a_1(z)|^2 + 2|a_3(z)|^2 = $ constant, (19.4-24) can be shown to have the solution

$$a_1(z) = a_1(0) \operatorname{sech} \frac{g a_1(0) z}{\sqrt{2}} \tag{19.4-25a}$$

$$a_3(z) = -\frac{j}{\sqrt{2}} a_1(0) \tanh \frac{g a_1(0) z}{\sqrt{2}}. \tag{19.4-25b}$$

Consequently, the photon flux densities $\phi_1(z) = |a_1(z)|^2$ and $\phi_3(z) = |a_3(z)|^2$ evolve in accordance with

$$\phi_1(z) = \phi_1(0) \operatorname{sech}^2 \frac{\gamma z}{2} \tag{19.4-26a}$$

$$\phi_3(z) = \frac{1}{2} \phi_1(0) \tanh^2 \frac{\gamma z}{2}, \tag{19.4-26b}$$

where $\gamma/2 = g a_1(0)/\sqrt{2}$, i.e.,

$$\gamma^2 = 2 g^2 a_1^2(0) = 2 g^2 \phi_1(0) = 8 d^2 \eta^3 \hbar \omega^3 \phi_1(0) = 8 d^2 \eta^3 \omega^2 I_1(0). \tag{19.4-27}$$

Since $\operatorname{sech}^2 + \tanh^2 = 1$, $\phi_1(z) + 2\phi_3(z) = \phi_1(0)$ is constant, indicating that at each position z, photons of wave 1 are converted to half as many photons of wave 3. The fall of $\phi_1(z)$ and the rise of $\phi_3(z)$ with z are shown in Fig. 19.4-1.

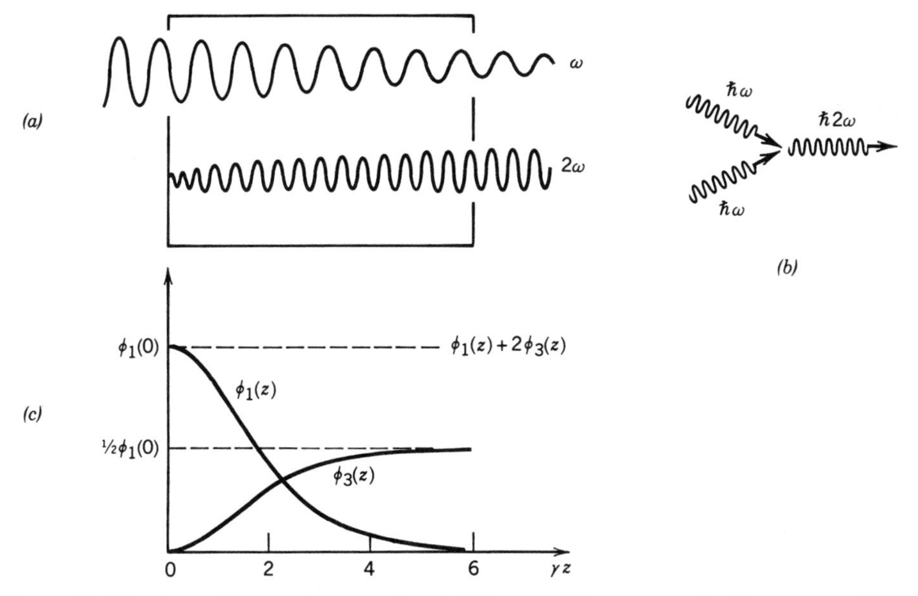

Figure 19.4-1 Second-harmonic generation. (*a*) A wave of frequency ω incident on a nonlinear crystal generates a wave of frequency 2ω. (*b*) Two photons of frequency ω combine to make one photon of frequency 2ω. (*c*) As the photon flux density $\phi_1(z)$ of the fundamental wave decreases, the photon flux density $\phi_3(z)$ of the second-harmonic wave increases. Since photon numbers are conserved, the sum $\phi_1(z) + 2\phi_3(z) = \phi_1(0)$ is a constant.

The efficiency of second-harmonic generation for an interaction region of length L is

$$\frac{I_3(L)}{I_1(0)} = \frac{\hbar\omega_3\phi_3(L)}{\hbar\omega_1\phi_1(0)} = \frac{2\phi_3(L)}{\phi_1(0)}$$

$$= \tanh^2\frac{\gamma L}{2}. \qquad (19.4\text{-}28)$$

For large γL (long cell, large input intensity, or large nonlinear parameter), the efficiency approaches one. This signifies that all the input power (at frequency ω) has been transformed into power at frequency 2ω; all input photons of frequency ω are converted into half as many photons of frequency 2ω.

For small γL [small device length L, small nonlinear parameter $\mathscr{d}$, or small input photon flux density $\phi_1(0)$], the argument of the tanh function is small and therefore the approximation $\tanh x \approx x$ may be used. The efficiency of second-harmonic generation is then

$$\frac{I_3(L)}{I_1(0)} \approx \tfrac{1}{4}\gamma^2L^2 = \tfrac{1}{2}\mathscr{g}^2L^2\phi_1(0) = 2\mathscr{d}^2\eta^3\hbar\omega^3L^2\phi_1(0) = 2\mathscr{d}^2\eta^3\omega^2L^2I_1(0),$$

so that

$$\boxed{\frac{I_3(L)}{I_1(0)} = 2\eta_o^3\omega^2\frac{\mathscr{d}^2}{n^3}\frac{L^2}{A}P,} \qquad (19.4\text{-}29)$$

Second-Harmonic Generation Efficiency

where $P = I_1(0)A$ is the incident optical power and A is the cross-sectional area. The efficiency is proportional to the input power P and the factor $\mathscr{d}^2/n^3$, which is a figure of merit used for comparing different nonlinear materials. For a fixed input power P, the efficiency is directly proportional to the geometrical factor L^2/A. To maximize the efficiency we must confine the wave to the smallest possible area A and the largest possible interaction length L. This is best accomplished with waveguides (planar or channel waveguides or fibers).

Effect of Phase Mismatch

To study the effect of phase (or momentum) mismatch, the general equations (19.4-22) are used with $\Delta k \neq 0$. For simplicity, we limit ourselves to the weak-coupling case for which $\gamma L \ll 1$. In this case, the amplitude of the fundamental wave $a_1(z)$ varies only slightly with z [see Fig. 19.4-1(*c*)], and may be assumed approximately constant. Substituting $a_1(z) \approx a_1(0)$ in (19.4-22b) and integrating, we obtain

$$a_3(L) = -j\frac{\mathscr{g}}{2}a_1^2(0)\int_0^L \exp(j\,\Delta k\,z')\,dz' = -\left(\frac{\mathscr{g}}{2\,\Delta k}\right)a_1^2(0)[\exp(j\,\Delta k\,L) - 1], \qquad (19.4\text{-}30)$$

from which $\phi_3(L) = |a_3(L)|^2 = (\mathscr{g}/\Delta k)^2\phi_1^2(0)\sin^2(\Delta k\,L/2)$, where $a_1(0)$ is assumed

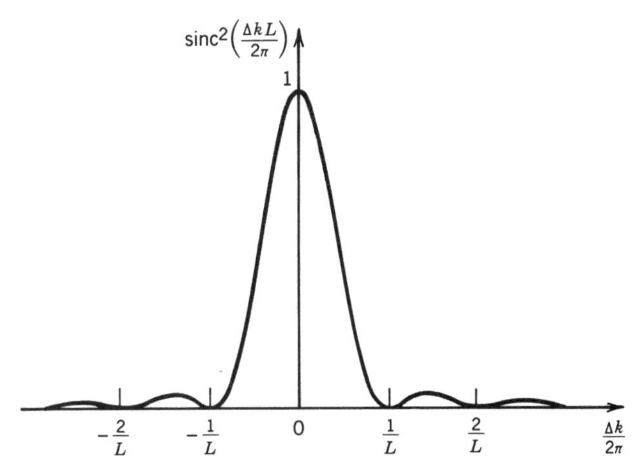

Figure 19.4-2 The factor by which the efficiency of second-harmonic generation is reduced as a result of a phase mismatch $\Delta k L$ between waves interacting within a distance L.

to be real. The efficiency of second-harmonic generation is therefore

$$\frac{I_3(L)}{I_1(0)} = \frac{2\phi_3(L)}{\phi_1(0)} = \tfrac{1}{2}\mathscr{g}^2 L^2 \phi_1(0)\,\operatorname{sinc}^2 \frac{\Delta k L}{2\pi}, \tag{19.4-31}$$

where $\operatorname{sinc}(x) = \sin(\pi x)/(\pi x)$.

The effect of phase mismatch is therefore to reduce the efficiency of second-harmonic generation by the factor $\operatorname{sinc}^2(\Delta k L/2\pi)$. This factor is unity for $\Delta k = 0$ and drops as Δk increases, reaching $(2/\pi)^2 \approx 0.4$ when $|\Delta k| = \pi/L$, and vanishing when $|\Delta k| = 2\pi/L$ (see Fig. 19.4-2). For a given L, the mismatch Δk corresponding to a prescribed efficiency reduction factor is inversely proportional to L, so that the phase matching requirement becomes more stringent as L increases. For a given mismatch Δk, the length $L_c = 2\pi/|\Delta k|$ is a measure of the maximum length within which second-harmonic generation is efficient; L_c is often called the **coherence length**. Since $|\Delta k| = 2(2\pi/\lambda_o)|n_3 - n_1|$, where λ_o is the free-space wavelength of the fundamental wave and n_1 and n_3 are the refractive indices of the fundamental and the second-harmonic waves, $L_c = \lambda_o/2|n_3 - n_1|$ is inversely proportional to $|n_3 - n_1|$, which is governed by the material dispersion.

The tolerance of the interaction process to the phase mismatch can be regarded as a result of the wavevector uncertainty $\Delta k \propto 1/L$ associated with confinement of the waves within a distance L [see Appendix A, (A.2-6)]. The corresponding momentum uncertainty $\Delta p = \hbar\,\Delta k \propto 1/L$, explains the apparent violation of the law of conservation of momentum in the wave-mixing process.

B. Frequency Conversion

A frequency up-converter (Fig. 19.4-3) converts a wave of frequency ω_1 into a wave of higher frequency ω_3 by use of an auxiliary wave at frequency ω_2, called the "*pump*." A photon $\hbar\omega_2$ from the pump is added to a photon $\hbar\omega_1$ from the *input signal* to form a photon $\hbar\omega_3$ of the *output signal* at an *up-converted* frequency $\omega_3 = \omega_1 + \omega_2$.

The conversion process is governed by the three coupled equations (19.4-15). For simplicity, assume that the three waves are phase matched ($\Delta k = 0$) and that the pump is sufficiently strong so that its amplitude does not change appreciably within the interaction distance of interest; i.e., $a_2(z) \approx a_2(0)$ for all z between 0 and L. The three equations (19.4-15) then reduce to two,

$$\frac{da_1}{dz} = -j\frac{\gamma}{2}a_3 \qquad (19.4\text{-}32a)$$

$$\frac{da_3}{dz} = -j\frac{\gamma}{2}a_1, \qquad (19.4\text{-}32b)$$

where $\gamma = 2\mathscr{g}a_2(0)$ and $a_2(0)$ is assumed real. These are simple differential equations with harmonic solutions

$$a_1(z) = a_1(0)\cos\frac{\gamma z}{2} \qquad (19.4\text{-}33a)$$

$$a_3(z) = -ja_1(0)\sin\frac{\gamma z}{2}. \qquad (19.4\text{-}33b)$$

The corresponding photon flux densities are

$$\phi_1(z) = \phi_1(0)\cos^2\frac{\gamma z}{2} \qquad (19.4\text{-}34a)$$

$$\phi_3(z) = \phi_1(0)\sin^2\frac{\gamma z}{2}. \qquad (19.4\text{-}34b)$$

Dependences of the photon flux densities ϕ_1 and ϕ_3 on z are sketched in Fig. 19.4-3(*c*). Photons are exchanged periodically between the two waves. In the region between $z = 0$ and $z = \pi/\gamma$, the input ω_1 photons combine with the pump ω_2 photons and generate the up-converted ω_3 photons. Wave 1 is therefore attenuated, whereas wave 3 is amplified. In the region $\gamma z = \pi$ to $\gamma z = 2\pi$, the ω_3 photons are more abundant; they disintegrate into ω_1 and ω_2 photons; so that wave 3 is attenuated and wave 1 amplified. The process is repeated periodically as the waves travel through the medium.

The efficiency of up-conversion for a device of length L is

$$\frac{I_3(L)}{I_1(0)} = \frac{\omega_3}{\omega_1}\sin^2\frac{\gamma L}{2}. \qquad (19.4\text{-}35)$$

For $\gamma L \ll 1$, and using (19.4-16), this is approximated by $I_3(L)/I_1(0) \approx (\omega_3/\omega_1)(\gamma L/2)^2 = (\omega_3/\omega_1)\mathscr{g}^2L^2\phi_2(0) = 2\omega_3^2L^2\mathscr{d}^2\eta^3I_2(0)$, from which

$$\frac{I_3(L)}{I_1(0)} = 2\eta_o^3\omega_3^2\frac{\mathscr{d}^2}{n^3}\frac{L^2}{A}P_2, \qquad (19.4\text{-}36)$$

Up-Conversion Efficiency

where A is the cross-sectional area and $P_2 = I_2(0)A$ is the pump power. The conversion efficiency is proportional to the pump power, the ratio L^2/A, and the material parameter $\mathscr{d}^2/n^3$.

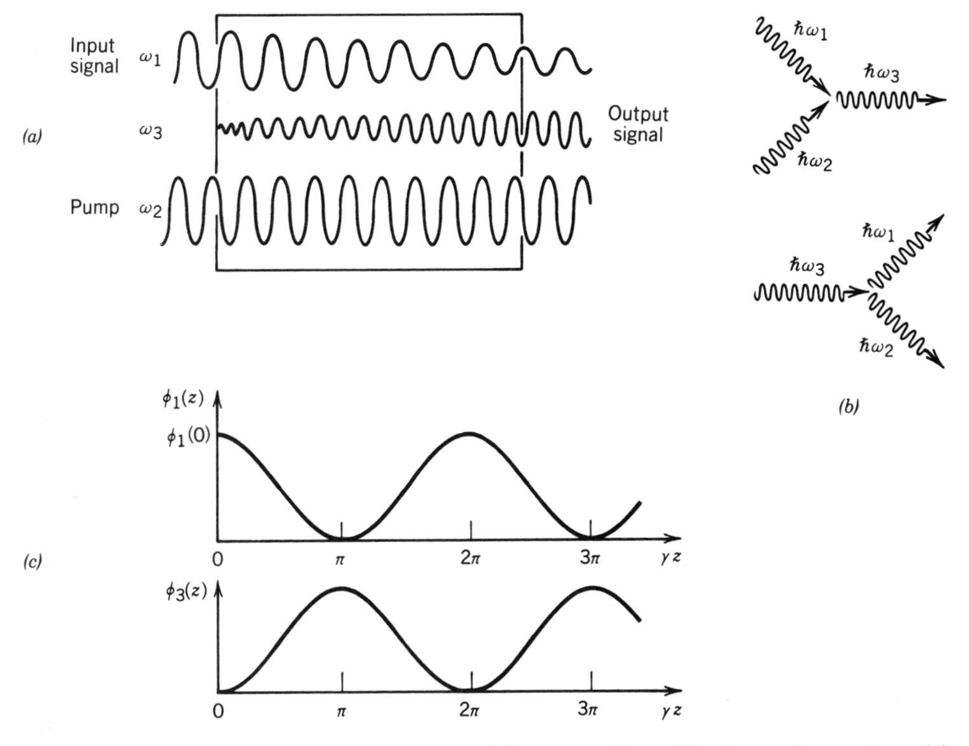

Figure 19.4-3 The frequency up-converter: (*a*) wave mixing; (*b*) photon interactions; (*c*) evolution of the photon flux densities of the input ω_1-wave and the up-converted ω_3-wave. The pump ω_2-wave is assumed constant.

EXERCISE 19.4-5

Infrared Up-Conversion. An up-converter uses a proustite crystal ($\mathscr{d} = 1.5 \times 10^{-22}$ MKS, $n = 2.6$). The input wave is obtained from a CO_2 laser of wavelength 10.6 μm, and the pump from a 1-W Nd^{3+}:YAG laser of wavelength 1.06 μm focused to a cross-sectional area 10^{-2} mm^2 (see Fig. 19.2-5). Determine the wavelength of the up-converted wave and the efficiency of up-conversion if the waves are collinear and the interaction length is 1 cm.

C. Parametric Amplification and Oscillation

Parametric Amplifiers

The parametric amplifier uses three-wave mixing in a nonlinear crystal to provide optical gain [Fig. 19.4-4(*a*)]. The process is governed by the same three coupled equations (19.4-15) with the waves identified as follows:

- Wave 1 is the "**signal**" to be amplified. It is incident on the crystal with a small intensity $I_1(0)$.
- Wave 3, called the "**pump**," is an intense wave that provides power to the amplifier.
- Wave 2, called the "**idler**," is an auxiliary wave created by the interaction process.

The basic idea is that a photon $\hbar\omega_3$ provided by the pump is split into a photon $\hbar\omega_1$, which amplifies the signal, and a photon $\hbar\omega_2$, which creates the idler [Fig. 19.4-4(*b*)].

Assuming perfect phase matching ($\Delta k = 0$), and an undepleted pump, $a_3(z) \approx a_3(0)$, the coupled-wave equations (19.4-15) give

$$\frac{da_1}{dz} = -j\frac{\gamma}{2}a_2^* \tag{19.4-37a}$$

$$\frac{da_2}{dz} = -j\frac{\gamma}{2}a_1^*, \tag{19.4-37b}$$

where $\gamma = 2g a_3(0)$. If $a_3(0)$ is real, γ is also real, and the differential equations have the solution

$$a_1(z) = a_1(0)\cosh\frac{\gamma z}{2} \tag{19.4-38a}$$

$$a_2(z) = -ja_1(0)\sinh\frac{\gamma z}{2}. \tag{19.4-38b}$$

The corresponding photon flux densities are

$$\phi_1(z) = \phi_1(0)\cosh^2\frac{\gamma z}{2} \tag{19.4-39a}$$

$$\phi_2(z) = \phi_1(0)\sinh^2\frac{\gamma z}{2}. \tag{19.4-39b}$$

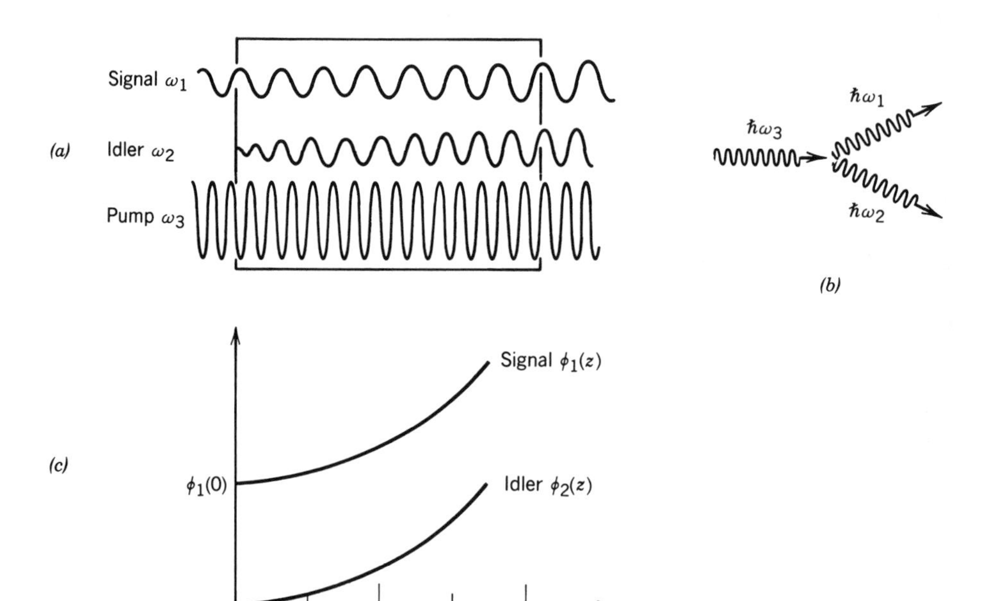

Figure 19.4-4 The parametric amplifier: (*a*) wave mixing; (*b*) photon mixing; (*c*) photon flux densities of the signal and the idler; the pump photon flux density is assumed constant.

Both $\phi_1(z)$ and $\phi_2(z)$ grow monotonically with z, as illustrated in Fig. 19.4-4(c). This growth saturates when sufficient energy is drawn from the pump so that the assumption of an undepleted pump no longer holds.

The total gain of an amplifier of length L is $G = \phi_1(L)/\phi_1(0) = \cosh^2(\gamma L/2)$. In the limit $\gamma L \gg 1$, $G = (e^{\gamma L/2} + e^{-\gamma L/2})^2/4 \approx e^{\gamma L}/4$, so that the gain increases exponentially with γL. The gain coefficient $\gamma = 2\mathscr{g}a_3(0) = 2\mathscr{d}(2\hbar\omega_1\omega_2\omega_3\eta^3)^{1/2}a_3(0)$, from which

$$\gamma = \left[8\eta_o^3\omega_1\omega_2\frac{\mathscr{d}^2}{n^3}\frac{P_3}{A}\right]^{1/2}, \tag{19.4-40}$$

Parametric Amplifier Gain Coefficient

where $P_3 = I_3(0)A$ and A is the cross-sectional area.

EXERCISE 19.4-6

Gain of a Parametric Amplifier. An 8-cm-long ADP crystal ($n = 1.5$, $\mathscr{d} = 7.7 \times 10^{-24}$ MKS) is used to amplify He–Ne laser light of wavelength 633 nm. The pump is an argon laser of wavelength 334 nm and intensity 2 MW/cm^2. Determine the gain of the amplifier.

Parametric Oscillators

A parametric oscillator is constructed by providing feedback at both the signal and the idler frequencies of a parametric amplifier, as illustrated in Fig. 19.4-5. Energy is supplied by the pump.

To determine the condition of oscillation, the gain of the amplifier is equated to the loss. Losses have not been included in the derivation of the coupled equations, (19.4-37), which describe the parametric amplifier. These equations can be modified by including phenomenological loss terms,

$$\frac{da_1}{dz} = -\frac{\alpha_1}{2}a_1 - j\frac{\gamma}{2}a_2^* \tag{19.4-41a}$$

$$\frac{da_2}{dz} = -\frac{\alpha_2}{2}a_2 - j\frac{\gamma}{2}a_1^*, \tag{19.4-41b}$$

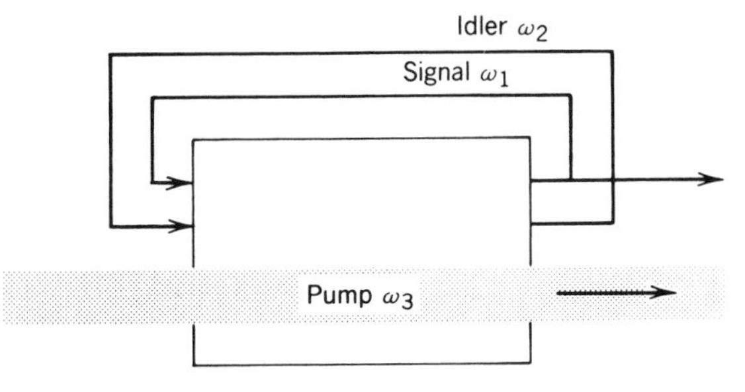

Figure 19.4-5 The parametric oscillator generates light at frequencies ω_1 and ω_2. A pump of frequency $\omega_3 = \omega_1 + \omega_2$ serves as the source of energy.

where α_1 and α_2 are power attenuation coefficients for the signal and idler waves, respectively. These terms represent scattering and absorption losses in the medium and losses at the mirrors of the resonator [see Fig. 19.2-7(*c*)] distributed along the length of the crystal as was done with the laser (see Sec. 14.1). In the absence of coupling ($\gamma = 0$), (19.4-41a) gives $a_1(z) = \exp(-\alpha_1 z/2) a_1(0)$, and $\phi_1(z) = \exp(-\alpha_1 z)\phi_1(0)$, so that the photon flux decays at a rate α_1. Equation (19.4-41b) gives a similar result.

The steady-state solution of (19.4-41) is obtained by equating the derivatives to zero,

$$0 = \alpha_1 a_1 + j\gamma a_2^* \tag{19.4-42a}$$

$$0 = \alpha_2 a_2 + j\gamma a_1^*. \tag{19.4-42b}$$

Equation (19.4-42a) gives $a_1/a_2^* = -j\gamma/\alpha_1$ and the conjugate of (19.4-42b) gives $a_1/a_2^* = \alpha_2/j\gamma$, so that for a nontrivial solution, $-j\gamma/\alpha_1 = \alpha_2/j\gamma$, from which

$$\gamma^2 = \alpha_1\alpha_2. \tag{19.4-43}$$

If $\alpha_1 = \alpha_2 = \alpha$, the condition of oscillation becomes $\gamma = \alpha$, meaning that the amplifier gain coefficient equals the loss coefficient. Since $\gamma = 2g a_3(0)$, the amplitude of the pump must be $a_3(0) \geq \alpha/2g$ and the corresponding photon flux density $\phi_3(0) \geq \alpha^2/4g^2$. Substituting from (19.4-16) for g, we obtain $\phi_3(0) \geq \alpha^2/8\hbar\omega_1\omega_2\omega_3\eta^3 d^2$. Thus the minimum pump intensity $\hbar\omega_3\phi_3(0)$ required for parametric oscillation is

$$I_3\big|_{\text{threshold}} = \frac{\alpha^2 n^3}{8\omega_1\omega_2\eta_o^3 d^2}. \tag{19.4-44}$$

Parametric Oscillation Threshold Pump Intensity

The oscillation frequencies ω_1 and ω_2 of the parametric oscillator are determined by the frequency- and phase-matching conditions, $\omega_1 + \omega_2 = \omega_3$ and $n_1\omega_1 + n_2\omega_2 = n_3\omega_3$. The solution of these two equations yields ω_1 and ω_2. Since the medium is always dispersive the refractive indices are frequency dependent (i.e., n_1 is a function of ω_1, n_2 is a function of ω_2, and n_3 is a function of ω_3). The oscillation frequencies may be tuned by varying the refractive indices using, for example, temperature control.

*19.5 COUPLED-WAVE THEORY OF FOUR-WAVE MIXING

We now derive the coupled differential equations that describe four-wave mixing in a third-order nonlinear medium, using an approach similar to that employed in the three-wave mixing case.

Coupled-Wave Equations

Four waves constituting a total field

$$\mathscr{E}(t) = \sum_{q=1,2,3,4} \operatorname{Re}\left[E_q \exp(j\omega_q t)\right]$$

$$= \sum_{q=\pm1,\pm2,\pm3,\pm4} \tfrac{1}{2} E_q \exp(j\omega_q t) \tag{19.5-1}$$

travel in a medium characterized by a nonlinear polarization density

$$\mathscr{P}_{\mathrm{NL}} = 4\chi^{(3)}\mathscr{E}^3. \tag{19.5-2}$$

The corresponding source of radiation, $\mathscr{S} = -\mu_o \partial^2 \mathscr{P}_{\mathrm{NL}}/\partial t^2$, is therefore a sum of $8^3 = 512$ terms,

$$\mathscr{S} = \tfrac{1}{2}\mu_o \chi^{(3)} \sum_{q,p,r=\pm 1, \pm 2, \pm 3, \pm 4} (\omega_q + \omega_p + \omega_r)^2 E_q E_p E_r \exp\left[j(\omega_q + \omega_p + \omega_r)t\right]. \tag{19.5-3}$$

Substituting (19.5-1) and (19.5-3) into the wave equation (19.4-1) and equating terms at each of the four frequencies ω_1, ω_2, ω_3, and ω_4, we obtain four Helmholtz equations with sources,

$$\left(\nabla^2 + k_q^2\right)E_q = -S_q, \qquad q = 1, 2, 3, 4, \tag{19.5-4}$$

where S_q is the amplitude of the component of $\mathscr{S}$ at frequency ω_q.

For the four waves to be coupled, their frequencies must be commensurate. Consider, for example, the case for which the sum of two frequencies equals the sum of the other two frequencies,

$$\boxed{\omega_3 + \omega_4 = \omega_1 + \omega_2.} \tag{19.5-5}$$

Frequency-Matching Condition

Three waves can then combine and create a source at the fourth frequency. Using (19.5-5), terms in (19.5-3) at each of the four frequencies are

$$S_1 = \mu_o \omega_1^2 \chi^{(3)} \left\{6E_3E_4E_2^* + 3E_1\left[|E_1|^2 + 2|E_2|^2 + 2|E_3|^2 + 2|E_4|^2\right]\right\} \tag{19.5-6a}$$

$$S_2 = \mu_o \omega_2^2 \chi^{(3)} \left\{6E_3E_4E_1^* + 3E_2\left[|E_2|^2 + 2|E_1|^2 + 2|E_3|^2 + 2|E_4|^2\right]\right\} \tag{19.5-6b}$$

$$S_3 = \mu_o \omega_3^2 \chi^{(3)} \left\{6E_1E_2E_4^* + 3E_3\left[|E_3|^2 + 2|E_2|^2 + 2|E_1|^2 + 2|E_4|^2\right]\right\} \tag{19.5-6c}$$

$$S_4 = \mu_o \omega_4^2 \chi^{(3)} \left\{6E_1E_2E_3^* + 3E_4\left[|E_4|^2 + 2|E_1|^2 + 2|E_2|^2 + 2|E_3|^2\right]\right\}. \tag{19.5-6d}$$

Each wave is therefore driven by a source with two components. The first is a result of mixing of the other three waves. The first term in S_1, for example, is proportional to $E_3E_4E_2^*$ and therefore represents the mixing of waves 2, 3, and 4 to create a source for wave 1. The second component is proportional to the complex amplitude of the wave itself. The second term of S_1, for example, is proportional to E_1, so that it plays the role of refractive-index modulation, and therefore represents the optical Kerr effect (see Exercise 19.3-4).

It is therefore convenient to separate the two contributions to these sources by defining

$$S_q = \bar{S}_q + (\omega_q/c_o)^2 \Delta\chi_q E_q, \qquad q = 1, 2, 3, 4 \tag{19.5-7}$$

where

$$\bar{S}_1 = 6\mu_o \omega_1^2 \chi^{(3)} E_3 E_4 E_2^* \tag{19.5-8a}$$

$$\bar{S}_2 = 6\mu_o \omega_2^2 \chi^{(3)} E_3 E_4 E_1^* \tag{19.5-8b}$$

$$\bar{S}_3 = 6\mu_o \omega_3^2 \chi^{(3)} E_1 E_2 E_4^* \tag{19.5-8c}$$

$$\bar{S}_4 = 6\mu_o \omega_4^2 \chi^{(3)} E_1 E_2 E_3^*, \tag{19.5-8d}$$

and

$$\Delta\chi_q = 6\frac{\eta}{\epsilon_o}\chi^{(3)}(2I - I_q), \qquad q = 1,2,3,4. \tag{19.5-9}$$

Here $I_q = |E_q|^2/2\eta$ are the intensities of the waves, $I = I_1 + I_2 + I_3 + I_4$ is the total intensity, and η is the impedance of the medium.

This enables us to rewrite the Helmholtz equations (19.5-4) as

$$\left(\nabla^2 + \bar{k}_q^2\right)E_q = -\bar{S}_q, \qquad q = 1,2,3,4, \tag{19.5-10}$$

where

$$\bar{k}_q = \bar{n}_q \frac{\omega_q}{c_o}$$

and

$$\bar{n}_q = \left[n^2 + \frac{6\eta}{\epsilon_o}\chi^{(3)}(2I - I_q)\right]^{1/2} = n\left[1 + \frac{6\eta}{\epsilon_o n^2}\chi^{(3)}(2I - I_q)\right]^{1/2}$$

$$\approx n\left[1 + \frac{3\eta}{\epsilon_o n^2}\chi^{(3)}(2I - I_q)\right],$$

from which

$$\boxed{\bar{n}_q \approx n + n_2(2I - I_q),} \tag{19.5-11a}$$

Optical Kerr Effect

where

$$n_2 = \frac{3\eta_o}{\epsilon_o n^2}\chi^{(3)}, \tag{19.5-11b}$$

which matches with (19.3-15).

The Helmholtz equation for each wave is modified in two ways:

- A source representing the combined effects of the other three waves is present. This may lead to the amplification of an existing wave, or the emission of a new wave at that frequency.
- The refractive index for each wave is altered, becoming a function of the intensities of the four waves.

Equations (19.5-10) and (19.5-8) yield four coupled differential equations which may be solved under the appropriate boundary conditions.

Degenerate Four-Wave Mixing

We now develop and solve the coupled-wave equations in the degenerate case for which all four waves have the same frequency, $\omega_1 = \omega_2 = \omega_3 = \omega_4 = \omega$. As was assumed in Sec. 19.3C, two of the waves (waves 3 and 4), called the pump waves, are plane waves propagating in opposite directions, with complex amplitudes $E_3(\mathbf{r}) = A_3 \exp(-j\mathbf{k}_3 \cdot \mathbf{r})$ and $E_4(\mathbf{r}) = A_4 \exp(-j\mathbf{k}_4 \cdot \mathbf{r})$, and wavevectors related by $\mathbf{k}_4 = -\mathbf{k}_3$. Their intensities are assumed much greater than those of waves 1 and 2, so that they are approximately undepleted by the interaction process, allowing us to assume that their complex envelopes A_3 and A_4 are constant. The total intensity of the four waves I is then also approximately constant, $I \approx [|A_3|^2 + |A_4|^2]/2\eta$. The terms $2I - I_1$ and $2I - I_2$, which govern the effective refractive index $\bar{n}$ for waves 1 and 2 in (19.5-11), are approximately equal to $2I$, and are therefore also constant, so that the optical Kerr effect amounts to a constant change of the refractive index. Its effect will therefore be ignored.

With these assumptions the problem is reduced to a problem of two coupled waves, 1 and 2. Equations (19.5-10) and (19.5-8) give

$$(\nabla^2 + k^2)E_1 = -\xi E_2^* \qquad (19.5\text{-}12a)$$

$$(\nabla^2 + k^2)E_2 = -\xi E_1^*, \qquad (19.5\text{-}12b)$$

where

$$\xi = 6\mu_o \omega^2 \chi^{(3)} E_3 E_4 = 6\mu_o \omega^2 \chi^{(3)} A_3 A_4 \qquad (19.5\text{-}13)$$

and $k = \bar{n}\omega/c_o$, where $\bar{n} \approx n + 2n_2 I$ is a constant.

The four nonlinear coupled differential equations have thus been reduced to two *linear* coupled equations, each of which takes the form of the Helmholtz equation with a source term. The source for wave 1 is proportional to the conjugate of the complex amplitude of wave 2, and similarly for wave 2.

Phase Conjugation

Assume that waves 1 and 2 are also plane waves propagating in opposite directions along the z axis, as illustrated in Fig. 19.5-1,

$$E_1 = A_1 \exp(-jkz), \qquad E_2 = A_2 \exp(jkz). \qquad (19.5\text{-}14)$$

This assumption is consistent with the phase-matching condition since $k_1 + k_2 = k_3 + k_4$.

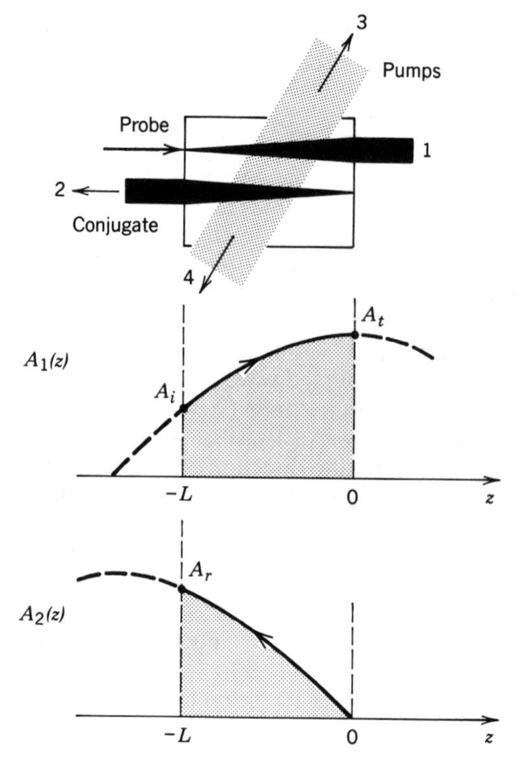

Figure 19.5-1 Degenerate four-wave mixing. Waves 3 and 4 are intense pump waves traveling in opposite directions. Wave 1, the probe wave, and wave 2, the conjugate wave, also travel in opposite directions and have increasing amplitudes.

Substituting (19.5-14) in (19.5-12) and using the slowly varying envelope approximation, (19.4-14), we reduce equations (19.5-12) to two first-order differential equations,

$$\frac{dA_1}{dz} = -j\gamma A_2^* \tag{19.5-15a}$$

$$\frac{dA_2}{dz} = j\gamma A_1^*, \tag{19.5-15b}$$

where

$$\gamma = \frac{\xi}{2k} = \frac{3\omega\eta_o}{\bar{n}}\chi^{(3)}A_3A_4 \tag{19.5-16}$$

is a coupling coefficient.

For simplicity, assume that A_3A_4 is real, so that γ is real. The solution of (19.5-15) is then two harmonic functions $A_1(z)$ and $A_2(z)$ with a 90° phase shift between them. If the nonlinear medium extends over a distance between the planes $z = -L$ to $z = 0$, as illustrated in Fig. 19.5-1, wave 1 has amplitude $A_1(-L) = A_i$ at the entrance plane, and wave 2 has zero amplitude at the exit plane, $A_2(0) = 0$. Under these boundary conditions the solution of (19.5-15) is

$$A_1(z) = \frac{A_i}{\cos \gamma L} \cos \gamma z \qquad (19.5\text{-}17)$$

$$A_2(z) = j \frac{A_i^*}{\cos \gamma L} \sin \gamma z. \qquad (19.5\text{-}18)$$

The amplitude of the reflected wave at the entrance plane, $A_r = A_2(-L)$, is

$$\boxed{A_r = -jA_i^* \tan \gamma L,} \qquad (19.5\text{-}19)$$

Reflected Wave Amplitude

whereas the amplitude of the transmitted wave, $A_t = A_1(0)$, is

$$\boxed{A_t = \frac{A_i}{\cos \gamma L}.} \qquad (19.5\text{-}20)$$

Transmitted Wave Amplitude

Equations (19.5-19) and (19.5-20) suggest a number of applications:

- The reflected wave is a conjugated version of the incident wave. The device acts as a **phase conjugator** (see Sec. 19.3C).
- The intensity reflectance, $|A_r|^2/|A_i|^2 = \tan^2 \gamma L$, may be smaller or greater than 1, corresponding to attenuation or gain, respectively. The medium can therefore act as a **reflection amplifier** (an "amplifying mirror").
- The transmittance $|A_t|^2/|A_i|^2 = 1/\cos^2 \gamma L$ is always greater than 1, so that the medium always acts as a **transmission amplifier**.
- When $\gamma L = \pi/2$, or odd multiples thereof, the reflectance and transmittance are infinite, indicating instability. The device is then an **oscillator**.

*19.6 ANISOTROPIC NONLINEAR MEDIA

In an anisotropic medium, each of the three components of the polarization vector $\mathcal{P} = (\mathcal{P}_1, \mathcal{P}_2, \mathcal{P}_3)$ is a function of the three components of the electric field vector $\mathcal{E} = (\mathcal{E}_1, \mathcal{E}_2, \mathcal{E}_3)$. These functions are linear for small magnitudes of $\mathcal{E}$ (see Sec. 6.3) but deviate slightly from linearity as $\mathcal{E}$ increases. Each of these three nonlinear functions may be expanded in a Taylor's series in terms of the three components $\mathcal{E}_1$, $\mathcal{E}_2$, and $\mathcal{E}_3$, as was done in (19.1-2) in the scalar analysis. Thus

$$\boxed{\mathcal{P}_i = \epsilon_o \sum_j \chi_{ij} \mathcal{E}_j + 2 \sum_{jk} d_{ijk} \mathcal{E}_j \mathcal{E}_k + 4 \sum_{jkl} \chi_{ijkl}^{(3)} \mathcal{E}_j \mathcal{E}_k \mathcal{E}_l,} \qquad i, j, k, l = 1, 2, 3. \qquad (19.6\text{-}1)$$

The coefficients χ_{ij}, d_{ijk}, and $\chi^{(3)}_{ijkl}$ are elements of tensors that correspond to the scalar coefficients χ, d, $\chi^{(3)}$, and (19.6-1) is a generalization of (19.1-2) applicable to the anisotropic case.

Symmetries

Because the coefficient d_{ijk} is a multiplier of the product $\mathscr{E}_j \mathscr{E}_k$, it must be invariant to exchange of j and k. Similarly, $\chi^{(3)}_{ijkl}$ is invariant to any permutations of j, k, and l. Equation (19.6-1) can be written in the form $\mathscr{P}_i = \epsilon_o \Sigma_j \chi^e_{ij} \mathscr{E}_j$, where χ^e_{ij} is an effective (field-dependent) tensor. By using an argument similar to that used for the linear lossless medium, it follows that χ^e_{ij} must be invariant to exchange of i and j. Thus the tensors χ_{ij}, d_{ijk}, and $\chi^{(3)}_{ijkl}$ are invariant to exchange of i and j. It follows that the three tensors are invariant to any permutations of their indices.

Elements of the tensors d_{ijk} and $\chi^{(3)}_{ijkl}$ are usually listed as 6×3 and 6×6 matrices $d_{Ik} = d_{iK}$ and $\chi^{(3)}_{IK}$, respectively, using the contracted notation defined in Table 18.2-1 on page 714, in which the single index $I = 1, \ldots, 6$ replaces the pair of indices (i, j), $i, j = 1, 2, 3$; and the index $K = 1, \ldots, 6$ replaces (k, l).

The tensors d_{ijk} and $\chi^{(3)}_{ijkl}$ are closely related to the Pockels and Kerr tensors $\mathfrak{r}_{ijk}$ and $\mathfrak{s}_{ijkl}$, respectively, as demonstrated in Problem 19.6-3, and they have the same symmetries. Tables 18.2-2 and 18.2-3 on pages 714 and 715, which list $\mathfrak{r}_{Ik}$ and $\mathfrak{s}_{IK}$, can be used to determine the symmetries of d_{Ik} and $\chi^{(3)}_{IK}$ for the different crystal groups. Table 19.6-1 provides values for the d_{Ik} coefficients for a number of crystals.

TABLE 19.6-1 Representative Magnitudes of Second-Order Nonlinear Optical Coefficients for Different Materials[a]

Crystal	d_{iK} (MKS units)[b]
Te	$d_{11} = 5.7 \times 10^{-21}$
GaAs	$d_{14} = 1.2 \times 10^{-21}$
Ag_3AsS_3 (proustite)	$d_{31} = 1.5 \times 10^{-22}$
	$d_{22} = 2.4 \times 10^{-22}$
	$d_{33} = 3.0 \times 10^{-22}$
$KNbO_3$	$d_{31} = 1.4 \times 10^{-22}$
	$d_{32} = 1.8 \times 10^{-22}$
$Ba_2NaNb_5O_{15}$ (bananas)	$d_{33} = 1.2 \times 10^{-22}$
	$d_{32} = 8.2 \times 10^{-23}$
$LiIO_3$	$d_{31} = 1.1 \times 10^{-22}$
	$d_{33} = 3.2 \times 10^{-23}$
$KTiOPO_4$ (KTP)	$d_{33} = 1.2 \times 10^{-22}$
	$d_{31} = 5.8 \times 10^{-23}$
	$d_{32} = 4.4 \times 10^{-23}$
$LiNbO_3$	$d_{31} = 4.3 \times 10^{-23}$
	$d_{22} = 2.3 \times 10^{-23}$
	$d_{33} = 3.9 \times 10^{-22}$
β-BaB_2O_4 (BBO)	$d_{22} = 1.4 \times 10^{-23}$
	$d_{31} = 7.1 \times 10^{-25}$
LiB_3O_5 (LBO)	$d_{32} = 1.1 \times 10^{-23}$
	$d_{31} = 1.0 \times 10^{-23}$
	$d_{33} = 5.6 \times 10^{-25}$
$NH_4H_2PO_4$ (ADP)	$d_{36} = 6.8 \times 10^{-24}$
KH_2PO_4 (KDP)	$d_{36} = 4.1 \times 10^{-24}$
	$d_{14} = 3.8 \times 10^{-24}$
Quartz	$d_{11} = 3.0 \times 10^{-24}$
	$d_{14} = 2.6 \times 10^{-26}$

[a] Actual values depend on the wavelength.
[b] The coefficients d/ϵ_o are often used in the literature (generally in units of pm/V). The coefficients in the table are readily converted to pm/V by dividing the tabulated values by $10^{-12}\epsilon_o = 8.85 \times 10^{-24}$.

EXERCISE 19.6-1

KDP. Use Table 18.2-2 on page 714 to verify that for crystals of $\bar{4}2m$ symmetry, such as potassium dihydrogen phosphate (KDP),

$$\mathscr{P}_1 = \epsilon_o \chi_{11} \mathscr{E}_1 + 2 d_{41} \mathscr{E}_2 \mathscr{E}_3 \tag{19.6-2}$$

$$\mathscr{P}_2 = \epsilon_o \chi_{22} \mathscr{E}_2 + 2 d_{41} \mathscr{E}_1 \mathscr{E}_3 \tag{19.6-3}$$

$$\mathscr{P}_3 = \epsilon_o \chi_{33} \mathscr{E}_3 + 2 d_{63} \mathscr{E}_1 \mathscr{E}_2, \tag{19.6-4}$$

where the axes 1, 2, 3 are the principal axes of the crystal. Determine the nonlinear polarization density vector for an electric field $\mathscr{E}$ in the x–y plane at an angle of 45° with the x and y axes.

Three-Wave Mixing in Anisotropic Second-Order Nonlinear Media

An optical field $\mathscr{E}(t)$ comprising two monochromatic linearly polarized waves of angular frequencies ω_1 and ω_2 and complex amplitudes $\mathbf{E}(\omega_1)$ and $\mathbf{E}(\omega_2)$ is applied to a second-order nonlinear crystal. The component of the polarization density vector at frequency $\omega_3 = \omega_1 + \omega_2$ may be determined by using the relation

$$P_i(\omega_3) = 2 \sum_{jk} d_{ijk} E_j(\omega_1) E_k(\omega_2), \qquad j, k = 1, 2, 3, \tag{19.6-5}$$

where $E_j(\omega_1)$, $E_k(\omega_2)$, and $P_i(\omega_3)$ are components of these vectors along the principal axes of the crystal. This equation is a generalization of (19.2-11d).

If $E_j(\omega_1) = E(\omega_1) \cos\theta_{1j}$ and $E_k(\omega_2) = E(\omega_2) \cos\theta_{2k}$, where θ_{1j} and θ_{2k} are the angles the vectors $\mathbf{E}(\omega_1)$ and $\mathbf{E}(\omega_2)$ make with the principal axes, then (19.6-5) may be written in the form

$$P_i(\omega_3) = 2 d_{\text{eff}} E(\omega_1) E(\omega_2), \tag{19.6-6}$$

where

$$d_{\text{eff}} = \sum_{jk} d_{ijk} \cos\theta_{1j} \cos\theta_{2k}, \qquad i, j, k = 1, 2, 3. \tag{19.6-7}$$

Equation (19.6-6) is in the form used in the scalar formulation in Secs. 19.2C and 19.4, where d_{eff} plays the role of the d coefficient.

Phase Matching in Three-Wave Mixing

As shown in Sec. 19.2C, the phase-matching condition $\mathbf{k}_3 = \mathbf{k}_1 + \mathbf{k}_2$ is necessary for efficient wave mixing. This condition is equivalent to $\omega_3 n_3 \hat{u}_3 = \omega_1 n_1 \hat{u}_1 + \omega_2 n_2 \hat{u}_2$, where $\hat{u}_1$, $\hat{u}_2$, and $\hat{u}_3$ are unit vectors in the directions of propagation of the waves. We assume that the three waves are normal modes of the crystal (see Sec. 6.3) with phase velocities c_o/n_a, c_o/n_b, and c_o/n_c. Note that n_a, n_b, and n_c depend on the directions of the waves, their polarizations, and on the frequencies. In a uniaxial crystal, n_a, n_b, and n_c may be the ordinary or extraordinary indices.

As an example, consider second-harmonic generation in a uniaxial crystal with waves traveling in the same direction. Assuming that waves 1 and 2 are identical, $\omega_1 = \omega_2 = \omega$, and $\omega_3 = 2\omega$, the phase-matching condition becomes $n_a = n_c$. It is then

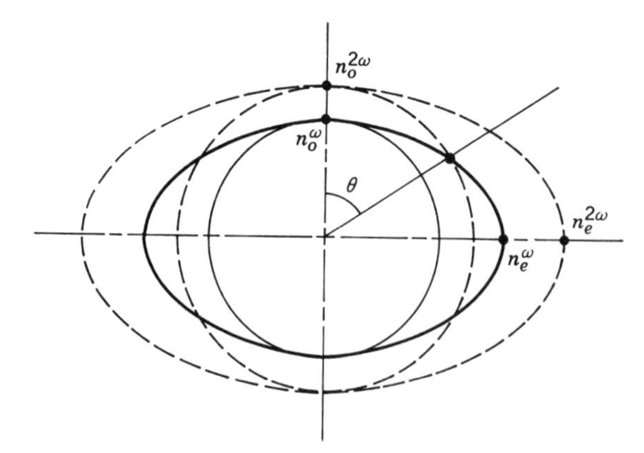

Figure 19.6-1 Matching the extraordinary refractive index of the fundamental wave to the ordinary refractive index of the second-harmonic wave.

necessary to find the direction and polarizations of the two waves such that the wave of frequency ω has the same refractive index as the wave of frequency 2ω.

As explained in Sec. 6.3, the normal modes for a wave traveling in a uniaxial crystal with refractive indices n_o and n_e are an ordinary wave with refractive index n_o (independent of direction) and an extraordinary wave with refractive index $n(\theta)$ satisfying $1/n^2(\theta) = \cos^2\theta/n_o^2 + \sin^2\theta/n_e^2$, where θ is the angle between the direction of the wave and the optic axis. The dependence of these two refractive indices on θ is illustrated by the ellipse and the circle in Fig. 19.6-1 (see also Fig. 6.3-11). Since n_o and n_e are frequency dependent, we denote them as n_o^{ω}, $n_o^{2\omega}$, n_e^{ω}, $n_e^{2\omega}$ and represent the ellipse/circle at the fundamental frequency ω by solid curves and the ellipse/circle at the second-harmonic 2ω by dashed curves. To match $n_a = n^{\omega}(\theta)$ to $n_b = n_o^{2\omega}$, a direction is found for which the circle at 2ω intersects the ellipse at ω, as illustrated in Fig. 19.6-1. This is achieved by selecting an angle θ for which

$$\frac{1}{n_o^{2\omega}} = \frac{\cos^2\theta}{\left(n_o^{\omega}\right)^2} + \frac{\sin^2\theta}{\left(n_e^{\omega}\right)^2}.$$

Thus the fundamental wave is an extraordinary wave and the second-harmonic wave is an ordinary wave.

*19.7 DISPERSIVE NONLINEAR MEDIA

This section provides a brief discussion of the origin of dispersion and its effect on nonlinear optical processes. For simplicity, anisotropic effects are not included. A dispersive medium is a medium with memory (see Sec. 5.2); the polarization density $\mathscr{P}(t)$ resulting from an applied electric field $\mathscr{E}(t)$ does not occur instantaneously. The response $\mathscr{P}(t)$ at time t is a function of the applied electric field $\mathscr{E}(t')$ at times $t' \le t$. When the medium is also nonlinear, the functional relation between $\mathscr{P}(t)$ and $\{\mathscr{E}(t'), t' \le t\}$ is nonlinear. There are two means for describing such nonlinear dynamical systems:

- A phenomenological integral relation between $\mathscr{P}(t)$ and $\mathscr{E}(t)$ based on an expansion, similar to the Taylor's series expansion, called the Volterra series expansion. The coefficients of the expansion characterize the medium phenomenologically. Coefficients similar to χ, $\mathscr{d}$, and $\chi^{(3)}$ are defined and turn out to be frequency dependent.

- A nonlinear differential equation for $\mathscr{P}(t)$, with $\mathscr{E}(t)$ as a driving force obtained by using a model that describes the physics of the polarization process.

Integral-Transform Description of Dispersive Nonlinear Media

If the deviation from linearity is small, a Volterra series expansion may be used to describe the relation between $\mathscr{P}(t)$ and $\mathscr{E}(t)$. The first term of the expansion is a linear combination of $\mathscr{E}(t')$ for all $t' \leq t$

$$\mathscr{P}(t) = \epsilon_o \int_{-\infty}^{\infty} x(t - t')\mathscr{E}(t')\, dt'. \tag{19.7-1}$$

This is a linear system with impulse-response function $\epsilon_o x(t)$ [see Section 5.2, in particular (5.2-17), and Appendix B].

The second term in the expansion is a superposition of the products $\mathscr{E}(t')\mathscr{E}(t'')$ at pairs of times $t' \leq t$ and $t'' \leq t$,

$$\mathscr{P}(t) = \epsilon_o \iint_{-\infty}^{\infty} x^{(2)}(t - t', t - t'')\mathscr{E}(t')\mathscr{E}(t'')\, dt'\, dt'', \tag{19.7-2}$$

where $x^{(2)}(t', t'')$ is a function of two variables that characterizes the second-order dispersive nonlinearity. The third term represents a third-order nonlinearity which can be characterized by a function $x^{(3)}(t', t'', t''')$ and a similar triple integral relation.

The *linear* dispersive contribution described by (19.7-1) can also be completely characterized by the response to monochromatic fields. If $\mathscr{E}(t) = \text{Re}\{E(\omega)\exp(j\omega t)\}$, then $\mathscr{P}(t) = \text{Re}\{P(\omega)\exp(j\omega t)\}$, where $P(\omega) = \epsilon_0 \chi(\omega) E(\omega)$ and $\chi(\omega)$ is the Fourier transform of $x(t)$ at $\nu = \omega/2\pi$. The medium is thus characterized completely by the frequency-dependent susceptibility $\chi(\omega)$.

The *second-order nonlinear* contribution described by (19.7-2) is characterized by the response to a superposition of *two* monochromatic waves of angular frequencies ω_1 and ω_2. Substituting

$$\mathscr{E}(t) = \text{Re}\{E(\omega_1)\exp(j\omega_1 t) + E(\omega_2)\exp(j\omega_2 t)\} \tag{19.7-3}$$

into (19.7-2), it can be shown that the polarization-density component of angular frequency $\omega_3 = \omega_1 + \omega_2$ has an amplitude

$$P(\omega_3) = 2\mathscr{d}(\omega_3; \omega_1, \omega_2)E(\omega_1)E(\omega_2). \tag{19.7-4}$$

The coefficient $\mathscr{d}(\omega_3; \omega_1, \omega_2)$ is a frequency-dependent version of the coefficient $\mathscr{d}$ in (19.2-11d). The relation between this coefficient and the response function $x^{(2)}(t', t'')$ is established by defining

$$\mathscr{X}^{(2)}(\omega_1, \omega_2) = \iint_{-\infty}^{\infty} x^{(2)}(t', t'')\exp[-j(\omega_1 t' + \omega_2 t'')]\, dt'\, dt'', \tag{19.7-5}$$

which is the two-dimensional Fourier transform of $x(t', t'')$ evaluated at $\nu_1 = -\omega_1/2\pi$ and $\nu_2 = -\omega_2/2\pi$ [see Appendix A, (A.3-2)]. Substituting (19.7-3) into (19.7-2) and using (19.7-5), we obtain

$$\mathscr{d}(\omega_3; \omega_1, \omega_2) = \epsilon_o \mathscr{X}^{(2)}(\omega_1, \omega_2). \tag{19.7-6a}$$

Thus the second-order nonlinear dispersive medium is completely characterized by either of the frequency-dependent functions, $\mathscr{X}^{(2)}(\omega_1, \omega_2)$ or $\mathscr{d}(\omega_3; \omega_1, \omega_2)$.

The degenerate case of second-harmonic generation in a second-order nonlinear medium is also readily described by substituting $\mathscr{E}(t) = \mathrm{Re}\{E(\omega)\exp(j\omega t)\}$ into (19.7-2) and using (19.7-5). The resultant polarization has a component at frequency 2ω with amplitude $P(2\omega) = \mathscr{d}(2\omega; \omega, \omega)E(\omega)E(\omega)$, where

$$\mathscr{d}(2\omega; \omega, \omega) = \tfrac{1}{2}\epsilon_o \mathscr{X}^{(2)}(\omega, \omega). \tag{19.7-6b}$$

Other $\mathscr{d}$ coefficients representing various wave mixing processes may similarly be related to the two-dimensional function $\mathscr{X}^{(2)}(\omega_1, \omega_2)$. The electro-optic effect, for example, is a result of interaction between a steady field ($\omega_1 = 0$) and an optical wave ($\omega_2 = \omega$) to generate a polarization density at $\omega_3 = \omega$. The pertinent coefficient for this interaction is $\mathscr{d}(\omega; 0, \omega) = 2\epsilon_0 \mathscr{X}^{(2)}(\omega, 0)$. This is the coefficient that determines the Pockels coefficient $\mathfrak{r}$ in accordance with (19.2-10).

In a *third-order nonlinear* medium, an electric field comprising three harmonic functions of angular frequencies ω_1, ω_2, and ω_3 creates a sum-frequency polarization density with a component at angular frequency $\omega_4 = \omega_1 + \omega_2 + \omega_3$ of amplitude $P(\omega_4) = 6\chi^{(3)}(\omega_4; \omega_1, \omega_2, \omega_3)E(\omega_1)E(\omega_2)E(\omega_3)$, where the function $\chi^{(3)}(\omega_4; \omega_1, \omega_2, \omega_3)$ replaces the coefficient $\chi^{(3)}$ which describes the nondispersive case. The function $\chi^{(3)}(\omega_4; \omega_1, \omega_2, \omega_3)$ can be determined from $\mathscr{x}^{(3)}(t', t'', t''')$ by relations similar to (19.7-6a).

In summary: As a consequence of dispersion, the second- and third-order nonlinear coefficients $\mathscr{d}$ and $\chi^{(3)}$ are dependent on the frequencies of the waves involved in the wave mixing process.

Differential-Equation Description of Dispersive Nonlinear Media

An example of a nonlinear dynamic relation between $\mathscr{P}(t)$ and $\mathscr{E}(t)$ described by a differential equation is the relation

$$\frac{d^2\mathscr{P}}{dt^2} + \sigma\frac{d\mathscr{P}}{dt} + \omega_0^2\mathscr{P} + \omega_0^2\epsilon_o\chi_0 b\mathscr{P}^2 = \omega_0^2\epsilon_o\chi_0\mathscr{E}, \tag{19.7-7}$$

where σ, ω_0, χ_0, and b are constants. Without the nonlinear term $\omega_0^2\epsilon_o\chi_0 b\mathscr{P}^2$, this equation describes a medium in which each atom is described by the harmonic-oscillator model of an electron of mass m subject to an electric-field force $e\mathscr{E}$, an elastic restraining force $-\kappa x$, and a frictional force $m\sigma\, dx/dt$, where x is the displacement of the electron from its equilibrium position and $\omega_0 = (\kappa/m)^{1/2}$ is the resonance angular frequency (see Sec. 5.5C). The medium is then linear and dispersive with a susceptibility

$$\boxed{\chi(\omega) = \chi_0\frac{\omega_0^2}{\left(\omega_0^2 - \omega^2\right) + j\omega\sigma}.} \tag{19.7-8}$$

Linear Susceptibility (Harmonic-Oscillator Model)

When the restraining force is a nonlinear function of displacement, $-\kappa x - \kappa_2 x^2$, where κ and κ_2 are constants, we have an anharmonic oscillator described by (19.7-7), where b is proportional to κ_2. The medium is then nonlinear.

EXERCISE 19.7-1

Polarization Density. Show that for a medium containing N atoms per unit volume, each modeled as an anharmonic (nonlinear) oscillator with restraining force $-\kappa x - \kappa_2 x^2$, the relation between $\mathscr{P}(t)$ and $\mathscr{E}(t)$ is the nonlinear differential equation (19.7-7), where $\chi_0 = Ne^2/m\omega_0^2\epsilon_o$ and $b = \kappa_2/e^3N^2$.

Equation (19.7-7) cannot be solved exactly. However, if the nonlinear term is small, an iterative approach provides an approximate solution. We write (19.7-7) in the form

$$\mathscr{L}\{\mathscr{P}\} = \mathscr{E} - b\mathscr{P}^2, \qquad (19.7\text{-}9)$$

where $\mathscr{L} = (\omega_0^2\epsilon_o\chi_0)^{-1}(d^2/dt^2 + \sigma d/dt + \omega_0^2)$ is a linear differential operator. The iterative solution of (19.7-9) is described by the following steps:

- Find a first-order approximation $\mathscr{P}_1$ by neglecting the nonlinear term $b\mathscr{P}^2$ in (19.7-9), and solving the *linear* equation

$$\mathscr{L}\{\mathscr{P}_1\} \approx \mathscr{E}. \qquad (19.7\text{-}10)$$

- Use this approximate solution to determine the small nonlinear term $b\mathscr{P}_1^2$.
- Obtain a second-order approximation by solving (19.7-9) with the term $b\mathscr{P}^2$ replaced by $b\mathscr{P}_1^2$. The solution of the resultant *linear* equation is denoted $\mathscr{P}_2$,

$$\mathscr{L}\{\mathscr{P}_2\} = \mathscr{E} - b\mathscr{P}_1^2. \qquad (19.7\text{-}11)$$

- Repeat the process to obtain a third-order approximation as illustrated by the block diagram of Fig. 19.7-1.

We first examine the special case of monochromatic light, $\mathscr{E} = \text{Re}\{E(\omega)\exp(j\omega t)\}$. In the first iteration $\mathscr{P}_1 = \text{Re}\{P_1(\omega)\exp(j\omega t)\}$, where $P_1(\omega) = \epsilon_o\chi(\omega)E(\omega)$ and $\chi(\omega)$ is given by (19.7-8). In the second iteration, the linear system is driven by a force

$$\begin{aligned}\mathscr{E} - b\mathscr{P}_1^2 &= \text{Re}\{E(\omega)e^{j\omega t}\} - b\left[\text{Re}\{\epsilon_o\chi(\omega)\,e^{j\omega t}\}\right]^2 \\ &= \text{Re}\{E(\omega)e^{j\omega t}\} - \tfrac{1}{2}b\,\text{Re}\{[\epsilon_o\chi(\omega)E(\omega)]^2e^{j2\omega t}\} - \tfrac{1}{2}b|\epsilon_o\chi(\omega)E(\omega)|^2.\end{aligned}$$

Since these three terms have frequencies ω, 2ω, and 0, the linear system responds with

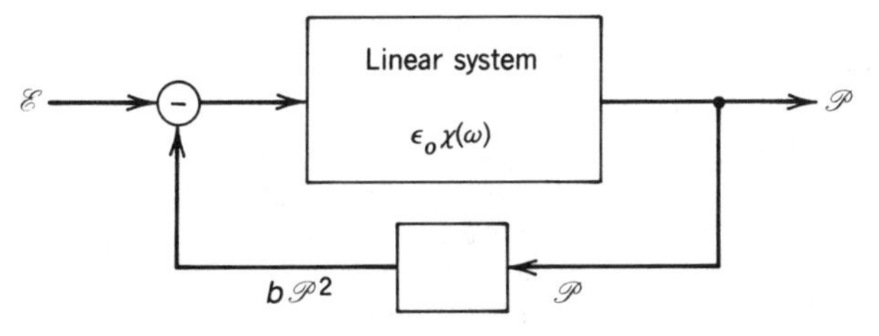

Figure 19.7-1 Block diagram representing the nonlinear differential equation (19.7-9). The linear system represented by the operator equation $\mathscr{L}\{\mathscr{P}\} = \mathscr{E}$ has a transfer function $\epsilon_o\chi(\omega)$.

susceptibilities $\chi(\omega)$, $\chi(2\omega)$, and $\chi(0)$, respectively. The component of $\mathscr{P}_2$ at frequency 2ω has an amplitude $P_2(2\omega) = \epsilon_o\chi(2\omega)\{-\frac{1}{2}b[\epsilon_0\chi(\omega)E(\omega)]^2\}$. Since $P(2\omega) = d(2\omega; \omega, \omega)E(\omega)E(\omega)$, we conclude that

$$d(2\omega; \omega, \omega) = -\tfrac{1}{2}b\epsilon_o^2[\chi(\omega)]^2\chi(2\omega). \tag{19.7-12}$$

EXERCISE 19.7-2

Miller's Rule. Show that for the nonlinear resonant medium described by (19.7-7) if the light is a superposition of two monochromatic waves of angular frequencies ω_1 and ω_2, then the second-order approximation described by (19.7-10) and (19.7-11) yields a component of polarization at frequency $\omega_3 = \omega_1 + \omega_2$ with amplitude $P_2(\omega_3) = 2d(\omega_3; \omega_1, \omega_2)E(\omega_1)E(\omega_2)$, where

$$d(\omega_3; \omega_1, \omega_2) = -\tfrac{1}{2}b\epsilon_o^2\chi(\omega_1)\chi(\omega_2)\chi(\omega_3). \tag{19.7-13}$$

Miller's Rule

Equation (19.7-13) is known as Miller's rule.

Miller's rule states that the coefficient of second-order nonlinearity for the generation of a wave of frequency $\omega_3 = \omega_1 + \omega_2$ from two waves of frequencies ω_1 and ω_2 is proportional to the product $\chi(\omega_1)\chi(\omega_2)\chi(\omega_3)$ of the linear susceptibilities at the three frequencies. The three frequencies must therefore lie within the optical transmission window of the medium (away from resonance). If these frequencies are much smaller than the resonance frequency ω_0, then (19.7-8) gives $\chi(\omega) \approx \chi_0$, and (19.7-13) yields $d(\omega_3; \omega_1, \omega_2) = -\frac{1}{2}b\epsilon_o^2\chi_0^3$, which is independent of frequency. The medium is then approximately nondispersive, and the results of the previous sections in which dispersion was neglected are applicable. Miller's rule also indicates that materials with large refractive indices (large χ_0) tend to have large d.

19.8 OPTICAL SOLITONS

When a pulse of light travels in a linear *dispersive* medium its shape changes continuously because its constituent frequency components travel at different group velocities and undergo different time delays [see Sec. 5.6 and Fig. 19.8-1(*a*)]. If the medium is also *nonlinear*, self-phase modulation (which results, for example, from the optical Kerr effect) alters the phase, and therefore the frequency, of the weak and intense parts of the pulse by unequal amounts. As a result of group-velocity dispersion, different parts of the pulse travel at different group velocities and the pulse shape is altered. The interplay between self-phase modulation and group-velocity dispersion can therefore result in an overall pulse spreading or pulse compression, depending on the magnitudes and signs of these two effects.

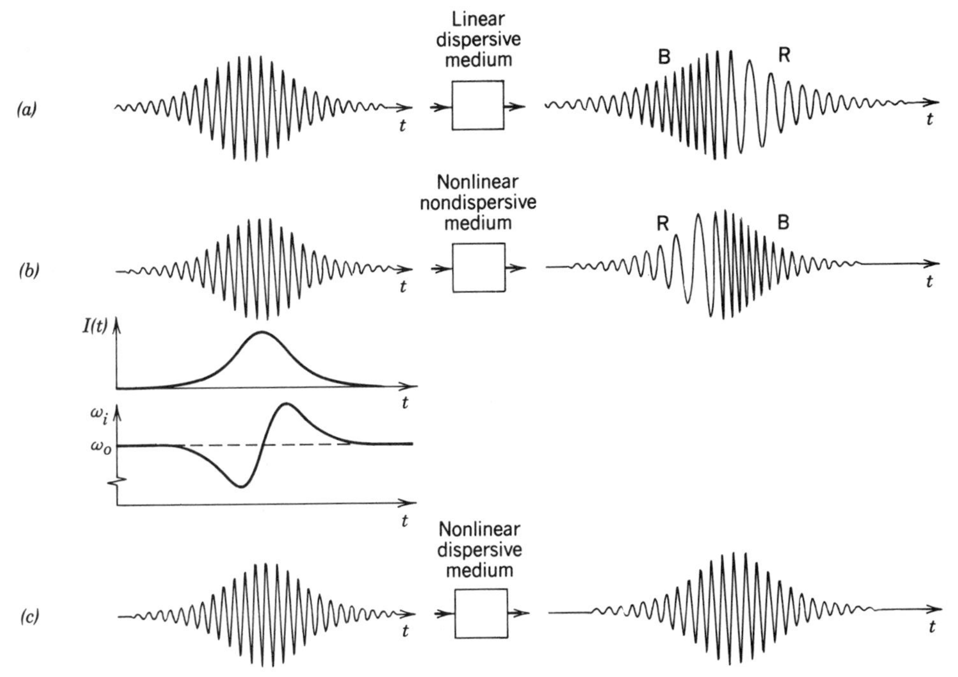

Figure 19.8-1 (a) Pulse spreading in a linear medium with anomalous dispersion; the shorter-wavelength component B has a larger group velocity and therefore travels faster than the longer-wavelength component R. (b) In a nonlinear medium, self-phase modulation ($n_2 > 0$) introduces a negative frequency shift in the leading half of the pulse (denoted R) and a positive-frequency shift in the trailing half (denoted B). The pulse is chirped, but its shape is not altered. If the chirped wave in (b) travels in the linear dispersive medium in (a), the pulse will be compressed. (c) If the medium is both nonlinear and dispersive, the pulse can be compressed, expanded, or maintained (creating a solitary wave), depending on the magnitudes and signs of the dispersion and nonlinear effects. This illustration shows a solitary wave.

Under certain conditions, an optical pulse of prescribed shape and intensity can travel in a *nonlinear dispersive* medium without ever altering its shape, as if it were traveling in an ideal *linear nondispersive* medium. This occurs when group-velocity dispersion fully compensates the effect of self-phase modulation. Such pulse-like stationary waves are called **solitary waves**. Optical **solitons** are special solitary waves that are orthogonal, in the sense that when two of these waves cross one another in the medium their intensity profiles are not altered (only phase shifts are imparted as a result of the interaction), so that each wave continues to travel as an independent entity.

The interplay between group-velocity dispersion and self-phase modulation may be understood by examining a pulse of intensity $I(z,t)$ and central angular frequency ω_0 traveling in the z direction in a nonlinear medium with refractive index $n = n_0 + n_2 I(z,t)$ [see Fig. 19.8-1(b)]. When the pulse travels a distance Δz it undergoes a phase shift $k_o[n_0 + n_2 I(z,t)]\,\Delta z$. The argument of the field is therefore $\varphi(t) = \omega_0 t - k_o[n_0 + n_2 I(z,t)]\,\Delta z$, so that the instantaneous angular frequency is $\omega_i = d\varphi/dt = \omega_0 - k_o n_2\,\Delta z\, dI(z,t)/dt$. If n_2 is positive, the frequency of the trailing half of the pulse (the right half) is increased (blue shifted) since $dI/dt < 0$, whereas the frequency of the leading half (the left half) is reduced (red shifted) since $dI/dt > 0$, as illustrated in Fig. 19.8-1(b). The pulse is therefore chirped (i.e., its instantaneous frequency varies with time). If the medium has anomalous dispersion (i.e., the dispersion coefficient D_λ is positive, or the coefficient $\beta'' = d^2\beta/d\omega^2$ [see (5.6-9)] is negative), the group velocity decreases with increasing wavelength. Thus the blue-shifted half of the pulse travels faster than the red-shifted half. As a result, the blue-shifted half catches up with the red-shifted half and the pulse is compressed (a related situation occurs in a medium with normal dispersion as shown in Fig. 5.6-4; this effect is used to generate ultrashort light pulses).

At a certain level of intensity and for certain pulse profiles, the effects of self-phase modulation and group-velocity dispersion are balanced so that a stable pulse, a soliton, travels without spread, as illustrated in Fig. 19.8-1(c). The chirping effect of self-phase modulation perfectly compensates the natural pulse expansion caused by the group-velocity dispersion. Any slight spreading of the pulse enhances the compression process, and any pulse narrowing reduces the compression process, so that the pulse shape and width are maintained. Solitons can be thought of as the modes (eigenfunctions) of a nonlinear dispersive system. A mathematical analysis of this phenomenon is based on solutions of the nonlinear wave equation that governs the propagation of the pulse envelope, as described subsequently.

The optical solitons described in this section are analogous to spatial solitons (self-guided beams). As explained in Sec. 19.3A, spatial solitons are monochromatic waves that are localized spatially in the transverse plane. They travel in a nonlinear medium without altering their spatial distribution, as a result of a balance between diffraction and self-phase modulation. Thus, spatial solitons are the transverse analogs of longitudinal (temporal) optical solitons. This analogy is not surprising since diffraction is the spatial equivalent of dispersion. The phenomena are described by the same differential equation, with space and time interchanged. In fact the term soliton refers to generic solutions describing pulses that propagate without change; they may be temporal or spatial.

Differential Equation for the Wave Envelope

To describe the propagation of an optical pulse in a nonlinear dispersive medium we start with the wave equation (19.1-3),

$$\nabla^2\mathscr{E} - \frac{1}{c_o^2}\frac{\partial^2\mathscr{E}}{\partial t^2} = \mu_o\frac{\partial^2}{\partial t^2}(\mathscr{P}_L + \mathscr{P}_{NL}), \tag{19.8-1}$$

where $\mathscr{P}_L$ and $\mathscr{P}_{NL}$ are the linear and the nonlinear components of the polarization density, respectively. Since the medium is dispersive, $\mathscr{P}_L(t)$ is related to $\mathscr{E}(t)$ by a time integral, the convolution in (5.2-17). The component $\mathscr{P}_{NL}$ is related to $\mathscr{E}$ by the nonlinear relation $\mathscr{P}_{NL} = 4\chi^{(3)}\mathscr{E}^3$, assumed here to be approximately instantaneous. Thus (19.8-1) gives a nonlinear integrodifferential equation in $\mathscr{E}$. Clearly, some approximations are necessary in order to solve this equation.

It is convenient to combine the linear terms in (19.8-1) and write

$$\nabla^2\mathscr{E} + \mathscr{F} = \mu_o\frac{\partial^2\mathscr{P}_{NL}}{\partial t^2}, \tag{19.8-2}$$

where

$$\mathscr{F} = -\mu_o\frac{\partial^2}{\partial t^2}(\epsilon_o\mathscr{E} + \mathscr{P}_L). \tag{19.8-3}$$

Since $\mathscr{P}_L$ is linearly related to $\mathscr{E}$, $\mathscr{F}$ must also be linearly related to $\mathscr{E}$. If $\mathscr{E} =$

$\mathrm{Re}\{E(\omega)\exp(j\omega t)\}$, then $\mathscr{F} = \mathrm{Re}\{F(\omega)\exp(j\omega t)\}$, where

$$F(\omega) = \beta^2(\omega)E(\omega). \tag{19.8-4}$$

The coefficient $\beta(\omega)$ is the propagation constant in the linear medium. In the absence of nonlinearity, (19.8-2) reproduces the Helmholtz equation $\nabla^2 E + \beta^2(\omega)E = 0$.

As in the analysis of pulse propagation in linear dispersive media (see Sec. 5.6), we consider a plane wave traveling in the z direction with central angular frequency ω_0 and central wavenumber $\beta_0 = \beta(\omega_0)$,

$$\mathscr{E} = \mathrm{Re}\{\mathscr{A}(z,t)\exp[j(\omega_0 t - \beta_0 z)]\}, \tag{19.8-5}$$

where the complex envelope $\mathscr{A}$ is assumed to be a slowly varying function of t and z (in comparison with the period $2\pi/\omega_0$ and the wavelength $\lambda = 2\pi/\beta_0$, respectively). Also, as in Sec. 5.6, for weak dispersion we approximate the propagation constant $\beta(\omega)$ by three terms of a Taylor's series expansion about ω_0, $\beta(\omega_0 + \Omega) = \beta_0 + \Omega\beta' + \frac{1}{2}\Omega^2\beta''$, where β_0, β', and β'' are the values of $\beta(\omega)$ and its first and second derivatives with respect to ω at $\omega = \omega_0$. The phase velocity c, the group velocity v, and the dispersion coefficient D_ν are related to the coefficients β_0, β', and β'' by $c = \omega_0/\beta_0$, $v = 1/\beta'$, and $D_\nu = 2\pi\beta''$, as defined in (5.6-8) and (5.6-9).

Using the three assumptions—slowly varying envelope, weak dispersion, and small nonlinear effect—it will subsequently be shown that the envelope $\mathscr{A}(z,t)$ satisfies the following differential equation:

$$\left(\frac{\partial}{\partial z} + \frac{1}{v}\frac{\partial}{\partial t}\right)\mathscr{A} - j\frac{\beta''}{2}\frac{\partial^2\mathscr{A}}{\partial t^2} - j\gamma|\mathscr{A}|^2\mathscr{A} = 0, \tag{19.8-6}$$

The Envelope Equation

where

$$\gamma = \tfrac{3}{2}\mu_o c\omega_0\chi^{(3)} = \frac{\omega_0}{2c_o}\frac{n_2}{\eta} \tag{19.8-7}$$

is a coefficient representing the nonlinear effect, $\eta = \eta_o/n$, $\eta_o = (\mu_o/\epsilon_o)^{1/2}$, and n_2 is the coefficient in the relation $n(I) = n + n_2 I$ defined by (19.3-6). For a linear medium ($\gamma = 0$) with no losses ($\alpha = 0$), and substituting $\beta'' = D_\nu/2\pi$ into (19.8-6), the envelope wave equation (5.6-17) is reproduced.

Derivation of the Envelope Equation

We begin with (19.8-2) and write

$$\mathscr{F} = \mathrm{Re}\{\mathscr{B}(z,t)\exp[j(\omega_0 t - \beta_0 z)]\} \tag{19.8-8a}$$

$$\mathscr{P}_{\mathrm{NL}} = \mathrm{Re}\{\mathscr{C}(z,t)\exp[j(\omega_0 t - \beta_0 z)]\}, \tag{19.8-8b}$$

where the complex envelopes $\mathscr{B}$ and $\mathscr{C}$ are assumed to be slowly varying functions of t and z. We will relate $\mathscr{B}$ to $\mathscr{A}$ in terms of the linear propagation constant $\beta(\omega)$, and relate $\mathscr{C}$ to $\mathscr{A}$ in terms of the nonlinear coefficient $\chi^{(3)}$, and ultimately substitute in (19.8-2) to obtain a differential equation for $\mathscr{A}$.

We now show that the envelopes $\mathscr{B}(z,t)$ and $\mathscr{A}(z,t)$ are related by

$$\mathscr{B} = \beta_0^2\mathscr{A} - j2\beta_0\beta'\frac{\partial\mathscr{A}}{\partial t} - \beta_0\beta''\frac{\partial^2\mathscr{A}}{\partial t^2}. \tag{19.8-9}$$

Writing $\mathscr{A}(z,t) = A(z,\Omega)\exp(j\Omega t)$ and $\mathscr{B}(z,t) = B(z,\Omega)\exp(j\Omega t)$ and using (19.8-4), (19.8-5), and (19.8-8a), we obtain

$$B(z,\Omega) = \beta^2(\omega_0 + \Omega)A(z,\Omega). \tag{19.8-10}$$

Substituting the approximation

$$\beta^2(\omega_0+\Omega) = \left(\beta_0 + \Omega\beta' + \tfrac{1}{2}\Omega^2\beta''\right)^2 \approx \beta_0^2 + 2\beta_0\left(\Omega\beta' + \tfrac{1}{2}\Omega^2\beta''\right)$$

into (19.8-10) gives

$$B(z,\Omega) = \beta_0^2 A(z,\Omega) + 2\beta_0\beta'\Omega A(z,\Omega) + \beta_0\beta''\Omega^2 A(z,\Omega). \tag{19.8-11}$$

Since $j\Omega A(z,\Omega)$ and $-\Omega^2 A(z,\Omega)$ are equivalent to $(\partial/\partial t)\mathscr{A}(z,t)$ and $(\partial^2/\partial t^2)\mathscr{A}(z,t)$, (19.8-11) yields (19.8-9).

The pertinent value of the nonlinear polarization density $\mathscr{P}_{\rm NL}$ is the component of $\mathscr{P}_{\rm NL} = 4\chi^{(3)}\mathscr{E}^3$ at frequency ω_0. This component has an envelope [see (19.3-3a)],

$$\mathscr{C} = 3\chi^{(3)}|\mathscr{A}|^2\mathscr{A}. \tag{19.8-12}$$

Substituting (19.8-9) and (19.8-12) into (19.8-8) and (19.8-2), we obtain a nonlinear partial differential equation for the envelope $\mathscr{A}$, which we simplify by using the slowly varying envelope approximation,

$$\frac{\partial^2}{\partial z^2}[\mathscr{A}\exp(-j\beta_0 z)] \approx \left[-2j\beta_0\frac{\partial\mathscr{A}}{\partial z} - \beta_0^2\mathscr{A}\right]\exp(-j\beta_0 z).$$

Since the nonlinearity is a small effect and the envelope $\mathscr{C}$ is slowly varying, we assume that $(\partial^2/\partial t^2)[\mathscr{C}\exp(j\omega_0 t)] \approx -\omega_0^2\mathscr{C}\exp(j\omega_0 t)$ and neglect higher-order terms. The resultant differential equation for $\mathscr{A}$ is (19.8-6).

Equation (19.8-6) may also be obtained if we assume that the nonlinear medium is approximately linear with a propagation constant $\beta(\omega) + \Delta\beta$, where $\Delta\beta = (\omega_0/c_o)n_2 I$. The intensity $I = |\mathscr{A}|^2/2\eta$ is assumed to be sufficiently slowly varying so that it may be regarded as time independent. The Fourier analysis which led to the differential equation for the linear medium, (5.6-17), is then simply modified by an added term proportional to $\Delta\beta\mathscr{A}$. This term produces the additional term $\gamma|\mathscr{A}|^2\mathscr{A}$, so that (19.8-6) is reproduced.

Solitons

Equation (19.8-6) governs the complex envelope $\mathscr{A}(z,t)$ of an optical pulse traveling in the z direction in an extended nonlinear dispersive medium with group velocity v, dispersion parameter β'', and nonlinear coefficient γ. A solitary-wave solution is possible if $\beta'' < 0$ (i.e., the medium exhibits anomalous group-velocity dispersion) and $\gamma > 0$ (i.e., the self-phase modulation coefficient $n_2 > 0$).

It is useful to standardize (19.8-6) by normalizing the time, the distance, and the amplitude to convenient scales τ_0, z_0, and $\mathscr{A}_0$, respectively:

- τ_0 is a constant representing the time duration of the pulse.
- The distance scale is taken to be

$$2z_0 = \frac{\tau_0^2}{|\beta''|}. \tag{19.8-13}$$

As shown in Sec. 5.6 [see (5.6-13) and (5.6-15)], if a Gaussian pulse of width τ_0 travels in a *linear* medium with dispersion parameter β'', its width increases by a factor of $\sqrt{2}$ after a distance $\tau_0^2/2|\beta''| = z_0$. The distance $2z_0$ is therefore called the dispersion distance (it is analogous to the depth of focus $2z_0$ in a Gaussian beam).

- The scale $\mathscr{A}_0$ is selected to be the amplitude at which the phase shift introduced by self-phase modulation for a propagation distance $2z_0$ is unity. Thus $(\omega_0/c_o)[n_2(\mathscr{A}_0^2/2\eta)]2z_0 = 1$. Since $\gamma = (\omega_0/2c_o)(n_2/\eta)$ and $2z_0 = \tau_0^2/|\beta''|$, this is equivalent to $\gamma\mathscr{A}_0^2\tau_0^2/|\beta''| = 1$, from which

$$\mathscr{A}_0 = \frac{(|\beta''|/\gamma)^{1/2}}{\tau_0}. \tag{19.8-14}$$

The corresponding intensity is $I_0 = \mathscr{A}_0^2/2\eta = (|\beta''|/2\gamma\eta)/\tau_0^2$. When the peak amplitude $\mathscr{A}$ of the incident pulse is much smaller than $\mathscr{A}_0$, the effect of group-velocity dispersion dominates and the nonlinear self-phase modulation is negligible. However, as we shall see subsequently, when $\mathscr{A} = \mathscr{A}_0$, these two effects compensate one another so that the pulse propagates without spread and becomes a soliton.

Using a coordinate system moving with a velocity v, and defining the dimensionless variables,

$$\mathscr{t} = \frac{(t - z/v)}{\tau_0} \tag{19.8-15a}$$

$$\mathscr{z} = \frac{z}{2z_0} = |\beta''|z/\tau_0^2 \tag{19.8-15b}$$

$$\psi = \frac{\mathscr{A}}{\mathscr{A}_0} = \tau_0\left(\frac{\gamma}{|\beta''|}\right)^{1/2}\mathscr{A}, \tag{19.8-15c}$$

(19.8-6) is converted into

$$j\frac{\partial\psi}{\partial\mathscr{z}} + \frac{1}{2}\frac{\partial^2\psi}{\partial\mathscr{t}^2} + |\psi|^2\psi = 0 \tag{19.8-16}$$

Nonlinear Schrödinger Equation

which is recognized as the nonlinear Schrödinger equation. The solution $\psi(\mathscr{z}, \mathscr{t})$ of (19.8-16) can be easily converted back into the physical complex envelope $\mathscr{A}(z, t)$ by use of (19.8-15).

The simplest solitary-wave solution of (19.8-16) is

$$\psi(\mathscr{z}, \mathscr{t}) = \operatorname{sech}(\mathscr{t}) \exp\left(j\frac{\mathscr{z}}{2}\right), \tag{19.8-17}$$

where $\operatorname{sech}(\cdot) = 1/\cosh(\cdot)$ is the hyperbolic-secant function. This solution is called the **fundamental soliton**. It corresponds to an envelope

$$\mathscr{A}(z,t) = \mathscr{A}_0 \operatorname{sech}\left(\frac{t - z/v}{\tau_0}\right) \exp\left(\frac{jz}{4z_0}\right), \tag{19.8-18}$$

Optical Soliton

which travels with velocity v without altering its shape. This solution is achieved if the incident pulse at $z = 0$ is

$$\mathscr{A}(0,t) = \mathscr{A}_0 \operatorname{sech}(t/\tau_0). \tag{19.8-19}$$

The envelope of the wave shown in Fig. 19.8-1(c) is a hyperbolic-secant function.

The envelope of the fundamental soliton is a symmetric bell-shaped function with peak value $\mathscr{A}(0,0) = \mathscr{A}_0$, width τ_0, and area $\int \psi(0,t)\,dt = 2\pi \mathscr{A}_0 \tau_0$. The intensity $I(0,t) = |\mathscr{A}(0,t)|^2/2\eta$ has a full width at half maximum $\tau_{\text{FWHM}} = 1.76\tau_0$. The width τ_0 may be arbitrarily selected by controlling the incident pulse, but the amplitude $\mathscr{A}_0$ must be adjusted such that $\mathscr{A}_0\tau_0 = (|\beta''|/\gamma)^{1/2}$. For a medium with prescribed parameters β'' and γ, therefore, the peak amplitude is inversely proportional to the width τ_0, and the peak power is inversely proportional to τ_0^2. The pulse energy $\int |\mathscr{A}|^2\,dt$ is directly proportional to $\mathscr{A}_0$, and therefore inversely proportional to τ_0. Thus a soliton of shorter duration must carry greater energy.

The fundamental soliton is only one of a family of solutions with solitary properties. For example, if the amplitude of the incident pulse $\psi(0, \mathscr{t}) = N \operatorname{sech}(\mathscr{t})$, where N is an integer, the solution, called the N-soliton wave, is a periodic function of z with period $\mathscr{z}_p = \pi/2$, called the soliton period. This corresponds to a physical distance $z_p = \pi z_0 = (\pi/2)\tau_0^2/|\beta''|$, which is directly proportional to τ_0^2. At $z = 0$ the envelope $\mathscr{A}(0,t)$ is a hyperbolic-secant function with peak amplitude $N\mathscr{A}_0$. As the pulse travels in the medium, it contracts initially, then splits into distinct pulses which merge subsequently and eventually reproduce the initial pulse at $z = z_p$. This pattern is repeated periodically. This periodic compression and expansion of the multi-soliton wave is accounted for by a periodic imbalance between the pulse compression, which results from the chirping introduced by self-phase modulation, and the pulse spreading caused by group-velocity dispersion. The initial compression has been used for generation of subpicosecond pulses.

To excite the fundamental soliton, the input pulse must have the hyperbolic-secant profile with the exact amplitude–width product $\mathscr{A}_0\tau_0$ in (19.8-14). A lower value of this product will excite an ordinary optical pulse, whereas a higher value will excite the fundamental soliton, or possibly a higher-order soliton, with the remaining energy diverted into a spurious ordinary pulse.

EXAMPLE 19.8-1. ***Solitons in Optical Fibers.*** Ultrashort solitons (several hundred femtoseconds to a few picoseconds) have been generated in glass fibers at wavelengths in the anomalous dispersion region ($\lambda_o > 1.3$ μm). They were first observed in a 700-m single-mode silica glass fiber using pulses from a mode-locked laser operating at a wavelength $\lambda_o = 1.55$ μm. The pulse shape closely approximated a hyperbolic-secant function of duration $\tau_0 = 4$ ps (corresponding to $\tau_{\text{FWHM}} = 1.76\tau_0 = 7$ ps). At this wavelength the dispersion coefficient $D_\lambda = 16$ ps/nm-km (see Fig. 8.3-5), corresponding to $\beta'' = D_\nu/2\pi = (-\lambda_o^2/c_o)D_\lambda/2\pi \approx -20$ ps^2/km. The refractive index $n = 1.45$ and the nonlinear coefficient $n_2 = 3.19 \times 10^{-16}$ cm^2/W correspond to $\gamma = (\pi/\lambda_o)(n_2/\eta) = 2.48 \times 10^{-16}$ m/V^2. The amplitude $\mathscr{A}_0 = (|\beta''|/\gamma)^{1/2}/\tau_0 \approx 2.25 \times 10^6$ V/m, corresponding to an intensity $I_0 = \mathscr{A}_0^2/2\eta \approx 10^6$ W/cm^2 (where $\eta = \eta_o/n = 260$ Ω). If the fiber area is 100 μm^2, this corresponds to a power of about 1 W. The soliton period $z_p = \pi z_0 = \pi\tau_0^2/2|\beta''| = 1.26$ km.

Soliton Lasers

Using Raman amplification (see Sec. 19.3A) to overcome absorption and scattering losses, optical solitons of a few tens of picoseconds duration have been successfully transmitted through many thousands of kilometers of optical fiber. Because of their unique property of maintaining their shape and width over long propagation distances, optical solitons have potential applications for the transmission of digital data through optical fibers at higher rates and for longer distances than presently possible with linear optics (see Sec. 22.1D).

Optical-fiber lasers have also been used to generate picosecond solitons. The laser is a single-mode fiber in a ring cavity configuration (Fig. 19.8-2). The fiber is a combination of an erbium-doped fiber amplifier (see Sec. 14.2E) and an undoped fiber providing the pulse shaping and soliton action. Pulses are obtained by using a phase modulator to achieve mode locking. A totally integrated system has been developed using an InGaAsP laser-diode pump and an integrated-optic phase modulator.

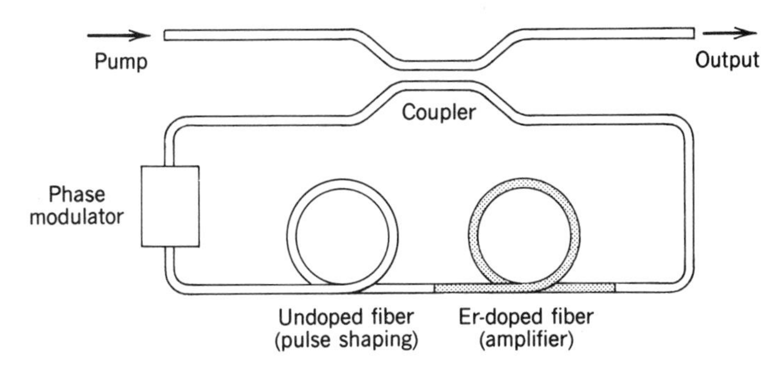

Figure 19.8-2 An optical-fiber soliton laser.

Dark solitons have also been observed. These are short-duration dips in the intensity of an otherwise continuous wave of light. They have properties similar to the "bright" solitons described earlier, but can be generated in the normal dispersion region ($\lambda_o < 1.3$ μm in silica optical fibers). They exhibit robust features that may be useful for optical switching.

READING LIST

General Books

H. M. Gibbs, G. Khitrova, and N. Peyghambarian, eds., *Nonlinear Photonics*, Springer-Verlag, New York, 1990.

P. N. Butcher and D. Cotter, *The Elements of Nonlinear Optics*, Cambridge University Press, New York, 1990.

A. Yariv, *Quantum Electronics*, Wiley, New York, 1967, 3rd ed. 1989.

V. S. Butylkin, A. E. Kaplan, Yu. G. Khronopulo, and E. I. Yakubovich, *Resonant Nonlinear Interactions of Light with Matter*, Springer-Verlag, Berlin, 1989.

R. A. Hann and D. Bloor, eds., *Organic Materials for Non-Linear Optics*, CRC Press, Boca Raton, FL, 1989.

P. W. Milonni and J. H. Eberly, *Lasers*, Wiley, New York, 1988, Chaps. 17 and 18.

N. B. Delone and V. P. Krainov, *Fundamentals of Nonlinear Optics of Atomic Gases*, Wiley, New York, 1988.

H. Haug, ed., *Optical Nonlinearities and Instabilities in Semiconductors*, Academic Press, Boston, 1988.

D. S. Chemla and J. Zyss, eds., *Nonlinear Optical Properties of Organic Molecules and Crystals*, vols. 1 and 2, Academic Press, Orlando, FL, 1987.

M. Schubert and B. Wilhelmi, *Nonlinear Optics and Quantum Electronics*, Wiley, New York, 1986.

F. A. Hopf and G. I. Stegeman, *Applied Classical Electrodynamics*, Vol. 2, *Nonlinear Optics*, Wiley, New York, 1986.

C. Flytzanis and J. L. Oudar, eds., *Nonlinear Optics: Materials and Devices*, Springer-Verlag, Berlin, 1986.

M. J. Weber, ed., *Handbook of Laser Science and Technology*, vol. III, *Optical Materials: Part 1, Nonlinear Optical Properties–Radiation Damage*, CRC Press, Boca Raton, FL, 1986.

B. B. Laud, *Lasers and Non-Linear Optics*, Wiley, New York, 1985.

A. Yariv and P. Yeh, *Optical Waves in Crystals*, Wiley, New York, 1984.

J. F. Reintjes, *Nonlinear Optical Parametric Processes in Liquids and Gases*, Academic Press, New York, 1984.

Y. R. Shen, *The Principles of Nonlinear Optics*, Wiley, New York, 1984.

M. S. Feld and V. S. Letokhov, eds., *Coherent Nonlinear Optics*, Springer-Verlag, New York, 1980.

D. C. Hanna, M. A. Yuratich, and D. Cotter, *Nonlinear Optics of Free Atoms and Molecules*, Springer-Verlag, New York, 1979.

H. Rabin and C. L. Tang, eds., *Quantum Electronics*, Academic Press, New York, 1975.

V. I. Karpman, *Nonlinear Waves in Dispersive Media*, Pergamon Press, Oxford, 1975.

I. P. Kaminow, *An Introduction to Electrooptic Devices*, Academic Press, New York, 1974.

G. B. Whitham, *Linear and Nonlinear Waves*, Wiley, New York, 1974.

F. Zernike and J. E. Midwinter, *Applied Nonlinear Optics*, Wiley, New York, 1973.

S. A. Akhmanov and R. V. Khokhlov, *Problems of Nonlinear Optics*, Gordon and Breach, New York, 1972.

R. H. Pantell and H. E. Puthoff, *Fundamentals of Quantum Electronics*, Wiley, New York, 1969.

G. C. Baldwin, *An Introduction to Nonlinear Optics*, Plenum Press, New York, 1969.

N. Bloembergen, *Nonlinear Optics*, W. A. Benjamin, Reading, MA, 1965, 1977.

Books on Ultrashort Pulses and Optical Solitons

P. J. Olver and D. H. Sattinger, eds., *Solitons in Physics, Mathematics, and Nonlinear Optics*, Springer-Verlag, New York, 1990.

A. Hasegawa, *Optical Solitons in Fibers*, Springer-Verlag, Berlin, 1989.

G. P. Agrawal, *Nonlinear Fiber Optics*, Academic Press, Boston, 1989.

P. G. Drazin and R. S. Johnson, *Solitons: An Introduction*, Cambridge University Press, New York, 1989.

E. M. Dianov, P. V. Mamyshev, A. M. Prokhorov, and V. N. Serkin, *Nonlinear Effects in Optical Fibers*, Harwood Academic Publishers, Chur, Switzerland, 1989.

W. Rudolph and B. Wilhelmi, *Light Pulse Compression*, Harwood Academic Publishers, Chur, Switzerland, 1989.

W. Kaiser, ed., *Ultrashort Laser Pulses and Applications*, Springer-Verlag, Berlin, 1988.

R. K. Dodd, J. C. Elbeck, J. D. Gibson, and H. C. Morris, *Solitons and Nonlinear Wave Equations*, Academic Press, New York, 1982.

G. L. Lamb, Jr., *Elements of Soliton Theory*, Wiley, New York, 1980.

K. Lonngren and A. Scott, eds., *Solitons in Action*, Academic Press, New York, 1978.

Special Journal Issues

Special issue on nonlinear optical phase conjugation, *IEEE Journal of Quantum Electronics*, vol. QE–25, no. 3, 1989.

Special issue on the quantum and nonlinear optics of single electrons, atoms, and ions, *IEEE Journal of Quantum Electronics*, vol. QE–24, no. 7, 1988.

Special issue on nonlinear guided-wave phenomena, *Journal of the Optical Society of America B*, vol. 5, no. 2, 1988.

Special issue on nonlinear optical processes in organic materials, *Journal of the Optical Society of America B*, vol. 4, no. 6, 1987.

Special issue on dynamic gratings and four-wave mixing, *IEEE Journal of Quantum Electronics*, vol. QE-22, no. 8, 1986.

Special issue on coherent optical transients, *Journal of the Optical Society of America B*, vol. 3, no. 4, 1986.

Special issue on stimulated Raman and Brillouin scattering for laser beam control, *Journal of the Optical Society of America B*, vol. 3, no. 10, 1986.

Special issue on excitonic optical nonlinearities, *Journal of the Optical Society of America B*, vol. 2, no. 7, 1985.

Articles

V. Mizrahi and J. E. Sipe, The Mystery of Frequency Doubling in Optical Fibers, *Optics and Photonics News*, vol. 2, no. 1, pp. 16–20, 1991.

G. I. Stegeman and R. Stolen, Nonlinear Guided Wave Phenomena, *Optics and Photonics News*, vol. 1, no. 12, pp. 34–36, 1990.

C. L. Tang, W. R. Bosenberg, T. Ukachi, R. J. Lane, and L. K. Cheng, NLO Materials Display Superior Performance, *Laser Focus World*, vol. 26, no. 9, pp. 87–97, 1990.

T. E. Bell, Light That Acts Like Natural Bits, *IEEE Spectrum*, vol. 27, no. 8, pp. 56–57, 1990.

W. P. Risk, Compact Blue Laser Devices, *Optics and Photonics News*, vol. 1, no. 5, pp. 10–15, 1990.

P. Thomas, Nonlinear Optical Materials, *Physics World*, vol. 3, no. 3, pp. 34–38, 1990.

M. de Micheli and D. Ostrowsky, Nonlinear Integrated Optics, *Physics World*, vol. 3, no. 3, pp. 56–60, 1990.

W. J. Tomlinson, Curious Features of Nonlinear Pulse Propagation in Single-Mode Optical Fibers, *Optics News*, vol. 15, no. 1, pp. 7–11, 1989.

J. Gratton and R. Delellis, An Elementary Introduction to Solitons, *American Journal of Physics*, vol. 57, pp. 683–687, 1989.

D. Marcuse, Selected Topics in the Theory of Telecommunications Fibers, in *Optical Fiber Telecommunications II*, S. E. Miller and I. P. Kaminow, eds., Academic Press, New York, 1988.

I. C. Khoo, Nonlinear Optics of Liquid Crystals, in *Progress in Optics*, E. Wolf, ed., vol. 26, North-Holland, Amsterdam, 1988.

D. M. Pepper, Applications of Optical Phase Conjugation, *Scientific American*, vol. 254, no. 1, pp. 74–83, 1986.

V. V. Shkunov and B. Ya. Zel'dovich, Optical Phase Conjugation, *Scientific American*, vol. 253, no. 6, pp. 54–59, 1985.

N. Bloembergen, Nonlinear Optics and Spectroscopy (Nobel lecture), *Reviews of Modern Physics*, vol. 54, pp. 685–695, 1982.

A. L. Mikaelian, Self-Focusing Media with Variable Index of Refraction, in *Progress in Optics*, E. Wolf, ed., vol. 17, North-Holland, Amsterdam, 1980.

W. Brunner and H. Paul, Theory of Optical Parametric Amplification and Oscillation, in *Progress in Optics*, E. Wolf, ed., vol. 15, North-Holland, Amsterdam, 1977.

J. A. Giordmaine, Nonlinear Optics, *Physics Today*, vol. 22, no. 1, pp. 39–53, 1969.

J. A. Giordmaine, The Interaction of Light with Light, *Scientific American*, vol. 210, no. 4, pp. 38–49, 1964.

PROBLEMS

19.2-1 **Frequency Up-Conversion.** A $LiNbO_3$ crystal of refractive index $n = 2.2$ is used to convert light of free-space wavelength 1.3 μm into light of free-space wavelength 0.5 μm, using a three-wave mixing process. The three waves are collinear plane waves traveling in the z direction. Determine the wavelength of the third wave (the pump). If the power of the 1.3-μm wave drops by 1 mW within an incremental distance Δz, what is the power gain of the up-converted wave and the power loss or gain of the pump within the same distance?

19.2-2 **Conditions for Three-Wave Mixing in a Dispersive Medium.** The refractive index of a nonlinear medium is a function of wavelength approximated by $n(\lambda_o) \approx n_0 - \xi\lambda_o$, where λ_o is the free-space wavelength and n_0 and ξ are constants. Show that three waves of wavelengths λ_{o1}, λ_{o2}, and λ_{o3} traveling in the same direction cannot be efficiently coupled by a second-order nonlinear effect. Is efficient coupling possible if one of the waves travels in the opposite direction?

*19.2-3 **Tolerance to Deviations from the Phase-Matching Condition.** (a) The Helmholtz equation with a source, $\nabla^2 E + k^2 E = -S$, has the solution

$$E(\mathbf{r}) = \int_V S(\mathbf{r}') \frac{\exp(-jk_o|\mathbf{r} - \mathbf{r}'|)}{4\pi|\mathbf{r} - \mathbf{r}'|} \, d\mathbf{r}',$$

where V is the volume of the source and $k_o = 2\pi/\lambda_o$. This equation can be used to determine the field emitted at a point $\mathbf{r}$, given the source at all points $\mathbf{r}'$ within the source volume. If the source is confined to a small region centered about the origin $\mathbf{r} = \mathbf{0}$ and $\mathbf{r}$ is a point sufficiently far from the source so that $r' \ll r$ for all $\mathbf{r}'$ within the source, then $|\mathbf{r} - \mathbf{r}'| = (r^2 + r'^2 - 2\mathbf{r} \cdot \mathbf{r}')^{1/2} \approx r(1 - \mathbf{r} \cdot \mathbf{r}'/r^2)$ and

$$E(\mathbf{r}) \approx \frac{\exp(-jk_o r)}{4\pi r} \int_V S(\mathbf{r}') \exp(jk_o \hat{\mathbf{r}} \cdot \mathbf{r}') \, d\mathbf{r}',$$

where $\hat{\mathbf{r}}$ is a unit vector in the direction of $\mathbf{r}$. Assuming that the volume V is a cube of width L and the source is a harmonic function $S(\mathbf{r}) = \exp(-j\mathbf{k}_s \cdot \mathbf{r})$, show that if $L \gg \lambda_o$, the emitted light is maximum when $k_o\hat{\mathbf{r}} = \mathbf{k}_s$ and drops sharply when this condition is not met. Thus a harmonic source of dimensions much greater than a wavelength emits a plane wave with approximately the same wavevector.
(b) Use the relation in part (a) and the first Born approximation to determine the scattered field, when the field incident on a second-order nonlinear medium is the sum of two waves of wavevectors $\mathbf{k}_1$ and $\mathbf{k}_2$. Derive the phase-matching condition $\mathbf{k}_3 = \mathbf{k}_1 + \mathbf{k}_2$ and determine the smallest magnitude of $\Delta\mathbf{k} = \mathbf{k}_3 - \mathbf{k}_1 - \mathbf{k}_2$ at which the scattered field E vanishes.

19.3-1 **Invariants in Four-Wave Mixing.** Derive equations for energy and photon-number conservation (the Manley–Rowe relation) for four-wave mixing.

19.3-2 **Power of a Spatial Soliton.** Determine an expression for the integrated intensity of the spatial soliton described by (19.3-10) and show that it is inversely proportional to the beam width W_0.

19.3-3 **An Opto-Optic Phase Modulator.** Design a system for modulating the phase of an optical beam of wavelength 546 nm and width $W = 0.1$ mm using a CS_2 Kerr cell of length $L = 10$ cm. The modulator is controlled by light from a pulsed laser of wavelength 694 nm. CS_2 has a refractive index $n = 1.6$ and a coefficient of third-order nonlinearity $\chi^{(3)} = 4.4 \times 10^{-32}$ (MKS units). Estimate the optical power P_π of the controlling light that is necessary for modulating the phase of the controlled light by π.

*19.4-1 **Gain of a Parametric Amplifier.** A parametric amplifier uses a 4-cm-long KDP crystal ($n \approx 1.49$, $\mathscr{d} = 8.3 \times 10^{-24}$ MKS units) to amplify light of wavelength 550 nm. The pump wavelength is 335 nm and its intensity is 10^6 W/cm^2. Assuming that the signal, idler, and pump waves are collinear, determine the amplifier gain coefficient and the overall gain.

*19.4-2 **Degenerate Parametric Down-Converter.** Write and solve the coupled equations that describe wave mixing in a parametric down-converter with a pump at frequency $\omega_3 = 2\omega$ and signals at $\omega_1 = \omega_2 = \omega$. All waves travel in the z direction. Derive an expression for the photon flux densities at 2ω and ω and the conversion efficiency for an interaction length L. Verify energy conservation and photon conservation.

*19.4-3 **Threshold Pump Intensity for Parametric Oscillation.** A parametric oscillator uses a 5-cm-long $LiNbO_3$ crystal with second-order nonlinear coefficient $\mathscr{d} = 4 \times 10^{-23}$ (MKS units) and refractive index $n = 2.2$ (assumed to be approximately constant at all frequencies of interest). The pump is obtained from a 1.06-μm Nd:YAG laser that is frequency doubled using a second-harmonic generator. The crystal is placed in a resonator using identical mirrors with reflectances 0.98. Phase matching is satisfied when the signal and idler of the parametric amplifier are of equal frequencies. Determine the minimum pump intensity for parametric oscillation.

*19.6-1 **Three-Wave Mixing in a Uniaxial Crystal.** Three waves travel at an angle θ with the optic axis (z axis) of a uniaxial crystal and an angle ϕ with the x axis, as illustrated in Fig. P19.6-1. Waves 1 and 2 are ordinary waves and wave 3 is an extraordinary wave. Show that the polarization density $P_{NL}(\omega_3)$ created by the electric fields of waves 1 and 2 is maximum if the angles are $\theta = 90°$ and $\phi = 45°$.

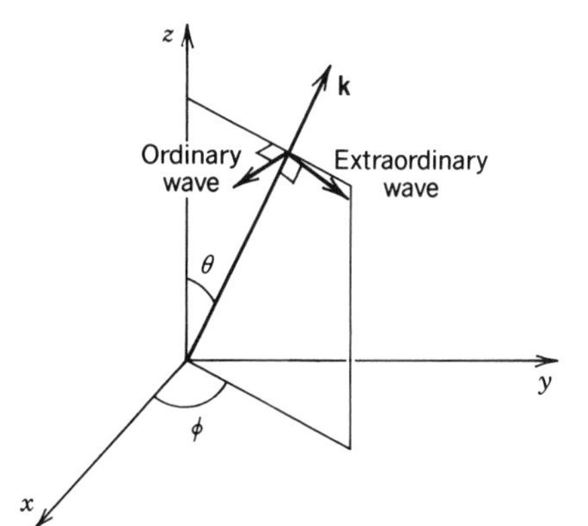

Figure P19.6-1 Three-wave mixing in a uniaxial crystal.

*19.6-2 **Phase Matching in a Degenerate Parametric Down-Converter.** A degenerate parametric down-converter uses a KDP crystal to down-convert light from 0.6 μm to 1.2 μm. If the two waves are collinear, what should the direction of propagation of the waves (in relation to the optic axis of the crystal) and their polarizations be so that the phase-matching condition is satisfied? KDP is a uniaxial crystal with the following refractive indices: at $\lambda_o = 0.6\ \mu m$, $n_o = 1.509$ and $n_e = 1.468$; at $\lambda_o = 1.2\ \mu$m, $n_o = 1.490$ and $n_e = 1.459$.

*19.6-3 **Relation Between Nonlinear Optical Coefficients and Electro-Optic Coefficients.** Show that the electro-optic coefficients are related to the coefficients of optical nonlinearity by $\mathfrak{r}_{ijk} = -4\epsilon_o \mathscr{d}_{ijk}/\epsilon_{ii}\epsilon_{jj}$ and $\mathfrak{s}_{ijkl} = -12\epsilon_o \chi^{(3)}_{ijkl}/\epsilon_{ii}\epsilon_{jj}$. These relations are generalizations of (19.2-10) and (19.3-2), respectively. *Hint:* If two matrices $\mathbf{A}$ and $\mathbf{B}$ are related by $\mathbf{B} = \mathbf{A}^{-1}$, the incremental matrices $\Delta\mathbf{A}$ and $\Delta\mathbf{B}$ are related by $\Delta\mathbf{B} = -\mathbf{A}^{-1}\,\Delta\mathbf{A}\,\mathbf{A}^{-1}$.

CHAPTER

20

ACOUSTO-OPTICS

20.1 INTERACTION OF LIGHT AND SOUND
- A. Bragg Diffraction
- B. Quantum Interpretation
- *C. Coupled-Wave Theory
- D. Bragg Diffraction of Beams

20.2 ACOUSTO-OPTIC DEVICES
- A. Modulators
- B. Scanners
- C. Interconnections
- D. Filters, Frequency Shifters, and Isolators

*20.3 ACOUSTO-OPTICS OF ANISOTROPIC MEDIA

Sir William Henry Bragg (1862–1942, left) and **Sir William Lawrence Bragg (1890–1971**, right), a father-and-son team, were awarded the Nobel Prize in 1915 for their studies of the diffraction of light from periodic structures, such as those created by sound.

The refractive index of an optical medium is altered by the presence of sound. Sound therefore modifies the effect of the medium on light; i.e., *sound can control light* (Fig. 20.0-1). Many useful devices make use of this **acousto-optic effect**; these include optical modulators, switches, deflectors, filters, isolators, frequency shifters, and spectrum analyzers.

Sound is a dynamic strain involving molecular vibrations that take the form of waves which travel at a velocity characteristic of the medium (the velocity of sound). As an example, a harmonic plane wave of compressions and rarefactions in a gas is pictured in Fig. 20.0-2. In those regions where the medium is compressed, the density is higher and the refractive index is larger; where the medium is rarefied, its density and refractive index are smaller. In solids, sound involves vibrations of the molecules about their equilibrium positions, which alter the optical polarizability and consequently the refractive index.

An acoustic wave creates a perturbation of the refractive index in the form of a wave. The medium becomes a *dynamic* graded-index medium—an inhomogeneous medium with a time-varying stratified refractive index. The theory of acousto-optics deals with the perturbation of the refractive index caused by sound, and with the propagation of light through this perturbed time-varying inhomogeneous medium.

The propagation of light in static (as opposed to time-varying) inhomogeneous (graded-index) media was discussed at several points in Chaps. 1 and 2 (Secs. 1.3 and 2.4C). Since optical frequencies are much greater than acoustic frequencies, the variations of the refractive index in a medium perturbed by sound are usually very slow in comparison with an optical period. There are therefore two significantly different time scales for light and sound. As a consequence, it is possible to use an adiabatic approach in which the optical propagation problem is solved separately at every instant of time during the relatively slow course of the acoustic cycle, always treating the material as if it were a static (frozen) inhomogeneous medium. In this quasi-stationary approximation, acousto-optics becomes the optics of an inhomogeneous medium (usually periodic) that is controlled by sound.

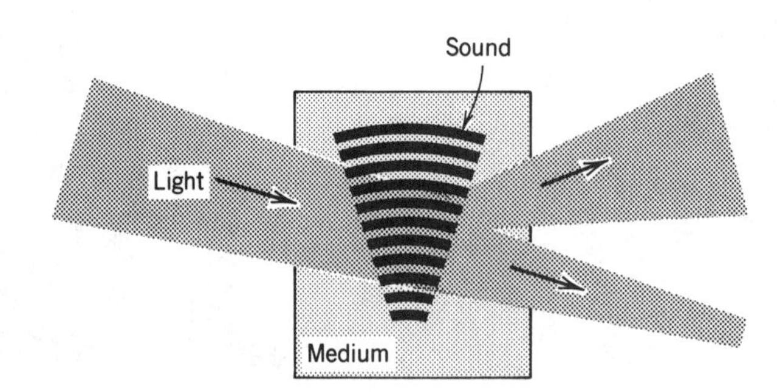

Figure 20.0-1 Sound modifies the effect of an optical medium on light.

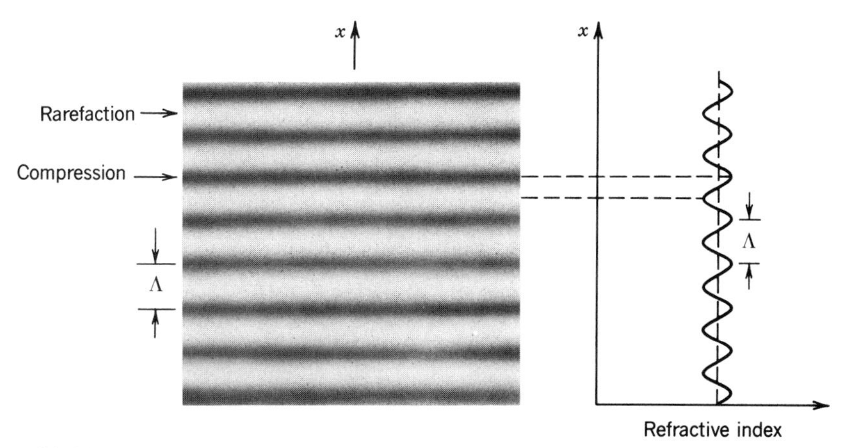

Figure 20.0-2 Variation of the refractive index accompanying a harmonic sound wave. The pattern has a period Λ, the wavelength of sound, and travels with the velocity of sound.

The simplest form of interaction of light and sound is the partial reflection of an optical plane wave from the stratified parallel planes representing the refractive-index variations created by an acoustic plane wave (Fig. 20.0-3). A set of parallel reflectors separated by the wavelength of sound Λ will reflect light if the angle of incidence θ satisfies the Bragg condition for constructive interference,

$$\boxed{\sin\theta = \frac{\lambda}{2\Lambda},} \qquad (20.0\text{-}1)$$

Bragg Condition

where λ is the wavelength of light in the medium (see Exercise 2.5-3). This form of light-sound interaction is known as **Bragg diffraction**, Bragg reflection, or Bragg scattering. The device that effects it is known as a Bragg reflector, a Bragg deflector, or a **Bragg cell**.

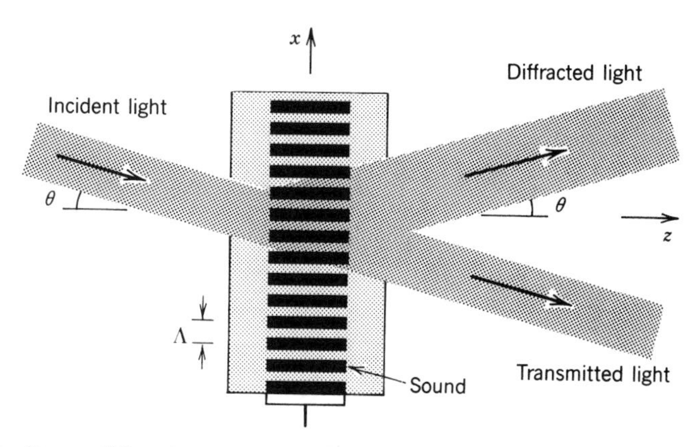

Figure 20.0-3 Bragg diffraction: an acoustic plane wave acts as a partial reflector of light (a beamsplitter) when the angle of incidence θ satisfies the Bragg condition.

Bragg cells have found numerous applications in photonics. This chapter is devoted to their properties. In Sec. 20.1, a simple theory of the optics of Bragg reflectors is presented for linear, nondispersive media. Anisotropic properties of the medium and the polarized nature of light and sound are ignored. Although the theory is based on wave optics, a simple quantum interpretation of the results is provided. In Sec. 20.2, the use of Bragg cells for light modulation and scanning is discussed. Section 20.3 provides a brief introduction to anisotropic and polarization effects in acousto-optics.

20.1 INTERACTION OF LIGHT AND SOUND

The effect of a scalar acoustic wave on a scalar optical wave is described in this section. We first consider optical and acoustic plane waves, and subsequently examine the interaction of optical and acoustic beams.

A. Bragg Diffraction

Consider an acoustic plane wave traveling in the x direction in a medium with velocity v_s, frequency f, and wavelength $\Lambda = v_s/f$. The strain (relative displacement) at position x and time t is

$$s(x,t) = S_0 \cos(\Omega t - qx), \tag{20.1-1}$$

where S_0 is the amplitude, $\Omega = 2\pi f$ is the angular frequency, and $q = 2\pi/\Lambda$ is the wavenumber. The acoustic intensity (W/m^2) is

$$I_s = \tfrac{1}{2}\varrho v_s^3 S_0^2, \tag{20.1-2}$$

where ϱ is the mass density of the medium.

The medium is assumed to be optically transparent and the refractive index in the absence of sound is n. The strain $s(x,t)$ creates a proportional perturbation of the refractive index, analogous to the Pockels effect in (18.1-4),

$$\Delta n(x,t) = -\tfrac{1}{2}\mathfrak{p} n^3 s(x,t), \tag{20.1-3}$$

where $\mathfrak{p}$ is a phenomenological coefficient known as the **photoelastic constant** (or strain-optic coefficient). The minus sign indicates that positive strain (dilation) leads to a reduction of the refractive index. As a consequence, the medium has a time-varying inhomogeneous refractive index in the form of a wave

$$\boxed{n(x,t) = n - \Delta n_0 \cos(\Omega t - qx),} \tag{20.1-4}$$

with amplitude

$$\Delta n_0 = \tfrac{1}{2}\mathfrak{p} n^3 S_0. \tag{20.1-5}$$

Substituting from (20.1-2) into (20.1-5), we find that the change of the refractive index is proportional to the square root of the acoustic intensity,

$$\boxed{\Delta n_0 = \left(\tfrac{1}{2}\mathscr{M} I_s\right)^{1/2},} \tag{20.1-6}$$

where

$$\mathscr{M} = \frac{\mathfrak{p}^2 n^6}{\varrho v_s^3} \tag{20.1-7}$$

is a material parameter representing the effectiveness of sound in altering the refractive index. $\mathscr{M}$ is a figure of merit for the strength of the acousto-optic effect in the material.

EXAMPLE 20.1-1. *Figure of Merit.* In extra-dense flint glass $\varrho = 6.3 \times 10^3$ kg/m^3, $v_s = 3.1$ km/s, $n = 1.92$, $\mathfrak{p} = 0.25$, so that $\mathscr{M} = 1.67 \times 10^{-14}$ m^2/W. An acoustic wave of intensity 10 W/cm^2 creates a refractive-index wave of amplitude $\Delta n_0 = 2.89 \times 10^{-5}$.

Consider now an optical plane wave traveling in this medium with frequency ν, angular frequency $\omega = 2\pi\nu$, free-space wavelength $\lambda_o = c_o/\nu$, wavelength in the unperturbed medium $\lambda = \lambda_o/n$ corresponding to a wavenumber $k = n\omega/c_o$, and wavevector $\mathbf{k}$ lying in the x–z plane and making an angle θ with the z axis, as illustrated in Fig. 20.0-3.

Because the acoustic frequency f is typically much smaller than the optical frequency ν (by at least five orders of magnitude), an adiabatic approach for studying light–sound interaction may be adopted: We regard the refractive index as a static "frozen" sinusoidal function

$$n(x) = n - \Delta n_0 \cos(qx - \varphi), \tag{20.1-8}$$

where φ is a fixed phase; we determine the reflected light from this inhomogeneous (graded-index) medium and track its slow variation with time by taking $\varphi = \Omega t$.

To determine the amplitude of the reflected wave we divide the medium into incremental planar layers orthogonal to the x axis. The incident optical plane wave is partially reflected at each layer because of the refractive-index change. We assume that the reflectance is sufficiently small so that the transmitted light from one layer approximately maintains its original magnitude (i.e., is not depleted) as it penetrates through the following layers of the medium.

If $\Delta r = (dr/dx)\,\Delta x$ is the incremental complex amplitude reflectance of the layer at position x, the total complex amplitude reflectance for an overall length L (see Fig. 20.1-1) is the sum of all incremental reflectances,

$$r = \int_{-L/2}^{L/2} e^{j2kx\sin\theta} \frac{dr}{dx}\, dx. \tag{20.1-9}$$

The phase factor $e^{j2kx\sin\theta}$ is included since the reflected wave at a position x is

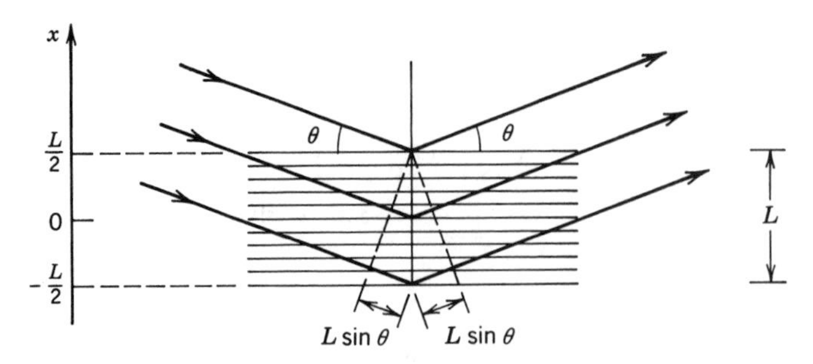

Figure 20.1-1 Reflections from layers of an inhomogeneous medium.

advanced by a distance $2x \sin \theta$ (corresponding to a phase shift $2kx \sin \theta$) relative to the reflected wave at $x = 0$ (see Fig. 20.1-1). The wavenumbers for the incident and reflected waves are taken to be the same, for reasons that will be explained later.

An expression for the incremental complex amplitude reflectance Δr in terms of the incremental refractive-index change Δn between two adjacent layers at a given position x may be determined by use of the Fresnel equations (see Sec. 6.2). For TE (orthogonal) polarization, (6.2-4) is used with $n_1 = n + \Delta n$, $n_2 = n$, $\theta_1 = 90° - \theta$, and Snell's law $n_1 \sin \theta_1 = n_2 \sin \theta_2$ is used to determine θ_2. When terms of second order in Δn are neglected, the result is

$$\Delta r = \frac{-1}{2n \sin^2 \theta} \Delta n. \qquad (20.1\text{-}10)$$

Equation (6.2-6) is similarly used for the TM (parallel) polarization, yielding

$$\Delta r = \frac{-\cos 2\theta}{2n \sin^2 \theta} \Delta n.$$

In most acousto-optic devices θ is very small, so that $\cos 2\theta \approx 1$, making (20.1-10) approximately applicable to both polarizations.

Using (20.1-8) and (20.1-10), we obtain

$$\frac{dr}{dx} = \frac{dr}{dn}\frac{dn}{dx} = \frac{-1}{2n \sin^2 \theta}\left[q \, \Delta n_0 \sin(qx - \varphi)\right] = r' \sin(qx - \varphi), \qquad (20.1\text{-}11)$$

where

$$r' = \frac{-q}{2n \sin^2 \theta} \Delta n_0. \qquad (20.1\text{-}12)$$

Finally, we substitute (20.1-11) into (20.1-9), and use complex notation to write $\sin(qx - \varphi) = [e^{j(qx-\varphi)} - e^{-j(qx-\varphi)}]/2j$, thereby obtaining

$$r = \tfrac{1}{2} j r' e^{j\varphi} \int_{-\frac{1}{2}L}^{\frac{1}{2}L} e^{j(2k \sin \theta - q)x}\, dx - \tfrac{1}{2} j r' e^{-j\varphi} \int_{-\frac{1}{2}L}^{\frac{1}{2}L} e^{j(2k \sin \theta + q)x}\, dx. \qquad (20.1\text{-}13)$$

The first term in (20.1-13) has its maximum value when $2k \sin \theta = q$, whereas the second is maximum when $2k \sin \theta = -q$. If L is sufficiently large, these maxima are sharp, so that any slight deviation from the angles $\theta = \pm \sin^{-1}(q/2k)$ makes the corresponding term negligible. Thus only one of these two terms may be significant at a time, depending on the angle θ. For reasons to become clear shortly, the conditions $2k \sin\theta \approx q$ and $2k \sin\theta \approx -q$ are called the upshifted and downshifted reflections, respectively. We first consider the upshifted condition, $2k \sin\theta \approx q$, for which the second term is negligible, and comment on the downshifted case subsequently. Performing the integral in the first term of (20.1-13) and substituting $\varphi = \Omega t$, we obtain

$$r = \tfrac{1}{2} j r' L \operatorname{sinc}\left[(q - 2k\sin\theta)\frac{L}{2\pi}\right] e^{j\Omega t}, \tag{20.1-14}$$

Amplitude Reflectance (Upshifted Case)

where $\operatorname{sinc}(x) = \sin(\pi x)/(\pi x)$. We proceed to discuss several important conclusions based on (20.1-14).

Bragg Condition

The sinc function in (20.1-14) has its maximum value of 1.0 when its argument is zero, i.e., when $q = 2k \sin\theta$. This occurs when $\theta = \theta_B$, where $\theta_B = \sin^{-1}(q/2k)$ is the **Bragg angle**. Since $q = 2\pi/\Lambda$ and $k = 2\pi/\lambda$,

$$\sin\theta_B = \frac{\lambda}{2\Lambda}. \tag{20.0-1}$$

Bragg Angle

The Bragg angle is the angle for which the incremental reflections from planes separated by an acoustic wavelength Λ have a phase shift of 2π so that they interfere constructively (see Exercise 2.5-3).

EXAMPLE 20.1-2. ***Bragg Angle.*** An acousto-optic cell is made of flint glass in which the sound velocity is $v_s = 3$ km/s and the refractive index is $n = 1.95$. The Bragg angle for reflection of an optical wave of free-space wavelength $\lambda_o = 633$ nm ($\lambda = \lambda_o/n = 325$ nm) from a sound wave of frequency $f = 100$ MHz ($\Lambda = v_s/f = 30$ μm) is $\theta_B = 5.4$ mrad $\approx$ 0.31°. This angle is internal (i.e., inside the medium). If the cell is placed in air, θ_B corresponds to an external angle $\theta'_B \approx n\theta_B = 0.61°$. A sound wave of 10 times greater frequency ($f = 1$ GHz) corresponds to a Bragg angle $\theta_B \approx 3.1°$.

The Bragg condition can also be stated as a simple relation between the wavevectors of the sound wave and the two optical waves. If $\mathbf{q} = (q, 0, 0)$, $\mathbf{k} = (-k\sin\theta, 0, k\cos\theta)$, and $\mathbf{k}_r = (k\sin\theta, 0, k\cos\theta)$ are the components of the wavevectors of the sound wave, the incident light wave, and the reflected light wave, respectively, the condition $q = 2k\sin\theta_B$ is equivalent to the vector relation

$$\mathbf{k}_r = \mathbf{k} + \mathbf{q}, \tag{20.1-15}$$

illustrated by the vector diagram in Fig. 20.1-2.

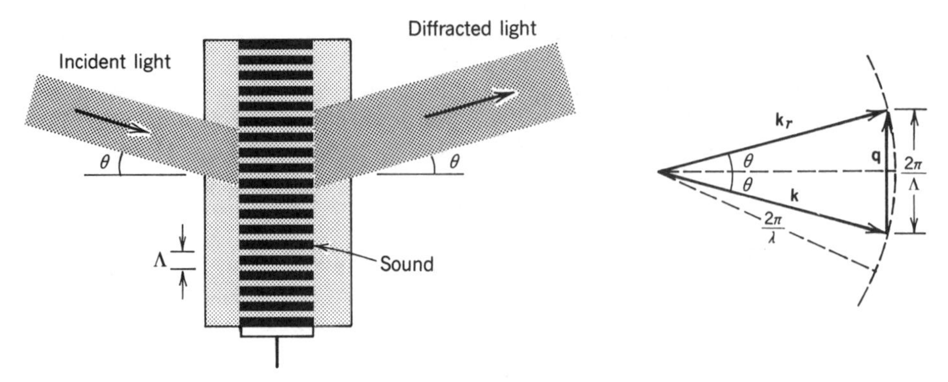

Figure 20.1-2 The Bragg condition $\sin\theta_B = q/2k$ is equivalent to the vector relation $\mathbf{k}_r = \mathbf{k} + \mathbf{q}$.

Tolerance in the Bragg Condition

The dependence of the complex amplitude reflectance on the angle θ is governed by the symmetric function $\operatorname{sinc}[(q - 2k\sin\theta)L/2\pi] = \operatorname{sinc}[(\sin\theta - \sin\theta_B)2L/\lambda]$ in (20.1-14). This function reaches its peak value when $\theta = \theta_B$ and drops sharply when θ differs slightly from θ_B. When $\sin\theta - \sin\theta_B = \lambda/2L$ the sinc function reaches its first zero and the reflectance vanishes (Fig. 20.1-3). Because θ_B is usually very small, $\sin\theta \approx \theta$, and the reflectance vanishes at an angular deviation from the Bragg angle of approximately $\theta - \theta_B \approx \lambda/2L$. Since L is typically much greater than λ, this is an extremely small angular width. This sharp reduction of the reflectance for slight deviations from the Bragg angle occurs as a result of the destructive interference between the incremental reflections from the sound wave.

Doppler Shift

In accordance with (20.1-14), the complex amplitude reflectance r is proportional to $\exp(j\Omega t)$. Since the angular frequency of the incident light is ω [i.e., $E \propto \exp(j\omega t)$], the reflected wave $E_r = rE \propto \exp[j(\omega + \Omega)t]$ has angular frequency

$$\omega_r = \omega + \Omega. \tag{20.1-16}$$

Doppler Shift

The process of reflection is therefore accompanied by a frequency shift equal to the

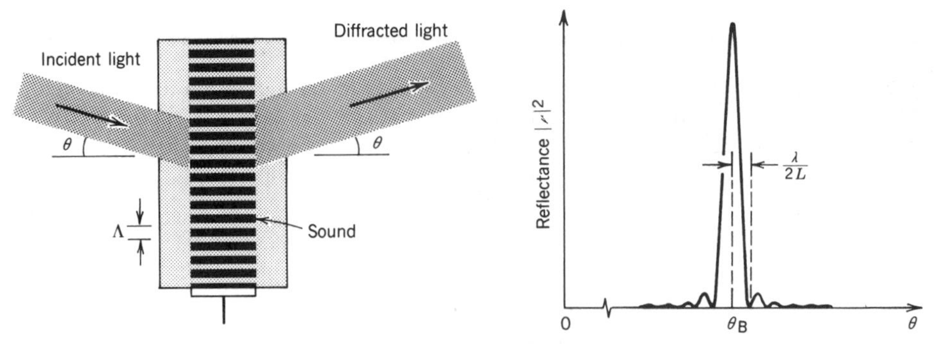

Figure 20.1-3 Dependence of the reflectance $|r|^2$ on the angle θ. Maximum reflection occurs at the Bragg angle $\theta_B = \sin^{-1}(\lambda/2\Lambda)$.

frequency of the sound. This can also be thought of as a Doppler shift (see Exercise 2.6-1 and Sec. 12.2D). The incident light is reflected from surfaces that move with a velocity v_s. Its Doppler-shifted angular frequency is therefore $\omega_r = \omega(1 + 2v_s \sin\theta/c)$, where $v_s \sin\theta$ is the component of velocity of these surfaces in the direction of the incident and the reflected waves. Using the relations $\sin\theta = \lambda/2\Lambda$, $v_s = \Lambda\Omega/2\pi$, and $c = \lambda\omega/2\pi$, (20.1-16) is reproduced. The Doppler shift equals the sound frequency.

Because $\Omega \ll \omega$, the frequencies of the incident and reflected waves are approximately equal (with an error typically smaller than 1 part in 10^5). The wavelengths of the two waves are therefore also approximately equal. In writing (20.1-9) we have implicitly used this assumption by using the same wavenumber k for the two waves. Also, in drawing the vector diagram in Fig. 20.1-2 it was assumed that the vectors $\mathbf{k}_r$ and $\mathbf{k}$ have approximately the same length $n\omega/c_o$.

Reflectance

The reflectance $\mathscr{R} = |r|^2$ is the ratio of the intensity of the reflected optical wave to that of the incident optical wave. At the Bragg angle $\theta = \theta_B$, (20.1-14) gives $\mathscr{R} = |r'|^2 L^2/4$. Substituting for r' from (20.1-12),

$$\mathscr{R} = \frac{\pi^2}{\lambda_o^2}\left(\frac{L}{\sin\theta}\right)^2 \Delta n_0^2, \qquad (20.1\text{-}17)$$

and using (20.1-6), we obtain

$$\mathscr{R} = \frac{\pi^2}{2\lambda_o^2}\left(\frac{L}{\sin\theta}\right)^2 \mathscr{M} I_s. \qquad (20.1\text{-}18)$$

Reflectance

The reflectance $\mathscr{R}$ is therefore proportional to the intensity of the acoustic wave I_s, to the material parameter $\mathscr{M}$ defined in (20.1-7) and to the square of the oblique distance $L/\sin\theta$ of penetration of light through the acoustic wave.

Substituting $\sin\theta = \lambda/2\Lambda$ into (20.1-18), we obtain

$$\mathscr{R} = 2\pi^2 n^2 \frac{L^2\Lambda^2}{\lambda_o^4} \mathscr{M} I_s.$$

Thus the reflectance is inversely proportional to λ_o^4 (or directly proportional to ω^4). The dependence of the efficiency of scattering on the fourth power of the optical frequency is typical of light-scattering phenomena.

The proportionality between the reflectance and the sound intensity poses a problem. As the sound intensity increases, $\mathscr{R}$ would eventually exceed unity, and the reflected light would be more intense than the incident light! This unacceptable result is a consequence of violating the assumptions of this approximate theory. It was assumed that the incremental reflection from each layer is too small to deplete the transmitted wave which reflects from subsequent layers. Clearly, this assumption does not hold when the sound wave is intense. In reality, a saturation process occurs, ensuring that $\mathscr{R}$ does not exceed unity. A more careful analysis (see Sec. 20.1C), in which depletion of the incident optical wave is included, leads to the following

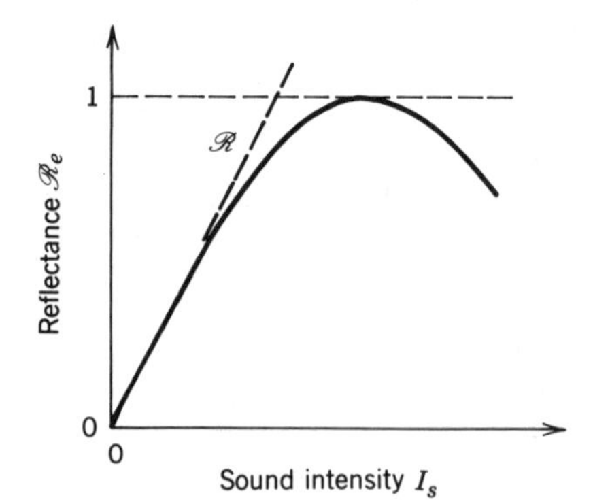

Figure 20.1-4 Dependence of the reflectance $\mathscr{R}_e$ of the Bragg reflector on the intensity of sound I_s. When I_s is small $\mathscr{R}_e \approx \mathscr{R}$, which is a linear function of I_s.

expression for the reflectance:

$$\boxed{\mathscr{R}_e = \sin^2 \sqrt{\mathscr{R}}\,,} \tag{20.1-19}$$

where $\mathscr{R}$ is the approximate expression (20.1-18) and $\mathscr{R}_e$ is the exact expression. This relation is illustrated in Fig. 20.1-4. Evidently, when $\mathscr{R} \ll 1$, $\sin\sqrt{\mathscr{R}} \approx \sqrt{\mathscr{R}}$, so that $\mathscr{R}_e \approx \mathscr{R}$.

EXAMPLE 20.1-3. ***Reflectance.*** A Bragg cell is made of extra-dense flint glass with material parameter $\mathscr{M} = 1.67 \times 10^{-14}$ m^2/W (see Example 20.1-1). If $\lambda_o = 633$ nm (wavelength of the He–Ne laser), the sound intensity $I_s = 10$ W/cm^2, and the length of penetration of the light through the sound is $L/\sin\theta = 1$ mm, then $\mathscr{R} = 0.0206$ and $\mathscr{R}_e = 0.0205$, so that approximately 2% of the light is reflected. If the sound intensity is increased to 100 W/cm^2, then $\mathscr{R} = 0.206$, $\mathscr{R}_e = 0.192$ (i.e., the reflectance increases to $\approx$ 19%).

Downshifted Bragg Diffraction

Another possible geometry for Bragg diffraction is that for which $2k \sin\theta = -q$. This is satisfied when the angle θ is negative; i.e., the incident optical wave makes an acute angle with the sound wave as illustrated in Fig. 20.1-5. In this case, the second term of (20.1-13) has its maximum value, whereas the first term is negligible. The complex amplitude reflectance is then given by

$$r = -\tfrac{1}{2} j r' L e^{-j\Omega t}. \tag{20.1-20}$$

In this geometry, the frequency of the reflected wave is downshifted, so that

$$\omega_s = \omega - \Omega \tag{20.1-21}$$

and the wavevectors of the light and sound waves satisfy the relation

$$\mathbf{k}_s = \mathbf{k} - \mathbf{q}, \tag{20.1-22}$$

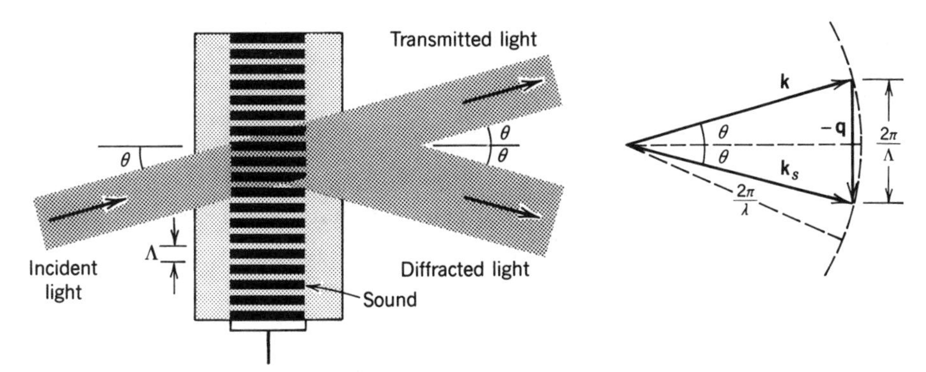

Figure 20.1-5 Geometry of downshifted reflection of light from sound. The frequency of the reflected wave is downshifted.

illustrated in Fig. 20.1-5. Equation (20.1-22) is a phase-matching condition, ensuring that the reflections of light add in phase. The frequency downshift in (20.1-21) is consistent with the Doppler shift since the light and sound waves travel in the same direction.

B. Quantum Interpretation

In accordance with the quantum theory of light (see Chap. 11), an optical wave of angular frequency ω and wavevector $\mathbf{k}$ is viewed as a stream of photons, each of energy $\hbar\omega$ and momentum $\hbar\mathbf{k}$. An acoustic wave of angular frequency Ω and wavevector $\mathbf{q}$ is similarly regarded as a stream of acoustic quanta, called **phonons**, each of energy $\hbar\Omega$ and momentum $\hbar\mathbf{q}$.

Interaction of light and sound occurs when a photon combines with a phonon to generate a new photon of the sum energy and momentum. An incident photon of frequency ω and wavevector $\mathbf{k}$ interacts with a phonon of frequency Ω and wavevector $\mathbf{q}$ to generate a new photon of frequency ω_r and wavevector $\mathbf{k}_r$, as illustrated in Fig. 20.1-6. Conservation of energy and momentum require that $\hbar\omega_r = \hbar\omega + \hbar\Omega$ and $\hbar\mathbf{k}_r = \hbar\mathbf{k} + \hbar\mathbf{q}$, from which the Doppler shift formula $\omega_r = \omega + \Omega$ and the Bragg condition, $\mathbf{k}_r = \mathbf{k} + \mathbf{q}$, are recovered.

*C. Coupled-Wave Theory

Bragg Diffraction as a Scattering Process

Light propagation through an inhomogeneous medium with dynamic refractive index perturbation $\Delta n(x, t)$ my also be regarded as a light-scattering process and the Born approximation (see Sec. 19.1) may be used to describe it. A perturbation $\Delta\mathscr{P}$ of the

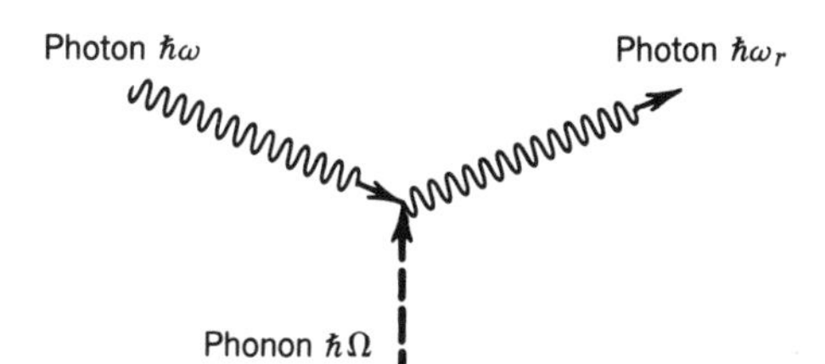

Figure 20.1-6 Bragg diffraction: a photon combines with a phonon to generate a new photon of different frequency and momentum.

electric polarization density acts as a source of light

$$\mathscr{S} = -\mu_o \frac{\partial^2 \Delta\mathscr{P}}{\partial t^2} \tag{20.1-23}$$

[see (5.2-19) and the discussion following (19.1-7)]. Since $\mathscr{P} = \epsilon_o \chi \mathscr{E} = \epsilon_o(\epsilon/\epsilon_o - 1)\mathscr{E} = \epsilon_o(n^2 - 1)\mathscr{E}$, where $\mathscr{E}$ is the electric field, the perturbation Δn corresponds to $\Delta\mathscr{P} = \epsilon_o\,\Delta(n^2 - 1)\mathscr{E} = 2\epsilon_o n\,\Delta n \mathscr{E}$, so that

$$\boxed{\mathscr{S} = -2\mu_o \epsilon_o n \frac{\partial^2}{\partial t^2}(\Delta n \mathscr{E}).} \tag{20.1-24}$$

Thus the source $\mathscr{S}$ is proportional to the second derivative of the product $\Delta n \mathscr{E}$. To determine the scattered field we solve the wave equation (19.1-6), $\nabla^2 \mathscr{E} - (1/c^2)\partial^2\mathscr{E}/\partial^2 t = -\mathscr{S}$, together with (20.1-24) and $\Delta n = -\Delta n_0 \cos(\Omega t - \mathbf{q}\cdot\mathbf{r})$.

The idea of the first Born approximation is to assume that the source $\mathscr{S}$ is created by the incident field only and to solve the wave equation for the scattered field. Substituting $\mathscr{E} = \mathrm{Re}\{A \exp[j(\omega t - \mathbf{k}\cdot\mathbf{r})]\}$ into (20.1-24), where A is a slowly varying envelope, we obtain

$$\mathscr{S} = -\left(\frac{\Delta n_0}{n}\right)\left(k_r^2 \,\mathrm{Re}\{A \exp[j(\omega_r t - \mathbf{k}_r\cdot\mathbf{r})]\} + k_s^2 \,\mathrm{Re}\{A \exp[j(\omega_s t - \mathbf{k}_s\cdot\mathbf{r})]\}\right), \tag{20.1-25}$$

where $\omega_r = \omega + \Omega$, $\mathbf{k}_r = \mathbf{k} + \mathbf{q}$, $k_r = \omega_r/c$; and $\omega_s = \omega - \Omega$, $\mathbf{k}_s = \mathbf{k} - \mathbf{q}$, $k_s = \omega_s/c$. We thus have two sources of light of frequencies $\omega \pm \Omega$, and wavevectors $\mathbf{k} \pm \mathbf{q}$, that may emit an upshifted or downshifted Bragg-reflected plane wave. Upshifted reflection occurs if the geometry is such that the magnitude of the vector $\mathbf{k} + \mathbf{q}$ equals $\omega_r/c \approx \omega/c$, as can easily be seen from the vector diagram in Fig. 20.1-2. Downshifted reflection occurs if the vector $\mathbf{k} - \mathbf{q}$ has magnitude $\omega_s/c \approx \omega/c$, as illustrated in Fig. 20.1-5. Obviously, these two conditions may not be met simultaneously.

We have thus independently proved the Bragg condition and Doppler-shift formula using a scattering approach. Equation (20.1-25) indicates that the intensity of the emitted light is proportional to $\omega_r^4 \approx \omega^4$, so that the efficiency of scattering is inversely proportional to the fourth power of the wavelength. This analysis can be pursued further to derive an expression for the reflectance by determining the intensity of the wave emitted by the scattering source (see Problem 20.1-2).

Coupled-Wave Equations

To go beyond the first Born approximation, we must include the contribution made by the scattered field to the source $\mathscr{S}$. Assuming that the geometry is that of upshifted Bragg diffraction, the field $\mathscr{E}$ is composed of the incident and Bragg-reflected waves: $\mathscr{E} = \mathrm{Re}\{E \exp(j\omega t)\} + \mathrm{Re}\{E_r \exp(j\omega_r t)\}$. With the help of the relation $\Delta n = -\Delta n_0 \cos(\Omega t - \mathbf{q}\cdot\mathbf{r})$, (20.1-24) gives

$$\mathscr{S} = \mathrm{Re}\{S \exp(j\omega t) + S_r \exp(j\omega_r t)\} + \text{terms of other frequencies},$$

where

$$S = -k^2 \frac{\Delta n_0}{n} E_r, \qquad S_r = -k_r^2 \frac{\Delta n_0}{n} E. \tag{20.1-26}$$

Comparing terms of equal frequencies on both sides of the wave equation, $\nabla^2\mathscr{E} - (1/c^2)\,\partial^2\mathscr{E}/\partial^2 t = -\mathscr{S}$, we obtain two coupled Helmholtz equations for the incident wave and the Bragg-reflected wave,

$$(\nabla^2 + k^2)E = -S, \qquad (\nabla^2 + k_r^2)E_r = -S_r. \tag{20.1-27}$$

These equations, together with (20.1-26), may be solved to determine E and E_r.

Consider, for example, the case of small-angle reflection ($\theta \ll 1$), so that the two waves travel approximately in the z direction. Assuming that $k \approx k_r$, the fields E and E_r are described by $E = A\exp(-jkz)$ and $E_r = A_r\exp(-jkz)$, where A and A_r are slowly varying functions of z. Using the slowly varying envelope approximation (see Sec. 2.2C), $(\nabla^2 + k^2)A\exp(-jkz) \approx -j2k(dA/dz)\exp(-jkz)$, (20.1-26) and (20.1-27) yield

$$\frac{dA}{dz} = j\tfrac{1}{2}\gamma A_r \tag{20.1-28a}$$

$$\frac{dA_r}{dz} = j\tfrac{1}{2}\gamma A, \tag{20.1-28b}$$

where

$$\gamma = k\frac{\Delta n_0}{n}. \tag{20.1-29}$$

If the cell extends between $z = 0$ and $z = d$, we use the boundary condition $A_r(0) = 0$, and find that equations (20.1-28) have the harmonic solution

$$A(z) = A(0)\cos\frac{\gamma z}{2} \tag{20.1-30a}$$

$$A_r(z) = jA(0)\sin\frac{\gamma z}{2}. \tag{20.1-30b}$$

These equations describe the rise of the reflected wave and the fall of the incident wave, as illustrated in Fig. 20.1-7. The reflectance $\mathscr{R}_e = |A_r(d)|^2/|A(0)|^2$ is therefore given by $\mathscr{R}_e = \sin^2(\gamma d/2)$, so that $\mathscr{R}_e = \sin^2\sqrt{\mathscr{R}}$, where $\mathscr{R} = (\gamma d/2)^2$. Using

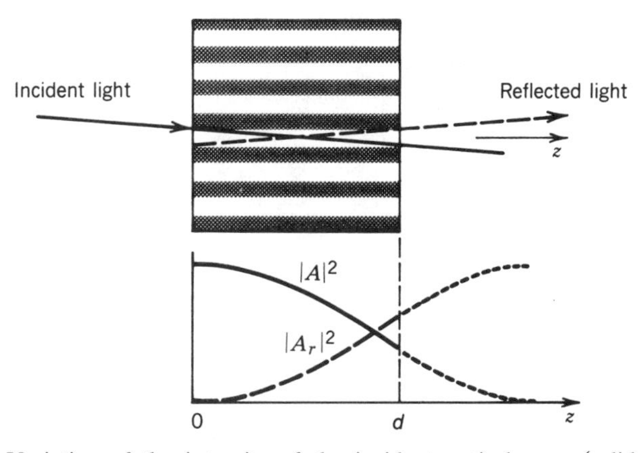

Figure 20.1-7 Variation of the intensity of the incident optical wave (solid curve) and the intensity of the Bragg-reflected wave (dashed curve) as functions of the distance traveled through the acoustic wave.

(20.1-29), we obtain $\mathscr{R} = (\pi^2/\lambda_o^2)\,\Delta n_0^2 d^2$. This is exactly the expression for the weak-sound reflectance in (20.1-17) with $d = L/\sin\theta$.

D. Bragg Diffraction of Beams

It has been shown so far that an optical *plane wave* of wavevector $\mathbf{k}$ interacts with an acoustic *plane wave* of wavevector $\mathbf{q}$ to produce an optical plane wave of wavevector $\mathbf{k}_r = \mathbf{k} + \mathbf{q}$, provided that the Bragg condition is satisfied (i.e., the angle between $\mathbf{k}$ and $\mathbf{q}$ is such that the magnitude $k_r = |\mathbf{k} + \mathbf{q}| \approx k = 2\pi/\lambda$). Interaction between a *beam* of light and a *beam* of sound can be understood if the beam is regarded as a superposition of plane waves traveling in different directions, each with its own wavevector (see the introduction to Chap. 4).

Diffraction of an Optical Beam from an Acoustic Plane Wave

Consider an optical *beam* of width D interacting with an acoustic *plane wave*. In accordance with Fourier optics (see Sec. 4.3A), the optical beam can be decomposed into plane waves with directions occupying a cone of half-angle

$$\delta\theta = \frac{\lambda}{D}. \tag{20.1-31}$$

There is some arbitrariness in the definition of the diameter D and the angle $\delta\theta$, and a multiplicative factor in (20.1-31) is taken to be 1.0. If the beam profile is rectangular of width D, the angular width from the peak to the first zero of the Fraunhofer diffraction pattern is $\delta\theta = \lambda/D$; for a circular beam of diameter D, $\delta\theta = 1.22\lambda/D$; for a Gaussian beam of waist diameter $D = 2W_0$, $\delta\theta = \lambda/\pi W_0 = (2/\pi)\lambda/D \approx 0.64\lambda/D$ [see (3.1-19)]. For simplicity, we shall use (20.1-31).

Although there is only one wavevector $\mathbf{q}$, there are many wavevectors $\mathbf{k}$ (all of the same length $2\pi/\lambda$) within a cone of angle $\delta\theta$. As Fig. 20.1-8 illustrates, there is only one direction of $\mathbf{k}$ for which the Bragg condition is satisfied. The reflected wave is then a plane wave with only one wavevector $\mathbf{k}_r$.

Diffraction of an Optical Beam from an Acoustic Beam

Suppose now that the acoustic wave itself is a beam of width D_s. If the sound frequency is sufficiently high so that the wavelength is much smaller than the width of

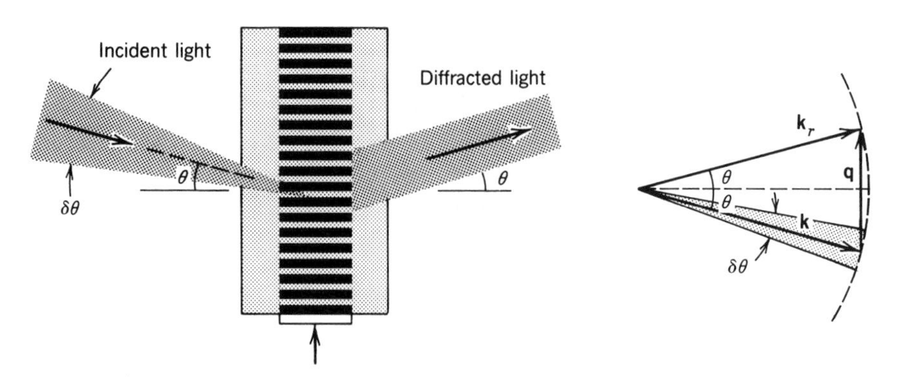

Figure 20.1-8 Diffraction of an optical *beam* from an acoustic *plane wave*. There is only one plane-wave component of the incident light beam that satisfies the Bragg condition. The diffracted light is a plane wave.

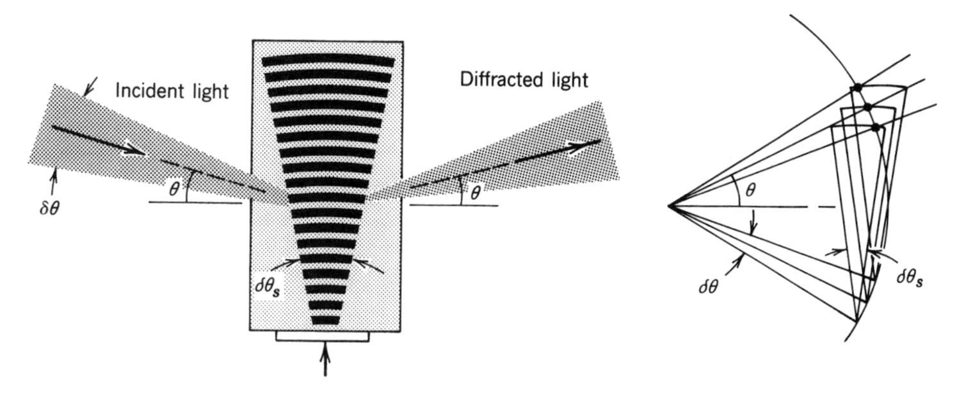

Figure 20.1-9 Diffraction of an optical *beam* from a sound *beam*.

the medium, sound propagates as an unguided (free-space) wave and has properties analogous to those of optical beams, with angular divergence

$$\delta\theta_s = \frac{\Lambda}{D_s}. \tag{20.1-32}$$

This is equivalent to many plane waves with directions lying within the divergence angle.

The reflection of an optical beam from this acoustic beam can be determined by finding matching pairs of optical and acoustic plane waves satisfying the Bragg condition. The sum of the reflected waves constitutes the reflected optical beam. There are many vectors **k** (all of the same length $2\pi/\lambda$) and many vectors **q** (all of the same length $2\pi/\Lambda$); only the pairs of vectors that form an isoceles triangle contribute, as illustrated in Fig. 20.1-9.

If the acoustic-beam divergence is greater than the optical-beam divergence ($\delta\theta_s \gg \delta\theta$) and if the central directions of the two beams satisfy the Bragg condition, every incident optical plane wave finds an acoustic match and the reflected light beam has the same angular divergence as the incident optical beam $\delta\theta$. The distribution of acoustic energy in the sound beam can thus be monitored as a function of direction, by using a probe light beam of much narrower divergence and measuring the reflected light as the angle of incidence is varied.

Diffraction of an Optical Plane Wave from a Thin Acoustic Beam; Raman–Nath Diffraction

Since a thin acoustic beam comprises plane waves traveling in many directions, it can diffract light at angles that are significantly different from the Bragg angle corresponding to the beam's principal direction. Consider, for example, the geometry in Fig. 20.1-10 in which the incident optical plane wave is perpendicular to the main direction of a thin acoustic beam. The Bragg condition is satisfied if the reflected wavevector $\mathbf{k}_r$ makes angles $\pm\theta$, where

$$\sin\frac{\theta}{2} = \frac{\lambda}{2\Lambda}. \tag{20.1-33}$$

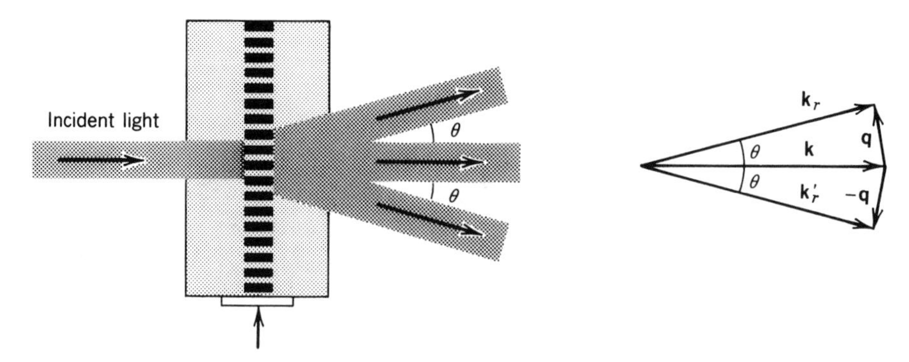

Figure 20.1-10 An optical plane wave incident normally on a thin-beam acoustic standing wave is partially deflected into two directions making angles $\approx \pm\lambda/\Lambda$.

If θ is small, $\sin(\theta/2) \approx \theta/2$ and

$$\boxed{\theta \approx \frac{\lambda}{\Lambda}.} \tag{20.1-34}$$

The incident beam is therefore deflected into either of the two directions making angles $\pm\theta$, depending on whether the acoustic beam is traveling upward or downward. For an acoustic standing-wave beam the optical wave is deflected in both directions.

The angle $\theta \approx \lambda/\Lambda$ is the angle by which a diffraction grating of period Λ deflects an incident plane wave (see Exercise 2.4-5). The thin acoustic beam in fact modulates the refractive index, creating a periodic pattern of period Λ confined to a thin planar layer. The medium therefore acts as a thin diffraction grating. This phase grating diffracts light also into higher diffraction orders, as illustrated in Fig. 20.1-11(a).

The higher-order diffracted waves generated by the phase grating at angles $\pm 2\theta$, $\pm 3\theta, \ldots$ may also be interpreted using a quantum picture of light–sound interaction. One incident photon combines with two phonons (acoustic quantum particles) to form a photon of the second-order reflected wave. Conservation of momentum requires that $\mathbf{k}_r = \mathbf{k} \pm 2\mathbf{q}$. This condition is satisfied for the geometry in Fig. 20.1-11(b). The second-order reflected light is frequency shifted to $\omega_r = \omega \pm 2\Omega$. Similar interpretations apply to higher orders of diffraction.

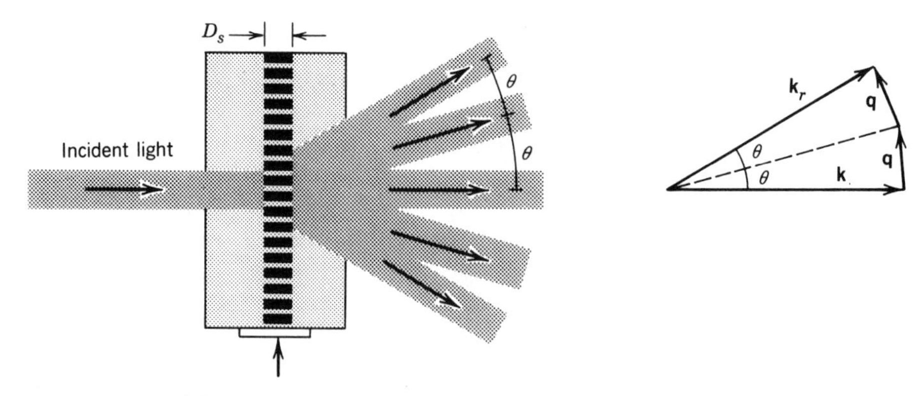

Figure 20.1-11 (a) A thin acoustic beam acts as a diffraction grating. (b) Conservation-of-momentum diagram for second-order acousto-optic diffraction.

The acousto-optic interaction of light with a *perpendicular thin* sound beam is known as **Raman–Nath** or **Debye–Sears scattering** of light by sound.[†]

20.2 ACOUSTO-OPTIC DEVICES

A. Modulators

The intensity of the reflected light in a Bragg cell is proportional to the intensity of sound, if the sound intensity is sufficiently weak. Using an electrically controlled acoustic transducer [Fig. 20.2-1(*a*)], the intensity of the reflected light can be varied proportionally. The device can be used as a linear analog modulator of light.

As the acoustic power increases, however, saturation occurs and almost total reflection can be achieved (see Fig. 20.1-4). The modulator then serves as an optical switch, which, by switching the sound on and off, turns the reflected light on and off, and the transmitted light off and on, as illustrated in Fig. 20.2-1(*b*).

Modulation Bandwidth

The bandwidth of the modulator is the maximum frequency at which it can efficiently modulate. When the amplitude of an acoustic wave of frequency f_0 is varied as a function of time by amplitude modulation with a signal of bandwidth B, the acoustic wave is no longer a single-frequency harmonic function; it has frequency components within a band $f_0 \pm B$ centered about the frequency f_0 (Fig. 20.2-2). How does monochromatic light interact with this multifrequency acoustic wave and what is the maximum value of B that can be handled by the acousto-optic modulator?

When both the incident optical wave and the acoustic wave are plane waves, the component of sound of frequency f corresponds to a Bragg angle,

$$\theta = \sin^{-1}\frac{\lambda}{2\Lambda} = \sin^{-1}\frac{f\lambda}{2v_s} \approx \frac{\lambda}{2v_s}f \tag{20.2-1}$$

(assumed to be small). For a fixed angle of incidence θ, an incident monochromatic optical plane wave of wavelength λ interacts with one and only one harmonic component of the acoustic wave, the component with frequency f satisfying (20.2-1), as illustrated in Fig. 20.2-3. The reflected wave is then *monochromatic* with frequency

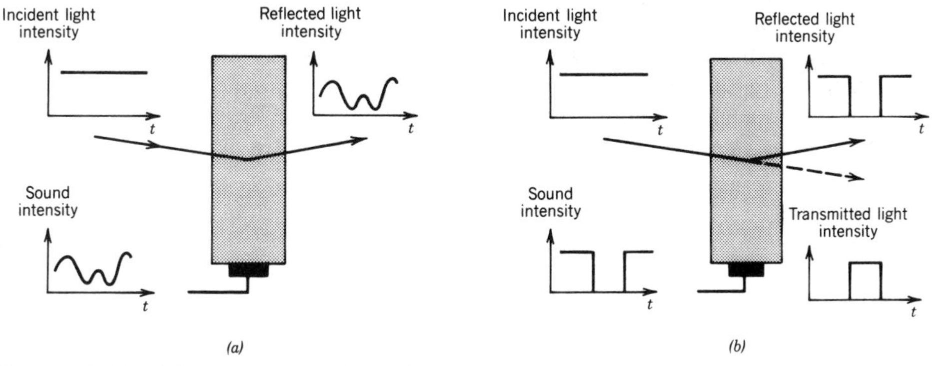

Figure 20.2-1 (*a*) An acousto-optic modulator. The intensity of the reflected light is proportional to the intensity of sound. (*b*) An acousto-optic switch.

[†]For further details, see, e.g., M. Born and E. Wolf, *Principles of Optics*, Pergamon Press, New York, 6th ed. 1980, Chap. 12.

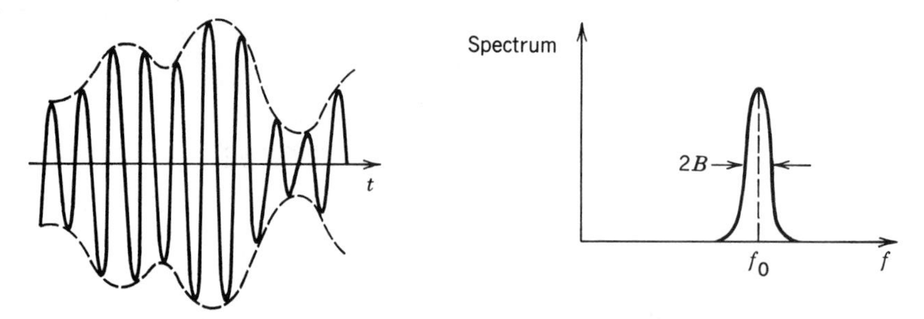

Figure 20.2-2 The waveform of an amplitude-modulated acoustic signal and its spectrum.

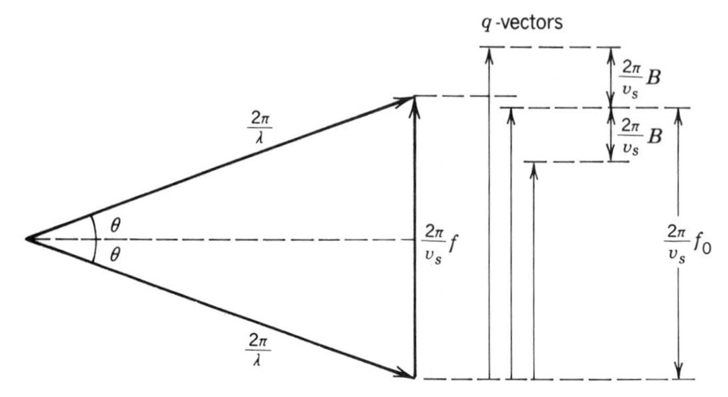

Figure 20.2-3 Interaction of an optical plane wave with a modulated (multiple frequency) acoustic plane wave. Only one frequency component of sound reflects the light wave. The reflected wave is monochromatic and not modulated.

$\nu + f$. Although the acoustic wave is modulated, the reflected optical wave is not. Evidently, under this idealized condition the bandwidth of the modulator is zero!

To achieve modulation with a bandwidth B, each of the acoustic frequency components within the band $f_0 \pm B$ must interact with the incident light wave. A more tolerant situation is therefore necessary. Suppose that the incident light is a beam of width D and angular divergence $\delta\theta = \lambda/D$ and assume that the modulated sound wave is planar. Each frequency component of sound interacts with the optical plane wave that has the matching Bragg angle (Fig. 20.2-4). The frequency band $f_0 \pm B$ is matched by an optical beam of angular divergence

$$\delta\theta \approx \frac{(2\pi/v_s)B}{2\pi/\lambda} = \frac{\lambda}{v_s}B.$$

The bandwidth of the modulator is therefore

$$B = v_s\frac{\delta\theta}{\lambda} = \frac{v_s}{D}, \tag{20.2-2}$$

or

$$B = \frac{1}{T}, \qquad T = \frac{D}{v_s}, \tag{20.2-3}$$

Bandwidth

where T is the transit time of sound across the waist of the light beam. This is an

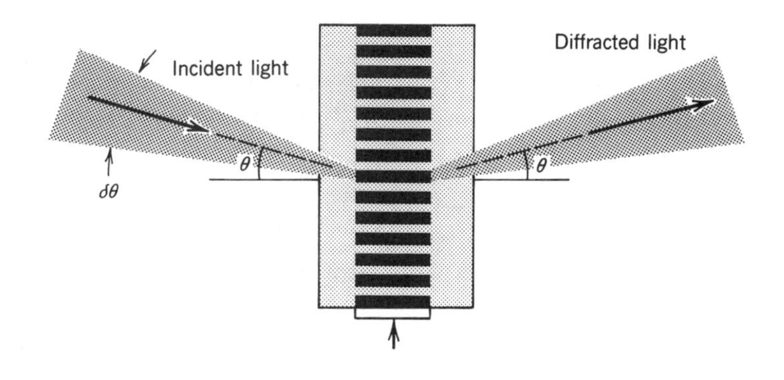

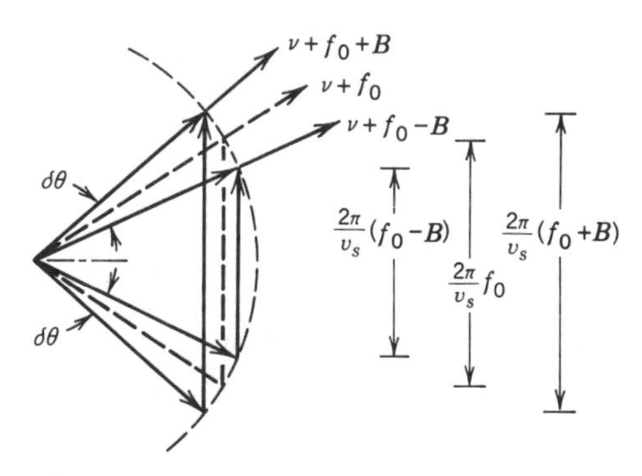

Figure 20.2-4 Interaction of an optical beam of angular divergence $\delta\theta$ with an acoustic plane wave of frequency in the band $f_0 \pm B$. There are many parallel **q** vectors of different lengths each matching a direction of the incident light.

expected result since it takes time T to change the amplitude of the sound wave at all points in the light–sound interaction region, so that the maximum rate of modulation is $1/T$ Hz. To increase the bandwidth of the modulator, the light beam should be focused to a small diameter.

EXERCISE 20.2-1

Parameters of Acousto-Optic Modulators. Determine the Bragg angle and the maximum bandwidth of the following acousto-optic modulators:

Modulator 1

Material: Fused quartz ($n = 1.46$, $v_s = 6$ km/s)
Sound: Frequency $f = 50$ MHz
Light: He–Ne laser, wavelength $\lambda_o = 633$ nm, angular divergence $\delta\theta = 1$ mrad

Modulator 2

Material: Tellurium ($n = 4.8$, $v_s = 2.2$ km/s)
Sound: Frequency $f = 100$ MHz
Light: CO_2 laser, wavelength $\lambda_o = 10.6$ μm, and beam width $D = 1$ mm

---

B. Scanners

The acousto-optic cell can be used as a scanner of light. The basic idea lies in the linear relation between the angle of deflection 2θ and the sound frequency f,

$$2\theta \approx \frac{\lambda}{v_s} f, \tag{20.2-4}$$

where θ is assumed sufficiently small so that $\sin\theta \approx \theta$. By changing the sound frequency f, the deflection angle 2θ can be varied.

One difficulty is that θ represents both the angle of reflection and the angle of incidence. To change the angle of reflection, both the angle of incidence and the sound frequency must be changed simultaneously. This may be accomplished by tilting the sound beam. Figure 20.2-5 illustrates this principle. Changing the sound frequency requires a frequency modulator (FM). Tilting the sound beam requires a sophisticated system that uses, for example, a phased array of acoustic transducers (several acoustic transducers driven at relative phases that are selected to impart a tilt to the overall generated sound wave). The angle of tilt must be synchronized with the FM driver.

The requirement to tilt the sound beam may be alleviated if we use a sound beam with an angular divergence equal to or greater than the entire range of directions to be scanned. As the sound frequency is changed, the Bragg angle is altered and the incoming light wave selects only the acoustic plane-wave component with the matching direction. The efficiency of the system is, of course, expected to be low. We proceed to examine some of the properties of this device.

Scan Angle

When the sound frequency is f, the incident light wave interacts with the sound component at an angle $\theta = (\lambda/2v_s)f$ and is deflected by an angle $2\theta = (\lambda/v_s)f$, as Fig. 20.2-6 illustrates. By varying the sound frequency from f_0 to $f_0 + B$, the deflection angle 2θ is swept over a scan angle

$$\Delta\theta = \frac{\lambda}{v_s} B. \tag{20.2-5}$$

Scan Angle

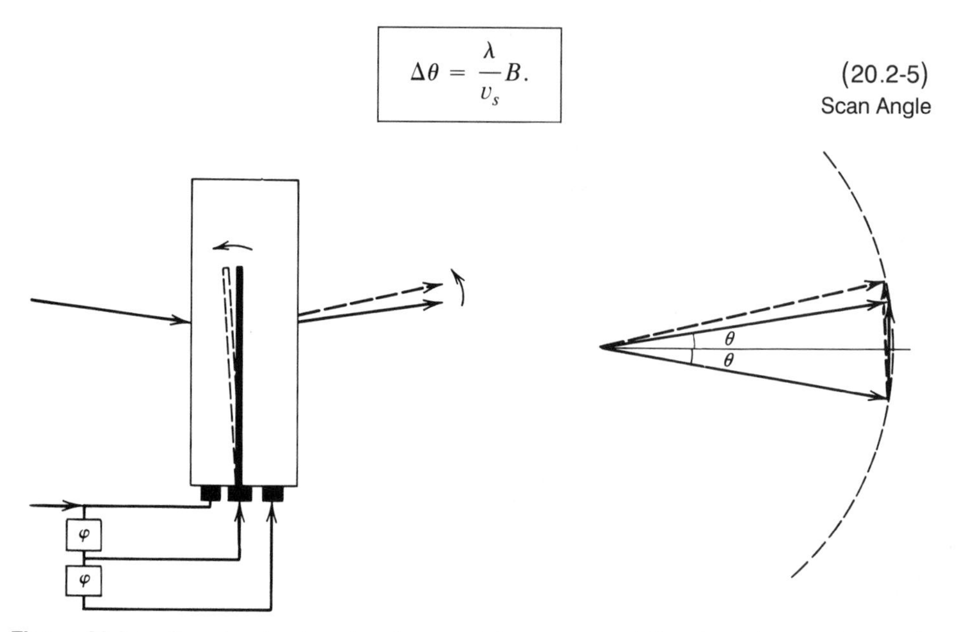

Figure 20.2-5 Scanning by changing the sound frequency *and* direction. The sound wave is tilted by use of an array of transducers driven by signals differing by a phase φ.

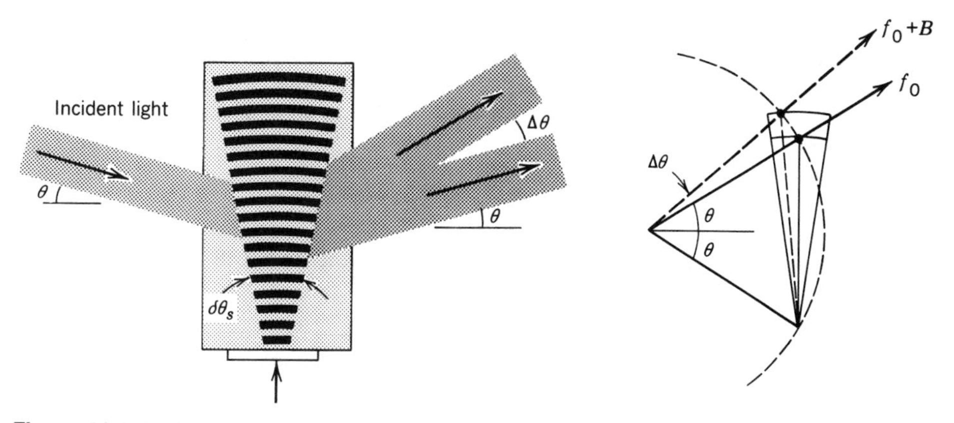

Figure 20.2-6 Scanning an optical wave by varying the frequency of a sound beam of angular divergence $\delta\theta_s$ over the frequency range $f_0 \le f \le f_0 + B$.

This, of course, assumes that the sound beam has an equal or greater angular width $\delta\theta_s = \Lambda/D_s \ge \Delta\theta$. Since the scan angle is inversely proportional to the speed of sound, larger scan angles are obtained by use of materials for which the sound velocity v_s is small.

Number of Resolvable Spots

If the optical wave itself has an angular width $\delta\theta = \lambda/D$, and assuming that $\delta\theta \ll \delta\theta_s$, the deflected beam also has a width $\delta\theta$. The number of resolvable spots of the scanner (the number of nonoverlapping angular widths within the scanning range) is therefore

$$N = \frac{\Delta\theta}{\delta\theta} = \frac{(\lambda/v_s)B}{\lambda/D} = \frac{D}{v_s}B,$$

or

$$\boxed{N = TB,} \tag{20.2-6}$$

Number of Resolvable Spots

where B is the bandwidth of the FM modulator used to generate the sound and $T = D/v_s$ is the transit time of sound through the light beam (Fig. 20.2-7).

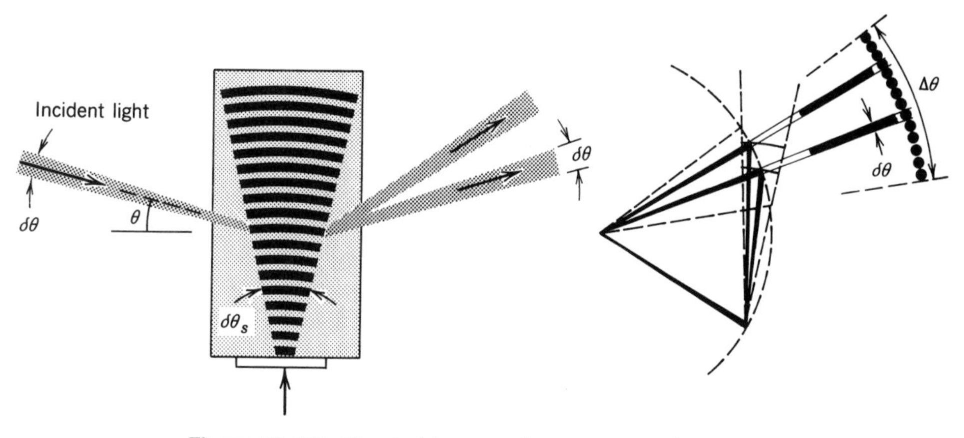

Figure 20.2-7 Resolvable spots of an acousto-optic scanner.

The number of resolvable spots is therefore equal to the time–bandwidth product. This number represents the degrees of freedom of the device and is a significant indicator of the capability of the scanner. To increase N, a large transit time T should be used. This is the opposite of the design requirement in an acousto-optic modulator, for which the modulation bandwidth $B = 1/T$ is made large by selecting a small T.

EXERCISE 20.2-2

Parameters of an Acousto-Optic Scanner. A fused-quartz acousto-optic scanner ($v_s = 6$ km/s, $n = 1.46$) is used to scan a He–Ne laser beam ($\lambda_o = 633$ nm). The sound frequency is scanned over the range 40 to 60 MHz. To what width should the laser beam be focused so that the number of resolvable points is $N = 100$? What is the scan angle $\Delta\theta$? What is the effect of using a material in which sound is slower, flint glass ($v_s = 3.1$ km/s), for example?

The Acousto-Optic Scanner as a Spectrum Analyzer

The proportionality between the angle of deflection and the sound frequency can be utilized to make an acoustic spectrum analyzer. A sound wave containing a spectrum of different frequencies disperses the light in different directions with the intensity of deflected light in a given direction proportional to the power of the sound component at the corresponding frequency (Fig. 20.2-8).

C. Interconnections

An acousto-optic cell can be used as an interconnection optical switch that routes information carried by one or more optical beams to one or more selected directions. Several interconnection schemes are possible:

- An acousto-optic cell in which the frequency of the acoustic wave is one of N possible values, $f_1, f_2, \ldots,$ or f_N, deflects an incident optical beam to one of N

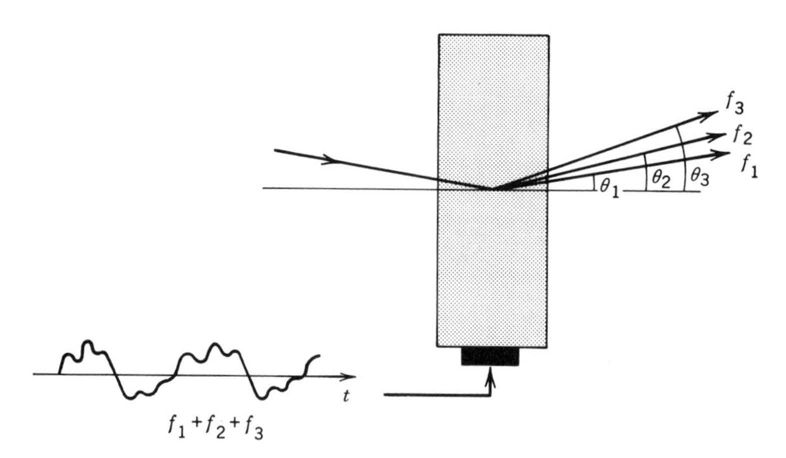

Figure 20.2-8 Each frequency component of the sound wave deflects light in a different direction. The acousto-optic cell serves as an acoustic spectrum analyzer.

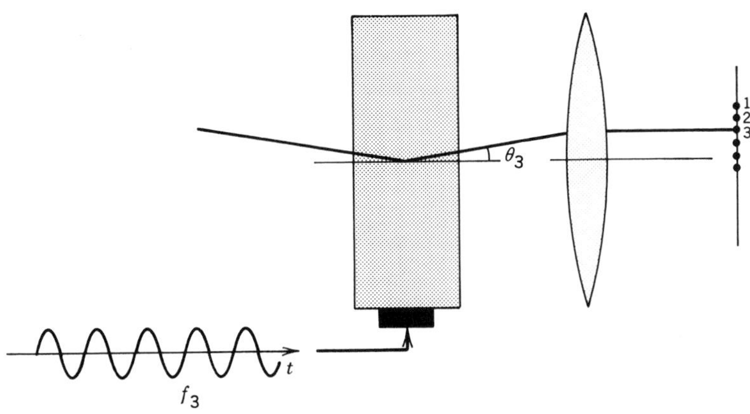

Figure 20.2-9 Routing an optical beam to one of N directions. By applying an acoustic wave of frequency f_3, for example, the optical beam is deflected by an angle θ_3 and routed to point 3.

corresponding directions, $\theta_1, \theta_2, \ldots,$ or θ_N, as illustrated in Fig. 20.2-9. The device routes one beam to any of N directions.

- By using an acoustic wave comprising two frequencies, f_1 and f_2, simultaneously, the incident optical beam is reflected in the two corresponding directions, θ_1 and θ_2, simultaneously. Thus one beam is connected to any pair of many possible directions as illustrated in Fig. 20.2-10. Similarly, by using an acoustic wave with M frequencies the incoming beam can be routed simultaneously to M directions. An example is the acoustic spectrum analyzer for which an incoming light beam is reflected from a sound wave carrying a spectrum of M frequencies. The light beam is routed to M points, with the intensity at each point proportional to the power of the corresponding sound-frequency component.
- The length of the acousto-optic cell may be divided into two segments. At a certain time, an acoustic wave of frequency f_1 is present in one segment and an acoustic wave of frequency f_2 is present in the other. This can be accomplished by generating the acoustic wave from a frequency-shift-keyed electric signal in the form of two pulses: a pulse of frequency f_1 followed by another of frequency f_2, each lasting a duration $T/2$, where $T = W/v_s$ is the transit time of sound through the cell length W (see Fig. 20.2-11). When the leading edge of the acoustic wave

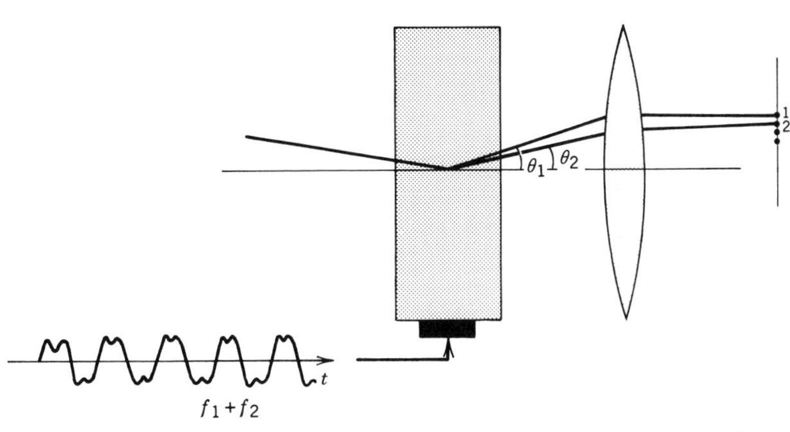

Figure 20.2-10 Routing a light beam simultaneously to a number of directions.

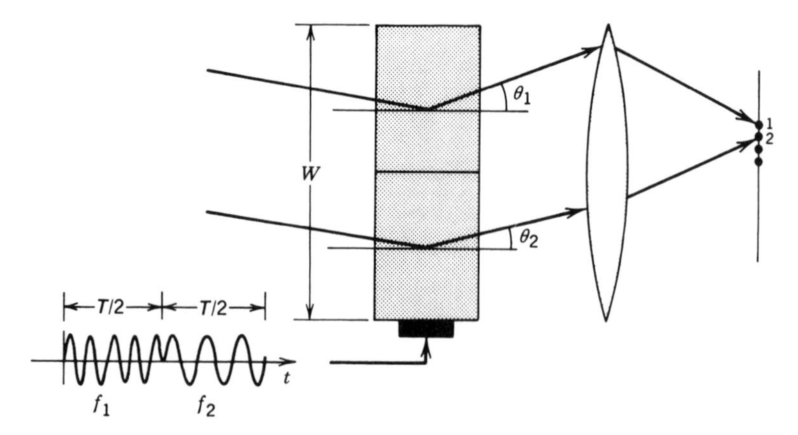

Figure 20.2-11 Routing each of two light beams to a set of specified directions. The acoustic wave is generated by a frequency-shift-keyed electric signal.

reaches the end of the cell, the cell processes two incoming optical beams by deflecting the top beam to the direction θ_1 corresponding to f_1, and the bottom beam to the direction θ_2 corresponding to f_2. This is a switch that connects each of two beams to any of many possible directions. By placing more than one frequency component in each segment, each of the two beams can itself be routed simultaneously to several directions.

- The cell may also be divided into N segments, each carrying a harmonic acoustic wave of the same frequency f but with a different amplitude. The result is a **spatial light modulator** that modulates the intensities of N input beams (Fig. 20.2-12). Spatial light modulators are useful in optical signal processing (see Sec. 21.5).
- The most general interconnection architecture is one for which the cell is divided into L segments, each of which carries an acoustic wave with M frequencies. The device acts as a random access switch that routes each of L incoming beams to M directions simultaneously (Fig. 20.2-13).

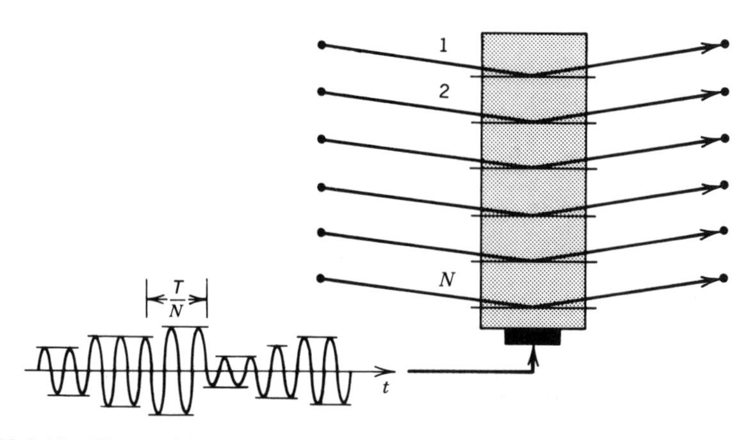

Figure 20.2-12 The spatial light modulator modulates N optical beams. The acoustic wave is driven by an amplitude-modulated electric signal.

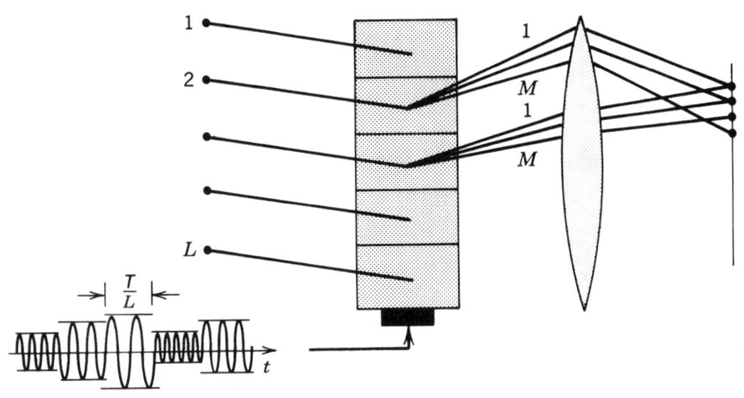

Figure 20.2-13 An arbitrary-interconnection switch routes each of L incoming light beams for the random access of M points.

Interconnection Capacity

There is an upper limit to the number of interconnections that may be established by an acousto-optic device, as will be shown subsequently. If an acousto-optic cell is used to route each of L incoming optical beams to a maximum of M directions simultaneously, then product ML cannot exceed the time–bandwidth product $N = TB$, where T is the transit time through the cell and B is the bandwidth of the acoustic wave,

$$ML \leq N. \tag{20.2-7}$$

Interconnection Capacity

This upper bound on the number of interconnections is called the interconnection capacity of the device.

An acousto-optic cell with L segments uses an acoustic wave composed of L segments each of time duration T/L. For each segment to address M independent points the acoustic wave must carry M independent frequency components per segment. For a signal of duration T/L there is an inherent frequency uncertainty of L/T hertz. The M frequency components must therefore be separated by at least that uncertainty. For the M components to be placed within the available bandwidth B, we must have $M(L/T) \leq B$, from which $ML \leq TB$, and hence (20.2-7) follows.

A single optical beam ($L = 1$), for example, can be connected to any of $N = TB$ points, but each of two beams can be connected to at most $N/2$ points, and so on. It is a question of dividing an available time–bandwidth product $N = TB$ in the form of L time segments each containing M independent frequencies. Examples of the possible choices are illustrated in the time–frequency diagram in Fig. 20.2-14.

D. Filters, Frequency Shifters, and Isolators

The acousto-optic cell is useful in a number of other applications, including filters, frequency shifters, and optical isolators.

Tunable Acousto-Optic Filters

The Bragg condition $\sin\theta = \lambda/2\Lambda$ relates the angle θ, the acoustic wavelength Λ, and the optical wavelength λ. If θ and Λ are specified, reflection can occur only for a single optical wavelength $\lambda = 2\Lambda\sin\theta$. This wavelength-selection property can be used to

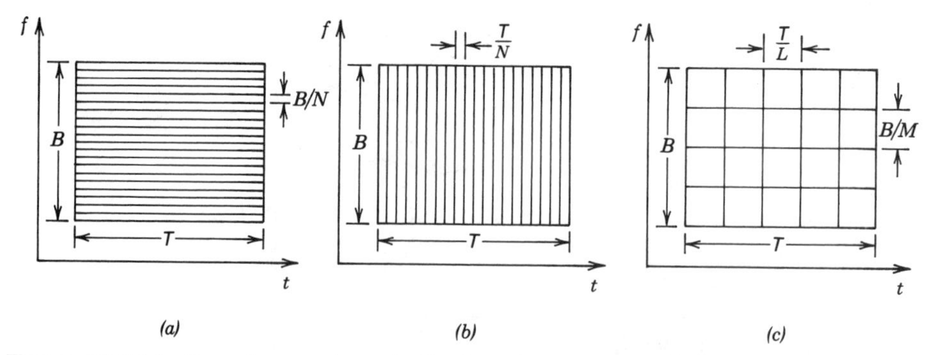

Figure 20.2-14 Several examples of dividing the time–bandwidth region TB in the time–frequency diagram into $N = TB$ subdivisions (in this diagram $N = 20$). (*a*) A scanner: a single time segment containing N frequency segments. (*b*) A spatial light modulator: N time segments each containing one frequency component. (*c*) An interconnection switch: L time segments each containing $M = N/L$ frequency segments (in this diagram, $N = 20$, $M = 4$, and $L = 5$).

filter an optical wave composed of a broad spectrum of wavelengths. The filter is tuned by changing the angle θ or the sound frequency f.

EXERCISE 20.2-3

Resolving Power of an Acousto-Optic Filter. Show that the spectral resolving power $\lambda/\Delta\lambda$ of an acousto-optic filter equals fT, where f is the sound frequency, T the transit time, and $\Delta\lambda$ the minimum resolvable wavelength difference.

Frequency Shifters

Optical frequency shifters are useful in many applications of photonics, including optical heterodyning, optical FM modulators, and laser Doppler velocimeters. The acousto-optic cell may be used as a tunable frequency shifter since the Bragg reflected light is frequency shifted (up or down) by the frequency of sound. In a heterodyne optical receiver, a received amplitude- or phase-modulated optical signal is mixed with a coherent optical wave from a local light source, acting as a local oscillator with a different frequency. The two optical waves beat (see Sec. 2.6B) and the detected signal varies at the frequency difference. Information about the amplitude and phase of the received signal can be extracted from the detected signal (see Sec. 22.5A). The acousto-optic cell offers a practical means for imparting the frequency shift required for the heterodyning process.

Optical Isolators

An optical isolator is a one-way optical valve often used to prevent reflected light from retracing its path back into the original light source (see Sec. 6.6C). Optical isolators are sometimes used with semiconductor lasers since the reflected light can interact with the laser process and create deleterious effects (noise). The acousto-optic cell can serve as an isolator. If part of the frequency-upshifted Bragg-diffracted light is reflected onto itself by a mirror and traces its path back into the cell, as illustrated in Fig. 20.2-15, it undergoes a second Bragg diffraction accompanied by a second frequency upshift.

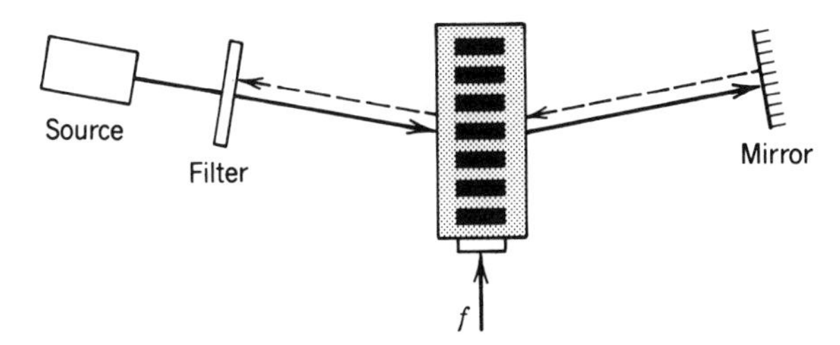

Figure 20.2-15 An acousto-optic isolator.

Since the frequency of the returning light differs from that of the original light by twice the sound frequency, a filter may be used to block it. Even without a filter, the laser process may be insensitive to the frequency-shifted light.

*20.3 ACOUSTO-OPTICS OF ANISOTROPIC MEDIA

The scalar theory of interaction of light and sound is generalized in this section to include the anisotropic properties of the medium and the effects of polarization of light and sound.

Acoustic Waves in Anisotropic Materials

An acoustic wave is a wave of material strain. Strain is defined in terms of the displacements of the molecules relative to their equilibrium positions. If $\mathbf{u} = (u_1, u_2, u_3)$ is the vector of displacement of the molecules located at position $\mathbf{x} = (x_1, x_2, x_3)$, the strain is a symmetrical tensor with components $s_{ij} = \frac{1}{2}(\partial u_i/\partial x_j + \partial u_j/\partial x_i)$, where the indices $i, j = 1, 2, 3$ denote the coordinates (x, y, z). The element $s_{33} = \partial u_3/\partial x_3$, for example, represents tensile strain (stretching) in the z direction [Fig. 20.3-1(a)], whereas s_{13} represents shear strain since $\partial u_1/\partial x_3$ is the relative movement in the x direction of two incrementally separated parallel planes normal to the z direction, as illustrated in Fig. 20.3-1(b).

An acoustic wave can be longitudinal or transverse, as illustrated in the following examples.

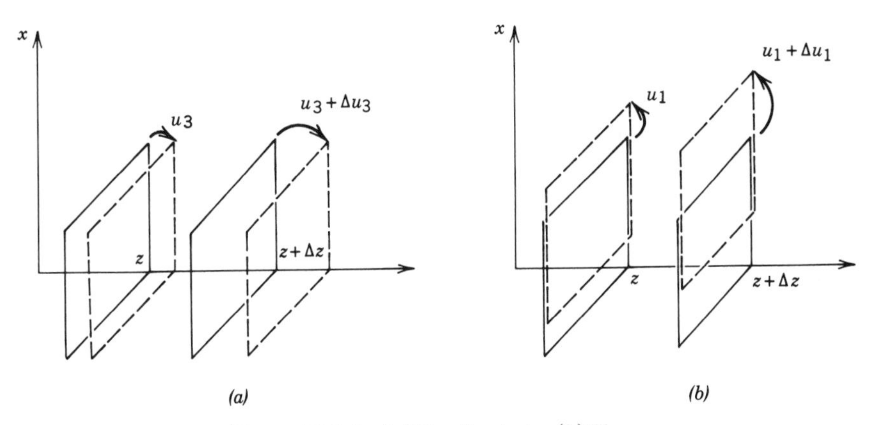

Figure 20.3-1 (a) Tensile strain. (b) Shear.

EXAMPLE 20.3-1. ***Longitudinal Wave.*** A wave with the displacement $u_1 = 0$, $u_2 = 0$, $u_3 = A_0 \sin(\Omega t - qz)$, where A_0 is a constant, corresponds to a strain tensor with all components vanishing except

$$s_{33} = S_0 \cos(\Omega t - qz), \tag{20.3-1}$$

where $S_0 = -qA_0$. This is a wave of stretching in the z direction traveling in the z direction. Since the vibrations are in the direction of wave propagation, the wave is longitudinal.

EXAMPLE 20.3-2. ***Transverse Wave.*** The displacement wave, $u_1 = A_0 \sin(\Omega t - qz)$, $u_2 = 0$, $u_3 = 0$, corresponds to a strain tensor with all components vanishing except

$$s_{13} = s_{31} = S_0 \cos(\Omega t - qz), \tag{20.3-2}$$

where $S_0 = -\frac{1}{2}qA_0$. This wave travels in the z direction but vibrates in the x direction. It is a transverse (shear) wave.

The velocities of the longitudinal and transverse acoustic waves are characteristics of the medium and generally depend on the direction of propagation.

The Photoelastic Effect

The optical properties of an anisotropic medium are characterized completely by the electric impermeability tensor $\boldsymbol{\eta} = \epsilon_0 \boldsymbol{\epsilon}^{-1}$ (see Sec. 6.3). Given $\boldsymbol{\eta}$, we can determine the index ellipsoid and hence the refractive indices for an optical wave traveling in an arbitrary direction with arbitrary polarization.

In the presence of strain, the electric impermeability tensor is modified so that η_{ij} becomes a function of the elements of the strain tensor, $\eta_{ij} = \eta_{ij}(s_{kl})$. This dependence is called the **photoelastic effect**. Each of the nine functions $\eta_{ij}(s_{kl})$ may be expanded in terms of the nine variables s_{kl} in a Taylor's series. Maintaining only the linear terms,

$$\eta_{ij}(s_{kl}) \approx \eta_{ij}(0) + \sum_{kl} \mathfrak{p}_{ijkl} s_{kl}, \qquad i, j, l, k = 1, 2, 3, \tag{20.3-3}$$

where $\mathfrak{p}_{ijkl} = \partial \eta_{ij} / \partial s_{kl}$ are constants forming a tensor of fourth rank known as the **strain-optic tensor**.

Since both $\{\eta_{ij}\}$ and $\{s_{kl}\}$ are symmetrical tensors, the coefficients $\{\mathfrak{p}_{ijkl}\}$ are invariant to permutations of i and j, and to permutations of k and l. There are therefore only six instead of nine independent values for the set (i, j) and six independent values for (k, l). The pair of indices (i, j) is usually contracted to a single index $I = 1, 2, \ldots, 6$ (see Table 18.2-1 on page 714). The indices (k, l) are similarly contracted and denoted by the index $K = 1, 2, \ldots, 6$. The fourth-rank tensor $\mathfrak{p}_{ijkl}$ is thus described by a 6×6 matrix $\mathfrak{p}_{IK}$.

Symmetry of the crystal requires that some of the coefficients $\mathfrak{p}_{IK}$ vanish and that certain coefficients are related. The matrix $\mathfrak{p}_{IK}$ of a cubic crystal, for example, has the structure

$$\mathfrak{p}_{IK} = \begin{bmatrix} \mathfrak{p}_{11} & \mathfrak{p}_{12} & \mathfrak{p}_{12} & 0 & 0 & 0 \\ \mathfrak{p}_{12} & \mathfrak{p}_{11} & \mathfrak{p}_{12} & 0 & 0 & 0 \\ \mathfrak{p}_{11} & \mathfrak{p}_{12} & \mathfrak{p}_{11} & 0 & 0 & 0 \\ 0 & 0 & 0 & \mathfrak{p}_{44} & 0 & 0 \\ 0 & 0 & 0 & 0 & \mathfrak{p}_{44} & 0 \\ 0 & 0 & 0 & 0 & 0 & \mathfrak{p}_{44} \end{bmatrix}. \tag{20.3-4}$$

Strain-Optic Matrix (Cubic Crystal)

This matrix is also applicable for isotropic media, with the additional constraint $\mathfrak{p}_{44} = \frac{1}{2}(\mathfrak{p}_{11} + \mathfrak{p}_{12})$, so that there are only two independent coefficients.

EXAMPLE 20.3-3. ***Longitudinal Acoustic Wave in a Cubic Crystal.*** The longitudinal acoustic wave described in Example 20.3-1 travels in a cubic crystal of refractive index n. By substitution of (20.3-1) and (20.3-4) into (20.3-3) we find that the associated strain results in an impermeability tensor with elements,

$$\eta_{11} = \eta_{22} = \frac{1}{n^2} + \mathfrak{p}_{12} S_0 \cos(\Omega t - qz)$$

$$\eta_{33} = \frac{1}{n^2} + \mathfrak{p}_{11} S_0 \cos(\Omega t - qz)$$

$$\eta_{ij} = 0, \qquad i \neq j.$$

Thus the initially optically isotropic cubic crystal becomes a uniaxial crystal with the optic axis in the direction of the acoustic wave (z direction) and with ordinary and extraordinary refractive indices, n_o and n_e, given by

$$\frac{1}{n_o^2} = \frac{1}{n^2} + \mathfrak{p}_{12} S_0 \cos(\Omega t - qz) \tag{20.3-5}$$

$$\frac{1}{n_e^2} = \frac{1}{n^2} + \mathfrak{p}_{11} S_0 \cos(\Omega t - qz). \tag{20.3-6}$$

The shape of the index ellipsoid is altered periodically in time and space in the form of a wave, but the principal axes remain unchanged (see Fig. 20.3-2). Since the change of the refractive indices is usually small, the second terms in (20.3-5) and (20.3-6) are small, so that the approximation $(1 + \Delta)^{-1/2} \approx 1 - \Delta/2$, when $|\Delta| \ll 1$, may be applied to approximate (20.3-5) and (20.3-6) by

$$n_o \approx n - \tfrac{1}{2} n^3 \mathfrak{p}_{12} S_0 \cos(\Omega t - qz) \tag{20.3-7}$$

$$n_e \approx n - \tfrac{1}{2} n^3 \mathfrak{p}_{11} S_0 \cos(\Omega t - qz). \tag{20.3-8}$$

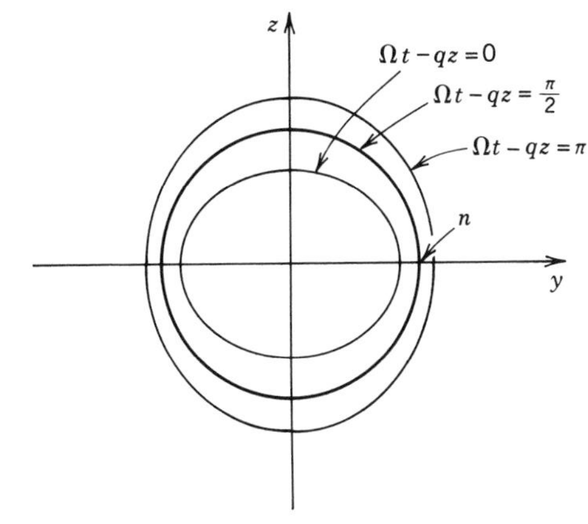

Figure 20.3-2 A longitudinal acoustic wave traveling in the z direction in a cubic crystal alters the shape of the index ellipsoid from a sphere into an ellipsoid of revolution with dimensions varying sinusoidally with time and an axis in the z direction.

EXERCISE 20.3-1

Transverse Acoustic Wave in a Cubic Crystal. The transverse acoustic wave described in Example 20.3-2 travels in a cubic crystal. Show that the crystal becomes biaxial with principal refractive indices

$$n_1 \approx n - \tfrac{1}{2} n^3 \mathfrak{p}_{44} S_0 \cos(\Omega t - qz) \tag{20.3-9}$$

$$n_2 \approx n \tag{20.3-10}$$

$$n_3 \approx n + \tfrac{1}{2} n^3 \mathfrak{p}_{44} S_0 \cos(\Omega t - qz). \tag{20.3-11}$$

In Example 20.3-3 and Exercise 20.3-1, the acoustic wave alters the index ellipsoid's principal values but not its principal directions, so that the ellipsoid maintains its orientation. Obviously, this is not always the case. Acoustic waves in other directions and polarizations relative to the crystal principal axes result in alteration of the principal refractive indices as well as the principal axes of the crystal.

Bragg Diffraction

The interaction of a linearly polarized optical wave with a longitudinal or transverse acoustic wave in an anisotropic medium can be described by the same principles discussed in Sec. 20.1. The incident optical wave is reflected from the acoustic wave if the Bragg condition of constructive interference is satisfied. The analysis is more complicated, in comparison with the scalar theory, since the incident and reflected waves travel with different velocities and, consequently, the angles of reflection and incidence need not be equal.

The condition for Bragg diffraction is the conservation-of-momentum (phase-matching) condition,

$$\mathbf{k}_r = \mathbf{k} + \mathbf{q}. \tag{20.3-12}$$

The magnitudes of these wavevectors are $k = (2\pi/\lambda_o)n$, $k_r = (2\pi/\lambda_o)n_r$, and $q = (2\pi/\Lambda)$, where λ_o and Λ are the optical and acoustic wavelengths and n and n_r are the refractive indices of the incident and reflected optical waves, respectively.

As illustrated in Fig. 20.3-3, if θ and θ_r are the angles of incidence and reflection, the vector equation (20.3-12) may be replaced with two scalar equations relating the z and x components of the wavevectors in the plane of incidence:

$$\frac{2\pi}{\lambda_o} n_r \cos\theta_r = \frac{2\pi}{\lambda_o} n \cos\theta$$

$$\frac{2\pi}{\lambda_o} n_r \sin\theta_r + \frac{2\pi}{\lambda_o} n \sin\theta = \frac{2\pi}{\Lambda},$$

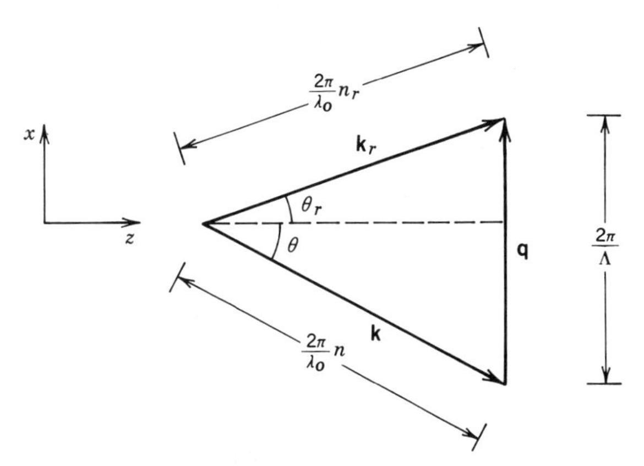

Figure 20.3-3 Conservation of momentum (phase-matching condition, or Bragg condition) in an anisotropic medium.

from which

$$n_r \cos \theta_r = n \cos \theta \tag{20.3-13a}$$

$$n_r \sin \theta_r + n \sin \theta = \frac{\lambda_o}{\Lambda}. \tag{20.3-13b}$$

Given the wavelengths λ_o and Λ, the angles θ and θ_r may be determined by solving equations (20.3-13). Note that n and n_r are generally functions of θ and θ_r that may be determined from the index ellipsoid of the unperturbed crystal.

Equations (20.3-13) can be easily solved when the acoustic and optical waves are collinear, so that $\theta = \pm\pi/2$ and $\theta_r = \pi/2$. The + and − signs correspond to back and front reflections, as illustrated in Fig. 20.3-4. The conditions (20.3-13) then reduce to one condition,

$$n_r \pm n = \frac{\lambda_o}{\Lambda}. \tag{20.3-14}$$

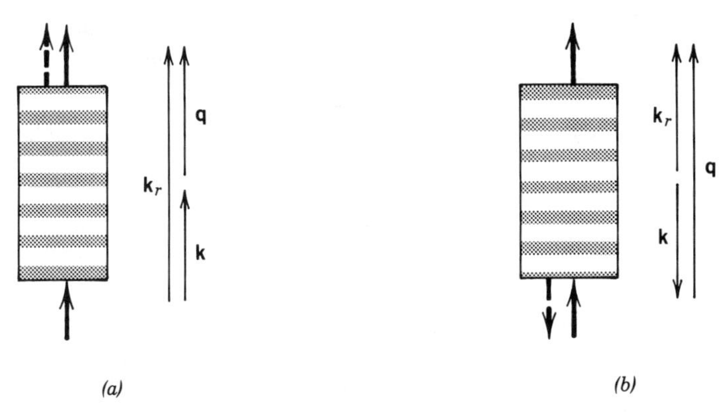

Figure 20.3-4 Wavevector diagram for front and back reflection of an optical wave from an acoustic wave.

For back reflection (+ sign), Λ must be smaller than λ_o, which is unlikely except for very high frequency acoustic waves. For front reflection (− sign), the incident and reflected waves must have different polarizations so that $n_r \neq n$.

READING LIST

Books

C. S. Tsai, *Guided-Wave Acoustooptics*, Springer-Verlag, Berlin, 1990.

L. N. Magdich and V. Ya. Molchanov, *Acoustooptic Devices and Their Applications*, Gordon and Breach, New York, 1989.

A. Korpel, *Acousto-Optics*, Marcel Dekker, New York, 1988.

A. Yariv, *Optical Electronics*, Holt, Rinehart and Winston, New York, 1971, 3rd ed. 1985.

A. Yariv and P. Yeh, *Optical Waves in Crystals*, Wiley, New York, 1984.

J. F. Nye, *Physical Properties of Crystals: Their Representation by Tensors and Matrices*, Clarendon Press, Oxford, 1957; Oxford University Press, New York, 1984.

M. Gottlieb, C. L. M. Ireland, and J. M. Ley, *Electro-Optic and Acousto-Optic Scanning and Deflection*, Marcel Dekker, New York, 1983.

N. J. Berg and J. N. Lee, eds., *Acousto-Optic Signal Processing*, Marcel Dekker, New York, 1983.

T. S. Narasimhamurty, *Photoelastic and Electro-Optic Properties of Crystals*, Plenum Press, New York, 1981.

M. Born and E. Wolf, *Principles of Optics*, Pergamon, New York, 1959, 6th ed. 1980, Chap. 12.

D. F. Nelson, *Electric, Optic, and Acoustic Interactions in Dielectrics*, Wiley, New York, 1979.

J. Sapriel, *Acousto-Optics*, Wiley, New York, 1979.

M. J. P. Musgrave, *Crystal Acoustics*, Holden-Day, San Francisco, 1970.

M. V. Berry, *The Diffraction of Light by Ultrasound*, Academic Press, New York, 1966.

Articles

A. C. Tam, Applications of Photoacoustic Sensing Techniques, *Reviews of Modern Physics*, vol. 58, pp. 381–431, 1986.

Special issue on acoustooptic signal processing, *Proceedings of the IEEE*, vol. 69, no. 1, 1981.

A. Korpel, Acousto-Optics, in *Applied Optics and Optical Engineering*, vol. 6, R. Kingslake and B. J. Thompson, eds., Academic Press, New York, 1980.

E. G. Lean, Interaction of Light and Acoustic Surface Waves, in *Progress in Optics*, vol. 11, E. Wolf, ed., North-Holland, Amsterdam, 1973.

E. K. Sittig, Elastooptic Light Modulation and Deflection, in *Progress in Optics*, vol. 10, E. Wolf, ed., North-Holland, Amsterdam, 1972.

R. W. Damon, W. T. Maloney, and D. H. McMahon, Interaction of Light with Ultrasound: Phenomena and Applications, in *Physical Acoustics: Principles and Methods*, vol. 7, W. P. Mason and R. N. Thurston, eds., Academic Press, New York, 1970.

R. Adler, Interaction between Light and Sound, *IEEE Spectrum*, vol. 4, no. 5, pp. 42–54, 1967.

E. I. Gordon, A Review of Acoustooptical Deflection and Modulation Devices, *Proceedings of the IEEE*, vol. 54, pp. 1391–1401, 1966.

PROBLEMS

20.1-1 **Diffraction of Light from Various Periodic Structures.** Discuss the diffraction of an optical plane wave of wavelength λ from the following periodic structures, indicating in each case the geometrical configuration and the frequency shift(s):
(a) An acoustic traveling wave of wavelength Λ.
(b) An acoustic standing wave of wavelength Λ.

(c) A graded-index transparent medium with refractive index varying sinusoidally with position (period Λ).
(d) A stratified medium made of parallel layers of two materials of different refractive indices, alternating to form a periodic structure of period Λ.

*20.1-2 **Bragg Diffraction as a Scattering Process.** An incident optical wave of angular frequency ω, wavevector $\mathbf{k}$, and complex envelope A interacts with a medium perturbed by an acoustic wave of angular frequency Ω and wavevector $\mathbf{q}$, and creates a light source $\mathscr{S}$ given by (20.1-25). The angle θ corresponds to upshifted Bragg diffraction, so that the scattering light source is $\mathscr{S} = \text{Re}\{S_r(\mathbf{r})\exp(j\omega_r t)\}$, where $S_r(\mathbf{r}) = -(\Delta n_0/n)k_r^2 A \exp(-j\mathbf{k}_r \cdot \mathbf{r})$, $\omega_r = \omega + \Omega$, and $\mathbf{k}_r = \mathbf{k} + \mathbf{q}$. This source emits a scattered field E. Assuming that the incident wave is undepleted by the acousto-optic interaction (first Born approximation, i.e., A remains approximately constant), the scattered light may be obtained by solving the Helmholtz equation $\nabla^2 E + k^2 E = -S$. This equation has the far-field solution (see Problem 19.2-3)

$$E(\mathbf{r}) \approx \frac{\exp(-jkr)}{4\pi r} \int_V S_r(\mathbf{r}') \exp(jk\hat{\mathbf{r}} \cdot \mathbf{r}')\, d\mathbf{r}',$$

where $\hat{\mathbf{r}}$ is a unit vector in the direction of $\mathbf{r}$, $k = 2\pi/\lambda$, and V is the volume of the source. Use this equation to determine an expression for the reflectance of the acousto-optic cell when the Bragg condition is satisfied. Compare the result with (20.1-18).

20.1-3 **Condition for Raman–Nath Diffraction.** Derive an expression for the maximum width D_s of an acoustic beam of wavelength Λ that permits Raman–Nath diffraction of light of wavelength λ (see Fig. 20.1-10).

20.1-4 **Combined Acousto-Optic and Electro-Optic Modulation.** One end of a lithium niobate ($LiNbO_3$) crystal is placed inside a microwave cavity with an electromagnetic field at 3 GHz. As a result of the piezoelectric effect (the electric field creating a strain in the material), an acoustic wave is launched. Light from a He–Ne laser ($\lambda_o = 633$ nm) is reflected from the acoustic wave. The refractive index is $n = 2.3$ and the velocity of sound is $v_s = 7.4$ km/s. Determine the Bragg angle. Since lithium niobate is also an electro-optic material, the applied electric field modulates the refractive index, which in turn modulates the phase of the incident light. Sketch the spectrum of the reflected light. If the microwave electric field is a pulse of short duration, sketch the spectrum of the reflected light at different times indicating the contributions of the electro-optic and acousto-optic effects.

20.2-1 **Acousto-Optic Modulation.** Devise a system for converting a monochromatic optical wave with complex wavefunction $U(t) = A\exp(j\omega t)$ into a modulated wave of complex wavefunction $A\cos(\Omega t)\exp(j\omega t)$ by use of an acousto-optic cell with an acoustic wave $s(x, t) = S_0 \cos(\Omega t - qx)$. *Hint*: Consider the use of upshifted and downshifted Bragg reflections.

20.2-2 **Frequency-Shift-Free Bragg Reflector.** Design an acousto-optic system that deflects light without frequency shifting it. *Hint*: Use two Bragg cells. (Reference: F. W. Freyre, *Applied Optics*, vol. 22, pp. 3896–3900, 1981.)

*20.3-1 **Front Bragg Diffraction.** A transverse acoustic wave of wavelength Λ travels in the x direction in a uniaxial crystal with refractive indices n_o and n_e and optic axis in the z direction. Derive an expression for the wavelength λ_o of an incident optical wave, traveling in the x direction and polarized in the z direction, that satisfies the condition of Bragg diffraction. What is the polarization of the front reflected wave? Determine Λ if $\lambda_o = 633$ nm, $n_e = 2.200$, and $n_o = 2.286$.

CHAPTER

21

PHOTONIC SWITCHING AND COMPUTING

21.1 PHOTONIC SWITCHES
- A. Switches
- B. Opto-Mechanical, Electro-Optic, Acousto-Optic, and Magneto-Optic Switches

21.2 ALL-OPTICAL SWITCHES

21.3 BISTABLE OPTICAL DEVICES
- A. Bistable Systems
- B. Principle of Optical Bistability
- C. Bistable Optical Devices
- D. Hybrid Bistable Optical Devices

21.4 OPTICAL INTERCONNECTIONS
- A. Holographic Interconnections
- B. Optical Interconnections in Microelectronics

21.5 OPTICAL COMPUTING
- A. Digital Optical Computing
- B. Analog Optical Processing

The ideas of **Johann (John) von Neumann (1903–1957)** had a major influence on the architecture of digital computers. He investigated the use of logic gates based on nonlinear dielectric constants. In 1953 he proposed that stimulated emission in a semiconductor material could be used to provide light amplification, which is the underlying principle for the operation of the semiconductor laser.

Switching is an essential operation in communication networks. It is also a basic operation in digital computers and signal processing systems. The current rapid development of high-data-rate fiber-optic communication systems has created a need for high-capacity repeaters and terminal systems for processing optical signals and, therefore, a need for high-speed photonic switches. Similarly, the potential for optical computing can only be realized if large arrays of fast photonic gates, switches, and memory elements are developed.

This chapter introduces the basic principles of the emerging technologies of photonic switching and optical signal processing. Many of the fundamental principles of photonics, which have been introduced in earlier chapters (Fourier optics and holography, guided-wave optics, electro-optics, acousto-optics, and nonlinear optics), find use here.

Section 21.1 provides a brief introduction to the general types and properties of switches and to photonic switching using opto-mechanical, acousto-optic, magneto-optic, and electro-optic devices. All-optical switches are described in Sec. 21.2. Section 21.3 is devoted to bistable optical devices. These are switches with memory—systems for which the output is one of only two states, depending on the current and previous values of the input. Section 21.4 covers optical interconnections and their applications in optical signal processing and in microelectronics. Finally, Sec. 21.5 outlines the basic features of optical processing and computing systems, both digital and analog.

21.1 PHOTONIC SWITCHES

A. Switches

A switch is a device that establishes and releases connections among transmission paths in a communication or signal-processing system. A control unit processes the commands for connections and sends a control signal to operate the switch in the desired manner. Examples of switches are shown in Fig. 21.1-1.

A 1×1 switch can be used as an elementary unit from which switches of larger sizes can be built. An $N \times N$ crosspoint-matrix (crossbar) switch, for example, may be constructed by using an array of N^2 1×1 switches organized at the points of an $N \times N$ matrix to connect or disconnect each of the N input lines to a free output line [see Fig. 21.1-1(*d*) and Fig. 21.1-2(*a*)]. The mth input reaches all elementary switches of the mth row, while the lth output is connected to outputs of all elementary switches of the lth column. A connection is made between the mth input and the lth output by activating the (m, l) 1×1 switch.

An $N \times N$ switch may also be built by use of 2×2 switches. An example is the 4×4 switch, made by the use of five 2×2 switches in the configuration shown in Fig. 21.1-2(*b*).

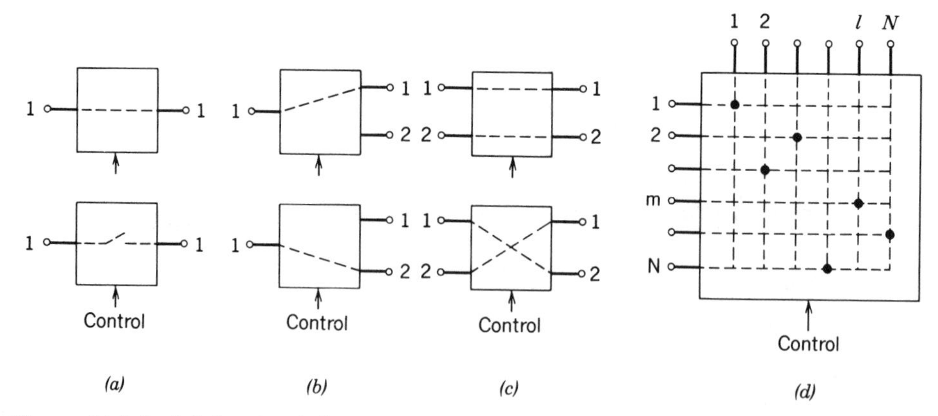

Figure 21.1-1 (*a*) 1×1 switch connects or disconnects two lines. It is an on–off switch. (*b*) 1×2 switch connects one line to either of two lines. (*c*) 2×2 crossbar switch connects two lines to two lines. It has two configurations: the bar state and the cross state. (*d*) $N \times N$ crossbar switch connects N lines to N lines. Any input line can always be connected to a free (unconnected) output line without blocking (i.e., without conflict).

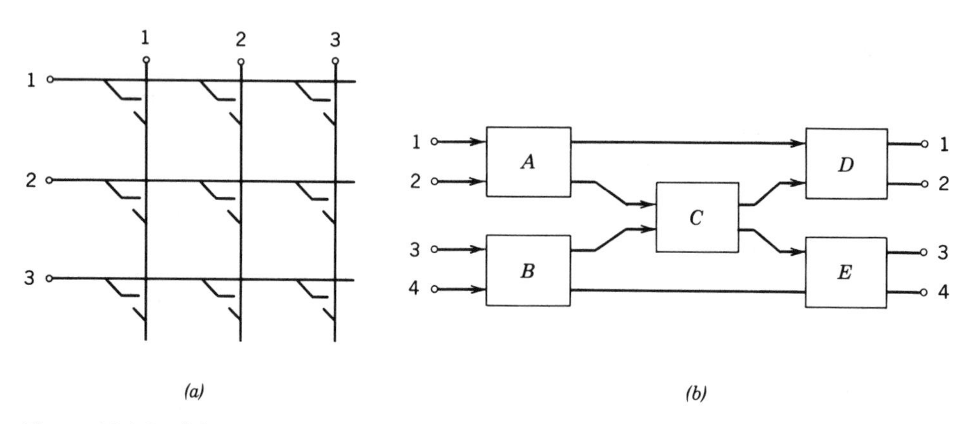

Figure 21.1-2 (*a*) A 3×3 switch made of nine 1×1 switches. (*b*) A 4×4 switch made of five 2×2 switches. Input line 1 is connected to output line 3, for example, if switches A and C are in the cross state and switch E is in the bar state.

A switch is characterized by the following parameters:

- *Size* (number of input and output lines) and *direction(s)*, i.e., whether data can be transferred in one or two directions.
- *Switching time* (time necessary for the switch to be reconfigured from one state to another).
- *Propagation delay time* (time taken by the signal to cross the switch).
- *Throughput* (maximum data rate that can flow through the switch when it is connected).
- *Switching energy* (energy needed to activate and deactivate the switch).
- *Power dissipation* (energy dissipated per second in the process of switching).
- *Insertion loss* (drop in signal power introduced by the connection).
- *Crosstalk* (undesired power leakage to other lines).
- *Physical dimensions*. This is important when large arrays of switches are to be built.

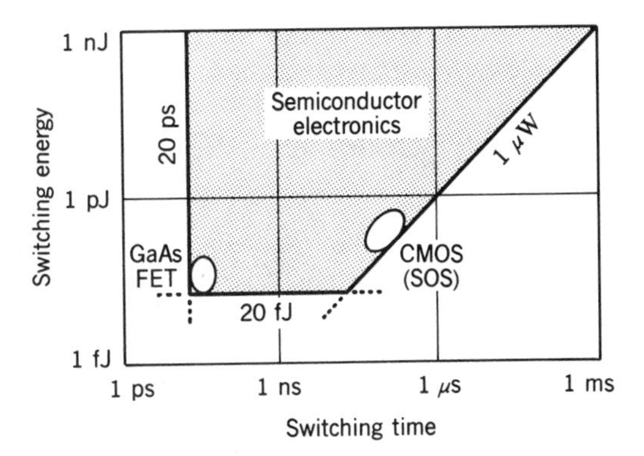

Figure 21.1-3 Limits of switching energy, switching time, and switching power for semiconductor devices. Both silicon-on-sapphire (SOS) complementary-symmetry metal-oxide-semiconductor (CMOS) and GaAs field-effect transistors (FET) are shown.

Electronic switches are used to switch electrical signals. The switch control is either electro-mechanical (using relays) or electronic (using semiconductor enabling logic circuits). Although it is difficult to provide precise limits on the minimum achievable switching time, switching energy, and switching power for semiconductor electronics technology, which continues to advance rapidly, the following bounds are representative of the orders of magnitude:

Minimum switching time	$= 10\text{–}20$ ps
Minimum energy per operation	$= 10\text{–}20$ fJ
Minimum switching power	$\approx 1\ \mu$W.

Limits of Semiconductor Electronic Switches

These limits are shown schematically in Fig. 21.1-3.

Josephson devices can operate at lower energies (tens of aJ; 1 aJ $= 10^{-18}$ J); a switching time of 1.5 ps has been demonstrated and subpicosecond operations are theoretically possible.

Optical signals may be switched by the use of electronic switches: the optical signals are converted into electrical signals using photodetectors, switched electronically, and then converted back into light using LEDs or lasers (Fig. 21.1-4). These optical/electrical/optical conversions introduce unnecessary time delays and power loss (in addition

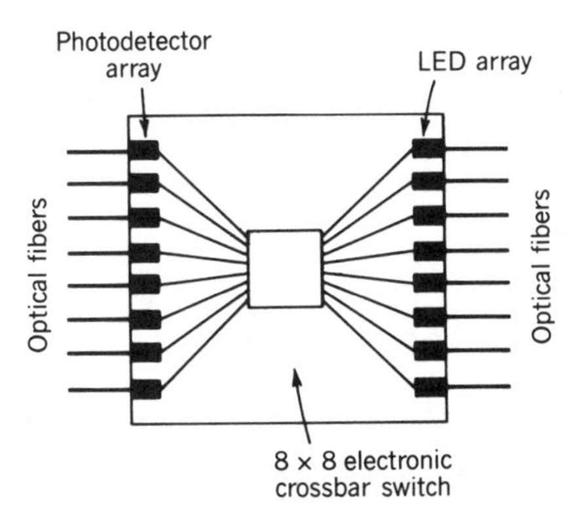

Figure 21.1-4 An optoelectronic 8 × 8 crossbar switch. Eight optical signals carried by eight optical fibers are detected by an array of photodetectors, switched using an 8 × 8 electronic crossbar switch, and regenerated using eight LEDs (or diode lasers) into eight outgoing optical fibers. The data rates that can be handled by silicon switches are currently a few hundred Mb/s, while GaAs switches can operate at rates exceeding 1 Gb/s.

to the loss of the optical phase caused by the process of detection). Direct optical switching is clearly preferable to electronic switching.

B. Opto-Mechanical, Electro-Optic, Acoustic-Optic, and Magneto-Optic Switches

Optical modulators and scanners can be used as switches. A modulator can be operated in the on–off mode as a 1×1 switch. A scanner that deflects an optical beam into N possible directions is a $1 \times N$ switch. These switches can be combined to make switches of higher dimensions.

Modulation and deflection of light can be achieved by the use of mechanical, electrical, acoustic, magnetic, or optical control; the switches are then called opto-mechanical (or mechano-optic), electro-optic, acousto-optic, magneto-optic, or opto-optic (all-optical), respectively. The remainder of this section provides a brief outline of opto-mechanical and magneto-optic switches, and a brief review of electro-optic and acousto-optic switches, which are discussed in Secs. 18.1B and 20.2, respectively. All-optical switches are covered in Sec. 21.2.

Opto-Mechanical Switches

Opto-mechanical switches use moving (rotating or alternating) mirrors, prisms, or holographic gratings to deflect light beams (Fig. 21.1-5). Piezoelectric elements may be

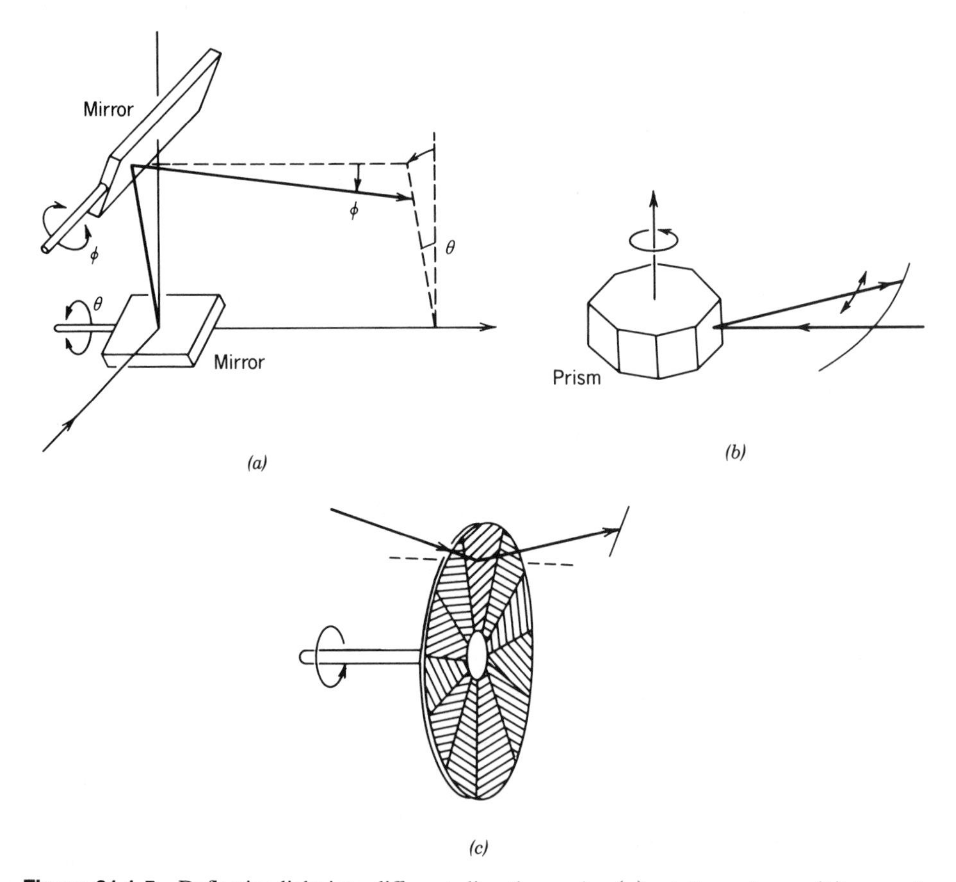

Figure 21.1-5 Deflecting light into different directions using (a) rotating mirrors; (b) a rotating prism; (c) a rotating holographic disk. Each sector of the holographic disk contains a grating whose orientation and period determine a scanning plane and scanning angle of the deflected light.

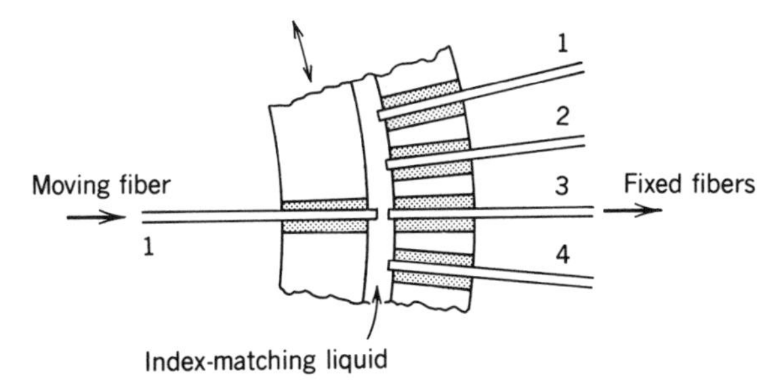

Figure 21.1-6 An optical fiber attached to a rotating wheel is aligned with one of a number of optical fibers attached to a fixed wheel. The fibers are placed in V-grooves. An index-matching liquid is used for better optical coupling.

used for fast mechanical action. A moving drop of mercury in a capillary cell can act as a moving mirror.

An optical fiber can be connected to any of a number of other optical fibers by mechanically moving the input fiber to align with the selected output fiber using a mechanism such as that illustrated in Fig. 21.1-6.

The major limitation of opto-mechanical switches is their low switching speeds (switching times are in the millisecond regime). Their major advantages are low insertion loss and low crosstalk.

Electro-Optic Switches

As discussed in Sec. 18.1, electro-optic materials alter their refractive indices in the presence of an electric field. They may be used as electrically controlled phase modulators or wave retarders. When placed in one arm of an interferometer, or between two crossed polarizers, the electro-optic cell serves as an electrically controlled light modulator or a 1×1 (on–off) switch (see Sec. 18.1B).

Since it is difficult to make large arrays of switches using bulk crystals, the most promising technology for electro-optic switching is integrated optics (see Chap. 7 and Sec. 18.1). Integrated-optic waveguides are fabricated using electro-optic dielectric substrates, such as $LiNbO_3$, with strips of slightly higher refractive index at the locations of the waveguides, created by diffusing titanium into the substrate.

An example of a 1×1 switch using an integrated-optic Mach–Zehnder interferometer is described in Sec. 18.1B and shown in Fig. 21.1-7(*a*). An example of a 2×2 switch is the directional coupler discussed in Sec. 18.1D and illustrated in Fig. 22.1-7(*b*). Two waveguides in close proximity are optically coupled; the refractive index is altered by applying an electric field adjusted so that the optical power either remains in the same waveguide or is transferred to the other waveguide. These switches operate at a few volts with speeds that can exceed 20 GHz.

An $N \times N$ integrated-optic switch can be built by use of a combination of 2×2 switches. A 4×4 switch is implemented by use of five 2×2 switches connected as in Fig. 21.1-2(*b*). This configuration can be built on a single substrate in the geometry shown in Fig. 21.1-8. An 8×8 switch is commercially available and larger switches are being developed.

The limit on the number of switches per unit area is governed by the relatively large physical dimensions of each directional coupler and the planar nature of the interconnections within the chip. To reduce the dimensions and increase the packing density of switches, intersecting (instead of parallel) waveguides are being investigated.

Because of the rectangular nature of integrated-optics technology, it is difficult to obtain efficient coupling to cylindrical waveguides (e.g., optical fibers). Relatively large insertion losses are encountered, especially when a single-mode fiber is connected to a directional coupler. Because the coupling coefficient is polarization dependent, the

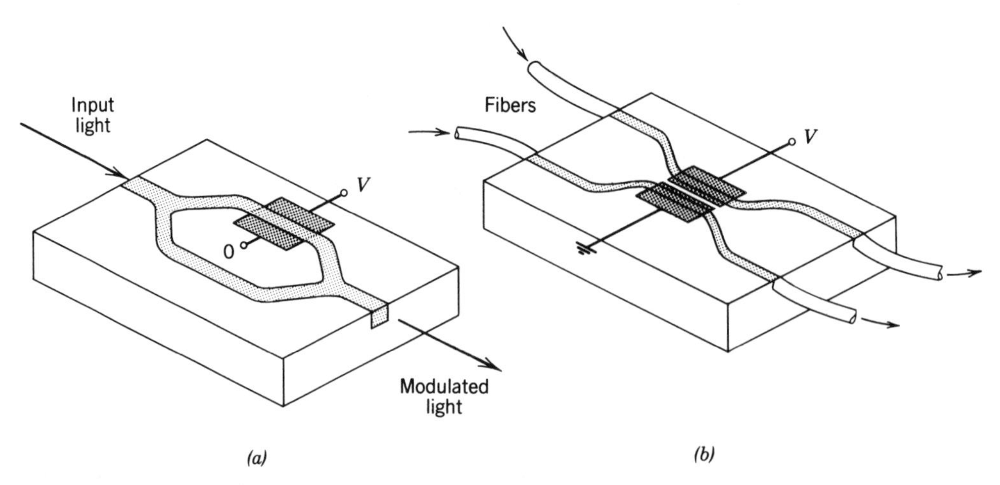

Figure 21.1-7 (*a*) A 1×1 switch using an integrated-optic Mach–Zehnder interferometer. (*b*) A 2×2 switch using an integrated electro-optic directional coupler.

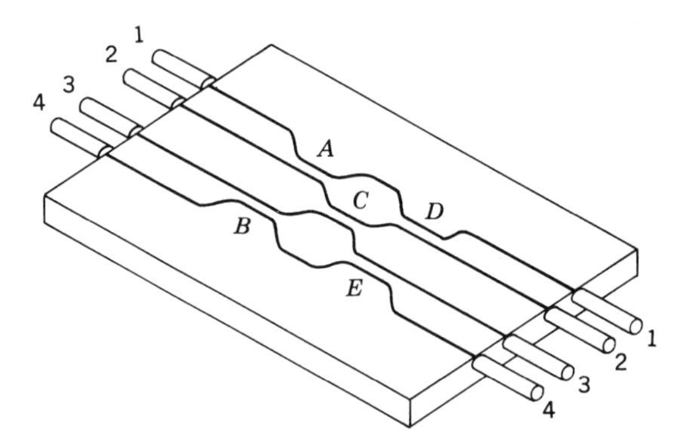

Figure 21.1-8 An integrated-optical 4×4 switch using five directional couplers A, B, C, D, and E on a single substrate.

polarization of the guided light must be properly selected. This imposes a restriction requiring that the input and output connecting fibers must be polarization maintaining (see Sec. 8.1C). Elaborate schemes are required to make polarization-independent switches.

Liquid crystals provide another technology that can be used to make electrically controlled optical switches (see Sec. 18.3). A large array of electrodes placed on a single liquid-crystal panel serves as a spatial light modulator or a set of 1×1 switches. The main limitation is the relatively low switching speed.

Acousto-Optic Switches

Acousto-optic switches use the property of Bragg deflection of light by sound (Chap. 20). The power of the deflected light is controlled by the intensity of the sound. The angle of deflection is controlled by the frequency of the sound. An acousto-optic modulator is a 1×1 switch. An acousto-optic scanner (Fig. 21.1-9) is a $1 \times N$ switch, where N is the number of resolvable spots of the scanner (see Sec. 20.2B). Acousto-optic cells with $N = 2000$ are available. If different parts of the acousto-optic cell carry sound waves of different frequencies, an $N \times M$ switch or interconnection device is

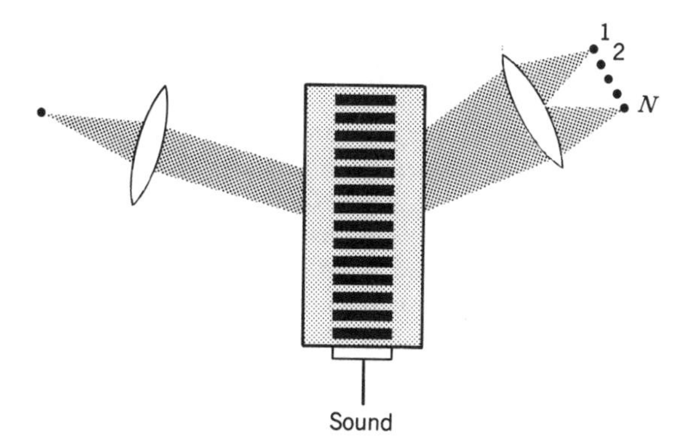

Figure 21.1-9 Acousto-optic switch.

obtained. Limitations on the maximum product NM achievable with acousto-optic cells have been discussed in Sec. 20.2C. Arrays of acousto-optic cells are also becoming available.

Magneto-Optic Switches

Magneto-optic materials alter their optical properties under the influence of a magnetic field. Materials exhibiting the Faraday effect, for example, act as polarization rotators in the presence of a magnetic flux density B (see Sec. 6.4B); the rotatory power ρ (angle per unit length) is proportional to the component of B in the direction of propagation. When the material is placed between two crossed polarizers, the optical power transmission $\mathscr{T} = \sin^2 \theta$ is dependent on the polarization rotation angle $\theta = \rho d$, where d is the thickness of the cell. The device is used as a 1×1 switch controlled by the magnetic field.

Magneto-optic materials have recently received more attention because of their use in optical-disk recording. In these systems, however, a thermomagnetic effect is used in which the magnetization is altered by heating with a strong focused laser. Weak linearly polarized light from a laser is used for readout.

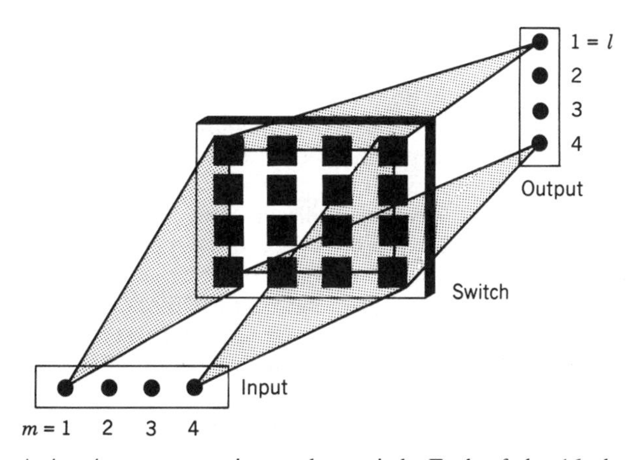

Figure 21.1-10 A 4×4 magneto-optic crossbar switch. Each of the 16 elements is a 1×1 switch transmitting or blocking light depending on the applied magnetic field. Light from the input mth point, $m = 1, 2, 3, 4$ is distributed to all switches in the mth column. Light from all switches of the lth row reaches the lth output point ($l = 1, 2, 3, 4$). The system is an implementation of the 4×4 switch depicted in Fig. 21.1-1(d).

The magneto-optic material is usually in the form of a film (e.g., bismuth-substituted iron garnet) grown on a nonmagnetic substrate. The magnetic field is applied by use of two intersecting conductors carrying electric current. The system operates in a binary mode by switching the direction of magnetization.

Arrays of magneto-optic switches can be constructed by etching isolated cells (each of size as small as 10×10 μm) on a single film. Conductors for the electric-current drive lines are subsequently deposited using usual photolithographic techniques. Large arrays of magneto-optic switches (1024×1024) have become available and the technology is advancing rapidly. Switching speeds of 100 ns are possible. Figure 21.1-10 illustrates the use of a 4×4 array of magneto-optic switches as a 4×4 switch.

21.2 ALL-OPTICAL SWITCHES

In an all-optical (or opto-optic) switch, light controls light with the help of a nonlinear optical material. Nonlinear optical effects may be direct or indirect. Direct effects occur at the atomic or molecular level when the presence of light alters the atomic susceptibility or the photon absorption rates of the medium. The optical Kerr effect (variation of the refractive index with the applied light intensity; see Sec. 19.3A) and saturable absorption (dependence of the absorption coefficient on the applied light intensity; see Sec. 13.3B) are examples of direct nonlinear optical effects.

Indirect nonlinear optical effects involve an intermediate process in which electric charges and/or electric fields play a role, as illustrated by the following two examples.

- In photorefractive materials (see Sec. 18.4), absorbed nonuniform light creates mobile charges that diffuse away from regions of high concentration and are trapped elsewhere, creating an internal space-charge electric field that modifies the optical properties of the medium by virtue of the electro-optic effect.
- In an optically-addressed liquid-crystal spatial light modulator (see Sec. 18.3B), the control light is absorbed by a photoconductive layer and the generated electric charges create an electric field that modifies the molecular orientation and therefore the indices of refraction of the material, thereby controlling the transmission of light.

In these two examples, optical nonlinear behavior is exhibited because of an intermediate effect: light creates an electric field that modifies the optical properties of the medium. Other indirect nonlinear optical effects will be discussed in Sec. 21.3 in connection with bistable optical devices.

Nonlinear optical effects (direct or indirect) may be used to make all-optical switches. The optical phase modulation in the Kerr medium (see Sec. 19.3A), for example, may be converted into intensity modulation by placing the medium in one leg of an interferometer, so that as the control light is turned on and off, the transmittance of the interferometer is switched between 1 and 0, as illustrated in Fig. 21.2-1.

The retardation between two polarizations in an anisotropic nonlinear medium may also be used for switching by placing the material between two crossed polarizers. Figure 21.2-2 illustrates an example of an all-optical switch using an anisotropic optical fiber exhibiting the optical Kerr effect.

An array of switches using an optically-addressed liquid-crystal spatial light modulator is illustrated in Fig. 21.2-3. The control light alters the electric field applied to the liquid-crystal layer and therefore alters its reflectance. Different points on the liquid-crystal surface have different reflectances and act as independent switches controlled

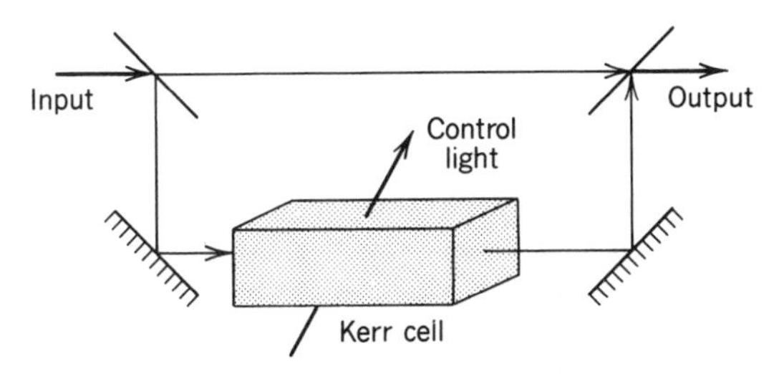

Figure 21.2.1 An all-optical on–off switch using a Mach–Zehnder interferometer and a material exhibiting the optical Kerr effect.

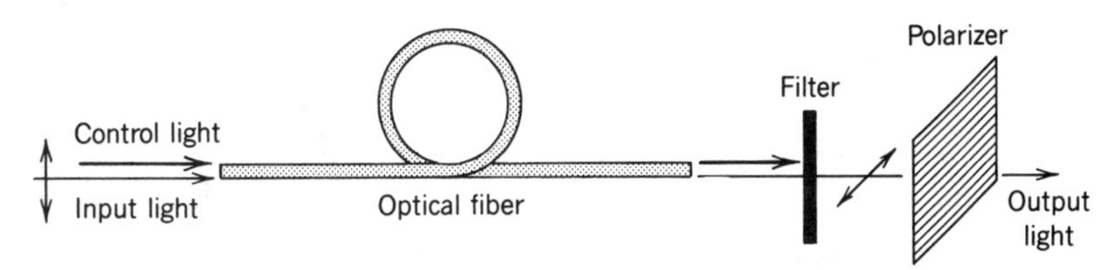

Figure 21.2-2 An anisotropic nonlinear optical fiber serving as an all-optical switch. In the presence of the control light, the fiber introduces a phase retardation π, so that the polarization of the linearly polarized input light rotates 90° and is transmitted by the output polarizer. In the absence of the control light, the fiber introduces no retardation and the light is blocked by the polarizer. The filter is used to transmit the signal light and block the control light, which has a different wavelength.

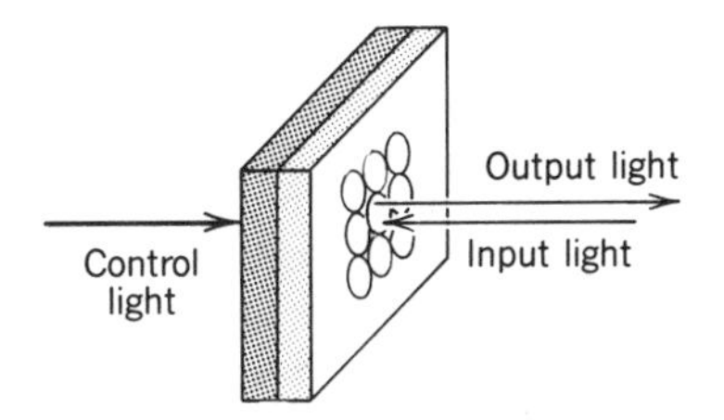

Figure 21.2-3 An all-optical array of switches using an optically addressed liquid-crystal spatial light modulator (light valve).

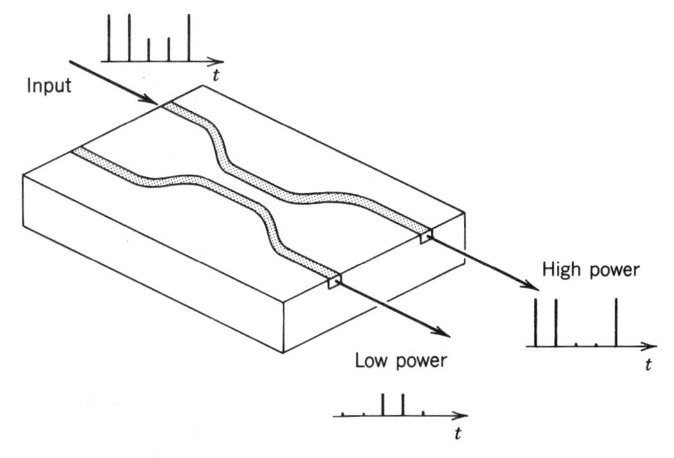

Figure 21.2-4 A directional coupler controlled by the optical Kerr effect. An input beam of low power entering one waveguide is channeled into the other waveguide; a beam of high power remains in the same waveguide.

by the input light beams. These devices can accommodate a large number of switches, but they are relatively slow.

It is not necessary that the control light and the controlled light be distinct. A single beam of light may control its own transmission. Consider, for example, the directional coupler illustrated in Fig. 21.2-4. The refractive indices and the dimensions may be selected so that when the input optical power is low, it is channeled into the other waveguide; when it is high the refractive indices are altered by virtue of the optical Kerr effect and the power remains in the same waveguide. The device serves as a self-controlled (self-addressed) switch. It can be used to sift a sequence of weak and strong pulses, separating them into the two output ports of the coupler. All-optical gates and optical-memory elements made of nonlinear optical materials will be discussed in Sec. 21.3.

Fundamental Limitations on All-Optical Switches

Minimum values of the switching energy E and the switching time T of all-optical switches are governed by the following fundamental physical limits.

Photon-Number Fluctuations. The minimum energy needed for switching is in principle one photon. However, since there is an inherent randomness in the number of photons emitted by a laser or light-emitting diode, a larger mean number of photons must be used to guarantee that the switching action almost always occurs whenever desired. For these light sources and under certain conditions (see Sec. 11.2C) the number of photons arriving within a fixed time interval is a Poisson-distributed random number n with probability distribution $p(n) = \bar{n}^n \exp(-\bar{n})/n!$, where $\bar{n}$ is the mean number of photons. If $\bar{n} = 21$ photons, the probability that no photons are delivered is $p(0) = e^{-21} \approx 10^{-9}$. An average of 21 photons is therefore the minimum number that guarantees delivery of at least one photon, with an average of 1 error every 10^9 trials. The corresponding energy is $E = 21h\nu$. For light of wavelength $\lambda_o = 1$ μm, $E = 21 \times 1.24 \approx 26$ eV $= 4.2$ aJ. This is regarded as a lower bound on the switching energy; it should be noted, however, that this is a practical bound, rather than a fundamental limit, inasmuch as sub-Poisson light (see Sec. 11.3B) may in principle be used. To be on the less optimistic side, a minimum of 100 photons may be used as a reference. This corresponds to a minimum switching energy of 20 aJ at $\lambda_o = 1$ μm. Note that, at optical frequencies, $h\nu$ is much greater than the thermal unit of energy $k_B T$ at room temperature (at $T = 300$ K, $k_B T = 0.026$ eV).

Energy–Time Uncertainty. Another fundamental quantum principle is the energy–time uncertainty relation $\sigma_E \sigma_T \geq h/4\pi$ [see (11.1-12)]. The product of the minimum switching energy E and the minimum switching time T must therefore be greater than $h/4\pi$ (i.e., $E \geq h/4\pi T = h\nu/4\pi\nu T$). This bound on energy is smaller than the energy of a photon $h\nu$ by a factor $4\pi\nu T$. Since the switching time T is not smaller than the duration of an optical cycle $1/\nu$, the term $4\pi\nu T$ is always greater than unity. Because E is chosen to be greater than the energy of one photon, $h\nu$, it follows that the energy–time uncertainty condition is always satisfied.

Switching Time. The only fundamental limit on the minimum switching time arises from energy–time uncertainty. In fact, optical pulses of a few femtoseconds (a few optical cycles) are readily generated. Such speeds cannot be attained by semiconductor electronic switches (and are also beyond the present capabilities of Josephson devices). Subpicosecond switching speeds have been demonstrated in a number of optical switching devices. Switching energies can also, in principle, be much smaller than in semiconductor electronics, as Fig. 21.2-5 illustrates.

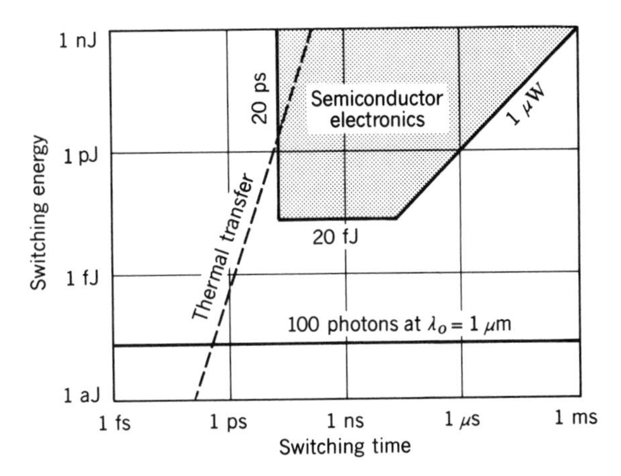

Figure 21.2-5 Limits on the switching energy and time for all-optical switches. Switching energy must be above the 100-photon line. If the switching is repetitive, points must lie to the right of the thermal-transfer line. Limits of semiconductor electronic devices are marked by the 1-μW, 20-fJ, and 20-ps lines.

Size. Limits on the size of photonic switches are governed by diffraction effects, which make it difficult to couple optical power to and from devices with dimensions smaller than a wavelength of light.

Practical Limitations

The primary limitation on all-optical switching is a result of the weakness of the nonlinear effects in currently available materials, which makes the required switching energy rather large. Another important practical limit is related to the difficulty of thermal transfer of the heat generated by the switching process. This limitation is particularly severe when the switching is performed repetitively. If a minimum switching energy E is used in each switching operation, a total energy E/T is used every second. For very short switching times this power can be quite large. The maximum rate at which the dissipated power must be removed sets a limit, making the combination of very short switching times and very high switching energies untenable. The thermal-transfer limit based on certain reasonable assumptions† is indicated on the diagram of Fig. 21.2-5. Note, however, that thermal effects are less restrictive if the device is operated at less than the maximum repetition rate; i.e., the energy of one switching operation has more than a bit time to be dissipated. The performance of a number of actual all-optical photonic switches is shown in Fig. 21.3-19 at the end of Sec. 21.3.

21.3 BISTABLE OPTICAL DEVICES

Highly sophisticated digital electronic systems (e.g., a digital computer) contain a large number of interconnected basic units: switches, gates, and memory elements (flip-flops). This section introduces bistable optical devices, which can be used as optical gates and

†See P. W. Smith and W. J. Tomlinson, Bistable Optical Devices Promise Subpicosecond Switching, *IEEE Spectrum*, vol. 18, no. 6, pp. 26–33, 1981.

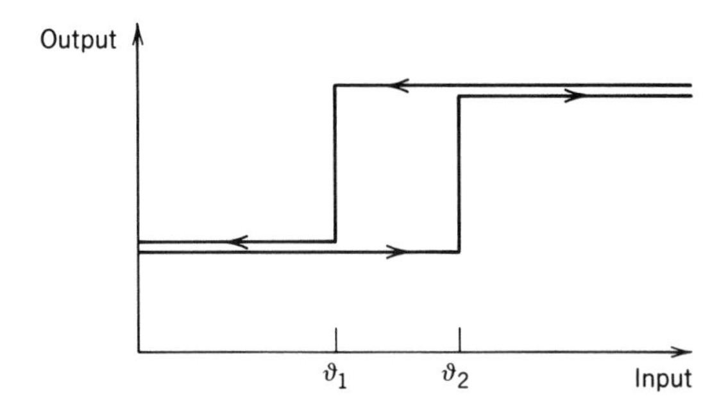

Figure 21.3-1 Input–output relation for a bistable system.

flip-flops. Potential applications in digital optical computing are discussed in Sec. 21.5A.

A. Bistable Systems

A bistable (or two-state) system has an output that can take only one of two distinct stable values, no matter what input is applied. Switching between these values may be achieved by a temporary change of the level of the input. In the system illustrated in Fig. 21.3-1, for example, the output takes its low value for small inputs and its high value for large inputs. When an increasing input exceeds a certain critical value (threshold) ϑ_2, the output jumps from the low to the high value. When the input is subsequently decreased, the output jumps back to the lower value when another critical value $\vartheta_1 < \vartheta_2$ is crossed, so that the input–output relation forms a hysteresis loop.

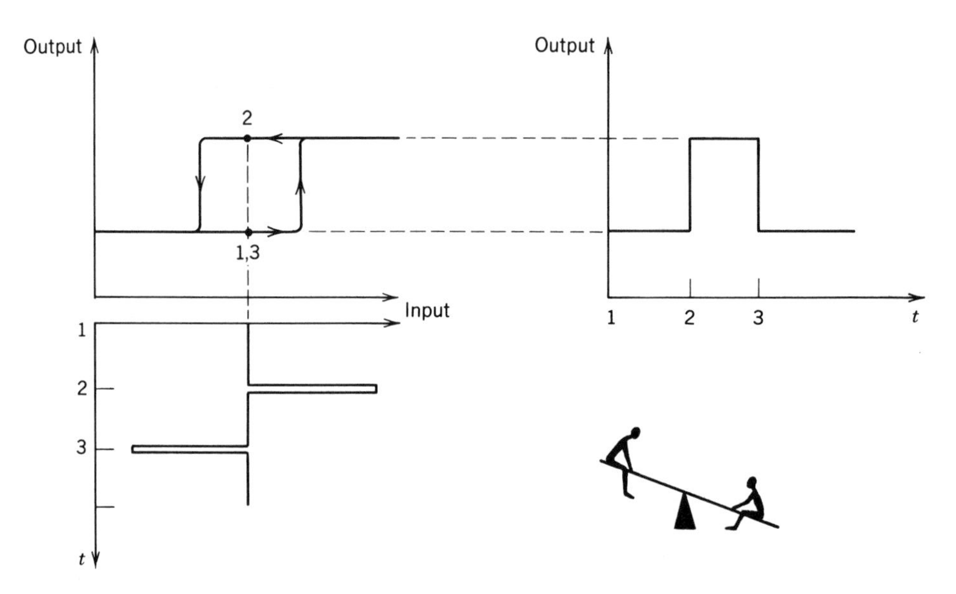

Figure 21.3-2 Flip-flopping of a bistable system. At time 1 the output is low. A positive input pulse at time 2 flips the system from low to high. The output remains in the high state until a negative pulse at time 3 flips it back to the low state. The system acts as a latching switch or a memory element.

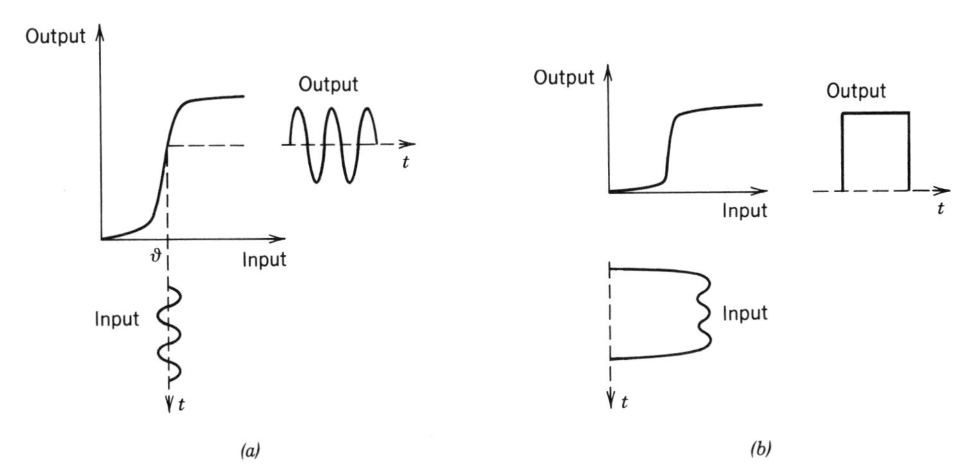

Figure 21.3-3 The bistable device as (*a*) an amplifier or (*b*) a thresholding device, pulse shaper, or limiter.

There is an intermediate range of input values (between ϑ_1 and ϑ_2) for which low or high outputs are possible, depending on the history of the input. Within this range, the system acts like a seesaw. If the output is low, a large positive input spike flips it to high. A large negative input spike flips it back to low. The system has a "flip-flop" behavior; its state depends on its history (whether the last spike was positive or negative; Fig. 21.3-2).

Bistable devices are important in the digital circuits used in communications, signal processing, and computing. They are used as switches, logic gates, and memory elements. The device parameters may be adjusted so that the two critical values (the thresholds ϑ_1 and ϑ_2) coalesce into a single value ϑ. The result is a single-threshold steep S-shaped nonlinear output–input relation. When biased appropriately the device can have large differential gain and can be used as an amplifier, like a transistor. It can

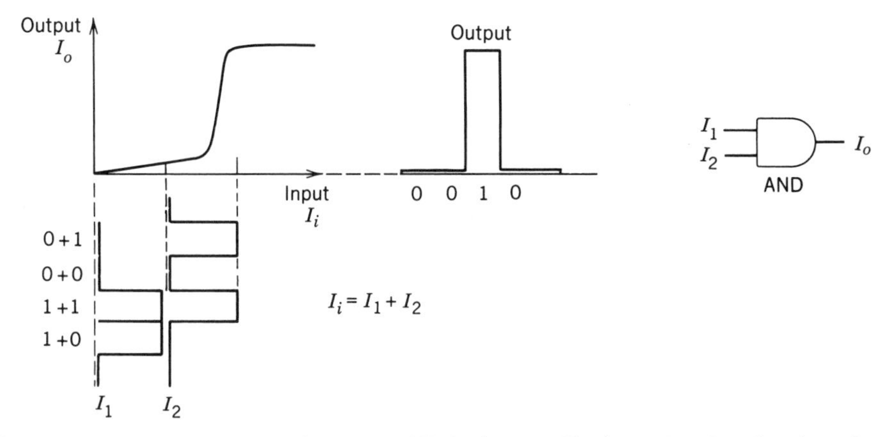

Figure 21.3-4 The bistable device as an AND logic gate. The input $I_i = I_1 + I_2$, where I_1 and I_2 are pulses representing the binary data. The output I_o is high if and only if both inputs are present.

also be used as a thresholding element in which the output switches between two values as the input exceeds a threshold, as a pulse shaper, or as a limiter (Fig. 21.3-3). A stable threshold and stable bias are necessary for these operations.

Bistable devices are also used as logic elements. The binary data are represented by pulses that are added and their sum used as input to the bistable device. With an appropriate choice of the pulse heights in relation to the threshold, the device can be made to switch to high only when both pulses are present, so that it acts as an AND gate, as illustrated in Fig. 21.3-4.

An *electronic* bistable (flip-flop) circuit is made by connecting the output of each of two transistors to the input of the other (see any textbook on digital electronic circuits). As will be explained subsequently, a photonic bistable system, on the other hand, uses a combination of a nonlinear optical material and optical feedback.

B. Principle of Optical Bistability

Two features are required for making a bistable device: *nonlinearity* and *feedback*. Both features are available in optics. If the output of a nonlinear optical element is fed back (by use of mirrors, for example) and used to control the transmission of light through the element itself, bistable behavior can be exhibited.

Consider the generic optical system illustrated in Fig. 21.3-5. By means of feedback the output intensity I_o is somehow made to control the transmittance $\mathscr{T}$ of the system, so that $\mathscr{T}$ is some nonlinear function $\mathscr{T} = \mathscr{T}(I_o)$. Since $I_o = \mathscr{T} I_i$,

$$I_i = \frac{I_o}{\mathscr{T}(I_o)}. \tag{21.3-1}$$

Input–Output Relation for a Bistable System

If $\mathscr{T}(I_o)$ is a nonmonotonic function, such as the bell-shaped function shown in Fig. 21.3-6(*a*), I_i will also be a nonmonotonic function of I_o, as illustrated in Fig. 21.3-6(*b*). Consequently, I_o must be a multivalued function of I_i; i.e., there are some values of I_i with more than one corresponding value of I_o, as illustrated in Fig. 21.3-6(*c*).

The system therefore exhibits bistable behavior. For small inputs ($I_i < \vartheta_1$) or large inputs ($I_i > \vartheta_2$), each input value has a single corresponding output value. In the intermediate range, $\vartheta_1 < I_i < \vartheta_2$, however, each input value corresponds to three possible output values. The upper and lower values are stable, but the intermediate value [the line joining points 1 and 2 in Fig. 21.3-6(*c*)] is unstable. Any slight perturbation added to the input forces the output to either the upper or the lower branch. Starting from small input values and increasing the input, when the threshold ϑ_2 is exceeded the output jumps to the upper state without passing through the

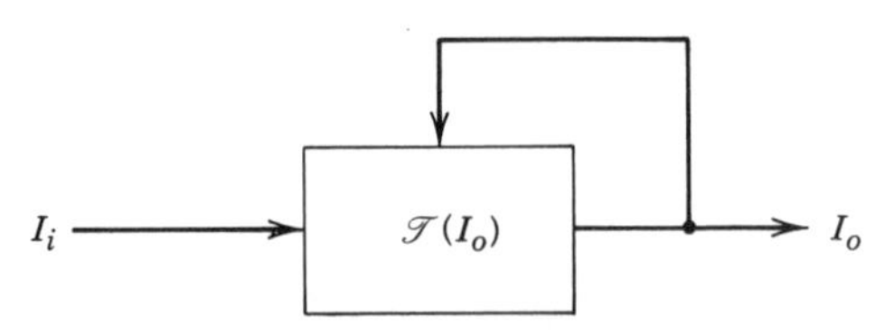

Figure 21.3-5 An optical system whose transmittance $\mathscr{T}$ is a function of its output I_o.

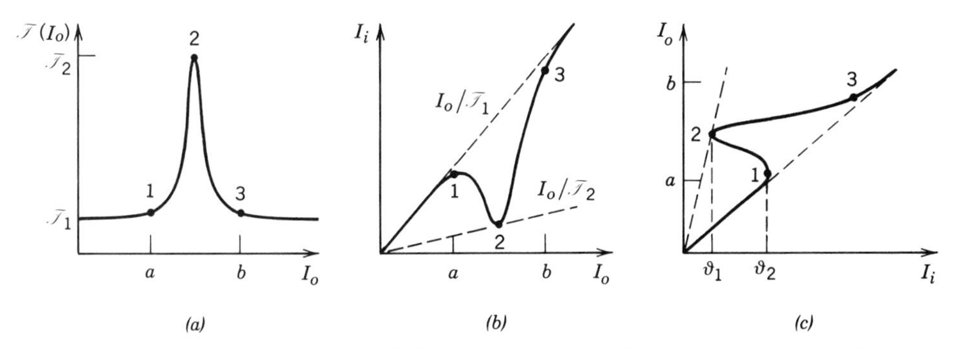

Figure 21.3-6 (*a*) Transmittance $\mathcal{T}(I_o)$ versus output I_o. (*b*) Input $I_i = I_o/\mathcal{T}(I_o)$ versus output I_o. For $I_o < a$ or $I_o > b$, $\mathcal{T}(I_o) = \mathcal{T}_1$ and $I_i = I_o/\mathcal{T}_1$ is a linear relation with slope $1/\mathcal{T}_1$. At the intermediate value of I_o for which $\mathcal{T}$ has its maximum value $\mathcal{T}_2$ (point 2), I_i dips below the line $I_i = I_o/\mathcal{T}_1$ and touches the lower line $I_i = I_o/\mathcal{T}_2$ at point 2. (*c*) The output I_o versus the input I_i is obtained simply by replotting the curve in (*b*) with the axes exchanged. (The diagram is rotated 90° in a counterclockwise direction and mirror imaged about the vertical axis.)

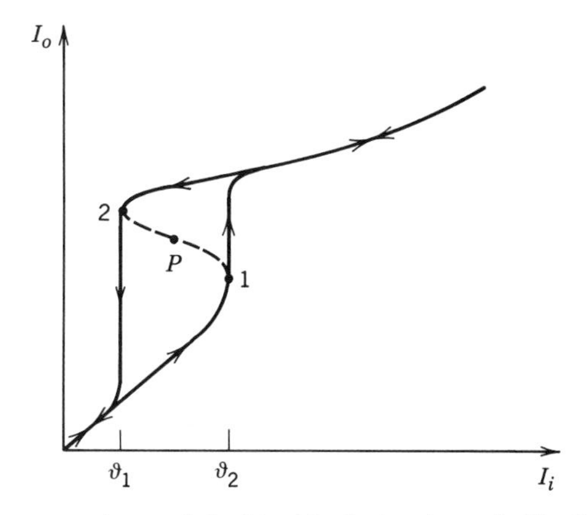

Figure 21.3-7 Output versus input of the bistable device shown in Fig. 21.3-5. The dashed line represents an unstable state.

unstable intermediate state. When the input is subsequently decreased, it follows the upper branch until it reaches ϑ_1 whereupon it jumps to the lower state, as illustrated in Fig. 21.3-7.

The instability of the intermediate state may be seen by considering point P in Fig. 21.3-7. A small increase of the output I_o causes a sharp increase of the transmittance $\mathcal{T}(I_o)$ since the slope of $\mathcal{T}(I_o)$ is positive and large [see Fig. 21.3-6(*a*) and note that P lies on the line joining points 1 and 2]. This, in turn, results in further increase of $\mathcal{T}(I_o)$, which increases I_o even more. The result is a transition to the upper stable state. Similarly, a small decrease in I_o causes a transition to the lower stable state.

The nonlinear bell-shaped function $\mathcal{T}(I_o)$ was used only for illustration. Many other *nonlinear* functions exhibit bistability (and possibly multistability, with more than two stable values of the output for a single value of the input).

EXERCISE 21.3-1

Examples of Nonlinear Functions Exhibiting Bistability. Use a computer to plot the relation between I_o and $I_i = I_o/\mathcal{T}(I_o)$, for each of the following functions:

(a) $\mathcal{T}(x) = 1/[(x-1)^2 + a^2]$
(b) $\mathcal{T}(x) = 1/[1 + a^2 \sin^2(x+\theta)]$
(c) $\mathcal{T}(x) = \frac{1}{2} + \frac{1}{2}\cos(x+\theta)$
(d) $\mathcal{T}(x) = \operatorname{sinc}^2[(a^2 + x^2)^{1/2}]$
(e) $\mathcal{T}(x) = (x+1)^2/(x+a)^2$.

Select appropriate values for the constants a and θ to generate a bistable relation. The functions in (b) to (e) apply to bistable systems that will be discussed subsequently.

C. Bistable Optical Devices

Numerous schemes can be used for the optical implementation of the foregoing basic principle. Two types of nonlinear optical elements can be used (Fig. 21.3-8):

- Dispersive nonlinear elements, for which the refractive index n is a function of the optical intensity.
- Dissipative nonlinear elements, for which the absorption coefficient α is a function of the optical intensity.

The optical element is placed within an optical system and the output light intensity I_o controls the system's transmittance in accordance with some nonlinear function $\mathcal{T}(I_o)$.

Dispersive Nonlinear Elements

A number of optical systems can be devised whose transmittance $\mathcal{T}$ is a nonmonotonic function of an intensity-dependent refractive index $n = n(I_o)$. Examples are interferometers, such as the Mach–Zehnder and the Fabry–Perot etalon, with a medium exhibiting the optical Kerr effect,

$$n = n_0 + n_2 I_o, \tag{21.3-2}$$

where n_0 and n_2 are constants.

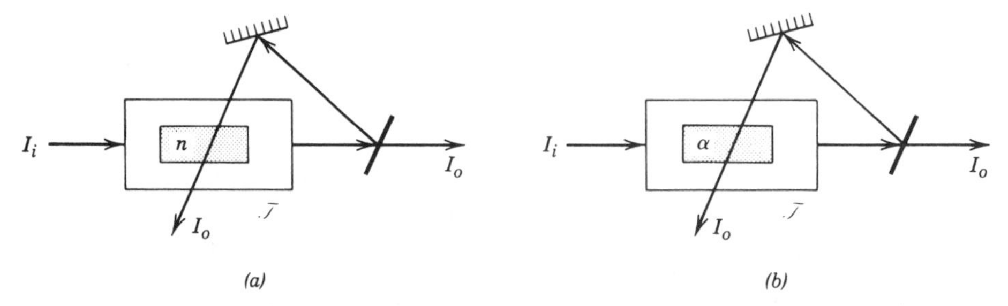

Figure 21.3-8 (*a*) Dispersive bistable optical system. The transmittance $\mathcal{T}$ is a function of the refractive index n, which is controlled by the output intensity I_o. (*b*) Dissipative bistable optical system. The transmittance $\mathcal{T}$ is a function of the absorption coefficient α, which is controlled by the output intensity I_o.

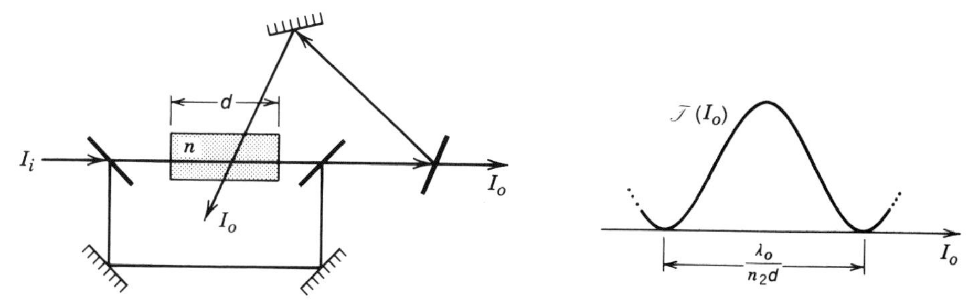

Figure 21.3-9 A Mach–Zehnder interferometer with a nonlinear medium of refractive index n controlled by the transmitted intensity I_o via the optical Kerr effect.

In the *Mach–Zehnder interferometer*, the nonlinear medium is placed in one branch, as illustrated in Fig. 21.3-9. The power transmittance of the system is (see Sec. 2.5A)

$$\mathcal{T} = \frac{1}{2} + \frac{1}{2}\cos\left(2\pi\frac{d}{\lambda_o}n + \varphi_0\right), \tag{21.3-3}$$

where d is the length of the active medium, λ_o the free-space wavelength, and φ_0 a constant. Substituting from (21.3-2), we obtain

$$\boxed{\mathcal{T}(I_o) = \frac{1}{2} + \frac{1}{2}\cos\left(2\pi\frac{d}{\lambda_o}n_2 I_o + \varphi\right),} \tag{21.3-4}$$

where $\varphi = \varphi_0 + (2\pi d/\lambda_o)n_0$ is another constant. As Fig. 21.3-9 shows, this is a nonlinear function comprising a periodic repetition of the generic bell-shaped function used earlier to demonstrate bistability [see Fig. 21.3-6(*a*)].

In a *Fabry–Perot etalon* with mirror separation d, the intensity transmittance is (see Sec. 2.5B)

$$\mathcal{T} = \frac{\mathcal{T}_{\max}}{1 + (2\mathcal{F}/\pi)^2 \sin^2[(2\pi d/\lambda_o)n + \varphi_0]}, \tag{21.3-5}$$

where $\mathcal{T}_{\max}$, $\mathcal{F}$, and φ_0 are constants and λ_o is the free-space wavelength. Substituting for n from (21.3-2) gives

$$\boxed{\mathcal{T}(I_o) = \frac{\mathcal{T}_{\max}}{1 + (2\mathcal{F}/\pi)^2 \sin^2[(2\pi d/\lambda_o)n_2 I_o + \varphi]},} \tag{21.3-6}$$

where φ is another constant. As illustrated in Fig. 21.3-10, this function is a periodic sequence of sharply peaked bell-shaped functions. The system is therefore bistable.

Intrinsic Bistable Optical Devices

The optical feedback required for bistability can be internal instead of external. The system shown in Fig. 21.3-11, for example, uses a resonator with an optically nonlinear medium whose refractive index n is controlled by the internal light intensity I within

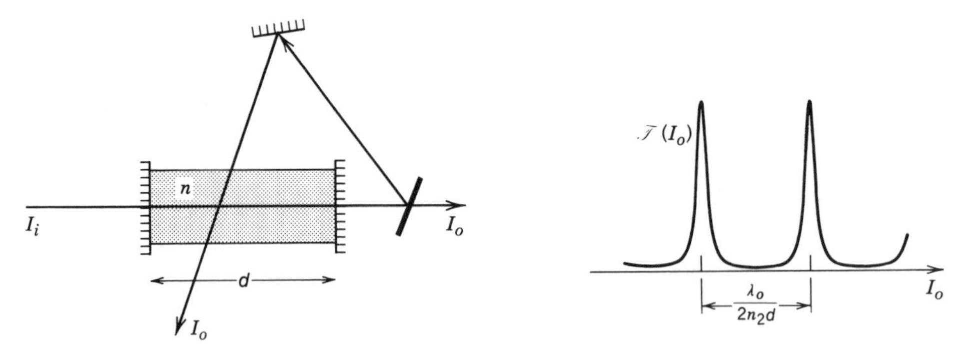

Figure 21.3-10 A Fabry–Perot interferometer containing a medium of refractive index n controlled by the transmitted light intensity I_o.

the resonator, instead of the output light intensity I_o. Since $I_o = \mathcal{T}_o I$, where $\mathcal{T}_o$ is the transmittance of the output mirror, the action of the internal intensity I has the same effect as that of the external intensity I_o, except for a constant factor. If the medium exhibits the optical Kerr effect, for example, the refractive index is a linear function of the optical intensity $n = n_0 + n_2 I$ and the transmittance of the Fabry–Perot etalon is

$$\mathcal{T}(I_o) = \frac{\mathcal{T}_{\max}}{1 + (2\mathcal{F}/\pi)^2 \sin^2[(2\pi d/\lambda_o) n_2 I_o/\mathcal{T}_o + \varphi]}. \tag{21.3-7}$$

Thus the device operates as a self-tuning system.

Dissipative Nonlinear Elements

A dissipative nonlinear material has an absorption coefficient that is dependent on the optical intensity I. The saturable absorber discussed in Sec. 13.3B is an example in which the absorption coefficient is a nonlinear function of I,

$$\alpha = \frac{\alpha_0}{1 + I/I_s}, \tag{21.3-8}$$

where α_0 is the small-signal absorption coefficient and I_s is the saturation intensity. If the absorber is placed inside a Fabry–Perot etalon of length d that is tuned for peak transmission (Fig. 21.3-12), then

$$\mathcal{T} = \frac{\mathcal{T}_1}{(1 - \mathcal{R}e^{-\alpha d})^2}, \tag{21.3-9}$$

where $\mathcal{R} = \sqrt{\mathcal{R}_1 \mathcal{R}_2}$ ($\mathcal{R}_1$ and $\mathcal{R}_2$ are the mirror reflectances) and $\mathcal{T}_1$ is a constant

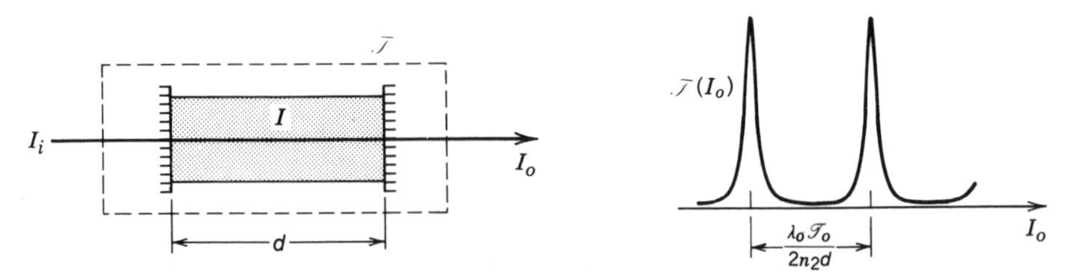

Figure 21.3-11 Intrinsic bistable device. The internal light intensity I controls the active medium and therefore the overall transmittance of the system $\mathcal{T}$.

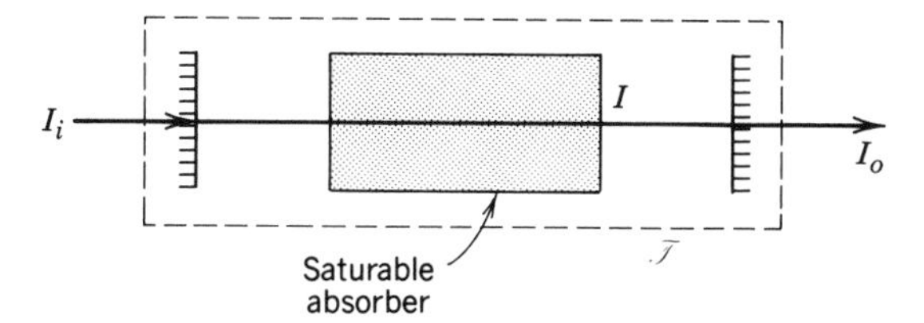

Figure 21.3-12 A bistable device consisting of a saturable absorber in a resonator.

(see Secs. 2.5B and 9.1A for details). If $\alpha d \ll 1$, i.e., the medium is optically thin, $e^{-\alpha d} \approx 1 - \alpha d$, and

$$\mathscr{T} \approx \frac{\mathscr{T}_1}{[1 - (1 - \alpha d)\mathscr{R}]^2}. \tag{21.3-10}$$

Because α is a nonlinear function of I, $\mathscr{T}$ is also a nonlinear function of I. Using the relation $I = I_o/\mathscr{T}_o$ and (21.3-8) and (21.3-10),

$$\mathscr{T}(I_o) = \mathscr{T}_2\left[\frac{I_o + I_{s1}}{I_o + (1 + a)I_{s1}}\right]^2, \tag{21.3-11}$$

where $\mathscr{T}_2 = \mathscr{T}_1/(1 - \mathscr{R})^2$, $a = \alpha_0 d\mathscr{R}/(1 - \mathscr{R})$, and $I_{s1} = I_s\mathscr{T}_o$. For certain values of a, the system is bistable [recall Exercise 21.3-1, example (e)].

Suppose now that the saturable absorber is replaced by an amplifying medium with saturable gain

$$\gamma = \frac{\gamma_0}{1 + I/I_s}. \tag{21.3-12}$$

The system is nothing but an optical amplifier with feedback, i.e., a laser. If $\mathscr{R}\exp(\gamma_0 d) < 1$, the laser is below threshold; but when $\mathscr{R}\exp(\gamma_0 d) > 1$, the system becomes unstable and we have laser oscillation. Lasers do exhibit bistable behavior. However, the theory of these phenomena is beyond the scope of this book.

In some sense, *the dispersive bistable optical system is the nonlinear-index-of-refraction (instead of nonlinear-gain) analog of the laser.*

Materials

Optical bistability has been observed in a number of materials exhibiting the optical Kerr effect (e.g., sodium vapor, carbon disulfide, and nitrobenzene). The coefficient of nonlinearity n_2 for these materials is very small. A long path length d is therefore required, and consequently the response time is large (nanosecond regime). The power requirement for switching is also high.

Semiconductors, such as GaAs, InSb, InAs, and CdS, exhibit a strong optical nonlinearity due to excitonic effects at wavelengths near the bandgap. A bistable device may simply be made of a layer of the semiconductor material with two parallel partially reflecting faces acting as the mirrors of a Fabry–Perot etalon (Fig. 21.3-13). Because of the large nonlinearity, the layer can be thin, allowing for a smaller response time.

GaAs switches based on this effect have been the most successful. Switch-on times of a few picoseconds have been measured, but the switch-off time, which is dominated by relatively slow carrier recombination, is much longer (a few nanoseconds). A switch-off time of 200 ps has been achieved by the use of specially prepared samples in

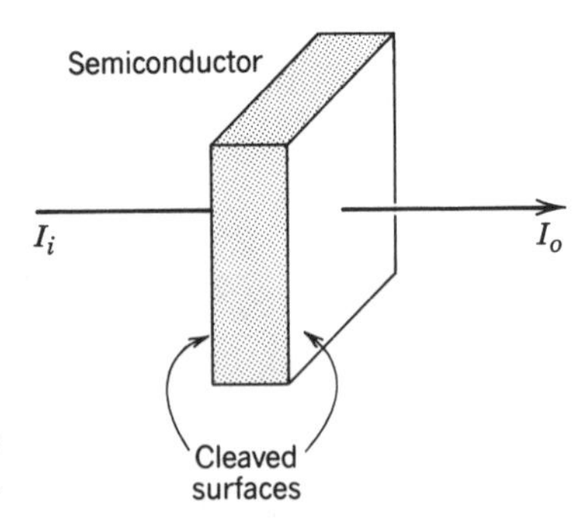

Figure 21.3-13 A thin layer of semiconductor with two parallel reflecting surfaces can serve as a bistable device.

which surface recombination is enhanced. The switching energy is 1 to 10 pJ. It is possible, in principle, to reduce the switching energy to the femtojoule regime. InAs and InSb have longer switch-off times (up to 200 ns). However, they can be speeded up at the expense of an increase of the switching energy. Semiconductor multiquantum-well structures (see Secs. 15.1G and 16.3G) are also being pursued as bistable devices, and so are organic materials.

The key condition for the usefulness of bistable optical devices, as opposed to semiconductor electronics technology, is the capability to make them in large arrays. Arrays of bistable elements can be placed on a single chip with the individual pixels defined by the light beams. Alternatively, reactive ion etching may be used to define the pixels. An array of 100×100 pixels on a 1-cm^2 GaAs chip is possible with existing technology. The main difficulty is heat dissipation. If the switching energy $E = 1$ pJ, and the switching time $T = 100$ ps, then for $N = 10^4$ pixels/cm^2 the heat load is $NE/T = 100$ W/cm^2. This is manageable with good thermal engineering. The device can perform 10^{14} bit operations per second, which is large in comparison with electronic supercomputers (which operate at a rate of about 10^{10} bit operations per second).

D. Hybrid Bistable Optical Devices

The bistable optical systems discussed so far are all-optical. Hybrid electrical/optical bistable systems in which electrical fields are involved have also been devised. An example is a system using a Pockels cell placed inside a Fabry–Perot etalon (Fig. 21.3-14). The output light is detected using a photodetector, and a voltage proportional to the detected optical intensity is applied to the cell, so that its refractive index variation is proportional to the output intensity. Using $LiNbO_3$ as the electro-optic material, 1-ns switching times have been achieved with $\approx$ 1-μW switching power and $\approx$ 1-fJ switching energy. An integrated optical version of this system [Fig. 21.3-14(*b*)] has also been implemented.

Another system uses an electro-optic modulator employing a Pockels cell wave retarder placed between two crossed polarizers (Fig. 21.3-15); see Sec. 18.1B. Again the output light intensity I_o is detected and a proportional voltage V is applied to the cell. The transmittance of the modulator is a nonlinear function of V, $\mathcal{T} = \sin^2(\Gamma_0/2 - \pi V/2V_\pi)$, where Γ_0 and V_π are constants. Because V is proportional to I_o, $\mathcal{T}(I_o)$ is a nonmonotonic function and the system exhibits bistability.

An integrated-optical directional coupler can also be used (Fig. 21.3-16). The input light I_i enters from one waveguide and the output I_o leaves from the other waveguide; the ratio $\mathcal{T} = I_o/I_i$ is the coupling efficiency (see Sec. 18.1D). Using (18.1-20) yields

$$\mathcal{T} = \left(\frac{\pi}{2}\right)^2 \operatorname{sinc}^2 \left\{ \frac{1}{2} \left[1 + 3\left(\frac{V}{V_0}\right)^2 \right]^{1/2} \right\}, \tag{21.3-13}$$

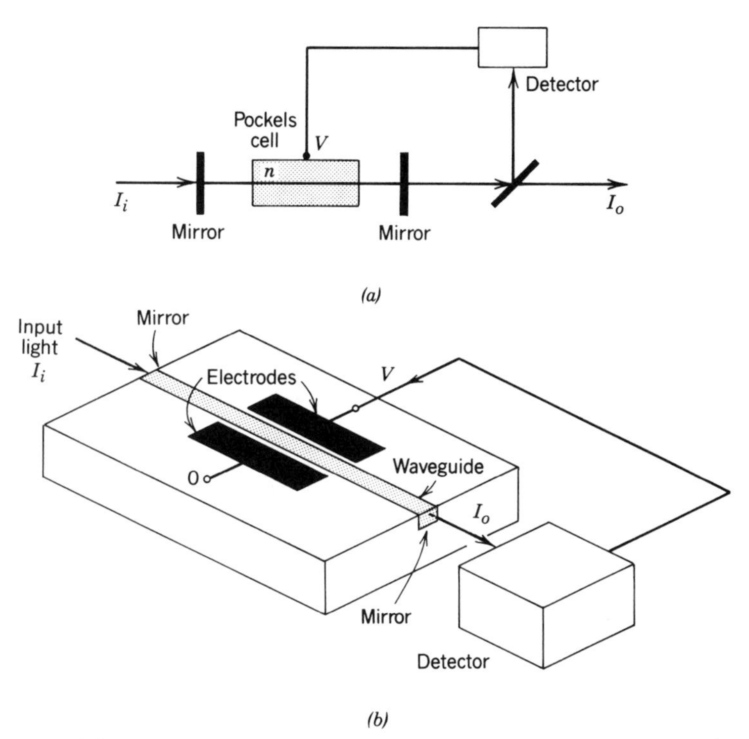

Figure 21.3-14 (*a*) A Fabry–Perot interferometer containing an electro-optic medium (Pockels cell). The output optical power is detected and a proportional electric field is applied to the medium to change its refractive index, thereby changing the transmittance of the interferometer. (*b*) An integrated-optical implementation.

where V is the applied voltage and V_0 is a constant. A bistable system is created by making V proportional to the output intensity I_o [see Exercise 21.3-1, example (d)].

Other nonlinear optical devices can also be used. An optically addressed liquid-crystal spatial light modulator (see Sec. 18.3B) can be used to create a large array of bistable elements (Fig. 21.3-17). The reflectance $\mathscr{R}$ of the modulator is proportional to the intensity of light illuminating its "write" side. The output reflected light is fed back

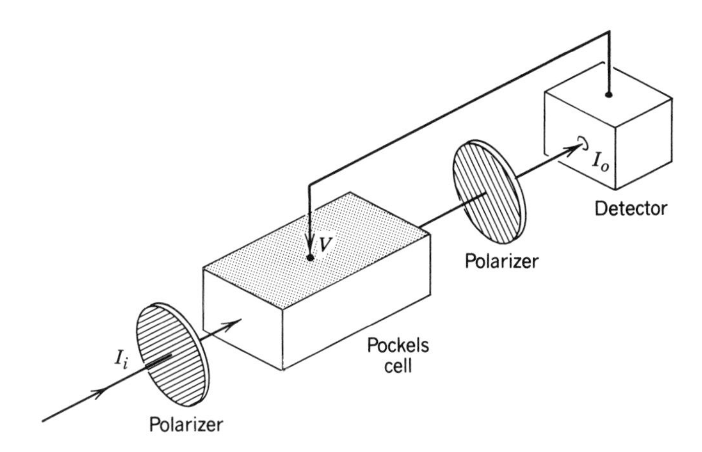

Figure 21.3-15 A hybrid bistable optical system uses an electro-optic modulator with electrical feedback.

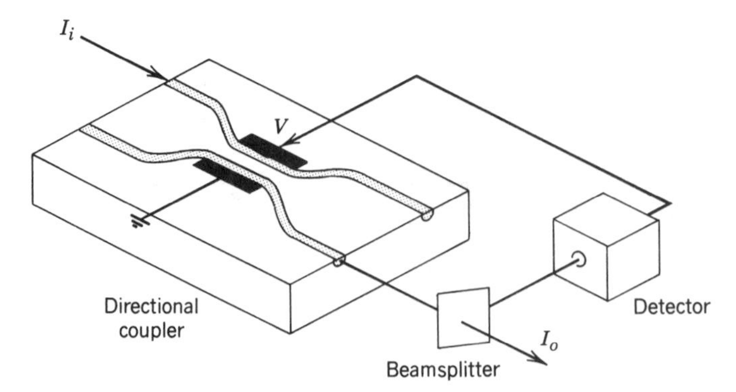

Figure 21.3-16 A bistable device uses a directional coupler with electrical feedback.

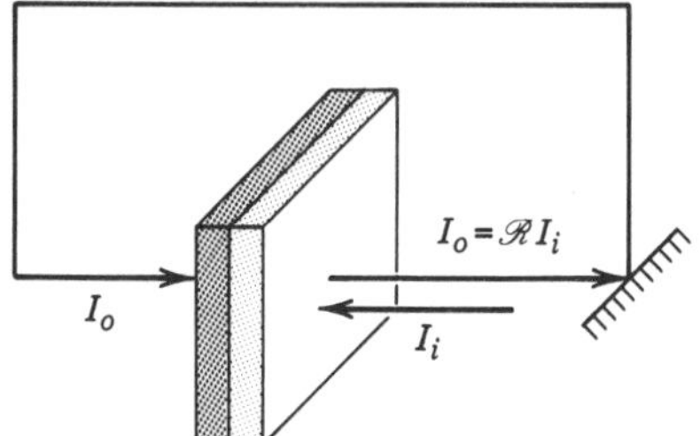

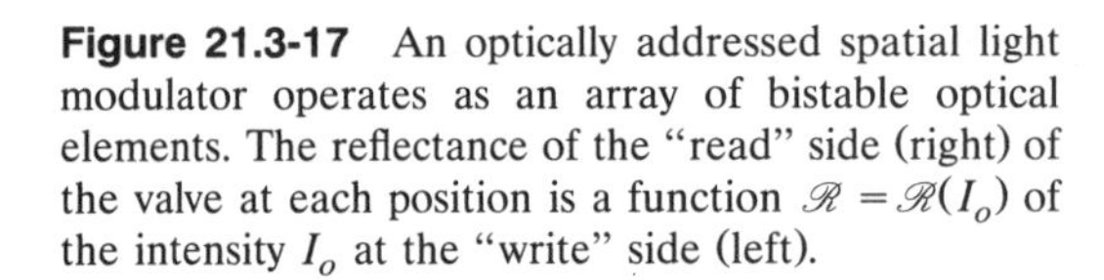
Figure 21.3-17 An optically addressed spatial light modulator operates as an array of bistable optical elements. The reflectance of the "read" side (right) of the valve at each position is a function $\mathscr{R} = \mathscr{R}(I_o)$ of the intensity I_o at the "write" side (left).

to "write" onto the device, so that $\mathscr{R} = \mathscr{R}(I_o)$. Since $\mathscr{R}(I_o)$ is a nonlinear function, bistable behavior is exhibited. Different points on the surface of the device can be addressed separately, so that the modulator serves as an array of bistable optical elements. Typical switching times are in the tens of milliseconds regime and switching powers are less than 1 μW.

The electro-optical properties of semiconductors offer many possibilities for making bistable optical devices. As mentioned earlier, the laser amplifier is an important example in which the nonlinearity is inherent in the saturation of the amplifier gain. InGaAsP laser-diode amplifiers have been operated as bistable switches with optical switching energy less than 1 fJ, and switching time less than 1 ns.

Self-Electro-Optic-Effect Device

Another electro-optic semiconductor device is the **self-electro-optic-effect device** (SEED). The SEED uses a heterostructure multiquantum-well semiconductor material made, for example, of alternating thin layers of GaAs and AlGaAs (Fig. 21.3-18). Because the bandgap of AlGaAs is greater than that of GaAs, quantum potential wells are formed (see Sec. 15.1G) which confine the electrons to the GaAs layers. An electric field is applied to the material using an external voltage source. The absorption coefficient is a nonlinear function $\alpha(V)$ of the voltage V at the wells. But V is dependent on the optical intensity I since the light absorbed by the material creates charge carriers which alter the conductance. Optical bistability is exhibited as a result of the dependence of the absorption $\alpha(V)$ on the internal optical intensity I.

This device operates without a resonator since the feedback is created internally by the optically generated electrons and holes. But it is not exactly an all-optical device since it involves electrical processes within the material and requires an external source of voltage. SEED devices can be fabricated in arrays operating at moderately high speeds and very low energies.

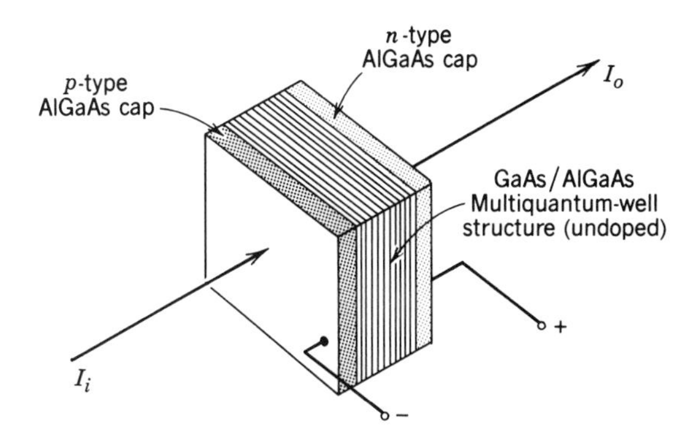

Figure 21.3-18 The self-electro-optic-effect device (SEED).

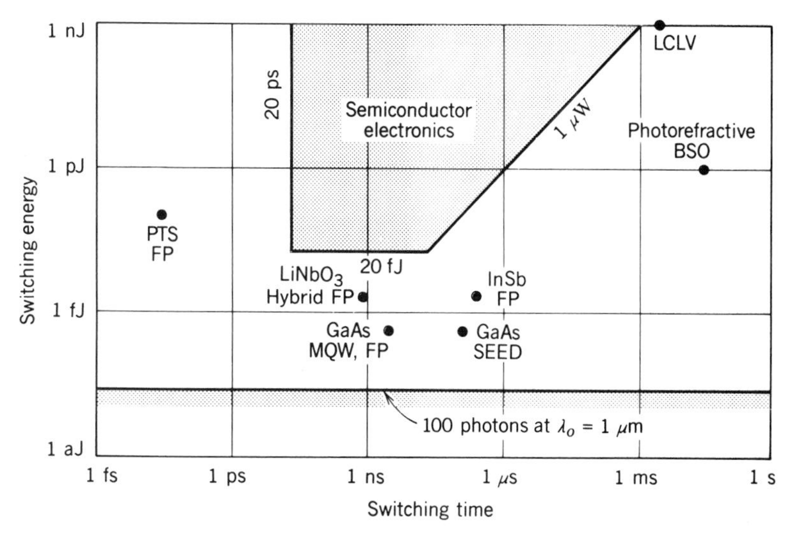

Figure 21.3-19 Switching energies and switching times of a number of optical bistable switches (LCLV = liquid-crystal light valve; FP = Fabry–Perot; SEED = self-electro-optic-effect device; MQW = multiquantum well; PTS = polymerized diacetylene, an organic material; BSO = bismuth silicon oxide). The photon-fluctuation limit on switching energy (100 photons of 1-μm wavelength) is marked. Limits of semiconductor electronic switches are also shown. (Data adapted from P. W. Smith and W. J. Tomlinson, Bistable Optical Devices Promise Subpicosecond Switching, *IEEE Spectrum*, vol. 18, no. 6, pp. 26–33, 1981 © IEEE.)

The performance of a number of bistable optical devices reported in the literature is summarized in Fig. 21.3-19.

21.4 OPTICAL INTERCONNECTIONS

Digital signal-processing and computing systems contain large numbers of interconnected gates, switches, and memory elements. In electronic systems the interconnections are made by use of conducting wires, coaxial cables, or conducting channels within semiconductor integrated circuits. Photonic interconnections may similarly be realized by use of optical waveguides with integrated-optic couplers (see Sec. 7.4B, and

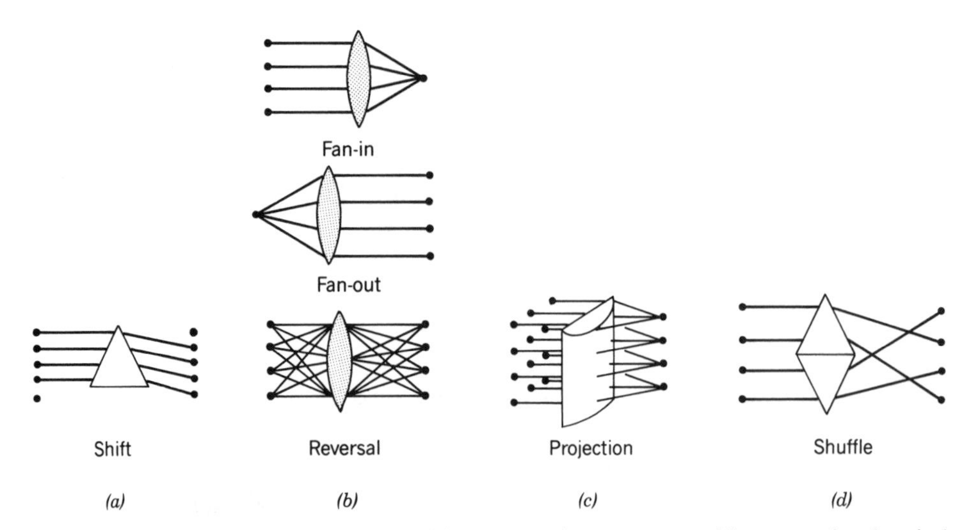

Figure 21.4-1 Examples of simple optical interconnection maps created by conventional optical components: (*a*) A prism bends parallel optical rays and establishes an ordered interconnection map with a shift. (*b*) A lens establishes a fan-in, a fan-out, or a reversal map. (*c*) An astigmatic optical system, such as a cylindrical lens, connects all points of each row in the input plane to a corresponding point in the output plane. (*d*) Two prisms are oriented to perform a perfect-shuffle interconnection map. The perfect shuffle is an operation used in sorting algorithms and in the fast Fourier transform (FFT).

Fig. 21.1-7, for example) or fiber-optic couplers and microlenses (see Sec. 22.2C and Fig. 22.2-12).

Free-space light beams may also be used for interconnections. This option is not available in electronic systems since electron beams must be in vacuum and cannot cross one another without mutual repulsion. This section is devoted to free-space optical interconnects.

Conventional optical components (mirrors, lenses, prisms, etc.) are used in numerous optical systems to establish optical interconnections, such as between points of the object and image planes of an imaging system. To appreciate the order of magnitude of the density of such interconnections, note that in a well-designed imaging system as many as 1000×1000 independent points per mm^2 in the object plane are connected optically by means of the lens to a corresponding 1000×1000 points per mm^2 in the image plane. For this to be implemented electrically, a million nonintersecting and properly insulated conducting channels per mm^2 would be required!

Conventional optical components may be used to create interconnection maps with simple patterns, such as shift, fan-in, fan-out, magnification, reduction, reversal, and shuffle, as Fig. 21.4-1 illustrates. Arbitrary optical interconnection maps, such as that illustrated in Fig. 21.4-2, require the design of custom optical components which may be quite complex and impractical. However, computer-generated holograms made of a

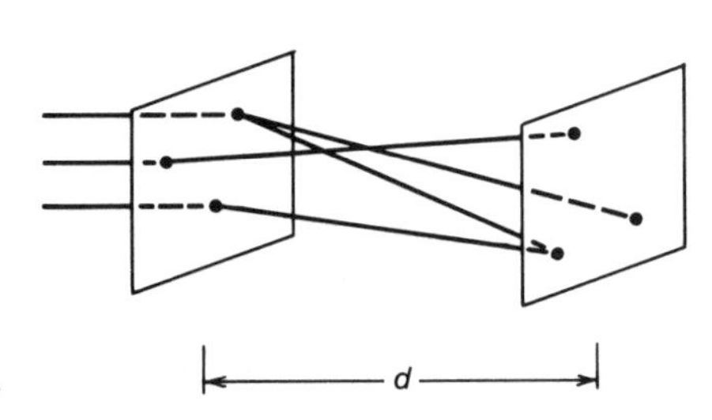

Figure 21.4-2 An arbitrary interconnection map.

large number of segments of phase gratings of different spatial frequencies and orientations have been used successfully to create high-density optical interconnections.

A. Holographic Interconnections

A phase grating is a thin optical element whose complex amplitude transmittance is a periodic function of unit amplitude, $t(x, y) = \exp[-j2\pi(\nu_x x + \nu_y y)]$, for example. The parameters ν_x and ν_y are the spatial frequencies in the x and y directions; they determine the period and orientation of the grating. It was shown in Secs. 2.4B and 4.1A that when a coherent optical beam of wavelength λ is transmitted through the grating, it undergoes a phase shift, causing it to tilt by angles $\sin^{-1}\lambda\nu_x \approx \lambda\nu_x$ and $\sin^{-1}\lambda\nu_y \approx \lambda\nu_y$, where $\lambda\nu_x \ll 1$ and $\lambda\nu_y \ll 1$, as illustrated in Fig. 21.4-3. By varying the spatial frequencies ν_x and ν_y (i.e., the periodicity and orientation of the grating) the tilt angles are altered.

As described in Sec. 4.1A, this principle may be used to make an arbitrary interconnection map by use of a phase grating made of a collection of segments of gratings of different spatial frequencies. Optical beams transmitted through the different segments undergo different tilts, in accordance with the desired interconnection map (Fig. 21.4-4). If the grating segment located at position (x, y) has frequencies

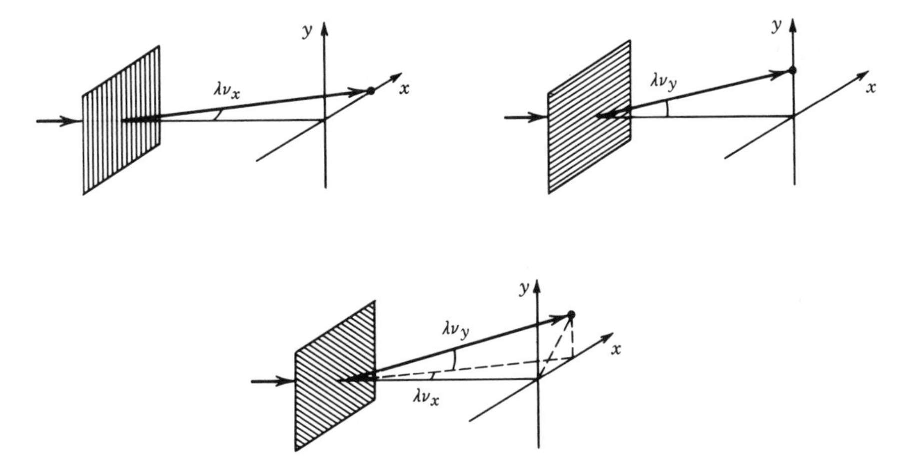

Figure 21.4-3 Bending of an optical wave as a result of transmission through a phase grating. The deflection angles, assumed to be small, depend on the spatial frequency and orientation of the grating.

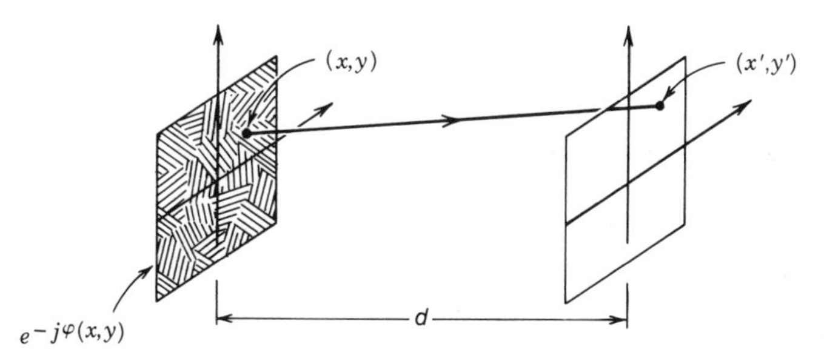

Figure 21.4-4 An interconnection map created by an array of phase gratings of different periodicities and orientations.

$\nu_x = \nu_x(x, y)$ and $\nu_y = \nu_y(x, y)$, the angles of tilt are approximately $\lambda\nu_x$ and $\lambda\nu_y$, and the beam hits the output plane at a point (x', y') satisfying

$$\frac{x' - x}{d} \approx \lambda\nu_x, \qquad \frac{y' - y}{d} \approx \lambda\nu_y, \tag{21.4-1}$$

where d is the distance between the hologram and the output plane and all angles are assumed to be small. Given the desired relation between (x', y') and (x, y), i.e., the interconnection map, the necessary spatial frequencies ν_x and ν_y may be determined at each position using (21.4-1).

In the limit in which the grating elements have infinitesimal areas, we have a continuous (instead of discrete) interconnection map: a geometric coordinate transformation rule that transforms each point (x, y) in the input plane into a corresponding point of the output plane (x', y'). If the desired transformation is defined by the two continuous functions

$$\boxed{x' = \psi_x(x, y), \qquad y' = \psi_y(x, y),} \tag{21.4-2}$$

the grating frequencies must vary continuously with x and y as in a frequency-modulated (FM) signal. Assuming that the grating has a transmittance $t(x, y) = \exp[-j\varphi(x, y)]$, the associated local (or instantaneous) frequencies are given by

$$2\pi\nu_x = \frac{\partial\varphi}{\partial x}, \qquad 2\pi\nu_y = \frac{\partial\varphi}{\partial y}. \tag{21.4-3}$$

(This is analogous to the instantaneous frequency of an FM signal.) Substituting into (21.4-1), we obtain

$$\boxed{\frac{\psi_x(x, y) - x}{d} = \frac{\lambda}{2\pi}\frac{\partial\varphi}{\partial x}, \qquad \frac{\psi_y(x, y) - y}{d} = \frac{\lambda}{2\pi}\frac{\partial\varphi}{\partial y}.} \tag{21.4-4}$$

These two partial differential equations may be solved to determine the grating phase function $\varphi(x, y)$.

EXAMPLE 21.4-1. ***Fan-In Map.*** Suppose that all points (x, y) in the input plane are to be steered to the point $(x', y') = (0, 0)$ in the output plane, so that a fan-in interconnection map is created. Substituting $\psi_x(x, y) = \psi_y(x, y) = 0$ in (21.4-4) and solving the two partial differential equations, we obtain $\varphi(x, y) = -\pi(x^2 + y^2)/\lambda d$. Not surprisingly, this is exactly the phase shift introduced by a lens of focal length d (see Sec. 2.4B).

EXERCISE 21.4-1

The Logarithmic Map. Show that the logarithmic coordinate transformation

$$x' = \psi_x(x, y) = \ln x$$

$$y' = \psi_y(x, y) = \ln y$$

is realized by a hologram with the phase function

$$\varphi(x, y) = \frac{2\pi}{\lambda d}\left(x \ln x - x - \tfrac{1}{2}x^2 + y \ln y - y - \tfrac{1}{2}y^2\right). \qquad (21.4\text{-}5)$$

Holographic interconnection devices are capable of establishing one-to-many or many-to-one interconnections (i.e., connecting one point to many points, or vice versa; Fig. 21.4-5). For example, if the grating centered at the location (x, y) is a superposition of two harmonic gratings so that its complex amplitude transmittance $t(x, y) = \exp[-j2\pi(\nu_{x1}x + \nu_{y1}y)] + \exp[-j2\pi(\nu_{x2}x + \nu_{y2}y)]$, the incident beam is split equally into two components, one tilted at angles $(\lambda\nu_{x1}, \lambda\nu_{y1})$ and the other at $(\lambda\nu_{x2}, \lambda\nu_{y2})$, where all angles are small. Weighted interconnections may be realized by assigning different weights to the different gratings. Arbitrary interconnection maps may therefore be created by appropriate selection of the grating spatial frequencies at each point of the hologram.

EXERCISE 21.4-2

Interconnection Capacity. The space–bandwidth product of a square hologram of size $d \times d$ is the product $N = (Bd)^2$, where B is the highest spatial frequencies that may be printed on the hologram. Show that if the hologram is used to direct each of L incoming beams to M directions, the product ML cannot exceed N,

$$\boxed{ML \leq N.}$$

Hint: Use an analysis similar to that presented in Sec. 20.2C in connection with acousto-optic interconnection devices [see (20.2-7)].

What is the maximum number of interconnections per mm^2 if the highest spatial frequency is 1000 lines/mm and if every point in the input plane is connected to every point in the output plane?

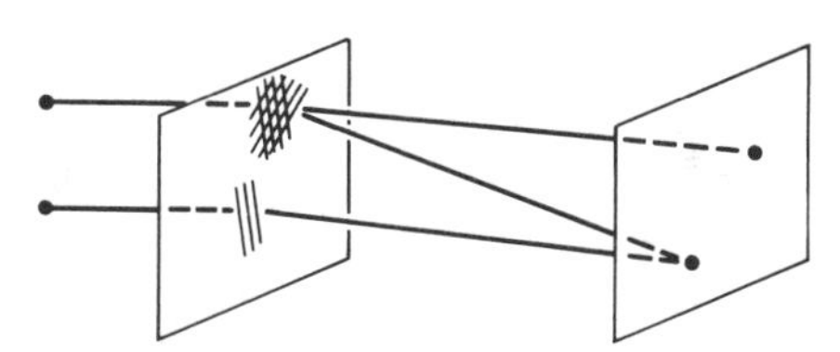

Figure 21.4-5 An arbitrary interconnection system containing one-to-many and many-to-one interconnections.

Once the appropriate phase $\varphi(x, y)$ is decided, the optical element is fabricated by using the techniques of **computer-generated holography**. This approach allows a complex function $\exp[-j\varphi(x, y)]$ to be encoded with the help of a binary function taking only two values, 1 and 0, or 1 and -1, for example. This is similar to encoding an image by use of black dots whose size or density vary in proportionality to the local gray value of the image (an example is the halftone process used for printing images in newspapers). With the help of a computer, the binary image is printed on a mask (a transparency) that plays the role of the hologram. The binary image may also be printed by etching grooves in a substrate, which modulate the phase of an incident coherent wave, a technology known as **surface-relief holography**. References discussing computer-generated holography are provided in the reading list.

Dynamic (reconfigurable) interconnections may be constructed using acousto-optic devices or magneto-optic devices. But the number of interconnection points is much smaller than is achievable by use of holographic gratings. Dynamic holographic interconnections may be achieved by use of nonlinear optical processes, such as four-wave mixing in photorefractive materials. Two waves interfere to create a grating from which a third wave is reflected. The angle between the two waves determines the spatial frequency of the grating, which determines the tilt of the reflected wave (Secs. 18.4 and 19.3C). These devices are the subject of current research.

B. Optical Interconnections in Microelectronics

The possibility of using optical interconnections to replace conventional electrical interconnections in microelectronics has led to a substantial research and development effort. With the successful use of fiber optics for computer-to-computer communications (in local area networks, for example) it is natural to consider the use of optical fibers for processor-to-processor, backplane-to-backplane, board-to-board, and chip-to-chip communications. However, the use of free-space optical communications at these different levels, and as well for intrachip interconnections, has also been explored.

Advances in high-speed high-density microelectronic circuitry and the emergence of parallel processing architectures have created communication bottlenecks so that interconnections have become a major problem. In very-large-scale integrated circuits (VLSI), interconnections occupy a large portion of the available chip area. To minimize the effect of interconnection time delays, which are becoming as long as, or even longer than, gate delays, considerable design effort is being devoted to the equalization of interconnect lengths. Optical interconnections have the potential for alleviating some of these problems.

Optical interconnections offer a number of basic advantages over electronic interconnections:

- *Density*. Electronic interconnections are planar or quasi-planar and cannot overlap or cross without proper insulation. Free-space optical interconnections can be three-dimensional. Optical beams can intersect (pass through one another) without mutual interference (provided that the medium is linear) and their size is limited only by optical diffraction. This allows for a much greater density of interference-free interconnections.
- *Delay*. Photons travel at the speed of light (0.3 mm/ps in free space). The propagation time delay is ≈ 3.3 ps/mm. By comparison, propagation delays of electrical signals in striplines fabricated on ceramics and polyimides are approximately 10.2 and 6.8 ps/mm, respectively. Whereas the velocity of light is independent of the number of interconnections branching from an interconnect, in electronic transmission lines the velocity is inversely proportional to the capacitance per unit length so that it depends on the total capacitive "load"; the

propagation delay time therefore increases with increase of the fan-outs. Optics offers a greater flexibility of fan-out and fan-in interconnections, limited only by the available optical power.

- *Bandwidth*. The density of optical interconnections is not affected by the bandwidth of the data carried by each connection. This is not the case in electronics for which the density of interconnections must be reduced sharply at high modulation frequencies to eliminate capacitive and inductive coupling effects between proximate interconnections. Optical interconnections have greater density–bandwidth products than those of electronic interconnections.
- *Power*. Electrical transmission lines must be terminated with their matched impedance to avoid reflections. This usually requires a larger expenditure of power. In optical interconnections, power requirements are limited by the sensitivity of photodetectors and the efficiencies of the electrical-to-optical and optical-to-electrical conversions as well as the power transmission efficiency of the routing elements (which also includes losses due to optical reflections).

Optical interconnections may be implemented within microelectronics by use of a number of electronic-optical transducers (light sources) acting as transmitters that beam the local electric signal to optical-electronic transducers (photodiodes) acting as receivers. A routing device (e.g., a reflection hologram) redirects the emitted light beams to the appropriate photodetector(s), as illustrated in Fig. 21.4-6. This idea can be applied to chip-to-chip or to intrachip interconnections.

There are a number of technical difficulties. Because light sources cannot be made using silicon (see Sec. 15.1D), another technology, GaAs for example, must be used. GaAs-on-Si technology (heteroepitaxy) must surmount the problems of lattice-parameter and thermal-expansion mismatch between the two materials. This is an area of ongoing research. Another difficulty is the design of light sources with sufficiently narrow beams. The design of efficient holograms and the problem of sensitivity to hologram misalignments are important, and must be addressed for this technology to become feasible.

A different approach is to replace the light sources with electro-optic modulators that modulate uniform light beams originating from an external source and reflect them onto a hologram, where they are routed back to the photodiodes on the chip (Fig. 21.4-7).

One-way optical interconnections may be achieved by use of an external light source that transmits information to a number of photodetectors on a silicon chip. One useful

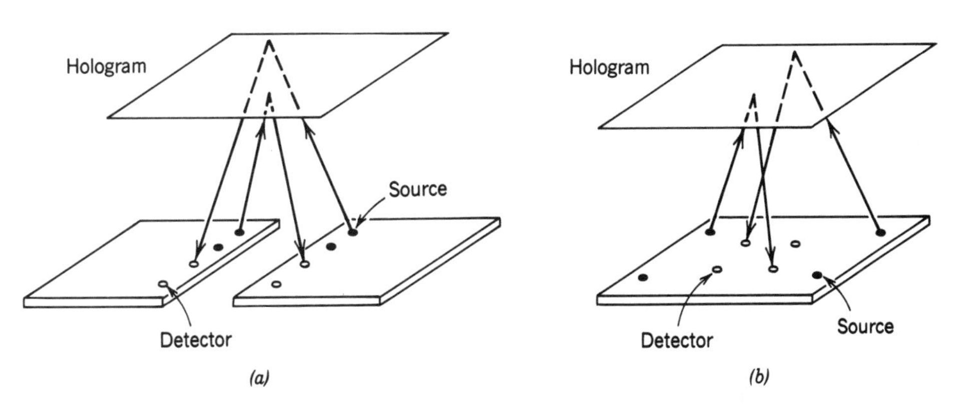

Figure 21.4-6 Optical interconnections using light sources (LEDs or diode lasers) connected optically to photodetectors by an external reflection hologram acting as a routing element: (*a*) chip-to-chip interconnections; (*b*) intrachip interconnections.

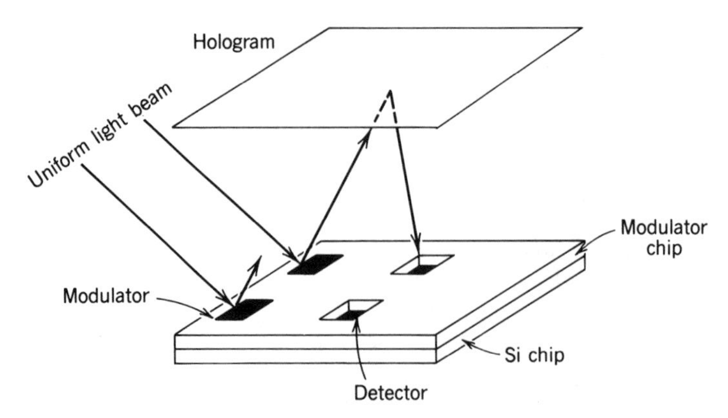

Figure 21.4-7 Interconnections using electro-optic modulators. Electrical signals are used to modulate light beams that are directed by a hologram onto photodiodes, where they are converted into electrical signals.

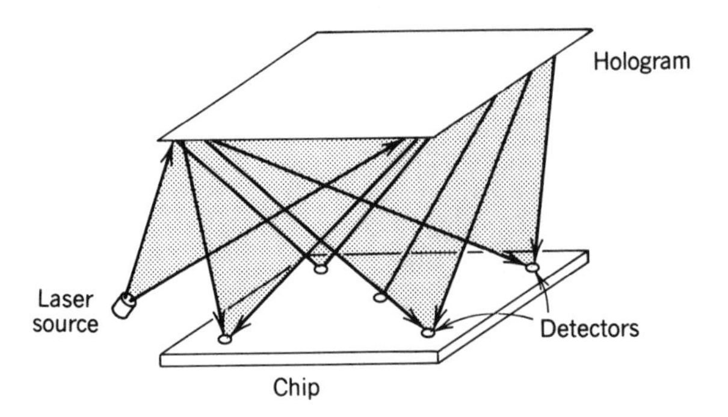

Figure 21.4-8 Clock pulses from an external light source are directed to photodetectors in a silicon chip. This reduces differential time delays and clock skew.

application is in optical clock distribution. This ensures accurate synchronization of high-speed synchronous circuits and alleviates the problem of clock skew that results from differential time delays (Fig. 21.4-8). The hologram may, of course, be eliminated and the light "broadcast" directly to all points on the chip. This creates a robust system that is insensitive to misalignment, but the power efficiency is low since a larger portion of the optical power is wasted.

Reprogrammable interconnections with dynamic holographic optical elements are also under investigation. Optical interconnections are likely to play an important future role in microelectronics.

21.5 OPTICAL COMPUTING

A. Digital Optical Computing

A digital electronic computer is made of a large number of interconnected logic gates, switches, and memory elements. Numbers are represented in a binary system and mathematical operations such as addition, subtraction, and multiplication are reduced to a set of logic operations. Instructions are encoded in the form of sequences of binary numbers. The binary numbers ("0" and "1") are represented physically by two values of the electric potential. The system operation is controlled by a clock that governs the flow of streams of "1" and "0" electrical pulses. Interconnections between the gates and switches are typically local or via a bus and the operation is sequential (i.e., time multiplexed).

It is natural to consider building an optical digital computer mimicking the electronic digital computer. The necessary optical hardware has already been introduced and discussed at length in this and earlier chapters. Electronic gates, switches, and memory elements are replaced by the corresponding optical devices; electrical interconnections within integrated circuits are replaced by waveguides in integrated optics; wires are replaced by optical fibers; the bits "1" and "0" are represented by two intensities of light, "bright" and "dark," for example; data enter the system in the form of light pulses at some clock rate; and the architecture is identical to that of the conventional electronic computer.

Although this straightforward duplication is possible (at least on a small scale), the size, speed, and switching energy and power of present state-of-the-art digital optical devices make the overall performance of the proposed optical computer significantly inferior to its electronic counterpart. As mentioned in Secs. 21.2 and 21.3, very fast optical switches are available, but not in large arrays, and the switching energy and power dissipation are prohibitively large. These limitations, however, are technological instead of fundamental. It is also important to note that the approach of mimicking the electronic computer does not exploit some basic differences between photonics and electronics, which, when properly utilized, could give photonics some important advantages.

Although it is necessary in electronic circuits to guide electrons within conduits (wires, microstrip lines, or planar conducting channels within planar integrated circuits), photons do not require such conduits and free-space three-dimensional global optical interconnections are possible, as described in Sec. 21.4. A large number of points in two parallel planes can be optically connected by a large three-dimensional network of free-space *global* interconnections established by use of a custom-made hologram. It is possible, for example, to have *each* of 10^4 points in the input plane interconnected to *all* 10^4 points of the output plane; or each point of 10^6 points in the input plane connected to an arbitrarily selected set of 100 points among 10^6 points in the output plane. This level of global interconnections is substantially greater than is possible in electronic circuits. A competitive optical computer can, and must, exploit this feature.

Consider, for example, the hypothetical computing system illustrated in Fig. 21.5-1, in which a two-dimensional array of N optical gates ($N = 10^6$, for example) are interconnected holographically. Each gate is connected locally or globally to a small or large number of other gates in accordance with a fixed wiring pattern representing, for example, arithmetic logic units, central processing units, or instruction decoders. The machine could be "programmed" or "reconfigured" by changing the interconnection hologram. This type of parallel architecture is significantly different from the usual bus-limited architecture typically used in very-large-scale integration (VLSI). The level of parallelism (i.e., size of global interconnections) envisioned in optical computers is also much higher than that possible in electronic array processors.

Such a system could, for example, be used to build an optical sequential logic machine. The gates are NOR gates. Each gate has two inputs and one output. The output optical beam from each gate is directed by the hologram to the appropriate input(s) of other gates. The electronic digital circuit is translated into a map of interconnections between output and input points in the gate plane. The interconnection map is coded on a fixed computer-generated hologram. Data arrive in the form of a number of optical beams connected directly to appropriate gate inputs and a number of gates deliver the final output of the processor.

If this parallelism of the processing elements and the interconnections were combined with high switching speeds, the result would yield staggering computational

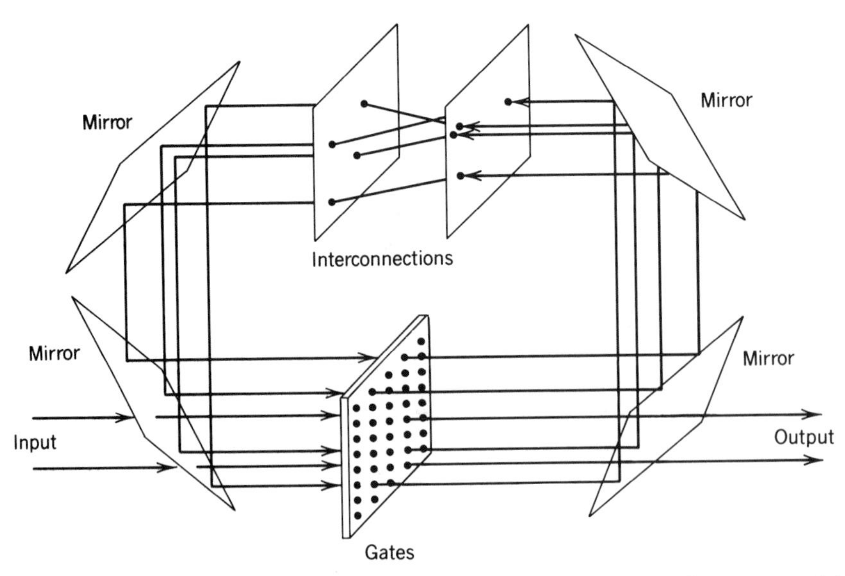

Figure 21.5-1 Possible architecture for all-optical digital computing. N gates are globally interconnected via a hologram.

power. Since the gates are operated in parallel, the data throughput is the product of the number of gates N and their switching speeds. If it were possible to have $N = 10^6$ optical gates operating with a switching time of 0.1 ns, the system could perform 10^{16} bit operations per second. This extremely high rate is approximately the same as that of the human brain and is orders of magnitude greater than the largest currently available electronic computer. As mentioned in Sec. 21.4, these numbers are, in principle, within the *fundamental* limits of photonics. The main technical difficulty remains: *the creation of large high-density arrays of fast optical gates that operate at sufficiently low switching energies and dissipate manageable powers*. Vigorous research toward achieving this goal is ongoing. If these optical machines become available, they are likely to be used in computational tasks suitable to this parallel architecture, e.g., digital image processing and artificial intelligence.

B. Analog Optical Processing

While effective application of the enormous capacity of optical global interconnections to *digital* computing awaits the development of large arrays of optical switches and gates, *analog* optical computing is presently a feasible technology with actual and potential applications in broadband signal processing, radar signal processing, image processing and machine vision, artificial intelligence, and associative memory operations in neural networks.

Most mathematical operations achievable with analog optical processors are combinations of the elementary operations of *addition* and *multiplication* performed many times in parallel by means of a large optical network of interconnections. Theoretically, all *linear* operations (weighted superpositions) can be implemented by use of these elementary operations. The routing elements used to establish the interconnections are usually conventional bulk optical component (lenses, for example), but holographic and acousto-optic devices are being used increasingly.

The variables on which the desired mathematical operation is to be performed are represented by physical (optical) quantities:

- In *incoherent optical processors*, the optical intensity, or the intensity reflectance or transmittance of a transparency or a spatial light modulator, may be used as the computation variable. These variables must be real values and cannot be negative.
- In *coherent optical processors*, the optical complex amplitude, or the complex amplitude transmittance or reflectance of a transparency or a modulator, may be used. These variables can be complex. Coherent optics permits the use of holograms as phase modulators and as interconnection elements.

Multiplication is achieved by transmitting the light through (or reflecting it from) a transparency or a modulator. In coherent processors the optical complex amplitude is multiplied by the amplitude transmittance of the transparency; in incoherent processors it is multiplied by the intensity transmittance. *Addition* is obtained when light beams are routed to the same point. In coherent processors, the complex amplitudes are added; in incoherent processors, the intensities are added.

Optical processors are inherently two-dimensional, so that data can be entered in the form of two-dimensional arrays, or images. This offers a great flexibility of interconnections and a variety of interesting signal-processing schemes. We shall illustrate these schemes by using examples from discrete processors operating on a finite number of variables. Examples of continuous processors operating on functions will then be presented.

Discrete Optical Processors

Summation. The operation $g = \Sigma_{lm} f_{lm}$ is performed by simple use of a fan-in interconnection map (implemented by a lens, for example; Fig. 21.5-2). The input variables f_{lm} $(l, m = 1, 2, \ldots, N)$ are represented by the intensities of N^2 optical beams, which are added to produce a light intensity g at the output.

Projection. The operation $g_l = \Sigma_m f_{lm}$ is similarly performed by ordering the input variables f_{lm} in the form of rows and columns in the input plane and using an interconnection map (implemented by a cylindrical lens, for example) that routes the beams of each row into a single point at the output plane where they are added (Fig. 21.5-3).

Inner and Outer Products. The inner product $g = \Sigma_m f_m h_m$ is a transformation of two input vectors f_m and h_m into a scalar g. It is basically a sum of products. The outer

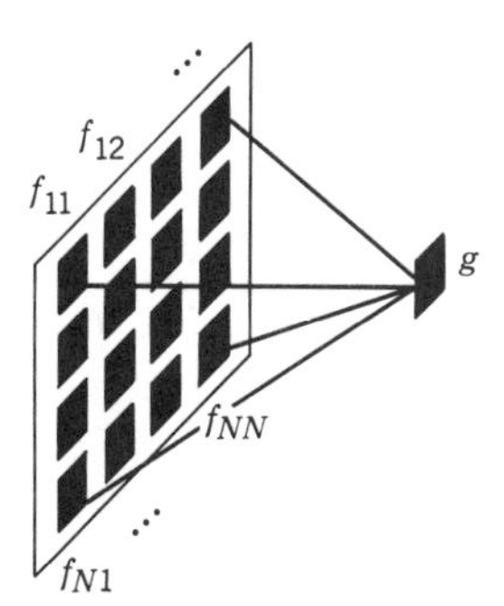

Figure 21.5-2 Optical summation.

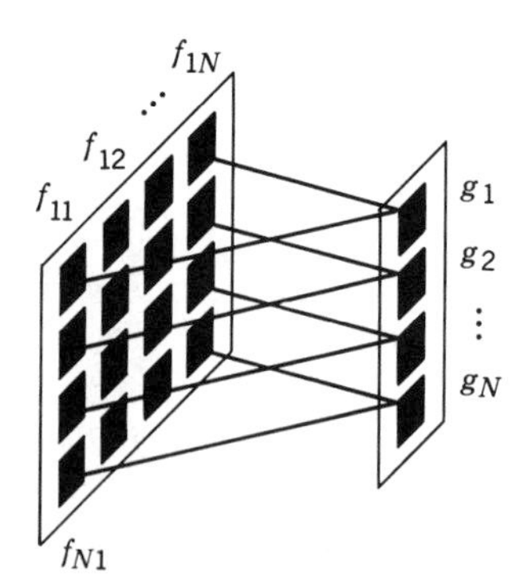

Figure 21.5-3 Optical projection.

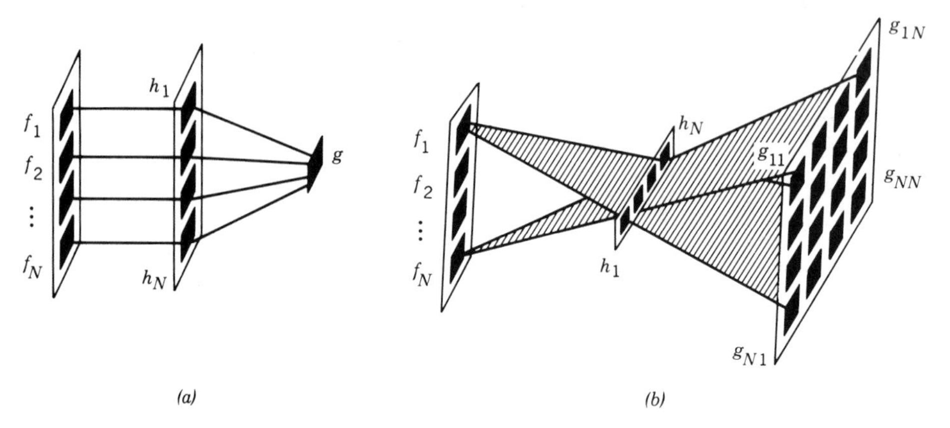

Figure 21.5-4 (*a*) Inner product. (*b*) Outer product.

product $g_{lm} = f_l h_m$ transforms two vectors into a matrix. These two operations are performed by use of a combination of a multiplication element and appropriate interconnections, fan-ins and fan-outs, for example (Fig. 21.5-4).

Matrix Multiplication. The operation $g_l = \sum_m A_{lm} f_m$ representing multiplication of a matrix of elements $\{A_{lm}\}$ by a vector of elements $\{f_m\}$ is a basic operation in linear algebra. It can be implemented by using a mask whose transmittances at an array of points are proportional to the elements $\{A_{lm}\}$ (Fig. 21.5-5). The elements are ordered

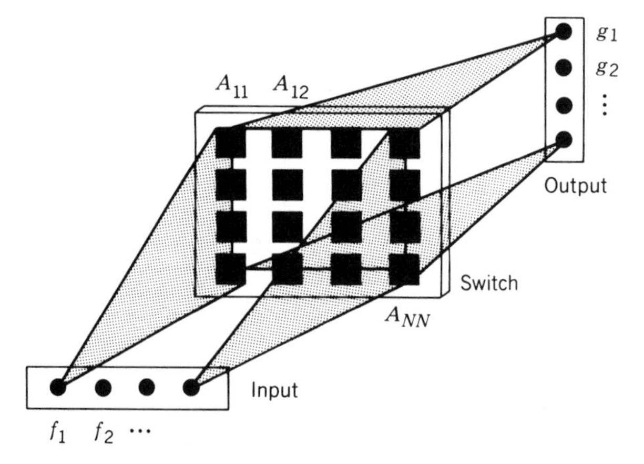

Figure 21.5-5 Optical matrix-vector multiplication.

in the form of a matrix. Two interconnection maps are used: One distributes (fans-out) the entry f_m to all elements of the mth column, where they are multiplied by $A_{1m}, A_{2m}, \ldots, A_{Nm}$; the second adds (fans-in) products in the lth row to obtain $g_l = \sum_m A_{lm} f_m$, for all $l = 1, 2, \ldots, N$. Fan-in and fan-out elements are implemented easily by use of conventional cylindrical lenses.

These five examples illustrate the flexibility of optical processors in performing the operations of linear algebra. Dynamic operations require the use of pulsed light sources. Dynamic transparencies are implemented by use of spatial light modulators.

Continuous Optical Processors

The generalization of these five operations to continuous functions is straightforward. The variables f_{lm}, g_l, and A_{lm} are replaced by the continuous functions $f(x, y)$, $g(x)$, and $A(x, y)$. The operations of integration, projection, inner product, outer product, and matrix-vector multiplication correspond to:

$$g = \iint f(x, y)\, dx\, dy \qquad \text{(integration)}$$

$$g(x) = \int f(x, y)\, dy \qquad \text{(projection)}$$

$$g = \int f(x) h(x)\, dx \qquad \text{(inner product)}$$

$$g(x, y) = f(x) h(y) \qquad \text{(outer product)}$$

$$g(x) = \int A(x, y) f(y)\, dy \qquad \text{(linear filtering).}$$

The Fourier Transform as an Interconnection Map

The Fourier transform is an important mathematical tool used in the analysis of linear systems and employed in numerous signal processing applications (see Appendices A and B). In Chap. 4 a theory of wave optics based on the Fourier transform was presented. In Sec. 4.2 it was shown that if a transparency of complex amplitude transmittance $f(x, y)$ is illuminated by a plane wave of coherent light, the transmitted light takes the form of plane waves traveling in different directions; the amplitude of the wave that makes an angle (θ_x, θ_y) is $F(\theta_x/\lambda, \theta_y/\lambda)$, where

$$F(\nu_x, \nu_y) = \int_{-\infty}^{\infty} \int_{-\infty}^{\infty} \exp\left[j2\pi(\nu_x x + \nu_y y) \right] f(x, y)\, dx\, dy$$

is the Fourier transform of $f(x, y)$. If these plane waves are focused by a lens of focal length f, the Fourier transform forms an image $F(x/\lambda f, y/\lambda f)$ in the focal plane of the lens, as illustrated in Fig. 21.5-6.

We can also think of a transparency with amplitude transmittance $f(x, y)$ as a holographic interconnection element, like the ones considered in Sec. 21.4A, connecting each point in the output plane to the entire input plane. The function $f(x, y)$ is decomposed into a sum of harmonic functions of different spatial frequencies (ν_x, ν_y) with amplitudes $F(\nu_x, \nu_y)$. As an interconnection element, the transparency "routes" the amplitude $F(\nu_x, \nu_y)$ in a direction at angles $\theta_x \approx \lambda\nu_x$ and $\theta_y \approx \lambda\nu_y$. The natural rules of wave propagation correspond to a Fourier-transform interconnection map! The lens merely funnels all the rays coming from each direction to a single point, i.e., acts as a fan-in interconnection element. The recognition of this natural property of

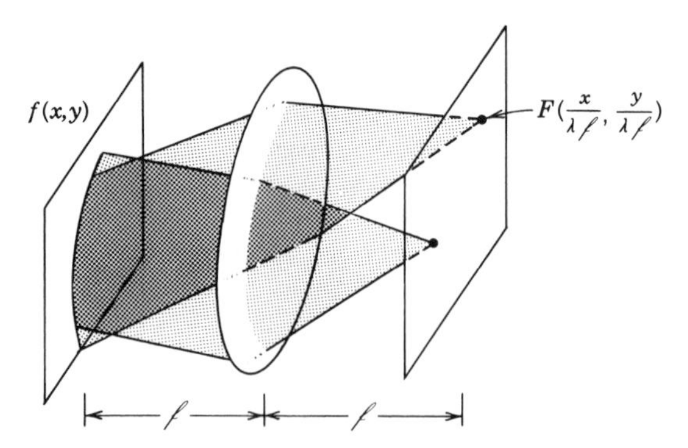

Figure 21.5-6 The optical Fourier transform as an interconnection map.

optical Fourier-transform generation has played an important historic role in motivating the use of optics for signal processing and computing.

Convolution and Correlation

The operation of convolution of two functions $f(x, y)$ and $h(x, y)$,

$$g(x, y) = \int_{-\infty}^{\infty}\int_{-\infty}^{\infty} f(x', y')h(x - x', y - y')\, dx'\, dy'$$

(see Appendix B), represents the action of a spatial filter of impulse-response function $h(x, y)$ on an input function $f(x, y)$. This operation may be implemented by exploiting the property that the Fourier transform of $g(x, y)$ is the product of the Fourier transforms of $f(x, y)$ and $h(x, y)$, i.e., $G(\nu_x, \nu_y) = F(\nu_x, \nu_y)\mathcal{H}(\nu_x, \nu_y)$. Optical implementation involves three steps: Fourier transforming $f(x, y)$ using a lens, multiplication with $\mathcal{H}(\nu_x, \nu_y)$ using an appropriate holographic mask (see Sec. 4.5), and inverse Fourier transforming the product $F(\nu_x, \nu_y)\mathcal{H}(\nu_x, \nu_y)$ using another lens (see Sec. 4.4B for details). Arbitrary two-dimensional shift-invariant spatial filters may thus be implemented optically. Filters of this type have numerous applications in image processing (image enhancement and image deblurring, for example).

The operation of cross-correlation between two functions $h(x, y)$ and $f(x, y)$ is defined by

$$g(x, y) = \int_{-\infty}^{\infty}\int_{-\infty}^{\infty} h^*(x', y')f(x' + x, y' + y)\, dx'\, dy'$$

[see (A.1-5)]. This operation may be implemented optically by exploiting the property that the Fourier transforms of $g(x, y)$, $f(x, y)$, and $h(x, y)$ are related by $G(\nu_x, \nu_y) = F(\nu_x, \nu_y)\mathcal{H}^*(\nu_x, \nu_y)$. The optical implementation is similar to that used in convolution. Cross-correlation is an important operation used in pattern recognition as a feature representing the degree of similarity between two images.

The multiplication operation $G = F\mathcal{H}^*$ may be implemented in real time by use of four-wave mixing in a nonlinear medium (see Sec. 19.3C). In accordance with (19.3-21), if the amplitude of waves 1 and 4 are proportional to $\mathcal{H}$ and F, respectively, and the amplitude of wave 3 is uniform, then the amplitude of wave 2 is proportional to the product $F\mathcal{H}^*$. As illustrated in Fig. 21.5-7, the Fourier transforms of $f(x, y)$ and

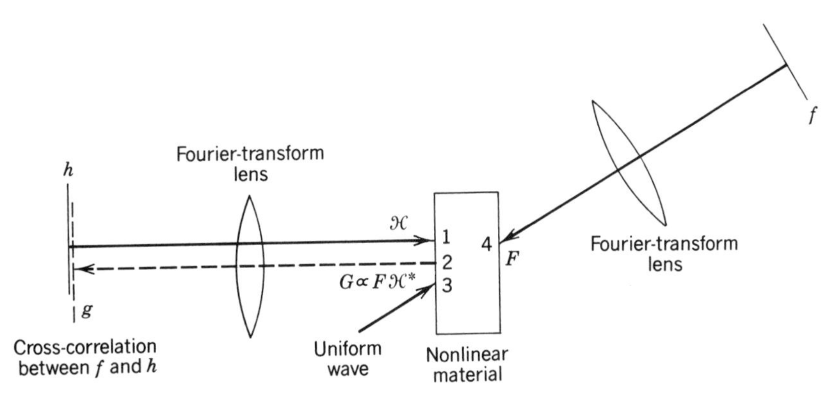

Figure 21.5-7 An optical system for performing the cross correlation between two spatial functions, $f(x, y)$ and $h(x, y)$, using two Fourier-transform lenses and four-wave mixing in a nonlinear optical material.

$h(x, y)$ are computed by use of Fourier transform lenses and the product $F(\nu_x, \nu_y)\mathcal{H}^*(\nu_x, \nu_y)$, which is generated by the mixing process, is inverse Fourier transformed by another Fourier transform lens, so that the cross correlation $g(x, y)$ is obtained in real time. The nonlinear material may be a Kerr medium or a photorefractive material (see Sec. 18.4).

Geometric Transformations

Another class of useful operations on two-dimensional signals (images) consists of geometric transformations. An image $f(x, y)$ is transformed into another image $g(x', y')$ by a change of the coordinate system $x' = \psi_x(x, y)$, $y' = \psi_y(x, y)$. These transformations include magnification, reduction, reversal, rotation, shift, perspective, and so on. The logarithmic transformation $\{x' = \ln x,\ y' = \ln y\}$ is useful in converting a change of scale in the original image into a displacement of the transformed image (because $\ln ax = \ln x + \ln a$). Similarly, the Cartesian-to-polar transformation maps a rotation of the original image into displacement in the transformed image. These operations are useful in scale-invariant and rotation-invariant pattern recognition. The optical implementation of geometric transformations using computer-generated holograms has been described in Sec. 21.4A.

Outlook

A large stock of discrete and continuous mathematical operations on arrays of variables and on two-dimensional functions may be implemented optically. Numerous other operations may be realized by serial and parallel combinations and cascades of these operations. The power of optical analog processors lies in the high degree of parallelism and the large size of the interconnection maps. However, analog computing has limited accuracy and dynamic range and is therefore suitable principally for computational tasks that are insensitive to error. A good example is the implementation of neural networks. These are networks with a high degree of global interconnection, involving simple operations of weighted superpositions and thresholding, that are cascaded and connected in a variety of forms. They implement algorithms which have an underlying redundancy, so that the limited accuracy of analog computing is tolerable.

The main challenge for optical processing lies in the development of high-resolution and fast-interface devices (spatial light modulators and array detectors) and in the design of robust and miniaturized optical systems.

READING LIST

PHOTONIC SWITCHING

Books

T. K. Gustafson and P. W. Smith, eds., *Photonic Switching*, Springer-Verlag, New York, 1988.

G. F. Marchall, ed., *Laser Beam Scanning: Opto-Mechanical Devices, Systems, and Data Storage Optics*, Marcel Dekker, New York, 1985.

H. A. Elion and V. N. Morozov, *Optoelectronic Switching Systems in Telecommunications and Computers*, Marcel Dekker, New York, 1984.

Articles and Special Issues of Journals and Proceedings

Nonlinear Optical Materials and Devices for Photonic Switching, *SPIE*, vol. 1216, 1990.

Y. Silberberg, Photonic Switching Devices, *Optics News*, vol. 15, no. 2, pp. 7–12, 1989.

Photonic Switching, *IEEE Journal of Selected Areas in Communications*, vol. 6, Aug. 1988.

S. F. Su, L. Jou, and J. Lenart, A Review on Classification of Optical Switching Systems, *IEEE Communications Magazine*, vol. 24, no. 5, pp. 50–55, 1986.

P. W. Smith, Applications of All-Optical Switching and Logic, *Philosophical Transactions of the Royal Society of London*, vol. A313, pp. 349–355, 1984.

OPTICAL BISTABILITY

Books

H. M. Gibbs, *Optical Bistability: Controlling Light with Light*, Academic Press, New York, 1985.

C. M. Bowden, M. Cifton, and H. R. Roble, eds., *Optical Bistability*, Plenum Press, New York, 1981.

Articles and Special Issues of Proceedings

High Speed Phenomena in Photonic Materials and Optical Bistability, *SPIE*, vol. 1280, 1990.

B. Chen, Integrated Optical Logic Devices, in *Integrated Optical Circuits and Components*, L. D. Hutcheson, ed., Marcel Dekker, New York, 1987, pp. 289–314.

H. M. Gibbs, Optical Bistability: Where Is It Headed? *Laser Focus*, vol. 21, Oct. 1985.

L. A. Lugiato, Theory of Optical Bistability, in *Progress in Optics*, vol. 21, E. Wolf, Ed., North-Holland, Amsterdam, 1984.

S. D. Smith and A. C. Walker, The Prospects for Optically Bistable Elements in Optical Computing, *SPIE Proceedings*, vol. 492, pp. 342–345, 1984.

P. W. Smith and W. J. Tomlinson, Bistable Optical Devices Promise Subpicosecond Switching, *IEEE Spectrum*, vol. 18, no. 6, pp. 26–33, 1981.

OPTICAL INTERCONNECTIONS

Articles and Special Issues of Journals and Proceedings

Optical Interconnects, *Applied Optics* (*Information Processing*), vol. 29, no. 8, 1990.

Optical Interconnections and Networks, *SPIE*, vol. 1281, 1990.

Optical Interconnects in the Computer Environment, *SPIE*, vol. 1178, 1990.

J. E. Midwinter, Digital Optics, Smart Interconnect or Optical Logic?, *Physics in Technology*, vol. 119, part I, pp. 101–108; part II, pp. 153–165, May 1988.

Optical Interconnections, *Optical Engineering*, vol. 25, Oct. 1986.

P. R. Haugen, S. Rychnovsky, A. Husain, and L. D. Hutcheson, Optical Interconnects for High Speed Computing, *Optical Engineering*, vol. 25, pp. 1076–1085, 1986.

A. A. Sawchuk and B. K. Jenkins, Dynamic Optical Interconnections for Parallel Processors, *SPIE Proceedings*, vol. 625, pp. 143–153, 1986.

D. H. Hartman, Digital High Speed Interconnects: A Study of the Optical Alternative, *Optical Engineering*, vol. 25, pp. 1086–1102, 1986.

A. Husain, Optical Interconnect of Digital Integrated Circuits and Systems, *SPIE Proceedings*, vol. 466, pp. 10–20, 1984.

J. W. Goodman, F. I. Leonberger, S. Y. Kung, and R. A. Athale, Optical Interconnections for VLSI Systems, *Proceedings of the IEEE*, vol. 72, pp. 850–866, 1984.

J. W. Goodman, Optical Interconnections in Microelectronics, *SPIE Proceedings*, vol. 456, pp. 72–85, 1984.

COMPUTER-GENERATED HOLOGRAPHY

W. J. Dallas, Computer-Generated Holograms, in *The Computer in Optical Research*, B. R. Frieden, ed., Springer-Verlag, New York, 1980, pp. 291–366.

W.-H. Lee, Computer-Generated Holograms: Techniques and Applications, in *Progress in Optics*, vol. 16, E. Wolf, ed., North-Holland, Amsterdam, 1978, pp. 119–232.

OPTICAL COMPUTING AND PROCESSING

Books

A. D. McAulay, *Optical Computer Architectures*, Wiley, New York, 1991.

P. K. Das, *Optical Signal Processing*, Springer-Verlag, New York, 1990.

R. Arrathoon, ed., *Optical Computing: Digital and Symbolic*, Marcel Dekker, New York, 1989.

H. Arsenault, T. Szoplik, and B. Macukow, eds., *Optical Processing and Computing*, Academic Press, Orlando, FL 1989.

D. G. Feitelson, *Optical Computing*, MIT Press, Cambridge, MA, 1988.

J. L. Horner, ed., *Optical Signal Processing*, Academic Press, New York, 1987.

F. T. S. Yu, *White-Light Optical Signal Processing*, Wiley, New York, 1985.

F. T. S. Yu, *Optical Information Processing*, Wiley, New York, 1983.

H. Stark, ed., *Applications of Optical Fourier Transforms*, Academic Press, New York, 1982.

S. H. Lee, ed., *Optical Information Processing: Fundamentals*, Springer-Verlag, Berlin, 1981.

M. Françon, *Optical Image Formation and Processing*, Academic Press, New York, 1979.

D. Casasent, ed., *Optical Data Processing: Applications*, Springer-Verlag, New York, 1978.

W. E. Kock, G. W. Stroke, and Yu. E. Nesterikhin, *Optical Information Processing*, Plenum Press, New York, 1976.

W. T. Cathey, *Optical Information Processing and Holography*, Wiley, New York, 1974.

A. R. Shulman, *Optical Data Processing*, Wiley, New York, 1970.

Special Issues of Journals and Proceedings

Digital Optical Computing, *SPIE Critical Reviews*, vol. CR35, 1990.

Advances in Optical Information Processing IV, *SPIE*, vol. 1296, 1990.

Digital Optical Computing II, *SPIE*, vol. 1215, 1990.

Selected Papers on Optical Computing, *SPIE*, vol. 1142, 1989.

Optical Computing '88, *SPIE*, vol. 963, 1989.

Optical Computing and Nonlinear Materials, *SPIE*, vol. 881, 1988.

Optical Computing, *Applied Optics*, vol. 27, May 1988.

Digital Optical Computing, *SPIE*, vol. 752, 1987.

Optical Information Processing II, *SPIE*, vol. 639, 1986.

Optical Computing, *SPIE*, vol. 625, 1986.

Digital Optical Computing, *Optical Engineering*, vol. 25, Jan. 1986.

Photonic Computing, *Applied Optics*, vol. 25, Sept. 15, 1986.
Optical and Hybrid Computing, *SPIE*, vol. 634, 1986.
Real Time Signal Processing VIII, *SPIE*, vol. 564, 1985.
Transformations in Optical Signal Processing, *SPIE*, vol. 373, 1984.
Optical Computing, *Proceedings of the IEEE*, vol. 72, July 1984.
Acoustooptic Signal Processing, *Proceedings of the IEEE*, vol. 69, Jan. 1981.
Optical Computing, *Proceedings of the IEEE*, vol. 65, Jan. 1977.

Articles

D. A. B. Miller, Optoelectronic Applications of Quantum Wells, *Optics and Photonics News*, vol. 1, no. 2, pp. 7–15, 1990.

B. S. Wherrett, The Many Facets of Optical Computing, *Computers in Physics*, vol. 2, pp. 24–27, Mar. 1988.

P. Batacan, Can Physics Make Optics Compute? *Computers in Physics*, vol. 2, pp. 9–15, Mar. 1988.

Y. S. Abumostafa and D. Psaltis, Optical Neural Computers, *Scientific American*, vol. 256, no. 3, pp. 88–95, 1987.

The Coming of the Age of Optical Computing, *Optics News*, Apr. 1986.

D. Casasent, Acoustooptic Linear Algebra Processors: Architectures, Algorithms, and Applications, *Proceedings of the IEEE*, vol. 72, pp. 831–849, 1984.

W. T. Rhodes and P. S. Guilfoyle, Acoustooptic Algebraic Processing Architectures, *Proceedings of the IEEE*, vol. 72, pp. 820–830, 1984.

E. Abraham, C. T. Seaton, and S. D. Smith, The Optical Digital Computer, *Scientific American*, vol. 248, no. 2, pp. 85–93, 1983.

J. Jahns, Concepts of Optical Digital Computing—A Survey, *Optik*, vol. 57, pp. 429–449, 1980.

J. W. Goodman, Operations Achievable with Coherent Optical Information Processing Systems, *Proceedings of the IEEE*, vol. 65, pp. 29–38, 1977.

L. J. Cutrona, E. N. Leith, C. J. Palermo, and L. J. Porcello, Optical Data Processing and Filtering Systems, *IRE Transactions on Information Theory*, vol. IT-6, pp. 386–400, 1960.

PROBLEMS

21.3-1 **Optical Logic.** Figure 21.3-4 illustrates how a nonlinear thresholding optical device may be used to make an AND gate. Show how a similar system may be used to make NAND, OR, and NOR gates. Is it possible to make an XOR (exclusive OR)? Can the same system be used to obtain the OR of N binary inputs?

21.3-2 **Bistable Interferometer.** A crystal exhibiting the optical Kerr effect is placed in one of the arms of a Mach–Zehnder interferometer. The transmitted intensity I_o is fed back and illuminates the crystal. Show that the intensity transmittance of the system is $I_o/I_i = \mathcal{T}(I_o) = \frac{1}{2} + \frac{1}{2}\cos(\pi I_o/I_\pi + \varphi)$, where I_π and φ are constants. Assuming that $\varphi = 0$, sketch I_o versus I_i and derive an expression for the maximum differential gain dI_o/dI_i.

21.4-1 **Interconnection Hologram for a Conformal Map.** Design a hologram to realize the geometric transformation defined by

$$x' = \psi_x(x, y) = \ln\sqrt{x^2 + y^2}$$

$$y' = \psi_y(x, y) = \tan^{-1}\frac{y}{x}.$$

This is a Cartesian-to-polar transformation followed by a logarithmic transformation of the polar coordinate $r = (x^2 + y^2)^{1/2}$. Determine an expression for the phase function $\varphi(x, y)$ of the hologram required.

21.5-1 **Optical Projection.** Design an optical system that implements the optical projection operation depicted in Fig. 21.5-3. Assume that the data $\{f_{lm}\}$ are entered by use of an array of LEDs. Use a spherical lens and a cylindrical lens, of appropriate focal lengths, to perform the necessary imaging in the vertical direction and focusing in the horizontal direction.

CHAPTER

22

FIBER-OPTIC COMMUNICATIONS

22.1 COMPONENTS OF THE OPTICAL FIBER LINK
- A. Optical Fibers
- B. Sources for Optical Transmitters
- C. Detectors for Optical Receivers
- D. Fiber-Optic Systems

22.2 MODULATION, MULTIPLEXING, AND COUPLING
- A. Modulation
- B. Multiplexing
- C. Couplers

22.3 SYSTEM PERFORMANCE
- A. Digital Communication System
- B. Analog Communication System

22.4 RECEIVER SENSITIVITY

22.5 COHERENT OPTICAL COMMUNICATIONS
- A. Heterodyne Detection
- B. Performance of the Analog Heterodyne Receiver
- C. Performance of the Digital Heterodyne Receiver
- D. Coherent Systems

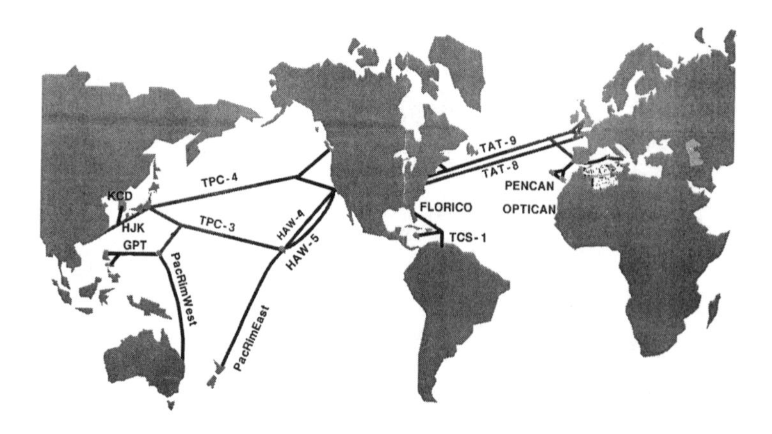

AT&T undersea fiber-optic communication network of the 1990s

Until recently, virtually all communication systems have relied on the transmission of information over electrical cables or have made use of radio-frequency and microwave electromagnetic radiation propagating in free space. It would appear that the use of light would have been a more natural choice for communications since, unlike electricity and radio waves, it did not have to be discovered. The reasons for the delay in the development of this technology are twofold: the difficulty of producing a light source that could be rapidly switched on and off and therefore could encode information at a high rate, and the fact that light is easily obstructed by opaque objects such as clouds, fog, smoke, and haze. Unlike radio-frequency and microwave radiation, light is rarely suitable for free-space communication.

Lightwave communications has recently come into its own, however, and indeed it is now the preferred technology in many applications. It is used for the transmission of voice, data, telemetry, and video in long-distance and local-area networks, and is suitable for a great diversity of other applications (e.g., cable television). Lightwave technology affords the user enormous transmission capacity, distant spacings of repeaters, immunity from electromagnetic interference, and relative ease of installation.

The spectacular successes of fiber-optic communications have their roots in two critical photonic inventions: the development of the light-emitting diode (LED) and the development of the low-loss optical fiber as a light conduit. Suitable detectors of light have been available for some time, although their performance has been improved dramatically in recent years. Interest in optical communications was initially stirred by the invention of the laser in the early 1960s. However, the first generation of fiber-optic communication systems made use of LED sources and indeed many present local-area commercial systems continue to do so. Nevertheless, most lightwave communication systems (such as long-haul single-mode fiber-optic systems and short-haul free-space systems) do benefit from the large optical power, narrow linewidth, and high directivity provided by the laser. The proposed extension of the fiber network to reach individual dwellings will rely on the use of diode lasers.

A fiber-optic communication system comprises three basic elements: a compact light source, a low-loss/low-dispersion optical fiber, and a photodetector. These optical components have been discussed in Chaps. 16, 8, and 17, respectively. In this chapter we examine their role in the context of the overall design, operation, and performance of an optical communication link. Optical accessories such as connectors, couplers, switches, and multiplexing devices, as well as splices, are also essential to the successful operation of fiber links and networks. Optical-fiber amplifiers have also proved themselves to be very valuable adjuncts to such systems. The principles of some of these devices have been discussed in Chap. 21 and in other parts of this book.

Although the waveguiding properties of different types of optical fibers have been discussed in detail in Chap. 8, this material is reviewed in Sec. 22.1 (in abbreviated form) to make this chapter self-contained. A brief summary of the properties of semiconductor photon sources and detectors suitable for fiber-optic communication systems is also provided in this section. This is followed, in Sec. 22.2, by an introduction to modulation, multiplexing, and coupling systems used in fiber-optic communications.

Section 22.3 introduces the basic design principles applicable to long-distance digital and analog fiber-optic communication systems. The maximum fiber length that can be used to transmit data (at a given rate and with a prescribed level of performance) is determined. Performance deteriorates if the data rate exceeds the fiber bandwidth, or if the received power is smaller than the receiver sensitivity (so that the signal cannot be distinguished from noise). The sensitivity of an optical receiver operating in a binary digital communication mode is evaluated in Sec. 22.4. It is of interest to compare these results with the sensitivity of an analog optical receiver, which was determined in Sec. 17.5D.

Coherent optical communication systems, which are introduced in Sec. 22.5, use light not as a source of controllable power but rather as an electromagnetic wave of controllable amplitude, phase, or frequency. Coherent optical systems are the natural extension to higher frequencies of conventional radio and microwave communications. They provide substantial gains in receiver sensitivity, permitting further spacings between repeaters and increased data rates.

22.1 COMPONENTS OF THE OPTICAL FIBER LINK

A. Optical Fibers

An optical fiber is a cylindrical dielectric waveguide made of low-loss materials, usually fused silica glass of high chemical purity. The core of the waveguide has a refractive index slightly higher than that of the outer medium, the cladding, so that light is guided along the fiber axis by total internal reflection. As described in Chap. 8, the transmission of light through the fiber may be studied by examining the trajectories of rays within the core. A more complete analysis makes use of electromagnetic theory. Light waves travel in the fiber in the form of modes, each with a distinct spatial distribution, polarization, propagation constant, group velocity, and attenuation coefficient. There is, however, a correspondence between each mode and a ray that bounces within the core in a distinct trajectory.

Step-Index Fibers

In a step-index fiber, the refractive index is n_1 in the core and abruptly decreases to n_2 in the cladding [Fig. 22.1-1(a)]. The fractional refractive index change $\Delta = (n_1 - n_2)/n_1$ is usually very small ($\Delta = 0.001$ to 0.02). Light rays making angles with the fiber axis smaller than the complement of the critical angle, $\bar{\theta}_c = \cos^{-1}(n_2/n_1)$, are guided within the core by multiple total internal reflections at the core–cladding boundary. The angle $\bar{\theta}_c$ in the fiber corresponds to an angle θ_a for rays incident from air into the fiber, where $\sin\theta_a = \text{NA}$ and $\text{NA} = (n_1^2 - n_2^2)^{1/2} \approx n_1(2\Delta)^{1/2}$ is called the numerical aperture. θ_a is the acceptance angle of the fiber.

The number of guided modes M is governed by the fiber V parameter, $V = 2\pi(a/\lambda_o)\text{NA}$, where a/λ_o is the ratio of the core radius a to the wavelength λ_o. In a fiber with $V \gg 1$, there are a large number of modes, $M \approx V^2/2$, and the minimum and maximum group velocities of the modes are $v_{\min} \approx c_1(1 - \Delta) = c_1(n_2/n_1)$ and $v_{\max} \approx c_1 = c_o/n_1$. When an impulse of light travels a distance L in the fiber, it undergoes different time delays, spreading over a time interval $2\sigma_\tau = L/c_1(1 - \Delta) - L/c_1 \approx (L/c_1)\Delta$. The result is a pulse of rms width

$$\sigma_\tau \approx \frac{L}{2c_1}\Delta. \tag{22.1-1}$$

Fiber Response Time
(Multimode Step-Index Fiber)

The overall pulse width is therefore proportional to the fiber length L and to the fractional refractive index change Δ. This effect is called **modal dispersion.**

Graded-Index Fibers

In a graded-index fiber, the refractive index of the core varies gradually from a maximum value n_1 on the fiber axis to a minimum value n_2 at the core–cladding boundary [Fig. 22.1-1(*b*)]. The fractional refractive index change $\Delta = (n_1 - n_2)/n_1 \ll 1$. Rays follow curved trajectories, with paths shorter than those in the step-index fiber. The axial ray travels the shortest distance at the smallest phase velocity (largest refractive index), whereas oblique rays travel longer distances at higher phase velocities (smaller refractive indices), so that the delay times are equalized. The maximum difference between the group velocities of the modes is therefore much smaller than in the step-index fiber.

When the fiber is graded optimally (using an approximately parabolic profile), the modes travel with almost equal group velocities. When the fiber V parameter, $V = 2\pi(a/\lambda_o)\text{NA}$, is large, the number of modes $M \approx V^2/4$; i.e., there are approximately half as many modes as in a step-index fiber with the same value of V. The group velocities then range between c_1 and $c_1(1 - \Delta^2/2)$, so that for a fiber of length L an input impulse of light spreads to a width

$$\sigma_\tau \approx \frac{L}{4c_1}\Delta^2. \tag{22.1-2}$$

Fiber Response Time (Graded-Index Fiber; Parabolic Profile)

This is a factor $\Delta/2$ smaller than in the equivalent step-index fiber. This reduction factor, however, is usually not fully met in practical graded-index fibers because of the difficulty of achieving ideal index profiles.

Single-Mode Fibers

When the core radius a and the numerical aperture NA of a step-index fiber are sufficiently small so that $V < 2.405$ (the smallest root of the Bessel function J_0), only a single mode is allowed. One advantage of using a single-mode fiber is the elimination of pulse spreading caused by modal dispersion. Pulse spreading occurs, nevertheless, since the initial pulse has a finite spectral linewidth and since the group velocities (and therefore the delay times) are wavelength dependent. This effect is called **chromatic dispersion**. There are two origins of chromatic dispersion: **material dispersion**, which results from the dependence of the refractive index on the wavelength, and **waveguide dispersion**, which is a consequence of the dependence of the group velocity of each mode on the ratio between the core radius and the wavelength. Material dispersion is usually larger than waveguide dispersion.

A short optical pulse of spectral width σ_λ spreads to a temporal width

$$\sigma_\tau = |D|\sigma_\lambda L, \tag{22.1-3}$$

Fiber Response Time (Material Dispersion)

proportional to the propagation distance L (km) and to the source linewidth σ_λ (nm), where D is the dispersion coefficient (ps/km-nm). The parameter D involves a combination of material and waveguide dispersion. For weakly guiding fibers ($\Delta \ll 1$), D may be separated into a sum $D_\lambda + D_w$ of the material and waveguide contributions. The geometries, refractive-index profiles, and pulse broadening in multimode step-

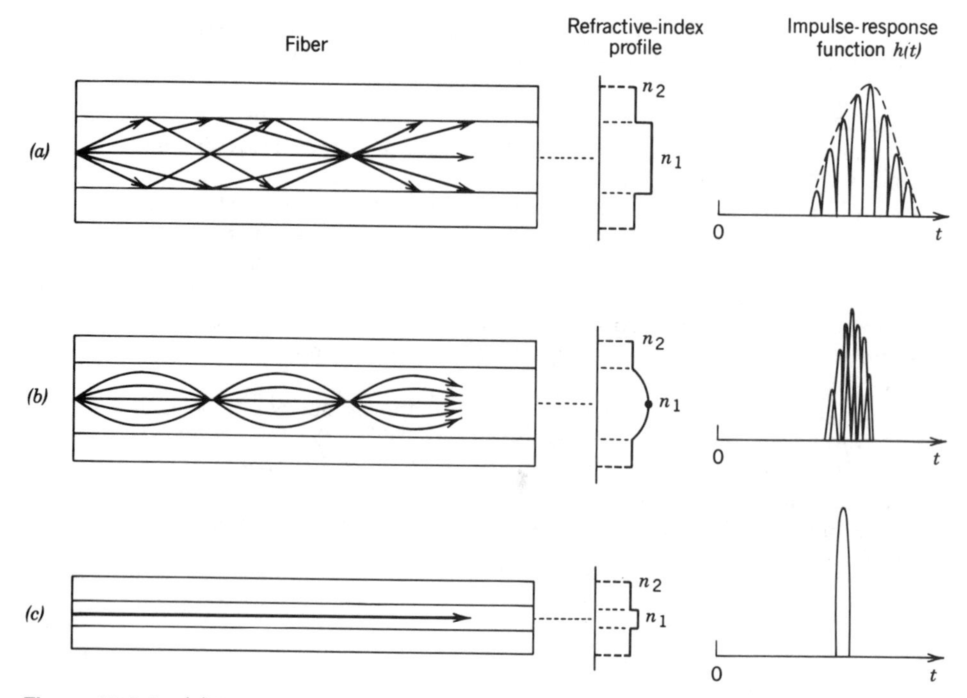

Figure 22.1-1 (*a*) Multimode step-index fibers: relatively large core diameter; uniform refractive indices in the core and cladding; large pulse spreading due to modal dispersion. (*b*) Graded-index fibers: refractive index of the core is graded; there are fewer modes; pulse broadening due to modal dispersion is reduced. (*c*) Single-mode fibers: small core diameter; no modal dispersion; pulse broadening is due only to material and waveguide dispersion.

index and graded-index fibers and in single-mode fibers are schematically compared in Fig. 22.1-1.

Material Attenuation and Dispersion

The wavelength dependence of the attenuation coefficients of different types of fused-silica-glass fibers are illustrated in Fig. 22.1-2. As the wavelength increases beyond the visible band, the attenuation drops to a minimum of approximately 0.3 dB/km at $\lambda_o = 1.3$ μm, increases slightly at 1.4 μm because of OH-ion absorption, and then drops again to its absolute minimum of ≈ 0.16 dB/km at $\lambda_o = 1.55$ μm, beyond which it rises sharply. The dispersion coefficient D_λ of fused silica glass is also wavelength dependent, as illustrated in Fig. 22.1-2. It is zero at $\lambda_o \approx 1.312$ μm.

Operating Wavelengths for Fiber-Optic Communications

As illustrated in Fig. 22.1-2, the minimum attenuation occurs at ≈ 1.55 μm, whereas the minimum material dispersion occurs at ≈ 1.312 μm. The choice between these two wavelengths depends on the relative importance of power loss versus pulse spreading, as explained in Sec. 22.3. However, the availability of an appropriate light source is also a factor. First-generation fiber-optic communication systems operated at ≈ 0.87 μm (the wavelength of AlGaAs light-emitting diodes and diode lasers), where both attenuation and material dispersion are relatively high. More advanced systems operate at 1.3 and 1.55 μm. A summary of the salient properties of silica-glass fibers at these three operating wavelengths is provided in Table 22.1-1.

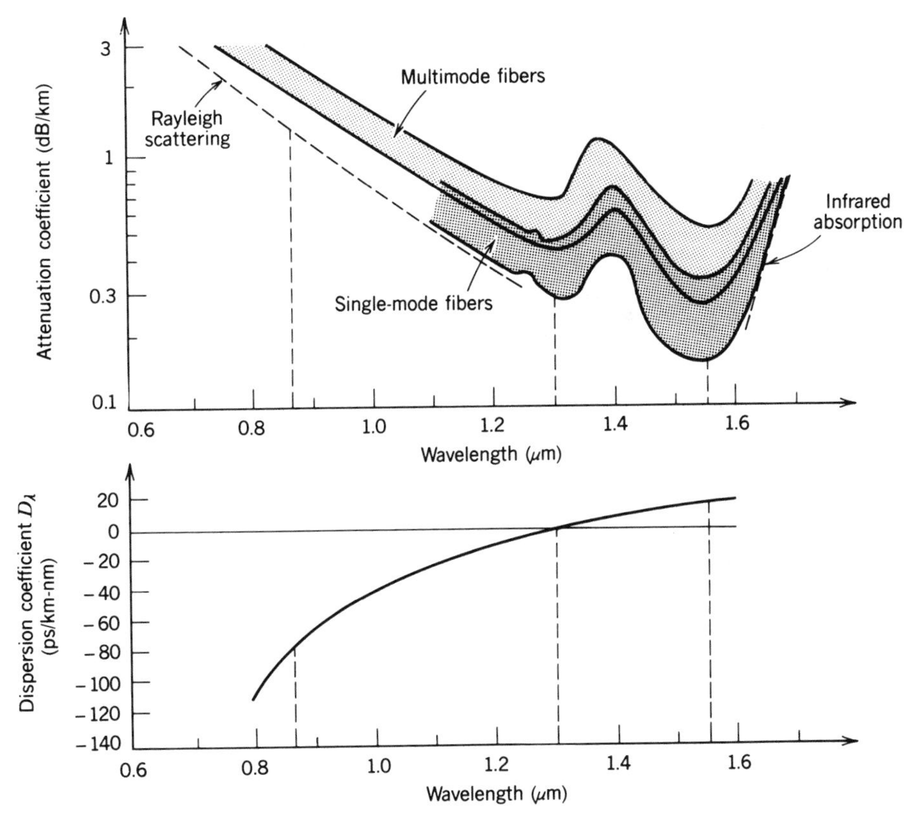

Figure 22.1-2 Wavelength dependence of the attenuation and material dispersion coefficients of silica-glass fibers, indicating three wavelengths at which fiber-optic communication systems typically operate: 0.87, 1.3, and 1.55 μm.

Advanced designs using graded-index single-mode fibers aim at balancing waveguide dispersion with material dispersion, so that the overall dispersion coefficient vanishes at $\lambda_o = 1.55$ μm rather than at 1.312 μm. This is achieved at the expense of a slight increase of the attenuation coefficient.

Transfer Function, Response Time, and Bandwidth

A communication channel is usually characterized by its impulse-response function $h(t)$. For the fiber-optic channel, this is the received power as a function of time when the input power at the transmitter side is an impulse function $\delta(t)$ [see Figs. 22.1-3(a) and 22.1-1]. An equivalent function that also characterizes the channel is the transfer function $\mathcal{H}(f)$. This is obtained, as illustrated in Fig. 22.1-3(b), by modulating the

TABLE 22.1-1 Minimum Attenuation and Material Dispersion Coefficients of Silica-Glass Fiber at Three Wavelengths[a]

λ_o (μm)	Attenuation (dB/km)	Dispersion (ps/km-nm)
0.87	1.5	−80
1.312	0.3	0
1.55	0.16	+17

[a]Actual values depend on the type of fiber and the dopants used.

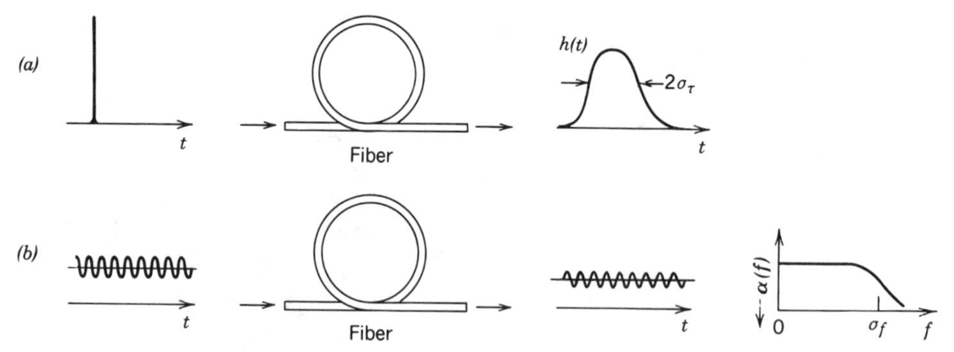

Figure 22.1-3 (*a*) Measurement of the impulse-response function $h(t)$. (*b*) Measurement of the transfer function $\mathcal{H}(f)$. The attenuation coefficient $\boldsymbol{\alpha}(f)$ is the negative of the absolute value of $\mathcal{H}(f)$ in dB units for $L = 1$ km.

input power ($P(z)$ at $z = 0$) sinusoidally at frequency f, $P(0) = P_0(0) + P_s(0)\cos(2\pi f t)$, where $P_s(0) < P_0(0)$, and measuring the output power after propagation a distance L through the fiber, $P(L) = P_0(L) + P_s(L)\cos(2\pi f t + \varphi)$. The transfer function is $\mathcal{H}(f) = [P_s(L)/P_s(0)]\exp(j\varphi)$. Clearly, $P_0(L) = \mathcal{H}(0)P_0(0)$, where $\mathcal{H}(0)$ is the transmittance $\mathcal{T}$.

The absolute value $|\mathcal{H}(f)|$ is the factor by which the amplitude of the modulated signal at frequency f is reduced as a result of propagation. The attenuation coefficient $\boldsymbol{\alpha}(f)$ is defined by

$$\boldsymbol{\alpha}(f) = \frac{-10\log_{10}(|\mathcal{H}(f)|)}{L} \qquad (22.1\text{-}4)$$

Fiber Attenuation Coefficient (dB/km) at Modulation Frequency f

and has units of dB/km. Thus $|\mathcal{H}(f)| = \exp[-\alpha(f)L]$, where $\alpha(f) \approx 0.23\boldsymbol{\alpha}(f)$ is the attenuation coefficient in units of km^{-1}. As shown in Appendix B, the transfer function $\mathcal{H}(f)$ is the Fourier transform of the impulse-response function $h(t)$, so that knowledge of one function is sufficient to determine the other.

Three important measures of the performance of the channel are determined from $h(t)$ or $\mathcal{H}(f)$:

- The *attenuation* of a steady (unmodulated) input optical power is determined by the transfer function $\mathcal{H}(f)$ at $f = 0$. Since $\mathcal{H}(f)$ is the Fourier transform of $h(t)$, $\mathcal{H}(0) = \int h(t)\,dt$ is the area under $h(t)$.
- The *response time* σ_τ is the width of $h(t)$. It limits the shortest time at which adjacent pulses may be spaced without significantly overlapping.
- The *bandwidth* σ_f (Hz) is the width of $|\mathcal{H}(f)|$. It serves as a measure of the maximum rate at which the input power may be modulated without significant increase of the attenuation. Since $\mathcal{H}(f)$ and $h(t)$ are related by a Fourier transform, the bandwidth σ_f is inversely proportional to the response time σ_τ.

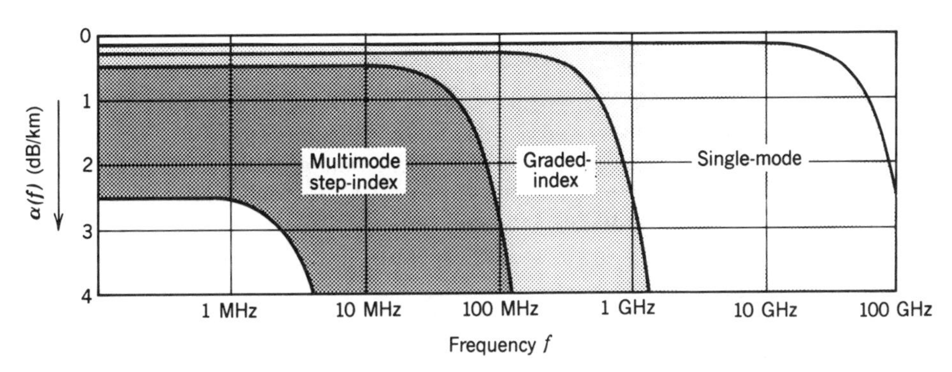

Figure 22.1-4 Typical attenuation coefficients (dB/km) as a function of the modulation frequency f for transmission through different types of optical fibers at various wavelengths. A wave whose power is modulated at frequency f is attenuated by $\boldsymbol{\alpha}(f)L$ dB upon propagation a distance L km. The unmodulated wave is attenuated at a rate $\boldsymbol{\alpha} = \boldsymbol{\alpha}(0)$ dB/km, where $\boldsymbol{\alpha}$ is the attenuation coefficient shown in Fig. 22.1-2.

The coefficient of proportionality depends on the actual profile of $h(t)$ (see Appendix A, Sec. A.2). We use the relation

$$\sigma_f = \frac{1}{2\pi\sigma_\tau}, \tag{22.1-5}$$

Fiber Bandwidth

for purposes of illustration.

The impulse-response function and the transfer function of the optical fiber depend on material attenuation, material and waveguide dispersion, and modal dispersion in the multimode case. The relative contribution of each of these factors depends on the type of fiber: step-index or graded-index, and multimode or single-mode, as illustrated in Fig. 22.1-1 (see also Fig. 8.3-8) and Fig. 22.1-4.

Examples

- In a *multimode step-index fiber* the impulse-response function is a sequence of pulses centered at the mode delay times $\tau_q = L/v_q$, $q = 1, \dots, M$, where v_q is the group velocity of mode q and M is the number of modes (see Fig. 8.3-7). The largest delay difference is $2\sigma_\tau = \tau_{\max} - \tau_{\min}$ where σ_τ is given by (22.1-1). The widths of these pulses are determined by material and waveguide dispersion and are usually much smaller than the delay difference $\tau_{\max} - \tau_{\min}$. A multimode step-index glass fiber with $n_1 = 1.46$ and fractional refractive index difference $\Delta = 0.01$, for example, has a response time $\sigma_\tau/L \approx \Delta/2c_1 \approx 24$ ns/km, corresponding to a bandwidth $\sigma_f L \approx L/2\pi\sigma_\tau \approx 6.5$ MHz-km. For a fiber of length $L = 10$ km, $\sigma_\tau \approx 240$ ns and $\sigma_f \approx 650$ kHz. In a 100-km fiber, an impulse spreads to a width of 2.4 μs and the bandwidth drops to 65 kHz.
- The response time of a multimode *graded-index fiber* with an optimal refractive-index profile, $n_1 = 1.46$, and $\Delta = 0.01$ under ideal conditions is, from (22.1-2), $\sigma_\tau/L \approx \Delta^2/4c_1 \approx 122$ ps/km. This corresponds to a bandwidth of 1.3 GHz-km.

Under these conditions, however, material dispersion may become important, depending on the spectral linewidth of the source.

- For a *single-mode fiber* with a light source of spectral linewidth $\sigma_\lambda = 1$ nm (from a typical single-mode laser) and a fiber dispersion coefficient $D_\lambda = 1$ ps/km-nm (for operation near $\lambda_o = 1.3$ μm), the response time given by (22.1-3) is $\sigma_\tau/L = 1$ ps/km, corresponding to a bandwidth $\sigma_f = 159$ GHz-km. A fiber of length 100 km has a response time 100 ps and bandwidth ≈ 1.6 GHz.

Advanced Materials

Several materials, with attenuation coefficients far smaller than that of silica glass, are being used in experimental optical systems in the mid-infrared region. These include heavy-metal fluoride glasses, halide-containing crystals, and chalcogenide glasses. For these materials, the infrared absorption band is located further in the infrared than in silica glass so that mid-infrared operation, with its attendant reduced Rayleigh scattering (which decreases as $1/\lambda_o^4$), is possible. Attenuations as small as 0.001 dB/km are expected to be achievable with fluoride-glass fibers operating at wavelengths in the 2 to 4 μm band. If these extremely low-loss materials are economically made into fibers, and if suitable semiconductor light sources are perfected for room-temperature operation in the mid-infrared band, repeaterless transmission over distances of several thousand, instead of hundreds, of kilometers would become routine.

Fiber Amplifiers

Erbium-doped silica fibers, serving as laser amplifiers (see Sec. 13.2C), are becoming increasingly important components of 1.55-μm fiber-optic communication systems. These devices offer high-gain amplification (30 to 45 dB), with low noise, near the wavelength of lowest loss in silica glass. They are pumped by InGaAsP diode lasers (usually at 1.48 μm), and exhibit low insertion loss (< 0.5 dB) and polarization insensitivity. They are usually operated in the saturated regime and exhibit minimal crosstalk between different signals that are simutaneously transmitted through them.

An Er^{3+}-doped fiber amplifier may be used as an optical-power amplifier placed directly at the output of the source laser, or as an optical preamplifier at the photodetector input (or both). It can also serve as an all-optical repeater, replacing the electronic repeaters that provide reshaping, retiming, and regeneration of the bits (e.g., those used in current long-haul undersea fiber-optic systems). All-optical repeaters are advantageous in that they offer increased gain and bandwidth, insensitivity to bit rate, and the ability to simultaneous amplify multiple optical channels.

Nonlinear Optical Properties of Fibers

At high levels of power (tens of milliwatts), optical fibers exhibit nonlinear properties, which have a number of undesirable effects such as an increase of the pulse spreading in single-mode fibers, crosstalk between counter-propagating waves used in two-way communications, and crosstalk between waves of different wavelengths used in wavelength-division multiplexing. However, the nonlinear properties of fibers may be harnessed for useful applications. Nonlinear dispersion (dependence of the phase velocity on the intensity) may be adjusted to compensate for chromatic dispersion in the fiber. The result is spreadless pulses known as optical solitons (see Sec. 19.8). The gain provided by a fiber amplifier can be used to compensate for the fiber attenuation so that ideally the pulses suffer no attenuation and no spreading. Nonlinear interactions can also be used to provide gain, but the properties of such amplifiers are generally inferior to those of laser amplifiers such as Er^{3+}:silica fiber.

B. Sources for Optical Transmitters

The basic requirements for the light sources used in optical communication systems depend on the nature of the intended application (long-haul communication, local-area network, etc.). The main features are:

- *Power.* The source power must be sufficiently high so that after transmission through the fiber the received signal is detectable with the required accuracy.
- *Speed.* It must be possible to modulate the source power at the desired rate.
- *Linewidth.* The source must have a narrow spectral linewidth so that the effect of chromatic dispersion in the fiber is minimized.
- *Noise.* The source must be free of random fluctuations. This requirement is particularly strict for coherent communication systems.
- Other features include ruggedness, insensitivity to environmental changes such as temperature, reliability, low cost, and long lifetime.

Both light-emitting diodes (LEDs) and laser diodes are used as sources in fiber-optic communication systems. These devices are discussed in Chap. 16.

Laser diodes have the advantages of high power (tens of mW), high speeds (in the GHz region), and narrow spectral width. However, they are sensitive to temperature variations. Multimode diode lasers suffer from partition noise, i.e., random distribution of the laser power among the modes. When combined with chromatic dispersion in the fiber, this leads to random intensity fluctuations and reshaping of the transmitted pulses. Laser diodes also suffer from frequency chirping, i.e., variation of the laser frequency as the optical power is modulated. Chirping results from changes of the refractive index that accompany changes of the charge-carrier concentrations as the injected current is altered. Significant advances in semiconductor laser technology in recent years have resulted in many improvements and in considerable increase of their reliability and lifetime.

Light-emitting diodes are fabricated in two basic structures: surface emitting and edge emitting. *Surface-emitting diodes* have the advantages of ruggedness, reliability, lower cost, long lifetime, and simplicity of design. However, their basic limitation is their relatively broader linewidth (more than 100 nm in the band 1.3 to 1.6 μm). When operated at their maximum power, modulation frequencies up to 100 Mb/s are possible, but higher speeds (up to 500 Mb/s) can only be achieved at reduced powers. The *edge-emitting diode* has a structure similar to the diode laser (with the reflectors removed). It produces more power output with relatively narrower spectral linewidth, at the expense of complexity.

Sources at 0.87 μm

AlGaAs light-emitting diodes and AlGaAs/GaAs double-heterostructure and quantum-well laser diodes have been used at this wavelength. Surface-emitting LEDs are used extensively.

Sources at 1.3 and 1.55 μm

InGaAsP LEDs have been used in this band with moderate speeds and powers. Single-mode systems make use of InGaAsP/InP double-heterostructure lasers together with single-mode fibers. The requirement for a narrow spectral linewidth is not as crucial at 1.3 μm since material dispersion is minimal. At 1.55 μm, however, it is important to use sources with narrow linewidths because of the presence of material dispersion. A number of technologies are available for providing single-longitudinal-mode lasers (single-frequency lasers) that are stable at high speeds of modulation (see

Sec. 16.3E). These include external-cavity lasers, distributed feedback (DFB) and distributed Bragg-reflector (DBR) lasers capable of providing spectral linewidths of 5 to 100 MHz at a few mW of output power with modulation rates exceeding 20 GHz, and cleaved-coupled-cavity (C^3) lasers which promise linewidths as low as 1 MHz (but are subject to thermal drift).

DFB lasers are probably the most commonly used. Current modulation can be employed since the frequency chirp can be made sufficiently small. DFB lasers with multiple sections and/or multiple electrodes are under development; these should provide further improvements in performance. Quantum-well lasers, in particular InGaAs strained-layer quantum-well lasers (see Sec. 16.3G), are highly promising. These devices offer lower thresholds and larger bandwidths than their lattice-matched cousins (theoretical calculations show that thresholds as low as 50 A/cm^2, and bandwidths as high as 100 GHz, are possible). The prospects for quantum-wire and quantum-dot lasers (see Sec. 15.1G) lie further in the future.

Sources at Longer Wavelengths

Interest in wavelengths longer than 1.55 μm is engendered by the development of low-loss fibers in the 2- to 4-μm wavelength band. Laser diodes that can be operated at room temperature at these wavelengths are being developed. Double-heterostructure InGaAsSb/AlGaAsSb lasers (lattice matched to a GaSb substrate), as an example, can be operated at $\lambda_o = 2.27$ μm at $T = 300$ K (so far only in the pulsed mode, however), with a threshold current density ≈ 1500 A/cm^2, differential quantum efficiency ≈ 0.5, and output power ≈ 2 W. Emission wavelengths from 1.8 to 4.4 μm can potentially be obtained for the range of InGaAsSb compositions that can be lattice matched to GaSb.

C. Detectors for Optical Receivers

A comprehensive discussion of semiconductor photon detectors is provided in Chap. 17. Two types of detectors are commonly used in optical communication systems: the *p-i-n* photodiode and the avalanche photodiode (APD). The APD has the advantage of providing gain before the first electronic amplification stage in the receiver, thereby reducing the detrimental effects of circuit noise. However, the gain mechanism itself introduces noise and has a finite response time, which may reduce the bandwidth of the receiver. Furthermore, APDs require a high-voltage supply and more complicated circuitry to compensate for their sensitivity to temperature fluctuations. The signal-to-noise ratio and the sensitivity of receivers using *p-i-n* photodiodes and APDs are discussed in Secs. 17.5 and 22.4.

Detectors at 0.87 μm

Silicon *p-i-n* photodiodes and APDs are used at these wavelengths. In state-of-the-art preamplifiers, silicon APDs enjoy a 10-to-15-dB sensitivity advantage over silicon *p-i-n* photodiodes because their internal gain makes the noise of the preamplifier relatively less important. The sensitivity of Si APDs at bit rates up to several hundred Mb/s corresponds to about 100 photons/bit. (For a discussion of receiver sensitivity, see Sec. 22.4.)

Detectors at 1.3 and 1.55 μm

Silicon is not usable in this region because its bandgap is greater than the photon energy. Germanium and InGaAs *p-i-n* photodiodes are both used; InGaAs is preferred because it has greater thermal stability and lower dark noise. Typical InGaAs *p-i-n* photodiodes have quantum efficiencies ranging from 0.5 to 0.9, responsivities ≈ 1 A/W, and response times that are in the tens of ps (corresponding to bandwidths up to

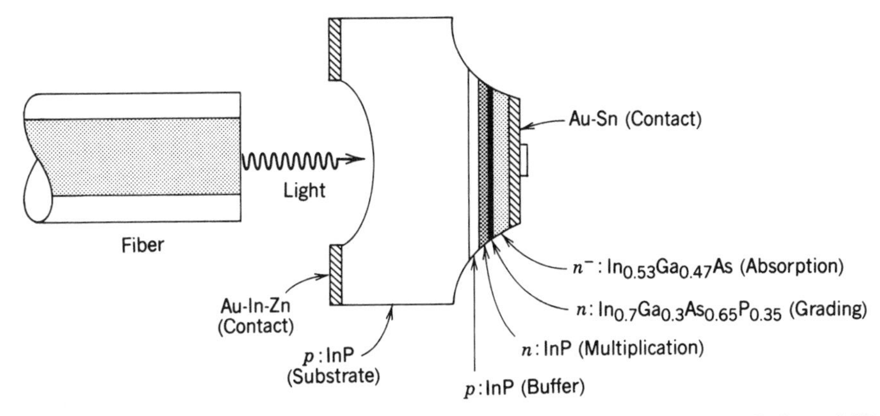

Figure 22.1-5 Structure of an SAGM APD. (Adapted from J. C. Campbell, A. G. Dentai, W. S. Holden, and B. L. Kasper, High-Performance Avalanche Photodiode with Separate Absorption, 'Grading', and Multiplication Regions, *Electronics Letters*, vol. 19, pp. 818–820, 1983.)

60 GHz). Some of these devices make use of waveguide structures. Schottky-barrier photodiodes are faster; their response times are in the ps regime, corresponding to bandwidths $\approx$ 100 GHz.

The development of low-noise APDs (for applications such as fiber-optic communications) has been a challenge. InGaAs APDs operating at speeds $\approx$ 2 Gb/s are widely available. However since the ionization ratio $\mathscr{k}$ is near unity, the gain noise is large. Furthermore, like all narrow bandgap materials, InGaAs suffers from large tunneling leakage currents when subjected to strong electric fields.

A solution to this latter problem makes use of a heterostructure of two materials—a small gap material for the absorption region, and a large-gap material for the multiplication region. Fig. 22.1-5 illustrates an SAGM (separate absorption, grading, multiplication) APD in which the absorption takes place in InGaAs and the multiplication in InP. The InGaAsP grading layer provides a smooth transition for the valence band edge which minimizes hole trapping and shortens the response time of the device. Holes multiply in this device. Quantum efficiencies are in the range 0.75 to 0.9, bandwidths extend up to $\approx$ 10 GHz, and gain-bandwidth products are as high as $\approx$ 75 GHz.

At longer wavelengths, junction photodiodes fabricated from II–VI materials (e.g., HgCdTe) and IV–VI materials (e.g., PbSnTe) are useful.

D. Fiber-Optic Systems

The various operating wavelengths and types of fibers, light sources, and detectors that may be used for building an optical link offer many possible combinations, some of which are summarized in Table 22.1-2.

Progress in the implementation of fiber-optic systems has generally followed a downward path along each of the columns of this table, toward longer wavelengths: from multimode to single-mode fibers, from LEDs to lasers, and from photodiodes to APDs. Appropriate materials for the longer wavelengths (e.g., quaternary sources and detectors) had to be developed to make this progress possible. Although there are many possible combinations of the different types of fibers, sources, and detectors, any number of which may be appropriate for certain applications, three systems are particularly noted:

System 1: Multimode Fibers at 0.87 μm. This is the early technology of the 1970s. Fibers are either step-index or graded-index. The light source is either an LED or a laser (AlGaAs). Both silicon *p-i-n* and APD photodiodes are used. The

TABLE 22.1-2 Operating Wavelengths and Frequently Used Components in Fiber-Optic Links

Wavelength λ_o (μm)	Fiber	Source		Detector	
0.87	Multimode step-index				Si
		LED	AlGaAs	*p-i-n*	
1.3	Multimode graded-index				Ge
		Laser	InGaAsP	APD	
1.55	Single-mode				InGaAs

performance of this system is limited by the fiber's high attenuation and modal dispersion.

System 2: Single-Mode Fibers at 1.3 μm. The move to single-mode fibers and a wavelength where material dispersion is minimal led to a substantial improvement in performance, limited by fiber attenuation. InGaAsP lasers are used with either InGaAs *p-i-n* or APD photodetectors (or Ge APDs).

System 3: Single-Mode Fibers at 1.55 μm. At this wavelength the fiber has its lowest attenuation. Performance is limited by material dispersion, which is reduced by the use of single-frequency lasers (InGaAsP).

These three systems, which are often referred to as the first three generations of fiber-optic systems, are used as examples in Sec. 22.3 and estimates of their expected performance are provided.

Most systems currently being installed belong to the third generation. As an example, the AT&T TAT-9 transatlantic fiber-optic cable (see page 874) makes use of single-mode fibers at 1.55 μm and low-chirp InGaAsP DFB single-frequency lasers. Information is transmitted at 560 Mb/s per fiber pair; some 80,000 simultaneous voice-communication channels are carried the approximately 6000 km from the U.S. and Canada to the U.K., France, and Spain. Repeaters, which are powered by high voltage sent along the length of the cable, are spaced more than 100 km apart.

Third-generation technology has been extended in a number of directions, and systems currently under development will incorporate many of the advances achieved in the laboratory. One relatively recent development of substantial significance is the Er^{3+}:silica-fiber amplifier (see Secs. 13.2C and 22.1A). This device will have a dramatic impact on the configuration of new systems. AT&T and KDD in Japan, for example, have joined together in the development of a transpacific fiber-optic link that will use fiber-amplifier repeaters spaced ≈ 40 km apart to carry some 600,000 simultaneous voice-communication channels. This is a dramatic improvement over the 80,000 simultaneous conversations supported by the electronically repeatered TAT-9 transatlantic cable put into service in 1991.

Optical soliton transmission is another area of high current interest and substantial promise. Solitons are short (typically 1 to 50 ps) optical pulses that can travel through long lengths of optical fiber without changing the shape of their pulse envelope. As discussed in Sec. 19.8, the effects of fiber dispersion and nonlinear self-phase modulation (arising, for example, from the optical Kerr effect) precisely cancel each other, so that the pulses act as if they were traveling through a linear nondispersive medium. Erbium-doped fiber amplifiers can be effectively used in conjunction with soliton transmission to overcome absorption and scattering losses. Prototype systems have already been operated at several Gb/s over fiber lengths in excess of 12,000 km. Soliton transmission at Tb/s rates is in the offing.

All of the systems described above make use of **direct detection**, in which only the signal light illuminates the photodetector. Fourth-generation systems make use of **coherent detection** (see Sec. 22.5), in which a locally generated source of light (the local oscillator) illuminates the photodetector along with the signal. Erbium-doped fiber amplifiers are also useful in conjunction with heterodyne systems. The use of coherent detection in a fiber-optic communication system improves system performance; however, this comes at the expense of increased complexity. As a result, the commercial implementation of coherent systems has lagged behind that of direct-detection systems.

22.2 MODULATION, MULTIPLEXING, AND COUPLING

A communication system (Fig. 22.2.1) is a link between two points in which a physical variable is *modulated* at one point and observed at the other point. In optical communication systems, this variable may be the optical intensity, field amplitude, frequency, phase, or polarization. To transmit more than one message on the same link, the messages may be marked by some physical attribute that identifies them at the receiver. This scheme is called *multiplexing*. A communication network is a link between multiple points. Messages are transmitted between the different points by a system of *couplers* and *switches* that route the messages to the desired locations. Modulation, multiplexing, coupling, and switching are therefore important aspects of communication systems. This section is a brief introduction to modulation, multiplexing, and coupling in fiber-optic communication systems. Photonic switches are considered in Chap. 21.

A. Modulation

Optical communication systems are classified in accordance with the optical variable that is modulated by the message:

Field Modulation. The optical field may serve as a carrier of very high frequency (2×10^{14} Hz at $\lambda_o = 1.5$ μm, for example). The amplitude, phase, or frequency may be modulated, much as the amplitude, phase, or frequency of electromagnetic fields of lower frequencies (such as radio waves) are varied in amplitude modulation (AM), phase modulation (PM), and frequency modulation (FM) systems (Fig. 22.2-2). Because of the extremely high frequency of the optical carrier, a very wide spectral band is available, and large amounts of information can, in principle, be transmitted.

Intensity Modulation. The optical intensity (or power) may be varied in accordance with a modulation rule by means of which the signal is coded (direct proportionality, for example, as illustrated in Fig. 22.2-3). The optical field oscillations at

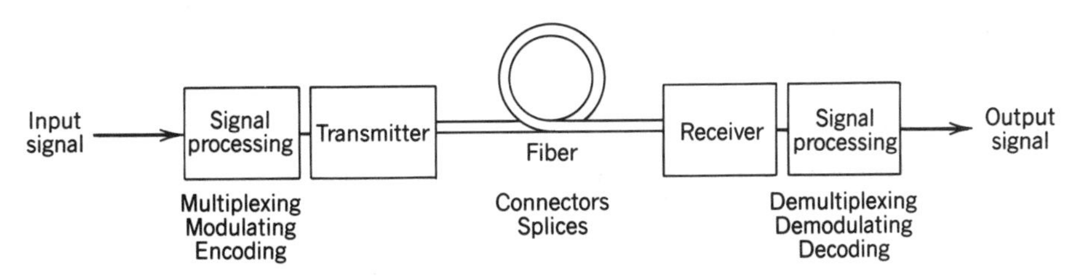

Figure 22.2-1 The fiber-optic communication system.

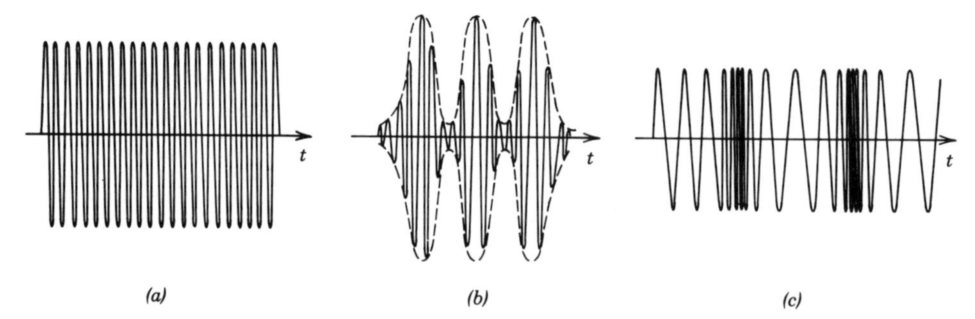

Figure 22.2-2 Amplitude and frequency modulation of the optical field: (*a*) unmodulated field; (*b*) amplitude-modulated field; (*c*) frequency-modulated field.

10^{14} to 10^{16} Hz are unrelated to the operations of modulation and demodulation; only *power* is varied at the transmitter and detected at the receiver. However, the wavelength of light may be used to mark different messages for the purpose of multiplexing.

Although modulation of the optical field is an obvious extension of conventional radio and microwave communication systems to the optical band, it is rather difficult to implement, for several reasons:

- It requires a source whose amplitude, frequency, and phase are stable and free from fluctuations, i.e., a highly coherent laser.
- Direct modulation of the phase or frequency of the laser is usually difficult to implement. An external modulator using the electro-optic effect, for example, may be necessary.
- Because of the assumed high degree of coherence of the source, multimode fibers exhibit large modal noise; a single-mode fiber is therefore necessary.
- Unless a polarization-maintaining fiber is used, a mechanism for monitoring and controlling the polarization is needed.
- The receiver must be capable of measuring the magnitude and phase of the optical field. This is usually accomplished by use of a heterodyne detection system.

Because of the requirement of coherence, optical communication systems using field modulation are called *coherent communication systems.* These systems are discussed in Sec. 22.5.

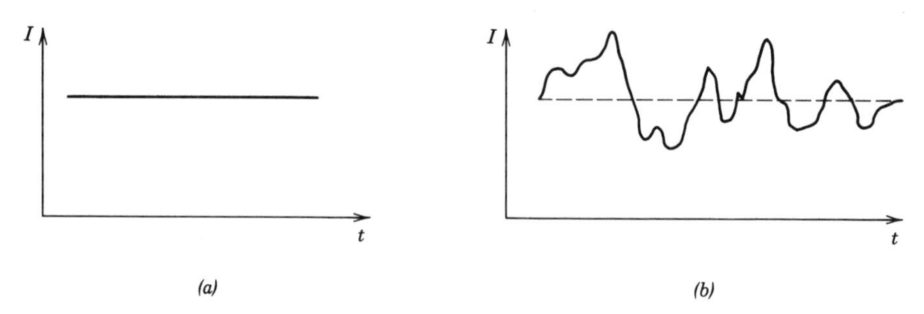

Figure 22.2-3 Intensity modulation: (*a*) unmodulated intensity; (*b*) modulated intensity.

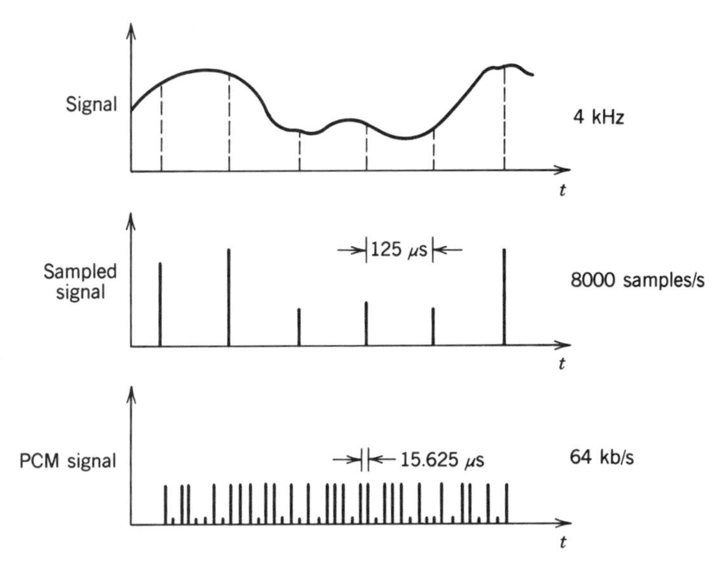

Figure 22.2-4 An example of PCM. A 4-kHz voice signal is sampled at a rate of 8×10^3 samples per second. Each sample is quantized to $2^8 = 256$ levels and represented by 8 bits, so that the signal is a sequence of bits transmitted at a rate of 64 kb/s.

The majority of commercial fiber communication systems at present use intensity modulation. The power of the source is modulated by varying the injected current in an LED or a diode laser. The fiber may be single-mode or multimode and the optical power received is measured by use of a *direct-detection receiver*.

Once the modulation variable is chosen (intensity, frequency, or phase), any of the conventional modulation formats (analog, pulse, or digital) can be used. An important example is **pulse code modulation** (PCM). In PCM the analog signal is sampled periodically at an appropriate rate and the samples are quantized to a discrete finite number of levels, each of which is binary coded and transmitted in the form of a sequence of binary bits, "1" and "0," represented by pulses transmitted within the time interval between two adjacent samples (Fig. 22.2-4).

If intensity modulation is adopted, each bit is represented by the presence or absence of a pulse of light. This type of modulation is called **on–off keying** (OOK). For frequency or phase modulation, the bits are represented by two values of frequency or phase. The modulation is then known as **frequency shift keying** (FSK) or **phase shift keying** (PSK). These modulation schemes are illustrated in Fig. 22.2-5. It is also possible to modulate the intensity of light with a harmonic function serving as a subcarrier whose amplitude, frequency, or phase is modulated by the signal (in the AM, FM, PM, FSK, or PSK format).

B. Multiplexing

Multiplexing is the transmission and retrieval of more than one signal through the same communication link, as illustrated in Fig. 22.2-6. This is usually accomplished by marking each signal with a physical label that is distinguishable at the receiver. Two standard multiplexing systems are in use: frequency-division multiplexing (FDM) and time-division multiplexing (TDM). In FDM, carriers of distinct frequencies are modulated by the different signals. At the receiver, the signals are identified by the use of filters tuned to the carrier frequencies. In TDM, different interleaved time slots are

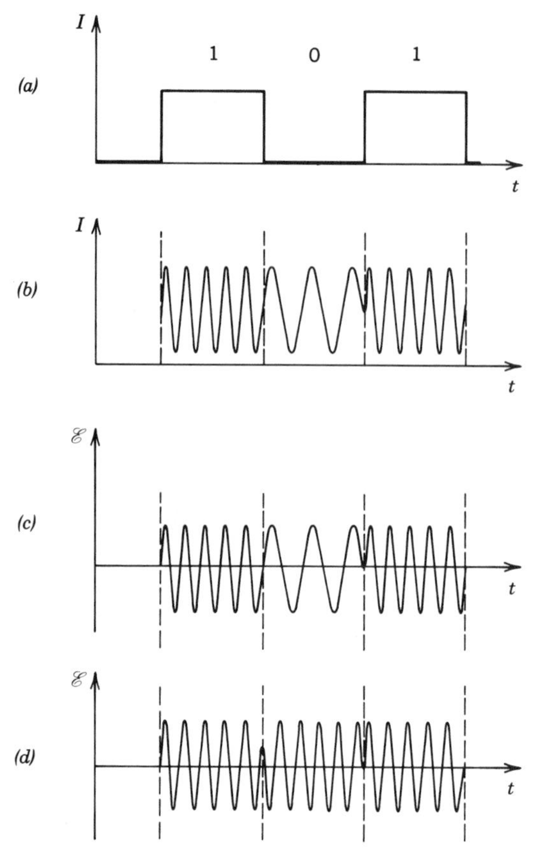

Figure 22.2-5 Examples of binary modulation of light: (*a*) on–off keying intensity modulation (OOK/IM); (*b*) frequency-shift-keying intensity modulation (FSK/IM); (*c*) frequency-shift-keying (FSK) field modulation; (*d*) phase-shift-keying (PSK) field modulation.

allotted to samples of the different signals. The receiver looks for samples of each signal in the appropriate time slots.

In optical communication systems based on intensity modulation, FDM may be implemented by use of subcarriers of different frequencies. The subcarriers are identified at the receiver by use of electronic filters sensitive to these frequencies, as illustrated in Fig. 22.2-7. It is also possible, and more sensible, to use the underlying optical frequency of light as a multiplexing "label" for FDM. When the frequencies of the carriers are widely spaced (say, greater than a few hundred GHz) this form of FDM is usually called **wavelength-division multiplexing** (WDM). A WDM system uses light sources of different wavelengths, each intensity modulated by a different signal. The

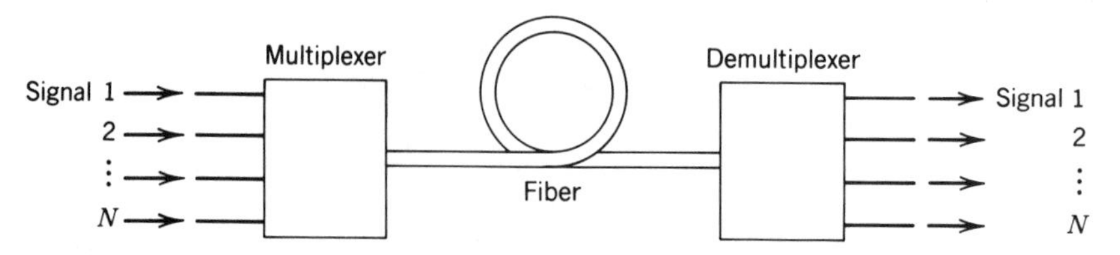

Figure 22.2-6 Transmission of N optical signals through the same fiber by use of multiplexing.

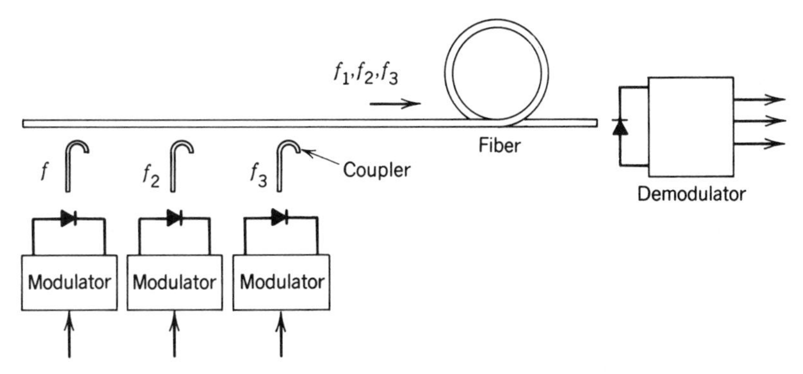

Figure 22.2-7 Frequency-division multiplexing using intensity modulation with subcarriers. Demultiplexing is accomplished by use of electronic filters.

modulated light beams are mixed into the fiber using optical couplers. Demultiplexing is implemented at the receiver end by use of optical (instead of electronic) filters that separate the different wavelengths and direct them to different detectors.

At $\lambda_o = 1.55\ \mu$m, for example, a frequency spacing of $\Delta\nu = 250$ GHz is equivalent to $|\Delta\lambda| = (\lambda_o^2/c_o)|\Delta\nu| = 2$ nm. Thus 10 channels cover a band of 20 nm. Since the carrier frequencies are widely spaced, each channel may be modulated at very high rates without crosstalk. However, from an optics perspective, a 2-nm spectral range is relatively narrow. The spectral linewidth of the light sources must be even narrower and their frequencies must be stable within this narrow spectral range.

Wavelength-division demultiplexers use optical filters to separate the different wavelengths. There are filters based on selective absorption, transmission, or reflection, such as thin-film interference filters. An optical fiber, with the two ends acting as reflectors, can serve as a Fabry–Perot etalon with spectral selectivity (see Sec. 2.5B). Other filters are based on angular dispersion, such as the diffraction grating. Examples of these filters are illustrated in Fig. 22.2-8. Another alternative is the use of hetero-

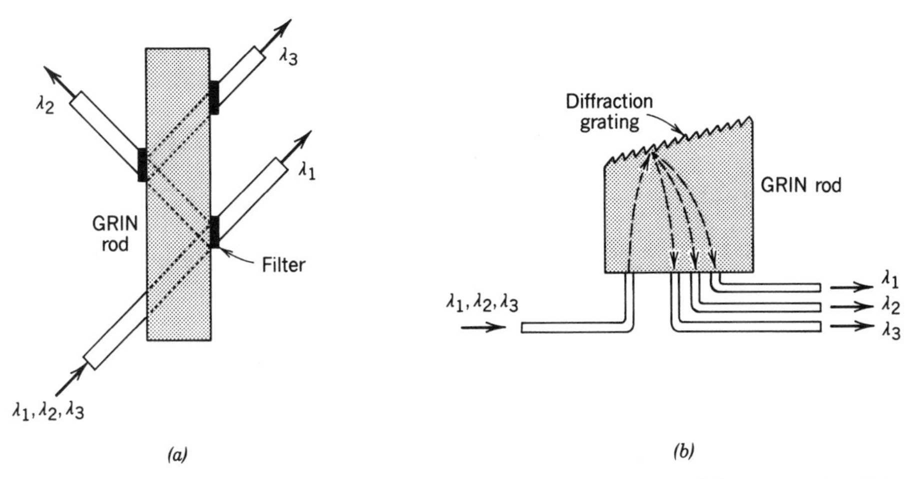

Figure 22.2-8 Wavelength-division demultiplexing using optical filters. (*a*) Each of the dielectric interference filters transmits only a single wavelength and reflects other wavelengths. A graded-index (GRIN) rod (see Sec. 1.3B) guides the waves between the filters. (*b*) A diffraction grating (Sec. 2.4B) separates the different wavelengths into different directions, and a graded-index (GRIN) rod guides the waves to the appropriate fibers.

dyne detection. A wavelength-multiplexed optical signal with carrier frequencies $\nu_1, \nu_2, \ldots$ is mixed with a local oscillator of frequency ν_L and detected. The photocurrent carries the signatures of the different carriers at the beat frequencies $f_1 = \nu_1 - \nu_L$, $f_2 = \nu_2 - \nu_L, \ldots$. These frequencies are then separated using electronic filters (see Sec. 22.5A).

C. Couplers

In addition to the transmitter, the fiber link, and the receiver, a communication system uses couplers and switches which direct the light beams that represent the various signals to their appropriate destinations. *Couplers* always operate on the incoming signals in the same manner. *Switches* are controllable couplers that can be modified by an external command. Photonic switches are described in Chap. 21.

Examples of couplers are shown schematically in Fig. 22.2-9. In the T-coupler, a signal at input point 1 reaches both output points 2 and 3; a signal at either point 2 or point 3 reaches point 1. In the star coupler, the signal at any of the input points reaches all output points. In the four-port directional coupler, a signal at any of the input points 1 or 2 reaches both output points 3 and 4; and a signal coming from any of the output points 3 or 4 in the opposite direction reaches both points 1 and 2. When operated as a switch, the four-port directional coupler is switched between the parallel state (1–3 and 2–4 connections) and the cross state (1–4 and 2–3 connections).

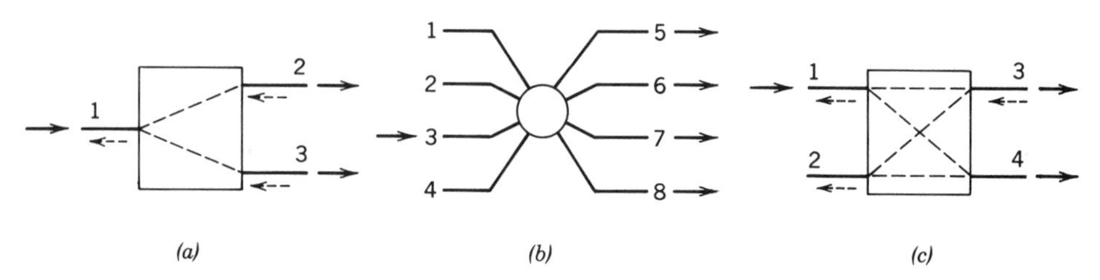

Figure 22.2-9 Examples of couplers: (a) T coupler; (b) star coupler; (c) directional coupler.

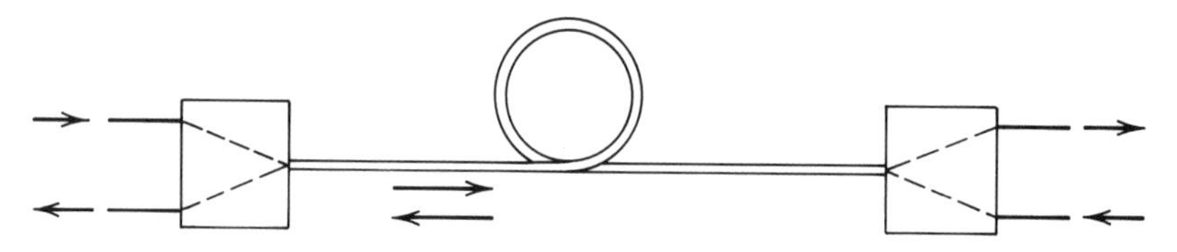

Figure 22.2-10 A duplex (two-way) communication system using two T couplers.

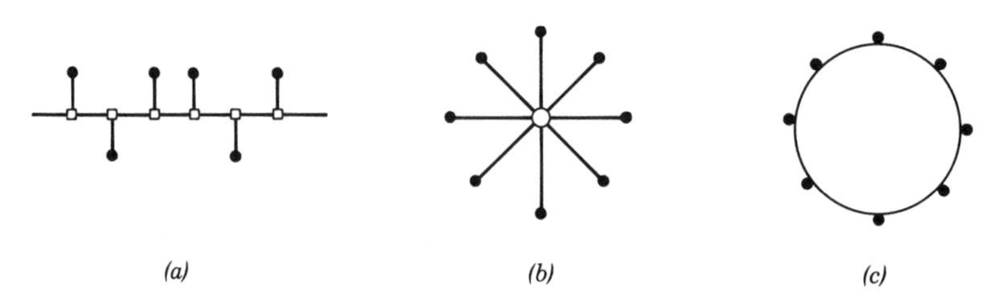

Figure 22.2-11 Examples of communication networks using couplers: (a) bus network; (b) star network; (c) ring network.

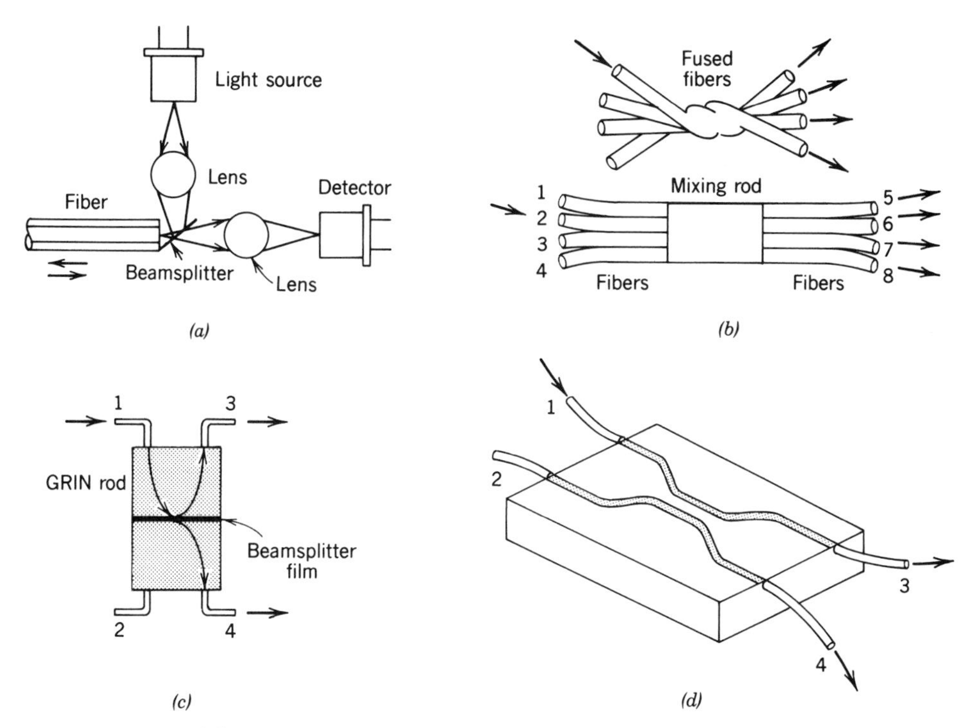

Figure 22.2-12 (*a*) A T coupler at one end of a duplex optical communication link using a beamsplitter and ball lenses (see Problem 1.2-4). (*b*) A star coupler using fused fibers and another using a mixing rod, a slab of glass through which light from one fiber is dispersed to reach all other fibers. (*c*) A four-port directional coupler using two GRIN-rod lenses separated by a beamsplitter film. (*d*) An integrated-optic four-port directional coupler (see Secs. 7.4B and 21.1B).

An important example illustrating the need for T-couplers is the duplex communication system used in two-way communications, as shown in Fig. 22.2-10. Couplers are essential to communication networks, as illustrated in Fig. 22.2-11. Optical couplers can be constructed by use of miniature beamsplitters, lenses, graded-index rods, prisms, filters, and gratings compatible with the small size of the optical beams transmitted by fibers. This new technology is called **micro-optics**. Integrated-optic devices (see Secs. 7.4B and 21.1B) may also be used as couplers; these are more suitable for single-mode guided light. Figure 22.2-12 shows some examples of optical couplers.

22.3 SYSTEM PERFORMANCE

In this section the basic concepts of design and performance analysis of fiber-optic communication systems are introduced using two examples: an on–off keying digital system and an analog system, both using intensity modulation.

A. Digital Communication System

Consider a fiber-optic communication system using an LED or a laser diode of power P_s (mW) and spectral width σ_λ (nm); an optical fiber of attenuation coefficient $\boldsymbol{\alpha}$ (dB/km), response time σ_τ/L (ns/km), and length L (km); and a *p-i-n* or APD

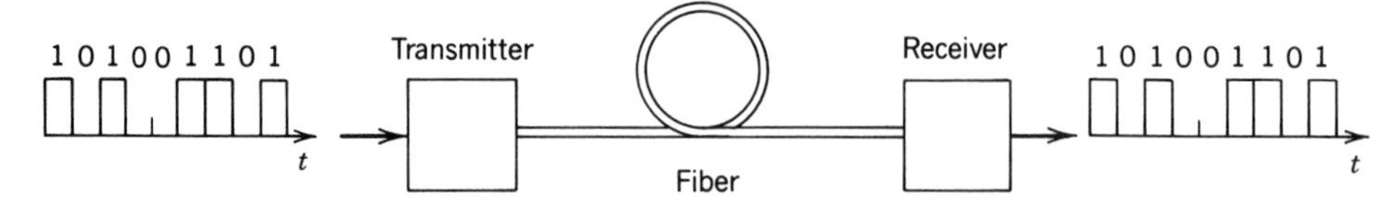

Figure 22.3-1 A binary on–off keying digital optical fiber link.

photodetector. The intensity of light is modulated in an on–off keying (OOK) system by turning the power on and off to represent bits "1" and "0," as illustrated in Fig. 22.3-1. The link transmits B_0 bits/s. Several of these links may be cascaded to form a longer link. An intermediate receiver-transmitter unit connecting two adjacent links is called a **regenerator** or **repeater**. Here we are concerned only with the design of a single link.

The purpose of the design is to determine the maximum distance L over which the link can transmit B_0 bits/s with a rate of errors smaller than a prescribed rate. Clearly, L decreases with increase of B_0. An equivalent problem is to determine the maximum bit rate B_0 a link of length L can transmit with an error rate not exceeding the allowable limit. The maximum bit-rate–distance product LB_0 serves as a single number that describes the capability of the link. We shall determine the typical dependence of L on B_0, and derive expressions for the maximum bit-rate–distance product LB_0 for various types of fibers.

The Bit Error Rate

The performance of a digital communication system is measured by the **probability of error** per bit, which we refer to as the **bit error rate** (BER). If p_1 is the probability of mistaking "1" for "0," and p_0 is the probability of mistaking "0" for "1," and if the two bits are equally likely to be transmitted, then BER $= \frac{1}{2}p_1 + \frac{1}{2}p_0$. A typical acceptable BER is 10^{-9} (i.e., an average of one error every 10^9 bits).

Receiver Sensitivity

The sensitivity of the receiver is defined as the minimum number of photons (or the corresponding optical energy) per bit necessary to guarantee that the rate of error (BER) is smaller than a prescribed rate (usually 10^{-9}). Errors occur because of the randomness of the number of photoelectrons detected during each bit, as well as the noise in the receiver circuit itself. The sensitivity of receivers using different photodetectors will be determined in Sec. 22.4. It will be shown that when the light source is a stabilized laser, the detector has unity quantum efficiency, and the receiver circuit is noise-free, an average of at least $\bar{n}_0 = 10$ photons per bit is required to ensure that BER $\leq 10^{-9}$. Therefore, *the sensitivity of the ideal receiver is 10 photons/bit*. This means that bit "1" should carry an average of at least 20 photons, since bit "0" carries no photons. In the presence of other forms of noise, the sensitivity may be significantly degraded.

A sensitivity of $\bar{n}_0$ photons corresponds to an optical energy $h\nu\bar{n}_0$ per bit and an optical power $P_r = (h\nu\bar{n}_0)/(1/B_0)$,

$$P_r = h\nu\bar{n}_0 B_0, \qquad (22.3\text{-}1)$$

which is proportional to the bit rate B_0. As the bit rate increases, a higher optical power is required to maintain the number of photons/bit (and therefore the BER) constant. It will be shown in Sec. 22.4 that when circuit noise is important, the receiver sensitivity $\bar{n}_0$ depends on the receiver bandwidth (i.e., on the data rate B_0). This behavior complicates the design problem. For simplicity, we shall assume here that the receiver sensitivity (photons per bit) is independent of B_0. For the purposes of illustration we shall use the nominal receiver sensitivities of $\bar{n}_0 = 300$ photons per bit for receivers operating at $\lambda_o = 0.87\ \mu$m and 1.3 μm, and $\bar{n}_0 = 1000$ photons per bit for receivers operating at $\lambda_o = 1.55\ \mu$m.

Design Strategy

Once we know the minimum power required at the receiver, the power of the source, and the fiber attenuation per kilometer, a power budget may be prepared from which the maximum fiber length is determined. We must also prepare a budget for the pulse spreading that results from dispersion in the fiber. If the width σ_τ of the received pulses exceeds the bit time interval $1/B_0$, adjacent pulses overlap and cause *intersymbol interference*, which increases the error rates. There are therefore two conditions for the acceptable operation of the link:

- The received power must be at least equal to the receiver power sensitivity P_r. A margin of 6 dB above P_r is usually specified.
- The received pulse width σ_τ must not exceed a prescribed fraction of the bit time interval $1/B_0$.

If the bit rate B_0 is fixed and the link length L is increased, two situations leading to performance degradation may occur: The received power becomes smaller than the receiver power sensitivity P_r, or the received pulses become wider than the bit time $1/B_0$. If the former situation occurs first, the link is said to be *attenuation limited.* If the latter occurs first, the link is said to be *dispersion limited.*

Attenuation-Limited Performance

Attenuation-limited performance is assessed by preparing a power budget. Since fiber attenuation is measured in dB units, it is convenient to also measure power in dB units. Using 1 mW as a reference, dBm units are defined by

$$\boxed{\mathscr{P} = 10 \log_{10} P,} \qquad P \text{ in mW}; \quad \mathscr{P} \text{ in dBm}.$$

For example, $P = 0.1$ mW, 1 mW, and 10 mW correspond to $\mathscr{P} = -10$ dBm, 0 dBm, and 10 dBm, respectively. In these logarithmic units, power losses are additive instead of multiplicative.

If $\mathscr{P}_s$ is the power of the source (dBm), $\boldsymbol{\alpha}$ is the fiber loss in dB/km, $\mathscr{P}_c$ is the splicing and coupling loss (dB), and L is the maximum fiber length such that the power delivered to the receiver is the receiver sensitivity $\mathscr{P}_r$ (dBm), then

$$\mathscr{P}_s - \mathscr{P}_c - \mathscr{P}_m - \boldsymbol{\alpha} L = \mathscr{P}_r \qquad \text{(dB units)}, \tag{22.3-2}$$

where $\mathscr{P}_m$ is a safety margin. The optical power is plotted schematically in Fig. 22.3-2 as a function of the distance from the transmitter.

The receiver power sensitivity $\mathscr{P}_r = 10 \log_{10} P_r$ (dBm) is obtained from (22.3-1),

$$\mathscr{P}_r = 10 \log \frac{\bar{n}_0 h\nu B_0}{10^{-3}} \text{ dBm}. \tag{22.3-3}$$

Thus $\mathscr{P}_r$ increases logarithmically with B_0, and the power budget must be adjusted for each B_0 as illustrated in Fig. 22.3-3.

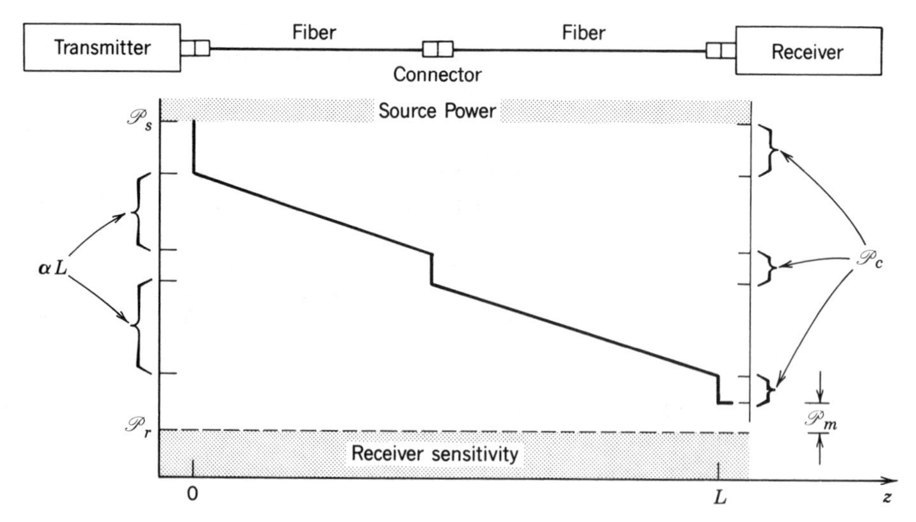

Figure 22.3-2 Power budget of an optical link.

The maximum length of the link is obtained by substituting (22.3-3) into (22.3-2),

$$L = \frac{1}{\alpha}\left(\mathscr{P}_s - \mathscr{P}_c - \mathscr{P}_m - 10\log\frac{\bar{n}_0 h\nu B_0}{10^{-3}}\right), \tag{22.3-4}$$

from which

$$L = L_0 - \frac{10}{\alpha}\log B_0, \tag{22.3-5}$$

Distance versus Bit Rate (Attenuation-Limited Fiber)

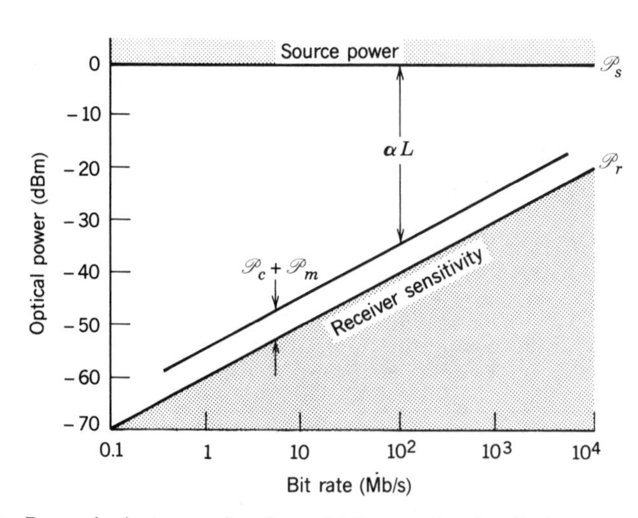

Figure 22.3-3 Power budget as a function of bit rate B_0. As B_0 increases, the power $\mathscr{P}_r$ required at the receiver increases (so that the energy per bit remains constant), and the maximum length L decreases.

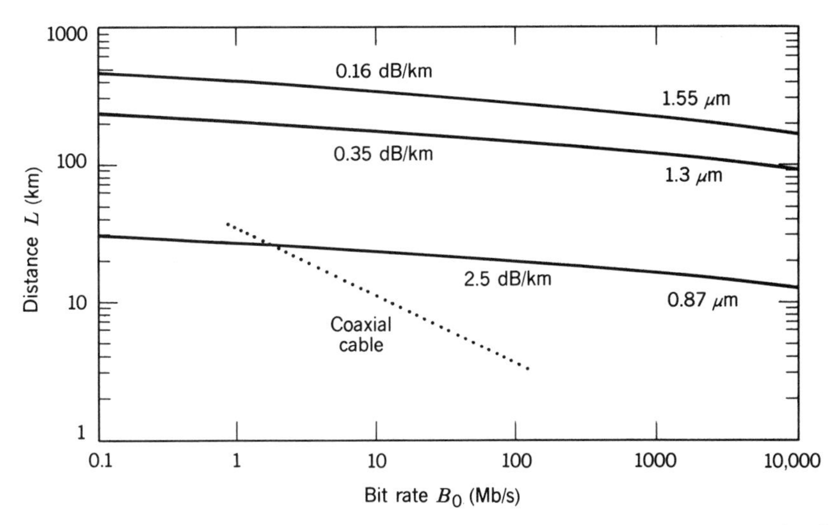

Figure 22.3-4 Maximum fiber length L as a function of bit rate B_0 under attenuation-limited conditions for a fused silica glass fiber operating at wavelengths $\lambda_o = 0.87$, 1.3, and 1.55 μm assuming fiber attenuation coefficients $\boldsymbol{\alpha} = 2.5$, 0.35, and 0.16 dB/km, respectively; source power $P_s = 1$ mW ($\mathscr{P}_s = 0$ dBm); receiver sensitivity $\bar{n}_0 = 300$ photons/bit for receivers operating at 0.87 and 1.3 μm and $\bar{n}_0 = 1000$ for the receiver operating at 1.55 μm; and $P_c = P_m = 0$. For comparison, the L–B_0 relation for a typical coaxial cable is also shown.

where $L_0 = [\mathscr{P}_s - \mathscr{P}_c - \mathscr{P}_m - 30 - 10\log(\bar{n}_0 h\nu)]/\boldsymbol{\alpha}$. The length drops with increase of B_0 at a logarithmic rate with slope $10/\boldsymbol{\alpha}$. Figure 22.3-4 is a plot of this relation for the operating wavelengths 0.87, 1.3, and 1.55 μm.

Dispersion-Limited Performance

The width σ_τ of the received pulse increases with increase of the fiber length L (see Sec. 22.1A). When σ_τ exceeds the bit time interval, $T = 1/B_0$, the performance begins to deteriorate as a result of intersymbol interference. We shall select the maximum allowed width to be one-fourth of the bit-time interval,

$$\sigma_\tau = \frac{T}{4} = \frac{1}{4B_0}. \tag{22.3-6}$$

The choice of the factor $\frac{1}{4}$ is clearly arbitrary and serves only to compare the different types of fibers:

- *Step-Index Fiber.* The width of the received pulse after propagation a distance L in a multimode step-index fiber is governed by modal dispersion. Substituting (22.1-1) into (22.3-6), we obtain the L–B_0 relation

$$\boxed{LB_0 = \frac{c_1}{2\Delta}.} \tag{22.3-7}$$

Bit-Rate – Distance Product (Modal-Dispersion-Limited Step-Index Fiber)

where $c_1 = c_o/n_1$ is the speed of light in the core material and $\Delta = (n_1 - n_2)/n_1$ is the fiber fractional index difference. For $n_1 = 1.46$ and $\Delta = 0.01$, the bit-rate–distance product $LB_0 \approx 10$ km-Mb/s.

- *Graded-Index Fiber.* In a multimode graded-index fiber of optimal (approximately parabolic) refractive index profile, the pulse width is given by (22.1-2). Using (22.3-6), we obtain

$$LB_0 = \frac{c_1}{\Delta^2}. \tag{22.3-8}$$

Bit-Rate – Distance Product (Modal-Dispersion-Limited Graded-Index Fiber)

For $n_1 = 1.46$ and $\Delta = 0.01$, the bit-rate–distance product $LB_0 \approx 2$ km-Gb/s.

- *Single-Mode Fiber.* Assuming that pulse broadening in a single-mode fiber results from material dispersion only (i.e., neglecting waveguide dispersion), then for a source of linewidth σ_λ the width of the received pulse is given by (22.1-3), so that

$$LB_0 = \frac{1}{4|D_\lambda|\sigma_\lambda}, \tag{22.3-9}$$

Bit-Rate – Distance Product (Material-Dispersion-Limited Single-Mode Fiber)

where D_λ is the dispersion coefficient of the fiber material. For operation near $\lambda_o = 1.3\ \mu$m, $|D_\lambda|$ may be as small as 1 ps/km-nm. Assuming that $\sigma_\lambda = 1$ nm (the linewidth of a single-mode laser), the bit-rate–distance product $LB_0 \approx 250$ km-Gb/s. For operation near $\lambda_o = 1.55\ \mu$m, $D_\lambda = 17$ ps/km-nm, and for the same source spectral width $\sigma_\lambda = 1$ nm, $LB_0 \approx 15$ km-Gb/s.

The distance versus bit-rate relations for these dispersion-limited examples are plotted in Fig. 22.3-5.

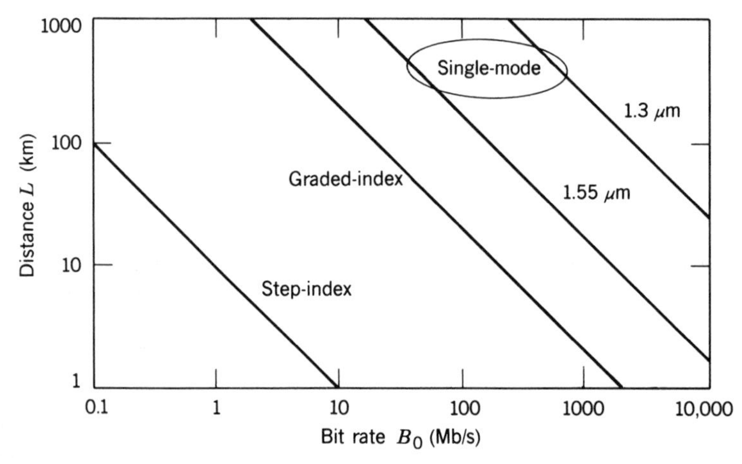

Figure 22.3-5 Dispersion-limited maximum fiber length L as a function of bit rate B_0 for: (*a*) multimode step-index fiber ($n_1 = 1.46$, $\Delta = 0.01$), $LB_0 = 10$ km-Mb/s; (*b*) multimode graded-index fiber with parabolic profile ($n_1 = 1.46$, $\Delta = 0.01$), $LB_0 = 2$ km-Gb/s; (*c*) single-mode fiber limited by material dispersion, operating at 1.3 μm with $|D_\lambda| = 1$ ps/km-nm and $\sigma_\lambda = 1$ nm, $LB_0 = 250$ km-Gb/s; (*d*) single-mode fiber limited by material dispersion, operating at 1.55 μm with $D_\lambda = 17$ ps/km-nm and $\sigma_\lambda = 1$ nm, $LB_0 \approx 15$ km-Gb/s.

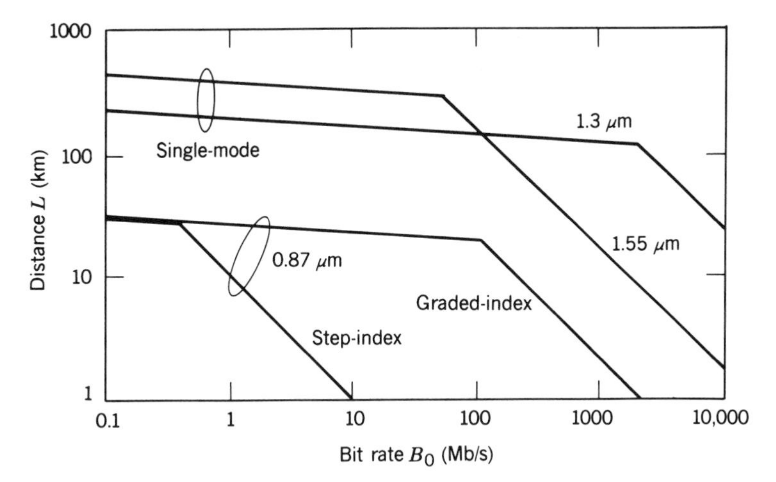

Figure 22.3-6 Maximum distance L versus bit rate B_0 for four examples of fibers. This graph is obtained by superposing the graphs in Figs. 22.3-4 and 22.3-5. Each curve represents the maximum distance L of the link at each bit rate B_0 that satisfies *both* the attenuation and dispersion limits, i.e., guarantees the reception of the required power *and* pulse width at the receiver. At low bit rates, the system is attenuation limited; L drops with B_0 logarithmically. At high bit rates, the system is dispersion limited and L is inversely proportional to B_0.

The attenuation-limited and dispersion-limited bit-rate–distance relations are combined in Fig. 22.3-6 by superposing Figs. 22.3-4 and 22.3-5. These relations describe the performance of three generations of optical fibers operating at $\lambda_o = 0.87$ μm (multimode step-index and graded-index), at 1.3 μm (single-mode), and at 1.55 μm (single-mode), respectively.

In creating these L–B_0 curves, many simplifying assumptions and arbitrary choices have been made. The values obtained should therefore be regarded as only indications of the order of magnitude of the relative performance of the different types of fibers.

The Best Possible Fiber-Optic Communication System

It is instructive to compare the performance of the practical systems shown in Fig. 22.3-7 with the "best" that can be achieved with silica glass fibers. The following assumptions are made:

- The fiber is a single-mode fiber operating at $\lambda_o = 1.55$ μm, where the attenuation coefficient is the absolute minimum $\boldsymbol{\alpha} \approx 0.16$ dB/km.
- The detector is assumed ideal (i.e., photon limited). This corresponds to a receiver sensitivity of 10 photons per bit, instead of 300 or 1000, which were assumed in the previous examples. Using (22.3-5) the attenuation-limited performance may be determined and is shown in Fig. 22.3-7.
- To reduce the material or waveguide dispersion, the spectral linewidth σ_λ of the source must be small. Spectral widths that are a small fraction of 1 nm are obtained with single-frequency lasers. However, an extremely narrow spectral width is incompatible with an extremely short pulse because of the Fourier transform relation between the spectral and temporal distributions. For a pulse of duration $T = 1/B_0$ the Fourier-transform limited spectral width is[†] $\sigma_\nu \approx 1/2T = B_0/2$. Since $\nu = c_o/\lambda_o$, σ_ν is related to σ_λ by $\sigma_\nu = |\partial\nu/\partial\lambda_o|\sigma_\lambda = (c_o/\lambda_o^2)\sigma_\lambda$. The

[†]This is the power-equivalent spectral width, which is defined by (A.2-10) in Appendix A and satisfies (A.2-12).

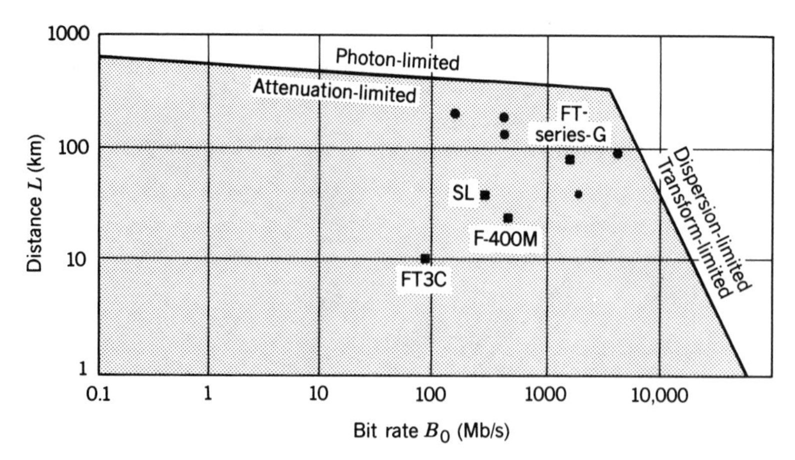

Figure 22.3-7 Distance versus bit rate for a fiber operating at $\lambda_o = 1.55\ \mu$m with attenuation coefficient $\boldsymbol{\alpha} = 0.16$ dB/km and dispersion coefficient $D_\lambda = 17$ ps/km-nm, using an ideal photon-limited receiver with a 10-photon/bit sensitivity and an ideal light source with Fourier-transform-limited spectral width. The squares represent the performance of commercial fiber-optic systems in operation. For example, the AT&T FT-series-G system operates at 1.7 Gb/s with a repeater distance of 90 km at $\lambda_o = 1.55\ \mu$m. The dots represent systems that have been tested in the laboratory. (See, e.g., P. S. Henry, R. A. Linke, and A. H. Gnauck, Introduction to Lightwave Systems, in *Optical Fiber Telecommunications II*, S. E. Miller and I. P. Kaminow, eds., Academic Press, New York, 1988.)

Fourier-transform-limited minimum value of σ_λ is therefore

$$\sigma_\lambda = \frac{\lambda_o^2}{2c_o} B_0, \tag{22.3-10}$$

which is directly proportional to the bit rate B_0. For $B_0 = 10$ Gb/s and $\lambda_o = 1.55$ μm, for example, $\sigma_\lambda = 0.16$ nm. When (22.3-10) is substituted in (22.3-9), we obtain the distance bit-rate relation

$$\boxed{LB_0^2 = \frac{c_o}{2|D_\lambda|\lambda_o^2},} \tag{22.3-11}$$

which is shown in Fig. 22.3-7 for $\lambda_o = 1.55\ \mu$m and $D_\lambda = 17$ ps/km-nm.

- By use of dispersion-shifted fibers it is possible to reduce the overall chromatic dispersion coefficient D at 1.55 μm by a factor of 10, for example. In this case the dispersion-limited line in Fig. 22.3-7 moves to the right to 10 times greater bit rates. However, this comes at the expense of some increase of attenuation, which results in moving the attenuation-limited line downward.

Dispersion as a Power Penalty

The assumption that the maximum acceptable width of the received pulses σ_τ is one-fourth of the bit time $T = 1/B_0$ is rather arbitrary. Wider pulses can in fact be tolerated, provided that the signal-to-noise ratio is improved by increasing the received power beyond the receiver sensitivity. The required increase, denoted $\mathscr{P}_{\text{ISI}}$ and called the intersymbol interference power penalty or the **dispersion power penalty**, is deter-

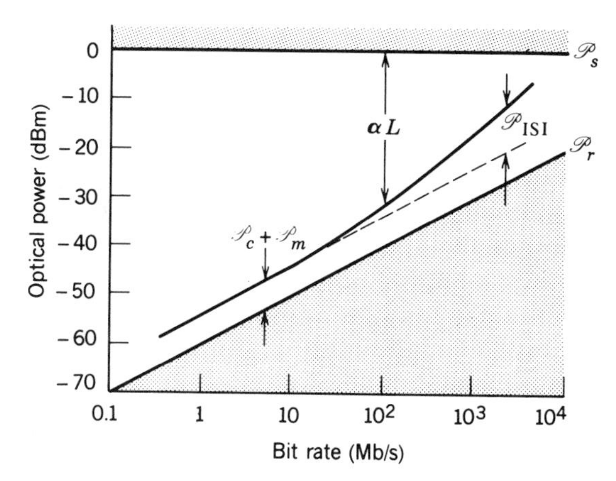

Figure 22.3-8 Power budget as a function of bit rate. $\mathcal{P}_s$ is the source power, $\mathcal{P}_c$ is the power loss at the couplers, $\mathcal{P}_r$ is the receiver sensitivity, $\mathcal{P}_m$ is the power safety margin, and $\mathcal{P}_{ISI}$ is the dispersion power penalty.

mined by ensuring that the error rate reaches the limit BER $= 10^{-9}$ when the received power is $\mathcal{P}_r + \mathcal{P}_{ISI}$ and the widths of the received pulses are σ_τ.

A rough estimate of $\mathcal{P}_{ISI}$ may be obtained by determining the attenuation coefficient $\boldsymbol{\alpha}(f)$ [see (22.1-4)] of the fiber at the modulation frequency $f = B_0/2$, which is the frequency of a periodic pulse train representing the bit sequence 101010. . . . The power penalty is then $\mathcal{P}_{ISI} = [\boldsymbol{\alpha}(f) - \boldsymbol{\alpha}(0)]L$ dB/km. Since σ_τ is the width of the fiber impulse-response function $h(t)$, the width of the transfer function $\mathcal{H}(f)$ is $\sigma_f = 1/2\pi\sigma_\tau$, so that the dispersion-limit condition $\sigma_\tau < 1/4B_0$ is equivalent to $(1/2\pi\sigma_f) < (1/8f)$, or $f < (\pi/4)\sigma_f$. At the prescribed limit $\sigma_\tau = 1/4B_0$, the modulation frequency $f = (\pi/4)\sigma_f$ is well within the fiber bandwidth σ_f so that the penalty is negligible. As σ_τ increases beyond the $1/4B_0$ limit, f eventually exceeds the fiber bandwidth σ_f, whereupon the dispersion power penalty increases sharply. If $h(t)$ is a Gaussian pulse, for example, $\mathcal{H}(f)$ is also Gaussian and its logarithm is proportional to f^2/σ_f^2, so that the dispersion power penalty in dB units is proportional to $(f/\sigma_f)^2$ or to $(\sigma_\tau/T)^2 = (\sigma_\tau B_0)^2$.

By treating dispersion as a power penalty, the attenuation-limited and dispersion-limited analyses are combined into one general design equation (Fig. 22.3-8)

$$\mathcal{P}_s = \mathcal{P}_c + \mathcal{P}_m + \boldsymbol{\alpha}(f)L + \mathcal{P}_r = \mathcal{P}_c + \mathcal{P}_m + \boldsymbol{\alpha}L + \mathcal{P}_{ISI} + \mathcal{P}_r. \quad (22.3\text{-}12)$$

Since $\mathcal{P}_{ISI}$ is a nonlinear function of B_0 and L, and $\mathcal{P}_r$ is a function of $\log B_0$, (22.3-12) is a nonlinear equation relating L to B_0. Its solution gives a smooth curve that joins the attenuation-limited and dispersion-limited curves determined earlier in the limits of small and large B_0, respectively, as illustrated in Fig. 22.3-9.

B. Analog Communication System

An analog fiber-optic communication system using intensity modulation is shown schematically in Fig. 22.3-10. The signal is a continuous function of time representing an audio, video, or data waveform. The power of the light source (usually an LED) is modulated by the signal and guided by the fiber to the receiver, where it is detected and amplified. Under ideal conditions, the original signal is reproduced.

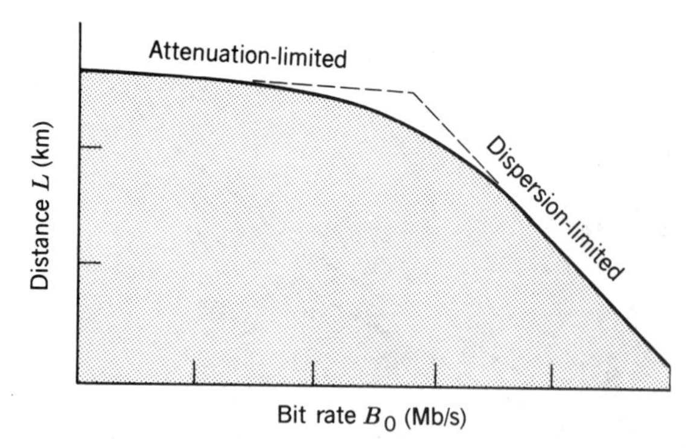

Figure 22.3-9 The L–B_0 relation obtained by treating dispersion as a power penalty.

There are two causes of signal distortion in this type of fiber-optic link:

- Because of the fiber attenuation, the received signal is weakened and may not be discernible from noise.
- Because of the fiber dispersion, the transmission bandwidth is limited and high frequencies are attenuated more than low frequencies, resulting in signal degradation.

Both of these deleterious effects increase with the increase of the fiber length L. The received optical power drops exponentially with L, whereas the fiber bandwidth is inversely proportional to L. The maximum allowable length of the link is determined by ensuring that two conditions are met:

- The fiber attenuation must be sufficiently small so that the received power is greater than the receiver power sensitivity P_r.
- The fiber bandwidth $\sigma_f = 1/2\pi\sigma_\tau$ must be greater than the bandwidth B at which the data are to be transmitted.

As discussed in Sec. 17.5, the sensitivity of an analog optical receiver is the smallest optical power necessary for the signal-to-noise ratio SNR of the photocurrent to exceed a prescribed value SNR_0. For an ideal receiver (with unity quantum efficiency and no circuit noise) $\text{SNR} = \bar{n} = (P/h\nu)/2B$, where B is the receiver bandwidth, P the optical power (watts), and $\bar{n}$ the average number of photons received in a time interval $1/2B$, regarded as the resolution time of the system. If SNR_0 is the minimum allowed signal-to-noise ratio, the receiver sensitivity becomes $\bar{n}_0 = \text{SNR}_0$ photons per resolution time and the corresponding power

$$P_r = h\nu\bar{n}_0(2B). \tag{22.3-13}$$

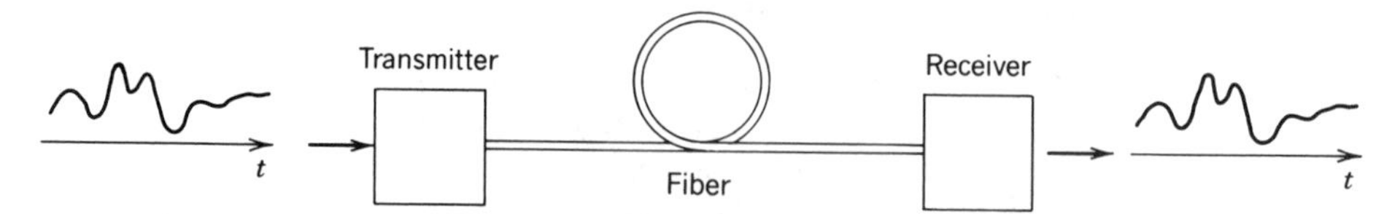

Figure 22.3-10 An analog optical fiber link.

This is identical to the expression (22.3-1) for the power sensitivity of the digital receiver if the resolution time $1/2B$ of the analog system is equated with the bit time $1/B_0$ of the digital system.

Because of the equivalence between (22.3-13) and (22.3-1) and because of the applicability of (22.3-12) to analog systems as well, the L–B_0 relations determined earlier for the binary digital system are applicable to the analog system, with B_0 replaced by $2B$, provided that the acceptable performance of the analog system is $\mathrm{SNR}_0 = 10$. As an example, a 1-km fiber link capable of transmitting digital data at a rate of 2 Gb/s with a BER not exceeding 10^{-9} can also be used to transmit analog data of bandwidth 1 GHz with a signal-to-noise ratio of at least 10.

In analog systems, however, the required signal-to-noise ratio is usually much greater than 10, so that the receiver sensitivity must be much greater than 10 photons per resolution time. For high-quality audio and video signals, for example, a 60-dB signal-to-noise ratio is often required. This corresponds to $\mathrm{SNR}_0 = 10^6$, or $\bar{n}_0 = 10^6$ photons per resolution time. Additional design considerations are particularly important in analog systems. For example, the nonlinear response of the light source and photodetector cause additional signal degradation and place restrictions on the dynamic range of the transmitted waveforms.

22.4 RECEIVER SENSITIVITY

The sensitivity of an analog receiver was defined in Sec. 17.5 as the minimum power of the received light (or the corresponding photon flux) necessary to achieve a prescribed signal-to-noise ratio SNR_0. In this section we discuss the sensitivity of the digital communication receiver. The sensitivity of a binary on–off keying system is defined as the minimum optical energy (or the corresponding mean number of photons) per bit necessary to obtain a prescribed bit error rate (BER). We first determine the sensitivity of the ideal detector and then consider the effects of circuit noise and detector gain noise. This section relies on the material in Sec. 17.5.

Sensitivity of the Ideal Optical Receiver

Assume that bits "1" and "0" of the on–off keying system described in Secs. 22.2A and 22.3A are represented by the presence and absence of optical energy, respectively (Fig. 22.4-1). During bit "1" an average of $\bar{n}$ photons is received. During bit "0" no photons are received. If the two bits are equally likely, the overall average number of photons per bit is $\bar{n}_a = \frac{1}{2}\bar{n}$. Since the actual number of detected photons is random, errors in bit identification occur.

For light generated by laser diodes, the probability of detecting n photons when an average of $\bar{n}$ photons is transmitted obeys the Poisson distribution $p(n) = \bar{n}^n \exp(-\bar{n})/n!$ (see Sec. 11.2). The receiver decides that "1" has been transmitted if it detects one or more photons. The probability p_1 of mistaking "1" for "0" is therefore equal to the probability of detecting no photons, i.e., $p_1 = p(0) = \exp(-\bar{n})$. When bit "0" is transmitted, there are no photons; the receiver decides correctly that bit "0" has been transmitted, so that $p_0 = 0$. The bit error rate is the average of the two error probabilities, $\mathrm{BER} = \frac{1}{2}(p_1 + p_0)$, from which

$$\mathrm{BER} = \tfrac{1}{2}\exp(-\bar{n}) = \tfrac{1}{2}\exp(-2\bar{n}_a). \qquad (22.4\text{-}1)$$

Figure 22.4-1 is a semilogarithmic plot of this relation.

The receiver sensitivity is defined as the average number of photons per bit required to achieve a certain BER (usually 10^{-9}). For $\mathrm{BER} = 10^{-9}$, (22.4-1) gives $\bar{n}_a \approx 10$

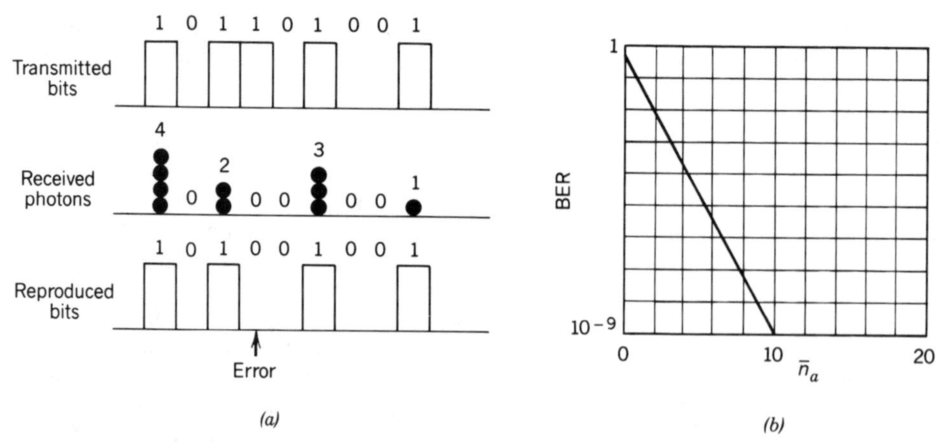

Figure 22.4-1 (*a*) Example of errors resulting from the random photon numbers. (*b*) Bit error rate BER versus the mean number of photons per bit $\bar{n}_a$ in an on–off keying system using an ideal receiver.

photons per bit. We conclude that:

> *The receiver sensitivity (for bit error rate* BER $= 10^{-9}$*) of an optical digital communication system using an ideal receiver is 10 photons per bit.*

EXERCISE 22.4-1

Effect of Quantum Efficiency and Background Noise on Receiver Sensitivity

(a) Show that for a receiver using a detector with quantum efficiency η, but otherwise ideal, BER $= \frac{1}{2}\exp(-2\eta\bar{n}_a)$, and that the sensitivity is $\bar{n}_a = 10/\eta$ photons per bit, corresponding to $\bar{m}_a = \eta\bar{n}_a = 10$ photoelectrons per bit.

(b) Assuming that bits "1" and "0" correspond to mean photon numbers $\bar{n}_1 = \bar{n} + \bar{n}_B$ and $\bar{n}_0 = \bar{n}_B$, where $\bar{n}$ is the mean number of signal photons and $\bar{n}_B$ is the mean of a Poisson-distributed background photon flux that is always present independently of the signal, determine an expression for the BER as a function of $\bar{n}$ and $\bar{n}_B$. Plot BER versus $\bar{n}_a = \frac{1}{2}\bar{n}$ for several values of $\bar{n}_B$. From this plot, determine the receiver sensitivity $\bar{n}_a$ as a function of $\bar{n}_B$. (*Hint:* The sum of two random numbers, each with a Poisson probability distribution, also has a Poisson distribution.)

It should be recognized that the ideal receiver sensitivity of 10 photons per bit is applicable only for light with a Poisson photon-number distribution. The sensitivity can, in principle, be improved by the use of photon-number-squeezed light (see Sec. 11.3B).

Sensitivity of a Receiver with Circuit Noise and Gain Noise

As explained in Sec. 17.5, a photodiode transforms an average fraction η of the received photons into photoelectron–hole pairs, each of which contributes a charge e to the electric current in the external circuit. The total charge accumulated in the bit time $T = 1/B_0$ is m (units of electrons). This number is random and has a Poisson distribution with mean $\bar{m} = \eta\bar{n}$ and variance $\bar{m}$.

Additional noise is introduced by the photodiode circuit in the form of a random electric current i_r of Gaussian probability distribution with zero mean and variance σ_r^2. Within the bit time interval $T = 1/B_0$, the accumulated charge $q = i_r T/e$ (in units of electrons) has an rms value $\sigma_q = \sigma_r T/e$. The parameter σ_q, called the circuit-noise parameter, depends on the receiver bandwidth B as described in Sec. 17.5C.

The total accumulated charge per bit $s = m + q$ (units of electrons) is the sum of a Poisson random variable m and an independent Gaussian random variable q. Its mean is

$$\mu = \bar{m} = \eta \bar{n} \tag{22.4-2}$$

and its variance is the sum of the variances,

$$\sigma^2 = \bar{m} + \sigma_q^2. \tag{22.4-3}$$

When $\bar{m}$ is large, the overall distribution may be approximated by a Gaussian distribution with mean μ and variance σ^2. We adopt this approximation in the present analysis.

For an avalanche photodiode (APD) of gain $\bar{G}$, the mean number of photoelectrons is amplified by a factor $\bar{G}$, but additional noise is introduced in the amplification process. The mean of the total collected charge per bit s (units of electrons) is

$$\mu = \bar{m}\bar{G} \tag{22.4-4}$$

and the variance is

$$\sigma^2 = \bar{m}\bar{G}^2 F + \sigma_q^2, \tag{22.4-5}$$

where $F = \langle G^2 \rangle / \langle G \rangle^2$ is the excess-noise factor of the APD (see Sec. 17.5B).

The receiver measures the charge s accumulated in each bit (by use of an integrator, for example) and compares it to a prescribed threshold ϑ. If $s > \vartheta$, bit "1" is selected; otherwise, bit "0" is selected. The probabilities of error p_1 and p_0 are determined by examining two Gaussian probability distributions of s that have

$$\begin{aligned} &\text{mean } \mu_0 = 0, \quad &&\text{variance } \sigma_0^2 = \sigma_q^2 \quad &&\text{for bit "0"} \\ &\text{mean } \mu_1 = \bar{m}\bar{G}, \quad &&\text{variance } \sigma_1^2 = \bar{m}\bar{G}^2 F + \sigma_q^2 \quad &&\text{for bit "1."} \end{aligned} \tag{22.4-6}$$

The probability p_0 of mistaking "0" for "1" is the integral of a Gaussian probability distribution $p(s)$ with mean μ_0 and variance σ_0^2 from $s = \vartheta$ to $s = \infty$. The probability p_1 of mistaking "1" for "0" is the integral of a Gaussian probability distribution with mean μ_1 and variance σ_1^2 from $s = -\infty$ to $s = \vartheta$. The threshold ϑ is selected such that the average probability of error $\text{BER} = \frac{1}{2}(p_0 + p_1)$ is minimized.

This type of analysis is the basis of the conventional theory of binary detection in the presence of Gaussian noise. If μ_0 and μ_1, and σ_0^2 and σ_1^2 are the means and variances associated with two Gaussian variables representing bits "0" and "1," and if σ_0 and σ_1 are much smaller than $\mu_1 - \mu_0$, the bit error rate for an optimal-threshold receiver is approximately

$$\text{BER} \approx \frac{1}{2}\left[1 - \text{erf}\left(\frac{Q}{\sqrt{2}}\right)\right], \tag{22.4-7}$$

where

$$Q = \frac{\mu_1 - \mu_0}{\sigma_0 + \sigma_1} \tag{22.4-8}$$

and

$$\text{erf}(z) = \frac{2}{\sqrt{\pi}} \int_0^z \exp(-x^2)\, dx \tag{22.4-9}$$

is the error function. When $Q = 6$, BER $\approx 10^{-9}$. The receiver sensitivity therefore corresponds to $Q = 6$, or

$$\boxed{\mu_1 - \mu_0 = 6(\sigma_0 + \sigma_1).} \tag{22.4-10}$$

Condition for BER $= 10^{-9}$ (Gaussian Approximation)

Substituting from (22.4-6) into (22.4-10) and defining $\overline{m}_a = \frac{1}{2}\overline{m}$ as the mean number of photoelectrons detected per bit, we obtain

$$\overline{m}_a = 18F + 6\frac{\sigma_q}{\overline{G}}. \tag{22.4-11}$$

Equation (22.4-11) relates the receiver sensitivity $\overline{m}_a$, which is the mean number of photoelectrons per bit required to make the BER $= 10^{-9}$, to the receiver parameters $\overline{G}$, F, and σ_q.

When the APD gain is sufficiently large so that $3\overline{G}F \gg \sigma_q$, the second term in (22.4-11) is negligible and

$$\boxed{\overline{m}_a \approx 18F.} \tag{22.4-12}$$

APD Receiver Sensitivity (No Circuit Noise)

For a receiver using a photodiode with no gain ($\overline{G} = 1$ and $F = 1$) and assuming that the circuit noise is negligible, $\overline{m}_a = 18$ photoelectrons per bit. This is different from the 10 photoelectrons per bit obtained earlier for this ideal receiver. The reason for the discrepancy is that the replacement of the Poisson distribution with the Gaussian distribution is not appropriate in this case. Typical sensitivities of several receivers are listed in Table 22.4-1. The actual values depend on the receiver circuit-noise parameter σ_q, which in turn depends on the bit rate B_0.

TABLE 22.4-1 Typical Sensitivities (Mean Number of Photons per Bit) of Some Optical Receivers Operating at Bit Rates in the Range 1 Mb/s to 2.5 Gb/s

Receiver	Receiver Sensitivity (photons/bit)
Photon-limited ideal detector	10
Si APD	125
Er-doped silica-fiber preamplifier/ InGaAs *p-i-n* photodiode	215
InGaAs APD	500
p-i-n photodiode	6000

22.5 COHERENT OPTICAL COMMUNICATIONS

Coherent optical communication systems may use field modulation (amplitude, phase, or frequency) instead of intensity modulation. They employ highly coherent light sources, single-mode fibers, and heterodyne receivers. In this section we examine the principles of operation of these systems, determine their performance advantage, and briefly discuss the requirements on the system components.

A. Heterodyne Detection

Photodetectors are responsive to the photon flux and, as such, are insensitive to the optical phase. It is possible, however, to measure the complex amplitude (both magnitude and phase) of the signal optical field by mixing it with a coherent reference optical field of stable phase, called the **local oscillator**, and detecting the superposition using a photodetector, as illustrated in Fig. 22.5-1. As a result of interference (beating) between the two fields, the detected electric current contains information about both the amplitude and phase of the signal field.

This detection technique is called **optical heterodyning, optical mixing, photomixing, light beating** (see Sec. 2.6B), or **coherent optical detection** (as opposed to **direct detection**). The coherent optical receiver is the optical equivalent of a superheterodyne radio receiver. The signal and local-oscillator waves usually have different frequencies (ν_s and ν_L). When $\nu_s = \nu_L$ the detector is said to be a **homodyne detector**.

Let $\mathscr{E}_s = \text{Re}\{A_s \exp(j2\pi\nu_s t)\}$ be the signal optical field, with $A_s = |A_s|\exp(j\varphi_s)$ its complex amplitude and ν_s its frequency. The magnitude $|A_s|$ or the phase φ_s are modulated with the information signal at a rate much slower than ν_s. The local oscillator field is described similarly by $\mathscr{E}_L$, A_L, ν_L, and φ_L. The two fields are mixed using a beamsplitter or an optical coupler, as illustrated in Fig. 22.5-1. If the incident fields are perfectly parallel plane waves and have precisely the same polarization, the total field is the sum of the two constituent fields $\mathscr{E} = \mathscr{E}_s + \mathscr{E}_L$. Taking the absolute square of the sum of the complex amplitudes, we obtain

$$|A_s \exp(j2\pi\nu_s t) + A_L \exp(j2\pi\nu_L t)|^2$$

$$= |A_s|^2 + |A_L|^2 + 2|A_s||A_L|\cos[2\pi(\nu_s - \nu_L)t + (\varphi_s - \varphi_L)].$$

Since the intensities I_s, I_L, and I are proportional to the absolute-square values of the

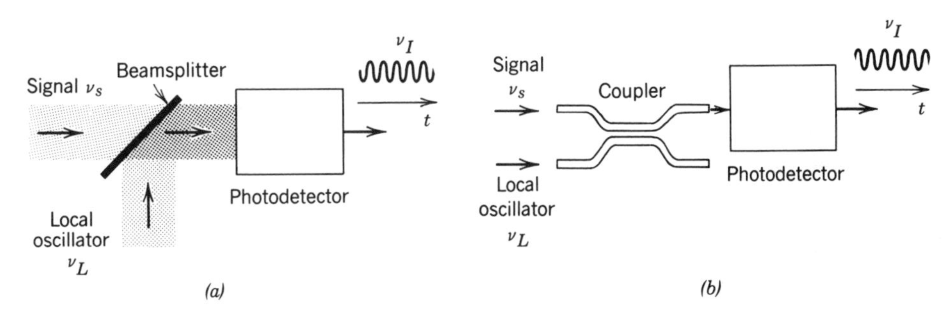

Figure 22.5-1 Optical heterodyne detection. A signal wave of frequency ν_s is mixed with a local oscillator wave of frequency ν_L using (*a*) a beamsplitter, and (*b*) an optical coupler. The photocurrent varies at the frequency difference $\nu_I = \nu_s - \nu_L$.

complex amplitudes,

$$I = I_s + I_L + 2(I_s I_L)^{1/2} \cos[2\pi\nu_I t + (\varphi_s - \varphi_L)],$$

where $\nu_I = \nu_s - \nu_L$ is the difference frequency.

The optical power collected by the detector is the product of the intensity and the detector area, so that

$$P = P_s + P_L + 2(P_s P_L)^{1/2} \cos[2\pi\nu_I t + (\varphi_s - \varphi_L)], \qquad (22.5\text{-}1)$$

where P_s and P_L are the powers of the signal and the local-oscillator beams, respectively. Slight misalignments between the directions of the two waves reduces or washes out the interference term [the third term of (22.5-1)], since the phase $\varphi_s - \varphi_L$ then varies sinusoidally with position within the area of the detector. The third term of (22.5-1) varies with time at the difference frequency ν_I with a phase $\varphi_s - \varphi_L$. If the signal and local oscillator beams are close in frequency, their difference ν_I can be far smaller than the individual frequencies.

The photocurrent i generated in a semiconductor photon detector is proportional to the incident photon flux Φ (see Sec. 17.1B). When ν_I is much smaller than ν_s and ν_L, the superposed light is quasi-monochromatic and the total photon flux $\Phi \approx P/h\bar{\nu}$ is proportional to the optical power, where $\bar{\nu} = \frac{1}{2}(\nu_s + \nu_L)$. The mean photocurrent is therefore $\bar{i} = \eta e \Phi = (\eta e / h\bar{\nu})P$, where e is the electron charge and η the detector's quantum efficiency, so that

$$\bar{i} = \bar{i}_s + \bar{i}_L + 2(\bar{i}_s \bar{i}_L)^{1/2} \cos[2\pi\nu_I t + (\varphi_s - \varphi_L)], \qquad (22.5\text{-}2)$$

where $\bar{i}_s = \eta e P_s / h\bar{\nu}$ and $\bar{i}_L = \eta e P_L / h\bar{\nu}$ are the photocurrents generated by the signal and local-oscillator individually. The local oscillator is usually much stronger than the signal, so that the first term in (22.5-2) is negligible and

$$\bar{i} \approx \bar{i}_L + 2(\bar{i}_s \bar{i}_L)^{1/2} \cos[2\pi\nu_I t + (\varphi_s - \varphi_L)]. \qquad (22.5\text{-}3)$$

Photomixing Current

The time dependence of the detected current $\bar{i}$ is sketched in Fig. 22.5-2(*a*). The second term in (22.5-3), which oscillates at the difference frequency ν_I, carries the

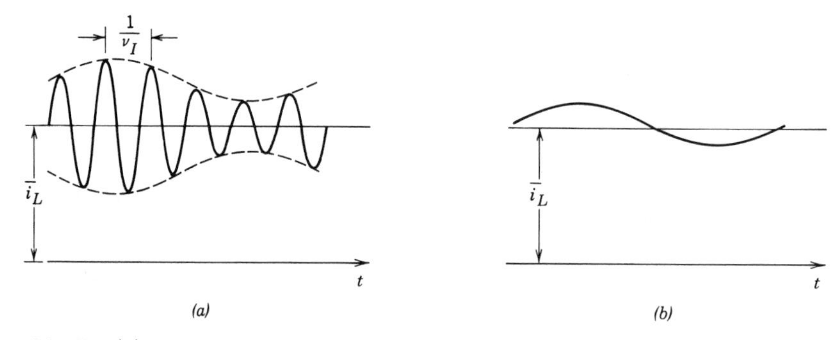

Figure 22.5-2 (*a*) Photocurrent generated by the heterodyne detector. The envelope and phase of the time-varying component carries complete information about the complex amplitude of the optical field representing the signal. (*b*) Photocurrent generated by the homodyne detector.

useful information. With knowledge of $\bar{i}_L$ and φ_L, the amplitude and phase of this term can be determined, and $\bar{i}_s$ and φ_s estimated, from which the intensity and phase (and hence the complex amplitude) of the measured optical signal can be inferred. The information-containing signal variables $\bar{i}_s$ or φ_s are usually slowly varying functions of time in comparison with the difference frequency ν_I, so they act as slow modulations of the amplitude and the phase of the harmonic function $2\bar{i}_L^{1/2}\cos(2\pi\nu_I t - \varphi_L)$. This amplitude- and phase-modulated current can be demodulated by drawing on the conventional techniques used in AM and FM radio receivers.

From a photon-optics point of view, this process can be understood in terms of the detection of polychromatic (two-frequency) photons (see Problem 11.1-7).

The homodyne system is a special case of the heterodyne system for which $\nu_s = \nu_L$ and $\nu_I = 0$. The demodulation process is different. A phase-locked loop is used to lock the phase of the local oscillator so that $\varphi_L = 0$ and (22.5-3) yields

$$\bar{i} = \bar{i}_L + 2\left(\bar{i}_s\bar{i}_L\right)^{1/2}\cos\varphi_s. \tag{22.5-4}$$

Amplitude and phase modulation is achieved by varying $\bar{i}_s$ or φ_s, respectively.

B. Performance of the Analog Heterodyne Receiver

Heterodyne detection is necessary whenever the phase of the optical field is to be measured. However, heterodyne detection can also be useful for measuring the optical intensity since it provides gain through the presence of the strong local oscillator. As such, it offers an alternative to both laser amplification (see Chap. 13 and Sec. 16.2) and APD amplification (see Sec. 17.4). This can provide a signal-to-noise ratio advantage over direct detection, as we show in this section.

The mean photocurrent $\bar{i}$ generated by a photodiode is accompanied by noise of variance

$$\sigma_i^2 = 2e\bar{i}B + \sigma_r^2, \tag{22.5-5}$$

where B is the receiver's bandwidth; the first term is due to photon noise and the second represents circuit noise (see Sec. 17.5). The intensity of the local oscillator can be made sufficiently large so that even if the signal is weak, the total current $\bar{i}$ is such that the circuit noise σ_r^2 is negligible in comparison with the photon noise $2e\bar{i}B$.

Assuming that $\bar{i}_L \gg \bar{i}_s$ and $2e\bar{i}_L B \gg \sigma_r^2$, we use (22.5-3) and approximate (22.5-5) by

$$\bar{i} \approx \bar{i}_L + 2\left(\bar{i}_s\bar{i}_L\right)^{1/2}\cos\left[2\pi\nu_I t + (\varphi_s - \varphi_L)\right] \tag{22.5-6a}$$

$$\sigma_i^2 \approx 2e\bar{i}_L B. \tag{22.5-6b}$$

In the case of amplitude modulation, the signal is represented by the rms value of the sinusoidal waveform in (22.5-6a), with the phase ignored. The electrical signal power is therefore $\frac{1}{2}[2(\bar{i}_s\bar{i}_L)^{1/2}]^2 = 2\bar{i}_s\bar{i}_L$ and the noise power is $\sigma_i^2 = 2e\bar{i}_L B$, so that the power signal-to-noise ratio is

$$\mathrm{SNR} = \frac{2\bar{i}_s\bar{i}_L}{2e\bar{i}_L B} = \frac{\bar{i}_s}{eB}. \tag{22.5-7}$$

If $\bar{m} = \bar{i}/2Be$ is the mean number of photoelectrons counted in the resolution time interval $T = 1/2B$, then

$$\boxed{\text{SNR} = 2\bar{m}.} \tag{22.5-8}$$

Signal-to-Noise Ratio of a Heterodyne Receiver

In comparison, the SNR of the direct-detection photodiode receiver measuring the same signal current $\bar{i}_s$ without the benefit of heterodyning is

$$\text{SNR} = \frac{\bar{i}_s^2}{2e\bar{i}_s B + \sigma_r^2} = \frac{\bar{m}^2}{\bar{m} + \sigma_q^2}, \tag{22.5-9}$$

where $\sigma_q^2 = (\sigma_r/2Be)^2$ is the circuit-noise parameter discussed in Sec. 17.5C. The principal advantage of the heterodyne system is now apparent. For strong light or low circuit noise ($\bar{m} \gg \sigma_q^2$), the direct-detection result is $\text{SNR} = \bar{m}$. The heterodyne receiver, which yields $\text{SNR} = 2\bar{m}$, offers a factor-of-2 improvement (3-dB advantage). But for weak light (or large circuit noise) the advantage can be even more substantial, since the heterodyne receiver has $\text{SNR} = 2\bar{m}$, whereas the SNR of the direct-detection receiver is reduced by circuit noise to $\text{SNR} = \bar{m}/(1 + \sigma_q^2/\bar{m})$.

The performance of a direct-detection *avalanche* photodiode receiver is also inferior to that of a heterodyne photodiode receiver. In accordance with (17.5-32), the SNR obtained when the APD gain is sufficiently large to overcome circuit noise is

$$\text{SNR} = \frac{\bar{m}}{F},$$

where F is the APD excess noise factor ($F > 1$). Therefore, even a noiseless APD receiver ($F = 1$) is a factor of 2 inferior to the heterodyne receiver.

Advantages of Heterodyne Receivers

In comparison with the direct-detection receiver, the heterodyne receiver has the following advantages:

- It is capable of measuring the optical phase and frequency.
- It permits the use of wavelength-division multiplexing (WDM) with smaller channel spacing ($\approx$ 100 MHz). In conventional direct-detection systems the channel spacing is of the order of 100 GHz.
- It permits the use of electronic equalization to compensate for pulse broadening in the fiber. Pulse broadening is a result of the dephasing of the different wavelength/frequency components because of differences in group velocities. Since the receiver monitors the phase, this dephasing may be removed by proper electronic filtering.
- By use of a strong reference field, the heterodyne receiver has an inherent noiseless gain conversion factor that effectively amplifies the signal above the circuit noise level.
- It provides a 3-dB advantage over even the noiseless direct-detection receiver.
- It is insensitive to unwanted background light with which the local oscillator does not mix. Heterodyning is one of the few ways of attaining photon-noise-limited detection in the infrared, where background noise is so prevalent.

The cost of these advantages is an increase in the system's complexity since heterodyning requires a stable local oscillator, an optical coupler in which the mixed fields are precisely aligned, and complex circuits for phase locking.

C. Performance of the Digital Heterodyne Receiver

In this section the performance and sensitivity of a digital coherent communication system are determined in the cases of amplitude and phase modulation.

On–Off Keying (OOK) Homodyne System

Consider an on–off keying (OOK) system transmitting data at a rate B_0 bits/s and using a homodyne receiver. Bits "1" and "0" are represented by the presence and absence of the signal $\bar{i}_s$ during the bit time $T = 1/B_0$, respectively. Assuming that $\varphi_s = \varphi_L = 0$ and $\nu_I = \nu_s - \nu_L = 0$, the measured current has the following means and variances obtained from (22.5-6a) and (22.5-6b):

$$\begin{aligned} &\text{mean } \mu_1 \approx \bar{i}_L + 2(\bar{i}_L \bar{i}_s)^{1/2}, \quad \text{variance } \sigma_1^2 \approx 2\bar{i}_L eB \quad \text{for bit "1"} \\ &\text{mean } \mu_0 \approx \bar{i}_L, \qquad\qquad\qquad\quad \text{variance } \sigma_0^2 \approx 2\bar{i}_L eB \quad \text{for bit "0."} \end{aligned} \tag{22.5-10}$$

The receiver bandwidth $B = B_0/2$ since the bit time $T = 1/B_0$ is the sampling time $1/2B$ for a signal of bandwidth B.

The performance of the binary communication system under the Gaussian approximation has been discussed in Sec. 22.4. The bit error rate is given by (22.4-7), where

$$Q = \frac{\mu_1 - \mu_0}{\sigma_1 + \sigma_0} = \left(\frac{\bar{i}_s}{2eB} \right)^{1/2} = \overline{m}^{1/2}, \tag{22.5-11}$$

and $\overline{m} = \bar{i}_s/2eB$ is the mean number of detected photoelectrons in bit 1. For a bit error rate BER $= 10^{-9}$, $Q \approx 6$ and therefore $\overline{m} = 36$, corresponding to a receiver sensitivity $\overline{m}_a = \frac{1}{2}\overline{m} = 18$ photoelectrons per bit (averaged over both bits).

Phase-Shift-Keying (PSK) Homodyne System

Here bits "1" and "0" are represented by a phase shift $\varphi_s = 0$ and π, respectively. Assuming that $\varphi_L = 0$, the means and variances of the photocurrent for bits "1" and "0" are, from (22.5-6),

$$\begin{aligned} &\text{mean } \mu_1 = \bar{i}_L + 2(\bar{i}_L \bar{i}_s)^{1/2}, \quad \text{variance } \sigma_1^2 = 2e\bar{i}_L B \quad \text{for bit "1"} \\ &\text{mean } \mu_0 = \bar{i}_L - 2(\bar{i}_L \bar{i}_s)^{1/2}, \quad \text{variance } \sigma_0^2 = 2e\bar{i}_L B \quad \text{for bit "0"} \end{aligned}$$

and therefore

$$Q = \frac{\mu_1 - \mu_0}{\sigma_1 + \sigma_0} = 2\left(\frac{\bar{i}_s}{2eB} \right)^{1/2} = 2(\overline{m})^{1/2}. \tag{22.5-12}$$

For a BER $= 10^{-9}$, $Q = 6$, from which $\overline{m} = 9$. Since *each* of the two bits must carry an average of nine photoelectrons in this case, the average number of photoelectrons per bit is $\overline{m}_a = \overline{m} = 9$. It follows that the receiver sensitivity is 9 photoelectrons/bit. The PSK homodyne receiver is twice as sensitive as the OOK homodyne receiver because it requires half the number of photoelectrons.

Comparison

The sensitivity of the heterodyne digital receiver can be determined by following a similar analysis. Table 22.5-1 lists the receiver sensitivities of several digital modulation systems, assuming $\eta = 1$. Although it appears that the direct-detection OOK system has about the same performance as the best coherent system (homodyne PSK), in

TABLE 22.5-1 Receiver Sensitivity of Different Receivers and Modulation Systems under Ideal Conditions (Photons per Bit)

	Direct Detection	Homodyne	Heterodyne
OOK	10	18	36
PSK	—	9	18
FSK	—	—	36

practice this is not so. In the homodyne system, circuit noise is overcome, whereas in the direct-detection system, circuit noise cannot be ignored, unless an APD is used. When an APD is used in a direct-detection receiver, circuit noise is overcome, but the APD gain noise raises the receiver sensitivity from 10 to at least $10F$, where F is the excess-noise factor. Direct-detection systems would have performance comparable to coherent-detection systems if a perfect APD with $F = 1$ (no excess noise) were available.

D. Coherent Systems

An essential condition for the proper mixing of the local oscillator field and the received optical field is that they must be locked in phase, be parallel, and have the same polarization in order to permit interference to take place. This places stringent requirements on the two lasers and on the fiber. The lasers must be single-frequency and have minimal phase and intensity fluctuations. The local oscillator is phase-locked to the received optical field by means of a control system that adjusts the phase and frequency of the local oscillator adaptively (using a phase-locked loop). The fiber must be single-mode (to avoid modal noise). The fiber must also be polarization-maintaining, or the receiver must contain an adaptive polarization-compensation system.

A schematic diagram of a coherent fiber-optic communication system using two lasers and phase modulation is shown in Fig. 22.5-3. The local oscillator field is mixed with the received optical field using an optical directional coupler. One branch of the coupler output contains the sum of the two optical fields and the other branch contains the difference. Using (22.5-2), the detected currents

$$\bar{i}_{\pm} = \bar{i}_s + \bar{i}_L \pm 2(\bar{i}_s \bar{i}_L)^{1/2} \cos[2\pi\nu_I t + (\varphi_s - \varphi_L)]$$

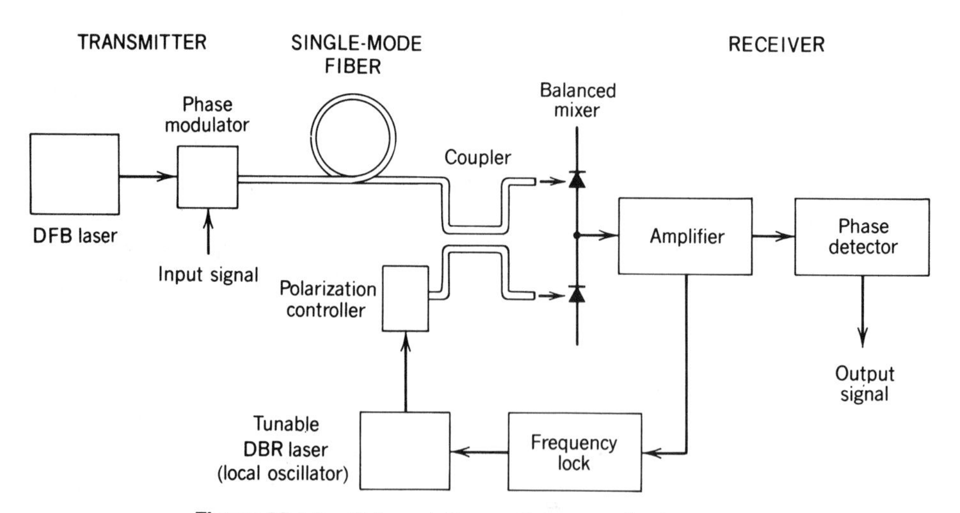

Figure 22.5-3 Coherent fiber-optic communication system.

are subtracted electronically, yielding $4(\bar{i}_s\bar{i}_L)^{1/2}\cos[2\pi\nu_I t + (\varphi_s - \varphi_L)]$, which is demodulated to recover the message. This type of coherent receiver is known as a **balanced mixer**. It has the advantage of canceling out intensity fluctuations of the local oscillator. A number of coherent fiber-optic communication systems have been implemented at $\lambda_o = 1.55\ \mu m$ (where fiber attenuation is minimal) with bit-rate–distance products matching theoretical expectations.

One example is provided by a system operating at a bit rate ≈ 4 Gb/s. A DFB laser with a 15-MHz CW linewidth was directly modulated in an FSK signal format. The local oscillator was a tunable DBR laser (see Sec. 16.3E). This system exhibited a receiver sensitivity ≈ 190 photons/bit and was used for transmission over a 160-km length of fiber.

READING LIST

See also the reading lists in Chapters 7, 8, 16, 17, and 21.

Books

T. Li, ed., *Optical Fiber Data Transmission*, Academic Press, Boston, 1991.

H. B. Killen, *Fiber Optic Communications*, Prentice Hall, Englewood Cliffs, NJ, 1991.

P. K. Cheo, *Fiber Optics and Optoelectronics*, Prentice Hall, Englewood Cliffs, NJ, 1990.

T. C. Edwards, *Fiber-Optic Systems: Network Applications*, Wiley, New York, 1989.

J. C. A. Chaimowicz, *Lightwave Technology*, Butterworths, Boston, 1989.

C. Lin, ed., *Optoelectronic Technology and Lightwave Communications Systems*, Van Nostrand Reinhold, New York, 1989.

S. E. Miller and I. P. Kaminow, eds., *Optical Fiber Telecommunications II*, Academic Press, New York, 1988.

S. Karp, R. Gagliardi, S. E. Moran, and A. Holland, *Optical Channels: Fibers, Clouds, Water, and the Atmosphere*, Plenum Press, New York, 1988.

W. B. Jones, Jr., *Introduction to Optical Fiber Communication Systems*, Holt, Rinehart and Winston, New York, 1988.

J. C. Palais, *Fiber Optic Communications*, Prentice Hall, Englewood Cliffs, NJ, 2nd ed. 1988.

T. Okoshi and K. Kikuchi, *Coherent Optical Fiber Communications*, Kluwer, Boston, 1988.

C. K. Kao, *Optical Fibre*, Peter Peregrinus, London, 1988.

C. D. Chaffee, *The Rewiring of America: The Fiber Optics Revolution*, Academic Press, Boston, 1988.

H. F. Taylor, ed., *Advances in Fiber Optics Communications*, Artech House, Norwood, MA, 1988.

S. Geckeler, *Optical Fiber Transmission Systems*, Artech House, Norwood, MA, 1987.

G. Mahlke and P. Gössing, *Fiber Optic Cables*, Wiley, New York, 1987.

P. K. Runge and P. R. Trischitta, eds., *Undersea Lightwave Communications*, IEEE Press, New York, 1986.

C. Baack, ed., *Optical Wideband Transmission Systems*, CRC Press, Boca Raton, FL, 1986.

S. D. Personick, *Fiber Optics: Technology and Applications*, Plenum Press, New York, 1985.

D. G. Baker, *Fiber Optic Design and Applications*, Reston Publishing, Reston, VA, 1985.

J. M. Senior, *Optical Fiber Communications*, Prentice-Hall, Englewood Cliffs, NJ, 1985.

J. Gowar, *Optical Communication Systems*, Prentice-Hall, Englewood Cliffs, NJ, 1984.

B. Culshaw, *Optical Fibre Sensing and Signal Processing*, Peter Peregrinus, London, 1984.

J. C. Daly, ed., *Fiber Optics*, CRC Press, Boca Raton, FL, 1984.

A. H. Cherin, *An Introduction to Optical Fibers*, McGraw-Hill, New York, 1983.

G. Keiser, *Optical Fiber Communications*, McGraw-Hill, New York, 1983.

D. J. Morris, *Pulse Code Formats for Fiber Optical Data Communication: Basic Principles and Applications*, Marcel Dekker, New York, 1983.

H. F. Taylor, ed., *Fiber Optics Communications*, Artech House, Dedham, MA, 1983.

Y. Suematsu and K. Iga, *Introduction to Optical Fiber Communications*, Wiley, New York, 1982.

H. Kressel, ed., *Semiconductor Devices for Optical Communication*, Springer-Verlag, New York, 2nd ed. 1982.

T. Okoshi, *Optical Fibers*, Academic Press, New York, 1982.

C. K. Kao, *Optical Fiber Systems*, McGraw-Hill, New York, 1982.

S. D. Personick, *Optical Fiber Transmission Systems*, Plenum Press, New York, 1981.

M. K. Barnoski, ed., *Fundamentals of Optical Fiber Communications*, Academic Press, New York, 2nd ed. 1981.

A. B. Sharma, S. J. Halme, and M. M. Butusov, *Optical Fiber Systems and Their Components*, Springer-Verlag, Berlin, 1981.

CSELT (Centro Studi e Laboratori Telecomunicazioni), *Optical Fibre Communications*, McGraw-Hill, New York, 1981.

C. P. Sandbank, ed., *Optical Fibre Communication Systems*, Wiley, New York, 1980.

M. J. Howes and D. V. Morgan, eds., *Optical Fibre Communications*, Wiley, New York, 1980.

J. E. Midwinter, *Optical Fibers for Transmission*, Wiley, New York, 1979.

S. E. Miller and A. G. Chynoweth, *Optical Fiber Telecommunications*, Academic Press, New York, 1979.

B. Saleh, *Photoelectron Statistics with Applications to Spectroscopy and Optical Communication*, Springer-Verlag, Berlin, 1978.

G. R. Elion and H. A. Elion, *Fiber Optics in Communication Systems*, Marcel Dekker, New York, 1978.

R. O. Harger, *Optical Communication Theory*, Dowden, Hutchinson & Ross, Stroudsburg, PA, 1977.

R. M. Gagliardi and S. Karp, *Optical Communications*, Wiley, New York, 1976.

W. K. Pratt, *Laser Communication Systems*, Wiley, New York, 1969.

Special Journal Issues

Special issue on optical fiber communication, *Optics and Photonics News*, vol. 1, no. 11, 1990.

Special issue on wide-band optical transmission technology and systems, *Journal of Lightwave Technology*, vol. LT-6, no. 11, 1988.

Special issue on fiber optic local and metropolitan area networks, *IEEE Journal of Selected Areas in Communications*, vol. SAC-6, no. 6, 1988.

Special issue on factors affecting data transmission quality, *Journal of Lightwave Technology*, vol. LT-6, no. 5, 1988.

Special issue on high speed technology for lightwave applications, *Journal of Lightwave Technology*, vol. LT-5, no. 10, 1987.

Special issue on coherent communications, *Journal of Lightwave Technology*, vol. LT-5, no. 4, 1987.

Special issue on fiber optic systems for terrestrial applications, *IEEE Journal of Selected Areas in Communications*, vol. SAC-4, no. 9, 1986.

Joint special issue on lightwave devices and subsystems, *Journal of Lightwave Technology*, vol. LT-3, no. 6; and *IEEE Transactions on Election Devices*, vol. ED-32, no. 12, 1985.

Special issue on fiber optics for local communications, *IEEE Journal of Selected Areas in Communications*, vol. SAC-3, no. 6, 1985.

Joint special issue on undersea lightwave communications, *Journal of Lightwave Technology*, vol. LT-2, no. 6; and *IEEE Journal of Selected Areas in Communications*, vol. SAC-2, no. 6, 1984.

Special issue on fiber optic systems, *IEEE Journal of Selected Areas in Communications*, vol. SAC-1, no. 3, 1983.

Special issue on communications aspects of single-mode optical fiber and integrated optical technology, *IEEE Journal of Quantum Electronics*, vol. QE-17, no. 6, 1981.

Special issue on optical-fiber communications, *Proceedings of the IEEE*, vol. 68, no. 10, 1980.
Special issue on quantum-electronic devices for optical-fiber communications, *IEEE Journal of Quantum Electronics*, vol. QE-14, no. 11, 1978.
Special issue on optical communication, *Proceedings of the IEEE*, vol. 58, no. 10, 1970.

Articles

E. Desurvire, Erbium-Doped Fiber Amplifiers for New Generations of Optical Communication Systems, *Optics & Photonics News*, vol. 2, no. 1, pp. 6–11, 1991.
K. Nakagawa and S. Shimada, Optical Amplifiers in Future Optical Communication Systems, *IEEE Lightwave Communication Systems Magazine*, vol. 1, no. 4, pp. 57–62, 1990.
P. E. Green and R. Ramaswami, Direct Detection Lightwave Systems: Why Pay More?, *IEEE Lightwave Communication Systems Magazine*, vol. 1, no. 4, pp. 36–49, 1990.
R. E. Wagner and R. A. Linke, Heterodyne Lightwave Systems: Moving Towards Commercial Use, *IEEE Lightwave Communication Systems Magazine*, vol. 1, no. 4, pp. 28–35, 1990.
J. A. Jay and E. M. Hopiavuori, Dispersion-Shifted Fiber Hits its Stride, *Photonics Spectra*, vol. 24, no. 9, pp. 153–158, 1990.
M. G. Drexhage and C. T. Moynihan, Infrared Optical Fibers, *Scientific American*, vol. 259, no. 5, pp. 110–116, 1988.
R. A. Linke and A. H. Gnauck, High-Capacity Coherent Lightwave Systems, *Journal of Lightwave Technology*, vol. 6, pp. 1750–1769, 1988.
S. F. Jacobs, Optical Heterodyne (Coherent) Detection, *American Journal of Physics*, vol. 56, pp. 235–245, 1988.
K. Nosu, Advanced Coherent Lightwave Technologies, *IEEE Communications Magazine*, vol. 26, no. 2, pp. 15–21, 1988.
W. J. Tomlinson and C. A. Brackett, Telecommunications Applications of Integrated Optics and Optoelectronics, *Proceedings of the IEEE*, vol. 75, pp. 1512–1523, 1987.
R. A. Linke and P. S. Henry, Coherent Optical Detection: A Thousand Calls on One Circuit, *IEEE Spectrum*, vol. 24, no. 2, pp. 52–57, 1987.
S. R. Nagel, Optical Fiber—the Expanding Medium, *IEEE Communications Magazine*, vol. 25, no. 4, pp. 33–43, 1987.
T. Li, Advances in Lightwave Systems Research, *AT & T Technical Journal*, vol. 66, no. 1, pp. 5–18, 1987.
H. Kogelnik, High-Speed Lightwave Transmission in Optical Fibers, *Science*, vol. 228, pp. 1043–1048, 1985.
M. C. Teich, Laser Heterodyning, *Optica Acta* (*Journal of Modern Optics*), vol. 32, pp. 1015–1021, 1985.
P. S. Henry, Lightwave Primer, *IEEE Journal of Quantum Electronics*, vol. QE-21, pp. 1862–1879, 1985.
I. W. Stanley, A Tutorial Review of Techniques for Coherent Optical Fiber Transmission Systems, *IEEE Communications Magazine*, vol. 23, no. 8, pp. 37–53, 1985.
T. Li, Lightwave Telecommunication, *Physics Today*, vol. 38, no. 5, pp. 24–31, 1985.
J. D. Crow, Computer Applications for Fiber Optics, *IEEE Communications Magazine*, vol. 23, no. 2, pp. 16–20, 1985.
Y. Suematsu, Long-Wavelength Optical Fiber Communication, *Proceedings of the IEEE*, vol. 71, pp. 692–721, 1983.

PROBLEMS

22.1-1 **Fiber-Optic Systems.** Discuss the validity of each of the following statements and indicate the conditions under which your conclusion is applicable.
(a) The wavelength $\lambda_o = 1.3\ \mu m$ is preferred to $\lambda_o = 0.87\ \mu m$ for all fiber-optic communication systems.

(b) The wavelength $\lambda_o = 1.55\ \mu m$ is preferred to $\lambda_o = 1.3$ for all fiber-optic communication systems.
(c) Single-mode fibers are superior to multimode fibers because they have lower attenuation coefficients.
(d) There is no pulse spreading at $\lambda_o \approx 1.312\ \mu m$ in silica glass fibers.
(e) Compound semiconductor devices are required for fiber-optic communication systems.
(f) APDs are noisier than *p-i-n* photodiodes and are therefore not useful for fiber-optic systems.

22.1-2 **Components for Fiber-Optic Systems.** The design of a fiber-optic communication system involves many choices, some of which are shown in Table 22.1-2 on page 886. Make reasonable choices for each of the applications listed below. More than one answer may be correct. Some choices, however, are incompatible.
(a) A transoceanic cable carrying data at a 100-Mb/s rate with 100-km repeater spacings.
(b) A 1-m cable transmitting analog data from a sensor at 1 kHz.
(c) A link for a computer local-area network operating at 500 Mb/s.
(d) A 1-km data link operating at 100 Mb/s with $\pm 50°C$ temperature variations.

22.3-1 **Performance of a Plastic Fiber Link.** A short-distance low-data-rate communication system uses a plastic fiber with attenuation coefficient 0.5 dB/m, an LED generating 1 mW at a wavelength of 0.87 μm, and a photodiode with receiver sensitivity -20 dBm. Assuming a power loss of 3 dB each at the input and output couplers, determine the maximum length of the link. Assume that the data rate is sufficiently low so that dispersion effects play no role.

22.3-2 **Power Budget.** A fiber-optic communication link is designed for operation at 10 Mb/s. The source is a 1-mW AlGaAs diode laser operating at 0.87 μm. The fiber is made of 1-km segments each with attenuation 3.5 dB/km. Connectors between segments have a loss of 2 dB each and input and output couplers each introduce a loss of 2 dB. The safety margin is 6 dB. Two receivers are available, a Si *p-i-n* photodiode receiver with sensitivity 5000 photons per bit, and a Si APD with sensitivity 125 photons per bit. Determine the maximum length of the link for each receiver.

22.4-1 **Dependence of Receiver Sensitivity on Wavelength.** The receiver sensitivity of an ideal receiver (with unity quantum efficiency and no circuit noise) operating at a wavelength 0.87 μm is -76 dBm. What is the sensitivity at 1.3 μm if the receiver is operated at the same data rate?

22.4-2 **Bit Error Rates.** A quantum-limited *p-i-n* photodiode (no noise other than photon noise) of quantum efficiency $\eta = 1$ mistakes a present $\lambda_o = 0.87\ \mu m$ optical signal of power P (bit 1) for an absent signal (bit 0) with probability 10^{-10}. What is the probability of error under each of the following new conditions?
(a) The wavelength is $\lambda_o = 1.3\ \mu m$.
(b) Original conditions, but now the power is doubled.
(c) Original conditions, but the efficiency is now $\eta = 0.5$.
(d) Original conditions, but an ideal APD with $\eta = 1$ and gain $G = 100$ (no gain noise) is used.
(e) As in part (d), but the APD has an excess noise factor $F = 2$ instead.

22.4-3 **Sensitivity of an AM Receiver.** A detector with responsivity $\Re$ (A/W), bandwidth B, and negligible circuit noise, measures a modulated optical power $P(t) = P_0 + P_s \cos(2\pi f t)$, with $f < B$. If $P_0 \gg P_s$, derive an expression for the minimum modulation power P_s that is measurable with signal-to-noise ratio $\mathrm{SNR}_0 = 30$ dB. What is the effect of the background power P_0 on the minimum observable signal P_s?

22.4-4 **Maximum Length of an Analog Link.** A fiber-optic communication link uses intensity modulation to transmit data at a bandwidth $B = 10$ MHz and signal-to-noise ratio of 40 dB. The source is a $\lambda_o = 0.87\ \mu$m light-emitting diode producing 100 μW average power with maximum modulation index of 0.5. The fiber is a multimode step-index fiber with attenuation coefficient 2.5 dB/km. The detector is an avalanche photodiode with mean gain $\overline{G} = 100$, excess noise factor $F = 5$, and responsivity of 0.5 A/W (not including the gain). Assuming that the circuit noise is negligible, calculate the optical power sensitivity of the receiver and the attenuation-limited maximum length L of the fiber.

22.4-5 **Sensitivity of a Photon-Counting Receiver.** A photodetector of quantum efficiency $\eta = 0.5$ counts photoelectrons received in successive time intervals of duration $T = 1\ \mu$s. Determine the receiver sensitivity (mean number of photons required to achieve SNR $= 10^3$) assuming a Poisson photon-number distribution. Assuming that the wavelength of the light is $\lambda_o = 0.87\ \mu$m, what is the corresponding optical power? If this optical power is received, what is the probability that the detector registers zero counts?

APPENDIX

A

FOURIER TRANSFORM

This appendix provides a brief review of the Fourier transform, and its properties, for functions of one and two variables.

A.1. One-Dimensional Fourier Transform

The harmonic function $F \exp(j2\pi\nu t)$ plays an important role in science and engineering. It has frequency ν and complex amplitude F. Its real part $|F|\cos(2\pi\nu t + \arg\{F\})$ is a cosine function with amplitude $|F|$ and phase $\arg\{F\}$. The variable t usually represents time; the frequency ν has units of cycles/s or Hz. The harmonic function is regarded as a building block from which other functions may be obtained by a simple superposition.

In accordance with the Fourier theorem, a complex-valued function $f(t)$, satisfying some rather unrestrictive conditions, may be decomposed as a superposition integral of harmonic functions of different frequencies and complex amplitudes,

$$f(t) = \int_{-\infty}^{\infty} F(\nu) \exp(j2\pi\nu t)\, d\nu. \tag{A.1-1}$$

Inverse Fourier Transform

The component with frequency ν has a complex amplitude $F(\nu)$ given by

$$F(\nu) = \int_{-\infty}^{\infty} f(t) \exp(-j2\pi\nu t)\, dt. \tag{A.1-2}$$

Fourier Transform

$F(\nu)$ is termed the **Fourier transform** of $f(t)$, and $f(t)$ is the **inverse Fourier transform** of $F(\nu)$. The functions $f(t)$ and $F(\nu)$ form a Fourier transform pair; if one is known, the other may be determined.

In this book we adopt the convention that $\exp(j2\pi\nu t)$ represents positive frequency, whereas $\exp(-j2\pi\nu t)$ is a harmonic function representing negative frequency. The opposite convention is used by some authors who define the Fourier transform in (A.1-2) with a positive sign in the exponent, and use a negative sign in the exponent of the inverse Fourier transform (A.1-1).

In communication theory, the functions $f(t)$ and $F(\nu)$ represent a signal, with $f(t)$ its time-domain representation and $F(\nu)$ its frequency-domain representation. The squared-absolute value $|f(t)|^2$ is called the **signal power**, and $|F(\nu)|^2$ is the energy spectral density. If $|F(\nu)|^2$ extends over a wide frequency range, the signal is said to have a wide bandwidth.

Properties of the Fourier Transform

Some important properties of the Fourier transform are provided below. These properties can be proved by direct application of the definitions (A.1-1) and (A.1-2) (see any of the books in the reading list).

- *Linearity.* The Fourier transform of the sum of two functions is the sum of their Fourier transforms.
- *Scaling.* If $f(t)$ has a Fourier transform $F(\nu)$, and τ is a real scaling factor, then $f(t/\tau)$ has a Fourier transform $|\tau|F(\tau\nu)$. This means that if $f(t)$ is scaled by a factor τ, its Fourier transform is scaled by a factor $1/\tau$. For example, if $\tau > 1$, then $f(t/\tau)$ is a stretched version of $f(t)$, whereas $F(\tau\nu)$ is a compressed version of $F(\nu)$. The Fourier transform of $f(-t)$ is $F(-\nu)$.
- *Time Translation.* If $f(t)$ has a Fourier transform $F(\nu)$, the Fourier transform of $f(t-\tau)$ is $\exp(-j2\pi\nu\tau)F(\nu)$. Thus delay by time τ is equivalent to multiplication of the Fourier transform by a phase factor $\exp(-j2\pi\nu\tau)$.
- *Frequency Translation.* If $F(\nu)$ is the Fourier transform of $f(t)$, the Fourier transform of $f(t)\exp(j2\pi\nu_0 t)$ is $F(\nu-\nu_0)$. Thus multiplication by a harmonic function of frequency ν_0 is equivalent to shifting the Fourier transform to a higher frequency ν_0.
- *Symmetry.* If $f(t)$ is real, then $F(\nu)$ has Hermitian symmetry [i.e., $F(-\nu) = F^*(\nu)$]. If $f(t)$ is real and symmetric, then $F(\nu)$ is also real and symmetric.
- *Convolution Theorem.* If the Fourier transforms of $f_1(t)$ and $f_2(t)$ are $F_1(\nu)$ and $F_2(\nu)$, respectively, the inverse Fourier transform of the product

$$F(\nu) = F_1(\nu)F_2(\nu) \tag{A.1-3}$$

is

$$f(t) = \int_{-\infty}^{\infty} f_1(\tau)f_2(t-\tau)\,d\tau. \tag{A.1-4}$$

Convolution

The operation defined in (A.1-4) is known as the convolution of $f_1(t)$ with $f_2(t)$. Convolution in the time domain is therefore equivalent to multiplication in the Fourier domain.

- *Correlation Theorem.* The correlation between two complex functions is defined as

$$f(t) = \int_{-\infty}^{\infty} f_1^*(\tau)f_2(t+\tau)\,d\tau. \tag{A.1-5}$$

Correlation

The Fourier transforms of $f_1(t)$, $f_2(t)$, and $f(t)$ are related by

$$F(\nu) = F_1^*(\nu)F_2(\nu). \tag{A.1-6}$$

- *Parseval's Theorem.* The signal energy, which is the integral of the signal power $|f(t)|^2$, equals the integral of the energy spectral density $|F(\nu)|^2$, so that

$$\int_{-\infty}^{\infty} |f(t)|^2 \, dt = \int_{-\infty}^{\infty} |F(\nu)|^2 \, d\nu. \tag{A.1-7}$$

Parseval's Theorem

TABLE A.1-1 Selected Functions and Their Fourier Transforms

Function	$f(t)$	$F(\nu)$
Uniform	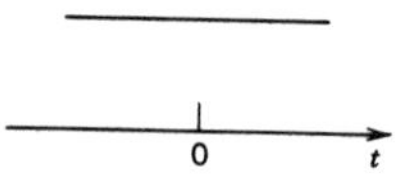 1	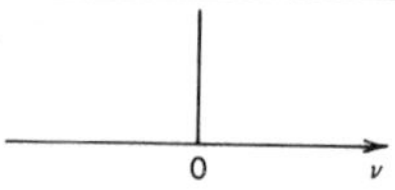 $\delta(\nu)$
Impulse	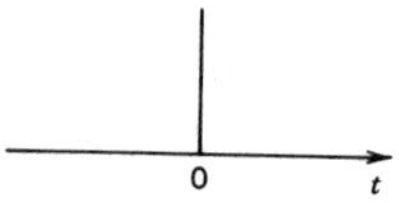 $\delta(t)$	1 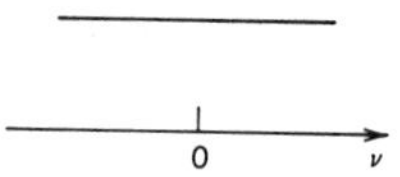
Rectangular	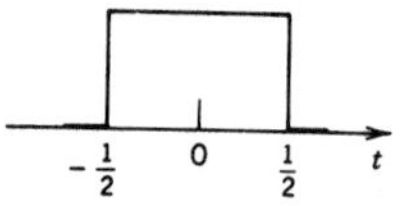 $\mathrm{rect}(t)$	$\mathrm{sinc}(\nu)$ 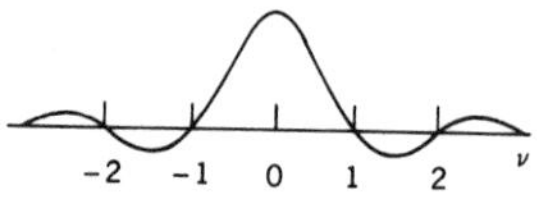
Exponential[a]	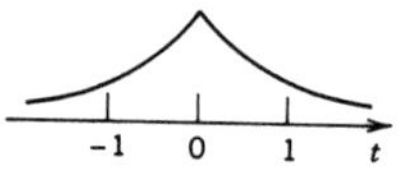 $\exp(-\lvert t\rvert)$	$\dfrac{2}{1+(2\pi\nu)^2}$ 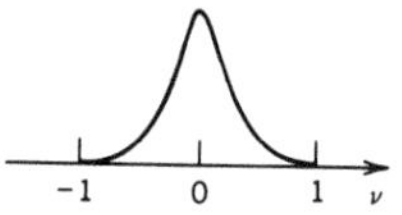
Gaussian	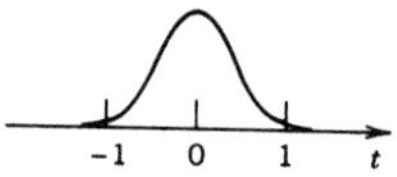 $\exp(-\pi t^2)$	$\exp(-\pi\nu^2)$ 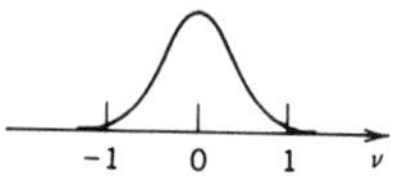
Chirp[b]	 $\exp(j\pi t^2)$	$e^{j\pi/4}\exp(-j\pi\nu^2)$ 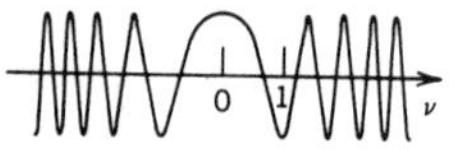
Sum of $M=2S+1$ impulses	 $\sum_{n=-S}^{S} \delta(t-n)$	$\dfrac{\sin(M\pi\nu)}{\sin(\pi\nu)}$ 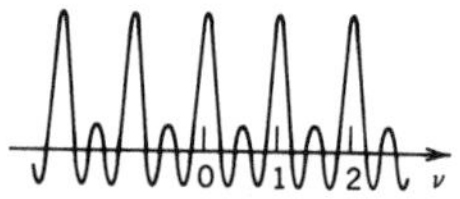
Infinite sum of impulses	 $\sum_{n=-\infty}^{\infty} \delta(t-n)$	$\sum_{n=-\infty}^{\infty} \delta(\nu-n)$ 

[a] The double-sided exponential function is shown. The Fourier transform of the single-sided exponential, $f(t) = \exp(-t)$ with $t \geq 0$, is $F(\nu) = 1/[1 + j2\pi\nu]$. Its magnitude is $1/[1 + (2\pi\nu)^2]^{1/2}$.
[b] The functions $\cos(\pi t^2)$ and $\cos(\pi\nu^2)$ are shown. The function $\sin(\pi t^2)$ is shown in Fig. 4.3-6.

Examples
The Fourier transforms of some important functions used in this book are listed in Table A.1-1. By use of the properties of linearity, scaling, delay, and frequency translation, the Fourier transforms of other functions may be readily obtained. In this table:

- $\mathrm{rect}(t) \equiv 1$ for $|t| \le \frac{1}{2}$, and is 0 elsewhere, i.e., it is a pulse of unit height and unit width centered about $t = 0$.
- $\delta(t)$ is the impulse function (Dirac delta function), defined as $\delta(t) = \lim_{\alpha \to \infty} \alpha\,\mathrm{rect}(\alpha t)$. It is the limit of a rectangular pulse of unit area as its width approaches zero (so that its height approaches infinity).
- $\mathrm{sinc}(t) = \sin(\pi t)/(\pi t)$ is a symmetric function with a peak value of 1.0 at $t = 0$ and zeros at $t = \pm 1, \pm 2, \dots$.

A.2. Time Duration and Spectral Width

It is often useful to have a measure of the width of a function. The width of a function of time $f(t)$ is its time duration and the width of its Fourier transform $F(\nu)$ is its spectral width (or bandwidth). Since there is no unique definition for the width, a plethora of definitions are in use. *All definitions, however, share the property that the spectral width is inversely proportional to the temporal width, in accordance with the scaling property of the Fourier transform.* The following definitions are used at different places in this book.

The Root-Mean-Square Width
The *root-mean-square (rms) width* σ_t of a nonnegative real function $f(t)$ is defined by

$$\sigma_t^2 = \frac{\int_{-\infty}^{\infty} (t - \bar{t})^2 f(t)\,dt}{\int_{-\infty}^{\infty} f(t)\,dt}, \qquad \text{where } \bar{t} = \frac{\int_{-\infty}^{\infty} t f(t)\,dt}{\int_{-\infty}^{\infty} f(t)\,dt}. \tag{A.2-1}$$

If $f(t)$ represents a mass distribution (t representing position), then $\bar{t}$ represents the centroid and σ_t the radius of gyration. If $f(t)$ is a probability density function, these quantities represent the mean and standard deviation, respectively. As an example, the *Gaussian function* $f(t) = \exp(-t^2/2\sigma_t^2)$ has an rms width σ_t. Its Fourier transform is given by $F(\nu) = (1/\sqrt{2\pi}\,\sigma_\nu)\exp(-\nu^2/2\sigma_\nu^2)$, where

$$\sigma_\nu = \frac{1}{2\pi\sigma_t} \tag{A.2-2}$$

is the rms spectral width.

This definition is not appropriate for functions with negative or complex values. For such functions the rms width of the squared-absolute value $|f(t)|^2$ is used,

$$\sigma_t^2 = \frac{\int_{-\infty}^{\infty} (t - \bar{t})^2 |f(t)|^2\,dt}{\int_{-\infty}^{\infty} |f(t)|^2\,dt}, \qquad \text{where } \bar{t} = \frac{\int_{-\infty}^{\infty} t|f(t)|^2\,dt}{\int_{-\infty}^{\infty} |f(t)|^2\,dt}.$$

We call this version of σ_t the *power-rms width.*

With the help of the Schwarz inequality, it can be shown that the product of the power rms widths of an arbitrary function $f(t)$ and its Fourier transform $F(\nu)$ must be greater than $1/4\pi$,

$$\sigma_t \sigma_\nu \geq \frac{1}{4\pi}, \tag{A.2-3}$$

Duration–Bandwidth Reciprocity Relation

where the spectral width σ_ν is defined by

$$\sigma_\nu^2 = \frac{\int_{-\infty}^{\infty} (\nu - \bar{\nu})^2 |F(\nu)|^2 \, d\nu}{\int_{-\infty}^{\infty} |F(\nu)|^2 \, d\nu}, \qquad \text{where } \bar{\nu} = \frac{\int_{-\infty}^{\infty} \nu |F(\nu)|^2 \, d\nu}{\int_{-\infty}^{\infty} |F(\nu)|^2 \, d\nu}.$$

Thus the time duration and the spectral width cannot simultaneously be made arbitrarily small.

The *Gaussian function* $f(t) = \exp(-t^2/4\sigma_t^2)$, for example, has a power-rms width σ_t. Its Fourier transform is also a Gaussian function, $F(\nu) = (1/2\sqrt{\pi}\,\sigma_\nu)\exp(-\nu^2/4\sigma_\nu^2)$, with power-rms width

$$\sigma_\nu = \frac{1}{4\pi\sigma_t}. \tag{A.2-4}$$

Since $\sigma_t \sigma_\nu = 1/4\pi$, the Gaussian function has the minimum permissible value of the duration–bandwidth product. In terms of the angular frequency $\omega = 2\pi\nu$,

$$\sigma_t \sigma_\omega \geq \tfrac{1}{2}. \tag{A.2-5}$$

If the variables t and ω, which usually describe time and angular frequency (rad/s), are replaced with the position variable x and the spatial angular frequency k (rad/m), respectively, then (A.2-5) translates to

$$\sigma_x \sigma_k \geq \tfrac{1}{2}. \tag{A.2-6}$$

In quantum mechanics, the position x of a particle is described by the wavefunction $\psi(x)$, and the wavenumber k is described by a function $\phi(k)$ which is the Fourier transform of $\psi(x)$. The uncertainties of x and k are the rms widths of the probability densities $|\psi(x)|^2$ and $|\phi(k)|^2$, respectively, so that σ_x and σ_k are interpreted as the uncertainties of position and wavenumber. Since the particle momentum is $p = \hbar k$ (where $\hbar = h/2\pi$ and h is Planck's constant), the position–momentum uncertainty product satisfies the inequality

$$\sigma_x \sigma_p \geq \frac{\hbar}{2}, \tag{A.2-7}$$

Heisenberg Uncertainty Relation

which is known as the **Heisenberg uncertainty relation**.

The Power-Equivalent Width

The power-equivalent width of a signal $f(t)$ is the signal energy divided by the peak signal power. If $f(t)$ has its peak value at $t = 0$, for example, then the power-equivalent width is

$$\tau = \int_{-\infty}^{\infty} \frac{|f(t)|^2}{|f(0)|^2}\, dt. \tag{A.2-8}$$

The *double-sided exponential function* $f(t) = \exp(-|t|/\tau)$, for example, has a power-equivalent width τ, as does the Gaussian function $f(t) = \exp(-\pi t^2/2\tau^2)$. This definition is used in Sec. 10.1, where the coherence time of light is defined as the power-equivalent width of the complex degree of temporal coherence.

The power-equivalent spectral width is similarly defined by

$$\mathscr{B} = \int_{-\infty}^{\infty} \frac{|F(\nu)|^2}{|F(0)|^2}\, d\nu. \tag{A.2-9}$$

If $f(t)$ is real, so that $|F(\nu)|^2$ is symmetric, and if it has its peak value at $\nu = 0$, the power-equivalent spectral width is usually defined as the positive-frequency width,

$$B = \int_{0}^{\infty} \frac{|F(\nu)|^2}{|F(0)|^2}\, d\nu. \tag{A.2-10}$$

In the case $F(\nu) = \tau/(1 + j2\pi\nu\tau)$, for example,

$$B = \frac{1}{4\tau}. \tag{A.2-11}$$

This definition is used in Sec. 17.5A to describe the bandwidth of photodetector circuits susceptible to photon and circuit noise (see also Problem 17.5-5).

Using Parseval's theorem (A.1-7) and the relation $F(0) = \int_{-\infty}^{\infty} f(t)\, dt$, (A.2-10) may be written in the form

$$B = \frac{1}{2T}, \tag{A.2-12}$$

where

$$T = \frac{\left[\int_{-\infty}^{\infty} f(t)\, dt\right]^2}{\int_{-\infty}^{\infty} f^2(t)\, dt} \tag{A.2-13}$$

is yet another definition of the time duration [the square of the area under $f(t)$ divided by the area under $f^2(t)$]. In this case, the duration–bandwidth product $BT = \frac{1}{2}$.

The $1/e$-, Half-Maximum, and 3-dB Widths

Another type of measure of the width of a function is its duration at a prescribed fraction of its maximum value ($1/\sqrt{2}$, $1/2$, $1/e$, or $1/e^2$, as examples). Either the half-width or the full width on both sides of the peak is used. Two commonly encountered measures are the full-width at half-maximum (FWHM) and the half-width at $1/\sqrt{2}$-maximum, called the 3-dB width. The following are three important examples:

- The *exponential function* $f(t) = \exp(-t/\tau)$ for $t \geq 0$ and $f(t) = 0$ for $t < 0$, which describes the response of a number of electrical and optical systems, has a $1/e$-maximum width $\Delta t_{1/e} = \tau$. The magnitude of its Fourier transform $F(\nu) = \tau/(1 + j2\pi\nu\tau)$ has a 3-dB width (half-width at $1/\sqrt{2}$-maximum)

$$\Delta\nu_{3\text{-dB}} = \frac{1}{2\pi\tau}. \tag{A.2-14}$$

- The *double-sided exponential function* $f(t) = \exp(-|t|/\tau)$ has a half-width at $1/e$-maximum $\Delta t_{1/e} = \tau$. Its Fourier transform $F(\nu) = 2\tau/[1 + (2\pi\nu\tau)^2]$, known as the *Lorentzian distribution*, has a full-width at half-maximum

$$\Delta\nu_{\text{FWHM}} = \frac{1}{\pi\tau}, \tag{A.2-15}$$

  and is usually written in the form $F(\nu) = (\Delta\nu/2\pi)/[\nu^2 + (\Delta\nu/2)^2]$ where $\Delta\nu = \Delta\nu_{\text{FWHM}}$. The Lorentzian distribution describes the spectrum of certain light emissions (see Sec. 12.2D).
- The *Gaussian function* $f(t) = \exp(-t^2/2\tau^2)$ has a full-width at $1/e$-maximum $\Delta t_{1/e} = 2\sqrt{2}\,\tau$. Its Fourier transform $F(\nu) = \sqrt{2\pi}\,\tau \exp(-2\pi^2\tau^2\nu^2)$ has a full-width at $1/e$-maximum

$$\Delta\nu_{1/e} = \frac{\sqrt{2}}{\pi\tau} \tag{A.2-16}$$

  and a full-width at half-maximum

$$\Delta\nu_{\text{FWHM}} = \frac{(2\ln 2)^{1/2}}{\pi\tau}, \tag{A.2-17}$$

  so that

$$\Delta\nu_{\text{FWHM}} = (\ln 2)^{1/2}\,\Delta\nu_{1/e} = 0.833\,\Delta\nu_{1/e}. \tag{A.2-18}$$

  The Gaussian function is also used to describe the spectrum of certain light emissions (see Sec. 12.2D) as well as to describe the spatial distribution of light beams (see Sec. 3.1).

A.3. Two-Dimensional Fourier Transform

We now consider a function of two variables $f(x, y)$. If x and y represent the coordinates of a point in a two-dimensional space, then $f(x, y)$ represents a spatial pattern (e.g., the optical field in a given plane). The harmonic function $F\exp[-j2\pi(\nu_x x + \nu_y y)]$ is regarded as a building block from which other functions may be composed by superposition. The variables ν_x and ν_y represent spatial frequencies in the x and y directions, respectively. Since x and y have units of length (mm), ν_x and ν_y have units of cycles/mm, or lines/mm. Examples of two-dimensional harmonic functions are illustrated in Fig. A.3-1.

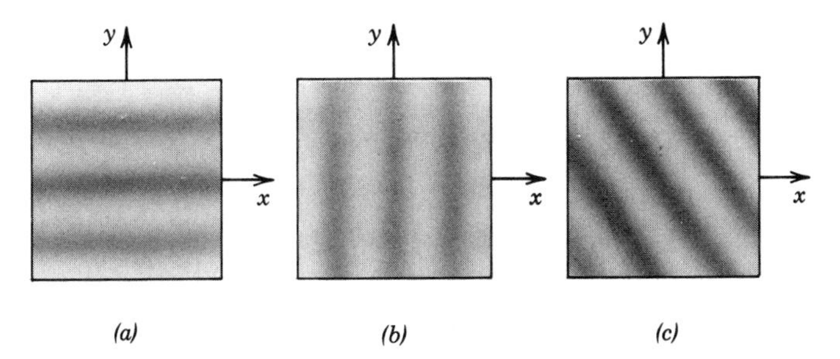

Figure A.3-1 The real part $|F|\cos[2\pi\nu_x x + 2\pi\nu_y y + \arg\{F\}]$ of a two-dimensional harmonic function: (*a*) $\nu_x = 0$; (*b*) $\nu_y = 0$; (*c*) arbitrary case. For this illustration we have assumed that $\arg\{F\} = 0$ so that dark and white points represent positive and negative values of the function, respectively.

The Fourier theorem may be generalized to functions of two variables. A function $f(x, y)$ may be decomposed as a superposition integral of harmonic functions of x and y,

$$f(x,y) = \iint_{-\infty}^{\infty} F(\nu_x,\nu_y)\exp\left[-j2\pi(\nu_x x + \nu_y y)\right] d\nu_x\, d\nu_y \tag{A.3-1}$$

Inverse Fourier Transform

where the coefficients $F(\nu_x, \nu_y)$ are determined by use of the two-dimensional Fourier transform

$$F(\nu_x,\nu_y) = \iint_{-\infty}^{\infty} f(x,y)\exp\left[j2\pi(\nu_x x + \nu_y y)\right] dx\, dy. \tag{A.3-2}$$

Fourier Transform

Our definitions of the two- and one-dimensional Fourier transforms, (A.3-2) and (A.1-2) respectively, differ in the sign of the exponent. The choice of this sign is, of course, arbitrary, as long as opposite signs are used in the Fourier and inverse Fourier transforms. In this book we have adopted the convention that $\exp(j2\pi\nu t)$ has positive temporal frequency ν, whereas $\exp[-j2\pi(\nu_x x + \nu_y y)]$ has positive spatial frequencies ν_x and ν_y. We have elected to use different signs in the spatial (two-dimensional) and temporal (one-dimensional) cases in order to simplify the notation used in Chap. 4 (Fourier optics), in which the traveling wave $\exp(+j2\pi\nu t)\exp[-j(k_x x + k_y y + k_z z)]$ has temporal and spatial dependences with opposite signs.

Properties

The two-dimensional Fourier transform has many properties that are obvious generalizations of those of the one-dimensional Fourier transform, and others that are unique to the two-dimensional case:

- *Convolution Theorem.* If $f(x, y)$ is the two-dimensional convolution of two functions $f_1(x, y)$ and $f_2(x, y)$ with Fourier transforms $F_1(\nu_x, \nu_y)$ and $F_2(\nu_x, \nu_y)$, respectively, so that

$$f(x, y) = \int_{-\infty}^{\infty}\int_{-\infty}^{\infty} f_1(x', y')f_2(x - x', y - y')\, dx'dy', \tag{A.3-3}$$

then the Fourier transform of $f(x, y)$ is

$$F(\nu_x, \nu_y) = F_1(\nu_x, \nu_y)F_2(\nu_x, \nu_y). \tag{A.3-4}$$

Thus, as in the one-dimensional case, convolution in the space domain is equivalent to multiplication in the Fourier domain.

- *Separable Functions.* If $f(x, y) = f_x(x)f_y(y)$ is the product of one function of x and another of y, then its two-dimensional Fourier transform is a product of one function of ν_x and another of ν_y. The two-dimensional Fourier transform of $f(x, y)$ is then related to the product of the one-dimensional Fourier transforms of $f_x(x)$ and $f_y(y)$ by $F(\nu_x, \nu_y) = F_x(-\nu_x)F_y(-\nu_y)$. For example, the Fourier transform of $\delta(x - x_0)\delta(y - y_0)$, which represents an impulse located at (x_0, y_0), is the harmonic function $\exp[j2\pi(\nu_x x_0 + \nu_y y_0)]$; and the Fourier transform of the Gaussian function $\exp[-\pi(x^2 + y^2)]$ is the Gaussian function $\exp[-\pi(\nu_x^2 + \nu_y^2)]$; and so on.
- *Circularly Symmetric Functions.* The Fourier transform of a circularly symmetric function is also circularly symmetric. For example, the Fourier transform of

$$f(x, y) = \begin{cases} 1, & (x^2 + y^2)^{1/2} \le 1 \\ 0, & \text{otherwise}, \end{cases} \tag{A.3-5}$$

denoted by the symbol $\text{circ}(x, y)$ and known as the *circ function*, is

$$F(\nu_x, \nu_y) = \frac{J_1(2\pi\nu_\rho)}{\nu_\rho}, \qquad \nu_\rho = (\nu_x^2 + \nu_y^2)^{1/2}, \tag{A.3-6}$$

where J_1 is the Bessel function of order 1. These functions are illustrated in Fig. A.3-2.

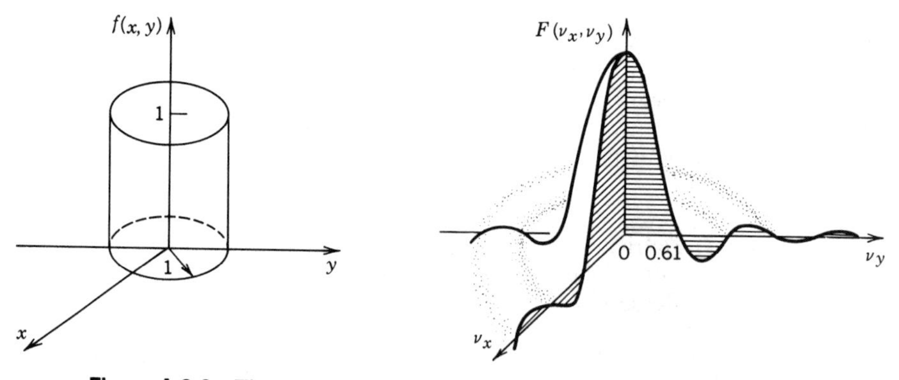

Figure A.3-2 The circ function and its two-dimensional Fourier transform.

READING LIST

E. Kamen, *Introduction to Signals and Systems*, Macmillan, New York, 1987, 2nd ed. 1990.

R. A. Gabel and R. A. Roberts, *Signals and Linear Systems*, Wiley, New York, 3rd ed. 1987.

C. D. McGillem and G. R. Cooper, *Continuous and Discrete Signal and System Analysis*, Holt, Rinehart and Winston, New York, 2nd ed. 1984.

A. V. Oppenheim and A. S. Willsky, *Signals and Systems*, Prentice-Hall, Englewood Cliffs, NJ, 1983.

R. N. Bracewell, *The Fourier Transform and Its Applications*, McGraw-Hill, New York, 2nd ed. 1978.

J. D. Gaskill, *Linear Systems, Fourier Transforms, and Optics*, Wiley, New York, 1978.

L. E. Franks, *Signal Theory*, Prentice-Hall, Englewood Cliffs, NJ, 1969.

A. Papoulis, *Systems and Transforms with Applications in Optics*, McGraw-Hill, New York, 1968.

A. Papoulis, *The Fourier Integral and Its Applications*, McGraw-Hill, New York, 1962.

APPENDIX

B

LINEAR SYSTEMS

This appendix provides a review of the basic characteristics of one- and two-dimensional linear systems.

B.1. One-Dimensional Linear Systems

Consider a system whose input and output are the functions $f_1(t)$ and $f_2(t)$, respectively. An example is a harmonic oscillator driven by a time-varying force $f_1(t)$ that responds by undergoing a displacement $f_2(t)$. The system is characterized by a rule that relates the output to the input. In general, the rule may take the form of a differential equation, an integral transform, or a simple mathematical operation such as $f_2(t) = \log f_1(t)$.

Linear Systems

A system is said to be *linear* if it satisfies the principle of superposition, i.e., if its response to the sum of any two inputs is the sum of its responses to each of the inputs separately. The output at time t is, in general, a weighted superposition of the input contributions at different times τ,

$$f_2(t) = \int_{-\infty}^{\infty} h(t;\tau) f_1(\tau)\, d\tau, \tag{B.1-1}$$

where $h(t;\tau)$ is a weighting function representing the contribution of the input at time τ to the output at time t. If the input is an impulse at τ, so that $f_1(t) = \delta(t-\tau)$, then (B.1-1) gives $f_2(t) = h(t;\tau)$. Thus $h(t;\tau)$ is the **impulse-response function** of the system (also known as the **Green's function**).

Linear Shift-Invariant Systems

A linear system is said to be **time-invariant** or **shift-invariant** if, when its input is shifted in time, its output shifts by an equal time, but otherwise remains the same. The impulse-response function is then a function of the time difference, $h(t;\tau) = h(t-\tau)$. Under these conditions, (B.1-1) becomes

$$\boxed{f_2(t) = \int_{-\infty}^{\infty} h(t-\tau) f_1(\tau)\, d\tau.} \tag{B.1-2}$$

Thus the output $f_2(t)$ is the convolution of the input $f_1(t)$ with the impulse-response

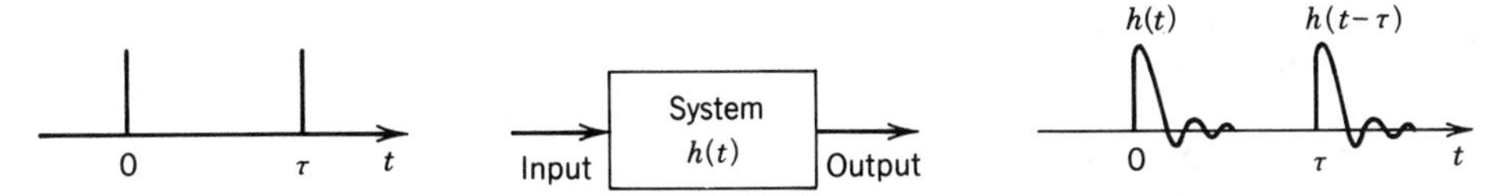

Figure B.1-1 Response of a linear shift-invariant system to impulses.

function $h(t)$ [see (A.1-4)]. If $f_1(t) = \delta(t)$, then $f_2(t) = h(t)$; and if $f_1(t) = \delta(t - \tau)$, then $f_2(t) = h(t - \tau)$, as illustrated in Fig. B.1-1.

The Transfer Function

In accordance with the convolution theorem discussed in Appendix A, the Fourier transforms $F_1(\nu)$, $F_2(\nu)$, and $\mathcal{H}(\nu)$, of $f_1(t)$, $f_2(t)$, and $h(t)$, respectively, are related by

$$F_2(\nu) = \mathcal{H}(\nu)F_1(\nu). \tag{B.1-3}$$

If the input $f_1(t)$ is a harmonic function $F_1(\nu)\exp(j2\pi\nu t)$, the output $f_2(t) = \mathcal{H}(\nu)F_1(\nu)\exp(j2\pi\nu t)$ is also a harmonic function of the same frequency but with a modified complex amplitude $F_2(\nu) = F_1(\nu)\mathcal{H}(\nu)$, as illustrated in Fig. B.1-2. The multiplicative factor $\mathcal{H}(\nu)$ is known as the system's **transfer function**. The transfer function is the Fourier transform of the impulse-response function. Equation (B.1-3) is the key to the usefulness of Fourier methods in the analysis of linear shift-invariant systems. To determine the output of a system for an arbitrary input, we simply decompose the input into its harmonic components, multiply the complex amplitude of each harmonic function by the transfer function at the appropriate frequency, and superpose the resultant harmonic functions.

Examples

- *Ideal system:* $\mathcal{H}(\nu) = 1$ and $h(t) = \delta(t)$; the output is a replica of the input.
- *Ideal system with delay:* $\mathcal{H}(\nu) = \exp(-j2\pi\nu\tau)$ and $h(t) = \delta(t - \tau)$; the output is a replica of the input delayed by time τ.
- *System with exponential response:* $\mathcal{H}(\nu) = \tau/(1 + j2\pi\nu\tau)$ and $h(t) = e^{-t/\tau}$ for $t \geq 0$, and $h(t) = 0$, otherwise; this represents the response of a system described by a first-order linear differential equation, e.g., that representing an *R-C* circuit with time constant τ. An impulse at the input results in an exponentially decaying response.
- *Chirped system:* $\mathcal{H}(\nu) = \exp(-j\pi\nu^2)$ and $h(t) = e^{-j\pi/4}\exp(j\pi t^2)$; the system distorts the input by imparting to it a phase shift proportional to ν^2. An input

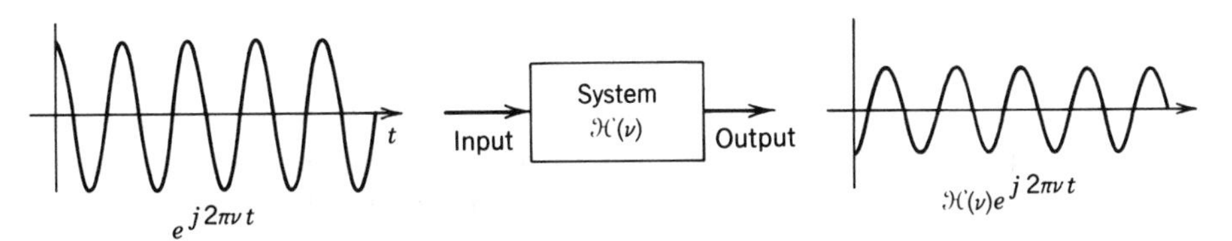

Figure B.1-2 Response of a linear shift-invariant system to a harmonic function.

impulse generates an output in the form of a chirped signal, i.e., a harmonic function whose instantaneous frequency (the derivative of the phase) increases linearly with time. This system describes the propagation of optical pulses through media with a frequency-dependent phase velocity (see Sec. 5.6). It also describes changes in the spatial distribution of light waves as they propagate through free space (see Sec. 4.1C).

Linear Shift-Invariant Causal Systems

The impulse response function $h(t)$ of a linear shift-invariant *causal* system must vanish for $t < 0$, since the system's response cannot begin before the application of the input. The function $h(t)$ is therefore not symmetric and its Fourier transform, the transfer function $\mathcal{H}(\nu)$, must be complex. It can be shown† that if $h(t) = 0$ for $t < 0$, then the real and imaginary parts of $\mathcal{H}(\nu)$, denoted $\mathcal{H}'(\nu)$ and $\mathcal{H}''(\nu)$ respectively, are related by

$$\mathcal{H}'(\nu) = \frac{1}{\pi}\int_{-\infty}^{\infty} \frac{\mathcal{H}''(s)}{s-\nu}\,ds \qquad \text{(B.1-4)}$$

$$\mathcal{H}''(\nu) = \frac{1}{\pi}\int_{-\infty}^{\infty} \frac{\mathcal{H}'(s)}{\nu-s}\,ds, \qquad \text{(B.1-5)}$$

Hilbert Transform

where the Cauchy principal values of the integrals are to be evaluated, i.e.,

$$\int_{-\infty}^{\infty} \equiv \lim_{\Delta\to 0}\left(\int_{-\infty}^{\nu-\Delta} + \int_{\nu+\Delta}^{\infty}\right), \qquad \Delta > 0.$$

Functions that satisfy (B.1-4) and (B.1-5) are said to form a Hilbert transform pair, $\mathcal{H}''(\nu)$ being the **Hilbert transform** of $\mathcal{H}'(\nu)$.

If the impulse response function $h(t)$ is also real, its Fourier transform must be symmetric, $\mathcal{H}(-\nu) = \mathcal{H}^*(\nu)$. The real part $\mathcal{H}'(\nu)$ then has even symmetry, and the imaginary part $\mathcal{H}''(\nu)$ has odd symmetry. The integrals in (B.1-4) and (B.1-5) may then be rewritten as integrals over the interval $(0, \infty)$. The resultant equations are known as the **Kramers–Kronig relations**

$$\mathcal{H}'(\nu) = \frac{2}{\pi}\int_{0}^{\infty} \frac{s\mathcal{H}''(s)}{s^2-\nu^2}\,ds \qquad \text{(B.1-6)}$$

$$\mathcal{H}''(\nu) = \frac{2}{\pi}\int_{0}^{\infty} \frac{\nu\mathcal{H}'(s)}{\nu^2-s^2}\,ds. \qquad \text{(B.1-7)}$$

Kramers–Kronig Relations

In summary, the Hilbert-transform relations, or the Kramers–Kronig relations, relate the real and imaginary parts of the transfer function of a linear shift-invariant

†See, e.g., L. E. Franks, *Signal Theory*, Prentice-Hall, Englewood Cliffs, NJ, 1969.

causal system, so that if one part is known at all frequencies, the other part may be determined.

Example: The Harmonic Oscillator

The linear system described by the differential equation

$$\left(\frac{d^2}{dt^2} + \sigma\frac{d}{dt} + \omega_0^2\right) f_2(t) = f_1(t) \tag{B.1-8}$$

describes a harmonic oscillator with displacement $f_2(t)$ under an applied force $f_1(t)$, where ω_0 is the resonance angular frequency and σ is a coefficient representing damping effects. The transfer function $\mathcal{H}(\nu)$ of this system may be obtained by substituting $f_1(t) = \exp(j2\pi\nu t)$ and $f_2(t) = \mathcal{H}(\nu)\exp(j2\pi\nu t)$ in (B.1-8), which yields

$$\mathcal{H}(\nu) = \frac{1}{(2\pi)^2}\frac{1}{\nu_0^2 - \nu^2 + j\nu\,\Delta\nu}, \tag{B.1-9}$$

where $\nu_0 = \omega_0/2\pi$ is the resonance frequency, and $\Delta\nu = \sigma/2\pi$. The real and imaginary parts of $\mathcal{H}(\nu)$ are therefore

$$\mathcal{H}'(\nu) = \frac{1}{(2\pi)^2}\frac{\nu_0^2 - \nu^2}{\left(\nu_0^2 - \nu^2\right)^2 + (\nu\,\Delta\nu)^2} \tag{B.1-10}$$

$$\mathcal{H}''(\nu) = -\frac{1}{(2\pi)^2}\frac{\nu\,\Delta\nu}{\left(\nu_0^2 - \nu^2\right)^2 + (\nu\,\Delta\nu)^2}. \tag{B.1-11}$$

Since the system is causal, $\mathcal{H}'(\nu)$ and $\mathcal{H}''(\nu)$ satisfy the Kramers-Kronig relations. When $\nu_0 \gg \Delta\nu$, $\mathcal{H}'(\nu)$ and $\mathcal{H}''(\nu)$ are narrow functions centered about ν_0. For $\nu \approx \nu_0$, $(\nu_0^2 - \nu^2) \approx 2\nu_0(\nu_0 - \nu)$ so that (B.1-10) and (B.1-11) may be approximated by

$$\mathcal{H}''(\nu) = -\frac{1}{(2\pi)^2}\frac{\Delta\nu/4\nu_0}{(\nu_0 - \nu)^2 + (\Delta\nu/2)^2} \tag{B.1-12}$$

$$\mathcal{H}'(\nu) = 2\frac{\nu - \nu_0}{\Delta\nu}\mathcal{H}''(\nu). \tag{B.1-13}$$

The transfer function of the harmonic-oscillator system is used in Chaps. 5 and 13 to describe dielectric and atomic systems. Equation (B.1-12) has a Lorentzian form.

B.2. Two-Dimensional Linear Systems

A two-dimensional system relates two two-dimensional functions $f_1(x, y)$ and $f_2(x, y)$, called the input and output functions. These functions may, for example, represent optical fields at two parallel planes, with (x, y) representing position variables; the system comprises the free space and optical components that lie between the two planes.

The concepts of linearity and shift invariance defined in the one-dimensional case are easily generalized to the two-dimensional case. The output $f_2(x, y)$ of a *linear*

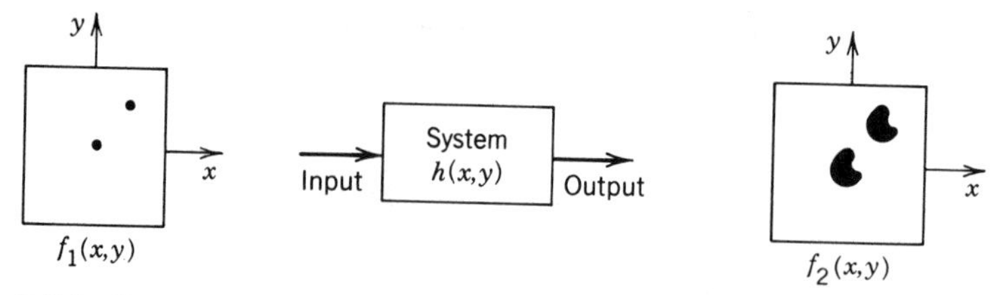

Figure B.2-1 Response of a two-dimensional linear shift-invariant system to two impulses.

system is related to its input $f_1(x, y)$ by a superposition integral

$$f_2(x, y) = \iint_{-\infty}^{\infty} h(x, y; x', y') f_1(x', y')\, dx'\, dy', \tag{B.2-1}$$

where $h(x, y; x', y')$ is a weighting function that represents the effect of the input at the point (x', y') on the output at the point (x, y). The function $h(x, y; x', y')$ is the **impulse-response function** of the system (also known as the **point-spread function**).

The system is said to be **shift-invariant** (or **isoplanatic**) if shifting its input in some direction shifts the output by the same distance and in the same direction without otherwise altering it (see Fig. B.2-1). The impulse response function is then a function of position differences $h(x, y; x', y') = h(x - x', y - y')$. Equation (B.2-1) then becomes the two-dimensional convolution of $h(x, y)$ with $f_1(x, y)$:

$$\boxed{f_2(x, y) = \iint_{-\infty}^{\infty} h(x - x', y - y') f_1(x', y')\, dx'\, dy'.} \tag{B.2-2}$$

Applying the two-dimensional convolution theorem discussed in Sec. A.3 of Appendix A, we obtain

$$\boxed{F_2(\nu_x, \nu_y) = \mathcal{H}(\nu_x, \nu_y) F_1(\nu_x, \nu_y),} \tag{B.2-3}$$

where $F_2(\nu_x, \nu_y)$, $\mathcal{H}(\nu_x, \nu_y)$, and $F_1(\nu_x, \nu_y)$ are the Fourier transforms of $f_2(x, y)$, $h(x, y)$, and $f_1(x, y)$, respectively.

A harmonic input of complex amplitude $F_1(\nu_x, \nu_y)$ therefore produces a harmonic output of the same spatial frequency but with complex amplitude $F_2(\nu_x, \nu_y) = \mathcal{H}(\nu_x, \nu_y)\, F_1(\nu_x, \nu_y)$, as illustrated in Fig. B.2-2. The multiplicative factor $\mathcal{H}(\nu_x, \nu_y)$ is

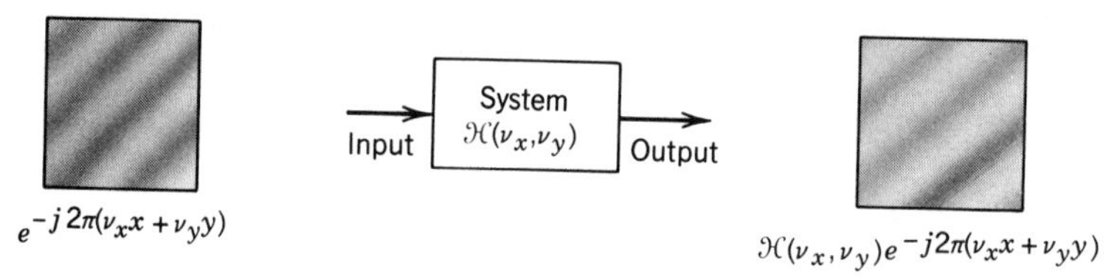

Figure B.2-2 Response of a two-dimensional linear shift-invariant system to harmonic functions.

the system's **transfer function**. The transfer function is the Fourier transform of the impulse-response function. Either of these functions characterizes the system completely and enables us to determine the output corresponding to an arbitrary input.

In summary, a two-dimensional linear shift-invariant system is characterized by its impulse-response function $h(x, y)$ or its transfer function $\mathcal{H}(\nu_x, \nu_y)$. For example, a system with $h(x, y) = \text{circ}(x/\rho_s, y/\rho_s)$ smears each point of the input into a patch in the form of a circle of radius ρ_s. It has a transfer function $\mathcal{H}(\nu_x, \nu_y) = \rho_s J_1(2\pi\rho_s\nu_\rho)/\nu_\rho$, where $\nu_\rho = (\nu_x^2 + \nu_y^2)^{1/2}$, which has the shape illustrated in Fig. A.3-2. The system severely attenuates spatial frequencies higher than $0.61/\rho_s$ lines/mm.

READING LIST

See the reading list in Appendix A.

APPENDIX

C

MODES OF LINEAR SYSTEMS

Every linear system is characterized by special inputs that are invariant to the system, i.e., inputs that are not altered (except for a multiplicative constant) upon passage through the system. These inputs are called the modes, or the eigenfunctions, of the system. The multiplicative constants are the eigenvalues; they are the attenuation or amplification factors of the modes.

A linear system is completely characterized by its eigenfunctions and eigenvalues. An arbitrary input function may be expanded as a combination of the eigenfunctions, each of which is multiplied by the corresponding eigenvalue upon transmission through the system, and the output is the sum of the resultant components. The modes are transmitted through the system without mixing among themselves.

The linear system that operates on the two-dimensional function $f(x, y)$ in accordance with (B.2-1), for example, is characterized by a number of modes satisfying the integral equation

$$\iint_{-\infty}^{\infty} h(x, y; x', y') f_q(x', y')\, dx'\, dy' = \lambda_q f_q(x, y), \qquad q = 1, 2, \ldots. \quad \text{(C.1-1)}$$

The functions $f_q(x, y)$ and the constants λ_q are the eigenfunctions and eigenvalues of the system, respectively. When $f_q(x, y)$ is the input to the system, the output is $\lambda_q f_q(x,y)$, which is identical to the input, except for the multiplicative factor λ_q.

An example (discussed in Sec. 9.2E) is light traveling a single round trip between two mirrors in a laser resonator. The distributions of light in the transverse plane at the beginning and at the end of the trip are the input and output to the system. The modes of the resonator are those light distributions that maintain their shape after one round trip, except for a multiplicative factor representing propagation and reflection losses. The modes are therefore the stationary distributions that remain unchanged after many round trips.

Another example (discussed in Chap. 7) is light traveling in an optical waveguide. The modes of the waveguide are those distributions in the transverse plane (the x–y plane) that are not altered as the light travels along the axis of the waveguide (the z direction). The eigenvalues are the phase factors $\exp(-j\beta_q z)$, where β_q is the propagation constant of mode q.

The concept of modes applies also to one-dimensional linear systems operating on functions $f(t)$. The modes of a linear shift-invariant system are the harmonic functions $\exp(j2\pi\nu t)$, since these functions maintain their harmonic nature (including the frequency) when they are transmitted through the system. The eigenvalue associated with the harmonic function of frequency ν is the system's transfer function $\mathcal{H}(\nu)$. In this case there is a continuum of modes indexed by the frequency ν.

Discrete linear systems are also important in optics. The linear system operating on vectors of size N (sets of numbers $X_1, X_2, \ldots, X_N$ arranged in a column matrix $\mathbf{X}$) is characterized by a square matrix $\mathbf{M}$ of size $N \times N$, which operates on an input vector $\mathbf{X}$ to generate an output vector $\mathbf{Y} = \mathbf{MX}$. The modes of such discrete systems are those input vectors that remain parallel to themselves upon transmission through the system, i.e., that obey $\mathbf{MX}_q = \lambda_q \mathbf{X}_q$, $q = 1, 2, \ldots, N$, where λ_q is a scalar. Thus the modes of the system are the eigenvectors $\mathbf{X}_q$ of the matrix $\mathbf{M}$, and the scalars λ_q are the corresponding eigenvalues.

The special case of discrete systems operating on vectors of size $N = 2$ is particularly important in optics. This system performs the matrix operation

$$\begin{bmatrix} Y_1 \\ Y_2 \end{bmatrix} = \begin{bmatrix} A & B \\ C & D \end{bmatrix} \begin{bmatrix} X_1 \\ X_2 \end{bmatrix},$$

with the input and output represented by vectors of size 2. There are two independent modes of the system, the eigenvectors of the *ABCD* matrix. Such systems describe the transformation of the polarization of light transmitted through an optical system (see Sec. 6.1B). The vector (X_1, X_2) represents the components of the input electric field in two orthogonal directions (the Jones vector), and (Y_1, Y_2) similarly describes the output electric field. The modes of the optical system are the vectors (X_1, X_2) (polarization states) that change only by a multiplicative factor on passing through the optical system. They represent the polarization states that are maintained as light travels through the system.

READING LIST

See the reading list in Appendix A.

SYMBOLS

Roman Symbols

a = Radius of an aperture or fiber [m]; also, Lattice constant [m]

a = Normalized complex amplitude of an optical field ($|\mathsf{a}|^2$ = photon number)

$\mathscr{a}$ = Amplitude (magnitude) of an optical wave; also, Normalized complex amplitude of an optical field ($|\mathscr{a}|^2$ = photon flux density)

A = Complex envelope of a monochromatic plane wave

$A(\mathbf{r})$ = Complex envelope of a monochromatic wave

A_ν = Complex envelope of the component of a wave at frequency ν

A = Area [m^2]; also, Element of the *ABCD* matrix

A_c = Coherence area [m^2]

$\mathscr{A}(\mathbf{r}, t)$ = Complex envelope of a polychromatic (e.g., pulsed) wave

$\mathbf{A}$ = Vector potential [$V \cdot s \cdot m^{-1}$]

$\mathbb{A}$ = Einstein A coefficient [s^{-1}]

ASE = Amplified spontaneous emission

B = Magnetic flux-density complex amplitude [Wb/m^2]; also, Bandwidth [Hz]

B_0 = Bit rate [bits/s]

B = Element of the *ABCD* matrix

$\mathscr{B}$ = Magnetic flux density [Wb/m^2]; also, Power-equivalent spectral width [Hz]

$\mathbb{B}$ = Einstein B coefficient [$m^3 \cdot J^{-1} \cdot s^{-2}$]

BER = Bit error rate

c = Speed of light; Phase velocity [m/s]

c_o = Speed of light in free space [m/s]

C = Electrical capacitance [F]

$C(\cdot)$ = Fresnel integral

C = Element of the *ABCD* matrix

$\mathscr{C}$ = Coupling coefficient in a directional coupler [m^{-1}]

d = Differential

$d\mathbf{r}$ = Incremental volume [m^3]

ds = Incremental length [m]
d = Distance; Length [m]
$\mathscr{d}$ = Coefficient of second-order optical nonlinearity [C · V^{-2}]
$\mathscr{d}_{ijk}$ = Element of the second-order optical nonlinearity tensor [C · V^{-2}]
$\mathscr{d}_{Ik}$ = Element of the second-order optical nonlinearity tensor (contracted indices) [C · V^{-2}]
$\mathscr{d}(\omega_3; \omega_1, \omega_2)$ = Coefficient of second-order optical nonlinearity (dispersive medium) [C · V^{-2}]
D = Diameter [m]; also, Electric flux-density complex amplitude [C/m^2]
D_w = Waveguide dispersion coefficient [s · m^{-2}]
D_x, D_y = Lateral widths [m]
D_λ = Material dispersion coefficient [s · m^{-2}]
D_ν = Material dispersion coefficient [s^2 · m^{-1}]
D = Element of the *ABCD* matrix
$\mathscr{D}$ = Electric flux density [C/m^2]

e = Magnitude of electron charge [C]
$\hat{\mathbf{e}}_x$ = Unit vector in the x direction
E = Electric-field complex amplitude [V/m]
E = Energy [J]
E_A = Acceptor energy level [J]
E_c = Energy at the bottom of the conduction band [J]
E_D = Donor energy level [J]
E_f = Fermi energy [J]
E_{fc} = Quasi-Fermi energy for the conduction band [J]
E_{fv} = Quasi-Fermi energy for the valence band [J]
E_g = Bandgap energy [J]
E_v = Energy at the top of the valence band [J]
E_ν = Energy spectral density [J · Hz^{-1}]
$\mathscr{E}$ = Electric field [V/m]

f = Focal length of a lens [m]; also, Frequency [Hz]
$f(E)$ = Fermi function
f_a = Probability that absorption condition is satisfied
$f_c(E)$ = Fermi function for the conduction band
f_{col} = Collision rate [s^{-1}]
f_e = Probability that emission condition is satisfied
f_g = Fermi inversion factor
$f_v(E)$ = Fermi function for the valence band
f = Frequency of sound [Hz]; also, Modulation frequency [Hz]
$\mathscr{f}$ = Focal length [m]
F = Excess-noise factor of an avalanche photodiode
$F_\#$ = F-number of a lens
$\mathscr{F}$ = Finesse of a resonator; also, Force [kg · m · s^{-2}]

g = Resonator g-parameter
$g(\mathbf{r}_1, \mathbf{r}_2)$ = Normalized mutual intensity

$g(\mathbf{r}_1, \mathbf{r}_2, \tau)$ = Complex degree of coherence
$g(\nu)$ = Lineshape function of a transition [Hz^{-1}]
$g(\tau)$ = Complex degree of temporal coherence
g_0 = Gain factor
$g_{\nu 0}(\nu)$ = Electron–phonon collisionally broadened lineshape function in a semiconductor [Hz^{-1}]
g = Degeneracy parameter
$\mathscr{g}$ = Coupling coefficient in a parametric interaction [m^{-3}]
G = Gain of an amplifier; also, Gain of a photon detector; also, Conductance [Ω^{-1}]
$G(\mathbf{r}_1, \mathbf{r}_2)$ = Mutual intensity [W/m^2]
$G(\mathbf{r}_1, \mathbf{r}_2, \tau)$ = Mutual coherence function [W/m^2]
$G(\nu)$ = Gain of an optical amplifier
$G(\tau)$ = Temporal coherence function [W/m^2]
G = Rate of photoionization in a photorefractive material [$m^{-3} \cdot s^{-1}$]
$\mathsf{G}_n(\cdot)$ = Hermite-Gaussian functions
G_0 = Rate of thermal electron-hole generation in a semiconductor [$m^{-3} \cdot s^{-1}$]
$\mathbf{G}$ = Coherency matrix [W/m^2]; also, Gyration vector of an optically active medium

h = Planck's constant [$J \cdot s$]
$h(t)$ = Impulse-response function of a linear system
$h(x, y)$ = Impulse-response function of a two-dimensional linear system
$\hbar = h/2\pi$ [$J \cdot s$]
H = Magnetic-field complex amplitude [A/m]
$\mathsf{H}_n(\cdot)$ = Hermite polynomials
$\mathscr{H}$ = Magnetic field [A/m]
$\mathcal{H}(\nu)$ = Transfer function of a linear system
$\mathcal{H}'(\nu)$ = Real part of the transfer function of a linear system
$\mathcal{H}''(\nu)$ = Imaginary part of the transfer function of a linear system
$\mathcal{H}(\nu_x, \nu_y)$ = Transfer function of a two-dimensional linear system

i = Electric current [A]; also, integer
i_e = Electron current [A]
i_h = Hole current [A]
i_p = Photoelectric current [A]
i_s = Reverse current in a semiconductor p–n diode [A]
i_t = Threshold current of a laser diode [A]
i_T = Transparency current for a laser-diode amplifier [A]
I = Optical intensity [W/m^2]
I_s = Saturation optical intensity of an amplifier or absorber [W/m^2]; also, Acoustic intensity [W/m^2]
I_ν = Intensity spectral density [$W \cdot m^{-2} \cdot Hz^{-1}$]
$\mathscr{I}$ = Moment of inertia [$kg \cdot m^2$]

$j = \sqrt{-1}$; also, integer
J = Electric current density [A/m^2]

J_e = Electron current density [A/m^2]
J_h = Hole current density [A/m^2]
$J_m(\cdot)$ = Bessel function of the first kind of order m
J_p = Photoelectric current density [A/m^2]
J_t = Threshold current density of a laser diode [A/m^2]
J_T = Transparency current density for a laser-diode amplifier [A/m^2]
$\mathbf{J}$ = Jones vector

k = Wavenumber [m^{-1}]; also, integer
k_B = Boltzmann's constant [J/K]
k_o = Free-space wavenumber [m^{-1}]
$k_T = (k_x^2 + k_y^2)^{1/2}$ = Lateral component of the wavevector [m^{-1}]
k_x, k_y = Components of the wavevector in the x and y directions [m^{-1}]
= Spatial angular frequencies in the x and y directions [rad/m]
$\mathfrak{k}$ = Ionization ratio for an avalanche photodiode
$\mathbf{k}$ = Wavevector [m^{-1}]
$\mathbf{k}_g$ = Grating wavevector [m^{-1}]
$K_m(\cdot)$ = Modified Bessel function of the second kind of order m

l = Length [m]; also, integer
l_c = Coherence length [m]
L = Length [m]; also, Electrical inductance [H]; also, Loss factor; also, integer
L_c = Coherence length in a parametric interaction [m]
$L_n(\cdot)$ = Laguerre polynomials
$L_0 = \pi/2\mathcal{C}$ = Coupling length (transfer distance) in a directional coupler [m]
LP = Linearly polarized mode

$m = m_0$ = Electron mass or atomic mass [kg]; also, integer; also, Contrast or modulation depth
m_c = Effective mass of a conduction-band electron [kg]
m_r = Reduced mass of an atom [kg]; also, Reduced mass of an electron–hole pair in a semiconductor [kg]
m_v = Effective mass of a hole [kg]
m = Photon number; also, Photoelectron number
M = Magnification in an image system; also, Number of modes; also, integer
M = Mass of an atom [kg]
$M(\nu)$ = Density of modes in a resonator or cavity [m$^{-3}\cdot$ Hz^{-1} for a 3-D resonator; m$^{-1}\cdot$ Hz^{-1} for a 1-D resonator]
$\mathscr{M}$ = Magnetization density [A/m]; also, Number of modes of thermal light; also, Figure of merit for the acousto-optic effect [m^2/W]
$\mathbf{M}$ = Ray-transfer matrix

n = Refractive index; also, integer
$n(\mathbf{r})$ = Refractive index of an inhomogeneous medium
$n(\theta)$ = Refractive index of the extraordinary wave with its wavevector at an angle θ with respect to the optic axis of a uniaxial crystal

n_e = Extraordinary refractive index
n_o = Ordinary refractive index
n_2 = Optical Kerr coefficient (nonlinear refractive index) [m^2/W]
n = Photon number
n = Photon-number density [m^{-3}]
n_s = Saturation photon-number density [m^{-3}]
n = Concentration of electrons in a semiconductor [m^{-3}]
n_i = Concentration of electrons/holes in an intrinsic semiconductor [m^{-3}]
N = Group index; also, integer; also, Number of atoms; also, Number of resolvable spots of a scanner
N_F = Fresnel number
N = Number density [m^{-3}]; also, $N = N_2 - N_1$ = Population density difference between energy levels 2 and 1 [m^{-3}]
N_a = Atomic number density [m^{-3}]
N_A = Number density of ionized acceptor atoms in a semiconductor [m^{-3}]
N_D = Number density of ionized donor atoms in a semiconductor [m^{-3}]
N_t = Laser threshold population difference [m^{-3}]
N_0 = Steady-state population difference in the absence of amplifier radiation [m^{-3}]
NA = Numerical aperture

p = Probability; also, Momentum [$kg \cdot m \cdot s^{-1}$]; also, Grade profile parameter of a graded-index fiber
$p(n)$ = Probability of n events
$p(x, y)$ = Aperture function or pupil function
p_{ab} = Probability density for absorption (mode containing one photon) [s^{-1}]
p_{sp} = Probability density for spontaneous emission (into one mode) [s^{-1}]
p_{st} = Probability density for stimulated emission (mode containing one photon) [s^{-1}]
p = Normalized electric-field quadrature component
p = Dipole moment [$C \cdot m$]
p = Concentration of holes in a semiconductor [m^{-3}]
p = Photoelastic constant (strain-optic coefficient)
p_{ijkl} = Element of the strain-optic tensor
p_{IK} = Element of the strain-optic tensor (contracted indices)
P = Electric polarization-density complex amplitude [C/m^2]
$P(\nu_x, \nu_y)$ = Fourier transform of the aperture function $p(x, y)$
P_{ab} = Probability density for absorption (mode containing many photons) [s^{-1}]
P_{NL} = Complex amplitude of the nonlinear component of the polarization density [C/m^2]
P_{sp} = Probability density for spontaneous emission (into any mode) [s^{-1}]
P_{st} = Probability density for stimulated emission (mode containing many photons) [s^{-1}]
P = Optical power [W]
P_ν = Power spectral density [$W \cdot Hz^{-1}$]

P_π = Half-wave optical power in a Kerr medium [W]
$\mathscr{P}$ = Electric polarization density [C/m^2]; also, Optical power [dBm]
$\mathscr{P}_L$ = Linear component of the polarization density [C/m^2]
$\mathscr{P}_{NL}$ = Nonlinear component of the polarization density [C/m^2]
$\mathcal{P}$ = Degree of polarization

q = Electric charge [C]; also, Wavenumber of an acoustic wave [m^{-1}]; also, integer (mode index, diffraction order)
$q(z)$ = Complex Gaussian-beam parameter [m]
$\mathbf{q}$ = Wavevector of an acoustic wave [m^{-1}]
Q = Electric charge [C]; also, Quality factor of an optical resonator

r = Radial distance in spherical coordinates [m]; also, Radial distance in a cylindrical coordinate system [m]
$\mathbf{r}$ = Position vector [m]
$r(\nu)$ = Rate of photon emission/absorption from a semiconductor [m^{-3}]
$\mathscr{r}$ = Complex amplitude reflectance; also, Round-trip (real) amplitude attenuation factor for a wave in a Fabry-Perot resonator
$\mathfrak{r}$ = Electron–hole recombination parameter [m^3/s]
$\mathfrak{r}_{nr}$ = Nonradiative electron–hole recombination parameter [m^3/s]
$\mathfrak{r}_r$ = Radiative electron–hole recombination parameter [m^3/s]
$\mathfrak{r}$ = Linear electro-optic (Pockels) coefficient [m/V]
$\mathfrak{r}_{ijk}$ = Element of the linear electro-optic tensor [m/V]
$\mathfrak{r}_{Ik}$ = Element of the linear electro-optic tensor (contracted indices) [m/V]
R = Radius of curvature [m]; also, Electrical resistance [Ω]
$R(z)$ = Radius of curvature of a Gaussian beam [m]
R = Pumping rate [s$^{-1}\cdot$ m^{-3}]; also, Recombination rate in a semiconductor [s$^{-1}\cdot$ m^{-3}]; also, Electron–hole injection rate in a semiconductor [s$^{-1}\cdot$ m^{-3}]
R_t = Laser threshold pumping rate [s$^{-1}\cdot$ m^{-3}]
$\mathscr{R}$ = Intensity or power reflectance
$\mathfrak{R}$ = Responsivity of a photon source [W/A]; also, Responsivity of a photon detector [A/W]
$\mathfrak{R}_d$ = Differential responsivity of a laser diode [W/A]
$\mathbf{R}(\theta)$ = Jones matrix for coordinate rotation by an angle θ

s = Length or distance [m]
$s(\mathbf{r}_1, \mathbf{r}_2, \nu)$ = Normalized cross-spectral density
$s(x, t)$ = Strain wavefunction
s_{ij} = Element of the strain tensor
$\mathfrak{s}$ = Quadratic electro-optic (Kerr) coefficient [m^2/V^2]
$\mathfrak{s}_{ijkl}$ = Element of the quadratic electro-optic tensor [m^2/V^2]
$\mathfrak{s}_{IK}$ = Element of the quadratic electro-optic tensor (contracted indices) [m^2/V^2]
S = Transition strength (oscillator strength) [m$^2\cdot$ Hz]
$S(\mathbf{r})$ = Complex amplitude for a radiation source [V/m^3]
$S(\cdot)$ = Fresnel integral

$\mathsf{S}(\mathbf{r})$ = Eikonal [m]

$\mathsf{S}(\mathbf{r}_1, \mathbf{r}_2, \nu)$ = Cross-spectral density [W/m^2 · Hz]

$\mathsf{S}(\nu)$ = Power spectral density [W/m^2 · Hz]

$\mathscr{S}$ = Poynting vector [W/m^2]

$\mathcal{S}$ = Photon spin [J · s]

SNR = Signal-to-noise ratio

t = Time [s]

t_{sp} = Spontaneous lifetime [s]

$\mathscr{t}$ = Complex amplitude transmittance

T = Temperature [K]

T = Transit time [s]; also, Counting time [s]; also, Switching time [s]; also, Bit time interval [s]; also, Resolution time ($\mathsf{T} = 1/2B$ where B = Bandwidth) [s]; also, Period of a wave ($\mathsf{T} = 1/\nu$ where ν = frequency) [s]

T_F = $1/\nu_F$ = Inverse of resonator-mode frequency spacing [s]; also, Period of a mode-locked laser pulse train [s]

$\mathscr{T}$ = Intensity or power transmittance; also, Power-transfer or power-transmission ratio

$\mathbf{T}$ = Jones matrix

TE = Transverse electric wave

TEM = Transverse electromagnetic wave

TM = Transverse magnetic wave

u = Displacement [m]

$u(\mathbf{r}, t)$ = Wavefunction of an optical wave

$\hat{\mathbf{u}}$ = Unit vector

$U(\mathbf{r})$ = Complex amplitude of a monochromatic optical wave

$U(\mathbf{r}, t)$ = Complex wavefunction of an optical wave

$U_\nu(\mathbf{r})$ = Fourier transform of the wavefunction of an optical wave

v = Group velocity of a wave [m/s]

v_s = Velocity of sound [m/s]

v = Velocity of an atom or object [m/s]

v_e = Velocity of an electron [m/s]

v_h = Velocity of a hole [m/s]

V = Volume [m^3]; also, Voltage [V]; also, Verdet constant [m/Wb]

V_c = Critical voltage for a liquid-crystal cell [V]

V_π = Half-wave voltage of an electro-optic retarder or modulator [V]

V_0 = Built-in potential difference in a p–n junction [V]; also Switching voltage of a directional coupler [V]

V = Fiber V parameter

$\mathsf{V}(\mathbf{r})$ = Potential energy [J]

$\mathscr{V}$ = Visibility

$\mathfrak{V}$ = $\mathfrak{V}$-number of a dispersive medium

w = Width [m]

w_d = Width of the absorption region in an avalanche photodiode [m]

w_m = Width of the multiplication region in an avalanche photodiode [m]
W = Work function [J]
$W(z)$ = Width or radius of a Gaussian beam at an axial distance z from the beam center [m]
W_0 = Waist radius of a Gaussian beam [m]
W = Probability density for absorption of pump light [s^{-1}]
W_i = Probability density for absorption and stimulated emission [s^{-1}]
$\mathscr{W}$ = Integrated optical power in units of photon number

x = Position coordinate; displacement [m]
x = Normalized electric-field quadrature component
$x(t)$ = Inverse Fourier transform of the susceptibility of a dispersive medium $\chi(\nu)$

y = Position coordinate [m]

z = Position coordinate (Cartesian or cylindrical coordinates) [m]
z_0 = Rayleigh range of a Gaussian beam [m]; also, Rayleigh range of a Gaussian pulse traveling in a dispersive medium [m]
Z = Atomic number

Greek Symbols

α = Attenuation or absorption coefficient [m^{-1}]; also, Apex angle of a prism; also, Twist coefficient of a twisted nematic liquid crystal [m^{-1}]
α_e = Electron ionization coefficient in a semiconductor [m^{-1}]
α_h = Hole ionization coefficient in a semiconductor [m^{-1}]
α_m = Loss coefficient of a resonator attributed to a mirror [m^{-1}]
α_r = Effective overall distributed loss coefficient [m^{-1}]
α_s = Loss coefficient of a laser medium [m^{-1}]
α_p = Mean value of p for a coherent state
α_x = Mean value of x for a coherent state
$\boldsymbol{\alpha}$ = Attenuation coefficient of an optical fiber [dB/km]

$\beta = k_z$ = Propagation constant [m^{-1}]
β' = First derivative of β with respect to ω [$m^{-1} \cdot s$]
β'' = Second derivative of β with respect to ω [$m^{-1} \cdot s^2$]
$\beta(\nu)$ = Propagation constant in a dispersive medium [m^{-1}]
$\beta_0 = \beta(\nu_0)$ = Propagation constant at the central frequency ν_0 [m^{-1}]

γ = Gain coefficient [m^{-1}]; also, Coupling coefficient in a parametric device [m^{-1}]; also, Nonlinear coefficient in soliton theory; also, Lateral decay coefficient in a waveguide [m^{-1}]; also, Magnetogyration coefficient [m^2/Wb]
$\gamma(\nu)$ = Gain coefficient of an optical amplifier [m^{-1}]
γ_p = Peak gain coefficient of a laser-diode amplifier [m^{-1}]
$\gamma_0(\nu)$ = Small-signal gain coefficient of an optical amplifier [m^{-1}]
Γ = Retardation; also, Confinement factor

$\delta(\cdot)$ = Delta function or impulse function
δx = Increment of x
$\delta \nu$ = Spectral width of resonator modes [Hz]
Δ = Thickness of a thin optical component [m]; also, Fractional refractive-index change in an optical fiber or waveguide
Δx = Increment of x
Δn = Concentration of excess electron–hole pairs [m^{-3}]
Δn_T = Transparency injected-carrier concentration for a laser-diode amplifier [m^{-3}]
$\Delta \nu$ = Spectral width or linewidth [Hz]
$\Delta \nu_c$ = $1/\tau_c$ = spectral width [Hz]
$\Delta \nu_D$ = Doppler linewidth [Hz]
$\Delta \nu_{\mathrm{FWHM}}$ = Full-width-at-half-maximum spectral width [Hz]
$\Delta \nu_s$ = Linewidth of a saturated amplifier [Hz]

ϵ = Electric permittivity of a medium [F/m]; also, Focusing error [m^{-1}]
ϵ_{ij} = Component of the electric permittivity tensor [F/m]
ϵ_o = Electric permittivity of free space [F/m]

$\zeta(z)$ = Excess axial phase of a Gaussian beam

η = Impedance of a dielectric medium [Ω]; also, Electric impermeability
η_{ij} = Component of the electric impermeability tensor
η_o = Impedance of free space [Ω]
η = Quantum efficiency; also, Efficiency of power transfer; also, Power-conversion (wall-plug) efficiency
η_d = External differential quantum efficiency
η_e = Emission efficiency; also, Overall transmission efficiency
η_{ex} = External quantum efficiency
η_i = Internal quantum efficiency

θ = Angle
$\bar{\theta}$ = $90° - \theta$ = Complement of angle θ
θ_a = Acceptance angle
θ_B = Brewster angle; also, Bragg angle
θ_c = Critical angle
$\bar{\theta}_c$ = Complementary critical angle
θ_d = Deflection angle of a prism
θ_s = Angle subtended by source
θ_0 = Divergence angle of a Gaussian beam
ϑ = Threshold

κ = Elastic constant of a harmonic oscillator [J/m^2]

λ = Wavelength [m]
λ_A = Acceptor long-wavelength limit [m]
λ_F = Wavelength spacing of adjacent resonator modes [m]
λ_g = Bandgap wavelength (long-wavelength limit) of a semiconductor [m]

λ_o = Free-space wavelength [m]
Λ = Spatial period of a grating or periodic structure [m]; also, Wavelength of an acoustic wave [m]

μ = Magnetic permeability [H/m]; also, Carrier mobility in a semiconductor $[m^2 \cdot s^{-1} \cdot V^{-1}]$
μ_e = Electron mobility $[m^2 \cdot s^{-1} \cdot V^{-1}]$
μ_h = Hole mobility $[m^2 \cdot s^{-1} \cdot V^{-1}]$
μ_o = Magnetic permeability of free space [H/m]

ν = Frequency [Hz]
ν_F = Frequency spacing of adjacent resonator modes; free spectral range of a Fabry-Perot spectrometer [Hz]
ν_s = Spatial bandwidth of an imaging system $[m^{-1}]$
ν_q = Frequency of mode q [Hz]
ν_x, ν_y = Spatial frequencies in the x and y directions $[m^{-1}]$
$\nu_\rho = (\nu_x^2 + \nu_y^2)^{1/2}$ = Radial component of the spatial frequency $[m^{-1}]$
ν_0 = Central frequency [Hz]

ξ = Coupling coefficient in four-wave mixing

ρ = Rotatory power of an optically active medium $[m^{-1}]$; also, $\rho = (x^2 + y^2)^{1/2}$ = Radial distance in a cylindrical coordinate system [m]
ρ_c = Coherence distance [m]
ρ_s = Radius of the Airy disk [m]; also, Radius of the blur spot of an imaging system [m]
ϱ = Mass density of a medium $[kg \cdot m^{-3}]$; also, Charge density $[C \cdot m^{-3}]$
$\varrho(k)$ = Wavenumber density of states $[m^{-2}]$
$\varrho(\nu)$ = Spectral energy density $[J \cdot m^{-3} \cdot Hz^{-1}]$; also, Optical joint density of states $[m^{-3} \cdot Hz^{-1}]$
$\varrho_c(E)$ = Density of states near the conduction band edge $[m^{-3} \cdot J^{-1}$ in a bulk semiconductor]
$\varrho_v(E)$ = Density of states near the valence band edge $[m^{-3} \cdot J^{-1}$ in a bulk semiconductor]

σ = Conductivity $[\Omega^{-1} \cdot m^{-1}]$; also, Damping coefficient of a harmonic oscillator $[s^{-1}]$
$\sigma(\nu)$ = Transition cross section $[m^2]$
σ_q = Circuit-noise parameter
σ_x = Standard deviation of a random variable x; rms width of a function of x
$\sigma_0 = \sigma(\nu_0)$ = Transition cross section at the central frequency ν_0 $[m^2]$

τ = Lifetime [s]; also, Decay time [s]; also, Width of a function of time [s]; also, Excess-carrier electron–hole recombination lifetime in a semiconductor [s]
τ_c = Coherence time [s]
τ_d = Delay time [s]
τ_e = Electron transit time [s]

τ_h = Hole transit time [s]
τ_m = Multiplication time in an avalanche photodiode [s]
τ_{nr} = Nonradiative electron–hole recombination lifetime [s]
τ_p = Resonator photon lifetime [s]
τ_r = Radiative electron–hole recombination lifetime [s]
τ_s = Saturation time constant of a laser transition [s]
τ_{21} = Lifetime of a transition between energy levels 2 and 1 [s]

ϕ = Angle in a cylindrical coordinate system; also, Photon flux density $[m^{-2} \cdot s^{-1}]$
$\phi(p)$ = Momentum wavefunction $[s^{1/2} \cdot kg^{-1/2} \cdot m^{-1/2}]$
ϕ_ν = Spectral photon flux density $[m^{-2} \cdot s^{-1} \cdot Hz^{-1}]$
$\phi_s(\nu)$ = Saturation photon-flux density $[m^{-2} \cdot s^{-1}]$
φ = Phase
$\varphi(\nu)$ = Phase-shift coefficient of an optical amplifier $[m^{-1}]$
Φ = Photon flux $[s^{-1}]$
Φ_ν = Spectral photon flux $[s^{-1} \cdot Hz^{-1}]$

χ = Electric susceptibility; also, Electron affinity [J]
χ' = Real part of the electric susceptibility χ
χ'' = Imaginary part of the electric susceptibility χ
$\chi(\nu)$ = Electric susceptibility of a dispersive medium
χ_{ij} = Component of the electric susceptibility tensor
$\chi^{(3)}$ = Coefficient of third-order optical nonlinearity $[C \cdot m \cdot V^{-3}]$
$\chi^{(3)}_{ijkl}$ = Element of the third-order optical nonlinearity tensor $[C \cdot m \cdot V^{-3}]$
$\chi^{(3)}_{IK}$ = Element of the third-order optical nonlinearity tensor (contracted indices) $[C \cdot m \cdot V^{-3}]$

$\psi(x)$ = Particle position wavefunction $[m^{-1/2}]$
$\Psi(\mathbf{r}, t)$ = Particle wavefunction $[m^{-3/2} \cdot s^{-1/2}]$

ω = Angular frequency [rad/s]
Ω = Angular frequency of an acoustic wave [rad/s]; also, Angular frequency of a harmonic electric signal [rad/s]; also, Solid angle

Mathematical Symbols

∂ = Partial differential
∇ = Gradient operator
$\nabla \cdot$ = Divergence operator
$\nabla \times$ = Curl operator
$\nabla^2 = \partial^2/\partial x^2 + \partial^2/\partial y^2 + \partial^2/\partial z^2$ = Laplacian operator
$\nabla_T^2 = \partial^2/\partial x^2 + \partial^2/\partial y^2$ = Transverse Laplacian operator

INDEX

ABCD law, 99
ABCD matrix, *see* Ray-transfer matrix
Aberration, 14, 16
Absorption, 436, 440–443
  band-to-band, 576–584, 586–587, 590
Absorption coefficient, 175–177, 181–183, 192, 318, 421, 465, 484, 586, 614, 620–622, 649. *See also* Attenuation coefficient, optical fiber
Absorption edge, 576
Acceptance angle, 17, 276–277, 876. *See also* Numerical aperture (NA)
Acceptor, 551, 656
Acoustic wave, 802, 825–826
  longitudinal, 826, 827
  transverse, 826, 828
Acousto-optic(s), 800–831
  anisotropic media, 825–830
  filter, 823–824
  frequency shifter, 824
  interconnections, 820–823
  isolator, 824–825
  modulator, 815–817, 831
  Raman-Nath diffraction, 813–815, 831
  scanner, 818–820
  spatial light modulator, 822
  spectrum analyzer, 820
  switch, 816, 838–839
ADP ($NH_4H_2PO_4$), 699, 714, 716, 720, 741, 773, 780
Airy disk, 130
Airy pattern, 130
Affinity, electron, 646, 662
AlAs (aluminum arsenide), 550, 576, 588
Alexandrite (Cr^{3+}:Al_2BeO_4), 519
AlGaAs (aluminum gallium arsenide), 549, 550, 569, 572, 576, 588, 590, 606, 619, 632, 633, 636, 637, 662, 669, 744, 854, 855, 883, 885, 886
AlP (aluminum phosphide), 550, 588
AlSb (aluminum antimonide), 550, 588
Amorphous solid, 210
Amplified spontaneous emission (ASE), 488–489, 493, 520
Amplifier, laser, *see* Laser amplifier
Amplifier, optical, *see* Laser amplifier
Amplitude, 44
Analytic signal, 73
Angular frequency, 44
Angular momentum, 393
Anisotropic media:
  acousto-optic, 825–830
  electro-optic, 712–721
  liquid crystal, 227–230, 721–727
  nonlinear-optic, 779–782, 841
  three-wave mixing, 781
  wave propagation, 210–220
Anode, 647
APD, *see* Avalanche photodiode
Aperture function, 128
Ar^+ (argon ion) laser, 480, 519, 521, 535, 538, 539
Array detector, 664–665
Atomic transition, 434–449. *See also* Laser transitions
AT&T, 874, 886
Attenuation coefficient, optical fiber, 296–298, 880–882
Avalanche buildup time, 671–673
Avalanche photodiode (APD), 666–673, 884–885
  excess noise factor, 679–681, 694
  gain, 669–671, 694
  gain, optimal, 688
  impact ionization, 666–667
  InGaAs, 884–885, 886
  ionization coefficient ratio, 667
  ionization coefficients, 666
  multilayer, 668
  noise, 678–681, 694, 905, 906
  quantum efficiency, 649–650, 694, 884–885
  reach-through, 668
  response time, 671–673, 884–885
  responsivity, 651, 669, 884
  separate absorption, grading, multiplication (SAGM), 885
  separate-absorption-multiplication (SAM), 667–668
  Si, 681, 694, 884, 885
  signal-to-noise ratio, 680–681, 686–687, 690, 695, 906

Balanced mixer, 912–913
Bandgap, direct and indirect, 547–548, 579–581
Bandgap energy, 544, 550, 551, 576–642
Bandgap wavelength, 550, 576, 605–606, 650, 662
Band offset, 568
Bandwidth, *see also* Fiber, optical, response time, Photodetector, response time; Spectral width
acousto-optic modulator, 815–817
definition, 921–924
electro-optic modulator, 700–701, 735
laser amplifier, 465, 480, 611–612, 641, 642
laser oscillator, 508–513, 521–522
optical fiber, 880–882
photodetector, 656, 657, 661, 663, 884–885
resonator modes, 318, 320
Bardeen, John, 542
Basov, Nikolai G., 460
$BaTiO_3$ (barium titanate), 729
Beam, acoustic, 812–815, 818–819
Beam, optical:
Bessel, 104–106
donut, 104
Gaussian, 51, 81–100, 134, 173, 188, 330–335, 341, 382, 389, 419, 420, 513–514, 791
Hermite–Gaussian, 100–104, 336–337, 513–515
Laguerre–Gaussian, 104
Beamsplitter, 12, 54–55, 389–390, 409–411, 420, 421
polarizing, 231–232, 711, 728
Beating:
light, 75–76, 907–909
single-photon, 419
Bernoulli distribution, 409
Bessel beam, 104–106
Bessel function, 104, 278, 493
Betaluminescence, 455
Biaxial crystal, 211
Binary semiconductor, 548, 550
Binomial distribution, 410
Bioluminescence, 455
Birefringence, 221
Bistable optical device, 843–855
dispersive, 848–850, 872
dissipative, 850–851
hybrid, 852–855
intrinsic, 849–850
self-electro-optic-effect (SEED), 854–855
Bistable system, 844–846
Bit error rate (BER), 894–906, 911, 916
Blackbody radiation, 452–454, 459, 683
Bloembergen, Nicolaas, 737
Blur spot, 136, 141–143
Blurred image, 136–143
Bohr, Niels, 423
Boltzmann constant, 405, 432
Boltzmann distribution, 405, 406, 432, 434, 452
Born, Max, 342
Born approximation, 742–743
Born postulate, 425
Bose–Einstein distribution, 406–407, 420, 421, 452, 489
Bragg, William Henry, 799
Bragg, William Lawrence,799
Bragg angle, 70, 801, 805
Bragg cell, 801
Bragg diffraction, 69–70, 801–815, 828–830, 831
coupled-wave theory, 810–812
Doppler shift, 806–807
downshifted, 808–809
optical and acoustic beams, 812–815
quantum interpretation, 809
Raman–Nath scattering, 813–815, 831
reflectance, 807–808
scattering theory, 809–810, 831
Bragg reflection, *see* Bragg diffraction
Bragg scattering, *see* Bragg diffraction
Brattain, Walter H., 542
Brewster angle, 207, 208, 231, 236
Brewster window, 208, 209, 516
Broadband light, 440–442
BSO (bismuth silicon oxide, $Bi_{12}SiO_{20}$), 711, 729, 855
Built-in field, 565
Buried heterostructure laser, 629–630
Burrus-type LED, 607

C^3 laser, *see* Cleaved-coupled-cavity (C^3) laser
Capacitance, diffusion, 566
Capacitance, junction, 566
Carrier concentration, 552–559
Carrier generation, 559–562
Carrier injection, 560, 566
Carrier mobility, 653
Cascade of optical components, 30–32
Cathodoluminescence, 455, 459
Causal system, 179, 466–468, 930
Caustic, 16
Cavity, *see* Resonator
Cavity dumping, 523, 541
CdS (cadmium sulfide), 551
CdSe (cadmium selenide), 551
CdTe (cadmium telluride), 175, 551, 662, 699, 714, 717
Chalcogenide glass, 882
Channel waveguide, 260–261
Charge-coupled device (CCD) detector, 664–665
Chemiluminescence, 455
Chirp function, 132, 920
Chirping, 132, 188, 787, 883, 929
Cholesteric liquid crystal, 227
Chromatic dispersion, 302, 877
Circular dichroism, 237
Circularly polarized light, 194, 196–198, 199, 201, 223–224, 236, 379, 393
Circular polarization, *see* Circularly polarized light
Circular waveguide, *see* Fiber, optical
Cladding, fiber, 17, 39, 273, 277, 876, 878
Cleaved-coupled-cavity (C^3) laser, 518, 631, 884

CO_2 (carbon dioxide), 426
CO_2 (carbon dioxide) laser, 477, 480, 519, 521, 535, 480, 539, 747, 771
Coherence:
average intensity, 345–346
complex degree of coherence, 354
complex degree of temporal coherence, 347
cross-spectral density, 356, 357
cross-spectral purity, 357
effect on image formation, 368–372
effect on interference, 360–366
effect of propagation, 367–368, 372–375
longitudinal, 357–359
mutual coherence function, 353, 355, 381
mutual intensity, 355, 367–375, 381
power spectral density, 349–350
spatial, 353–357, 362–376, 381
spectral width, 351–352
temporal, 346–353, 361–362
temporal coherence function, 346, 347
Coherence area, 356
Coherence distance, 364, 365, 375
Coherence length, 349, 352, 358, 359, 381
Coherence time, 348, 349, 351, 352
Coherency matrix, 377
Coherent detection, 887, 888, 907–913
Coherent imaging, 135–143, 371–372
Coherent light, 344, 347, 354
Coherent optical communications, 888, 907–913
Coherent-state light, 414
Collision broadening, 446, 583
Color, 350
Communications, optical, *see* Fiber-optic communications
Complex amplitude, 45
Complex amplitude transmittance, *see* Transmittance, complex amplitude
Complex analytic signal, 73
Complex degree of coherence, 354
Complex envelope, 47
Complex representation, 73
Complex wavefunction, 45
Compound semiconductor, 548–551
Computer-generated holography, 860
Computing, optical, 136–139, 862–869
analog, 136–139, 864–869. *See also* Processing, optical
continuous, 867
digital, 862–864
discrete, 865–867
logic, 845, 848–855, 872
matrix operations, 866–867
Concentration, electron and hole, 552–559
Conduction band, 429, 431, 543–545, 668
Conductivity, 192, 655, 693
Confinement:
carriers, in semiconductor, 567–568, 618–619
photons, in waveguide, 254, 569, 618–619
rays, in resonator, 327–330
Confinement factor, 271, 621
Confocal parameter, beam, 86
Confocal resonator, 329–330, 334, 337, 339, 341
Conjugate holographic image, 145, 146–147
Conjugate wave, 78, 51, 758–760
Conjugation, phase, 758–761, 777–779
Convolution, 120, 919
optical, 868–869
Convolution theorem, 919, 925–926
Cooling, laser, 449–450
Core, fiber, 17, 39, 274, 876–878
Corning, Inc., 272
Correlation, 919
optical, 156, 868–869
Coupled waves:
acousto-optic, 810–812
degenerate four-wave mixing, 777
degenerate three-wave mixing, 764
directional coupler, 264–269, 707–709, 837–838, 841–842, 852–854, 892–893
four-wave mixing, 774–779
frequency conversion, 769–771
parametric amplifier, 771–773
parametric oscillator, 773–774
phase conjugation, 777–779
second-harmonic generation, 766–769
three-wave mixing, 762–774
up-conversion, 771
Couplers, 892–893
Coupling:
between modes, 304–305
into waveguide, 261–264
between waveguides, 264–269
Critical angle, 11, 17, 206–208, 249, 260, 276–277, 602–603, 876
Cross-correlation, optical, 155, 868–869
Cross-spectral density, 356, 357
Cross-spectral purity, 357
Crystal lattice constant, 550, 551, 578, 637
Crystal optics, 194–237
CS_2 (carbon disulfide), 741, 797
Cutoff condition:
dielectric waveguide, 252, 271
optical-fiber waveguide, 282
planar-mirror waveguide, 245
Cylindrical lens, 40, 115, 116, 856
Cylindrical wave, 78

Dark current, 674
Dark soliton, 793
Debye-Sears diffraction, 813–815
Decay time, 435
Decibel (dB) units, 296–297, 880
Deflector, *see* Scanner
Defocused imaging system, 155
Degeneracy parameters, 433
Degenerate four-wave mixing, 758–760, 777
Degenerate semiconductor, 558
Degree of polarization, 379
Delta function, 920, 921

Density:
of resonator modes, 315–316, 324–326, 452, 459, 683
of states, 552–553, 571–573, 597
optical joint density of states, 579, 610, 634
Depletion layer, 563
Depth of focus, 87
Detector, *see* Photodetector
Diatomic molecule, 426
Dichroism, 231
Dielectric constant, 163, 169
Dielectric medium, 162–167, 168–169, 179–182, 191, 192
anisotropic, 165, 210–223, 227–230, 712–718, 721, 779–782
dispersive, 165, 169, 176–191, 255–258, 285–286, 294–295, 298–306, 308, 309, 587, 782–788, 876–882
inhomogeneous, 164, 169, 800. *See also* Graded-index fiber; Graded-index optics; Graded-index slab
nonlinear, 166–167, 739–743, 848–852, 872
Differential quantum efficiency, 625
Diffraction:
Bragg, *see* Bragg diffraction
Debye-Sears, *see* Diffraction, Raman-Nath
Fraunhofer, *see* Fraunhofer diffraction
Fresnel, *see* Fresnel diffraction
Raman-Nath, 813–815, 831
Diffraction grating, 60–61, 78–79, 112, 145, 150, 154, 830
Diffusion capacitance, 566
Digital communications, 886, 889, 893–901, 911–912, 916
Digital optical computing, 862–864
Diode junction, 563–567
Diode laser, *see* Laser diode
Dipole moment, 161, 739
Direct-bandgap semiconductor, *see* Bandgap, direct and indirect
Direct detection, 887
Directional coupler, *see* Coupled waves, directional coupler
Dispersion, 176–179. *See also* Dielectric medium, dispersive
angular, 178
anomalous, 186
coefficient, 179, 185–191
normal, 186
in optical fibers, *see* Fiber, optical, dispersion
Dispersion relation, propagation in a crystal, 215–216
Dispersive medium, 165, 169, 466–467. *See also* Dielectric medium, dispersive
Dispersive nonlinear medium, 782–786, 796, 798
Distributed Bragg reflector (DBR) laser, 631–632, 884, 912
Distributed feedback (DFB) laser, 631–632, 884, 912
Divergence, angular, 86, 93, 106, 129, 130, 134, 608–609, 630–631, 812–813
Donor, 551, 656
Donut beam, 104
Doped semiconductor, 551–552
Doppler broadened lineshape function, see Lineshape function, Doppler-broadened
Doppler broadening, 447–449, 486, 510–513
Doppler linewidth, 448, 480, 538
Doppler radar, 76
Doppler shift, 76, 806–807
Double heterostructure, 567–569, 618–619, 623–624, 661–662, 883–884
Double refraction, 221–223, 236, 706
Duration-bandwidth reciprocity relation, 922
Dye laser, 428, 480, 519–520, 521, 535
Dynode, 646–647

Edge-emitting LED, 606–608, 883
Edge enhancement, 139
Eigenvalue, 934–935
Eikonal, 25, 26, 52, 289–290
Eikonal equation, 25–26, 53
Einstein, Albert, 384, 423
Einstein A and B coefficients, 441–443
Elasto-optic effect, 735
Electric dipole, 161, 739
Electric displacement, 160–161
Electric field, 159
Electric flux density, 160–161
Electric permittivity, *see* Permittivity
Electric susceptibility, *see* Susceptibility
Electroluminescence, 455
Electromagnetic optics, 158–192
Electromagnetic wave, 169–174
Electron mobility, 655
Electro-optic directional coupler, 707–709, 837–838
Electro-optic effect, 697–700, 712–719, 721, 745–746
Electro-optic modulator, 700–705, 710–712, 719–727
double-refraction, 706, 736
half-wave voltage, 700
integrated-optic, 702, 704, 736
intensity modulator, 702–705, 720–721, 735, 736
longitudinal, 700–701
phase modulator, 700–701, 719–720, 735
transverse, 700–701
traveling wave, 701
Electro-optic scanner, 705–707
Electro-optic switch, 702–705, 837–838
Electro-optic wave retarder, 701–702
Elliptical mirror, 7
Elliptical polarization, 194
Energy, optical, 44, 168, 386–388, 400–401, 407, 411, 413
in anisotropic media, 215, 218–220
Energy bands, 429–431
AlGaAs, 431
GaAs, 430, 545
Si, 430, 545

Energy conservation, 765
Energy levels:
 AlGaAs/GaAs multiquantum-well, 431, 590
 C^{6+}, 427
 CO_2, 426
 diatomic molecule, 425–427
 dye molecule, 427, 428
 H, 427
 He, 428
 $LiNbO_3$ (photorefractive), 729
 molecular rotation, 426
 molecular vibration, 425–426
 N_2, 425, 426
 Nd^{3+}:YAG, 479
 Ne, 428
 quantum well, 431, 432
 ruby, 429–430, 477
Energy-momentum relations:
 electrons/holes, 432, 545–547, 569–570, 572, 578
 photons, 390, 419, 750, 757, 809
Energy per mode, average, 452, 459, 683
Energy-time uncertainty, 396, 444, 842
Epitaxy, 569
Er^{3+}:silica fiber, 476, 477, 479–480, 519, 535, 609, 793, 882, 886
Etalon, *see* Resonator
Evanescent wave, 253
Excess noise factor, 679–681, 694
Excimer laser, 519, 521
Exciton, 574
Extinction coefficient, 175, 253
Extraordinary refractive index, 211, 218–220
Extraordinary wave, 218–220
Extrinsic semiconductor, 551–552, 656

Fabry, Charles, 310
Fabry–Perot etalon, *see* Resonator
Fabry–Perot filter, *see* Resonator
Fabry–Perot interferometer, *see* Resonator
Fabry–Perot resonator, *see* Resonator
Faraday effect, 225–227, 233–234, 839–840
Faraday rotator, 225, 234, 839–840
Feedback, 314, 495–496, 498, 620, 773, 848–855
Fermat, Pierre de, 1
Fermat's Principle, 4
Fermi-Dirac distribution, 434, 554
Fermi energy, 434, 554, 590
Fermi function, 554
Ferroelectric liquid crystals, 726–727
Fiber, optical, 17, 272–309
 attenuation, 296–298, 880–882
 bandwidth, 880–882
 characteristic equation, 280, 281
 cladding, 17, 39, 273, 277, 876, 878
 core, 17, 39, 274, 876–878
 coupling, 39
 coupling efficiency, 307
 dispersion, 189–190, 298–304, 308, 309, 876–882
  chromatic, 302, 877
  flattened, 302, 303
  material, 300, 877–878
  modal, 299, 308, 877
  nonlinear, 303, 882
  shifted, 302, 303
  waveguide, 301–302, 877
 erbium-doped, 476, 477, 479–480, 519, 535, 609, 793, 882, 886
 extrinsic losses, 298
 field distribution, 277–279
 graded-index, 23–25, 40, 273–274, 287–296, 877–882
 grade profile parameter, 288
 group velocities, 285–286, 294–295, 308
 impulse response function, 304–306, 879–880
 materials, 274, 882
 modal noise, 286
 modes, 280–284, 292–296, 308
 nonlinear effects, 744, 792–793, 882
 number of modes, 282–284, 292–293, 296
 numerical aperture, 17–18, 24–25, 39, 275–277, 308, 876
 polarization-maintaining, 287
 propagation constants, 284–285, 294, 308
 pulse propagation, 182–192, 304–306, 309, 792–793, 878
 quasi-plane wave, 289–291
 rare-earth doped, 476, 477, 479–480, 518–519, 609, 793, 882
 Rayleigh scattering, 297–298, 308
 rays in, 24, 275–277, 288–289
 resonator, 311
 response time, 299–306, 308, 309, 876–877
 single-mode, 273, 274, 286–287, 298, 877–882, 886
 soliton laser, 793
 solitons, 792–793
 speckle, 286
 step-index, 274–287, 308, 876, 878
 transfer function, 879–882
 V parameter, 279–280, 876
 weakly guiding, 280
Fiber-optic communications, 874–917
 attenuation-limited, 895–897, 902
 bit error rate, 894–905, 916
 coherent, 907–913
 couplers, 892–893
 detectors, 884–885
 dispersion-limited, 897–898, 902
 dispersion power penalty, 900–901
 distance *vs.* bit rate, 897–900, 902
 Er^{3+}:silica-fiber amplifiers, 875, 882, 886
 fibers, 876–882
 modulation, 887–889
 multiplexing, 889–892
 power budget, 895–897
 receiver sensitivity:
  analog, 689–690
  coherent, 912
  digital, 894, 903–906
 soliton, 886
 sources, 883–884
 switches, 833–843

Fiber-optic communications (*Continued*)
system performance, 893–903
systems, 885–887
undersea network, 874, 886
Finesse, 71, 316, 319, 320, 321, 499–500
Flint glass, 177, 803, 805, 808
Fluctuations, *see* Coherence; Noise
Fluorescence, 456
Fluoride glass, 882
Flux, photon, 398–403, 420
F-number of a lens, 95, 141–143, 371
Focal length:
lens, 15
mirror, 9
Focal plane, 31
Focal point, 31
4-f system, 136–139
Fourier, Jean-Baptiste Joseph, 108
Fourier optics, 108–156
Fourier plane, 137
Fourier transform:
one-dimensional, 918–921
optical, 121–127, 153, 382, 867
Table, 920
two-dimensional, 153, 924–926
Fourier-transform holography, 147
Fourier-transform spectroscopy, 362
Four-level laser, 472–474, 476, 478–480, 492
Four-wave mixing, 756–760, 774–779, 796
Frauenhofer, Josef von, 108
Fraunhofer approximation, 122, 123, 374
Fraunhofer diffraction, 128–131, 154
circular aperture, 130, 131, 812
diffraction grating, 154
oblique wave illumination, 154
rectangular aperture, 129–130, 812
Free-carrier transitions, 574
Free electron laser, 520–521
Free spectral range, spectrum analyzer, 322
Frequency:
instantaneous, 114, 787
of light, 42, 44, 158
of resonator modes, 313
pulling, 502–503
spacing of adjacent resonator modes, 313
spatial, 109
Frequency conversion, 456–457, 746–747, 769–771, 796
Frequency-division multiplexing (FDM), 889–891
Frequency-shift keying (FSK), 889–890
Fresnel, Augustin Jean, 193
Fresnel approximation, 49, 50, 118–121, 123, 363
Fresnel diffraction, 131–134, 188
Gaussian aperture, 133–134
slit, 132–133
two pinholes, 68, 154, 362–366, 394
Fresnel equations, 205
Fresnel integrals, 133
Fresnel number, 50, 119, 123, 132–134
Fresnel zone plate, 116
Fringes, 64–65, 67–68, 382, 394
Fused silica, 175, 177, 190, 274, 297–298, 300–301, 744, 878–882

GaAs (gallium arsenide), 18, 175, 430, 431, 545–548, 550, 557, 562, 563, 569, 572, 575, 576, 586–588, 590, 591, 594, 596, 602, 605, 606, 636, 638, 640–642, 662, 692, 714, 717, 729, 735, 780, 851, 852, 855, 883
Gabor, Dennis, 108
Gain:
avalanche photodiode, 670–673, 688, 694, 695, 885
laser, 462–465, 482–484, 491, 493, 520, 642, 651
photoconductor, 655, 657, 694
Gain coefficient, 464–467, 480–487, 497, 510, 585, 611–617, 620, 634–636, 641
saturated, 481, 492–493
Gain-guided laser diode, 621–624
Gain noise, APD, 678–681, 694, 695
Gain switching, 522, 526–527, 540
GaP (gallium phosphide), 550, 575, 576, 588
GaSb (gallium antimonide), 550, 576, 588
Gas laser, 480, 519, 521, 538, 539
Gauss, Karl Friedrich, 80
Gaussian beam, 51, 81–106, 121, 133–134, 173, 255, 331–337, 382, 389, 420
collimation, 96
complex amplitude, 83
complex envelope, 82
confocal parameter, 86
depth of focus, 86
divergence, 86
elliptic, 107
expansion, 97
focusing, 94, 107
intensity, 83
partially coherent, 382
phase, 87, 107
power, 84, 107
q parameter, 82, 90
radius, 85
radius of curvature, 88
Rayleigh range, 82
reflection from mirror, 97
refraction, 107
relaying, 96
shaping, 94
spot size, 85, 107
transmission through arbitrary system, 98–100
transmission through GRIN slab, 107
transmission through lens, 92–97
waist radius, 85
wavefront, 87
Gaussian lineshape function, *see* Lineshape function, Gaussian
Gaussian mutual intensity, 381
Gaussian probability distribution, 905
Gaussian pulse, 187, 396

Gaussian spectrum, 349, 351, 448
Ge (germanium), 175, 177, 548, 550, 574–576, 588, 656–657, 694, 886
General Electric Corporation, 592
Generalized pupil function, 140–141
Generation, carrier, 559–560
Geometrical optics, *see* Ray optics
Glass, 175, 177, 178, 803, 805, 808
Graded-index fiber:
  group velocities, 294
  modes, 292
  number of modes, 296
  numerical aperture, 24
  optimal index profile, 295
  propagation constants, 294
  quasi-plane waves, 289
  rays, 23, 40
  V parameter, 293
Graded-index lens, 63
Graded-index (GRIN) optics, 18–26
Graded-index slab, 20–23, 39, 62, 78
Grating, *see* Diffraction grating
Grating equation, 61
Grating spectrometer, 62
GRIN, *see* Graded-index (GRIN) optics
Group index, 179, 189–190
Group velocity, 179, 185, 186, 189, 190, 192, 245, 255–256, 285, 294, 301, 308
Group-velocity dispersion, 257, 299
Guided-wave optics, 238–271
Guoy phase shift, 87, 89
Gyration vector, 224

H (hydrogen), 427
Half-wave plate, *see* Retarder, wave
Harmonic oscillator:
  classical, 180, 931
  nonlinear, 784–786
  quantum, 412–414
He (helium), 428
Heisenberg uncertainty relation, 413, 922
Helmholtz equation, 46, 168
  paraxial, 50, 78, 189
He-Ne (helium-neon) laser, 480, 519, 521, 535, 539
Hermite-Gaussian beam, 100–104, 107, 336–337, 514
Hermite polynomials, 102
Hero's principle, 4
Hertz, Heinrich, 644
Heterodyne detection, 907–913
Heterojunction, 567–569
HgCdTe (mercury cadmium telluride), 551, 662, 633
HgTe (mercury telluride), 551
Hilbert transform, 467, 930
Hole burning, 487
Hole mobility, 655
Holes in semiconductors, 544
Hologram, *see* Holography
Holographic interconnections, 857–858
Holographic scanner, 115
Holographic spatial filter, 148
Holography, 143–151
  computer-generated, 860
  Fourier transform, 147
  off-axis, 146
  rainbow, 151
  real-time, 759
  reflection, 151
  spherical reference wave, 155
  surface-relief, 860
  volume, 149–151
  white light, 149–151
Homodyne detection, 907–913
Homojunction, *see* Junction
Huygens, Christiaan, 41
Huygens-Fresnel principle, 121
Hysteresis, 844, 847

IBM Corporation, 592
Idler wave, 749, 771
Image correlation, 868
Image detectors, 647, 664–665
Image formation:
  coherent light, 135–143, 371–372
  4-*f* lens system, 137–139
  imaging equation, 15
  impulse-response function, 136, 141–142
  incoherent light, 368–372
  lens, 30, 31, 60, 135, 136, 139–143
  mirror, 10
  partially coherent light, 366–372
  resolution, 371–372
  spherical boundary, 14, 15
  transfer function, 138–143
Image intensifier, 646
Image magnification, 15, 142
Image processing, 138–139, 869
Impact ionization, 666
Impedance:
  dielectric medium, 170
  free space, 171
Impermeability tensor, 211
Impulse-response function:
  dispersive medium, 186
  free space, 120
  imaging system:
    coherent, 369–372
    defocused, 136, 155
    4-*f*, 138
    incoherent, 369–372
    single-lens, 141
  linear system, 828, 832
InAs (indium arsenide), 550, 575, 576, 588, 714, 717
Incoherent light, image formation, 368–372
Incoherent-to-coherent converter, 712
Index ellipsoid, 212–215
Index-guided laser diode, 621–624

Index of refraction, *see* Refractive index
Indicatrix, optical, 212–215
Indirect-bandgap semiconductor, *see* Bandgap, direct and indirect
Induced emission, *see* Stimulated emission
Inelastic collisions, 446
Infrared, 158
InGaAs (indium gallium arsenide), 550, 633, 638, 658, 663
InGaAsP (indium gallium arsenide phosphide), 549, 550, 576, 588, 605, 613, 616, 617, 619, 623, 626, 628, 632, 633, 637, 640–642, 658, 662, 663
Inhibited spontaneous emission, 459
Inhomogeneous broadening, 446
Inhomogeneous medium, 164
Injection:
  carrier, 560, 566
  minority carrier, 560–562
Injection electroluminescence, 455
Injection laser diode, *see* Laser diode
InP (indium phosphide), 550, 575, 576, 588, 619
InSb (indium antimonide), 550, 575, 588, 663, 692
Instantaneous frequency, 114, 787
Instantaneous intensity, 345
Integrated optics, 238–271
Intensity, average, 345–346
Intensity, optical, 44, 161, 168
  instantaneous, 345
  monochromatic light, 46
  quasi-monochromatic light, 74
Intensity modulation, 887
Interconnections:
  acousto-optic, 820–823
  capacity, 823, 859
  coordinate transformations, 869, 872
  holographic, 114–116, 153, 857–858
  in microelectronics, 860–862
Interference:
  effect of spatial coherence, 362-365
  effect of temporal coherence, 361-362, 365–366
  interference equation, 64
  multiple waves, 68, 70, 76
  partially coherent light, 360–366
  plane wave and spherical wave, 67
  single-photon, 394–395, 419
  two oblique plane waves, 65–67
  two spherical waves, 67
  two waves, 63
Interferogram, 362
Interferometer:
  Mach–Zehnder, 65, 66, 395, 703, 704, 736, 841, 849
  Michelson, 65, 66, 79, 362
  Michelson stellar, 375–376
  Sagnac, 65, 66
Internal reflection, total, 11
Intersymbol interference, 889–902
Intrinsic semiconductor, 548–551
Invariants, three-wave mixing, 765
Inverse Fourier transform, 918, 925
Inversion, population, 464, 468–476
Ionization ratio, 667
Isolator, optical, 233–234, 236, 824–825

Johnson noise, *see* Noise, optical receiver, thermal noise
Joint density of states, *see* Optical joint density of states
Jones matrix, 199–203
  coordinate transformation, 202
  linear polarizer, 200
  polarization rotator, 201
  wave retarder, 200, 201
Jones vector, 197
Junction:
  p–i–n, 567, 593, 601, 657, 659
  p–n, 563–567, 661
Junction capacitance, 566

KDP (KH_2PO_4), 699, 714, 716, 720, 735, 744, 780, 781, 797, 798
Kerr, John, 696
Kerr coefficients, 700, 713, 715, 718–719
Kerr effect, 697–700, 751
  optical, 752, 754, 757, 769
$KNbO_3$ (potassium niobate), 729
Kramers–Kronig relations, 179, 466–468, 930
k selection rule, 578
k space, 324, 325
k surface, 216, 217, 219

Laguerre–Gaussian Beam, 104
Lamb dip, 513
Lambertian source, 608
Laser, *see also* Laser amplifier
  alexandrite, 519
  Ar^+ (argon ion), 480, 519, 521, 535, 538, 539
  ArF, 521
  cavity dumped, 523, 541
  cleaved-coupled-cavity (C^3), 518, 631, 884
  CO_2 (carbon dioxide), 477, 480, 519, 521, 535, 539, 747, 771
  colliding pulse mode, 535
  color center, 521
  distributed Bragg reflector, 631–632, 884, 912
  distributed feedback, 631–632, 884, 912
  dye, 428, 480, 519–520, 521, 535
  Er^{3+}:silica fiber, 519, 535
  Er^{3+}:YAG, 519
  etalon, 539
  excimer, 519, 521
  four-level, 472–474, 492
  free electron, 520–521
  frequencies, 501–502, 539
  frequency pulling, 502–503
  gain switched, 522, 526–527, 540
  gas, 480, 519, 521, 538, 539
  HCN, 521
  H_2O (water vapor), 519, 521

He-Cd (helium cadmium), 519, 521
He-Ne (helium neon), 480, 519, 521, 535, 539
internal photon flux density, 503
internal photon-number density, 507
Kr^+ (krypton ion), 519, 521
KrF, 519, 521
liquid, 519–520
mode-locked, 524, 531–536
modes:
lateral or transverse, 516
longitudinal, 509–513, 516–518, 538
selection, 515–518
multiline, 515–516
multiquantum-well, 636–637
Nd^{3+}:glass (neodymium glass), 478–480, 518, 519, 521, 535
Nd^{3+}:selenium oxychloride, 519
Nd^{3+}:YAG (neodymium YAG), 478–480, 518, 519, 521, 535
Nd^{3+}:YLF, 519
Nd^{3+}:YSGG, 519
oscillation threshold, 500–501
plasma, 520
polarization, 515, 516
power, 503–508, 539
pulsed, 522–536
Q-switched, 523, 527–531, 540, 541
resonator, *see* Resonator
ruby, 477–478, 480, 521, 531, 535
semiconductor, *see* Laser diode
single-mode, 516–518, 631–632
solid-state, 518–519
soliton, 793
spatial distribution, 513–515
spectral distribution, 508–513
threshold population difference, 500, 539
transients, 524–526, 540
Ti^{3+}:Al_2O_3 (Ti:sapphire), 480, 519, 521, 535
three-level, 474–476
transversely excited atmospheric (TEA), 477
wavelengths, 521
x-ray, 520
Laser amplifier, 460–493. *See also* Semiconductor laser amplifier
amplified spontaneous emission (ASE), 488–489, 493, 520
bandwidth, 465–466
C^{6+}, 427, 520–521
Doppler broadened, 486–487
dye, 480
Er^{3+}:silica fiber, 476, 477, 479–480, 609, 793, 882, 886
gain, 462–465, 491, 493, 520, 642, 651
saturated, 482–484, 492–493
gain coefficient, 464–466, 480–487, 497, 510
hole burning, 487
inhomogeneously broadened, 446–449
Nd^{3+}:glass, 478–480
Nd^{3+}:YAG, 478–480
noise, 488–489
phase shift, 466–468
power source, 468–480
pumping, 472–480
four-level, 472–474
three-level, 474–476, 492
two-level, 492
ruby, 477–478, 480
saturation intensity, 492
saturation photon-flux density, 481, 482, 492
saturation time constant, 471, 472–476
semiconductor, *see* Semiconductor laser amplifier
spectral broadening, 482
Laser cooling, 449–450
Laser diode, 619–638. *See also* Semiconductor laser amplifier
AlGaAs (aluminum gallium arsenide), 632, 633, 637
arrays, 637–638
cleaved-coupled-cavity (C^3), 631, 884
differential quantum efficiency, 625
distributed Bragg reflector (DBR), 631–632, 884, 912
distributed feedback (DFB), 631–632, 884, 912
double heterostructure, 626, 629–630
efficiency:
emission, 624
overall, 626
gain coefficient, 611–617, 620, 634–636, 641
InGaAs (indium gallium arsenide), 638
InGaAsP (indium gallium arsenide phosphide), 623–624, 626, 632, 633, 637
light-current curve, 625
modes, 629–631
multiquantum-well (MQW), 636
power, 624–625
quantum-well, 632–636
radiation pattern, 629–631
resonator, 620–622
responsivity, 626
single-frequency, 631–632
spatial distribution, 629–631
spectral distribution, 627–629, 642–643
strained-layer, 637
surface-emitting (SELD), 632, 637–638
threshold current density, 622–624, 642
transparency current density, 616
Laser diode amplifier, *see* Semiconductor laser amplifier
Laser transitions, 480, 518–522, 535, 632, 633
Laser trapping of atoms, 449–450
Lattice constant, 550–551
LED (light-emitting diode), 594–609
circuit, 608–609
coupling to a fiber, 640
edge-emitting, 606–607
external quantum efficiency, 604, 640
injection electroluminescence, 594–600
internal quantum efficiency, 602
materials, 605–606

LED (light-emitting diode) (*Continued*)
overall quantum efficiency, 640
photon flux, 600–603
power, 603
response time, 606
responsivity, 604
spatial distribution, 608
spectral distribution, 599–600, 605
spectral linewidth, 600, 640
superluminescent, 627
surface-emitting, 606–608
trapped light, 18
Lens, 14
complex amplitude transmittance, 58
convex, 14
cylindrical, 40
double-convex, 14, 59
F-number, 95, 141–143
focal length, 15
lens law, 15
plano-convex, 58
thick, 31
$LiNbO_3$ (lithium niobate), 699, 701, 704, 709, 714, 715, 719, 720, 729, 736, 780, 796, 797, 831, 852, 855
$LiTaO_3$ (lithium tantalate), 699, 714, 715, 719
Lifetime broadening, 444
Light emitting diode, *see* LED
Light guide, *see* Waveguide
Light mixing, 75
Light pressure, 391
Light valve, liquid-crystal, 728, 855
Lightwave communications, *see* Fiber-optic communications
Lincoln Laboratory, M.I.T., 592
Linearly polarized light, 194, 196–198
Linear system, 928–935
causal, 930
impulse-response function, 928, 932
modes, 934–935
one-dimensional, 928–931
point spread function, 932
shift-invariant, 928, 932
transfer function, 929, 932–933
two-dimensional, 931–933
Line broadening, 444–449
collision, 446, 583
Doppler, 447–449, 486–487, 512–513
homogeneous, 446, 480, 510–511, 583
inhomogeneous, 446–449, 480, 511–512
lifetime, 444–446, 583
Lineshape function, 437
average, 446
Doppler-broadened, 447–449
Gaussian, 448–449
Lorentzian, 180–181, 444, 465, 583, 931
Linewidth, *see* Spectral width
Liquid crystal, 227–230, 235
cholesteric, 227
display, 727
light valve, 728, 855
modulator, 721–727
ferroelectric, 726–727
nematic, 721–724
twisted nematic, 724–726
nematic, 227
retarder, 721–727
smectic, 227
spatial light modulators, 727–728
twisted nematic, 227–230
Liquid laser, 519–520
Local oscillator, 907, 912
Logic, optical, 845, 848–867, 872
Longitudinal coherence, 357–359
Lorentzian lineshape function, *see* Lineshape, function, Lorentzian
Losses:
in fibers, 296–298
in resonators, 316–321
LP modes, fiber, 280
Luminescence, 454–457

Mach–Zehnder interferometer, 65, 66, 395, 703, 704, 735, 838, 849
Magnetic field, 159
Magnetic flux density, 160
Magnetic permeability, 159
Magnetization density, 161
Magnetogyration coefficient, 226
Magneto-optic effect, 225–227
Magneto-optic modulator, 735
Maiman, Theodore H., 494
Mandel's formula, 408
Manley–Rowe relations, 750, 765, 796
Mass, effective, 546–547
Mass action, law of, 557
Material dispersion, 300
Matrix:
ABCD, 28
coherency, 377
Jones, 199–203
ray-transfer, 28
Matrix optics, 26–37
Maxwell, James Clerk, 157
Maxwell's equations:
dielectric medium, 163
free space, 159
monochromatic fields, 167, 168
Memory element, optical, 846–855
MgF_2 (magnesium fluoride), 175
Michelson interferometer, 65, 66, 79
Michelson stellar interferometer, 375–376
Microchannel plate, 646–647
Miller's rule, 786
Minority carrier injection, 560–562
Mirror:
concave, 8
convex, 8
elliptical, 7
focal length, 6, 9
paraboloidal, 6, 8
planar, 6
spherical, 8–10

Mirror waveguide, *see* Waveguide, planar-mirror
Mobility, 655
Modal noise, 286
Mode density, 324, 326
Mode locking, 524, 531–536
Modes:
  fiber, 280–286
  laser, *see* Resonator, modes
  linear system, 934–935
  optically active medium, 224
  planar-dielectric waveguide, 249–258
  planar-mirror waveguide, 242–248
  polarization system, 203
  propagation in a crystal, 213
  rectangular dielectric waveguide, 259–260
  rectangular mirror waveguide, 259
  resonator, *see* Resonator, modes
Modulation:
  field, 887
  frequency shift keying (FSK), 889, 890
  intensity modulation, 887
  on–off keying (OOK), 889, 890, 903, 911–912
  phase shift keying (PSK), 889, 890, 911–912
  pulse code (PCM), 889
Modulator:
  acousto-optic, 815–817, 831
  electro-optic, 700–705, 710–712, 719–721
  liquid crystal, 721–727
  magneto-optic, 839
  opto-optic, 797, 840–843
Momentum, photon, 390–391, 419, 420
Momentum of electron/hole, 545
Momentum wavefunction, 412
Monochromatic light, 44
Multilayer photodetectors, 688–689
Multiplexing:
  frequency, 889–890
  time, 889–890
  wavelength, 890–892
Multiquantum well, 569–573, 854, 855
Multiquantum-well laser, 636–637
Mutual coherence function, 353, 355, 381
Mutual intensity, 355, 367–375, 381

N_2 (nitrogen), 425
Nd^{3+}:glass (neodymium glass) laser, 478–480, 518, 519, 521, 535
Nd^{3+}:YAG (neodymium YAG) laser, 478–480, 518, 519, 521, 535
Nd^{3+}:YLF (neodymium YLF) laser, 519
Nd^{3+}:YSGG (neodymium YSGG) laser, 519
Ne (neon), 428
Negative-binomial distribution, 420
Nematic liquid crystal, 227
Network, star, 892
Newton, Isaac, 1
Neyman type-A distribution, 459
Noise:
  laser amplifier, 488–489
  optical fiber, 286–287
  optical field, 411–415
  optical receiver:
    background noise, 674
    bipolar transistor amplifier noise, 690
    circuit noise, 681–685
    circuit noise parameter, 683–685
    FET amplifier noise, 690
    Johnson noise, *see* Noise, optical receiver, thermal noise
    minimum detectable signal, 674
    Nyquist noise, *see* Noise, optical receiver, thermal noise
    receiver sensitivity, 674, 689–690, 695
    resistance-limited-amplifier noise, 683–684, 688–690
    signal-to-noise ratio, 674, 685–689, 694, 695
    thermal noise, 682–683
    transistor amplifier noise, 684–685, 690
  photodetector:
    avalanche photodiode, 679–681, 694
    dark current noise, 674
    excess noise factor, 679–681, 694
    gain noise, 678–681, 694, 695
    photocurrent noise, 676–678
    photoelectron noise, 675
    photon noise, 403, 409, 675
  photon number, 403, 409
  photon partition, 409–411
Noise factor, excess, 679–681, 694, 905, 906
Nonlinear optical coefficients, 740, 743, 751, 779, 780
Nonlinear optics:
  anisotropic effects, 779–782
  dispersive effects, 782–786
  fibers, 792–793
  photorefractive effect, 729–733
  pulse propagation, 786–793
  second-order effects, 743–751, 762–774
  third-order effects, 751–761, 774–779
Nonlinear wave equation, 741
Normal modes, *see* Modes
Normal surface, 216
Numerical aperture (NA), *see also* Acceptance angle
  graded-index fiber, 24, 25, 308
  step-index fiber, 17, 39, 275–277
Nyquist noise, *see* Noise, thermal

Occupancy of energy levels, 553–555
On–off keying (OOK), *see* Modulation
Optical activity, 223–225
Optical bistability, 846–855
Optical communications, *see* Fiber-optic communications
Optical computing, *see* Computing, optical
Optical Doppler radar, 76
Optical fiber, *see* Fiber, optical
Optical Fourier transform, 121–127
Optical indicatrix, 212–215
Optical isolator, 233–234, 236
Optical joint density of states, 579, 610, 634
Optical Kerr effect, 752, 754, 757, 769

Optical logic, 845, 848–867, 872
Optical materials, 175, 177
Optical path length, 3, 78
Optical processing, *see* Processing, optical
Optical receiver, *see* Receiver sensitivity
Optical rectification, 744
Optical resonator, *see* Resonator
Optic axis, 211
Optoelectronic integrated circuits, 240
Ordinary refractive index, 211, 218–220
Ordinary wave, 218–220
Orthogonal polarizations, 198
Oscillation condition, 500
Oscillation threshold, 500–501
Oscillator strength, 437

Parabolic index profile, 21, 23, 288
Paraboloidal mirror, 6, 8
Paraboloidal wave, 49
Parametric amplifier, 749, 771–773, 797
  coupled-wave equations, 771–773
  gain coefficient, 773, 797
  idler, 749
  pump, 749
  signal, 749
Parametric conversion, 749, 769–771, 797, 798
Parametric interactions, 748–751
Parametric oscillator, 749, 773–774, 797
Paraxial approximation, 8
Paraxial Helmholtz equation, 50, 51
Paraxial optics, 8
Paraxial ray, 8
Paraxial ray equation, 20
Paraxial wave, 50–52
Parseval's theorem, 920
Partial coherence, *see* Coherence
Partially coherent imaging, 366–372
Partially coherent light, 343–383
Partially coherent plane wave, 357–359
Partially coherent spherical wave, 359
Partially polarized light, 376–379, 383
Partial polarization, *see* Partially polarized light
Path length, optical, 3
Pattern recognition, optical, 868
Pauli exclusion principle, 433, 544
Periodic optical system, 32–37
  sequence of lenses, 35
  resonator, 36
Periodic table of elements, 548
Permeability, magnetic, 159
Permittivity:
  dielectric medium, 163
  free space, 159
  relative, 163
  tensor, 210
Perot, Alfred, 310
Phase, 44
Phase conjugate resonator, 761
Phase conjugation, 758–761, 777–779
Phase matching:
  directional couplers, 267
  four-wave mixing, 757
  second-harmonic generation, 768–769, 782
  three-wave mixing, 747, 781, 796
Phase modulator, 797
Phase object, 154
Phase shift keying, 889, 890, 911–912
Phase velocity, 48
Phosphorescence, 456
Photocathode, 647
Photoconductivity, 654–657
Photoconductor, 654–657
  circuit, 693
  excess noise factor, 694
  extrinsic, 656
  gain, 655
  response time, 657
  spectral response, 656
Photodetector:
  gain, 651
  linear dynamic range, 650
  long-wavelength limit, 650. *See also* Bandgap wavelength
  noise, *see* Noise, photodetector
  quantum efficiency, 649–650
  response time, 652–654, 657, 658, 661, 663, 671–673, 884–885
  responsivity, 650–651
  thermal detectors, 645
  two-photon, 693
Photodiode, 648
  array, 664–665
  avalanche, *see* Avalanche photodiode (APD)
  bias circuits, 658–660
  heterostructure, 661–663
  metal-semiconductor, 662–665
  photoconductive, 659
  photovoltaic, 659
  p-i-n, 660–661
  p-n, 657–660
  quantum efficiency, 663, 693
  response time, 658
  Schottky-barrier, 662–665
Photoeffect:
  external, 645
  internal, 647
Photoelastic constant, 802
Photoelastic effect, 802, 826–827
Photoelectric detector, *see* Photodetector
Photoelectron emission, 645–647
Photoemissive detector, 645–647
Photoluminescence, 455
Photomultiplier tube, 646
Photon, 386
  absorption and emission, 434–443
  counting, 403
  detector, *see* Photodetector
  energy, 387–388, 418
  flux, 398–411, 420
    partitioning, 409–411, 421

flux density, 399
interference, 394–395, 419
lifetime, 320, 340
momentum, 390–391, 419, 420
noise, 403, 409, 675
number, 388, 400
polarization, 391–394
position, 388–390, 418
radiation pressure, 391
spin, 393–394
stream, 398–411
random partitioning, 409–411, 421
time, 395–396
time-energy uncertainty, 396
Photon-number conservation, 750, 765
Photon-number noise, 842
Photon-number-squeezed light, 415–416
Photon-number statistics, 403–409, 420, 422
binomial, 421
Boltzmann, 405–406
Bose–Einstein, 406–407, 420, 452, 489
Laguerre-polynomial, 489, 493
Mandel's formula, 408
negative-binomial, 420
partioned photons, 409–411, 421
Poisson, 403–405, 420
Photorefractive effect, 729–733
Phototube, 646
Photovoltaic detector, 659
p–i–n junction, 567
Planar dielectric waveguide, *see* Waveguides, planar dielectric
Planar mirror, 6
Planar-mirror resonator, 311–327, 329, 340
Planck, Max, 384
Planck's constant, 387
Plane of incidence, 5
Plane wave, 47, 170
Plasma laser, 520
p–n junction, 563–567
Pockels, Friedrich, 696
Pockels coefficients, 699, 713–718
Pockels effect, 697–699
Pockels readout optical modulator (PROM), 711–712
Point-spread function, *see* Impulse-response function, imaging system
Poisson, Siméon, 644
Poisson distribution, 403–405, 420
Polarization, 193–237
circular, 194, 196–198, 236
degree of, 379
ellipse, 195
elliptical, 194
linear, 194, 196–198
normal modes, 203
partial, 376–379, 383
rotator, 201, 203, 233–234, 235
TE, 204–209
TM, 204–209
Polarization density, 161
Polarized light, 194–203, 378
Polarizer, 200, 203, 230–232, 237
Polarizing beamsplitter, 231, 232
Polychromatic light, 72
Power, optical, 44, 161, 168
Power spectral density, 349–350
Poynting vector, 161
Principal axes, 211
Principal point, 31
Principal refractive indices, 211
Prism, 11, 12, 178
polarizing, 232
Rochon, 232
Sénarmont, 232
Wollaston, 232
Prism coupler, 263
Probability Bernoulli distribution, 409
binomial, 410
Boltzmann, 405–406
Bose–Einstein, 406–407, 420, 489
exponential, 409
Gaussian, 905
geometric, 406
negative-binomial, 420
Neyman type-A, 459
noncentral-chi-square, 493
Poisson, 403–405, 420
Probability of error, 894, 903–906
Probability of energy-level occupancy, 432–434
Processing, optical:
analog, 864–869
coherent, 121–127, 136–139, 865, 867–869
convolution and correlation, 868–869
digital, 862–864
discrete, 865–867
Fourier-transform, 121–127, 867–868
geometric transformations, 869, 872
incoherent, 865–867
matrix operations, 866–867
Prokhorov, Aleksandr M., 460
Propagation in anisotropic crystal, 210–223
Propagation constant, 175
Propagation of partially coherent light, 366–376
Proustite (Ag_3AsS_3), 771, 780
PtSi (platinum silicide), 662, 664–665
Pulse code modulation, 889
Pulse compression, 188
Pulsed laser, 522–536
cavity dumping, 523, 541
gain switching, 522, 526–527, 540
mode locking, 524, 531–536
Q-switching, 523, 527–531, 540, 541
Pulsed light:
complex wavefunction, 73
in dispersive linear medium, 182–189
in dispersive nonlinear medium, 754–755
in fibers, 792
plane wave, 74
solitons, 754–755, 786–793

Pulsed light (*Continued*)
  spherical wave, 79
Pulse spreading, 182–189, 192
Pulse width, 187
Pumping, 468–480
Pupil function, 135
Purity, cross-spectral, 357

Q-switching, 523, 527–531, 540, 541
Quadrature components, field, 411
Quadrature-squeezed light, 414–415
Quadric representation of tensor, 212
Quality factor *Q*, resonator, 321
Quantum dot, 572–573
Quantum efficiency:
  differential, 625
  external, 604, 640
  internal, 562–563, 640
  overall, 640
Quantum electrodynamics, 385
Quantum of light, *see* Photon
Quantum noise, *see* Photon, noise
Quantum optics, 385
Quantum states of light, 411–416
Quantum well, 569–571, 573, 590
Quantum-well lasers, 632–636
Quantum wire, 572–573
Quarter-wave plate, *see* Retarder, wave
Quartz, 175–177, 780
  fused, 817, 820
Quasi-equilibrium, semiconductor, 558
Quasi-Fermi energy, 558–559
Quasi-monochromatic light, 73, 355, 364
Quasi-plane wave, 174
Quaternary semiconductor, 549

Radiation pattern, 608, 629–631
Radiation pressure, 391
Radiative transitions, 434–437, 576–581
Radius of curvature, Gaussian-beam, 88, 90, 91
Rainbow holography, 151
Raman gain, 755
Raman–Nath diffraction, 813–815, 831
Ramo's theorem, 652
Random light, *see* Coherence
Random partitioning, photon, 409–411, 421
Rare-earth-doped fibers, 479–480
Rate equations, 451, 459
Ray, 3
  angle, 27
  in graded index fiber, 23–25
  in graded index slab, 20–23
  height, 27
  meridional, 275
  paraxial, 8
  in periodic system, 32–37
  skewed, 275
  in step-index fiber, 275–277
Ray equation, 19–20
Rayleigh, Lord (John W. Strutt), 80
Rayleigh range, 82
Rayleigh scattering, 297–298
Ray optics, 1–40, 52–53
  paraxial, 20
Ray-transfer matrix, 28–30
  cascaded components, 30
  cylindrical lens, 40
  free space, 28
  GRIN plate, 40
  lens system, 40
  planar boundary, 28
  planar mirror, 29
  spherical boundary, 29
  spherical mirror, 29
  thick lens, 31
  thin lens, 29
Receiver sensitivity:
  analog, 689–690
  digital, 894, 903–906
  frequency shift keying, 913
  heterodyne detection, 912
  homodyne detection, 911–912
  ideal (photon-limited), 904, 906
  on–off keying:
    coherent, 911–912
    direct-detection, 906, 912
  phase-shift keying, coherent, 911, 912
Reciprocity, optical, 761
Recombination, 559–563, 590, 591
  lifetime, 561–563
Rectification, optical, 744
Reference wave, *see* Holography
Reflectance:
  complex amplitude:
    external, 206–208
    internal, 206–208
    planar boundary, 205–209
    spherical mirror, 79
  power, planar boundary, 209
Reflection, 5, 7–11, 53–53, 203–209, 236
  law of, 5
  total internal, 11
Reflection grating, 61, 151, 760
Reflection hologram, 151
Refraction, 5, 6, 54, 55, 203–209
  conical, 237
  double, 221–223, 236
  external, 10
  internal, 10
  law of, 6
Refractive index, 3, 43, 163, 164, 169, 176, 177, 181, 183, 587–588
  anisotropic medium, 211–218
  extraordinary, 218–220
  graded, 19–26, 39, 40, 288, 303
    quadratic profile, 21, 23, 40, 288
  inhomogeneous medium, 164. *See also* Refractive index, graded
  ordinary, 218–220
  principal, 211

semiconductor, 587–588
silica glass, 190, 300
Resolution:
acousto-optic scanner, 818–820
electro-optic scanner, 705–706
imaging system, 136, 141–143, 155, 371
Resonance frequencies, 317, 318
Resonant medium, 179–183, 192
Resonator, 36, 72, 310–341, 419
concentric, 329–330
confinement, 327–330
confocal, 329–330, 334, 337, 339, 341
diffraction loss, 337–339, 341
fiber, 311
finesse, 316
finite aperture, 337–339
free spectral range, 332
Fresnel number, 339
g-parameters, 329
loss, 316–321
modes:
density of, 315, 324, 326
frequencies, 313, 315, 335–337, 340
frequency spacing, 313
Gaussian, 330–336, 341
Hermite–Gaussian, 336–337
longitudinal, 336
transverse, 336
phase conjugate, 761
photon lifetime, 320, 340
planar-mirror, 311–327, 329, 340
quality factor *Q*, 321
ray confinement, 327–330
ring, 311, 315
spectral response, 317, 340
spectral width, 318
spectrum analyzer, 321–322
spherical-mirror, 327–339, 341
stability, 327–330
symmetrical, 329, 333
three-dimensional, 324–327
two-dimensional, 323
unstable, 341, 515
Response time, optical fiber, *see* Fiber, optical, response time
Response time, photodetector, *see* Photodetector, response time
Responsivity, 604–605, 626, 650–651
Retarder, wave, 200, 201, 203, 232–233, 235, 236
electro-optic, 701–702
liquid crystal, 721–727
Ring aperture, 155
Ring resonator, 311, 315
rms width, 921–922
Rochon prism, 232
Rotational energy levels of diatomic molecule, 425
Rotatory power:
Faraday rotator, 225–227
optically active medium, 223
Ruby, 429–430
laser, 477–478, 480, 521, 531

Saturable absorber, 484–485, 850
Saturated gain, 481, 492–493
Saturation intensity, 492
Saturation photon-flux density, 481, 482, 492
Scalar wave, 43, 174
Scalar wave equation, 43
Scanner:
acousto-optic, 818–820
electro-optic, 705–706
holographic, 115
mechanical, 836
Schawlow, Arthur L., 494
Shockley, William P., 542
Schottky-barrier photodiode, 662–665
Schrödinger's equation:
nonlinear, 754, 791
time-dependent, 425
time-independent, 425
Secondary emission, 646–647
Second-harmonic generation, 541, 743–744
Self-focusing, 753
Self-guided beam, 754–755, 797
SELFOC lens, 21–23, 40, 63, 78, 107
Self-phase modulation, 753, 787
Semiconductor:
absorption, 576–584, 586–587, 590
absorption edge, 576
acceptor, 551
bandgap energy, 544, 550, 551
bandgap wavelength, 550, 576, 650
band-to-band transitions, 559–560, 574–587
binary, 548, 550
carrier concentration, 552–559
degenerate, 558
density of states, 552–553, 571–573
donor, 551
doped, 551–552
effective mass of electron/hole, 546–547
elemental, 548, 550
energy–momentum relation, 546–547
excitonic transitions, 574
extrinsic, 551–552
Fermi energy, 554, 590
Fermi function, 554
free-carrier transitions, 574
generation of electron–hole pairs, 559–560
heterostructure, 567–569
injection, 560, 566
intrinsic, 548–551
lattice constant, 550, 551
law of mass action, 557
occupancy probability, 553–555
quantum efficiency, internal, 562–563
quantum well, 569–571
quasi-equilibrium, 558
quasi-Fermi energy, 558–559
quaternary, 549

Semiconductor (*Continued*)
 recombination, 559–563, 590, 591
 recombination lifetime, 561–563
 refractive index, 587–588
 spontaneous emission, 584–585, 590
 stimulated emission, 576–585, 590
 ternary, 549
Semiconductor laser, *see* Laser diode
Semiconductor laser amplifier, 609–619
 bandwidth, 611, 641, 642
 GaAs/AlGaAs, 619
 gain coefficient, 585–586, 610–616, 620, 641, 642
 heterostructure, 617–619
 InGaAsP, 613, 615, 616, 617, 619, 641
 pumping, 612–617
Sénarmont prism, 232
Shot noise, 676
Si (silicon), 175, 177, 430, 545–548, 550, 557, 563, 575, 576, 588, 656, 661–665, 673, 681, 692, 693, 694, 884–886
Signal-to-noise ratio, 674–689
Silica glass, 175, 177, 876
sinc function, 921
Single-mode:
 fiber, 273, 274, 286
 laser, 516–518
 waveguide, 252, 271
Slab waveguide, *see* Waveguide, planar dielectric
Slowly varying envelope approximation, 51, 184
Smectic liquid crystal, 227
Snell's law, 6
SNR, *see* Signal-to-noise ratio
Solar cell, 659
Solid-state laser, 518–519
Solitary wave, 787
Soliton, 754–755, 786–793
 dark, 793
 envelope equation, 788–790
 fibers, 792
 fundamental, 792
 laser, 793
 spatial, 754–755, 797
Sonoluminescence, 455
Sound, *see* Acoustic wave
Space-charge field, 729–732
Spatial amplitude modulation, 113
Spatial bandwidth, 139, 143
Spatial coherence, 353–357, 362–376, 381
Spatial filter, optical, 136–143, 154
 high-pass, 139
 low-pass, 138, 139
 Vander Lugt, 148
 vertical-pass, 139
Spatial frequency, 109, 111
Spatial frequency modulation, 114
Spatial frequency multiplexing, 114
Spatial harmonic function, 111
Spatial light modulator, 709–712
Spatially incoherent light, 368–376
Spatial soliton, 754–755, 797
Spatial spectral analysis, 112
Speckle, 286
Spectral distribution, see Bandwidth; Line broadening; Spectral width
Spectral linewidth, *see* Spectral width
Spectral photon-flux density, 401
Spectral width, 351–352, 381, 382, 921–924. *See also* Bandwidth
 electroluminescence, 600
 laser diode, 627–628, 631
 LED, 352, 605, 640
 multimode laser, 352, 627–628
 resonator modes, 318, 320
 single-mode laser, 352, 510–511, 521–522, 631
 sodium lamp, 352
 sunlight, 352
Spectrometer, *see* Spectrum analyzer
Spectrum analyzer:
 acousto-optic, 820
 diffraction grating, 62
 Fabry–Perot etalon, 72, 321–322
 Fourier-transform, 362
Speed of light, 3, 43, 160, 163, 164
Spherical boundaries, 13
Spherical mirror, 8–10, 79
Spherical wave, 48, 78, 79, 171
Spin, photon, 393
Spontaneous emission, 435, 438, 442, 458, 459, 584–585, 590
 inhibited, 459
Spontaneous lifetime, 439
Squeezed-state light, 414–416
Stability, resonator, 327–330
Standing wave, 79, 191
Star network, 892
Stationary random light, 345
Statistical average, 345
Statistical optics, *see* Coherence
Stellar interferometer, Michelson, 375–376
Step-index fiber:
 characteristic equation, 281
 group velocities, 285
 LP modes, 280
 mode cutoff, 282
 number of modes, 282
 numerical aperture, 17, 39, 275–277
 propagation constants, 284, 285
 rays, 275–277
 V parameter, 279
Stimulated emission, 436, 440–443, 458, 459, 576–585, 590
Strain, 802
 tensor, 825–826
Strained-layer laser diode, 637
Strain-optic tensor, 826
Strip waveguide, 261
Superlattice, 572
Superluminescent LED, 627
Superposition, 43

Surface-emitting laser diode (SELD), 632, 637–638
Surface-emitting LED, 606–608, 883
Susceptibility, 162, 166, 169, 175–176, 179–181, 739–741
Susceptibility tensor, 165
Switches, 833–843
 acousto-optic, 838–839
 all-optical, 840–843
 electro-optic, 837–838
 magneto-optic, 839–840
 optoelectronic, 835
 opto-mechanical, 836–837
 opto-optic, 840–843
Switching energy, 834, 835, 843, 855
Switching power, 834, 835, 843, 855
Switching time, 834, 835, 842, 843, 855
Symmetrical resonator, 329, 333

Te (tellurium), 780, 817
Temporal coherence, 346–353, 361–362
Temporal coherence function, 346, 347
TEM wave, 170
TE ("s") polarization, 204–205, 246, 255
Ternary semiconductor, 549
Thermal light, 405–407, 424, 450–454
Thermal noise, 682–683
Thin optical component, 56
Third-harmonic generation, 751–752
Three-level laser, 474–476
Three-wave mixing, 746–750, 762–774, 781–782, 796, 797
Threshold, laser, 500, 539, 622–624, 636
Threshold, parametric oscillator, 774
Threshold current density, laser diode, 622–624, 642
$Ti^{3+}:Al_2O_3$ (Ti:sapphire) laser, 480, 519, 521, 535
Time-dependent Schrödinger equation, 425
Time-division multiplexing, 889–890
Time-energy uncertainty, 396, 397, 842
Time-independent Schrödinger equation, 425
Time reversal, 759
Time-varying light, *see* Pulsed light
TM ("p") polarization, 204–205, 246, 255
Total internal reflection, 11, 16
Townes, Charles H., 460
Transformation, coordinate, 202
Transients, laser, 524–526, 540
Transition cross section, 435
Transition linewidth, 437
Transition strength, 437, 439
Transmission grating, 61, 151, 760
Transmittance, complex amplitude:
 diffraction grating, 61
 graded-index thin plate, 63
 prism, 57
 thin lens, 58
 transparent plate, 55–57
Transverse electric (TE) mode, 204–209
Transverse electromagnetic (TEM) wave, 170
Transverse magnetic (TM) mode, 204–209
Trapping of atoms, 449–450
Trapping of light, 18, 39
Twisted nematic liquid crystal, 724–726
Two-dimensional Fourier transform, 153, 924–926
Two-dimensional linear system, 931
Two-photon absorption, 693
Two-wave mixing:
 in photorefractive material, 733
 in third-order nonlinear medium, 756
Tyndall, John, 238

Ultraviolet, 158
Uncertainty relation, Heisenberg, 922
Undersea fiber-optic network, 874, 886
Uniaxial crystal, 211
 negative, 211
 positive, 211
Unpolarized light, 378
Unstable resonator, 341, 515

Valence band, 429
Van Cittert–Zernike theorem, 372–373
Vander Lugt filter, 148
Vector potential, 172
Velocity of light, *see* Speed of light
Verdet constant, 225–227
Visibility, fringe, 79, 360, 361, 382
Visible light, 42, 158
V number:
 of dispersive material, 178
 of optical fiber, 279
Volume hologram, 149–151
von Neumann, Johann (John), 832

Wave equation:
 free space, 160
 inhomogeneous medium, 164
 linear homogeneous medium, 43, 163
 nonlinear medium, 167
 partial coherence, 355
Wavefront, 46
Wavefunction, 43
Waveguides:
 channel, 260–261
 circular, *see* Fiber, optical
 couplers, 262–269
 lens, 16
 planar dielectric, 248–269
  asymmetric, 258
  confinement factor, 254
  dispersion relation, 256
  extinction coefficient, 253
  field distributions, 252–254, 271
  group velocities, 255, 258
  mode cutoff, 252, 271
  mode excitation, 261–263
  number of modes, 251
  numerical aperture, 251

Waveguides (*Continued*)
propagation constants, 250
rectangular, 259–260
single-mode, 252, 271
symmetric, 240–257
TE and TM modes, 255
two-dimensional, 259–261
planar-mirror, 240–248
cut-off, 245
dispersion relation, 245
field distributions, 242–244, 270
group velocities, 245
modal dispersion, 270
number of modes, 244
propagation constants, 242
single-mode, 245
TE modes, 241–245
TM modes, 246
Waveguides, coupling between, 264–269, 271
Wavelength, 42, 47, 158
Wavelength-division multiplexing (WDM), 890–892
Wavenumber, 46
Wave optics, 41–79, 174
Wavepacket, 75
Wave–particle duality, 389
Wave plate, *see* Retarder, wave
Wave restoration, 761
Wave retarder, *see* Retarder, wave
Wavevector, 47
White-light hologram, 149–151
Width of a function:
1/*e*, 923–924
3-dB, 923–924
full-width at half-maximum (FWHM), 923–924
half-maximum, 923–924
power-equivalent, 922–923
root-mean-square (rms), 921–922
Wien's law, 459
Wiener–Khinchin theorem, 350, 381
Wolf, Emil, 342
Wollaston prism, 232
Work function, photoelectric, 546

X-ray, 158
X-ray laser, 520

YAG (yttrium aluminum garnet) laser, 478–480, 518, 519, 521, 535
Y branch, 261
YLF (yttrium lithium fluoride) laser, 519
Young, Thomas, 41
Young's two-pinhole interference experiment, 68, 154, 362–366, 394
YSGG (yttrium scandium gallium garnet) laser, 519

ZnSe (zinc selenide), 175
Zone plate, 116